FUNDAMENTALS OF ELECTRICAL ENGINEERING AND ELECTRONICS

FUNDAMENTALS OF
ELECTRICAL ENGINEERING AND ELECTRONICS

IN

INTERNATIONAL SYSTEM (SI) OF UNITS
(Incorporating rationalized M.K.S.A. system)

[For the Examinations of B.Sc. (Engg.); B.E. (Electronics); Sec. A of A.M.I.E.(I), A.M.Ae. S.I. ; City & Guilds (London); I.E.R.E. (London); Diploma (Elect. & Mech.) and Diploma (Electronics & Commn.) etc.]

B.L. THERAJA

(Thoroughly Revised Edition)

Dear Students,
Beware of fake /pirated editions. Many of our best selling titles have been unlawfully printed by unscrupulous persons. Your sincere effort in this direction may stop piracy and save intellectuals' rights.
For the genuine book check the 3-D hologram which gives a ***rainbow*** *effect.*

2000

S. CHAND & COMPANY LTD.

RAM NAGAR, NEW DELHI-110 055

S. CHAND & COMPANY LTD.
Head Office : 7361, RAM NAGAR, NEW DELHI - 110 055
Phones : 3672080-81-82; Fax : 91-11-3677446
Shop at: **schandgroup.com**
E-mail: **schand@vsnl.com**

Branches :
No. 6, Ahuja Chambers, 1st Cross, Kumara Krupa Road, **Bangalore**-560 001. Ph : 2268048
285/J, Bipin Bihari Ganguli Street, **Calcutta**-700 012. Ph : 2367459, 2373914
152, Anna Salai, **Chennai**-600 002. Ph : 8522026
Pan Bazar, **Guwahati**-781 001. Ph : 522155
Sultan Bazar, **Hyderabad**-500 195. Ph : 4651135, 4744815
Mai Hiran Gate, **Jalandhar** - 144008. Ph. 401630
613-7, M.G. Road, Ernakulam, **Kochi**-682 035. Ph : 381740
Mahabeer Market, 25 Gwynne Road, **Lucknow**-226 018. Ph : 226801, 284815
Blackie House, 103/5, Walchand Hirachand Marg , Opp. G.P.O., **Mumbai**-400 001. Ph : 2690881, 2610885
3, Gandhi Sagar East, **Nagpur**-440 002. Ph : 723901
104, Citicentre Ashok, Govind Mitra Road, **Patna**-800 004. Ph : 671366

First Edition 1961
Subsequent Editions and Reprints 1965, 66, 68, 72, 74, 76, 77, 78, 79, 80, 81, 82 (Twice), 83, 84 (Twice), 85, 86 (Twice), 87, 88, 89 (Twice), 90, 91 (Twice), 92 (Twice), 93, 95, 96, 97 (Twice), 98,
Reprint 2000

ISBN : 81-219-0099-9

PRINTED IN INDIA

By Rajendra Ravindra Printers (Pvt.) Ltd., 7361, Ram Nagar, New Delhi-110 055 and published by S. Chand & Company Ltd., 7361, Ram Nagar, New Delhi-110 055.

Dedicated
to
My wife

PREFACE TO THE TWENTY-EIGHTH EDITION

Time, he explained, is the prime factor
Governing human advancement:
The heart of measurement is time;
Measurement is the blood of Science

—Cal Clothier in *Knowledge is Power*

It is a pleasure to present the twenty-eighth revised edition of this popular text-book which has been enthusiastically received by numerous teachers and students both in India and abroad. In the *Electrical Engineering* portion of the book, lot of fresh material has been added in the form of articles on *Source Conversion, Swinburn's Test, Battery Charging from AC Source, Speed of Rotor Field in Three-phase Motors, Neutral Current in Unbalanced Star Connection and Balanced Y/Δ and Δ/Y connections.*

In the *Electronics* portion of the book, following four new chapters have been added: **Optoelectronic Devices, Digital Electronic Devices, Analog and Digital Communication and Electron Ballistics.** Lot of new material concerning *tungsten filament lamps, discharge lamps, sodium vapour lamps, starters and stroboscopic effect* has been added in the chapter on **Illumination.**

There has been extensive pruning of the Solved Examples in the text. Majority of the old examples have been replaced by questions set in the latest examination papers of different engineering colleges and technical institutions.

It is sincerely hoped that with these extensive additions and revisions, the present edition will prove even more useful to my esteemed readers than the earlier ones.

As ever before, I take this opportunity to thank my publishers particularly Sh. Ravinder Kumar Gupta for the personal interest he took in the printing of this book. I would also love to record my sincere appreciation for Ms. Shweta Bhardwaj for cheerfully rendering secretarial services in the preparation of this edition and to Mr. Cal Clothier, Lecturer, Leads Polytechnic, U.K. for borrowing a few beautiful lines from his famous poem "Knowledge is Power."

New Delhi
March, 1997

B.L. THERAJA

CONTENTS

1 ELECTRIC CURRENT AND OHM'S LAW

1.1. Modern Electron Theory

Modern research has established that all matter whether solid, liquid or gaseous, consists of minute particles called **'molecules'** which are themselves made up of still minute particles known as **atoms.** Those substances whose molecules consist of similar atoms are known as **elements** and those whose molecules consist of dissimilar atoms are called **compounds.** The number of elements so far discovered is 106, whereas the number of compounds is unlimited.

An atom is taken to consist of the following:

1. It has a hard central core, known as **nucleus**. It contains two types of particles*; one is known as **proton** and carries positive charge, the other is **neutron** (discovered by Chadwick in 1932) which is electrically neutral *i.e.* it carries no charge, though it is as heavy as a proton. The protons and neutrons are very closely held together with tremendous nuclear forces.
2. Revolving round the relatively massive nucleus, in more or less elliptical orbits (or shells), are infinitesimally small particles known as **electrons.** These electrons carry the smallest negative charge and have a negligible mass. The mass of electron is approximately 1/1840 that of a proton.

Such a view of an atom, known as Bohr-Rutherford model, is shown in Fig. 1.1. It has been found that an atom is like a miniature solar system, a heavy positively charged nucleus taking the place of the Sun at the centre, with orbital electrons acting like planets. The planetary electrons revolve at distances which are much greater than the size of nucleus or electrons themselves; hence most of the space occupied by an atom is empty! In fact, a solid may be thought of as a sponge-like structure, in which nucleus and electrons occupy but little of the space taken up by the solid.

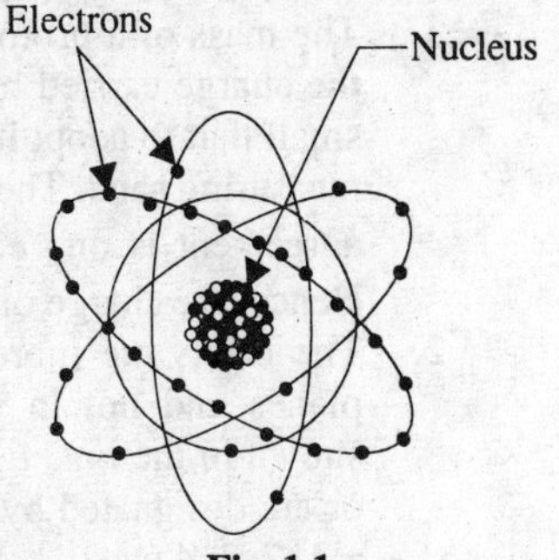

Fig. 1.1

It has also been found that the effective diameters of the atoms are of the order of 10^{-10} metre and the diameters of their nuclei of the order of 10^{-15} metre. Hence, the diameter of an atom is roughly 100,000 times greater than the diameter of the nucleus. Some rough idea of the vast emptiness existing within an atom can be got by imagining that the model of hydrogen atom may consist of a cricket ball, with a small soap bubble revolving round it at a distance of 3 km or so!

The particles discussed above form the fundamental bricks of which all matter is made. Atoms of all substances consist of identical protons, neutrons and electrons etc., the only difference being in their number and relative configuration.

It has been found that the positive charge on a proton is numerically equal to the negative charge of an electron. Normally, an atom is electrically neutral, because it consists of as many protons as electrons. The number of protons in the nucleus of an atom gives the atomic number (*Z*) of the substance whose atom it is. The total weight of a nucleus (*i.e.* protons plus neutrons)

---

*In addition to protons and neutrons, there have been discovered other particles like mesons and neutrino etc. and electrons inside the nucleus. All these particles within the nucleus are known as *nucleons.*

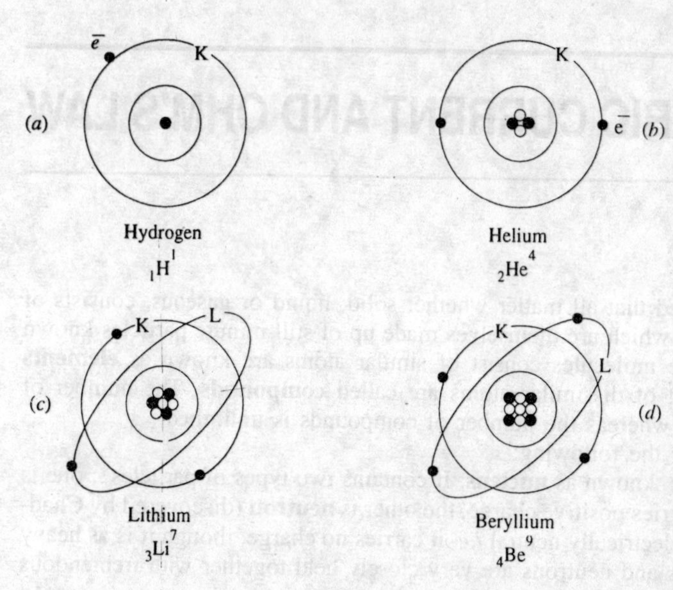

Fig. 1.2

is called the atomic mass number (A). If the number of protons in a nucleus is changed, then transmutation of one element into another can be achieved.

The simplest atom is that of hydrogen. It consists of one positive proton and one orbital electron, as shown in Fig. 1.2(*a*).

Next is the helium atom which has two planetary electrons and whose nucleus contains two protons and two uncharged neutrons as shown in Fig. 1.2(*b*). Similarly, in Fig. 1.2(*c*) and (*d*) are shown atoms of lithium and beryllium. Under each figure is given the symbol used in Nuclear Physics. The subscript is the atomic number and the superscript represents the atomic weight. Oxygen atom ($_8O^{16}$) has 8 orbital electrons and its nucleus consists of 8 protons and 8 neutrons. Heaviest atom is that of a newly-discovered element (still un-named) with Z = 106.

The following important points about atomic structure should be understood clearly:—

1. The mass of a proton is 1.66×10^{-27} kg and that of an electron 9.1×10^{-31} kg. Though the charge carried by an electron is the natural unit of electricity, but it is so extremely small that to adopt it as a unit of electricity would be like adopting the grain as a unit for measuring sand. The practical unit of charge or quantity of electricity adopted by general agreement is one coulomb* which is equal to the charge of 6.242×10^{18} electrons. Hence, the charge of a single electron is $1/6.242 \times 10^{18}$ *i.e.* 1.602×10^{-19} coulomb.
2. The orbits are more or less elliptical in shape and lie in all planes and not in one, though for convenience they are so shown in the Fig. 1.3. These orbits are marked *K, L, M, N* etc., or are designated by their principal quantum number *n* where *n* = 1, 2, 3, 4 etc.
3. The maximum number of electrons possible in any extra-nuclear orbit or shell is fixed and is given by $2n^2$ subject to the condition that maximum number of electrons in any orbit will not exceed 32. Counting these orbits from nucleus outwards, the first orbit can have a maximum number of 2 electrons; second orbit 8 electrons; third orbit 18 electrons and so on. For example, a copper atom has 29 electrons which will be distributed as follows (Fig. 1.3).
   1st orbit : 2 electrons; 2nd orbit : 8 electrons; 3rd orbit : 18 electrons; 4th orbit : 1 electron
   Similarly, silver atom has 47 electrons which will be grouped as 2, 8, 18, 18, 1.

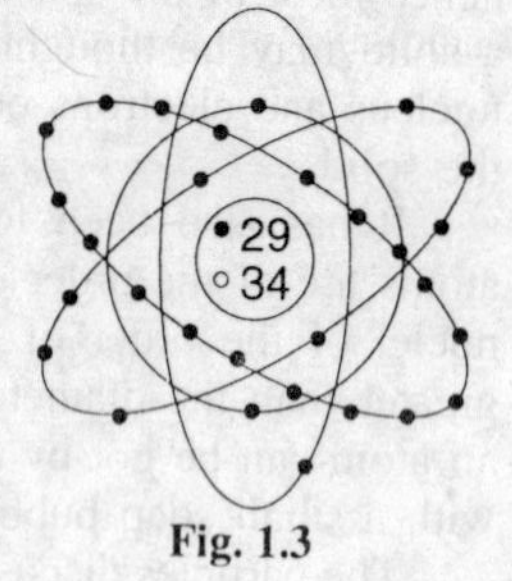

Fig. 1.3

4. The centripetal force necessary to keep electrons rotating in elliptical orbits round the nucleus is supplied by the force of attraction between their charges, as given by Coulomb's laws. It is obvious that nearer the electron is to the nucleus, greater is the force with which it is bound to it.

*After Charles A. Coulomb (1736-1806), French philosopher, distinguished for his investigations in electricity and mathematics.

The electrons in the outermost orbit experience a very weak force of attraction for two reasons: (*a*) force varies inversely as the square of the distance between two charges (*b*) the presence of a large number of electrons in the intermediate orbits acts as a partial screen between the nucleus and the outermost electrons. This screening or shielding action results in reduced attraction between the two. It is found that in metals, the outermost electrons are very loosely attached to the atom. In fact, they can be hardly said to be attached to one parent atom, they very freely move from one atom to another and behave very much like the molecules of a gas in a draught-free room. They wander about with random motion between atoms, continuously colliding with one another, but not moving in any particular direction. These free-moving and unattached electrons form what is known as *electron gas*. The condition of these electrons in a copper wire is shown diagrammatically in Fig. 1.4. The atoms are arranged in a particular pattern called "crystal lattice". The + signs indicate that they are charged positively (because of having lost some electrons) *i.e.* they are now ions. Although, these ions can oscillate about their mean position, yet for the present, they are shown stationary.

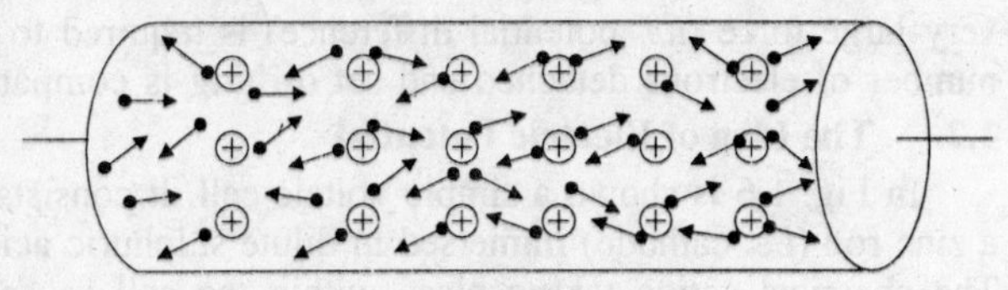

Fig. 1.4

The electrons, shown as black dots with arrows, wander about in all sorts of manner and directions. When this wire is joined across the terminals of a battery, the electrons experience an attractive force due to anode and a repulsive force due to cathode, with the result that they start drifting from cathode to anode, as shown in Fig. 1.5. When some external force (*i.e.* potential difference) is applied to these atoms, the outermost one or more electrons get easily detached from the parent atom and start drifting along and so give rise to flow of electrons.

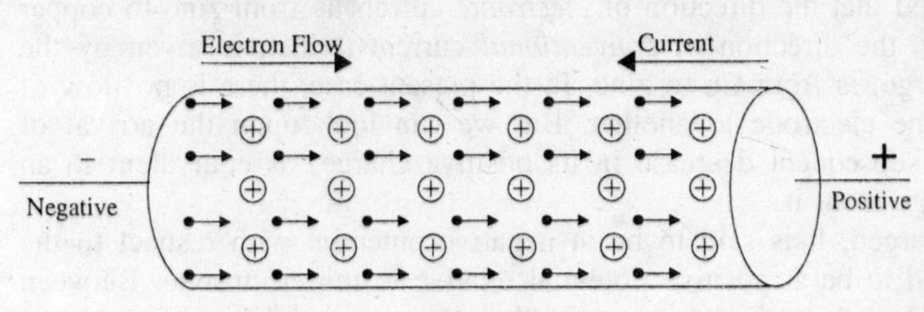

Fig. 1.5

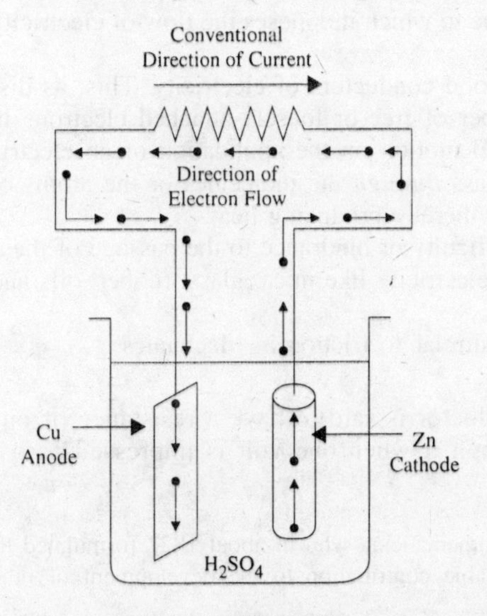

Fig. 1.6

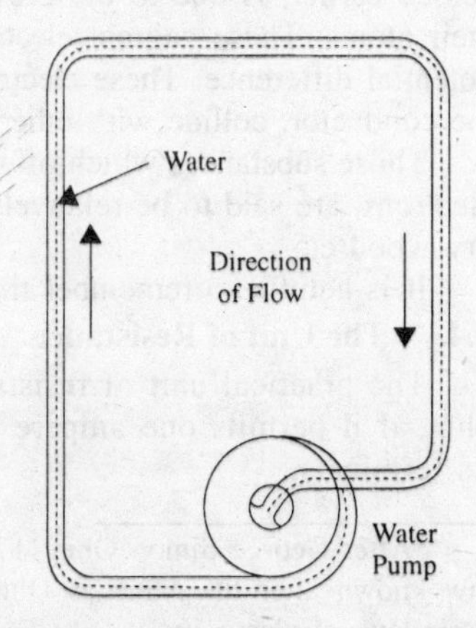

Fig. 1.7

This continuous flow of electrons constitutes an electric current. It is found that those substances, whose atoms have their outer-most orbits incomplete, act as good conductors of electricity *i.e.*, they permit an easy detachment of their outer-most electrons and offer very little hindrance to their flow 'through' their atoms. Such substances are known as *good conductors*. But substances whose electrons are rigidly held to their atoms are termed as *bad conductors*. In their case, a very large force (*i.e.* potential difference) is required to detach their electrons and even then the number of electrons detached and set drifting is comparatively small.

1.2. The Idea of Electric Potential

In Fig. 1.6 is shown a simple voltaic cell. It consists of a copper plate (known as anode) and a zinc rod (*i.e.* cathode) immersed in dilute sulphuric acid (H_2SO_4) contained in a suitable vessel. The chemical action taking place within the cell causes the electrons to be removed from Cu plate and to be deposited on the zinc rod (for details see Art. 2.4). This transfer of electrons is accomplished through the agency of the diluted H_2SO_4, which is known as an electrolyte. The result is that the zinc rod becomes negative, due to the deposition of electrons on it and the Cu plate becomes positive, due to the departure of electrons from it. The large number of electrons collected on the zinc rod is being attracted by anode, but is prevented from returning to it by the force set up by the chemical action within the cell. But if the two electrodes are joined by a wire *externally*, then electrons rush to anode, thereby equalizing the charges of the two electrodes. However, due to the continuity of chemical action, a continuous difference in the number of electrons on the two electrodes is maintained, which keeps up a continuous flow of current through the external circuit. The action of an electric cell is similar to that of a water pump which, while working, maintains a continuous flow of water *i.e.* water current through the pipe (Fig. 1.7).

It should be particularly noted that the direction of *electronic* current is from zinc to copper in the external circuit. However, the direction of *conventional* current (which is given by the direction of flow of *positive* charge) is from Cu to zinc. In the present case, there is no flow of positive charge as such from one electrode to another. But we can look upon the arrival of electrons on copper plate (with subsequent decrease in its positive charge) as equivalent to an actual departure of positive charge from it.

When zinc is negatively charged, it is said to be at negative potential with respect to the electrolyte, whereas anode is said to be at positive potential relative to the electrolyte. Between themselves, Cu plate is assumed to be at a higher potential than the zinc rod. This difference in potential is continuously maintained by the chemical action going on in the cell, which supplies energy to establish this potential difference.

1.3. Resistance

It may be defined as the property of a substance, due to which it opposes the flow of electricity (*i.e.* electrons) through it.

Metals (as a class), acids and salt solutions are good conductors of electricity. This, as discussed earlier, is due to the presence of a large number of free or loosely-attached electrons in their atoms. These vagrant electrons assume a directed motion on the application of an electric potential difference. These electrons, while flowing, pass *through* the molecules or the atoms of the conductor, collide with other atoms and electrons, thereby producing heat.

Those substances which offer relatively greater difficulty or hindrance to the passage of these electrons, are said to be relatively poor conductors of electricity like mica, glass, rubber, oils and dry wood etc.

It is helpful to remember that electric friction is similar to friction in Mechanics.

1.4. The Unit of Resistance

The practical unit of resistance is ohm*. A conductor is said to have a resistance of one ohm, if it permits one ampere current to flow through it when one volt is impressed across its terminals.

---

*After George Simon Ohm (1787-1854), a German mathematician who in about 1827 formulated the law known after his name as Ohm's Law. He made notable contribution to the development of other basic laws of electricity.

TABLE No. 1.1
Multiples and Submultiples of Ohm

Prefix	*Its Meaning*	*Symbol*	*Equal to*
Mega —	One million	MΩ	10^6 Ω
Kilo —	One thousand	KΩ	10^3 Ω
Centi —	One hundredth	—	—
Milli —	One thousandth	mΩ	10^{-3} Ω
Micro —	One millionth	μΩ	10^{-6} Ω

For insulators, whose resistances are very high, a much bigger unit is used *i.e.* megohm = 10^6 ohm (the perfix 'mega' or 'mego' meaning a million) or kilohm = 10^3 ohm (kilo means thousand). In the case of very small resistances, smaller units like milliohm = 10^{-3} ohm or microhm = 10^{-6} ohm are used. The symbol for ohm is Ω.

1.5. Laws of Resistance

The resistance R offered by a conductor depends on the following factors :

1. It varies directly as its length.

If, for example, one metre of a copper wire has a resistance of 0.04 Ω, then two metres will have a resistance of 0.08 Ω and 50 metres will have a resistance of 2.0 Ω.

2. It varies inversely as the cross-section of the conductor.
3. It depends on the nature of the material.
4. It also depends on the temperature of the conductor.

Neglecting the last factor for the time being, we can say that $R \propto \frac{l}{A}$ or $R = \rho \frac{l}{A}$ ...(*i*)

where l is the length, A the area of cross-section of the conductor and ρ is a constant depending on the nature of the material of the conductor and is known as the *specific resistance* or *resistivity.*

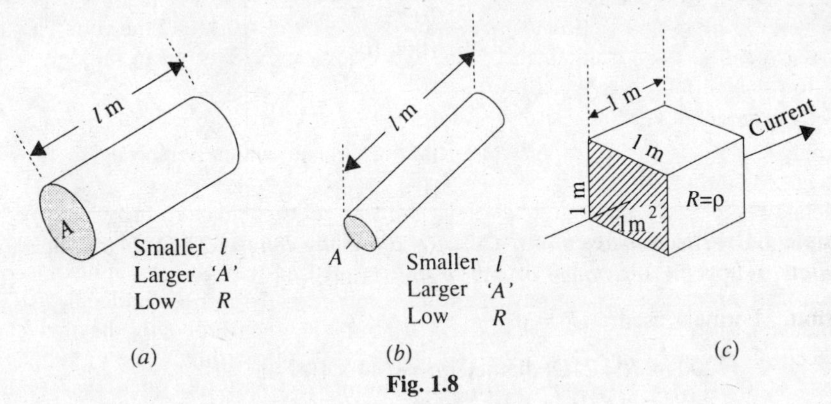

Fig. 1.8

If in Eq. (*i*) above, we put

l = 1 m and A = 1 m² (*i.e.* 1 m³ of a material) then, $R = \rho$ [Fig. 1.8 (*c*)]

Hence, specific resistance of a material may be defined as :

the resistance between the opposite faces of a metre cube of that material.

1.6. Unit of Resistivity

From Eq. (*i*), we have $\rho = \frac{AR}{l}$

In the S.I. system of units, $\rho = \frac{A \text{ metre}^2 \times R \text{ ohm}}{l \text{ metre}} = \frac{AR}{l}$ ohm- metre (Ω-m)

Values of resistivity and temperature coefficients for various materials are given in Table No. 1.2. The resistivities of commercial materials may differ by several per cent due to impurities etc.

TABLE No. 1.2
Resistivities and Temperature Coefficients

Material		*Resistivity in ohm-metre at 20°C × 10^{-8}*	*Temperature coefficient at 20°C × 10^{-4}*
Aluminium	...	2.69	40.3
Brass	...	6–8	20
Carbon	...	7000	–5
Constantan or Eureka	...	49	+ 0.1 to – 0.4
Copper	...	1.72	39.3
German Silver (84 Cu ; 2 Ni ; 1 Zn)	...	20.2	2.7
Iron	...	9.8	65
Manganin (84 Cu ; 12 Mn ; 4 Ni)	...	42– 44	0.25
Mercury		95.8	8.9
Nichrome (80 Cu ; 20 Cr)	...	108.5	1.4
Platinum	...	9 – 15.5	36.7
Silver	...	1.64	38
Tungsten	...	5.5	47
Amber	...	5×10^{14}	
Bakelite	...	10^{10}	
Glass	...	10^{10}–10^{12}	
Mica	...	10^{15}	
Rubber	...	10^{16}	
Shellac	...	10^{14}	
Sulphur	...	10^{15}	

Example 1.1. *The resistance of a Cu wire 200 metre long is 21 Ω. If its thickness is 0.44 mm, calculate its specific resistance in ohm-metre* (Ω-m).

Solution. Formula used : $R = \rho \dfrac{l}{A}$

Here l = 200 m, R = 21Ω, diameter $d = 0.44 \times 10^{-3}$ m³

∴ $A = \pi d^2/4 = p \times (0.44 \times 10^{-3})^2/4 \text{m}^2$

Now, $\rho = \dfrac{AR}{l} = \dfrac{\pi(0.44 \times 10^{-3})^2 \times 21}{4 \times 200} = \mathbf{1.597 \times 10^{-8}}$ **ohm-m**

Example 1.2. *A current of 0.2A is passed through a coil of iron wire which has a cross-sectional area of 0.01 cm². If the resistivity of iron is 14 ×10^{-8} Ω–m and the p.d. across the ends of the coil is 21 volts, what is the length of the wire? What is the conductivity of the wire and the rate of generation of heat in the coil? Take J = 4200 J/kcal.*

Solution. Resistance of the coil R = 21/ 0.2 = 105Ω

$A = 0.01 \text{ cm}^2 = 0.01 \times 10^{-4} \text{ m}^2$; $\rho = 14 \times 10^{-8}$ Ω–m

Now $R = \rho \frac{l}{A}$

$\therefore \quad l = \frac{AR}{\rho} = \frac{105 \times 0.01 \times 10^{-4}}{14 \times 10^{-8}} = \mathbf{750\ m}$

Conductivity $\sigma = \frac{1}{\rho} = \frac{1}{14 \times 10^{-8}} = \frac{10^8}{14} \cdot \frac{1}{\text{ohm} - \text{m}} = \mathbf{7.14 \times 10^6}$ **siemens /metre**

Heat produced/second $= \frac{VI}{4200} = \frac{21 \times 0.2}{4200} = \mathbf{10^{-3}}$ **kcal /s.**

Example 1.3. *Two wires, having equal lengths and made of the same material, have resistances of 25 Ω and 49 Ω respectively. Find their relative diameters.*

Solution $R = \rho \frac{l}{A} = \rho \frac{l}{\pi d^2/4}$

Since the lengths of the wires and their material are the same, ρ and l are constant,

$\therefore \quad R \propto \frac{1}{d^2}$ —where d is the diameter of the wire.

$\therefore \quad R_1 \propto \frac{1}{d_1^2}$ and $R_2 \propto \frac{1}{d_2^2} \quad \therefore \frac{R_2}{R_1} = \frac{d_1^2}{d_2^2}$

$\therefore \quad \frac{d_1}{d_2} = \sqrt{\frac{R_2}{R_1}} = \sqrt{\frac{49}{25}} = \frac{7}{5}$; $\quad \mathbf{d_1 = 1.4\ d_2}$

1.7. Conductance (G) and Conductivity (σ)

Conductance (G) is reciprocal of resistance.* Whereas resistance of a conductor measures the *opposition* which it offers to the flow of current, the conductance measures the *inducement* which it offers to its flow.

From Eq. (*i*) of Art. 1.5, we have $R = \rho \frac{l}{A}$

$$\therefore \quad G = \frac{1}{\rho} \cdot \frac{A}{l} = \frac{\sigma A}{l} \quad ...(i)$$

where σ is called the *conductivity* or *specific conductance* of a conductor. Its unit is Siemens/metre (S/m). The unit of conductance is Siemens (S) whereas the old unit was mho.

1.8. Effect of Temperature on Resistance

The effect of rise in temperature is :

1. to *increase* the resistance of the pure metals. The increase is large and fairly regular for normal ranges of temperature. The temperature / resistance graph is a straight line (Fig. 1.9). As would be presently clarified, metals have a positive temperature-coefficient of resistance.
2. to *increase* the resistance of alloys, though, in their case, the increase is relatively small and irregular. For some high resistance alloys like Eureka (60% Cu and 40% Ni) and manganin, the increase in resistance is (or can be made) negligible over a considerable range of temperature.
3. to *decrease* the resistance of electroytes, insulators (such as paper, rubber, glass, mica etc.) and partial conductors, such as carbon. Hence, insulators are said to possess a *negative* temperature-coefficient of resistance.

1.9. Temperature-coefficient of Resistance

Let a metallic conductor having a resistance of R_0 at 0°C be heated to t°C and let its resistance at this temperature be R_t. Then, considering normal ranges of temperature, it is found that the increase in resistance $\Delta R = R_t - R_0$ depends

*In a.c. circuits, it has a slightly different meaning.

1. directly on its initial resistance,
2. directly on the rise in temperature,
3. on the nature of the material of the conductor.

or $R_t - R_0 \propto R_0 \times t$ or $R_t - R_0 = \alpha R_0 t$ ...(*i*)

where α (alpha) is a constant and is known as the *temperature coefficient of resistance* of that conductor.

Rearranging Eq. (*i*) above, we get

$$\alpha = \frac{R_t - R_0}{R_0 \times t} = \frac{\Delta R}{R_0 \times t}$$

If $R_0 = 1\Omega$, $t = 1°C$, then $\alpha = \Delta R$

Hence, temperature-coefficient at 0°C may be defined as

the change in resistance per ohm per degree change in temperature from 0°C.

Now, $\Delta R / R_0$ represents the *fractional* change in resistance *i.e.*, the change in resistance expressed as fraction of the original resistance. Hence, temperature-coefficient of resistance may also be defined as :

the fractional change in resistance per degree change in temperature from 0°C.

From Eq. (*i*) above, we find that $R_t = R_0 (1 + \alpha t)$

Since this formula is true both for rise and fall in temperature, in general, it may be written as

$$R_t = R_0 (1 \pm \alpha t)$$

It should be remembered that the above equation holds good both for rise and fall in temperature. As temperature of a conductor is decreased, its resistance is also decreased. In Fig. 1.9 is shown the temperature/resistance graph for copper, which is practically a straight line. If this line was extended backwards, it would cut the temperature-axis at a point where temperature is –234.5°C (a number quite easy to remember). It means that, theoretically, the resistance of a copper conductor will become zero at this point though as shown by solid line, in practice, the curve departs from a straight line at very low temperatures. From the two similar triangles of Fig. 1.9, it is seen that

$$\frac{R_t}{R_0} = \frac{t + 234.5}{234.5} = \left(1 + \frac{t}{234.5}\right)$$

$$R_t = R_0\left(1 + \frac{t}{234.5}\right) \text{ or } R_t = R_0 (1 + \alpha t)$$

where, $\alpha = 1/234.5$ for copper.

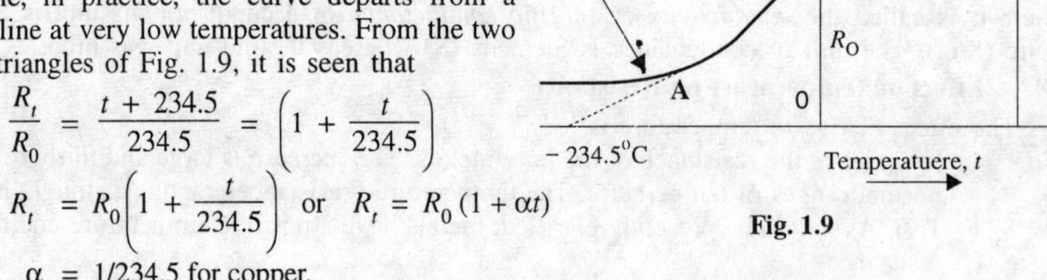

Fig. 1.9

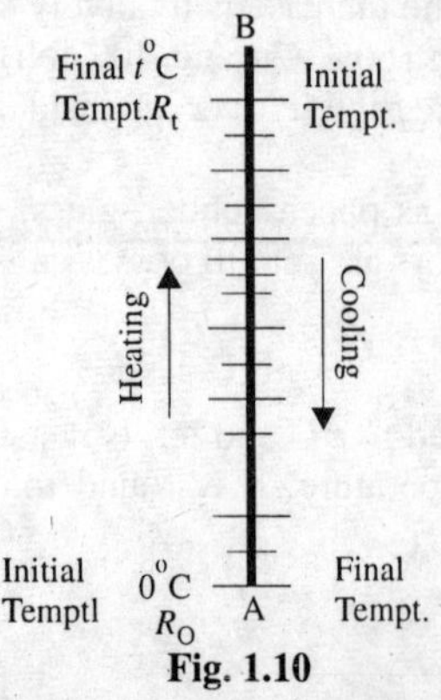

Fig. 1.10

1.10. Value of α at Different Temperatures

So far we did not make any distinction between values of α at different temperatures. But it is found that value of α itself is not constant, but depends on the initial temperature on which the increment in resistance is based. When the increment is based on the resistance measured at 0°C, then it has the value of α_0. At any other initial temperature t°C, value of α is α_t and so on. It should be remembered that, for any conductor, α_0 has the maximum value.

With reference to Fig. 1.10, the value of R_0 in terms of R_t during cooling is given by

$$R_0 = R_t [1 + \alpha_t (-t)] = R_t(1 - \alpha_t t) \quad ...(i)$$

where α_t is the temperature coefficient at t°C. It should be noted that here the initial starting point is B and temperature is decreased from t°C to 0°C at point A.
Now, taking point A as the initial point, we get

$$R_t = R_0(1 + \alpha_0 t) \quad ...(ii)$$

where α_0 is temperature -coefficient as referred to 0°C.

From Eq. (*i*) above, we have

$$\alpha_t = \frac{R_t - R_0}{R_t \times t}$$

Substituting the value of R_t fromEq. (*ii*) above, we get

$$\alpha_t = \frac{R_0(1 + \alpha_0 t) - R_0}{R_0(1 + \alpha_0 t) \times t} = \frac{\alpha_0}{1 + \alpha_0 t}$$

$$\therefore \quad \alpha_t = \frac{\alpha_0}{1 + \alpha_0 t} \quad ...(iii)$$

In general, let

α_1 = tempt-coefficient at t_1°C ; α_2 = tempt-coefficient at t_2°C

Then, from Eq. (*iii*) above, we get

$$= \frac{\alpha_0}{1 + \alpha_0 t_1} \text{ or } \frac{1}{\alpha_1} = \frac{1 + \alpha_0 t_1}{\alpha_0}$$

Similarly, $$\frac{1}{\alpha_1} = \frac{1 + \alpha_0 t_1}{\alpha_o}$$

Substracting one from the other, we get

$$\frac{1}{\alpha_2} - \frac{1}{\alpha_1} = (t_2 - t_1) \text{ or } \frac{1}{\alpha_2} = \frac{1}{\alpha_1} + (t_2 - t_1)$$

$$\therefore \quad \alpha_2 = \frac{1}{1/\alpha_1 + (t_2 - t_1)}$$

Values of α for copper at different temperatures are given in Table No. 1.3.

TABLE No. 1.3

Different Values of α for Copper

Tempt. in °C	0	5	10	20	30	40	50
α	0.00427	0.00418	0.00409	0.00393	0.00378	0.00364	0.00352

In view of the dependence of α on the initial temperature, we may define the temperature coefficient of resistance at a given temperature as the change in resistance per ohm per degree centigrade change in temperature from the given temperature.

In case R_0 is not given, then relation between the known resistance R_1 at t_1°C and the unknown resistance R_2 at t_2°C can be found as follows :

$$R_2 = R_0(1 + \alpha_0 t_2) \quad \text{and} \quad R_1 = R_0(1 + \alpha_0 t_1)$$

$$\frac{R_2}{R_1} = \frac{1 + \alpha_0 t_2}{1 + \alpha_0 t_1} \quad ...(iv)$$

The above expression can be simplified by a little approximation as follows :

$$\frac{R_2}{R_1} = (1 + \alpha_0 t_2)(1 + \alpha_0 t_1)^{-1}$$

$= (1 + \alpha_0 t_2)(1 - \alpha_0 t_1)$ [Using Binomial Theorem for expansion and neglecting squares and higher powers of $(\alpha_0 t_1)$].

$= 1 + \alpha_0 (t_2 - t_1)$ [Neglecting product $\alpha_{02} t_1 t_2$]

$\therefore \quad R_2 = R_1[1 + \alpha_0(t_2 - t_1)$ ...(*v*)

For more accurate calculations, however, Eq. (*iv*) should be used.

Note. We could also use the equation $R_2 = R_1[1 + \alpha_1(t_2 - t_1)]$

1.11. Variation of Resistivity with Temperature

Not only resistance but specific resistance or resistivity of metallic conductors also increases with rise in temperature and *vice versa*. The relation between ρ_0 and ρ_t within normal ranges of temperature is given by

$$\rho_t = \rho_0(1 + \alpha_0 t)$$

where ρ_0 = resistivity at 0°C

ρ_t = resistivity at t°C.

Example 1.4. *The base of an incandescent lamp with a tungsten filament is marked 120-V, 60-W. Measurement on a Wheatstone bridge of the resistance of the lamp at 20°C indicates 20Ω. What is the normal temperature of incandescence, if the resistance-temperature coefficient of tungsten is 5 × 10⁻³ per °C at 20°C ?*

Solution. Working current of the lamp is = 60/120 = 0.5 A

Hot resistance of the lamp = 120/0.5 = 240 Ω

Let t°C be the normal working temperature of the lamp.

Then $R_t = 240\Omega$; $R_{20} = 20\Omega$; $\alpha_{20} = 5 \times 10^{-3}/°C$

Now $R_t = R_{20}[1 + \alpha_{20}(t - 20)]$

$\therefore \quad 240 = 20[1 + 5 \times 10^{-3}(t - 20)]$; $t =$ **2220°C**

Example 1.5. A conductor has a cross-section of 10 cm² and specific resistance of 7.5 μΩ–cm at 0°C. What will be its resistance in ohm per km when the temperature is 40°C? Take the temp-coeff. of the material = 0.005 per °C.

Solution. It will be assumed that tempt. coeff. of 0.005 per °C represents α_0.

$$\rho_{40} = \rho_0(1 + 40\,\alpha_0) = 7.5\,(1 + 40 \times 0.005) = 9\mu\Omega\text{–cm} = 9 \times 10^{-5}\ \Omega\text{–m}$$

$$R_{40} = \rho_{40}\frac{l}{A}$$

Here, $l = 1\text{ km} = 10^3\text{ m}$: $A = 10\text{ cm}^2 = 10 \times 10^{-4} = 10^{-3}\text{ m}^2$

$R_{40} = 9 \times 10^{-8} \times 10^3/10^{-3} =$ **0.09 Ω**

Example 1.6. *A platinum coil has a resistance of 3.146 Ω at 40°C and 3.767 Ω at 100°C. Find the resistance at 0°C and the temperature coefficient of resistance at 40°C.*

(Electric Circuits–I, Punjab Univ. 1994)

Solution. $R_{100} = R_0(1 + 100\,\alpha_0)$ ...(*i*)

$R_{40} = R_0(1 + 40\,\alpha_0)$ ...(*ii*)

$$\therefore \quad \frac{3.767}{3.146} = \frac{1 + 100\alpha_0}{1 + 40\alpha_0}, \qquad \alpha_0 = 0.00379 \text{ or } 1.240 \text{ per °C}$$

From (*i*) we have

$$3.767 = R_0(1 + 100 \times 0.00379), \qquad \therefore \quad R_0 = 2.732\ \Omega$$

Now, $$\alpha_{40} = \frac{\alpha_0}{1 + 40\alpha_0} = \frac{0.00379}{1 + 40 \times 0.00379} + \frac{1}{280} \text{ per °C}$$

Example 1.7. *An aluminium resistor has a resistance of 43.6 ohms at 20°C and 47.2 ohms at 40°C. Calculate the temperature coefficients of resistance.* **(A.M.I.E. Summer 1990)**

Solution. Formula used : $R_t = R_0\,(1 + \alpha_0\, t)$

$$R_{20} = 43.6 = R_0\,(1 + 20\,\alpha_0)$$

$$R_{40} = 47.2 = R_0\,(1 + 40\,\alpha_0)$$

$$\therefore \quad \frac{47.2}{43.6} = \frac{1 + 40\,\alpha_0}{1 + 20\,\alpha_0} \;; \quad \alpha_0 = 0.045 \text{ per °C.}$$

Example 1.8. *Two coils connected in series have resistances of 600Ω and 300Ω and tempt. coeff. of 0.1% and 0.4% respectively at 20°C. Find the resistance of the combination at a tempt. of 50°C. What is the effective temperature coeff. of combination ?*

Solution. Resistance of 600 Ω resistor at 50°C is $= 600[1 + 0.001\,(50 - 20)] = 618\ \Omega$

Similarly, the resistance of 300W resistor at 50°C is $= 300[1 + 0.004\,(50 - 20)] = 336\Omega$

Hence, total resistance of combination at 50° C is $= 618 + 336 =$ **954 Ω**

Let β = resistance-temperature coefficient at 20°C

Now, combination resistance at 20°C = 900 Ω

Combination resistance at 50°C = 954 Ω

$\therefore \quad 954 = 900[1 + \beta(50 - 20)] \quad \therefore \quad \beta =$ **0.002 per ° C**

Example 1.9. *By what percent will the resistance of a copper conductor 100 m long, cross-section 0.75 mm² increase when heated from 20°C to 100°C. Take α_0 = 0.00393 per °C and specific resistance at 20°C = 1.724 × 10⁻⁸ Ω –m for Copper.*

(Elements of Elect. Engg.-II, Punjab Univ. 1993)

Solution. $R_{20} = \rho_{20}\ l/A = 1.724 \times 10^{-8} \times 1000/0.75 \times 10^{6} = 23\Omega,\ \rho_{100} = \rho_{20}\,[1 + \alpha_0\,(100 - 20)]$

$$= 1.724 \times 10^{-8}\,(1 + 0.00393 \times 80) = 2.266 \times 10^{-8}$$

$$R_{100} = \rho_{100}\ \rho/A = 2.226 \times 10^{-8} \times 1000/0.75 \times 10^{-6} = 30.2\Omega$$

% change $= (30.2 - 23)/23 = 0.313$ or **31.3%**

Example 1.10. *A coil of copper wire has resistance of 90 ohm at 20°C and is connected to a 230 V supply. By how much should the voltage be increased to keep the current constant if the temperature of the coil rises to 60°C? Take the coefficient of resistance of copper as 0.00428 per °C at 0°C.* **(Electric Circuits-I, Punjab Univ. 1993)**

Solution. Current at 20°C = 230/90 = 2.555A

The resistance of the Cu wire at 60°C can be found from the relation

$$R_{60} = R_{20}\,[1 + \alpha_0\,(60 - 20)] = 90\,(1 + 0.00428 \times 40) = 105.4\ \Omega$$

Voltage required at 60°C is

$$V = I \times R_{60} = 2.55 \times 105.4 = 269.3\text{V}$$

$\therefore$ increase in voltage $= 269.3 - 230 =$ **39.3V**

Example 1.11. *Two materials have resistance-temperature coefficient of 0.004 and 0.0002 respectively at a given temperature. In what ratio should the wires made of these materials be connected in series so as to have an overall tempt-coefficient of 0.002?*

Solution. Let R_1 and R_2 be the resistances of wires of the two given materials which are to be joined in series. Their ratio may be found by simple technique illustrated in Fig. 1.11.

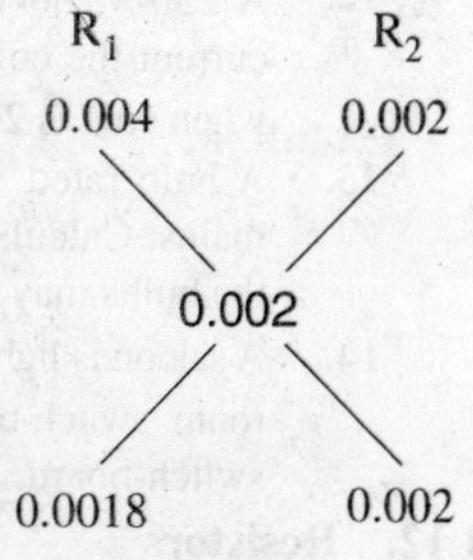

$$R_1/R_2 = 0.0018/0.002 = \mathbf{9/10}$$

Fig. 1.11

Hence, R_1 is 9/10 of R_2. If resistance per unit length of the two wires was given, we could find the ratio of their respective lengths also.

Tutorial Problems No. 1.1

1. Calculate the resistance of a 100 m length of wire having a uniform cross-section of 0.1 mm^2 if the wire is made of manganin having a resistivity of $50 \times 10^{-8} \Omega$ –m.
If the wire is drawn out to three times its original length, by how many times would you expect its resistance to be increased? **[500 Ω ; 9 times]**
2. A cube of material of side one cm has a resistance of 0.001 Ω between its opposite aces. If the same volume of the material has a length of 8 cm and a uniform cross-section, what will be the resistance of the length? **[0.064 Ω]**
3. A lead wire and an aluminium wire are connected together in parallel. The currents flowing in the respective wires are in the ratio of 39 : 40.The lead wire is 65 per cent longer than the aluminium wire and the ratio of their specific resistances is 98 : 13. Find the ratio of their cross-sectional areas. **[12.22 : 1]**
4. A coil of Cu wire has a resistance of 25 Ω at a temperature of 15°C, a coil of carbon wire has a resistance of 25.5 Ω at a temperature of 15°C. At what temperature will the resistance be the same in both coils? α for Cu = 4.28×10^{-3} per °C, α for carbon = – 0.52×10^{-3} per °C. **[19.4°C]**
5. An electric radiator is required to dissipate 1 kW when connected to a 230 V supply. If the coils of the radiator are of wire 0.5 mm in diameter having resistivity of 60×10^{-8} Ω –m, calculate the necessary length of wire. **[17.32 m]**
6. Ten cm^3 of copper are (*a*) drawn into a wire 100 m long (*b*) rolled into a square sheet of 10 cm side. Find the resistance of the wire and the resistance between opposite faces of the plate if the specific resistance of the copper is 1.7×10^{-8} Ω –μ. **[(a) 17 Ω (b) 1.7 × 10–9 Ω]**
7. Two conductors, one of copper and the other of iron, are connected in parallel and at 20°C carry equal currents. What proportion of current will pass through each if the temperature is raised to 100°C? For copper α = 0.0043/°C and for iron α = 0.0063/°C. **[52.8% ; 47.2%]**
8. An aluminium wire 5 metre long and 2 mm diameter is connected in parallel with a copper wire 3 metre long. The total current is 4 A and that in the aluminium wire is 2.5A. Find the diameter of the copper wire. The respective resistivities of copper and aluminium are 1.7 and 2.6 μΩ-cm. **[0.97 mm]**
9. A potential difference of 200 V is applied to a copper field coil at a temperature of 15°C and the current is 10A. What will be the mean temperature of the coil when the current has fallen to 5A, the applied voltage being the same as before? Given a = 1/ 234.5 per °C at 0°C. **[264.5°C]**
10. The current through an electriccal conductor is 1A when the temperature of the conductor is 0°C and 0.7A when the temperature is 100°C. What would be the current when the temperature of the conductor is 1200°C and what is the temperature coefficient of resistance of the conductor ? **[α = 0.0043/°C ; 0.16A]**
11. The shunt winding of a certain d.c. generator has a resistance of 135 Ω at a temperature of 15°C. Given that α for copper = 4.28×10^{-3} per °C, calculate the resistance of the winding when the temperature has reached 52°C. **[155 Ω]**
12. A 230-V, 60-W lamp has a filament whose normal working temperature is 2000°C. Find the current the cold lamp will take when first switched on. Assume that the temperature of the lamp when cold is 20°C and α for the filament is 0.0045 per °C. **[2.58A]**
13. A bulb rated 110-V, 60-W is connected with another bulb rated 110-V, 100-W across a 220-V mains. Calculate the resistance which should be joined in parallel with the first bulb so that both the bulbs may take their rated power. **[302.5 Ω]**
14. A saloon is lighted by 30 lamps marked 110-V, 100-W. The distance between the saloon and the engine room switch-board is 55 m. If cable resistance is 1.2 Ω /km, find the voltage necessary at switch-board. **[113.6V]**

1.12. Resistors

Different resistors, whether fixed or of variable type, are commonly made of metals and alloys. The resistance of an incandescent lamp is due to its coil, which is made of tungsten wire. The heating elements of toasters, iron etc. are made of nickel-chromium alloy in wire or ribbon form. The small-sized colour-banded resistors employed in radio and TV receivers use a carbon composition as the resistance material.

A rheostat is a variable wire resistance having two connections, one fixed and the other movable.

Those variable resistors which have three terminals, two at the ends and one movable contact are called potentiometers.

1.13. Linear and Non-linear Resistors

Linear resistors are those in which current produced is *directly* proportional to the applied voltage. Their current versus voltage graph is straight and linear. In other words, their resistance remains constant.

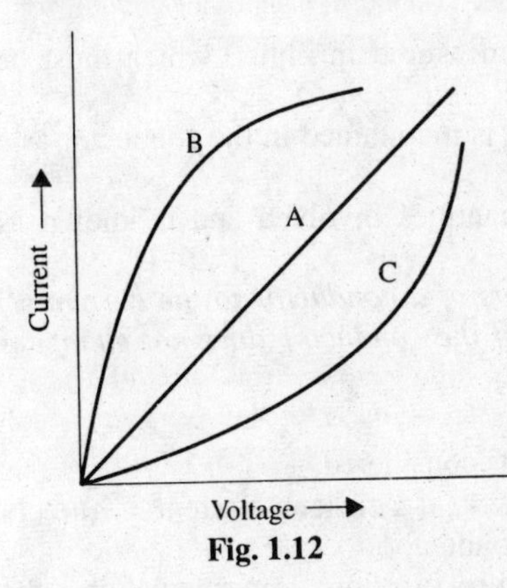

Fig. 1.12

Non-linear resistors are those whose current does not change linearly with changes in applied voltage. It is so because current flow always results in the production of heat, which either increases (as in metals) or decreases their resistance (as in insulators). Because of this change in resistance, the current through such a resistor is not directly proportional to the impressed voltage. Circuits of this type are called *non-linear circuits* and the current-carrying elements are known as *non-linear resistors*. In Fig. 1.12, *A* represents the case of a linear resistor, whereas *B* and *C* represent non-linear resistors. In case of *B*, current increases more than proportionately with applied voltage, which means that for this material, resistance decreases with rise in temperature. It is examplified by Thyrite, which is a non-metallic non-linear resistor.

Curve *C* also represents a non-linear resistor but one whose resistance increases more than proportionately with applied voltage. Example of such a resistor is a semi-conductor material chemically similar to ceramic oxide.

1.14. Uses of Non-linear Resistors

Mostly, they are used as protective devices in various electrical circuits, as explained below.

(*i*) **Thyrite.** It is non-metallic and is made by combining silicon carbide with a binder under high pressure and temperature. Electric contacts are made to the material by spraying a metal coating on its both surfaces. At normal temperatures, the resistance of Thyrite is extremely high. However, when impressed voltage is doubled, its current increases 10 to 100 times. Because of this property, a Thyrite resistor is connected directly across the winding terminals of a motor or a generator in order to limit the induced discharge e.m.f. and thereby protect the winding insulation against failure, when the impressed voltage is suddenly withdrawn.

(*ii*) **Thermistor.** Its resistance is comparatively low at normal temperatures and remains nearly constant up to a certain critical point, beyond which it rises sharply. Usually, a thermistor (about the size of an aspirin tablet) is kept in direct contact with a motor winding and is connected into a control circuit. For normal operating temperature, its resistance is low enough to permit motor operation. But as soon as temperature exceeds permissible limit, its resistance increases sharply which results in opening the control circuit, thereby stopping the motor.

1.15. Stranded Wires and Cables

Stranding of wires and cables is done for achieving greater flexibility. Following strandings are employed :

1. **Bunch Stranding.** A bunch-stranded wire is nothing else but a collection of ordinary wires without any particular geometrical pattern.

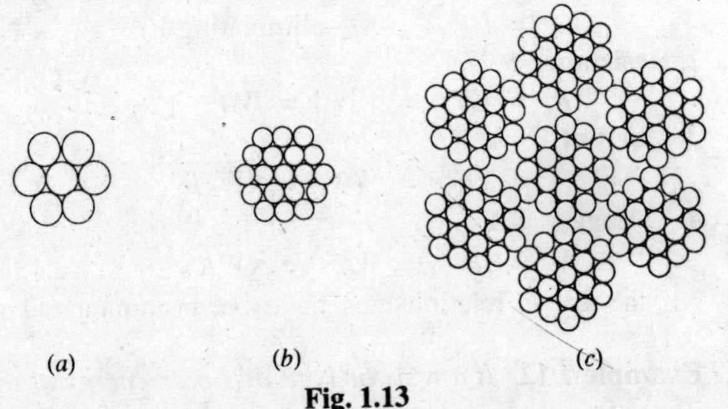

Fig. 1.13

2. Concentric Stranding. In this case, a centre wire is stranded by one or several definite layers of wires. It is so arranged that each layer contains six more wires than the layer immediately below it [Fig. 1.13 (*a*)]. When several layers are employed, each layer has a twist opposite to that of the layer beneath it [Fig. 1.13 (*b*)].

3. Rope-stranded Cable. It is a concentric-stranded assembly made up of several concentric cables twisted together [Fig. 1.13 (*c*)].

1.16. Ohm's Law

Whenever electric current flows through a conductor, the following three factors are present:

1. The pressure or potential difference V across the conductor (measured in volts) causing the current to flow.
2. The opposition or resistance R of the conductor (measured in ohms) which must be overcome.
3. The current strength I (measured in amperes) which is maintained in the conductor as a result of pressure overcoming the resistance.

There exists a definite relationship between the three quantities involved and is known as **Ohm's Law**. It may be stated thus :

the ratio of potential difference (V) between any two points of a conductor to the current (I) flowing between them is constant, provided the temperature of the conductor does not change.

In other words, $\frac{V}{I}$ = constant or $\frac{V}{I} = R$

where R is the resistance of the conductor between the two points considered.

Put in another way, it simply means that provided R is kept constant, current is direcly proportional to the potential difference across the ends of a conductor.

If V is measured in volts and I in amperes, then R is given in ohms. For calculations, the following three forms of the Ohm's law should be remembered.

1. $I = \frac{V}{R}$ *i.e.*, current $= \frac{\text{potential difference}}{\text{resistance}}$
2. $V = IR$ *i.e.*, p.d. = current ö resistance
3. $R = \frac{V}{I}$ *i.e.*, resistance $= \frac{\text{p.d.}}{\text{current}}$

This law is applicable not only to d.c. circuit but to a.c. circuit as well, provided account is taken of the induced e.m.f. resulting from the self-inductance of the circuit and of the distribution of current in the cross-section of the circuit.

1.17. Relations Derived from Ohm's Law

Following additional relationships connected directly or indirectly with Ohm's law are worth nothing.

1. Power. It is given by the product of voltage (V) and current (I)

$$W = VI$$

Its unit is watt.

Other forms of the above formula are

$$W = V^2/R \quad \text{—eliminating } I$$
$$= I^2R \quad \text{—eliminating } V$$

2. Resistance

$$R = V/I = V^2/W = W/I^2$$

3. Current

$$I = V/R = W/V = \sqrt{W/R}$$

4. Voltage

$$V = IR = W/I = \sqrt{WR}$$

All the above relationships have been summarized in Fig. 1.14.

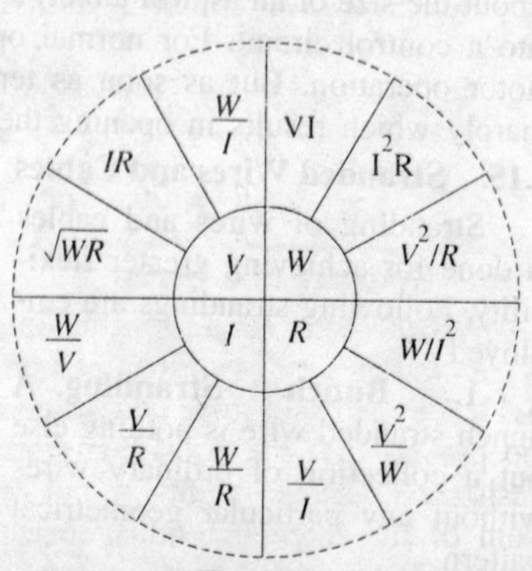

Fig. 1.14

Example 1.12. *If a resistor is to dissipate energy at the* *rate of 250 W, find ıts resistance for a terminal voltage of 100 V.* **(A.M.I.E. Winter 1986)**

Solution. Here ; $W = 250$ watt, $V = 100$ volt ; $R = ?$

Now, $W = \frac{V^2}{I}$ or $R = \frac{V^2}{I} = \frac{100^2}{250} = \mathbf{40\Omega}$

Example 1.13. *In the bridge circuit of Fig. 1.15 calculate the reading of a voltmeter connected across (i) AB (ii) BC (iii) AD (iv) DC and (v) BD.*

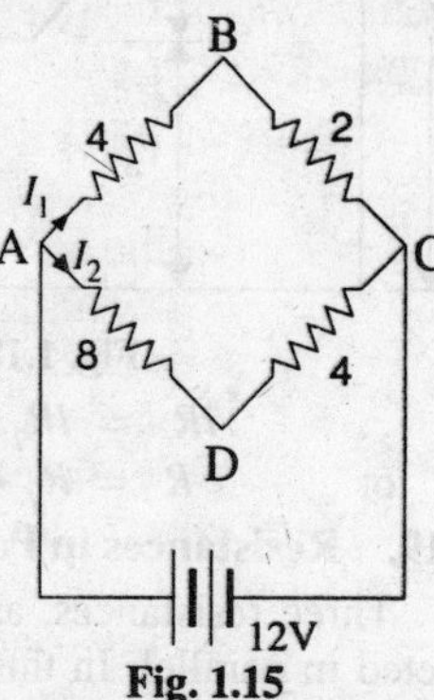

Fig. 1.15

Solution. The two branches *ABC* and *ADC* consist of two resistors connected in series. These two branches are connected in parallel across the 12-V battery. Taking branch *ABC* and applying Ohm's law, we have

$I_1 = 12/6 = 2\text{A}$

(*i*) $V_{AB} = I_1 \times R_{AB} = 2 \times 4 = \mathbf{8V}$

(*ii*) $V_{BC} = I_1 \times R_{BC} = 2 \times 2 = \mathbf{4V}$

$I_2 = 12/12 = 1\text{A}$

(*iii*) $V_{AD} = I_2 \times R_{AD} = 1 \times 9 = \mathbf{8V}$

(*iv*) $V_{DC} = I_2 \times R_{DC} = 1 \times 4 = \mathbf{4V}$

(*v*) Since $V_{AB} = V_{AD}$, the two points *B* and *D* are at the same potential. Hence, a voltmeter connected across them will read zero.

Example 1.14. *An incandescent projector lamp has the rated voltage of 60 volts and hot resistance of 20Ω. Find the series resistance required to operate the lamp from a 75 volt supply.*

Solution. Rated value of lamp current = 60/20 = 3A

Let *R* be the required series resistance [Fig. 1.16 (*a*)].

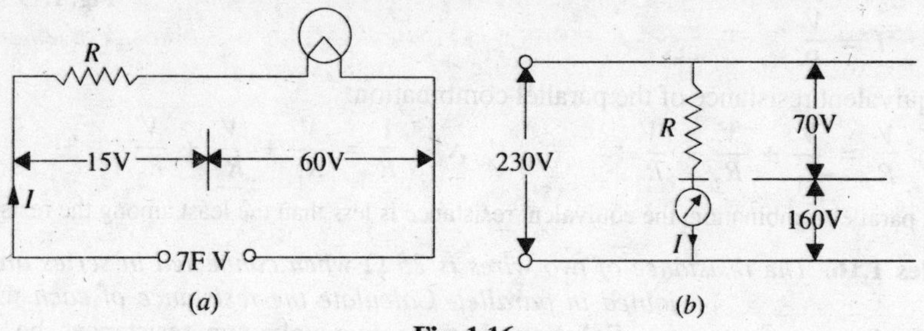

Fig. 1.16

Then, the excess voltage of (75-60) = 15V is to be dropped on *A*.

$\therefore \quad R = 15/3 = \mathbf{5\Omega}$

Example 1.15. *A voltmeter has a resistance of 20,000 Ω. When connected in series with an external resistance across a 230-V supply, the instrument reads 160V. What is the value of external resistance?*

Solution. The circuit is shown in Fig. 1.16 (*b*). The voltage drop across external resistance $R = 230 - 160 = 70$ *V*.

Circuit current $I = 160/20{,}000 = 1/125$ A

Now $IR = 70$

$\therefore \quad \frac{1}{125} \times R = 70$ or $R = \mathbf{8{,}750\ \Omega}$

1.18. Resistances in Series

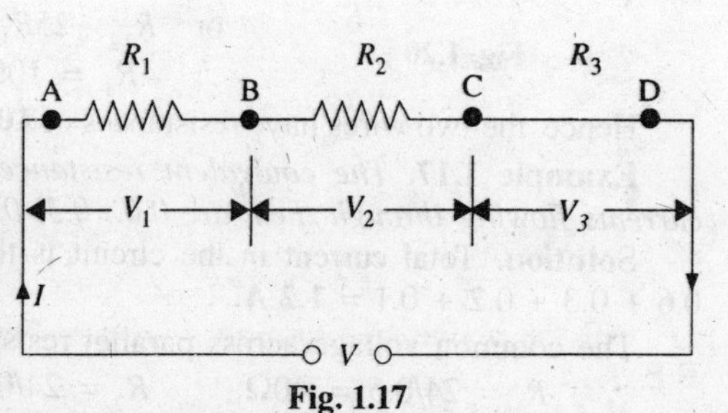

Fig. 1.17

When some conductors having resistances of R_1, R_2 and R_3 etc., are joined end-on-end, as in Fig. 1.17, they are said to be connected in series. It can be proved that the equivalent resistance or total resistance between points. *A* and *D* is equal to the sum of the three individual resistances. Being a

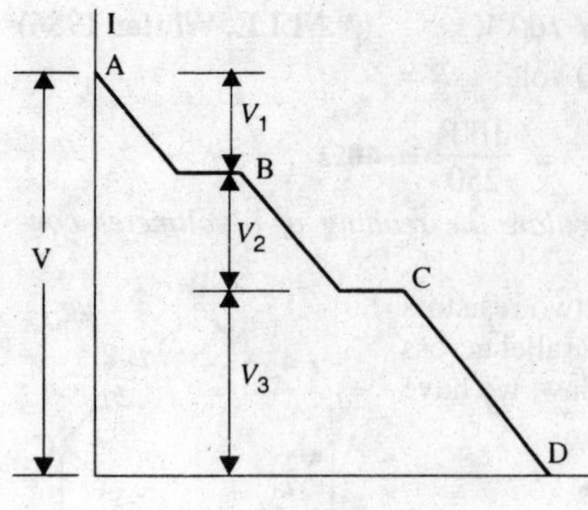

Fig. 1.18

Series circuit, it should be remembered that (*i*) same current passes through all the three conductors (*ii*) but voltage drop across each is different due to its resistance being different and is given by Ohm's law and (*iii*) sum of the three voltage drops is equal to the voltage applied across the three conductors (Fig. 1.18). There is a progressive fall in potential as we go from point *A* to *D* as shown in Fig. 1.19.

$$\therefore \quad V = V_1 + V_2 + V_3 = IR_1 + IR_2 + IR_3 \quad \text{...Ohm's Law}$$

But, $V = IR$

where *R* is the equivalent resistance of the series combination.

$$\therefore \quad IR = IR_1 + IR_2 + IR_3$$

$$\text{or} \quad R = R_1 + R_2 + R_3$$

1.19. Resistances in Parallel

Three resistances, as joined in Fig. 1.19, are said to be connected in parallel. In this case (*i*) p.d. across all resistances is the same (*ii*) current in each resistor is different and is given by Ohm's law and (*iii*) the total current is the sum of the three separate branch currents.

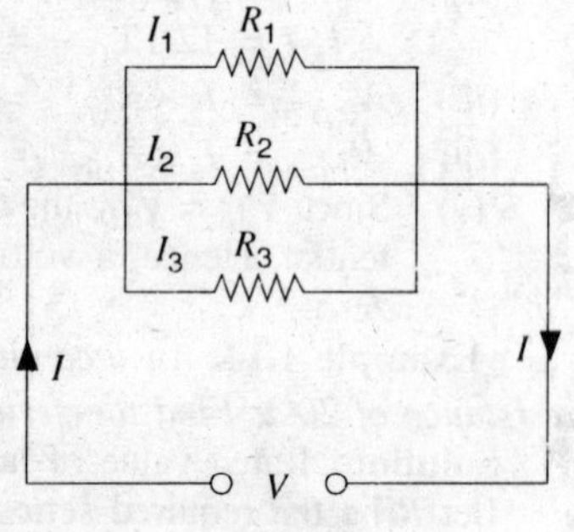

Fig. 1.19

$$\therefore \quad I = I_1 + I_2 + I_3 = \frac{V}{R_1} + \frac{V}{R_2} + \frac{V}{R_3}$$

Now, $I = \frac{V}{R}$

where R = equivalent resistance of the parallel combination.

$$\therefore \quad \frac{V}{R} = \frac{V}{R_1} + \frac{V}{R_2} + \frac{V}{R_3}; \qquad \therefore \quad \frac{1}{R} = \frac{V}{R_1} + \frac{V}{R_2} + \frac{V}{R_3}$$

Note : In parallel combination, the equivalent resistance is less than the least among the resistors.

Examples 1.16. *The resistance of two wires is 25 Ω when connected in series and 6Ω when joined in parallel. Calculate the resistance of each wire.*

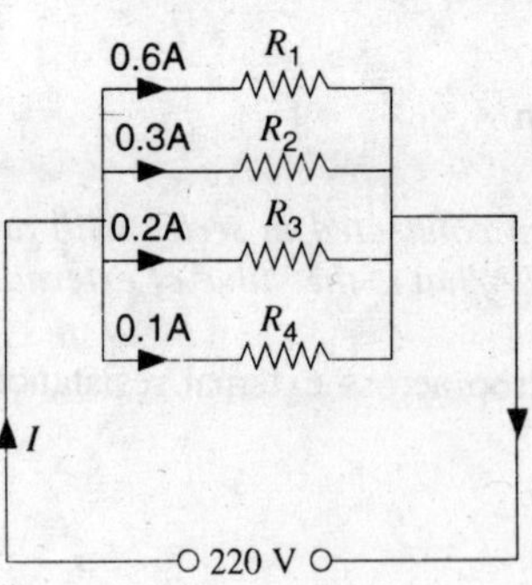

Fig. 1.20

Solution. Let the two unknown resistances be R_1 and R_2. Then, when in series

$$R_1 + R_2 = 25 \quad \text{...(i)}$$

When joined in parallel

$$\frac{1}{R_1} + \frac{1}{R_2} = \frac{1}{6} \quad \text{or, } 6 = \frac{R_1 R_2}{R_1 + R_2} \quad \text{...(ii)}$$

Putting the value of R_2 from Eq. (*i*) in Eq. (*ii*), we have

$$6 = \frac{R_1(25 - R_1)}{25}$$

$$\text{or} \quad R_{12} - 25R_1 + 150 = 0 \qquad \text{or} \quad (R_1 - 15)(R_1 - 10) = 0$$

$$\therefore \quad R_1 = 10\Omega, \text{ so } R_2 = 15\Omega \text{ or } R_1 = 15\Omega, \text{ so } R_2 = 10\Omega$$

Hence, the two wires have resistances of **10Ω** and **15Ω.**

Example 1.17. *The equivalent resistance of four resistors joined in parallel is 20Ω. The currents flowing through them are 0.6., 0.3, 0.2 and 0.1 A. Find the value of each resistor.*

Solution. Total current in the circuit is the sum of the four branch currents. Its value is = 0.6 + 0.3 + 0.2 + 0.1 = **1.2 A.**

The common voltage across parallel resistors = 20 × 1.2 = 24 V (Fig. 1.20)

$$\therefore \quad R_1 = 24/0.6 = 40\Omega; \qquad R_2 = 24/0.3 = 80\Omega$$

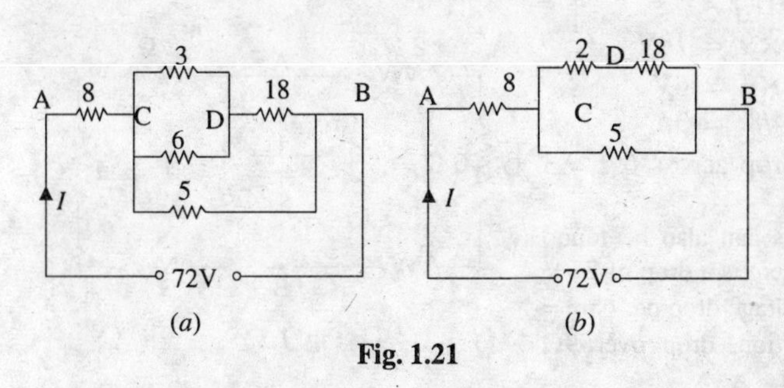

Fig. 1.21

R_3 = 24/0.2 = 120Ω;
R_4 = 24/0.1= 2.40Ω

Example 1.18. *Find the total resistance between points A and B of the circuit shown in Fig. 1.21 (a). Also, calculate the voltage drop across each resistor if a p.d. of 72 V is applied across the two terminals of the circuit. The numerical values represent resistances in ohm.*

Solution. Since, the two resistances of 3 Ω and 6 Ω are in parallel, their combined resistance = 3 || 6 = 3 × 6/ (3 + 6) = 2 Ω as shown in Fig. 1.21 (*b*)

The two series-connected resistances of 2Ω and 18Ω can be reduced to a single resistance of 20Ω as shown in Fig. 1.22 (*a*)

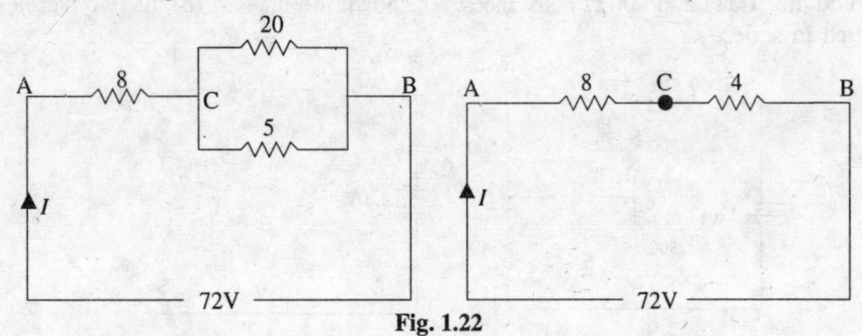

Fig. 1.22

Since the 20Ω and 5Ω resistances are in parallel, their equivalent resistance = 20 || 5 = **4 Ω** as shown in Fig. 1.22 (*b*). Obviously, total resistance between points *A* and *B* = 8 + 4 = 12 Ω

Total circuit current I = 72/12 = 6A

We would use the voltage divider formulae of Art. 1.20 to find the voltage drop across the two resistances of Fig. 1.22 (*b*).

Drop across 8Ω resistance = 72 × 8/ (8 + 4) = **48 V**

Drop across 4 Ω resistance = 72 – 48 = 24 V

(or drop = 72 × 4/12 = 24 V)

Now, this 4 Ω resistance represents the parallel combination of 20Ω and 5Ω resistances in Fig. 1.22 (*a*)

∴ drop across 5Ω resistance = **48 V**

Now, drop across 20Ω resistance is also 48Ω. However, this 20Ω resistance represents two series-connected resistances of 2Ω and 18Ω as shown in Fig. 1.21(*b*). Again, applying voltage divider formula, we get

Drop across 18Ω resistance = 48 × 18/(2 + 18) = **43.2 V**

Drop across 2Ω resistance = 48 – 43.2 = **4.8 V**

(or drop = 48 × 2 /20 = 4.8 V)

Now, the 2Ω resistance represents a parallel combination of 3Ω and 6Ω resistances in Fig. 1.21 (*a*). Hence, drop across each is the same *i.e.*, **4.8 V.**

Example 1.19. *In the circuit shown in Fig. 1.23, find the voltage across and the current in each element.*

Solution. As seen, there are two parallel paths between points *A* and *B*; one path is *ACB* and has a total resistance of (9 + 3) total resistance of (2 + 6) = 8Ω. Both paths have a p.d. of 24 V applied across them.

Current through path *ACB* is = 24/12 = **2A**

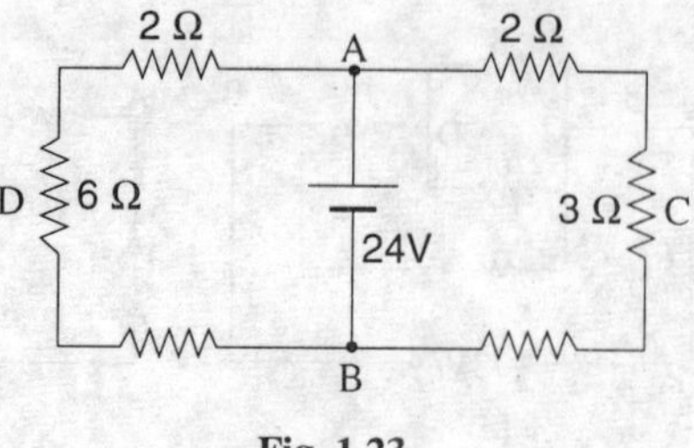

Fig. 1.23

Drop across 9Ω = 2 × 9 = **18V**
Drop across 3Ω = 2 × 3 = **6V**
Current through path *ADB* is = 24/8 = **3A**
Drop across 2Ω = 3 × 2 = 6V ; Drop across 6Ω = 3 × 6 = **18V**

Note : Drop across different elements can also be found without the help of current. For example, there is a drop of 24 V over 12Ω resistance in path *ACB*. Hence, voltage drop per ohm = 24/12 = 2 V/Ω. Accordingly, by proportion, drop over 9Ω resistance = 9 × 2 = 18 V and so on.

Example. 1.20. *What is the current through S [Fig. 1.24 (a)] when closed ?*
(Electrical Science, AMIE Winter 1994)

Solution. When switch *S* is closed, the 200 and 500 Ω resistances become connected in parallel. So do the 300 Ω and 400 Ω resistances. As shown in Fig. 1.24 (*b*) the two parallel groups are connected in series.

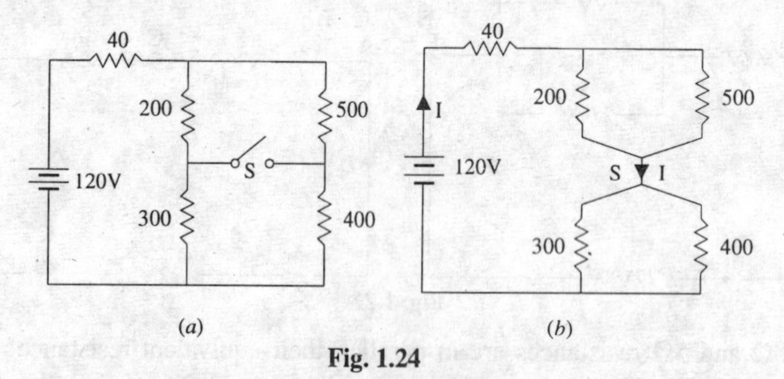

Fig. 1.24

The combined resistances of the two series-connected parallel groups is

$$= 200 \parallel 500 + 300 \parallel 400 = 2200/7\ \Omega$$

Hence, total resistance of the circuit is

$$= 40 + (2200/7) = 2480/7\Omega$$

$$I = 120/(2480/7) = \mathbf{0.339\ A}$$

Example 1.21. *What is the equivalent resistance of the ladder network shown in Fig. 1.25 (i) with the 75Ω load resistor connected, as shown and (ii) with the load resistor disconnected?*
(Electrical Science, AMIE Winter 1994)

Solution. (*i*) With reference to the ladder network of Fig. 1.25, we have

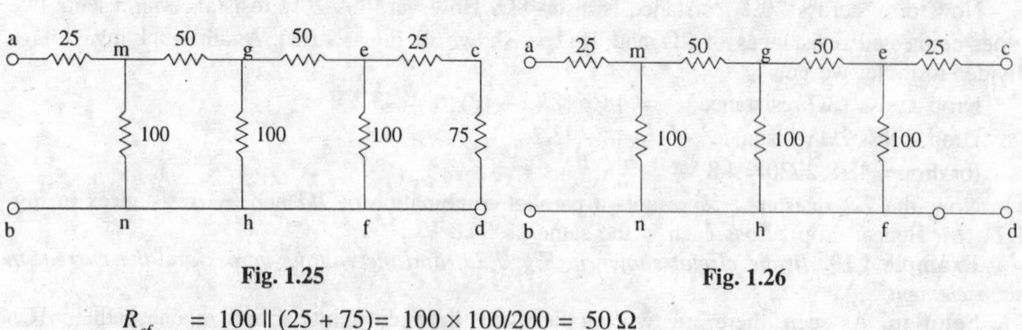

Fig. 1.25 Fig. 1.26

$$R_{ef} = 100 \parallel (25 + 75) = 100 \times 100/200 = 50\ \Omega$$

$$R_{gh} = 100 \parallel (50 + 50) = 50\ \Omega$$

$$R_{mn} = 100 \parallel (50 + 50) = 50\ \Omega$$

$$\therefore\ R_{ab} = (25 + 50) = 75\ \Omega$$

(ii) With Load Resistor Disconnected

When the load resistor is disconnected, the output terminals *c, d* become open-circuited. In that case, the 25Ω resistance has no effect since no current flows through that part of the circuit. Hence, resistance to the right part of the circuit. Hence, resistance to the right of nodes *g* and *h* is given by

$$R_{gh} = 100 \parallel (50 + 100) \quad = 100 \times 150.250 = 60\ \Omega$$

Resistance to the right of terminals *m* and *n* is given by

$$R_{mn} = 100 \parallel (50 + 60) \quad = 100 \times 110/210 = 52.4\ \Omega$$

This resistance is in series with the 25 Ω resistance connected to the terminal *a*. Hence, R_{ab} = (25 + 52.4) = **77.4 Ω.**

Example 1.22. *A current of 20 A flows through two ammeters A and B joined in series. Across A, the potential difference is 0.2 V and across B it is 0.3 V. Find how the same current will divide between A and B when they are joined in parallel.* **(A.M.I.E. Winter 1940)**

Solution. The two ammeters are connected in series in Fig. 1.25 (*a*).

$$R_A = 0.2/20 = 0.01\ \Omega \quad ; \quad R_B = 0.3/20 = 0.015\Omega$$

The same two ammeters are connected in parallel in Fig. 1.27 (*b*).

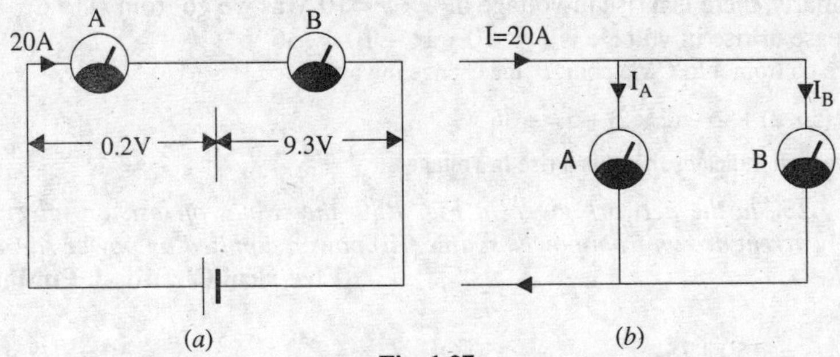

Fig. 1.27

$$I_A = I \times \frac{R_B}{R_A + R_B} = 20 \times \frac{0.015}{0.025} = \mathbf{12\ A}$$

$$I_B = I \times \frac{R_A}{R_A + R_B} = 20 \times \frac{0.01}{0.025} = \mathbf{8\ A}$$

Example 1.23. *Calculate the current through R in the network shown in Fig. 1.28. All the values of resistors are in ohms.* **(Elements of Elect, Engg., Punjab Univercity 1993)**

Solution. Between points *A* and *B* there are two parallel paths, one of resistance 10Ω and the other of (10 + 5) = 15Ω. Hence, *RAB* = 15 || 10 = 6Ω

$$R_{AC} = 10 \parallel (6 + 4) = 5\Omega$$

Hence, resistance as seen by the battery = 15 + 5 = 20Ω

Battery current, *I* = 40/20 = 2*A*

This current divides equally into two equal parts at point *A*, one ampere flowing along 10Ω resistance and the other along (6 + 4) = 10Ω path. Obviously, current through *R* is = 1/2 *A* = 0.5 *A*.

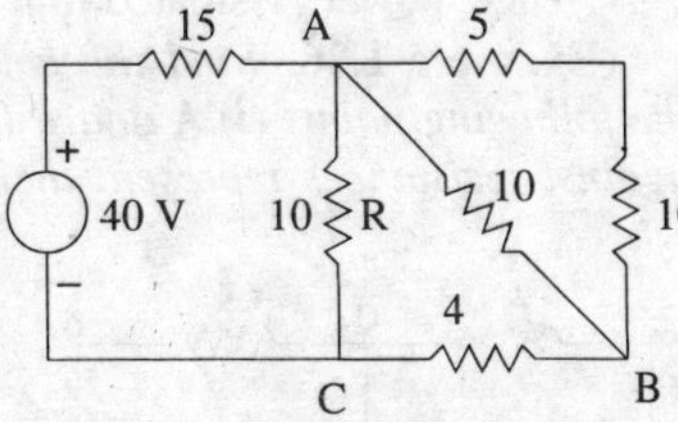

Fig. 1.28

Example 1.24. In the circuit shown in Fig. 1.29 (*a*), determine the voltage rise from *A* and *C* and the power absorbed by the portion *AD*.

Solution. It should be noted that 45-V and 5-V batteries are connected in additive series whereas 10-V battery is connected in opposition to them *i.e.*, in subtractive series.

Hence, net driving voltage around the circuit = (45 + 5) – 10 = 40 V. Total resistance = (2 + 5 + 8 + 5) = 20 Ω. Circuit current = 40/20 = 2A. As shown, it flows clockwise round the circuit. Since current flows from C to D to A, it is obvious that C is at a higher electrical potential than A.

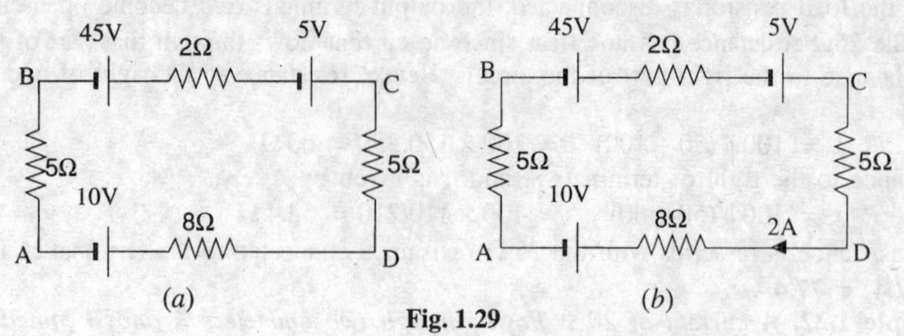

Fig. 1.29

As we go from A to C via point D, we meet the following voltages:

(*i*) Since we go from –ve to +ve terminal of the 10–v battery, there is an *increase* of 10V.

(*ii*) There is a voltage drop of $8 \times 2 = 16$V over the 8-Ω resistance. This represents an *increase* or rise in voltage, because we are going upstream *i.e.* opposite to the direction of flow of current.

(*iii*) Similarly, there is a rise in voltage of $2 \times 5 = 10$ V as we go from D to C.

Total increase or rise in voltage is $= 10 + 16 + 10 = 36$ V

Note : If we go from A to C via point B, the change in voltage is

$$= -(5 \times 2) + 45 - (2 \times 2) + 5 = +36 \text{ V}$$

The positive sign indicates that it is a rise in voltage.

Example 1.25. *In the network given in Fig. 1.30, the values of resistors given are in ohms. Determine : (i) current drawn from power source, (ii) power supplied by power source, (iii) power dissipated in R.* **(Electrical Circuits-I, Punjab Univ. 1993)**

Solution.

$$R_{BC} = 150 \parallel 75 = 50\ \Omega$$

$$R_{AD} = 200 \parallel (150 + 50) = 100\ \Omega$$

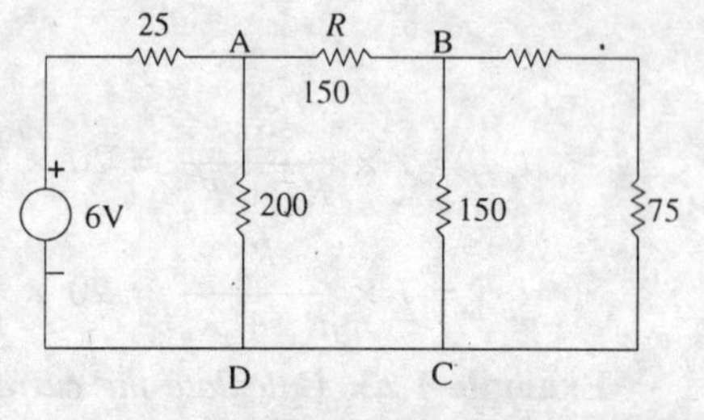

Fig. 1.30

The circuit resistance as seen by the battery = (100 + 25) = 125 Ω

(*i*) current $I = 6/125 = 0.048$ A= **48 mA**

(*ii*) power supplied by the battery = 6×0.048 = **0.288 Ω**

(*iii*) The battery current of 0.048 A is divided into two equal parts at point A. Hence, current passing through $R = 0.048/2 = 0.024$A

Hence, power dissipated in R of 150 Ω is $= 0.0242 \times 150 = 0.0864\ \Omega =$ **86.4 mΩ**

Example 1.26. *Find the equivalent resistance of the circuit given in Fig. 1.31 (a) between the following points (i) A and B (ii) C and D (iii) E and F (iv) A and F and (v) A and C. Figures against resistances represent their values in ohms.*

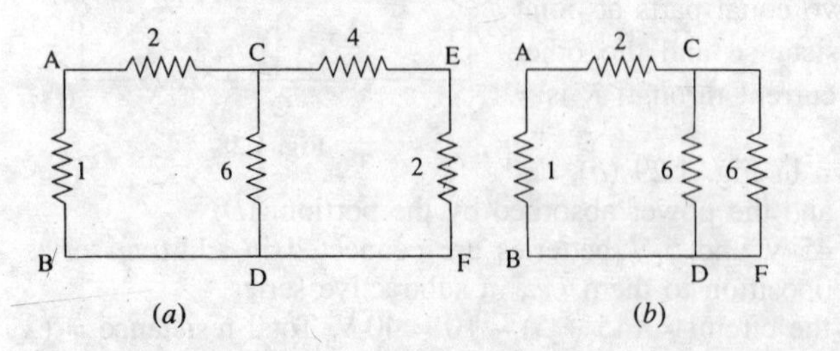

Fig. 1.31

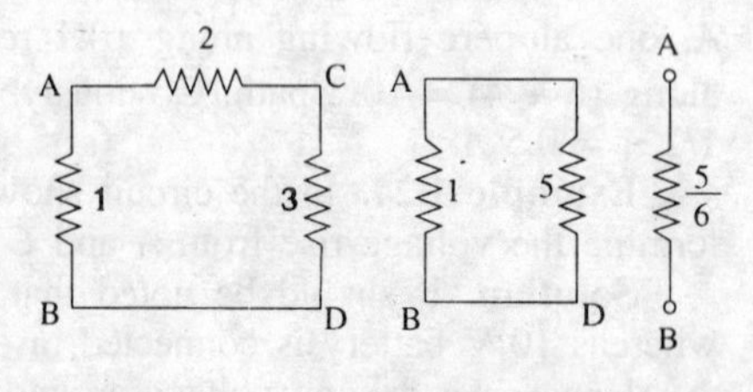

Fig. 1.32

Solution.

(*i*) Resistance Between A and B

In this case, the circuit to the right side of *AB* is in parallel with 1Ω resistance connected directly across points *A* and *B*.

The different steps to simplify the circuit are shown in Fig. 1.31 (*b*) and Fig. 1.32. The equivalent resistance R_{AB} = **5/6 ohm.**

(*ii*) Resistance Between C and D

In this case, there are three parallel paths between points *C* and *D*. One is *CD* itself, the second is *CFD* and the third is *CBD* [Fig. 1.33 (*a*)].

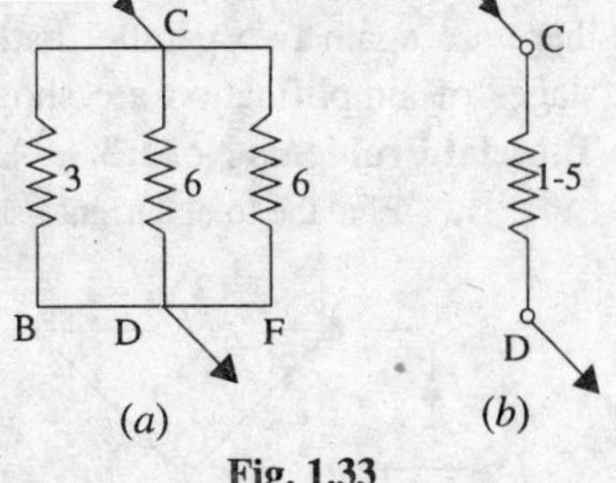

Fig. 1.33

It is so, because a current entering at point *C* will have three paths to go to *D*, as shown. The equivalent resistance R_{CD}, as shown in Fig. 1.33 (*b*), equals 1.5Ω.

(*iii*) Resistance Between E and F

As seen from Fig. 1.34 (*a*), in this case, the whole circuit to the left of *EF* is in parallel with the 2Ω resistance connected directly across *EF*.

After various simplifications, as shown in Fig. 1.34,

$$R_{EF} = \mathbf{1.5\ \Omega}$$

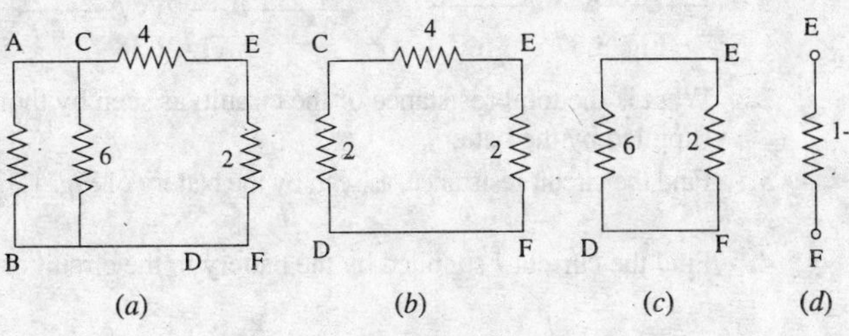

Fig. 1.34

(*iv*) Resistance Between A and F

As shown in Fig. 1.35 (*a*), if a current enters the circuit at point *A*, it has two choices : partly it goes along *AC* and partly along *ABDF*. At point *C*, there are again two parallel paths: one along *CEF* and the other along *CDF*. As shown in Fig. 1.35 (*b*), equivalent resistance between points *C* and *F* is = 6/2 = 3Ω. Fig. 1.35 (*c*) shows that there are two parallel paths between *A* and *F*. Their equivalent resistance = 5 ‖ 1 = 5/6 Ω as shown in Fig. 1.35 (*d*).

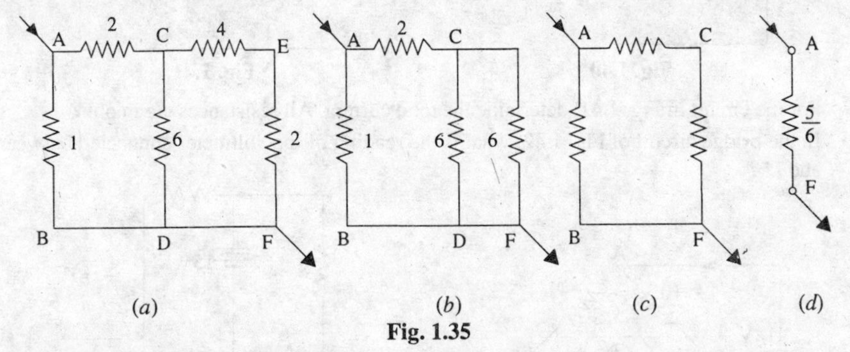

Fig. 1.35

(*v*) Resistance Between A and C

In this case, as shown in Fig. 1.36 (*a*), if a current enters the circuit at point *A* (and leaves

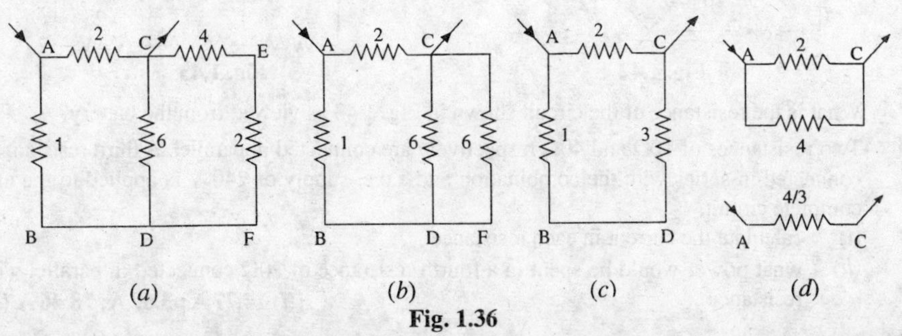

Fig. 1.36

at C) it has two choices : one is directly from A to C and the other is along ABD. At point D, there are again two parallel paths : one directly along DC and other along $DFEC$. The different stages of simplification are shown in Fig. 1.36. It is seen that $R_{AC} = 4/3\Omega$.

Tutorial Problems No. 1.2

1. Find the total current I supplied by the 12-V battery of Fig. 1.37 **[3A]**

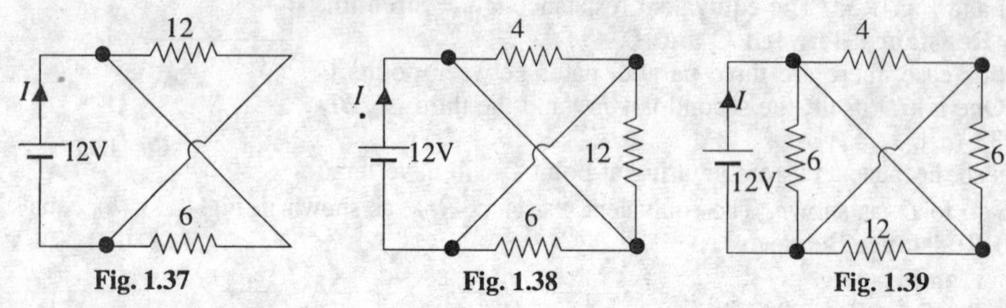

Fig. 1.37 **Fig. 1.38** **Fig. 1.39**

2. What is the total resistance of the circuit, as seen by the battery in Fig. 1.38 ? Calculate the power supplied by the battery. **[2Ω ; 72Ω]**
3. Find the circuit resistance, as seen by the battery of Fig. 1.39. What is the value of the total current I? **[1.5 Ω ; 8A]**
4. Find the current I supplied by the battery in the circuit of Fig. 1.40. All resistances are in ohm. **[9A]**

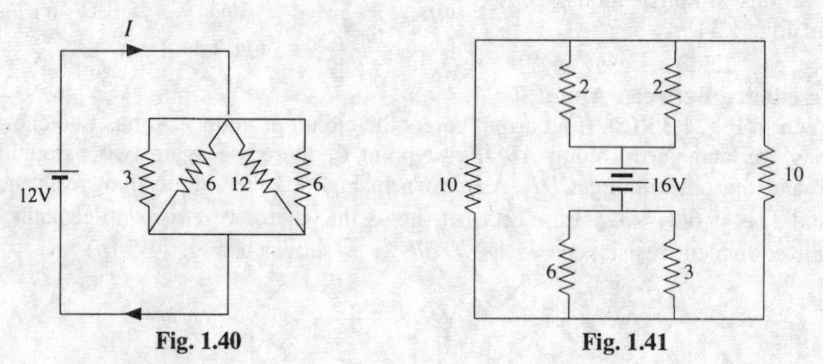

Fig. 1.40 **Fig. 1.41**

5. For the circuit in Fig. 1.41, determine the total current. All resistances are in ohm. **[8Ω ; 2A]**
6. In the bridge circuit of Fig. 1.42, what is the reading of the voltmeter connected between points B and D ? **[IV]**

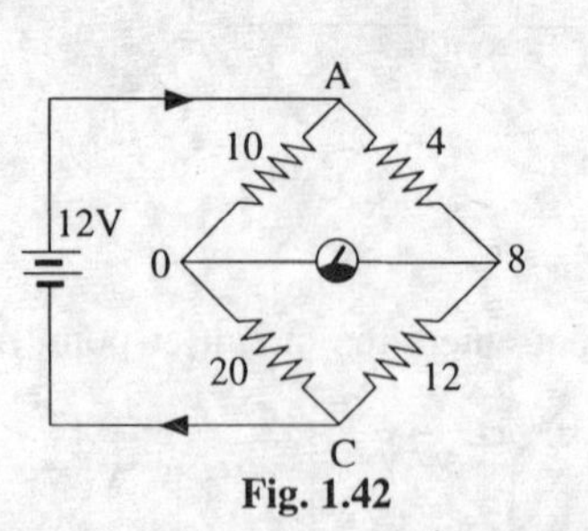

Fig. 1.42

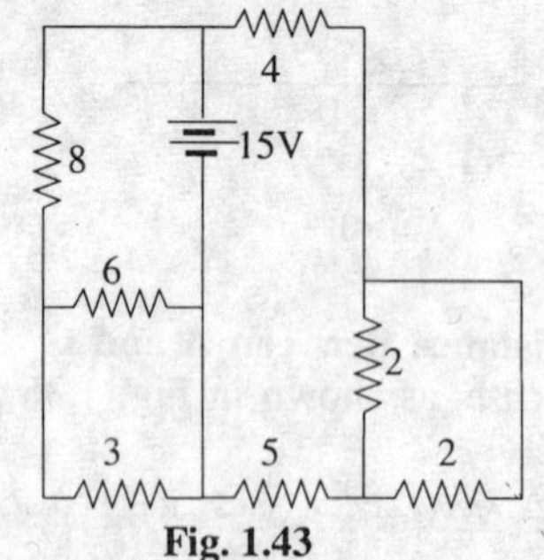

Fig. 1.43

7. What is the resistance of the circuit shown in Fig. 1.43 as viewed from the battery? **[5Ω]**
8. Two resistances of 10Ω and 40Ω respectively are connected in parallel. A third resistance of 5Ω is connected in series with the combination and a.d.c. supply of 240 V is applied to the ends of the complete circuit.
   (*i*) calculate the current in each resistance
   (*ii*) what power would be spent in a fourth resistance of 20Ω connected in parallel with the 5Ω resistance. **[(*i*) 14.77 A ; 3.69 A; 18.46 A (*ii*) 320 Ω]**

9. Two resistances of 4Ω and 12Ω respectively are connected in parallel with each other. Another resistance of 10Ω is connected in series with the combination.
Calculate the respective d.c. voltages which should be applied across the whole circuit
(*i*) to pass 6*A* through the 10Ω resistance, (*ii*) to pass 6 *A* through the 12Ω resistance.
Calculate also the total power from the supply in each case. **[(*i*) 78V; 468Ω (*ii*) 312V ; 7488 Ω]**

10. For the circuit shown in Fig. 1.44, find the equivalent resistances between points *A* and *B*. **[5Ω]**

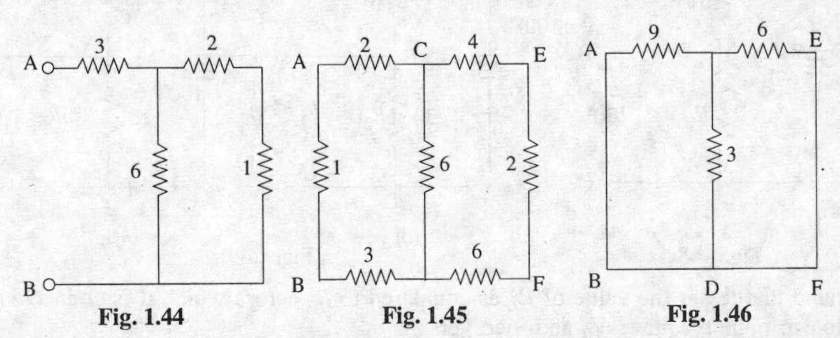

Fig. 1.44 **Fig. 1.45** **Fig. 1.46**

11. In the circuit shown in Fig. 1.45, the figures against resistances represent their values in ohms. Find the equivalent resistance of this circuit between points (*i*) *A* and *B* (*ii*) *C* and *D* (*iii*) *E* and *F* (*iv*) *A* and *D* (*v*) *A* and *F* (*vi*) *A* and *C* (*vii*) *C* and *B* (*viii*) *E* and *D* (*ix*) *C* and *F* (*x*) *C* and *E* (*xi*) *B* and *D* and (*xii*) *B* and *E*.

Hint. For finding R_{AE} and R_{BE}, delta/star transformation will have to be used.

[(*i*) 9/10Ω (*ii*) 12/5Ω (*iii*) 26/15Ω (*iv*) 2.4Ω (*v*) 22/5Ω (*vi*) 8.5Ω (*vii*) 21/10Ω (*viii*) 56/15Ω (*ix*) 18/5Ω (*x*) 44/15Ω (*xi*) 21/10Ω (*rii*) 381/90Ω]

12. In Fig. 1.46, find the resistance between points (*i*) *C* and *D* (*ii*) *C* and *F* (*iii*) *A* and *B* (*iv*) *A* and *C* (*v*) *A* and *E* (*vi*) *E* and *F*. **[(*i*) 18/11Ω (*ii*) 18/11Ω (*iii*) zero (iv) 18/11Ω (*v*) zero (*vi*) zero]**

1.20. Voltage Divider Formula

It is used for finding voltage drops across different resistors connected in series. Since in a series circuit, same current flows through each resistor, voltage drops are directly proportional to their ohmic values. If, in Fig. 1.47, R_2 is twice R_1, then $V_2 = 2V_1$ and so on. Now, total resistance of the series circuit is

$$R = R_1 + R_2 + R_3 = 12\,\Omega$$

According to the voltage divider formula, various voltage drops are

$$V_1 = V.\frac{R_1}{R} = 24 \times \frac{2}{12} = 4\text{V}$$

$$V_2 = V.\frac{R_2}{R} = 24 \times \frac{4}{12} = 8\text{V}$$

$$V_3 = V.\frac{R_3}{R} = 24 \times \frac{6}{12} = 12\text{ V}$$

As expected, $V_1 + V_2 + V_3 = 4 + 8 + 12 = 24$ V

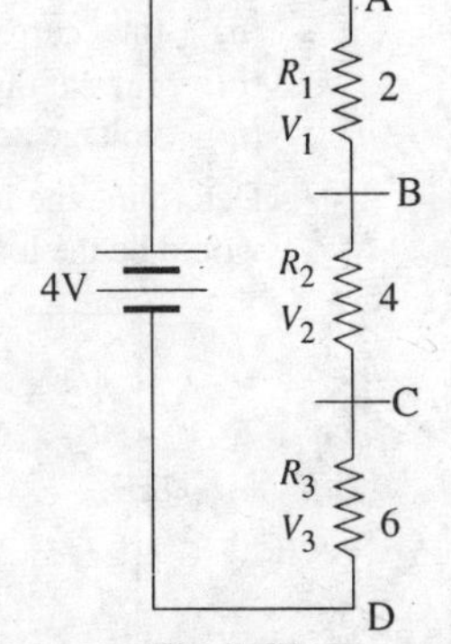

Fig. 1.47

1.21. Unloaded and Loaded Voltage Divider

An unloaded voltage divider is a series combination of a few resistors connected across a single voltage source, say, a battery as shown in Fig. 1.47. Such dividers are often used for obtaining different voltages from a given single source for feeding different networks. Fig. 1.48 shows a simple unloaded voltage divider which produced an output voltage $V_0 = 100 \times 30/(20 + 30) = 60$V. However, as shown in Fig. 1.49 (*a*), if a load resistor R_L is connected across the output, V_0 would be reduced by an amount depending on R_L.

It is so because resistance across B and C is reduced from R_2 to $R_2 \parallel R_L$. As seen from Fig. 1.48 (*b*),

$$R_{BC} = 30 \parallel 15 = 10\Omega \quad \therefore \quad V_0 = 100 \times 10/(10+20) = 33.3\text{V}$$

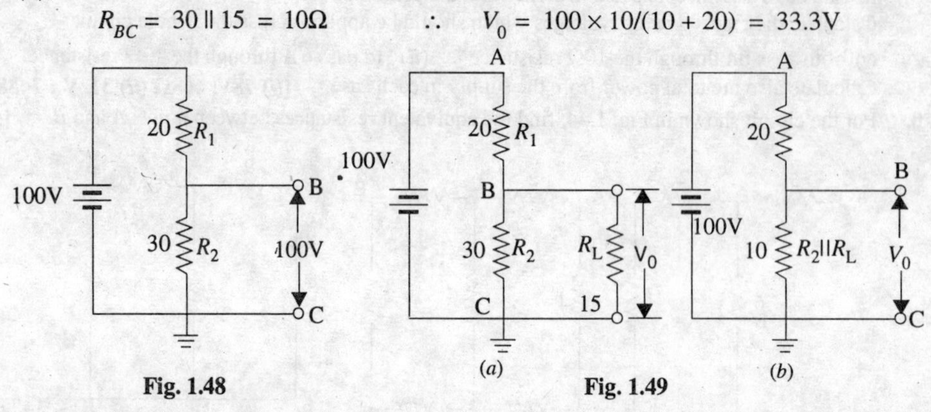

Fig. 1.48 **Fig. 1.49**

It is found that lesser the value of R_L as compared to R_2 across which it is connected, greater the reduction in output voltage V_0 and vice versa.

Tutorial Problems No. 1.3

1. Find voltages at points *A, B, C, D* and E in Fig. 1.59. **[V_A = 100 V; V_B = 83.3V ; V_C = 66.6 V ; V_D = 33.3 V ; V_E = 0]**

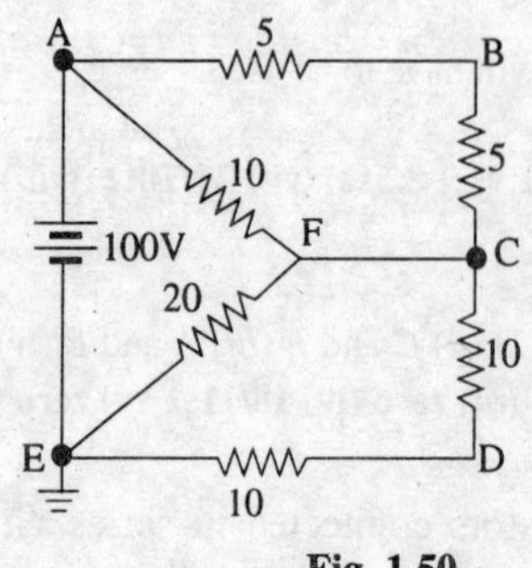

Fig. 1.50

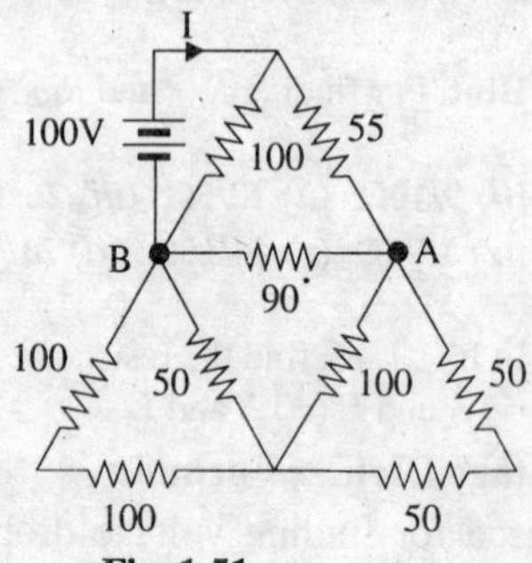

Fig. 1.51

2. For the circuit shown in Fig. 1.51, calculate the following :
   (*i*) effective resistance as seen by the source
   (*ii*) total current taken from the source
   (*iii*) current through 90Ω resistor
   (*iv*) voltage across *A* and *B* **[(i) 50Ω (*ii*) 2A (*iii*) 1A (*iv*) 45V]**
3. Determine the unloaded output voltage V_0 in Fig. 1.52. If a 60 Ω load resistor is connected, what would be the loaded output voltage? **[18V ; 16V]**

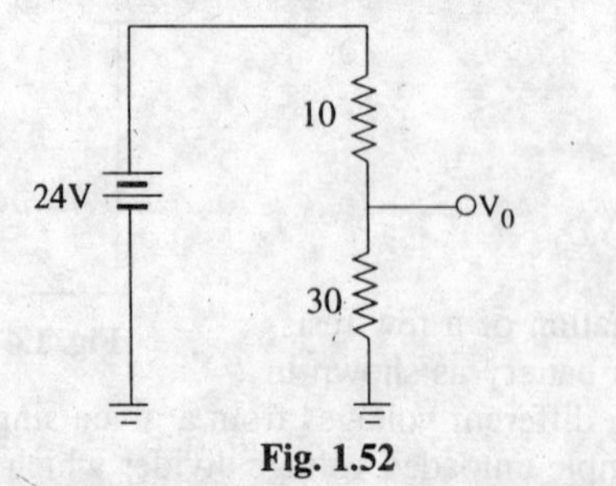

Fig. 1.52

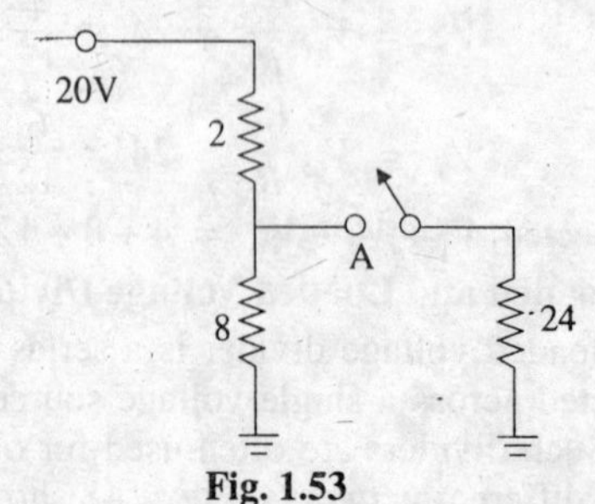

Fig. 1.53

4. In Fig. 1.53, find the voltage at point *A* when the switch is open. Also, find the value of this voltage when the switch is closed. **[16V ; 15V]**

1.22. Circuit Ground

Voltage is always relative *i.e.*, voltage at any point in a circuit is always measured relative to another point called the *reference* point. It is usually called *ground*. The term *ground* is derived from the method used in ac power lines in which one side of the line is neutralized by connecting it to a water pipe or a metal rod driven deep into the ground. This method of grounding is called *earth ground.*

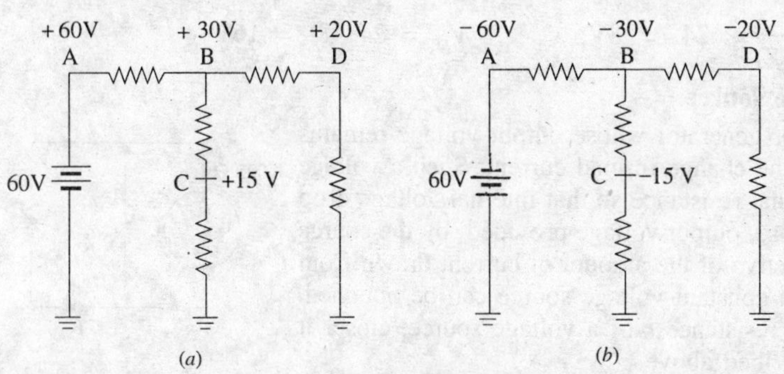

Fig. 1.54

However, in electronic circuits, the metal chassis or frame or cabinet that houses the assembly is used as *common* or *reference* point and is called the *chassis* or *circuit ground.* All circuit voltages (whether positive or negative) are measured with respect to this ground, which is supposed to be at 0V.

However, the chassis ground has not to be necessarily connected to the earth ground, though it is mostly earth-grounded in order to prevent shock hazard due to potential difference between the chassis and earth ground.

In Fig. 1.51 (*a*), negative side of the battery has been earthed so that all circuit voltages are positive with respect to the ground. However, Fig. 1.54 (*b*) shows the case where positive side of the battery has been grounded thereby making all circuit potentials negative with respect to the ground.

Example 1.27. *Determine the voltages of the points A, B, C, D and E in the circuit of Fig. 1.55(a) and (b). What is the value of V_{CE} in Fig. 1.55(a) and of V_{BD} in Fig. 1.55(b). Assume a drop of 15V across each resistor.*

Solution. In Fig. 1.55 (*a*), all voltages will be with respect to point *D* which has been grounded and hence taken to be at 0V. The different point voltages with their polarities are as under :

V_C = **+ 15V** ; V_B = 15 + 15 = **+ 30V**

V_A = 30 + 15 = **45V**

V_D = **0V** ; V_E = −15V, V_{CE} = $V_C - V_E$ = 15 − (−15) = + 30V

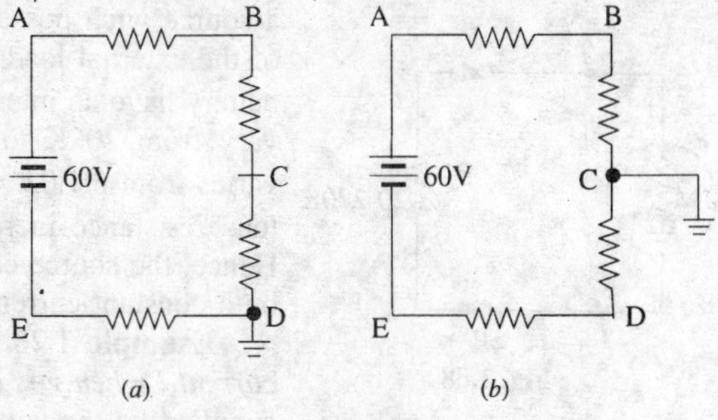

It shows that potential of point *C* with respect to point *E* is + 30 V though it is +15 V with respect to point *D*.

Fig. 1.55

In Fig. 1.55(*b*), point *C* is the ground. Hence,

V_A = **+ 30V;** V_B = **+ 15V;** V_C = 0V ; V_D = − 15V, V_E = **− 30V**

Similarly, V_{BD} = $V_B - B_D$ = +15 − (−15) = **+ 30V**

Example 1.25. *Determine the voltage at each point with respect to ground in Fig. 1.56. All resistances are in ohm.*

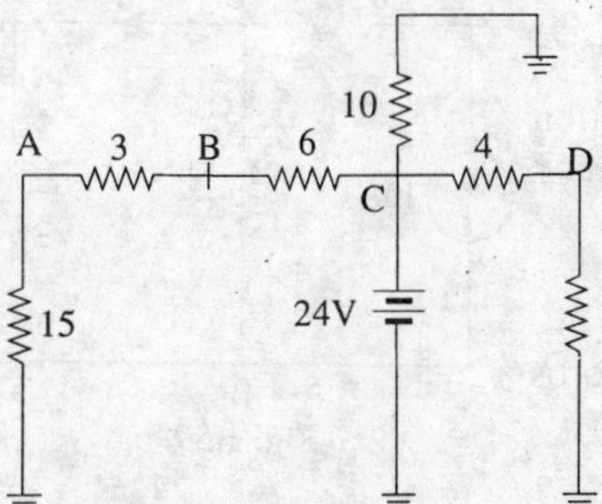

Solution. It should be noted that the three branches meeting at point *C* in the figure are in parallel with each other across the 24–V battery. It is so because they are all connected between the positive terminal of the battery and the ground. We would use the voltage divider formula (Art. 1.21) for finding drops across different series-connected resistors.

Fig. 1.56

Since point C is directly connected across the positive battery terminal, $V_C = +\ 24V$

$$V_B = \text{drop across } (15 + 3) = 18\ \Omega \text{ or}$$
$$= 24\text{ V} - \text{drop across } 6\ \Omega$$

$$\text{Drop across } 18\ \Omega = \frac{18}{(15 + 3 + 6)} \times 24 = \mathbf{18V}$$

$$V_D = \text{drop across } 8\Omega \text{ or } = 24\text{V} - \text{drop across } 4\ \Omega$$

$$\text{Drop across } 4\Omega = \frac{4}{(4 + 8)} \times 24 = \mathbf{8V}; \quad \therefore \quad V_D = 24 - 8 = \mathbf{16V}$$

1.23. Ideal Constant-Voltage Source

It is that voltage source or generator whose output voltage remains absolutely constant whatever the change in load current. Such a voltage source must possess zero internal resistance so that internal voltage drop in the source is zero. In that case, output voltage provided by the source would remain constant irrespective of the amount of current drawn from it. In practice, none such ideal constant-voltage source can be obtained. However, smaller the internal resistance r of a voltage source, closer it comes to an ideal source described above.

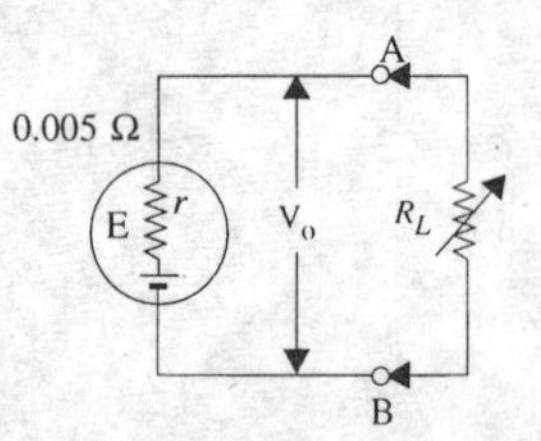

Fig. 1.57

Suppose, a 6-V battery has an internal resistance of 0.005Ω (Fig. 1.57). When it supplies no current *i.e.*, it is on no load, V_0 = 6V *i.e.*, output voltage provided by it at its output terminals A and B is 6V. If load current increases to 100 A, internal drop = 100 × 0.005 = 0.5. Hence,

$$V_0 = 6 - 0.5 = 5.5\text{ V.}$$

Obvisously, an output voltage of 5.5-6 V can be considered constant as compared to wide variation in load current from 0_A to 100 A.

1.24. Ideal Constant-Current Source

It is that voltage source whose internal resistance is infinity. In practice, it is approached by a source wich possesses very high resistance as compared to that of the external load resistance. As shown in Fig. 1.58, let the 6-V battery have an internal resistance of 1 MΩ and let load resistance vary from 20 K to 200 K. The current supplied by the source varies from 6/1.02 = 5.9 μA to 6/1.2 = 5 μA. As seen, even when load resistance increases 10 times, current decreases by 0.9 μA. Hence, the source can be considered, for all practical purposes, to be a constant-current source.

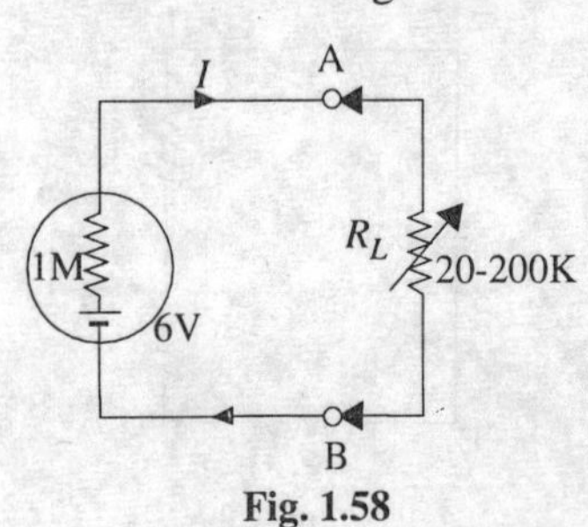

Fig. 1.58

Example 1.26. *In the circuit of Fig. 1.59, find the value of current I when v is (i) 1V (ii) 2 V and (iii) 3 V. What is the power supplied by the voltage source in each case?*

Solution. As seen, the circuit contains an ideal current source and a voltage source. Also, current through the 0.2Ω resistor is determined by the value of v connected directly across it.

***(i)* When V = 1 V**

In this case, I_{AB} = 1/0.2 = **5A**

∴ I = 10 – 5 = 5A

Since I comes out to be positive, it flows in the same direction as shown. Obviously, the voltage source does not supply any power. Rather, it absorbs a power of 1 × 5 = 5 Ω.

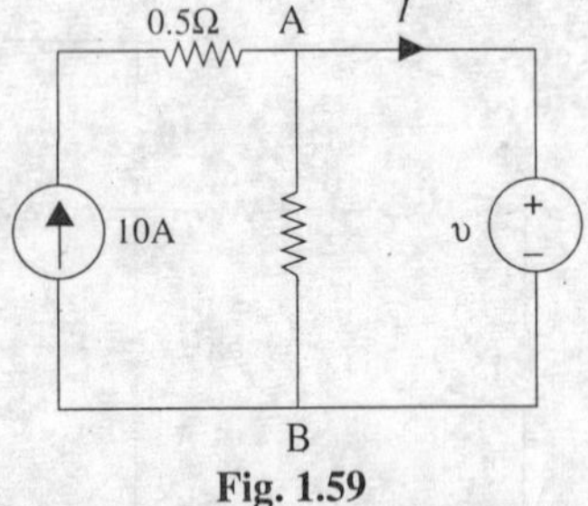

Fig. 1.59

***(ii)* When V = 2V**

Here, I_{AB} = 2/0.2 = 10A, I = 10 – 10 = **0A**

Since I = 0, the voltage source neither supplies nor absorbs any power from the source.

***(iii)* When V = 3V**

$$I_{AB} = 3/0.2 = 15\text{A} \qquad \therefore \quad I = 10 - 15 = \mathbf{-5A}$$

Since I comes out to be negative, it means that it flows in a direction opposite to that shown in Fig. 1.56. Since I flows out of the voltage source, power *supplied* by it = 3 × 5 = **15Ω**

Example 1.27. *Find the values of v_1 and v_2 in the circuit of Fig. 1.60.*

Solution. It is obvious that both current sources feed the common resistance of 3Ω. Since current through 3Ω resistor is

$$(3+1) = 4\text{A}$$

$$\therefore \quad v_2 = 4 \times 3 = 12\text{V}$$

Now, v_1 equals the sum of drops over 2Ω and 3Ω resistors. The drop over 2Ω resistor = 2 × 3 = 6V

$$\therefore \quad v_1 = 6\text{V} + 12\text{V} = \mathbf{18V}$$

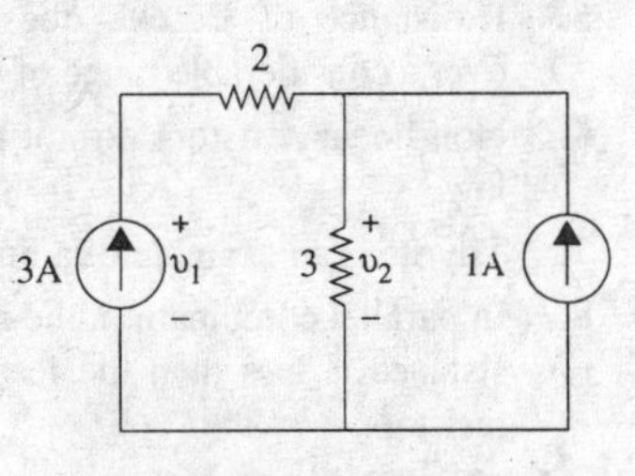

Fig. 1.60

HIGHLIGHTS

1. Laws of resistance are summed up as $R = \rho \dfrac{1}{A}$ ohm
2. Specific resistance or resistivity is defined as the resistance between the opposite faces of a metre cube of that material.
3. Conductance is reciprocal of resistance (R). Its unit is Siemens (S).
4. Conductivity (σ) is reciprocal of resistivity (ρ). Its unit is Siemens/metre.
5. Temperature coefficient is generally defined as the change in resistance per ohm per degree change in temperature.
6. In the case of good conductors, especially metals, the variations of resistance with temperature (within normal range) are given by
$$R_t = R_0(I \pm a_0 t)$$
7. It is found that α is not constant even for a given material. Its value depends on the initial temperature on which the increment in resistance is based. The values of α for various temperatures can be found from the relations.
$$\alpha_t = \frac{\alpha_0}{1 + \alpha_0 \tau} \quad \text{and} \quad \alpha_2 = \frac{1}{1/\alpha_1 + (t_2 - t_1)}$$
8. In the same way, resistivity (ρ) of a given material is also not the same at all temperatures. It varies as
$$\rho_t = \rho_0(1 + \alpha_0 t)$$
9. Ohm's law can be written as $I = V/R$. This law is applicable both to d.c. as well as a.c. circuits.

OBJECTIVE TESTS —1

A. Fill in the following blanks :

1. Electric resistance is similar to .... in Mechanics.
2. The resistivity of a material is the resistance between the opposite faces of a ....cube of that material.
3. Reciprocal of resistivity is ....
4. Temperature coefficient may be defined as the ....change in resistance per degree change in temperature.
5. Linear resistors are those in which current produced is .... proportional to the applied voltage.
6. Thermistor is an example of a .... resistor.
7. Temperature coefficient of a metal conductor increases with .... in temperature.
8. The variable resistor which has two fixed terminals and one movable contact is called a............... .

B. Answer True or False :

1. Resistivity of a material depends inversely on its length.
2. Resistivities of commercial materials differ by several percent due to impurities.
3. Reciprocal of resistance is called conductance.

4. As its temperature is increased, the resistance of an insulator is increased.

5. Resistance of Eureka does not change over a considerable range of temperature.

6. Non-linear resistors do not obey Ohm's law.

7. Thyrite is an example of a linear resistor.

8. In parallel combination, the equivalent resistance is less than the least among the resistors.

C. Multiple Choice Questions

1. Which of the following material has nearly zero temperature coefficient of resistance?

(*a*) copper (*b*) carbon
(*c*) manganin (*d*) mica

2. A coil has a resistance of 100Ω at $90°C$. At $100°C$, its resistance is 101Ω. The temperature-coefficient of the wire at $90°C$ is

(*a*) 0.01 (*b*) 0.1
(*c*) 0.0001 (*d*) 0.001

3. You have to replace a 1500Ω resistor in a radio. You have no 1500Ω resistor but have several 1000Ω ones which you would connect

(*a*) three in parallel
(*d*) three in series
(*c*) two in parallel and one in series
(*d*) two in parallel

4. When two resistances are connected in series, they have

(*a*) same resistance values
(*b*) same voltage across them
(*c*) same current passing through them
(*d*) different resistance values

5. Two resistors are said to be connected in series when

(*a*) both carry equal current
(*b*) total current equals the sum of branch currents
(*c*) sum of *IR* drops equals battery emf
(*d*) same current passes through both

6. In a parallel circuit, all components must

(*a*) have the same p.d. across them
(*b*) have same value
(*c*) carry equal currents
(*d*) carry same current

7. The resistivity of a material is 1.78×10^{-8} ohm-m. This material is

(*a*) conductor
(*b*) insulator
(*c*) semi-conductor
(*d*) none of these

(Elect. Engg. A.M.Ae. S.I. June 1994)

8. If the diameter of a copper wire is doubled, its current carrying capacity becomes

(*a*) twice (*b*) four times
(*c*) half (*d*) none of these

(Elect. Engg. A.M.Ae. S.I. June 1994)

9. The current rating of a 1 k Ω, 0.5 watt resistor is

(*a*) 2.23 A
(*b*) 1 A
(*c*) 22.36 mA
(*d*) none of these

(Elect. Engg. A.M.Ae. S.I. June 1994)

10. The resistance of a human body is around .... ohm.

(*a*) 100 (*b*) 1000
(*c*) 25 (*d*) none of these

(Elect. Engg. A.M.Ae. S.I. June 1994)

ANSWERS

A. 1. friction 2. mete 3. conductivity 4. fractional 5. directly 6. non-linear 7. rise/increase 8. potentiometer

B. 1. F 2. T 3. T 4. F 5. T 6. T 7. F 8. T

C. 1. *c* 2. *d* 3. *c* 4. *c* 5. *d* 6. *a*

2 DIVISION OF CURRENT

2.1. Primary Cell

It essentially consists of two dissimilar conducting eletrodes (one anode and the other cathode) immersed in a liquid called electrolyte, which acts chemically on one of the two electrodes more readily than on the other. By using the energy released by chemical action, electrons are shifted from one electrode to another, thereby creating a potential difference between the two electrodes. The value of *total* potential difference created between the electrodes, when *the cell is not connected* to an external circuit, is known as its *electromotive force* (*E.M.F.*)

Now, every cell has some internal resistance, which depends upon the construction and condition of the cell. The internal resistance depends on the area of electrodes, the distance between the electrodes, the temperature, strength and density of the electrolyte. When the cell supplies current to an external circuit, there is always some internal voltage drop due to this internal resistance. Hence, the voltage available for external circuit is decreased by this much amount. The net voltage available at the terminals for external circuit is known as the *terminal potential difference* (*T.P.D.*)

It may be noted that a primary cell cannot be recharged.

2.2. Cell and Battery

The word 'cell', means one unit or a combination of materials, for converting chemical energy into electrical energy. A 'battery' means a combination of these units or cells.

2.3. E.M.F. and Terminal Potential Difference

The e.m.f. of the cell, as said earlier, is the total potential difference established within the cell between the two electrodes, when the cell is not supplying any current (so that there is no internal voltage drop). The e.m.f. can be measured by connecting a suitable voltmeter across the electrodes. But the terminal potential difference is equal to the e.m.f. *minus* the internal voltage drop. In other words, the *T.P.D.* is the potential difference available for external circuit. If 'i' is the current supplied by the cell and r its internal resistance, then.

Terminal potential difference, V = e.m.f.$-ir$ or $V = E - ir$

where V = terminal potential difference and
E = e.m.f. generated within the cell.

It should be noted that whereas e.m.f. E is constant, the terminal potential difference V is not, as it depends on load current supplied by the cell.

Also, the term e.m.f. has local significance *i.e.* it is spoken of with reference to the generator itself, but potential difference is distributive. For example, we cannot say that an e.m.f. of 220 volts exists across an electric bulb. It is the potential difference of 220 V. In other words, p.d. can be distributed away from its source of generation.

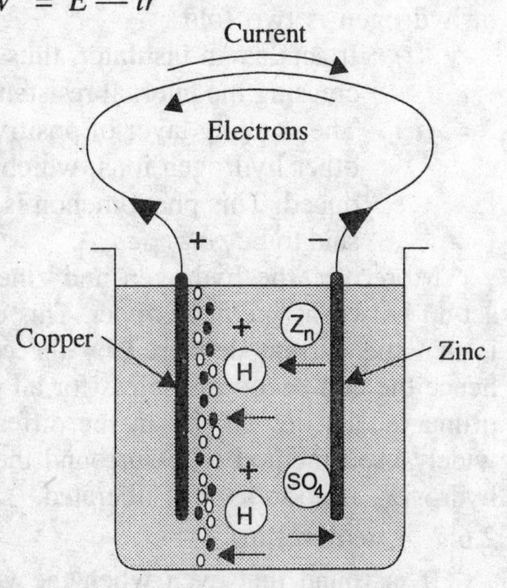

Fig. 2.1

2.4. The Simple Voltaic Cell

As shown in Fig. 2.1, such a cell consists of a copper (Cu) plate and a zinc (Zn) rod placed in dilute sulphuric acid (H_2SO_4). When terminals of the cell are joined together by a wire, hydrogen is given off at the Cu plate and a current is found to flow from Cu to Zn in the external circuit and from Zn to Cu within the cell. Cu plate forms the positive electrode *i.e.* anode and zinc the negative electrode *i.e.* cathode. While the current is flowing, some of the zinc is found to dissolve in the acid.

The production of the e.m.f. can be explained with the help of contact difference of potential. Whenever a metal plate is brought in contact with an electrolyte, there always develops a tendency either for some of the metal to go into the solution in the form of positive ions, thus leaving the plate negatively charged or for some of the positive ions in the solution to be deposited on the plate, which consequently becomes positively charged. In the case of a simple voltaic cell, when Zn is immersed in dilute H_2SO_4, some of the Zn enters the electrolyte in the form of positive ions which further combine with the negative sulphions in the electrolyte to form zinc sulphate ($\overset{++}{Zn} + \overset{--}{SO_4} = ZnSO_4$). However, this action stops when the contact potential between Zn and electrolytic solution reaches the value of 0.62 V, zinc being at the lower potential *i.e.*, negative with respect to the solution film adjacent to it (Fig. 2.2).

Similarly, when Cu plate is placed in contact with the electrolyte, the positive hydrogen ions in the solution have a tendency to get deposited on it until its potential rises nearly to 0.46 V above that of the solution. Hence, a total potential difference of 0.62 – (– 0.46) = 1.08 V is developed between the two electrodes (Fig. 2.2).

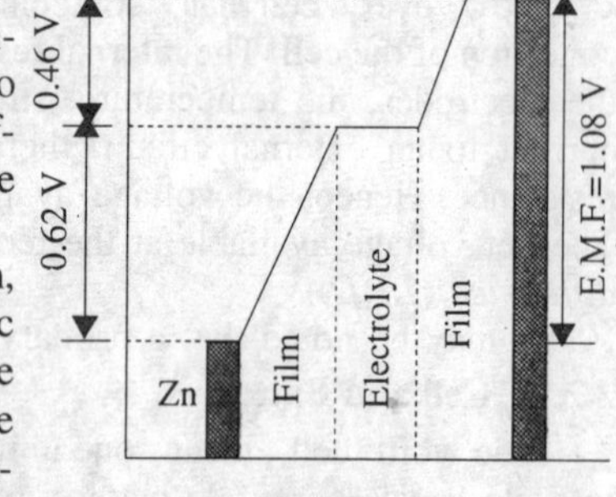

Fig. 2.2

As more and more of positive zinc ions pass into solution, an increasing number of electrons becomes available at the zinc electrode. These electrons run towards the copper anode via the external wire. There these electrons combine with the positive hydrogens ions, thus converting them into electrically neutral atoms. These atoms then combine in pairs to form molecules of hydrogen which bubbles off from the copper plate. The chemical action taking place may be represented by the following equation

$$Zn + H_2SO_4 = ZnSO_4 + H_2$$

2.5. Polarisation

If an ammeter is included in the external circuit of the voltaic cell, it indicates a gradual decrease in the current flowing. After some time, the current may cease altogether. The decrease is due to the collection of hydrogen bubbles on the surface of Cu plate. The effect of this layer of hydrogen is two fold :

(*i*) It acts as an insulator, thus reducing the effective area of the Cu plate and thereby increasing the internal resistance of the cell.

(*ii*) The sticking layer of positive hydrogen ions on the Cu plate exerts a repulsive force on other hydrogen ions, which are approaching the copper plate. Hence, the current is reduced. This phenomenon is called polarisation and the cell which is in this condition is said to be *polarised.*

Moreover, the hydrogen and zinc set up between them an e.m.f. which is *opposite* to that set up between zinc and copper. This e.m.f. is known as *back e.m.f.* The result of all these factors is that the current delivered by the cell diminishes to a very low value after a short time and hence the cell becomes useless for all practical purposes. The essential difference between various primary cells lies chiefly in the different methods employed to overcome this defect. The most widely-used method is to surround the anode by a solid or liquid depolariser, which oxidizes the hydrogen as soon as it is liberated.

2.6. Local Action

It is found that even when the voltaic cell is not supplying any load current, zinc goes on continuously dissolving in the electrolyte. This is due to the fact that some traces of impurities like iron and lead in the commercial zinc form tiny local cells, which are short-circuited by the main body of zinc. The action of these parasitic cells cannot be controlled, so that there is some wastage of zinc. This phenomenon is known as local action and can be prevented by amalgamating the zinc plate *i.e.* by rubbing mercury over the zinc plate. Mercury is supposed to cover the impurities and maintain a film of zinc dissolved in mercury.

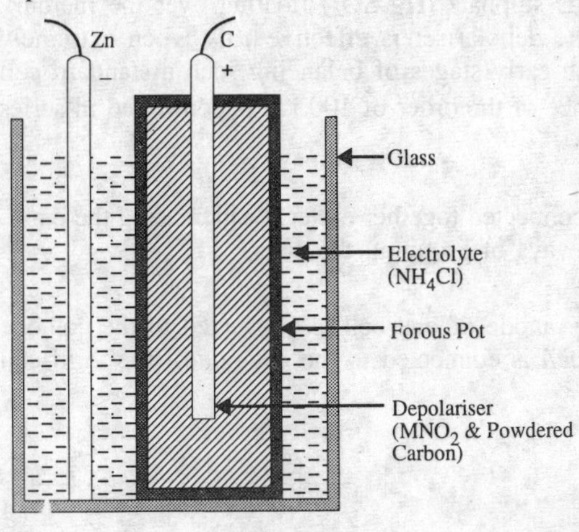

Fig. 2.3

2.7. Leclanche Cell

It is the only primary cell of importance and is now-a-days mainly used in commercial practice. It is used, in large numbers, for domestic bell and telephone circuits and also in the 'dry' form for flash lamps and wireless batteries etc. A section of the cell is shown in Fig. 2.3. The cathode consists of a zinc rod placed inside a solution of ammonium chloride (sal ammoniac) NH_4Cl contained in a glass jar. The anode is a carbon rod which is surrounded by solid manganese dioxide (MnO_2) and crushed carbon in a porous pot. Manganese dioxide is rich in oxygen which easily combines with the hydrogen approaching the carbon anode. Crushed carbon is mixed with MnO_2 for improving its conductivity and hence the internal resistance of the cell.

Chemical Action. When the anode and cathode are joined by a wire, zinc dissolves in the electrolyte thereby producing zinc chloride ($ZnCl_2$), ammonia (NH_3) and hydrogen. The hydrogen released is in the form of H+ ions :

$$Zn + 2NH_4Cl = ZnCl_2 + 2NH_3 + H_2$$

Conversion of zinc into zinc chloride is the source of energy produced by the cell. The $ZnCl_2$ and NH_3 dissolve in the water present in the electrolyte. Then H+ ions migrate towards the carbon rod through the porous pot and are converted into hydrogen atoms with the electrons reaching the carbon rod from the zinc rod *via* the external wire. Hydrogen attacks MnO_2 thus forming water

$$H_2 + 2MnO_2 = H_2O + Mn_2O_3$$

The main advantages of this cell are that the materials required for its construction are relatively cheap and its e.m.f. is 1.5 V. But its only serious disadvantage is that the solid depolariser (MnO_2) is slow in action *i.e.* hydrogen is evolved too fast to be oxidised completely by MnO_2. The result is that polarisation does take place if current is drawn from the cell for too long a time. However, if the cell is disconnected from the circuit for some time, the hydrogen present is oxidised by MnO_2 and the cell recovers. Hence, this cell is useful in circuits where current is required intermittently for short period, as in bell circuits or in a torch.

2.8. Standard Cell

A standard cell is one whose e.m.f. varies very little with time and temperature. It is not intended as a source of energy but for use as a secondary standard of voltage in potentiometer circuits for laboratory calibration of voltmeters, ammeters etc. Such a cell maintains a very constant voltage over long periods of time, provided no appreciable current is ever drawn from the cell. Most commonly used standard cell is the saturated Weston Cadmium cell. If the current ever drawn from it does not exceed 0.1 mA, then its e.m.f. which is about 1.0183 V can be trusted to remain constant to about 1 part in 100,000.

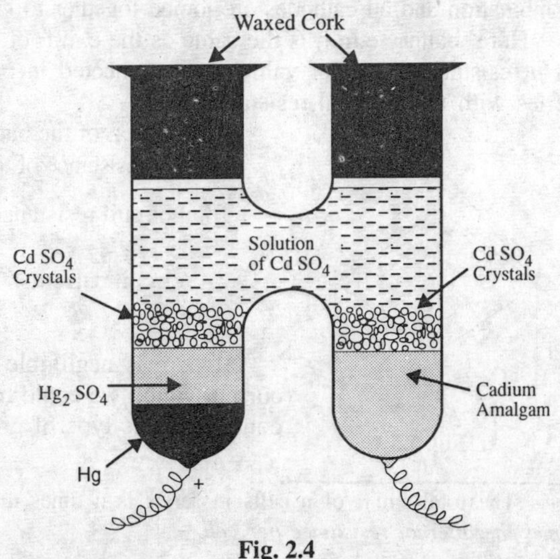

Fig. 2.4

Construction. The cell is housed in a H-shaped glass tube because its electrodes are liquid or semi-liquid (Fig. 2.4). Its anode is mercury and cathode is mercury-cadmium amalgam (*i.e.* 10 to 15% of cadmium dissolved in mercury). The electrolyte is a solution of cadmium sulphate ($CdSO_4$). The depolariser is a paste of mercurous sulphate (Hg_2SO_4) floating over the mercury anode and is an almot insoluble compound. The depolariser is effective only when extremely small currents are drawn from the cell. Hence, in early stages of balancing such a standard cell against a potentiometer wire, a protective resistance of the order of 100 kΩ is connected in series with the cell.

2.9. Grouping of Cells

A given number of cells may be grouped or connected together either for increasing the e.m.f. or the current. Following are the three different ways of grouping the cells.

2.10. Series Grouping

In series grouping, the positive electrode *i.e.* anode of one cell is connected to the cathode of the second cell and the anode of the second cell is connected to the cathode of the third cell and so on ,as shown in Fig. 2.5.

Let n = number of cells connected in series
r = internal resistance/cell
R = external load resistance
E = the e.m.f./cell

Then, total e.m.f. of the battery of n cells is $= nE$
Internal resistance of the battery $= nr$*
Total circuit resistance $= R + nr$

circuit current $$I = \frac{nE}{R + nr}$$

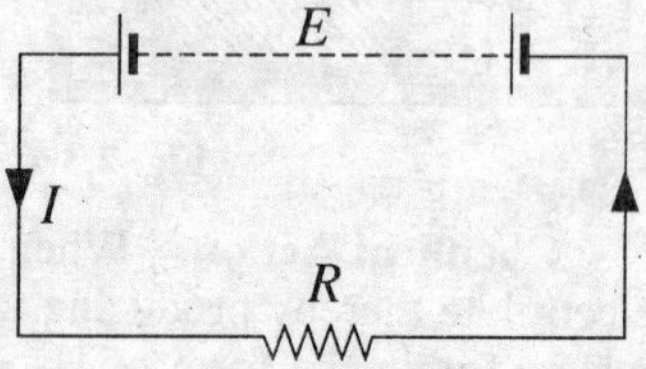

Fig. 2.5

If R is negligible as compared to 'nr', then

$$I = nE/nr = E/r = \text{current of one cell}$$

Hence, in this case, there is no increase of the current by joining n cells in series.

If, however, 'nr' is negligible relative to the external or load resistance R, then

$$I = nE/R = n \times E/R = n \times \text{current due to one cell}$$

In this case, we find that current delivered by the battery has increased n times.

Hence, we conclude that series grouping results in maximum current when internal resistance of battery is negligible as compared to the external resistance.

2.11. Parallel Grouping

In this method of grouping, anodes of all the cells are joined together to give the anode of the combination and all cathodes are joined together to give the cathode of the combination (Fig. 2.6).

Here, battery e.m.f. is the same as the e.m.f. of each cell. Moreover, the equivalent resistance of n resistances each of value r all connected in parallel is r/n. This equivalent resistance is in series with the external resistance R.

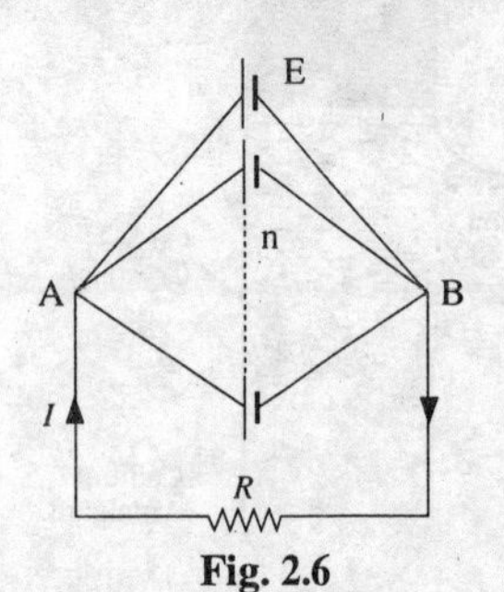

Fig. 2.6

Now, e.m.f. of the battery $= E$
Internal resistance of the battery $= r/n$
Total circuit resistance $= R + \frac{r}{n}$

$\therefore$ circuit current $$I = \frac{E}{R + \frac{r}{n}}$$

If 'r/n' is negligible relative to R, then $I = E/R$ *i.e.* equal to the current which we could have got even from one cell. Hence, under this condition, this type of grouping is not useful.

*The total e.m.f. of n cells in series is n times the e.m.f. per cell and total internal resistance is n *times the internal resistance per cell.*

However, if R is negligible relative to r/n, then $I = nE/r$. This current is n times the current we could get from one cell.

Hence, this grouping is useful when external resistance is small as compared to the internal resistance of the battery.

2.12. Mixed Grouping

In this case, a few cells are connected in series and some such series groups are connected in parallel as shown in Fig. 2.7.

Let the number of parallel rows be m and the number of cells, joined in series, in each row be n.

The internal resistance of this series parallel group can be found thus (Fig. 2.8).

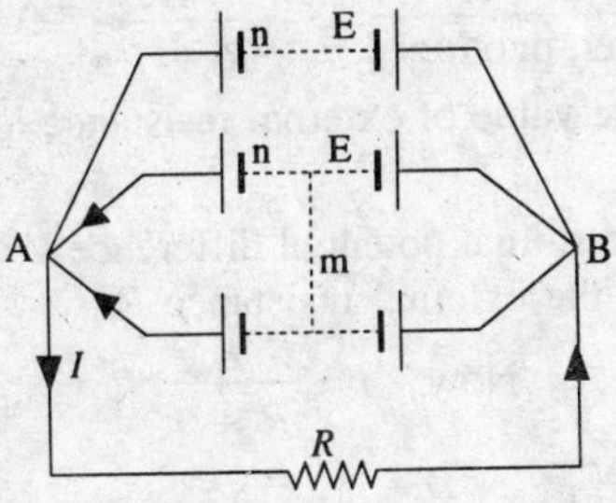

Fig. 2.7

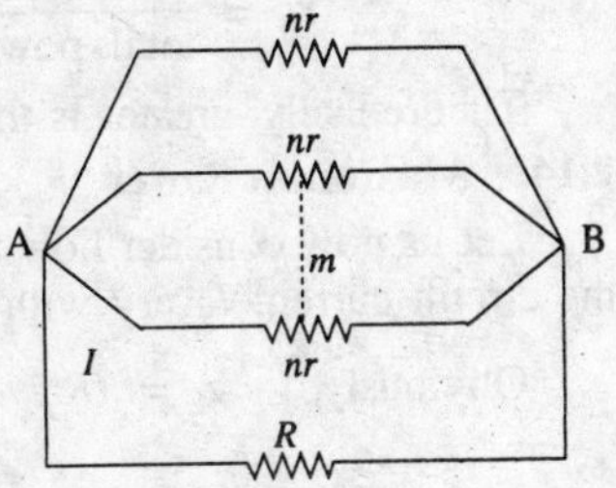

Fig. 2.8

Resistance of each row $= nr$

Equivalent resistance of such m rows connected in parallel $= nr/m$

$$\therefore \quad \text{total circuit resistance} = R + \frac{nr}{m}$$

$$\text{e.m.f. of the battery} = \text{e.m.f. of one row} = nE$$

$$\therefore \quad I = \frac{nE}{R + \frac{nr}{m}} = \frac{mnE}{mR + nr} = \frac{NE}{mR + nr}$$

where, $N = m \times n$ = total number of given cells.

Now numerator being constant, the circuit current will be maximum only when the denominator is minimum. The denominator will have minimum value* only when

$$mR = nr \qquad \text{or} \qquad R = \frac{nr}{m}$$

i.e. external resistance = internal resistance of the battery

Hence, series-parallel combination will yield maximum current when external or load resistance equals internal resistance of the battery. It should, however, be noted that this arrangement has an electrical efficiency of 50% only, because internal and external resistances being equal, half the power developed is wasted within the battery itself.

Note. For finding a suitable arrangement for obtaining maximum current from a given number of cells, the following two equations should be used to find the number of parallel rows and the number of cells joined in series in each row.

(*i*) $mn = N$ and (*ii*) $mR = nr$

If m turns out to be unity or less, then it means that all the N cells should be joined in series.

If, on the other hand, n comes out to be unity, then all the cells should be joined in parallel.

2.13. Efficiency of a Cell

In general, the efficiency of any system may be defined by the relation,

$$\text{efficiency} = \frac{\text{output}}{\text{input}} \times 100 \text{ percent}$$

---

* It may be proved thus : Let it be equal to y.

Then $y = mR + nr = \sqrt{(mR)^2} + \sqrt{(nr)^2} + (\sqrt{mR} - \sqrt{nr})^2 + \sqrt{mR}\sqrt{nr}$

Now, y will be minimum when right-hand side expression has a minimum value. The minimum value a squared quantity can have is zero.

$\therefore$ y is minimum when $(\sqrt{mR} - \sqrt{nr})^2 = 0$ or $mR = nr$.

Consider a cell having an e.m.f. of E, an internal resistance of r and delivering power to an external load resistance of R. If I is the circuit current, then useful power developed is I^2R watts and power lost within the cell is I^2r watts. Total power developed is $I^2R + I^2r = EI$ watts

$$\therefore \quad \eta = \frac{\text{useful power}}{\text{total power produced}} = \frac{I^2R}{I^2R + I^2r} = \frac{R}{R + r}$$

Theoretically, greater is the value of external resistance R, higher is the efficiency of the cell.

2.14. Maximum Power

Let us now consider how terminal potential difference V and the circuit current I vary with the external resistance R.

Obviously, $V = IR$; Now, $I = \dfrac{E}{R + r}$

$$\therefore \quad V = \frac{E}{R + r}R = E\left(1 - \frac{r}{R + r}\right)$$

As shown in Fig. 2.9, V increases as R increases. Similarly, $I = \dfrac{E}{R + r}$. As R increases, I decreases (Fig. 2.9). The useful power or power delivered to external resistance R is $P = VI$ watt.

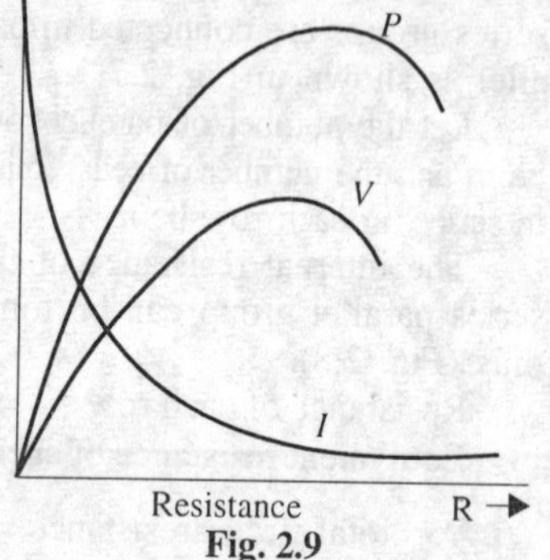

Fig. 2.9

By multiplying the ordinates for V and I, the third curve for power P can be drawn. It is seen from Fig. 2.9 that power P is maximum for a certain value of R. If external resistance is made more or less than this value, the power delivered is decreased.

Example 2.1. *A battery of 24 cells is required to be formed into a group in such a manner that it will send a maximum current through an external resistance of 5Ω. Each cell has an e.m.f. of 2.1 V and an external resistance of 2Ω. Find the best grouping and the value of the current through the external resistance.*

Solution. Let m = No. of rows in parallel ; n = No. of cells in each row

then, $N = n \times m = 24$ ...(*i*)

Now, for maximum current, $mR = nr$

or, $m \times 5 = n \times 2$ or $n = \dfrac{5m}{2}$ ...(*ii*)

Putting value of n from (*ii*) in (*i*), we get

$$\frac{5m^2}{2} = 24 \quad \text{or} \quad m^2 = \frac{48}{5} = 9.6$$

or m = **3 (to the nearest whole number)** $\therefore$ n = **8**

Hence, cells should be arranged in 3 rows, each containing 8 cells.

$$\text{Current through 5-}\Omega \text{ resistor} = \frac{2.1 \times 8}{5 + 16/3} = \mathbf{1.63\ A\ (approx)}$$

Example 2.2. *Find the minimum number of cells connected in two rows in parallel required to pass a current of 6 A through an external resistance of 0.7 Ω. Take the electromotive force of each cell as 2.1 volts and internal resistance of each cell as 0.5 Ω.*

Solution. Let 'n' be the number of cells in each of the two parallel rows.

E.M.F. of the battery $= 2.1\,n$ volt

Internal resistance of the battery $= 0.5\,n/2\ \Omega$

Total circuit resistance $= 0.7 + (0.5\,n/2)\ \Omega$

$$\text{Circuit current} = \frac{2.1\,n}{0.7 + (0.5\,n/2)}$$

$$\therefore \quad 6 = \frac{2.1\,n}{0.7 + (0.5\,n/2)} \quad \therefore \quad n = 7$$

Hence, minimum number of cells is $= 2 \times 7 = 14$.

Example 2.3. (*a*) *Five batteries are connected in parallel to supply current to a load resistance* *for heating. Each battery consists of 12 cells connected in series. The e.m.f. per cell is 2 volts and the individual cell resistance is 0.15 Ω. What should be the resistance of the load so that power consumed in the load is 384 watts ? For what load will the power consumption be maximum ?*

(b) Find the value of this maximum power.

Solution. (*a*) Total e.m.f. of the group $= 12 \times 2 = 24$ V

Internal resistance of one row $= 12 \times 0.15 = 1.8\ \Omega$

Combined internal resistance of 5 rows $= \dfrac{12 \times 0.15}{5} = 0.36\Omega$

Let external resistance be R Ω.

Then, current $I = \dfrac{24}{R + 0.36} A$

Power consumption in $R = I_2R$ watts $= \left(\dfrac{24}{R + 0.36}\right)^2 \times R$ watts

$$\therefore \quad \left(\frac{24}{R + 0.36}\right)^2 R = 384 \quad \text{or} \quad R^2 - 0.78\,R + 0.1296 = 0$$

$$\text{or} \quad R = \frac{0.78 \pm \sqrt{0.78^2 - 4 \times 0.1296}}{2} = \frac{0.78 \pm 0.3}{2} = 0.54\ \Omega \quad \text{or} \quad 0.24\ \Omega$$

Power consumption is maximum when external resistance = internal resistance

i.e., $R = \mathbf{0.36\ \Omega}$

(*b*) In this case, $I = \dfrac{24}{0.36 + 0.36} = \dfrac{100}{3}$ A

∴ maximum power consumed $= (100/3)2 \times 0.36 =$ **400 W**

Tutorial Problems No. 2.1

1. A battery of nine primary cells is connected

(*a*) all cells in series, (*b*) all cells in parallel,

(c) three sets in parallel each consisting of three cells in series. Each cell has an e.m.f. of 1.4 V and an internal resistance of 0.45Ω. The battery terminals are connected to a circuit of 7.2Ω. Calculate in each case :

(*i*) the current in the 7.2Ω resistance. (*ii*) the voltage drop across the resistance.

[(*a*) 1.12A ; 8.07V (*b*) 0.193A ; 1.39V (*c*) 0.549A; 3.95V]

2. Thirty-six cells of 2 V/cell and of internal resistance 1.5Ω/cell are given. Find how should they be joined as to send maximum current through an external resistance of 6 Ω. Find also the value of this current. **[3 rows of 12 cells each ; 2A]**

3. Eight cells are connected in series to an external resistance of 4W but one of the cells is connected in opposition. If the e.m.f. and internal resistance/cell are 2 volts and 3W respectively, find the current in the 4 ohm resistor. **[3/7 A]**

2.15. Division of Current in Parallel Circuits

In Fig. 2.10, two resistances are joined in parallel, across a voltage V. The current in each branch, as given by Ohm's Law, is

$$I_1 = V/R_1 \quad \text{and} \quad I_2 = V/R_2 \quad \therefore \quad \frac{I_1}{I_2} = \frac{R_2}{R_1}$$

$$\text{As} \quad \frac{1}{R_1} = G_1 \quad \text{and} \quad \frac{1}{R_2} = G_2$$

$$\therefore \quad \frac{I_1}{I_2} = \frac{G_1}{G_2}$$

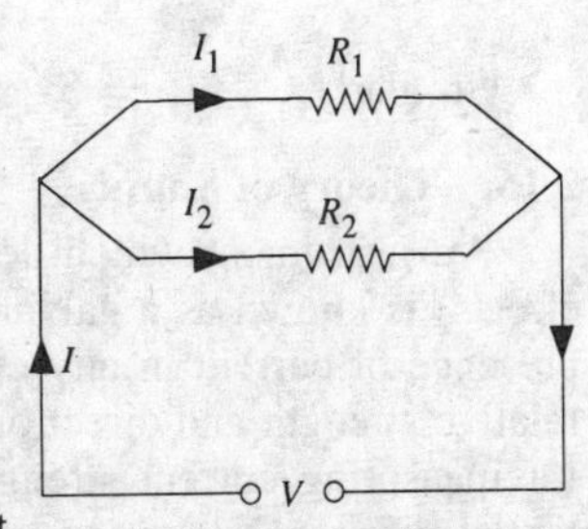

Fig. 2.10

Hence, the division of current in the branches of a parallel circuit is directly proportional to the conductance of the branches or inversely proportional to their resistances. We may also express the branch currents in the terms of the total circuit current thus :

Now $I_1 + I_2 = I \quad \therefore \quad I_2 = I - I_1$

Hence $\dfrac{I_1}{I - I_1} = \dfrac{R_2}{R_1} \quad \therefore \quad I_1 R_1 = R_2(I - I_1)$ or $I_1 = I\dfrac{R_2}{R_1 + R_2}$

Similarly, $I_2 = I\dfrac{R_2}{R_1 + R_2}$

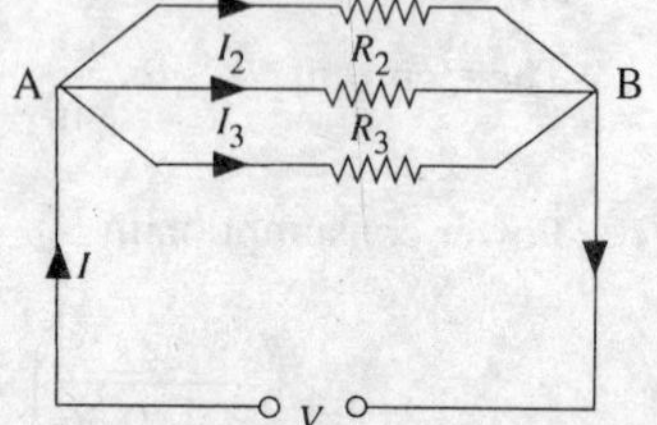

Fig. 2.11

Take the case of three resistors connected in parallel across a voltage V (Fig. 2.11). The total current $I = I_1 + I_2 + I_3$. Let the equivalent resistance be R. Then

$$V = IR$$

Also $V = I_1 R_1$

$\therefore$ $IR = I_1 R_1$

or $\dfrac{I}{I_1} = \dfrac{R_1}{R}$

or $I_1 = IR/R_1$ ...(i)

Now $\dfrac{1}{R} = \dfrac{1}{R_1} + \dfrac{1}{R_2} + \dfrac{1}{R_3}$ or $R = \dfrac{R_1 R_2 R_3}{R_2 R_3 + R_3 R_1 + R_1 R_2}$

From (*i*) above, $I_1 = I\left(\dfrac{R_2 R_3}{R_1 R_2 + R_2 R_3 + R_3 R_1}\right)$

Similarly, $I_2 = I\left(\dfrac{R_1 R_3}{R_1 R_2 + R_2 R_3 + R_3 R_1}\right)$ and $I_3 = I\left(\dfrac{R_1 R_2}{R_1 R_2 + R_2 Rsub3 + R_3 R_1}\right)$

Example 2.4. *An electric current of 5A is divided into three branches, the lengths of the wires in the three branches being proportional to 1, 2, and 3. Find the current in each branch (the wires are of the same material and cross-section).*

Solution. Since resistance is directly proportional to the length of a wire, the resistances of three branches are in the ratio 1 : 2 : 3. Let

$R_1 = 1\Omega$, then $R_2 = 2\Omega$ and $R_3 = 3\Omega$

As seen from Art. 2.15

$$I_1 = I\left(\frac{R_2 R_3}{R_1 R_2 + R_2 R_3 + R_3 R_1}\right)$$

Now $R_1 R_2 + R_2 R_3 + R_3 R_1 = 1 \times 2 + 2 \times 3 + 3 \times 1 = 11$

$\therefore$ $I_1 = 5 \times \dfrac{2 \times 3}{11} = \mathbf{\dfrac{30}{11}\ A}$

Similarly, $I_2 = 5 \times \dfrac{3 \times 1}{11} = \mathbf{\dfrac{15}{11}\ A}$ and $I_3 = 5 \times \dfrac{2}{11} = \mathbf{\dfrac{10}{11}\ A}$

2.16. Theory of Shunt

An instrument which detects electric current (by utilizing one of its effects) is known as a galvanometer. It is used for not only detecting the presence of current in any circuit but can also be used for measuring its relative strength and direction. The same galvanometer can be used either for measuring current strength *i.e.* as an ammeter or for measuring potential difference between two points *i.e.*, as a voltmeter.

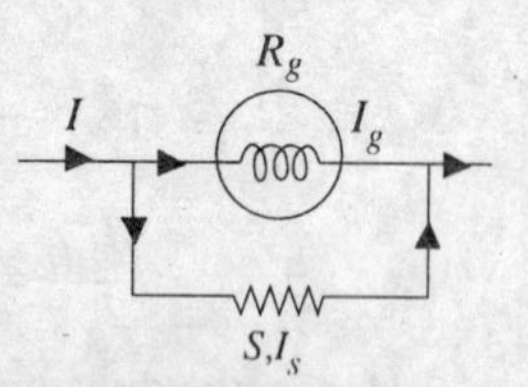

Fig. 2.12

Galvanometer itself can usually carry small current. If the current is too large, then only a part of the total current should be passed through the galvanometer, the remainder passing through a small resistance S connected across the galvanometer as shown in Fig. 2.12. The low resistance S provides a 'bypath' for the current and is called a shunt and the galvanometer is then said to be 'shunted'. Shunted galvanometers can be made to record currents many times greater than their normal full-scale deflection currents. The ratio of the maximum current (with shunt) to the full-scale deflection current (without shunt) is known as the 'multiplying power' or 'multiplying factor' of the shunt.

Let R_g = galvanometer resistance ; S = shunt resistance
I = line current ; I_g = full-scale deflection current of the galvanometer.

As seen from Fig. 2.12, the voltage across the galvanometer and the shunt resistance is the same as both are joined in parallel.

$$\therefore \quad R_g \times I_g = S \times (I - I_g)$$

because current through shunt is $I_s = (I - I_g)$

$$\therefore \quad S = \frac{R_g I_g}{I - I_g} \; ; \quad \text{Also,} \quad \frac{I}{I_g} = \left(1 + \frac{R_g}{S}\right)$$

$$\therefore \quad \text{multiplying power of shunt} = \left(1 + \frac{R_g}{S}\right)$$

Obviously, lower the value of shunt, greater is its multiplying power.

2.17. Ammeter Shunt

The range of ammeters can also be extended by suitably adjusting the value of shunt resistance.

Example 2.5. *A 1.0 A, 0.1Ω internal resistance ammeter is required for measuring current upto 10 amperes. How is it to be done? Explain giving connection diagram.*

(A.M.I.E. Winter 1987)

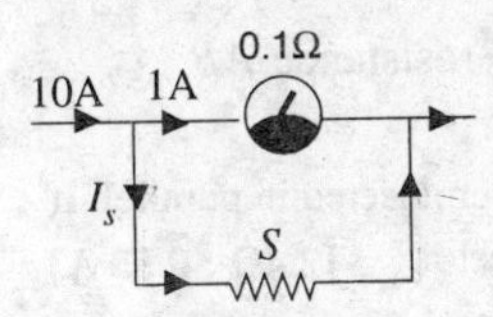

Fig. 2.13

Solution. The current range can be extended by using a suitable shunt, as shown in Fig. 2.13. Since line current is 10A, current passing through the shunt is (10 – 1) = 9A

$$\therefore \quad 0.1 \times 1 = 9 \times S \qquad \therefore \qquad S = \frac{\mathbf{1}}{\mathbf{90}}\,\Omega$$

Example 2.6. *In the circuit of Fig. 2.14 (a), the resistance of ammeter A is 0.15 Ω and that of voltmeter V is 16 kΩ. The voltmeter reads 120V and ammeter 0.35A. What is (i) the apparent power read by the meters and (ii) actual power taken by the lamp.*

(A.M.I.E. Summer 1988)

Solution. The given circuit can be re-drawn as shown in Fig. 2.14 (*b*).

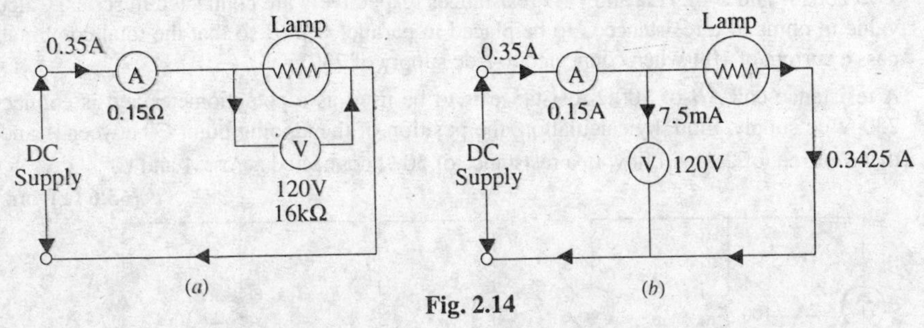

Fig. 2.14

Current drawn by voltmeter = 120/16K = 7.5 mA

(*i*) apparent power read by the instruments $= 120 \times 0.35$ = **42 W**

(*ii*) current actually passing through the lamp $= 0.35 - 0.0075 = 0.3425$ A

Actual power drawn by the 40–W lamp $= 120 \times 0.3425 = \mathbf{41\ W}$

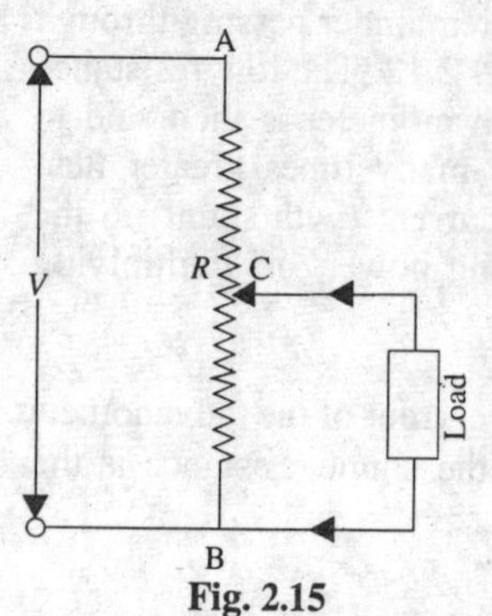

Fig. 2.15

2.18. Simple Potentiometer

A simple potentiometer or as it is sometimes called, a potential divider, is used for obtaining a variable voltage from a constant-voltage supply. The whole of supply voltage is dropped across a resistance *AB* as shown in Fig. 2.15 and by changing the position of the sliding contact *C* over the potentiometer resistance *AB*; any voltage from zero upto that of the supply can be obtained for applying it across the load. The volume control of radio or TV receiver is a common application of a potentiometer used as a voltage divider.

Example 2.7. *A potentiometer of resistance 80 Ω is connected across a supply of 120 volts. A current of 2A is required in the 10 Ω resistor. Find the position of the tapping point C and calculate the resistance between points A and C.*

Solution. Let the resistance of potentiometer between points *C* and $B = R\ \Omega$ as shown in Fig. 2.16. The point *C* is such that for that p.d. across the coil, a current of 2A passes through it. Now, p.d. across coil $= 10 \times 2 = 20$V. Same is the p.d. across resistance *R*. Hence, current through $CB = 20/R$ amperes.

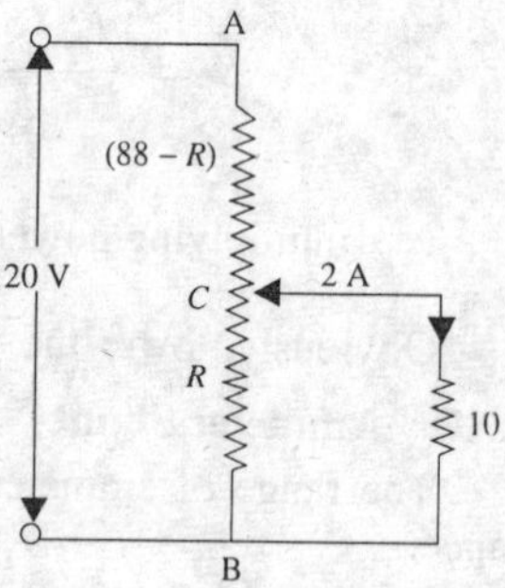

$\therefore$ total current entering at point $A = 2 + (20/R)$ amperes.

This current while passing through portion *AC* produces a voltage drop of $(80 - R)(2 + 20/R)$

However, as voltage across $CB = 20$V, voltage drop across $AC = 120 - 20 = 100$V

$\therefore \quad 100 = (80 - R)(2 + 20/R)$

or $\quad R_2 - 20R - 800 = 0$

or $\quad (R - 40)(R + 20) = 0 \quad \therefore \quad R = 40\ \Omega$

Fig. 2.16

Hence, the point *C* is situated at the middle point of the potentiometer resistance *AB*.

Tutorial Problems No. 2.2

1. Calculate the combined resistance of four conductors of 6, 9, 12 and 18 Ω connected in parallel. If current flowing in the circuit is 2 *A*, find the current flowing in the 9 Ω resistor. **[2.4Ω ; 0.53 A]**
2. *A* 150 Ω resistance coil *AB* is connected to terminals at 240V dc. Calculate the value of a further resistance coil which when connected between the mid-point of AB and the end *A*, will carry a current of 0.8 *A*. **[112.5 Ω]**
3. Two lamps are connected in parallel to a constant dc supply of 110 V. The resistances of the lamps are as 11 is to 9. The current flowing in the circuit is 4.5 *A*. Find the resistance of each lamp. **[54.72Ω ; 44.4 Ω]**
4. Two coils *A* and *B* of 11Ω and 14Ω resistances respectively are connected in series. Calculate the value in ohms of a resistance *C* to be placed in parallel with *B* so that the total combination will pass a current of 10*A* when connected to a dc supply of 230V. **[84 Ω]**
5. A resistance coil *AB* of 100Ω resistance is to be used as a potentiometer and is connected to a 230-V dc supply. Find, by calculation, the position of the tapping point *C* between *A* and *B* such that a current of 2 *A* will flow in a resistance of 50 Ω connected across *A* and *C*. **[63.6 Ω from A to C]**

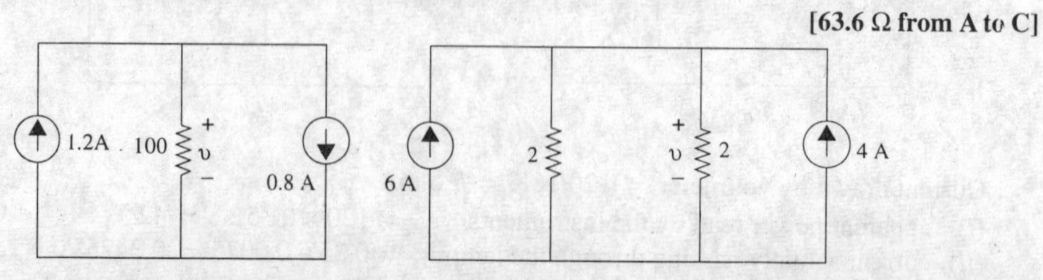

Fig. 2.17 Fig. 2.18

6. For the circuit of Fig. 2.17, find the value of v. **[40 V]**
7. In the circuit of Fig. 2.18, find the value of v. **[10 V]**
8. Find the values of i_2 and i_3 in Fig. 2.19. All resistance value are in ohm. **[2mA ; 5mA]**

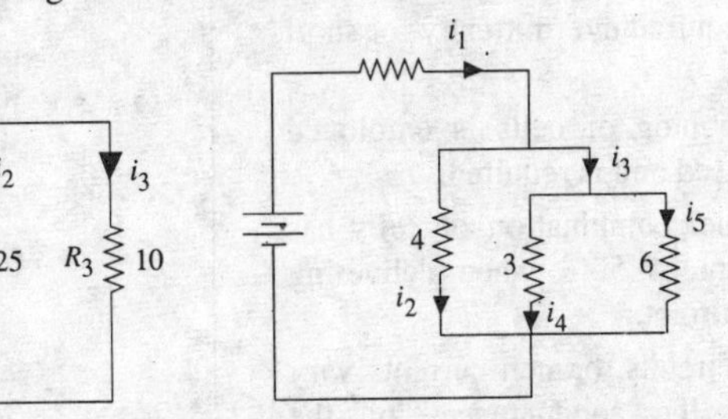

Fig. 2.19 **Fig. 2.20**

9. For the circuit shown in Fig. 2.20, find the values of i_1, i_2, i_3 and i_5 if $i_4 = 6$A. **[13.5 A; 4.5A, 9A, 3A]**

HIGHLIGHTS

1. The e.m.f. (E) of a cell is connected to its terminal p.d. V by the relation $E = V + ir$
2. A given number of cells may be joined (*i*) in series (*ii*) in parallel and (*iii*) in series-parallel combination.
3. For finding the best arrangement for grouping of cells, the following two equations may be used :

   (*i*) $N = m \times n$ (*ii*) $mR = nr.$
4. In parallel combination of various resistances, the circuit current divides inversely in the ratio of the branch resistances.

(*a*) $$I_1 = I\frac{R_2}{R_1 + R_2}, \quad I_2 = I\frac{R_1}{R_1 + R_2}$$ for 2.branches circuit

(*b*) $$I_1 = I\frac{R_2R_3}{\Sigma R_1R_2}, \quad I_2 = I\frac{R_3R_1}{\Sigma R_1R_2}, \quad I_3 = I\frac{R_1R_2}{\Sigma R_1R_2}$$ for 2.branches circuit

5. Shunt is a low resistance connected in parallel with a galvanometer for increasing its current range. Smaller the value of shunt resistance, greater is the extension in its range.
6. A simple potentiometer or potential divider is a suitable resistance from which any percentage of the applied voltage can be obtained :

OBJECTIVE TESTS —2

A. Fill in the following blanks :

1. A........... cell is one that cannot be recharged.
2. A battery consists of a number of ....
3. EMF of a cell is always .... than its terminal voltage.
4. A standard cell is one whose emf varies very little with time and ....
5. Series grouping of cells gives maximum current only when internal resistance of the battery is ....as compared to external resistance.
6. Series parallel grouping of cells yields maximum current when external resistance .... internal resistance of the battery.
7. Division of current in a parallel circuit is directly proportional to the............ of the branches.
8. The main purpose of using shunt across a galvanometer is to.......... its current range.

B. Answer True or False

1. A primary cell essentially consists of two similar electrodes immersed in an electrolyte.

2. Whereas emf of a given cell is constant, its terminal potential difference is not.
3. Leclanche cell is useful in circuits where current is required intermittently for short periods.
4. Parallel grouping of cells is employed when increased emf is required.
5. Series-parallel combination of cells has an efficiency of 50% when delivering maximum current.
6. In parallel circuits, branch currents vary directly as the conductances of the branches.
7. A shunted galvanometer has higher current range than an unshunted one.
8. A potentiometer is used for obtaining variable voltages from a given voltage supply.

C. Multiple Choice Questions

1. The emf of a given cell is
   (*a*) distributive
   (*b*) variable
   (*c*) constant
   (*d*) less than terminal p.d.
2. A Lechlanche cell has
   (*a*) a depolariser
   (*b*) a cathode of carbon
   (*c*) an electrolyte of H_2SO_4
   (*d*) large current capacity.
3. A standard cell is primarily used
   (*a*) for supplying large currents
   (*b*) as a source of energy
   (*c*) for providiing variable voltage
   (*d*) as a secondary standard of voltage
4. Parallel grouping of cells yields maximum current when external resistance
   (*a*) equals internal resistance per cell
   (*b*) is greater than internal resistance of the battery
   (*c*) equals internal resistance of the battery
   (*d*) is less than the internal resistance of the battery.
5. Parallel grouping of cells yields maximum current when external resistance.
   (*a*) reduce its resistance
   (*b*) increase its current range
   (*c*) increase its voltage range
   (*d*) decrease its loading effect.
6. A simple potentiometer is correctly called a
   (*a*) current divider
   (*b*) voltage stablizer
   (*c*) variable resister
   (*d*) voltage dividor.
7. Four 1.5 V cells are connected in parallel. The output voltage is
   (*a*) 1.0 V (*b*) 1.5 V
   (*c*) 6.0 V (*d*) none of these

(Elect. Engg. A.M.Ae. S.I. June 1994)

ANSWERS

A. 1. primary 2. cells 3. greater 4. temperature 5. negligible 6. equals 7. conductance 8. extend/increase

B. 1. F 2. T 3. T 4. F 5. T 6. T 7. T 8. T

C. 1. *c* 2. *a* 3. *d* 4. *c* 5. *b* 6. *d*

3 NETWORK ANALYSIS

3.1. General

Many laws and theorems are available for solving networks of conductors, both active and passive. The main advantage of these theorems is that they save time and considerably reduce the laborious mathematical work involved in solving networks, thereby minimising chances of error. The different laws and theorems discussed in the text are :

1. Kirchhoff's Laws 2. Maxwell's Loop Current Theorems 3. Superposition Theorems 4. Thevenin's Theorem 5. Norton's Theorem 6. Maximum Power Transfer Theorem 7. Delta / Star Transformation. 8. Star / Delta Transformation.

3.2. Kirchhoff's Laws*

These laws are more comprehensive than Ohm's law and are used for solving electrical networks which may not be readily solved by the latter. The two laws are :

(1) Kirchhoff's First Law or Point Law or Current Law (KCL)

It states that :

in any network of conductors, the algebraic sum of the currents meeting at a point (or junction) is zero.

Put in another way, it simply means that the total current *leaving* a junction is equal to the total current *entering* the junction. It is obviously true because there is no accumulation or depletion of current at any junction of the network.

Explanation

Consider the case of a few conductors meeting at a point A as in Fig. 3.1. Some conductors have currents leading to point A whereas others have currents leading away from point A. Assuming the incoming currents to be positive and the outgoing currents negative, we have

$$I_1 + (-I_2) + (-I_3) + I_4 + (-I_5) = 0$$

or, $$I_1 + I_4 - I_2 - I_3 - I_5 = 0$$

or, $$I_2 + I_4 = I_2 + I_3 + I_5$$

or incoming currents = outgoing currents

We can express the above conclusion thus $\Sigma I = 0$ ... at a junction

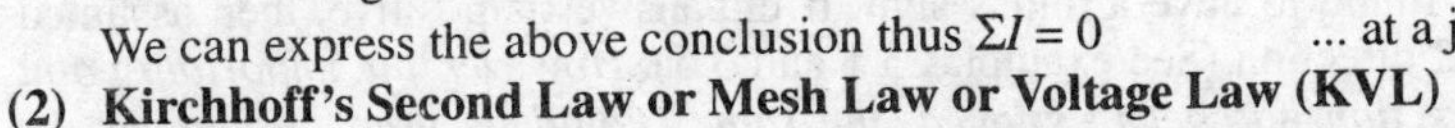

Fig. 3.1

(2) Kirchhoff's Second Law or Mesh Law or Voltage Law (KVL)

It states that :

the algebraic sum of the product of current and resistance in each of the conductors in any closed mesh (or path) in a network plus the algebraic sum of the e.m.f.s in that path is zero.

$$\Sigma IR + \Sigma \text{e.m.f.} = 0 \qquad \text{...round a mesh}$$

The basis of this law is this : If one starts from a particular junction and goes round the mesh till one comes back to the starting point, then one must be at the same potential with which one started. Hence, it means that all the sources of e.m.f. met on the way, must necessary be equal to the voltage drops in the resistances, every being given its proper sign, plus or minus.

3.3. Determination of Sign

In applying Kirchhoff's laws to specific problems, particular attention should be paid to the algebraic signs of voltage drops and e.m.f.s; otherwise results will come out to be wrong. Following sign convention is suggested :

*After Gustav Robert Kirchhoff (1824-1827), a German physicist.

A *rise* in voltage should be given a +ve sign and a *fall* in voltage a –ve sign. Keeping this in mind, it is clear that as we go from the –ve terminal of a battery to its +ve terminal, there is a rise in potential, hence this voltage should be given a +ve sign [Fig. 3.2 (*a*)]. If, on the other hand, we go from +ve terminal to –ve termina l [Fig. 3.2 (*b*)], then there is a *fall* in potential, hence this voltage should be preceded by a –ve sign. *It is important to note that the sign of the battery e.m.f. is independent of the direction of the current through that branch.*

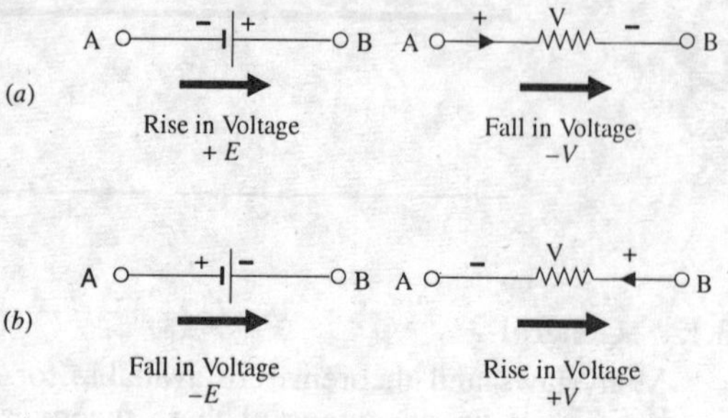

(*a*) Rise in Voltage $+E$ — Fall in Voltage $-V$

(*b*) Fall in Voltage $-E$ — Rise in Voltage $+V$

Fig. 3.2

Now, take the case of a resistor. If we go through a resistor in the same direction as the current, then there is a *fall* in potential, because current flows from higher to lower potential. Hence, this voltage fall should be taken –ve. However, if we go in a direction opposite to that of the current, then there is a rise in voltage. Hence, this voltage rise should be given a positive sign.

It is clear that the sign of voltage drop across a resistor depends on the direction of current through that resistor.

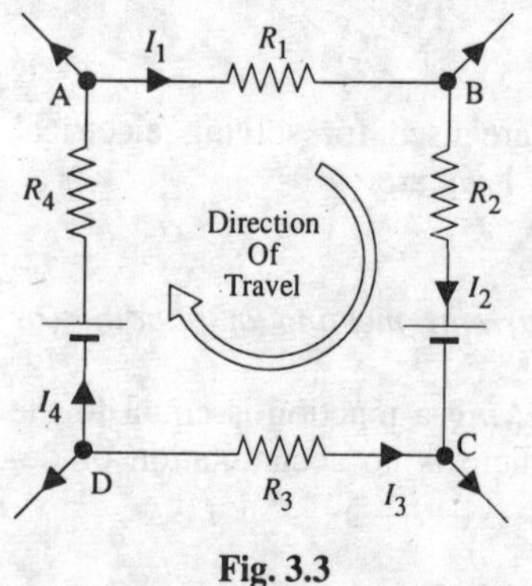

Fig. 3.3

Consider the closed path *ABCDA* in Fig. 3.3. Different voltage drops will have the following signs :

$I_1 R_1$ is –ve (fall in potential)

$I_2 R_2$ is –ve . " "

$I_3 R_3$ is +ve (rise in potential)

$I_4 R_4$ is –ve (fall in potential)

E_2 is –ve (fall in potential)

E_1 is +ve (rise in potential)

Using Kirchoff's Second Law, we get

$$-I_1 R_1 - I_2 R_2 + I_3 R_3 - I_4 R_4 - E_2 + E_1 = 0 \quad \text{or} \quad I_1 R_1 + I_2 R_2 - I_3 R_3 + I_4 R_4 = E_1 - E_2$$

3.4. Assumed Direction of Current

In applying Kirchhoff's laws to electrical networks, the question of assuming proper direction of current usually arises. The direction of current flow may be assumed either clockwise or anticlockwise. If the assumed direction of current is not the actual direction, then on solving the question, this current will be found to have a minus sign. If the answer is positive, then assumed direction is the same as actual direction (See examples 3.1 and 3.5). *However, the important point is that once a particular direction has been assumed, the same should be used throughout the solution of the question.*

Note. It should be noted that Kirchhorff's laws are applicable both to d.c. and a.c. voltages and currents. However, in the case of alternating currents and voltages, any e.m.f. of self-inductance or that existing across a capacitor should be also taken into account.

Example 3.1. *A bridge network ABCD is arranged as follows: resistances between terminals A-B, B-C, C-D, D-A and B-D are 10, 30, 15, 20 and 40 Ω respectively. A 2-V battery of negligible internal resistance is connected between terminals A and C. Determine the value and direction of the current in 40 Ω resistor.*

(A.M.I.E. Winter 1993)

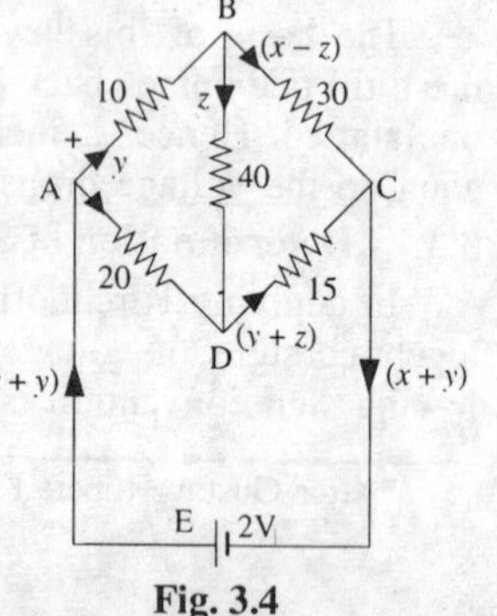

Fig. 3.4

Solution: The directions of various currents are as shown in Fig.3.4. Applying Kirchhoff's Second Law to various closed circuits, we have

Circuit ABDA

$$-10x - 40z + 20y = 0 \quad \text{or} \quad x - 2y + 4z = 0 \qquad ...(i)$$

Circuit BCDB

$$-30(x-z)+15(y+z)+40z=0 \quad \text{or} \quad 6x-3y-17z=0 \qquad ...(ii)$$

Circuit ADCEA

$$-20y-15(y+z)+2=0 \quad \text{or} \quad 35y+15z=2 \qquad ...(iii)$$

Multiplying Eq. (*i*) by 6 and subtracting Eq. (*ii*) from it, we have

$$9y-41z=0 \quad ...(iv)$$

Multiplying Eq.(*iii*) by 9 and then subtracting Eq. (*iv*) from it after multiplying it by 35, we have,

$$1570z = 18, z = 18/1570 = \mathbf{9/785\ A.} = 0.0115\ \text{A}$$

Since z turns out to be positive, its actual direction of flow is the same as assumed in Fig.3.4.

Example 3.2. *Find the current distribution in the network shown in Fig. 3.5.*

Electrical Science AMIE Winter 1993)

Solution. Let the current distribution in the different branches of the circuit be as shown in Fig. 3.5 will apply *KVL* to the closed circuit *ABC* and *BDC*.

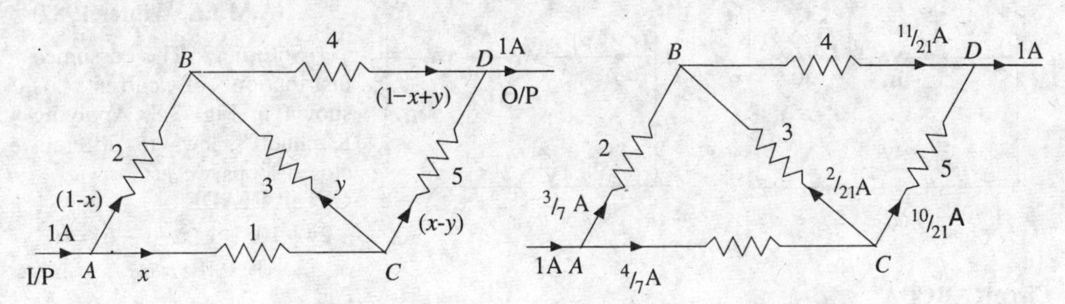

Fig. 3.5 **Fig. 3.6**

Circuit ABCA

Starting from point A and goiin clockwise we have

$$-2(1-x)+3y+x\times 1 = 0 \quad \text{or} \quad 3x+3y = 2 \qquad ...(i)$$

Circuit BDCB

$$-4(1-x+y)+5(x-y)-3y = 0 \quad \text{or} \quad 9x-12y = 4 \qquad ...(ii)$$

Multiplying (*i*) by 3 and substracting (*ii*) from it, we get

$$20y = 2 \quad \text{or} \quad y = 2/21\ \text{A}$$

Since y comes to be the positive, it means that its actual direction of flow is the same as shown in Fig. 3.5.

Substituting this value of y in (*i*) above, we get

$$3x = 12/7 \quad \text{or} \quad x = 4/7\ \text{A}$$

Hence, current distribution in the various branches of the network is as shown in Fig. 3.6.

Example 3.3. *Two batteries A and B are connected in parallel and a load of 10 Ω is connected across their terminals. A has an e.m.f. of 12 V and internal resistance of 2 Ω; B has an e.m.f. of 8 V and an internal resistance of 1 Ω. Using Kirchhoff's laws, determine the values and directions of the currents flowing in each of the batteries and in the external resistance. Also, determine the potential difference across the external resistance.* **(A.M.I.E. Winter 1988)**

Solution: The two batteries are eonnected in parallel as shown in Fig.3.7. It should be noted that unless stated otherwise, *the similar ends of the two batteries are assumed to be joined together.* Let the directions of the two branch currents be as shown.

Applying Kirchhoff's law to the two loops, we get the following equations:–

Loop CEDAC. Starting from point *C* anticlockwise, we get

$$-10(x+y)-2x+12 = 0 \quad \text{or} \quad 6x+5y=6 \qquad ...(i)$$

Loop CEDBC. Starting from point *C*, we get

$$-10(x+y)-1\times y+8= 0 \quad \text{or} \quad 10x+11y=8 \qquad ..(ii)$$

Solving for x and y from (*i*) and (*ii*), we get

$x = 13/8 = \mathbf{1.625\ A\ (discharge)},$

$y = \mathbf{-0.75\ A\ (charge)}$

Since y turns out to be *negative,* its actual directino of flow is opposite to that shown in Fig.3.6. It is a *charging* current and not discharging one.

Current in 10 Ω resistor = (1.625 − 0.75) = **0.875 A**

P.D. across external resistor = 10 × 0.875 = **8.75 V.**

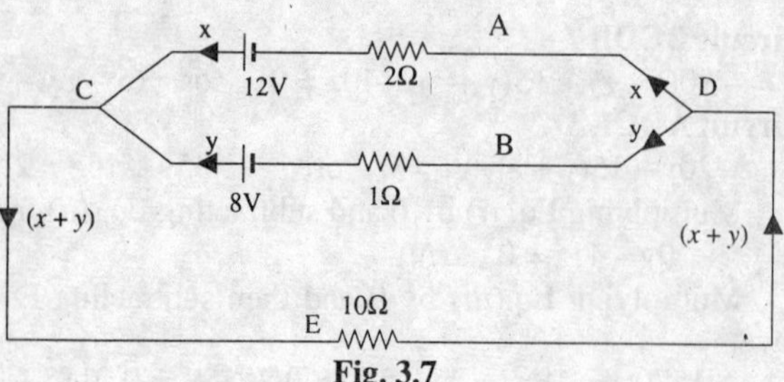

Fig. 3.7

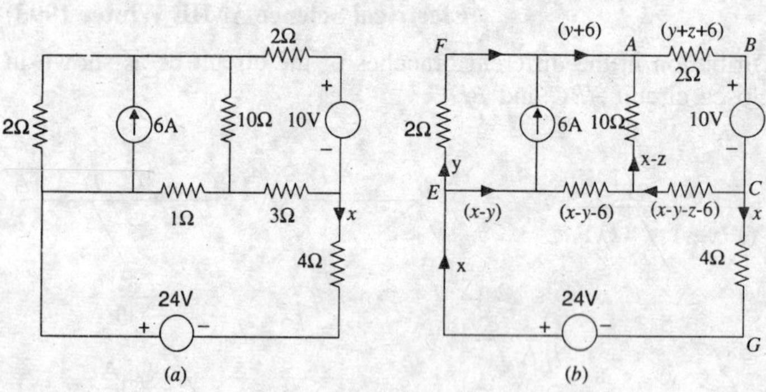

Fig. 3.8

Example 3.4. *Determine the current x in the 4-chm resistance in the circuit shown in Fig.3.8 (a) below.*

(A.M.I.E. Winter 1987)

Solution: The assumed distribution of currents is shown in Fig. 3.8. Applying Kirchhoff's laws to different closed loops we get

Circuit EFADE

$-2y + 10z + 1(x - y - 6) = 0$

or $x - 3y + 10z = 6$ ...(*i*)

Circuit ABCDA

$-2(y + z + 6) - 10 + 3(x - y - z - 6) - 10z = 0$ or $3x - 5y - 15z = 40$ ...(*ii*)

Circuit EDCGE

$-1(x - y - 6) - 3(x - y - z - 6) - 4x + 24 = 0$ or $8x - 4y - 3z = 48$ ...(*iii*)

Multiplying Eq.(*i*) by 5 and Eq.(*ii*) by 3 and then subtracting Eq.(*ii*) from Eq.(*i*), we get

$-4x + 95z = -90$ or $4x - 95z = 90$ ...(*iv*)

Next, multiplying Eq.(*ii*) by 4 and Eq.(*iii*) by 5 and subtracting Eq.(*iii*) from Eq.(*ii*), we get

$-28x - 45z = -80$ or $28x + 45z = 80$ ...(*v*)

Multiplying Eq.(*iv*) by 45 and Eq.(*v*) by 95 and adding the two, we get

$284x = 1165$ or $x = 1165/284 = \mathbf{4.1\ A}$

Example 3.5. *Formulate the Kirchhoff voltage law equations for the circuit of Fig.3.9 and find the values of* I_1, I_2 and I_3. **(A.M.I.E. Winter 1984)**

Solution. Circuit ABFGHA

$-I_1 \times 1 - 3(I_1 + I_2) - 6(I_1 + I_2 + I_3) - 0.5 I_1 + I_2 = 0$

$\therefore\ 7I_1 + 6I_2 + 4I_3 = 8$ ...(*i*)

Circuit BCDFB. Starting from point *B*, we get

$+0.4I_2 - 10 - 1.5I_3 + 3(I_1 + I_2) = 0$

$\therefore\ 30I_1 + 34I_2 - 15I_3 = 100$ ...(*ii*)

Circuit FDEGF

$+0.3(I_2 + I_3) - 6 + 6(I_1 + I_2 + I_3) + 1.5I_3 = 0$

$20I_1 + 21I_2 + 26I_3 = 20$ ...(*iii*)

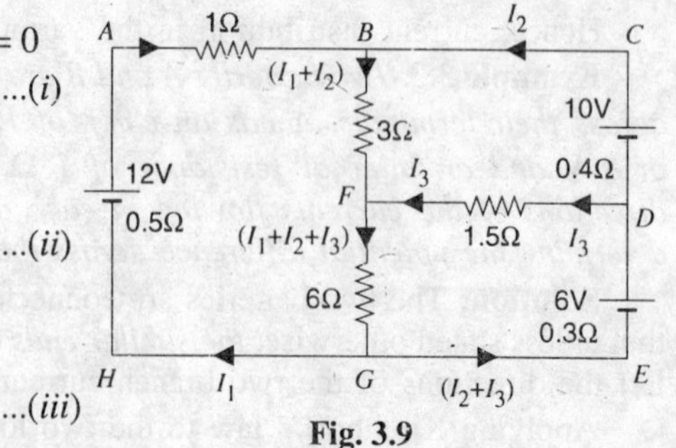

Fig. 3.9

Solving for the three currents, we get

$I_1 = \mathbf{-1.25\ A};\quad I_2 = \mathbf{3.54\ A};\quad I_3 = \mathbf{-1.13\ A}$

The negative signs mean that actual directions of flow of I_1 and I_3 are opposite to those shown in Fig. 3.9.

Example 3.6. *Use KCL to find the current supplied by the voltage–controlled current source in Fig.3.10.*

Solution: We will apply *KCL* to node *A*. Currents coming towards *A* would be taken positive and those going away from it is negative.

$\therefore \quad -2 - i_1 + 2v - i_2 = 0 \qquad \text{or} \qquad 2v = i_1 + i_2 + 2$

Now, $i_1 = v/3$ and $i_2 = v/6$ $\therefore$ $2v = v/3 + v/6 + 2$ or $v = 4/3$

Hence, value of current source = $2 \times 4/3 =$ **8/3 A**

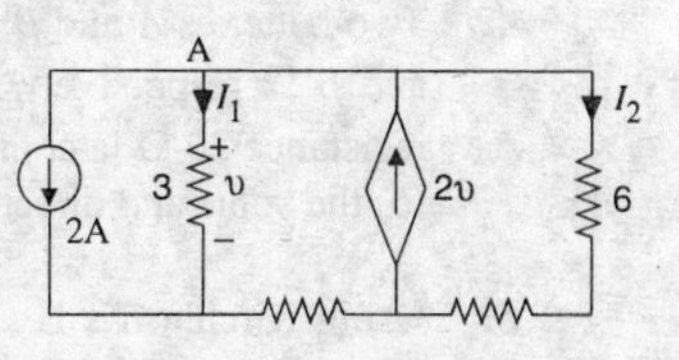

Fig. 3.10

Example 3.7. *Using KVL, find the value of* v *in the circuit of Fig. 3.11.*

Solution: As seen, in this circuit, the value of the dependent current source depends on the current i_1 through the 4 Ω resistor. It is obvious that current through 2 Ω resistor is $10i_1$. Let us apply **KVL** to the closed circuti ***ABCDA***. Starting from point *A* in the clockwise direction, we have

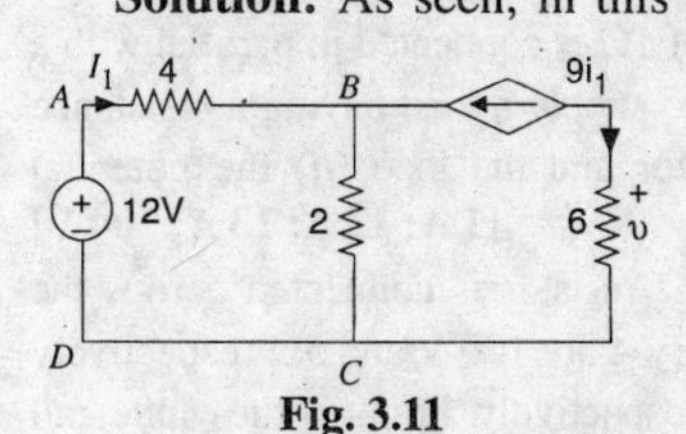

Fig. 3.11

$$-4i_1 - 10i_1 \times 2 + 12 = 0 \quad \text{or} \quad i_1 = 0.5 \text{ A}$$

Now, current through 6 Ω resistor is $9\,i_1 = 9 \times 0.5 = 4.5$ **A.**

Hence, $v = 6 \times 4.5 =$ **27 V.**

Example 3.8. *Find the current in each branch of the given network in Fig.3.12.*

(A.M.I.E. Summer 1990)

Solution: Starting from point *B* and applying *KVL* to circuit *BAFEB,* we get

$$-5\,(x - y) - 15 - 15\,(x - y) + 8y + 15 = 0$$

$\therefore \qquad 5x = 7y \qquad \ldots(i)$

From circuit *CBEDC*, we get

$$-10\,x - 15 - 8y - 10x + 25 = 0$$

$10x + 4y = 5 \qquad \ldots(ii)$

From (*i*) and (*ii*), we get

$x =$ **7/18 A;** $\qquad y =$ **5/18 A**

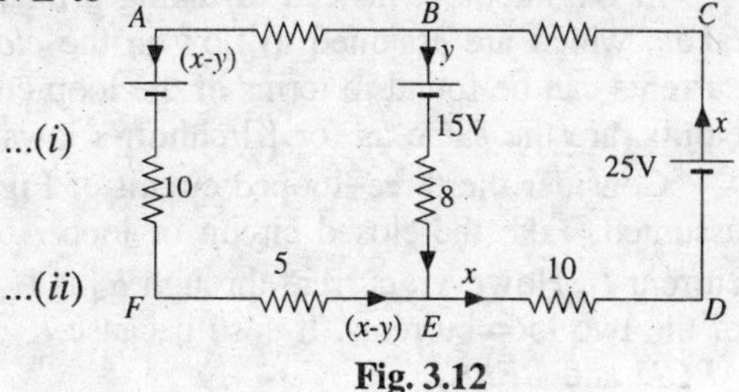

Fig. 3.12

Tutorial Problems No. 3.1

1. Find the value of current *i* in Fig. 3.13 by using *KCL* and Ohm's law. **[1A]**

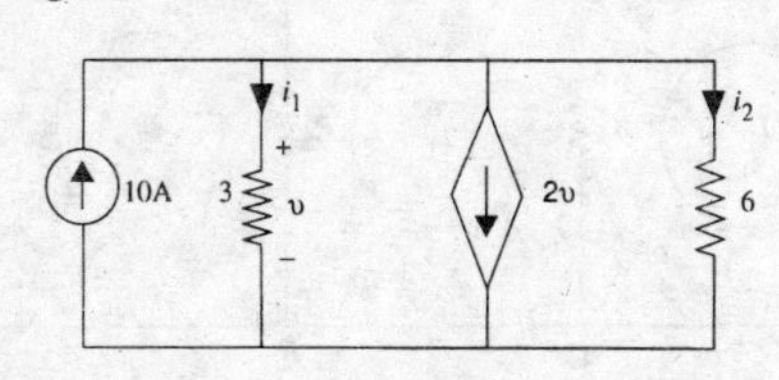

Fig. 3.13 **Fig. 3.14**

2. By using *KCL*, compute the values of currents i_1, i_2 and the values of the voltage–dependent current source in the circuit of Fig. 3.14. **[4/3 A, 2/3 A, 8 A]**

3. A Wheatstone bridge *ABCD* is arranged as follows: resistance between *A–B, B–C, C–D, D–A* and *B–D* are 10, 20, 15, 5 and 40 Ω respectively. A 20–V battery of negligible internal resistance is

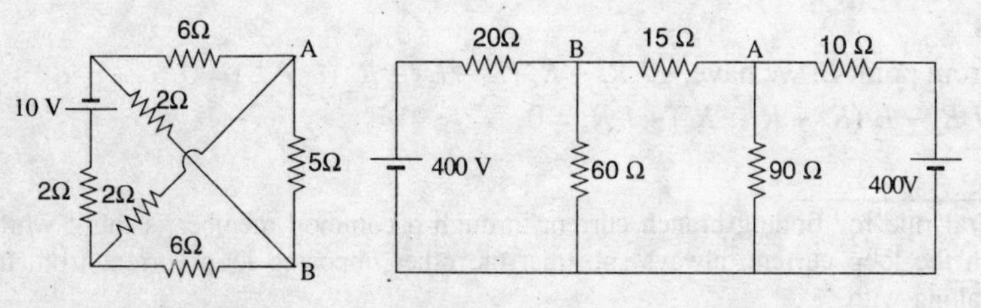

Fig. 3.15 **Fig. 3.16**

connected between terminals A and C. Determine the current in each resistor.

[AB = 0.645; BC = 0.678 A; AD = 1.025 A; DC = 0.992 A; DB = 0.033 A]

4. Two batteries A and B are connected in parallel. The e.m.f. and internal resistance of A are 110 V and 6 Ω respectively and the corresponding values for B are 130 V and 4 Ω respectively. A resistance 20 Ω is connected across the battery terminals. Calculate (*a*), the terminal voltage and (*b*) the value and direction of the current in each battery.

   [(*a*) 108.9 V (*b*) 0.1785 A discharge; 5.27 A discharge]

5. Using Kirchhoff's laws, calculate the voltage across AB in Fig.3.15 and indicate the polarity of this voltage. **[Point B is 2 V above point A]**

6. Using Kirchhoff's laws or otherwise, find he magnitude and direction of current flowing through the 15–Ω resistance of Fig.3.16. **[1 A from A to B]**

7. A battery having an e.m.f. of 105 V and an internal resistance of 1Ω is connected in parallel with a d.c. generator of e.m.f. 110 V and internal resistance of 0.5Ω to supply a load having a resistance of 8Ω. Calculate (*i*) the currents in the battery, the generator and the load (*ii*) the potential difference across the load. **[1 A; 12 A; 13 A; 104 V]**

8. Two batteries, A and B, are connected in parallel and an 80Ω resistor is connected acroos the battery terminals. The e.m.f. and the internal resistance of battery A are 100 V and 5Ω respectively and the corresponding values for battery B are 95 V and 3Ω respectively. Find (*a*) the value and direction of the current in each battery and (*b*) the terminal voltage.

   [1.069 A discharge; 0.1145 a discharge; 94.66 V]

3.5. Maxwell's Loop Current Method or Theorem

In this method, instead of taking *branch* currents (as in Kirchhoff's laws) *loop* currents are taken, which are assumed to flow in the *clockwise* direction. Of course, where required, branch currents can be found in terms of the loop currents. Sign conventions for the IR drops and bettery e.m.fs. are the same as for Kirchhoff's laws.

Consider the three–looped circuit of Fig. 3.17 where loop currents of I_1, I_2 and I_3 have been assumed. Take the closed circuit or loop $ABCDA$. As seen, current through R_1 is the same loop current I_1. However, current through R_4 is $(I_1 - I_2)$. Here, branch current is equal to the *difference* of the two loop currents. It is so because R_4 happens to be the common member of the two loops $ABCD$ and $BEFC$.

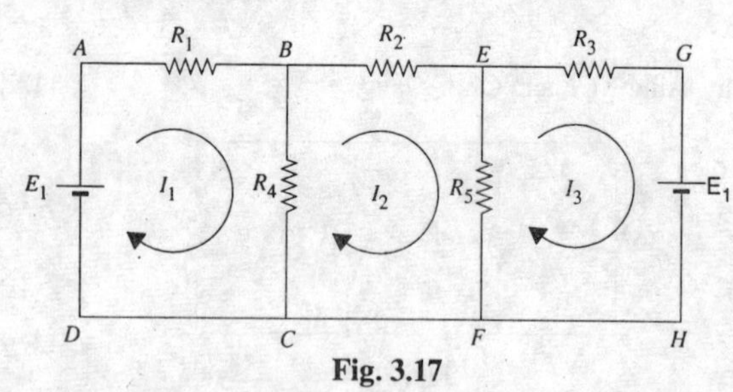

Fig. 3.17

Consider the second loop $BEFCB$. Current through R_2 is I_2. But current through R_5 is $(I_2 - I_3)$ and that through R_4 is $(I_2 - I_1)$ and *not* $(I_1 - I_2)$ *as was the case when we went round* the first loop a little earlier.*

Similarly, current through R_2 is I_3 and that through R_5 is $(I_3 - I_2)$ and not $(I_2 - I_3)$ *as was the case for the second loop.*

The voltage equations for the three loops are as under:

Loop ABCDA

Starting from point A, we have, $-I_1R_1 - R_4(I_1 - I_2) + E_1 = 0$

Simplifying it further, we have, $I_1(R_1 + R_4) - I_2R_4 = E_1$ ...(*i*)

Loop BEFCB

Starting from point B, we have, $-I_2R_2 - R_5(I_2 - I_3) - R_4(I_2 - I_1) = 0$

or $I_1R_4 - I_2(R_2 + R_4 + R_5) + I_3R_5 = 0$ ...(*ii*)

---

*The general rule for finding branch current through a common member is this : while going roud a loop along with the loop current, always substract the other opposing loop current from the loop current you are going along with.

Loop EGHFE

Starting from point *E*, we get, $I_3R_3 - E_2 - R_5(I_3 - I_2) = 0$

or $\quad I_2R_5 - I_3(R_3 + R_5) = E_2$ ...(*iii*)

The three equations given above may be used for finding the values and directions of flow of loop currents I_1, I_2 and I_3.

Example 3.9. *Find I_1, I_2 and I_3 in the network shown in Fig.3.18. using loop-cvrrent method. Figures against resistances indicate their values in ohms.*

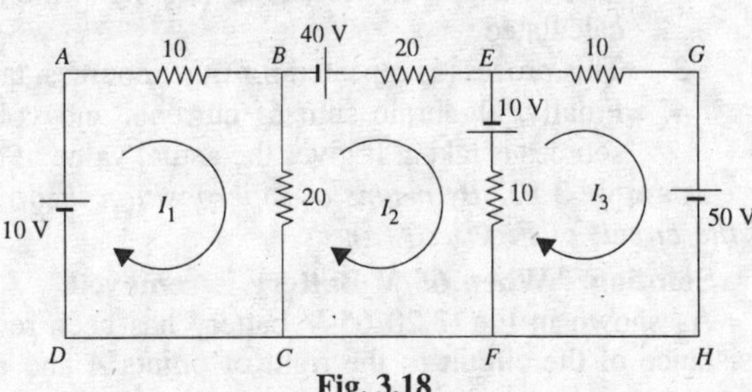

Fig. 3.18

Solution: Different loops would be taken one after another.

Loop ABCDA

$-10I_1 - 20(I_1 - I_2) - 10 = 0$

or $\quad 3I_1 - 2I_2 = -1$ ...(*i*)

Loop BEFCB

$40 - 20I_2 + 10 - 10(I_2 - I_3) - 20(I_2 - I_1) = 0$ or $\quad 2I_1 - 5I_2 + I_3 = -5$ ...(*ii*)

Loop EGHFE

$-10I_2 + 50 - 10(I_3 - I_2) - 10 = 0$ or $\quad I_2 - 2I_3 = -4$ ...(*iii*)

Multiplying Eq.(*ii*) by 2 and adding it to Eq.(*iii*), we get $\quad 4I_1 - 9I_2 = -14$ ...(*iv*)

Solving for I_1 and I_2 from Eq.(*i*) and (*iv*), we get

$I_1 - \mathbf{1A}$ and $I_2 = \mathbf{2A}$

Substituting these values in Eq. (*iii*), we have, $I_3 = \mathbf{3A}$

3.6. Source Conversion

A given voltage source, with a series resistance, can be converted into an equivalent current source with a paralledl resistance. Conversely, a current source with a parallel resistance, can be converted into an equivalent voltage source with a series resistance.

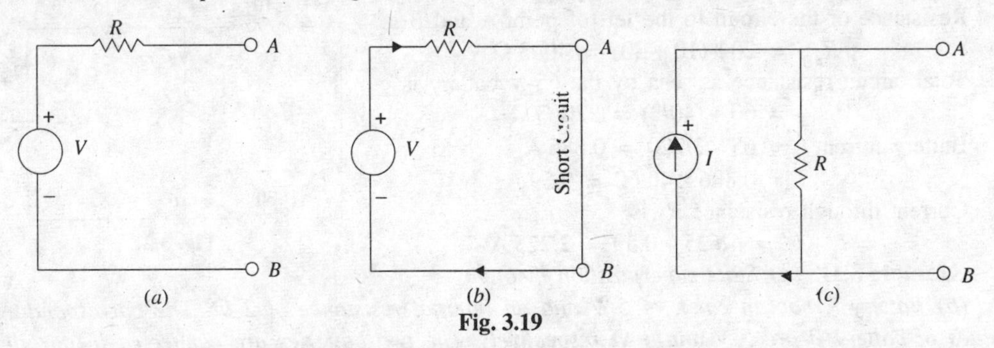

Fig. 3.19

Suppose, we want to convert the voltage source of Fig. 3.19 (*a*) into an equivalent current source.First, we will find the value of current supplied by the source when a short is put across its terminals *A* and *B*, as shwon in Fig. 3.19 (*b*). This current is $I = V/R$.

A current source, supplying this current *I* and having the same resistance *R*, connected in parallel with it, represents the equivalent source. It is shown in Fig. 3.19 (*c*). Similarly, a current source of *I* and a parallel resistance *R* can be converted into a voltage source of $V = I/R$ and a resistance *R* in series iwth it.

3.7. Superposition Theorem

According to this theorem, if there are a number of voltage or current sources acting simultaneously in a network, then each source can be treated as if it acts inpependently of the other.

Hence, we can calculate the effect of one source at a time and then superimpose *i.e.* algebrically add the results of all the other sources.

Following steps are taken while applying this theorem to the solution of networks which contain more than one voltage or current source.

1. First, all sources except the one under consideration are removed. While removing the voltage sources, their internal resistances (if any) are left behind. While removing current sources, they are replaced by an open circuit, since their internal resistance (by definition) is infinite.
2. Next, currents in various resistors and their voltage drops due to this single source are calculated.
3. This process is repeated for other sources, taken one at a time.
4. Finally, algebraic sum of currents and voltage drops over a resistor due to different sources is taken. It gives the actual value of current and voltage drop in that resistor.

Example 3.10. *By means of superposition theorem, find the current which flows through R2 in the circuit of Fig. 3.20 (a).* **(Electric Circuits-I, Punjab Univ. 1994)**

Solution. When 65 V Battery is removed.

As shown in Fig. 3.20 65-V battery has been removed and replaced by a short circuit. The resistance of the circuit to the right of points A and B is

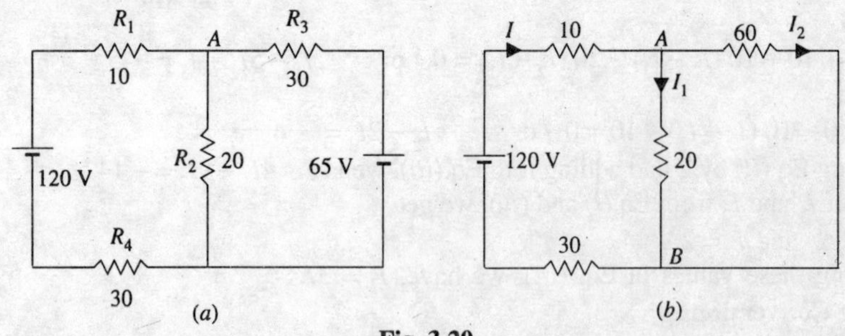

Fig. 3.20

Using current divider formula, we get $I_1 = 2.18 \times 60/80 = 1.635$ A

When 120 V Battery is Removed (Fig. 3.21)

Resistance of the circuit to the left of point A and B is

$$R_{AB} = 20 \parallel (10 + 30) = 40/3\ \Omega$$

Total circuit resistance as seen by the 65-V battery is

$$= 60 + (40/3) = 220/3\ \Omega$$

Battery current $I_2 = 65 \times 3/220 = 0.886$ A

$I_1 = 0.886 \times 40/60 = 0.59$ A

Current through resistance R_2 is

$$= I_1 + I_1' = 1.635 + 0.59 = 2.225 \text{ A}$$

Fig. 3.21

Example 3.11. *(a) State superposition theorem.*

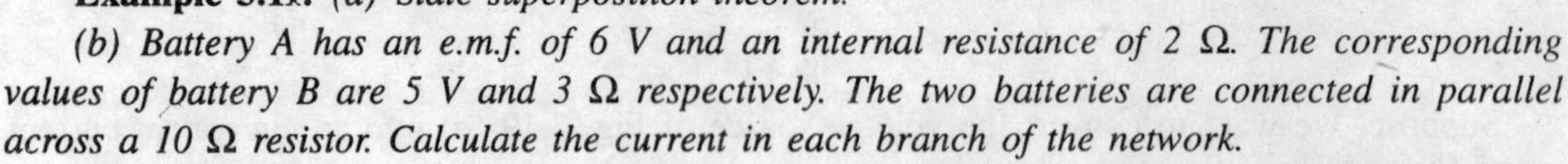

(b) Battery A has an e.m.f. of 6 V and an internal resistance of 2 Ω. The corresponding values of battery B are 5 V and 3 Ω respectively. The two batteries are connected in parallel across a 10 Ω resistor. Calculate the current in each branch of the network.

(Elect. Engg.-I, Punjab Univ. 1994)

Solution. When 5V Battery is Removed

As shown in Fig. 3.22 (*b*), the 5V battery has been removed and replaced by a short circuit.

$R_{AB} = 3 \parallel 10 = 2.3\ \Omega$

Resistance, as seen by 6-V battery = 2 + 2.3 = 4.3 Ω

$I_1' = 6/4.3 = 1.395$ A

$I_2' = 1.395 \times 10/(10 + 3) = 1.07$ A

$I_3' = 1.395 - 1.07 = 0.325$ A

When 6-V Battery is Removed [Fig. 3.23]

The total resistance as seen by the 5-V battery is

$I_1'' = 3 + 10 \parallel 2 = 4.67\ \Omega$

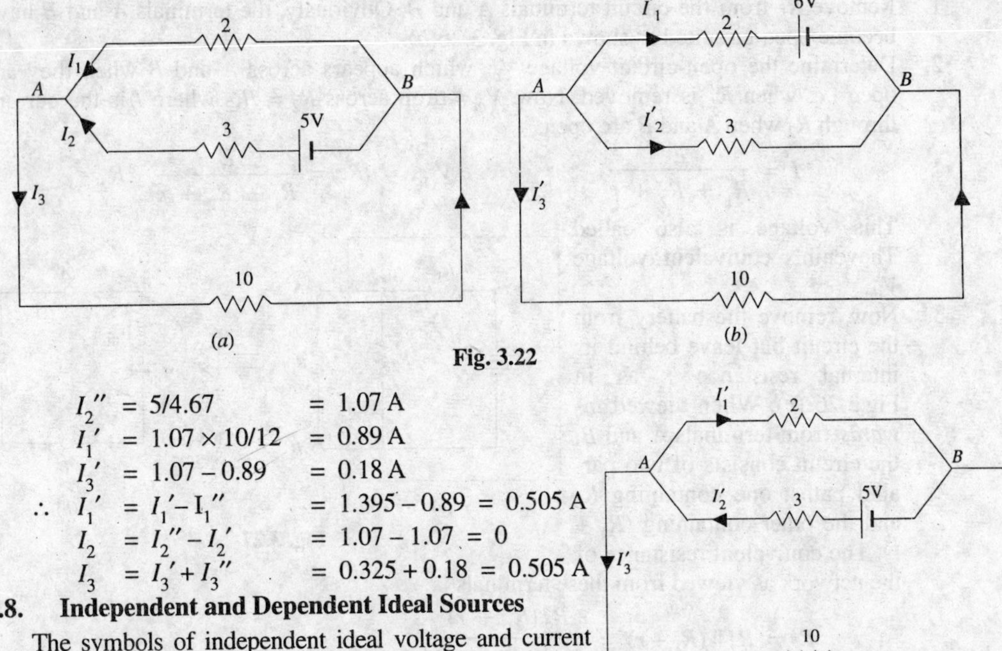

Fig. 3.22

$$
\begin{aligned}
I_2'' &= 5/4.67 && = 1.07\text{ A}\\
I_1' &= 1.07 \times 10/12 && = 0.89\text{ A}\\
I_3' &= 1.07 - 0.89 && = 0.18\text{ A}\\
\therefore\quad I_1' &= I_1' - I_1'' && = 1.395 - 0.89 = 0.505\text{ A}\\
I_2 &= I_2'' - I_2' && = 1.07 - 1.07 = 0\\
I_3 &= I_3' + I_3'' && = 0.325 + 0.18 = 0.505\text{ A}
\end{aligned}
$$

Fig. 3.23

3.8. Independent and Dependent Ideal Sources

The symbols of independent ideal voltage and current sources are shown in Fig.3.24. Since these sources are unaffected by the behaviour of the circuit in which they are connected, they are called *independent sources.*

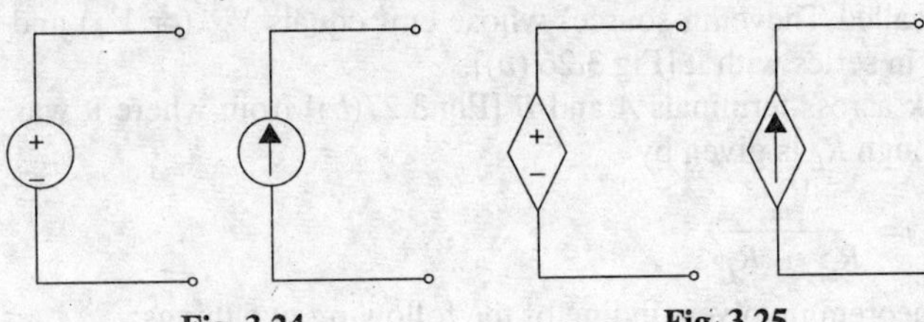

Fig. 3.24 Fig. 3.25

An ideal voltage or current source, whose value depends on a voltage or current existing at some other location in the circuit, is called a *dependent* (*or controlled*) source. It is represented by a diamond–shaped symbol (Fig.3.25) so as not to confuse it with an independent source.

If a dependent voltage source is controlled by a voltage, it is called voltage–controlled voltage source (VCVS) but if it is controlled by current them it is called current-controlled voltage source (CCVS).

Similarly, we have a voltage-controlled current source (VCCS) and a current-controlled current source (CCCS).

3.9. Thevenin's Theorem

According to this theorem, a given network, when viewed from its any two terminal points, can be replaced by a *single voltage source in series with a single resistance*. The equivalent voltage source is designated V_{th} and the equivalent resistance, R_{th}.

Suppose, it is required to find current flowing through load resistance R_L connected across terminals A and B as shown in Fig.3.25(a). As seen, the circuit contains a battery of emf E and internal resistance, r. We will proceed as follows:

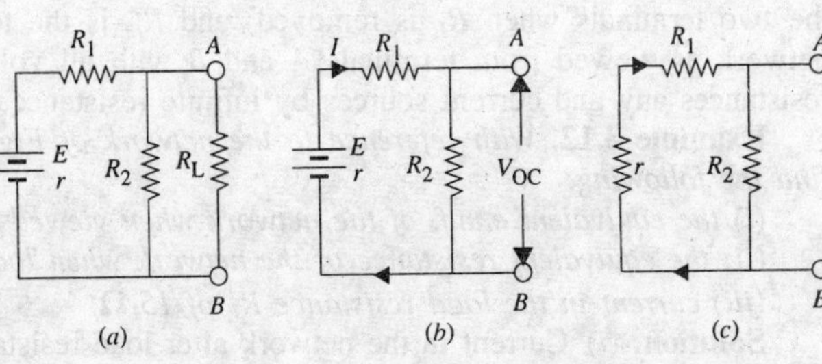

Fig. 3.26

1. Remove R_L from the circuit terminals *A* and *B*. Obviously, the terminals *A* and *B* have become open-circuited as shown in Fig. 3.26(*b*).
2. Determine the open-circuit voltage V_{oc} which appears across *A* and *B* when they are open *i.e.* when R_L is removed. Now, V_{oc}= drop across $R_2 = IR_2$ where *I* is the current through R_2 when *A* and *B* are open.

$$I = \frac{E}{R_1 + R_2 + r} \qquad \therefore \quad V_{OC} = IR_2 = \frac{E}{R_1 + R_2 + r} \cdot R_2$$

This voltage is also called Thevenin's equivalent voltage V_{th}.

3. Now, remove the battery from the circuit but leave behind its internal resistance *r* as in Fig.3.26 (*c*). When viewed *inwards* from terminals *A* and *B*, the circuit consists of two parallel paths: one containing R_2 and the other containing $(R_1 + r)$. The equivalent resistance of the network as viewed from these terminals is

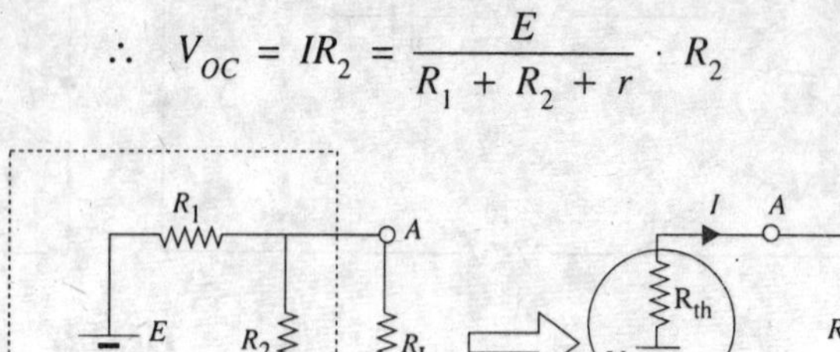

Fig. 3.27

$$R = R_2 \parallel (R_1 + r) = \frac{R_2 (R_1 + r)}{R_2 + (R_1 + r)}$$

This is Thevenin's equivalent resistance, R_{th}.

Hence, as viewed from terminals *A* and *B*, whole network (excluding R_L) can be reduced to a single voltage source (called Thevenin source) whose emf equals V_{oc} (or V_{th}) and which has a resistance of R_{th} in series with it [Fig.3.26 (*b*)].

4. Finally, R_L is connected back across terminals *A* and *B* [Fig.3.27(*b*)] from where it was earlier removed. Current through R_L is given by

$$I = \frac{V_{th}}{R_{th} + R_L}$$

As seen, use of Thevenin's theorem involves finding of the following two things:

(*i*) open-circuit voltage V_{th} (or V_{oc}) and

(*ii*) internal or equivalent resistance *R* of the network as viewed from open-circuited terminals *A* and *B*.

After this replacement of the whole network by a single source of e.m.f. and a single resistance has been accomplished, it is easy to find current in the load resistance joined across terminals *A* and *B*.

Hence, Thevenin's theorem, as applicablee to d.c. circuits, may be stated as under-

The current flowing through a load resistance R_L connected across any two terminals *A* and *B* of a network is given by $V_{th} /(R + R_L)$ where V_{th} is the open-circuit voltage (*i.e.* voltage across the two terminals when R_L is removed) and R_{th} is the terminal or equivalent resistance of the network as viewed from terminals *A* and *B* with all voltage sources replaced by their internal resistances any and current sources by infinite resistance.

Example 3.12. *With reference to the network of Fig.3.28, by applying Thevenin's theorem, find the following:*

(*i*) *the equivalent e.m.f. of the network when viewed from terminals A and B.*

(*ii*) *the equivalent resistance of the network when looked into from terminals A and B.*

(*iii*) *current in the load resistance R_L of 15 Ω.*

Solution: (*i*) Current in the network after load resistance has been removed [Fig. 3.28(*b*)] = 24/ (12 + 3 + 1) = 1.5 A.

∴ voltage across terminals

$AB = V_{th} = 12 \times 1.5 = \mathbf{18\ V}$

Hence, so far as terminals *A* and *B* are concerned, the network has an e.m.f. of 18 volt (and not 24 V).

(*ii*) There are two parallel paths between points *A* and *B* [Fig.3.29 (*a*)]. Imagine that battery of 24 V is removed but not its internal resistance. Then, equivalent resistance of the circuit as looked into from points *A* and *B* is

(*a*) (*b*)

Fig. 3.28

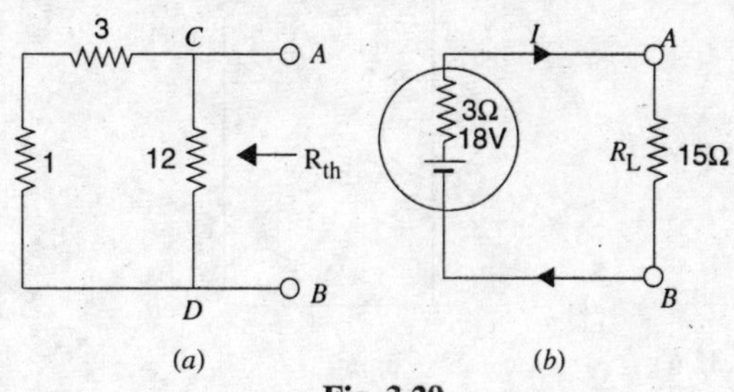

(*a*) (*b*)

Fig. 3.29

$$R_{th} = \frac{12 \times 4}{12 + 4} = 3\ \Omega$$

(*iii*) When load resistance of 15 Ω is connected across the terminals, then the network is reduced to the structure shown in Fig.3.29 (*b*).

$$I = 18/(15 + 3) = \mathbf{1A}$$

Example 3.13. *The arms of an unbalanced Wheatstone bridge network AB, BC, CD and DA comprise of resistance 10 Ω, 30 Ω, 15 Ω and 20 Ω respectively. A d.c. battery of 2 volts (having negligible internal resistance) is connected across terminals AC and a galvanometer of internal resistance 40 Ω is connected across terminals BD. Calculate the value and direction of the current in the galvonometer circuit BD applying Thevenin's Theorem. (assume the polarity of the battery voltage at the terminal A to be positive.* **Electrical Science AMIE Summer 1993).**

Solution. We will apply Thevenin's theorem to the given circuit as shown in Fig. 3.30 (*a*).

(*i*) Remove the galvanometer connected between the points *B* and *D*, thereby obtaining the circuit of Fig. 3.30 (*b*).

(*ii*) Now, let us find the p.d. between points *B* and *D*. It would be seen that with galvanometer removed, *ABC* and *ADC* become potential dividers connected in parallel across the 2V battery.

Potential of point *B* with respect to the negative terminal of the battery *i.e.* ground is

2V (*a*)

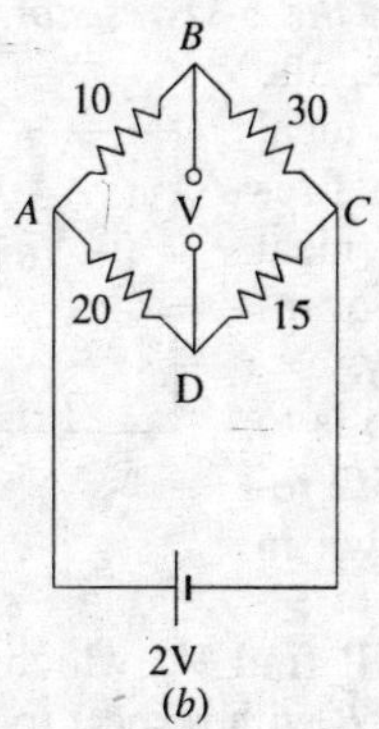

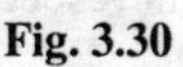

(*b*)

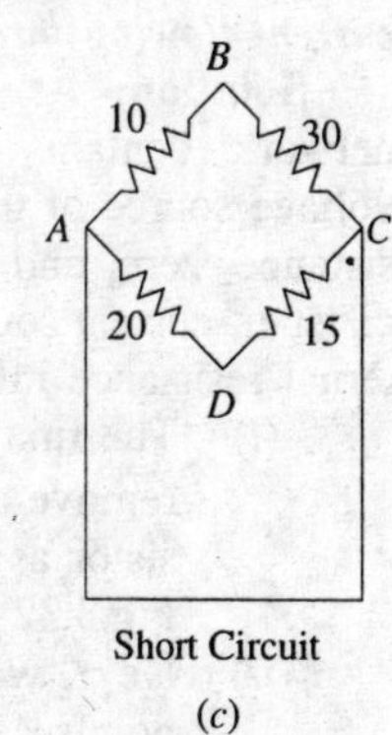

(*c*)

Fig. 3.30

$$V_B = 2 \times 30/(30 + 10) = 1.5\ \text{V}$$

$$V_D = 2 \times 15/(20 + 15) = 0.86\ \text{V}$$

$$\therefore\ V_{BD} = 1.5 - 0.86 = 0.64\ \text{V}$$

(*iii*) Now, remove the battery from the circuit leaving its internal resistance behind which, in this case, is zero. Hence, it amount to placing a short circuit between points *A* and *C* as shown in Fig. 3.30 (*c*).

(*iv*) Next, let us find the internal resistance of the whole network, when viewed from open points *B* and *D*. It should be kept in mind that path *BA* is an parallel with path *BC* and path *AD* is parallel with *CD*, as shown in Fig. 3.31 (*a*). The equivalent resistance between points *B* and *D* is (7.5 + 8.6) = 16.1 Ω. Hence, R_{th} = 16.1 Ω.

(*v*) Hence, so far as point *B* and *D* are concerned, the network has a source of e.m.f. of 0.64 V and an equivalent resistance of 16.1 Ω. Thevenin's source along with its internal resistance is shown in Fig. 3.31 (*c*).

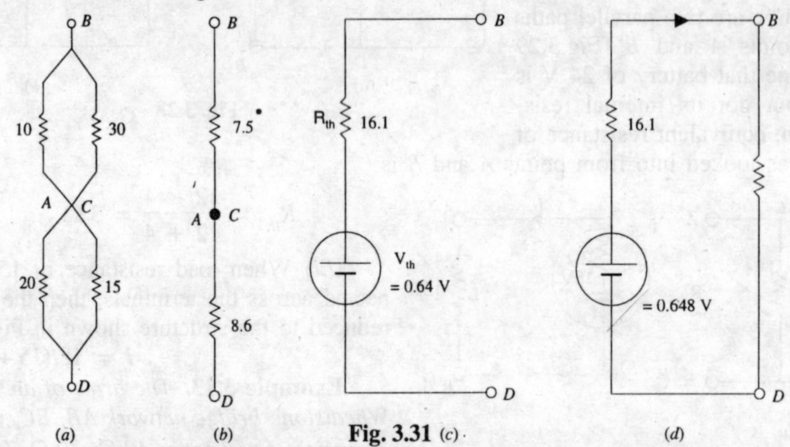

Fig. 3.31 (*c*)

(*vi*) Connected the 40 Ω galanometer (initially removed) to this Thevenin's source, as shown in Fig. 3.30 (*d*) and calculate the current flowing through it.

$$I = \frac{0.64}{16.1 + 40} = \mathbf{0.011\ A}$$

As seen, this current flows from point *B* to *D*.

It may be noted that this question is similar to that given in Example 3.1 which was solved by using Kirchhoff's laws.

Example 3.14. *With the help of Thevenin's theorem, find the magnitude and direction of the current flowing through the 5-Ω resistor in Fig. 3.32 (a).*

Solution: As seen, the network contains an ideal voltage source of internal resistance zero and an ideal constant-current source of internal resistance infinity.

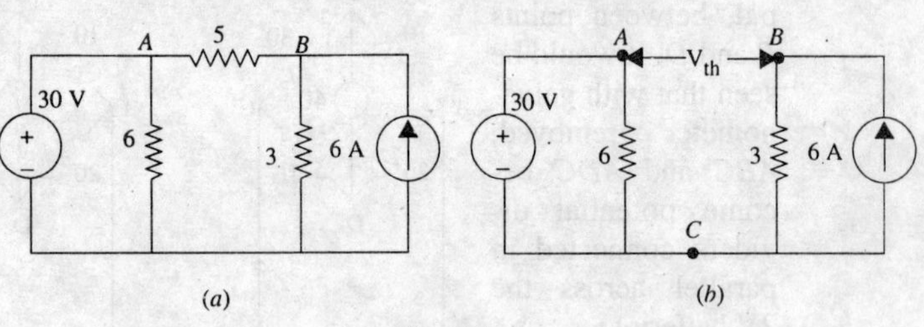

Fig. 3.32

(*i*) The first step is to remove the 5Ω resistor as shown in Fig.3.32 (*b*).

(*ii*) Next, we will find V_{th} which equals V_{AB}. For this purpose, we will find voltages of points *A* and *B* with respect to common point *C*. Since the 30-V voltage sour-ce is connected ac-ross the 6-Ω re- sistor, voltage of *A* with respect to point *C* is 30 V. As whole of 6A current passes through 3Ω resistor, drop across it is = 6×3 = 18 V. Hence, V_B = 18 V

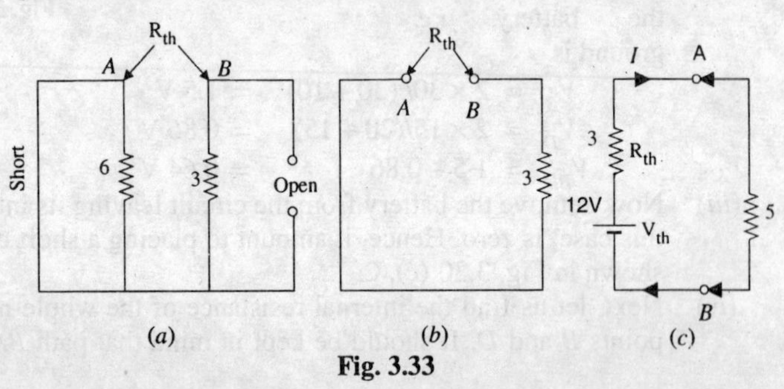

Fig. 3.33

with respect to point *C*. Obviously, $V_{AB} = V_{th} = 30 - 18 = 12$ V. Since $V_A > V_B$, current flows from *A* to *B*.

(*iii*) Now, the voltage source is removed leaving behind its internal resistance which is zero in the present case. Hence, the voltage source is replaced by a short as shown in Fig.3.30(*a*). The 6A current source is replaced by an open *i.e.* infinite resistance as shown. As seen from Fig.3.32(*b*), $R_{th} = 3\Omega$ because 6Ω resistor is shorted out.

(*iv*) The equivalent Thevenin source alongwith its resistance is shown in Fig. 3.32 (*c*). For finding *I*, the 5Ω resistor has been connected across this source.

$$I = 12/(5+3) = \mathbf{1.5A}$$

Example 3.15. *Find the current through 60 ohm resitance in the circuit. [Fig. 3.34 (a)].*

(Electrical Engg. A.M.A.e. S.I. June 1994)

Solution. We will solve this question first by using Kirchhoff's laws and then by Thevenin's theorem.

(*a*) Using Kirchhoff's Laws

Starting from point *A* in Fig. 3.34 (*b*) and applying *KVL* to the closed circuit *ABCA*, we get

$$-10x - 60(x-z) + 50y = 0 \quad \text{or} \quad 7x - 5y - 6z = 0 \quad \text{...(i)}$$

Similarly, from the closed circuit BCDB, we get

$$-60(x-z) - 10(x+y-z) + 50z = 0 \quad \text{or} \quad 7x + y - 12z = 0 \quad \text{...(ii)}$$

Now, taking the closed circuit *ACDA*, we obtain

$$-50y - 10(x+y-z) + 80 = 0 \quad \text{or} \quad x + 6y - z = 8 \quad \text{...(iii)}$$

Substracting (*ii*) from (*i*), we get

$$-6y + 6z = 0 \quad \text{or} \quad y = z \quad \text{...(iv)}$$

Putting $y = z$ in (*i*) above, we get $x = 11y/7$

Substituting this value of *x* and replacing *z* by *y* in (*iii*) we get

$$(11y/7) + 5y = 8,$$

or $y = 28/23\ A = z$

$$\therefore \quad x = 11y/7 = \frac{11 \times 28}{23 \times 7}$$

$$= 44/23\ \text{A}$$

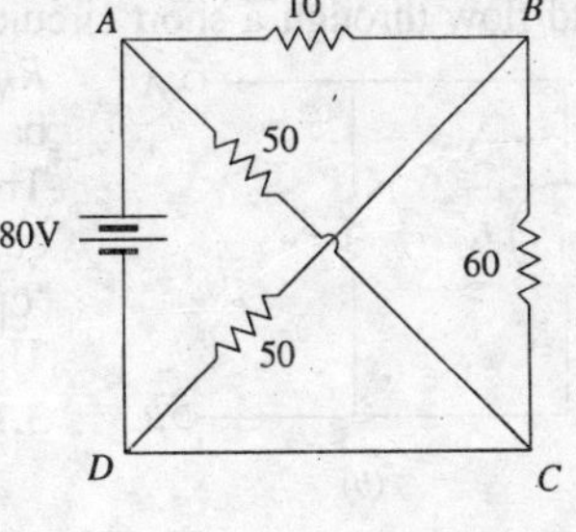

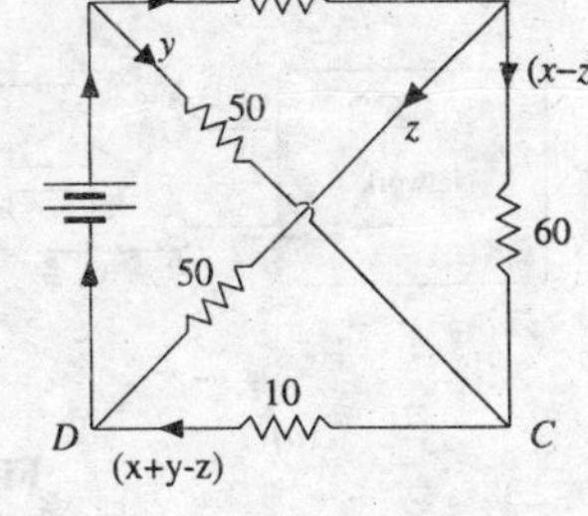

Fig. 3.34

∴ current through 60 Ω resistance $= x - z = (44/23) - (28/23) = 16/23 = \mathbf{0.6956\ A}$

(*b*) Using Thevenin's Theorem

The 60 Ω resistance has been removed from the terminals *B* and *C* as shown in Fig. 3.35 (*a*). Now, we have to find the open-circuit voltage or Thevenin voltage V_{th} across open terminals *B* and *C*. For this purpose, we will find V_B and V_C with respect to the common ground point *d*. It will be seen that there are two parallel path across the 80 V battery, one is *ABD* and the other is *ACD*. Applying the voltage divider formula (Art.) to the path *ABD*, we get

$$V_B = 80 \times 50/(50+10) = 66.7\ \text{V}$$

Similarly, from the parallel path ACD, we get

$$V_C = 80 \times 10/60 = 13.3\ \text{V}$$

$$\therefore \quad V_{BC} = 66.7 - 13.3 = 53.4\ \text{V. It represents } V_{th}$$

For finding Thevenin resistance R_{th}, the 80 V battery has been removed and replaced by a short-circuit, as shown in Fig. 3.35 (*b*). As we go from point *B* to point *E*, there are two parallel resistances of 10 Ω and 50 Ω, giving the combined resistance of 10 × 50/60 = 25/3 Ω.

Similary, between point *E* and *C* there are again two parallel resistances, giving a combined resistances of 10 × 50/60 = 25/3Ω. As we go from point *B* to *C via* point *E*, there are two resistances of 25/3Ω each in series giving a total resistance of 2 × 25/3 = 50/3 Ω .

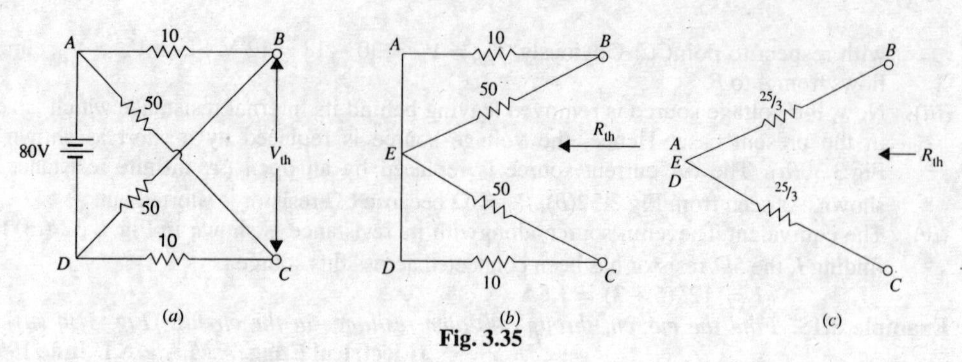

Fig. 3.35

Hence, $R_{BC} = R_{th} = 50/3\ \Omega$

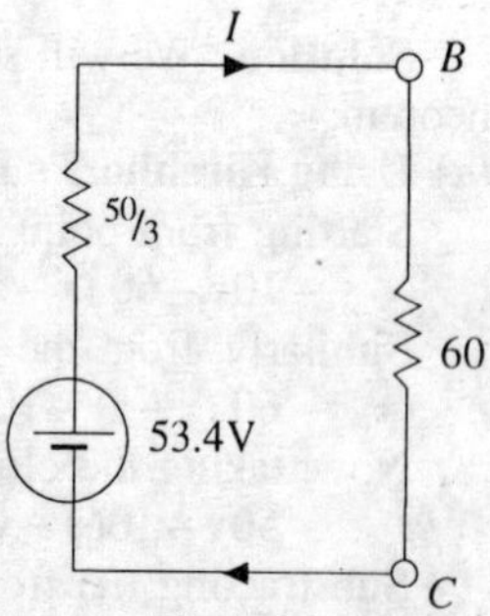

The equivalent Thevenin's source is shown in Fig. 3.36. The 60 Ω resistance (which had been removed earlier) is brought back and joined across terminals B and C as shown. The circuit $I = 53.4/60 + (50/3) = 0.6965$ A — as before.

Fig. 3.36

3.10. Norton's Theorem

This theorem is used where it is easier to simplify a network in term of currents instead of voltages. This theorem reduces a normally complicated network to a simple parallel circuit consisting of

(*a*) an ideal current source I_N of infinite internal resistance and

(*b*) a resistance R_N (or conductance $G_N = 1/R_N$) in parallel with it is shown in Fig.3.37 (*b*).

Here, I_N isthe current which would flow through a short circuit placed across terminals A and B. R_N is the circuit resistance looking back from the open A-B terminals. These terminals are not short-circuited for finding R_N but are kept 'open' as for calculating R_{th} for Thevenin's Theorem.*

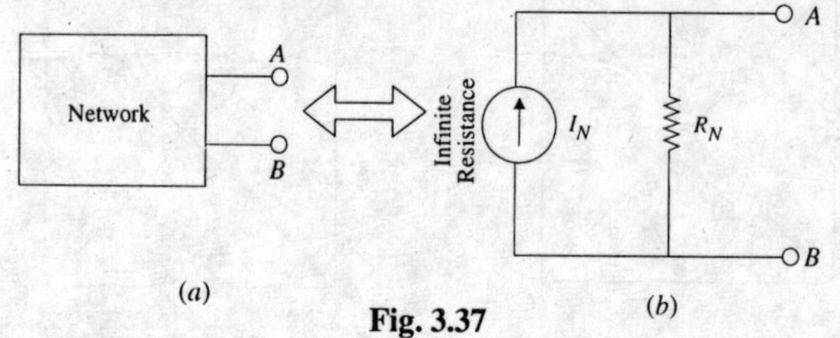

Fig. 3.37

3.11. How to Nortonise a Given Circuit ?

Suppose we want to Nortonise the circuit shown in Fig.3.38(*a*) *i.e.,* we want to find Norton's equivalent of this circuit between terminals A and B.

1. First Step. Put a short terminals A and B [Fig.3.38(*b*)]. As seen, it results in shorting out 12 Ω resistor as shown separately in Fig. 3.38 (*c*).

$$I_{SC} = 24/4 = 6\text{A}$$

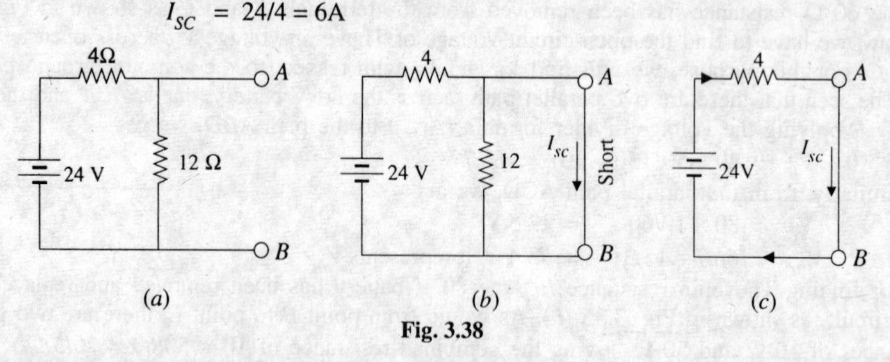

Fig. 3.38

*In fact, this resistance is the same, both for Thevenin and Norton equivalent circuits. In Norton's case, this resistor is in parallel with the current, whereas in Thevenin's case, it is in series with V_{th}.

The current is usually called Norton current I_N.

2. Second Step. Remove the short from terminals A and B so that they are again open.

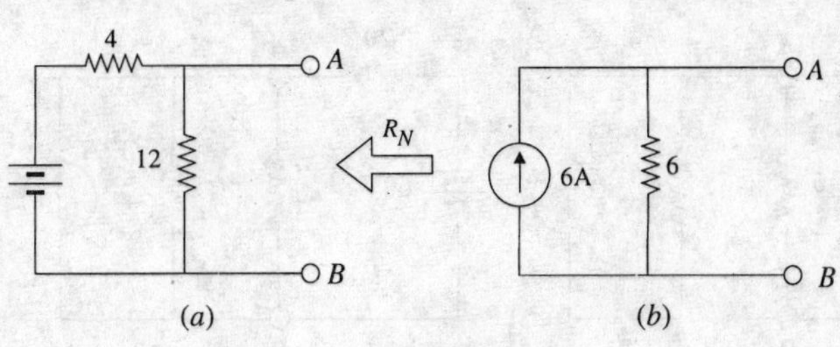

Fig. 3.39

3. Third Step. Remove the battery and replace it by its internal resistance which, in the present case, is zero. The resistance R_N of the circuit as viewed back or looked into from open terminals A and B is

$$R_N = 12 \parallel 4 = 3\Omega$$

Hence, Norton's equivalent of the given circuit in Fig. 3.38(a) becomes that shown in Fig. 3.39(b). It consists of a 6-A constant current source (or infinite resistance) in parallel with a 3Ω resistance.

Example 3.16. *Using Norton's theorem, calculate the current flowing through the 12-Ω resistor shown in Fig.3.40.*

Solution: The given circuit is shown in Fig.3.40(a). When applying Norton's theorem, the procedure would be as under:

1. First Step. Remove 12 Ω resistor from terminals A and B and then put a short-circuit across them as shown in Fig. 3.40(b). The current passing through 4 Ω resistor is also the short-circuit current I_{SC} (also written as I_N). For finding this current, we have first to rent I_{SC} (also written as I_N). For finding this current, we have first to find I by simplifying the circuit. As seen, total circuit resistance.

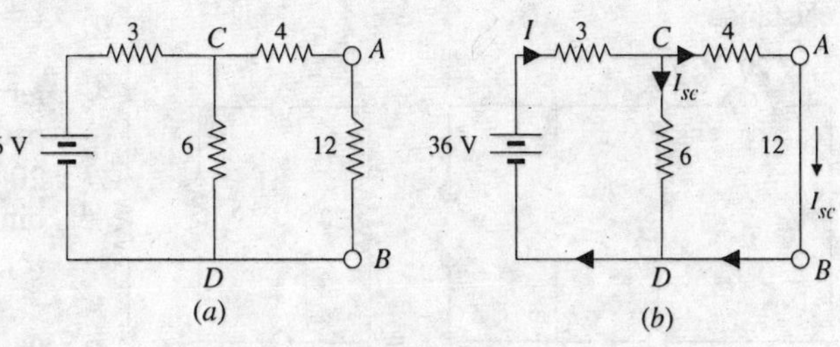

Fig. 3.40

$$= 3 + 6 \parallel 4 = 3 + 2.4 = 5.4\ \Omega$$

$$\therefore \quad I = 36/5.4 = 20/3\ \text{A}$$

The current divides at point C into two unequal parts.

$$I_{SC} = \frac{20}{3} \times \frac{6}{10} = 4\ \text{A}$$

2. Second Step. Remove the short circuit, thereby leaving terminals A and B open. Also remove the battery. Since its internal resistance is zero, it is replaced by a resistanceless piece of connecting wire thereby closing [Fig.3.41(a)]. The resistance R_N of the circuit as viewed back from terminals A and B is = 4 + 6 ∥ 3 = 6 Ω.

$$\therefore \ I_L = 4 \times \frac{6}{(6+12)} = 1.33\ \text{A}$$

from A to B

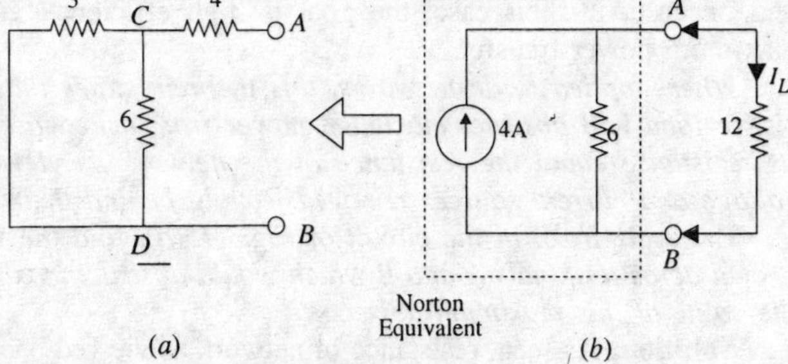

Fig. 3.41

Hence, the Norton's equivalent of the given circuit is as shown in Fig. 3.41(b). Load current I_L through 12 Ω resistor can be found by using the Current Division Formula (Art.2-15)

Example 3.17. *The circuit of Fig.3.42(a) is excited by a voltage source and a current source. Using Norton's theorem, calculate current through the 6-Ω resistor.*

Solution: In Fig. 3.42 (b), 6-Ω resistor has been removed and a short placed across terminals A and B. Short-circuit current I_{SC} (or I_N) is

$= \text{current from voltage source} + \text{current of current source} = \dfrac{60}{20} + 12 = 15 \text{ A}$

It should be noted that short across AB also shorts out 5-Ω resistor. Hence, all the 12 A current passes through short-circuit and none through 5-Ω resistor.

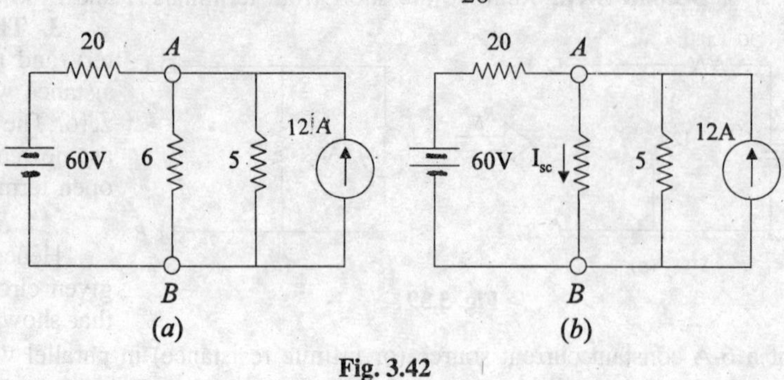

Fig. 3.42

In Fig.3.43(*a*), 'short' has been removed leaving terminals A and B open. Voltage source has been replaced by a connecting wire of zero resistance (since its internal resistance is zero). Current source has been replaced by an 'open' since it has infinite resistance.

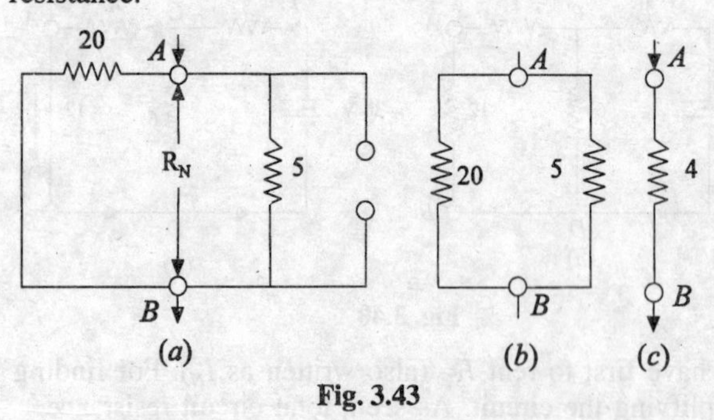

Fig. 3.43

When looked into the circuit from terminals A and B, there are two parallel paths between them having resistance of 20 Ω and 5 Ω [Fig.3.43(*b*)]. Their combined resistance, as seen from Fig. 3.43 (*c*), is 20 ‖ 5 = 4 Ω. Hence, $R_N = 4$ Ω.

The Norton's equivalent of the original circuit with respect to terminals A and B is shown in Fig.3.44. The 6-Ω resistance has been connected back to the terminals A and B (from where it was removed earlier).

$$I_L = 15 \times 4/(6+4) = \mathbf{6A}$$

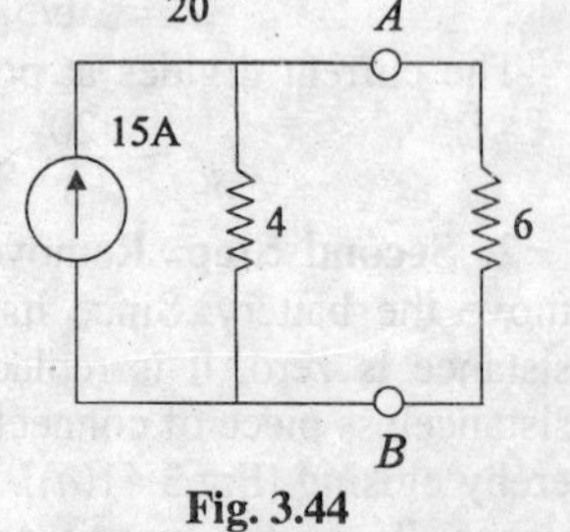

Fig. 3.44

3.12. Maximum Power Transfer Theorem

This theorem is very useful for analysing electronic and communication networks where main consideration is to transfer maximum power to the load irrespective of the efficiency. Its application to power transmission and distribution networks is limited because, in their case, the goal is high efficiency and not maximum power transfer.

When applied to dc networks, this theorem states that a resistive load will abstract maximum power from a network when its resistance equal the resistance of the network as viewed from the output terminals with all voltage and current sources removed leaving behind their internal resistances.

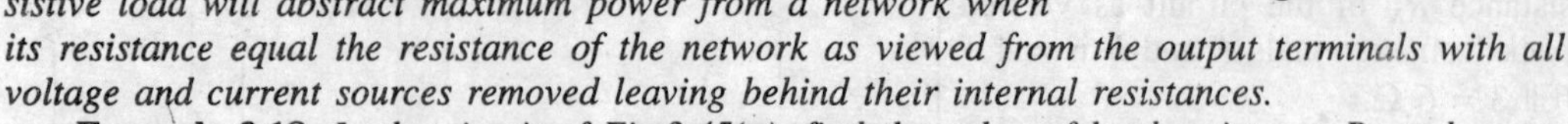

Example 3.18. *In the circuit of Fig.3.45(a), find the value of load resistance R_L to be connected acros terminals A and B which would abstract maximum power from the circuit. Also, find the value of this maximum power.*

Solution: As seen, resistance of network as viewed back from terminals A and B (with battery removed) is

$$= 4 + 6 \parallel 3 = 6\,\Omega$$

Hence, R_L should be equal to 6 Ω.

Let us now find power developed in R_L for which purpose we have to find I_L.

In Fig. 3.45(*b*), total circuit resistance = 3 + 6 ‖ 10 = 27/4 Ω.

$$I = 36 \div 27/4 = 16/3 \text{ A} \quad \therefore \quad I_L = \frac{16}{3} \times \frac{6}{(6 + 10)} = 2 \text{ A}$$

Max.power possible in load resistance $R_L = 2^2 \times 6 =$ **24 W**

It can be verified that if we have any other value of R_L, power drawn will be less than 24Ω.

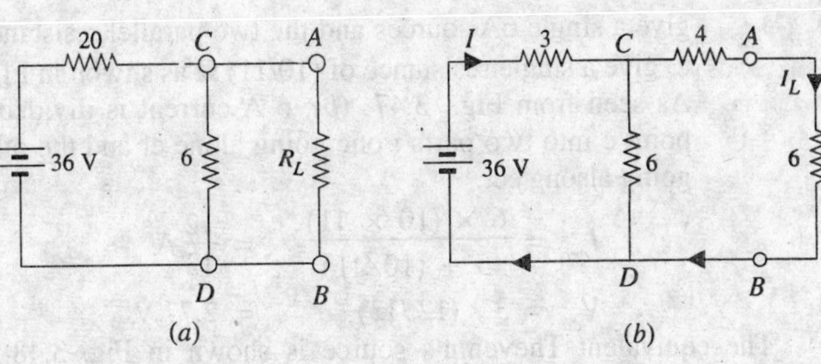

Fig. 3.45

Example 3.19. *In the network shown in Fig. 3.46 (a), find resistance R_L connected between terminals a and b so that maximum power is developed across R_L. What is the maximum power?*

(Electrical Science AMIE Winter 1993)

Solution. For finding the maximum power absorbed by R_L, we will find Thevevin's equivalents circuit for the given network.

(*i*) For finding the R_{th}, we will replace the voltage source by short-circuit and the current source by an eopen-circuit as shown in Fig. 3.46 (*b*). We will simplify the circuit by using series and parallel laws of resistance.

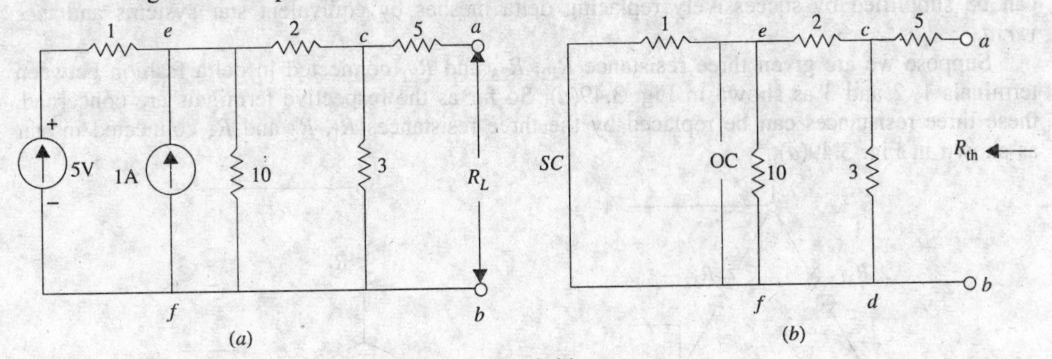

Fig. 3.46

$$R_{ef} = 10 \parallel 1 = 10/11\ \Omega$$
$$R_{cd} = 3 \parallel (2 + 10/11) = 1.477\ \Omega$$
$$R_{th} = R_{ab} = 5 + 1.477 = 6.477\ \Omega$$

(*ii*) V_{th}, Since there is no flow of current through 5Ω resistance, $V_{th} = V_{cd}$ *i.e.*, drop across 3 Ω resistance, For this purpose, we will use source conversion technique (Art.) to convert the 5V source and its 1Ω resistance connected across terminals e and *f* in Fig. 3.46 (*a*) into Norton's equivalent source. Short-circuit current = 5/1 = 5A. Hence, 5V source with 1Ω series resistance is equivalent to a Norton's source of 5A with a 1Ω resistance in parallel as shown in Fig. 3.47 (*a*). The two current sources are lumped together to

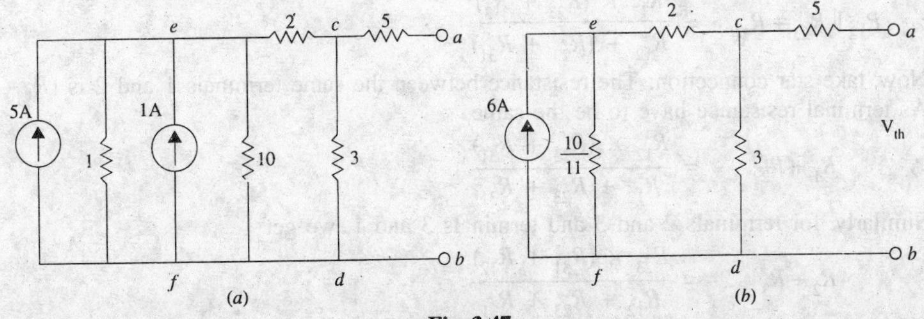

Fig. 3.47

give a single 6A sources and the two parallel resistances of 1 Ω and 10 Ω are combined to give a single resistance of (10/11) Ω as shwon in Fig. 3.47 (*b*).

As seen from Fig. 3.47 (*b*), 6 A current is divided at point e into two parts : one going alone ef and the other going alsong ect.

$$I_{ecd} = \frac{6 \times (10 \times 11)}{5 + (10/11)} = \frac{12}{13} \text{A}$$

$$\therefore \quad V_{cd} = 3 \times (12/13) = 2.77 \text{ V.}$$

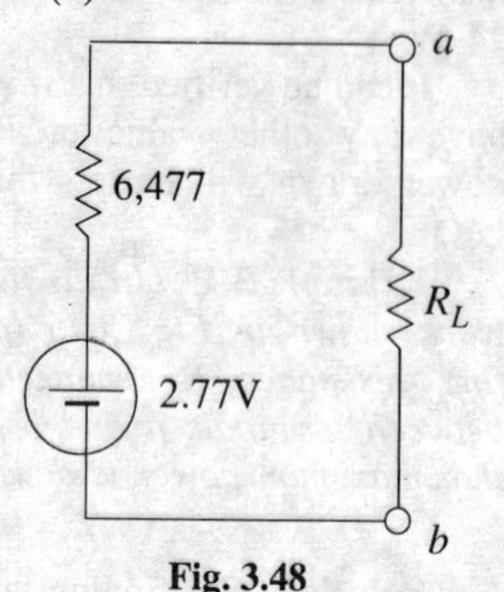

Fig. 3.48

The equivalent Thevenin's source is shown in Fig. 3.48 R_L will absorb maximum power when it equals R_{th} *i.e.*, when R_L = 6.477 Ω.

Current passing through R_L = 277/2 × 6.477 = 0.214 A

Maximum power absorbed by R_L = 0.2142 × 6.477 = 0.3 W

3.13. Delta/Star Transformation

In solving networks (having considerable number of branches) by the application of Kirchhoff's laws or by Loop Current method, one sometimes experiences great difficulty due to a large number of simultaneous equations that have to be solved. However, such complicated networks can be simplified by successively replacing delta meshes by equivalent star systems and *vice versa.*

Suppose we are given three resistance R_{12}, R_{23} and R_{31} connected in delta fashion between terminals 1, 2 and 3 as shown in Fig. 3.49(*a*). So far as the respective terminals are concerned, these three resistances can be replaced by the three resistances R_1, R_2 and R_3 connected in star as shown in Fig. 3.49(*a*).

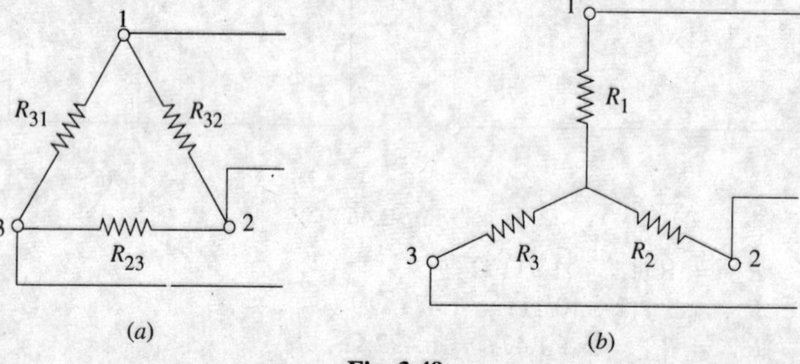

Fig. 3.49

These two arrangements will be electrically equivalent if the resistance as measured between any pair of terminals is the same in both the arrangements. Let us find this condition.

First, take delta connection: Between terminals 1 and 2, there are two parallel paths: one having a resistance of R_{12} and the other having a resistance of $(R_{23} + R_{31})$.

∴ resistance between terminals 1 and 2 is

$$R_{12} \parallel R_{23} + R_{31} = \frac{R_{12} \times (R_{23} + R_3)}{R_{12} + (R_{23} + R_{31})}$$

Now, take star connection: The resistance between the same terminals 1 and 2 is $(R_1 + R_2)$. As terminal resistance have to be the same

$$\therefore \quad R_1 + R_2 = \frac{R_{12} \times (R_{23} + R_{31})}{R_{12} + R_{23} + R_{31}} \quad ...(i)$$

Similarly, for terminals 2 and 3 and terminals 3 and 1, we get

$$R_2 + R_3 = \frac{R_{23} \times (R_{31} + R_{12})}{R_{12} + R_{23} + R_{31}} \quad ...(ii)$$

and $$R_3 + R_1 = \frac{R_{31} \times (R_{12} + R_{23})}{R_{12} + R_{23} + R_{31}} \quad ...(iii)$$

Now, subtracing (*ii*) from (*i*) and adding the result to (*iii*), we get

$$R_1 = \frac{R_{12} R_{31}}{R_{12} + R_{23} + R_{31}}; \quad R_3 = \frac{R_{31} R_{23}}{R_{12} + R_{23} + R_{31}}$$

and $$R_3 = \frac{R_{31} R_{23}}{R_{12} + R_{23} + R_{31}}$$

How to remember?

It is seen from above that each numerator is the product of the two sides of the delta which meet at the point in star terminals. Hence, it should be remembered that: *resistance of each arm of the star is given by the product of the resistance of the two delta sides that meet at its ends divided by the sum of the three delta resistances.*

3.14. Star / Delta Transformation

This transformation can be easily done by using equations (*i*), (*ii*) and (*iii*) given above. Multiplying (*i*) and (*ii*), (*ii*) and (*iii*), (*iii*) and (*i*) and adding them together and then simplifying them, we get

$$R_{12} = \frac{R_1 R_2 + R_2 R_3 + R_3 R_1}{R_3} = R_1 + R_2 + \frac{R_1 R_2}{R_3}$$

$$R_{12} = \frac{R_1 R_2 + R_2 R_3 + R_3 R_1}{R_1} = R_2 + R_3 + \frac{R_2 R_3}{R_1}$$

$$R_{31} = \frac{R_1 R_2 + R_2 R_3 + R_3 R_1}{R_2} = R_1 + R_3 + \frac{R_1 R_3}{R_2}$$

How to Remember?

The equivalent delta resistance between any two terminals is given by the sum of star resistances between those terminals plus the product of these two star resistances divided by the third star resistance.

Example 3.20. *Three resistance R, 2R and 3R are connected in delta [Fig.3.50(a)]. Determine the resistance for an equivalent star connection.*

In Fig. 3.50, 160 volts are applied to the terminal AB. Determine the resistance (a) between the terminals A and B and (b) the current I.

Solution. The three resistances are joined to delta in Fig. 3.50(*a*).

Remembering the rule given in Art. 3.13, we have in Fig. 3.50(*b*)

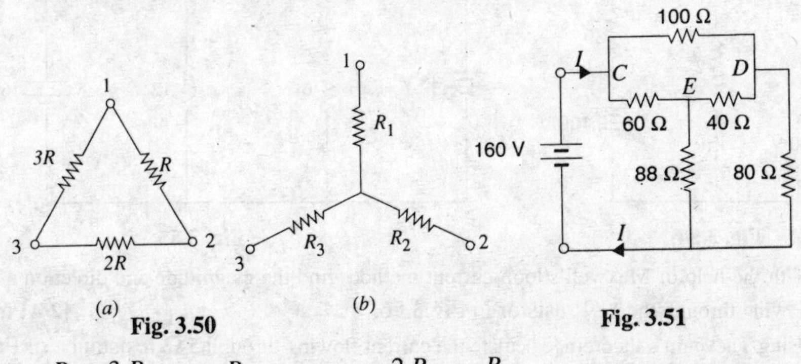

Fig. 3.50

Fig. 3.51

$$R_1 = \frac{R \times 3R}{R + 2R + 3R} = \frac{R}{2}; \; R_2 = R \times \frac{2R}{6R} = \frac{R}{3}; \; R_3 = 2R \times 3R/6R = R$$

Take the network of Fig. 3.51. The three resistances of 100 Ω, 60 Ω and 40 are delta-connected between terminal points C, D and E as shown in Fig. 3.52 (a). They can be converted into equivalent star connection as shown in Fig. 3.52 (b).

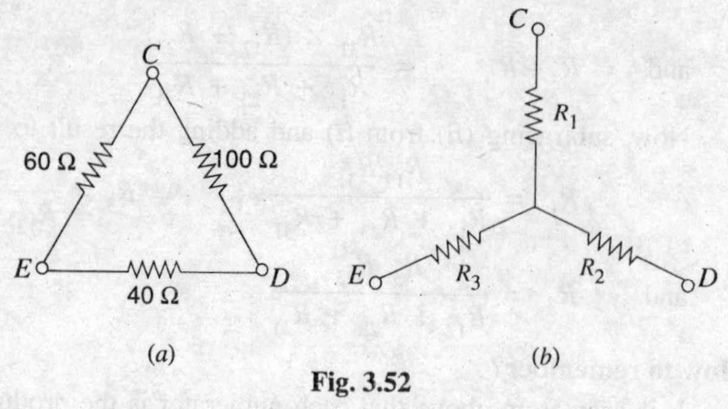

Fig. 3.52

$$R_1 = \frac{60 \times 100}{100 + 60 + 40} = 30\ \Omega;$$

$R_2 = 100 \times 40/200 = 20\ \Omega$;

$R_3 = 40 \times 60/200 = 12\ \Omega$

Then the network of Fig. 3.51 is reduced to a simple structure of Fig. 3.53(a).

As seen, there are two parallel paths between point S and B: one of resistance $(20 + 80) = 100\Omega$. Hence, equivalent resistance between points S and B.

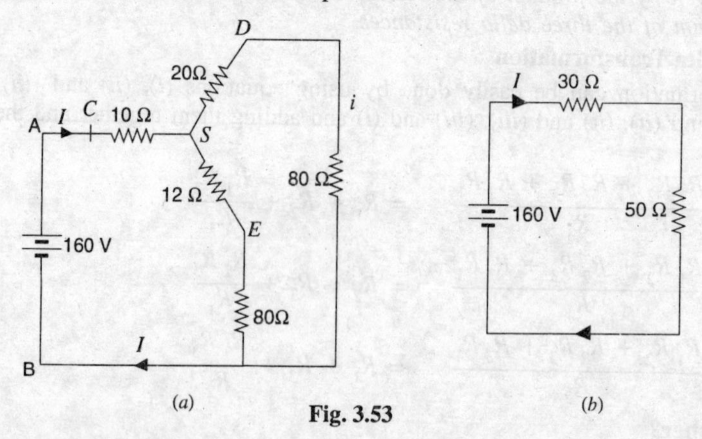

Fig. 3.53

$= 100 \times 100/200 = 50\ \Omega$

The whole network is reduced to a simple circuit of Fig.3.53(b).

(a) $\therefore$ resistance between points A and B $= 30 + 50 =$ **80 Ω**

(b) current I $= 160/80 =$ **2 A**

Tutorial Problems No. 3.2

1. Use Loop Current method for finding the current flowing through the 2-Ω resistor in Fig. 3.54. **[5 A]**

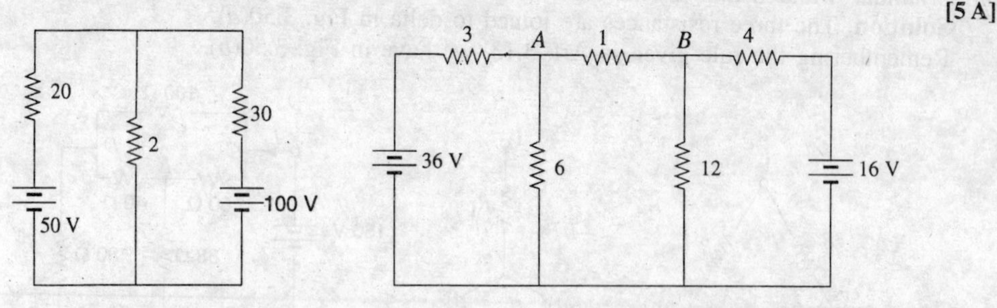

Fig. 3.54 Fig. 3.55

2. With the help of Maxwell's loop-current method, find the magnitude and direction of the current flowing through the 1-Ω resistor in Fig.3.56. **[2 A; from A to B]**
3. Using Thevenin's theorem, calculate the current flowing through 5 Ω resistance of Fig. 3.56.
4. With the help of Thevenin's theorem, find current through 8-Ω resistor of Fig.3.57. **[0.3 A]**
5. Calculate the magnitude and direction of flow of curent through the 12 ohm resistance of Fig.3.58. Use Thevenin's theorem for the purpose. What would be the current if 100-V battery

connections were reversed? Also, if 120-V battery is reversed?

[2.25 A from A to B; 9.75 A same direction; 9.75 A from B to A]

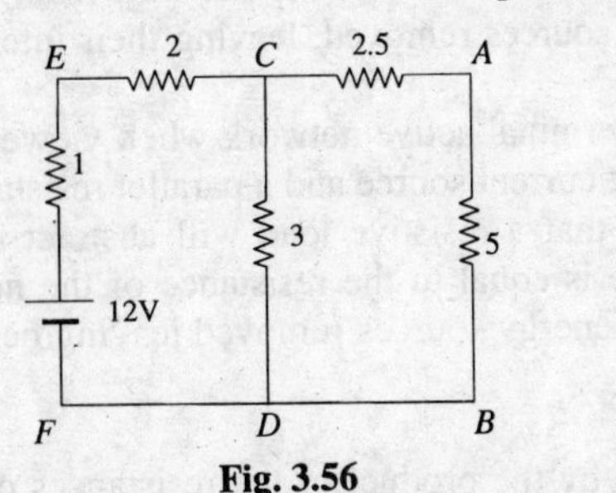

Fig. 3.56

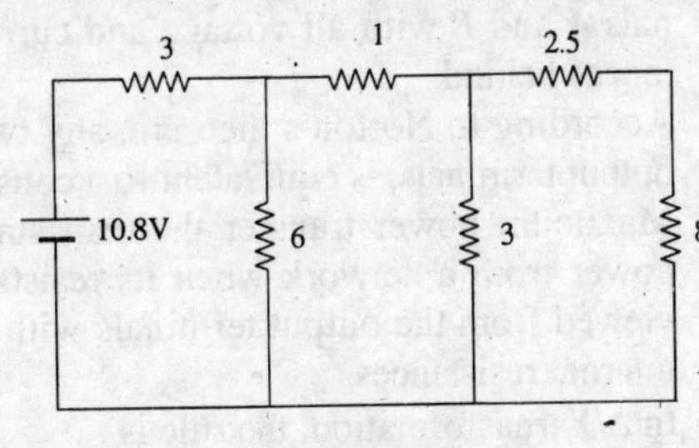

Fig. 3.57

6. Using Thevenin's theorem, find the current flowing through the 10-Ω resistor of Fig.3.59. Also, calculate the current when (*i*) 2-V battery connection is reversed and (*ii*) 8-V battery connection is reversed. **[0.135 A from A to B (*i*) 0.3 A from A to B (*ii*) 0.3 A from B to A]**

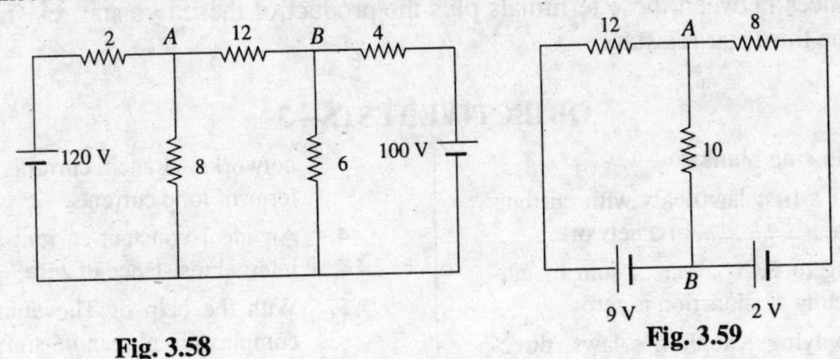

Fig. 3.58

Fig. 3.59

7. Using Δ/γ transformation, find the value of the equivalent resistance between points *A* and *B* of the circuit shown in Fig.3.60. **[6Ω]**

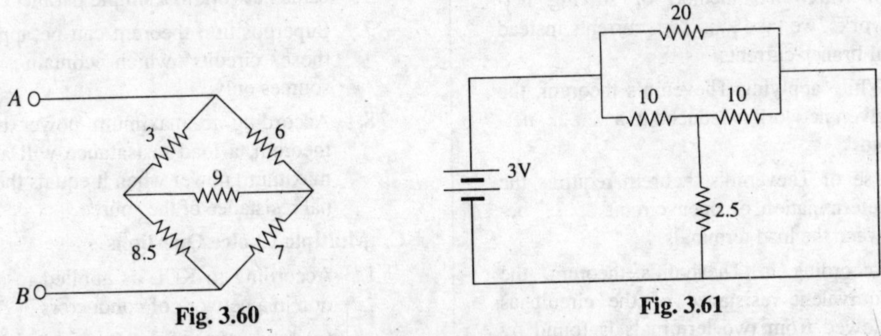

Fig. 3.60

Fig. 3.61

8. Calculate the current supplied by the 3.V battery in Fig. 3.61. **[0.4 A]**

HIGHLIGHTS

1. Two laws of Kirchhoff can be briefly stated as below:—
   (*a*) Point Law or Current Law
   $\Sigma I = 0$ — at a point or junction
   (*b*) Mesh Law or Voltage Law
   $\Sigma IR + \Sigma\, e.m.f. = 0$ — round a mesh.
2. According to Superposition theorem if there are a number of voltage and current sources acting simultaneously in any linear bilateral network, then each source acts independently of the others.
3. Thevenin's theorem may be stated as follows:
   the current flowing through a load resistance R_L connected across any two terminals *A*

and B of a network is given in $V_{th}/(R+R_L)$ where V_{th} is the open-circuit voltage and R_{th} is the internal resistance of the network as viewed back into the network from the terminals A and B with all voltage and current sources removed, leaving their internal resistances behind.

4. According to Norton's theorem, any two-terminal active network when viewed from its output terminals, is equivalent to a constant current source and a parallel resistance.
5. Maximum power transfer theorem states that a resistive load will abstract maximum power from a network when its resistance is equal to the resistance of the network as viewed from the output terminals with all energy sources removed leaving behind their internal resistances.
6. In Δ/Y trnasformation, the rule is :
resistance of each arm of the star is given by the product of the resistances of the two delta sides that meet at its end divided by the sum of the three delta resistances.
7. In Y/Δ transformation, the rule is :
the equivarent delta resistance between any two terminals is given by the sum of star resistances between these terminals plus the product of these two star resistances divided by the third star resistance.

OBJECTIVE TESTS—3

A. Fill in the following blanks:

1. Kirchhoff's first law deals with currents meeting at a .... of a network.
2. According to KCl .... sum of currents meeting at a junction is zero.
3. While applying Kirchhoff's laws, direction of current flow may be assumed either clockwise or ....
4. In Maxwell's method of solving networks, we take .... currents instead of branch currents.
5. While applying Thevenin's theorem, the given network is reduced to a .... network.
6. Use of Thevenin's theorem requires the determination of open circuit .... between the load terminals.
7. According to Thevenin's theorem, the equivalent resistance of the circuit as viewed from two terminals is found by replacing the current sources by .... circuits.
8. Norton's theorem is used where it is easier to simplify a network in terms of .... instead of voltages.

B. Answer True or False

1. Kirchhoff's laws make use of branch currents.
2. While applying Kirchhoff's laws, flow of branch currents may be assumed in any direction we like.
3. In Maxwell's method of solving eletrical networks, branch currents are found in termsof loop currents.
4. An ideal constant-current source has an internal resistance of zero.
5. With the help of Thevenin's theorem, a complex circuit can be simplified to a series circuit.
6. Norton's theorem reduces a given complicated network to a simple parallel circuit.
7. Superposition theorem can be applied to those circuits which contain voltage sources only,
8. According to maximum power transfer theorem, a load resisatance will abstract maximum power when it equals the internal resistance of the source.

C. Multiple Choice Questions

1. According to KCL as applied to a junction in a network of conductors
(a) total sum of currents meeting at a junction is zero.
(b) no current can leave the junction without some entering it.
(c) net current flow at the junction is positive
(d) algebraic sum of the currents meeting at a junction is zero.
2. Kirchhoff's voltage law is concerned with
(a) IR drops
(b) battery EMFs
(c) junction voltages

(*d*) both (*a*) and (*b*)

3. According to the commonly-used sign convention for voltages

(*a*) a fall in voltage is considered positive

(*b*) a rise in voltage is considered positive

(*c*) *IR* drop is taken as negative

(*d*) battery EMFs are taken as positive.

4. While Thevenizing a circuit between two terminals, V_{th} equals

(*a*) short-circuit terminal voltage

(*b*) open-circuit terminal voltage

(*c*) EMF of the battery nearest to the terminals

(*d*) net voltage available in the circuit.

5. Norton's equivalent of a circuit is

(*a*) constant-current source with a conductance in parallel

(*b*) constant-current source in series with an infinite resistance

(*c*) constant-voltage source in parallel with a high resistance

(*d*) single current source and a single voltage source

6. While calculating R_{th}, constant-current sources in the circuit are

(*a*) replaced by 'opens'

(*b*) replaced by 'shorts'

(*c*) treated in parallel with other voltage sources

(*d*) converted into equivalent voltage sources.

ANSWERS

A. 1. junction 2. algebraic 3. anti-clockwise 4. loop 5. series 6. voltage 7. open 8. currents

B. 1. T 2. T 3. T 4. F 5. T 6. T 7. F 8. T

C. 1. *d* 2. *d* 3. *b* 4. *b* 5. *a* 6. *a*

4 WORK, POWER AND ENERGY

4.1. Heating Effect of Electric Current

It is a matter of common experience that a conductor, when carrying current, becomes hot after some time. As explained earlier, an electric current is just a directed flow or drift of electrons through a substance. The moving electrons, as they pass '*through*' the molecules or atoms of that substance, collide with other electrons. This electronic collision results in the production of heat. This explains why passage of current is always accomplanied by generation of heat. the heat so produced is measured in the following units.

4.2. Unit of Heat

The unit generally employed in scientific work is calorie. A calorie is defined as the quantity of heat that will raise the temperature of 1 gram of water through one degree centigrade.

For very accurate work, the particular degree has been specified as from 14.5°C to 15.5°C.

In the SI system, the unit of heat is kilo/calorie (k/cal) whichis defined as the amount of heat required to heat one kg of water through 1°C or 1°K.

4.3. Joule's Law of Electric Heating

The amout of work required to maintain a current of I amperes through a resistance of R ohm (Fig. 4.1) for t seconds is

$$\text{W.D} = I^2 Rt \text{ joules}$$

$$= VIt \text{ joules} \qquad (\because R = V/I)$$

$$= Wt \text{ joules} \qquad (\because Watt = V \times I)$$

$$= \frac{V_2 t}{R} \text{ joules} \qquad (\because I = V/R)$$

This work is converted into heat and is dissipated away. Heat produced is

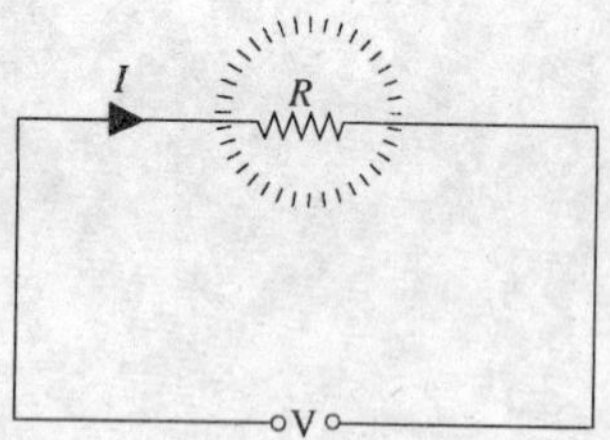

Fig. 4.1

$$H = \frac{\text{work done (W.D.)}}{\text{mechanicalequivalentof heat}} = \frac{\text{W.D.}}{J}$$

where
$$J = 4.186 \text{ joules /cal}$$
$$= 4.2 \text{ joules /cal (approx)}$$
$$= 4{,}186 \text{ joules/kcal}$$
$$= 4{,}200 \text{ joules /kcl (approx)}$$

$$\therefore \quad H = \frac{I^2Rt}{4.2} \text{ cal} = \frac{I^2Rt}{4{,}200} \text{ kcal}$$

Other expressions for heat produced are

$$H = \frac{VIt}{4{,}200} \text{kcal} = \frac{Wt}{4{,}200} \text{kcal} = \frac{V^2 . t}{4{,}200R} \text{ kcal}$$

Example 4.1. *A potential difference of 10 V is applied across 2.5 Ω resistor. Calculate the current, the power dissipated and the energy transformed into heat in 5 minutes.*

A.M.I.E Winter 1986)

Solution. $I = V/R = 10/2.5 = \mathbf{4A}$ $\quad W = VI = 10 \times 4 = \mathbf{40\ \Omega}$

$$H = \frac{Wt}{4200} = \frac{40 \times (5 \times 60)}{4200} = \textbf{2.86 kcal}$$

4.4. Thermal Efficiency

It is defined as the ratio of the heat *actually* utilised to the total heat produced electrically. Consider the case of the electric kettle used for boiling water. Out of the total heat produced (*i*) some goes to heat the apparatus itself *i.e.* kettle (*ii*) some is lost by radiation and convection etc. and (*iii*) the rest is utilised for heating the water. Out of these, the heat utilised for useful purpose is that in (*iii*). Hence, thermal efficiency of this electric apparatus is the ratio of heat utilized to heat water to the total heat produced.

4.5. General Formula

Suppose m kg of water contained in a vessel of water equivalent W kg is to be heated electrically from t_1° C to t_2° C in t seconds. The heat required is

$$Q = (m + W)(t_2 - t_1) \text{ kacl}$$

Assuming an efficiency of η for the electric heater or other energy conversion apparatus used, we get

$$\frac{VIt}{4200} \times \eta = (m + W)(t_2 - t_1); \qquad \frac{Wt}{4200} \times \eta = (m + W)(t_2 - t_1)$$

$$\frac{V^2 t}{4200R} \times \eta = (m + W)(t_2 - t_1)$$

Example 4.2. *An electric lamp consumes 100 watts of power. If the supply voltage is 220V, determine the current through the filament, its resistance and the energy consumed in 20 minutes.*

(Elect. Engg.-I, Punjab Univ. 1993)

Solution. $P = VI$ or $100 = 220 \times I$, $I = \textbf{0.454 A}$

Also, $P = V^2/R$ or $R = 2202/100 = \textbf{484 }\Omega$

Energy consumed $= VIt = P.t = 100 \times (20 \times 60) = \textbf{120,000 J}$

Example 4.3. *An electric kettle was marked 500-W, 230-V and was found to take 15 minutes to bring 1 kg of water at 15°C to boiling point. Determine the heat efficiency of the kettle.*

(A.M.I.E Summer 1988)

Solution. $H = \frac{Wt}{4200}$ kcal $\quad \therefore \quad H = 500 \times 15 \times 60/4200$ kcal

Heat actually absorbed by water

= Mass × Sp. Heat × Rise in temperature $= 1 \times 1 \times (100 - 15) = 85$ kcal

$$\text{Heat efficiency} = \frac{\text{heat absorbed}}{\text{heat produced}} = \frac{85 \times 4{,}200}{500 \times 15 \times 60} = 0.79 \text{ or } 70\%$$

Example 4.4. *A heating coil of resistance 50Ω is immersed in a vessel containing 500 gm of water. If on passing a current of 2A the temperature rise at the end of 10 minutes is observed to be 57.2°C, neglecting losses, calculate Joule's equivalent.*

(Electrical Engg. A.M.Ae.S.I. June 1992)

Solution. Work done electrically $= I^2Rt = 2^2 \times 50 \times (10 \times 60) = 120{,}000$ J

Neglecting losses, heat produced is $= H = ms\,(t_2 - t_1) = 0.5 \times 1 \times 57.2 = 28.6$ kcal.

$$\text{Joule's equivalent} = \frac{\text{work done}}{\text{heat produced}} = \frac{120{,}000}{28.6} = \textbf{4196 J/kcal.}$$

Example 4.5. *Determine the kW input necessary for an electric heater to raise the temprature of 5 litres of water from 15°C to 90°C in 10 minutes. Take the heat loss in radiation during the period as 65000 J and water equivalent of the heater as 130 g. Calculate approximately, the efficiency η of the operation. How long will it take to heat the water through the same range, if the input kW were halved? Take 1 kcal = 4180 J.*

(Electric Engg. A.M.Ae.S.I. June 1993)

Solution. Mass of water = 5.0 kg ; water equivalent of heater = 0.13 kg

∴ effective mass of water heated = 5.13 kg

Temperature rise = 90° – 15° = 75°C

Heat absorbed in 10 minutes = 5.13 × 75 = 384.75 kcal

= 384.75 × 4180 = 1, 608, 255 J

Radiation loss in 10 minutes = 65,000 J

∴ total energy consumed by water and calorimeter plus radiation loss

= 1 608,255 + 65, 000 = 1,673,255 J

Rate of energy consumption = 1,673,255/10 × 60 = 2,789 W

Rating of the heater = 2, 789/1000 = 2.789 kW = **3 kW (approx)**

Efficiency

Heat employed usefully = heat absorbed by water

= 5 × 75 = 5 × 75 × 4180 J = 1,567,500 J

∴ η of operation = 1,567,500/1,673,255 = 0.937 = **93.7%**

Example 4.6. *An immersion heater rated at 3 kW is used to heat a copper tank weighing 20 kg and holding 120 litres of water. How long will it take to raise the temperature of the water from 10°C to 60°C if 20 per cent of the energy supplied is wased in heat losses ?*

Assume specific heat of copper = 0.095 ; J = 4.2 joules/calorie.

Solution. Water equivalent of the copper tank = 20 × 0.095 = 1.9 kg of water

Effective amount of water = 120 + 1.9 = 121.9 kg

Rise in temperature = $(t_2 - t_1)$ = (60 – 10) = 50°C

Amount of heat reqd. Q = 121.9 × 50 = 6,095 kcal

Since thermal efficiency is given as 80% or 0.8,

$$\therefore \quad \frac{Wt}{4200} \times \eta = Q \quad \text{or} \quad \frac{3000t}{4200} \times 0.8 = 6095$$

∴ t = 10,666 second = **2 h 58 min.**

4.6. Quantity of Electricity

The difference between the rate of flow of electricity and the total quantity of electricity passing through a circuit in a given time, should be understood clearly. The rate of flow of electricity gives the current strength and is usually measured in amperes. The total *quantity* of electricity or charge is given by the product of current strength and time.

$$\therefore \quad Q = I \times t$$

If I is one ampere and it flows for one second, then

$$Q = 1 \text{ ampere} \times 1 \text{ second} = 1 \text{ A-s} = 1 \text{ coulomb}$$

Hence, one columb is that *quantity of electricity which flows per second past any point in a conductor, when current of one ampere flows through it.*

The bigger unit generally employed is ampere-hour (Ah)

The ampere-hour is the charge transferred by a current of one ampere while flowing for one hour through a given circuit.

I Ah = 3600 coluombs

For example, a current of 5 ampere strength, when flowing for 3 hours will caryy over a charge of 5 × 3 = 15 Ah or 15 × 3600 = 54,000 coulombs.

4.7. Electric Power

Power is the rate of doing work and is independent of the tota amount of work to be done. The rate of working (or power) is found by dividing the work done by the time required to do it.

$$\therefore \quad \text{electric power} = \frac{\text{electric work done}}{\text{time taken}}$$

We have seen that work done electrically in time t seconds = VIt joules

$$\therefore \quad \text{power} = \frac{VIt}{t} = VI$$

If V is in volts and I in amperes, then product VI is in watts.

$\therefore$ power in watts = volt × amperes

One watt may be defined as *the rate of doing one joule of work per second.*

$\therefore$ 1 watt = 1 joule/second

Bigger units are :—

1 kW = 1000 W = 10^3 W ; 1 MW = 10^6 W

4.8. Electric Energy

The unit of energy is Joule. Other units are :

1 watt-hour (Wh) = 1 watt × 1 hour = (1 J/s) × 3600 s = 3600 J

1 kilowatt-hour (kWh) = 3600 × 1000 = 36×10^5 J

The bill for electric charges is based on the number of kWh consumed.

4.9. Some Units in SI System

The SI system has only absolute units. It does not allow gravitional units. However, till switch-over to the SI system is complete, gravitional units have also been given where they are still commonly used.

(*a*) **Mass.** It is the the quantity of matter contained in a body. Its basic unit is kilogram (kg).

Other multiples commonly used are : 1 quintal = 100 kg ; 1 tonne = 1000 kg

(*b*) **Force**

(*i*) The absolute unit of force is newton (*N*). Its definition may be obtained from Newton's Second Law of Motion *i.e.* $F = ma$

If m = 1 kg and a = 1 m/s², then F = 1 newton.

Hence, one newton is that force which can give an acceleration of 1 m/s² to a mass of 1 kg.

(*ii*) Gravitational unit of force is kg-wt. It may be defined as follows—

It is the force which can impart an acceleration of 9.8 m/s² to a mass of 1 kg.

Obviously, 1 kg–wt = 9.8 N

In engineering literature, gravitational unit of force is written as kgf (instead of kg-wt), the letter '*f*' being added in order to distinguish it from the unit of mass which is kg.

(*c*) **Weight.** It is the force with which earth pulls a body down-wards. Its units are the same as for force.

(*i*) Absolute unit of weight is newton. (*ii*) Gravitational unit of weight is kg-wt.

Note. If a body has a mass of *m*. kg. then its weight is = mg = $m \times 9.8$ N

Now 1 kg-wt = 9.8 N

$\therefore$ weight = m × 9.8/9.8 = m kg-wt

Obviously, the mass of a body in *absolute* unit *i.e.* kg is *numerically* equal to its weight in gravitional unit *i.e.*, kg-wt.

It should be clearly understood that a body weighing 500 kg *i.e.*, 500 kg-wt has a mass of 500 kg.

(*d*) **Work.** If a force of F is applied in overcoming certain opposition or in moving a body through a distance S along the direction of force, then work done is

$$\text{W.D.} = F \times S$$

(*i*) Absolute unit of work is joule, which is defined below :

If F = 1N ; S = 1 m ; W.D. = 1 m–N or joule

Hence, one joule is the work done by a force of one newton when applied over a distance of one metre.

(*ii*) Gravitational unit of work is m-kg wt.

If F = 1 kg-wt ; S = 1 m ; W.D. = 1 m-kg wt or m-kg

Hence, one m-kg is the work done by a force of one kg-wt, when applied over a distance of one metre.

Obviously, 1 kg-wt = 9.8 joules.

Note. Usually the word 'wt' is omitted in which case unit of work becomes m-kg.

(*e*) **Torque.** It is defined as the turning moment of a force about an axis. It is given by the product of force and the radius at which it acts.

$\therefore \quad T = F \times R$

(*i*) *Absolute unit*

If F = 1 newton and R = 1 metre, then T = 1N × 1m = 1N–m

(*ii*) *Gravitational unit*

If F = 1 kg-wt and R = 1 m, then T = 1 kg-wt × 1 m = 1 kg-wt-m

Usually, the word 'wt' is omitted, hence the above unit is commonly written as kg-m.

Obviously, 1 kg-m = 9.8 N-m

(*f*) **Kilowatt-hour and Kilocalorie**

1 kWh = 36 × 105 joules

Also 1 kcal = 4,186 joules, $\therefore$ 1kWh = 36 × 10^5/4,186 = 860 kcal

(*g*) **Power.** If a force of F newton imparts a velocity of v metre/second to a body, then power developed is

$$P = F \times v \text{ watt}$$

If velocity v is in km/s, then, power = $f \times v$ kilowatt

Also, power = Tω watts

where, T = torque in N-m

ω = angular velocity in radian/second = 2πN rad/s

Here, N is motor speed in revolutions per second (rps).

Note. A velocity of 72 km/h = 72 × 1000/3600 = 20 m/s. As seen, the conversion factor is = 1000/3600 = 5/18.

Example 4.7. *The demand for the lighting of a small village is 50 A at 200 V. This is supplied from a dynamo at a distant station, the generated voltage being 220 V. Find (i) the resistance of the leads from the dynamo to the village (ii) the energy consumed in 10 hours in the village in kWh (iii) the number of kWh wasted in the leads in the same time.* **(A.M.I.E. Winter 19189)**

Solution. (*i*) Voltage drop over the leads (come and go) = 220 – 200 = 20 V

If R is the resistance of one conductor, then

$$2IR = 20 \quad \text{or} \quad 2 \times 50 \times R = 20 \text{ ; } R = 0.2\ \Omega$$

(*ii*) Energy consumed $= VIt = 200 \times 50 \times 10 \text{ Wh} = \dfrac{200 \times 50 \times 10}{1000} = \mathbf{100\ kWh}$

(*iii*) Energy wasted = 20 V × 50 A × 10 h = 10,000 Wh = **10 kWh**

Example 4.8. *An electric kettle contains 1.5 kg of water at 15°C. It takes 15 minutes to raise the temperature to 95°C. Assuming the heat losses due to radiation and heating the ketttle to be 14 kcal, find the current taken. The supply voltage is 100 V.*

Solution. The amount of heat required to raise the temperature of 1.5 kg of water from 15°C to 95°C is

= 1.5 (95 – 15) = 120 kcal

Total losses = 14 kcal

Hence, total amount of energy taken from the supply is = 120 + 14 = 134 kcal

Now, 1kWh = 860 kcal or 1 kcal = 1/860 kWh

$\therefore$ electric energy consumed = 134 × 1/860 = 67/430 kWh = 67 × 1000/430 Wh

Let I be the current drawn from the supply. Since time taken is 15 minutes *i.e.*, 1/4 hour, energy consumed is

VIt = 100 I × 1/4 = 25 I watt-hours $\therefore$ 25 I = 6700/43 ; I = **6.23 A**

Example 4.9. *Two heaters A and B are in parallel across supply voltage V. Heater A produces 500 kcal in 20 minutes and B produces 1000 kcal in 10 minutes. The resistance of A is 10 Ω.*

What is the resistance of B ? If the same heaters are connected in series across the same voltage, how much heat will be produced in 5 minutes ? **(A.M.I.E. Summer 1991)**

Solution. Heat produced $H = V^2\, t/R$

For heater A : 500 $= V^2 \times 20/10$ ...(*i*)

For heater B : $1000 = V^2 \times 10/R_2$ ...(*ii*)

Dividing (*i*) by (*ii*), we get $R_2 = 2.5\ \Omega$

Total resistance when connected in series = 12.5 Ω

Heat produced, $H = V^2 \times 5/12.5$ ...(*iii*)

From (*i*) and (*iii*) we get, H = **100 kcal.**

Example 4.10. *Two coils are connected in parallel and a voltage of 200 V is aapplied to the terminals. The total current taken is 15 A and the power dissipated in one of the coils is 1,500 watts. What is the resistance of each coil ?* **(A.M.I.E. Summer, 1990)**

Solution. Since the two coils are connected in parallel, voltage across each is the same *i.e.* 200 V. If I_1 is the current drawn by one coil, then

$VI_1 = W_1$ or $200 \times I_1 = 1500$; $I_1 = 7.5$ A

$R_1 = V/I_1 = 200/7.5 = \mathbf{26.7\ \Omega}$

Now, $I_2 = 15 - I_1 = 15 - 7.5 = 7.5$ A

Since, current drawn by the second coil is the same, hence, R_2 = **26.7. Ω.**

Example 4.11. *An electric boiler has two heating elements each of 230-V, 3.5 kW rating and contains 8 litres of water at 30°C. Assuming 10 per cent loss of heat from the boiler, find how long, after switching on the heater circuit, will the water boil at the atmospheric pressure (a) if the two elements are in parallel (b) if the two elements are in series. The supply voltage is 230 V?*

Mechanical equivalent of heat is 4,200 joules/kcal and 1 litre of water holds one kilogram of water.

Solution. Rating of each element = 3.5 kW = 3500 W

Combined rating of the two identical elements connected in parallel = 3500 × 2 = 7000 W

Combined rating of two identical elements when connected in series

= 3500/2 = 1750 W (Why ?)

Heat required $Q = 8(100 - 30)$ = 560 kcal

(*i*) **Elements in Parallel**

$$\frac{Wt}{4200} \times \eta = Q \quad \text{or} \quad \frac{7000 \times t}{4200} \times 0.9 = 560, \qquad t = \mathbf{373.3} \text{ second}$$

(*ii*) **Element in Series**

$$\frac{1750t}{4200} \times 0.9 = 560 \text{ or } t = \mathbf{1493.2 \text{ second}}$$

Note. Since power rating is one-fourth, time taken would be four times that in the first case.

Example 4.12. *An immersion heater, which is to operate on 230-V supply has to raise the temperature of 45.36 kg of water from 20°C to 95°C in 1 hour. Taking the efficiency of operation as 86%, determine (a) the rating of the heater and (b) the resistance of the heating elements in the heater. Calculate also the cost of the heating operation if energy costs 50 paise per k Wh.*

Solution. Mass of water = 45.36 kg

Temp. rise = 95 – 20 = 75°C

Heat utilized Q = 75 × 45.36 kcal

Efficiency = 0.86

Let W be the rating of the heater in watts.

Now, $$\frac{Wt}{4186} \times \eta = 75 \times 45.36$$

$\therefore \quad \dfrac{W \times 3600}{4186} \times 0.86 \quad = \ 75 \times 45.36$

(*a*) $W = 4600$ W $= \mathbf{4.6\ kW}$

(*b*) Now watts $= V^2/R$

$\therefore$ $4600 = 230^2/R$ or $R = 230^2/4600 = \mathbf{11.5\ \Omega}$

Energy input in one hour = 4.6 × 1 = 4.6 kWh

Cost = 4.6 × 50 = 230 paise = **Rs. 2 and 30 paise**

Example 4.13. *Find the amout of electrical energy expended in raising the temperature of 45 litres of water by 75°C. To what height could a weight of 5 tonnes be raised with the expenditure of the same energy? Assume efficiencies of the heating equipment and lifting equipment to be 90% and 70% respectively.* **(Electrical Engg., A.M.Ae.S.I. Dec. 1991.)**

Solution. Mass of water heated = 45 kg ; Heat reqd = 45 × 75 = 3.375 kcal

Heat produced electrically = 3,375/0.9 = 3,750 kcal.

Now, 1 kcal = 4,186 J

Electrical energy expended = 3,750 × 4,186 J = 15.6975 MJ

Energy available for lifting the load is = 0.7 × 15.6915 = 10.98825 MJ

If h meteres is the height through which the load of 5 tonnes can be lifted, then

potential energy of the load = mgh joules = 5 × 1000 × 9.81 h joules

$\therefore \quad 500 \times 9.81 \times h = 10.98825 \times 10^6$

$h = \mathbf{224\ metres}$

Example 4.14. *Find the current per line in a 3-phase arc-furnace to melt 5 metric tonne of steel in one hour at an overall efficiency of 50% if the arc is 115 V, initial temperature 30°C, melting point of steel 1370°C, specific heat of steel 278 J/kg/°C, latent heat of steel 37,000 J/kg.* **(Electrical Engg. A.M.Ae.S.I. Dec. 1994.)**

Solution. Total heat energy required to heat and melt steel $= mL + ms\,(t_2 - t_1)$

$mL = 5000 \times 37{,}000 = 1850 \times 10^5$ J

$ms\,(t_2 - t_1) = 5000 \times 278 \times 1340 = 18{,}626 \times 10^5$ J

Total heat energy reqd. = $18{,}626 \times 10^5 + 1850 \times 10^5 = 20{,}476 \times 10^5$ J

Since efficiency of operation is 50%, input energy = $20{,}476 \times 10^5/0.5 = 40{,}952 \times 10^5$J

Time taken is 1 hr = 3600 second

$\therefore$ input $= 40{,}952 \times 10^5/3600 = 11.375 \times 10^5$ W

If I_L is the line current, then

$\sqrt{3} \times 150 \times I_L \times 1 = 11{,}37{,}500$ or $I_L = \mathbf{5{,}710\ A}$

A power factor of unity has been assumed.

Example 4.15. *A diesel-electric generating set supplies an output of 50 kW. The calorific value of the fuel oil used is 12,500 kcal/kg. If the overall efficiency of the unit is 35% (i) calculate the mass of oil required per hour and (ii) the electrical energy generated per tonne of the fuel.* **(A.M.I.E. Summer 1988)**

Solution. Output = 50 kW ; Overall efficiency η = 0.35

Input = 50/0.35 = 142.86 kW

Input per hour = 142.86 × 860 = 122,960 kcal

Since 1 kg of fuel oil produces 12,500 kcal

(*a*) mass of oil reqd. = 122,860/12,500 = **9.83 kg**

(*b*) 1 tonne of fuel = 1000 kg

Heat content = 1000 × 12,500 = 12.5 × 106 kcal

= 12.5 × 106/860 = 14,535 kWh

Overall η = 0.35

Energy output $= 14,535 \times 0.35 =$ **5087 kWh**

Example 4.16. *Calculate the current required by a 1,500 volt d.c. locomotive when drawing 100 tonne load at 45 km/h with atractive resistance of 5 kg/tonne along (a) level track (b) agradient of 1 in 50. Assume a motor efficiency of 90 per cent.*

Solution. (*a*) As shown in Fig. 4.2 (*a*), in this case, only tractive resistance is to be overcome.

Tractive resistance $F = 5 \times 100 = 500$ kg-wt $= 500 \times 9.81 = 4,905$ N

Velocity $= 45$ km/h $= 45 \times 5/18 = 12.5$ m/s

Power required $= 4,905 \times 12.5 = 61,310$ W

$\therefore$ lccomotor power output $= 61,310$ W; $\eta = 0.9$

Motor power input $= 61,310/0.9 = 68,125$ W

Current drawn $= 69,125/1500 =$ **45.4 A**

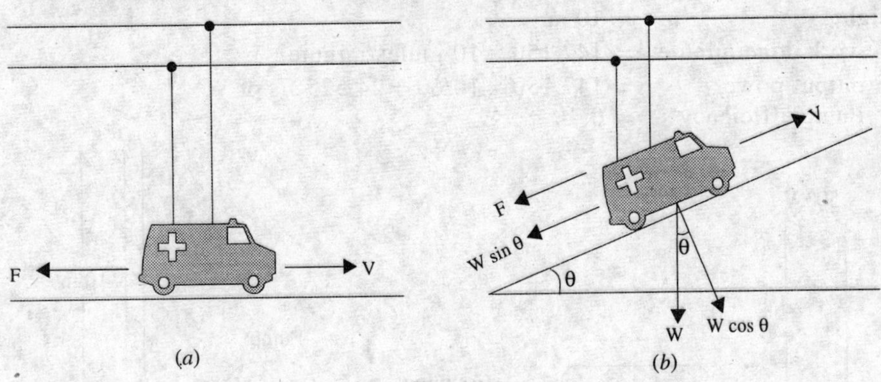

Fig. 4.2

(*b*) In this case, downward component of load is also to be taken into consideration [Fig. 4.2 (*b*)]

Load component acting downward along the gradient

$= W \sin\theta = 100 \times 1/50$ tonne $= 2$ tonne $= 2,000$ kg-wt

$= 2000 \times 9.81 = 19,620$ N

Tractive resistance $= 4905$ N (as before).

Total force required $= 19,620 + 4905 = 24,525$ N

Power output required $=$ force $\times$ velocity $= 24,525 \times 12.5$ W

Power input $= 24,525 \times 12.5/0.9$ W

Current drawn $= \dfrac{24,525 \times 12.5}{1500 \times 0.9} =$ **227 A**

Example 4.17. *In an electric furnace, 100 kg of tin are to be melted in an hour from an initial temperature of 15°C. Find the power required and the cost of the melting with energy at 50 paise per unit. (Sp. heat of tin = 0.056 ; melting point of tin = 235 °C ; latent heat of fusion of tin = 13.31 kilocalorie per kg ; thermal efficienty = 70 percent ; J = 4200 joules / kcal).*

Solution. Heat required to bring 100 kg of tin from 15°C to melting point of 235°C

$= 100 \times 0.056 \times (235 - 15) = 1232$ kcal

Heat of fusion $= 100 \times 13.31 = 1331$ kcal

Total heat needed $= 1232 + 1331 = 2,563$ kcal

Thermal efficiency $= 0.7$

$\therefore$ total heat produced electrically

$H = 2,563/0.7 = 3661.4$ kcal

$J = 4200$ joules/kcal

Also, $W = JH$

$\therefore$ total energy produced/hour

$= 3661.4 \times 4200$ J/h $= 153.78 \times 10^5$ J/h

Rate of energy production = $153.78 \times 105/3600$ J/s = **4278 watt**

Energy produced = $\dfrac{152.56 \times 10^5}{36 \times 10^5}$ = 4.272 kWh

Cost = $4.272 \times 50 = 214$ P = **Rs. 2 and 14 P**

Example 4.18. *An electrically-driven pump lifts 15 tonne of water in a minute to a height of 10 m. Assuming an efficiency of 70% for the pump and 90% for the motor, calculate (i) motor output in kW (ii) input current to the motor if the supply voltage is 500 V.*

Also, find the cost of running the pump for 3 hours a day for 30 days if the price of electrical energy is 90 paise/kWh.

Solution. The motor-pump arrangement is shown in Fig. 4.3.

Wt. of water lifted = 15,000 kg-wt = $15,000 \times 9.81$ = 147,150 N

Height = 10 m

∴ work done/minute = $147,150 \times 10$ joules/minute

∴ output power = $147,150 \times 10/60$ = 24,525 J/s or W

Pump efficiency = 0.7

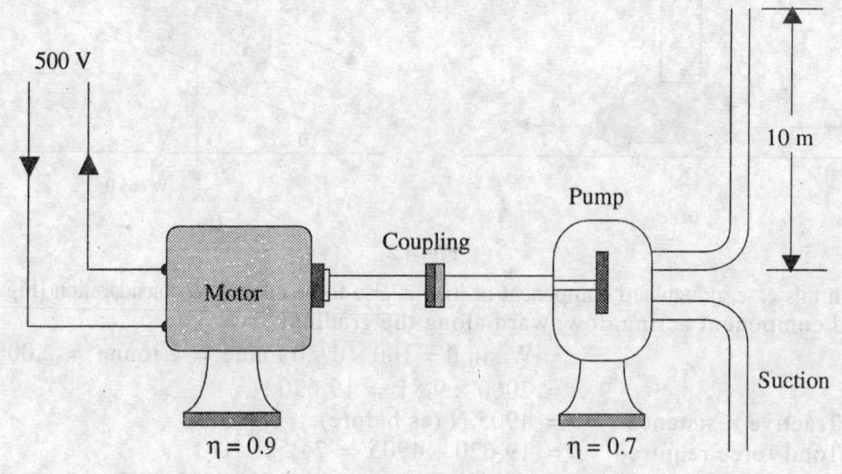

Fig. 4.3

(*i*) Pump input or motor power input = 24,524/0.7 = 35,036 W = 35.036 kW

(*ii*) Motor efficiency = 0.9

Motor power input = 35,036/0.9 = 38,930 W

Current drawn by motor = 38,930/50 = 77.86 A

Cost

Running time of the pump = 30×3 = 90 h

Energy consumed by the motor = 38.93×90 = 3504 kWh

Cost = Rs. 3504×0.9 = **Rs. 3154**

Example 4.19. *Calculate the number of kWh of electric energy obtained per hour from a generating plant whose overall efficiency is 26 per cent, given :*

Amout of coal burnt/hour = 4 tonne; calorific value of coal = 6,000 kcal/kg

Solution. Total amount of heat produced = $4,000 \times 6,000 = 24 \times 10^6$ kcal/h; $\eta = 0.26$

Amount of heat or thermal energy utilized = $0.26 \times 24 \times 10^6 = 624 \times 10^4$ kcal

Now 1 kWh = 860 kcal

∴ electric energy produced/ η is = $624 \times 10^4/860$ = **7256 kWh/h**

Example 4.20. *An electric lift with a cage (or body) weighing 1/2 tonne and a balance weight of 2 tonne takes a load of 3 tonne to a height of 20 metre in 30 seconds returning empty in the same time. The efficiency of the motor and gearing is 75% in either direction. If it makes*

15 double journeys per hour, how many kWh will be consumed in an hour?

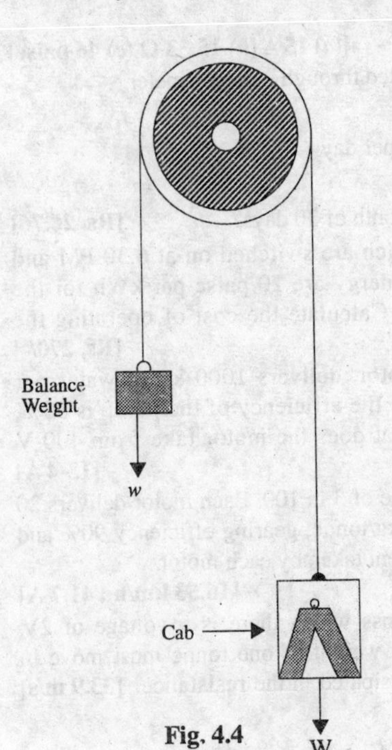

Fig. 4.4

Solution. The lift is shown in Fig. 4.4

Wt. lifted during upward journey

$= 3 + \frac{1}{2} - 2 = 1\frac{1}{2}$ tonne-wt = 1500 kg-wt

Wt. pulled down during downward journey

$= 2 - \frac{1}{2} = 1\frac{1}{2}$ tonne-wt = 1500 kg-wt

Distance travelled either way = 20/m

Work done per double journey

$= 2 \times (1500 \times 20) = 60,000$ m-kg wt

$= 60,000 \times 9.81 = 588,600$ J

Motor and gearing $\eta = 0.75$

Energy consumed per double journey = 588,600/0.75

= 784,800 J

Number of double journeys made per hour = 15

Energy consumed/h = $15 \times 784,800$ J

Now 1kWh $= 36 \times 10^5$J

$\therefore$ energy consumed/h

$= 15 \times 784,800/36 \times 10^5 = 3.27$ kWh

Example 4.21. *A dc motor develops a full-load torque of 25 N-m when running at 1200 rpm. If motor efficiency is 85%, calculate the current drawn by it from a 200-V supply.*

Solution. Power developed $P = T\omega = T \times 2\pi N$ newton-metre where N is motor speed in rps.

Now, $T = 25$ N-m, $N = 1200/60 = 20$ rps

$P = 25 \times 2\pi \times 20 = 3,142$ W

Since motor $\eta = 0.85$, power drawn from the supply is $= 3142/0.85 = 3,696$ W

$V.I = 3696$ or $I = 3696/200 =$ **18.48 A**

Tutorial Problems No. 4.1

1. Calculate the length of wire of resistance 6 Ω / m required to construct a heating coil capable of producing 200 cal/min when connected to the 100-V supply mains. **[119 m]**
2. An electric kettle is required to heat 0.5 kg of water from 10°C to boiling point in 5 minutes, the supply voltage being 230-V. If the efficiency of the kettle is 80 per cent, calculate the resistance of the heating element. Assume the specific heat capacity of water to be 4.2 kJ/kg. °K. **[67.2 Ω]**
3. A coil of resistance 100 Ω is immersed in a vessel containing 0.5 kg of water at 16°C and is connected to a 220-V electric supply. Calculate the time required to boil away all the water. Given J = 4,200 joules/kcal ; latent heat of steam = 536 kcal/kg. **[44 min 50 second]**
4. An electric fire takes a power of 1kW when connected directly across a 200-V mains. If long leads having a total resistance of 1Ω , are used to connect the fire to a 200-V mains, calculate (*a*) the power taken from the mains and (*b*) the power used in the fire itself. It may be assumed that the resistance of the fire remains constant. **[(a) 976 W (b) 952 W]**
5. An aluminium kettle, weighing 2 kg, holds 2 litres of water and consumes electric power at the rate of 2 kW. If 40% of the heat supplied is wasted, find the time taken to bring the kettle of water to boiling point from an initial temperature of 20°C. Specific heat of aluminium = 0.2 kcal/kg/°C, J = 4,200 joules/kcal. **[11.2 min]**
6. An electric cooker is rated at a maximum of 3.45 kW at 230 V. Determine (*a*) the current when used at maximum rating (*b*) the resistance of the heating element (*c*) the cost of using the cooker

for three hours, the first half hour being at maximum rating and the rest of the time at one-third maximum rating. Electric energy costs 10 paise/ kWh.

[(*a*) 15 A (*b*) 15.33 Ω (*c*) 46 paise]

7. The following are the details of the load on a circuit connected through a supply meter.

(*a*) six lights of 40 W each working for 4 hours / day.

(*b*) two fluorescent tubes of 125W each working 2 hours per day.

(*c*) one 1-kW heater working 3 hours / day

If one kWh costs 20 paise, what will be the total cost in a month of 30 days? **[Rs. 26.76]**

8. A floodlit sign is provided with four 1000-W lamps which are switched on at 6.30 PM and off at 11-30 PM every day. The charges for electrical energy are 20 paise per kWh for the first 100 kWh plus 50 paise for every additional kWh. Calculate the cost of operating the floodlighting for one month of 30 days. **[Rs. 270/-]**

9. A centrifugal pump, which is gear-driven by a d.c. motor, delivers 1000 kg of water per minute to a tank 20 metre above the level of the sump. If the efficiency of the pump is 80%, that of the gearing 90% and that of the motor 85%, what does the motor take from 400-V supply ? **[13-4 A]**

10. A 20-tonne tram-car driven by two motors ascends an incline of 3 in 100. Each motor delivers 20 kW when running on 600 V. The tractive resistance is 10 kg/tonne, gearing efficiency 90% and motor efficiency 80%. Find the speed of the car and the current taken by each motor.

[16.53 km/h ; 41.7 A]

11. A current of 80A flows for one hour in a resistance across which there is a voltage of 2V. Determine the velocity in meters per second with which a weight of one tonne must move in order that its kinetic energy shall be equal to the energy dissipated in the resistance. **[33.9 m/s]**

HIGHLIGHTS

1. One kilocalorie (kcal) is the amount of heat required to heat one kg of water through one °C.
2. According to Joule's Law of electric heating, the amount of heat produced is

$$H = \frac{I^2 Rt}{J}; \quad \text{where } J = 4186 \text{ joules/kcal} \cong 4200 \text{ joules/kcal}$$

3. General formula for heating is

$$\frac{I^2 Rt}{J} \times \eta = (m + W)(t_2 - t_1); \qquad \frac{Wt}{4200} \times \eta = (m + w)(t_2 - t_1)$$

$$\frac{V^2 t}{4000R} \times \eta = (m + W)(t_2 - t_1)$$

4. Quantity of electricity or charge is measured in terms of coulomb (C). One columb of charge is the charge carried over by a current of one ampere flowing for one second. Larger unit of charge is ampere-hour (Ah).

$$1 \text{ Ah} = 3600 \text{ coulomb}$$

5. Electric power is given in watts

$$\text{volts} \times \text{amperes} = \text{watts}$$

6. Energy is measured in kWh.

$$1 \text{ kWh} = 36 \times 10^5 \text{ joules} = 860 \text{ kcal}$$

OBJECTIVE TESTS—4

A. Fill in the following blanks :

1. Ampere-hour is the unit of ....
2. Kilowatt-hour is the unit of ....
3. Force is measured in ....
4. Power is given by the product of force and ....
5. One kWh equals nearly .... kcal.
6. If two identical 220-V, 4.kW heating elements are connected in series across a 220-V supply, their combined rating becomes ....kW.

B. Answer True or false.

1. The unit of heat is kilocalorie.

2. Kolowatt-hour is the unit of power.
3. Power is given by the product of torque and angular velocity.
4. More heat is produced by connecting two identical heaters in parallel across their rated supply than in series.
5. Torque is given by the product of force and the radius at which this force acts.
6. Metre-newton is the unit of torque.

C. Multiple Choice Questions.

1. Newton-metre is the unit of
   (*a*) torque (*b*) energy
   (*c*) power (*d*) work
2. One kWh equals nearly .... kcals.
   (*a*) 3600 (*b*) 4200
   (*c*) 860 (*d*) 9800
3. The cost of running 2 kW heater for 10 hours at 50 paise/kWh is Rs. ....
   (*a*) 5 (*b*) 10
   (*c*) 1 (*d*) 2
4. If a 220-V heater is used on 110-v supply, heat produced by it would be—as much.
   (*a*) one-half (*b*) twice
   (*c*) one-fourth (*d*) four times
5. For a given line voltage, four heating elements will produce maximum heat when connected
   (*a*) all in parallel
   (*b*) all in series
   (*c*) two parallel pairs in series
   (*d*) one pair in parallel with the other two in series.
6. A torque of 50 N-m driving a rotor at 600 rpm produces a power of .... watts.
   (*a*) 500 (*b*) 3140
   (*c*) 1570 (*d*) 30,000
7. The electrical energy required to heat a bucket of water to a certain temperature is 2 kWh. If the heat losses are 25%, the energy input is .... kWh.
   (*a*) 1.5 (*b*) 2
   (*c*) 2.5 (*d*) 2.67

(Elect. Engg. A.M.Ae.S.I. Dec. 1993)

ANSWERS

A. 1. charge 2. energy 3. newton 4. velocity 5. 860 6. 2
B. 1. T 2. F 3. T 4. T 5. T 6. F
C. 1. *a* 2. *c* 3. *b* 4. *c* 5. *a* 6. *b*

5 ELECTROSTATICS

5.1. Static Electricity

In the preceding chapters, we concerned ourselves exclusively with current electricity *i.e.* electricity in motion. Now, we will discuss the behaviour of static electricity and the laws governing it. In fact, electrostatics is that branch of engineering which deals with the phenomena associated with electricity *at rest.*

It has been already discussed that generally an atom is electrically neutral *i.e.* in a normal atom, the sum total of positive charge of the protons is exactly equal to the sum total of negative charge of the electrons.

If, somehow, some negatively-charged electrons are removed from the atoms of a body, it is left with a preponderance of positive charge. Then, it is said to be positively charged. If, on the other hand, some electrons are added to it, then negative charge out-balances the positive charge and the body is said to be negatively charged.

In brief, we can say that positive electrification of a body results from a deficiency of electrons whereas negative electrification results from an excess of electrons.

The total deficiency or excess of electrons in a body is known as its charge.

5.2. Absolute and Relative Permittivities of a Medium

While discussing electrostatic phenomenon, a certain property of the medium called its *permittivity* plays a very important role. Every medium is supposed to possess two permittivities.

(*i*) absolute permittivity ($\in$) and (*ii*) relative permittivity ($\in_r$).

For measuring relative permittivity, vacuum or free space is chosen as the reference medium. It is allotted an absolute permittivity of 8.854×10^{-12} farad/metre (F/m). Its symbol is $\in_0$. Obviously, the relative permittivity of vacuum with reference to itself is unity. Hence, for *free space*

absolute permittivity, $\in_0 = 8.854 \times 10^{-12}$ F/m

relative permittivity $\in_r = 1$

Now, take any other medium. If its permittivity, as compared to vacuum is $\in_r$, then its absolute permittivity is $\in = \in_0 \in_r$ farad/metre

If, for example, relative permittivity of mica is 5, then

$$\in_r = 5 \qquad \text{(being a ratio, it has no units)}$$

$$\in = \in_0 \in_r = 8.854 \times 10^{-12} \times 5 \text{ F/m} = 44.27 \times 10^{-12} \text{ F/m}$$

5.3. Laws of Electrostatics

First Law. Like charges of electricity repel each other, whereas unlike charges attract each other.

Second Law. According to this law, the force exerted between two *point* charges (*i*) is directly proportional to the product of their strength (*ii*) is inversely proportional to the square of the distance between them and (*iii*) is inversely proportional to the absolute permittivity ($\in$) of the surrounding medium.

This law is known as Coulomb's Law and can be expressed mathematically as

$$F \propto \frac{Q_1 Q_2}{\in d_2} \quad \text{or} \quad F = k\frac{Q_1 Q_2}{\in d^2}$$

where k is the constant of proportionality whose value depends on the system of units employed. In the S.I. system of units, $k = 1/4\pi$.

$$\therefore \quad F = \frac{Q_1 Q_2}{4\pi \in d^2}$$

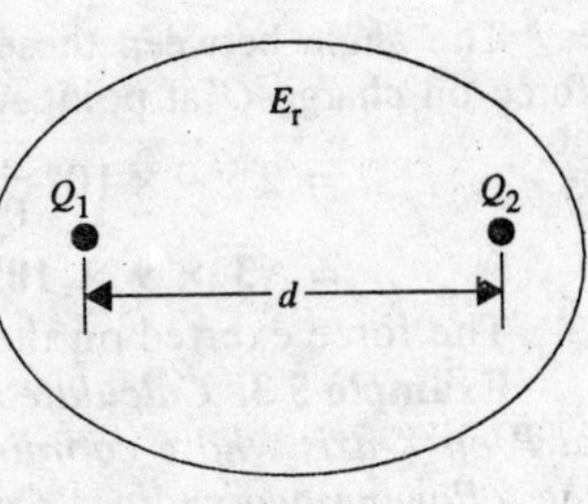

Fig. 5.1

If Q_1 and Q_2 are in coulombs, d in metres and $\in$ in farad/metre. then F is in newtons.

The above equation may be written as

$$F = \frac{Q_1 Q_2}{4\pi \in_0 \in_r d^2} \text{N} \quad \text{— in a medium}$$

$$= \frac{Q_1 Q_2}{4\pi \in_0 d^2} \text{N} \quad \text{— in air}$$

Now, $\dfrac{1}{4\pi \in_0} = \dfrac{1}{4\pi \times 8.854 \times 10^{-12}} = 9 \times 10^9$ (approx)

In that case, Coulomb's law for electric force becomes

$$F = 9 \times 10^9 \frac{Q_1 Q_2}{\in_r d^2} \text{N} \quad \text{— in a medium}$$

$$= 9 \times 10^9 \frac{Q_1 Q_2}{d^2} \text{N} \quad \text{— in air} \quad \ldots(i)$$

If in Eq (i) above,

$Q_1 = Q_2 = Q$ (say);

$d = 1$ metre ; $F = 9 \times 10^9$ N

then, $Q_2 = 1$ or $Q = \pm 1$ coulomb

Hence, one coulomb of charge may be defined as :

That charge which when placed in air (strictly vacuum) from an equal and similar charge at a distance of one metre repels it with a force of 9×10^9 N.

Although coulomb is found to be a unit of convenient size in dealing with electric current, yet from the standpoint of electrostatics, it is an enormous unit. Hence, its submultiples like microcoulomb (μC) and micro-microcoulomb (μμC) are generally used.

$1 \mu C = 10^{-6}$ C and $1 \mu\mu C = 10^{-12}$ C

It may be noted here that *relative* permittivity of air is one, of Teflon is 2, of water 81, of paper between 2 and 3, of glass between 5 and 10, of ceramic 1200 and of mica between 2.5 and 6.

It may be noted that Coulomb's law is true for atomic as well as macroscopic field.

Example 5.1. *Find the force of interaction between two charges spaced 10 cm apart in vacuum. The charges are 4×10^{-5} and 6×10^{-8} coulomb respectively. If the same charges are separated by the same distance in kerosene ($\in_r = 2$),* what is the corresponding force of interaction ? **(A.M.I.E. Summer 1988)**

Solution. In vacuum $F = 9 \times 10^9 \; Q_1 . Q_2/d^2$ newton

$\therefore \quad F = 9 \times 10^9 \times 4 \times 10^{-5} \times 6 \times 10^{-8}/(0.1) = $ **2.16 N**

In a medium, $F = 9 \times 10^9 \; Q_1 \; Q_2/\in_r d^2$ newton

$\therefore \quad F = 9 \times 10^9 \times 4 \times 10^{-5} \times 6 \times 10^{-8}/(0.1)^2 \times 2 = $ **1.08 N**

Example 5.2. *Three identical point charges Q coulombs each are placed at the vertices of an equilateral triangle 10 cm. apart. Calculate the force on each charge.*

Solution. Let us consider the force on the charge Q placed at point A (Fig. 5.2)

Force along BA due to charge at point B is F

$$= 9 \times 10^9 \frac{Q^2}{0.1^2} \text{N}$$

Force along *CA* due to charge at point *C* is $F = 9 \times 10^9 \dfrac{Q^2}{0.1^2}$ N

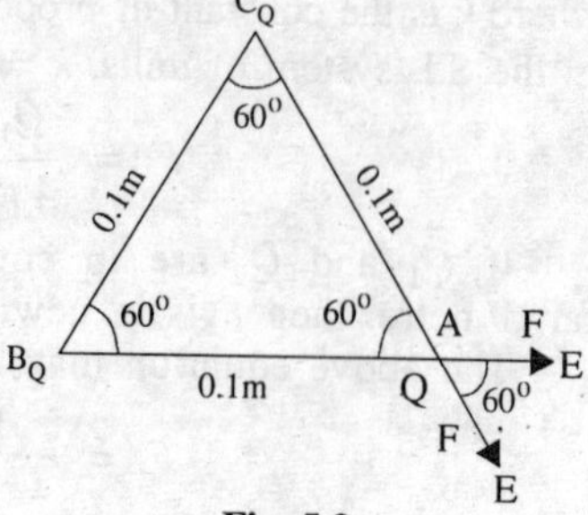

Fig. 5.2

The angle between these two forces is 60°. The resultant force on charge *Q* at point *A*.

$$= 2 \times 9 \times 10^9 \frac{q^2}{0.1^2} \times \cos 60°/2$$

$$= \sqrt{3} \times 9 \times 10^{11} Q^2 \text{ newton}$$

The force exerted on the other charges is also the same.

Example 5.3. *Calculate the force on a unit positive charge at P on X-axis whose coordinates are (x = 2m, y = 0) due to the following two charges : a positive charge of 10^{-9} coulomb situated inair at the origin (x = 0, y = 0) and a negative charge of -2×10^{-9} coulomb situated on the X-axis (x = 1m, y = 0).* **(A.M.I.E. Summer 1988)**

Solution. Repulsive force on unit charge at *P* due to the charge at the origin

$$= \quad 9 \times 10^9 \times 1 \times 10^{-9}/2^2 = 2.25 \text{ N}$$

Attractive force on the unit charge at *P* due to the negative charge

$$= \quad 9 \times 10^9 \times 1 \times 2 \times 10^{-9}/1^2 = 18 \text{ N}$$

∴ net attractive force on the unit charge

$$= \quad 18 - 2.25 = \mathbf{15.75\ N}$$

Example 5.4. *Calculate the distance of separation between two electrons (in vacuum) for which the electric force between them is equal to the gravitational force on one of them at the earth surface.*

Given : Mass of electron $= 9.1 \times 10^{-31}$ *Kg*

Charge of electron $= 1.6 \times 10^{-19} C$ **(Electrical Engg. A.M.Ae.S.I. June 1993)**

Solution : Gravitational force on one electron = mg newton = $9.1 \times 10^{-31} \times 9.81$ N

Electrostatic force between the two electrons

$$= \quad 9 \times 10^9 \frac{Q^2}{d^2} = \frac{9 \times 10^9 \times (1.6 \times 10^{-19})^2}{d^2} \text{ N}$$

Equating the two forces we get

$$9 \times 10^9 \times 2.56 \times 10^{-38}/d^2$$

$$= 9.1 \times 10^{-31} \times 9.81, \quad d = \mathbf{5.08\ m}$$

5.4. Electric Field

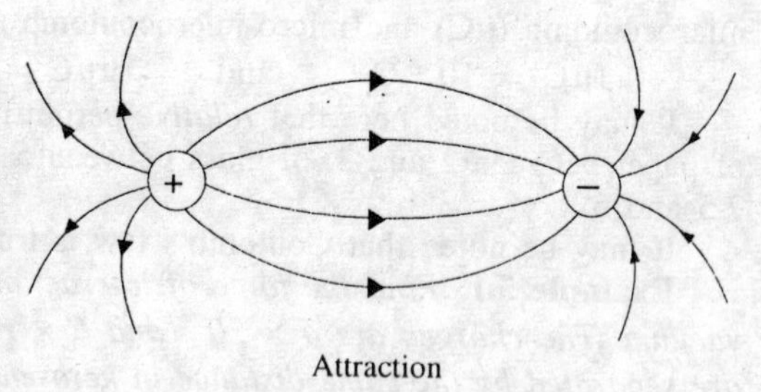

Attraction

It is found that the medium around a charge is always under stress and that a force acts on a positive or negative charge when placed in that medium. If the charge is sufficiently large, then it may create such a huge stress as to cause the mechanical rupture of the medium, followed by the passage of an arc discharge.

The region in which the stress exists or in which electric forces act, is called an electric field or dielectric field or electrostatic field.

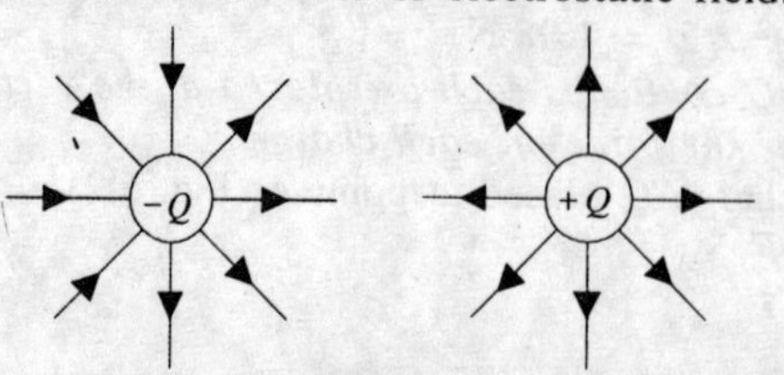

Fig. 5.4

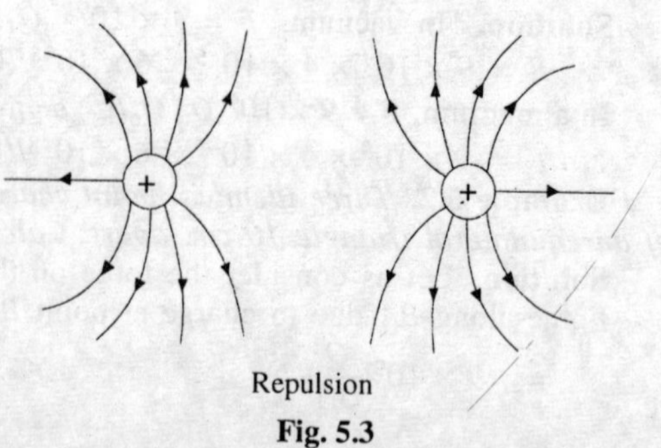

Repulsion

Fig. 5.3

The stress is represented by imaginary *lines of force*. The direction of the line of force at any point is the direction along which a unit positive charge placed at that point would move if free to do so. It was suggested by Faraday that the electric field should be imagined to be divided into *tubes of force* containing a fixed number of lines of force. He assumed these tubes to be elastic and having the property of contracting longitudinally and repelling laterally. With the help of these properties, it becomes easy to explain (*i*) why unlike charges attract each other and try to come nearer to each other and (*ii*) why like charges repel each other (Fig. 5.3).

However, it is more common to use the term *lines of force*. These lines are supposed to emanate from a positive charge and end on a negative charge (Fig. 5.4). These lines always leave or enter a conducting surface normally. The number of lines of force emanating from a charge of + Q coulomb are $Q/\in$ where $\in$ is the absolute permittivity.

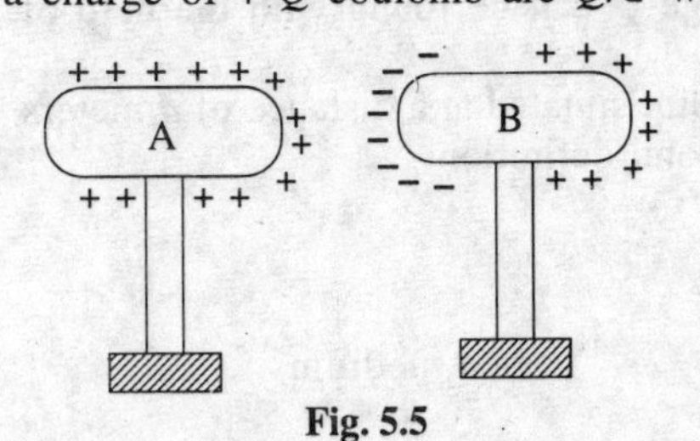

Fig. 5.5

5.5. Electrostatic Induction

It is found that when an uncharged body is brought near a charged body, then it also acquires some charge. This phenomenon of an uncharged body getting charged merely by the nearness of a charged body is known as *induction.* In Fig. 5.5, a positively-charged body A is brought close to a perfectly insulated uncharged body B. It is found that the end of B nearer to A gets negatively charged whereas the farther end becomes positively charged. The negative and positive charges of B are known as *induced charges*. The negative charge of B is called *bound charge* because it must remain on B so long as positive charge of A remains there. However, the positive charge on the farther end of B is called free charge. In Fig. 5.6, the body B has been earthed by a wire. The positive charge flows to earth leaving negative charge behind. If A is removed, then this negative charge will also go to earth, leaving B uncharged. It is found that :

(*i*) a positive charge induces a negative charge and *vice versa.*

(*ii*) each of the two induced charges is equal to the inducing charge.

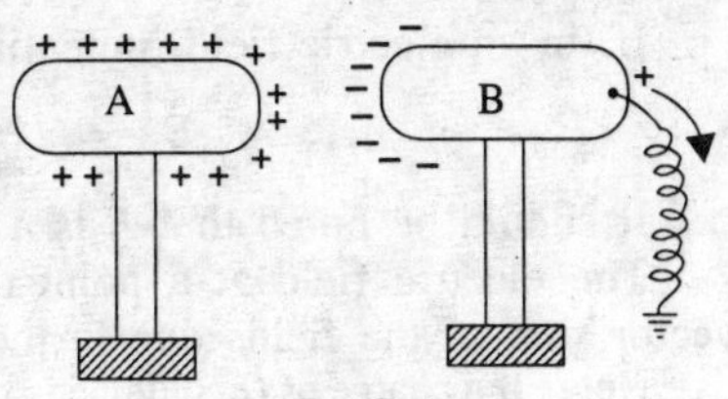

Fig. 5.6

5.6. Flux Per Unit Charge

In the S.I. System of Units, one *tube* of flux is supposed to emanate from a charge of one coulomb. Hence, if the charge is Q coulombs, then flux is given by

$$\Psi = Q \text{ coulombs}$$

In other words, flux is also measured in coulombs.

5.7. Electric Flux Density (D)

Flux density is defined as the flux per unit area held perpendicular to the tubes of flux. If Ψ coulomb is the flux passing normally through an area of A metre^2, then

$$D = \frac{\Psi}{A} \text{ C/m}^2$$

5.8. Field Strength or Field Intensity or Electric Intensity E.

Field strength or intensity of electric field at any point situated within an electrostatic field may be defined in any one of the following three ways :

(*a*) It is equal to the lines of *force passing normally through a unit cross-section at that point.*

Suppose there is a point charge of Q coulomb. The number of lines of force produced by it is $Q/\in$. If these lines fall on an area of A m^2, then

$$E = \frac{Q/\in}{A} = \frac{Q}{\in A} \text{ V/m}$$

$$\therefore \quad E = \frac{Q}{\epsilon_0 \epsilon_r A} \text{ V/m} \quad \text{... in a medium}$$

$$= \frac{Q}{\epsilon_0 A} \text{ V/m} \quad \text{... in air}$$

(*b*) It is numerically equal to the *force experience by a unit positive charge placed at the point in question.*

In this case, the unit of E is N/ C.

For example, if a charge of q coulomb placed at a particular point within an electric field experiences a force of F newton, the electric field at that point is given by

$$E = F/q \quad \text{N/C}$$

The value of E within the field due to a point charge Q can be found with the help of Coulomb's laws.

Suppose it is required to find the electric field at a point situated at a distance of d meters from a charge of Q coulombs, then according to the second definition

$$F = \frac{Q \times 1}{4\pi \epsilon_0 \epsilon_r d^2} \text{ N/C}$$

$$= \frac{Q}{4\pi \epsilon_0 \epsilon_r d^2} \text{ N/C} \quad \text{... in a medium}$$

$$= \frac{Q}{4\pi \epsilon_0 d^2} \quad \text{... in air}$$

(*c*) Electric intensity at any point in an electric field is *equal to the potential gradient at that point.*

If, in an electric field, potential falls by an amount—dV over a distnace of dx, then

$$E = \frac{-dv}{dx} \text{ V/m}$$

It should be noted that E is a vector quantity, having both magnitude and direction.

The electric field at a point due to more than one charge can be found by taking the vector sum of the fields due to individual charges.

Note. If a charge of Q coulomb is placed in an electric field of E V/m, then force acting on the charge is $F = EQ$ newton.

Example 5.5. *Three point charges of $+16 \times 10^{-9}$ C, $+ 64 \times 10^{-9}$ C and $- 48 \times 10^{-9}$ C, are placed at the corners of a square of 4 cm sides. Calculate the electric field at the fourth corner.*

Solution. The square is shown in Fig. 5.7. Values of electric field at S along SL due to charge at P is

$$= 9 \times 10^9 \times \frac{Q}{d^2} \text{ N/C}$$

$$= 9 \times 10^9 \times \frac{16 \times 10^{-9}}{0.04^2} = 9 \times 10^4 \text{ N/C}$$

Similarly, value of electric field at S along SM due to charge at Q is

$$= 9 \times 10^2 \times \frac{64 \times 10^{-9}}{(\sqrt{0.0032})^2} = 18 \times 10^4 \text{ N/C}$$

Also, value of electric field at S along SR due to charge at R is

$$= 9 \times 10^9 \times \frac{49 \times 10^{-9}}{0.04^2} = 27 \times 10^4 \text{ N/C}$$

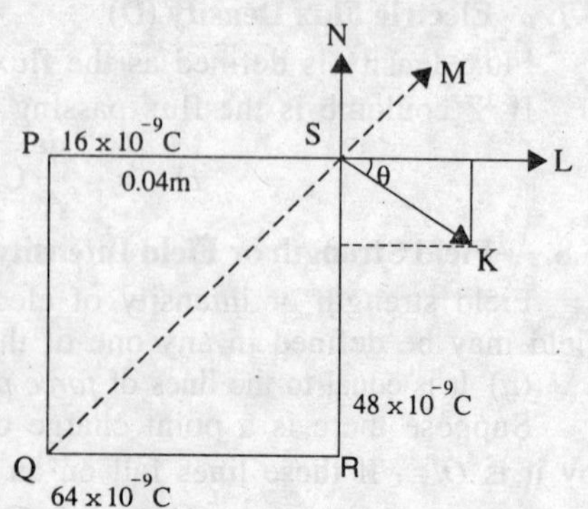

Fig. 5.7

Resolving these electric fields into their X- and Y-components, we have

Total *X*-component along

$$SL = 9 \times 10^4 + 18 \times 10^4 \cos 45° = 9 \times 10^4 + \frac{18 \times 10^4}{\sqrt{2}}$$

$$= 21.73 \times 10^4 \text{ N/C}$$

Total *Y*-component along $SR = 27 \times 10^4 - 18 \times 10^4 \sin 45° = 14.27 \times 10^4$ N/C

Hence, resultant field along $SK = \sqrt{(21.73 \times 10^4)^2 + (14.27 \times 10^4)^2} = 26 \times 10^4$ N/C

$$\tan\theta = \frac{Y\text{–component}}{X\text{–component}} = 14.27/21.73 = 0.656$$

$$\theta = \mathbf{33.3°}$$

5.9. Electric Field Intensity E and Electric Displacement D

$$E = \frac{Q}{\in A} \qquad \text{... Art. 5.8}$$

$$D = \frac{\Psi}{A} = \frac{Q}{A} \qquad \text{... Art. 5.7}$$

$$\therefore \quad E = \frac{D}{\in}$$

or $\quad D = \in E \text{ C/m}^2 = \in_0 E_r\, E \text{ C/m}^2$

In other words, the product of electric intensity *E* at any point within a dielectric medium and the absolute permittivity $\in = \in_0 \in_r$ at the same point is called the "displacement" at that point.

Like electric intensity *E*, displacement *D* is also a vector quantity whose direction at every point is the same as that of *E* but whose magnitude is $\in_0 \in_r$ times *E*. As *E* is represented by lines of force, similarly *D* may also be represented by tubes called *tubes of displacement* or *tubes of flux*. The tangent to these tubes at any point gives the direction of *D* at that point and the number of tubes per unit area taken perpendicular to their direction is numerically equal to the displacement at that point. Hence, the number of displacement per unit area (*D*) is $\in_0 \in_r$ times the number of *lines* of force per unit area (*E*) at that point.

One useful property of *D* is that its surface integral over any closed surface equals the enclosed charge.

Let us find the value of *D* at a point distant *r* metres from a point chagte of *Q* coulombs. Imagine a sphere of radius *r* metres surrounding the charge. Total flux is *Q* coulombs and it falls on a surface area of $4\pi r^2$ metres. Hence, electric flux density is

$$D = \frac{Q}{4\pi r^2} \text{ C/m}^2$$

5.10. Electric Potential and Energy

We know that a body raised above the ground level has a certain amount of mechanical potential energy which, by definition, is given by the amount of work done in raising it to that height. If, for example, a weight of 5 kg is raised against gravity through 10 m, then the potential energy of the body is 5 × 10 = 50 m-kg. The body falls because there is attraction due to gravity and always proceeds from a place of higher potential energy to one of lower potential energy. So, we speak of gravitational 'potential' energy or briefly 'potential' at different points in the earth's gravitational field.

Now, consider an electric field. Imagine an isolated positive charge *Q* placed in air (Fig. 5.8). Like earth's gravitational field, it has its own electrostatic field which theoretically extends up to infinity. If the charge *X* is very far away from *Q*, say, at infinity, then force on it is practically zero. As *X* is brought nearer to *Q*, a force of repulsion acts on it. Because similar charges repel each other, work or energy is required to bring it to a point like *A* in the electric field. Hence, when at

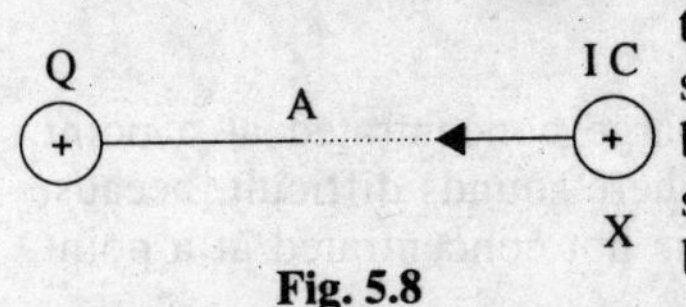

Fig. 5.8

point A, X has some amount of electric potential energy. Similar other points in the field will also have some potential energy.

In the gravitational field, usually 'sea level' is chosen as the place of 'zero' potential. In an electric field, infinity is chosen as the theoretical place of 'zero' potential although, in practice, earth is chosen as 'zero' potential, because earth is such a large conductor that its potential remains practically constant although it keeps on losing and gaining electricity every day.

5.11. Potential and Potential Difference

As explained above, since the force on a charge at infinity is zero, hence 'infinity' is chosen as the theoretical place of zero electric potential. Therefore, potential at a point in an electric field may be defined as *numerically equal to the work done in bringing a positive charge of one coulomb from infinity to that point against the electric field.*

The unit of this potential will depend on the unit of charge taken and the work done.

If, in shifting one coulomb from infinity to a certain point in the electric field, the work done is one joule, then potential of that point is one volt.

Obviously, potential is work per unit charge.

$$\therefore \qquad 1 \text{ volt} = \frac{1 \text{ joule}}{1 \text{ coulomb}}$$

Similarly, potential difference (p.d.) of one volt exists between two points if one joule of work is done in shifting a charge of one coulomb from one point to the other.

5.12. Potential at a Point

Consider a positive point charge of Q coulombs placed in air. At a point x metres from it, the force on one coulomb positive charge is $Q/4\pi \in_0 x^2$ (Fig. 5.9). Suppose, this one coulomb charge is moved towards Q through a small distance dx. Then, work done

$$= \frac{Q}{4\pi\in_0 x^2} \times (-dx)$$

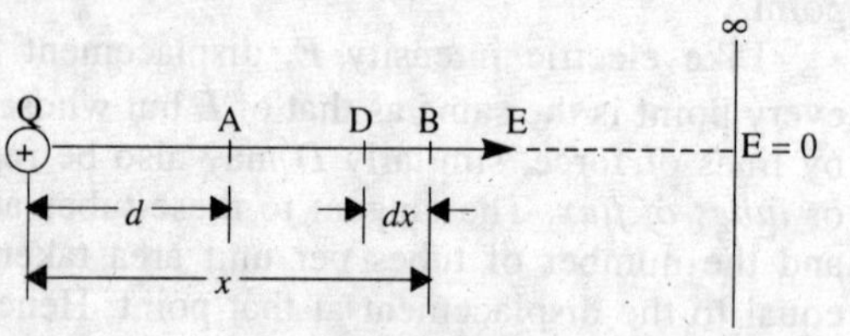

Fig. 5.9

The negative sign is taken because dx is measured along the negative direction of x.

The total work done in bringing this coulomb of positive charge from infinity to any point d metres from Q is given by

$$W = \int_{x=\infty}^{x=d} -Q \cdot \frac{dx}{4\pi\in_0 x^2} = -\frac{Q}{4\pi\in_0}\int_{\infty}^{d}\frac{dx}{x^2} = -\frac{Q}{4\pi\in_0}\left[-\frac{1}{x}\right]_{\infty}^{d}$$

$$= -\frac{Q}{4\pi\in_0}\left[-\frac{1}{d} - \left(-\frac{1}{\infty}\right)\right] = -\frac{Q}{4\pi\in_0 d} \text{ joules}$$

By definition, this work in joules is numerically equal to the potential of that point in volts.

$$\therefore \qquad V = \frac{Q}{4\pi\in_0 d} = 9 \times 10^9 \frac{Q}{d} \text{ volts} \qquad \text{.. in air}$$

$$\text{and} \qquad V = \frac{Q}{4\pi\in_0\in_r d} = 9 \times 10^9 \frac{Q}{\in_r d} \text{ volt} \qquad \text{.. in medium}$$

We find that as d increases, V decreases, till it becomes zero at infinity.

Note. If a charge of Q coulomb is shifted against a p.d. of V volts, work done is W.D. = QV joules.

5.13. Potential Due to a Charged Sphere

The above formula $V = Q/4\pi\in_0\in_r d$ applies only to a charge concentrated at a point. The problem of finding potential at a point outside a charged sphere sounds difficult, because the charge on the sphere is distributed over its surface and so, is not concentrated at a point.

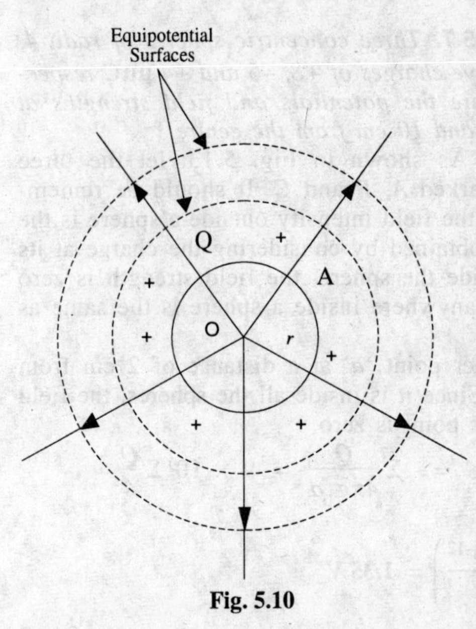

Fig. 5.10

But the problem is easily solved by noting that the lines of force of a charged sphere, like A in Fig. 5.10, spread out normally from its surface. If produced backwards, they meet at the centre of A. Hence, for finding the potentials at points outside the sphere, we can imaginve the charge on the sphere as concentrated at its centre. If r is the radius of the sphere in metres and Q its charge in coulombs, then potential of its surface is $Q/4\pi\epsilon_0 r$ volts and electric intensity is $Q/4\pi\epsilon_0 r^2$. At any other point 'd' metres from the centre of the sphere, the corresponding values are $Q/4\pi\epsilon_0 d$ and $Q/4\pi\epsilon_0 d^2$ respectively. The variations of the potential and electric intensity with distance for a charged sphere are shown in Fig. 5.11.

5.14. Equipotential Surfaces

An equipotential surface is a surface in an electric field such that all points on it are at the same potential. For example, different spherical surfaces around a charged sphere are equipotential surfaces. One important property of an equipotential surface is that lines of force are always normal to it.

5.15. Potential and Electric Intensity Inside a Conducting Sphere

It has been experimentally found that when charge is given to a conducting body, say, a sphere, then it resides entirely on its outer surface *i.e.* within a conducting body (whether hollow or solid), the charge is zero. Hence (*i*) flux is zero (*ii*) field intensity is zero (*iii*) all points within the conductor are at the same potential as its surface (Fig. 5.12).

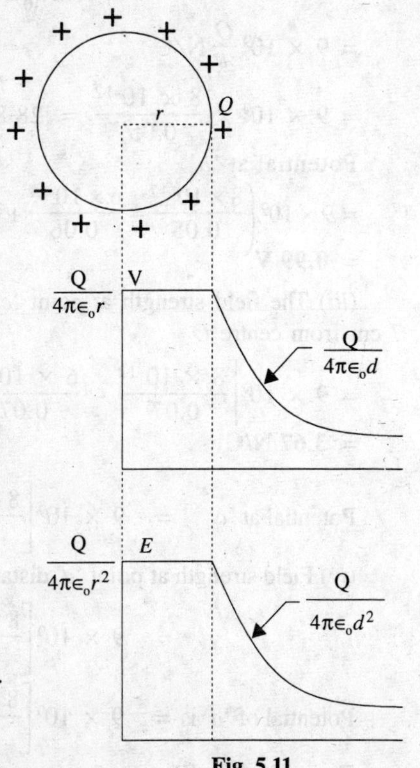

Fig. 5.11

Example 5.6. *A hollow sphere is charged to 12 μC of electricity. Find the potential :*

(a) at its surface

(b) inside the sphere

(c) at a distnace of 0.3 metre from the surface.

The radius of the sphere is 0.1 metre.

Solution. The potential at the surface of the sphere is $\dfrac{Q}{4\pi\epsilon_0\epsilon_r d}$ where ϵ_r is the relative permittivity of the surrounding medium. Potential anywhere inside the sphere is the same. For calculations, charge on the sphere is supposed to be concentrated at its centre.

(*a*) $V = \dfrac{Q}{4\pi\epsilon_0 d} = 9 \times 10^9 \dfrac{Q}{d}$ volt

$= 9 \times 10^9 \times \dfrac{12 \times 10^{-6}}{0.1} = 108 \times 10^4$ volt.

(*b*) Inside also V = $\mathbf{108 \times 10^4}$ **V**

(*c*) Distance of the point from the centre is 0.4 m

$\therefore$ potential $= 9 \times 10^9 \times \dfrac{12 \times 10^{-6}}{0.4} =$ $\mathbf{27 \times 10^4}$ **V**

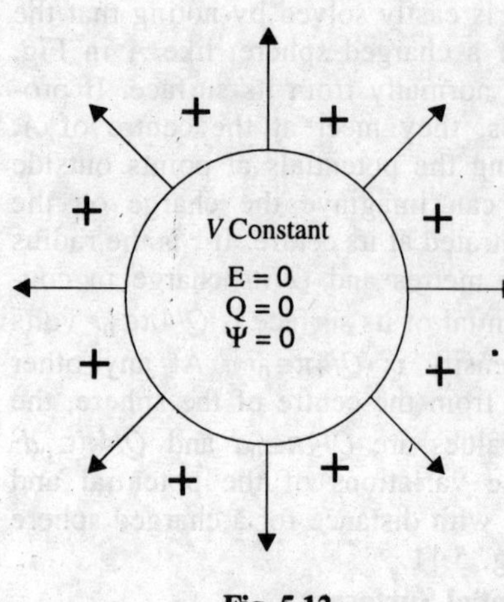

Fig. 5.12

Example 5.7. *Three concentric spheres of radii 4, 6 and 8 cm have charges of +8, –6 and +4 μμC respectively. What are the potentials and field strengths at points 2, 5, 7 and 10 cm from the centre ?*

Solution. As shown in Fig. 5.13, let the three spheres be marked *A*, *B* and *C*. It should be remembered that (*i*) the field intensity outside a sphere is the same as that obtained by considering the charge at its centre (*ii*) inside the sphere, the field strength is zero (*iii*) potential anywhere inside a sphere is the same as at its surface.

(*i*) Consider point '*a*' at a distance of 2 cm from the centre *O*. Since it is inside all the spheres, the field strength at this point is zero.

$$\text{Potential at } `a' = \Sigma \frac{Q}{4\pi\epsilon_0 d} = 9 \times 10^9 \Sigma \frac{Q}{d}$$

$$= 9 \times 10^9 \left(\frac{8 \times 10^{-12}}{0.04} - \frac{6 \times 10^{-12}}{0.06} + \frac{4 \times 10^{-12}}{0.08} \right) = \mathbf{1.35\ V}$$

(*ii*) Since point '*b*' is outside sphere *A*, but inside *B* and *C*

$$\therefore \quad \text{electric field} = \frac{Q}{4\pi\epsilon_0 d^2}$$

$$= 9 \times 10^9 \frac{Q}{d^2} \text{ N/C}$$

$$= 9 \times 10^9 \times \frac{8 \times 10^{-12}}{0.05^2} = 28.8 \text{ N/C}$$

Potential at '*b*'

$$= 9 \times 10^9 \left(\frac{8 \times 10^{-12}}{0.05} - \frac{6 \times 10^{-12}}{0.06} + \frac{4 \times 10^{-12}}{0.08} \right)$$

$$= \mathbf{0.99\ V}$$

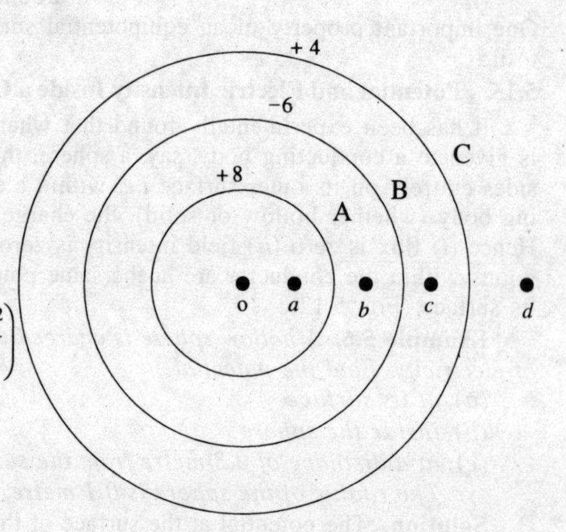

Fig. 5.13

(*iii*) The field strength at point '*c*' distant 7 cm from centre *O*

$$= 9 \times 10^9 \left[\frac{8 \times 10^{-12}}{0.07^2} - \frac{6 \times 10^{-12}}{0.07} \right]$$

$$= \mathbf{3.67\ N/C}$$

$$\text{Potential at } `c' = 9 \times 10^9 \left[\frac{8 \times 10^{-12}}{0.07} - \frac{6 \times 10^{-12}}{0.07} + \frac{4 \times 10^{-12}}{0.08} \right] = \mathbf{0.707}$$

(*iv*) Field strength at point '*d*' distant 10 cm from point *O* is

$$= 9 \times 10^9 \left[\frac{8 \times 10^{-12}}{0.1^2} - \frac{6 \times 10^{-12}}{0.1^2} + \frac{4 \times 10^{-12}}{0.1^2} \right] = \mathbf{5.4\ N/C}$$

$$\text{Potential of } `d' \text{ is} = 9 \times 10^9 \left[\frac{8 \times 10^{-12}}{0.1} - \frac{6 \times 10^{-12}}{0.01} + \frac{4 \times 10^{-12}}{0.01} \right] = \mathbf{5.4\ N/C}$$

Example 5.8. *What is the potential at the centre of a square having charges of +2, +1, –2 and +3 in terms of 10^{-8} C each if side of the square is 1.41 metre.*

Solution. It should be remembered that potential is a scalar quantity. The total potential at the centre *O* (Fig. 5.14) is equal to the *algebraic* sum of the potentials due to the four charges. As seen, $OB = OD = OC = OA = 2/2 = 1$ metre. Potential at point *O* due to charge at

(*i*) point $A = 9 \times 10^9 \times 2 \times 10^{-8}/1$
$= 180$ V

(*ii*) point $B = 9 \times 10^9 \times 1 \times 10^{-8}/1$
$= 90$ V

(*iii*) point $C = -9 \times 10^9 \times 2 \times 10^{-8}/1 = -180$ V

(*iv*) point $D = 9 \times 10^9 \times 3 \times 10^{-8}/1 = 270$ V

$\therefore$ potential of point $O = 180 + 90 - 180 + 270 =$ **360 V**

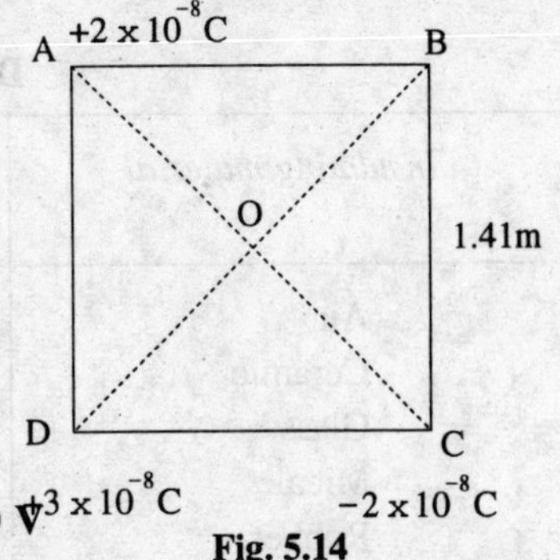

Fig. 5.14

5.16. Potential Gradient (g)

It is defined as *the rate of change of potential with distance measured in the direction of electric force.*

i.e.
$$g = \frac{dV}{dx}$$

Its unit is volt/metre although its derivative volt/cm or kV/mm is generally used in practice. Suppose, in an electric field of strength *E*, there are two points *dx* metre apart. The p.d. between them is

$$dV = E(-dx) = -E.dx$$

$$\therefore \quad E = -\frac{dV}{dx} = -g \quad ...(i)$$

Hence, it means that electric intensity at a point is equal to the negative potential gradient at that point.

Now, electric intensity is measured in newton/coulomb (Art. 5.8) but equation (*i*) above shows that its unit may also be volt/metre, the same as that of potential gradient. It can be shown that N/C is the same as V/m.

$$1 \text{ volt} = 1\,\frac{\text{joule}}{\text{coulomb}} \quad \text{(Art. 5.11)}$$

$$= 1\,\frac{\text{metre} - \text{newton}}{\text{coulomb}}$$

$$\therefore \quad 1\,\frac{\text{volt}}{\text{metre}} = 1\,\frac{\text{metre} - \text{newton}}{\text{coulomb} - \text{metre}} = 1\,\frac{\text{newton}}{\text{coulomb}}$$

$$\therefore \quad 1 \text{ V/m} = 1 \text{ N/C.}$$

5.17. Dielectric Strength of a Medium

It is measured in terms of the potential difference which when applied across 1 metre (or 1 mm) thickness of a dielectric medium or insulating medium will break down its insulation. It is also expressed in V/m. For example, when we say, the breakdown potential of air is 3×10^6 V/m, then we mean that the maximum p.d. which 1 metre thickness of air can withstand across it, is 3×10^6 volts. If the voltage exceeds this value, then air insulation breaks down and a spark is produced which punctures the medium.

It is obvious that dielectric strength of a medium represents the maximum value of the electric intensity or potential gradient that can be safely established in that medium.

TABLE No. 5.1
Dielectric Constant and Strength

Insulating material	*Dielectric constant or relative permittivity*	*Dielectric strength in kV/mm*
Air	neraly 1	3.2
Ceramic	1200	40
Glass	5–12	12 – 20
Mica	4–8	20 – 60
Rubber	2.5	—
Wood	2.5– 6.8	—
Micanite	4.5–6	25 – 35
Paper	1.8– 2.6	—
Paraffin Wax	1.7– 2.3	30
Porcelain	5 – 6.7	15
Quartz	4.5– 4.7	8
Sulphur	3.6– 4.1	
Teflon	2.0	60

Example 5.9. *Find the radius of an isolated sphere capable of being charged to 1 million volt potential before sparking into the air. Given that breakdown voltage of air is 30,000 V/cm.*

Solution. Let r metre be the radius of the sphere.

Then $$V = \frac{Q}{4\pi\epsilon_0 r} = 10^6 \text{ V} \quad \ldots(i)$$

Breakdown voltage = 30,000 V/cm = 3,000,000 V/m = 3×10^6 V/m

Since electric intensity equals breakdwon voltage

$$E = \frac{Q}{4\pi\epsilon_0 r^2} = 3 \times 10^6 \text{ V/m} \quad \ldots(ii)$$

Dividing (*i*) by (*ii*), we get

$$r = \frac{1}{3}\text{ m} = \mathbf{0.33 \text{ metre}}$$

Example 5.10. *It is found experimentally that an electric field of 3×10^6 V/m in air will cause electrical breakdown of the air. What is the greatest charge that can be placed on a sphere of 1 metre diameter? What is the potential of the sphere for the charge ?*

Solution. Breakdown voltage = 3×10^6 V/m

$$\therefore \quad E = \frac{Q}{4\pi\epsilon_0 r^2} = 3 \times 10^9 \text{ V/m}$$

$$\text{or} \quad 9 \times 10^9 \frac{Q}{0.5^2} = 3 \times 10^6$$

$$\therefore \quad Q = \frac{3 \times 10^6 \times 0.5^2}{9 \times 10^9} = \mathbf{8.33 \times 10^{-5} \text{ C}}$$

Now, potential $$V = \frac{Q}{4\pi\epsilon_0 r} = 9 \times 10^9 \times \frac{Q}{r}$$

$$= 9 \times 10^9 \frac{8.33 \times 10^{-5}}{0.5} = 15 \times 10^5 \text{ volt} = \mathbf{1{,}500 \text{ kV}}$$

Example 5.11. *Two brass plates are arranged horizontally, one 2 cm above the other and the lower plate is earthed. The plates are charged to a difference of potential of 6,000 volts. A drop of oil with an electric charge of 1.6 × 10^–19^ C is in equilibrium between the plates so that it neither rises nor falls. What is the mass of the drop ?*

Solution. The elctric intensity is equal to the potential gradient between the plates.

$$g = 6000/2 = 3000 \text{ volt/cm} = 3 \times 10^5 \text{ V/m}$$
$$E = 3 \times 10^5 \text{ V/m or N/C}$$
$$\therefore \text{ force on drop} = E \times Q = 3 \times 10^5 \times 1.6 \times 10^{-19}$$
$$= 4.8 \times 10^{-14} \text{ N} = \text{wt of drop} = mg \text{ newton}$$
$$m \times 9.81 = 4.8 \times 10^{-14}$$
$$m = 4.89 \times 10^{-15} \text{ kg}$$

Example 5.12. *In a parallel plate capacitor with solid dielectric, the plates are 0.015 cm apart and when charged, the surface charge density is 10^{-9} coulomb per sq cm. Calculate the electric flux density. If the relative permittivity of the solid dielectric is 6, calculate the potential difference between the plates.* **(A.M.I.E. Winter 1987)**

Solution. As seen from Art. 5.9

$$D = \frac{\Psi}{A} = \frac{C}{A} = \sigma$$

Hence, electric flux density = charge density = 10^{-9} C/cm² = 10^{-5} C/m²

$$\text{Now} \quad E = \frac{D}{\epsilon_0 \epsilon_r} = \frac{V}{d} \quad \therefore \quad V = \frac{d^D}{\epsilon_0 \epsilon_r}$$

$$= 0.015 \times 10^{-2} \times \frac{10^{-5}}{8.854 \times 10^{-12} \times 6} = \mathbf{28.2\ V}$$

Example 5.13. A parallel-plate capacitor has its plates 0.2 mm apart, with a solid dielectric between having $\epsilon_r = 3.3$. The capacitor is charged so that surface charge density is 4 × 10^{-3} m μC/cm². Find the electric flux density and the voltage between the plates.

Solution. Electric flux density = charge density

$$\therefore \quad D = 4 \times 10^{-2} \ \mu\text{C/cm}^2 = 4 \times 10^{-5} \text{ C/m}^2$$

$$\text{Now} \quad E = \frac{dV}{dx}$$

where dV = p.d. between plates; dx = distnace between plates

Here, dx = 0.2 mm = 2×10^{-4} m

$$E = \frac{D}{\epsilon_0 \epsilon_r} = \frac{4 \times 10^{-5}}{8.854 \times 10^{-12} \times 3.3}$$
$$= 1.37 \times 10^6 \text{ V/m}$$
$$dV = E.dx = 1.37 \times 10^6 \times 2 \times 10^{-4} = \mathbf{274V}$$

Example 5.14. *A 0.23 cm thick sheet of fibre, with a relative permittivity of 5, is inserted between two parallel metal plates 0.25 cm apart. A p.d. of 1,500 V is applied between the plates. Determine the electric field intensity in the fibre and in the air film between the fibre and the metal plates. Will the air break down?* **(A.M.I.E. Nov. 1990)**

Solution. The arrangement is shown in Fig. 5.15.

Let E_1 and E_2 be the electric intensities or potential gradients in air and fibre sheet respectively.

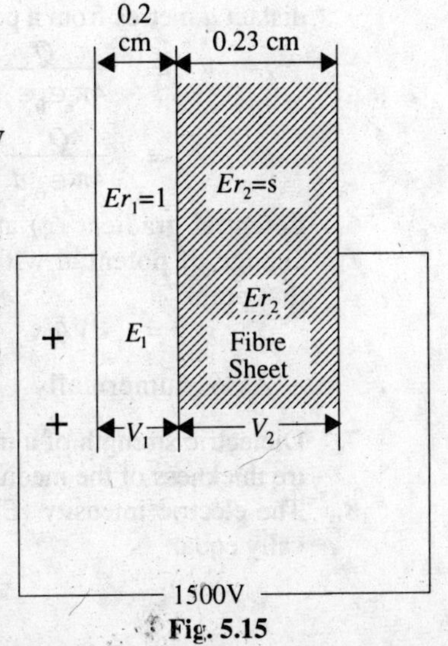

Fig. 5.15

$$\text{Now,} \quad E_1 = V_1/x_1$$
$$\text{or} \quad V_1 = E_1.x_1 = 0.02 \times 10^{-2} E_1$$
$$V_2 = E_2.x_2 = 0.23 \times 10^{-2} E_2$$
$$\text{Also,} \quad V = V_1 + V_2$$

$1500 = 10^{-2}(0.02\,E_1 + 0.23\,E_2)$ or $15 \times 10^6 = 2E_1 + 23E_2$ ...(*i*)

Now, value of electric flux density D is the same in the two media.

$\therefore\ D = \epsilon_0 \epsilon r_1 E_1 = \epsilon_0 \epsilon r_2 E_2$ or $E_1 = 5E_2$ ...(*ii*)

From Eq. (*i*) and (*ii*), we get

$33\,E_2 = 15 \times 10^6$ or $E_2 = \mathbf{0.455 \times 10^6}$ **V/m**; $E_1 = \mathbf{2.275 \times 10^6}$ **V/m**

Since dielectric strength of air is 3×10^6 V/m, it is obvious that there will be no breakdwon of the air film.

HIGHLIGHTS

1. Coulomb's Law of electrostatic force between two point charges is

$$F = \frac{Q_1 Q_2}{4\pi \epsilon_0 \epsilon_r d^2} \text{ newton}; \quad \text{where } \epsilon_0 = 8.854 \times 10^{-12} \text{ F/m}$$

2. Electric flux is measured in coulombs and is equal to the charge

$$\therefore \quad \Psi = Q \text{ coulombs}$$

3. Electric intensity at a point within a field is equal to the force in newtons experienced by a positive charge of 1C placed at that point. In other words, electric intensity is force per unit charge.

$$E = F/q \text{ newton/coulomb}$$

Other expressions for E are :

$$E = \frac{D}{\epsilon_0 \epsilon_r} \text{ V/m}$$

In the field due to a point charge Q, its vlue is

$$E = \frac{Q}{4\pi \epsilon_0 \epsilon_r d^2} \text{ V/m}$$

4. Electric displacement or flux density D is defined as

$$D = \Psi/A \text{ C/m}^2$$

It is related to E by the equations

$$D = \epsilon \text{ C/m}^2$$

$$= \epsilon_0 \epsilon_r E \text{ C/m}^2$$

Also, $D = \sigma \text{ C/m}^2$ where σ is charge density

5. Potential at any point within a field is numerically equal to the work done in shifting one coulomb from infinity to that point against the electric force. The potential at any point distant d metres from a point charge of Q coulombs is given by

$$V = \frac{Q}{4\pi \epsilon_0 \epsilon_r d} \text{ volts} \quad \text{...in a medium}$$

$$= \frac{Q}{4\pi \epsilon_0 d} \text{ volts} \quad \text{...in air}$$

6. Potential gradient (g) at any point within an electric field is defined as the rate of change of potential with distance measured in the direction of the electric force at that point

$$g = dV/dx$$

Also, numerically $E = g = -\dfrac{dV}{dx}$ V/m

7. Dielectric strength of a medium is given by the p.d. which when applied across one metre thickness of the medium will break down its insulation. It is measured in V/m.
8. The electric intensity (E), potential gradient (g) and dielectric strength are all numerically equal.

OBJECTIVE TESTS—5

A. Fill in the following blanks :

1. Coulomb's inverse square law is .... for atomic as well as macroscopic field. **(A.M.I.E. Winter 1984)**
2. The absolute permittivity of vacuum is .... farad/metre.
3. The unit of electric flux is ....
4. The electrostatic potential inside a hollow conductor is .... **(A.M.I.E. Winter 1984)**
5. In an electric field, electric intensity at any point is numerically equal to the .... gradient at the point.
6. Dielectric strength of a medium is expressed in ....
7. The absolute permittivity of a dielectric medium is given by the ratio of electric .... and electric intensity.
8. Coulomb is the unit of electric ....

B. Answer True or False.

1. Coulomb's inverse square law is true only for electric charges.
2. Absolute permittivity of free space is unity.
3. Electric field is represented by lines of force.
4. Electric flux is numerically equal to the electric charge.
5. Electric field has the unit of newton/weber.
6. No flux can exist inside a hollow charged sphere.
7. Electric field inside a charged hollow sphere is zero.
8. Dielectric strength of a medium represents the maximum value of the electric intensity that can be safely established in that medium.

C. Multiple Choice Questions

1. A coulomb is that charge which placed in vacuum from an equal and similar charge at a distance of one metre repels it with a force of .... newton.
   (*a*) $\neq 1$ (*b*) 9×10^{9}
   (*c*) 8.854×10^{-12} (*d*) 10^{-12}
2. A charge of + Q coulomb placed in a medium of relative permittivity $\in_r$ will give out a flux of .... coulomb.
   (*a*) Q (*b*) $\in_0 Q$
   (*c*) $\in_0 \in_r Q$ (*d*) $Q/\in_0$
3. The relation between electric displacement D and electric intensity E in a medium is given by
   (*a*) $D = E/\in_0$ (*b*) $D = \in E$
   (*c*) $D = E/\in$ (*d*) $D = \in_r E$
4. The electrostatic potential inside a positively-charged sphere is
   (*a*) maximum (*b*) minimum
   (*c*) zero (*d*) constant
5. Potential gradient has the same unit as that of
   (*a*) electric flux
   (*b*) electric field
   (*c*) dieleetric strength
   (*d*) charge density
   (*e*) both (*b*) and (*c*)
6. The space between two charged metallic plates is filled partly by air and party by fibre sheet of $\in_r = 5$.
   The electric intensity in fibre sheet is .... times greater than that in air,
   (*a*) 5 (*b*) 25
   (*c*) 1/25 (*d*) 1/5
7. The force between two charges is 60 N. If the *x* distnace between the charges is doubled, the force will be
   (*a*) 60 newton. (*b*) 30 newton.
   (*c*) 15 newton. (*d*) 10 newton.
   (Elect. Engg. A.M.Ae. S.I. Dec. 1993.)
8. The electric field intensity at a point situated 4 metres from the point charge is 500 N/C. If the distance is reduced to 2 metres, the field intensity will be
   (*a*) 1000 newton/coulomb.
   (*b*) 2000 newton/coulomb.
   (*c*) 250 newton/coulomb.
   (*d*) none of these.
   (Elect. Engg. A.M.Ae.S.I. June 1994)

ANSWERS

A. 1. true 2. 8.854×10^{-12} 3. coulomb 4. constant 5. potential 6. volt/metre 7. displacement 8. charge/flux

B. 1. F 2. F 3. T 4. T 5. T 6. T 7. F 8. T

C. 1. *b* 2. *a* 3. *b* 4. *d* 5. *e* 6. *a*

6 CAPACITANCE

6.1. Capacitor

A capacitor essentially consists of two conducting surfaces separated by a layer of an insulating medium called *dielectric*. The conducting surfaces may be in the form of either circular (or rectangular) plates or of spherical or cylindrical shape. The purpose of a capacitor is to store electrical energy by electrostatic stress in the dielectric.

A parallel-plate capacitor is shown in Fig. 6.1. One plate is joined to the positive end of the supply and the other to the negative end or is earthed. It is experimentally found that in the presence of an earthed plate *B*, plate *A* is capable of withholding more charge than when *B* is not there. When such a capacitor is put across a battery, there is a momentary flow of electrons form *A* to *B*. As negatively-charged electrons are withdrawn from *A*, it becomes positive and as these electrons collect on *B*, it becomes negative. Hence, p.d. is established between plates *A* and *B*. This transient flow of electrons gives rise to a charging current. The strength of the charging current is maximum when the two plates are uncharged, but it then decreases and finally ceases when p.d. across the plates becomes slowly and slowly equal and opposite to the battery e.m.f.

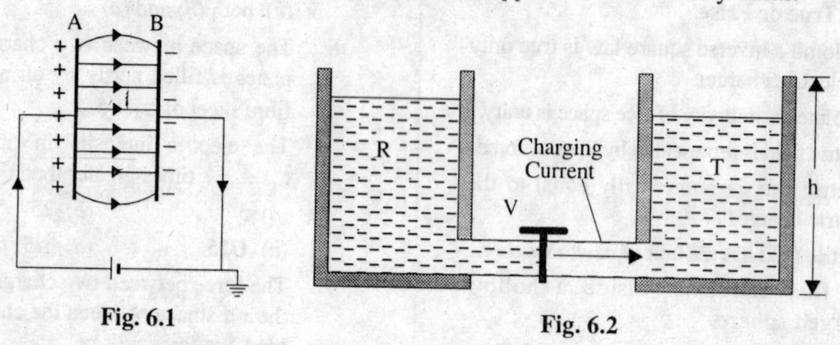

Fig. 6.1

Fig. 6.2

In fact, the charging of a capacitor is in many respects similar to the charging of a water tank *T* from a reservoir *R* (Fig. 6.2).

On opening the valve, the water rushes through the connecting pipe from *R* to *T*. At first, the rush would be great and then flow of water will go on decreasing as the level of water in *T* goes on increasing. Obviously, water flow will stop when the water levels in *R* and *T* become equal.

In the same way, when capacitor is fully charged *i.e.* when p.d. across its plates becomes the same in magnitude as the charging e.m.f., then no current flows. Two points should be mentioned here :—

(*i*) The battery itself does not create electricity on plates *A* and *B*, its function is merely to cause the transfer of electrons from *A* to *B*

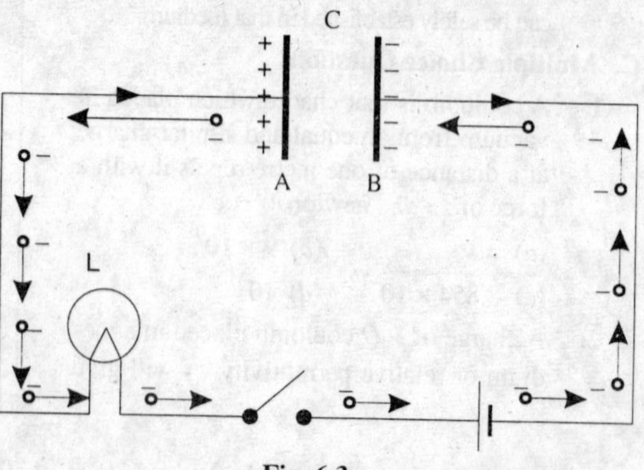

Fig. 6.3

and hence create a p.d. between them. Its functions is analogous to that of a pump.

(ii) When capacitor is connected across a battery, no continuous current can flow 'through' the capacitor. There is only a momentary shift of electrons from one plate to another. If a suitable lamp is included in the circuit as shown in Fig. 6.3, it will glow for a very short while *i.e.* so long as capacitor is being charged. After that, it does not glow. A capacitor blocks the passage of direct current.

It has been found that the capacitor plates is under a state of strain and energy is stored in it. If the p.d. across the plates is increased, the strain in the dielectric increases, till an electrical breakdown of the medium, accompanied by a spark across the plates, takes place. The maximum voltage per metre thickness which a medium can withstand without a rupture or breakdown is called its dielectric strength.

6.2. Capacitance

The property of a capacitor is 'store electricity' may be called its capacitance.

The capacitance of a capacitor may be defined as

the amount of charge required to create a unit potential difference between its plates.

Suppose, we give Q coulombs of charge to one of the two plates of a capacitor and it a p.d. of V volts is etstablished between the two, then its capacitance is

$$C = \frac{Q}{V} \frac{\text{coulomb}}{\text{volt}}$$

By definition, the unit of capacitance is coulomb/volt which is also called *farad* (in honour of Michael Faraday).

$$\therefore \quad 1 \text{ farad} = 1 \text{ coulomb/volt}$$

One farad is defined as the capacitance of a capacitor which requires a charge of one coulomb to establish a p.d. of one volt between its plates.

One farad is actually too large for practical purposes. Hence, much smaller units like micro-farad (μF) and micro-microfarad (μμF) or picofarad (pF) are generally employed

$$1\,\mu = 10^{-6} \text{ F}$$

$$1\,\mu\mu\text{F or pF} = 10^{-12} \text{ F}$$

6.3. Capacitance of an Isolated Sphere

Consider a charged sphere of radius r metres having a charge of Q coulombs placed in air as shown in Fig. 6.4.

It has been proved in Art. 5.12 that the free surface potential V of such a sphere with respect to infinity (in practice, earth) is given by

$$V = \frac{Q}{4\pi\epsilon_0 r}$$

$$\therefore \quad \frac{Q}{V} = 4\pi\epsilon_0 r$$

By definition $Q/V =$ capacitance C

$$\therefore \quad C = 4\pi\epsilon_0 r \text{ F} \quad \text{...in air}$$

$$= 4\pi\epsilon_0\epsilon_r r \text{ F} \text{ ...in a medium}$$

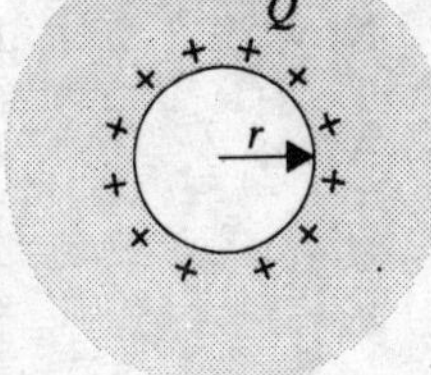

Fig. 6.4

Note. It is sometimes felt surprising that an isolated sphere can act as a capacitor because, at first sight, it appears to have one plate only. The question arises as to which is the second plate or surface. But if we remember that the surface potential V is with respect to infinity (actually, earth), then it is obvious that the other surface is earth. The capacitance $4\pi\epsilon_0 r$ exists between the surface of the sphere and earth.

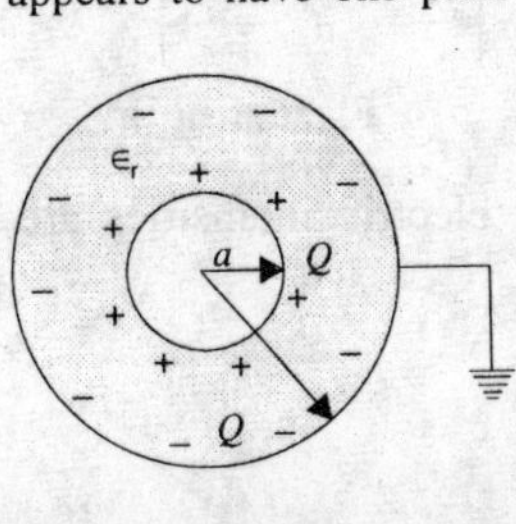

Fig. 6.5

6.4. Spherical Capacitor

Consider a spherical capacitor consisting of two concentric spheres of radii a and b metres as shown in Fig. 6.5. Suppose, the inner sphere is given a charge of $+Q$ coulombs. It will induce a charge of $-Q$ coulombs on the inner surface of the outer sphere and a charge of $+Q$ coulombs on its outer surface which will go to earth. If the dielectric medium between the two spheres has a relative permittivity of $\in_r$, then the free surface potential of the inner sphere due to its own charge is $Q/4\pi\in_0\in_r a$ volts. The potential of the inner sphere due to $-Q$ charge on the inner surface of the outer sphere is $-Q/4\pi\in_0\in_r b$ (remembering that potential anywhere inside a sphere is the same as at its surface).

$\therefore$ total potential difference between the two surfaces is

$$V = \frac{Q}{4\pi\in_0\in_r a} - \frac{Q}{4\pi\in_0\in_r b} = \frac{Q}{4\pi\in_0\in_r}\left(\frac{1}{a}-\frac{1}{b}\right) = \frac{Q}{4\pi\in_0\in_r}\left(\frac{b-a}{ab}\right)$$

$$\frac{Q}{V} = \frac{4\pi\in_0\in_r ab}{b-a} \qquad \therefore \quad C = 4\pi\in_0\in_r \frac{ab}{b-a}\ \text{F}$$

Example 6.1. *The radii of two spheres differ by 4 cm and the capacitance of this spherical capacitor is 53-33 pF. If outer sphere is earthed, calculate the radii assuming air as the dielectric.*

Solution. $C = 4\pi\in_0 \dfrac{ab}{b-a}$ F

Here $b = (a + 0.04)$ metre ; $C = 53.33$ pF $= 53.33 \times 10^{-12}$ F

$\in_0 = 8.854 \times 10^{-12}$ F/m

$$\therefore \quad 53.33 \times 10^{-12} = \frac{a\,(a+0.04)}{0.04} \times 4\pi \times 8.854 \times 10^{-12}$$

$$\therefore \quad \frac{a(a+0.04)}{0.04} = \frac{53.33 \times 10^{-12}}{4\pi \times 8.854 \times 10^{-12}} = 0.48$$

$\therefore \quad a^2 + 0.04\,a = 0.0192$ or $a^2 + 0.04a - 0.0192 = 0$

or $(a+0.16)(a-0.12) = 0$ or $a = 0.12$ m $=$ **12 cm**

$\therefore \quad b = 0.16$ m $=$ **16 cm**

6.5. Parallel-plate Capacitor

(i) Uniform Dielectric Medium

A parallel-plate capacitor consisting of two plates M and N each of area A m² separated by a thickness d metres of a medium of relative permittivity $\in_r$ is shown in Fig. 6.6. If a charge of $+Q$ coulombs is given to plate M, then flux between the two plates is Q coulomb and flux density in the dielectric medium is

$$D = \Psi/A = Q/A\ \text{C/m}^2$$

Now $$E = \frac{D}{\in} = \frac{D}{\in_0\in_r} = \frac{Q/A}{\in_0\in_r}$$

Also, E = potential gradient $= V/d$

$$\therefore \quad \frac{V}{d} = \frac{Q}{\in_0\in_r A} \quad \text{or} \quad \frac{Q}{V} = \frac{\in_0\in_r A}{d}$$

$$\therefore \quad C = \frac{\in_0\in_r A}{d}\ \text{F} \quad \text{...in a medium}$$

$$= \frac{\in_0 A}{d}\ \text{F} \quad \text{... in air}$$

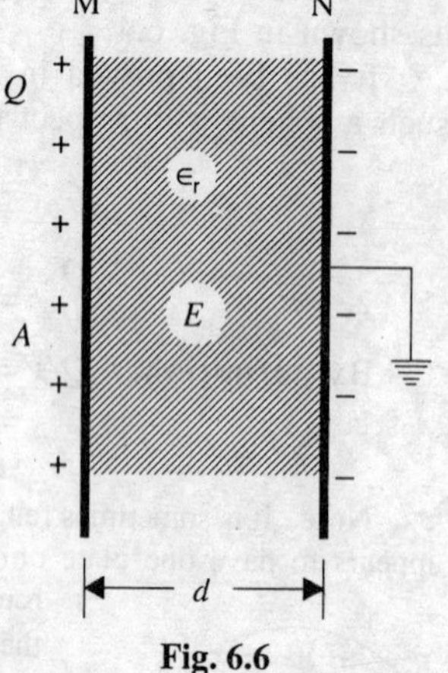

Fig. 6.6

(ii) Medium Partly Air

As shown in Fig. 6.7, the medium consists partly of air and partly of a parallel-sided dielectric slab of thickness t and relative permittivity $\in_r$. The electric flux density $D = Q/A$ is the same in both media. But electric intensities are different.

In air $$E_1 = \frac{D}{\in_0} = \frac{Q}{\in_0 A}$$

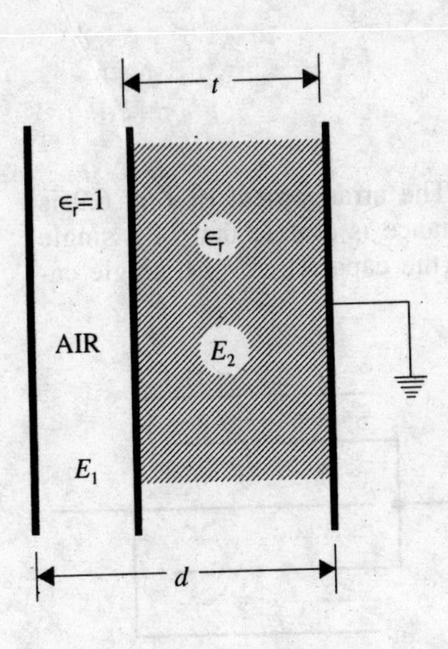

Fig. 6.7

In the medium $E_2 = \dfrac{D}{\epsilon_0 \epsilon_r} = \dfrac{Q}{\epsilon_0 \epsilon_r A}$

P.D. between plates $V = E_1 . (d - t) + E_2$

$$= \frac{D}{\epsilon_0}(d-t) + \frac{D}{\epsilon_0 \epsilon_r} t = \frac{D}{\epsilon_0}\left(\frac{t}{\epsilon_r} + d - t\right)$$

$$= \frac{Q}{\epsilon_0 A}\left[d - \left(t - \frac{t}{\epsilon_r}\right)\right]$$

$$\therefore \quad \frac{Q}{V} = \frac{\epsilon_0 A}{\left[d - \left(t - \dfrac{t}{\epsilon_r}\right)\right]}$$

$$\text{or } C = \frac{\epsilon_0 A}{\left[d - \left(t - \dfrac{t}{\epsilon_r}\right)\right]} \text{ farad} \qquad ...(i)$$

If the medium were totally air, the capacitance would have been

$$C = \frac{\epsilon_0 A}{d} \text{ farad} \qquad ...(ii)$$

From (*i*) and (*ii*) it is obvious that when a dielectric slab of thickness t and relative permittivity ϵ_r is introduced between the plates of an air capacitor, its capacitance increases because as seen from (*i*), the denominator decreases. The distance between the plates is effectively reduced by $\left(t - \dfrac{t}{\epsilon_r}\right)$. To bring the capacitace back to its original value, the capacitor plates will have to be further separated by that much distance in air. Obviously, the new separation between the two plates would be

$$= d + \left(t - \frac{t}{\epsilon_r}\right)$$

(*iii*) Composite Medium

A parallel-plate capacitor having three slabs of different dieletric media is shown in Fig. 6.8. As before

$$V = V_1 + V_2 + V_3$$

$$= E_1 d_1 + E_2 d_2 + E_3 d_3$$

$$= \frac{D}{\epsilon_0 \epsilon_{r1}} d_1 + \frac{D}{\epsilon_0 \epsilon_{r2}} d_2 + \frac{D}{\epsilon_0 \epsilon_{r3}} d_3$$

$$= \frac{D}{\epsilon_0}\left(\frac{d_1}{\epsilon_{r1}} + \frac{d_2}{\epsilon_{r2}} + \frac{d_3}{\epsilon_{r3}}\right)$$

$$= \frac{Q}{\epsilon_0 A}\left(\frac{d_1}{\epsilon_{r1}} + \frac{d_2}{\epsilon_{r2}} + \frac{d_3}{\epsilon_{r3}}\right)$$

$$\therefore \quad \frac{Q}{V} = C = \frac{\epsilon_0 A}{\left(\dfrac{d_1}{\epsilon_{r1}} + \dfrac{d_2}{\epsilon_{r2}} + \dfrac{d_3}{\epsilon_{r3}}\right)}$$

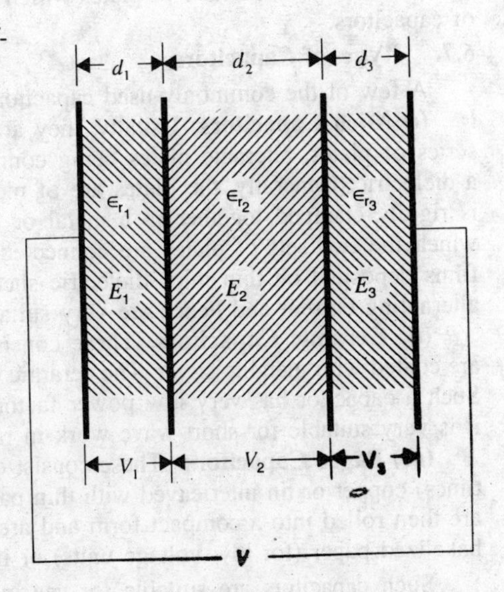

Fig. 6.8

In general $$C = \frac{\epsilon_0 A}{\Sigma d/\epsilon_r} \text{ F}$$

6.6. Multiplate and Variable Capacitors

Multiplate capacitors are shown in Fig. 6.9 and Fig. 6.10. The arrangement of Fig. 6.9 is equivalent to two capacitors joined in parallel. Hence, its capacitance is double that of a single capacitor. Similarly, the arrangement of Fig. 6.10 has four times the capacitance of a single capacitor.

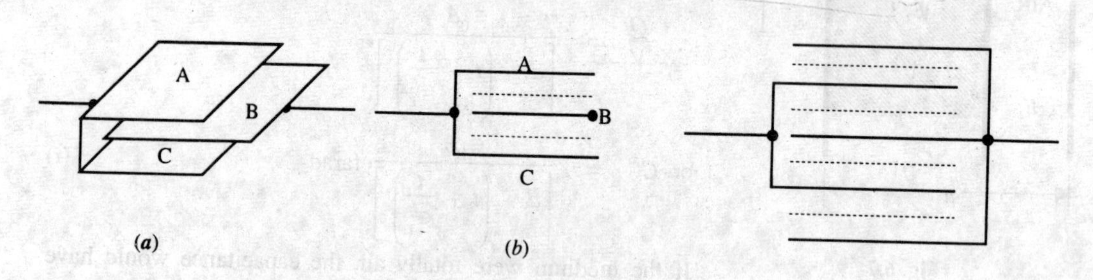

Fig. 6.9 **Fig. 6.10**

If one set of plates is fixed and the other is capable of rotation, then capacitance of such a multiple capacitor can be varied. Such variable-capacitance air capacitors are widely used in radio work (Fig. 6.11). The set of fixed plates F is insulated from the other set R which can be rotated by turning the knob K. The common area between the two sets is varied by rotating K, hence the capacitance between the two is altered. Minimum capacitance is obtained when R is completely rotated out of F and maximum when R is completely rotated in *i.e.* when the two sets of plates completely overlap each other.

The capacitance of such a capacitor is

$$= \frac{(n-1)\epsilon_0 \epsilon_r A}{d} \text{ F}$$

when n is the total number of plates which means that $(n - 1)$ is the number of capacitors.

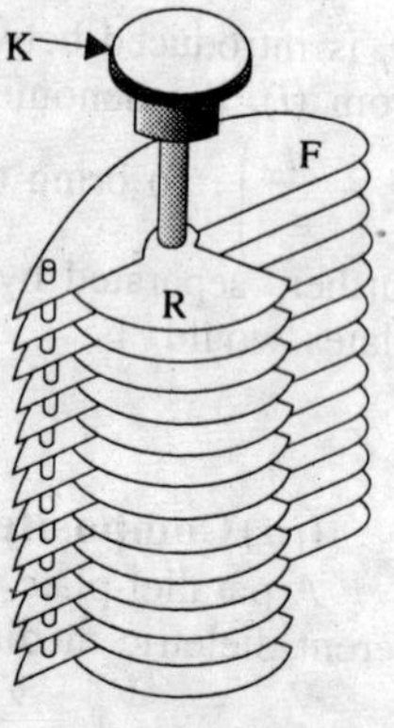

Fig. 6.11

6.7. Types of Capacitors

A few of the commonly-used capacitors are as follows :

(i) Mica Capacitors. Usually, they are multiplate type consisting of a series of plates, alternate plates being connected together and separated by a dielectric. Generally, the plates are of metal foil separated by thin mica sheets. The assembly is rigidly clamped together in a metal or ebonite case having two metal tugs or wire ends to which the two sets of plates are connected. The silver-mica type of capacitor consists of silver films deposited on thin mica dielectric sheets. Mica capacitors have very low power factors on alternating current and hence are very suitable for use at radio frequencies.

(ii) Ceramic Capacitors. These consist of discs of ceramic material whose opposite faces are coated with metallic silver. The ceramic disc acts as the dielectric and silver coatings as plates. Such a capacitor has very low power factor which decreases with increase in frequency. Hence, it is very suitable for short-wave work in radio.

(iii) Paper Capacitors. These consist of sheets of metal foils usually aluminium (but sometimes) copper or tin interleaved with thin paper impregnated with oil or wax. The foils and paper are then rolled into a compact form and are housed in hermetically sealed tubes of cardboard or bakelized paper (for low voltage units) or in oil-filled metal boxes (for high-voltage units).

Such capacitors are suitable for use in audio-frequency (A.F.) stages of radio receivers as by-pass and coupling capacitors.

(*iv*) **Electrolytic Capacitors.** These are either wet type or dry type. The wet type consists of an aluminium anode which is centrally mounted in a metal cylinder filled with an electrolytic solution usually ammonium borate which acts as a cathode. Two connections, one from the anode and the other from the cathode are brought out. When current is passed between the anode and cathode, a very thin film of aluminium oxide (Al_2O_3) is formed on the anode surface. This film has insulating properties and the anode-cathode combination, thereafter, becomes a capacitor of very large capacitance because dielectric film is extremely thin. For successful operation, such capacitors must always be used with proper polarity.

The dry type capacitor consists of positive and negative electrodes of aluminium foil which are separated from each other by a porous paper or gauze saturated with either a very viscous liquid or paste containing the electrolyte. The whole arrangement is formed into a roll and housed in a waxed carboard tube. Due to their high capacitance, such capacitors are generally used in smoothing circuits in radio work.

Example 6.2. *A capacitor consists of two similiar square aluminium plates, each 10 cm × 10 cm mounted parallel and opposite each other. What is their capacitance in μμF when the distance between them is 1 cm and the dielectric air? If the capacitor is given a charge of 500 μμC, what will be the difference of potential between plates ? How will this be affected if the space between the plates is filled with wax which has a relative permittivity of 4 ?*

Solution. $C = \dfrac{\epsilon_0 A}{d}$ F

Here $\epsilon_0 = 8.854 \times 10^{-12}$ F/m ; $d = 1$ cm $= 10^{-2}$ m;

$A = 10 \times 10 = 100 \text{ cm}^2 = 10^{-2} \text{ m}^2$

$$\therefore \quad C = \frac{8.854 \times 10^{-12} \times 10^{-2}}{10^{-2}} = 8.854 \times 10^{-12} \text{ F} = 8.854 \ \mu\mu\text{F}$$

Now $C = \dfrac{Q}{V} \quad \therefore \quad V = \dfrac{Q}{C}$

$$\therefore \quad V = \frac{500 \times 10^{-12}}{8.854 \times 10^{-12}} = \textbf{56.5 volts}$$

Now, when was is introduced, their capacitance is increased four times because

$$C = \frac{\epsilon_0 \epsilon_r A}{d} \text{ F} = 4 \times 8.854 = 35.4 \ \mu\mu\text{F}$$

The p.d. will, obviously, decrease to one-fourth value.

$\therefore \quad V = 56.5/4 = \textbf{14.1 volts}$

Example 6.3. *A parallel-plate capacitor has plates 0.15 mm apart, a plate area of 0.1 m² and a dielectric of relative permittivity 3. Find electric flux density, electric field intensity and the p.d. between the plates if capacitor charge is 0.5 mC* **(A.M.I.E. Summer 1989)**

Solution. $C = \epsilon_0 \epsilon_r \ A/d = 8.854 \times 10^{-12} \times 3 \times 0.1/0.15 \times 10^{-3}$

$= 1.77 \times 10^{-8}$ F

$D = Q/A = 0.5 \times 10^{-6}/0.1 = \mathbf{0.5 \times 10^{-5}\ C/m^2}$

$Q = CV \quad \therefore \quad V = Q/C = 0.5 \times 10^{-6}/1.77 \times 10^{-8} = \mathbf{28\ V}$

$E = V/d = 28/0.15 \times 10^{-3} = \mathbf{186{,}667\ V/m}$

Example 6.4. *A parallel-plate capacitor has plates of area 2m² spaced by three slabs of difference dielectric materials. The relative permittivities are 2, 3, 6 and the thickness are 0.4, 0.6 and 1.2 mm respectively. Calculate the combined capacitance and the electric stress in each material when the applied voltage is 1000 V.* **(A.M.I.E. Summer 1990)**

Solution. $$C = \frac{\epsilon_0 A}{\dfrac{d_1}{\epsilon_{r1}} + \dfrac{d_2}{\epsilon_{r2}} + \dfrac{d_3}{\epsilon_{r3}}} \text{ F}$$

$$= \frac{8.854 \times 10^{-12} \times 2}{\frac{0.4 \times 10^{-3}}{2} + \frac{0.6 \times 10^{-3}}{3} + \frac{1.2 \times 10^{-3}}{6}} \text{ F} = \mathbf{0.0295 \times 10^{-6} \text{ F}}$$

Charge on plate,

$$Q = CV = 0.0295 \times 10^{-6} \times 1000 \text{ C} = 29.5 \times 10^{-6} \text{ C}$$

Electric flux density

$$D = \frac{29.5 \times 10^{-6}}{2} = 14.75 \times 10^{-6} \text{ C/m}^2$$

$$\therefore \quad E_1 = \frac{D}{\epsilon_0 \epsilon_{r1}} = \frac{14.75 \times 10^{-6}}{8.854 \times 10^{-12} \times 2} = \mathbf{833.3 \text{ kV/m}}$$

$$E_2 = \frac{14.75 \times 10^{-6}}{8.854 \times 10^{-12} \times 3} = 555.4 \text{ kV/m}$$

and $$E_3 = \frac{14.75 \times 10{-6}}{8.854 \times 10^{-12} \times 6} = 277.7 \text{ kV/m}$$

Example 6.5. *A parallel-plate capacitor has plates of 1500 cm² separated by 5 mm with air dielectric. If a layer of dielectric 2 mm thick and relative permittivity 3 is now introduced between the plates, what must be the separation between the plates to bring the capacitance to the original value.?*

Solution. By introducing a layer of a dielectric, the effective distance between the plates is reduced by $(t - t/\epsilon_r)$

$$\therefore \text{ decrease in separation} = t - \frac{t}{\epsilon_r} = 2 - \frac{2}{3} = 1.33 \text{ mm}$$

Hence, the plates have to be further separated by a distance of 1.33 mm. The total distance or the new separation between the plates should be = 5 + 1.33 = **6.33 mm.**

Example 6.6. *A capacitor is composed of two plates separated by a sheet of insulating material 3 mm thick and of relative permittivity 4. The distance between the plates is increased to allow the insertion of a second sheet 5 mm thick of relative permittivity ϵ_r. If the capacitance of the capacitor so formed is $\frac{1}{3}$ of the original capacitance, find ϵ_r*

Solution. Let the plate area be A m².

In the first case, $$C_1 = \frac{\epsilon_0 \epsilon_r A}{d} = \frac{4\epsilon_0 A}{3 \times 10^{-3}} \text{ F}$$

In the second case, $$C_2 = \frac{\epsilon_0 A}{(d_1/\epsilon_{r1} + d_2/\epsilon_{r2})} \text{ F} = \frac{\epsilon_0 A}{\frac{3 \times 10^{-3}}{4} + \frac{5 \times 10^{-3}}{\epsilon_{r2}}} \text{ F}$$

Since $$C_2 = \frac{C_1}{3}$$

$$\therefore \quad \frac{\epsilon_0 A}{\frac{3 \times 10^{-3}}{4} + \frac{5 \times 10^{-3}}{\epsilon_{r2}}} = \frac{4\epsilon_0 A}{3 \times 3 \times 10^{-3}} \quad \therefore \epsilon_{r2} = \mathbf{3.3}$$

Example 6.7. *Calculate the capacitance and energy stored in a parallel-plate capacitor which consists of two metal plates, each 60 cm² separated by a dielectric of 1.5 mm thickness and of $\epsilon_r = 3.5$ if a p.d. of 1000 V is applied across it.* **(A.M.I.E. Winter 1980)**

Given : $\epsilon_0 = 8.854 \times 10^{-12}$ F/m

Solution. $C = \dfrac{\epsilon_0 \epsilon_r A}{d}$ farad

Here, $\epsilon_r = 3.5, A = 60 \text{ cm}^2 = 6 \times 10^{-3} \text{ m}^2, d = 1.5 \times 10^{-3}$ m

$\therefore \quad C = \dfrac{8.854 \times 10^{-12} \times 3.5 \times 6 \times 10^{-3}}{1.5 \times 10^{-3}} = 124 \times 10^{-12} \text{ F} = \mathbf{124\ pF}$

Energy stored $= \dfrac{1}{2} CV^2 = \dfrac{1}{2} \times 124 \times 10^{-12} \times 1000^2 = 62 \times 10^{-6} \text{ J} = \mathbf{62\ \mu J.}$

Tutorial Problems No. 6.1

1. The plates of a 0.0005 μF parallel-plate capacitor are spaced 5 mm apart in air. What is the area of each of the capacitor plates?
   If two dielectrics, each 2.5 mm thick and of relative permittivities 2 and 3 were placed between the plates, what would be the new capacitance ? **[0.28 m^2 ; 0.0012 μF]**
2. A parallel-plate capacitor has an effective plate area of each plate of 100 cm^2, the plates being separated by a dielectric 0.5 mm thick. Its capacitance is 442 μμF and it is charged to a p.d. of 10 kV. Calculate :
   (*a*) the potential gradient in the dielectric, (*b*) the electric flux density in the dielectric
   (*c*) the relative permittivity of the dielectric material **[(*a*) 2 × 10^7 V/m (*b*) 442 mC/m^2 (*c*) 2.5]**
3. A p.d. of 10 kV is applied across a capacitor formed by two parallel metal sheets, each of area 100 cm^2 separated by a dielectric 3 mm thick having a capacitance of 4 × 10^{-4} μF. Find :
   (*a*) total flux (*b*) potential gradient
   (*c*) the relative permittivity of the dielectric material
   (*d*) the electric flux density. **[(a) 4μC (b) 3.3 kV/mm (c) 13.6 (d) 0.4 mC/m^2]**
4. Calculate the capacitance of a capacitor which has five plates spaced 1 mm apart in air. Each plate has an area of 20 cm^2. What would be the capacitance of this capacitor if it were immersed in oil of relative permittivity 2.2 ? **[70.7 pF ; 155-5 pF]**
5. A p.d. of 15 kV is applied across the terminals of a capacitor consisting of two circular plates, each having an area of 200 cm^2 and separated by 1 mm thickness of dielectric. The capacitance of this capacitor is 4.5 × 10^{-4} μF. Calculate the dielectric constant (ϵ_r) and the electric flux density. **[337.5 μC/m^2 ; 2.54]**
6. A capacitor consists of two metal plates each 40 cm × 40 cm, spaced 6 mm apart. The space between the metal plates is filled with a glass plate 5 mm thick and a layer of paper 1 mm thick. The relative permittivities of glass and paper are 8 and 2 respectively. Calculate (*i*) the capacitance (*ii*) the potential gradient in each dielectric in kV/mm due to a p.d. of 10 kV between metal plates (neglect any fringing flux). *(A.M.I.E. Nov. 1972)*
   [(*i*) 1.257 × 10^{-9} F (*ii*) 1.11 kV/mm, 1.44 kV/mm**]**
7. A sheet of mica, 1 mm thick and of relative permittivity 6, is interposed between two parallel brass plates 3 mm apart. The remainder of the space between the plates is occupied by air. Calculate the area of each plate if the capacitance between them is 0.001 μF. Assuming that air can withstand a potential gradient of 3 MV/m, show that a p.d. of 5 kV between the plates will not cause a flashover. **[0.245 m^2 ; 2.3 MV/m in air gap]**

6.8. Cylindrical Capacitor

A single-core cable or cylindrical capacitor consisting of two co-axial cylinders of radii a and b metres, is shown in Fig. 6.12. Let the charge per metre length of the cable on the outer surface of the inner cylinder be $+Q$ coulombs and on the inner surface of the outer cylinder be $-Q$ coulombs (by induction). For all practical purposes, the charge $+Q$ coulombs/metre on the surface of the inner cylinder can be sup-

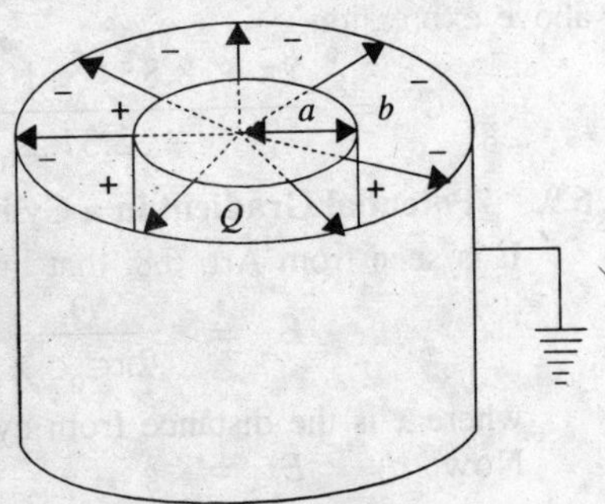

Fig. 6.12

posed to be located along its axis. Let $\in_r$ be the relative permittivity of the medium between the two cylinders. The outer cylinder is earthed.

Now, let us find the value of electric intensity at any point distant x metres from the axis of the inner cylinder. As shown in Fig. 6.13, consider an imaginary co-axial cylinder of radius x metres and length one metre between the two given cylinders. The electric field between the two cylinders is radial as shown. Total flux coming out radially from the curved surface of this imaginary cylinder is Q coulombs. Area of the curved surface

$$= 2\pi x \times 1 = 2\pi x \text{ m}^2$$

Hence, the value of the electric flux density on the surface of the imaginary cylinder is

$$D = \frac{\text{flux in coulombs}}{\text{area in m}^2} = \frac{\Psi}{2} = \frac{Q}{2\pi x} \text{ C/m}^2$$

The value of electric intensity

$$E = \frac{D}{\in_0 \in_r} \text{ V/m} = \frac{Q}{2\pi \in_0 \in_r x} \text{ V/m}$$

Now $dV = -E.dx$

$$V = \int_{x=b}^{x=a} -E.dx = \int_b^a \frac{-Q.dx}{2\pi \in_0 \in rx}$$

$$= \frac{-Q}{2\pi \in_0 \in_r} \int_b^a \frac{dx}{x} = \frac{-Q}{2\pi \in_0 \in_r} |\log x|_b^a$$

$$= \frac{-Q}{2\pi \in_0 \in_r} (\log_e a - \log_e b)$$

$$= \frac{-Q}{2\pi \in_0 \in_r} \log_e a/b = \frac{Q}{2\pi \in_0 \in_r} \log_e b/a$$

$$\frac{Q}{V} = \frac{2\pi \in_0 \in_r}{\log_e b/a}$$

$$\therefore C = \frac{2\pi \in_0 \in_r l}{2.3 \log_{10} b/a} \text{ F/m} \qquad (\because \log_e b/a = 2.3 \log_{10} b/a)$$

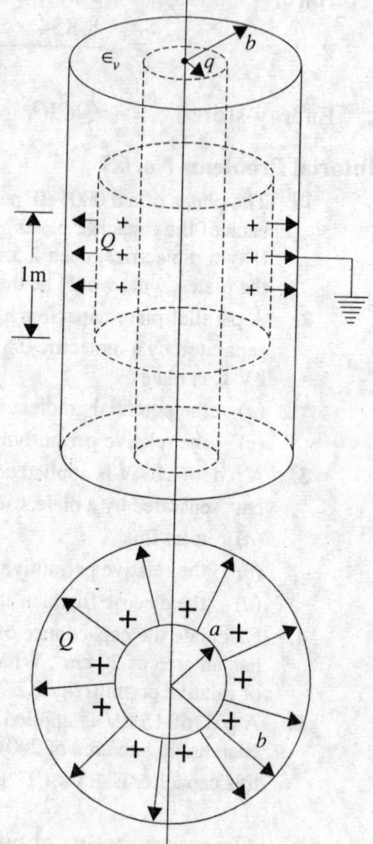

Fig. 6.13

The capacitance of l metres length of this cable is

$$C = \frac{2\pi \in_0 \in_r}{2.3 \log_{10} b/a} \text{ F}$$

The capacitance of 1 km length of the cable in μF can be found by putting l = 1 km in the above expression.

$$C = \frac{2\pi \times 8.854 \times 10^{-12} \times \in_r 1000}{2.3 \log_{10} b/a} \text{ F/km} = \frac{0.024 \in_r}{\log_{10} b/a} \text{ μF}/km$$

6.9. Potential Gradient in a Cylindrical Capacitor

It is seen from Art. 6.8. that in a cable capacitor.

$$E = \frac{Q}{2\pi \in_0 \in_r x} \text{ V/m}$$

where x is the distance from cylinder axis to the point under consideration

Now $\quad E = g$

$$\therefore \quad g = \frac{Q}{2\pi\epsilon_0\epsilon_r x} \text{ V/m} \quad ...(i)$$

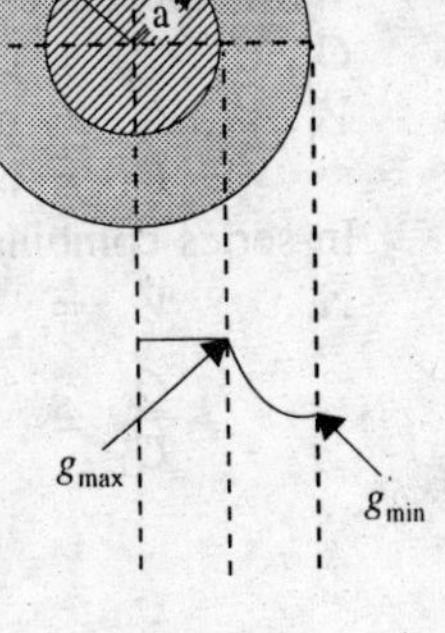

Fig. 6.14

From Art. 6.8, we find that

$$V = \frac{Q}{2\pi\epsilon_0\epsilon_r} \log_e b/a$$

$$\therefore \quad Q = \frac{2\pi\epsilon_0\epsilon_r V}{\log_e b/a}$$

Subsituting this value of Q in (i), above, we get

$$g = \frac{2\pi\epsilon_0\epsilon_r V}{\log_e b/a \times 2\pi\epsilon_0\epsilon_r x} \text{ V/m} \quad \text{or} \quad g = \frac{V}{x \log_e b/a} \text{ volt/metre}$$

$$\text{or} \quad g = \frac{V}{2.3 x \log_{10} b/a} \text{ volt/metre}$$

Obviously, potential gradient varies inversely as x.

Minimum value of $x = a$, hence maximum value of potential gradient is

$$g_{max} = \frac{V}{2.3 a \log_{10} b/a} \text{ V/m}$$

$$\text{Similarly,} \quad g_{min} = \frac{V}{2.3 b \log_{10} b/a} \text{ V/M}$$

Example 6.8. *A concentric cable has a core diameter of 2.36 cm. It is insulated with impregnated paper of thickness 1.65 cm and a relative permittivity of 3.5 and covered with lead sheath. If working voltage is 660 V d.c., calculate the capacitance of the cable in μF/km and maximum electric stress (potential gradient).* **(A.M.I.E. Winter, 1988)**

Solution. Given :

$a = 2.36/2 = 1.18$ cm ; $b = 1.18 + 1.65 = 2.83$ cm

$b/a = 2.83/1.18 = 2.4$; $\epsilon_r = 3.5$

$$C = \frac{0.024\epsilon_r}{\log_{10} b/a} \,\mu\text{F/km} = 0.024 \times 3.5/\log_{10} 2.4 = 0.084/0.38 = \mathbf{0.22 \text{ mF/km}}$$

$$g_{max} = \frac{V}{2.3 a \log_{10} b/a} = \frac{660}{2.3 \times 1.18 \times 10^{-5} \times 0.38} = \mathbf{64 \text{ kV/m}}$$

Example 6.9. *Calculate (a) the capacitance and (b) the resistance of a cable 5 km long having an inner core of diameter 1 cm., and an impregnated insulation of thickness 0.7 cm and relative permittivity 4. Specific resistance of copper may be taken as* 2μ Ω-cm.

Solution. (a) $C = \dfrac{0.024\epsilon_r}{\log_{10} b/a}$ μF/km

Here $a = 0.5$ cm ; $b = 0.5 + 0.7 = 1.2$ cm

$b/a =$ 12/5 2.4 ; $\log_{10} 2.4 = 0.3802$

$C = 0.024 \times 4/0.3802$ μF/km

Capacitance for 5 km length

$= 5 \times 0.024 \times 4/0.3802 =$ **1.263 μF.**

(b) Conductor resistance $R = \rho \dfrac{l}{A}$

Here $\rho = 2 \times 10^{-6}$ Ω-cm $l = 5$ km $= 10^5$ cm

$A = \pi d^2/4 \text{ cm}^2 = \pi/4 \text{ cm}^2$

$$R = \frac{2 \times 10^{-6} \times 5 \times 10^5}{\pi/4} = \mathbf{1.27\ \Omega}$$

6.10. Capacitors in Series

With reference to Fig. 6.15, let

C_1, C_2, C_3 = capacitances of three capacitors
V_1, V_2, V_3 = p.ds. across three capacitors ; V = applied voltage across combination ;
C = combined or equivalent capacitance.

In series combination, charge on all capacitors is the same but p.d. across each is different.

$$\therefore \quad V = V_1 + V_2 + V_3$$

$$\frac{Q}{C} = \frac{Q}{C_1} + \frac{Q}{C_2} + \frac{Q}{C_3} \quad \text{or} \quad \frac{1}{C} = \frac{1}{C_1} + \frac{1}{C_2} + \frac{1}{C_3}$$

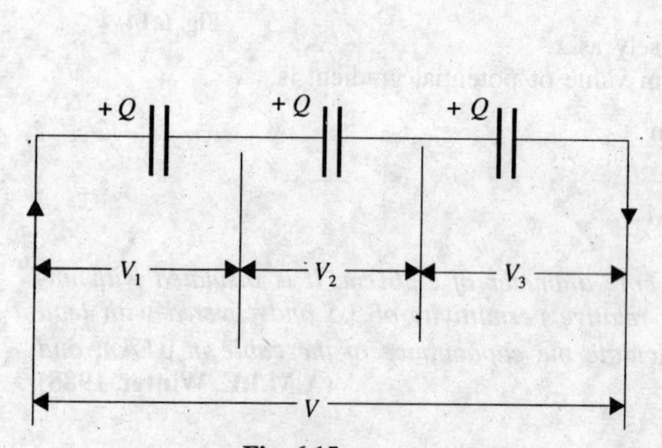

Fig. 6.15

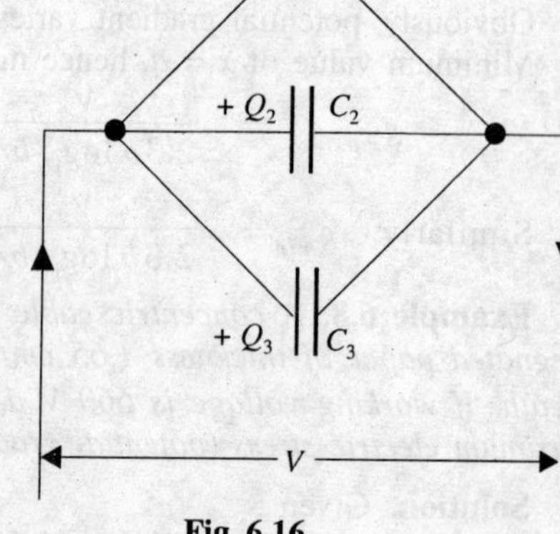

Fig. 6.16

6.11. Capacitors in Parallel

In this case, p.d. across each is the same but charge on each is different (Fig. 6.16).

$$\therefore \quad Q = Q_1 + Q_2 + Q_3 \quad \text{or} \quad CV = C_1V + C_2V + C_3V$$

$$\text{or} \quad C = C_1 + C_2 + C_3$$

6.12. Voltage Across Series-connected Capacitors

First consider the case when two capacitors of capacitances C_1 and C_2 are connected in series across a supply voltage of V as shown in Fig. 6.17 (*a*). If V_1 and V_2 are the voltage developed across them, then

$$V = V_1 + V_2 \qquad \ldots(i)$$

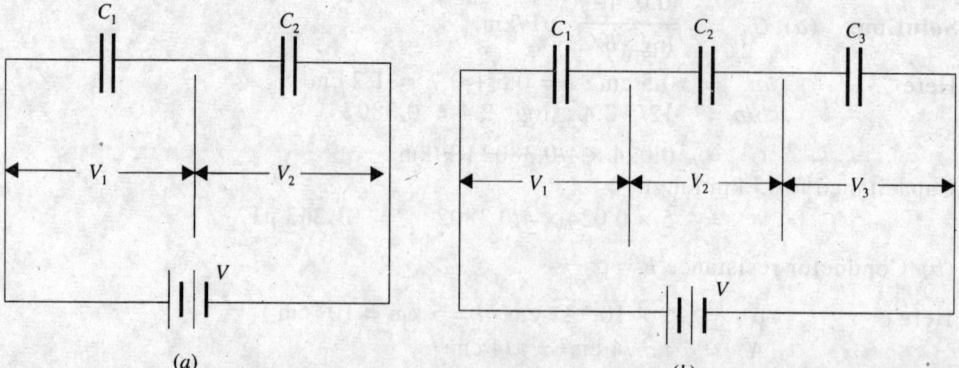

(*a*) (*b*)

Fig. 6.17

Since charge across each is the same,

$$\therefore \quad Q = C_1V_1 = C_2V_2 \quad \text{or} \quad V_2 = V_1C_1/C_2$$

Substituting this value in Eq. (*i*) above, we have

$$V = V_1 + V_1\frac{C_1}{C_2} \quad \text{or} \quad V_1 = V\frac{C_2}{C_1 + C_2}$$

$$\text{Similarly,} \quad V_2 = V\frac{C_1}{C_1 + C_2}$$

Now, consider the case shown in Fig. 6.17 (*b*). Here

$$V = V_1 + V_2 + V_3 \qquad (ii)$$

$$\text{Also,} \quad Q = C_1V_1 = C_2V_2 = C_3V_3$$

$$\therefore \quad V_2 = V_1\frac{C_1}{C_2} \text{ and } V_3 = V_1\frac{C_1}{C_3}$$

Substituting these values in Eq. (*i*) above, we have

$$V = V_1 + V_1\frac{C_1}{C_2} + V_1\frac{C_1}{C_3} = V_1\left(1 + \frac{C_1}{C_2} + \frac{C_1}{C_3}\right)$$

$$\therefore \quad V_1 = V\frac{C_2C_3}{C_1C_2 + C_2C_3 + C_3C_1}$$

$$\text{Also,} \quad V_2 = V\frac{C_1C_3}{C_1C_2 + C_2C_3 + C_3C_1} \quad \text{and} \quad V_3 = V\frac{C_1C_2}{C_1C_2 + C_2C_3 + C_3C_1}$$

Example 6.10. *The total capacitance of two capacitors is 0.03 μF when joind in series and 0.16 μF when connected in parallel. Find the capacitance of each capacitor.*

Solution. Let C_1 and C_2 be the unknown capacitances.

Then $C_1 + C_2 = 0.16$ ...(*i*) when in parallel

$$\frac{C_1C_2}{C_1 + C_2} = 0.03 \qquad \text{...(ii) when in series}$$

Eliminating C_2 from (*i*) and (*ii*), we get

$$\frac{C_1(0.16 - C_1)}{0.16} = 0.03$$

$$\text{or} \quad C_{12} - 0.16C_1 + 0.0048 = 0$$

$$\text{or} \quad (C_1 - 0.12)(C_1 - 0.04) = 0$$

$$\therefore \quad C_1 = 0.12 \quad \text{or} \quad 0.04$$

$$\text{Hence} \quad C_2 = 0.04 \quad \text{or} \quad 0.12$$

So, one capacitor is of **0.12 μF** and the other of **0.04 μF** capacitance.

Example 6.11. *Three capacitors are connected in series across 135-V supply. The voltages across them are 30, 45, and 60 and charge on each is 4500 microcoulombs. Find the capacitance of each capacitor and that of the combination.*

Solution. $C_1 = \dfrac{Q}{V_1} = \dfrac{4500\mu C}{30\,V} = 150\ \mu F$

$$C_2 = \frac{4500\mu\,C}{45\,V} = 100\ \mu F \quad \text{and} \quad C_2 = \frac{4500\mu C}{60\,V} = 75\ \mu F$$

$$\text{But,} \quad \frac{1}{C} = \frac{1}{C_1} + \frac{1}{C_2} + \frac{1}{C_3} = \frac{1}{150} + \frac{1}{100} + \frac{1}{75} = \frac{9}{300} = \frac{3}{100}$$

$$\therefore \quad \boldsymbol{C = \frac{100}{3} = 33.4\ \mu F}$$

or $$C = \frac{Q}{V} = \frac{4500}{135} = \mathbf{33.3\ \mu F}$$

Example 6.12. *A 10 μF, a 20 μF and a 40 μF capacitor are connected in series to a 399 volts source of e.m.f.*

(i) What is the equivalent capacitance?

(ii) What is the magnitude of charge across each capacitor ?

(iii) What is the potential difference across each capacitor ?

(Electrical Science, A.M.I.E. Winter 1994.)

Solution. The three capacitors are connected in series, as shown in Fig. 6.18. The equivalent capacitance is given by :

(*i*) $$\frac{1}{C} = \frac{1}{10} + \frac{1}{20} + \frac{1}{40} \quad \therefore \quad C = \frac{40}{7}\ \mu F$$

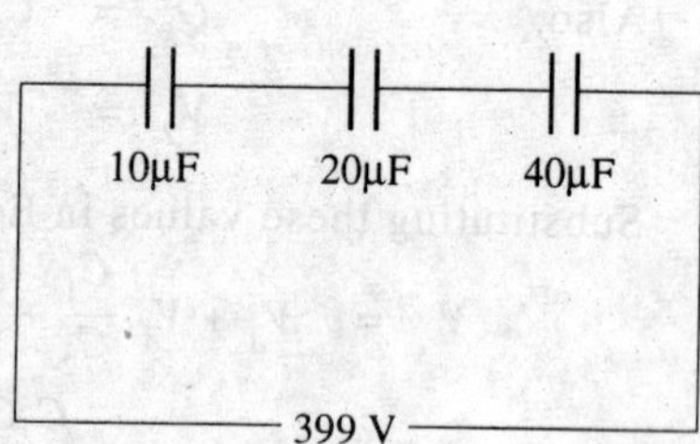

Fig. 6.18

(*ii*) The total charge across the combined capacitance is given by

$$Q = CV = (40/7) \times 399$$
$$= 15{,}960/7 = 2280\ \mu C$$

Charge across each capacitor would be the same *i.e.* **2280 μF.**

(*iii*) $V_1 = Q/C_1 = 2280/10 = \mathbf{228\ V}$

$V_2 = Q/C_2 = 2280/20 = \mathbf{114\ V}$

$V_3 = Q/C_3 = 2280/40 = \mathbf{57\ V}$

(*iv*) Alternative Solution

We may also use the formulae given in Art. for finding V_1, V_2 and V_3.

$$V_1 = V.\frac{C_2 C_3}{C_1 C_2 + C_2 C_3 + C_3 C_1}$$

$$= 399 \times \frac{20 \times 40}{200 + 800 + 400} = 228\ V$$

$$V_2 = 399 \times \frac{10 \times 40}{1400} = 114\ V \quad \text{and} \quad V_3 = 399 \times \frac{10 \times 20}{1400} = 57\ V$$

Example 6.13. *A 9 μF capacitor is connected in series with a parallel combination of two capacitors of values of 4μF and 2μF respectively (i) Determine the capacitance of the combination (ii) If a p.d. of 20 V is maintained across the combination, determine the charge on the 9 μF capacitor and the energy stored in the 4μF capacitor.* **(A.M.I.E. Winter 1990)**

Solution. (*i*) Capacitance of parallel combination $= 4 + 2 = 6\ \mu F$

Total capacitance $= 9 \parallel 6 = 3.6\ \mu F$

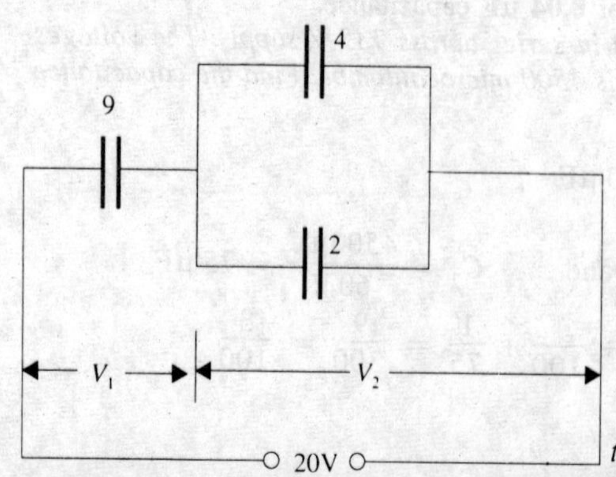

Fig. 6.19

$$(ii)\ V_1 = V\frac{C_2}{C_1 + C_2}$$

$$= 20 \times \frac{6}{9 + 6} = 8\ V$$

$$V_2 = 20 - 8 = 12\ V$$

Charge on 9μF $= 9 \times 8 = 72\ \mu C$

Energy stored in 4μF capacitor

$$= \frac{1}{2} \times 4 \times 12^2 = \mathbf{288\ \mu J}$$

Example 6.14. *Two capacitors have capacitances of 6μF and 10μF respectively. (i) Find the total capacitance when they are connected (1) in* *parallel (2) in series. (ii) When the above two capacitors are connected in series across a 200 V supply, find the potential difference across each capacitor and the charge on each capacitor.* **(A.M.I.E. Summer 1990)**

Solution. (*i*) (1) $C = C_1 + C_2 = 6 + 10 = 16\ \mu F$

(2) $C = C_1C_2/(C_1 + C_2) = 6 \times 10/16 = 3.75\ \mu F$

(*ii*) $V_1 = V\dfrac{C_2}{C_1 + C_2} = 200 \times \dfrac{10}{16} = \mathbf{125\ V}$

$V_2 = V\dfrac{C_1}{C_1 + C_2} = 200 \times \dfrac{6}{16} = \mathbf{75\ V}$

or $V_2 = V - V_1 = 200 - 125 = \mathbf{75\ V}$

Since the two capacitors are in series, charge on each is the same. It would be the same charge as on the equivalent capacitor

$\therefore \quad Q_1 = Q_2 = Q = CV = 3.75 \times 200 = 750\ \mu C$

Example 6.15. *In the circuit shown in the figure below, find the charge stored in each capacitor at steady-state condition. Find also the steady-state branch currents.* **(A.M.I.E. Summer 1987)**

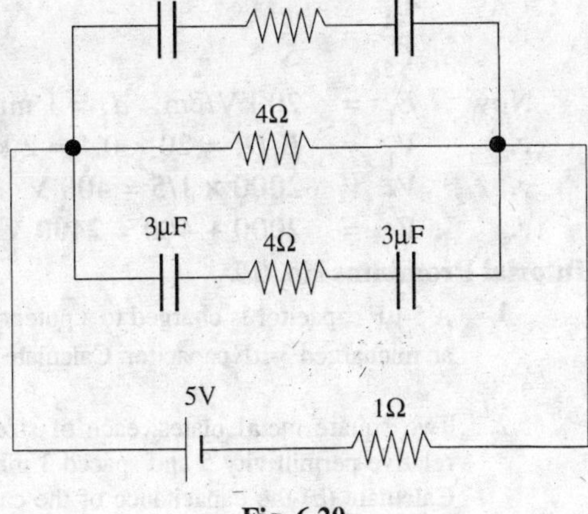

Fig. 6.20

Solution. Under steady-state conditions, there would be no current flow through branch No. 1 and 3. Consequently, there would be no voltage drop across the 4Ω resistors. Since the two series connected capacitors are equal, the voltage across each 3 μF capacitor would be 5/2 = 2.5 V.

$\therefore$ charge stored in each 3μF capacitor is

$Q = CV = 3 \times 2.5 = 7.5\ \mu C$

Under steady-state conditions, current will flow through branch No. 2 only.

$\therefore \quad I = 5/(4 + 1) = \mathbf{1A}$

Example 6.16. *Find the charges on capacitors in Fig. 6.21 and the p.ds. across them.*

Solution. Equivalent capacitance between points *A* and *B* is $C_1 + C_2 = 5 + 3 = 8\ \mu F$.

Capacitance of the whole combination

$C = \dfrac{8 \times 2}{8 + 2} = 1.6\ \mu F$

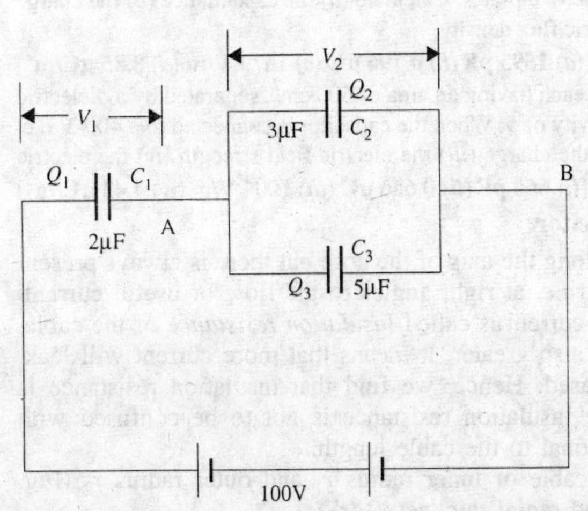

Fig. 6.21

Charge on the combination is

$Q_1 = CV = 100 \times 1.6 = \mathbf{160\ \mu C}$

$V1 = \dfrac{Q_1}{C_1} = \dfrac{160}{2} = \mathbf{80\ V}$

and $V_2 = 100 - 80 = \mathbf{20\ V}$

$Q_2 = C_2V_2 = 3 \times 10^{-6} \times 20$

$= 60 \times 10^{-6}\ C = \mathbf{60\ \mu C}$

$Q_3 = C_3V_2 = 5 \times 10^{-6} \times 20 = \mathbf{100\ \mu C}$

Example 6.17. *Two perfectly-insulated capacitors are connected in series. One is an air capacitor with a plate area of 100 cm², the plates being 1 mm apart ; the other has a plate area of 10 cm², the plates being separated by a solid dielectric 0.1 mm apart with a relative* *permittivity of 5. Find the voltage across the combination if the potential gradient in the air capacitor is 20 kV/cm.*

Solution. Let C_1 and C_2 be capacitances and V_1 and V_2 the potentials across the two capacitors in series as in Fig. 6.22.

Since the two capacitors are in series, charge across each is the same.

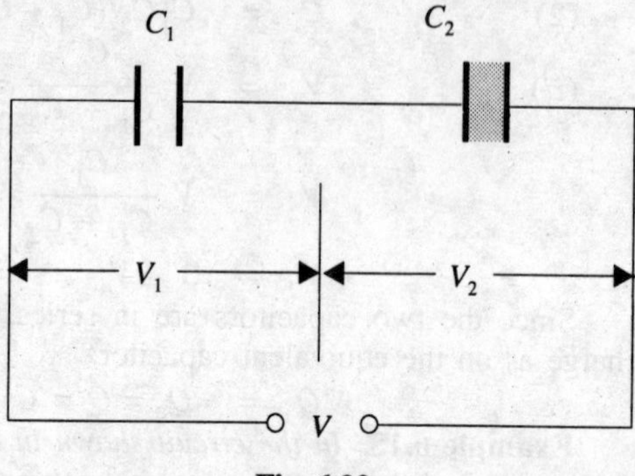

Fig. 6.22

$\therefore \quad Q = C_1V_1 = C_2V_2$

or $\quad C_1V_1 = C_2V_2$

or $\quad V_2 = V_1 \dfrac{C_1}{C_2}$

Now, $\quad C_1 = \dfrac{\epsilon_0 \times 100 \times 10^{-4}}{1 \times 10^{-3}}$

and $\quad C_2 = \dfrac{5\epsilon_0 \times 10 \times 10^{-4}}{0.1 \times 10^{-3}}$

$\therefore \quad \dfrac{C_1}{C_2} = \dfrac{1}{5}$

Now $\quad E_1 = 20$ kV/cm; $d_1 = 1$ mm $= 0.1$ cm

$\therefore \quad V_1 = E_1D_1 = 20 \times 0.1 = 2$ kV $= 2000$ V

$\therefore \quad V_2 = 2000 \times 1/5 = 400$ V

$\therefore \quad V = 2000 + 400 =$ **2400 V**

Tutorial Promblms No. 6.2

1. A 5-μF capacitor is charged to a potential difference of 100 V and then connected in parallel with an uncharged 3-μF capacitor. Calculate the potential difference across the parallel capacitors. **[62.5 V]**

2. Two square metal plates, each of size 200 mm, are completely immersed in insulating oil of relative permittivity 5 and spaced 3 mm apart. A p.d. of 600 V is maintained between the plates. Calculate (*a*) the capacitance of the capacitor (*b*) the charge stored on the plates (*c*) the electric field strength in the dielectric (*d*) the electric flux density. **[(*a*) 590 pF (*b*) 0.354 μC (*c*) 200 kV/m (*d*) 8.85 μC/m^2]**

3. A capacitor consists of two metal plates, each having an area of 900 cm^2, spaced 3 mm apart. The whole of the space between the plates is filled with a dielectric having a relative permittivity of 6. A p.d. of 500 V is maintained between the two plates. Calculate (*a*) the capacitance (*b*) the charge (*c*) the electric field strength (*d*) the electric flux density. **[(*a*) 1593 pF (*b*) 0.796 μC (*c*) 167 kV/m (*d*) 8.85 μC/m^2]**

4. A capacitor consists of two metal plates, each having an area of 600 cm^2, separated by a dielectric 4 mm thick which has a relative permittivity of 5. When the capacitor is connected to a 400-V d.c. supply, calculate (*i*) the capacitance (*ii*) the charge (*iii*) the electric field strength (*iv*) the electric flux density. **[(*i*) 664 pF (*ii*) 0.656 μC (*iii*) 100 kV/m (iv) 4.43 μC/m^2]**

6.13. Insulation Resistance of a Cable Capacitor

In a cable capacitor, useful current flows along the axis of the core but there is always present some leakage of current. This leakage is radial *i.e.* at right angles to the flow of useful current. The resistance offered to this radial leakage of current is called *insulation resistance* of the cable. If cable length is greater, then leakage area is also greater. It means that more current will leak. In other words, insulation resistance is decreased. Hence, we find that insulation resistance is inversly proportional to the cable length. The insulation resistance is not to be confused with conductor resistance which is directly proportional to the cable length.

Consider l metre length of a single-core cable of inner radius r_1 and outer radius r_2 (Fig. 6.23). Imagine an annular ring of radius 'r' and radial thickness 'dr'.

If resistivity of insulating material is ρ, then resistance of this narrow ring is

$$dR = \frac{\rho\, dr}{2\pi r \times l}$$

$\therefore$ insulation resistance of l metre length of the cable is

$$\int_0^R dR = \int_{r_1} \frac{\rho\, dr}{2\pi\, rl}$$

or
$$R = \frac{\rho}{2\pi rl} \left| \log_e r \right|_{r_1}^{r_2}$$

$$R = \frac{\rho}{2\pi l} \log_e r_2/r_1$$

$$\therefore \quad R = \frac{2.3\rho}{2\pi l} \log_{10} r_2/r_1 \ \Omega$$

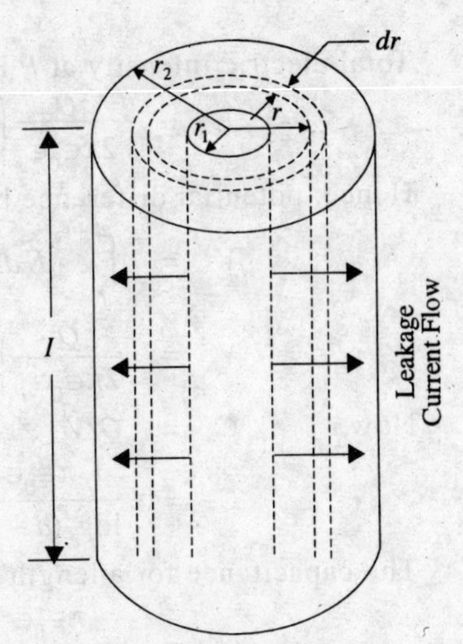

Fig. 6.23

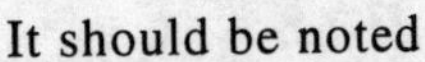

It should be noted

(i) that R is inversely proportional to the cable length.

(ii) that R depends upon the ratio r_2/r_1 and not on the thickness of the insulator itself.

Example 6.18. *Find the insulation resistance per metre of a single-core cable from the following data :—*

Core radius = 1.3 cm

Outer diameter of cable = 8 cm

Specific resistance of insulation material = 8×10^{14} ohm-metre.

Solution. Insulation resistance

$$R = \frac{2.3\,\rho}{2\pi l} \log_{10} r_2/r_1 \text{ ohm}$$

Here $\rho = 8 \times 10^{14}\ \Omega\text{-m}$; l = 1m

$r_1 = 1.3$ cm ; $r_2 = 8/2 = 4$ cm

$$\therefore \quad R = \frac{2.3 \times 8 \times 10^{14}}{2\pi \times 1} \log_{10} 4/1.3 = \mathbf{1.43 \times 10^{14}\ \Omega}$$

6.14. Capacitance Between two Parallel Wires

This case is of practical importance in overhead transmission lines. The simples system is 2-were system (either d.c. or a.c.) In the case of a.c. system, if the transmission line is long and voltage high, the charging current drawn by the line due to the capacitance between conductors is appreciable and affects its performance considerably.

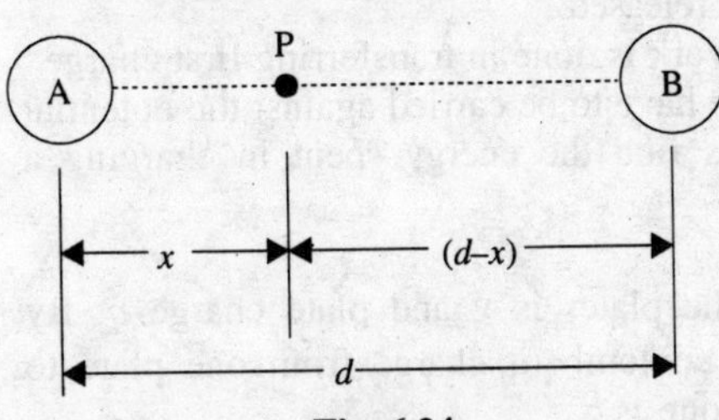

Fig. 6.24

With reference to Fig. 6.24, let

d = distance between centres of the wires A and B

r = radius of each wire ($\ll d$)

Q = charge in coulomb/metre of each wire*

Now, let us consider electric intensity at any point P between conductors A and B.

Electric intensity at P due to charge $+Q$ coulomb/metre² on A is

$$= \frac{Q}{2\pi\epsilon_0\epsilon_r x} \text{ V/m} \qquad \text{—towards } B$$

Electric intensity at P due to charge $-Q$ coulomb/metre² on B

$$= \frac{Q}{2\pi\epsilon_0\epsilon_r (d - x)} \text{ V/m} \qquad \text{—towards } B$$

*If charge on A is $+Q$ then on B it will be $-Q$.

Total electric intensity at P is

$$E = \frac{Q}{2\pi\epsilon_0\epsilon_r}\left(\frac{1}{x} - \frac{1}{(d-x)}\right)$$

Hence, potential difference between the two wires is

$$V = \int_r^{d-r} E.dx = \frac{Q}{2\pi\epsilon_0\epsilon_r}\int_r^{d-r}\left(\frac{1}{x} + \frac{1}{d-x}\right)dx$$

$$V = \frac{Q}{2\pi\epsilon_0\epsilon_r}\Big|\log_e x - \log_e d - x\Big|_r^{d-r}$$

Now, $C = Q/V$

$$\therefore \quad C = \frac{\pi\epsilon_0\epsilon_r}{\log_e (d-r)/r} = \frac{\pi\epsilon_r\epsilon_0}{2.3\log_{10}(d-r)/r} \cong \frac{\pi\epsilon_0\epsilon_r}{2.3\log_{10} d/r}\ \text{F/m}$$

The capacitance for a length of l metres is

$$C = \frac{\pi\epsilon_0\epsilon_r l}{2.3\log_{10} d/r}\ \text{F}$$

The capacitance per kilometre is

$$C = \frac{\pi \times 8.854\times 10^{-12} \times \epsilon_r \times 100 \times 10^6}{2.3\log_{10} d/r} = \frac{0.0121\,\epsilon_r}{\log_{10} d/r}\ \text{F/km}$$

Example 6.19. *The conductors of a two-wire transmission line (4 km long) are spaced 45 cm between centres. Calculate the capacity of the line if each conductor has a diameter of 1.5 cm*

(A.M.I.E. Winter 1988)

Solution $\quad C = \dfrac{0.0121\epsilon_r}{\log_{10} d/r}\ \mu\text{F/km}$

Here, $\quad d = 45$ cm; $\quad r = 1.5/2 = 0.75$ cm, $d/r = 45/0.75 = 60$

$\epsilon_r = 1$

$\therefore \quad C = 4 \times 0.0121 \times 1/\log_{10}60 = \mathbf{0.0272\ \mu F}$

6.15. Energy Stored in a Capacitor

Charging of a capacitor always involves some expenditure of energy by the charging agency. This energy is stored up in the electrostatic field set up in the dielectric medium. On discharging the capacitor, the field collapses and the stored energy is released.

To begin with when the capacitor is uncharged, little work is done in transferring first charge* from one plate to another. But further instalments of charge have to be carried against the potential difference already established between the plates. Let us find the energy spent in charging a capacitor of capacitance C to a voltage V.

(*a*) First Method

Suppose at any stage of charging, the p.d. across the plates is v and plate charge q. By definition, v is equal to the work done in shifting one coulomb of charge from one plate to another. If 'dq' is the charge next transferred, the work done is

$$dw = v.dq$$

Now $\quad q = Cv \qquad \therefore \quad dq = C.dv$

$$\therefore \quad dw = Cv\,dv$$

Total work done in giving V units of potential is

$$W = \int_0^V C\,v\,dv = C\left|\frac{v^2}{2}\right|_0^V$$

*It charge on A is +Q then on B it will be –Q

$$\therefore \quad W = \frac{1}{2} CV^2 \text{ joules}$$

provided C is in farads and V is in volts.

Also $$W = \frac{1}{2} QV \text{ joules} \qquad (\because C = Q/V)$$

$$= \frac{Q^2}{2C} \text{ joules} \qquad (\because V = Q/C)$$

If Q is in coulombs and C is in farads, then energy stored is given in joules.

(*b*) Second Method

For constant capacitance C, $V \propto Q$, hence the graph between V and Q is a straight line passing through the origin as shown in Fig. 6.25. The energy required to charge the capacitor with Q coulombs for a p.d. of V volts is given by the area of the triangle OAB which is equal to $\frac{1}{2}$ (base × height).

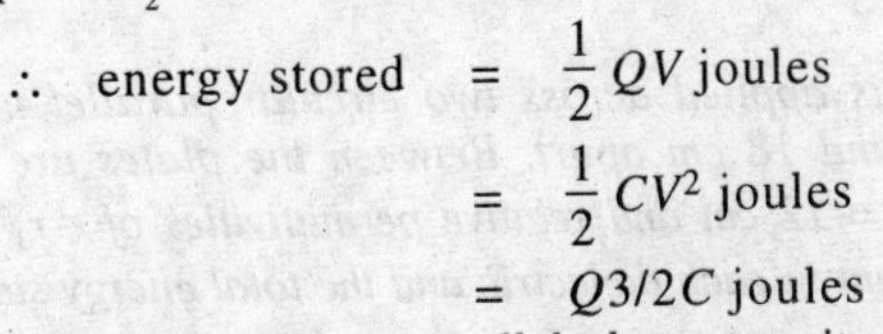

$$\therefore \text{ energy stored} = \frac{1}{2} QV \text{ joules}$$

$$= \frac{1}{2} CV^2 \text{ joules}$$

$$= Q3/2C \text{ joules}$$

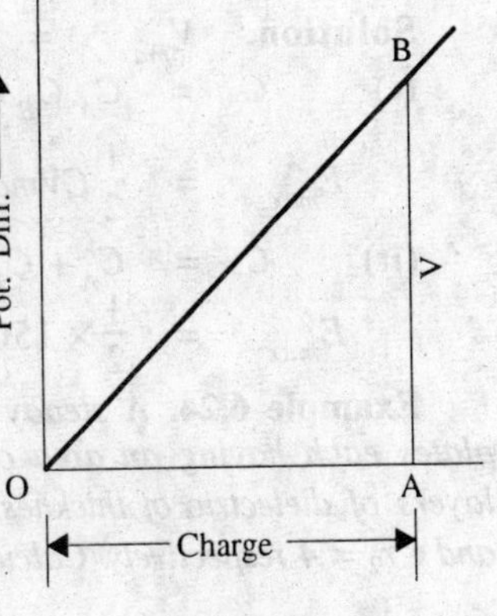

Fig. 6.25

Example 6.20. *A parallel-plate capacitor has 20 plates each of 0.5 m² area, the separation between the plates being 0.2 cm with a dielectric of relative permittivity 2.5.*

Calculate (i) the capacitance (ii) the value of the stored energy if it is charged to 1000 volts. **(A.M.I.E. Winter 1988)**

Solution. $$C = \frac{(n-1)\in_0 \in_r A}{d} \qquad \text{—Art 6.6}$$

$$= \frac{(20-1) \times 8.854 \times 10^{-12} \times 2.5 \times 0.5}{2 \times 10^{-3}} = \mathbf{105 \times 10^{-9}\ F}$$

Energy stored $$= \frac{1}{2} CV^2 = \frac{1}{2} \times 105 \times 10^{-9} \times 1000^2 = \mathbf{0.0524\ J}$$

Example 6.21. *A capacitor of capacitance 1 mF is charged to 10 kV and then discharged through a wire. Find the heat produced in the wire in calorie.* **(A.M.I.E. Winter 1991)**

Solution. $E = \frac{1}{2} CV^2 = \frac{1}{2} \times 1 \times 10^{-6} \times (10 \times 10^3)^2 = 50$ J

This stored energy is converted into heat during discharge. Now, 1 kcal = 4200 J.

$\therefore$ $H = 50/4200 = 0.019$ kcal. = **11.9 calorie.**

Example 6.22. *Two capacitors of capacitances 8 μF and 2 μF are connected in series across a 100–V d.c. supply. Now, if the supply voltage is removed and capacitors are then connected in parallel, what will be the final charge on each capacitor ?* **(A.M.I.E. Summer 1990)**

Solution. Since the two capacitors are in series (Art 6.12)

$$V_1 = V\frac{C_2}{C_1 + C_2} = 100 \times \frac{2}{8+2} = 20 \text{ V}$$

$$V_2 = V\frac{C_1}{C_1 + C_2} = 100 \times \frac{8}{10} = 80 \text{ V}$$

$$Q_1 = C_1V_1 = 8 \times 20 = 160\ \mu\text{C and } Q_2 = C_2V_2 = 2 \times 80\ 160\ \mu\text{C}$$

When the two capacitors are connected in parallel, they attain a common potential, say V, which is given by

$$V = \frac{\text{total charge}}{\text{total capacitance}} = \frac{2 \times 160}{8 + 2} = 32 \text{ V}$$

$$\therefore \quad Q_1 = C_1V = 8 \times 32 = \mathbf{256\ \mu C}$$

$$Q_2 = C_2V = 2 \times 32 = \mathbf{64\ \mu C}$$

Example 6.23. *Two capacitors A and B are placed (i) in series (ii) in parallel. Capacitor C_A = 100 μF and C_B = 50 μF. Find the maximum stored energy in the circuit when 240-V, 50-Hz supply is applied.* **(A.M.I.E. Winter 1989)**

Solution. $V_{max} = \sqrt{2}.V = 240\sqrt{2}.V$ Art. 16.13

(*i*) $C = C_A.C_B/(C_A + C_B) = 100 \times 50/150 = 100/3\ \mu F$

$$E_{max} = \frac{1}{2}CVmax^2 = \frac{1}{2}(100/3)(240\sqrt{2})^2 = \mathbf{1.92\ J}$$

(*ii*) $C = C_A + C_B = 150\ \mu F$,

$$E_{max} = \frac{1}{2} \times 150 \times (240\sqrt{2})^2 = \mathbf{8.64\ J}$$

Example 6.24. *A steady voltage of 5,000 V is applied across two circular parallel metal plates each having an area of one square metre and 18 cm apart. Between the plates are two layers of dielectric of thicknesses t_1 = 6 cm and t_2 = 12 cm and relative permittivities of $\in r_1 = 3$ and $\in r_2 = 4$ respectively. Calculate the voltage gradient in each dielectric and the total energy stored.*

Solution. $$C = \frac{\in_0 A}{\left(\frac{t_1}{\in_{r1}} + \frac{t_2}{\in_{r2}}\right)} = \frac{8.854 \times 10^{-12} \times 1}{\left(\frac{0.06}{3} + \frac{0.12}{4}\right)} = 1.77 \times 10^{-10}\text{ F}$$

$$\text{Total energy stored} = \frac{1}{2} \times 1.77 \times 10^{-10} \times 5,000^2 \text{ J} = \mathbf{22.1 \times 10^{-4}\ J}$$

Now, $D = \in_0 \in_{r1} E_1 = \in_0 \in_{r2} E_2$

$\therefore \quad 3E_1 = 4E_2$ ...(*i*)

Also $V = V_1 + V_2 = E_1t_1 + E_2t_2$

$\therefore \quad 5 \times 10^3 = 0.06\ E_1 + 0.12\ E_2$

or $6E_1 + 6E_2 = 5 \times 10^5$

$3E_1 + 6E_2 = 2.5 \times 10^5$ ...(*ii*)

From (*i*) and (*ii*), we get

$$E_2 = 2.5 \times 10^4 \text{ V/m} ; E_1 = \mathbf{3.33 \times 10^4\ V/m}$$

6.16. Energy Stored Per Unit Volume of the Dielectric

It has been shown in Art. 6.15 that energy stored in a capacitor is

$$E = \frac{1}{2}CV^2$$

In the case of a parallel-plate capacitor

$$C = \in A/d$$

$$\therefore \quad E = \frac{1}{2} \times \frac{\in A}{d}.V^2 = \frac{1}{2}\in A.D.\left(\frac{V}{d}\right)^2 = \frac{1}{2}\in E^2 \times Ad$$

Now $A \times d$ = volume of the dielectric medium.

$$\therefore \text{ energy/volume} = \frac{1}{2}\in E^2 = \frac{D^2}{2\in} \text{ J/m}^2$$

The above expression represents the energy required for creating a unit volume of an electric field having a flux denisty of D C/m².

6.17. Force of Attraction Between Oppositely-charged Plates

Fig. 6.24 shows a parallel-plate capacitor having a plate area of A m² and plate separation of d metres. Let the force of attraction between the two oppositely-charged plates M and N be F newton. Let plate N be pulled further apart against force F through a small distance dx. Work down is $F \times dx$. This energy goes to supply the necessary energy for the additional volume of the electric field produced.

Additional volume of electric field produced is

$$= A \times dx \text{ metre}^2$$

$$\text{Energy required} = A\,dx \times \frac{D^2}{2\epsilon}$$

$$\therefore \quad F \times dx = \frac{D^2}{2\epsilon} A\,dx \quad \text{or} \quad F = \frac{D^2 A}{2\epsilon} \text{ newton}$$

Fig. 6.26

Force per unit area of either plate is

$$= \frac{D^2}{2\epsilon} \text{ N/m}^2 = \frac{1}{2} \epsilon E^2 \text{ N/m}^2$$

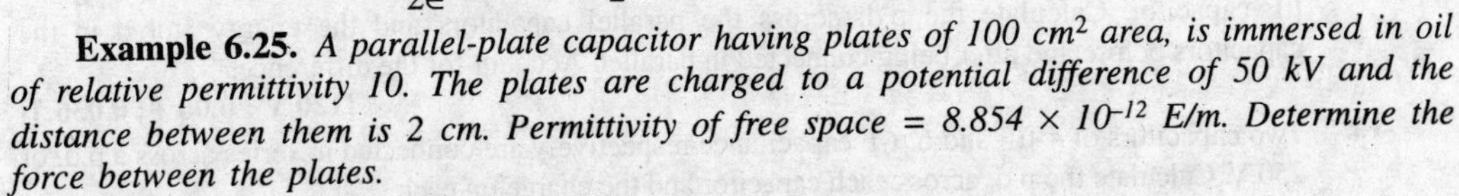

Example 6.25. *A parallel-plate capacitor having plates of 100 cm² area, is immersed in oil of relative permittivity 10. The plates are charged to a potential difference of 50 kV and the distance between them is 2 cm. Permittivity of free space = 8.854 × 10⁻¹² E/m. Determine the force between the plates.*

Solution. $F = \frac{1}{2} \epsilon_0 \epsilon_r A E^2$ newton

Here, $E = 50 \times 10^3/2 \times 10^{-2} = 25 \times 10^5$ V/m ; $\epsilon_r = 10$

$A = 100 \text{ cm}^2 = 10^{-2} \text{ m}^2$; $\epsilon_0 = 8.854 \times 10^{-12}$ F/m

$$\therefore \quad F = \frac{1}{2} \times 8.854 \times 10^{-12} \times 10 \times 10^{-2} \times (25 \times 10^5)^2 = \mathbf{2.77\ N}$$

Tutorial Problems No. 6.3

1. A submarine cable is 3,000 km long. The inner core is of 0.4 cm diameter. The outside diameter of the gutta-percha is 1 cm. If relative permittivity of gutta-percha is 4.05, find the capacitance of the cable. **[739 μF]**
2. A circuit comprises of a 2 μF capacitor in parallel with a 3 μF capacitor and a 5 μF capacitor is in series with this group. Calculate the effective capacitance of the circuit.
   A direct voltage of 100 V is applied to the above circuit. Calculate the p.d. across each capacitor and the charge stored by each capacitor. Show how the capacitors could be connected to give an effective capacitance of 6.2 μF. **[2.5 μF ; 250 μC, 50 V; 100 μC ; 150 ; μC]**
3. Two initially uncharged capacitors, of 1 and 2 μF are connected in series across a 100 V source. Calculate :
   (*a*) the capacitance of the two in series, (*b*) the voltage across each capacitance
   (*c*) the energy stored in each. **[(*a*) $\frac{2}{3}$ μF (*b*) 66.7 V ; 33.3 V (*c*) 2.22 mJ ; 1.11 mJ]**
4. Two perfectly-insulated capacitors are connected in series ; one is an air capacitor with a plate area 100 cm^2, the plates being 1 mm apart. The other has a plate area of 10 cm^2, the plates being separated by a dielectric 0.1 mm thick with a relative permittivity of 5. Find theh p.d. across the combination if the potential gradient in the air capacitor is 20 kV/cm. **[2,4000 V]**
5. Four perfect capacitors are connected as shown to the three terminals *A, B* and *C* (Fig. 6.25).
   (*a*) Calculate the value of capacitance that would be measured across terminals *BC*.
   (*b*) Terminals *B* and *C* are now connected together. What capacitance would be measured across *AB?*
   (*c*) Find the energy stored in the whole circuit in the second condition when a 100-V battery is connected across *AB*. **[(*a*) $\frac{30}{61}$ μF (*b*) $\frac{61}{30}$ μF (*C*) 10.2 mj]**
6. Two parallel plates spaced 1.0 cm apart in air, each have an area of 1.0 m^2. If the p.d. between them is 1,000 V, calculate
   (*a*) the energy stored in joules , (*b*) the force between them

If the dielectric were mica with $\in_r = 7$ instead of air, calculate the new values of energy and force.
[(*a*) 442 mJ (*b*) 0.309 N]

7. A parallel-plate capacitor is immersed in alcohol of $\in_r = 25$. The plates are charged to a p.d. of 20 kV and the distance between the plates is 15mm. Determine per m^2 of plate area the force between the plates and the energy stored.
[197.9 N/m^2 ; 2.95 J] (*A.M.I.E. May 1979*)

8. Two capacitors of capacitance 2 μF and 4μF respectively are connected in series. A potential difference of 900 volts is applied between the extreme terminals. Find the p.d. across each capacitor.
[600 V, 300 V] (*A.M.I.E. Winter 1983*)

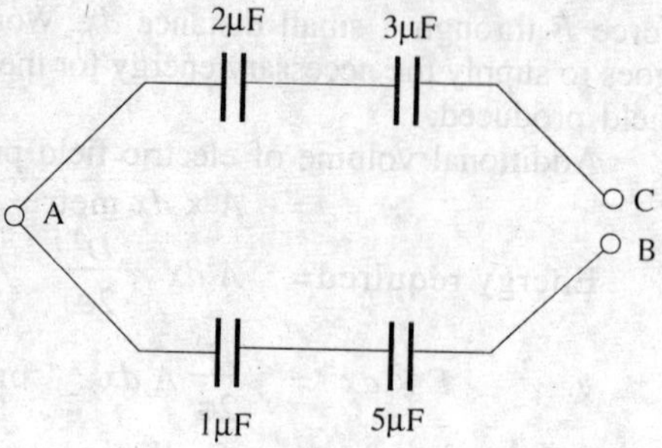

2μF 3μF A C B 1μF 5μF

Fig. 6.27

9. A 3-μF capacitor is charged to a p.d. of 200 V and then connected in parallel with an uncharged 2 μF capacitor. Calculate the p.d. across the parallel capacitors and the energy stored in the capacitors before and after being connected in parallel. Account for the difference.
[120 V ; 0.06 J ; 0.036.J]

10. Two capacitors of 4-μF and 6.μ-F capacitance respectively are connected in series across a p.d. of 250 V. Calculate the p.d. across each capacitor and the charge on each.
The capacitors are disconnected from the supply p.d. and reconnected in parallel with each other, with terminals of similar polarity being joined together. Calculate the new p.d. and charge for each capacitor.
What would have happened if, in making the parallel connection, the connections of one the capacitors had been reserved ?
[150 V ; 100 V ; 600 μC ; 120 V, 480 μC ; 720 μC]

11. A 9-μF capacitor is connected in series with two capacitors, 4-μF and 2-μF respectively, which are connected in parallel. Determine the capacitance of the combination.
If a p.d. of 20 V is maintained across the combination, determine the charge on the 9-μF capacitor and the energy stored in the 4-μF capacitor.
[3.6 mF ; 72 mC, 288 mJ]

12. A circuit consists of two capacitors *A* and *B* in parallel connected in series with another capacitor *C*. The capacitances of *A, B* and *C* are 6 μF, 10μF and 16 μF respectively. When the circuit is connected across a 400-V d.c. supply, calculate : (*i*) the potential difference across each capacitor (*ii*) the charge on each capacitor.
[(*i*) 200 V (*ii*) 1.2 μC ; 2μC ; 3.2 μC]

13. Two capacitors, *A* and *B*, having capacitances of 20 μF and 30 μF respectively, are connected in series to a 600-V d.c. supply. Determine the p.d. across each capacitor.
If a third capacitor *C* is connected in parallel with *A* and it is then found that the p.d. across *B* is 400 V, calculate the value of *C* and the energy stored in it.
[360 V ; 240 V ; 40 μF ; 0.8 J]

6.18. Charging of a Capacitor

In Fig. 6.28 is shown an arrangement by which a capacitor *C* may be charged through a high resistance *R* from a battery of *V* volts. The voltage across *C* can be measured by a suitable voltmeter. When switch *S* is connected to '*a*', *C* is charged but when it is connected to '*b*", *C* is short-circuited through *R* and is thus discharged. The voltage across *C* does not rise to *V* instantaneously but builds up slowly. Charging current is maximum at the start *i.e.* when *C* is uncharged, then it gradually decreases and finally ceases when p.d. across capacitor plates becomes equal and opposie to that of the battery. At the time of switching on the circuit, charging current is maximum. Hence, whole of *V* drops across *R* and none across *C*. When *C* becomes fully charged , *i* is zero. Hence, there is no drop across *R*. The whole of battery voltage appears aacross *C*. At any instant, let

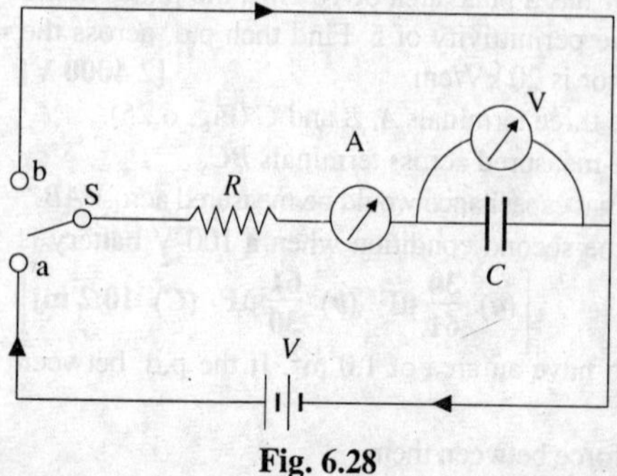

Fig. 6.28

v = p.d. across C; i = charging current; q = charge on capacitor plates

The applied voltage V is always equal to the sum of

(*i*) resistive drop (iR), (*ii*) voltage across capacitor (v)

$$\therefore \quad V = iR + n \qquad ...(i)$$

$$\text{Now} \quad i = \frac{dq}{dt} = \frac{d}{dt}(Cv) = C\frac{dv}{dt}$$

$$\therefore \quad V = v + CR\frac{dv}{dt} \qquad ...(ii)$$

$$\therefore \quad \frac{dv}{V - v} = -\frac{dt}{CR}$$

Intergrating both sides, we get $-\int d\frac{v}{V-v} = -\frac{t}{CR}\int dt$

$$\therefore \quad \log_e (V - v) = -\frac{t}{RC} + K \qquad ...(iii)$$

where K is the constant of integration whose value can be found from initial known conditions. We know that at the start of charging when $t = 0$, $v = 0$.

Substituting these values in Eq. (*ii*), we get $\log_e V = K$

Hence, Eq. (*iii*) becomes

$$\log_e(V - v) = \frac{-t}{CR} + \log_e V$$

$$\therefore \quad \log_e \frac{V - v}{V} = \frac{-t}{CR} = -\frac{t}{\lambda}$$

where $\lambda = CR$ = time constant

$$\therefore \quad \frac{V - v}{V} = e^{-t/\lambda} \quad \text{or} \quad v = V(1 - e^{-t/\lambda}) \qquad ...(iv)$$

This gives variation with time of voltage across the capacitor plates and is shown in Fig. 6.29.

Since $v = \frac{q}{C}$ and $V = \frac{Q}{C}$

$\therefore$ Equation (iv) becomes

$$\frac{q}{C} = \frac{Q}{C}(1 - e^{-e/\lambda})$$

$$\therefore \quad q = Q(1 - e^{-t/\lambda}) \qquad ...(v)$$

We find that the increase of charge, like growth of potential, follows an exponential law in which the steady value is reached after infinite time.

$$\text{Now} \quad i = \frac{dq}{dt}$$

$\therefore$ from equation (v), differentiating both sides, we get

$$\frac{dq}{dt} = i = Q\frac{d}{dt}(1 - e^{-t/\lambda})$$

$$= Q\left(+\frac{1}{\lambda}e^{-t/\lambda}\right) = \frac{Q}{\lambda}e^{-t/\lambda}$$

$$= \frac{CV}{CR}e^{-t/\lambda}$$

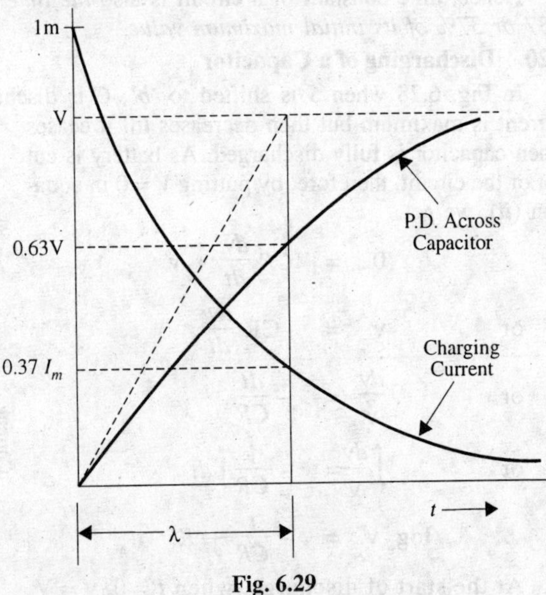

Fig. 6.29

($\because Q = CV$ and $\lambda = CR$)

$\therefore \quad t = \dfrac{V}{R} e^{-t/\lambda}$ or $i = I_m e^{-t/\lambda}$ ...(*vi*)

where I_m = maximum current = V/R

As charging continues, charging current decreases according to equation (*vi*) as shown in Fig. 6.27.

It may be noted that current reaches its maximum value in about five time constants of the circuit.

6.19. Time Constnat

(*a*) Just at the start of charging, p.d. across capacitor is zero, hence from Eq. (*ii*), putting v = 0, we get $V = CR\dfrac{dv}{dt}$

$\therefore$ initial rate of rise of voltage $\dfrac{dv}{dt} = \dfrac{V}{CR}$ volt/second

If this rate of rise were **maintained,** then time taken to reach voltage *V* would **have been** = $V \div V/CR = CR$. This time is known as time constant (λ) of the circuit.

(*b*) From equation (iv), we find that if $t = \lambda$, then

$$v = V(1 - e^{-t/\lambda}) = V(1 - e^{-l/\lambda}) = V(1 - e^{-1})$$

$$= V\left(1 - \frac{1}{e}\right) = V\left(1 - \frac{1}{2.718}\right)$$

$$\therefore \quad v = 0.632\text{ V}$$

Hence, time constant may be defined as *the time during which capacitor voltage rises to 0.632 or 63.2% of its final steady value.*

(*c*) From equation (*vi*) by putting $t = \lambda$, we get

$$i = I_m e^{-\lambda/\lambda} = I_m e^{-1}$$

$$i = \frac{I_m}{2.718} = 0.37\, I_m \;;\; i = 0.37\, I_m$$

Hence, time constant of a circuit is also *the time during which the charging current falls to 0.37 or 37% of its initial maximum value.*

6.20. Discharging of a Capacitor

In Fig. 6.28 when *S* is shifted to '*b*', *C* is discharged through *R*. To begin with, discharge current is maximum but then decreases till it ceases when capacitor is fully discharged. As battery is cut out of the circuit, therefore, by putting *V* = 0 in equation (*ii*), we get

$$0 = CR\frac{dv}{dt} + v$$

or $$v = -CR\frac{dv}{dt}$$

or $$\frac{dv}{v} = -\frac{dt}{CR}$$

or $$\int\frac{dv}{v} = -\frac{1}{CR}\int dt$$

$$\therefore \quad \log_e v = -\frac{t}{CR} + K$$

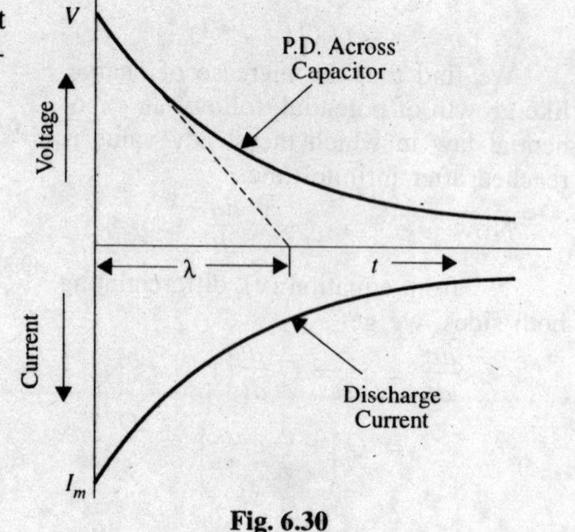

Fig. 6.30

At the start of discharge, when $t = 0$, v = V

$\therefore \quad \log_e V = 0 + K$ or $\log_e V = K$

Putting this value above, we get

$$\log_e V = -\frac{t}{\lambda} + \log_e V$$

or $\log_e v/V = -t/\lambda$

or $\frac{v}{V} = e^{-t/\lambda}$ or $v = V e^{-t\backslash\lambda}$

Similarly $q = Q e^{-t/\lambda}$ and $i = -I_m e^{-t/\lambda}$

The fall of potential and the discharging current are shown in Fig. 6.30.

Example 6.26. *A capacitor is charged from a d.c. source through a resistor of 0.5 megohm. If the p.d. across it reaches 75% of its initial value in half a second, find its capacitance.*

Solution. Using the relation $v = V(1 - e^{-t/\lambda})$

we have $v = 0.75\ V$ — *V* being the applied voltage

$t = 0.5$ second

$\therefore\ 0.75\ V = V(1 - e^{-1/2\lambda})$ or $e^{-1/2\lambda} = 0.25$

$\therefore\ \frac{-1}{2\lambda} = \log_e 0.25$ or $\lambda = 0.3612$ second (approx)

Now $\lambda = CR$

When *C* is in farads and *R* in ohms, they λ is in seconds.

$\therefore\ C \times 0.5 \times 10^6 = 0.3612$

$C = 0.3612/500{,}000$ farad

$= 0.7224 \times 10^{-6}$ farad = **0.7224 μF**

Example 6.27. *An 8 μF capacitor is connected in series with a 0.5 MΩ resistor across a 200 V d.c. supply. Calculate (i) the time constant (ii) the initial charging current (iii) the time taken for the p.d. across the capacitor to grow to 160 V and (iv) the current and the p.d. across the capacitor in 4 seconds after it is connected to the supply.*

(Electrical Science, AMIE Summer, 1993)

Solution. (*i*) Time constant, $\lambda = CR = 8 \times 10^{-6} \times 0.5 \times 10^6 = 4$ second

(*ii*) Initial charging current is maximum because at the start, there is no voltage across the capacitor to oppose the applied voltage. Initial current is restricted by *R* alone.

$\therefore\ I_m = V/R = 200/0.5 \times 10^6 = 0.4 \times 10^{-3}$ A = **0.4 mA**

(*iii*) $\Omega = V(1 - e^{-t/4})$ or $160 = 200(1 - e^{-t/4})$

$\therefore\ 1 - e^{-t/4} = 0.8$ or $e^{t/4} = 5$ or $(t/4)\log_e e = \log_e 5$

or $t/4 = 1.61$ or t = **6.44 seconds.**

(*iv*) Since the given time of 4 seconds represents the time constant of the circuit. Hence, capacitor voltage will raise to 63.2% of its maximum voltage and its charging current will fall to 37% of its maximum initial value. Hence, after 4 seconds.

$v = 0.632 \times 200 =$ **126.4 V**

$i = 0.37\ I_m = 0.37 \times 0.4 =$ **0.1448 mA**

Example 6.28. *A resistor R and a 2-μF capacitor are connected in series across a 200–V d.c. supply. Across the capacitor is connected a neon lamp that strikes at 120 V. Calculate R to make the lamp strike 5 second after the switch is closed.*

Solution. The circuit is shown in Fig. 6.31 (*a*). The neon lamp lights up as soon as voltage across the capacitor rises to 120 V. Now, the equation for the rise of voltage across the capacitor is

$v = V(1 - e^{-t/l})$ where $\lambda = CR$

Here $t = 5$ second ; $V = 200$ V, $v = 120$ V

$\therefore\ 120 = 200(2 - e^{-5/\lambda})$

or $e^{-5/\lambda} = 0.4$ or $e^{5/\lambda} = 2.5$

or $5/l = \log_e 2.5 = 2.3 \times \log_{10} 2.5 = 2.3 \times 0.3979$

$\therefore\ \lambda = 5.464$ second

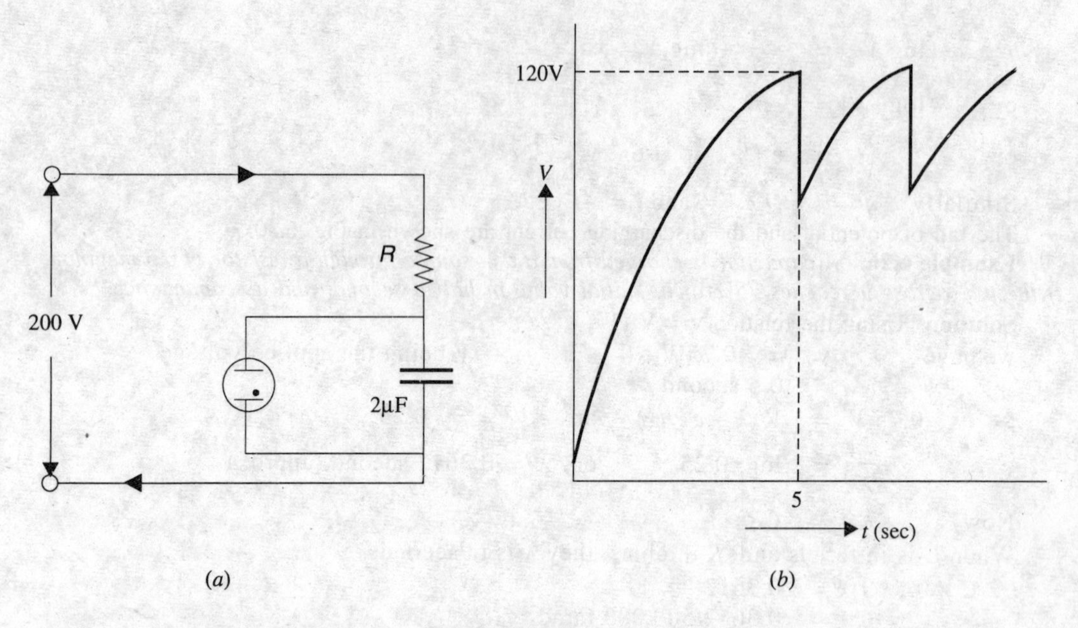

Fig. 6.31

$\therefore \quad R = 5.464/2 \times 10^{-6} = 2.73 \times 10^6\ \Omega$ = **2.73 MΩ**

Example 6.29. *If a capacitor of 1μF and resistance of 82 kΩ are connected in series with an e.m.f. of 100 V, calculate the magnitude of energy and the time in which the energy stored in the capacitor will reach half of its equilibrium value.* **(A.M.I.E. Summer 1989)**

Solution. Equilibrium value of stored energy

$$= \frac{1}{2}CV^2 = \frac{1}{2} \times 1 \times 10^{-6} \times 100^2 = 0.005 \text{ J}$$

Half of this energy = 0.0025 J

Since stored energy varies as V^2, half of the equlibrium energy would be stored in that time during which capacitor voltage rises to $V/\sqrt{2}$ or $100/\sqrt{2}$ or 70.7 V

Now, $\quad v = V(1 - e^{-t/\lambda})$

Now, $\quad v = 70.7$ V; V = 100 V ; $\lambda = CR = 0.082$ second s

$\therefore \quad 70.7 = 100\,(1 - e^{-t/0.082})$

$\therefore \quad t$ = **0.1 second.**

6.21. Leakage in a Capacitor

An ideal capacitor is one whose dielectric medium is a perfect insulator *i.e.* has infinite resistance. Once charged, it remains in this full-charged condition for ever. Actually, no insulating material is perfect. Hence, dielectric of every capacitor conducts a small amount of current called leakage current. The magnitude of the leakage current depends on the insulation resistance of the dielectric. That is why the charge of an acutal or non-ideal capacitor leaks off after some time. Such a non-ideal capacitor can be represented by an ideal capacitor connected in parallel with an extremely high resistor (Fig. 6.32). This parallel resistor represents the resistance of the dielectric through which leakage current flows.

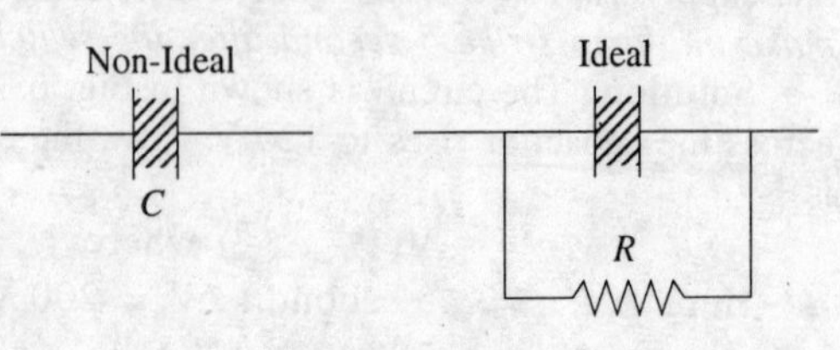

Fig. 6.32

Example 6.30. *A 20-μF capacitor is found to have an insulation resistance of 50 MΩ measured between the terminals. If this capacitor is charged off a dc supply of 230-V, find the time*

required after disconnection from the supply, for the p.d. across the capacitor to fall to 60V.

Solution. Obviously, the fully-charged capacitor discharges through its insulation resistance.

Here $V = 230$ V ; $v = 60$ V, $\lambda = CR = 20\times 10^{-6} \times 50 \times 10^6$

$= 1000$ second

Now, $v = V_e - t/\lambda$ or $60 = 230\, e^{-t/1000}$

or $ct/1000 = \frac{23}{6}$ or $\frac{t}{1000} \cdot \log_e e = \log_e 23/4 = 1.343$

$t =$ **1343 second**

Tutorial Problems No. 6.4

1. A capacitor of 100 μF is connected in series with an 8 kΩ resistor. Estimate the time constant of the circuit. If the combination is connected suddenly to a 100V d.c. supply, find (*i*) the initial rate of rise of the p.d. across the capacitor (*ii*) the initial charging current and (*iii*) the ultimate charge on the capacitor. **[0.8 second (i) 125 V/s (*ii*) 12.5 mA (*iii*) 0.01 C]**
2. A capacitor of 2 μF capacitance is joined in series with a 2-MΩ resistance to a d.c. supply of 100V. Calculate the current flowing and the energy stored in the capacitor at the end of an interval of 4 seconds from the start. **[18.4 μA ; 4 mJ]**
3. A capacitor of 100 μF is charged through a 500Ω resistance from a d.c. supply of 1000 V. What period after switching on will elapse before the capacitor acquires 50% of its final charge? What will be the maximum rate of change of current ? **[34.6 ms ; 40 A/s decreasing]**
4. A 10–μ F capacitor in series with a 10-kΩ resistor is connected across a 500-V d.c. supply. The fully charged capacitor is disconnected from the supply and discharged by connecting a 1000-Ω resistor across its terminals. Calculate : (*a*) the initial value of the charging current (*b*) the initial value of the discharge current and (*c*) the amount of heat, in joules, dissipated in the 1000-Ω resistor. **[50 mA ; 500 mA ; 1.25 J]**

HIGHLIGHTS

1. Capacitance of a capacitor is given by the ratio of its charge to the p.d. between its plates *i.e.*

$$C = Q/V \text{ farad}$$

2. Capacitance of a spherical capacitor is

$$C = 4\pi\epsilon_0\epsilon_r \cdot \frac{ab}{b-a} \text{ F}$$

3. Capacitance of a parallel-plate capacitor is given by

$$C = \frac{\epsilon_0\epsilon_r A}{d} \quad \text{— with a medium}$$

$$= \frac{\epsilon_0 A}{d} \quad \text{— with air}$$

4. For a composite medium, the capacitance is given by

$$C = \frac{\epsilon_0 A}{\Sigma d/\epsilon_r} \text{ farad}$$

5. Capacitance of a multiple parallel-plate capacitor having *n* plates is given by

$$C = \frac{(n-1)\,\epsilon_0\epsilon_r A}{d} \text{ farad}$$

6. Capacitance of a cable capacitor is given by

$$C = \frac{2\pi\epsilon_0\epsilon_r l}{2.3\log_{10} b/a}$$

Also, $$C = \frac{0.024\epsilon_r}{\log_{10} b/a} \mu\text{F/km}$$

7. Electric stress or potential gradient in a cable capacitor is given by

$$g = \frac{V}{2.3x \log_{10} b/a} \text{volt/metre}$$

8. Capacitance of combined capacitors is given by

$$C = C_1 + C_2 \quad \text{—when in parallel}$$

$$C = \frac{C_1 C_2}{C_1 + C_2} \quad \text{—when in series}$$

or, in general $$\frac{1}{C} = \frac{1}{C_1} + \frac{1}{C_2} + \frac{1}{C_3} + \quad \text{—when in series}$$

9. Insulation resistance of a cable capacitor is given by

$$C = \frac{2.3\rho}{2\pi l} \log_{10} r_2/r_1 \ \Omega$$

10. Energy stored in a capacitor is

$$E = \frac{1}{2} CV^2 \text{ joules}$$

11. Energy stored per unit volume of a dielectric is

$$E = \frac{\epsilon E^2}{2} = \frac{D^2}{2\epsilon} \text{ J/m}^2$$

12. Force is attraction between two oppositely-charged plates is

$$F = \frac{D^2 A}{2\epsilon} \text{ newton}$$

13. Time constant of an R-C circuit is defined as

(*i*) the time during which capacitor voltae rises to 63.2% of its final steady value.

Or

(*ii*) the time during which the chaging current falls to 37% of its initial maximum value.

14. During the chaging of a capacitor

(*a*) voltage across it rises according to the equation

$$v = V(1 - e^{-t/\lambda})$$

(*b*) charge increase according to the relation

$$q = Q(1 - e^{-t/\lambda})$$

(*c*) the charging current decreases according to the equation

$$i = I_m e^{-t/\lambda}$$

15. During the discharging of a capacitor

(*i*) voltage across it decreases according to the equation

$$v = Ve^{-t/\lambda}$$

(*ii*) charge decreases according to the equation

$$q = Q e^{-t/\lambda}$$

(*iii*) current decreases according to the equation.

$$i = -I_m e^{-t/\lambda}$$

OBJECTIVE TESTS—6

A. Fill in the following blanks :

1. The unit of capacitance is ....
2. The capacitance of a parallel-plate capacitor is given by the amount of charge required to create a .... potential difference between its plates.
3. Electrolytic capacitors are generally used in .... circuits in radio work.

4. In a cylindrical capacitor, maximum potential gradient occurs at the ....surface of the inner core.
5. Increased capacitance can be obtained by connecting the given capacitors in ....
6. The insulation resistance of a cable capacitor .... with increase in its length.
7. During the time constant, the voltage across the capacitor being charged rises to about .... percent of its final steady value.
8. A capacitor becomes almost discharged in about .... time constants.

B. Answer True or False :

1. An isolated single sphere possesses no capacitance.
2. Capacitance of a parallel-plate capacitor increases with increase in its plate separation.
3. The capacitance of a multiplate capacitor having *n* plates is *n* times the capacitance of a single capacitor.
4. Ceramic capacitors are suitable for short-wave work in radio.
5. In a cylindrical capacitor, minimum potential gradient occurs at the outer surface of its outer cylinder.
6. Total capacitance may be increased by joining the given capacitors in series.
7. Energy is required for charging a capacitor.
8. Rate of rise of voltage during charging of a capacitor is the same as its rate of decrease during discharging.

C. Multiple Choice Questions

1. The unit of capacitance is
   (*a*) farad (*b*) coulomb
   (*c*) volt (*d*) metre.
2. Paper capacitors are suitable for
   (*a*) short-wave work in radio
   (*b*) smoothing circuits in radio work
   (*c*) audio-frequency stages of radio receivers
   (*d*) use at radio frequencies.
3. In a cylindrical capacitor, maximum potential gradient occurs at the
   (*a*) outer surface of outer cylinder
   (*b*) outer surface of inner cylinder
   (*c*) cylinder axis
   (*d*) inner surface of the outer-cylinder.
4. When a 2-μF capacitor is series-connected with a parallel combination of 6 and 3 μF capacitors, the total capacitance is .... μF.
   (*a*) 2 (*b*) 11
   (*c*) 4 (*d*) 1
5. The insulation resistance per km of a single-core cable capacitor is 4×10^{12} Ω. Its value for 2 km length would be .... $\times 10^{12}$ Ω.
   (*a*) 8 (*b*) 16
   (*c*) 1 (*d*) 2
6. The time constant of an R-C circuit is defined as the time during which capacitor charging current becomes............percent of its............value.
   (*a*) 37, final (*b*) 63, final
   (*c*) 63, initial (*d*) 37, initial

ANSWERS

A. 1. Farad 2. unit 3. smoothing/filtering 4. outer 5. parallel 6. decreases 7. 63.2 8. five

B. 1. F 2. F 3. F 4. T 5. T 6. F 7. T 8. T

C. 1. *a* 2. *c* 3. *b* 4. *c* 5. *d* 6. *d*

7 MAGNETISM AND ELECTROMAGNETISM

7.1. Magnetic Field

The space or region around a magnet which is permeated by the lines of force and within which conductors carrying electric current are perceptibly influenced, is conventionally called the magnetic field of force or simply a magnetic field. It is assumed that lines of force emanate from a *N*-pole, pass through the surrounding medium, re-enter the *S*-pole and complete their path from *S* to *N* pole through the body of the magnet. Since every line of force must have a complete circuit, it is impossible to get a magnet having only one pole. These lines of force complete their paths independently and never cut or cross or merge into each other.

7.2. Pole Strength

It is well-known that a magnet can be made by stroking an iron piece with a permanent magnet or by placing an iron bar inside a long coil of insulated wire (*i.e.* a solenoid) carrying a strong current. It is found that the magnet made by the electrical method (known as electromagnet) is capable of raising a heavier load than the magnet made by stroking, showing that the poles are stronger in the first case.

In order to deal with pole strength, we need a unit of the strength or a 'unit pole' just as we need the definition of the mass of one kg for measuring the masses of different objects. As discussed in the next article, a unit pole-strength is defined as the strength of that pole which when placed in vacuum at a distance of 1 meter from a similar and equal pole repels it with a force of 1/4 μ_0 newton where

$$\mu_0 = \text{permeability of free space} = 4\pi \times 10^{-7} \text{ H/m}$$

7.3. Laws of Magnetic Force

Coulomb was the first to determine experimentally the quantitative expression for the magnetic force between two isolated point poles. It may be noted here that, in view of the fact that magnetic poles always exist in pairs, it is impossible, in practice, to get an isolated pole. The concept of an isolated pole is purely theoretical. However, poles of a thin but long magnet may be assumed to be isolated point poles for all practical purposes (Fig. 7.1). By using a torsion balance, he found that the force between two magnetic poles placed in a medium is

(*i*) directly proportional to their pole strengths,

(*ii*) inversely proportional to the square of the distance between them,

(*iii*) inversely proportional to the absolute permeability of the surrounding medium.

For example, if m_1 and m_2 represent the magnetic strengths of the two poles (their unit as yet being undefined) and *d* the distance in metres between them (Fig. 7.2), then the force *F* is given by the expression :

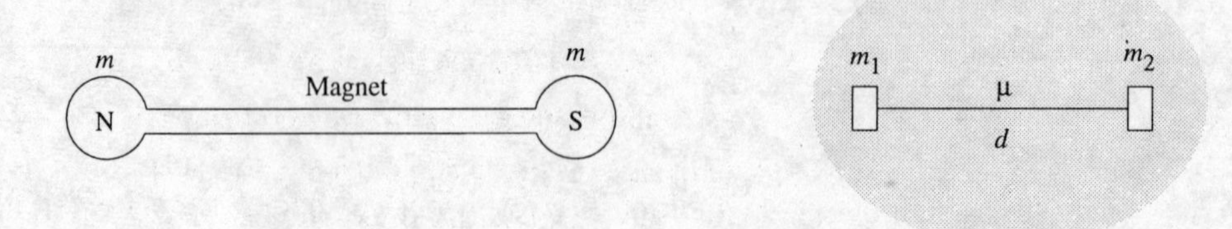

Fig. 7.1 Fig. 7.2

$$F \propto \frac{m_1 m_2}{\mu d^2} = k \frac{m_1 m_2}{\mu d^2}$$

where k is a constant of proportionality whose value depends on the system of units employed. In the M.K.S.A. as well as SI system, value of $k = 1/4\pi$

$$\therefore \quad F = \frac{m_1 m_2}{4\pi \mu d^2} = \frac{m_1 m_2}{4\pi \mu_0 \mu_4 d^2}$$

If the two poles are placed in vacuum or air, then

$$F = \frac{m_1 m_2}{4\pi \mu_0 d^2}$$

If, in the above equation,

$$m_1 = m_2 = m \text{ (say)} ; \quad r = 1 \text{ metre}$$

$$F = \frac{1}{4\pi \mu_0} \text{ newton}$$

Then $m^2 = 1$ or $m = \pm 1$ weber*

Hence, a unit magnetic pole may be defined as that pole which when placed in air at a distance of one metre from a similar and equal pole repels it with a force of $1/4\pi\mu_0$ newtons.**

7.4. Field Intensity or Field Strength or Magnetising Force (H)

Field strength at any point within a magnetic field is measured by the force experienced by a *N*-pole of one weber placed at that point. Hence, unit of H is N/Wb.***

Suppose it is required to find the field intensity at a point A distant d metres from a pole of strength m webers lying in air. For that purpose, imagine a pole of one weber placed at that point A. That force experienced by this pole is

$$F = \frac{m \times 1}{4\pi \mu_0 d^2} \text{ newton}$$

$$\therefore \quad H = \frac{m}{4\pi \mu_0 d^2} \text{ N/Wb (or AT/m)}$$

Also, if a pole of m Wb is placed in a uniform field or strength H N/Wb, then force experienced by the pole is = mH newtons.

It should be noted that field strength is a vector quantity having both magnitude and direction.

7.5. Magnetic Potential (M)

The magnetic potential at any point within a magnetic field is measured by the work done in shifting a *N*-pole of one weber from infinity to that point against the force of magnetic field. Its value is

$$M = \frac{m}{4\pi \mu_0 d} \text{ J/Wb}$$

It is a scalar quanity.

7.6. Magnetic Flux (Φ)

In the SI system of units, a unit *N*-pole is supposed to radiate out flux of one weber. Its symbol is Φ. Therefore, the flux coming out of a *N*-pole of m weber is given by

$$\Phi = m \text{ Wb}$$

---

*To commemorate the memory of German physicist Wilthelm Edward Weber (1804-1891).

**A unit magnetic pole is also defined as that magnetic pole which when placed at a distance of one metre from a very long straight conductor carrying a current of one ampere experiences a force of $1/2\pi$ newtons.

***It should be noted that N/Wb is the same thing as ampere-turn/m or ampere/metre.

7.7. Flux Density (B)

Flux density at a point is given by the magnetic flux passing per unit cross-section at that point.

It is usually designated by the capital letter B and is measured in weber/metre^2. The recommended name for the unit weber/metre^2 is tesla (T).

If Φ is the total flux passing through an area of A m^2, then

$$B = \Phi/A \quad \text{Wb/m}^2 \text{ or tesla (T)}$$

7.8. Magnetic Induction

A magnet can impart magnetism to a magnetic substance without actually coming in physical contact with it. For example, if an unmagnetised piece of soft iron be placed in the field of a magnet, it also becomes magnetised. The iron piece is under induction, the magnet is the inducing body and this phenomenon is known as *magnetic induction*. When magnetism is developed by induction, at least two poles are produced in the body. The two poles, one of the body and the other of the inducing magnet which are nearest to each other, are unlike poles (Fig. 7.3). Magnetic induction can be easily explained on Molecular Theory of Magnetism (Art. 7.13).

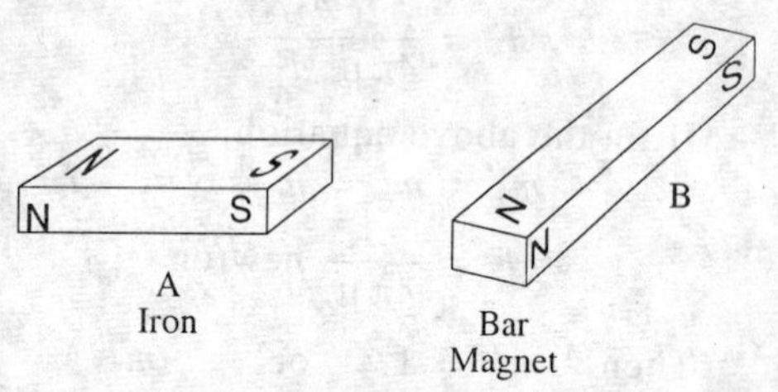

Fig. 7.3

7.9. Absolute Permeability (μ) and Relative Permeability (μ_r)

In Fig. 7.4 is shown a bar of magnetic material, say, iron placed in a uniform field of strength H N/Wb (or AT/m). The iron bar is magnetised by induction. Suppose it develops a polarity of m webers. Then, its own flux is m webers. These lines of flux emanate from its N-pole, re-enter its S-pole and continue from S to N-pole within the magnet. These lines are seen to be in opposition to the lines of force of the main field H outside the magnet but in the same direction within it. The resultant field is shown in Fig. 7.5.

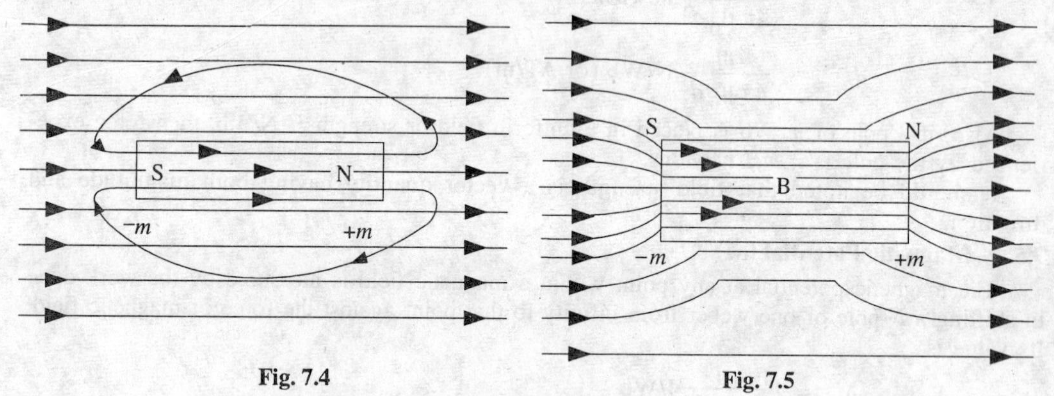

Fig. 7.4 Fig. 7.5

Let Φ = total flux through the bar in Wb

A = pole area of the bar in m^2

The flux density established through the bar is

$$B = \Phi/A \text{ Wb/m}^2$$

The absolute permeability of a medium is given by the ratio

$$\mu = \frac{B}{H} = \frac{\text{flux density}}{\text{magnetisingforce}}$$

$$B = \mu H$$

Now, suppose a magnetising force of H is set up first in vacuum (or air) and then in some other medium. The flux density established by it is

$$B_0 = \mu_0 H \qquad \text{—in vacuum}$$

and $B = \mu_0 \mu_r H$ —in that medium

The relative permeability of that medium is given by

$$\mu_r = \frac{\text{flux density in the medium}}{\text{flux density in vacuum}} \quad \text{—for same } H$$

$$= \frac{B}{B_0} \quad \text{— for same } H$$

Being a mere ratio, it has no units.

It is found that for ferromagnetic materials μ_r is considerably greater than unity, for paramagnetic materials is slightly greater than unity and for diamagnetic materials slightly less than unity.

7.10. Intensity of Magnetisation (J or I)

It is defined as the pole strength developed per unit area of the bar. Also, it is the **magnetic** moment developed per unit volume of the bar.

Let m = pole strength induced in the bar in Wb

A = face or pole area of the bar in m^2

Then $I = \frac{m}{A}$ Wb/m^2

Now, a pole of m webers, produces a flux of m webers. Hence, intensity of magnetisation of a substance may alo be defined as *the flux density produced in it due to its own magnetism.* It is known as induction flux density B_i.

If l is the magnetic length of the bar, then the product $(l \times m)$ is known as its magnetic moment M.

$$\therefore \quad I = \frac{m}{A} = \frac{m \times l}{A \times l} = \frac{M}{V} = \frac{\text{magnetic moment}}{\text{Volume}}$$

7.11. Susceptibility (K)

Susceptibility is defined as the ratio of intensity of magnetisation I to the magnetising force H.

$\therefore \quad K = I/H$ henry/metre.

7.12. Relation Between B, H, I and K

It is obvious from the above discussion that flux density B in a material is given by

$$B = B_0 + B_i = B_0 + m/A \qquad \therefore \quad B = \mu_0 H + I$$

Now, absolute permeability

$$\mu = \frac{B}{H} = \frac{\mu_0 H + I}{H} = \mu_0 + \frac{I}{H}$$

$$\therefore \quad \mu = \mu_0 + K; \quad \text{Also} \quad \mu = \mu_0 \mu_r$$

$$\therefore \quad \mu_0 \mu_r = \mu_0 + K$$

$$\therefore \quad \mu_r = 1 + \frac{K}{\mu_0}$$

For para-magnetic substances, K is positive, for non-magnetic substances, is zero and for diamagnetic substances, it is negative. For para-magnetic substances $\mu_r > 1$, for non-magnetic substances, $\mu_r = 1$ and for diamagnetic substances, $\mu_r < 1$.

7.13. Magnetic Screening

There is no known substance which will insulate against magnetic flux. In other words, magnetic flux can pass through all substances including even glass, rubber and wood etc. Hence, watches, compasses and precision electrical instruments get adversely affected by earth's field and from stray magnetic fields produced by generators and current-carrying conductors in their vicinity. Obviously, it is desirable to protect these instruments from such fields. As shown in Fig. 7.6, it can be done by enclosing the instruments in a soft-iron case or shell. Since soft-iron has very high permeability, magnetic flux prefers to pass through it instead of the instrument. The shell acts as a magnetic shield or screen.

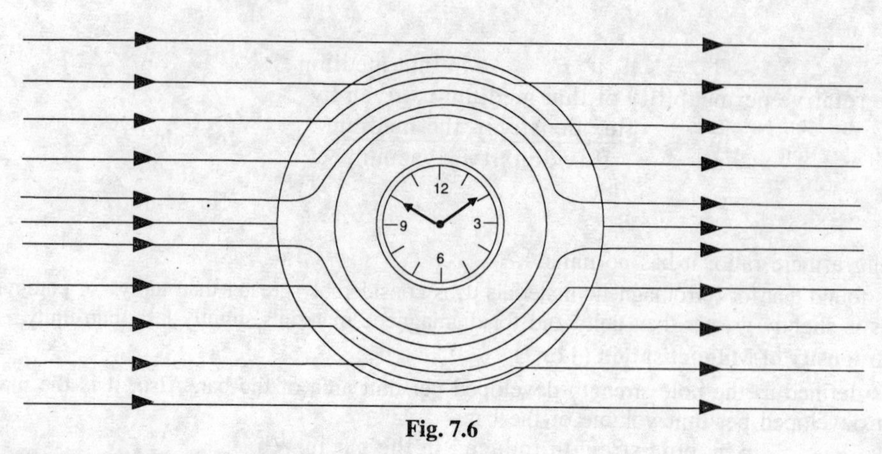

Fig. 7.6

7.14. Weber and Ewing's Molecular Theory

This theory was first advanced by Weber in 1852 and was, later on, further developed by Ewing in 1890. The basic assumption of this theory is *that molecules of all substances are inherently magnets in themselves, each having a N- and S-pole.* In an unmagnetised state it is supposed tht these small molecular magnets lie in all sorts of haphazard manner forming more or less closed loops (Fig. 7.7).

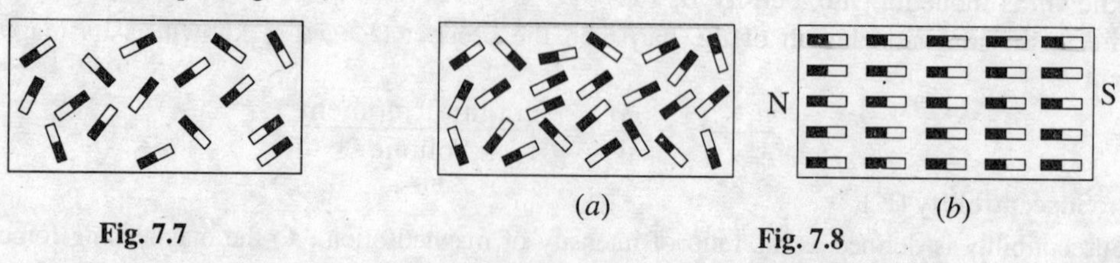

Fig. 7.7 Fig. 7.8

According to the laws of attraction and repulsion, these closed magnetic circuits are satisfied internally, hence there is no resultant external magnetism exhibited by the iron bar. But, when such an iron bar is placed in a magnetic field or under the influence of a magnetising force, then these molecular magnets start turning round their axes and orientate themselves more or less along straight lines parallel to the direction of the magnetising force. This linear arrangement of the molecular magnets results in *N*-polarity at one end of the bar and *S*-polarity at the other (Fig. 7.8). As the small magnets turn more nearly in the direction of the magnetising force, it requires more and more of this force to produce a given turning moment, thus accounting for the magnetic saturation. On this theory, the hysteresis loss is supposed to be due to molecular friction of these turning magnets.

Because of the limited knowledge of molecular structure available at the time of Weber, it was not possible to explain firstly as to why the molecules themselves are magnets and secondly, why it is impossible to magnetise certain substances like wood etc. The first objection was explained by Ampere who maintained that orbital movement of the electrons round the nucleus of an atom constituted a flow of current which, due to its associated magnetic effect, made the molecule a magnet.

Later on, it became difficult to explain the phenomenon of diamagnetism, erratic behaviour of ferromagnetic (intensely magnetisable) substances like iron, steel etc. and the paramagnetic (weakly magnetisable) substances like aluminium and chromium etc. Moreover, it was asked : if molecules of all substances are magnets, then why does not wood or air etc. become magnetised?

All this has been explained satisfactorily by the atom-domain theory which has superseded the molecular theory. It is beyond the scope of this book to go into the details of this theory. The interested reader is advised to refer to some standard book on magnetism. However, it may just be mentioned that this theory takes into account not only the planetary motion of the electron, but its rotation about its own axis as well. This latter rotation is called 'electron spin'. The gyroscopic behaviour of an electron gives rise to a magnetic moment which may be either positive or negative. A substance is ferromagnetic or diamagnetic accordingly as there is an excess of unbalanced positive spins or negative spins. Substances like wood or air are non-magnetisable because in their case, the positive and negative electron spins are equal hence they cancel each other out.

7.15. Curie Point

As the temperature of a magnetic material is increased, the agitation of its molecules is increased. As a result, the alignment of molecular magnets is disturbed and the substance loses some of its magnetism. Fig. 7.9 shows the decrease of magnetic strength with increase in temperature. While heating a magnetic substance, a certain temperature is reached when it loses all its magnetism. This temperature is called Curie point. Its value is about 750°C.

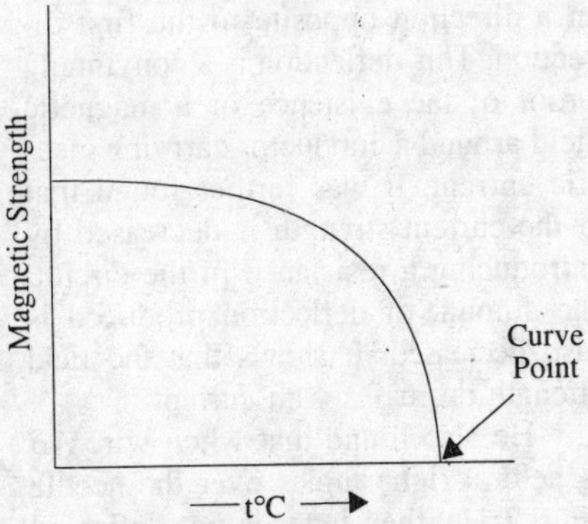

Fig. 7.9

7.16. Magnetic and Non-magnetic Substances

Different substances can be divided into either (*i*) magnetic substances or (*ii*) non-magnetic substances.

Materials that cannot be magnetised are called non-magnetic substances. Examples are : wood, rubber, paper, wax and plastic etc.

Those materials that can be magnetised are called *magnetic* substances. These can be further subdivided as under:

(*i*) **ferromagnetic substances**—which can be strongly magnetised by a magnetic field. Examples are:iron, steel, nickel, cobalt and alloys such as Alnico. Their relative permeability is very high (upto 100,000 or so) but varies with the magnetising force.

(*ii*) **paramagnetic substances**—which are only slightly attracted by a magnetic field. Examples are : aluminium, chromium, sodium and oxygen etc. Their relative permeability is *slightly* greater than unity. For example, μ_r for aluminium is 1.000022.

(*iii*) **diamagnetic substances**—which are slightly repelled by magnetic fields. Examples are: bismuth, zinc, sliver, gold, glass, water, hydrogen and nitrogen etc. Their relative permeability is only *slightly* less than unity.

Based on different considerations, magnetic materials may also be divided into (*i*) soft magnetic materials and (*ii*) hard magnetic materials. The soft magnetic materials like low-carbon electrical steel have high permeability, low coercivity and small hysteresis loss which makes them suitable for transformer cores, electro-magnets and measuring devices etc. The hard magnetic materials have high coercivity, relatively low permeability and large hysteresis loss and are suitable for making permanent magnets.

7.17. Ferrites

These are ferromagnetic ceramics and are generally produced by powdered metallurgy techniques using iron oxide. Ferrites have a fairly constant permeability and low hysteresis loss. Because of their extremely high electrical resistivity, they do not have eddy current loss. For this reason, they are ideally suited for high frequency applications. Examples are : as cores for radio frequency transformers and for switching and memory devices in high-speed computers.

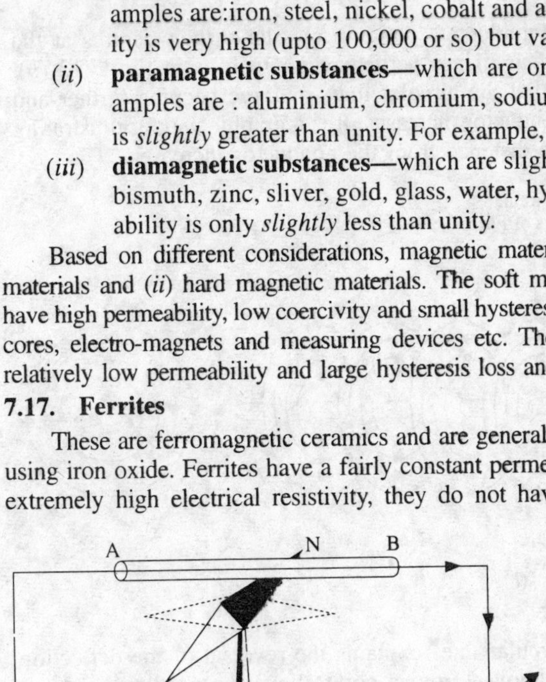

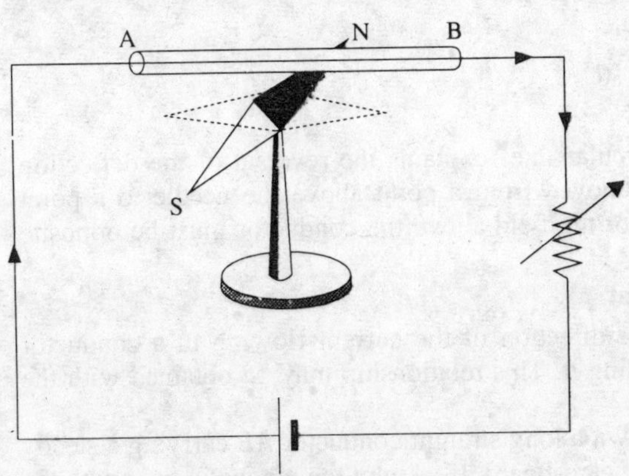

Fig. 7.10

ELECTROMAGNETISM

7.18. Magnetic Effects of Electric Current

In 1819, about two decades after the invention of voltaic cell,Oersted of Copenhagen discovered that an electric current is always accompanied by a magnetic field. He verified the existence of magnetic effects by the following simple experiments.

When Oersted placed a current-carrying conductor *AB* near and parallel to a magnetic needle, as in Fig. 7.10, he found that the needle was deflected in a certain direction. When the wire AB was placed beneath the needle then it was deflected in a direction opposite to the first direction. This deflection is a convincing proof of the existence of a magnetic field around a conductor carrying electric current. It was further found that if the current strength is decreased by introducing a resistance in the circuit, the amount of deflection produced is also decreased. It shows that the field strength decreases with current.

He also found that when wire AB is held at right angles over the needle (Fig. 7.11), then latter is not deflected at all. It shows that magnetic field of the current is at right angles to it, hence, in this case, parallel to the needle which explains the absence of deflection.

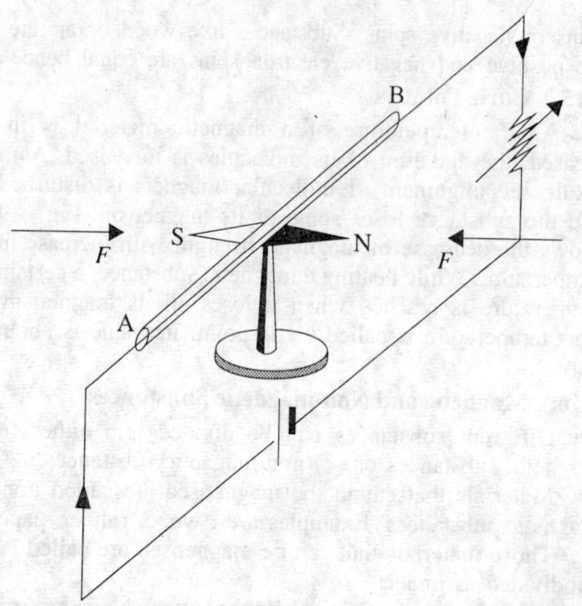

Fig. 7.11

Further investigation showed that the magnetic field consists of lines of force which form complete circles round the conductor. These circles have their centres at the centre of the current-carrying conductor and their planes perpendicular to it as in Fig. 7.12 (*b*). It should be noted from Fig. 7.12(*a*) and (*c*) that the circular lines of force become further and further apart as the distance from each to the conductor increases. It is a graphic way of illustrating the fact that the field strength decreases as the distance from the conductor increases.

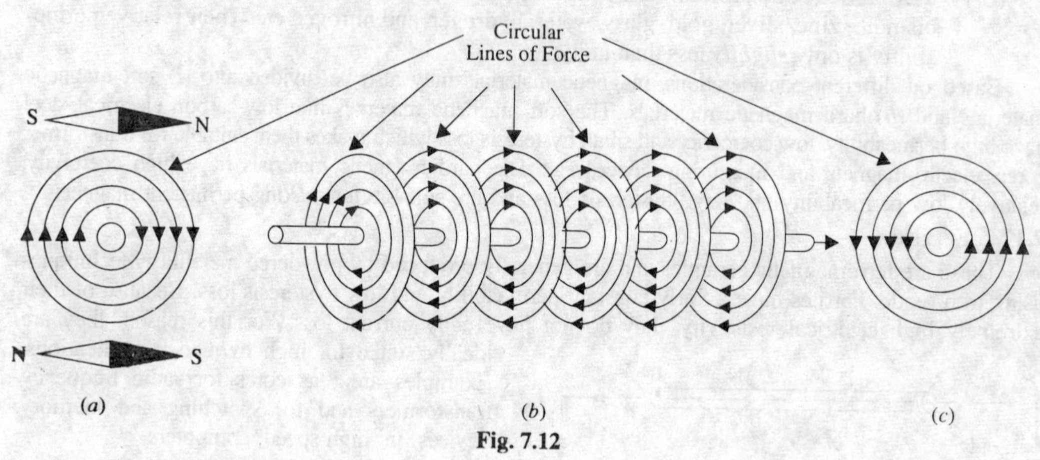

Fig. 7.12

The fact that magnetic field consists of circular lines explains the reversal of the deflection of the magnetic needle when conductor AB is moved from a point above the needle to a point beneath it [Fig. 7.12 (*a*)] because the direction of the field above the conductor must be opposite to that beneath the conductor.

7.19. Direction of Magnetic Field and Current

There exists a definite relation between the direction of the current flowing in a conductor and the direction of the magnetic field surrounding it. This relationship may be obtained with the help of the following simple rules:

(1) **Right-hand Rule.** In Fig. 7.13 is shown a long straight conductor AB carrying a steady current and passing through a piece of card-board. If we plot the magnetic field with the

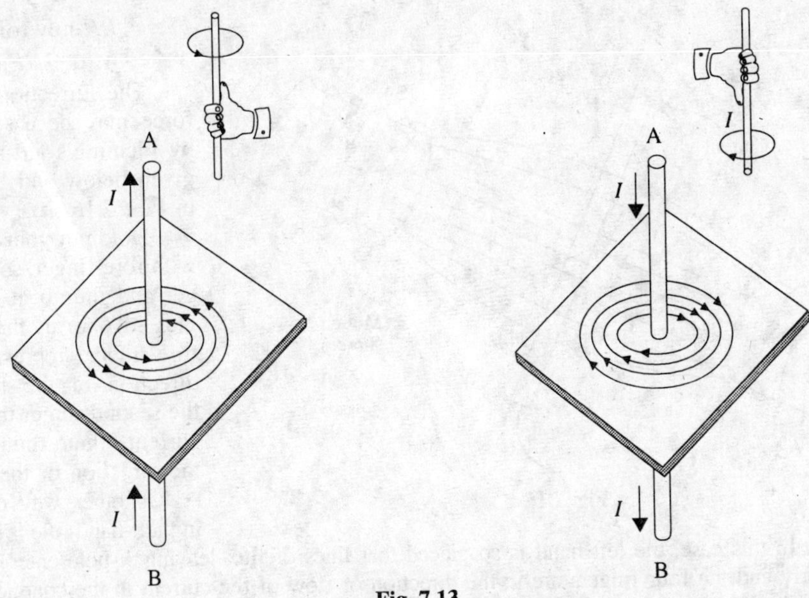

Fig. 7.13

help of a compass needle, we obtain the pattern shown in Fig. 7.13 (*a*) when current flows upwards from *B* to *A*. The lines of force are circular in shape and their direction is the direction in which an isolated *N* pole would move if free to do so. In Fig. 7.13 (*b*) is shown the field when current flows downwards from *A* to *B*. The direction of the field in each case can be found by the right-hand rule thus:

Grasp the wire or conductor in the *right hand* with the thumb pointing in the direction of the current. The fingers then curl round in the direction of the lines of force.

(2) **Maxwell's Corkscrew Rule.** The direction of the current and that of the magnetic field are related to each other as the forward travel of a corkscrew and the direction in which it is rotated (Fig. 7.14).

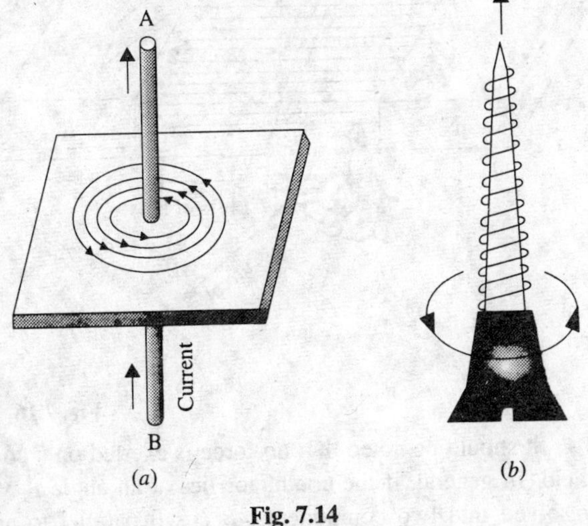

Fig. 7.14

Imagine a right-handed corkscrew held with its axis parallel to the conductor and pointing in the direction in which current is flowing. The direction of the magnetic lines of force is the same as that in which corkscrew must be rotated to cause it to advance in the direction of the current [Fig. 7.14 (*b*)].

7.20. Force on a Current-carrying Conductor Lying in a Magnetic Field

It is found that whenever a current-carrying conductor is placed in a magnetic field, it experiences a force which acts in a direction perpendicular both to the direction of the current and the field. In Fig. 7.15 is shown a conductor *XY* lying vertically in a uniform horizontal field of flux density *B* produced by two solenoids *A* and *B*. If *l* is the length of the conductor lying within this field and *I* ampere the current carried by it, then the magnitude of the force exerted on the conductor is

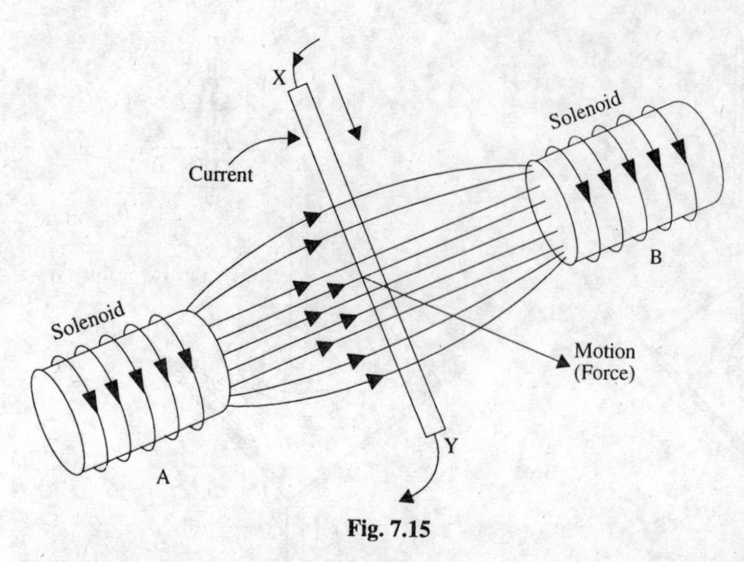

Fig. 7.15

$$F = BIl \text{ newton}$$
$$= \mu_0 \mu_r HIl \text{ newton}$$

The direction of this force may be easily found by Fleming's left-hand rule given below and illustrated in Fig. 7.16 (*a*).

Hold out your left hand with fore-finger, second finger and thumb at right angles to one another. If the fore-finger represents the direction of the field and the second finger that of the current, then thumb gives the direction of the motion.

Another way of applying left-hand rule is shown in Fig. 7.16(*b*). In this case, the left hand is so placed that lines of flux leaving *N*-pole enter the palm perpendicularly and the four fingers are in the direction of flow of the current in the conductor, then thumb gives the direction of the force.

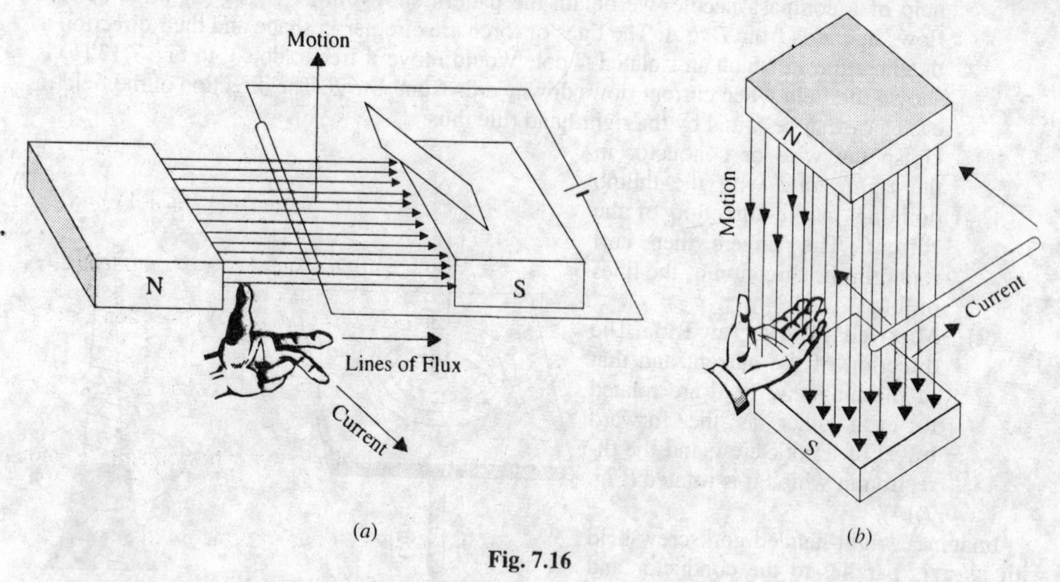

Fig. 7.16

It should be noted that no force is exerted on a conductor when it lies parallel to the magnetic field. In general, if the conductor lies at an angle θ with a field of flux density B, then B can be resolved into two components, $B \cos \theta$ parallel to and $B \sin \theta$ perpendicular to the conductor. The former produces no effect whereas the latter is responsible for the motion observed. In that case,

$$F = BIl \sin \theta \text{ newton}$$

7.21. Work Law and its Applications

If a unit *N*-pole is moved in a magnetic field, it experiences a magnetic force which depends on the field strength. If the unit pole is moved in opposition to the lines of force, then work is done against the magnetic field. Conversely, if the unit pole is moved in the direction of the magnetic field, then work is done by the magnetic force on whatever is restraining the movement of the pole. Consider N straight conductors, each carrying I amperes as shown in Fig. 7.17. The product 'NI' is called the 'ampere-turns'. According to Work Law.

The net work done on or by a unit N-pole in moving once round any single closed path in a magnetic field is numerically equal to the ampere-turns linked with that path.

Mathematically, $\oint H_r \, dr = NI$

where H_r is the magnetising force at a distance of r metres.

Explanation. In Fig. 7.17, path a and b are both closed paths which link N conductors once. Hence, work done in moving a unit N-pole once round either of these two paths in 'NI'. Even though path 'c' is also a closed path, it does not link any current-carrying conductor, hence work done in this case would be zero. Work law will now be used to find the magnetising force (H) due to (a) a long straight current-carrying conductor (b) a long solenoid.

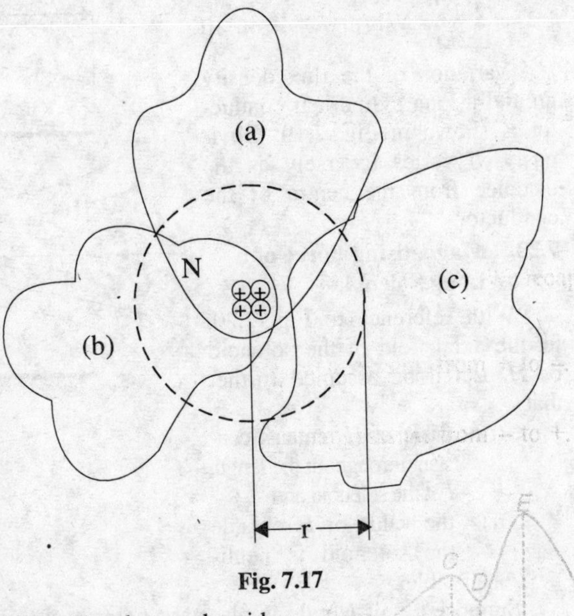

Fig. 7.17

7.22. Magnetising Force of a Long Straight Conductor

Fig. 7.18 shows a bundle of N long straight conductors each carrying I amperes. It is required to find the value of the magnetising force at point C distant r metres from the centre of the N

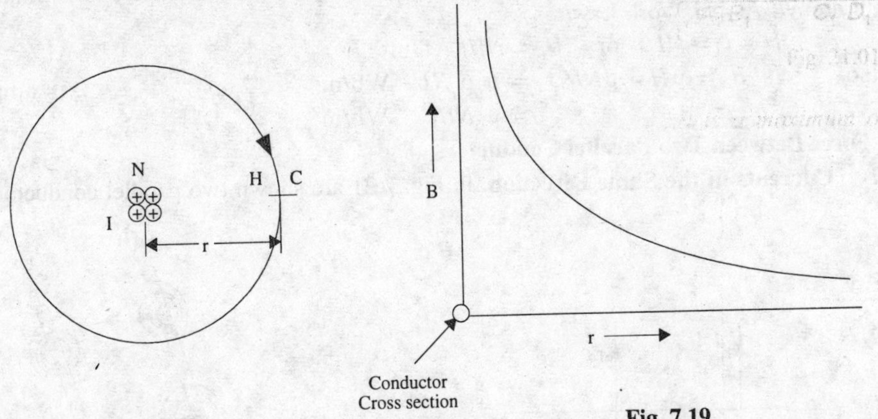

Fig. 7.18

Fig. 7.19

conductors. As shown, the conductors are surrounded by circular lines of force or flux. If H is the field strength a point C, then it means that a unit N-pole placed at C would experience a force of H newton. The direction of this force would be tangential to the circular lines of force passing through C. If the unit N-pole is moved once round this circular path against H, then work done on it is = force × distance = $H \times 2\pi r$ joules

Ampere-turns linked with closed path = NI

According to Work Law, the two must be equal.

$$\therefore \quad H \times 2\pi r = NI \quad \text{or} \quad H = \frac{NI}{2\pi r} \text{ AT/m}$$

$$\therefore \quad B = \frac{\mu NI}{2\pi r} \text{ Wb/m}^2 = \frac{\mu_0 \mu_r NI}{2\pi r} \text{ Wb/m}^2 \quad \text{... in a medium}$$

$$= \frac{\mu_0 NI}{2\pi r} \text{ Wb/m}^2 \quad \text{... in air}$$

Variation of the flux density around a long cylindrical conductor is shown in Fig. 7.19. Obviously, B varies inversely as the distance from the centre of the conductor.

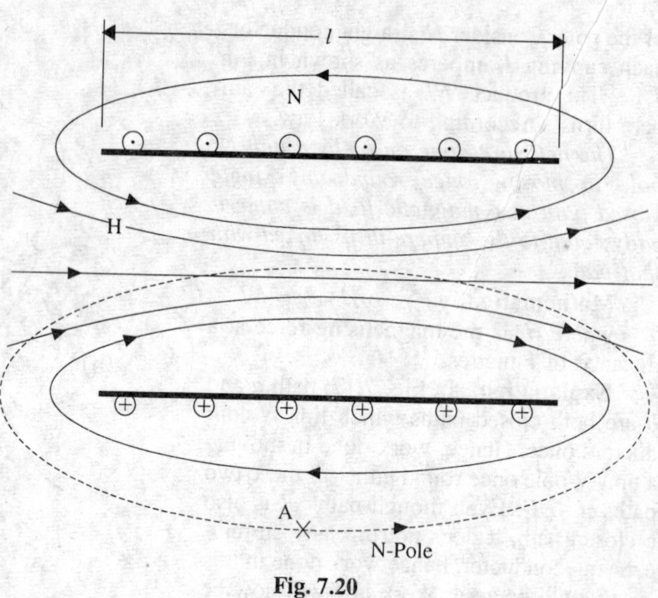

Fig. 7.20

7.23. Magnetising Force of a Long Solenoid

With reference to Fig. 7.20, let the axial field of the solenoid be H. Let it be assumed further that

(*i*) value of H remains constant throughout the length l of the solenoid and

(*ii*) the value of H outside the solenoid is negligible.

Suppose a unit *N*-pole is placed at point A outside the solenoid and is taken once round the dotted path in opposition to H. Remembering that a force of H newton acts on this unit *N*-pole only over the length l of the journey (it being negligible elsewhere), work done in one round is

$$= H \times l \text{ joules}$$

If N is the number of turns of the solenoid and I its current, then, ampere-turns linked with this path are NI. As per Work Law,

$$H \times l = NI \quad \text{or} \quad H = NI/l \text{ AT/m}$$

Also $$B = \mu H = \mu NI/l = \mu_0 \mu_r NI/l \text{ Wb/m}^2 \quad \text{...in a medium}$$

$$= \mu_0 NI/l \text{ Wb/m}^2 \quad \text{...in air}$$

7.24. Force Between Two Parallel Conductors

(*i*) **Currents in the Same Direction.** In Fig. 7.21 are shown two parallel conductor P and

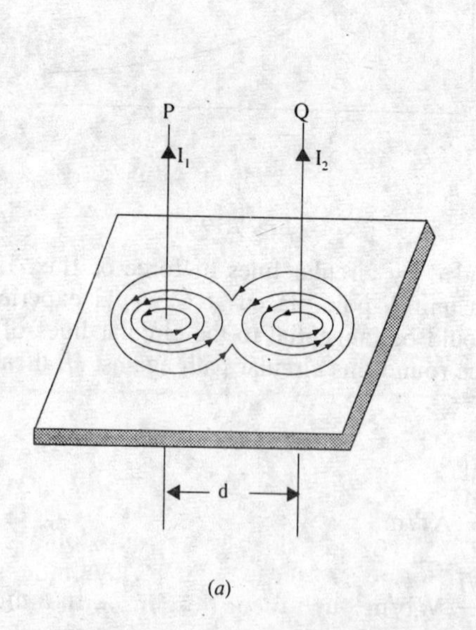

(*a*)

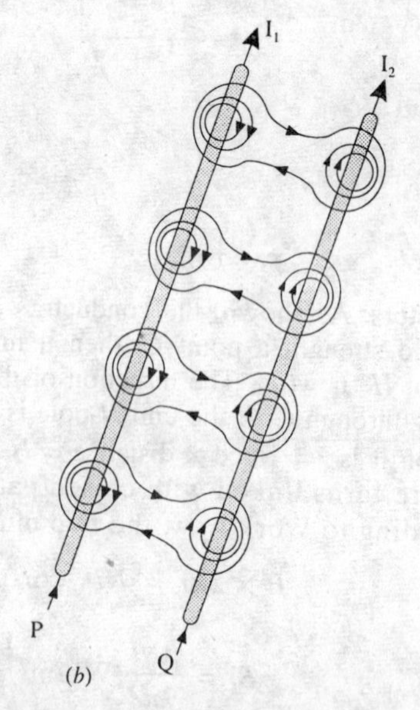

(*b*)

Fig. 7.21

Q carrying currents I_1 and I_2 amperes in the same direction *i.e.* upwards. The field strength in the space between the two conductors is decreased due to the two fields there being in opposition to each other. Hence, the resultant field is as shown in the figure. Obviously, the two conductors are attracted towards each other.

(*ii*) **Currents in Opposite Direction.** If, as shown in Fig. 7.22, the parallel conductors carry currents in the opposite directions, then field strength is increased in the space between the two conductors due to the two fields being in the same direction there. Because of the lateral repulsion of the lines of force, the two conductors experience a mutual force of repulsion as shown.

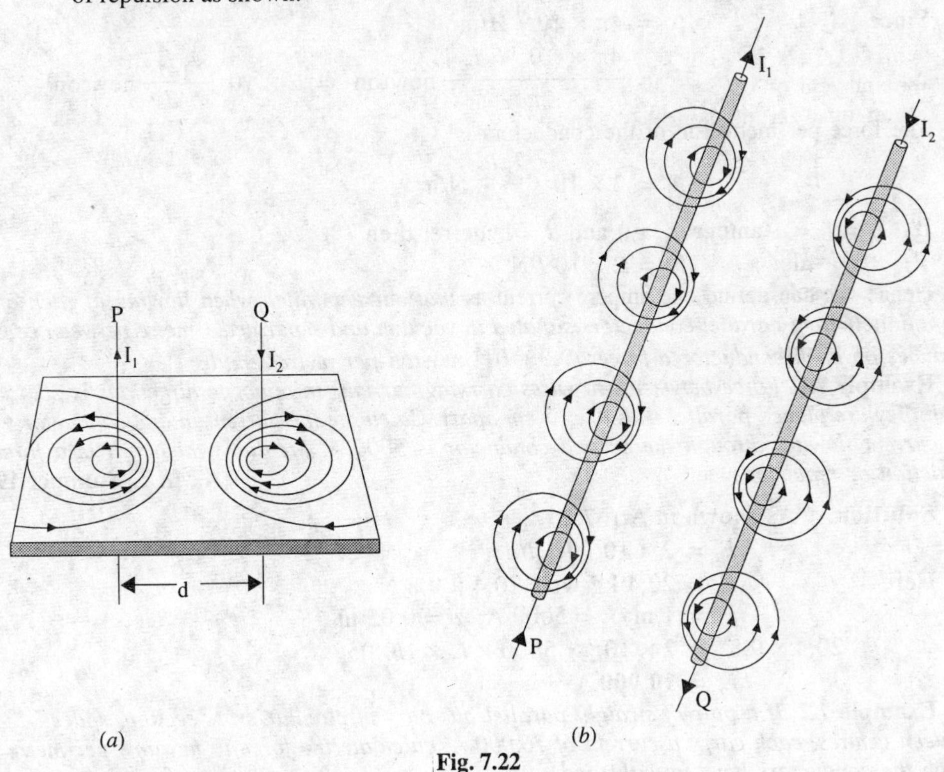

Fig. 7.22

7.25. Magnitude of Mutual Force

It is obvious that each of the two parallel conductors lies in the magnetic field of the other conductor. For example, conductor Q lies in the magnetic field of P and P lies in the field of Q. If 'd' metres is the distance between them, then flux density at Q due to P is (Art. 7.22)

$$B = \frac{\mu_0 I_1}{2\pi d} \text{ Wb/m}^2$$

If l is the length of the conductor Q lying in this flux density, then force (either of attraction or repulsion) as given in Art. 7.20 is

$$F = BI_2 \, l \text{ newton} \qquad \text{or} \qquad F = \frac{\mu_0 I_1 I_2 l}{2\pi d} \text{ newton}$$

Obviously, conductor P will experience an equal force in the opposite direction.

The above facts are known as Laws of Parallel Currents and may be stated as follows:

(*i*) Two parallel conductors attract each other if currents through them flow in the *same* direction and repel each other if the currents through them flow in *opposite* directions.

(*ii*) The force between two such parallel conductors is proportional to the product of the current strengths and to the lengths of the conductors considered and varies inversely as the distance between them.

7.26. Definition of Ampere

It has been proved in Art. 7.25 above that the force between two infinitely long parallel current-carrying conductors is given by the expression

$$F = \frac{\mu_0 I_1 I_2 l}{2\pi d} \text{ newton}$$

Since $\mu_0 = 4\pi \times 10^{-7}$ H/m

$$\therefore \quad F = \frac{4\pi \times 10^{-7} I_1 I_2 l}{2\pi d} \text{ newton} = 2 \times 10^{-7} \frac{I_1 I_2 l}{d} \text{ newton}$$

The force per metre run of the conductors is

$$F = 2 \times 10^{-7} \frac{I_1 I_2}{d} \text{ N/m}$$

If $I_1 = I_2 = 1$ ampere (say) and $d = 1$ metre, then

$$F = 2 \times 10^{-7} \text{ N}$$

Hence, we can define one ampere current as *that current which when flowing in each of the two infinitely long parallel conductors situated in vacuum and separated 1 metre between centres, produces on each conductor a force of 2 × 10⁻⁷ newton per metre length.*

Example 7.1. *Force between two wires carrying currents in opposite directions is 20.4 kg/m when they are placed parallel with axes 5 cm apart. Calculate the current in one conductor when the current flowing through the second conductor is 5000 A. Mention whether it is a force of attraction or repulsion.* **(A.M.I.E. Summer 1989)**

Solution. As shown in Art. 7.25

$$F = 2 \times 10^{-7} I_1 I_2 l/d$$

Here $F = 20.4$ kg-wt $= 20.4 \times 9.8$ N

$l = 1$ m; $I_1 = 5000$ A; $d = 0.05$ m

$$\therefore \quad 20.4 \times 9.8 = 2 \times 10^{-7} \times 5000 \times I_2 \times 1/0.05$$

$$\therefore \quad I_2 = \mathbf{10{,}000\ A}$$

Example 7.2. *If a pair of straight parallel bus-bars of circular cross-section, spaced 23 cm between centres, each carry a current of 70,000 A, calculate the force in newtons per metre run which the conductors have to withstand.*

Solution. $F = \frac{\mu_0 I_1 I_2 l}{2\pi d} = 2 \times 10^{-7} \frac{I_1 I_2 l}{d}$ newton

Here, $I_1 = I_2 = 70{,}000$ A ; $d = 0.23$ m ; $l = 1$ m

$$F = 2 \times 10^{-7} \times 70{,}000 \times 70{,}000/0.23 = \mathbf{4{,}260\ N/m}$$

Note. It is a force of repulsion since currents flow in opposite directions in the bus-bars

7.27. Magnetic Circuit

It may be defined as the route or path which is followed by magnetic flux. The laws of magnetic circuit are quite similar to (but not the same as) those of the electric circuit.

Consider a solenoid or a toroidal iron ring having a magnetic path of l metre, area of cross-section A m² and a coil of N turns carrying I amperes wound anywhere on it as in Fig. 7.23.

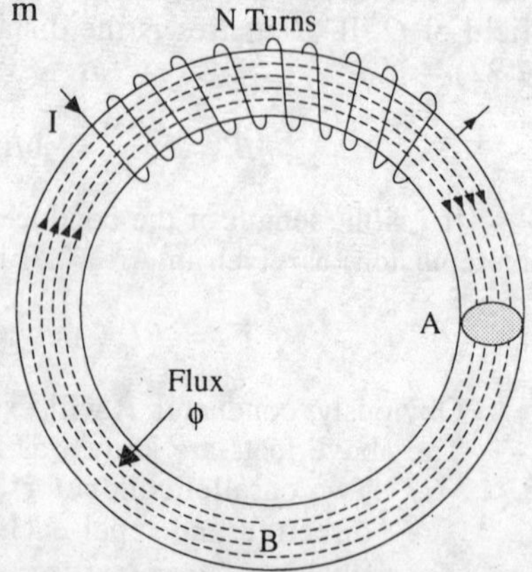

Fig. 7.23

Then, as seen from Art. 7.23, field-strength inside the solenoid is

$$H = \frac{NI}{l} \text{ AT/m}$$

Now, $B = \mu_0 \mu_r H \quad \therefore \quad B = \frac{\mu_0 \mu_r NI}{l} \text{ Wb/m}^2$

Total flux produced

$$\Phi = B \times A = \frac{\mu_0 \mu_r ANI}{l} \text{ Wb}$$

$$\Phi = \frac{NI}{l/\mu_0 \mu_r A} = \frac{NI}{S}$$

The numerator '*NI*' which produces magnetisation in the magnetic circuit is known as magnetomotive force (m.m.f.). Obviously, its unit is ampere-turn.

It is analogous to e.m.f. in an electric circuit.

The denominator $\frac{1}{\mu A}$ or $\frac{l}{\mu_0 \mu_r A}$ is called the *reluctance* of the circuit and is analogous to resistance in electric circuit.

$$\therefore \quad \text{Flux} = \frac{\text{mmf}}{\text{reluctance}}$$

Sometimes, the above equation is called the "Ohm's Law of Magnetic Circuit" because it resembles a similar expression in electric circuit *i.e.*

$$\text{current} = \frac{\text{emf}}{\text{resistance}}$$

7.28. Definitions

1. Magnetomotive force (m.m.f.). It drives or tends to drive flux through a magnetic circuit and corresponds to electromotive force (e.m.f.)in an electric circuit. It is given by the product '*NI*'.

M.M.F. is equal to the work done in joules in carrying a unit magnetic pole once through the entire magnetic circuit. It is measured in ampere-turns.

In fact, as p.d. between any two points is measured by the work done in carrying a unit charge from one point to another, similarly, m.m.f. between two points is measured by work done in joules in carrying a unit magnetic pole from one point to another.

2. Ampere-turns. It is the unit of magnetomotive force (m.m.f.).

3. Reluctance

It is the name given to that property of a material which opposes the creation of magnetic flux in it. It, in fact, measures the resistance offered to the passage of magnetic flux through a material and is analogous to resistance in an electric circuit even in form. Its unit is AT/Wb*

$$\text{Reluctance} = \frac{1}{\mu_0 \mu_r A} \text{ or } \frac{1}{\mu A}$$

$$\text{Resistance} = \rho \frac{l}{A} = \frac{l}{\sigma A}$$

In other words, the reluctance of a magnetic circuit is the number of ampere-turns required per weber of magnetic flux in the circuit. Since 1 AT/Wb = 1/henry, hence unit of reluctance is also 'reciprocal' henry.

4. Flux. It is equal to the total number of lines of induction existing in a magnetic circuit and is analogous to current in an electric circuit. It is measured in webers.

---

* From the relation $\Phi = \frac{\text{mmf}}{\text{reluctance}}$ it is obvious that reluctance = mmf/Φ. Since m.m.f. is in ampere-turns and flux in webers, unit of reluctance is ampere-turn/Wb or ampere/Wb since turn has no units.

5. Permeance. It is reciprocal of reluctance and implies the ease or readiness with which magnetic flux is developed. It is analogous to conductance in electric circuits. It is measured in terms of Wb/AT or henry.

6. Reluctivity. It is specific reluctance and corresponds to resistivity which is 'specific resistance'.

7.29. Composite Magnetic Circuit

In Fig 7.24 is shown a composite magnetic circuit consisting of three different magnetic materials of different permeabilities and lengths and one air gap ($\mu_r = 1$). Each path will have its own reluctance. The total reluctance is the sum of individual reluctances as they are joined in series.

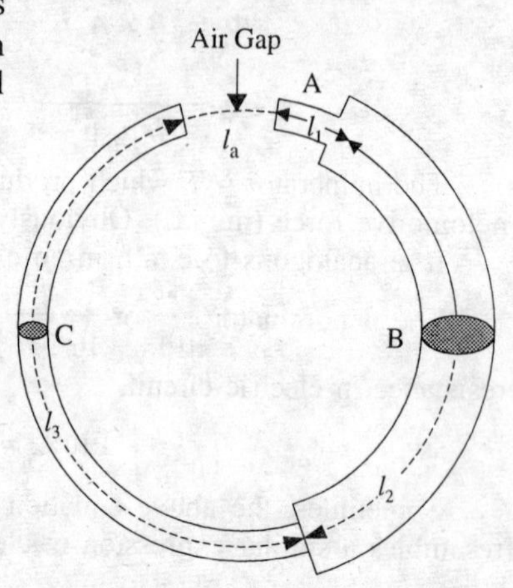

Fig. 7.24

$$\therefore \quad \text{total reluctance} = \Sigma \frac{l}{\mu_0 \mu_r A}$$

$$= \frac{l_1}{\mu_0 \mu_{r1} A_1} + \frac{l_2}{\mu_0 \mu_{r2} A_2} + \frac{l_3}{\mu_0 \mu_{r3} A_3} + \frac{l_a}{\mu_0 A_g}$$

$$\text{flux } \Phi = \frac{\text{m.m.f.}}{\Sigma \dfrac{l}{\mu_0 \mu_r A}}$$

7.30. How to Find Ampere-turns?

It has already been shown in Art. 7.19 that $H = NI/l$

$$\therefore \quad NI = H \times l$$

$$\therefore \quad \text{ampere-turns or AT} = H \times l$$

Hence, following procedure should be adopted for calculating the total ampere-turns of a composite magnetic path.

1. Find H for each portion of the composite circuit. For air, $H = B/\mu_0$, otherwise $H = B/\mu_0\mu_r$
2. Find ampere-turns for each path separately by using the relation $AT = H \times l$.
3. Add up these ampere-turns to get the total ampere-turns for the entire circuit.

7.31. Comparison Between Magnetic and Electric Circuits

SIMILARITIES

Magnetic Circuit	Electric Circuit

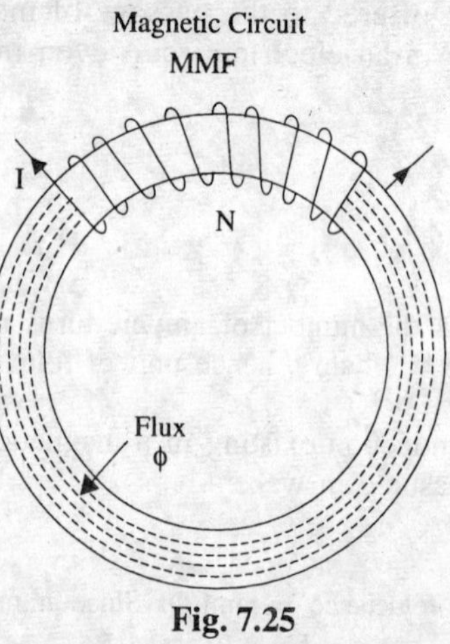

Magnetic Circuit
MMF
I
N
Flux
ϕ

Fig. 7.25

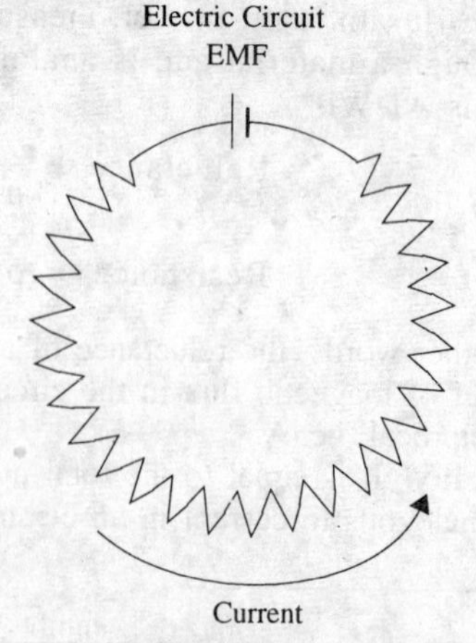

Fig. 7.26

1. Flux $= \dfrac{\text{m.m.f.}}{\text{reluctance}}$	Current $= \dfrac{\text{e.m.f.}}{\text{resistance}}$
2. M.M.F. (ampere-turns)	E.M.F. (in volts)
3. Flux (in webers)	Current I (in amperes)
4. Flux density B (in Wb/m^2)	Current density (A/m^2)
5. Reluctance	Resistance
$S = \dfrac{l}{\mu A} = \dfrac{l}{\mu_0 \mu_r A}$	$R = \rho \dfrac{l}{A} = \dfrac{l}{\sigma A}$
6. Permeance (= 1/reluctance)	Conductance (= 1/resistance)
7. Reluctivity	Resistivity
8. Permeability (= 1/reluctivity)	Conductivity (= 1/resistivity)

DIFFERENCES

1. Strictly speaking, flux does not actually 'flow' in the sense in which an electric current flows.
2. If temperature is kept constant, then conductivity (or resistivity) of an electric circuit is constant and is independent of current strength but the magnetic conductivity *i.e.* permeability depends on the total flux or flux density established through the material (Art. 7.34).
3. Flow of current in an electric circuit involves continuous expenditure of energy but in a magnetic circuit, energy is needed only for creating the flux initially but not for maintaining it.

7.32. Equivalent Electrical Circuits

Every magnetic circuit can be represented by its equivalent electrical circuit. While drawing it, it should be remembered that current corresponds to flux, resistance to reluctance and e.m.f. to m.m.f. Moreover, laws of series and parallel resistances also apply to a magnetic circuit.

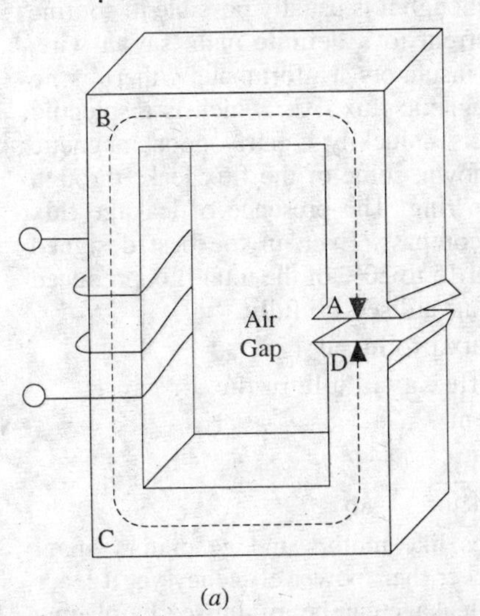

(*a*)

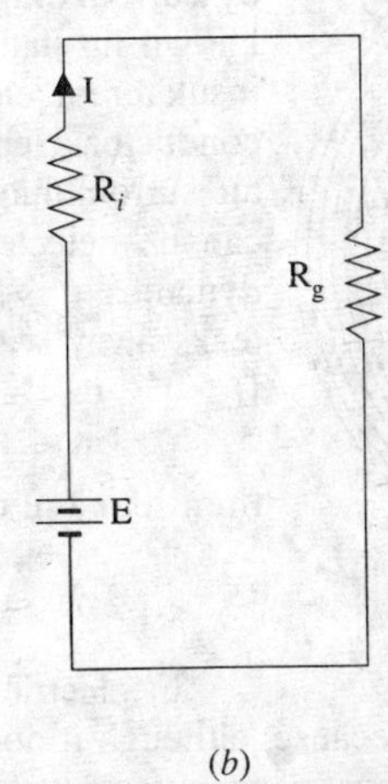

(*b*)

Fig. 7.27

Consider the circuit shown in Fig. 7.27 (*a*). The coil provides the mmf whereas the reluctances of the iron path and air-gap are in series with each other. The equivalent electrical circuit is shown in Fig. 7.27 (*b*). Here, R_i represents resistance of the iron path *ABCD* whereas R_g represents that of the air-gap *AD*. Same curent passes through both. The battery emf must equal sum of drop over R_i and R_g. Similarly, total mmf required is equal to the sum of those required for iron path and air-gap.

Now, consider the circuit shown in Fig. 7.28 (*a*). The reluctance of the iron path *AB* is represented by R_{AB} which is in series with the source of emf *E*. The two iron paths *ACB* and *ADB* are in parallel with respect to points *A* and *B*. The total flux is divided inversely as the reluctances of the two parallel paths. Similarly, in Fig. 7.28 (*b*), current I is divided at point A inversely as the resistances of the two parallel paths ACB and ADB.

$$E = \text{drop over } R_{AB} + \text{drop over either } ACB \text{ or } ADB$$

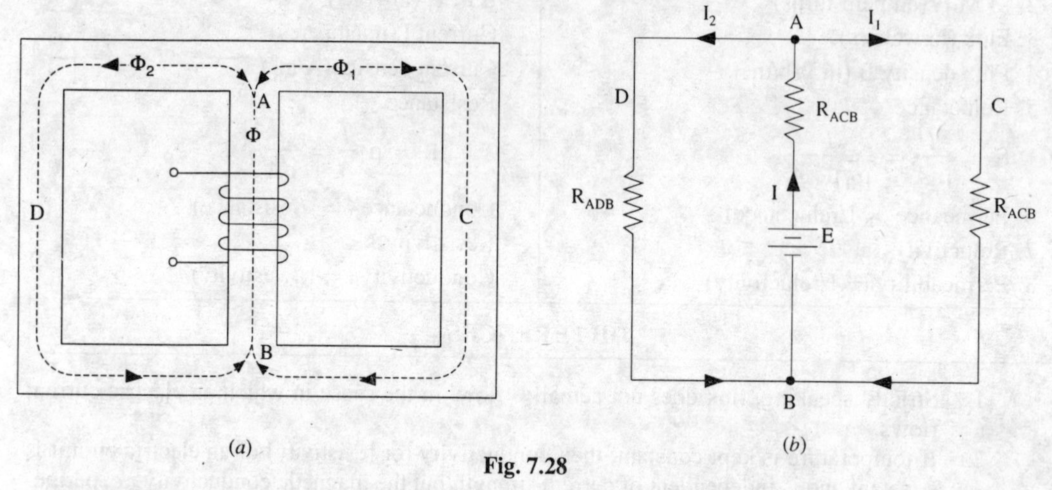

Fig. 7.28

Similarly, total mmf required in Fig. 7.28 (*a*) is the sum of that required for iron path AB and that required for either ACB or ADB (but not both). Same mmf which produces Φ_1 also produces Φ_2.

7.33. Leakage Flux and Hopkinson's Leakage Coefficient

Leakage flux is the flux which follows a path not intended for it. In Fig. 7.29 is shown an iron ring wound with a coil and having an air-gap. The flux in the air-gap is known as the useful flux because it is only this flux which can be utilised for various useful purposes.

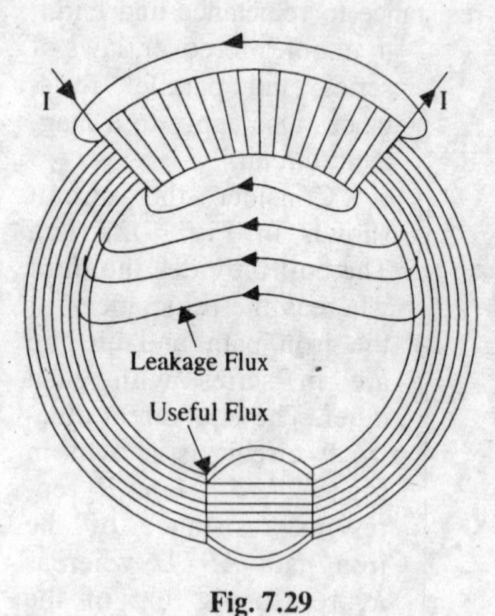

Fig. 7.29

It is found that it is impossible to confine all the flux to the iron path only, although it is usually possible to confine most of the electric current to a definite path, say, a wire, by surrounding it with insulators. Unfortunately, there is no known insulator for magnetic flux. Air, which is a splendid insulator of electricity, is unluckily a fairly good magnetic conductor. Hence, as shown, some of the flux leaks through air surrounding the iron ring. The presence of leakage flux can be detected by a compass. Even in the best designed dynamos, it is found that 15 to 20% of the total flux produced leaks away without being utilised usefully.

If Φ_t = total flux produced

Φ = useful flux available in the air gap

then leakage coefficient

$$\lambda = \frac{\text{total flux}}{\text{useful flux}} = \frac{\Phi_t}{\Phi}$$

In electric machines like motors and generators, magnetic leakage is undesirable, because, although it does not lower their power efficency yet it leads to their increased weight and cost of manufacture. Magnetic leakage can be minimised by placing the exciting coils or windings as closely as possible to the air-gap or to the points in the magnetic circuit where flux is to be utilised for useful purposes.

It is also seen from Fig. 7.29 that there is fringing or spreading of lines of flux at the edges of the air-gap. This fringing increases the effective area of the air-gap.

The value of λ for modern electric machines varies between 1.1 and 1.25.

7.34. Magnetisation Curves

The approximate magnetisation or B/H curves of few magnetic materials are shown in Fig. 7.30.

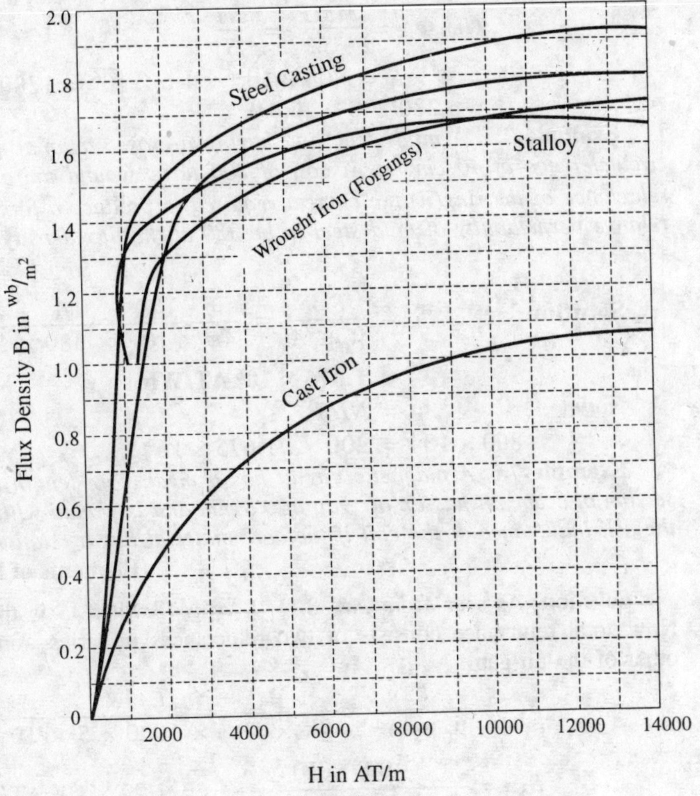

Fig. 7.30

These curves can be determined by the following methods provided the material is in the form of a ring:

(*a*) by means of a Ballistic Galvanometer and

(*b*) by means of a Fluxmeter

It is worth noting that the slope of a magnetisation curve is not constant. It is maximum at the beginning and minimum towards the end. The slope of a B/H curve represents μ or μ_r. Hence, it means that at low values of B, μ_r is high and at high values of B, μ_r is low. Since $S = l/\mu_0 \mu_r A$, it means that reluctance offered by the material also varies with B because μ_r depends on B. At high flux densities, μ_r is less, so that S is more. At low flux densities, μ_r is large, hence S is small.

Example 7.3 *Calculate the reluctance and permeance of a ferromagnetic circuit of mean length 20 cm, having area 10 cm² and a relative permeability 100. If the circuit is uniformly wound with 1000 turns of wire carrying 0.1 A, determine the magnetic flux in the core.*

(Elements of Elect. Engg-II, Punjab Univ. 1993.)

Solution. The reluctance of the circuit is given by

$$S = \frac{1}{\mu_0 \mu_r A} = \frac{0.2}{4\pi \times 10^{-7} \times 100 \times 10 \times 10^{-4}}$$

$$= 1.59 \times 10^6 \text{ H}^{-1}$$

$$\text{Permeance} = 1/S = 1/1.59 \times 10^6 = 6.29 \times 10^{-7} \text{ H}$$

$$\Phi = \frac{MMF}{S} = \frac{NI}{S} = \frac{1000 \times 0.1}{1.59 \times 10} = \mathbf{62.9 \times 10^{-6} \text{ Wb}}$$

Example 7.4 *A mild steel ring having a cross-sectional area of 500 mm and a mean circumference of 400 mm has a coil of 200 turns wound uniformly around it. Calculate (i) the reluctance of the ring and (ii) the current required to produce a flux of 800 μWb in the ring. Take relative permeability of mild steel as 400 at the given flux density.*

(Elect. Engg.-1, Punjab Univ. 1994)

Solution. (*i*) $S = \dfrac{l}{\mu_0 \mu_r A} = \dfrac{400 \times 10^{-3}}{4\pi \times 10^{-7} \times 400 \times 500 \times 10^{-6}}$

$$= \mathbf{1.6 \times 10^6 \text{ AT/Wb}}$$

$$(ii)\ \Phi = \frac{MMF}{S} = \frac{NI}{S}$$

$\therefore \quad NI = \Phi \times S = 800 \times 10^{-6} \times 1.6 \times 10^{6} = 1280$

$I = 1280/200 =$ **6.4 A**

Example 7.5. *A mild steel ring having a cross-sectional area of 5 cm² and a and a mean circumference of 40 cm has a coil of 200 turns wound uniformly around it. Calculate (i) the reluctance of the ring(ii) the current required to produce a flux of 800 μWb in the ring. Assume relative permeability of mild steel to be 380 at the flux density developed in the core.*

(Elect. Engg.-I, Punjab Univ. 1993)

Solution. $$S = \frac{1}{\mu_0 \mu_r A} = \frac{0.4}{4\pi \times 10^{-7} \times 380 \times 5 \times 10^{-4}}$$

$= \mathbf{1.675 \times 10^{6}\ AT/Wb}$

Now, $\Phi = NI/S$

$\therefore \quad 800 \times 10^{-6} = 200 \times I/1.675 \times 10^{4} \qquad \therefore \quad I =$ **6.7 A**

Example 7.6. *A magnetic circuit has a mean core length of 100 cm and a uniform cross-section of 5 cm². It has an air gap of 0.8 mm and is wound with a coil of 1200 turns. Determine the self inductance of the coil if the core material has a relative permeability of 1000.*

(Elements of Elect.Engg. Punjab Univ. 1994.)

Solution. As seen from Art. 8.11, $L = N^2/S$ where S is the total reluctance of the circuit. Now, total reluctance consists of two reluctances in series—one of the magnetic core and the other of the air-gap.

$$S_1 = \frac{l}{\mu_0 \mu_r A} = \frac{l}{4\pi \times 10^{-7} \times 1000 \times 5 \times 10^{-5}} = 6.37 \times 10^{6}\ \text{AT/Wb}$$

$$S_2 = \frac{-0.8 \times 10^{-3}}{4\pi \times 10^{-7} \times 5 \times 10^{-4}} = 1.27 \times 10^{6}\ \text{AT/Wb}$$

$\therefore$ total reluctance, $S = S_1 + S_2 = (6.37 + 1.27) \times 10^{6} = 7.64 \times 10^{6}$ AT/Wb

$\therefore \quad L = 1200^2/7.64 \times 10^{6} = 0.188\ \text{H} =$ **188 mH**

Example 7.7. *Calculate the relative permittivity of an iron ring when the exciting current taken by the 600-turn coil is 1.2 A and the total flux produced is 1 mWb. The mean circumference of the ring is 0.5 m and the area of the cross-section is 10 cm².* **(A.M.I.E. Winter 1990)**

Solution. $\Phi = MMF/S$ or $S = MMF/\Phi$; $\quad MMF = NI = 600 \times 1.2 = 720$

$\Phi = 1\ \text{mWb} = 1 \times 10^{-3}$ Wb $\qquad \therefore \quad S = 720/1 \times 10^{-3} = 720{,}000$

Now, $S = l/\mu_0\mu_r A$ or $720{,}000 = 0.5/4\pi \times 10^{-7} \times \mu_r \times 10 \times 10^{-4}$;

$\mu_r =$ **5.50**

Example 7.8. *An iron ring of mean length 50 cm has an air-gap of 1 mm and a winding of 200 turns. If the permeability of the iron is 300 when a current of 1.5 A flows through the coil, find the flux density.*

(A.M.I.E. Winter 1989)

Solution. Let A be the area of cross-section of the iron path as well as that of the air-gap.

Now, $$\Phi = \frac{MMF}{S} = \frac{MMF}{S_1 + S_2}$$

$$\therefore \quad BA = \frac{MMF}{\dfrac{l_1}{\mu_0 \mu_r A} + \dfrac{l_2}{\mu_0 A}} = \mu_0 \mu_r A \times \frac{MMF}{l_1 + \mu_r l_2}$$

$$\therefore \quad B = \mu_0 \mu_r \times \frac{MMF}{l_1 + \mu_r l_2}$$

$$= 4\pi \times 10^{-7} \times 300 \times \frac{200 \times 1.5}{0.5 + 300 \times 1 \times 10^{-3}} = \mathbf{0.14\ Wb/m^2}.$$

Example 7.9. *Estimate the number of ampere-turns necessary to produce a flux of 0.5 milli-weber round an iron ring of 5 cm² cross-section and 30 cm mean diameter, having an air-gap of 1 mm wide across it. Assume the relative permeability of iron as 1000. Neglect the leakage flux outside the air-gap.* **(A.M.I.E. Summer 1990)**

Solution. Iron Path

$$S_1 = \frac{l_1}{\mu_0 \mu_r A} = \frac{\pi \times 0.3}{4\pi \times 10^{-7} \times 1000 \times 5 \times 10^{-4}} = 1.5 \times 10^6 \text{ AT/Wb}$$

$$MMF_2 = \Phi S_1 = 0.5 \times 10^{-3} \times 1.5 \times 10^6 = 750 \text{ AT}$$

Air Gap

$$S_2 = \frac{l_2}{\mu_0 A} = \frac{1 \times 10^{-3}}{4\pi \times 10^{-7} \times 5 \times 10^{-4}} = 1.6 \times 10^6 \text{ AT/Wb}$$

$$MMF_2 = \Phi S_2 = 0.5 \times 10^{-3} \times 1.6 \times 10^6 = 800 \text{ AT}$$

Total MMF required = 750 + 800 = **1550 AT**

Example 7.10. *A cast-steel ring of mean circumference 50 cm has a cross-section of 0.52 cm². It has a saw-cut of 1 mm at one place. Given the following data:*

B (Wb/m²):	*1.0*	*1.25*	*1.46*	*1.60*
μ_r	*714*	*520*	*360*	*247*

Calculate how many ampere-turns are required to produce a flux of 0.052 mWb if leakage factor is 1.2 **(A.M.I.E. Winter 1988)**

Solution. Air Gap

Φ = 0 .052 mWb = 0.052×10^{-3} Wb = 52×10^{-6} Wb

A = 0.52 cm² = 0.52×10^{-4} m² = 52×10^{-6} m²

$B = \Phi/A = 52 \times 10^{-6}/52 \times 10^{-6} = 1$ Wb/m²

$H = B/\mu_0 = 1/4\pi \times 10^{-7} = 795 \times 10^3$ AT/m

l = 1 mm = 10^{-3} m

AT required = $H \times l = 795 \times 10^3 \times 10^{-3} = 795$ AT

Cast-steel Path

$B = 1.2 \times B_{air} = 1.2 \times 1 = 1.2$ Wb/m²

If we draw a graph between flux density and relative permeability (μ_r) from the given data, then value of μ_r for B = 1.2 Wb/m² is = 560

$\therefore$ $H = B/\mu_0\mu_r = 1.2/4\pi \times 10^{-7} \times 560 = 1705$ AT/m

l = 50 cm = 0.5 m

AT reqd. = $1705 \times 0.5 = 853$

Total AT reqd. = 795 + 853 = **1648**

Example 7.11. *An iron ring has a X-section of 3 cm² and a mean diameter of 25 cm. An air gap of 0.4 mm has been cut across the section of the ring. The ring is wound with a coil of 200 turns through which a current of 2 A is passed. If the total magnetic flux is 0.24 mWb, find the relative permeability of iron, assuming no magnetic leakage.*

(Electrical Engg. A.M.Ae. S.I. June 1992)

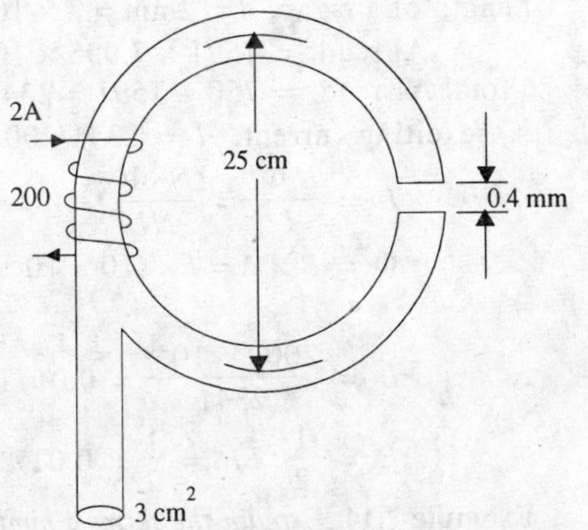

Fig. 7.31

Solution. Air Gap (Fig. 7.31)

$\Phi = 0.24 \times 10^{-3}$ Wb, $A = 3 \times 10^{-4}$m²

$B = 0.24 \times 10^{-3}/3 \times 10^{-4} = 0.8$ Wb/m²

$H = B/\mu_0 = 0.8/4\pi \times 10^{-7} = 6.367 \times 10^5$ AT/m

Air-gap length $= 0.4 \times 10^{-3}$ m

AT reqd. $= 6.367 \times 10^5 \times 0.4 \times 10^{-3} = 255$

Total AT provided $= 200 \times 2 = 400$

AT available for iron path $= 400 - 255 = 145$

H for iron path $= AT/l = 145/\pi \times 0.25 = 185$ AT/m

Now, $B = \mu_0 \mu_r$ H or $\mu_r = 0.8/4\pi \times 10^{-7} \times 185 =$ **3440.**

Example 7.12. *A magnetic ring has a mean circumference of 1.5 metre and is of 0.01 m² in cross-section and is wound with 175 turns. A saw cut of 4 mm wide is made in the ring. Calculate the magnetising current required to produce flux of 0.8 mWb in the air gap. Assume permeability of iron as 400 and leakage factor as 1.25.*

(Electrical Engg., A.M.Ae. S.I. Dec. 1991.)

Solution. Air Gap

$\Phi = 0.8$ mWb $= 0.8 \times 10^{-3}$ Wb $= 8 \times 10^{-4}$ Wb, $A = 0.01$ m² $= 1 \times 10^{-2}$ m².

$B = \Phi/A = 8 \times 10^{-4}/1 \times 10^{-2} = 0.08$ Wb/m²

$H = B/\mu_0 = 0.08/4\pi \times 10^{-7} = 6.37 \times 10^4$ AT/m

$l = 4$ mm $= 4 \times 10^{-3}$ m

AT reqd. $= H \times l = 6.37 \times 10^4 \times 4 \times 10^{-3} = 254$

Iron Path

$B = 1.25 \times B_{air} = 1.25 \times 0.08 = 0.1$ Wb/m²

$\therefore$ $H = B/\mu_0\mu_r = 0.1/4\pi \times 10^{-7} \times 400 = 1989$ AT/m, $l = 1.5$ m

AT reqd. $= 1989 \times 1.5 = 2983$

Total AT reqd. $= 254 + 2983 = 3237$

$\therefore$ magnetizing current reqd. $= 3237/175 =$ **18.5 A**

Example 7.13. *An iron ring of mean diameter 15 cm and 10 cm² in cross-section is wound with 200 turns of wire. There is an air gap of 2 mm cut in the ring. For a flux density of 1 Wb/m² and a relative permeability of 500, find the exciting current, the inductance and the stored energy.*

(Electrical Science AMIE Winter 1993)

Solution. Iron Path

$H = B/\mu_0\mu_r = 1/500 \times 4\pi \times 10^{-7} = 1591$ AT

Length of the iron path $= \pi D = 0.15\pi$ metre

AT reqd. $= H \times l = 1591 \times 0.15\,\pi = 750$

Air Gap

$H = B/\mu_0 = 1/4\pi \times 10^{-7} = 7.955 \times 10^5$ AT

Length of air-gap, $l = 2$mm $= 2 \times 10^{-3}$ m

At reqd. $= 1591 \times 7.995 \times 10^5 = 1591$

Total AT reqd. $= 750 + 1591 = 2341$

$\therefore$ exciting current, $I = 2341/200 =$ **11.7 A**

$$L = \frac{N\Phi}{I} = \frac{N^2\Phi}{NI}$$

$\Phi = B \times A = l \times (10 \times 10^{-4}) = 10^{-3}$ Wb

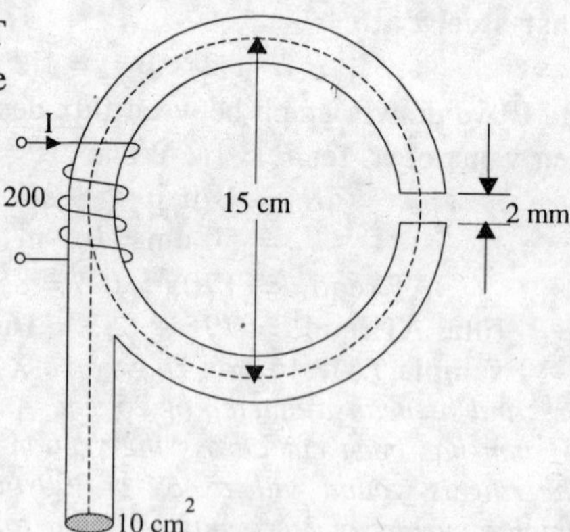

Fig. 7.32

$$\therefore \quad L = \frac{200^2 \times 10^{-3}}{2341} = 0.0171 \text{ H} = \mathbf{17.1\ mH}$$

$$E = \frac{1}{2}.LI^2 = \frac{1}{2} \times 0.0171 \times (11.7)^2 = \mathbf{1.17\ J}$$

Example 7.14. *Explain the terms : mmf, magnetic flux, reluctance and flux density. Estimate their respective values in the following case: A steel ring 30 cm mean diameter and of circular* *section 2 cm in diameter has an air gap 1 mm long. It is wound uniformly with 600 turns of wire carrying a currunt of 2.5 A. Neglect magnetic leakage. The iron path takes 40% of the total magnetomotive force.* **(A.M.I.E. Winter 1986)**

Solution. Total $MMF = NI = 600 \times 2.5 =$ **1500 AT**

Let M_1 and M_2 be the *MMFs* of the iron path and air-gap respectively and S_1 and S_2 their reluctances. Since the two reluctances are in series with the same flux passing through them

$$\frac{M_1}{M_2} = \frac{S_1}{S_2} \quad \text{or} \quad \frac{M_1 + M_2}{M_2} = \frac{S_1 + S_2}{S_2} \quad \text{or} \quad \frac{M}{M_2} = \frac{S}{S_2}$$

$$M = 1500 \text{ AT} \; ; \; M_2 = 60\% \text{ of } 1500 = 900 \text{ AT}$$

$$S_2 = \frac{l_2}{\mu_0 A} = \frac{1 \times 10^{-3}}{4\pi \times 10^{-7} \times \pi(1 \times 10^{-2})^2} = 2.5 \times 10^6 \text{ AT/Wb}$$

$$\therefore \quad \frac{1500}{900} = \frac{S}{2.5 \times 10^6} \; ; \quad S = \mathbf{4.17 \times 10^6 \text{ AT/Wb}}$$

Now, $M_2 = \Phi S_2 \quad \therefore \; \Phi = M_2/S_2 = 900/2.5 \times 10^6 =$ **0.36 mWb**

$$B = \frac{\Phi}{A} = \frac{0.36 \times 10^{-3}}{\pi \times 10^{-4}} = \mathbf{1.146 \text{ Wb/m}^2}$$

Example 7.15. *A magnetic ring has a mean circumference of 1.5m and is of 0.01 m^2 in cross-section, and is wound with 250 turns. A saw-cut of 4 mm wide is made in the ring. Calculate the magnetizing current required to produce a flux of 0.8 × 10^{-3} Weber in the air-gap. Assume relative permeability of iron as 400 and leakage factor as 1.25.*

(Magnetic Ckts & Materials, Punjab Univ. 1993)

Solution. Air Gap (Fig. 7.33)

$B = 0.8 \times 10^{-3}/0.01 = 0.08$ T

$H = B/\mu_0 = 0.08/4\pi \times 10^{-7} = 6.36 \times 10^4$ AT/m

AT reqd. $= H \times l = 6.36 \times 10^4 \times 4 \times 10^{-3} = 254.4$

Magnetic Ring

$B = 1.25 \times 0.08 = 0.1$ T

$H = B/\mu_0\mu_r = 0.1/4\pi \times 10^{-7} \times 400 = 199$

AT reqd. $= H \times l = 199 \times 1.5 = 298.5$

Total AT reqd. $= 254.4 + 298.5 = 552.9$

$\therefore$ magnetizing current reqd. $= 552.9/250 =$ **2.21 A**

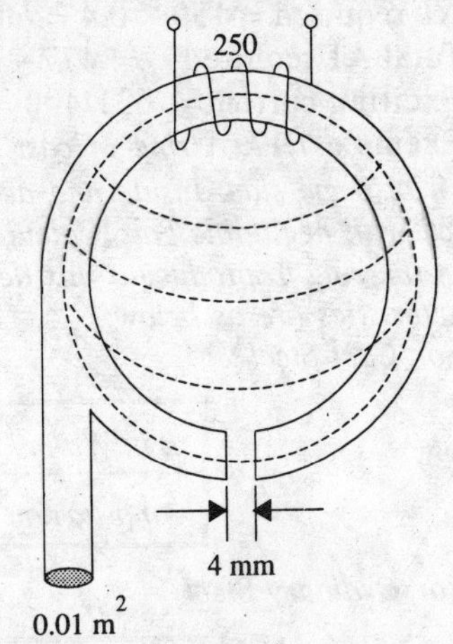

Fig. 7.33

Example 7.16. *A shell-type magnetic core made of silicon steel has the dimensions as shown in Fig. 7.33 below. The area of cross-section of the central limb is 10 cm^2 and that of each side limb is 8 cm^2. The central limb is wound with a coil of 400 turns. Estimate the ampere-turns required for (i) air-gap (ii) central limb and (iii) the rest of the magnetic circuit when the air-gap flux is 1.2 mWb. Assume no leakage. Aslo, calculate the current required in the coil to produce the above flux. The data for the B-H curve for silicon steel are :*

B (Wb/m²)	*0.5*	*0.75*	*1.0*	*1.2*	*1.4*
H(AT/m)	*100*	*150*	*240*	*470*	*900*

Solution. As seen from Fig. 7.34, the magnetic path consists of the cnetral limb with the air-gap *i.e.* path *EN* and two parallel paths between *E* and *N i.e.* path *EFN* and *EGN*. The total m.m.f. *i.e.* ampere-turns required are the sum of those required for (*i*) path *MN* (*ii*) air-gap and (*iii*) either of the two parallel paths *EFN* or *EGN*.

(i) Air Gap

Φ = 1.2 mWb, A = 10 cm^2

= $10 \times 10^{-4} = 10^{-3}$ m^2

$$B = \frac{1.2 \times 10^{-3}}{10^{-3}} = 1.2 \text{ Wb/m}^2$$

l = 0.5 mm = 0.5×10^{-3} m

$H = B/\mu_0 = 1.2/4\pi \times 10^{-7}$

= 95.45×10^4 AT/m

AT required = $H \times l = 95.45 \times 10^4 \times 0.5 \times 10^{-3}$ = **477**

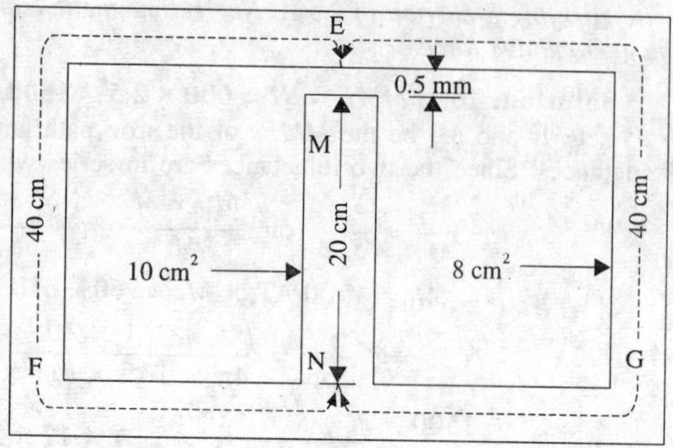

Fig. 7.34

(ii) Central limb MN

B = 1.2 Wb/m^2 —as in the air-gap

Value of H corresponding to the above flux density as found from the given data is = 470 AT/m. Since MN = 20 cm = 0.2 m

∴ AT required = 470 × 0.2 = **94.**

(iii) Side limb EFN

At point E, the flux is divided into two equal parts. Flux through either side limb is = 1.2/2 = 0.6 mWb.

∴ $B = 0.6 \times 10^{-3}/8 \times 10^{-4} = 0.75$ Wb/m^2

From the given data, corresponding value of H = 150 AT/m

AT required = 150 × 0.4 = **60**

Total AT required = 477 + 94 + 60 = **631**

Exciting current= 631/400 = **1.58 A**

Example 7.17. *A ring of cast steel has an external diameter of 24 cm and a square cross section of 3 cm side. Inside and across the ring, an ordinary steel bar 18 cm × 3 cm × 0.4 cm is fitted with negligible gap. Calculate the number of ampere-turns required to be applied to one half of the ring to produce a flux density of 1 Wb/m^2 in the other half. Neglect leakage. The B-H characteristics are as below:*

For Cast Steel

B in Wb/m^2	*1.0*	*1.1*	*1.2*
Amp-turn/m	*900*	*1020*	*1220*

For Ordinary Steel

B in Wb/m^2	*1.2*	*1.4*	*1.45*
Amp-turn/m	*590*	*1200*	*1650*

Solution. The magnetic circuit is shown in Fig. 7.35

The m.m.f. (or AT) produced on the half A acts across the parallel magnetic circuits C and D. First, total AT across C is calculated and since these ampturns are also applied across D, the flux density B in D can be estimated. Next, flux density in A is calculated and, therefore, the AT required for this flux density. In fact, the total AT (or m.m.f) required is the sum of that required for A and that for any one of the two parallel paths C or D.

Value of flux density in C = 1.0 Wb/m^2

Mean diameter of the ring = (24 + 18)/2 = 21 cm

∴ mean circumference = $\pi \times 21$ = 66 cm

Length of a path A or $C = 66/2 = 33$ cm = 0.33 m

Value of AT/m for a flux density of 1.0 Wb/m^2 as seen from the given B-H characteristics = 900 AT/m

$\therefore$ total AT for path $C = 900 \times 0.33 = 297$

The same ATs are applied across path D.

Length of path D = 18 cm = 0.18 m

$\therefore$ AT/m for D = 297/0.18 = 1650

Value of B corresponding to this value of AT/m from the given table is = 1.45 Wb/m^2

Flux through path $C = B \times A$

$= 1.0 \times 9 \times 10^{-4} = 9 \times 10^{-4}$ Wb

Flux through path $D = 1.45 \times (3 \times 0.4 \times 10^{-4})$

$= 1.74 \times 10^{-4}$ Wb

$\therefore$ total flux through $A = 9 \times 10^{-4} + 1.74 \times 10^{-4}$

$= 10.74 \times 10^{-4}$ Wb

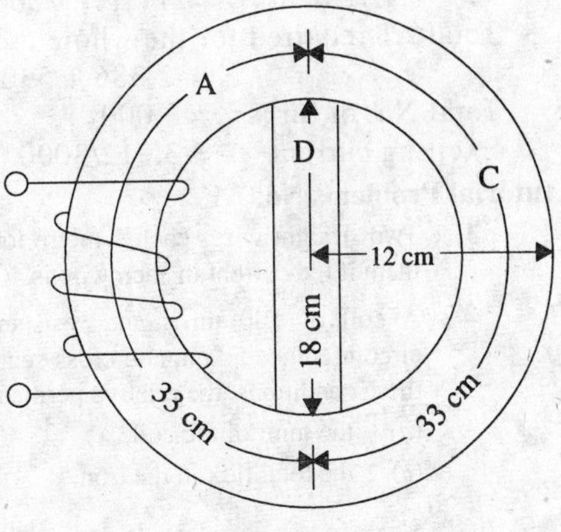

Flux density through $A = 10.74 \times 9 \times 10^{-4}$

$= 1.193$ Wb/m^2

Fig. 7.35

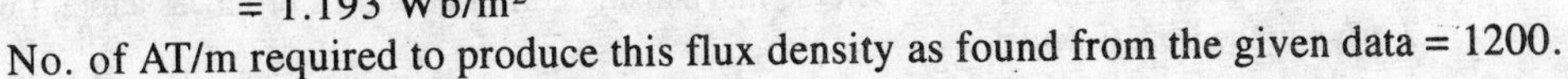

No. of AT/m required to produce this flux density as found from the given data = 1200.

$\therefore$ Ampere-turns required for path $A = 1200 \times 0.33 = 396$

Total AT required = 396 + 297 = **693.**

Example 7.18. *Determine the exciting current required to establish a flux of 1.6 mWb in each air gap in the magnetic circuit shown in Fig. 7.36. Dimensions given are in mm. Cross-section of A, B, C is square 40 mm side and that of iron armature D is 30 mm × 40 mm. The relative permeabilities of A, B, and C are 900 and that of D is 750. There are 1500 turns in each of the exciting coils connected in series. Neglect fringing and leakage.*

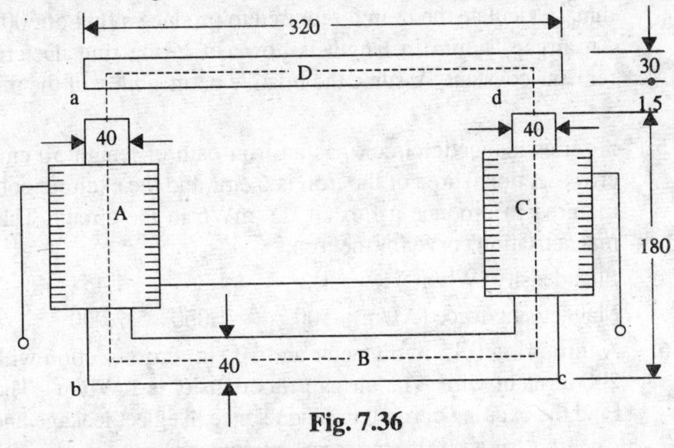

Fig. 7.36

Solution. The mean flux path is shown by the dotted line in Fig. 7.36. Let us find AT required for each path separately.

Air Gaps

$\Phi = 1.6 \times 10^{-3}$ Wb; $A = (40 \times 40) \times 10^{-6} = 16 \times 10^{-4}$ m^2

$B = 1.6 \times 10^{-3}/16 \times 10^{-4} = 1$ Wb/m^2

$H = B/\mu_0 = 1/4\pi \times 10^{-7} = 79.54 \times 10^4$ AT/m

Length of the two air-gaps is = 2 × 1.5 = 3 mm

AT reqd. $= 79.54 \times 10^4 \times 3 \times 10^{-3} = 2{,}386$

Path A, B and C

$B = 1$ Wb/m^2

$H = 1/4\pi \times 10^{-7} \times 900 = 884$ AT/m

Mean length of the path '*abcd*'= 160 + 280 + 160 = 600 mm

AT reqd. $= 884 \times 600 \times 10^{-3} = 530$

Iron Armature D

$\Phi = 1.6$ mWb; $A = (30 \times 40) \times 10^{-6} = 12 \times 10^{-4}$ m^2

$B = 1.6 \times 10^{-3}/12 \times 10^{-4} = 4/3$ Wb/m^2

$H = 4/3 \times 4\pi \times 10^{-7} \times 750 = 1415$ AT/m

Mean flux length $= 320 - (2 \times 20) = 280$ mm

AT reqd. $= 1415 \times 280 \times 10^{-3} = 396$

Total AT required for the whole circuit

$= 2,386 + 530 + 396 = 3,312$

Total No. of turns $= 3,000$

Exciting current $= 3,312/3000 =$ **1.104 A**

Tutorial Problems No. 7.1

1. Two straight wires each 1 metre long are parallel and 0.02 m apart. Calculate the force between them if the current in each wire is 40 A. **[16 × 10^{-3} N]** *(A.M.I.E. Summer 1975)*

2. A coil of 500 turns and resistance 20 Ω is wound uniformly on an iron ring of mean circumference 50 cm and cross-sectional area 4 cm^2. It is connected to a 24-V d.c. supply. Under these conditions, the relative permeability of iron is 800. Calculate the values of :

   (*a*) the mmf of the coil, (*b*) the magnetising force,

   (*c*) the total flux in the iron, (*d*) the reluctance of the ring

   [(*a*) 600 AT (*b*) 1200 AT/m (*c*) 0.483 mWb (*d*) 1.24 × 10^6 AT/Wb]

3. A magnetic circuit consists of an iron ring of mean circumference 80 cm with cross-sectional area 12 cm^2 throughout. A current of 2 A in the magnetising coil of 200 turns produces a total flux of 1.2 mWb in the iron. Calculate:

   (*a*) the flux density in the iron

   (*b*) the absolute and relative permeability of iron

   (*c*) the reluctance of the circuit. **[(*a*) 1 Wb/m^2 (*b*) 0.002; 1590 (*c*) 3.33 × 10^5 AT/Wb]**

4. A mild steel ring has a mean circumference of 500 mm and a uniform cross-sectional area of 300 mm^2. Calculate the m.m.f. required to produce a flux of 500 μWb.

   An airgap, 1 mm in length is now cut in the ring. Determine the flux produced if the m.m.f. remains constant. Assume the relative permeability of the mild steel to remain constant at 1200.

   [553 A; 147 μWb]

5. A series magnetic circuit has an iron path of length 50 cm and an air-gap of length 1 mm. The cross-sectional area of the iron is 6 cm^2 and the exciting coil has 400 turns. Determine the current required to produce a flux of 0.9 mWb in the circuit. The following points are taken from the magnetisation curve for the iron.

Flux density (Wb/m^2) :	1.2	1.35	1.45	1.55
Magnetising force (AT/m):	500	1,000	2,000	4,000

   [6.35 A]

6. An iron ring 0.15 m diameter and 10 cm^2 cross-section with a sawcut 2 mm wide is wound with 200 turns of wire. The air gap flux density is 1 Wb/m^2. The relative permeability of iron is 500. Find the exciting current and inductance. Neglect leakage and fringing. **[11.7 A; 17.1 mH]**

7. An iron ring 100 cm mean circumference and of circular cross-section 5 cm^2 has a saw-cut 2 mm in length. It is wound with 500 turns of wire. If 0.5 mWb flux exists across the air-gap, what will be the value of exciting current? Take leakage coefficient = 1.26 and relative permeability of iron = 500.

   [3.6 A] *(A.M.I.E. Summer 1976)*

8. A steel ring 15 cm mean radius and of circular section 1 cm in radius has an air gap of 1 mm length. It is wound uniformly with 500 turns of wire carrying a current of 3 A. Neglect magnetic leakage. The air gap takes 60% of the total M.M.F. Find reluctance.

   [4.17 × 10^{-6} AT/Wb] *(A.M.I.E. Summer 1987)*

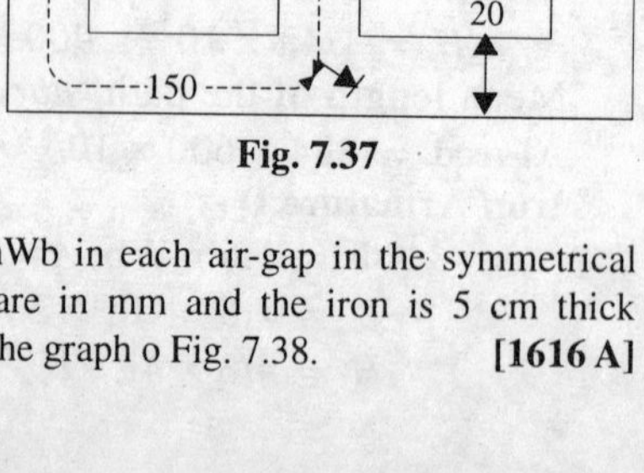

Fig. 7.37

9. Calculate the mmf required to produce a flux of 0.8 mWb in each air-gap in the symmetrical magnetic circuit shown in Fig. 7.37. All dimensions are in mm and the iron is 5 cm thick throughout. The material is cast steel with data shown in the graph o Fig. 7.38. **[1616 A]**

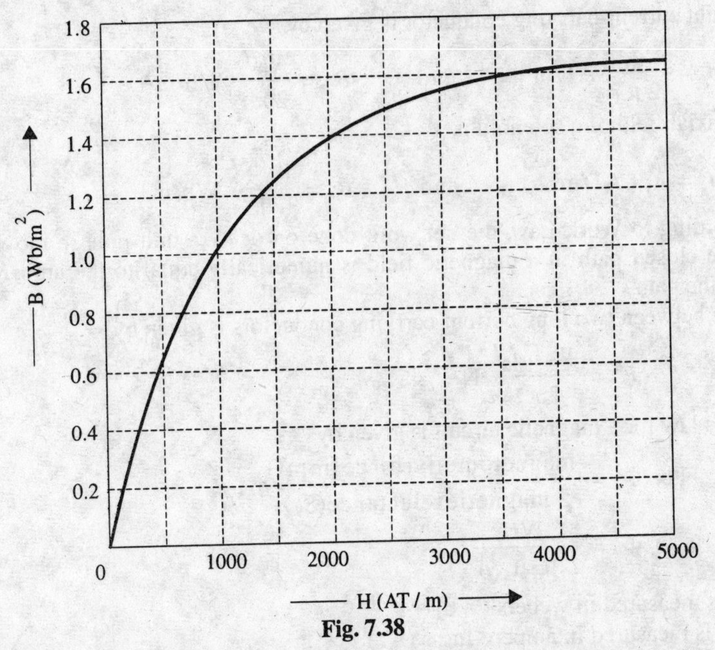

Fig. 7.38

10. Calculate the approximate value of current needed to produce a flux of 2 mWb in the air gap of the magnetic circuit shown in Fig. 7.39. The material is cast steel (for data refer to Fig. 7.38) and is 10 cm. thick.The exciting coil has 1000 turns and is wound on the right hand limb. Neglect leakage and fringing. All dimensions are in cm. **[1.92 A]**

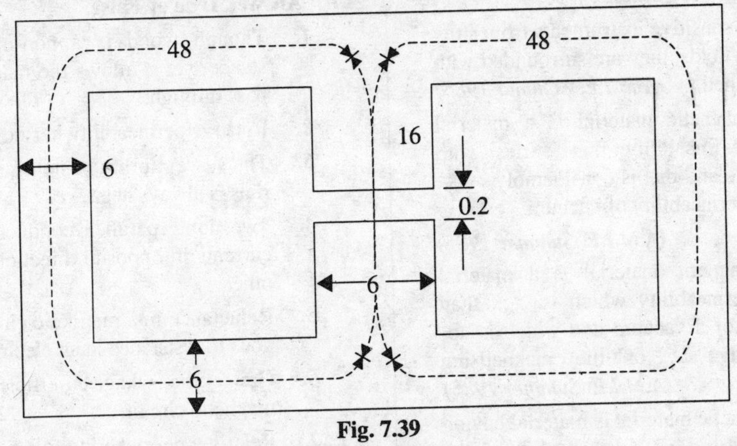

Fig. 7.39

HIGHLIGHTS

1. The force between two magnetic poles is given by

$$F = \frac{m_1 \ m_2}{4\pi \ \mu_0 \ \mu_r \ d^2} \text{ N}$$

2. Flux density is given by

$$B = \mu_0 \mu_r \ H \text{ Wb/m}^2$$

3. Force acting on a current-carrying conductor lying in and at right angles to a magnetic field is given by

$$F = BIl = \mu_0 \ \mu_r \ HIl \text{ newton}$$

4. The field strength at a point lying at a perpendicular distance of r metres from a long

straight current-carrying conductor is given by

$$H = \frac{I}{2\pi r}\ \text{AT/m} \quad \text{and} \quad B = \frac{\mu_0 I}{2\pi r}\ \text{Wb/m}^2 \quad \text{— in air}$$

5. The axial field of a solenoid is

$$H = \frac{NI}{l}\ \text{AT/m} \quad \text{and} \quad B = \frac{\mu_0 NI}{l}\ \text{Wb/m}^2 \quad \text{— in air}$$

6. According to Work Law, the net work done on or by a unit pole in moving round any single closed path in a magnetic field is numerically equal to the ampere-turns linked with the path.
7. Force between two long current-carrying conductors is given by

$$F = \frac{\mu_0 I_1 I_2 l}{2\pi d}\ \text{N}$$

8. Ohm's law for a magnetic circuit is given by

$$\text{Flux } (\Phi) = \frac{\text{magnetomotiveforce (mmf)}}{\text{magnetic reluctance(S)}}$$

$$\Phi = \frac{NI}{l/\mu_0 \mu_r A}$$

Flux is measured in webers (Wb)
MMF is measured in ampere-turns (AT)
Reluctance is measured in amp-turns/weber (AT/Wb)

OBJECTIVE TESTS—7

A. Fill in the following blanks.

1. To shield sensitive instruments from stray magnetic fields, they are surrounded with an ....shell. *(A.M.I.E. Summer 1985)*
2. A ferromagnetic material is a material having permeability that varies with ....force and that is considerably .... than the permeability of vacuum. *(A.M.I.E. Summer 1985)*
3. A paramagnetic material is a material having permeability which is ....than that ....of a vacuum and which is approximately ....of the magnetising force. *(A.M.I.E. Summer 1985)*
4. A diamagnetic material is material having permeability .... than that of the vacuum. *(A.M.I.E. Summer 1985)*
5. Ferromagnetic materials lose their.........property at a ....known as Curie point. *(A.M.I.E. Summer 1985)*
6. Two closely-lying long parallel conductors carrying currents in opposite directions.........each other.
7. Permeability in a magnetic circuit is analogous to .... in an electric circuit.
8. Iron is used for magnetic circuit as .... is used for electric circuits.

B. Answer True or False.

1. Though wood is a non-magnetic substance, yet it allows the magnetic flux to pass through.
2. Relative permeability has no units.
3. The susceptibility of non-magnetic materials is always negative.
4. Two long parallel conductors carrying currents in opposite directions repel each other.
5. Reluctance in a magnetic circuit is analogous to resistance in an electric circuit.
6. Greater the leakage co-efficient, better the magnetic circuit.
7. Relative permeability of a material increases with increase in the flux density established in it.
8. Relative permeability of vacuum is zero.

C. Multiple Choice Questions.

1. The unit of magnetising force is
   (*a*) henry/metre (*b*) weber/metre^2
   (*c*) ampere/metre (*d*) joule/weber.
2. The magnetising force produced by a solenoid depends on
   (*a*) the number of its turns
   (*b*) the current carried by it

(*c*) its length
(*d*) all of the above.

3. Aluminium can be classified as a ...... material.
(*a*) paramagnetic (*b*) ferromagnetic
(*c*) diamagnetic (*d*) soft magnetic.

4. Ferrites are ferromagnetic ceramics which have a fairly constant ....
(*a*) hysteresis loss
(*b*) relative permeability
(*c*) frequency response
(*d*) magnetic susceptibility.

5. Permeance of a magnetic circuit is given by the reciprocal of its ....
(*a*) reluctance (*b*) permeability
(*c*) flux density (*d*) reluctivity.

6. Magnetic leakage in electric machines leads to their
(*a*) increased weight
(*b*) increased cost
(*c*) reduced efficiency
(*d*) all of the above
(*e*) only (*a*) and (*b*).

7. The inductance of the coil wound with 200 turns on an iron ring with relatively long air gap and found to produce a flux of 130 μWb, when a current of 3A is passed through it is given by .... milli-henry.
(*a*) 8.67 (*b*) 86.7
(*c*) 867 (*d*) 1 .

(Elect. Engg. A.M.Ae. S.I. Dec. 1994)

ANSWERS

A. 1. iron 2. magnetising, greater 3. greater, independent 4. lesser 5. magnetic, temperature 6. repel 7. conductivity 8. copper.

B 1. T 2. T 3. F 4. T 5. T 6. F 7. F 8. F

C. 1. *c* 2. *d* 3. *a* 4. *b* 5. *a* 6. *e*

8 ELECTROMAGNETIC INDUCTION

8.1. Relation Between Magnetism and Electricity

It is well known that whenever an electric current flows through a conductor, a magnetic field is immediately brought into existence in the space surrounding the conductor. We can say that when electrons are in motion, they produce a magnetic field. The converse of this is also true *i.e.* when a magnetic field embracing a conductor moves relative to the conductor, it produces a flow of electrons. This phenomenon whereby an e.m.f. and hence current (*i.e.* flow of electrons) is induced in any conductor that is cut across or is cut by a magnetic flux is known as *electromagnetic induction.* The historical background of this phenomenon is this:

After the discovery by Oersted that electric current produces a magnetic field, scientists began to search for the reverse phenomenon from about 1821 onwards. The problem they put to themselves was how to 'convert' magnetism into 'electricity'. It is recorded that Michael Faraday* was in the habit of walking about with magnets in his pockets so as to keep himself constantly reminded of the problem. After nine years of continuous research and experimentation, he succeeded in producing electricity by 'converting' magnetism. In 1831, he formulated the basic laws underlying the phenomenon of electromagnetic induction (known after his name), upon which is based the operation of most of the commercial apparatus like motors, generators and transformers etc.

8.2. Production of Induced E.M.F. and Current

In Fig. 8.1 is shown an insulated coil whose terminals are connected to a sensitive galvanometer *G*. It is placed close to a stationary bar magnet initially at position *AB* (shown dotted). As it is, some lines of flux from the *N*-pole of the magnet are linked with or thread through the coil but, as yet, there is not deflection of the galvanometer. Now, suppose that the magnet is *suddenly* brought closer to the coil in position *CD* (Fig. 8.1). Then, it is found that there is a jerk or a sudden but momentary deflection in the galvanometer and that this *lasts so long as the magnet is in motion relative to the coil, not otherwise.*

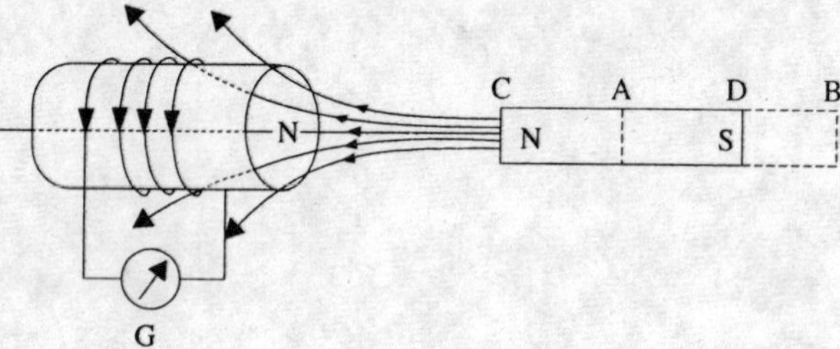

Fig. 8.1

The deflection is reduced to zero when the magnet becomes again stationary at its new position *CD*. It should be noted that due to the approach of the magnet, flux linked with the coil is increased.

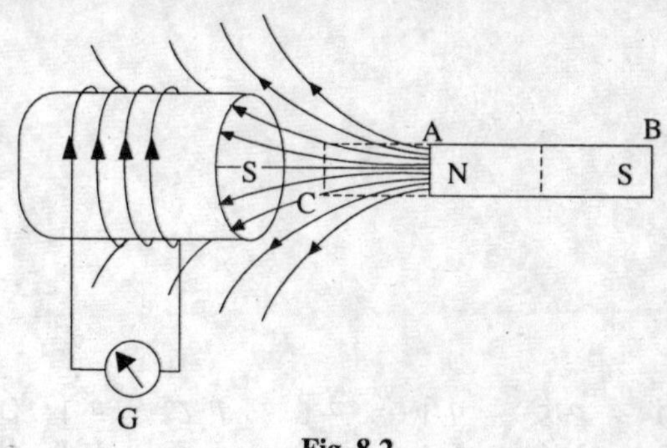

Fig. 8.2

Next, the magnet is *suddenly* withdrawn away from the coil as shown in Fig. 8.2. It is found that again there is a *momentary* deflection in the galvanometer and it persists so

*Michael Faraday (1791-1867), an English physicist.

long as the magnet is in motion, not when it becomes stationary. It is important to note that this deflection is in a direction opposite to that of Fig. 8.1. Obviously, due to the withdrawal of the magnet, flux linked with the coil is decreased.

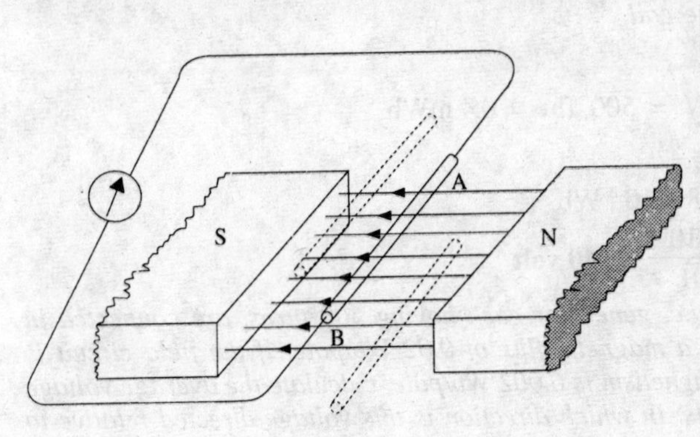

Fig. 8.3

The deflection of the galvanometer indicates the production of e.m.f. in the coil. The only cause of the production can be the sudden approach or withdrawal of the magnet from the coil. It is found that the actual cause of this e.m.f. is the change of the flux linking with the coil. This e.m.f. exists so long as the change in flux exists. Stationary flux, however strong, will never induce any e.m.f. in a stationary conductor. In fact, the same results can be obtained by keeping the bar magnet stationary and moving the coil suddenly away or towards the magnet.

The direction of current set up by the induced e.m.f. is as shown in the two figures given above.

The production of this electromagnetically-induced e.m.f. is further illustrated by considering a conductor *AB* lying within a magnetic field and connected to a galvanometer as shown in Fig. 8.3. It is found that whenever this conductor is moved up or down, a *momentary* deflection is produced in the galvanometer. It means that some transient e.m.f. is induced in *AB*. The magnitude of this induced e.m.f. (and hence the amount of deflection in the galvanometer) *depends on the quickness of the movement of AB.*

From this experiment we conclude that whenever a conductor cuts or *shears* the magnetic lines of flux, an e.m.f. is always induced in it.

It is also found that if the conductor is moved parallel to the direction of the lines of flux (so that it cuts none of these lines), then no e.m.f. is induced.

8.3. Faraday's Laws of Electromagnetic Induction

Faraday summed up the above facts into two laws known as Faraday's laws of electromagnetic Induction.

First Law. It states :—

'When the magnetic flux linked with a circuit changes, an e.m.f. is always induced in it.' *or*

'Whenever a conductor cuts across magnetic lines of flux, an e.m.f. is induced in that conductor.

Second Law. It states :—

'The magnitude of the induced e.m.f. is equal to the rate of change of *flux linkages*'.

Explanation. Suppose a coil has *N* turns and flux through it changes from an initial value of Φ_1 Wb to the final value of Φ_2 Wb in time *t* seconds. Then, remembering that by flux linkages is meant the product of number of turns by the flux linked with coil, we have

Initial flux linkages $= N\Phi_1$; Final flux linkages $= N\Phi_2$

$$\therefore \text{ induced e.m.f. } e = \frac{N\Phi_2 - N\Phi_1}{t} \text{ volt} \quad \text{or} \quad e = N\frac{\Phi_2 - \Phi_1}{t} \text{ volt}$$

Putting the above expression in its differential from, we get

$$e = \frac{d}{dt}(N\Phi) \quad \text{or} \quad e = N\frac{d\Phi}{dt} \text{ volt}$$

Usually, a minus sign is given to the right-hand side expression to signify the fact that the induced e.m.f. sets up current in such a direction that magnetic effect produced by it opposes the very cause producing (Art. 8.6).

$$\therefore \qquad e = -N\frac{d\Phi}{dt} \text{ volt}$$

Example 8.1. *A coil of 500 turns is linked by a flux of 0.4 mWb. If the flux is reversed in 0.01 second, find the e.m.f. induced in the coil.*

Solution. Induced e.m.f. is

$$e = N\frac{d\Phi}{dt} \text{ volt}; \qquad N = 500, \text{ flux} = 0.4 \text{ mWb}$$

$$dt = 0.01 \text{ second}; \qquad e = ?$$

$$d\Phi = 0.4-(-0.4) = 0.8 \text{ mWb} = 8\times10^{-4} \text{ Wb}$$

$$e = 500 \times \frac{8\times10^{-4}}{0.01} = \mathbf{40\ volt}$$

Example 8.2. *The field of a 6-pole d.c. generator each having 500 turns, are connected in series. When the field is excited, there is a magnetic flux of 0.02 Wb/pole. If the field circuit is opened in 0.02 second and the residual magnetism is 0.002 Wb/pole, calculate the average voltage which is induced across the field terminals. In which direction is this voltage directed relative to the direction of the current?*

Solution. Total number of the turns

$$N = 6\times500 = 3000$$

Total initial flux $= 6\times0.02 = 0.12$ Wb

Total residual flux $= 6\times0.002 = 0.012$ Wb

Change in flux, $d\Phi = 0.12 - 0.012 = 0.108$ Wb

Time of opening the circuit

$$dt = 0.02 \text{ second}$$

$$\therefore \quad \text{induced e.m.f.} = N\frac{d\Phi}{dt} = 3000\times\frac{0.108}{0.02} = \mathbf{16{,}200\ V}$$

The direction of this induced e.m.f. is the same as that of the *original* direction of the exciting current.

Example 8.3. *A coil of resistance 100 Ω is placed in a magnetic field of 1 mWb. The coil has 100 turns and a galvanometer of 400Ω resistance is connected in series with it. Find the average e.m.f. and the current if the coil is moved in 1/10th second from the given field to a field strength of 0.2 mWb.*

Solution. Induced e.m.f. $e = N.\frac{d\Phi}{dt}$ volt

Here, $d\Phi = 1 - 0.2 = 0.8 \text{ mWb} = 0.8\times10^{-3}$ Wb

$dt = 1/10 \text{ second} = 0.1$ s; $\quad N = 100$

$\therefore \quad e = 100\times0.8\times10^{-3}/0.1 = \mathbf{0.8\ V}$

Total circuit resistance $= 100 + 400 = 500\ \Omega$

$\therefore$ induced current $= 0.8/500 = 1.6\times10^{-3}$ A $=$ **1.6 mA**

8.4. Direction of induced E.M.F. and Current

There exists a definite relationship between the direction of the induced current, the direction of the lines of flux and the direction of motion of the conductor. The direction of the induced current may be found easily by applying either Fleming's Right hand Rule or Lenz's Law.

8.5. Fleming's Right-hand Rule

Hold out the right hand with the first finger, second finger and thumb at right angles to each other [Fig 8.4 (*a*)].

If Forefinger represents the direction of the lines of Flux, the thuMb points in the direction of Motion, then seCond finger points in the direction of the induced Current.

Another way of applying right-hand rule is shown in Fig. 8.4 (*b*). According to this method, when the right hand is so placed in a magnetic field along the conductor that flux coming from

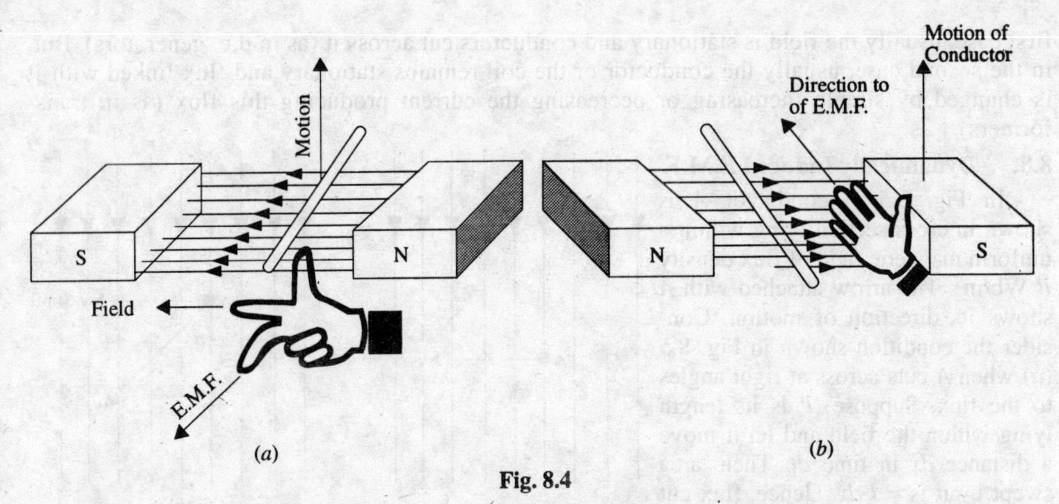

Fig. 8.4

N-pole *enters* the palm perpendicularly and the thumb points in the direction in which the conductor moves, the other four fingers give the direction of the induced e.m.f.

8.6. Lenz's Law*

The direction of the induced current may also be found by this law which was formulated by Lenz in 1835. This law states, in effect, that electromagnetically-induced current always flow in such a direction that the action of the magnetic field set up by it tends to oppose the very cause which produces it.

This statement will be clarified with reference to Fig. 8.1 and 8.2. It is found that when *N*-pole of the bar magnet approaches the coil, the induced current set up by the induced e.m.f. flows in the *anti-clockwise* direction in the coil as seen from the magnet side. The result is that that face of the coil becomes a *N*-pole and so tends to retard the onward approach of the *N*-pole of the magnet (like poles repel each other). The mechanical energy spent in overcomming this repulsive force is converted into electrical energy which appears in the coil.

When the magnet is withdrawn, as in Fig. 8.2, the induced current flows in the *clockwise* direction, thus making the face of the coil (facing the magnet) a *S*-pole. Therefore, the *N*-pole of the magnet has to be withdrawn against the attractive force of the *S*-pole of the coil. Again, the mechanical energy required to overcome this force of attraction is converted into electric energy.

It can be shown that Lenz's law is a direct consequence of Law of Conservation of Energy. Imagine for a moment that when *N*-pole of the magnet (see Fig. 8.1) approaches the coil, induced current flows in such a direction as to make the coil face a *S*-pole. Then, due to inherent attraction between unlike poles, the magnet would be automatically pulled towards the coil without the expenditure of any mechanical energy. It means that we would be able to create electric energy out of nothing which is denied by the inviolable Law of Conservation of Energy. In fact, to maintain the sanctity of this law, it is imperative for the induced current to flow in such a direction that the magnetic effect produced by it tends to oppose the very cause which produces it. In the present case, it is the relative motion of the magnet with respect to the coil which is the cause of the production of the induced current. Hence, the induced current always flows in such a direction as to tend to oppose this relative motion *i.e.* approach or withdrawal of magnet in the present case.

8.7. Induced E.M.F.

Induced e.m.f. can be either (*i*) *dynamically induced* or (*ii*) *statically induced* . In the

*After the Russian-born geologist and physicist Heinrich Friedrich Emil Lenz (1808 - 1865). He did lot of investigation on electromagnetism himself and carefully studied the experiments conducted by Faraday. The familiar statement afterwards known as Lenz's law that "the electrodynamic action of an induced current opposes equally the mechanical action inducing it" was contained in one of his papers "On the Direction of Galvanic Currents which are Excited through Electro-dynamic Action" read on 29th Nov. 1833.

first case, usually the field is stationary and conductors cut across it (as in d.c. generators). But, in the second case, usually the conductor or the coil remains stationary and flux linked with it is changed by simple increasing or decreasing the current producing this flux (as in transformers).

8.8. Dynamically Induced E.M.F.

In Fig. 8.5, a conductor A is shown in cross-section lying within a uniform magnetic field of flux density B Wb/m^2. The arrow attached with A shows its direction of motion. Consider the condition shown in Fig. 8.5 (*a*) when A cuts across at right angles to the flux. Suppose 'l' is its length lying within the field and let it move a distance dx in time dt. Then, area swept by it is $= l\ dx$. Hence, flux cut $= l.dx \times B$

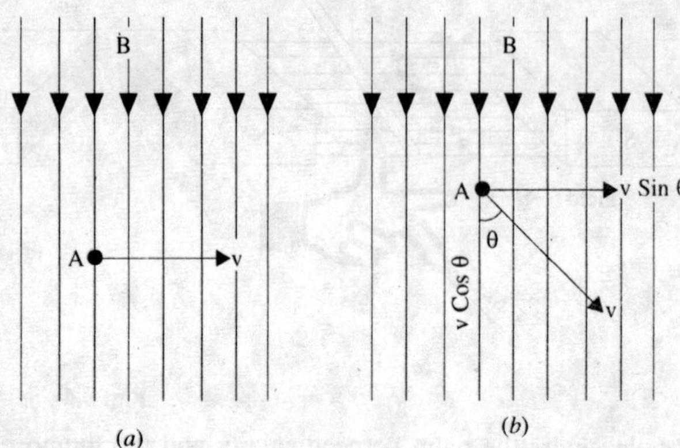

Fig. 8.5

$\therefore$ change in flux, $d\Phi = B\, l\, dx$

Time taken $= dt$

Hence, according to Faraday's Laws (Art. 8.3) the e.m.f. induced in it (known as dynamically induced e.m.f.) is

= rate of change of flux linkages

$$= \frac{d\Phi}{dt} = \frac{Bldx}{dt} = Bl\frac{dx}{dt} = Blv \text{ volt}$$

where $\frac{dx}{dt}$ = conductor velocity v

If the conductor A moves at an angle θ with the direction of the lines of flux [Fig. 8.5 (*b*) then the induced e.m.f. is

$$e = Blv \sin\theta \text{ volt}$$

The direction of the induced e.m.f. is given by Fleming's Right-hand rule (Art. 8.5).

It should be noted that generators work on the production of dynamically induced e.m.f. in the conductors housed in a revolving armature lying within a strong magnetic field.

Example 8.4. *A straight conductor has an active length of 20 cm and moves in a uniform field of 0.5 T at the rate of 5 ms. Find the generated e.m.f. if the motion of the conductor is (a) parallel (b) perpendicular (c) at an angle of 30° to the direction of the field.*

(Electrical Engg. A.M.Ae. S.I. Dec. 1993)

Solution. (*a*) Since the conductor moves parallel to the field, it does not cut it. Hence, no dynamically induced e.m.f. is produced. Alternatively,

$$e = \text{Blv} \sin\theta = 0.5 \times 0.2 \times 5 \times \sin 0° = \mathbf{0}$$

(*b*) Here, $\theta = 90°$

$$e = \text{Blv} \sin\theta = 0.5 \times 0.2 \times 5 \times \sin 90° = \mathbf{0.5\ V}$$

(*c*) Here, $\theta = 30°$, $e = 0.5 \times 0.2 \times 5 \times \sin 30° = \mathbf{0.25\ V}$

Example 8.5. *Calculate the e.m.f. generated in the axle of a car travelling at 72 km/h assuming the length of the axle to be 2 m and vertical component of the earth's magnetic field to be 40 μWb/m^2.*

Solution. Car velocity = 72 km/h = 72 × 5/18 = 20 m/s

Area of the magnetic field swept by the axle

= axle length × car velocity

$\therefore$ $A = 2 \times 20 = 40\ \text{m}^2$

flux cut, $d\Phi = BA = 40 \times 10^{-6} \times 40 = 1600 \times 10^{-6}$ Wb

time taken = 1 second

$$e = N\frac{d\Phi}{dt} = 1 \times \frac{1600 \times 10^{-6}}{1} = 1600 \times 10^{-6} \text{ V}$$

$= \mathbf{1.6\ mV}$

Example 8.6. *A conductor 30 cm long rotates about one end at 1000 r.p.m. in a plane perpendicular to a magnetic field of strength 0.5 Wb/m². Find the emf induced in it.*

(A.M.I.E. Winter 1981)

Solution. As shown in Fig. 8.6, as the rod *MN* rotates about an axis passing through its end *M*, cuts the flux perpendicularly.

Now, $e = Bl\upsilon$ volt

The velocity of outer end of the rod is maximum and that of the inner end is zero. Linear velocity of outer end is

$v = \omega r = (1000/60) \times 0.3$

$= 5$ m/s

Velocity of the inner end is zero because it forms part of the axis so that its $r = 0$.

$\therefore$ mean velocity of the revolving rod

$= (5 + 0)/2 = 2.5$ m/s

$\therefore \quad e = 0.5 \times 0.3 \times 2.5 = \mathbf{0.375\ V.}$

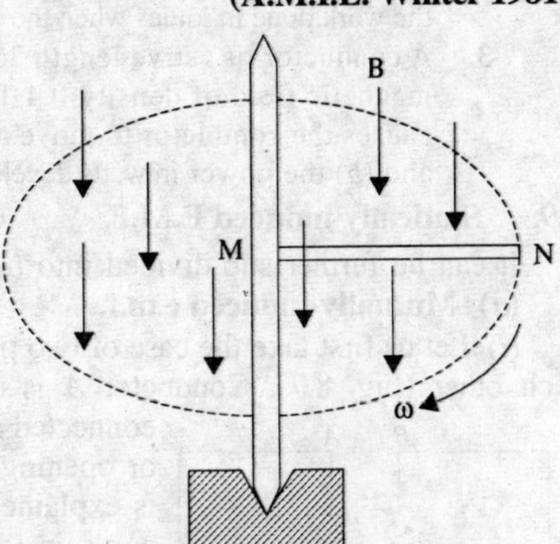

Fig. 8.6

Example 8.7. *A conductor 10 cm long and carrying a current of 50 A lies perpendicular to a field of strength 1000 AT/m, Calculate*

(*i*) *the force acting on the conductor,*

(*ii*) *the mechanical power to move this conductor against this force with a speed of 1 m/s and*

(*iii*) *e.m.f. induced in the conductor.*

(Magnetic Circuits & Materials, Punjab Univ. 1993)

Solution. (*i*) $F = Bll$ *newton*

Now, $H = 100$ AT/m

$\therefore \quad B = \mu_0 H = 4\pi \times 10^{-7} \times 10^3 = 4\pi \times 10^{-4}$ Wb/m²

$F = 4\pi \times 10^{-4} \times 50 \times 0.1 = \mathbf{6.28 \times 10^{-3}\ N}$

(*ii*) $P = F \times \upsilon = 6.28 \times 10^{-3} \times 1 = \mathbf{6.28 \times 10^{-3}\ W}$

(*iii*) $e = Bl\upsilon = 4\pi \times 10^{-4} \times 0.1 \times 1 = \mathbf{4\pi \times 10^{-5}\ V}$

Note. Incidentally, electric power produced

$= eI = 4\pi \times 10^{-5} \times 50 = 6.28 \times 10^{-3}$ W

the same as mechanical power input.

Example 8.8. *In a 4-pole dynamo, the flux/pole is 15 mWb. Calculatethe average e.m.f. induced in one of the armature conductors if armature is driven at 600 r.p.m.*

Solution. It should be noted that each time a conductor passes under a pole (whether *N* or *S*) it cuts a flux of 15 mWb. Hence, total flux cut in one revolution is 15 × 4 = 60 mWb. Since conductor is rotating at 600/60 = 10 r.p.s., time taken for one revoltution is 1/10 second = 0.1 second.

$\therefore$ average e.m.f. generated $= N\frac{d\Phi}{dt}$ volt

Here $N = 1$; $d\Phi = 60 \text{ mWb} = 6 \times 10^{-2}$ Wb

$dt = 0.1$ second

$e = 1 \times 6 \times 10^{-2}/0.1 = \mathbf{0.6\ V}$

Tutorial Problem No. 8.1

1. A conductor, 0.6 m long, is carrying a current of 75 A and is placed at right-angles to a magnetic field of uniform flux density. Calculate the value of the flux density if the mechanical force on the conductor is 30 N. **[0.667 Wb/m^2]**
2. A conductor, 500 mm long, is moved at a uniform speed at right angles to its length and to a uniform magnetic field having a density of 0.4 T. If the e.m.f. generated in the conductor is 2V and the conductor forms part of a closed circuit having resistance of 0.5 Ω, calculate (*i*) velocity of the conductor in metres/second, (*ii*) the force acting acting on the conductor in newtons, (*iii*) the work done in joules when the conductor has moved 600 mm. **[10m/s ; 0.8 N; 0.48 J]**
3. A conductor of active length 30 cm carries a current of 100 A and lies at right-angles to a magnetic field of density 0.4 T. Calculate the force in newtons exerted on it. If the force causes the conductor to move at a velocity of 10 m/s, calculate (*a*) the e.m.f. induced in it and (*b*) the power in watts developed by it. **[12 N; 1.2 V; 120 W]**

8.9. Statically-induced E.M.F.

It can be further sub-divided into (*a*) *mutually-induced e.m.f. and* (*b*) *self-induced e.m.f.*

(*a*) Mutually-induced e.m.f.

(*i*) Let us first take the case of two parallel conductors *A* and *B* lying close to but not touching each other (Fig. 8.7). Conductor *A* is connected to a battery through a key *K*, whereas *B* is connected across a voltmeter. It is found that at the instant of closing or opening *K*, there is a momentary deflection in the voltmeter. As is explained below, it is due to the production of mutually-induced e.m.f. in *B* which is produced because of change in flux linked with it.

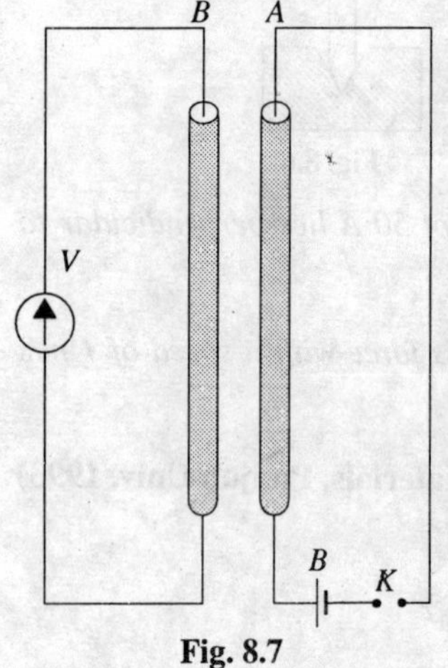

Fig. 8.7

Explanation It has been shown before that every current-carrying conductor is enshrouded by circular magnetic lines of force. But it should be remembered that this field is not established to its full strength *instantaneously* on closing the key *K*. It needs both time and energy (supplied by the battery) to bring this field to its full-value because current always takes some time (though very small) to come to its maximum steady value (see Art. 9.7 and Fig. 9.8). The gradual increase of current strength and hence of field strength around *A* is shown in Fig. 8.8.

In Fig. 8.8 (*a*), *K* has been closed and only a few lines of flux encircle *A*. Fig. 8.8 (*b*), (*c*) and (*d*) show in a rough diagrammatic way the increasing number of lines of flux that might be produced at successive instants as the current in *A* increases. In Fig. 8.8 (*d*), it is assumed that both current and flux around *A* have reached their maximum steady values and will continue at these values till *K* is opened.

As the flux increases, these lines of flux may be thought of as emanating *i.e.* expanding outwards from the axis of the conductor *A* just as ripples emanate from the point where a stone piece is dropped in a pond. These circular lines of flux expand very rapidly and, moreover, they expand at right angles to the conductor.

It will be seen from Fig. 8.8 (*c*) that some flux produced by current in *A* has cut conductor *B*. The flux that cuts *B* induces an e.m.f. in *B* which is called *mutually-induced e.m.f.* The induced e.m.f. is so called because it is produced in a conductor due to the changes of flux in a neighbouring conductor. In Fig. 8.8 (*d*) maximum flux has been established. Hence, *B* will not be cut further by more flux with the result that no further e.m.f. will be induced in it, because e.m.f. is induced only when the flux linked with a coil or a conductor is changing.

When *K* is opened, the current in *A* will rapidly but not *instantaneously* decrease to zero value. Hence, flux will also decrease to zero. The circular lines of flux will, therefore, contract, return to the axis of *A* and vanish. These lines while returning to *A* will cut *B* in a direction opposite to that in which they cut it while expanding outwards from *A*. Hence, they will again induce an e.m.f. in *B*, though this time in the opposite direction.

So, we conclude that :—

1. *A momentary e.m.f. is induced in B whenever K is closed or opened. This is shown by the momentary deflection of the voltmeter.*
2. *No. e.m.f. is induced in B when current through A is constant.*
3. *The direction of the induced e.m.f. in B in opposite to that in A.*

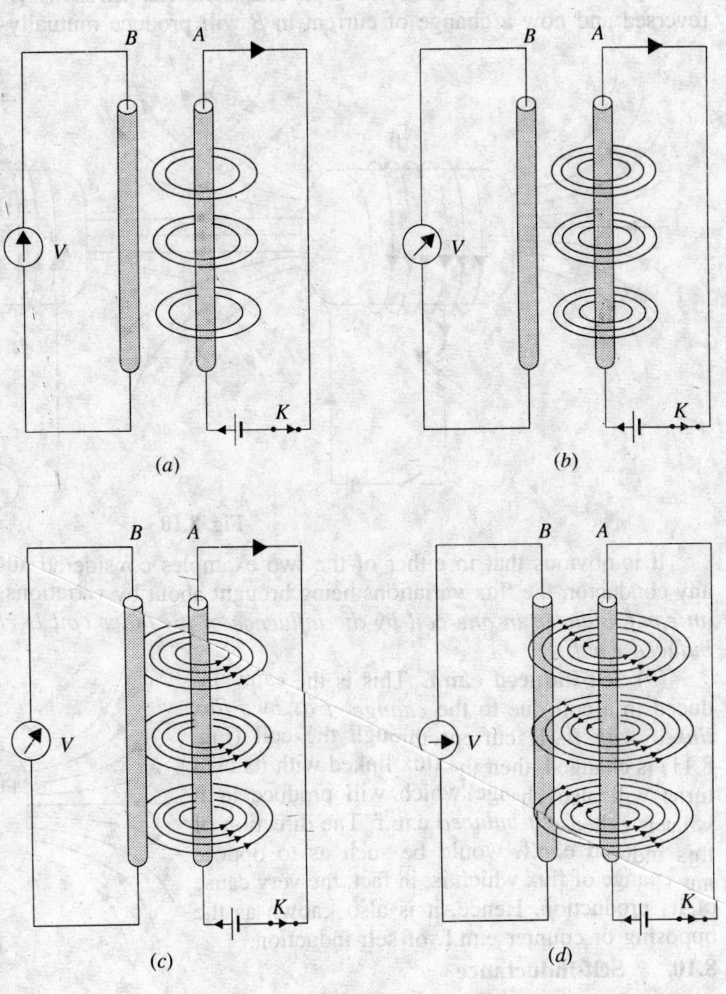

Fig. 8.8

(*ii*) Let us now consider two coils *A* and *B* lying close to each other (Fig. 8.9).

Coil *A* is joined to a battery, a switch and a variable resistance *R* whereas coil *B* is connected to a sensitive voltmeter *V*. When current through *A* is established by closing the key, its magnetic flux is set up which partly links with or threads through the coil *B*. As current through *A* is changed, the flux linked with *B* is also changed. Hence, mutually-induced e.m.f. is produced in *B* whose magnitude is given by Faraday's Laws (Art. 8.3) and direction by Lenz's Law (Art. 8.6).

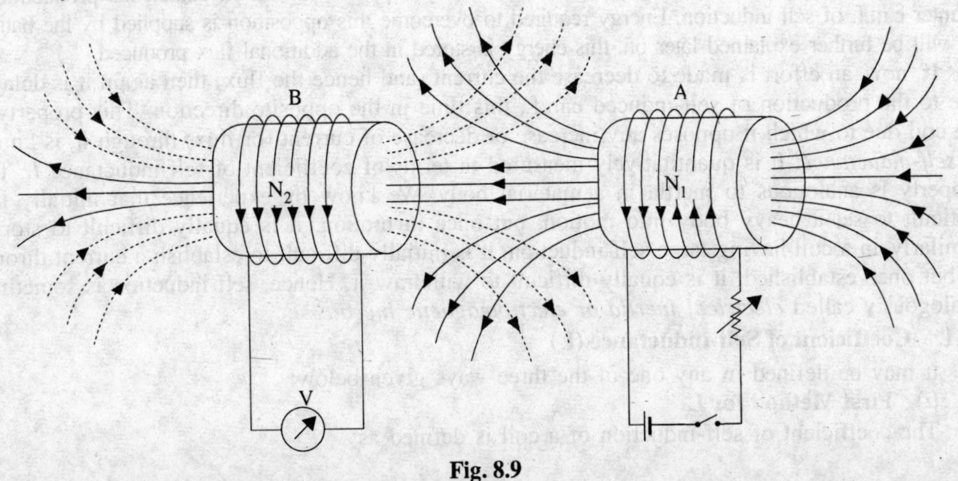

Fig. 8.9

If, now battery is connected to B and voltmeter across A [Fig. 8.10] then the situation is reversed and now a change of current in B will produce mutually-induced e.m.f. in A.

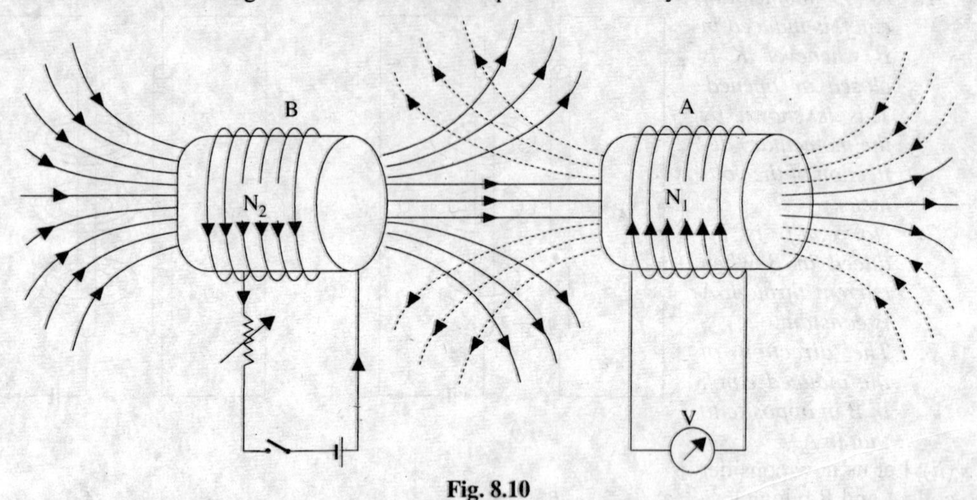

Fig. 8.10

It is obvious that in either of the two examples considered above, there is no movement of any conductor, the flux variations being brought about by variations in current strength only. *Such an e.m.f. induced in one coil by the influence of the other coil is called (statically but) mutually induced e.m.f.*

(*b*) **Self-induced e.m.f.** This is the e.m.f. induced in a coil due to the *change of its own flux linked with it.* If current through the coil (Fig. 8.11) is changed, then the flux linked with its own turns will also change which will produce in it what is called *self-induced* e.m.f. The direction of this induced e.m.f. would be such as to oppose any change of flux which is, in fact, the very cause of its production. Hence, it is also known as the opposing or counter e.m.f. of self-induction.

Fig. 8.11

8.10. Self-inductance

Imagine a coil of wire, similar to the one shown in Fig. 8.11, connected to a battery through a rheostat. It is found that whenever an effort is made to increase current (and hence flux) through it, it is always opposed by the instantaneous production of counter e.m.f. of self-induction. Energy required to overcome this opposition is supplied by the battery. As will be further explained later on, this energy is stored in the additional flux produced.

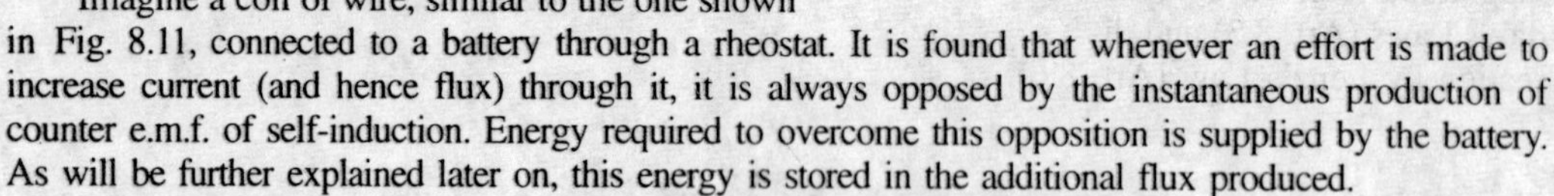

If, now, an effort is made to decrease the current (and hence the flux) then again it is delayed due to the production of self-induced e.m.f., this time in the opposite direction. This property of the coil due to which it opposes any increase or decrease of current (or flux) through it, is known as *self-inductance.* It is quantitatively measured in terms of coefficient of self-inductance L. This property is analogous to inertia in a material body. We know by experience that initially it is difficult to set a heavy body into motion, but once in motion, it is equally difficult to stop it. Similarly, in a coil having large self-induction, it is initially difficult to establish a current through it, but once established, it is equally difficult to withdraw it. Hence, self-induction is sometimes analogously called *electrical inertia or electromagnetic inertia.*

8.11. Coefficient of Self-inductance (L)

It may be defined in any one of the three ways given below:

(*i*) First Method for L

The coefficient of self-induction of a coil is defined as

"the weber-turns per ampere in the coil".

By 'weber-turns' is meant the product of flux in webers and the number of turns with which the flux is linked. In other words, it is the flux linkages of the coil.

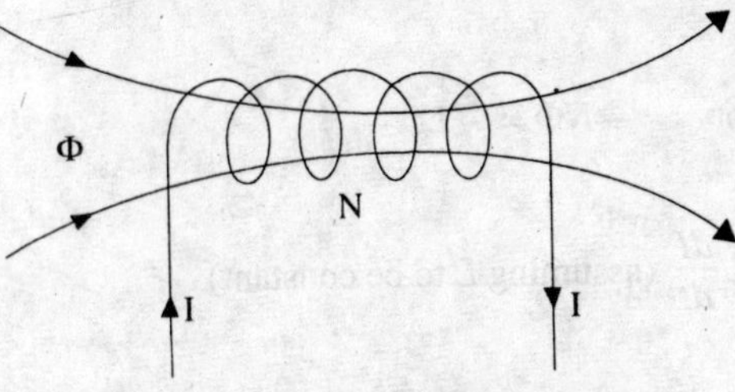

Fig. 8.12

Consider a solenoid having N turns and carrying a current of I amperes. (Fig. 8.12) If the flux produced is Φ webers, then weber-turns are $N\Phi$. Hence, weber-turns per ampere are $N\Phi/I$.

By definition, $L = \dfrac{N\Phi}{I}$

The unit of self-induction is henry*. If in the above relation,

$N\Phi$ = 1 Wb-turn, I = 1 ampere

then L = 1 henry (H)

Hence, a coil is said to have a self-inductance of one henry if a current of one ampere flowing through it produces flux linkages of one Wb-turn in it.

Therefore, the above relation becomes

$$L = \frac{N\Phi}{I} \text{ henry}$$

Example 8.9. *The field winding of a d.c. electromagnet is wound with 960 turns and has resistance of 50 Ω. When the exciting voltage is 230 V, the magnetic flux linking the coil is 0.005 Wb. Calculate the self-inductance of the coil and the energy stored in the magnetic field.*

Solution. Formula used : $L = \dfrac{N\Phi}{I}$ henry

Current through coil = 230/50 = 4.6 A

Φ = 0.005 Wb; N = 960

$$\therefore \quad L = \frac{960 \times 0.005}{4.6} = 1.0435 \text{ H}$$

$$\text{Energy stored} = \frac{1}{2}LI^2 = \frac{1}{2} \times 1.0435 \times 4.6^2 = \mathbf{11.04\ J}$$

(*ii*) **Second Method for L**

We have seen in Art. 7-23 that flux produced in a solenoid is

$$\Phi = \frac{NI}{l/\mu_0\mu_r A} \qquad \therefore \quad \frac{\Phi}{I} = \frac{N}{l/\mu_0\mu_r A}$$

Now $$L = N.\frac{\Phi}{I} = N.\frac{N}{l/\mu_0\mu_r A} \text{ henry}$$

$$\therefore \quad L = \frac{N^2}{l/\mu_0\mu_4 A} = \frac{N^2}{S} \text{ henry}$$

or $$L = \frac{\mu_0\mu_r A N^2}{l} \text{ henry}$$

It gives the value of self-induction in terms of the dimensions of the solenoid.

Example 8.10. *A magnetic field is produced by a coil of 300 turns which is wound on a closed iron ring. The ring has a cross-section of 20 cm² and mean length of 120 cm. The permeability of the iron is 800. If the current in the coil is 10 A, find the energy stored in the magnetic field.*

Solution. $L = \mu_0\mu_r AN^2/l$ henry

$= 4\pi \times 10^{-7} \times 800 \times 20 \times 10^{-4} \times 300^2/1.2 = 0.151$ H

*After the American scientist Joseph Henry (1707 - 1876), a contemporary of Faraday.

$$E = \frac{1}{2}LI^2 = \frac{1}{2} \times 0.151 \times 10^2 = \mathbf{7.55\ J}$$

(iii) Third Method for L

As seen from above,

$$L = \frac{n\,\Phi}{I} \quad \therefore \quad N\Phi = LI \quad \text{or} \quad -N\Phi = -LI$$

Differentiating both sides, we get

$$-\frac{d}{dt}(N\Phi) = -\frac{d}{dt}(LI) \qquad \therefore\ -N\,.\frac{d\Phi}{dt} = -L\frac{dI}{dt} \text{ (assuming } L \text{ to be constant)}$$

As seen from Art. 8.3

$$-N\,.\frac{d\Phi}{dt} = \text{self-induced e.m.f. } e_L$$

$$\therefore \qquad e_L = -L\frac{dI}{dt} \quad \text{or} \quad L = \frac{-e_L}{dI/dt}$$

If $\frac{dI}{dt} = 1$ ampere/second and $e_L = 1$ volt, then $L = 1$ henry

Hence, a coil has a self-inductance of one henry if one volt is induced in it when current through it changes at the rate of one ampere per second.

Example 8.11. *The inductance of a coil is 0.15 H. The coil has 100 turns. Find the following: (i) total magnetic flux through the coil when the current is 4 A, (ii) energy stored in the magnetic field, (iii) voltage induced in the coil when the current is reduced to zero in 0.01 second.*

(A.M.I.E. Winter 1989)

Solution. (*i*) $L = N\Phi/I$; $0.15 = 100\ \Phi/4$, $\Phi = 6$ **mH**

(*ii*) $E = \frac{1}{2} \times (6 \times 10^{-3}) \times 4^2 = \mathbf{0.048\ J}$; (*iii*) $e_L = L.dI/dt = 0.15 \times 4/0.01 = \mathbf{60\ V}$

Example 8.12. *Calculate the inductance of an air-cored toroid 25 cm mean diameter and 6.25 cm² circular cross section wound uniformly with 1000 turns of wire. Also, find the e.m.f. induced when a current increasing at the rate of 200 A per sec flows in the winding.*

(A.M.I.E. Winter 1992)

Solution. $l = \pi d = \pi \times 0.25 = 0.785$ m; $A = 6.25 \times 10^{-4}$ m²

$$L = \frac{N^2}{l/\mu_0\,\mu_r\,A} = \frac{\mu_0\,\mu_r\,AN^2}{l}$$

$$= 4\pi \times 10^{-7} \times 1 \times 6.25 \times 10^{-4} \times 1000^2/0.785 = \mathbf{10\ \mu H}$$

$$e_L = L dI/dt = 10 \times 10^{-6} \times 200 = \mathbf{2\ mV}$$

8.12. Mutual Inductance

In Art. 8.9 (Fig. 8.9) we have seen that any change of current in coil A is always accompanied by the production of mutually-induced e.m.f. in coil B. Mutual inductance may, therefore, be defined as the ability of one coil (or circuit) to produce an e.m.f. in a nearby coil by induction when the current in the first coil changes. This action being reciprocal, the second coil can also induce an e.m.f. in the first when current in the second coil changes. This ability of reciprocal induction is measured in terms of the coefficient of mutual induction M.

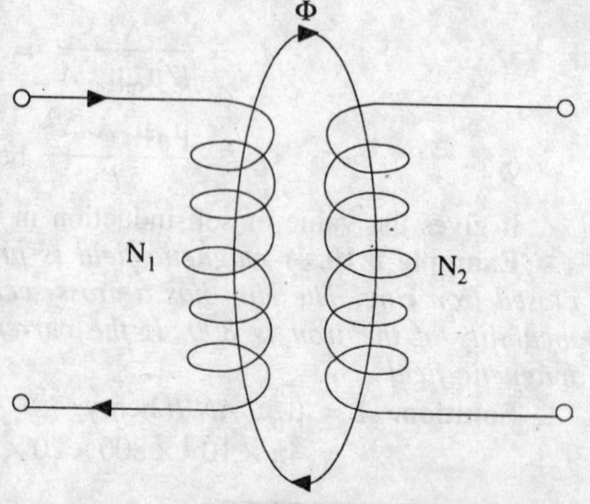

Fig. 8.13

8.13. Coefficient of Mutual Inductance (M)

It can also be defined in three ways as given below:

(*i*) First Method for M

Let there be two magnetically-coupled coils having N_1 and N_2 turns respectively (Fig. 8.13).

Coefficient of mutual inductance between the two coils is defined as *'the weber-turns in one coil due to one ampere current in the other'*.

Let a current of I_1 amperes when flowing in the first coil produce a flux of Φ_1 weber in it. *It is supposed that whole of this flux links with the turns of the second coil**. Then, flux-linkage *i.e.* weber-turns in the second coil for unit current in the first coil are $N_2\Phi_1/I_1$. Hence, by definition.

$$M = \frac{N_2 \Phi_1}{I_1}$$

If weber-turns in *second* coil due to one ampere current in the first coil *i.e.* $\frac{N_2 \Phi_1}{I_1} = 1$, then, as seen from above, $M = 1$ H.

Hence, two coils are said to have a mutual inductance of 1 henry if one ampere current flowing in one coil produces flux-linkages of one Wb-turn in the other.

In general, $M = N_2\Phi_2/I_1$ *where* Φ_2 *is the flux actually linked with the second coil.*

Example 8.13. *Two identical coils X and Y of 1,000 turns each lie in parallel planes such that 80% of flux produced by one coil links with the other. If a current of 5A flowing in X produces a flux of 0.05 mWb* in it,. *Find the mutual inductance between X and Y.*

Solution. Formula used : $M = \frac{N_2 \Phi_2}{I_1}$ henry

Flux produced in X = 0.05 mWb = 0.05×10^{-3} Wb

Flux linked with Y = $0.05 \times 10^{-3} \times 0.8 = 0.04 \times 10^{-3}$ Wb

$$\therefore \quad M = \frac{1000 \times 0.04 \times 10^{-3}}{5} = \mathbf{8 \times 10^{-3}\ H}$$

Example 8.14. *Two coupled coils have a coefficient of coupling 0.85,* $N_1 = 100$ *and* $N_2 = 800$, *with coil 1 open and a current of 5A in coil 2, the flux* Φ_2 *is 0.35 mWb. Find* L_1 L_2 *and M.*

(Electrical Engg. A.M.Ae. S.I. June 1994.)

Solution. $$L_2 = \frac{N_2 \Phi_2}{I_2} = \frac{800 \times 0.35 \times 10^{-3}}{5} = \mathbf{56\ mH}$$

Flux linked with coil No. 1, $\Phi_1 = .35 \times 0.85 = 0.2975$ mWb

$$\therefore \quad M = \frac{N_1 \Phi_1}{I_2} = \frac{100 \times 0.2975 \times 10^{-3}}{5} = \mathbf{5.95\ mH}$$

Now, $$k = \frac{M}{\sqrt{L_1 L_2}} \text{ or } 0.85 = 5.95 \times \frac{10^{-3}}{\sqrt{L_1 \times 56 \times 10^{-3}}}$$

$$\therefore \quad L_1 = \mathbf{0.875\ mH}$$

(*ii*) Second Method for M

We will now deduce an expression for the coefficient of mutual inductance in terms of the dimensions of the two coils.

$$\text{Flux in the first coil } \Phi_1 = \frac{N_1 I_1}{l/\mu_0 \mu_r A} \text{ Wb}$$

$$\text{Flux/ampere} = \frac{\Phi_1}{I_1} = \frac{N_1}{l/\mu_0 \mu_r A}$$

Assuming that whole of this flux (it usually is some percentage of it) is linked with the other coil having N_2 turns, then weber-turns in it due to flux/ampere in the first coil is

*In other words, $\Phi_2 = \Phi_1$. In case it is not so, then only that part of the flux of the first coil is used which is linked with the second coil.

$$M = \frac{N_2 \Phi_1}{I_1} = \frac{N_2 N_1}{l/\mu_0 \mu_r A} \text{ henry}$$

$$\therefore \quad M = \frac{\mu_0 \mu_r A N_1 N_2}{l} \text{ henry}$$

Also, $$M = \frac{N_1 N_2}{l/\mu_0 \mu_r A} = \frac{N_1 N_2}{\text{reluctance}} = \frac{N_1 N_2}{S} \text{ H}$$

Example 8.15. *A solenoid 70 cm in length and of 2100 turns has a radius of 4.5 cm. A second coil of 750 turns is wound upon the middle part of the solenoid. Find the self-inductance of the solenoid and the mutual inductance of the two coils.* **(A.M.I.E. Summer 1988)**

Solution. Since μ_r is not given, it would be presumed to be unity.

$$L = \frac{\mu_0 \mu_r A N^2}{l} = \frac{4\pi \times 10^{-7} \times 1 \times \pi (4.5 \times 10^{-2})^2 \times 2100^2}{0.7} = \mathbf{51 \text{ mH}}$$

Since the second coil is wound on the middle part of the solenoid, whole of the flux produced by the solenoid is linked with the coil *i.e.* coefficient of coupling is unity.

$$M = \mu_r \mu_0 A N_1 N_2 / l = 4\pi \times 10^{-7} \times \pi (4.5 \times 10^{-2})^2 \times 2100 \times 750/0.7 = \mathbf{18 \text{ mH}}$$

(*iii*) Third Method for M

As seen from above

$$M = \frac{N_2 \Phi_2}{I_1} = \frac{N_2 \Phi_1}{I_1} \qquad (\because \Phi_2 = \Phi_1)$$

$$\therefore \quad N_2 \Phi_1 = M I_1 \quad \text{or} \quad -N_2 \Phi_1 = -M I_1$$

Differentiating both sides, we get

$$-\frac{d}{dt}(N_2 \Phi_1) = -M \cdot \frac{dI_1}{dt} \qquad \text{(assuming } M \text{ to be constant)}$$

Now, $-\frac{d}{dt}(N_2 \Phi_1)$ = mutually-induced e.m.f. e_M in the second coil.

$$\therefore \quad e_M = -M \frac{dI_1}{dt}$$

If $dI_1/dt = 1$ ampere/second and $e_M = 1$ volt, then $M = 1$ H

Hence, two coils are said to have a mutual inductance of one henry if current changing at the rate of 1 ampere/second in one coil induces an e.m.f. of one volt in the other .

Example 8.16. *Two coils having 30 and 60 turns are wound side-by-side on a closed iron circuit of cross-section 0.01 m² and of mean length 1.5 m, (i) calculate the mutual inductance between the coils if the relative permeability of iron is 2000, (ii) a current in the first coil grows steadily from 0 to 10A in 0.01 second, Find the e.m.f. induced in the coil.*

(Magnetic circuits and Materials, Punjab Univ. 1993.)

Solution. (*i*) $$M = \frac{N_1 N_2}{l/\mu_0 \mu_r A} = \frac{\mu_0 \mu_r A N_1 N_2}{l}$$

$$\therefore \quad M = \frac{4\pi \times 10^{-7} \times 2000 \times 0.01 \times 30 \times 60}{1.5} = \mathbf{0.03 \text{ H}}$$

(*ii*) $$e_M = M dI_1/\text{dt.} = 0.03 \times 10/0.01 = \mathbf{30V}$$

Example 8.17. *Two coils, A of 5,000 turns and B of 3,000 turns lie in parallel planes. A current of 6 A in coil A produces a flux of 0.1 mWb. If 60% of the flux produced by coil A links with the turns of coil B, calculate the e.m.f. inducted in coil B when the current in coil A changes from 5 A to –5A in 0.01 second.*

Solution. Flux/ampere in coil A $= \Phi_1/I_1 = 0.1 \times 10^{-3}/6 = 1.667 \times 10^{-5}$ Wb

Flux linked with coil B $= 0.6 \times 1.667 \times 10^{-5}$ Wb

Now, $M = \dfrac{N_2 \Phi_2}{I_1} = 3000 \times 10^{-5} = 0.03$ H

Mutually-induced e.m.f. in coil B

$$e_M = M\frac{dI_1}{dt} = 0.03 \times 10/0.01 = \mathbf{30\ V}$$

Example 8.18. *Two identical coils A and B each having 1000 turns lie in parallel planes such that 65% of the flux produced by one coil links with the other coil and vice versa. A current of 10A in coil A produces in it a flux of 10^{-4} Wb. If the current in coil A changes from + 12 A to –12 A in 0.02 second, what would be the magnitude of the emf induced in coil B ? Calculate also the self-inductance of each coil and mutual inductance.*

(Magnetic Circuits & Materials, Punjabi Univ. 1994.)

Solution. Flux linked with coil $B = 0.65 \times 10^{-4}$ Wb

$$\therefore \quad M = \frac{N_2 \Phi_2}{I_1} = \frac{1000 \times 0.65 \times 10^{-4}}{10} = 6.5 \text{ mH}; \quad e_M = MdI_1/dt$$

$$dI_1 = 12 - (-12) = 24 \text{ A}, \; dt = 0.02 \text{ second}$$

$$\therefore \; e_M = 6.5 \times 10^{-3} \times 24/0.02 = \mathbf{7.8\ V}$$

$$L = N_1\Phi_1/I_1 = 1000 \times 10^{-4}/10 = \mathbf{10\ mH}$$

Example 8.19. *Two identical 2000-turn coils A and B lie in parallel plane such that 70% of flux produced by one links with the other. A current of 4 A in coil A produces in it a flux of 0.06 mWb. If the current in the coil A changes from +5A to –5A in 0.01 second, what e.m.f. is induced in the coil B?*

Calculate the self-inductance of each coil and their mutual inductance.

(Elements of Elect. Engg., Punjab Univ. 1993)

Solution. The self inductance of coil A is given by

$$L_1 = \frac{N_1 \Phi_1}{I_1} = \frac{2000 \times 0.06 \times 10^{-3}}{4} = \mathbf{0.03\ H}$$

Flux linked with coil B, $\Phi_2 = 70\%$ of $0.06 = 0.042$ mWb

$$\therefore \quad M = \frac{N_2 \Phi_2}{I_1} = \frac{2000 \times 0.042 \times 10^{-3}}{4} = \mathbf{0.021\ H}$$

Change in current, $dI_1 = 5 - (-5) = 10$A

The e.m.f. induced in coil B is given by

$$e_2 = \frac{MdI_1}{dt} = \frac{0.021 \times 10}{0.01} = \mathbf{21\ V}$$

Example 8.20. *The self-inductance of a coil of 500 turns is 0.25 H. If 60% of the flux is linked with a second coil of 10,000 turns, calculate (i) the mutual inductance of the two coils, (ii) e.m.f. induced in the second coil when current in the first coil changes at the rate of 100 A/second..*

Solution. $L_1 = \dfrac{N_1 \Phi_1}{I_1} = 0.25$

$\therefore$ Flux/ampere in first coil $= \dfrac{\Phi_1}{I_1} = \dfrac{0.25}{N_1} = 0.25/500 = 5 \times 10^{-4}$ Wb

Flux linked with second coil $= 0.6 \times 5 \times 10^{-4} = 3 \times 10^{-4}$ Wb

(*i*) $\therefore \quad M = N_2\dfrac{\Phi_1}{I_1} = 10{,}000 \times 3 \times 10^{-4}$ Wb $= \mathbf{3\ H}$

(*ii*) $\quad e_M = M\, dI_1/dt = 3 \times 100 = \mathbf{300\ V}$

8.14. Coefficient of Magnetic Coupling

Two coils are said to be magnetically coupled if full or part of the flux produced by one links with the other. Let L_1 and L_2 be the self-inductances of the two coils and M their mutual inductance, then

$$k = \frac{M}{\sqrt{L_1 L_2}}$$

When all the flux produced by one coil links with the other, then mutual inductance between the two is maximum and is given by

$$M = \sqrt{L_1 L_2}$$

In that case, $k = 1$

When there is no common flux between the two coils, they are said to be magnetically isolated. Since, in that case, $M = 0$, $k = 0$ also.

Hence, coefficient of coupling may be defined as the ratio of actual mutual inductance present between the two coils to the maximum possible value.

Example. 8.21. *Two identical coils X and Y of 1000 turns each lie in parallel planes. A current of 5 A in coil X produces 0.5 × 10⁻³ weber of flux in it. The coupling coefficient between the coils is 0.8. Determine the self-inductance and mutual inductance of the arrangement.*

(Electrical Engg., A.M.Ae.S.I. Dec. 1989)

Solution.

Self-inductance of coil $X = N_1\Phi_1/I_1 = 1000 \times 0.5 \times 10^{-3}/5 =$ **0.1 H**

Since the two coils are identical, self-inductance of Y is the same *i.e.* 0.1 H.

Now, $M = \frac{k}{\sqrt{L_1 L_2}} = 0.8 \times 0.1 = \mathbf{0.08\ H}$

Example 8.22. *The windings of a transformer have an inductance $L_1 = 6$ H, $L_2 = 0.06$ H and a coefficient of coupling $k = 0.9$. Find the e.m.f. induced in both windings when the primary current increases at the rate of 1000 A/s.*

Solution. It is obvious that self-induced e.m.f. e_L is produced in primary but mutually-induced e.m.f. e_M is produced in secondary.

$$e_L = L.dI_1/dt = \times 1000 = \mathbf{6000\ V}$$

Now, $M = k\sqrt{L_1 L_2}$ —Art. 8.14

$$= 0.9\sqrt{6 \times 0.06} = 0.54$$

$\therefore$ $e_M = M.dI_1/dt = 0.54 \times 1000 = \mathbf{540\ V}$

Example 8.23. *Two long single-layer solenoids have the same length and the same number of turns but are placed co-axially one within the each other. The diameter of the inner coil is 8 cm and that of the outer coil 10 cm. Calculate the coefficient of coupling between the coils.*

Solution. Coefficient of coupling

$$k = M/\sqrt{L_1 L_2}$$

Since N and l are the same for the two solenoids,

$$L_1 = \mu A_1 N^2/l \text{ and } L_2 = \mu A_2 N^2/l$$

$$\therefore \sqrt{L_1 L_2} = \frac{\mu N^2}{l}\sqrt{A_1 A_2} \quad ...(i)$$

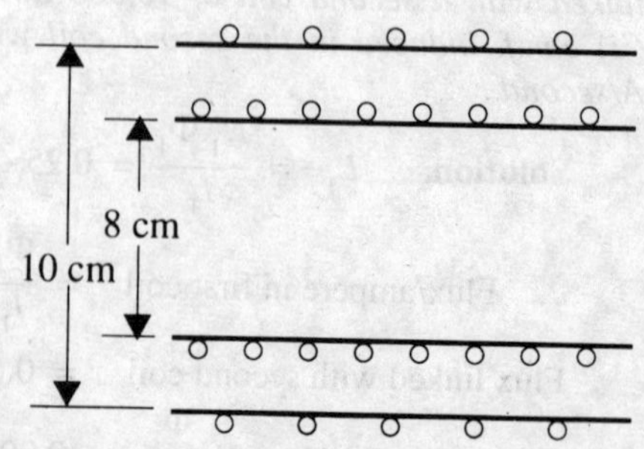

Fig. 8.14

Now, let us find the value of M between the solenoids by using the relation

$$M = N_2\Phi_2/I_1 \quad \text{(Art. 8.13)}$$

Flux density produced by the outer solenoid is

$$B = \mu H = \mu N I_1/l$$

Flux linked with the inner solenoid of cross-section A_2 is

$$\Phi = BA_2 = \pi NI_1A_2/l$$

$$\therefore \quad M = \frac{N_2\,\Phi_2}{I_1} = \frac{N\Phi_2}{I_2} = \mu N^2A_2/l$$

From (*i*) and (*ii*) above, we get,

$$k = \frac{M}{\sqrt{L_1\,L_2}} = \sqrt{\frac{A_2}{A_1}} = \frac{d_2}{d_1} = \frac{8}{10} \quad \text{or} \quad k = 0.8 \quad \text{or} \quad \mathbf{80\%}$$

Example 8.24. *A solenoid 70 cm in length and of 2100 turns has a radius of 4.5 cm. A second coil of 750 turns is wound upon the middle part of the solenoid. Find the self-inductance of the solenoid and the mutual inductance of the two coils.* **(A.M.I.E. Summer 1987)**

Solution. $A = \pi r^2 = \pi(4.5 \times 10^{-2})^2 = 20.25\pi \times 10^{-4}$ m²

$\mu_r = 1$ (assumed), $l = 70$ cm $= 0.7$ m

$M = 4\pi \times 10^{-7} \times 20.25\pi \times 10^{-4} \times 2100 \times 750/0.7$ H= **18mh**

Example 8.25. *The number of turns in a coil is 250. when a current of 2 A flows in this coil, the flux in the coil is 0.3 mWb. When this current is reduced to zero in 2 milliseconds, the voltage induced in a coil lying in the vicinity of coil is 63.75 volts. If the coefficient of coupling between the coils is 0.85, find self inductances of the two coils, mutual inductance and the number of turns in the second coil.* **(Electrical Engg. A.M.Ae.S.I. Dec. 1993)**

Solution. $L_1 = \dfrac{N_1\;\Phi_1}{I_1} = \dfrac{250 \times 0.3 \times 10^{-3}}{2} = \mathbf{37.5 \times 10^{-3}\ H}$

$e_M = M.dI_1/dt$ or $63.75 = M \times (2 - 0)/2 \times 10^{-3}$,

$\therefore \quad M = 63.75$ mH

$k = M/\sqrt{L_1\,L_2}$ or $L_1\,L_2 = M^2/k^2$ or $L_2 = (63.75 \times 10^{-3})^2/0.85^2 \times 37.5 \times 10^{-3}$

$=$ **150 mH**

Now, $\Phi_2 = k\Phi_1 = 0.85 \times 0.3 = 0.255$ mWb

$M = N_2\Phi_2/I_1$ or $63.75 \times 10^{-3} = N_2 \times 0.255 \times 10^{-3}/2$ or $N_2 = \mathbf{500}$

8.15. Inductances in Series

(*i*) When the coils are so joined in series such that their fluxes (or m.m.fs) are additive *i.e.* in the same direction (Fig. 8.15).

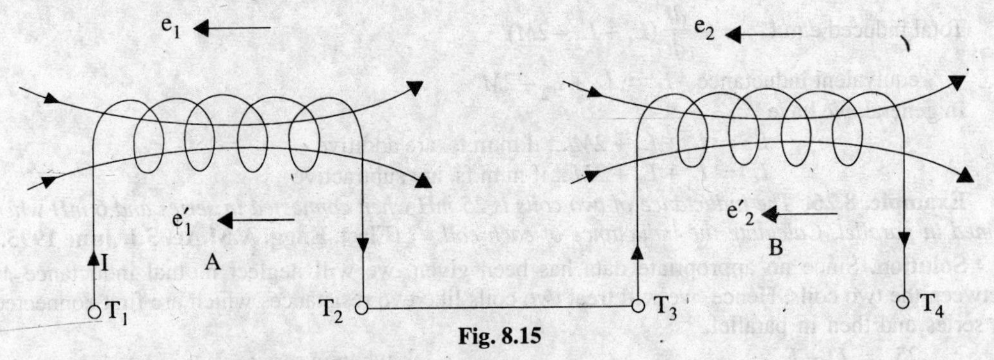

Fig. 8.15

Let M = coefficient of mutual inductance

L_1 = coefficient of self-inductance of 1st-coil

L_2 = coefficient of self-inductance of 2nd coil

Then, self-induced e.m.f. in A is $= e_1 = -L_1 \cdot \dfrac{dI}{dt}$

Mutually-induced e.m.f. in A due to change of current in B $= e_1' = -M \cdot \dfrac{dI}{dt}$

Self-induced e.m.f. in B is $= e_2 = -L_2 \cdot \dfrac{dI}{dt}$

Mutually-induced e.m.f. in B due to change of current in A is $= e_2' = -M \cdot \dfrac{dI}{dt}$

(All have -ve sign because both self and mutually-induced e.m.f.s' are in opposition to the applied e.m.f.)

Total induced e.m.f. in the combination $= -\dfrac{dI}{dt}(L_1 + L_2 + 2M)$ ...(*i*)

If L is the equivalent inductance, then total induced e.m.f. in that single coil would have been

$$= -L\frac{dI}{dt} \qquad ...(ii)$$

Equating (*i*) and(*ii*), we have

$$L = L_1 + L_2 + 2M$$

(*ii*) When the coils are so joined that their fluxes are in opposite directions (Fig. 8.16)

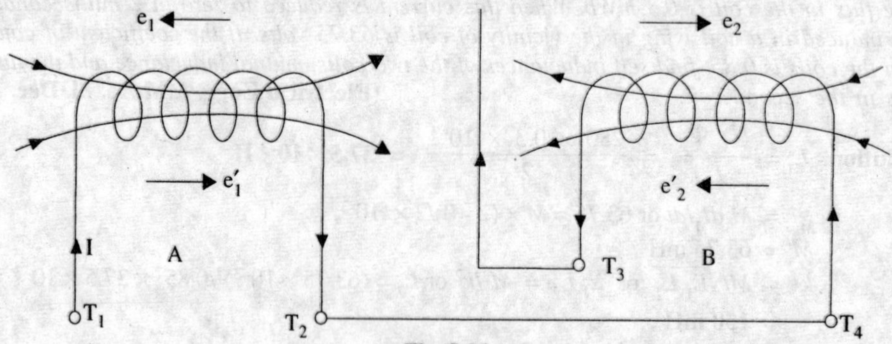

Fig. 8.16

As before, $e_1 = -L_1\dfrac{dI}{dt}$; $e_1' = +M\dfrac{dI}{dt}$ (mark the direction)

$e_2 = -L_2\dfrac{dI}{dt}$; $e_2' = +M\dfrac{dI}{dt}$

Total induced e.m.f. $= -\dfrac{dI}{dt}(L_1 + L_2 - 2M)$

$\therefore$ equivalent inductance $L = L_1 + L_2 - 2M$

In general, we have

$$L = L_1 + L_2 + 2M \text{ ... if m.m.fs. are additive}$$

$$L = L_1 + L_2 + 2M \text{ ...if m.m.fs. are subtractive}$$

Example. 8.26. *The inductance of two coils is 25 mH when connected in series and 6 mH when joined in parallel. Calculate the inductance of each coil.* **(Elect. Engg. A.M.Ae. S.I. June 1995.)**

Solution. Since no appropriate data has been given, we will neglect mutual inductance M between the two coils. Hence, we will treat two coils like two resistances which are first connected in series and then in parallel.

$\therefore\ 25 = L_1 + L_2$ ...when in series

$$6 = \frac{L_1\ L_2}{L_1 + L_2} = \frac{L_1\ L_2}{25} \qquad \text{... when in parallel}$$

$\therefore\ L_1L_2 = 6 \times 25 = 150$ or $L_2 = 150/L_1$

Substituting this value of L_2 in the above equation, we get

$25 = L_1 + 150/L_1$ or $L_1^2 - 25\,L_1 + 150 = 0$

Solving the above quadratic equation, we get

$$l_1 = \frac{25 \pm \sqrt{25^2 - 4 \times 150}}{2} = \frac{25 \pm 5}{2} = 15 \text{ or } 10$$

It means that L_1 is either 15 mH or 10 mH

If, $L_1 = 15$ mH, then, $L_2 =$ **10 mH**

When, $L_1 = 10$ mH, then, $L_2 =$ **15 mH**

Example 8.27. *A wooden ring has a mean length of magentic path of 100 cm and a cross-section of 10 cm². A coil of 2000 turns of wire is wound over it uniformly. Closely wound over this is a second coil of 4000 turns. The terminals of each coil are brought out separately.*

(a) What is the inductance of the first coil alone?

(b) What is the inductance of the second coil alone?

(c) If the two coils were connected electrically in series so that their m.m.fs. are additive, what would be the inductance of the combination?

Solution. $N_1 = 2000$; $N_2 = 4000$

$l = 1$ m ; $A = 10^{-3}$ m²

$\mu_0 = 4\pi \times 10^{-7}$ H/m; $\mu_r = 1$

(a) $L_1 = \mu_0 \mu_r A N_1^2/l$ H $= 10^{-3} \times 4\pi \times 10^{-7} \times 2000^2 =$ **5.03 mH**

(b) $L_2 = 10^{-3} \times 4\pi \times 10^{-7} \times 4000^2 =$ **20.11 mH**

(c) As one coil is completely wound over the other, the coefficient of coupling $k = 1$

$\therefore$ $M = \sqrt{L_1 L_2} = \sqrt{5.03 \times 20.11} = 10.06$ mH

$L = L_1 + L_2 + 2M = 5.03 + 20.11 + 2 \times 10.06 =$ **45.26 mH**

Example 8.28. *The combined inductance of two coils when connected in series are 0.42 H and 0.096 H for series-aiding and series-opposing connections respectively. If one of the coils when isolated has an inductance 0.1 H, calculate (i) the self-inductance of the other coil, (ii) mutual inductance between the coils, (iii) the coupling coefficient.*

(Electrical Engg., A.M.Ae.S.I. Dec. 1992)

Solution. (*ii*) For series-aiding

$L = L_1 + L_2 + 2M$ or $0.42 = L = L_1 + L_2 + 2M$

For series-opposing, $L = L_1 + L_2 - 2M$ or $0.096 = L = L_1 + L_2 - 2M$

From the above, we get, $M =$ **0.081 H**

(*i*) Let $L_1 = 0.1$ *H*. Then, substituting this value above,

we get, $L_2 =$ **0.482 H**

(iii) Coupling coefficient, $k = \frac{M}{\sqrt{L_1 L_2}} = \frac{0.081}{\sqrt{0.1 \times 0.482}} =$ **0.4 H**

Example 8.29. *Two coils with terminals T_1, T_2, and T_3, T_4 respectively are placed side by side. Measured separately, the inductance of the first coil is 1200 mH and of the second 800 mH. With T_2 joined to T_3,* the inductance between T_1 and T_4 *is 2500 mH. What is the mutual inductance between the two coils? And what would be the inductance between T_1 and T_3* with T_2 joined to T_4 ?

Solution. Formula used : $L = L_1 + L_2 \pm 2M$

$L_1 = 1200$ mH, $L_2 = 800$ mH

When joined additively, $L = L_1 + L_2 + 2M$

or $2500 = 1200 + 800 + 2M$

$\therefore$ $M =$ **250 mH**

When T_2 is joined to T_4, fluxes are in opposition

$L = 1200 + 800 - 2 \times 250 =$ **1500 mH**

Example 8.30. *Two coils are placed side by side. Their combined inductance when connected in series is 1 H or 0.2 H depending on the relative directions of the currents in the coils. Calculate the mutual inductance between the two coils. If the self-inductance of one of the coils when isolated is 0.2 H, find the self-inductance of the other coil.*

Solution. Let L_1 and L_2 be the coefficients of self-inductance of the two coils and M their coefficient of mutual inductance. Obviously, when the two coils are connected in series such that their fluxes are in opposite directions, their combined inductance is 0.2 H and when their fluxes are in the same direction, their combined inductance is 1 H.

$\therefore \quad 1 = L_1 + L_2 + 2M$ ...(*i*)

and $\quad 0.2 = L_1 + L_2 - 2M$ ...(*ii*)

Adding (*i*) and (*ii*), we get

$1.2 = 2(L_1 + L_2)$; Now, $L_1 = 0.2$ H

$\therefore \quad 1.2 = 2(0.2 + L_2)$

$\therefore \quad 2L_2 = 0.8$ H $\quad \therefore \; L_2 = \mathbf{0.4\ H}$

Putting this value of L_2 in (*i*) above, we get

$1 = 0.2 + 0.4 + 2M$

$\therefore \quad 2M = 0.4$ or $M = \mathbf{0.2\ H}$

Tutorial Problems No. 8.2

1. A d.c. motor field pole is wound with 800 turns and carries a flux of 20 mWb when excited. The exciting current is then switched off and the flux reduced to 0.2 mWb in 50 milli-seconds. Calculate the average e.m.f. induced in the coil. **[316.8 V]**

2. An iron-cored toroidal coil has 100 turns, a cross-sectional area of 10cm^2 and a mean length of 314 cm. If the relative permittivity of iron is 1000, calculate the inductance of the coil. **[4 mH]**

3. A flux of 0.5 mWb is produced in a coil of 900 turns wound on a wooden ring by a current of 3 A. Calculate (*a*) the inductance of the coil, (*b*) the average e.m.f. induced in the coil when a current of 5 A is switched off assuming the current to fall to zero in 1 ms, (*c*) the mutual inductance between the coils, if a second coil of 600 turns was uniformly wound over the first coil. **[(*a*) 0.15 H (*b*) 750 V (*c*) 0.1H]**

4. An iron ring, having a mean circumference of 250 mm and a cross-sectional area of 400 mm^2, is wound with a coil of 70 turns. From the following data, calculate the current required to set up a magnetic flux of 510 μWb.

B(Wb/m^2)	1.0	1.2	1.4
H(AT/m)	350	600	1250

Calculate also (*a*) the inductance of the coil at this current, (*b*) the self-induced e.m.f. if this current is switched off in 0.005 s. **[(*a*) 2.68 A (*b*) 13.32 mH (*c*) 7.14 V]**

5. A coil consists of 750 turns and a current of 10 A in the coil gives rise to magnetic flux of 1200 μWb. Calculate the inductance of the coil and determine the average e.m.f. induced in the coil when this current is reversed in 0.01 s. **[0.09 H ; 180 V]**

6. A non-magnetic ring having a mean diameter of 300 mm and a cross-sectional area of 500 mm^2 is unformly wound with a coil of 200 turns. Calculate from first principles the inductance of the winding. **[26.67 μH]**

7. If a coil of 150 turns is linked with a flux of 0.01 Wb when carrying a current of 10 A, calculate the inductance of the coil. If this current is uniformly reversed in 0.1 second, calculate the induced e.m.f. **[0.15 H ; 30 V]**

8. Two coils *A* and *B* are coupled together, coil *A* having 400 turns and coil *B* 600 turns. When 5A flows in coil *A*, a flux of 5 mWb links with coil *B*. Calculate the self-inductance of coil A and the mutual inductance between the two coils. **[0.4 H ; 0.6 H]**

9. A wooden ring has a mean diameter of 40 cm and a cross-sectional area of 3 cm^2 and is uniformly wound with two coils *A* and *B*. Coil *A* has 400 turns and coil *B* 500 turns. Calculate the mutual inductance between the two coils. Neglect any magnetic leakage.

If the current in coil *A* is 2A and this is reversed in 5 milli-seconds, calculate the value of the e.m.f. induced in coil *B*. **[60 μH ; 48 mV]**

10. A sample of cast steel in the shape of a ring has a mean diameter of 0.15 m and a cross-sectional area of 0.182×10^{-3} m^2. Determine μ_r of the sample when a coil of 330 turns carrying a current of 2.4 A produces a flux of 0.22 mWb in the ring. What is the inductance of the coil under this condition? **[572 ; 30.3 mH]**

11. A non-magnetic ring having a mean diameter of 30 cm and a cross-sectional area of 4 cm^2 is uniformly wound with two coils *A* and *B*, one over the other. *A* has 90 turns and *B* has 240 turns. Calculate the mutual inductance between the coils.
Also, calculate the e.m.f. induced in *B* when a current of 6 A in coil *A* reversed in 0.02 second. Indicate the direction of this e.m.f. with reference to the initial direction of the current.
[11.52 μA ; 6.9 mV ; same]

12. A coil of 50 turns having a mean diameter of 3 cm is placed coaxially at the centre of a solenoid 60 cm long wound with 2500 turns and carrying a current of 2A.
Determine the mutual inductance of the arrangement. **[0.185 mH]**

13. The total inductance of two coils, *A* and *B* when connected in series, is 0.5 H or 0.2 H depending on the relative directions of the current in the coils. Coil *A* when isolated from coil *B*, has a self-inductance of 0.2 H. Calculate:
(*a*) the mutual inductance between the two coils,
(*b*) the self-inductance of coil *B*,
(*c*) the coupling factor between the coils,
(*d*) the two possible values of the induced e.m.f. in coil *A* when the current is decreasing at 1000 A/s in the series circuit. **[(*a*)0.075 H (*b*) 0.15 H (*c*) 0.433 H (*d*) 275 or 125 V]**

14. When two identical coupled coils are connected in series the inductance of the combination is found to be 80 mH. When the connections to one of the coils are reversed, a similar measurement indicates 20 mH. Find the coupling coefficient between the two coils. **[0.6]**

HIGHLIGHTS

1. Faraday's Laws of Electromagnetic Induction are :
(*i*) Whenever the magnetic flux linked with a circuit changes, an e.m.f. is induced in it.
or
Whenever a conductor cuts across magnetic flux, an e.m.f. is induced in that conductor.
(*ii*) The magnitude of the induced e.m.f. is equal to the rate of change of flux linkages.

$$e = \frac{d}{dt}(N\Phi) \text{ volts} = N\frac{d\Phi}{dt} \text{ volts}$$

2. The direction of the induced e.m.f. and hence current may be found either by applying Fleming's Right-hand rule or Lenz's Law which states that the electromagnetically induced current always flows in such a direction that the action of the magnetic field set up by it is such as to oppose the very cause which produces it.

3. The induced e.m.f. may be subdivided as follows :

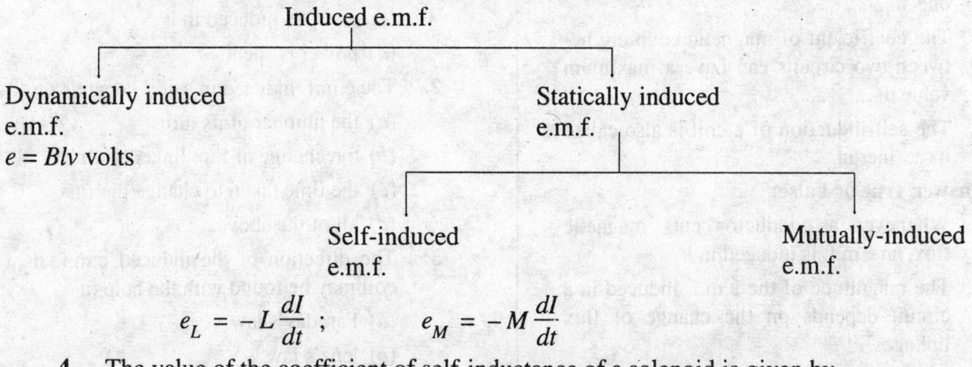

$$e_L = -L\frac{dI}{dt} \; ; \qquad e_M = -M\frac{dI}{dt}$$

4. The value of the coefficient of self-inductance of a solenoid is given by

$$L = \frac{N\Phi}{I} H \text{ i.e. weber-turns/ampere} = \mu_0 \mu_r \, A N^2/l \, H$$

$$= \frac{e_L}{dI/dt} H$$ *i.e.* numerically equal to the induced e.m.f. per unit rate of change of current.

5. Similarly, the coefficient of mutual inductance of two coils is given by

$$M = \frac{N_2 \Phi_2}{I_1} H$$ *i.e.* flux linkages in one coil per ampere current in the other

$$= \mu_0 \mu_r A N_1 N_2 / l \text{ H}.$$

$$= \frac{e_M}{dI_1/dt}$$ *i.e.* numerically equal to the induced e.m.f. in one coil per unit rate of change of current in the other.

6. Coefficient of coupling between two magnetically-coupled coils is given by

$$k = \frac{M}{\sqrt{L_1 L_2}}$$

7. The equivalent inductance of two series-connected inductances is

$$L = L_1 + L_2 + 2M$$ ...when two fluxes assist each other

$$= L_1 + L_2 - 2M$$ ...when two fiuxes oppose each other.

OBJECTIVE TESTS—8

A. Fill in the following blanks:

1. Whenever flux linked with a circuit changes, an .... is always induced in it.
2. According to Lenz's law, the current induced in a coil always .... the cause producing it.
3. A generator works on the production of .... induced e.m.f in its armature conductors.
4. Transformers work on the production of .... induced e.m.f.
5. The coefficient of self-induction of a coil is given by the weber-turns per .... in it.
6. When current changing at the rate of one ampere per second induces one volt in a coil, it is said to have a self-inductance of one ....
7. The coefficient of magnetic coupling between two circuits can have a maximum value of ....
8. The self-induction of a coil is also called its .... inertia.

B. Answer True or False:

1. Whenever a conductor cuts magnetic flux, an e.m.f. is induced in it.
2. The magnitude of the e.m.f. induced in a circuit depends on the change of flux linkages.
3. The direction of the e.m.f. induced in a conductor is given by Fleming's Left-hand rule.
4. Induced e.m.f. can be either mutually-induced or statically-induced.
5. The property of a coil due to which it opposes any increase or decrease of the current flowing through it is called self-inductance.
6. Transformers work on the principle of mutual induction.
7. Mutual induction between two coils is dependent on the number of turns of both coils.
8. Coefficient of magnetic coupling between two coils depends on their coefficient of mutual inductance only.

C. Multiple Choice Questions:

1. Whenever a magnet is quickly brought towards an open-circuited stationary coil
   (*a*) a current is induced in it
   (*b*) work has to be done
   (*c*) e.m.f. is induced in it
   (*d*) power is spent.
2. The e.m.f. induced in a coil depends on
   (*a*) the number of its turns
   (*b*) the change of flux linked with it
   (*c*) the time taken to change the flux
   (*d*) all of the above.
3. The direction of the induced e.m.f. in a coil may be found with the help of
   (*a*) Faraday's law
   (*b*) lenz's law
   (*c*) Fleming's left-hand rule
   (*d*) Steinmetz law.
4. The dynamically-induced e.m.f. in a conductor does NOT depend on

(*a*) flux density
(*b*) its active length
(*c*) its conductivity
(*d*) its velocity.

5. An e.m.f. is induced in the coil whenever flux through it
(*a*) is decreased
(*b*) is increased
(*c*) is abruptly reduced to zero
(*d*) all of the above.

6. Current changingn at the rate of 0.5 A/s induces an e.m.f. of 2 V in a coil. The self-inductance of the coil is .... henry.
(*a*) 4 (*b*) 2
(*3*) 1 (*d*) 3

7. A 100 turn coil has an inductance of 6 mH. If the number of turns is increased to 200, all other quantities remaining the same, the inductance will be .... milli-henry.
(*a*) 1.5 (*b*) 3
(*c*) 12 (*d*) 24

(Elect. Engg. A.M.Ae. S.I. Dec. 1993.)

8. The self inductances of two coils are 4 mH and 9 mH. If the coefficient of coupling is 0.5, the mutual inductance between the coils is ...... mH.
(*a*) 1 (*b*) 2
(*c*) 3 (*d*) 4

(Elect,. Engg. A.M.Ae. S.I. June 1994.)

ANSWERS

A. 1. e.m.f. 2. opposes 3. dynamically 4. mutually/statically 5. ampere
6. henry 7. unity 8. electrical/electromagnetic

B. 1. T 2. F 3. F 4. T 5. T 6. T 7. T 8. F

C. 1. *c* 2. *d* 3. *b* 4. *c* 5. *d* 6. *a* 7. *b* 8. *a*

9

MAGNETIC HYSTERESIS

9.1. Magnetic Hysteresis

It may be defined as the lagging of magnetisation or induction flux density (B) behind the magnetising force (H). Alternatively, it may be defined as that quality of a magnetic substance due to which energy is dissipated in it on the reversal of its magnetism.

Let us take an unmagnetised bar of iron AB and magnetise it by placing it within the field of a solenoid (Fig. 9.1). The field H produced by the solenoid is called the magnetising field or force. The value of this magnetising force H can be increased or decreased by increasing or decreasing current through it. Let H be increased in steps from zero up to a certain maximum value and the corresponding values of induction flux density (B) be noted. If we plot the relation between H and B, a curve like OA, as shown in Fig. 9.1, is obtained. The material becomes magnetically saturated for $H = OM$ and has, at that time, a maximum flux density of B_m established through it.

If H is now decreased gradually (by decreasing solenoid current) flux density B will not decrease along AO (as might be expected) but will decrease less rapidly along AC. When H is zero, B is not but has a definite value $= OC$. It means that on removing the magnetising force H, the iron bar is not completely demagnetised. This value of $B(= OC)$ measures *retentivity or remanence* of the material.

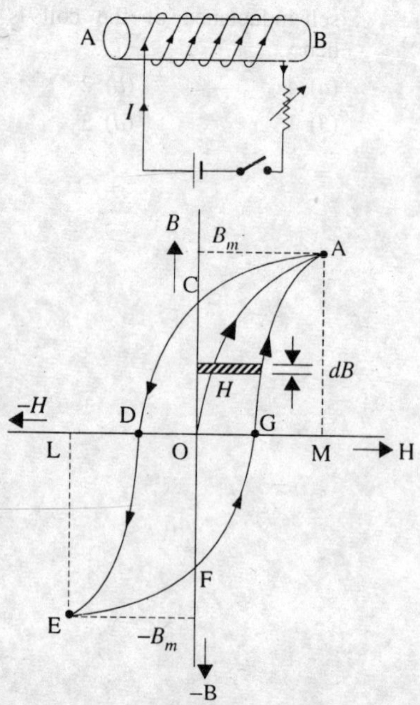

To demagnetise the iron bar we have to apply the magnetising force H in the reverse direction. When H is reversed by reversing current through the solenoid, then B is reduced to zero at point D where $H = OD$. This value of H required to wipe off residual magnetism is known as *coercive force* and is a measure of the *coercivity* of the material *i.e.* the 'tenacity' with which it holds on to its magnetism. Its value varies considerably for different materials being about 40,000 AT/m for Alnico and nearly 3 AT/m for Mumetal.

Fig. 9.1

If, after the magnetisation has been reduced to zero, value of H is further increased in the 'negative' *i.e.* reverse direction, the iron bar again reaches a state of magnetic saturation represented by point E. By taking H back from its value corresponding to negative saturation ($=OL$) to its value for positive saturation ($=OM$), a similar curve $EFGA$ is obtained; If we again start from G, the same curve $GACDEFG$ is obtained once again.

It is seen that B always lags behind H. The two never attain zero value simultaneously. This lagging of B behind H is given the name 'hysteresis' which literally means '*to lag behind*'. The closed loop $ACDEFGA$ which is obtained when iron bar is taken through one complete cycle of reversal of magnetisation is known as 'hysteresis loop'.

By one cycle of reversal of magnetisation of a magnetic material is meant its being first magnetised in one direction, then demagnetised and being magnetised again but in opposite direction as shown in Fig. 9.2.

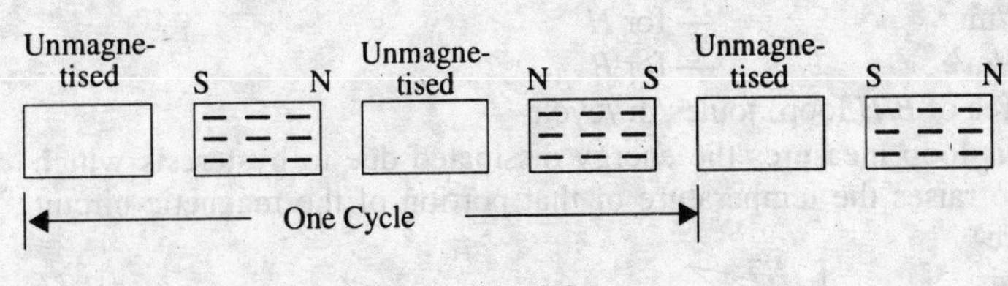

Fig. 9.2

9.2. Area of Hysteresis Loop

Just as the area of an indicator diagram measures the energy made available in a machine when taken through a cyclic operation, so also the area of the hysteresis loop represents the net energy spent in taking the iron bar through one cycle of magnetisation.

According to *Weber's Molecular Theory* of magnetism, when a magnetic material is magnetised, its molecules are forced along a straight line. So energy is spent in this process. Now, if iron has no retentivity, then energy spent in straightening the molecules could be recovered by reducing H to zero in the same way as the energy stored up in a spring can be recovered by allowing the spring to release its energy by driving some kind of load. Hence, in the case of magnetisation of a material of *high retentivity*, all the energy put into it originally for straightening the molecules is not recovered when H is reduced to zero. We will now proceed to find this loss of energy per cycle of magnetisation.

Let l = mean length of the iron bar, A = its area of cross-section

N = No. of turns of wire in the solenoid

If B is the flux density at any instant, then $\Phi = BA$

When current through the solenoid changes, flux also changes and so produces an induced e.m.f. whose value is

$$e = N\frac{d\Phi}{dt} \text{ volts} \qquad \text{(neglecting –ve sign)}$$

$$= N\frac{d}{dt}(BA) = NA\frac{dB}{dt} \text{ volts}$$

Now, $H = \dfrac{NI}{l}$ or $I = \dfrac{Hl}{N}$

The power or rate of expenditure of energy in maintaining the current 'i' against induced e.m.f. 'e' is

$$= ei \text{ watt}$$

$$= NA\frac{dB}{dt} \times \frac{Hl}{N} \text{ watt}$$

$$= Al\,H\frac{dB}{dt} \text{ watt}$$

Energy spent in time 'dt' during which flux has changed is

$$= AlH\frac{dB}{dt} \times dt \text{ joules} \quad = AlH.\,dB \text{ joules}$$

Total *net* work done for one cycle of magnetisation is

$$W = Al \oint H\,dB \text{ joules}$$

where $\oint$ stands for integration over whole cycle. Now '$H\,dB$' stands for the area of horizontal shaded strip in Fig. 9.1. Hence, $\oint H\,dB$ = area of the loop *i.e.* the area between the B/H curve and the B-axis.

$\therefore$ work done/cycle $= Al \times$ (area of the loop) joules

Now Al = volume of the material

$\therefore$ net work done/cycle/m^3 = (loop area) joules

$\therefore$ W = (area of B/H loop) joules/m^3/cycle

Precaution

Scales of B and H should be taken into consideration while calculating the actual loop area. For example, if the scales are,

	1 cm	=	x AT/m	— for H
and	1 cm	=	y Wb/m^2	— for B
then	W	=	xy (area of B/H loop) joules/m^3/cycle	

As seen from above, hysteresis loop measures the energy dissipated due to hysteresis which appears in the form of heat and so raises the temperature of that portion of the magnetic circuit which is subjected to magnetic reversals. The shape of the hysteresis loop depends on the nature of the magnetic substance (Fig. 9.3).

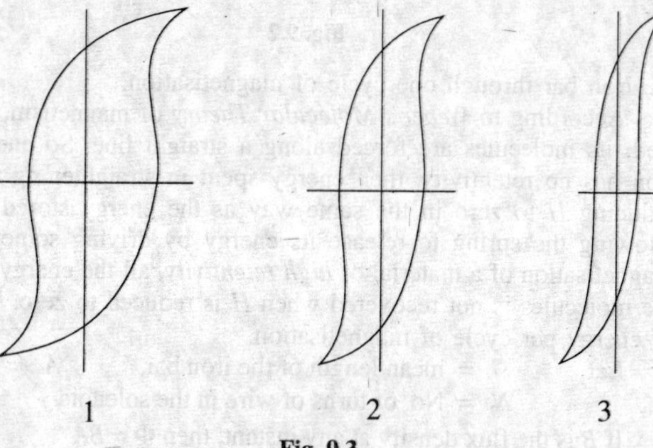

Fig. 9.3

Loop 1 is for hard steel. Due to its high retentivity and coercivity, it is well suited for making permanent magnets. But due to large hysteresis loss (as shown by large loop area) it is not suitable for rapid reversals of magnetisation. Certain alloys of aluminium, nickel and steel called alnico alloys have been found extremely suitable for making permanent magnets because they have high coercivity and high magnetic stability against the effect of heat, vibration and demagnetising fields but mechanically they are very hard, impossible to forge and difficult to machine except by grinding.

Loop 2 is for wrought iron and cast steel. It shows that these materials have high permeability and fairly good coercivity, hence making them suitable for cores of electromagnets.

Loop 3 is for alloyed sheet steel and it shows high permeability and low hysteresis loss. Hence, such materials are most suited for making armatures and transformer cores which are subjected to rapid reversals of magnetisation.

Example 9.1. *The hysteresis loop of a sample of sheet steel subjected to a maximum flux density of 1.3Wb/m^2 has an area of 93 cm^2, the scales being 1 cm = 0.1 Wb/m^2 and 1 cm = 50 AT/m. Calculate the hysteresis loss in watts when 1500 cm^2 of the same material is subjected to an alternating flux density of 1.3 Wb/m^2 peak value at a frequency of 65 Hz.*

(Magnetic Ckts & Materials, Punjab Univ. 1994)

Solution. Loss = xy (area of B/H loop) J/m^3/cycle

$= 0.1 \times 50 \times 93 = 465$ J/m^3/cycle

Volume = 1500 cm^3 = 15×10^{-4} m^3

Number of reversals per second = 65

$W_h = 465 \times 15 \times 10^{-4} \times 65$ J/s = **45.3 W**

Example 9.2. *The area of hysteresis loop for a certain magnetic material is 60 cm^2 and the scales of the graph on which it is drawn are*

H-axis, 1 cm = 200 AT/m;   B-axis, 1 cm = 0.2 Wb/m^2

Calculate the loss in watt/kg. The frequency is 50 Hz and the density of the material is 0.8 g/cm^3.

Solution. loss = xy(area of B/H loop) J/m^3/cycle

Here $x = 200$, $y = 0.2$

Loss $= 200 \times 0.2 \times 60 = 2400$ J/m^3/cycle

Volume of 1 kg $= \dfrac{1000}{8} \times 10^{-6}$ m^2 $= 10^{-3}/8$ m^3

$\therefore$ loss per kg $= 2400 \times 10^{-3}/8 = 0.3$ J/cycle

No. of reversals/second = 50

$\therefore$ hysteresis loss = 0.3×50 = **15 J/s or watts**

Example 9.3. *Calculate the hourly loss of energy in kWh in a specimen of iron, the hysteresis* *loop of which is equivalent in area to 150 J/m³. Frequency 50 Hz, specific gravity of iron 7.5, weight of specimen 10 kg.* **(Magnetic Ckt & Materials Punjab Univ. 1994)**

Solution. Hysteresis loss $= 250 \text{ J/m}^3\text{/cycle}$

Now, $\text{specific gravity} = \dfrac{\text{density of iron}}{\text{density of water}}$

$\therefore \quad 7.5 = \dfrac{\text{density of iron}}{1000 \text{ kg/m}^3}$

$\therefore \quad \text{density of iron} = 7.5 \times 10^3 \text{ kg/m}^3$

$\text{Volume of iron} = 10/7.5 \times 10^3 = 10^{-2}/7.5$

Number of cycles of magnetic reversals/hour $= 60 \times 50 = 3000$

$\therefore \quad \text{hysteresis loss/hr} = 250 \times (10^{-2}/7.5) \times 3000 = 1000 \text{ J}$

$= 1000/36 \times 10^5 = \mathbf{27.8 \times 10^{-5}}$ **kWh**

9.3. Steinmetz Law

It was experimentally found by Steinmetz* that hysteresis loss per m³ per cycle of magnetisation of a magnetic material depends on (*i*) the maximum flux density established in it *i.e.* B_{max} and (*ii*) the magnetic quality of the material.

$\therefore$ hysteresis loss $W_h \propto B_{max}^{1.6}$ joules/m³/cycle

$= \eta B_{max}^{1.6}$ joules/m³/cycle

where η is a constant depending on the nature of the magnetic material and is known as **Steinmetz hysteresis coefficient.** The index 1.6 is empirical and holds good if the value of B_{max} lies between 0.1 and 1.2 Wb/m². If B_{max} is either lesser than 0.1 Wb/m² or greater than 1.2 Wb/m², then index is greater than 1.6.

If V is the volume of the magnetic material in m³ and is subjected to f reversals/second, then loss is given by

$$W_h = \eta B_{max}^{1.6} f V \text{ watt}$$

The armatures of electric motors and generators and transformer cores etc. which are subjected to rapid reversals of magnetisation should, obviously, be made of substances having low hysteresis coefficients in order to reduce hysteresis loss. Values of hysteresis coefficients for some of the most commonly used materials are given as follows:

TABLE NO. 9.1
Hysteresis Coefficients

Material		*Hysteresis Coefficient* η *(joules/m³)* $\times 10^2$
Cast iron	...	27.63 to 40.2
Sheet iron	...	10.05
Cast steel	...	7.54 to 30.14
Hard cast steel	...	63 to 70.34
Silicon steel (4.8% Si)	...	1.91
Hard tungsten steel	...	145.1
Good dynamo sheet steel	...	5.02
Mild steel castings	...	7.54 5o 22.61
Nickel	...	32.66 to 100.5
Permalloy	...	0.25

Example 9.4. *A cylinder of iron of volume* 8×10^{-3} m³ *revolves for 20 minutes at a speed of 3000 r.p.m. in a two pole field of field-strength 0.8 Wb/m². If the hysteresis coefficient of iron is 753.6 joules/m³, specific heat of iron is 0.11, the loss due to eddy current is equal to that due*

*C.P. Steinmetz (1865 – 1923) – an American electrical engineer.

to hysteresis and 25% of the heat produced is lost by radiation, find the temperature rise of iron. Take density of iron as 7.8 g/cm³

Solution. An armature revolving in a multipolar field under-goes one magnetic reversal after passing under one pair of poles. In other words, number of magnetic reversals is the same as the number of *pair* of poles. If P is the number of poles, then magnetic reversals in one revolution are $P/2$. If speed of armature rotation is N r.p.m., then number of revolutions/second = $N/60$.

∴ No. of reversals/second = reversals in one revolution × revolutions/second

$$= \frac{P}{2} \times \frac{N}{60} = \frac{PN}{120} \text{ reversal/second}$$

Here, $N = 3000$ r.p.m. $P = 2$

$$\therefore \quad f = \frac{3000 \times 2}{120} = 50 \text{ reversals/second}$$

According to Steinmetz, hysteresis loss

$$W_h = \eta B_{max}^{1.6} f V \text{ watt}$$

It may be noted that f here stands for magnetic reversals/second and not for frequency of armature rotation.

$$\therefore \quad W_h = 753.6 \times (0.8)^{1.6} \times 50 \times 8 \times 10^{-3} \text{ J/s}$$

Loss in 20 minutes $= 753.6 \times (0.8)^{1.6} \times 50 \times 8 \times 10^{-3} \times 1200$ J

$= 253.2 \times 10^3$ J

Eddy current loss = 253.2×10^3 J

Total loss $= 506.4 \times 10^3$ J

Heat produced $= 506.4 \times 10^3/4.2 = 120{,}570$ cal

Heat utilized $= 120{,}570 \times 0.75 = 90{,}430$ cal

Heat absorbed by iron = $(8000 \times 7.8) \times 0.11 \times t$ cal ∴ $(8000 \times 7.8) \times 0.11t$

$= 90{,}430$ ∴ $t =$ **13.17° C**

Example 9.5. *Calculate the energy lost per hour by hysteresis in an iron armature weighing 50.86 kg in a 4-pole dynamo running at 1500 r.p.m., the limits of induction being ± 0.3 Wb/m², the hysteresis coefficient being 376.8 joules/m³ and specific gravity of the armature iron being 7.75.*

Solution. Assuming that for this B_{max}, the index of 1.6 is applicable, we have

$$W_h = \eta B_{max}^{1.6} f V \text{ watt}$$

Here $f = 500 \times 4/120 = 50$ Hz

$$V = \frac{50.86 \times 1000}{7.75} \times 10^{-6} = 6.561 \times 10^{-3} \text{ m}^3$$

$$\therefore \quad W_h = 376.8 \times (0.3)^{1.6} \times 50 \times 6.561 \times 10^{-3} \text{ J/s} = 18 \text{ J/s}$$

Loss/hour $= 18 \times 3600 = 64{,}800$ J

$= 64{,}800/36 \times 10^5 =$ **0.018 kWh**

Example 9.6. *In a certain transformer, the hysteresis loss is 300 W when the maximum flux density is 0.9 Wb/m² and the frequency 50 Hz. What would be the hysteresis loss if the maximum flux density were increased to 1.1 Wb/m² and the frequency reduced to 40 Hz.*

Assume the hysteresis loss over this range to be proportional to $B_{max}^{1.7}$

Solution. Hysteresis loss, $W_h = \eta_{max}^{1.7} f V$ joules/second

$$\therefore \quad W_h \propto B_{max}^{1.7} f$$

In the first case, $W_h = 300$ W; $B_{max} = 0.9$ Wb/m², $f = 50$ Hz

In the second case

$$B_{max} = 1.1 \text{ Wb/m}^2 \,; f = 40 \text{ Hz} \,; W_h = ?$$

$$\frac{300}{W_h} = \frac{(0.9)^{1.7} \times 50}{(1.1)^{1.7} \times 40}; \quad W_h = 300 \times (1.1/0.9)^{1.7} \times 4/5 = \mathbf{337.7 \text{ W}}$$

9.4. Energy Stored in a Magnetic Field

For establishing a magnetic field, energy must be spent, though no energy is required to *maintain* it. Take the example of the exciting coils of an electromagnet. The energy supplied to it is spent in two ways (*i*) part of it goes to meet I^2R loss and is lost once for all, (*ii*) part of it goes to create flux and is stored in the magnetic field as potential energy and is similar to the potential energy of a raised weight. When a mass m is raised through a height of h, then potential energy stored in it is mgh. Work is done in *raising* this mass but once raised to a certain height, no further expenditure of energy is required to *maintain* it at that position. This mechanical potential energy can be recovered, so can the electrical energy stored in a magnetic field.

When current through an inductive coil is gradually changed from zero to a maximum value I, then every change of it is opposed by the self-induced e.m.f. produced due to this change. Energy is needed to overcome this opposition. This energy is stored in the magnetic field of the coil and is, later on, recovered when that field collapses. The value of this stored energy can be found in the following two ways.

(*i*) First Method

Let i = instantaneous value of current

e = induced e.m.f. at the instant = $L\, di/dt$

Then, work done in time dt in overcoming this opposition is

$$dW = e\,i\,dt$$

$$= L\frac{di}{dt} \times i \times dt \qquad (\because\ e = L\,di/dt)$$

$$= L\,i\,di$$

Total work done in establishing the maximum steady current of I is

$$\int_0^W dW = \int_0^I L\,i\,dt = \frac{1}{2}LI^2$$

or $$W = \frac{1}{2}LI^2$$

If L is in henrys and I in amperes, then this value is in joules.

$$\therefore \quad W = \frac{1}{2}LI^2 \text{ Joules}$$

(*ii*) Second Method

If current grows uniformly from zero value to this maximum value I, then average current is $I/2$. If L is the inductance of the circuit, then self-induced e.m.f is $e = L.I/t$ where 't' is the time for the change from zero to I.

∴ average power absorbed = induced e.m.f. × average current

$$= L.\frac{I}{t} \times \frac{1}{2}I = \frac{1}{2}\frac{LI^2}{t}$$

Total energy absorbed = power × time

$$= \frac{1}{2}\frac{LI^2}{t} \times t = \frac{1}{2}LI^2$$

$$\therefore \text{ energy stored} = \frac{1}{2}LI^2 \text{ joules}$$

Example 9.7. *An iron core choking coil has a length 1 metre, a mean diameter of 0.06 m and is wound uniformly with 1000 turns of wire. Find the energy stored in joules when the coil is carrying a current of 10 A. Take relative permeability = 1*

(Magnetic Ckts. and Materials, Punjab Univ. 1993)

Solution. The self inductance of the choking coil is given by

$$L = \mu_0 \mu_r AN^2/l$$

$$A = \pi d^2/4 = \pi \times 0.06^2/4 = 0.0028 \text{ m}^2$$

$$\therefore \quad L = 4\pi \times 10^{-7} \times 1 \times 0.0028 \times 1000^2/1 = 0.00352 \text{ H}$$

Energy stored in the coil is $= \frac{1}{2} L I^2$

$\therefore \quad E = \frac{1}{2} \times 0.00352 \times 10^2 = \mathbf{0.176\ J}$

9.5. Energy Stored Per Unit Volume of a Magnetic Field

It has already been shown that the energy stored in a magnetic field of length l metre and of cross-section A m^2 is

$$E = \frac{1}{2} L I^2 \text{ joules}$$

$$= \frac{1}{2} \times \frac{\mu_0 \mu_r A N^2}{l} I^2 \text{ joules}$$

Now, $\quad H = \frac{NI}{l}$ AT/m

$$E = \left(\frac{NI}{l}\right)^2 \times \frac{1}{2} \mu_0 \mu_r A l \text{ joules}$$

$$= \frac{1}{2} \mu_0 \mu_r H^2 \times A l \text{ joules}$$

Now $\quad Al =$ volume of the magnetic field in m^3

$\therefore$ energy stored/m^3 $= \frac{1}{2} \mu_0 \mu_r H^2$

$$= \frac{1}{2} BH \text{ joules} \qquad (\because \mu_0 \mu_r H = B)$$

$$= \frac{B^2}{2 \mu_0 \mu_r} \text{ joules} \qquad \text{— in a medium}$$

$$= \frac{B^2}{2\mu_0} \text{ joules} \qquad \text{— in air}$$

or $$= \frac{\mu_0 H^2}{2} \text{ joules} \qquad \text{— in air}$$

9.6. Lifting power of a Magnet

In Fig. 9.4, let

P = force in newtons between two poles

A = pole area in m^2

If one of the poles (say, upper one) is pulled apart against this attractive force through a distance of dx metres (dx being so small that neither P nor B are altered appreciably) then work done

$$= P \times dx \text{ joules} \qquad ...(i)$$

This work goes to provide energy for the additional volume of the magnetic field created.

Additional volume of magnetic field created is

$$= A \times dx \text{ m}^3$$

Energy required per m^3

$$= \frac{B^2}{2\mu_0} \text{ joules} \qquad \text{... Art. 9.5}$$

$\therefore$ energy required for the new volume

$$= \frac{B^2}{2\mu_0} \times A\, dx \qquad ...(ii)$$

Equating (i) and (ii), we get

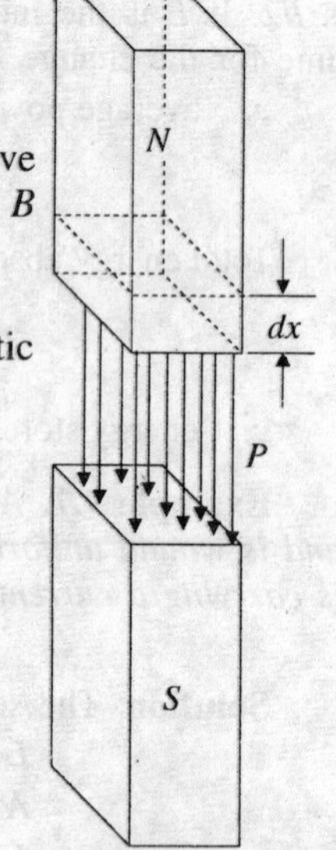

Fig. 9.4

$$P.dx = \frac{B^2 \times A \cdot dx}{2\mu_0}$$

$$\therefore \quad P = \frac{B^2 A}{2\mu_0} \text{ newton}$$

or $$P = \frac{B^2 A}{9.81 \times 2\mu_0} = \frac{B^2 A}{19.62\,\mu_0} \text{ kg-wt}$$

Also, $$P = \frac{B^2}{2\mu_0} \text{ N/m}^2$$

Example 9.8. *In a telephone receiver, the size of each of the poles of the electromagnet is 1.2 cm × 0.2 cm and the flux between each pole and the diaphragm is 4 × 10^{-4} Wb. With what force is the diaphragm attracted to the poles ?* **(Magnetic Ckts. & Materials, Punjab Univ. 1994)**

Solution. Pulling force between each pole and the diaphragm is

$$P = \frac{B^2 A}{2\mu_0} \text{ newton}$$

Since there are two poles, the total force applied on the diaphragm is

$$= 2 \times \frac{B^2 A}{2\mu_0} = \frac{B^2 A}{\mu_0}$$

$$A = 1.2 \times 0.2 = 0.24 \text{ cm}^2 = 0.24 \times 10^{-4} \text{ m}^2$$

$$B = \frac{\Phi}{A} = \frac{4 \times 10^{-4}}{0.24 \times 10^{-4}} = 16.6 \text{ Wb/m}^2$$

$\therefore$ total force on the diaphragm is

$$= \frac{16.6^2 \times 0.24 \times 10^{-4}}{4\pi \times 10^{-7}} = \mathbf{5.26 \times 10^3\ N}$$

Example 9.9. *A horse-shoe magnet is formed out of a bar of wrought iron 45.72 cm long, having a cross-section of 6.45 cm². Exciting coils of 500 turns are placed on each limb and connected in series. Find the exciting current necessary for the magnet to lift a load of 64 kg assuming that the load has negligibe reluctance and makes close contact with the magnet. Relative permeability of iron = 700.* **(Electrical Engg. A.M.Ae. S.I June 1992)**

Solution. Horse-shoe magnet is shown in Fig. 9.5.

Force of attraction at each pole = 64/2 = 32 kg-wt

$$= 32 \times 9.81 \text{ N} = 314 \text{ N}$$

$$A = 6.45 \times 10^{-4} \text{ m}^2$$

$$F = \frac{B^2 A}{2\mu_0}$$

$$\therefore \quad 314 = \frac{B^2 \times (6.45 \times 10^{-4})}{2 \times 4\pi \times 10^{-7}}$$

$$\therefore \quad B = 1.22 \text{ Wb/m}^2$$

Now, $H = B/\mu_0 \mu_r$

$= 1.1/4\pi \times 10^{-7} \times 700 = 1387$ AT/m

Length of the path $= 0.4572$ m

$\therefore$ A.T. required $= 1387 \times 0.4572 = 634$

No. of turns $= 500 \times 2 = 1000$

$\therefore$ current required $= 634/1000 =$ **0.634**

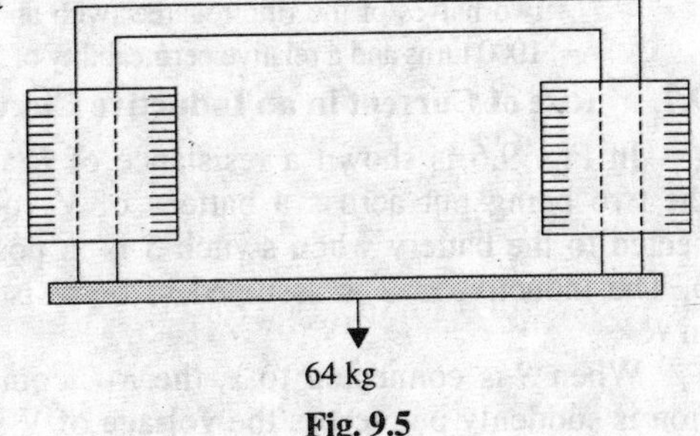

Fig. 9.5

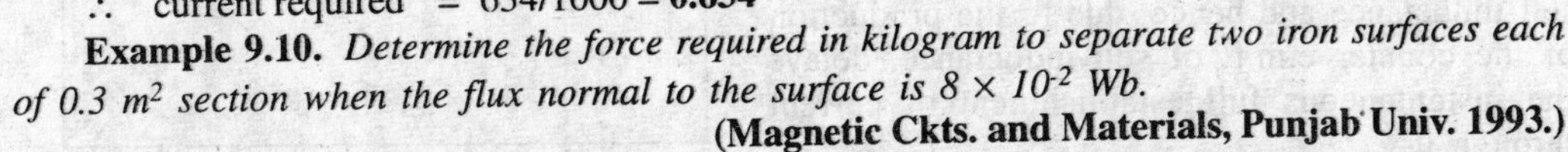

Example 9.10. *Determine the force required in kilogram to separate two iron surfaces each of 0.3 m² section when the flux normal to the surface is 8 × 10^{-2} Wb.*

(Magnetic Ckts. and Materials, Punjab Univ. 1993.)

Solution. The pulling force required is given by

$$P = \frac{B^2 A}{2\mu_0} \text{ newton} = \frac{B^2 A}{19.62\,\mu_0} \text{ kg-Wt}$$

$$B = 8 \times 10^{-2}/0.3 = 0.267 \text{ Wb/m}^2$$

$$\therefore \quad P = \frac{(0.267)^2 \times 0.3}{19.62 \times 4\pi \times 10^{-7}} = \mathbf{867.4 \text{ kg-Wt}}$$

Tutorial Problems No. 9.1

1. The armature of a 4-pole d.c. motor has a volume of 0.012 m^3. In a test on the sheet iron used in the armature, carried out to the same value of maximum flux density as exists in the armature, the area of the hysteresis loop obtained represented a loss of 200 J/m^3. Determine the hysteresis loss in watt when the armature rotates at a speed of 900 r.p.m. **[72 W]**
2. The hysteresis loop for a specimen weighing 12 kg is equivalent to 300 J/m^3. Find the loss of energy per hour at 50 Hz. Density of iron is 7.5 g/cm^3. **[0.024 kWh]**
3. The area of a hysteresis loop plotted for a sample of iron is 67.1 cm^2, the maximum flux density being 1.06 Wb/m^2. The scales of B and H are such that 1 cm = 0.12 Wb/m^2 and 1 cm = 7.07 A/m. Find the loss due to hysteresis if 750 g of this iron were subjected to an alternating magnetic field of maximum flux density 1.06 Wb/m^2 at a frequency of 60 Hz. The density of the iron is 7700 kg/m^3 **[0.333 W]**
4. A hysteresis loop is plotted against a horizontal axis which scales 1000 A/m = 1 cm and a vertical axis which scales 1 Wb/m^2 = 5 cm. If the area of the loop is 9 cm^2 and the overall height is 14 cm, calculate (*a*) the hysteresis loss in joules per cubic metre per cycle, (*b*) the maximum flux density, (*c*) the hysteresis loss in watts per kilogram, assuming the density of the material to be 7800 kg/m^3. **[800 J ; 1.4 Wb/m^2 ; 11.55 W at 50 Hz]**
5. A 4-pole, 500-V shunt-wound d.c. motor has field resistance of 500 ohm and a flux of 0.025 Wb per pole. Calculate the inductance of the field circuit and the energy stored if each field coil has 1200 turns. **[120 H ; 60 J]**
6. A lifting magnet is required to raise a load of one tonne with a factor of safety of 1.5. If the flux density across the pole faces is 0.8 Wb/m^2, calculate the area of each pole face. **[577 cm^2]**
7. A ring of iron has a mean diameter of 30 cm and a circular cross-sectional area of 6 cm^2. It is wound with 1000 turns of wire carrying a current of 3 A. The ring is split across a diameter and an air-gap of 0.25 mm is formed each side. Assuming μ_r = 500 for iron, determine the approximate pull in kg-wt required to separate the two parts. **[122 kg]**
8. The two halves of a cast-iron ring of mean circumference 50 cm and mean cross-sectional area 10 cm^2 are separated by brass plates 2.5 mm thick. Calculate the flux density and current required to hold the two halves of the ring together with an attractive force of 25 kg. The ring is uniformly wound with 1000 turns and a relative permeability of 2500 may be assumed. **[0.556 Wb/m^2 ; 2.3 A]**

9.7. Rise of Current in an Inductive Circuit

In Fig. 9.6 is shown a resistance of R Ω in series with a coil of self-inductance L henrys, the two being put across a battery of V volts. The R-L combination becomes suddenly connected to the battery when switch S is in position a and is short-circuited when S is in position b. The inductive coil is assumed to be resistanceless, its actual small resistance being included in R.

When S is connected to a, the R-L combination is suddenly put across the voltage of V volts. Let us take the instant of closing S as the starting zero time. It is found that current does not reach its maximum value instantaneously but takes some finite time. It is easily explained by recalling that the coil possesses electrical inertia *i.e.* self-inductance and hence, due to the production of the counter e.m.f. of self-inductance, delays the instantaneous full establishment of current through it.

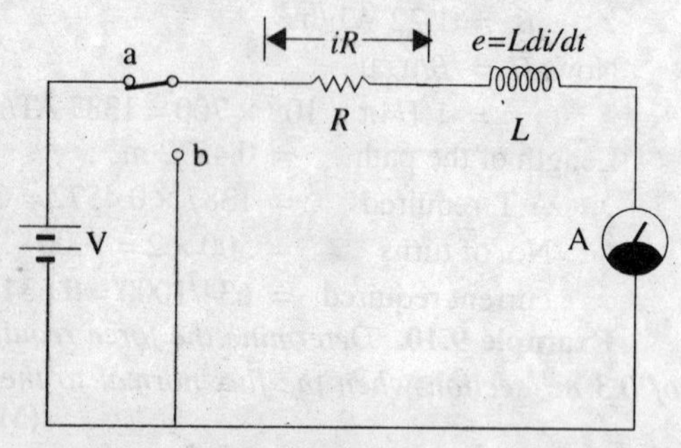

Fig. 9.6

At the *instant* of switching on, i is zero but its rate of change (*i.e.* di/dt) is maximum. Hence, drop across R is zero and the entire battery voltage is used to oppose $e = L\,di/dt$. However, after sometime when steady conditions are established, current is maximum (= V/R) and steady (*i.e.* $di/dt = 0$). Hence, $e = 0$. It means that under steady conditions, all the battery voltage drops over R because none is required for opposing self-induced e.m.f. in the coil. However, at any time after switching on the circuit and before the establishment of steady condition, V is partly dropped across R and partly across L.

We will now investigate the growth of current i through such an inductive circuit.

The applied voltage V must, at any instant, supply not only the ohmic drop iR over the resistance R but must also overcome the e.m.f. of self-inductance *i.e.* $L\,di/dt$.

$$\therefore \quad V = iR + L\,di/dt \qquad \text{...}(i)$$

$$\text{or} \quad (V - iR) = L\,di/dt$$

$$\therefore \quad \frac{di}{V - iR} = \frac{dt}{L}$$

Multiplying both sides by ($-R$), we get

$$(-R)\frac{di}{(V - iR)} = -\frac{R}{L}dt$$

Integrating both sides, we get

$$\int \frac{(-R)\,di}{(V - iR)} = -\frac{R}{L}\int dt$$

$$\therefore \quad \log_e^{V-iR} = -\frac{R}{L}t + K \qquad \text{...}(ii)$$

where e is the Napierian logarithmic base = 2.718 and K is constant of integration whose value can be found from the initial known conditions.

To begin with, when $t = 0$, $i = 0$, hence putting these values in Eq. (*ii*) above, we get

$$\log_e^V = K$$

Substituting this value of K in the above given equation, we have

$$\log_e^{V-iR} = -\frac{R}{L}t \log_e^V$$

$$\text{or} \quad \log_e^{V-iR} - \log_e^V = -\frac{R}{L}t$$

$$\text{or} \quad \log_e \frac{V - iR}{V} = -\frac{R}{L}t = -\frac{i}{\lambda}$$

$$\text{where,} \quad L/R = \lambda \qquad \text{... time constant}$$

$$\therefore \quad \frac{V - iR}{V} = e^{-t/\lambda}$$

$$\text{or} \quad i = \frac{V}{R}(1 - e^{-t/\lambda}) \qquad \text{...}(iii)$$

Now, V/R represents the maximum steady value I_m of the current that would eventually be established through the $R - L$ circuit.

$$i = I_m(1 - e^{-t/\lambda}) \qquad \text{...}(iv)$$

This is an exponential equation whose graph is shown in Fig. 9.7. It is seen from it that current rise is rapid at first and then decreases until at $t = \infty$, it becomes zero. Theoretically, current does not reach its maximum steady value I_m until infinite time. However, in practice, it reaches this value in a relatively short time of about five time constants of the circuit.

The rate of rise of current *i.e.* di/dt at any stage can be found by differentiating Eq. (*iv*) above w.r.t. time. However, the initial rate of rise of current can be obtained by putting $t = 0$ and $i = 0$ in Eq. (*i*) above.*

---

*Initial value of di/dt can also be found by differentiating Eq. (*iv*) and putting $t = 0$ in it.

$\therefore \quad V = 0 \times R + L\dfrac{di}{dt}$

$\therefore \quad \dfrac{di}{dt} = \dfrac{V}{L}$ ... initial value

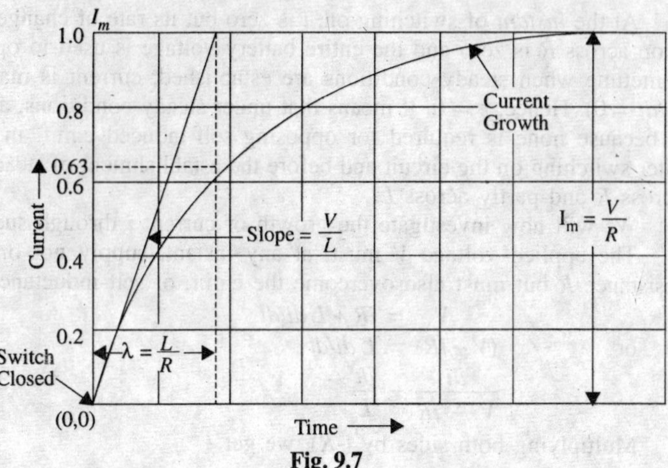

Fig. 9.7

The constant $\lambda = L/R$ is known as the *time-constant* of circuit. It can be variously defined as under :

(*i*) *It is the time during which current would have reached its maximum value of $I_m(= V/R)$ had it maintained its initial rate of rise.*

Time taken =

$$\frac{I_m}{\text{initial rate of rise}} = \frac{V/R}{V/L} = \frac{L}{R}$$

(*ii*) But actually the current takes more time because its rate of the rise decreases gradually. In actual practice, in a time equal to the time constant, it merely reaches 0.632 of its maximum value as shown below :

Putting $t = L/R = \lambda$ in Eq. (*iv*) above, we get

$$i = I_m(1 - e^{-\lambda/\lambda}) = I_m\left(1 - \frac{1}{e}\right)$$

$$= I_m\left(1 - \frac{1}{2.718}\right) = 0.632\, I_m$$

Hence, the time-constant λ of an *R-L* circuit may also be defined *as the time during which the current actually rises to 0.632 of its maximum steady value (Fig. 9.7).*

This delayed rise of current in an inductive circuit is utilised in providing time-lag in the operation of electric relays and trip coils etc.

9.8. Decay of Current in an Inductive Circuit

When the switch *S* (Fig. 9.6) is put in position *b*, then the *R-L* circuit is short-circuited. It is found that the current does not cease immediately (as it would do in a non-inductive circuit) but continues to flow and is reduced to zero only after an appreciable time has elapsed since the instant of short-circuit.

The equation for decay of current with time is found by putting $V = 0$ in Eq. (*i*) of Art. 9.7.

$$0 = iR + L\frac{di}{dt}$$

or $$\frac{di}{i} = -\frac{R}{L}\,dt$$

Integrating both sides, we have

$$\int \frac{di}{i} = -\frac{R}{L}\int dt$$

$$\therefore \quad \log_e i = -\frac{R}{L}t + K \qquad ...(i)$$

Now, at the instant of switching off the current, $i = I_m$ and if time is counted from this instant, then $t = 0$.

$$\therefore \quad \log_e I_m = 0 + K$$

Putting this value of *K* in eq. (*i*) above, we get,

$$\log_e i = -\frac{t}{\lambda} + \log_e I_m$$

$$\therefore \quad \log_e i/I_m = -\frac{t}{\lambda} \quad \text{or} \quad \frac{i}{I_m} = e^{-t/\lambda}$$

or $$i = I_m e^{-t/\lambda} \qquad ...(ii)$$

It is a decaying exponential function and is plotted in Fig. 9.8. It can be shown again that, theoretically, current should take infinite time to reach zero value although in actual practice it does so in a relatively short time.

Again, putting $t = \lambda$ in Eq. (*ii*) above, we get

$$i = \frac{I_m}{e} = \frac{I_m}{2.178} = 0.37\, I_m$$

Hence, time constant (λ) of an *R-L* circuit may be defined as *the time during which current falls to 0.37 of its maximum steady value while decaying (Fig. 9.8)*

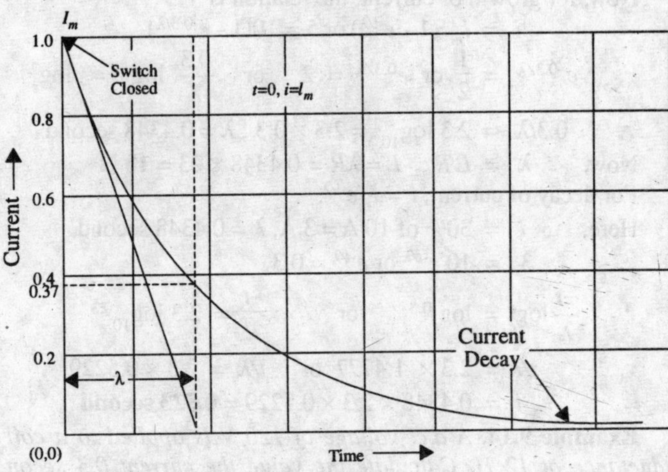

Fig. 9.8

Example 9.11. *A choke with an inductance of 20 H and a resistance of 16Ω is connected through a switch to a 24 volt source of e.m.f.*

(*i*) *What is the initial current when switch is closed ?*

(*ii*) *What is the initial rate of change of current when the switch is closed?*

(*iii*) *What is the final steady-state current?*

(*iv*) *What is the current at the instant of 1.25 second after switching?*

(Electrical Science, AMIE Winter 1994)

Solution.

(*i*) Since inductance does not permit instantaneous change in current, hence initial current in the circuit is **zero**.

(*ii*) Initial rate of change of current = V/L = 24/20 = **1.2 A/s**

(*iii*) The final steady current = V/R = 24/16 = **1.5 A**

(*iv*) Since 1.25 second represents the time constant of the curcuit, hence current at the instant of 1.25 s will be 63.2% of the maximum value.

$$i = 0.632 \times 1.5 = \mathbf{0.948\ A}$$

Example 9.12. *A coil of resistance 20 ohms and inductance 0.8 henry is connected to a 200 volt d.c. supply. Find (i) rate of change of current at the instant of closing the switch, (ii) final steady value of current, (iii) time constant of circuit, (iv) time taken for current to rise to half its final value and (v) energy stored in the field when steady state is reached.* **(A.M.I.E. Summer 1988)**

Solution. (*i*) Initial rate of change of current at the instant of closing the switch is $(di/dt)_t = 0$ V/L = 200/0.8 = 250 **A/s**, (*ii*) Final steady value of the current is $I_m = V/R$ = 200/20 = **10 A**, (*iii*) Time constant, λ - L/R = 0.8/20 = **0.04 second** (*iv*) Here, $i = I_m/2$. Also, for growth, $i = I_m (1 - e^{-t/\lambda})$ $\therefore$ $I_m/2 = I_m (1 - e^{-t/0.04})$; t = 0.0277 **second** (*v*) $E = \frac{1}{2} Li^2 = \frac{1}{2} \times 0.8 \times 10^2$ = **40 J**

Example 9.13. *The field winding of a d.c. machine takes a steady current of 10 A when connected to a d.c. source of 230 volts. On switching the voltage to the field, it is observed* *that it takes 0.3 second to reach 5 A. Find the value of the inductance of the field. Find also the time for the current to fall to 30% of the final steady value if the yield terminals are short-circuited.* **(A.M.I.E. Winter 1989)**

Solution.

Here, $i = 5\text{A}, I_m = 10\text{ A}, t = 0.3$ second ; $R = 230/10 = 23\ \Omega$

Now, for growth of current, the relation is

$$i = I_m (1 - e^{-t/\lambda}) \text{ or } 5 = 10(1 - e^{-0.3/\lambda})$$

$$\therefore \quad e^{-0.3/\lambda} = \frac{1}{2} \text{ or } e^{0.3/\lambda} = 2 \quad \text{or} \quad \frac{0.3}{\lambda} \log_e^e = \log_e^2$$

$$\therefore \quad 0.3/\lambda = 2.3 \log_{10}{}^2 = 2.3 \times 0.3 \text{ ; } \lambda = 0.4348 \text{ second}$$

Now, $\lambda = L/R$; $L = \lambda R = 0.4348 \times 23 = \mathbf{10\ H}$

For decay of current, $i = I_m e^{-t/\lambda}$

Here, i = 30% of 10 A = 3 A, $\lambda = 0.4348$ second

$$\therefore \quad 3 = 10\, e^{-t/\lambda} \text{ or } e^{-t/\lambda} = 0.3$$

$$\therefore \quad \frac{-t}{\lambda} \log_e^e = \log_e{}^{0.3} \quad \text{or} \quad \frac{-t}{\lambda} = 2.3 \log_{10}{}^{0.3}$$

$$\therefore \quad -t/\lambda = 2.3 \times \bar{1}.4777 \quad \text{or} \quad t/\lambda = 2.3 \times 0.5229$$

$$t = 0.4348 \times 2.3 \times 0.5229 = \mathbf{0.523\ second}$$

Example 9.14. *A d.c. voltage of 120 V is applied to a coil haivng a resistance of 8 Ω and inductance of 12 H. Calculate the value the current 0.3 second after switching on the supply. With the current having reached the final steady value, how much time would it take for the current to reach a value of 6 A after switching off the supply?*

Solution. For Growth of Current

$$i = I_m (1 - e^{-t/\lambda})$$

$$I_m = V/R = 120/8 = 15 \text{ A}$$

$$\lambda = L/R = 12/8 = 1.5 \text{ second}$$

$$t = 0.3 \text{ second}$$

$$\therefore \quad i = 15\,(1 - e^{-0.3/1.5}) = 15(1 - e^{-02})$$

$$= 15(1 - 0.8485) = \mathbf{2.72\ A}$$

For Decay of Current

$$i = I_m e^{-t/\lambda}$$

$$6 = 15\, e^{-t/1.5} \text{ or } e^{-t/1.5} = 0.4$$

$$\text{or} \quad e^{t/1.5} = 2.5 \qquad \therefore \quad t = \mathbf{1.374\ second}$$

Example 9.15. *What is meant by time-constant of a circuit?*

A 50 V battery, a 100-Ω resistance and a coil having an inductance of 4H and a resistance of 100 Ω are connected in series. Find (a) the current after a period equivalent to the time constant of the coil and (b) the rate at which the current will then be increasing.

Solution. $i = I_m (1 - e^{-R/L.t})$

Total resistance $R = 100 + 100 = 200\ \Omega$; $R/L = 200/4 = 50$

$$I_m = 50/200 = 0.25 \text{ A}$$

Hence, the above equation becomes

$$i = 0.25\,(1 - e^{-R/L.t})$$

Time constant of the *coil*

$$= 4/100 = 0.04 \text{ second}$$

Hence, putting $t = 0.04$ second in the above equation, we get

(*a*) $i = 0.25\,(1 - e^{-50 \times 0.04}) = 0.25(1 - e^{-2}) = \mathbf{0.216\ A}$

(*b*) $L\, di/dt = V - iR$

$$\therefore \quad 4.di/dt = 50 - 0.216 \times 200 = 6.8$$

$\therefore$ *di/dt* at that time $= 6.8/4 = \mathbf{1.7\ A/s}$

Example 9.16. *A relay coil has a time constant of 2 millisecond, an operating current of 0.25 A and a hold-on current of 0.1 A. When a 50-V d.c. supply is suddenly switched on to the coil, there is a delay of 2 millisecond before the relay operates. Calculate*

(a) the resistance of the coil,

(b) the inductance of the coil,

(c) the delay before the relay releases when the supply is suddenly disconnected from the coil.

Solution. After switching on, it takes 2 milliseconds to operate the coil *i.e.* for the current to reach a value of 0.25 A. Using the equation

$\therefore \quad i = I_m(1 - e^{-t/\lambda})$, we get

$0.25 = I_m(1 - e^{-2/2})$

Now, $I_m = V/R = 50/R$

$\therefore \quad 0.25 = \frac{50}{R}(1 - e^{-1}) = \frac{50}{R}\left(1 - \frac{1}{2.718}\right)$

(a) $R = \mathbf{126.1\ \Omega}$

(b) Now $\lambda = \frac{L}{R} = 2 \times 10^{-3} \qquad \therefore \quad L = \mathbf{252.2\ mH}$

The moment current decreases to 0.1 A or less, the hold-on coil would no longer work. Using the relation

$i = I_m e^{-t/\lambda}$ we get

$0.1 = \frac{50}{126.1} e^{-t/2} \quad \text{or} \quad e^{-t/2} = 0.252$

$\therefore \quad -\frac{t}{2} = 2 \log_{10}^{0.252}$

$= 2.3 \times \bar{1}.4014 = 2.3 \times -0.5986$

$\therefore \quad t = \mathbf{2.75\ millisecond.}$

Example 9.17. *A telephone operating at a current of 120 mA, has an inductance of 10 henrys and resistance of 100 ohm. If a 24-volt battery having negligible internal resistance be suddenly applied to the system, calculate the operating time.* **(A.M.I.E. Summer 1987)**

Solution. Maximum steady current $I_m = 24/100 = 0.24$ A = 240 mA. The operating time is equal to the time required by the current to rise from 0 to 120 mA

Now, $i = I_m(1 - e^{-t/\lambda})$

$I_m = 240$ mA, $i = 120$ mA, $\lambda = L/R = 10/100 = 0.1$ s

$\therefore \quad 120 = 240\,(1 - e^{-t/0.1})\,; \qquad t = \mathbf{0.069\ s}$

Example 9.18. *Consider a series R-L circuit connected to a battery source of 1 volt. If R = 1 ohm and L = 1 henry, find the current i (t) through the inductance. Find the voltage across resistance and the inductance separately at the time of switching the supply and when sufficient time has elapsed after switching on the source.* **(A.M.I.E. Summer 1994)**

Solution. $\lambda = L/R = 1/1 = 1$ s ; $I_m = V/R = 1/1 = 1$ A

The current *i*(*t*) through the coil at any instant is given by

$i(t) = I_m(1 - e^{-t/\lambda}) = 1(1 - e^{-t/1}) = (1 - e^{-t})$

As explained in Art. 9.7

(*i*) At the time of switching on, drop across *R* is zero but that across *L* is maximum *i.e.* **1 V.**

(*ii*) Under steady conditions, current is maximum so that drop across *R* is maximum *i.e.* **1 V** and that across *L* is **zero.**

Tutorial Problems No. 9.2

1. A coil with a self-inductance of 2.4 H and resistance of 12 Ω is suddenly switched across a 120-V d.c. supply of negligible internal resistance. Determine the time-constant of the coil, the instantaneous value of the current after 0.1 second, the final steady value of the current and the time taken for the current to reach 5 A. **[0.2 second ; 3.94 A ; 10 A ; 0.139 second]**

2. A relay has a resistance of 300 Ω and is switched on to a 100-V d.c. supply. If the current reaches 63.2 per cent of its final steady value oin 0.002 second, determine.
   (*a*) the time constant of the circuit,
   (*b*) the inductance of the circuit,
   (*c*) the final steady value of the current,
   (*d*) the initial rate of rise of current. **[(*a*) 0.002 second (*b*) 0.6 H (*c*) 0.366 A (*d*) 183 A/s]**
3. A coil of 200 turns has a resistance of 50 Ω and produces a magnetic flux of 5 mWb when a steady current of 4 A is passed through it. The coil is connected across a d.c. supply of 200 V. Calculate
   (*a*) time constant of the coil,
   (*b*) the induced e.m.f. in the coil at the instant when the current has risen to 3 A,
   (*c*) the rate of growth of the current at this instant
   (*d*) the energy stored in the magnetic field when the current has reached its steady value. **[(*a*) 50 ms (*b*) 50 V (*c*) 20 A/s (*d*) 20 J]**
4. A coil has a self-inductance of 0.75 H and a resistance of 25 Ω. It is connected to a d.c. supply of 50 V. If the supply is suddenly removed and replaced by a short-circuit, calculate the time taken for the current to decay from 1.5 A to 0.5 A. **[33 ms]**
5. A solenoid of inductance 30 H and of resistance 30 Ω is connected in series with a battery of 100 V. In how much time, after connecting the battery, will the current reach half of its final value. *(A.M.I.E. May 1975)*
6. A coil having a resistance of 25 Ω and an inductance of 2.5 H is connected across a 50-V d.c. supply. Determine (*a*) initial rate of growth of current, (*b*) value of current after 0.15 second and (*c*) time required for the current to grow to 1.8 A **[(*a*) 20 A/s (*b*) 1.55 A (*c*) 0.23 s]**
7. A coil, of inductance 5 H and resistance 100 Ω, carries a steady current of 2 A. Calculate the initial rate of fall of current in the coil after a short-circuit switch connected across its terminals has been closed. What was the energy stored in the coil and in what form was it dissipated ? **[40 A/s ; 10 J, heat]**

HIGHLIGHTS

1. By magnetic hysteresis is meant the lagging of induction flux density (*B*) behind magnetising force (*H*).
   It results in some net loss (known as hysteresis loss) when a magnetic material is carried through one complete cycle of reversal of magnetisation.
2. Hysteresis loss is
   (*a*) = *xy* (area of hysteresis loop) J/m³/cycle
   (*b*) $= \eta B_{max}^{1.6} fV$ watt —Steinmetz Law
3. Energy stored in an inductive coil is
   $$E = \frac{1}{2} LI^2 \text{ joules}$$
4. Energy stored per unit volume of a magnetic field is
   $$= \frac{B^2}{2\mu_0} \text{ J/m}^3$$
5. Lifting power of a magnet is
   $$= \frac{B^2 A}{2\mu_0} \text{ N}$$
6. In an inductive circuit, the current rises according to the relation
   $$i = I_m(1 - e^{-t/\lambda})$$
   Similarly, the current decays according to the relation
   $$i = I_m e^{-t/\lambda}$$
   The time constant of the circuit is
   $$\lambda = L/R \text{ second.}$$

OBJECTIVE TESTS — 9

A. Fill in the following blanks.

1. The *B/H* loop area represents .... loss of a magnetic material.
2. Magnetic hysteresis is primarily due to the .... of the magnetic material.
3. The hysteresis loss in a magnetic material is converted into ....
4. Hard steel is well-suited for making permanent magnets because it has high coercivity and high ....
5. High .... and low hysteresis loss of alloy sheet steel make it most suited for making transformer cores.
6. The experimental law for the hysteresis loss of a magnetic material was found by ....
7. Energy stored in a coil depends on the .... of the current passing through it.
8. Lifting power of an electromagnet varies as the square of its ....

B. Answer True or False.

1. Hysteresis loss cannot occur in non-magnetic materials.
2. The area of the hysteresis loop represents hysteresis loss in J/s.
3. For a given current, energy stored in the magnetic field of a coil depends on the square of its turns.
4. The lifting power of an electromagnet varies directly as the area of its poles.
5. The initial rate of rise of current through a coil is the same as its initial rate of decay.
6. The current through a coil becomes steady after about five time constants.
7. During decay, the current through a coil falls to 63.2% of its initial value in one time constant.
8. The time constant of a circuit is given by the ratio of resistance over inductance.

C. Multiple Choice Questions.

1. A magnetic material having high retentivity and high coercivity is most suited for making
   (*a*) electromagnets
   (*b*) armatures
   (*c*) permanent magnets
   (*d*) transformer cores.
2. Hysteresis loss in a magnetic material is due to its
   (*a*) retentivity
   (*b*) coercivity
   (*c*) flux density
   (*d*) both (*a*) and (*b*) .
3. According to Steinmetz law, hysteresis loss in a magnetic material depends on its.
   (*a*) frequency of magnetic reversals
   (*b*) flux density
   (*c*) volume
   (*d*) all of the above.
5. The delay in the establishment of steady current through a coil depends on the value of
   (*a*) applied voltage
   (*b*) self-inductance
   (*c*) resistance
   (*d*) (*b*) and (*c*).
6. The initial rate of rise of current through a coil is
   (*a*) the minimum
   (*b*) the same as rate of decay
   (*c*) independent of its inductance
   (*d*) inversely dependent on the applied voltage .

ANSWERS

A. 1. hysteresis 2. coercivity 3. heat 4. retentivity 5. permeability 6. Steinmetz 7. square 8. flux density

B. 1. T 2. F 3. T 4. T 5. T 6. T 7. F 8. F

10 D.C. GENERATORS

10.1. Generator Principle

An electrical generator is a machine which converts mechanical energy (or power) into electrical energy (or power).

This energy conversion is based on the principle of the production of dynamically (or motionally) induced e.m.f. As seen from Fig. 8-3 whenever a conductor cuts magnetic flux, dynamically induced e.m.f. is produced in it according to Faraday's Laws of Electromagnetic Induction. This e.m.f. will cause a current to flow if the conductor circuit is closed.

Hence, the basic essential parts of an electrical generator are (*i*) a magnetic field and (*ii*) a conductor or conductors which can move so as to cut the flux.

10.2. Simple Loop Generator

(*a*) Construction

In Fig. 10.1 is shown a single-turn rectangular copper coil *ABCD* rotating about its own axis in a magnetic field provided by either permanent magnets or electromagnets. The two ends of the coil are joined to two sliprings or discs '*a*' and '*b*' which are insulated from each other and from the central shaft. Two collecting brushes (or carbon) press against the sliprings. Their function is to collect the current induced in the coil and to convey it to the external load resistance *R*.

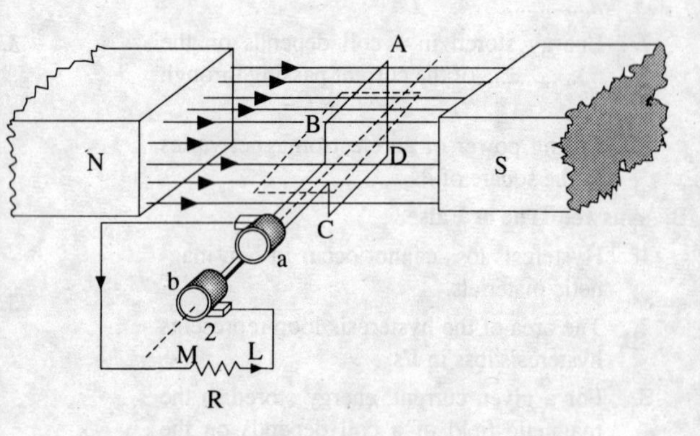

Fig. 10.1

The rotating coil may be called armature and the magnets as field magnets.

(*b*) Working

Imagine the coil to be rotating in clockwise direction (Fig. 10.2). As the coil assumes successive positions in the field, the flux linked with it changes. Hence, an e.m.f. is induced in it which is proportional to the rate of change of flux linkages ($e = -Nd\Phi/dt$). When the plane of the coil is at right angles to the lines of flux *i.e.* when it is in position 1, the flux linked with the coil is maximum but *rate of change of flux linkages is minimum.* This is so because in this position, the coil sides *AB* and *CD* do not cut or shear the lines of flux; rather they slide along them *i.e.* they move parallel to them. Hence, there is no induced e.m.f. in the coil. Let us take this no-e.m.f. or vertical position of the coil as the starting position. The angle of rotation or time will be measured from this position.

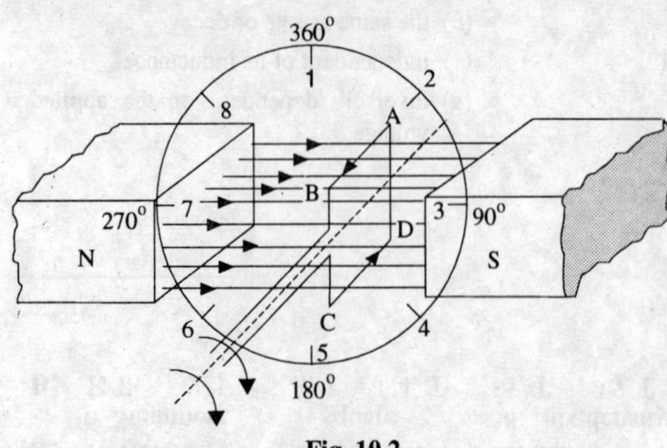

Fig. 10.2

As the coil continues rotating further, the rate of change of flux linkages (and hence induced e.m.f. in it) increases, till position 3 is reached where $\theta = 90°$. Here, the coil plane is horizontal *i.e.* parallel to the lines of flux. As seen, the flux linked with the coil is minimum but *rate of change of flux linkages or rate of flux cutting is maximum.* Hence, maximum e.m.f. is induced in the coil when in this position (Fig. 10.3).

In the next quarter revolution *i.e.* from 90° to 180°, the flux linked with the coil gradually *increases,* but rate of change of flux *decreases.* Hence, the induced e.m.f. decreases gradually till in position 5 of the coil, it is reduced to zero value (Fig. 10.3).

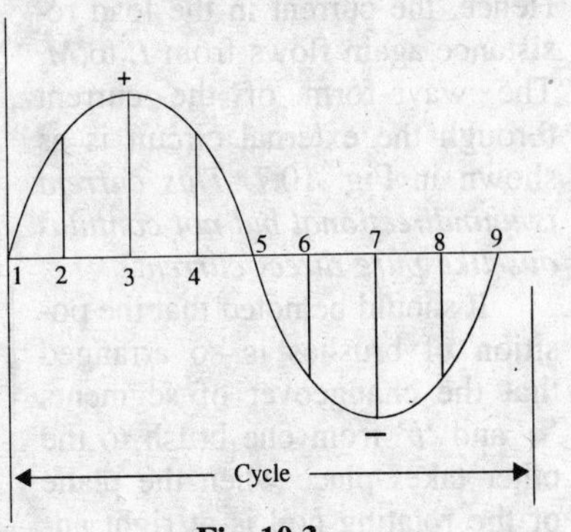

So, we find that in the first revolution of the coil, no (or minimum) e.m.f. is induced in it when in position 1, maximum e.m.f. is induced when in position 3 and no e.m.f. is induced when in position 5. The direction of this induced e.m.f. can be found by applying Fleming's Right-hand rule which gives its direction from *A* to *B* and *C* to *D*. Hence, the direction of current flow is *ABMLCD* (Fig. 10.1). The current through the load resistance *R* flows from *M* to *L* during the first half revolution of the coil.

Fig. 10.3

In the next half revolution *i.e.* from 180°, the variations in the magnitude of e.m.f. are similar to those in the first half revolution. Its value is maximum when coil is in position 7 and minimum when it is in position 1. But it will be found that the direction of the induced current is from *D* to *C* and *B* to *A*. Hence, the path of current flow is along *DCLMBA* which is just the reverse of the previous direction of flow.

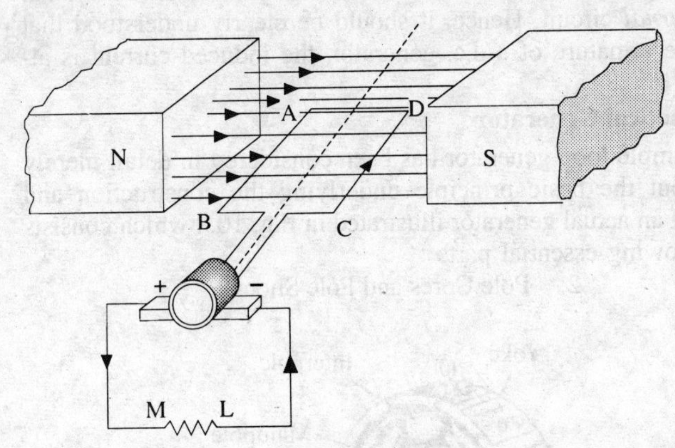

Fig. 10.4

Therefore, we find that the current which we obtain from such a simple generator reverses its direction after every half revolution. Such a current undergoing periodic reversals is known as alternating current (A.C). It is, obviously, different from a direct current (D.C.) which continuously flows in one and the same direction. It should be noted that A.C. not only reverses its direction, it does not even keep its magnitude constant while flowing in any one direction. The two half-cycles may be called positive and negative half cycles respectively (Fig. 10.3).

For making the flow of current unidirectional in the external circuit, the slip-rings are replaced by split-rings (Fig. 10.4). The split-rings are made out of a conducting cylinder which is cut into two halves or segments insulated from each other by a thin sheet of mica or some other insulating material (Fig. 10.5).

As before, the coil ends are joined to these segments on which rest the carbon brushes.

It is seen [Fig. 10.6 (*a*)] that in the first half revolution, current flows along *ABLMCD i.e.* brush No. 1 which is in contact with segment '*a*', acts as the positive end of the supply and brush No. 2 and '*b*' as the negative end. In the next half revolution [Fig. 10.6 (*b*)], the direction of the induced current in the coil is reversed. But at the same time, the positions of segments '*a*' and '*b*' are also reversed with the result that brush No. 1

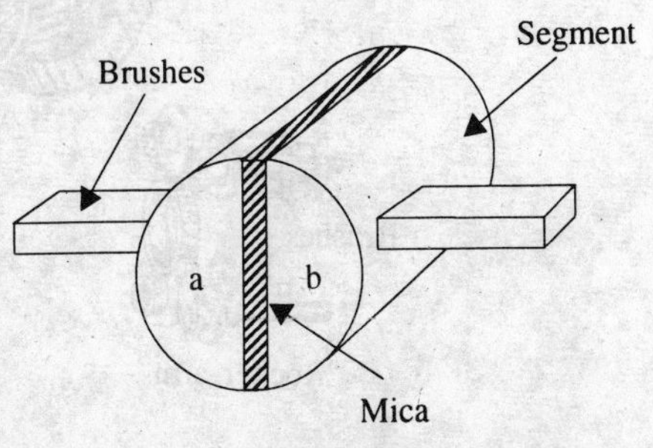

Fig. 10.5

comes in touch with that segment which is positive *i.e.* segment '*b*'. Hence, the current in the load resistance again flows from *L* to *M*. The wave-form of the current through the external circuit is as shown in Fig. 10.7. *This current is unindirectional but not continuous like pure direct current.*

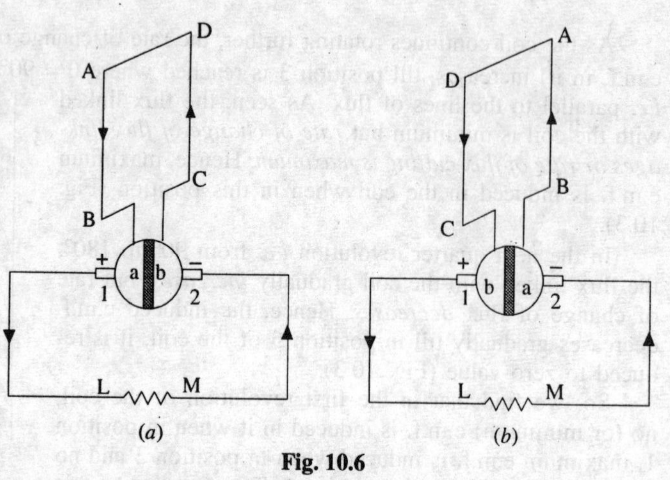

Fig. 10.6

It should be noted that the position of brushes is so arranged that the changeover of segments '*a*' and '*b*' from one brush to the other takes place when the plane of the rotating coil is at right angles to the plane of the lines of flux because in that position, the induced e.m.f. in the coil is zero.

Another important point to remember is that even now the current induced in the coil sides is alternating as before. It is only due to the rectifying action of the split-rings (also called commutator) that it becomes unidirectional in the *external* circuit. Hence, it should be clearly understood that even in the armature of a d.c. generator, the induced current is alternating.

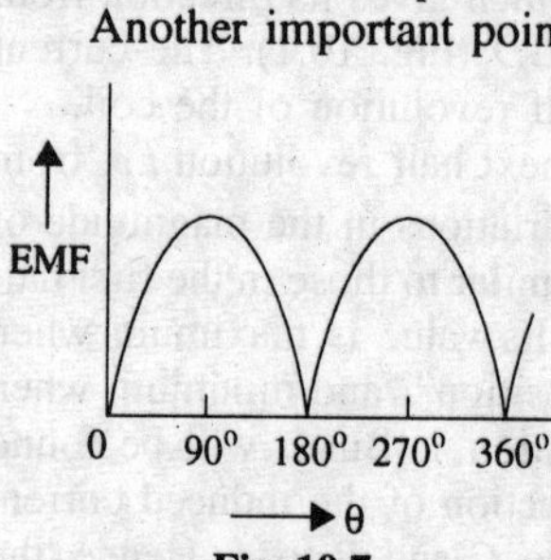

EMF

0 90° 180° 270° 360°

θ

Fig. 10.7

10.3. Practical Generator

The simple loop generator has been considered in detail merely to bring out the basic principle underlying the construction and working of an actual generator illustrated in Fig. 10.8 which consists of the following essential parts :

1. Magnetic Frame or Yoke.
2. Pole Cores and Pole Shoes.

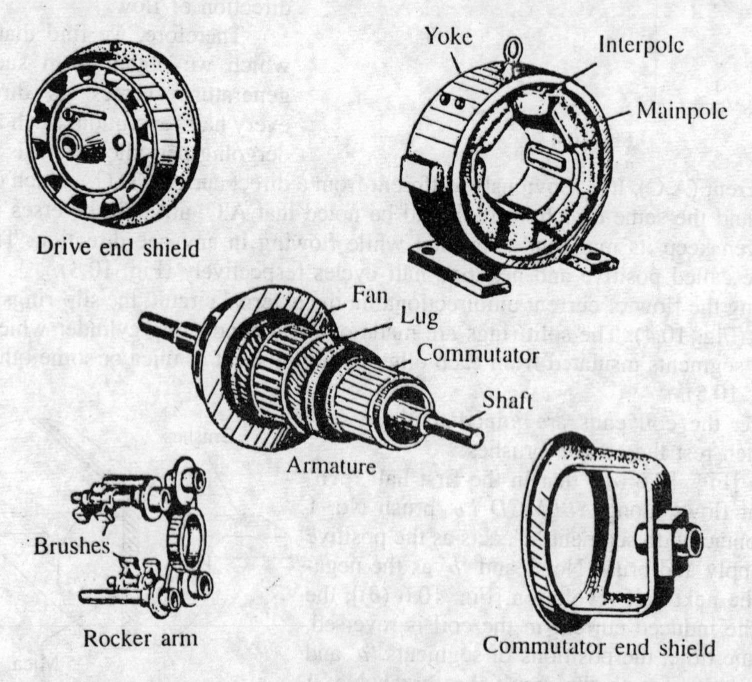

Fig. 10.8

3. Pole Coils or Field Coils.
4. Armature Core.
5. Armature Windings or Conductors.
6. Commutator.
7. Brushes and Bearings.

Of these, the yoke, the pole cores, the armature core and air gaps between the poles and the armature core form the magnetic circuit whereas the rest form the electrical circuit.

Fig. 10.8 gives the view of a d.c. machine taken apart. Full details of each component are given below:

10.4. Yoke

The outer frame or yoke serves double purpose:

(*i*) it provides mechanical support for the poles and acts as a protecting cover for the whole machine and

(*ii*) it carries the magnetic flux produced by the poles.

In small generators where cheapness rather than weight is the main consideration, yokes are made of cast iron. But for large machines, usually cast steel or rolled steel is employed.

10.5. Pole Cores and Pole Shoes

The field magnets consist of pole cores and pole shoes. The pole shoes serve two purposes. (*i*) they spread out the flux in the air-gaps and also, being of larger cross-section, reduce the reluctance of the magnetic path and (*ii*) they support the exciting coils.

10.6. Pole Coils

The field coils or pole coils, which consist of copper wire or strip, are former-wound for the correct dimension. Then, the former is removed and the wound coil put into place over the core.

When current is passed through these coils, they electromagnetise the poles which produce the necessary flux that is cut by the revolving armature conductors.

10.7. Armature Core

It houses the armature conductors or coils and causes them to rotate and hence cut the magnetic flux of the field magnets. In addition to this, its most important function is to provide a path of very low reluctance to the flux passing through the armature from a *N*-pole to a *S*-pole.

It is cylindrical or drumshaped and is built up of usually circular sheet steel discs or laminations approximately 0.06 mm thick (Fig. 10.9)

The purpose of using laminations is to reduce the loss due to eddy currents. Thinner the laminations, greater is the resistance offered to the induced e.m.f., smaller the current and hence less the I^2R loss in the core.

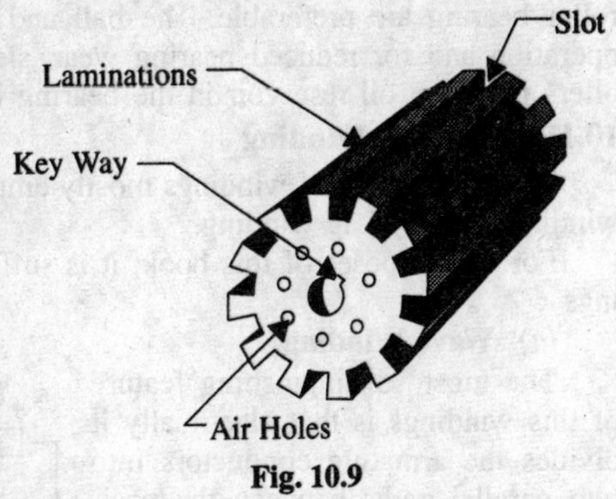

Fig. 10.9

10.8. Armature Windings

The armature windings are usually former-wound. These are first wound in the form of flat rectangular coils and are then pulled into their proper shape in a coil puller. Various conductors of the coils are insulated from each other. The conductors are placed in the armature slots which are lined with tough insulating material. This slot insulation is folded over above the armature conductors placed in the slot and is secured in place by special hard wooden or fibre wedges.

10.9. Commutator

The function of the commutator is to facilitate the collection of current from the armature conductors. As shown in Art 10.2, it rectifies *i.e.* converts the alternating current induced in the armature into unidirectional current. It is of cylindrical structure and is built up of wedge-shaped segments of high-conductivity hard-drawn or drop-forged copper. These segments are insulated from each other by thin layers of mica. Each commutator segment is connected to the armature conductors by means of a copper lug or strip (or riser).

10.10. Brushes and Bearings

The brushes whose function is to collect current from comutators, are usually made of carbon and are in the shape of a rectangular block. These brushes are housed in brush-holders (usually of the box-type variety) which are mounted on brush-holder studs or brackets.

In turn, the brush-holder studs are mounted on a brush yoke or rocker arm. The brush-holder studs are insulated from the brush yoke by means of insulation seelves and discs. The brush yoke, brush-holders and brushes make up the brush gear which is shown in Fig. 10.10.

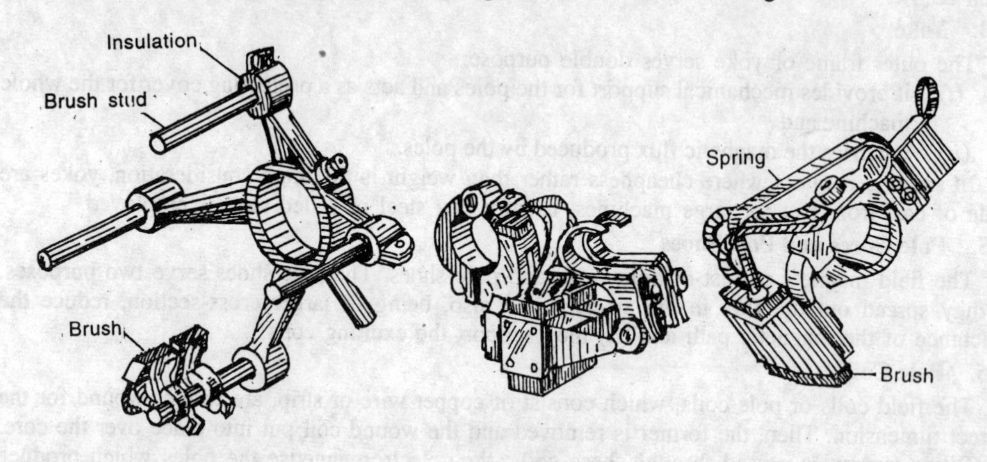

Fig. 10.10

Because of their reliability, ball bearings are frequently employed ; though for heavy duties, roller bearing are preferable. The ball and rollers are generally packed in hard oil for quieter operation and for reduced bearing wear, sleeve bearings are used which are lubricated by ring oilers fed from oil reservoir in the bearing bracket.

10.11. Armature Winding

Two basic types of windings mostly employed for drum-type armature are known as (*i*) Wave winding and (*ii*) Lap winding.

For the purposes of this book, it is sufficient to know the following facts about these windings :

(*a*) Wave Winding

The most distinguishing feature of this windings is that electrically it divides the armature conductors into two parallel paths between the positive and negative brushes irrespective of the number of poles of the machine.

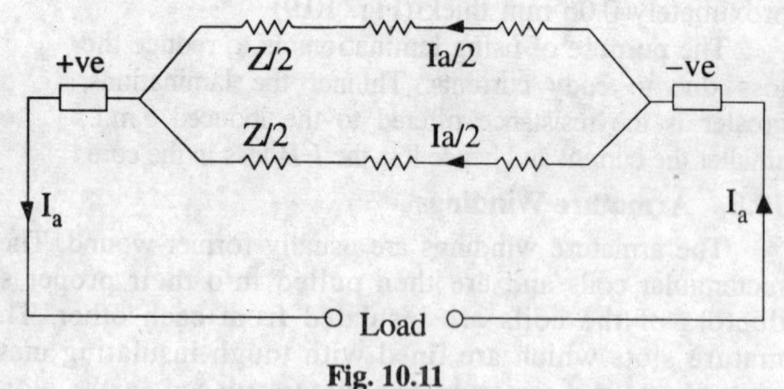

Fig. 10.11

As shown in Fig. 10.11, as the armature current enters the negative brush, it finds two parallel paths of equal resistance available for going to the positive brush. Hence, it divides equally into two parts. Each path consists of $Z/2$ conductors connected in series (Z–being the total number of armature conductors) and each carries a current of $I_a/2$ where I_a is the total armature current.

For example, in the case of a 4-pole, wave-wound generator having 30 armature conductors, each of the two parallel paths will have 15 conductors.

(*b*) Lap Winding

In this case, the armature conductors are divided into as *many parallel paths as the number of poles of the generator.* If there are *P*-poles and *Z* armature conductors, then there are *P* parallel paths, each consisting of *Z/P* conductors connected in series between the positive and negative set of brushes.

Fig. 10.12 shows the case of a 4-pole machine. As armature current enters the negative brush, it has four parallel paths available for going to the positive brush. Each path has *Z*/4 conductors and carries a current of I_a /4.

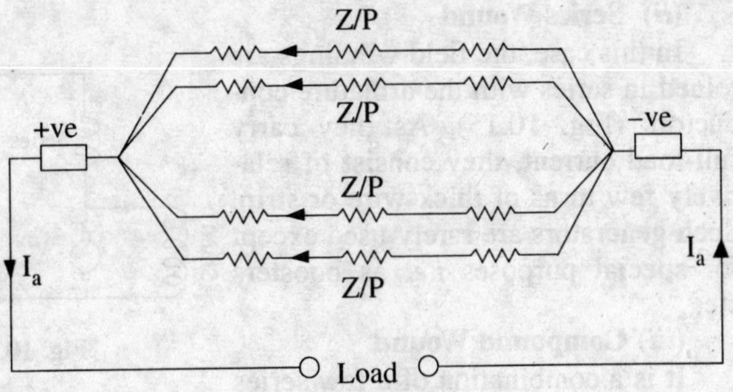

Fig. 10.12

10.12. Armature Resistance

Let, l = length of each armature conductor

S = its cross-section

A = No. of parallel paths

= 2 —for wave-winding

= P —for lap-winding

R = resistance of the *whole* winding

$$\text{Then,} \quad R = \frac{\rho\, l}{S} \times Z$$

$$\text{Resistance of each parallel path} = \frac{\rho\, lZ}{SA}$$

There are such *A* paths in parallel, hence equivalent resistance called armature resistance is

$$R_a = \frac{1}{A}\,\frac{\rho\, LZ}{SA} \quad \text{or} \quad R_a = \frac{\rho\, LZ}{SA^2}$$

10.13. Types of Generators

Generators are usually classified according to the way in which their fields are excited. Generators may be divided into (*a*) separately-excited generators and (*b*) self-excited generators.

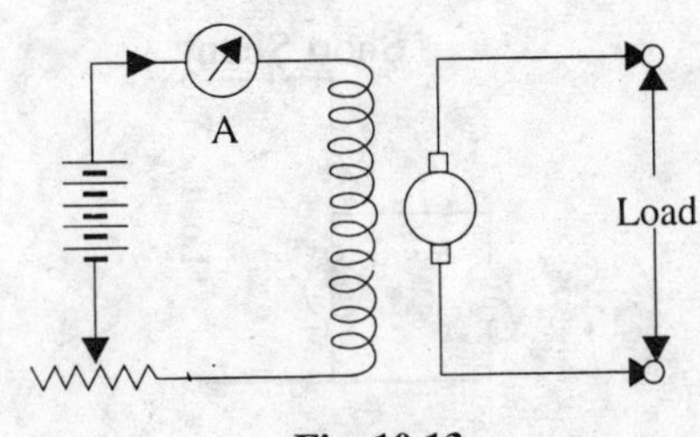

Fig. 10.13

(*a*) **Separately-excited** generators are those whose field magnets are energised from an independent external source of direct current. It is shown diagrammatically in Fig. 10.13.

(*b*) **Self-excited** generators are those whose field magnets are energised by the current produced by the generators themselves. Due to residual magnetism, there is always present some flux in the poles. When the armature is rotated, some e.m.f. and hence some induced current is produced which is partly or fully passed through the field coils thereby strengthening the residual pole flux further.

There are three types of self-excited generators named according to the manner in which their field coils (or windings) are connected to the armature.

(*i*) Shunt Wound

The field windings are connected across or in parallel with the armature conductors and have the full voltage of the generator applied across them (Fig. 10.14).

The field coil consists of many turns of fine gauge copper wire. Such generators are in much common use.

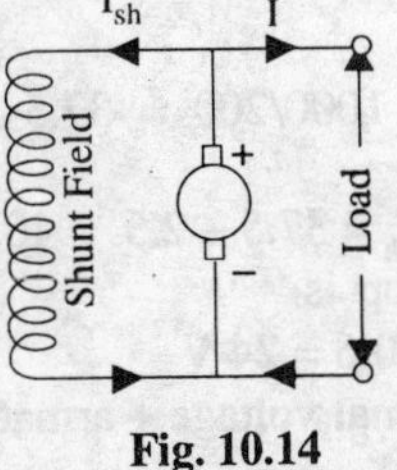

Fig. 10.14

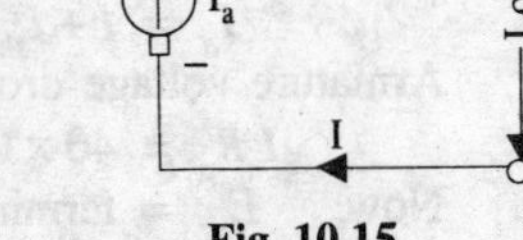

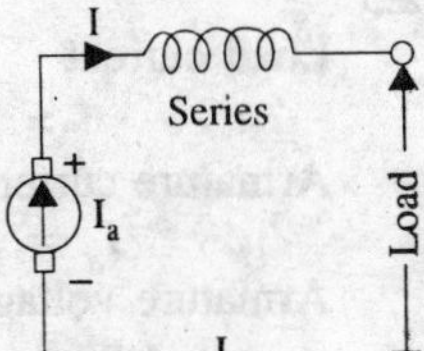

Fig. 10.15

***(ii)* Series Wound**

In this case, the field windings are joined in series with the armature conductors (Fig. 10.15). As they carry full-load current, they consist of relatively few turns of thick wire or strip. Such generators are rarely used except for special purposes *i.e.* as boosters etc.

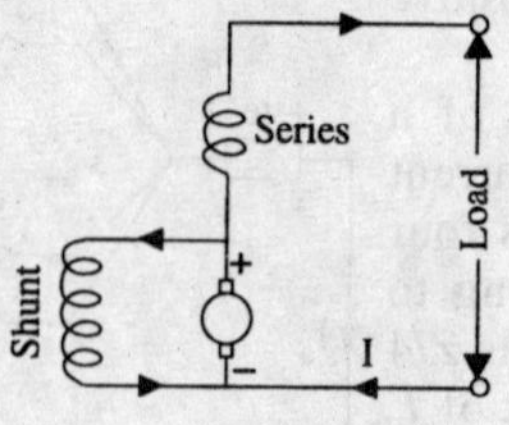

Fig. 10.16

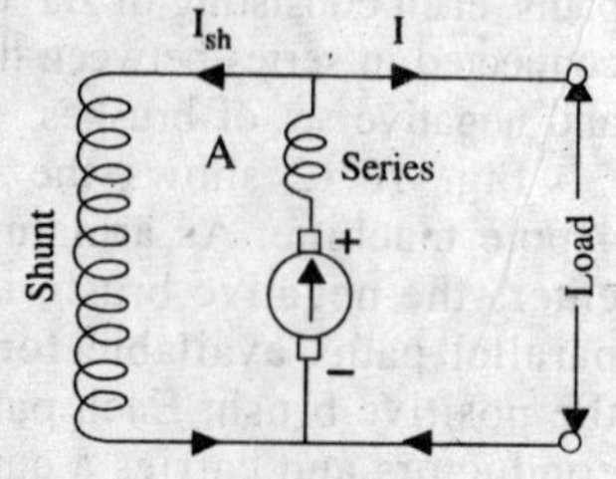

Fig. 10.17

***(iii)* Compound Wound**

It is a combination of a few series and a few shunt windings and can be either short-shunt or long-shunt as shown in Fig. 10.16 and Fig. 10.17 respectively. Various types of d.c. generators have been shown separately in Fig. 10.18.

VARIOUS TYPES OF D.C. GENERATORS

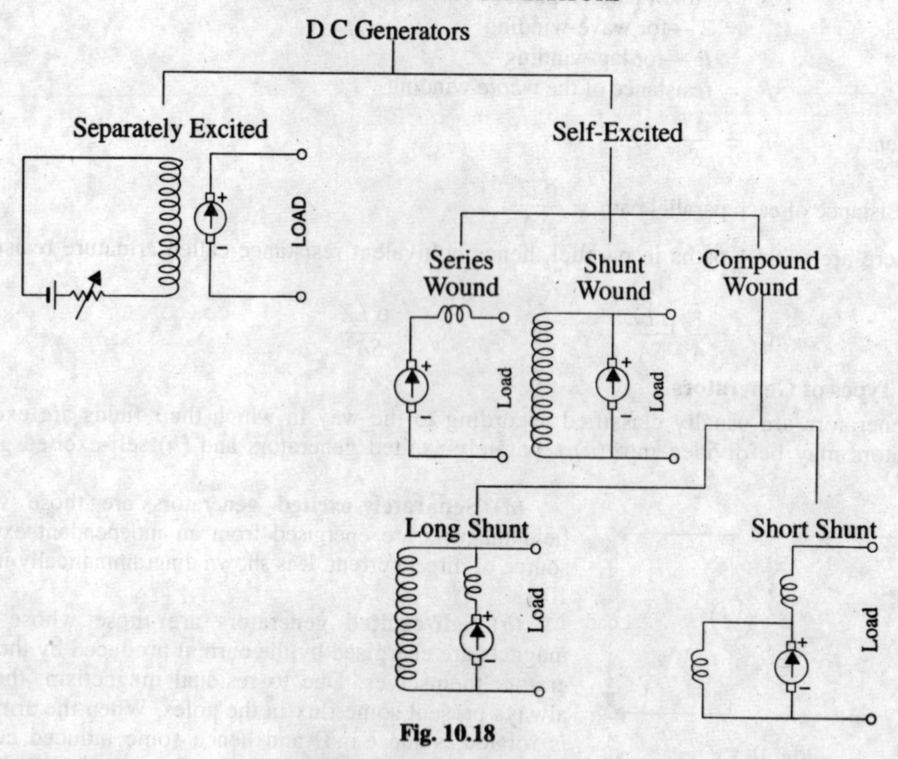

Fig. 10.18

Example 10.1. *A d.c. shunt generator supplies a load of 7.5 kW at 200 V. Calculate the induced e.m.f. if armature resistance is 0.6 ohm and field resistance is 80 ohms.* **(A.M.I.E. Summer 1991)**

Solution. Generator circuit is shown in Fig. 10.19.

Current through shunt field winding is I_{sh} = 200/80 = 2.5

Load current

$$= 7.5 \times 1000/200 = 37.5 \text{ A}$$

Armature current is

$$I_a = I + I_{sh} = 37.5 + 2.5 = 40 \text{ A}$$

Armature voltage drop is

$$I_a R_a = 40 \times 0.6 = \mathbf{24\ V}$$

Now, E_g = terminal voltage + armature drop

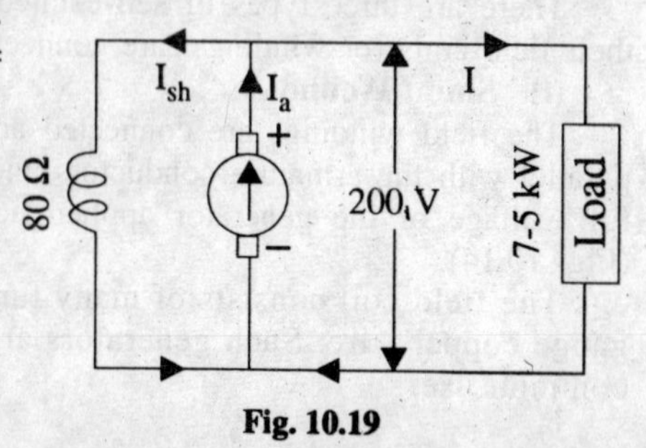

Fig. 10.19

∴ e.m.f. generated in armature

E_g = 200 + 24 = **224 V.**

Example 10.2. *A 4-pole, long-shunt compound generator suplies 100 A at a terminal voltage of 500 V. If armature resistance is 0.02 Ω, series field resistance is 0.04 Ω and shunt field resistance 100 Ω, find the generated e.m.f. Take drop per brush as 1 V. Neglect, armature reaction.*

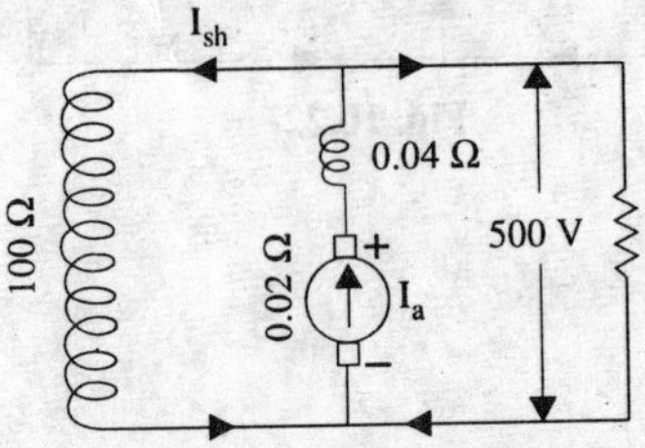

Fig. 10.20

Solution. Generator circuit is shown in Fig. 10.20.

I_{sh} = 500/100 = 5 A

Current through armature and series field windings

= 100 + 5 = 105 A

Voltage drop on series field winding

= 105 × 0.04 = 4.2 V

Armature voltage drop

= 105 × 0.02 = 2.1 volt

Drop at brushes = 2 × 1 = 2 V

Now $E_g = V + I_a R_a$ + brush drop

∴ generated e.m.f. E_g = 500 + 4.2 + 2.1 + 2 = **508.3 V**

Example 10.3. *A short-shunt cumulative compound d.c. generator supplies 7.5 kW at 230 V. The shunt field, series field and armature resistances are 100, 0.3 and 0.4 ohms respectively. Calculate the induced e.m.f. and the load resistance.* **(A.M.I.E. Winter 1994)**

Solution. Generator circuit is shown in Fig. 10.21. Load current

= 7.5 × 1000/230 = 32.6 A

Voltage drop in series winding

= 0.3 × 32.6 = 9.78 V

Voltage across shunt winding

= 230 + 9.78 = 239.78 V

I_{sh} = 239.78/100 = 2.4 A

I_a = 32.6 + 2.4 = 35 A

$I_a R_a$ = 35 × 0.4 = 14 V

Induced e.m.f. = 230 + 9.78 + 14 = **253.78 V**

If R is load resistance, then, $W = V^2/R$

∴ $R = V^2/W = 230^2/7.5 \times 1000$ = **7.05 Ω**

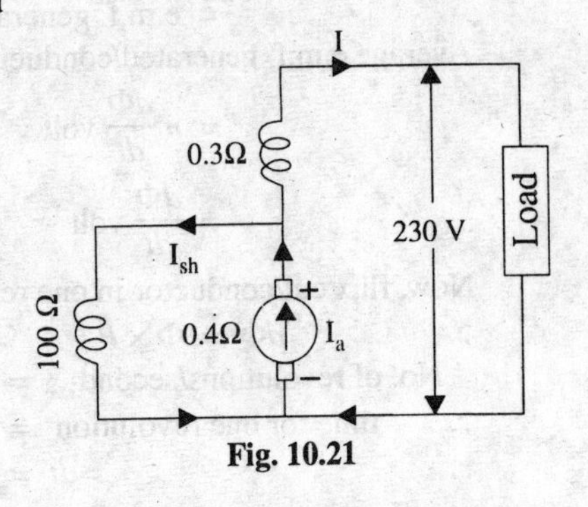

Fig. 10.21

Example 10.4. *The resistance of the field circuit of a shunt-excited d.c. generator is 200 Ω. When the output of the generator is 100 kW, the terminal voltage is 500 V and the generated e.m. f. 525 V. Calculate (a) the armature resistance and (b) the value of the generated e.m.f. when the output is 60 kW, if the terminal voltage then is 520 V.*

Solution. (*a*) Output current is

I = 100,000/500 = 200 A

I_{sh} = 500/200 = 2.5 A ; I_a = 202.5 A

Armature drop = 525 – 500 = 25 V

∴ $I_a R_{sh}$ = 25 or 202.5 × R_a = 25 ; R_a = **0.123** Ω

(*b*) Output current = 60,000/520 = 115.4 A

I_{sh} = 520/200 = 2.6 A ; I = 115.4 + 2.6 = 118 A

∴ $I_a R_a$ = 118 × 0.123 = 14.56 V

∴ generated e.m.f. E_g = 520 + 14.56 = **534.56 V**

Example 10.5. *A 30-kW, 300-V, d.c. shunt generator has armature and field resistances of 0.05 ohm and 100 ohm respectively. Calculate the total power developed by the armature when it delivers full output power.* **(A.M.I.E. Summer 1988)**

Solution. The generator circuit is shown in Fig. 10.22.

Load current

$I = 30{,}000/300 = 100$ A

$I_{sh} = 300/100 = 3$ A

$I_a = 100 + 3 = 103$ A

$I_a R_a = 103 \times 0.05 = 5.15$ V

$E_g = V + I_a R_a = 300 + 5.15$

$= 305.15$ V

Power developed in armature

$= E_g I_a = 305.15 \times 103/1000 =$ **31.43 kW**

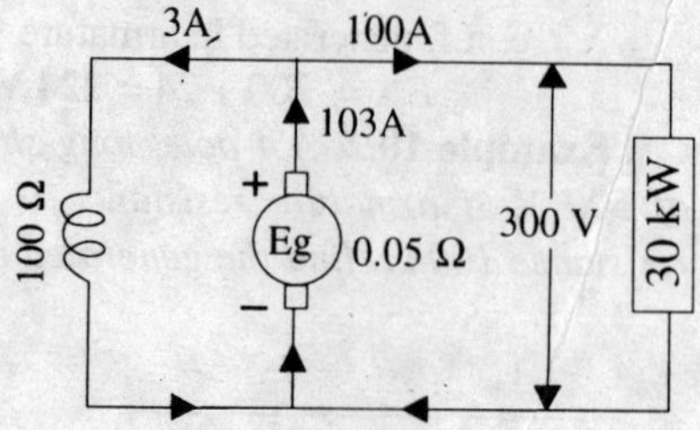

Fig. 10.22

10.14. Generated E.M.F. or E.M.F. Equation of a D.C. Generator

Let Φ = flux/pole in webers

Z = total number of armature conductors

= No. of slots × conductors/slot

P = No. of poles

A = No. of parallel paths in armature

N = armature rotation in r.p.m.

E_g = e.m.f. induced in any parallel path in the armature.

Generated e.m.f. E_g

= e.m.f. generated in any one of the armature parallel paths.

Average e.m.f. generated/conductor

$$= n\frac{d\Phi}{dt} \text{ volt} \qquad \text{(Ref. Art. 8.3)}$$

$$= \frac{d\Phi}{dt} \text{ volt} \qquad (\because n = 1)$$

Now, flux cut/conductor in one revolution = ΦP weber

$\therefore \quad d\Phi = \Phi \times P$

No. of revolutions/second $= N/60$

$\therefore$ time for one revolution $= 60/N$ second

$\therefore \quad dt = 60/N$ second

$\therefore$ flux cut per conductor/second

$$= \frac{d\Phi}{dt} = \frac{\Phi P}{60/N} = \frac{\Phi PN}{60} \text{ Wb/s}$$

$\therefore$ e.m.f. generated /conductor $= \dfrac{\Phi PN}{60}$ volt

For a wave-wound generator

No. of parallel paths = 2

No. of conductors (in series) in one path = $Z/2$

$$\therefore \text{ E.M.F. generated/path} = \frac{\Phi PN}{60} \times \frac{Z}{2} = \frac{\Phi ZPN}{120} \text{ volt}$$

For a lap-wound generator

No. of parallel paths = P

No. of conductors (in series) in one path = Z/P

$$\therefore \text{ e.m.f. generated/path} = \frac{\Phi PN}{60} \times \frac{Z}{P} = \frac{\Phi ZN}{60} \text{ volt}$$

In general, generated e.m.f. is $E_g = \dfrac{\Phi ZN}{60} \times \left(\dfrac{P}{A}\right)$ volt where $A = 2$ — for wave-winding

$= P$ — for lap-winding

Example 10.6. *Calculate the e.m.f. generated by a 4-pole, wave-wound armature having 45 slots with 18 conductors per slot when driven at 1200 r.p.m. The flux per pole is 0.016 Wb.*

Solution. $E = \dfrac{\Phi ZN}{60} \times \left(\dfrac{P}{A}\right)$ volt

Here $\Phi = 0.016$ Wb; $N = 200$ r.p.m.

$Z = 45 \times 18 = 810$; $P = 4$

$A = 2$ (wave-winding)

$$E_g = \frac{0.016 \times 810 \times 1200}{60}\left(\frac{4}{2}\right) = \mathbf{518.4\ volt}$$

Example 10.7. *An 8-pole lap-connected armature of a d.c. machine has 960 conductors, a flux of 40 mWb per pole and a speed of 400 r.p.m. Calculate the e.m.f. generated an open circuit. If the above armature were wave-connected, at what speed must it be driven to generate 400 volts?* **(A.M.I.E. Winter 1989)**

Solution. (*i*) $E_g = \dfrac{(40 \times 10^{-3}) \times 960 \times 400}{60} \times \left(\dfrac{8}{8}\right) = \mathbf{256\ V}$

(*ii*) $400 = \dfrac{(40 \times 10^{-3}) \times 960\, N}{60} \times \left(\dfrac{8}{2}\right)$; **N = 156.2 pm**

Example 10.8. *A d.c. generator develops an e.m.f. of 200 V when driven at 1000 r.p.m. with a flux per pole of 0.02 Wb. It is desired that this e.m.f. be increased to 210 V at 1100 r.p.m. What should be the value of the flux per pole under the new circumstances ?*

Solution. $E_g = \dfrac{\Phi ZN}{60}\left(\dfrac{P}{A}\right)$ volt

$E \propto \Phi N$

In the first case, $200 \propto 0.02 \times 1000$ ..(*i*)

In the second case,

$210 \propto \Phi \times 1100$ ...(*ii*)

$\therefore \dfrac{210}{200} = \dfrac{1100\ \Phi}{1000 \times 0.02}$; $\Phi = \mathbf{0.019\ Wb}$

Example 10.9. *The armature of a 4-pole d.c. generator is required to generate an e.m.f. of 520 V on open circuit when revolving at a speed of 660 r.p.m. Calculate the magnetic flux per pole required if the armature has 144 slots with 2 coil sides per slot, each coil consisting of three turns. The armature is wave-wound.*

Solution. Formula used :

$$E_g = \frac{\Phi ZN}{60} \times \left(\frac{P}{A}\right) \text{volt}$$

Here $E_g = 520$ V; $N = 660$ r.p.m.

$Z = 144 \times 2 \times 3 = 864$; $P = 4$; $A = 2$

$$\therefore\ 520 = \frac{\Phi \times 864 \times 660}{60} \times \left(\frac{4}{2}\right)$$

$$\therefore\ \Phi = \frac{60 \times 520}{2 \times 864 \times 660} = \mathbf{0.0274\ Wb}$$

Example 10.10. *A 4-pole, lap-wound, d.c. shunt generator has a useful flux per pole of 0.07 Wb. The armature winding consists of 220 turns each of 0.004 Ω resistance. Calculate the terminal voltage when running at 900 r.p.m. if the armature current is 50 A.*

Solution. Since each turn has two sides

$Z = 220 \times 2 = 440$; $N = 900$ r.p.m.

$\Phi = 0.07$ Wb ; $P = A = 4$

$$E_g = \frac{\Phi ZN}{60} \times \left(\frac{P}{A}\right) \text{volt}$$

$$= \frac{0.07 \times 440 \times 900}{60} \times \left(\frac{4}{4}\right) = 462 \text{ volt}$$

Total resistance of 220 turns or 440 conductors

$= 220 \times 0.004 = 0.88\ \Omega$

Since there are 4 parallel paths in armature,

∴ resistance of each path = 0.88/4 = 0.22 Ω

Now, there are four such resistances in parallel each of value 0.22 Ω.

∴ armature resistance $R_a = 0.22/4 = 0.055\ \Omega$

Armature drop $= I_a R_a = 50 \times 0.055 = 2.75$ volt

Now, terminal voltage $V = E_g - I_a R_a = 462 - 2.75 =$ **459.25 V**

Example 10.11. *A 4-pole d.c. shunt generator with a wave-wound armature having 390 conductors has to supply a load of 500 lamps each of 100 W at 250 V. Allowing 10 V for the voltage drop in the connecting leads between the generator and the load and a contact drop of 1 volt per brush, calculate the speed at which the generator should be driven. The flux per pole is 30 mWb and the armature and the shunt field resistances are respectively 0.05 Ω and 65 Ω.*

(A.M.I.E. Winter 1989

Solution. Load current $= 500 \times 100/250 = 200$ A

Voltage across shunt $= 250 + 10 = 260$ V

$I_{sh} = 260/65 = 4$ A

$I_a = 200 + 4 = 204$ A

$I_a R_a = 204 \times 0.05 = 10.2$ V

Brush drop $= 2 \times 1 = 2$ V

Drop in connecting leads $= 10$ V

Generated e.m.f. $= 250 + 10 + 2 + 10.2 = 272.2$ V

Now, $$E_g = \frac{\Phi ZN}{60} \times \left(\frac{P}{A}\right) \text{volt}$$

∴ $$272.2 = \frac{30 \times 10^{-3} \times 390 \times N}{60} \times \left(\frac{4}{2}\right)$$

∴ $N =$ **698 r.p.m.**

Example 10.12. *A 6-pole generator armature has 1000 conductors and is wave-wound. If the flux per pole is 20 milliweber and the speed is 500 rpm, calculate the e.m.f. generated. If the above machine is self-excited and the armature and the field resistances are 0.5 Ω and 250 Ω respectively, calculate the output current when the armature current is 40 A. Neglect brush contact drop and armature reaction.* **(Electrical Science, AMIE Winter 1993)**

Solution. Using the given data, the generated e.m.f. can be found as under :

$$E_g = \frac{\Phi ZN}{60} \left(\frac{P}{A}\right)$$

$$= \frac{20 \times 10^{-3} \times 1000 \times 500}{60} \times \left(\frac{6}{2}\right) = \mathbf{500\ V}$$

When the machine is run as a self-excited generator,

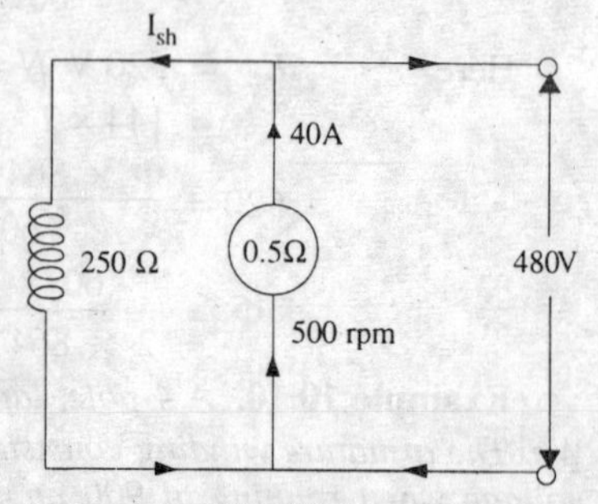

Fig. 10.23

the current and the voltage values are as shown in Fig. 10.23. The output voltage will be given by the generated voltage minus the armature drop.

$\therefore \quad V = E_g - I_a R_a = 500 - 40 \times 0.5 = 480$ V

$I_{sh} = 480/250 = 1.92$ A

$\therefore$ output current, $I = 40 - 1.92 =$ **38.1 A**

Example 10.13. *A commutator machine is rated at 5 kW, 250 V, 2000 r.p.m. The armature resistance. R_a is 1 Ω. Driven from the electrical end at 2000 r.p.m. the no-load power input to the armature is I_a = 1.2 A at 250 V with the field winding (R_f = 250 Ω) excited by I_f = 1 A. Estimate the efficiency of this machine as a 5 kW generator.*

(Electrical Science, AMIE Summer 1994.)

Solution. As shown in Fig. 10.24 (*a*) when the machine is driven as a motor, it takes a supply current of (1 + 1.2) = 2.2 A on no-load.

No-load input = $250 \times 2.2 = 550$ W

No-load arm. Cu. loss = $1.2^2 \times 1.44$ W $\cong$ 1.5 W

$\therefore$ constant losses = $550 - 1.5 = 548.5$ W

The same machine runs as a generator as shown in Fig. 10.24 (*b*).

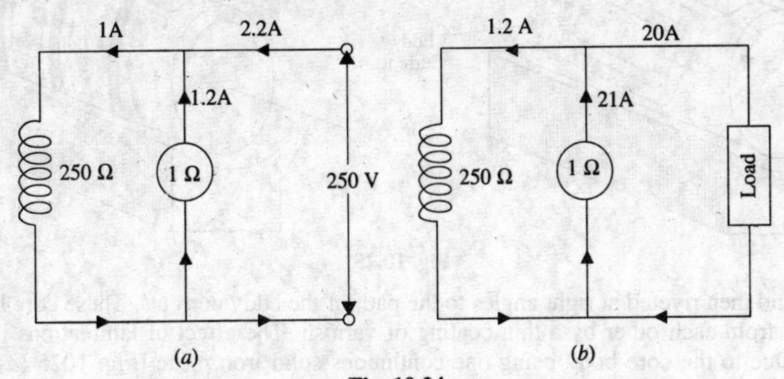

Fig. 10.24

F.L output current = 5000/250 = 20 A

I_{sh} = 1 A. $\quad I = 20 + 1 = 21$ A

F.L. arm. Cu. loss = $21^2 \times 1 = 441$ W

Total losses $= 441 + 548.5 = 989.5$ W

$\eta = 5000/(5000 + 989.5) = 0.835$ or **83.5%**

10.15. Iron Loss in Armature

Due to the rotation of the iron core of the armature in the magnetic flux of the field poles, there are some losses taking place continuously in the core and are known as Iron losses or Core losses. Iron loss consists of (*i*) **Hysteresis loss** and (*i*) **Eddy Current loss.**

(*i*) Hysteresis Loss (W_h).

This loss is due to the reversal of magnetism of the armature core. Every portion of the rotating core passes under *N*-and *S*-pole alternately, thereby attaining *S*-and *N*-polarity respectively. The core undergoes one complete cycle of magnetic reversal after passing under one *pair* of poles. If *P* is the number of poles and *N* the armature speed in r.p.m, then frequency of magnetic reversals is

$$f = PN/120 \text{ reversals/second}$$

The loss depends upon the volume and grade of iron, maximum value of flux density B_{max} and frequency of magnetic reversals. For normal flux densities (*i.e.* up to 1.5 Wb/m²), hysteresis loss is given by **Steinmetz** formula. According to this formula

$$W_h = \eta B_{max}^{1.6} \, F \, V \text{ watt}$$

where V = volume of the core in m^3

η = Steinmetz hysteresis coefficient

Value of η for

Good dynamo sheet steel = 502.4, silicon steel = 191

***(ii)* Eddy Current Loss**

When the armature core rotates, then it also cuts the flux. Hence, an e.m.f is induced in the body of the core according to the laws of electromagnetic induction. This e.m.f., though small, sets up a large current in the body of the core due to its small resistance. This current is known as eddy current. The power loss due to the flow of this current is known as 'eddy current' loss. This loss would be considerable if solid iron core were used. In order to reduce this loss and the consequent heating of the core to a small value, the core is built up of thin laminations which

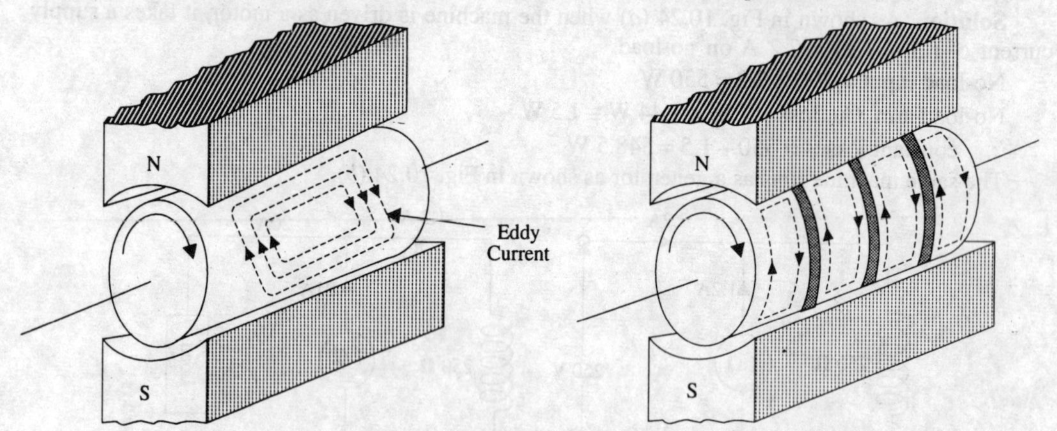

Fig. 10.25

are stacked and then riveted at right angles to the path of the eddy currents. These core laminations are insulated from each other by a thin coating of varnish. The effect of laminations is shown in Fig. 10.25. Due to the core body being one continuous solid iron piece [Fig. 1025 *(a)*] the magnitude of eddy current is large. As armature cross-sectional area is large, its resistance is very small, hence eddy current as well as the loss is large. In Fig. 10.5 *(b)* the same core has been split up into thin circular discs insulated from each other. It is seen that now each current path, being of much less cross-section, has a very high resistance. Hence, magnitude of eddy currents is reduced very much, thereby drastically reducing eddy current loss.

It is found that eddy current loss W_e is given by the following relation :

$$We = kB_{max}^2 f^2 t^2 \text{ V watt}$$

where B_{max} = maximum flux density in the core

f = frequency of magnetic reversals

t = thickness of each lamination

V = volume of armature core

It is seen from above that this loss varies directly as the square of the thickness of laminations, hence it should be kept as small as possible. Another point to note is that $W_h \propto f$ but $W_e \propto f^2$. This fact makes it possible to separate the two losses experimentally, if so desired.

As said earlier, these iron losses, if allowed to take place unchecked, not only reduce the efficiency of the generator but also raise the temperature of the core. As the output of the machine is limited in most cases, by the temperature rise, these losses have to be kept as small as is economically possible.

Eddy current loss is reduced by using a laminated core but hysteresis loss cannot be reduced this way. For reducing the hysteresis loss, those metals are chosen for the armature core which have a low hysteresis coefficient. Generally, special silicon steels, such as stalloys are used whih not only have a low hysteresis coefficient, but which possess high electrical resistivity.

10.16. Total Loss in a D.C. Generator

The various losses occuring in a generator can be subdivided as follows:

(*a*) **Copper Losses** (or I^2R loss)

(*i*) Armature copper loss = $I_a^2R_a$ (not $E_g\,I_a$)

where R_a = resistance of armature and interpoles and series field winding etc.

This loss is about 30 to 40% of full load losses.

(*ii*) Field copper loss : In the case of shunt generators, it is practically constant and = $I_{sh}^2R_{sh}$ (or VI_{sh}). In the case of series generators, it is = $I_{sc}^2\,R_{sc}$ where R_{sc} is resistance of the series field winding.

This loss is about 20 to 30% of F.L. losses.

(*iii*) The loss due to brush contact resistance. It is usually included in the armature copper loss.

(*b*) **Magnetic Losses** (also known as iron or core losses).

(*i*) Hysteresis loss, $W_h \propto B_{max}^{1.6}\,f$

(*ii*) Eddy current loss, $W_e \propto B_{max}^{2}\,f^2$

These losses are practically constant for shunt and compound-wound generators, because field current, in their case, is approximately constant.

Both these losses total up to about 20 to 30% of F.L. losses.

(*c*) **Mechanical Losses**

These consist of

(*i*) friction loss at bearings and commutator.

(*ii*) air-friction or windage loss of rotating armature.

These are about 10 to 20% of F.L. losses.

The total losses in a d.c. generator are summed up below.

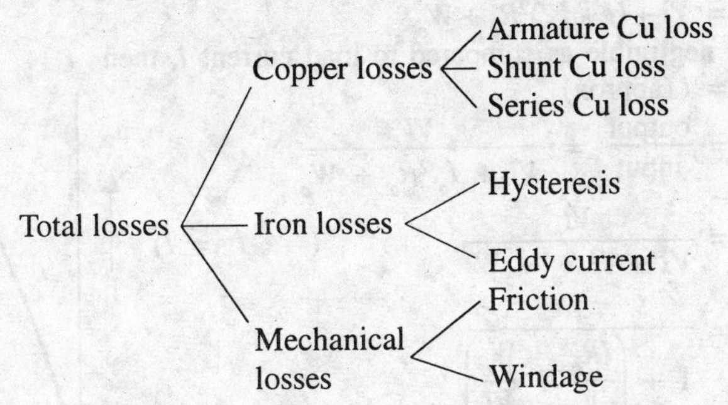

10.17. Stray Losses

Usually magnetic and mechanical losses are collectively known as **stray losses.**

10.18. Constant and Standing Losses

As said above, field Cu loss is constant for shunt and compound generators. Hence, stray losses and shunt Cu losses are constant in their case. These losses are known as *standing or constant* losses W_c.

Hence, for shunt and compound generators,

$$\text{Total losses} = \text{armature copper loss} + W_c$$
$$= I_a^2R_a + W_c = (I + I_{sh})^2\,R_a + W_c$$

Armature Cu loss, $I_a^2R_a$ is known as *variable* loss because it varies with the load current.

$\therefore$ total losses = variable losses + constant losses W_c

10.19. Power Stages

Various power stages in the case of a d.c. generator are shown below:

Following are the three generator efficiencies :

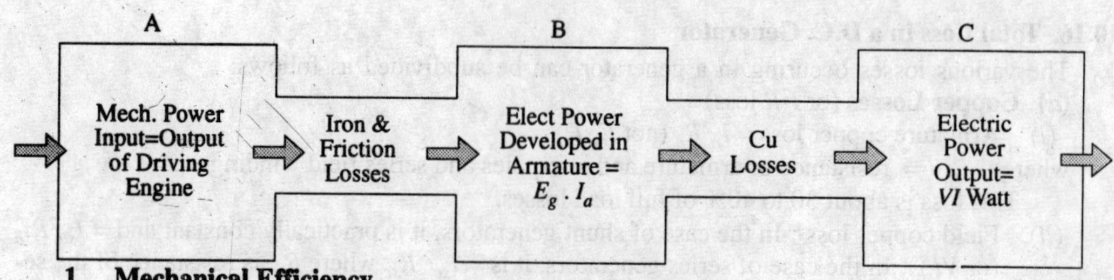

1. Mechanical Efficiency

$$\eta_m = \frac{B}{A} = \frac{\text{total watts generated in armature}}{\text{mechanical power supplied}}$$

2. Electrical Efficiency

$$\eta_e = \frac{C}{B} = \frac{\text{watts available in load circuit}}{\text{total watts generated}} = \frac{VI}{E_g I_a}$$

3. Overall or Commercial Efficiency

$$\eta_c = \frac{C}{A} = \frac{\text{watts available in load circuit}}{\text{mechanical power supplied}}$$

It is obvious that overall efficiency $\eta_c = \eta_m \times \eta_e$. For good generators, its value may be as high as 95%.

Note. Unless specified otherwise, commercial efficiency is always to be understood.

10.20. Condition for Maximum Efficiency

$$\text{Generator output} = VI \text{ watts}$$

$$\text{Generator input} = \text{output} + \text{losses}$$

$$= VI + I_a^2 R_a + W_c$$

$$= VI + (I + I_{sh})^2 R_a + W_c \qquad (\because \; I_a = I + I_{sh})$$

However, if I_{sh} is negligible as compared to load current I, then

$$I_a = I \text{ (approx)}$$

$$\eta = \frac{\text{output}}{\text{input}} = \frac{VI}{VI + I_a^2 R_a + W_c}$$

$$= \frac{VI}{VI + I^2 R_a + W_c} \qquad (\because \; I_a = I)$$

$$= \frac{1}{1 + \left(\frac{IR_a}{V} + \frac{W_c}{VI}\right)}$$

Now, efficiency is maximum when denominator is minimum

i.e. when $\frac{d}{dI}\left(\frac{IR_a}{V} + \frac{W_c}{VI}\right) = 0$

or $\frac{R_a}{V} - \frac{W_c}{VI^2} = 0$ or $I^2 R_a = W_c$

Hence, efficiency is maximum when

variable loss = constant loss

The load current corresponding to maximum efficiency is given by relation

$$I^2 R_a = W_c$$

or

$$I = \sqrt{\frac{W_c}{R_a}}$$

η

I

Fig. 10.26

Variation of η with load current is shown in Fig. 10.26.

Example 10.14. *A shunt generator delivers 195 A at a terminal p.d. of 250 V. The armature* *resistance and shunt field resistance are 0.02 Ω and 50 Ω respectively. The iron and friction losses equal 950 W. Find*

(a) e.m.f. generated, (b) Cu losses,
(c) output of the prime mover in kW, (d) commercial, mechanical and electrical efficiencies.

Solution. (*a*) I_{sh} = 250/50 = 5 A, I_a = 195 + 5 = 200 A

Armature voltage drop = I_aR_a = 200 × 0.02 = 4 V

∴ generated e.m.f. = 250 + 4 = **254 V**

(*b*) Armature Cu loss = $I_a^2R_a = 200^2 \times 0.02$ = 800 W

Shunt Cu loss = $V.I_{sh}$ = 250 × 5 = 1250 W

∴ total Cu loss = 1250 + 800 = **2050 W**

(*c*) Stray losses = 950 W

Total losses = 2050 + 950 = 3,000 W

output = 250 × 195 = 48,750 W

Input = 48,750 + 3,000 = 51,750 W

∴ output of the prime mover = 51,750 W = 51.75 kW

(*d*) Input = 51,750 W; Stray loss = 950 W

Electrical power produced in armature

= 51,750 – 950 = 50,800 W

$$\eta_m = \frac{50,800}{51,750} \times 100 = \mathbf{98.2\%}$$

Electrical or Cu losses

= 2050 W

$$\therefore \quad \eta_e = \frac{48,750}{48,750 + 2,050} \times 100 = \mathbf{95.9\%}$$

$$\eta_c = \frac{48,750}{51,750} \times 100 = \mathbf{94.2\%}$$

Example 10.15. *A 400-V shunt generator has a full-load current of 200 A, its armature resistance is 0.06 Ω and field resistance 100 Ω; the stray losses are 2,000 W. Find kW output of prime mover when it is delivering full load and find the load for which the efficiency of the generator is maximum.* **(Elect. Science, AMIE 1994)**

Solution. Full-load output current = 200 A

∴ F.L. output = 400 × 200 = 80,000 W

I_{sh} = 400/100 = 4 A

∴ I_a = 200 + 4 = 204 A

Armature Cu loss = $204^2 \times 0.06$ = 2,497 W

Shunt Cu loss = 400 × 4 = 1,600 W

Stray losses = 2,000 W

∴ total losses in the generator

= 2,000 + 2,497 + 1,600 = 6,097 W

∴ total input = 80,000 + 6, 097 = 86,097 W

∴ kW output of the prime mover

= 86,097/1000 = **86.1 kW**

The load current for which efficiency is maximum is given by

$$I = \sqrt{W_c/R_a}$$

Now, constant loss = stray losses + shunt Cu loss

= 2,000 + 1,600 = 3,600 W

$$I = \sqrt{\frac{3600}{0.06}} = \mathbf{245\ A}$$

Example 10.16. *A shunt generator has a F.L. current of 196 A at 220 V. The stray losses are 720 W and the shunt field coil resistance is 55 Ω. It has a F.L. efficiency of 88%, find the armature resistance. Also, find the load current corresponding to maximum efficiency.*

Solution. Output = 220 × 196 = 43,120 W

η = 88% (overall efficiency)

∴ electrical input = 43,120/0.88 = 49,000 W

∴ total losses = 49,000 – 43,120 = 5,880 W

Shunt field current = 220/55 = 4 A

∴ I_a = 196 + 4 = 200 A

∴ shunt Cu loss = 220 × 4 = 880 W

Stray losses = 720 W

Constant losses = 880 + 720 = 1,600 W

∴ armature Cu loss = 5,880 - 1,600 = 4,280 W

∴ $I_a^2 R_a$ = 4,280 W

$200^2 R_a = 4,280$ or $R_a = \dfrac{4,280}{200 \times 200} = \mathbf{0.107\ \Omega}$

For maximum efficiency,

$I^2 R_a$ = constant losses = 1,600 W

$$I = \sqrt{\frac{1600}{0.107}} = \mathbf{122.3\ A}$$

Example 10.17. *A 100-kW, 460-V shunt generator was run as a motor on no-load at its rated voltage and speed. The total current taken was 9.8 A including a shunt current of 2.7 A. The resistance of the armature circuit at normal working temperature was 0.11 ohm. Calculate the efficiencies at (a) full-load and (b) half-load.* **(Electrical Engg. A.M.Ae.S.I.Dec. 1994)**

Solution. No-load input of a shunt motor represents its losses only which consist of (*i*) variable armature Cu. loss (*ii*) constant losses.

No-load input = 460 × 9.8 = 4508 W

No-load arm. current = 9.8 – 2.7 = 7.1 A

No-load arm. Cu. loss = $7.1^2 \times 0.11$ = 5.5 W

∴ constant losses = 4508 – 5.5 = 4502.5 W

(*a*) Full Load

F.L. input current = 100,000/460 = 217.4 A

Arm. current = 217.4 - 2.7 = 214.7 A

F.L arm. Cu. loss = $(214.7)^2 \times 0.11$ = 5070.5 W

Total losses = 5070.5 + 4502.5 = 9573 W = 9.573 kW

F.L. output = 100 – 9.573 = 90.427 kW

F.L.η = 90.427/100 = 0.904 or **90.4%**

(*b*) Half Load

Input current = 217.4/2 = 108.7 A

Arm. current = 108.7 - 2.7 = 106 A

Arm. Cu. loss = $106^2 \times 0.11$ = 1236 W

Total losses = 1236 + 4502.5 = 5738.5 W = 5.738 kW

Output = 50 – 5.738 = 44.262 kW

Half-load η = 44.262/50 = 0.885 or **88.5%**

Example 10.18. *A shunt generator gives a full-load output of 6 kW at a terminal potential of 200 V. The armature and field resistances are 0.5 Ω and 50 Ω respectively. If the mechanical and iron losses combined amount to 500 W, calculate the kW input required at the driving shaft on full load and the full-load efficiency.*

Solution. I_{sh} = 200/50 = 4A

Output current $= 6{,}000/200 = 30$ A

$I_a = 30 + 4 = 34$ A

$\therefore \quad I_a^2R_a = 34^2 \times 0.5 = 578$ W

Shunt field Cu loss $= 200 \times 4 = 800$ W

Total losses $= 578 + 800 + 500 = 1{,}878$ W

$\therefore$ input $= 6{,}000 + 1{,}878 = 7{,}878$ W $=$ **7.878 kW**

$$\text{Full load } \eta = \frac{6{,}000 \times 100}{7{,}878} = \mathbf{76.2\%}$$

Tutorial Problems No. 10.1

1. Calculate the flux per pole required on full-load for a 50-kW, 400-V, 8-pole, 600 r.p.m. d.c. shunt generator with 256 conductors arranged in a lap–connected winding. The armature winding resistance is 0.1 Ω, the shunt field resistance is 200 Ω and there is a brush contact voltage drop of 1 volt at each brush on full-load. **[162 mWb]**

2. A 4-pole, short-shunt compound generator has armature, shunt field and series field resistance of 0.4 Ω, 160 Ω and 0.2 Ω respectively.
The armature is lap-wound with 440 conductors and is driven at 600 r.p.m. Calculate the flux per pole when the machine is delivering 120 A at 400 V. **[108 mWb]**

3. A 6-pole, wave-wound shunt generator delivers 75 kW at 120 V. The armature has 270 conductors and a total resistance of 0.02 Ω. The flux per pole is 0.01 Wb. Calculate the speed of the machine in r.p.m. if the field resistance is 60 Ω and the total voltage drop across the brushes is 2 V. **[1000 r.p.m.]**

4. A d.c., six-pole shunt generator has a wave-wound armature with 412 conductors. The flux per pole is 20 mWb and it is driven at a speed of 1250 r.p.m. The armature resistance is 0.15 Ω, the field resistance is 125 Ω and the total brush contact drop is 2.4 V. If the terminal p.d. is maintained at 500 V, calculate the current supplied to the load. **[80 A]**

5. A 4-pole machine has a lap-wound armature with 90 slots each containing six conductors. If the machine runs at 1500 r.p.m. and the flux per pole is 0.03 Wb, calculate the e.m.f. generated.
If the machine is run as a shunt generator with the same field flux, the armature and field resistances being 1.0 Ω and 200 Ω respectively, calculate the output current when the armature current is 25 A. If due to a fall in speed, the e.m.f. becomes 380 V, calculate the load current in a 40 Ω load. **[405 V ; 23.1 A ; 9.2 A]**

6. In a d.c. motor-generator set, the generator supplies a load of 198 A at 60 V. The resistance of the generator armature is 0.05 Ω and the shunt field resistance is 30 Ω. The iron, friction and windage losses in the generator amount to 1.2 kW. The 500-V d.c. motor which drives the generator at 600 r.p.m. has an efficiency of 80%. Calculate (*a*) the output of the motor in kW (*b*) current taken by the motor. **[(*a*) 15.12 kW (*b*) 38 A]**

7. A 20-kW, 440-V, short-shunt compound d.c. generator has a full-load efficiency of 87%. If the resistance of the armature is 0.4 Ω and that of the series and shunt fields is 0.25 Ω and 240 W respectively, calculate the combined bearing, windage and core loss of the machine. **[725 W]**

8. A short-shunt compound d.c. generator supplies a current of 100 A at a voltage of 220 V. If the resistance of the shunt field is 50 Ω, of the series field 0.025 Ω, of the armature 0.05, the total brush drop is 2 V and the iron and friction loss amount 1 kW, find

(*a*) the generated e.m.f.,

(*b*) the copper losses ,

(*c*) the output of the prime mover driving the generator,

(*d*) generator efficiency. **[(*a*) 229.7 V (*b*) 1.995 kW (*c*) 24.99 kW (*d*) 0.88]**

10.21. Armature Reaction

By armature reaction is meant the effect of magnetic field set up by armature current on the distribution of flux under main poles of a generator (or motor). This armature field has two effects:

1. it demagnetises (or weakens) the main pole flux and

Accordingly, there are two components of armature reaction, one is called the 'demagnetising component' and the other 'distorting component'. Both these components increase with increase in the armature current, that is, with increase in the load on the generator. Under severe overloads or short circuit, the demagnetising component of armature reaction may become so strong as to reverse the polarity of the main poles. In general, when the main flux Φ is decreased due to armature reaction, the e.m.f. induced in the armature of a d.c. generator is also decreased because $E \propto \Phi$.

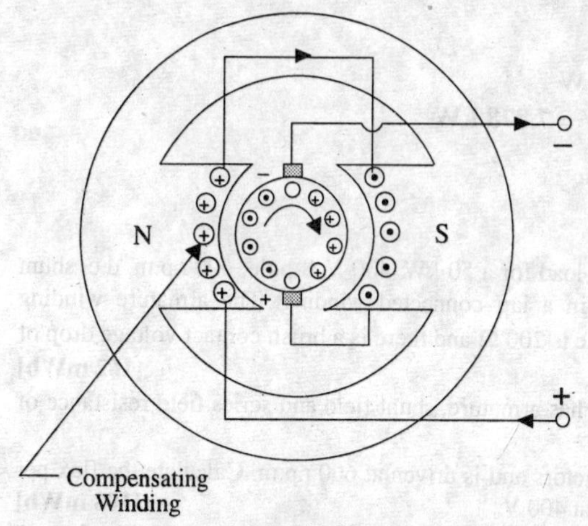

Fig. 10.27

The demagnetising effect of armature reaction is neutralized by adding a few extra ampere-turns to the main field winding. The distorting effect is neutralized by using 'compensating windings'. These windings are embedded in slots in the pole-shoes and are connected in series with armature in such a way that current in them flows in a direction opposite to that of the armature current flowing in the armature conductors directly below the pole shoes.(Fig. 10.27).

10.22. Commutation

As briefly explained in Art. 10.2, the induced currents in the armature conductors of a d.c. generator are alternating currents *i.e.* these currents flow in one direction when conductors are under *N*-pole and in exactly opposite direction when they are under *S*-pole. As conductors pass out of the influence of a *N*-pole and enter that of a *S*-pole, the current in them is reversed. This reversal of current takes place along magnetic neutral axis (*M.N.A.*) or brush axis *i.e.* when the brush-spans and hence short-circuits the particular coil undergoing reversal of current through it.

This process by which current in the short-circuited armature coil is reversed while it crosses the *M.N.A.* is called 'commutation'.

The brief period during which coil remains short-circuited is known as commutation period T_c.

If current reversal is completed within the time T_c, then commutation is ideal. If not, then sparking is produced between the brush and the commutator which damages both.

The main factor which does not allow the armature current to completely reverse its direction within the specified period of T_c is the production of the self-induced e.m.f. called *reactance voltage* in the conductors through which the current is reversing. If current changes from $+ I$ to $- I$ in time T_c, then reactance voltage is $L\ 2I/T_c$ where L is the inductance of the armature conductors. Commutation can be improved *i.e.* current reversal can be made sparkless by using interpoles. These are small poles fixed to the yoke and spaced in between the main poles as shown in Fig. 10.28. They are wound with a few thick copper-wire turns and are connected in series with the armature so that they carry full armature current. Their polarity is the same as that of the main pole *ahead* in the direction of rotation.

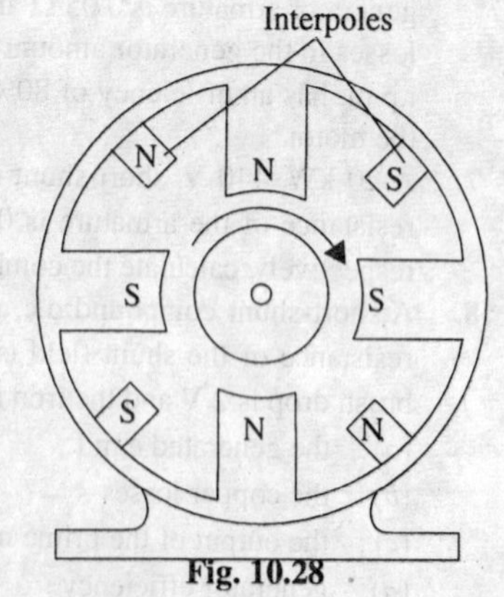

Fig. 10.28

HIGHLIGHTS

HIGHLIGHTS

1. The e.m.f. generated within the armature of a d.c. generator is given by

$$E_g = \frac{\Phi ZN}{60} \times \left(\frac{P}{A}\right) \text{ volts} \qquad \text{... when } N \text{ is in r.p.m.}$$

$$= \Phi ZN\,(P/A) \text{ volts} \qquad \text{... when } N \text{ is in r.p.s.}$$

2. The losses in a generator are (*i*) Cu or electrical losses, (*ii*) iron or magnetic losses and (*iii*) mechanical losses.
3. Magnetic and mechanical losses are collectively known as stray losses.
4. In the case of a shunt generator, the shunt Cu loss is practically constant. Hence, stray losses plus Cu losses are collectively referred to as constant losses (W_c).
5. The condition for maximum efficiency of a d.c. shunt generator is

armature Cu loss = constant losses

6. The load current corresponding to maximum efficiency is given by

$$I = \sqrt{\frac{W_c}{R_a}}$$

7. By 'armature reaction' is meant the effect of the magnetic field set up by armature current on the main pole field. There are two components of armature reaction:
(*i*) demagnetising component which weakens the main pole flux and
(*ii*) distorting component which distorts it.
8. The process by which current in the short-circuited armature coil is reversed while it crosses the magnetic neutral axis is called 'commutation'.

OBJECTIVE TESTS—10

A. Fill in the following blanks.

1. In a d.c. generator, the armature core provides a path of very .... reluctance to the pole flux.
2. The main purpose of laminating the armature core of a d.c. generator is to reduce .... loss.
3. The armature winding of a d.c. generator which gives more parallel paths in the armature is called .... winding.
4. Those d.c. generators whose field magnets are energised by their own generated current are called .... generators.
5. In a d.c. generator, magnetic and mechanical losses are collectively known as .... losses.
6. The efficiency of a d.c. generator is maximum when its variable loss equals .... loss.
7. The compensating windings in a d.c. generator are used for neutralizing the .... effect of armature reaction.
8. The interpoles of a d.c. generator help to improve its ....

B. Answer True or False.

1. The commutator of a d.c. generator functions as a rectifier.
2. Wave-wound d.c. generators provide less current but more voltage.
3. Hysteresis loss in a d.c. generator may be reduced by laminating the armature core.
4. The efficiency of a d.c. generator is maximum when its copper loss equals iron loss.
5. Armature reaction can be increased by increasing the field current of a d.c. generator.
6. Due to armature reaction, the main pole flux of a d.c. generator is weakened as well as distorted.
7. Delayed commutation produces sparking between brush and commutator of a d.c. generator.
8. Both compensating windings and interpoles help in improving the commutation in a d.c. generator.

C. Multiple Choice Questions.

1. The armature conductors of a 6-pole, lap-wound d.c generator are divided into .... parallel paths.
(*a*) 2 (*b*) 3
(*c*) 6 (*d*) 4
2. In a long-shunt compound-wound gene-

arator, the shunt field is connected in parallel with

(*a*) armature (*b*) series field

(*c*) parallel combination of armature and series field

(*d*) series combination of armature and series field .

3. If the flux/pole of a d.c. generator is halved but its speed is doubled, its generated e.m.f. will

(*a*) be halved

(*b*) remain the same

(c) be doubled

(*d*) be quadrupled.

4. Stray losses in a d.c. generator consist of .... losses.

(*a*) magnetic and mechanical

(*b*) magnetic and electrical

(*c*) electrical and mechanical

(*d*) copper and iron.

5. The overall efficiency of a d.c. shunt generator is maximum when its variable loss equals .... loss.

(*a*) stray (*b*) iron

(*c*) constant (*d*) mechanical.

6. In a d.c. generator, the main function of compensating windings is to

(*a*) assist in commutation

(*b*) reduce demagnetising effect of armature reaction

(*c*) reduce distorting effect of armature reaction

(*d*) eliminate reactance voltage.

7. The armature of a dc machine has a resistance of 0.1 ohm and is connected to a 230 V supply. If it is running as a generater giving 80 A, then the generated emf is given by .... volt.

(*a*) 238 (*b*) 224

(*c*) 230 (*d*) 400.

(Elect.Engg. A.M.Ae. S.I. Dec. 1994.)

ANSWERS.

A. 1. low 2. eddy current 3. lap 4. self-excited 5. stray 6. constant 7. distorting 8. commutation

B. 1. T 2. T 3. F 4. F 5. F 6. T 7. T 8. F

C. 1. *c* 2. *d* 3. *b* 4. *a* 5. *c* 6. *c*

11 GENERATOR CHARACTERISTICS

11.1. Characteristics of D.C. Generators

Following are the three *most* important characteristics or curves of a d.c. generator :

1. Open Circuit Characteristic (O.C.C)

It is also known as Magnetic Characteristic or No-load Saturation Characteristic. It shows the relation between the no-load generated e.m.f. in the armature E_0 and the field or exciting current I_f at a given fixed speed. In fact, it is just the magnetisation curve for the material of the electromagnets. Its shape is practically the same for all generators whether separately-excited or self-excited.

2. Internal or Total Characteristic (E/I_a)

It gives the relation between e.m.f. E actually induced in the armature (after allowing for the demagnetising effect of armature reaction) and the armature current I_a. This characteristic is of interest mainly to the designer.

3. External Characteristic (V/I)

It is also referred to as performance characteristic or sometimes voltage regulation curve.

It gives relation between the terminal voltage V and the load current I. This curve lies below the internal characteristic because it takes into account the voltage drop over the armature circuit resistance. The values of V are obtained by subtracting I_aR_a from corresponding values of E. This characteristic is of great importance in judging the suitability of a generator for a particular purpose. It may be obtained in two ways : (*i*) by making simultaneous measurements with a suitable voltmeter and ammeter on a loaded generator or (*ii*) graphically from the *O.C.C.* provided the armature and field resistances are known and also if the demagnetising effect, under rated load conditions, of the armature reaction (from the short-circuit test) is known.

11.2. Open Circuit Characteristic

The *O.C.C.* for d.c. generators is found as described below:

The field winding of the generator (whether shunt or series wound) is disconnected from the machine and, instead, connected to an external separate d.c. source like a battery as shown in Fig. 11.1. The exciting current is changed rheostatically and its value read on the ammeter. The armature is driven at a constant speed, say, N and the generated e.m.f. on no-load E_0 is measured by the voltmeter connected across the armature. The field (or exciting) current I_f is increased by suitable steps (starting from zero) and the corresponding values of E_0 are measured. On plotting the relation between E_0 and I_f, a curve of the form shown in Fig. 11.1 is obtained.

Due to residual magnetism in the poles, some e.m.f. is generated even when exciting current is zero. Hence, the curve starts a little way up instead of starting from the origin. The length OA represents the e.m.f. induced due to residual flux only. It is seen that the initial portion of the curve is practically straight. This is due to the fact that at low flux densities, the reluctance of the magnetic path is almost constant. The magnetic path (in the generator) lies partly through air (clearance between the pole-shoes and the armature) and partly through iron (*i.e.* poles). At low flux densities, the reluctance of iron path is negligible because of its high permeability, hence the total reluctance is very nearly represented by the reluctance of the air-gap alone. Therefore, the flux (and hence E_0) increases linearly with exciting current*. But at higher flux densities, where

*The initial slope of the O.C.C. is determined by the reluctance of the air-gap *i.e.* by the air-gap width.

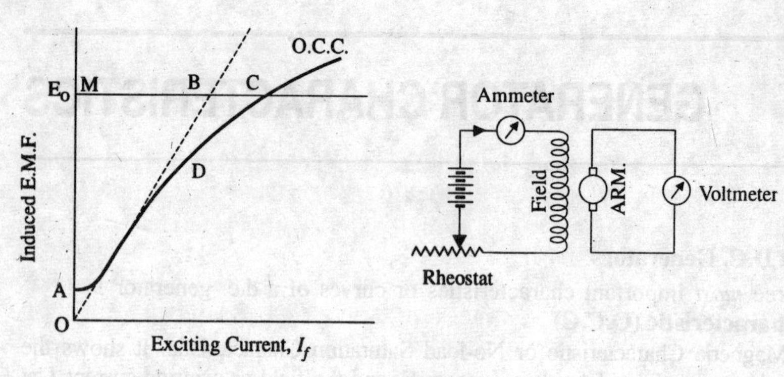

Fig. 11.1

permeability of iron is less, the reluctance of iron path is no longer negligible, rather it goes on increasing as flux density is increased*. The result is that the total reluctance is no longer constant which means that the straight line relationship between E_0 and I_f no longer holds good.

At point *D*, the magnetic saturation of the poles sets in and hence the curve bends over. Over the saturated portion of the curve, a greater increase in exciting current is required for producing a given increase in the generated e.m.f.

It may be pointed out here that in all subsequent figures, the modification of *O.C.C.* due to the effects of residual magnetism would be considered negligibly small and hence the *O.C.C.* would be drawn starting from the origin.

11.3. Critical Resistance for a Series Generator

Now, connect the field windings back to the armature and in series with it and run the machine as a series generator. Suppose the load on the generator consists of a variable resistance. This resistance can be represented by a straight line passing through the origin in Fig. 11.2 because the axes of the graph are volts and amperes. Suppose the load resistance is 20 Ω. Then, corresponding to a voltage of 100 volt, current is 100/20 = 5A. Hence, point *B* (5 A, 100 V) is the point which lies on the 20-Ω resistance line. Join point *O* to *B* and produce further. This line is known as the resistance line. The slope of this line, obviously, represents 20 Ω. Resistance line *OA* represents a resistance of value less than 20 Ω. But lines *OT* and *OD* represent resistances of higher value. Obviously, steeper the line is, greater is the resistance.

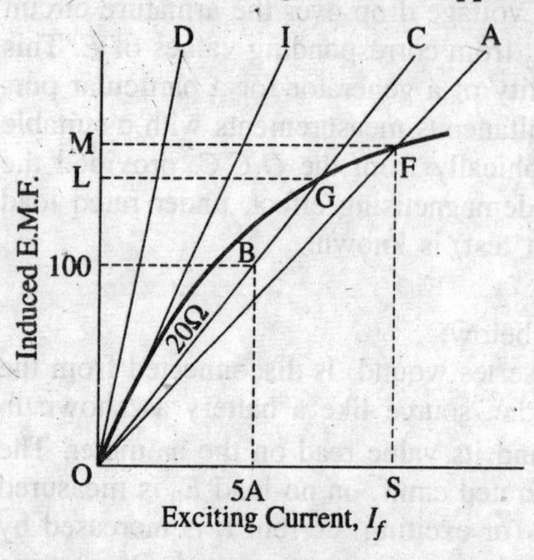

Fig. 11.2

Suppose the load resistance is 20 Ω as represented by resistance line *OBC*. Let the armature of this series generator be set into rotation at *N* r.p.m. Due to residual flux, a small e.m.f. would be produced initially. If the load resistance is not large, this e.m.f. will be able to set up a small current which will increase the field ampere-turns, which in turn, will increase the pole flux. Increased pole flux will lead to more induce e.m.f. which will result in more field ampere-turns and so on. In this way, the cumulative building up action of the generator will start till equilibrium is reached at some point *G* where the resistance line cuts the *O.C.C.* The voltage to which the machine will build up = *OL*.

If the load resistance is decreased (say, equal to that represented by resistance line *OA*) then the machine will build up to a maximum voltage of *OM* which is somewhat greater than the previous value of *OL*.

On the other hand, if the load resistance is increased, then the maximum voltage to which the machine builds up is decreased. If the resistance is increased too much, in fact, so much that

*For example for point *C* in Fig. 11.1.

MB = ampere-turns required by the air-gap.

BC = ampere-turns required by the iron path.

its resistance line does not intersect $O.C.C.$ at all (as line OD), then with this load resistance, the machine will never start building up *i.e.* the generator will fail to excite with this high resistance. It is so because under these conditions the electrical power induced in the armature is not sufficient to increase the pole flux after supplying the I^2R loss. Hence, the cumulative building-up does not commence.

The maximum possible resistance with which the generator would *just* excite, at a particular speed, is given by the slope of the line OT which is tangent to the initial straight portion of the $O.C.C.$ as shown in Fig. 11.2. This resistance is known as the *critical resistance* for that particular speed.

Hence, we conclude that a series generator will build up the voltage only when the total circuit resistance (*i.e.* load resistance and field winding resistance etc.) is less than the critical resistance. Because, it is only then the mutual reinforcement of induced e.m.f., exciting current and flux takes place.

11.4. Critical Resistance for a Shunt Generator

Now, connect the field windings back to armature and run the machine as a shunt generator. Due to residual magnetism, some initial e.m.f. and hence current, would be generated. This current while passing through the field coils, will strengthen the magnetism of the poles (provided field coils are properly connected). This will increase the pole flux which will further increase the generated e.m.f. Increased e.m.f. means more current which further increases the flux and so on. This mutual reinforcement of e.m.f. and flux proceeds on till equilibrium is reached at some point like P (Fig. 11.3). This point lies on the resistance line OA of the field winding. Let R be the resistance of the field winding. Line OA is drawn such that its slope equals the field winding resistance R *i.e.* every point on this curve is such that

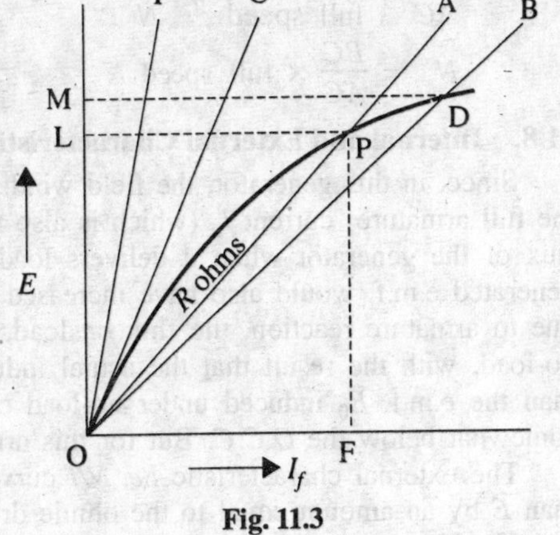

Fig. 11.3

$$\frac{\text{volt}}{\text{amperes}} = R$$

The voltage OL, corresponding to point p, represents the maximum voltage to which the machine will build up with R as field resistance. OB represents smaller resistance and the corresponding voltage OM is slightly greater than OL. If field resistance is increased, then slope of the resistance line increases and hence the maximum voltage to which the generator will build up at a given speed, decreases. If R is increased so much that the resistance line does not cut the $O.C.C$ at all (like OT), the machine will fail to excite *i.e.* there will be no 'build up' of the voltage. If the resistance line just lies along the slope, then with that value of field resistance the machine will just excite. The value of resistance represented by the tangent to the curve is known as *critical resistance* R_c for a *given speed.*

11.5. How to Find Critical Resistance R_c?

First, $O.C.C.$ is plotted from the given data. Then, tangent is drawn to its initial part. The slope of this curve gives the critical resistance for the speed at which the data were obtained.

11.6. How to Draw O.C.C. at Different Speeds?

Suppose we are given the data for $O.C.C.$ of a generator run at a fixed speed, say, N_1. It will be shown that $O.C.C.$ at any other constant speed N_2 can be deduced from the $O.C.C.$ for N_1. In Fig. 11.4, curve OA represents $O.C.C.$ at N_1.

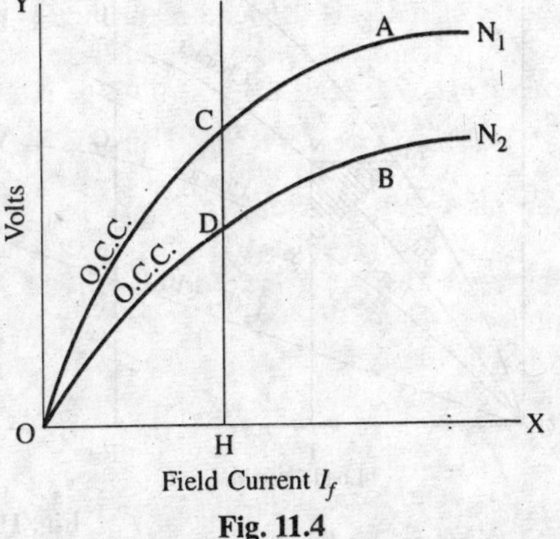

Fig. 11.4

Since $E \propto N$ for any fixed excitation, hence

$$\frac{E_2}{E_1} = \frac{N_2}{N_1} \quad \text{or} \quad E_2 = E_1 \times \frac{N_2}{N_1}$$

As seen, for $I_f = OH$, $E_1 = HC$. The value of new voltage for the same I_f but at N_2 is

$$E_2 = HC \times \frac{N_2}{N_1} = HD$$

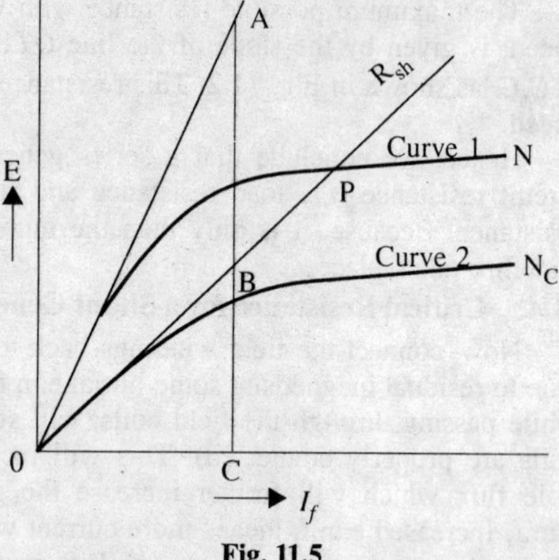

Fig. 11.5

In this way, point D is located. In a similar way, other such points can be found and the new *O.C.C.* drawn (see Ex. 11.1).

11.7. Critical speed N_c

Critical speed of shunt generator is that speed for which the *given* shunt field resistance represents critical resistance. In Fig. 11.5, curve 2 corresponds to critical speed because R_{sh} line is tangential to it. Obviously

$$\frac{BC}{AC} = \frac{N_c}{\text{full speed}} = \frac{N_c}{N}$$

$$\therefore \quad N_c = \frac{BC}{AC} \times \text{full speed } N$$

11.8. Internal and External Characteristics of a Series Generator

Since, in this generator, the field windings are in series with the armature, hence they carry the full armature current I_a (which is also the load current). If there were nothing to reduce the flux of the generator when it delivers load, then as I_a were increased, the flux and hence the generated e.m.f. would also have increased as given by the *O.C.C.* characteristic *OA*. However, due to armature reaction, the flux on load, for a given current, would be less than the flux on no-load, with the result that the actual induced e.m.f. E (under load conditions) is a little less than the e.m.f. E_0 induced under no-load conditions. Hence, the internal characteristic *OB* lies somewhat below the *O.C.C.* But for this armature reaction, the two would have coincided.

The external characteristic *i.e. V/I* curve lies still lower because the terminal p.d. V is less than E by an amount equal to the ohmic drop in the generator.

If R_a = armature resistance, R_{se} = resistance of series field winding

Then, ohmic drop is $= I_a(R_a + R_{se})$

$$\therefore \quad V = E - I_a(R_a + R_{se})$$

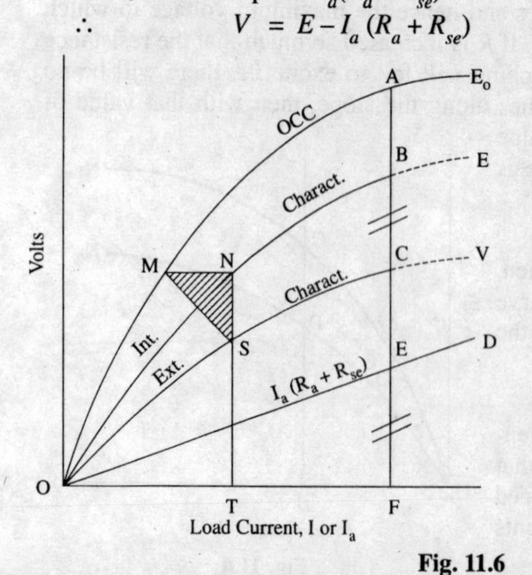

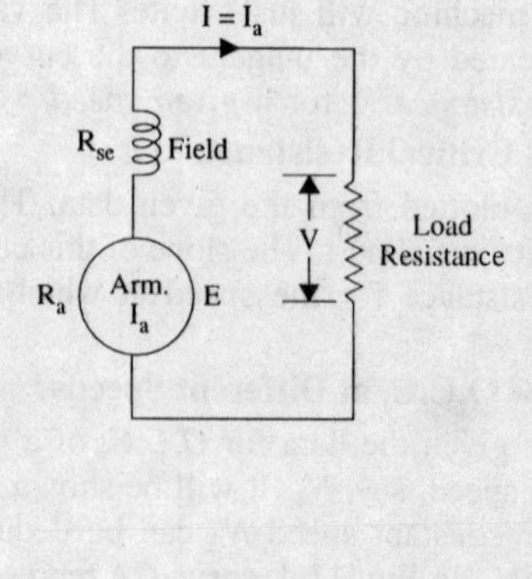

Fig. 11.6

The ohmic voltage drop is represented by the straight line *OD* in Fig. 11.6. If ordinates of curve *OD* are subtracted from those of internal characteristic *OB*, then we get the external characteristic *OC*. For example, for a load current = *OF*, the ohmic drop is *EF*. If we mark off *BC* =

EF, then, we get point *C* on the external characteristic. By obtaining a number of similar points, we can draw the external characteristic.

It will be noticed that a series generator is a variable voltage generator *i.e.* its voltage increases with load current.

It would also be seen that with increase in load current, the demagnetising effect of armature reaction is also increased. In the shaded right-angled triangle '*MNS*', for a load current of *OT*,

MN = effect of armature reaction in terms of current

NS = internal voltage drop.

If '*n*' is the number of turns per pole then the product '$n \times MN$' represents the total demagnetising effect of armature reaction for a load current = *OT*.

Beyond point *C*, when load current is increased, then due to excessive demagnetising effect of armature reaction, terminal p.d. *V* is decreased instead of increasing because the induced e.m.f. *E* itself is reduced. The portion beyond point *C* represents unstable conditions.

It is seen that the initial portion of the external characteristic is straight *i.e.* over a restricted range, the terminal p.d. *V* is proportional to the load current. This proportionality makes a series generator suitable for being used as a booster *i.e.* for 'boosting up' the supply voltage. We know that in a supply system, the voltage at the consumer end is always less than that at the supply end by an amount equal to the drop in the cable. This voltage drop due to line resistance is directly proportional to the current. Hence, if a series generator is connected in the line (as shown in Fig. 11.7) and is so adjusted that at all load currents it supplies an additional voltage which is just equal to the line drop then the voltage at the consumer's end can be maintained constant. The booster of Fig. 11.7 is driven by a shunt motor.

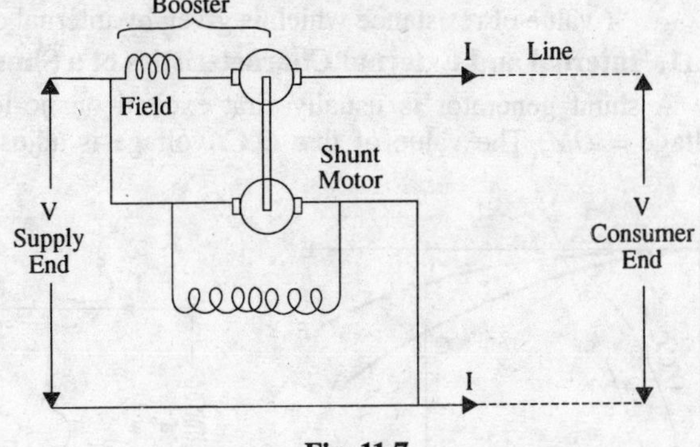

Fig. 11.7

11.9. Voltage Build up of a Shunt Gnerator

Before loading a shunt generator, it is allowed to build up its voltage, As, usually, there is always present some residual magnetism in the poles, a small e.m.f. is produced initially. This e.m.f. circulates a small current in the field circuit which increases the pole flux (provided field circuit is properly connected to armature, otherwise this current may wipe off the residual magnetism). When flux is increased, generated e.m.f. is increased which further increases the flux and so on.

Now, the generated e.m.f. in the armature has to

(*a*) supply the ohmic drop $I_f R_{sh}$ in the winding and (*b*) to overcome the opposing self-induced e.m.f. in the field coils *i.e.* $L\, dI/dt$, because field coils have appreciable self-inductance.

If (and so long as) the generated e.m.f. is in excess of the ohmic drop $I_f\, R_{sh}$, energy would continue being stored in the pole fields. For example, as shown in Fig. 11.8 corresponding to field current *OA*, the generated e.m.f. is *AC*. Of this, *AB* goes to supply ohmic drop $I_f\, R_{sh}$ and *BC* goes to overcome self-induced e.m.f. in the coil. Corresponding to $I_f = OF$, whole of the generated e.m.f. is used to overcome ohmic drop. None is left to overcome $L.dI/dt$. Hence, no further energy is stored in the pole fields. Subsequently, there is no further increase in pole flux and the generated e.m.f. With the given shunt field resistance represented by line OP, the maximum voltage to which the machine will build up is *OE*. If resistance is

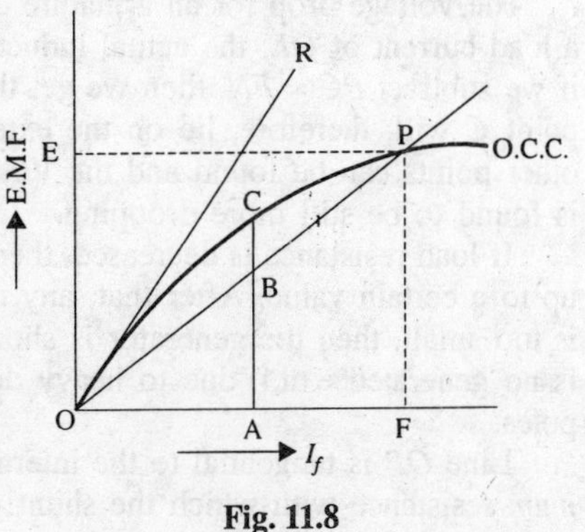

Fig. 11.8

decreased, it will build up to somewhat higher voltage. Line *OR* represents the resistance known as *critical resistance.* If shunt field resistance is greater than this value, the generator will fail to excite.

11.10. Conditions for Build up of a Shunt Generator

We may summarise the conditions necessary for the build up of a (self-excited) shunt generator as follows :

1. There must be some residual magnetism in the poles.
2. For the given direction of rotation, the shunt field coils should be correctly connected to the armature *i.e.* they should be so connected that the induced current reinforces the e.m.f. produced initially due to residual magnetism.
3. If excited on open circuit, its shunt field resistance should be less than the critical resistance (which can be found from its *O.C.C.)*
4. If excited on the load, its shunt field resistance should be more than a certain minimum value of resistance which is given by internal characteristic (Fig. 11.10).

11.11. Internal and External Characteristics of a Shunt Generator

A shunt generator is usually first excited on no-load, so that it gives its full open-circuit voltage = *OM*. The value of this *O.C.* voltage is adjusted with the help of field rheostat *Rh* as shown in Fig. 11.9. Then, the load is switched on. If there were no demagnetising effect of armature reaction and no ohmic drop in armature circuit, then the voltage would have remained constant as shown by the horizontal line *MA*. But, in actual practice, due to armature reaction, especially at higher loads, the pole flux and hence the *actual* induced e.m.f. is reduced. Therefore, the internal (or E/I_a) characteristic droops down slightly. The external characteristic can be obtained by further subtracting ohmic voltage drop (for various load currents) from the internal characteristic.

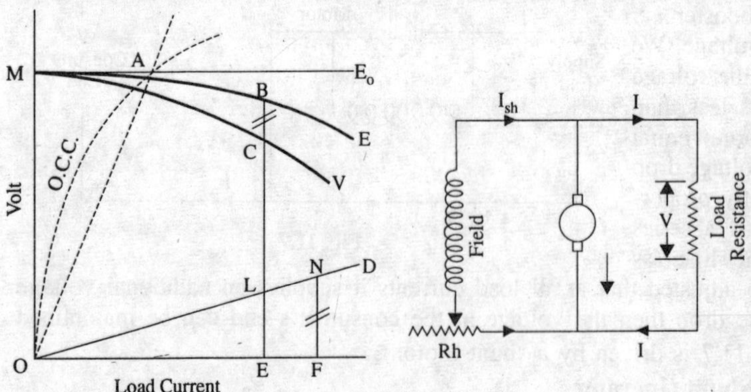

Fig. 11.9

We know that for a shunt generator

$$I_a = I + I_{sh}$$

The resistance line *OD* represents the shunt field resistance (and the interpole winding resistance). For a load current *OE,* the armature current is *OE* + *EF* where *EF* is the shunt field current I_{sh}. The voltage drop for an armature current *OE* is *FN*. Now, for a load current of *OE,* the actual induced e.m.f. is *BE*. Out of this if we subtract *BC* = *FN*, then we get the terminal p.d. of *EC*. The point *C* will, therefore, lie on the external characteristic. Similar other points can be found and the *V*/*I* graph can be drawn which is found to be still more drooping.

If load resistance is decreased, then armature current increases up to a certain value. After that, any decrease in load resistances is too small, then the generator is short-circuited and hence there is no generated e.m.f. due to heavy demagnetisation of the main poles.

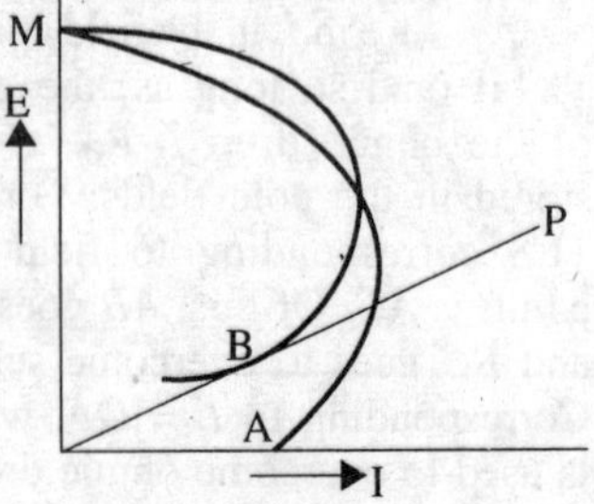

Fig. 11.10

Line *OP* is tangential to the internal characteristic *MB* and its slope gives the value of *minimum* resistance with which the shunt generator will excite if excited on load.

It will be seen from Fig. 11.9 that a shunt generator gives its greatest voltage at no-load, the voltage V falling off as output current is increased. However, the fall in voltage from no-load to full-load is small and the terminal p.d. can always be maintained constant by adjusting the shunt field regulator Rh.

11.12. Voltage Regulation

By voltage regulation of a generator is meant the change in its terminal voltage with the change in load current when it is run at a constant speed. If the change in voltage between no-load and full-load is small, then the generator is said to have good regulation, but if the change in voltage is large, then it has poor regulation. The voltage regulation of a d.c. generator is defined as the change in voltage when the load is reduced from rated value to zero, expressed as a percentage of the rated load voltage.

If no-load voltage is 240 V and rated-load voltage is 220 V,

then $$\text{regulation} = \frac{240 - 220}{220} = 0.091 \quad \text{or} \quad \mathbf{9.1\%}$$

Example 11.1. *The O.C.C. of a d.c. shunt generator running at 1000 r.p.m. is as follows:*

O.C. volts :	52.5	107.5	155	196.5	231	256.4	275	287.5
I_f *in amps* :	*1*	2	3	4	5	6	7	8

Estimate the voltage to which machine will build up if the speed is 800 r.p.m. and field circuit resistance is 30 Ω.

Solution. The *O.C.C.* for 1000 r.p.m. has been plotted in Fig. 11.11. Resistance line *OA* for 30 Ω is drawn as follows :

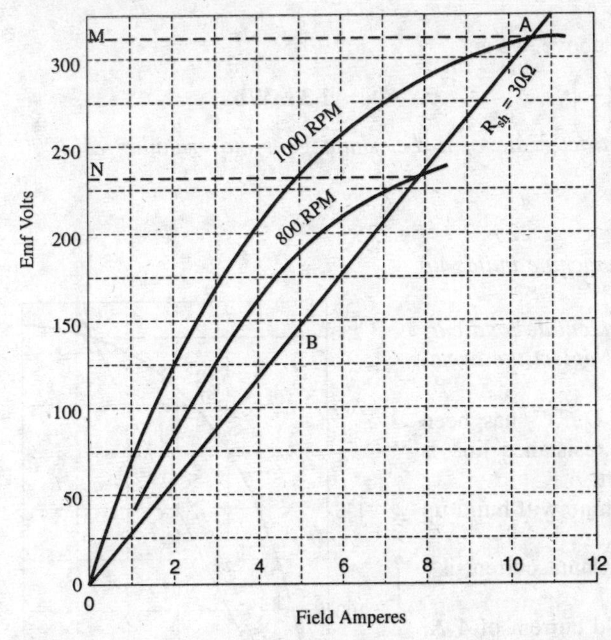

Fig. 11.11

Take 5 ampere current. Multiplying it by 30 Ω, we get 150 V. Hence, join the ordinates of 5 A and 150 V. We get the point B (5 A, 150 V). Now, draw a line joining the origin (0, 0) and B. This gives 30 Ω resistance line. It cuts the *O.C.C.* at A. Draw a horizontal line from A cutting the Y-axis at M. OM gives the maximum voltage to which the machine will build up with 30 Ω resistance in the shunt field and at 1000 r.p.m. It will be found that OM = **310 V.**

For finding the e.m.f. induced at 800 r.p.m., a new curve is drawn. Since e.m.f. is proportional to the speed, hence voltage values at 1000 r.p.m. are multiplied by 800/1000 = 0.8 and *O.C.C.* is plotted as before. With R_{sh} = 30 Ω, e.m.f. is = ON = **230 V.**

Example 11.2. *The magnetisation curve at 800 r.p.m. of a 4-pole, 220-V, shunt generator with 576 conductors lap-wound armature is as follows:*

Field current (Amp.)	0	0.5	*1*	2	3	4	5
E.M.F. (Volts)	10	50	100	175	220	245	262

(a) *If field circuit resistance is 75 Ω, calculate (i) the speed at which the machine just fails to excite (ii) the flux per pole if the open circuit p.d. is 225 V.*

(*b*) *What is the residual flux?*

Solution. *O.C.C.* has been drawn from the given data in Fig. 11.12.

OA is 75-Ω line and *OT* is tangential to the curve. Any line *BC* is drawn perpendicular to the base, cutting lines *OA* and *OT* at points *D* and *B* respectively.

(*a*) (*i*) $\frac{N_c}{N} = \frac{DC}{BC}$

or $\frac{N_c}{800} = \frac{150}{194}$

$\therefore \quad N_c = 800 \times \frac{150}{194} = \mathbf{620 \ r.p.m.}$

Fig. 11.12

(*ii*) Using the relation

$$E = \frac{\Phi Z N}{60}\left(\frac{P}{A}\right) \text{volts} \qquad ...(i)$$

we get $225 = \frac{\Phi \times 576 \times 800}{60} \times \left(\frac{4}{4}\right)$ $\qquad \therefore \ \Phi = \mathbf{29.3 \ mWb}$

(*b*) With no exciting current, the e.m.f. induced due to residual flux is 10 V (as given in the table).

Hence, using again equation (*i*) above, we get

$10 = \frac{\Phi \times 576 \times 800}{60} \times \left(\frac{4}{4}\right)$ $\qquad \therefore \ \Phi = \mathbf{1.3 \ mWb}$

Example 11.3. *The open-circuit characteristic of a d.c. shunt generator running at 1000 r.p.m. is as follows :*

Field Current Amps :	*2*	*4*	*6*
Open Circuit Voltage:	*180*	*225*	*235*

Determine the voltage to which the machine builds up if the field circuit resistance is 50 Ω.

If the armature resistance is 0.1 Ω, calculate load current for a terminal voltage of 200 volts. Neglect the armature reaction.

Solution. As shown in Fig. 11.13, *O.C.C.* has been plotted from the given data. The shunt resistance line is drawn as usual. It cuts the *O.C.C.* at point *A*

Hence, the voltage to which the machine will build up is **230 V.**

When terminal voltage is 200 V, the shunt current is

$I_{sh} = 200/50 = 4 \text{ A}$

As seen from *O.C.C.*, with shunt field current of 4 A, the total e.m.f. induced in the armature (neglecting armature reaction) is 225V. As armature reaction is negligible, the difference of 225 - 200 = 25 V is the drop over the armature resistance alone.

$\therefore \quad I_a R_a = 25$

or $\quad I_a = 25/0.1 = 250 \text{ A}$

As load current $I = I_a - I_{sh}$

$\therefore \quad I = 250 - 4 = \mathbf{246 \ A}$

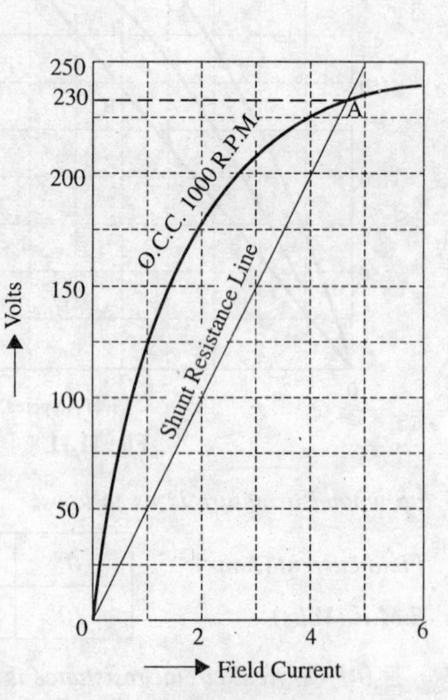

Fig. 11.13

11.13. The Compound Generator

It is a generator which has series as well as parallel field windings. When the shunt field is connected in parallel with the series combination of the armature and series field (Fig. 10-17), the compound generator is said to be connected in long shunt. However, when the shunt field is connected in parallel with only the armature, the compound generator is said to be connected in short shunt (Fig. 10.16). It is found that output (*i.e.* external) characteristics of long and short shunt compound generators are almost identical.

Though connected differently, both the shunt and series field widings are located on the same pole pieces of the generator. If the series and shunt fluxes assist each other, the generator is called cumulatively-compounded. If the two fluxes oppose each other, we get differentially-compounded generator which, however, is not used in practice.

11.14 Degree of Compounding

The level of compounding in a cumulatively-compound generator can be altered by changing the amount of current passing through the series field winding with the help of a by-pass rheostat *RH* as shown in Fig. 11.14.

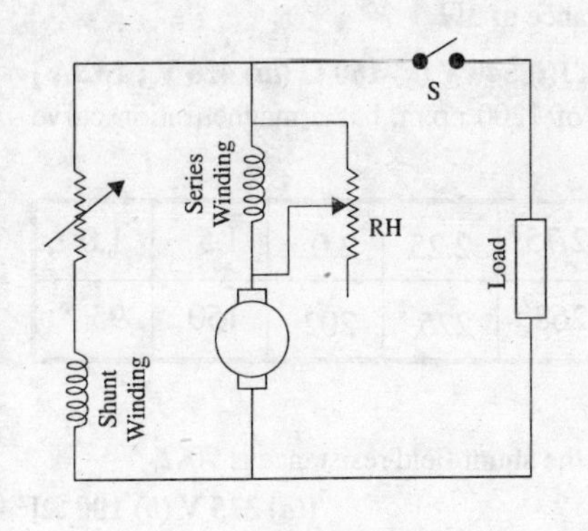

Fig. 11.14

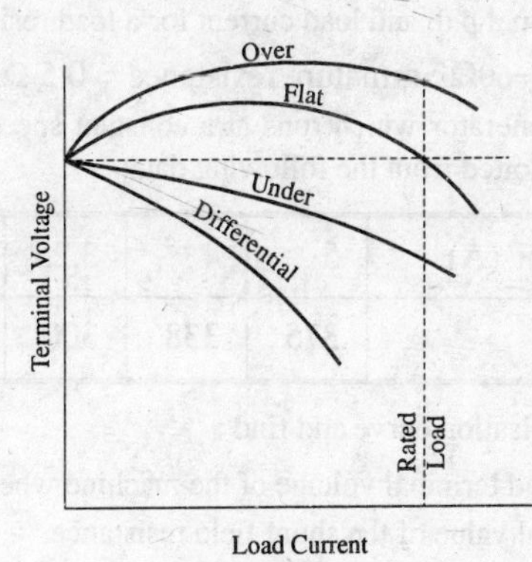

Fig. 11.15

When the series field current is such that it causes the full-load terminal voltage to be about the same as the no-load voltage, the generator is said to be flat-compounded. If the full-load voltage is greater than the no-load voltage, the generator is said to be over-compounded. However, when full-load voltage is less than the no-load voltage, the generator is under-compounded (Fig. 11.15).

However, it should be noted that even in the case of a flat-compounded generator, the terminal voltage is not constant from no-load to full-load. As seen from Fig. 11.15 at half-load, the voltage is greater than the full-load voltage.

Compound-wound gneerators are more extensively used because they can be designed to have a wide variety of characteristics. For example, an over-compounded generator may be used to compensate for the voltage drop in a feeder system.

Example 11.4. *A shunt generator is to be converted into a level compounded generator by the addition of a series field winding. From a test on the machine with shunt excitation only, it is found that the shunt current is 4. 1 A to give 440 V on no-load and 5.8 A to give the same voltage when the machine supplies its full-load of 200 A. The shunt winding has 1200 turns per pole. Find the number of series turns required per pole.* **(Elect. Engg. A.M.Ae. S.I. June 1995)**

Solution. It is seen that for obtaining the same output voltage of 440 V, the excitation of the shunt generator has to be increased to compensate for the voltage drops. In the present case, the extra shunt excitation required is = 1200 (5.8 – 4.1) = 2040 AT. The same excitation can be provided by connecting a few series turns in the armature circuit while keeping the shunt excitation constant at its no-load value. If N_{se} is the number of series turns required then, since armatere current is 200 A

$$N_{se} \times 200 = 2040, N_{se} = 10.4 \cong \mathbf{11}$$

Tutorial Problems No. 11.1

1. A d.c. shunt-wound generator has the following open-circuit magnetisation curve at its rated speed.

Field current (A)	0.5	1.0	1.5	2.0	3.0	4.0
E.M.F. (V)	180	340	450	500	550	570

The resistance of the field circuit is 200 Ω and that of the armature circuit is 0.5 W. If the generator is driven at its rated speed, find the terminal voltage (*i*) on open-circuit (*ii*) when the armature current is 60 A. **[540 V; 505 V]**

2. A d.c. generator has the following magnetisation characteristic at 1200 r.p.m.

Field current (A)	1	2	4	6	8	10
Generated E.M.F. (V)	192	312	468	566	626	660

If the generator is shunt-excited and driven at 1,000 r.p.m., determine

(*i*) The voltage to which it will excite on open-circuit.

(*ii*) The approximate value of the critical resistance of the shunt field circuit.

(*iii*) The terminal p.d. and load current for a load resistance of 3Ω.

Field resistance=60Ω ; armature resistance = 0.5 Ω **[(*i*) 540 V (*ii*) 160 Ω (*iii*) 426 V ; 142 A]**

3. A d.c. shunt generator which runs at a constant speed of 1200 r.p.m. has a magnetisation curve which can be plotted from the following data:

Field current (A)	5	4	3.25	2.75	2.25	2.0	1.5	1.0
E.M.F. (V)	375	338	300	268	225	202	150	95

Plot the magnetisation curve and find :

(*a*) the no-load terminal voltage of the machine when the shunt field resistance is 90Ω,

(*b*) the critical value of the shunt field resistance. **[(*a*) 315 V (*b*) 100 Ω]**

4. The *O.C.C.* of a d.c. generator driven at 1000 r.p.m. is given by the following data :

Field current (A)	0.2	0.4	0.6	0.8	1.0	1.2
E.M.F. (V)	46	88	126	152	165	173

Estimate :

(*a*) the critical field resistance when the generator is run as a shunt machine at 1200 r.p.m.,

(*b*) the *O.C.* voltage at 1200 r.p.m. with a field resistance of 190 Ω,

(*c*) the resistance to be inserted in series with the shunt field to reduce the *O.C.* voltage to 180 V at 1200 r.p.m. **[(*a*) 266 Ω (*b*) 202 Ω (*c*) 44 Ω**

HIGHLIGHTS

1. The three main important characteristics of a d.c. generator are as follows :

(*a*) No-load Saturation Characteristic (E_0 / I_f). It is also known as Magnetic Characteristic or Open Circuit Characteristic. It is, in fact, magnetisation curve of the material of the electromagnets and is the same for the generators.

(*b*) Internal or Total Characteristic (E/I_a).
It gives relation between the e.m.f. *E actually* induced in the armature and the armature current I_a.

(*c*) External Characteristic (*V/I*).
It gives relation between terminal voltage *V* and the load current *I*. It is also known as performance characteristic or voltage regulation curve.

2. The maximum possible resistance with which the generator would just excite, at a particular speed, is called the critical resistance of the generator at that particular speed.

3. A series generator has got rising voltage characteristic *i.e.* its voltage rises with current.
4. Conditions for the build-up of a shunt generator are :
   (*i*) there must be some residual magnetism in the poles,
   (*ii*) for the given direction of rotation, the shunt field coils should be correctly connected to the armature,
   (*iii*) its shunt resistance must be less than the critical resistance.
5. For a shunt generator, the voltage equation is

$$E_g = V + I_a R_a + \text{brush drop}$$

OBJECTIVE TESTS—11

A. Fill in the following blanks :

1. The curve showing the relation between the terminal voltage *V* and the load current of a d.c. generator is called .... characteristic.
2. The value of resistance represented by the tangent to the *O.C.C.* of a d.c. generator is called its .... resistance for the given speed.
3. There must be some .... magnetism in the poles for the build-up of a d.c. generator.
4. The speed for which the given .... field resistance of a shunt generator represents critical resistance is called critical speed.
5. The rising *V/I* characteristic of a series generator makes it suitable for use as a ....
6. A shunt generator has slightly .... external characteristics.
7. In a cumulatively-compounded d.c. generator, the shunt and series fields .... each other.
8. The level of compounding in a cumulatively-compound d.c. generator is generally altered by changing the amount of current passing through the .... field winding.

B. Answer True or False.

1. The open-circuit characteristic is similar for all types of d.c. generators.
2. The O.C.C. of a d.c. generator is just the magnetisation curve for the material of the electromagnets.
3. A d.c. series generator will not excite if its load resistance line does not intersect its O.C.C.
4. The shunt field resistance of a d.c. generator represents its critical resistance.
5. The only condition for the build-up of d.c generator is that its poles should have some residual magnetism.
6. A shunt generator gives its greatest voltage at no-load.
7. A shunt generator has negative voltage regulation.
8. A flat-compound d.c. generator has constant voltage from no-load to full-load.

C. Multiple Choice Questions

1. The O.C.C. of a d.c. generator is also called its .... characteristic.
   (*a*) internal (*b*) magnetic
   (*c*) external (*d*) performance
2. The line representing the critical resistance of a d.c. generator .... its O.C.C.
   (*a*) intersects
   (*b*) does not intersect
   (*c*) just touches
   (*d*) runs parallel to
3. Which of the following will NOT prevent a self-excited shunt generator from building upto its full voltage ?
   (*a*) wrong direction of rotation
   (*b*) open field
   (*c*) no residual magnetism
   (*d*) speed too high
4. The level of compounding in a cumulatively-compound d.c. generator is usually adjusted by
   (*a*) altering series field current
   (*b*) changing shunt field current
   (*c*) connecting it long-shunt
   (*d*) connecting it short-shunt

5. Which of the following generator provides approximately constant voltage from no-load to full-load?

(*a*) series (*b*) shunt

(*c*) flat-compound

(*d*) over-compound

6. The...............d.c. generator has the poorest voltage regulation.

(*a*) over-compound

(*b*) flat-compound

(*c*) shunt (*d*) series

12 D.C. MOTOR

12.1. Motor principle

An electric motor is a machine which converts electrical energy into mechanical energy. Its action is based on the principle that when a current-carrying conductor is placed in a magnetic field, it experiences a mechanical force whose direction is given by Fleming's left-hand rule (Art. 7.19) and whose magnitude is given by

$$F = BIl \text{ newton} \qquad \text{(Art. 7.20)}$$

Constructionally, there is no basic difference between a d.c. generator and a d.c. motor (Fig. 12.1). In fact, the same d.c. machine can be used interchangeably as a generator or as a motor. D.C. motors are also, like generators, shunt-wound or series-wound or compound-wound.

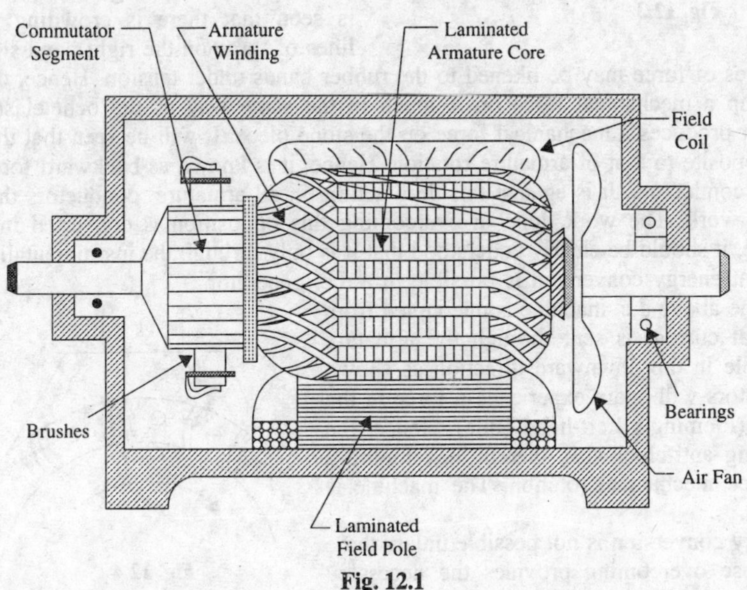

Fig. 12.1

In Fig. 12.2 is shown a part of a multipolar d.c motor. When its field magnets are excited and its armature conductors are supplied with current from the supply mains, they experience a force tending to rotate the armature. Armature conductors under *N*-pole are assumed to carry current downwards (crosses) and those under *S*-poles to carry current upwards (dots). By applying Fleming's Left-hand rule, the direction of the force on each conductor can be-found. It is shown by small arrows placed above each conductor. It will be seen that each conductor experiences a force *F* which tends to rotate the armature in anti-clockwise direction.These forces collectively produce a driving torque which sets the armature rotating.

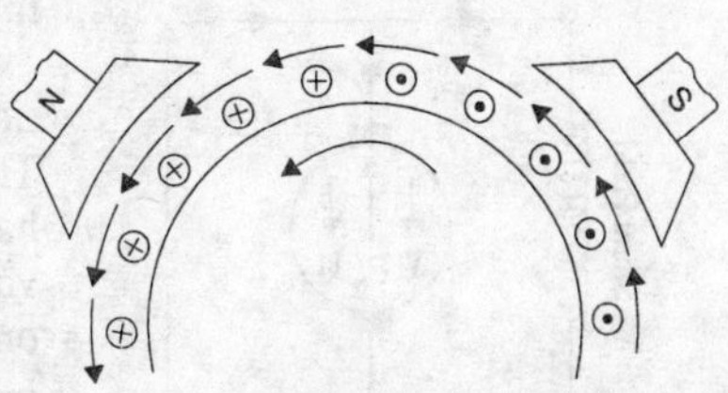

Fig. 12.2

It should be noted that the function of the commutator in the motor is the same as in a generator. By reversing current in each conductor as it passes from one pole to another, it helps to develop a continuous and unidirectional torque.

12.2. Comparison of Generator and Motor Action

As said above, the same d.c. machine can be used, at least theoretically, as a generator or as a motor. When operating as a generator, it is driven by a mechanical machine and it develops voltage which in turn produces a current flow in an electric circuit. When operating as a motor, it is supplied by electric current and it develops torque which, in turn, produces mechanical rotation.

Let us first consider its operation as a generator and see how exactly and through which agency, mechanical power is converted into electric power.

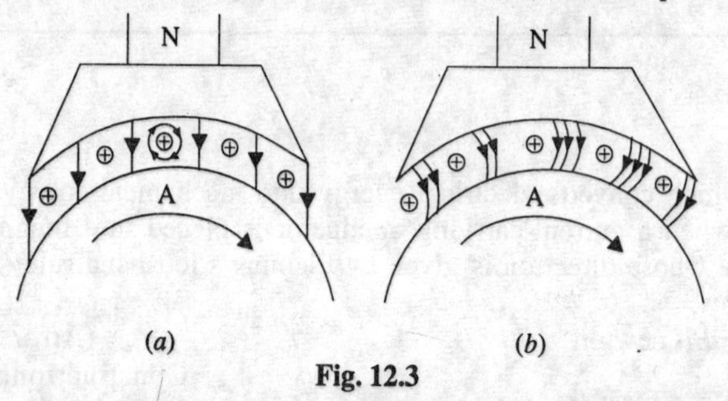

Fig. 12.3

In Fig. 12.3 is shown part of a generator whose armature is being driven clockwise by its prime mover.

Fig. 12.3 (*a*) represents the fields set up independently by the main poles and the armature conductors like *A* in the figure. The resultant field or magnetic lines of force are shown in Fig. 12.3 (*b*). It is seen that there is crowding of lines of force on the right-hand side of *A*. These magnetic lines of force may be likened to the rubber bands under tension. Hence, the bent lines of force set up a mechanical force on *A* much in the same way as the bent elastic rubber band of a catapult produces a mechanical force on the stone piece. It will be seen that this force is in a direction opposite to that of armature rotation. Hence, it is known as backward force or magnetic drag on the conductors. It is against this drag acting on all armature conductors that the prime mover has to work. The work done in overcoming this opposition is converted into electric energy. Therefore, it should be clearly understood that it is only through the instrumentality of this magnetic drag that energy conversion is possible in a d.c. generator.

Next, suppose that the above d.c. machine is uncoupled from its prime mover and that current is sent through the armature conductors under a *N*-pole in the downward direction as shown in Fig. 12.4. The conductors will again experience a force in the anti-clockwise direction (Fleming's Left-hand Rule). Hence, the machine will start rotating anti-clockwise, thereby developing a torque which can produce mechanical rotation. The machine is then said to be motoring.

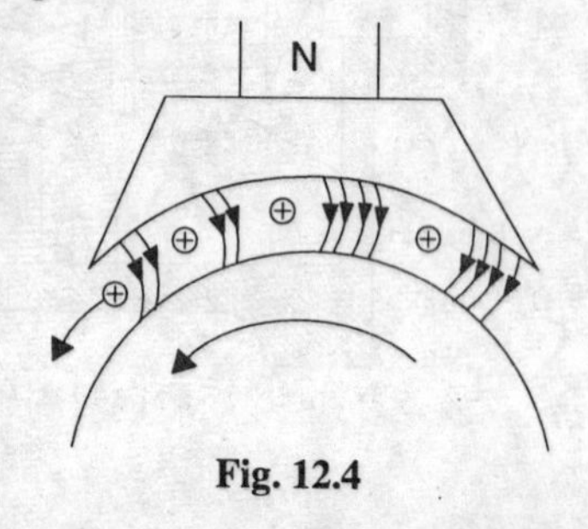

Fig. 12.4

As said above, energy conversion is not possible unless there is some opposition whose overcoming provides the necessary means for such conversion. In the case of a generator, it was the magnetic drag which provided the necessary opposition. But what is the equivalent of that in the case of a motor. Well, it is the back e.m.f. It is explained in this manner ?

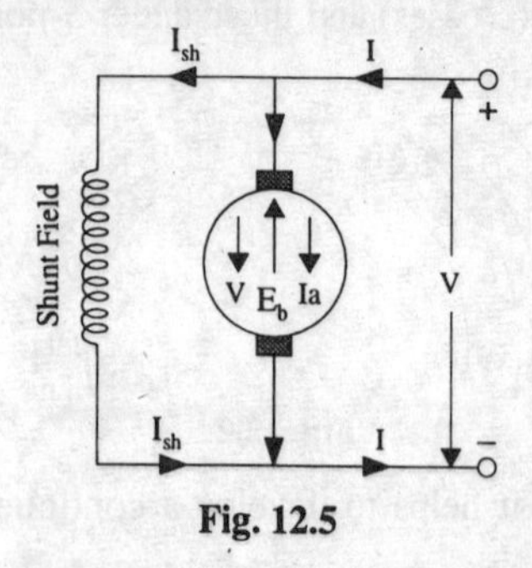

Fig. 12.5

As soon as the armature starts rotating, dynamically (or motionally) induced e.m.f. is produced in the armature conductors. The direction of this induced e.m.f. as found by Fleming's Right-hand Rule, is outward *i.e.* in direct opposition to the applied voltage (Fig. 12.4). That is why it is known as back e.m.f. E_b or counter e.m.f. Its value is the same as for the motionally induced e.m.f in the generator ; $E_b = (\Phi ZN) \times (P/A)$ volts. The applied voltage *V* has to force current through the armature against this back e.m.f. E_b. The electric work done in overcoming this opposition is converted into mechanical energy developed in the armature. Therefore, it is obvious that but for the production of this opposing e.m.f., energy conversion would not have been possible.

Now, before leaving this topic, let it be pointed out that in an actual motor with slotted armature, the torque is not due to mechancial force on the conductors themselves, but due to the tangential pull on the armature teeth as shown in Fig. 12.6.

It is seen in Fig. 12.6 (*a*) that the main flux is concentrated in the form of tufts at the armature teeth while the armature flux is shown by the dotted lines embracing the armature slots. The effect of armature flux on the main flux, as shown in Fig. 12.6 (*b*), is two-fold:

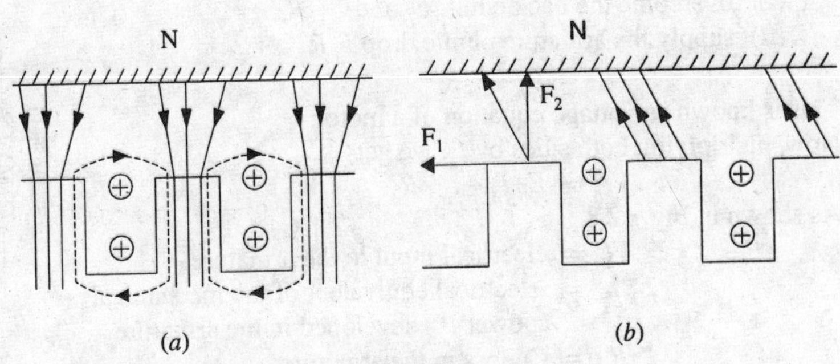

Fig. 12.6

1. It increases the flux on the left-hand side of the teeth and decreases it on the right-hand side, thus making the distribution of the flux density across the tooth section unequal.
2. It inclines the direction of lines of force in the air-gap so that they are not radial but are disposed in a manner shown in Fig. 12.6 (*b*). The pull exerted by the poles on the teeth can now be resolved into two components. One is the tangential component F_1 and the other vertical component F_2. The vertical component F_2, when considered for all the teeth round the armature, adds up to zero. But the component F_1 is not cancelled and it is this tangential component which, acting on all the teeth, gives rise to the torque.

12.3 Significance of the Back E.M.F.

As explained in Art 12.2, when the motor armature rotates, the conductors also rotate and hence cut the flux. In accordance with the laws of electromagnetic induction, e.m.f. is induced in them whose direction, as found by Fleming's Right-hand Rule, is in opposition to the applied voltage (Fig. 12.7). Because of its opposite direction, it is referred to as back e.m.f. E_b. The equivalent circuit of a motor is shown in Fig. 12.7 (*b*). The rotating armature generating the back e.m.f. E_b is like a battery of e.m.f. E_b put across a supply mains of V volts. Obviously, V has to drive I_a against the opposition of E_b. The power required to overcome this opposition is $E_b I_a$ watts.

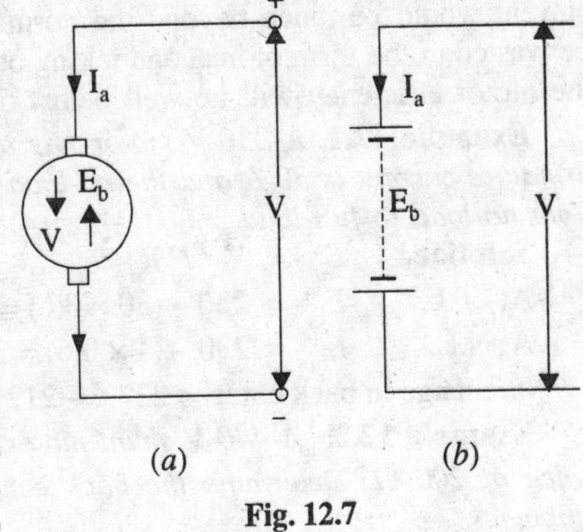

Fig. 12.7

In the case of a battery, this power over an interval of time is converted into chemical energy, but in the present case, it is converted into mechanical energy.

It will be seen that $I_a = \dfrac{\text{net voltage}}{\text{resistance}} = \dfrac{V - E_b}{R_a}$ where R_a is the resistance of the armature circuit.

As pointed out above

$$E_b = \Phi ZN \times (P/A) \text{ volts} \qquad \text{... } N \text{ in r.p.s.}$$

$$= \frac{\Phi ZN}{60}\left(\frac{P}{A}\right) \qquad \text{... } N \text{ in r.p.m.}$$

Back e.m.f. depends, among other factors, upon the armature speed. If speed is high, E_b is large, hence armature current I_a, as seen from the above equation, is small. If the speed is less, then E_b is less, hence more current flows which develops more torque (Art. 12.7). So, we find that E_b acts like a governor *i.e.* it makes a motor self-regulating so that it draws as much current as is just necessary.

12.4. Voltage Equation of the Motor

The voltage V applied across the motor armature (Fig. 12.8) has to (*i*) overcome the back e.m.f. E_b and

(*ii*) supply the armature ohmic drop $I_a R_a$

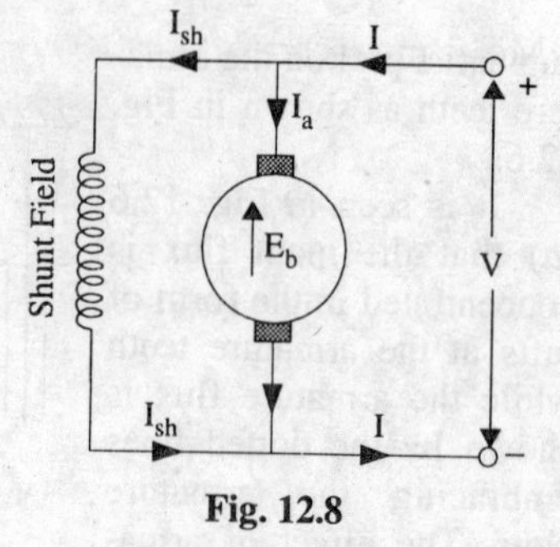

Fig. 12.8

$$\therefore \quad V = E_b + I_a R_a$$

This is known as voltage equation of a motor.

Now, multiplying both sides by I_a, we get

$$VI_a = E_b I_a + I_a^2 R_a$$

As shown in Fig. 12.8,

VI_a = electrical input to the armature.

$E_b I_a$ = electrical equivalent of the mechanical power P_m developed in the armature.

$I_a^2 R_a$ = Cu loss in the armature.

Hence, out of the armature input, some is wasted as I^2R loss and the rest is converted into mechanical work within the armature.

12.5. Condition for Maximum Power

The mechanical power developed by a motor is

$$P_m = VI_a - I_a^2 R_a$$

Differentiating both sides with respect to I_a, we get

$$dP_m / dI_a = V - 2 I_a R_a = 0 \qquad \therefore \quad I_a R_a = V/2$$

As $\quad V = E_b + I_a R_a$ and $I_a R_a = V/2 \qquad \therefore \quad E_b = V/2$

The mechanical power developed by a motor is maximum when the back e.m.f. is equal to half the applied voltage. This condition is, however, not realized in practice because in that case, current would be much beyond the normal current of the motor. Moreover, half the input would be wasted in the form of heat and taking other losses (mechanical and magnetic) into consideration, the motor efficiency will be well below 50 per cent.

Example 12.1. *A 230-V motor has an armature circuit resistance of 0.6 Ω. If the full-load armature current is 30 A and the no-load armature current is 4 A, find the change in back e.m.f. from no-load to full-load.*

Solution. $E_b = V - I_a R_a$

At F.L. $\quad E_b = 230 - (30 \times 0.6) = 212$ V

At N.L. $\quad E_{bo} = 230 - (4 \times 0.6) = 227.6$ V

∴ change in back e.m.f. = 227.6 – 212=**15.6 V**

Example 12.2. *A 440-V shunt motor has an armature resistance of 0.8 Ω and a field resistance of 200 Ω. Determine the back e.m.f. when giving an output of 7.46 kW at 85 per cent efficiency.*

Solution. Motor input power = 7.46×1000/0.85 W

Motor input current = 7460/0.85 × 440 = 19.95 A

I_{sh} = 440/200 = 2.2 A

I_a = 19.95– 2.2= 17.75 A

Now $\quad E_b = V - I_a R_a = 440 - (17.75 \times 0.8) =$ **425.8 V**

Example 12.3. *A 25-kW, 250-V, d.c. shunt generator has armature and field resistances of 0.06 Ω and 100 Ω respectively. Determine the total armature power developed when working (i) as a generator delivering 25 kW output and (ii) as a motor taking 25 kW input.*

Solution. As generator (Fig. 12.9)

Output current = 25,000/250 = 100 A

I_{sh} = 250/100 = 2.5 A; $\quad I_a$ = 102.5 A

Generated e.m.f = $250 + I_a R_a$ = 250+6.15 = 256.15 V

Power developed in armature

$$= E_g I_a = \frac{256.15 \times 102.5}{1000} = \textbf{26.25 kW}$$

As Motor (Fig. 12.10)

Motor input current = 100 A

I_{sh} = 2.5 A;

I_a = 97.5 A

E_b = 250 – (97.5 × 0.06)

= 250 – 5.85 = 244.15 V

Power developed in armature

= $E_b I_a$ watt

$= \dfrac{244.15 \times 97.5}{1000}$

= **23.8 kW**

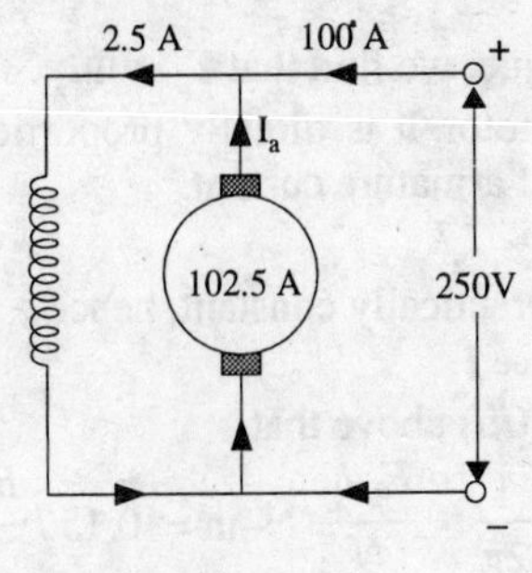

Fig. 12.9

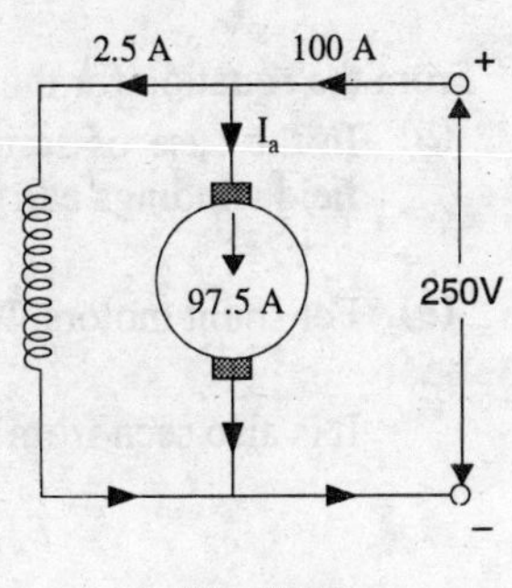

Fig. 12.10

Tutorial Problems No. 12.1

1. What do you understand by the term back e.m.f.? A. d.c. motor connected to 460-V supply has an armature resistance of 0.15Ω. Calculate
   (*a*) the value of back e.m.f. when the armature current is 120 A.
   (*b*) the value of armature current when the back e.m.f. is 447.4 V. **[(a) 442 V (b) 84 A]**
2. A d.c. motor connected to a 460-V supply takes an armature current of 120 A on full-load. If the armature circuit has a resistance of 0.25 Ω, calculate the value of the back e.m.f at this load. **[430 V]**

12.6 Torque

By the term torque is meant the turning or twisting moment of a force about an axis. It is measured by the product of the force and the radius at which this force acts.

Consider a pulley of radius *R* metres acted upon by a circumferential force of *F* newton which causes it to rotate at *N* r.p.s. (Fig. 12.11).

Then, torque $T = F \times R$ newton-metre

Work done by this force in one revolution

= Force × distance = $F \times 2\pi R$ joules

Work done per second

$W = F \times 2\pi R \times N$ joules/second

$= (F \times R) \times 2\pi N$ joules/second

Now $2\pi N$ = angular velocity ω in radian/second

and $F \times R$ = torque *T*

∴ work done/second = $T \times \omega$ joules/second

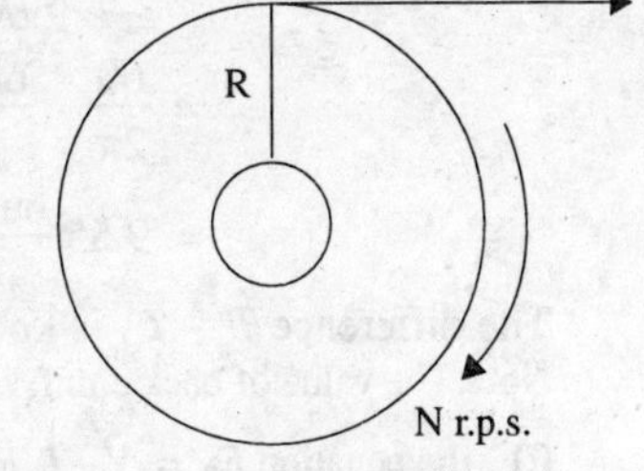

Fig. 12.11

It is also obvious from above that if *T* is the torque in N-m and ω the angular velocity in radian/second, then

Power developed = $T \times \omega$ watt

12.7 Armature Torque of a Motor

Let T_a be the torque developed by the armature of a motor running at *N* r.p.s. *i.e.* with an angular velocity of $\omega = 2\pi N$ rad/s. If T_a is in N-m, then

$$\text{Power developed} = T_a \times \omega = T_a \times 2\pi N \text{ watt} \quad ...(i)$$

We also know that electrical power converted into mechanical power in the armature is (Art. 12.4)

$$= E_b I_a \text{ watt} \quad ...(ii)$$

Equating (*i*) and (*ii*), we get

$$T_a \times 2\pi N = E_b I_a \quad ...(iii)$$

Since $E_b = \Phi ZN \times (P/A)$ volt

∴ $T_a \times 2\pi N = \Phi ZN\,(P/A).\,I_a$

or $T_a = \dfrac{1}{2\pi}\,\Phi Z I_a\,(P/A)$ N–m = $0.159\Phi Z I_a\,(P/A)$ N-m

$= (0.159/9.81)\,\Phi Z I_a (P/A)$kg-m = $0.0162\,\Phi\, Z I_a\,(P/A)$ kg-m

From the equation for the torque, we find that $T_a \propto \Phi I_a$

(*a*) In the case of series motor, Φ is directly proportional to I_a (before saturation) because field windings carry full armature current.

$\therefore \quad T_a \propto I_a^2$

(*b*) For shunt motors, Φ is practically constant, hence

$T_a \propto I_a$

It is also seen from Eq. (*iii*) above that

$$T_a = \frac{1}{2\pi} \cdot \frac{E_b I_a}{N} \text{ N–m} = 0.159 \frac{E_b I_a}{N} \text{ N–m} = 0.0162 \frac{E_b I_a}{N} \text{ kg–m}$$

If N is in rpm, then as seen from Eq. (*iii*) above,

$$T_a = \frac{E_b I_a}{2\pi N/60} = \frac{60}{2\pi} \cdot \frac{E_b I_a}{N} = 9.55 \frac{E_b I_a}{N} \text{ N–m}$$

12.8 Shaft Torque (T_{sh})

The whole of the armature torque, as calculated above, is not available for doing useful work because a certain percentage of this is required for supplying iron and friction losses in the motor.

The torque which is available for doing useful work is known as shaft torque T_{sh}. It is so called because it is available at the shaft.

Output $= T_{sh} \times 2\pi N$ watt

where T_{sh} is in N-m and N in rps

$$\therefore \quad T_{sh} = \frac{\text{output in watts}}{2\pi N} \text{ N–m} \qquad \text{— } N \text{ in r.p.s.}$$

$$= \frac{\text{output in watts}}{2\pi N/60} \text{ N–m} \qquad \text{— } N \text{ in r.p.m.}$$

$$= \frac{60}{2\pi} \cdot \frac{\text{output}}{\text{N}}$$

$$= 9.55 \frac{\text{output}}{N} \text{ N–m} \qquad \text{— } N \text{ in r.p.m.}$$

The difference $T_a - T_{sh}$ is known as lost torque

Note. The value of back e.m.f. E_b can be found from

(*i*) the equation $E_b = V - I_a R_a$ (*ii*) the formula $E_b = \Phi ZN \times (P/A)$ volt

Example 12.4. *A d.c. motor takes 40 A at 220 V and runs at 800 r.p.m. If the armature and field resistances are 0.2* Ω *and 0.1* Ω *respectively, find the torque developed in the armature.*

Solution. $T_a = 9.55\, E_b I_a / N$ newton - metre

Now, $E_b = V - I_a (R_a + R_{se}) = 220 - 40(0.2 + 0.1) = 208$ V

$\therefore \quad T_a = 9.55 \times 208 \times 40/800 =$ **99.3 N-m.**

Example 12.5. *Determine the value of torque in N – m units established by the armature of a 4 - pole motor having 774 conductors, two paths in parallel, 24 mWb per pole when the total armature current is 50 amperes.*

Solution. Formula used:

$T_a = 0.159\, \Phi Z I_a (P/A)$ N – m

Here $Z = 774; \quad I_a = 50$A

$\Phi = 24 \times 10^{-3}$ Wb, $P = 4, A = 2$

$$T_a = 0.159 \times 24 \times 10^{-3} \times 774 \times 50 \times \frac{4}{2} \text{ N–m} = \textbf{295.3 N–m}$$

Example 12.6. *A cutting tool exerts a tangential force of 400 newtons on a steel bar of diameter 10 cm, which is being turned in a simple lathe. The lathe is driven by chain at 840 r.p.m. from a 220 – V, d.c. motor which runs at 1800 r.p.m.*

Calculate the current taken by the motor if its efficiency is 80%. What size is the motor pulley if the lathe pulley has a diameter of 24 cm?

Solution. Torque = tangential force × radius = $400 \times 0.05 = 20$ N – m

Output power $= T_a \times 2\pi N$ watt $= 20 \times 2\pi \times (840/60) = 1760$ **W**

Motor $\eta = 0.8$

∴ motor input = 1760/0.8 = 2,200 **W**

Current drawn by motor = 2,200/220 = **10A**

Let N_1 and D_1 be the speed and diameter of the driver pulley respectively and N_2 and D_2 the respective speed and diameter of the lathe pulley.

Then $N_1 \times D_1 = N_2 \times D_2$

∴ $1800 \times D_1 = 840 \times 0.24$

∴ $D_1 = 840 \times 0.24/1800 = 0.112$ m = **11.2 cm**

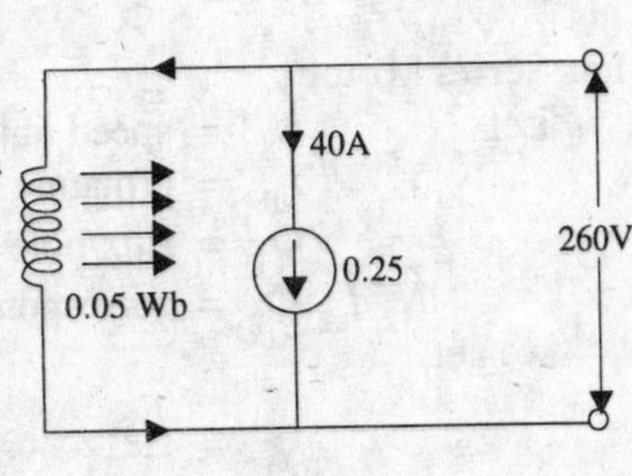

Fig. 12.12

Example 12.7. *A D.C. motor has 6 poles, flux per pole of 0.05 Wb with lap wound armature of 600 conductors. Motor speed is 500 r.p.m. Determine the applied voltages and back e.m.f. given:*

armature resistance is 0.25 Ω, armature current 40 amperes. Also, determine the torque developed by the motor in N-m.

(Elements of Elect. Engg. Punjab Univ. 1993)

Solution. $E_b = \frac{\Phi ZN}{60}\left(\frac{P}{A}\right) = \frac{0.05 \times 600 \times 500}{60} =$ **250 V**

Arm. drop $= I_a R_a = 40 \times 0.25 = 10$ V

∴ applied voltage, $V = E_b + I_a R_a = 250 + 10 =$ **260 V**

$T_a = 0.159\, E_b I_a / N$ ... N in rps

∴ $T_a = 0.159 \times 250 \times 40/(500/60) =$ **1908 N – m**

Example 12.8. *A six pole 250 V series motor is wave connected. There are 240 slots and each slot has four conductors. The flux per pole is 1.75×10^{-2} webers when the motor is taking 80 A. The field resistance is 0.05 Ω, the armature resistance is 0.1 Ω and the iron and fricitional loss is 0.1 KW. Calculate (a) speed (b) BHP (c) shaft torque (d) the pull in newtons at the rim of the pulley of dia 25 cm.* ***(Electrical Engg. A.M.Ae.S.I. June 1993)***

Solution. $E_b = 250 - 80\,(0.05 + 0.1) = 238$ V

Now, $E_b = \frac{\Phi ZN}{60}\left(\frac{P}{A}\right)$

or, $238 = \frac{17.5 \times 10^{-3} \times (240 \times 4) \times N}{60}\left(\frac{6}{3}\right)$

(*a*) ∴ $N =$ **283.3 rpm**

(*b*) Total input $= 250 \times 80 = 20{,}000$ V = 20 kW

Cu loss $= 80^2 \times 0.15 = 960$ W = 0.96 kW

Total loss = 0.1 + 0.96 = 1.06 kW

Output = 20 – 1.06 = 18.94 kW = (18.94/0.746) = **25.4 BHP**

(*c*) $T_{sh} = 9.55 \times$ output/N $= 9.55 \times 18{,}940/283.3 =$ **638.5 N – m**

(*d*) If F is the pull in newtons at the rim of the pulley, then

$F \times (0.25/2) = 638.5$ or $F =$ **5110 N**

12.9. Speed of a D.C.Motor

From the voltage equation of a motor (Art. 12.4), we get

$E_b = V - I_a R_a$

or $\Phi ZN(P/A) = V - I_a R_a$.

$$\therefore \qquad N = \frac{V - I_a R_a}{\Phi} \times \left(\frac{A}{ZP}\right) \text{r.p.s.}$$

Now $V - I_a R_a = E_b$

$$\therefore \qquad N = \frac{E_b}{\Phi} \times \left(\frac{A}{ZP}\right) \text{ r.p.s.} \qquad \text{or} \qquad N = \frac{kE_b}{\Phi}$$

It shows that speed is directly proportional to back e.m.f. E_b and inversely to the flux Φ.

or $$N \propto \frac{E_b}{\Phi}$$

For Series Motor

Let N_1 = speed in the 1st case

I_{a1} = armature current in the 1st case

Φ_1 = flux/pole in the 1st case

N_2, I_{a2}, Φ_2 = corresponding values in the second case. Then, using the above relation, we get,

$$N_1 \propto \frac{E_{b1}}{\Phi_1} \qquad \text{where} \quad E_{b1} = V - I_{a1}R_a$$

$$N_2 \propto \frac{E_{b2}}{\Phi_2} \qquad \text{where} \quad E_{b2} = V - I_{a2}R_a$$

$$\therefore \qquad \frac{N_2}{N_1} = \frac{E_{b2}}{E_{b1}} \times \frac{\Phi_1}{\Phi_2}$$

Prior to saturation of magnetic poles, $\Phi \propto I_a$

$$\therefore \qquad \frac{N_2}{N_1} = \frac{E_{b2}}{E_{b1}} \times \frac{I_{a1}}{I_{a2}}$$

For Shunt Motor

In this case, the same equation applies,

$$\therefore \qquad \frac{N_2}{N_1} = \frac{E_{b2}}{E_{b1}} \times \frac{\Phi_1}{\Phi_2}$$

If $\Phi_2 = \Phi_1$, then $\dfrac{N_2}{N_1} = \dfrac{E_{b2}}{E_{b1}}$

12.10. Speed Regulation

The term speed regulation refers to the change in the speed of a motor with change in applied load torque, other conditions remaining constant. By change in speed here is meant the change which occurs under these conditions due to inherent properties of the motor itself and not those changes which are affected through manipulation of rheostats or other speed controlling devices.

The speed regulation is defined as *the change in speed when the load on the motor is reduced from rated value to zero, expressed as per cent of the rated load speed.*

$$\therefore \quad \% \text{ speed regulation} = \frac{N.L. \text{ speed} - F.L. \text{ speed}}{F.L. \text{ speed}} \times 100$$

$$= \frac{N_0 - N}{N} \times 100 = \frac{dN}{N} \times 100$$

Example 12.9. *A 6-pole, lap-wound shunt motor has 500 conductors in the armature. The resistance of the armature path is 0.05 Ω. The resistance of the shunt field is 25 Ω. Find the speed of the motor when it takes 120 A from a d.c. main of 100 V supply. Flux per pole = 20 mWb.*

Solution. $$N = \frac{E_b}{\Phi}\left(\frac{A}{ZP}\right) \text{ r.p.s.}$$

$A = 6, P = 6, Z = 500; \quad E_b = V - I_a R_a$

Now, $I = 120$ A (given); $\quad I_{sh} = 100/25 = 4$A

$I_a = 120 - 4 = 116$A

$\therefore \quad E_b = 100 - (116 \times 0.05) = 94.2$ V

$\therefore \quad N = \dfrac{94.2}{20 \times 10^{-3}} \left(\dfrac{6}{500 \times 6}\right) = 9.42$ r.p.s. $=$ **565 r.p.m.**

Examp1e 12.10. *A d.c. shunt generator delivers 50 kW at 250 V when running at 400 r.p.m. The armature and field resistances are 0.02 Ω and 50 Ω respectively. Calculate the speed of the same machine when running as a shunt motor and taking 50 kW input at 250 V. Allow 1 volt per brush for contact drop.*

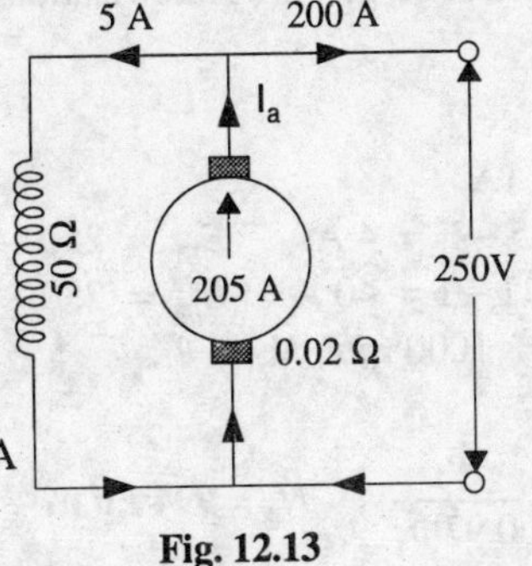

Fig. 12.13

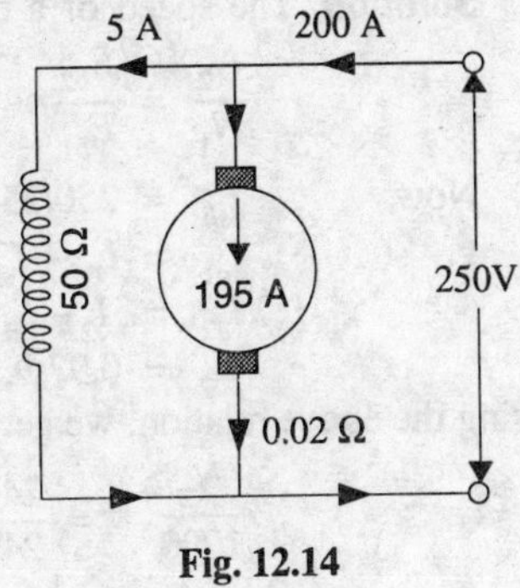

Fig. 12.14

Solution. As Generator (Fig. 12.13)

$I_{sh} = 250/50 = 5$A

$\therefore \quad I_a = I + I_{sh}$

$= 200 + 5 = 205$ A

$I_a R_a = 205 \times 0.02$

$= 4.1$ V

Brush drop $= 2 \times 1 = 2$V

$\therefore \quad E_{b1} = 250 + 4.1 + 2 = 256.1$ V; $\quad N_1 = 400$ r.p.m.

As Motor (Fig. 12.14)

$E_b = V - I_a R_a -$ brush drop

$= 250 - (195 \times 0.02) - 2 = 244.1$ V

$$\frac{N_2}{N_1} = \frac{E_{b2}}{E_{b1}} \times \frac{\Phi_1}{\Phi_2}$$

$\therefore \quad \dfrac{N_2}{400} = \dfrac{244.1}{256.1} \qquad (\because \; \Phi_1 = \Phi_2)$

$\therefore \quad N_2 = 400 \times 244.1/256.1 =$ **381.4 r.p.m.**

Example 12.11. *A 4-pole, 250 V, 7.46 kW d.c. shunt motor with a full-load efficiency of 85 per cent has 382 wave-connected armature conductors. The shunt field resistance is 125 Ω and the armature resistance is 1 Ω. If the motor takes a current of 4 A from mains on no - load and the flux per pole is 20 mWb, calculate its no-load speed.*

At what speed will the motor run on full-load?

Solution. As seen from Art. 12.9

$$N = \frac{E_b}{\Phi}\left(\frac{60\,A}{ZP}\right) \text{ r.p.m.}$$

Now $\quad I_{sh} = 250/125 = 2$ A; $\quad I_a = 4 - 2 = 2$ A

$\therefore \quad E_{bo} = V - I_{a0} R_a = 250 - (2 \times 1) = 248$ V

$\Phi = 20$ mWb $= 20 \times 10^{-3}$ Wb

$Z = 382; A = 2; P = 4$

$\therefore$ No-load speed $N_0 = \dfrac{248}{20 \times 10^{-3}} \left(\dfrac{60 \times 2}{382 \times 4}\right) = 973.8$ r.p.m.

Full-load motor input current

$I = 7.46 \times 1000/0.85 \times 250 = 35.1$ A

$I_{sh} = 2$ A (same as before)

Full-load $I_a = 35.1 - 2 = 33.1$ A

$E_b = 250 - (33.1 \times 1) = 216.9$ V

Now, $\frac{N}{N_0} = \frac{E_b}{E_{b0}}$ $N = 973.8 \times 216.9/248 =$ **851.7 r.p.m.**

Example 12.12. *A 250-V shunt motor on no-load runs at 1000 rpm and takes 5 A. The field and armature resistances are 250 Ω and 0.25 Ω respectively. Calculate the speed when the motor is loaded such that it takes 41 A if the armature reaction weakens the field by 3 %.*

(Electrical Engg., A.M.Ae.S.I. Dec. 1993)

Solution. The speed of a d.c. motor is given by the relation

$$\frac{N_2}{N_1} = \frac{E_{b2}}{E_{b1}} \times \frac{\Phi_1}{\Phi_2}$$

Now, $I_{sh} = 250/250 = 1$A

∴ $I_{a1} = I_1 - I_{sh} = 5 - 1 = 4$ A $E_{b1} = 250 - 4 \times 0.25 = 249$ V

$I_{a2} = I_2 - I_{sh} = 41 - 1 = 40$ A $E_{b2} = 250 - 40 \times 0.25 = 240$ V

$\Phi_2 = 0.97\Phi$, $N_1 = 1000$ rpm, $N_2 = ?$

Using the above relation, we get

$$\frac{N_2}{1000} = \frac{240}{249} \times \frac{\Phi_1}{0.97\Phi_1}, \quad N_2 = \mathbf{994\ r.p.m.}$$

Example 12.13. *The armature resistance of 220 V d.c. shunt motor is 0.4 Ω and it takes a no-load armature current of 2 A and runs at 1350 rpm. Find the speed when taking an armature current of 50 A if armature reaction weakens the flux by 2 %.*

(Electrical Engg. A.M.Ae.S.I. Dec. 1991.)

Solution. The speed relationship of a d.c. motor is given by

$$\frac{N}{N_0} = \frac{E_b}{E_{b0}} \times \frac{\Phi_0}{\Phi}$$

$E_{bo} = 220 - 2 \times 0.4 = 219.2$ V; $N_o = 1350$ rpm

$E_b = 220 - 50 \times 0.4 = 200$ V; $\Phi = 0.98\,\Phi_o$

$$\therefore \frac{N}{1350} = \frac{200}{219.2} \times \frac{\Phi_0}{0.98\,\Phi_0} \qquad \therefore N = \mathbf{1257\ r.p.m.}$$

Example 12.14. *A 250-volt series motor has an armature resistance of 0.08 Ω and field resistance of 0.02 Ω and produces full-load torque when running at 500 r.p.m. taking a current of 40 A.*

Calculate : (a) armature current (b) the speed when producing full-load torque. Neglect effect of armature reaction and saturation.

Solution. (*a*) Since saturation is neglected

$T \propto \Phi I_a \propto I_a^2$ $T_{a1} \propto I_{a1}^2 \propto 40^2$ and $T_{a2} \propto I_{a2}^2$

$T_{a2}/T_{a1} = I_{a2}^2/40^2$

Now $T_{a2}/T_{a1} = 1/2$ or $I_{a2}^2/40^2 = 1/2$

or $I_{a2} = 40/\sqrt{2} =$ **28.28 A**

(*b*) $$\frac{N_2}{N_1} = \frac{E_{b2}}{E_{b1}} \times \frac{I_{a1}}{I_{a2}}$$

$E_{b1} = 250 - (40 \times 0.1) = 246$ V $E_{b2} = 250 - (28.28 \times 0.1) = 247.2$ V

$$\therefore \frac{N_2}{500} = \frac{247.2}{246} \times \frac{40}{28.28} \qquad \therefore N_2 = \mathbf{711\ r.p.m.}$$

Tutorial Problems No. 12.2

1. A d.c. machine has the following details:
   Poles = 4; flux per pole = 58.7 mWb
   Armature winding = wave; No. of slots = 48
   Conductors per slot = 4; Armature resistance = 0.2Ω
   When the machine was connected to a 400 - V supply, it ran as a motor at a speed of 1050 r.p.m. Calculate the value of back e.m.f. and hence find the value of the gross torque developed by the motor. **[395 V; 90 N–m]**
2. The armature of a 4-pole, d.c. shunt motor has a lap-connected armature winding with 740 conductors. The no-load flux per pole is 30 mWb. If the armature current is 40 A and the effect of armature reaction at this current is to reduce the air-gap flux per pole by 4 %, determine the torque developed. **[135.7 N–m]**
3. A shunt machine has an armature resistance of 0.05 Ω and a field resistance of 50 W. When connected across a 200-V supply, it runs as a motor at 500 r.p.m. and takes a total input of 100 kW. At what speed must the machine be driven as a generator to supply an output of 100 kW at a terminal voltage of 200 V? **[642 r.p.m.]**
4. A shunt motor supplied at 460 V runs light at 750 r.p.m. If it is then connected to a 230-V supply, at what speed will it run; assuming that flux is proportional to field current. **[750 r.p.m.]**
5. A 37-3 kW, 500-V shunt motor has a full-load efficiency of 84%; it has six poles, the flux per pole is 40 mWb; the armature is wave-wound and has 492 conductors; the resistance of the armature circuit is 0.12 Ω and the resistance of the shunt field is 250 Ω.
   Calculate (i) full-load speed, (*ii*) the useful load torque. **[497 r.p.m. (*ii*) 719.2 N-m]**
6. A d.c. shunt-excited machine run^ at 750 r.p.m. and supplies a load of 45 kW to 225-V bus-bars. Determine the speed at which the machine will run as a shunt motor when taking 36 kW from the 225-V bus-bars.
   Field resistance = 45Ω; Armature resistance = 0.1Ω. Neglect armature reaction effect.
   [640 r.p.m.]
7. A four-pole d.c motor is connected to a 500-V d.c. supply and takes an armature current of 80 A. The resistance of the armature circuit is 0.4 Ω. The armature is wave-wound with 522 conductors and the useful flux per pole is 0.025 Wb. Calculate (*a*) the back e.m.f. of the motor, (*b*) the speed of the motor, (*c*) the torque in newton-metres developed by the armature.
   [(*a*)468 V (*b*)1075 r.p.m. (*c*)333 N-m]
8. A four-pole, 460-V shunt motor has its armature wave-wound with 888 conductors. The useful flux per pole is 0.02 Wb and the resistance of the armature circuit is 0.7 Ω. If the armature current is 40 A, calculate (*a*) the speed and (*b*) the torque in newton-metres. **[(*a*) 730r.p.m. (*b*) 226 N-m]**
9. A four-pole motor has its armature lap-wound with 1040 conductors and runs at 1000 r.p.m. When taking an armature current of 50 A from a 250-V d.c. supply. The resistance of the armature circuit is 0.2 Ω. Calculate (*a*) the useful flux per pole (*b*) the torque developed by the armature in newton-metres. **[(*a*) 13.85 mWb (*b*) 114.7 N-m]**
10. The armature of a 6-pole, d.c. motor is (i) wave-wound (*ii*) lap-wound. If the armature has 61 slots with 8 conductors per slot, calculate the torque developed in the armature for both the above types of windings when the input current to the armature is 840 A and the flux per pole is 35 mWb.
    [(*i*) 1140 N-m (*ii*) 380 N-m]

12.11. Motor Characteristics

The characteristic curves of a motor are those curves which show relation between the following quantities:

1. Torque and armature current *i.e.* T_a/I_a characteristic. It is also known as *electrical characteristic.*

2. Speed and armature current *i.e.* N/I_a characteristic.

3. Speed and torque *i.e.* N/T_a characteristic. It is also known as *mechanical characteristic.* It can be found from (1) and (2) above.

While discussing motor characteristics, the following relations should always be kept in mind.

$$(i)\ T_a \propto \Phi I_a \quad (ii)\ I_a = \frac{V - E_b}{R_a} \quad \text{and} \quad (iii)\ N \propto \frac{E_b}{\Phi}$$

12.12 Characteristics of Series Motors

1. T_a / I_a characteristic. We have seen that $T_a \propto \Phi\ I_a$. In this case, as field windings also carry the armature current, hence $\Phi \propto I_a$ upto the point of magnetic saturation. Hence, before saturation.

$$T_a \propto \Phi I_a \quad \text{or} \quad T_a \propto I_a^2$$

At light loads, I_a and hence Φ is small. But as I_a increases, T_a increases as the square of the current. Hence, T_a/I_a curve is a parabola as shown in Fig. 12.15. After saturation, Φ is almost independent of I_a, hence $T_a \propto I_a$ only. So, the characteristic becomes a straight line. The shaft torque T_{sh} is less than armature torque due to stray losses. It is shown dotted in the figure. So, we conclude that (prior to magnetic saturation) on heavy loads, a series motor exerts a torque proportional to the square of armature current. Hence, in case where huge starting torque is required for accelerating heavy masses quickly as in hoists and electric trains etc., series motors are used.

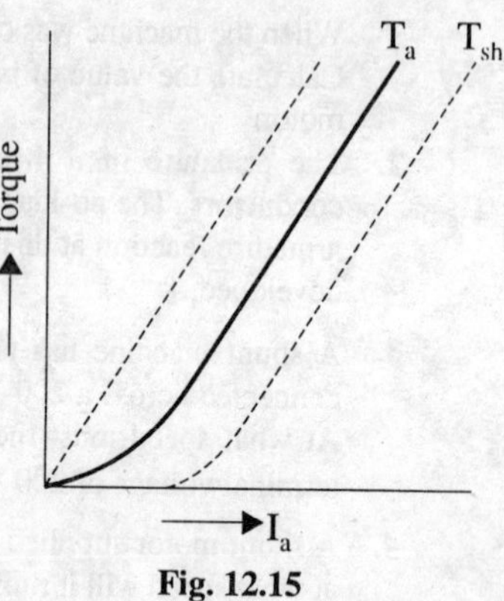

Fig. 12.15

2. N/I_a Characteristic: Variations of speed can be deduced from the formula:

$$N \propto \frac{E_b}{\Phi}$$

Change in E_b for various load currents is small and hence may be neglected for the time being. With increased I_a, Φ also increases. Hence. speed varies inversely as armature current as shown in Fig. 12.16.

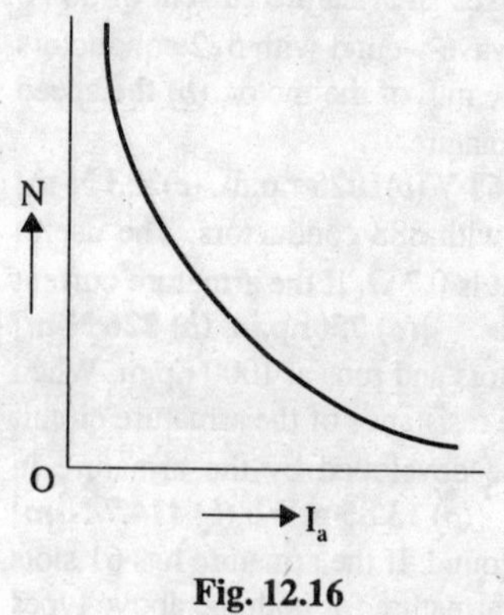

Fig. 12.16

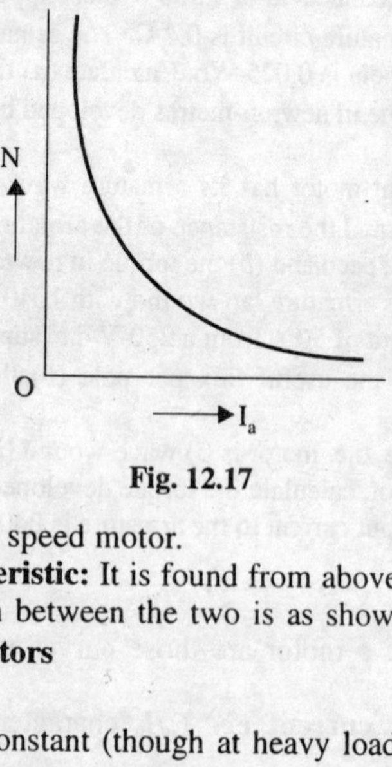

Fig. 12.17

When load is heavy, I_a is large. Hence, speed is low (this decreases E_b and allows more armature current to flow). But when load and hence I_a falls to a small value, speed becomes dangerously high. Hence, a series motor should never be started without some mechanical (not belt-driven) load on it, otherwise it may develop excessive speed and get damaged due to heavy centrifugal forces so produced. It should be noted that series motor is a variable speed motor.

3. N/T_a or **mechanical characteristic:** It is found from above that when speed is high, torque is small and *vice-versa.* The relation between the two is as shown in Fig. 12.17.

12.13. Characteristics of Shunt Motors

1. T_a / I_a Characteristic

Assuming Φ to be practically constant (though at heavy loads, Φ decreases somewhat due to increased armature reaction) we find that $T_a \propto \Phi$.

Hence, the electrical characteristic as shown in Fig. 12.18 is practically a straight line through the origin. Shaft torque is shown dotted. Since a heavy starting load will need a heavy starting current, shunt motor should never be started on (heavy) load.

2. N/I_a Characteristic

If Φ is assumed constant, then $N \propto E_b$. As E_b is also practically constant, speed, for most purposes, is constant (Fig. 12.19).

But strictly speaking, both E_b and Φ decrease with increasing load. The decrease in E_b is obvious but decrease in Φ is due to increased armature reaction. However, E_b decreases slightly more than Φ hence, on the whole, there is some decrease in speed. Therefore, the actual speed curve is slightly drooping as shown in Fig. 12.19. But for all practical purposes, shunt motor is taken as a constant-speed motor.

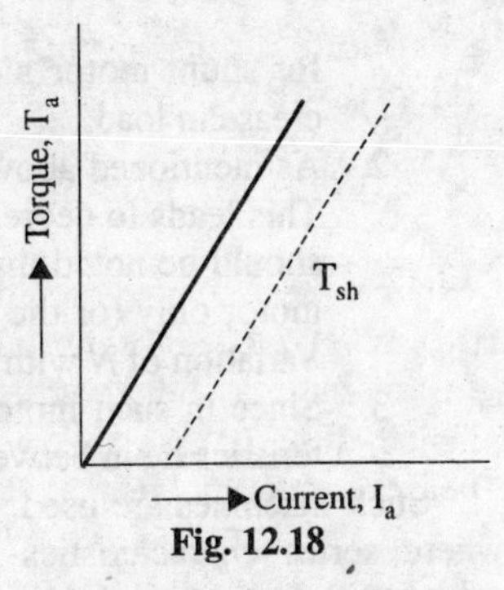

Fig. 12.18

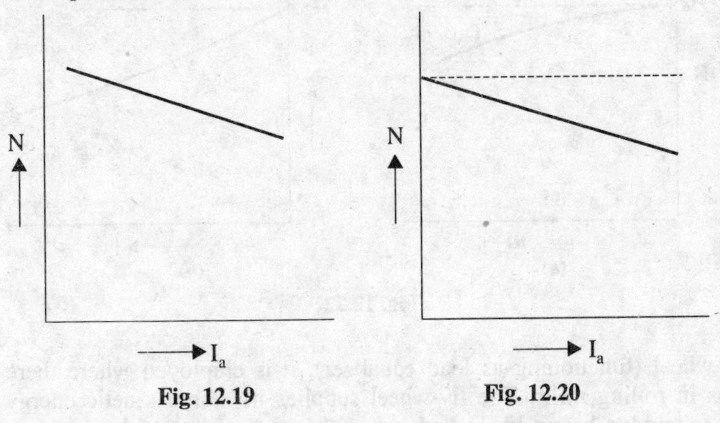

Fig. 12.19 **Fig. 12.20**

Because there is no appreciable change in the speed of a shunt motor from no-load to full-load, it may be connected to loads which are totally and suddenly thrown off without any fear of excessive speed resulting. Due to the constancy of their speed, shunt motors are suitable for driving shafting, machine tools, lathes, wood-working machines and for all other purposes where an approximately constant speed is required.

3. N/T_a characteristic can be deduced from (1) and (2) above and is shown in Fig. 12.20.

12.14 Compound Motors

In such motors, a part of the field winding is in parallel with the armature and the other part is in series with it. Compound motors are of two types:

1. *Cumulative-compound* motors. In their case, series field winding helps or aids the shunt winding as shown in Fig. 12.21 (*a*).
2. *Differential-compound motors.* In their case, the fluxes of the two windings oppose each other as shown in Fig. 12.22 (*a*).

12.15. Characteristics of Cumulative Compound Motors

1. It is seen from Fig. 12.21 (*b*) that as armature current increases, series flux increases thereby increasing the total flux of the motor. Consequently, torque is increased. However, one point worth noting is that this increase in T_a with I_a is greater than it would be

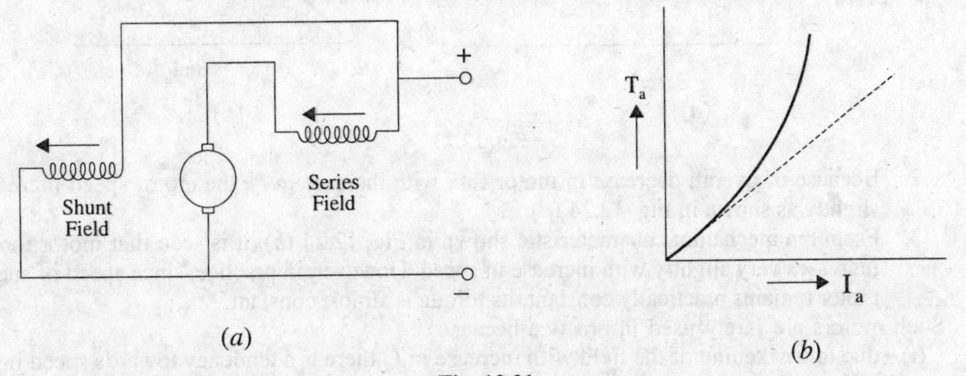

Fig. 12.21

for shunt motor alone. Obviously, such motors develop a high torque with sudden increase in load.

2. As mentioned above, as I_a increases, the series flux and hence total motor flux increases. This leads to decrease in motor speed which has a definite value N_o at no-load. Again, it should be noted that decrease in speed is more rapid than would be the case for a shunt motor only (or the decrease is less rapid, than would be the case for series motor only). Variation of N with I_a is shown in Fig. 12.22 (*a*).
3. Since in such motors, series excitation helps shunt excitation, their mechanical characteristics lie in between those of shunt and series motors as shown in Fig. 12.22 (*b*).

Such machines are used where series characteristics are required and where, in addition, the load is likely to be removed totally such as in some types of coal-cutting machines or for driving heavy machine tools which have to take sudden deep cuts quite often. Due to shunt windings, speed will not become excessively high but due to series windings, it will be able to take heavy load. In conjunction with fly-wheel (functioning as load equalizer), it is employed where there are sudden temporary loads as in rolling mills. The fly-wheel supplies its stored kinetic energy when motor slows down due to sudden heavy load. And, when due to removal of load, motor speeds up, it gathers up its kinetic energy.

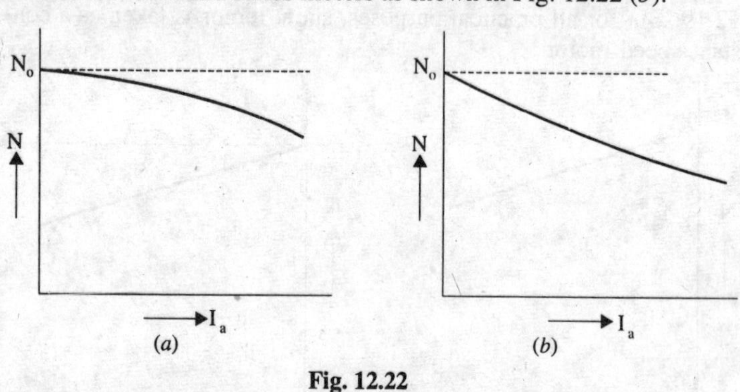

Fig. 12.22

12.16 Characteristics of Differential Compound Motors

1. Since in their case, series field opposes the shunt field, total motor flux is somewhat decreased with increase in I_a. Hence, T_a does increase with I_a but not as rapidly as would be the case for a series motor only as shown iin Fig. 12.23 (*b*).

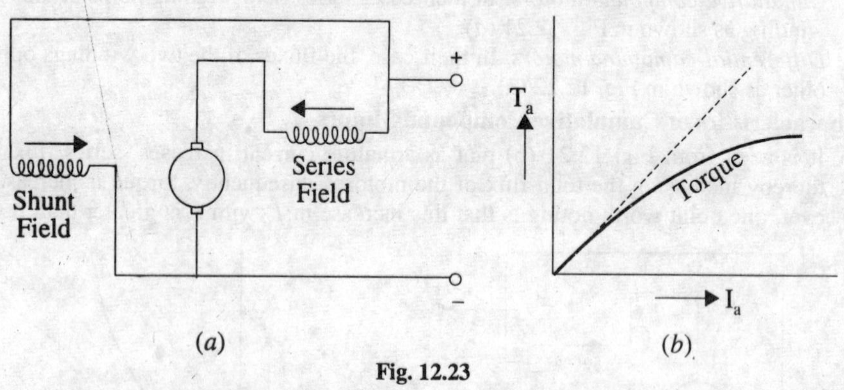

Fig. 12.23

2. Because of overall decrease in motor flux with increase in I_a, the motor speed increases slightly as shown in Fig. 12.24 (*a*).
3. From the mechanical characteristic shown in Fig. 12.24 (*b*), it is seen that motor torque increases very slightly with increase in speed. However, in practice, since speed of such a motor remains practically constant, its torque is almost constant.

Such motors are rarely used in practice because

(*i*) due to weakening of the field with increase in I_a, there is a tendency towards speed **instability** with a consequent possibility of motor running away;

(*ii*) unless the series field is short-circuited during starting, there is a strong possibility of the motor starting in the wrong direction. This happens when the starting current in the series field is so large that series ampere-turns overbalance the shunt field turns.

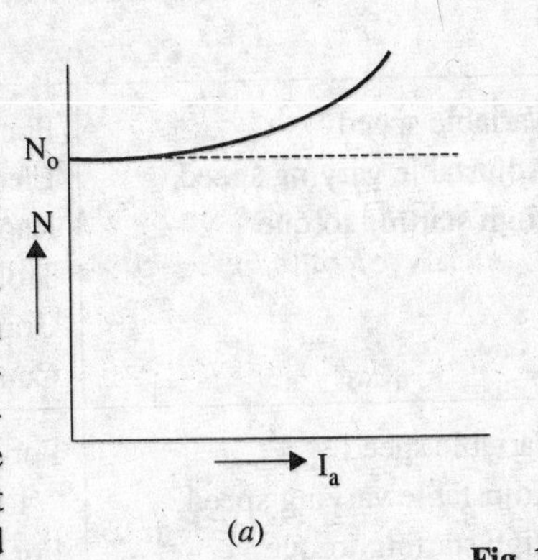

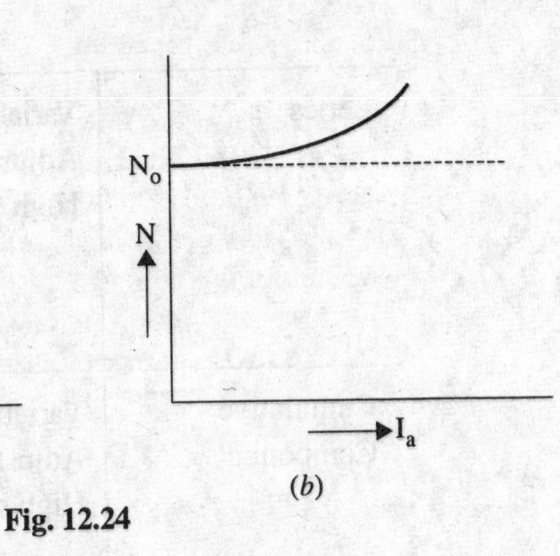

Fig. 12.24

12.17 Comparison of Shunt and Series Motors

1. Shunt Motors

The different characteristics have been discussed in Art. 12.13. It is clear that

(*a*) speed of a shunt motor is sufficiently constant,

(*b*) for the same current input, its starting torque is not as high as that of a series motor.

Hence, shunt motors are used

(*i*) when the speed has to be maintained approximately constant from *N.L.* to *F.L. i.e.* for driving a line of shafting etc.,

(*ii*) when it is required to drive the load at various speeds, any one speed being kept constant for a relatively long period *i.e.* for individual driving of such machines as lathes. The shunt regulator enables the required speed control to be obtained easily and economically.

2. Series Motors

The operating characteristics have been discussed in Art. 12.12. These motors

(*a*) have a relatively huge starting torque,

(*b*) have low speed at high loads and dangerously high speed at low loads.

Hence, such a motor is used

(*i*) when a large starting torque is required *i.e.* for driving hoists, cranes, trams etc.,

(*ii*) when the motor can be directly coupled to a load such as a fan whose torque increases with speed,

(*iii*) if constancy of speed is not essential, then, in fact, the decrease of speed with increase in load, has the advantage that the power absorbed by the motor does not increase as rapidly as the torque. For instance, when torque is doubled, the power approximately increases by about 50 to 60% only, ($\therefore I_a \propto \sqrt{T_a}$)

(*iv*) a series motor should not be used where there is a possibility of the load decreasing to a very small value. Thus, it should not be used for driving centrifugal pumps or for a belt drive of any kind.

Summary of Applications

Type of Motor	*Characteristics*	*Applications*
Shunt	Approximately constant speed Adjustable speed Medium starting troque (up to 1.5 F.L. torque)	For driving constant-speed line shafting Lathes Centrifugal pumps Machine tools Blowers and fans Reciprocating pumps

Series	Variable speed Adjustable varying speed High starting torque	For traction work i.e. Electric Locomotives Rapid Transit systems Trolley Cars etc. Crances and hoists Conveyors
Cumulative Compound	Variable speed Adjustable varying speed High starting torque	For intermittent high torque loads For shears and punches Elevators Conveyors Heavy machine tools Rolling mills Large presses

12.18 Losses and Efficiency

The losses taking place in the motor are the same as in generators (Art. 10.16). These are (*i*) Copper losses, (*ii*) Magnetic losses and (*iii*) Mechanical losses.

The condition for maximum *power* developed by the motor is

$$I_a R_a = V/2 = E_b \quad \text{(Art.12.5)}$$

The condition for maximum *efficiency* is that armature *Cu* losses should be equal to constant losses (Art. 10-20).

12.19. Power Stages of a D.C. Motor

The various stages of energy transformation in a motor and also the various losses occurring in it are shown in the power flow diagram of Fig. 12.25.

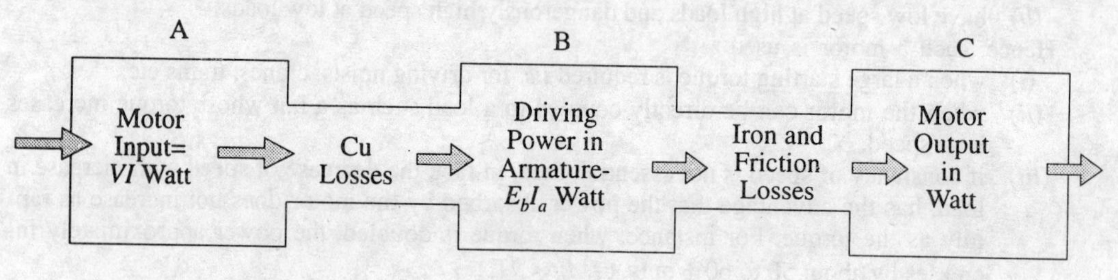

Fig. 12.25

Overall or commercial efficiency $\eta_c = \dfrac{C}{A}$

Electrical efficiency $\eta_e = \dfrac{B}{A}$

Mechanical efficiency $\eta_m = \dfrac{C}{B}$

The efficiency curve for a motor is similar in shape to that for a generator (Art. 10-20).

12.20 Swinburne's Test or No Load Test

This method gives the no-load losses of d.c. motors and is particularly applicable to shunt and compound-wound motors.

Knowing the losses, one can find the efficiency of the machine. For this purpose, the motor is run at no-load at its rated voltage and speed. The no-load input current I_o is measured by an ammeter. If I_{sh} is the shunt current the no-load armature current $I_{ao} = (I_o - I_{sh})$

As seen, the no-load motor input is $= VI_0$, armature input $= VI_{ao} = V(I_o - I_{sh})$ and power input to shunt $= VI_{sh}$

No-load power input to armature, supplies the following:

(*i*) iron losses in the armature, (*ii*) friction loss,

(*iii*) windage loss, (*iv*) armature *Cu* loss.

If we substract from the total input the no-load armature *Cu* loss, than we get constant losses. Knowing the constant losses, we can determine the motor efficiency at any other load.

Example 12.15. *A 250-volt, d.c. shunt motor takes 4 A at rated voltage on no-load. Armature and field resistances are 0.5 Ω and 250 Ω respectively. Calculate the efficiency of the motor for a load current of 40 A.*

Solution. No-load motor input $= 250 \times 4 = 1000$ **W**

This input goes to meet no-load losses which consist of (*i*) armature Cu loss and (*ii*) constant losses. This no-load data can be used to find the constant losses of the motor.

$I_{sh} = 250/250 = 1\text{A}; \quad I_a = 4 - 1 = 3$ **A**

Armature Cu loss $= I_a^2 R_a = 3^2 \times 0.5 = 4.5$ W

Constant losses of motor $= 1000 - 4.5 = 995.5$ W

With load current of 40 A

$I_a = 40 - 1 = 39\text{A}$

Armature Cu loss $= 39^2 \times 0.5 = 760.5$ W

Motor input $= 250 \times 40 = 10{,}000$ W

Total losses $= 760.5 + 995.5 = 1{,}756$ W

Motor output $= 10{,}000 - 1{,}756 = 8{,}244$ W

Motor $\eta = 8{,}244/10{,}000 = 0.8244$ or **82.44%**

Example 12.16. *A 4-pole, 500-V, 74.6kW wave-connected motor has 492 conductors. The flux per pole is 50 mWb and full-load efficiency is 92%. Its field resistance is 250 Ω amd armature circuit resistance is 0.1 Ω. Calculate for full-load (a) motor speed and (b) the useful torque developed.*

Solution. Motor input $= 74.6 \times 1000/0.92$

Motor input current $= \dfrac{74{,}600}{0.92 \times 500} = 162.1$ A

$I_{sh} = 500/250 = 2\text{A}$

$I_a = 162.1 - 2 = 160.1$ A

$E_b = 500 - (160.1 \times 0.1) = 484$ V

(*a*) $E_b = \Phi ZN \times \left(\dfrac{P}{A}\right)$ volt

$\therefore \quad 484 = 50 \times 10^{-3} \times 492 \times N \times (4/2)$

$\therefore \quad N = 9.83$r.p.s. $=$ **590 r.p.m.**

(*b*) $T_{sh} = 9.55 \cdot \dfrac{74.6 \times 1000}{590} =$ **1207 N-m**

Example 12.17. *The armature of a 220-volt, 4-pole series motor is lap-wound. There are 280 slots and each slot has 4 conductors. The current is 45 A and the flux per pole is 0.018 weber. The field resistance is 0.3 Ω, the armature resistance is 0.5 Ω and iron and frictional losses total 800 watts. The pulley diameter is 40 cm. Find the pull at the rim of the pulley.*

(Electrical Science, AMIE Winter 1993.)

Solution. We will first find the back e.m.f. E_b from the given data and thence find the motor speed.

$E_b = V - I_a (R_a + R_{se})$

$= 220 - 45\,(0.5 + 0.3) = 184$ V

$E_b = \dfrac{\Phi ZN}{60}\left(\dfrac{P}{A}\right)$

or $184 = \dfrac{0.018 \times (280 \times 4) \times N}{60} \times \left(\dfrac{4}{4}\right)$, $N = 548$ rpm.

Motor input = 220 × 45 = 9900 W

Arm. Cu. loss = $45^2 \times 0.8 = 1620$ W

Iron and friction losses = 800 W

∴ total loss = 1620 + 800 = 2420 W

Motor output = 9900 – 2420 = 7480 W

T_{sh} = 9.55 × 7480/448 = 135.34 N–m

Let F be the pull in newton at the rim of the pulley.

Then, $F \times 0.4 = 135.34$ F = **338.4 N**

Tutorial Problems No. 12.3

1. A 7.46-kW, 230-V, shunt motor has a full-load speed of 1200 r.p.m. The resistances of the armature and thee field circuit are 0.3 Ω and 115 Ω respectively. The full-load efficiency of the motor is 85%. If the total input to the motor is 620 W when it runs light, determine the no-load speed of the motor.
   Neglect brush contact drop and the effect of the armature reaction. **[1260 r.p.m.]**
2. A d.c. shunt motor has the following particulars:
   8.952 kW output; 440V, armature resistance = 1.5 Ω, brush contact drop = 2V; shunt resistance = 700 Ω. Iron and friction losses on full-load = 450 W. Calculate the efficiency when taking the full-load current of 24 A. **[85%]**
3. A d.c. series motor on full-load takes 50A from 230-V d.c. mains. The total resistance of the motor is 0.22 Ω. If the iron and friction losses together amount to 5% of the input, calculate the output furnished by the motor. Total voltage drop due to brush contact is 2V. **[10.26 kW]**
4. A 250-V shunt motor takes a current of 80 A and runs at 850 r.p.m. The resistance of the shunt field is 125 Ω and the armature circuit resistance including the brushes is 0.1 Ω. If the iron and friction losses are 1.5 kW, calculate (i) the output (*ii*) shaft torque and (*iii*) the efficiency.
   [(*i*) 17.35kW (*ii*) 195.4 N–m (*iii*) 86.95%]
5. A 250-V, d.c. compound motor, connected long-shunt, with its series field directly in series with the armature, takes an armature current of 80 A at full-load. Estimate the efficiency at full-load, given
   Armature resistance = 0.09 Ω
   Shunt field resistance = 125 Ω
   Series field resistance = 0.04 Ω
   Friction and iron losses at full-load = 750 W **[89.8%]**
6. A 2-pole d.c. shunt motor operating from a 200-V supply takes a full-load current of 35 A, the no-load current being 2 A. The field resistance is 500 Ω and the armature has a resistance of 0.6 Ω. Calculate the full-load efficiency of the motor. Take the brush drop as being equal to 1.5 V per brush arm. Neglect temperature rise. **[82.6]**

HIGHLIGHTS

1. Armature current of a d.c. motor is given by $I_a = \dfrac{V - E_b}{R_a}$
2. Voltage equation of the motor is $V = E_b + I_a R_a$ + brush drop
3. Condition for maximum power developed in armature is
   $E_b = V/2$
4. Armature torque of a motor is given by
   $T_a = 0.159\, \Phi Z I_a (P/A)$ N–m
   $= 0.0162\, \Phi Z I_a (P/A)$ kg–m

Also, $T_a = 0.159 \dfrac{E_b I_a}{N}$ N-m — N in r.p.s

$$= 0.0162 \frac{E_b I_a}{N} \text{ Kg-m} \qquad \text{— } N \text{ in r.p.s.}$$

5. Also, $$T_a = 9.55 \frac{E_b I_a}{N} \text{ N–m} \qquad \text{— } N \text{ in r.p.m.}$$

6. Shaft or useful torque is given by

$$T_{sh} = \frac{\text{output in watts}}{2\pi N} \text{ N–m} \qquad \text{— } N \text{ in r.p.s.}$$

$$= 9.55 \frac{\text{output in watts}}{N} \text{ N–m} \qquad \text{— } N \text{ in r.p.m}$$

7. Speed of a d.c. motor is given by

$$N = \frac{E_b}{\Phi}\left(\frac{60\,A}{ZP}\right) \text{ r.p.m.}$$

8. Speed regulation of a motor is given by

$$\% \text{ speed regn.} = \frac{N.L. \text{ speed} - F.L. \text{ speed}}{F.L. \text{ speed}} \times 100$$

OBJECTIVE TESTS—12

A. Fill in the following blanks:

1. An electric motor converts electrical energy into .... energy.
2. The function of a commutator in a d.c. motor is to make its armature torque.........
3. An electric motor achieves energy conversion due to the presence of .... e.m.f. induced in its armature conductors.
4. An electric motor develops maximum power when its back e.m.f equals .... the applied voltage.
5. The power developed by a motor is given by the product of its ...., and angular velocity.
6. The speed of a d.c. motor varies directly as its back e.m.f. and inversely as the .... per pole.
7. Prior to magnetic saturation, the armature torque of a series motor is proportional to the .... of its armature current.
8. Whereas shunt motor is called an approximately constant-speed motor, a series motor is called a .... speed motor.

B. Answer True or False:

1. A d.c. generator and a d.c. motor work on the same basic principle.
2. Constructionally, there is no basic difference between a d.c. generator and a d.c. motor.
3. Back e.m.f. regulates the speed of a d.c. motor.
4. Torque of a motor is determined by its speed.
5. The electrical characteristic of a series d.c. motor is a parabola.
6. In a d.c. motor, it is the speed which depends on torque and not *vice versa.*
7. Cumulatively-compound motors are well-suited for driving heavy machine tools.
8. When heavy load comes on a d.c. shunt motor, it adjusts by reducing speed.

C. Multiple Choice Questions:

1. An electric motor is used for
   (*a*) generating power
   (*b*) changing mechanical energy to electrical
   (*c*) changing electrical energy to mechanical
   (*d*) increasing the energy put into it.
2. .... represents the electrical energy converted into machanical energy by a d.c. motor.
   (*a*) $E_b I_a$ (*b*) $V I_a$
   (*c*) $E_g I_a$ (*d*) $V I_{sh}$
3. For doubing the torque developed by a d.c. series motor, its armature current has to increase by about .... per cent.
   (*a*) 100 (*b*) 75
   (*c*) 25 (*d*) 45.
4. When load is removed, ... motor will run at the highest speed.

(*a*) shunt (*b*) series
(*c*) cumulative-compound
(*d*) differential-compound.

5. The armature torque of a d.c. motor is a function of its
(*a*) pole flux
(*b*) armature current
(*c*) speed
(*d*) both (*a*) and (*b*)
(*e*) both (*a*) and (*c*).

6. A d.c. series motor is best suited for driving
(*a*) lathes
(*b*) heavy machine tools
(*c*) cranes and hoists
(*d*) shears and punches.

7. In an identical condition, the ratio of starting torque in a series dc motor to shunt dc motor is given by
(*a*) 1.0
(*b*) less than 1.0
(*c*) more than 1.0
(*d*) equal to zero.

(Elect. Engg. A.M.Ae.S.I. Dec. 1994.)

8. Fleming's left hand rule is used to find
(*a*) direction of force on a current carrying conductor in a magnetic field.
(*b*) direction of flux in a solenoid
(*c*) direction of magnetic field due to current carrying conductor
(*d*) none of these.

(Elect. Engg. A.M.Ae.S.I. Dec. 1993)

ANSWERS

A. 1. mechanical 2. unidirectional 3. back counter 4. half 5. torque 6. flux 7. square 8. variable

B. 1. F 2. T 3. T 4. F 5. T 6. T 7. T 8. F.

C. 1. *c* 2. *a* 3. *d* 4. *b* 5. *d* 6. *c*

13 SPEED CONTROL OF D.C. MOTORS

13.1. Factors Controlling the Speed

It has been shown earlier that the speed of a motor is given by the relation

$$N = \frac{V - I_a R_a}{\Phi} \times \frac{60\,A}{ZP} = K \frac{V - I_a R_a}{\Phi} \text{ r.p.m.}$$

where R_a = armature *circuit* resistance.

It is obvious that the speed can be controlled by varying (*i*) flux per pole, Φ (flux control), (*ii*) resistance R_a of armature circuit (rheostatic control). These methods as applied to shunt and series motors will be discussed below.

13.2. Speed Control of Shunt Motors

(i) Variation of Flux or Flux Control Method

It is seen from above that $N \propto 1/\Phi$. By decreasing the flux, the speed can be increased and *vice versa*. Hence, the name *flux* or *field control* method. The flux of a d.c. motor can be changed by changing I_{sh} with the help of a shunt field rheostat (Fig. 13.1). Since I_{sh} is relatively small, shunt field rheostat has to carry only small current which means I^2R loss is small, so that rheostat is small in size. This method is, therefore, very efficient. In non-interpolar machines, the speed can be increased by this method in the ratio 2:1. Any further weakening of flux Φ adversely affects the commutation and hence puts a limit to the maximum speed obtainable with this method. In machines fitted with interpoles, a ratio of maximum to minimum speeds of 6:1 is fairly common.

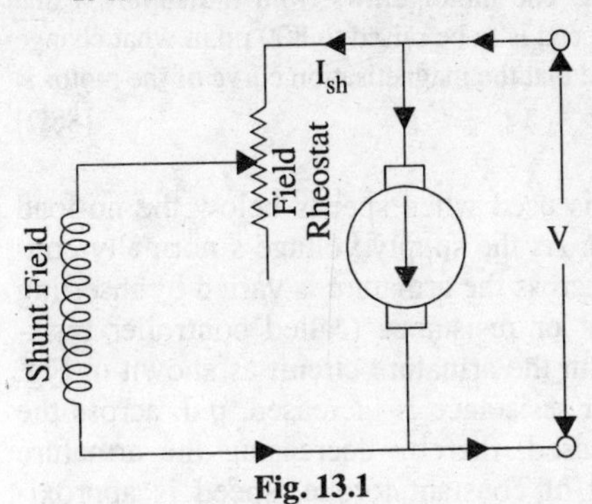

Fig. 13.1

Example 13.1. *A 500-V shunt motor runs at its normal speed of 250 rpm when the armature current is 200 A. The resistance of armature is 0.12 Ω. Calculate the speed when a resistance is inserted in the field, reducing the shunt field to 80% of normal value, and the armature current is 100 A.* ***(Electrical Engg. A.M.Ae.S.I. June 1992.)***

Solution. $E_{b1} = V - I_{a1} R_a = 500 - 200 \times 0.12 = 476$ V

$E_{b2} = V - I_{a2} R_a = 500 - 100 \times 0.12 = 488$ V

$\Phi_2 = 0.8\,\Phi_1$

Now, $\dfrac{N_2}{N_1} = \dfrac{E_{b2}}{E_{b1}} \times \dfrac{\Phi_1}{\Phi_2}$

$\dfrac{N_2}{250} = \dfrac{488}{476} \times \dfrac{1}{0.8}$, $\quad N_2 = \mathbf{320.4\ r.p.m.}$

Example 13.2. *Explain, in detail, what happens when the flux is reduced by 10% in a 200-V d.c. shunt motor, having an armature resistance of 0.2 Ω, carrying a current of 50 A and running at 960 r.p.m. prior to weakening the field. The total torque may be assumed to remain constant, iron and friction losses may be neglected.*

Solution. By decreasing the flux, speed can be increased as shown by the following relation

$$\frac{N_2}{N_1} = \frac{E_{b2}}{E_{b1}} \times \frac{\Phi_1}{\Phi_2}$$

Now, $T_a \propto \Phi I_a$

Since torque is the same in both cases,

$$\Phi_1 I_{a1} = \Phi_2 I_{a2} \quad \text{or} \quad \Phi_1 \times 50 = 0.9\,\Phi_1 \times I_{a2}$$

$\therefore$ $I_{a2} = 50/0.9 = 55.55$ A; $E_{b1} = 200 - (0.2\times50) = 190$ V

$E_{b2} = 200 - (0.2\times55.55) = 188.9$ V; $\Phi_2 = 0.9\,\Phi_1$; $N_1 = 960$ r.p.m.

$$\therefore \quad \frac{N_2}{960} = \frac{188.9}{190} \times \frac{\Phi_1}{0.9\,\Phi_1} \qquad \therefore \quad N_2 = \mathbf{1058\ r.p.m.}$$

Tutorial Problems No. 13.1

1. A 250-V d.c. shunt has an armature circuit resistance of 0.5 Ω and a field circuit resistance of 125 Ω. It drives a load at 1000 r.p.m. and takes 30 A. The field circuit resistance is then slowly increased to 150 Ω. If the flux and field current can be assumed to bee proportional and if the load torque remains constant, calculate the final speed and armature current. **[1186 r.p.m.; 33.6 A]**
2. A 250-V shunt motor with an armature resistance of 0.5 Ω and a shunt field resistance of 250 Ω drives a load the torque of which remains constant. The motor draws from the supply a line current of 21 A when the speed is 600 r.p.m. If the speed is to be raised to 800 r.p.m what change must be affected in the shunt field resistance? Assume that the magnetisation curve of the motor is a straight line. **[88Ω]**

***(ii)* Armature or Rheostatic Control Method**

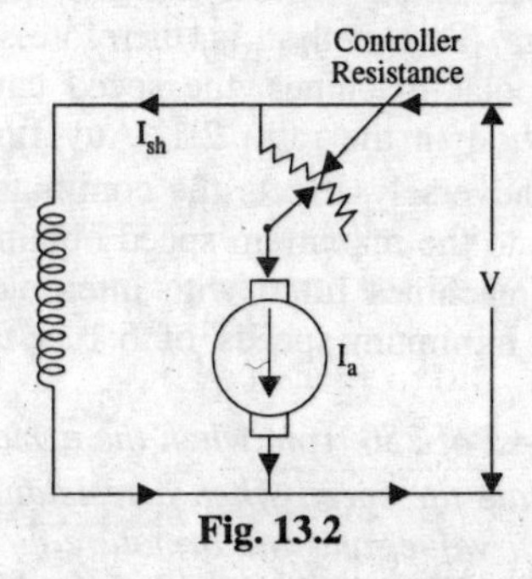

Fig. 13.2

This method is used when speeds below the no-load speed are required. As the supply voltage s normally constant, the voltage across the armature is varied by inserting a variable rheostat or resistance (called controller resistance) in series with the armature circuit as shown in Fig. 13.2. As controller resistance is increased, p.d. across the armature is decreased, thereby decreasing the armature speed. For a load of constant torque, speed is approximately proportional to the p.d across the armature. From the speed *Vs* armature current characteristic (Fig.13.3) it is seen that greater the resistance in the armature circuit, greater is the fall in speed.

Let I_{a1} = armature current in the first case

I_{a2} = armature current in the second case

(If I_{a1}=I_{a2}, then the load is of constant torque)

N_1, N_2 = corresponding speeds

V = supply voltage

Then $N_1 \propto V - I_{a1}R_a$

Now $V - I_{a1} \quad R_a = E_{b1}$

$\therefore \quad N_1 \propto E_{b1}$

Let some controller resistance of value R be added to the armature circuit so that total armature circuit resistance becomes $(R + R_a) = R_t$

Then, $\quad N_2 \propto V - I_{a2}R_t \propto E_{b2} \qquad \therefore \quad \dfrac{N_2}{N_1} = \dfrac{E_{b2}}{E_{b1}}$

(In fact, it is a simplified form of the relation given in Art. 12–9 becauce here $\Phi_1 = \Phi_2$).

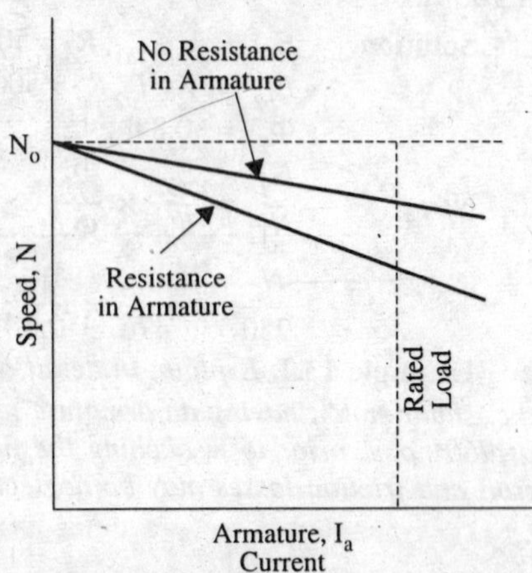

Fig. 13.3

Considering no-load speed, we have

$$\frac{N}{N_o} = \frac{V - I_a R_t}{V - I_{ao} R_a}$$

Neglecting $I_{ao}R_a$ with respect to V, we get

$$N = N_o \left(1 - \frac{I_a R_t}{V}\right)$$

It is seen that fr a given resistance R_t, the speed is a linear function of armature current I_a as shown in Fig. 13.4.

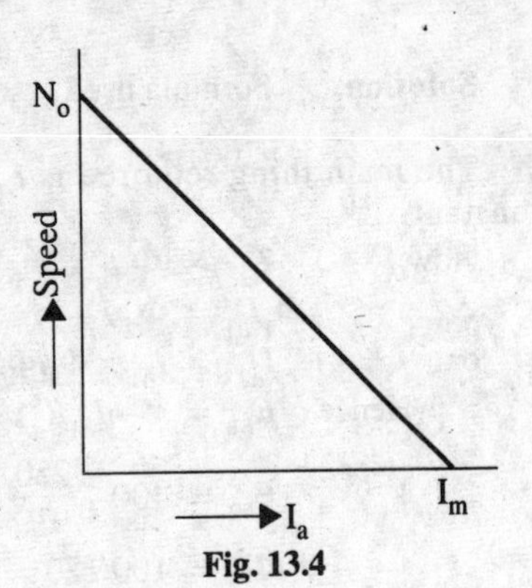

Fig. 13.4

The load current for which the speed would be zero is found by putting $N = 0$ in the above relation.

$$\therefore \quad 0 = N_o \left(1 - \frac{I_a R_t}{V}\right) \quad \text{or} \quad I_a = V/R_t$$

This is the maxium current and is known as stalling current.

This method is very wasteful, expensive and unsuitable for rapidly changing loads, because for a given value of R_t, speed will change with load.

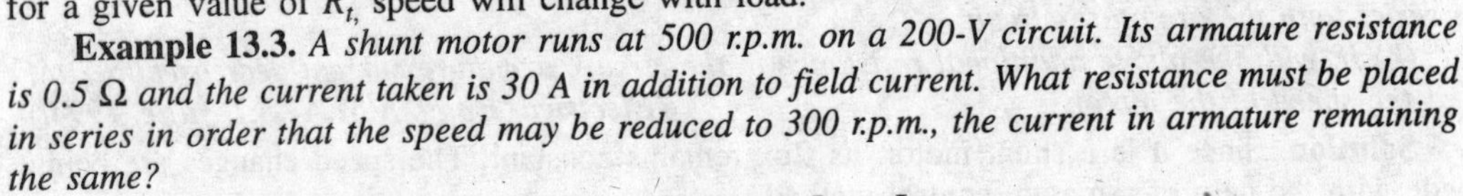

Example 13.3. *A shunt motor runs at 500 r.p.m. on a 200-V circuit. Its armature resistance is 0.5 Ω and the current taken is 30 A in addition to field current. What resistance must be placed in series in order that the speed may be reduced to 300 r.p.m., the current in armature remaining the same?*

Solution. Since field current remains the same, $\Phi_2 = \Phi_1$

$$\therefore \quad \frac{N_2}{N_1} = \frac{E_{b2}}{E_{b1}}$$

$$E_{b1} = 200 - (0.5 \times 30) = 185\,; \qquad E_{b2} = (200\text{–}30\,R_t) \text{ volt}$$

$$\therefore \quad \frac{300}{500} = \frac{200 - 30\,R_t}{185} \qquad \therefore \quad R_t = 2.97\ \Omega$$

Additional resistance required $= 2.97 - 0.5 = \mathbf{2.47\ \Omega}$

Example 13.4. *A 240 volt shunt motor runs at 850 r.p.m. when the armature current is 70 A. The armature resistance is 0.1 Ω. Calculate the required resistance to be placed in series with the armature to reduce the speed to 650 r.p.m. when the armature current is 50 A.*

(Electrical Science, AMIE Winter 1994.)

Solution. Since shunt current and hence flux remains the same, the speed relationship is given by

$$\frac{N_2}{N_1} = \frac{E_{b2}}{E_{b1}}$$

$$E_{b1} = V - I_{a1} R_a = 240 - 70 \times 0.1 = 233 \text{ V}$$

Let R be the resistance connected in series with the motor armature. Putting $(R_a + R) = R_t$ we get

$$E_{b2} = 240 - 50\,R_t, \quad N_1 = 850 \text{ r.p.m.,} \quad N_2 = 650 \text{ r.p.m.}$$

$$\therefore \quad \frac{650}{850} = \frac{240 - 50\,R_t}{233}, \qquad \therefore \quad R_t = 1.236\ \Omega$$

Hence, $R = 1.236 - 0.1 = \mathbf{1.136\ \Omega}$

Example 13.5. *A shunt motor, fed from a 400–V direct current supply, takes an armature current of 100 A when running at 800 r.p.m. If the total torque developed remains unchanged, find the speed at which the motor will run if the flux is increased to 120% of its original value and a resistance of 0.8 Ω is connected in series with the armature. The armature resistance is 0.2 Ω.*

Solution. Formula used : $\dfrac{N_2}{N_1} = \dfrac{E_{b2}}{E_{b1}} \times \dfrac{\Phi_1}{\Phi_2}$

The main thing required is I_{a2}. It can be found from the fact that motor torque remains constant.

Now, $T_a \propto \Phi I_a$

$\therefore$ $\Phi_1 I_{a1} = \Phi_2 I_{a2}$

or $I_{a2} = I_{a1} \times \Phi_1/\Phi_2 = 100/1.2 = 250/3$ A

Hence $E_{b1} = V - I_{a1} R_a = 400 - (100 \times 0.2) = 380$ V

$$E_{b2} = 400 - \frac{250}{3}\ (0.8 + 0.2) = \frac{950}{3}\ \text{V}$$

$\Phi_1/\Phi_2 = 100/120 = 1/1.2;\ N_1 = 800$ r.p.m.

$$\frac{N_2}{800} = \frac{950/3}{380} \times \frac{1}{1.2} \qquad \therefore\ N_2 = \mathbf{555\ r.p.m.}$$

Example 13.6. *A 240-V d.c. shunt motor has an armature resistance of 0.25 ohm and runs at 1000 rpm taking an armature current of 40 A. It is desired to reduce the speed to 800 rpm.*

(a) If the armature current remains the same, find the additional resistance to be connected in series with the armature circuit.

(b) If, with the above additional resistance in the circuit, armature current decreases to 20A, find the speed of the motor. ***(Electrical Engg. A.M.Ae.S.I. June. 1994.)***

Solution. Since it is a shunt motor, its flux remains constant. The speed changes are being made with the help of armature control method.

The speed formula used is $N_2/N_1 = E_{b2}/E_{b1}$

(*a*) $N_1 = 1000$ rpm, $E_{b1} = 240 - 40 \times 0.25 = 230$ V

$N_2 = 800$ rpm, $E_{b2} = 240 - 40\,R_t$, where R_t is the total armature circuit resistance.

$$\therefore\ \frac{800}{1000} = \frac{240 - 40\,R_t}{230} \quad \text{or} \quad R_t = 1.4\ \Omega$$

If R is the additional resistance connected in the armature circuit, then

$R + 0.25 = 1.4\ \Omega$ or $R = 1.4 - 0.25 = \mathbf{1.15\ \Omega}$

(*b*) Here, $N_1 = 1000$ rpm, $E_{b1} = 230$ V — as before

$N_2 = ?$, $E_{b2} = 240 - 20\,R_t = 240 - 20 \times 1.4 = 212$ V

$$\therefore\ \frac{N_2}{1000} = \frac{212}{230} \qquad \therefore\ N_2 = \mathbf{922\ r.p.m.}$$

13.3. Speed Control of Series Motors

(*i*) Flux Control Method

Variations in the flux of a series motor can be brought about in any one of the following ways:

(a) Field Divertor

The series windings are shunted by a variable resistance known as field divertor (Fig. 13.5). Any desired amount of current can be passed through the divertor by adjusting its resistance. Hence, the flux can be decreased and consequently, the speed of the motor increased.

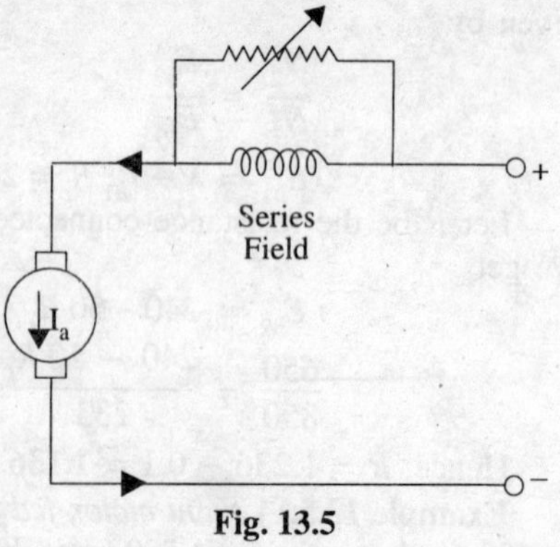

(b) Armature Divertor

A divertor across the armature can be used for giving speeds lower than the normal speeds (Fig. 13.6). For a *given constant load torque, if I_a is reduced due to armature divertor*, then Φ must increase ($\because T_a \propto \Phi I_a$). This results in an increase in current taken from the supply (which increases the flux) and a fall in speed ($\because N \propto 1/\Phi$). The variations in speed

Divertor

Series Field

Fig. 13.5

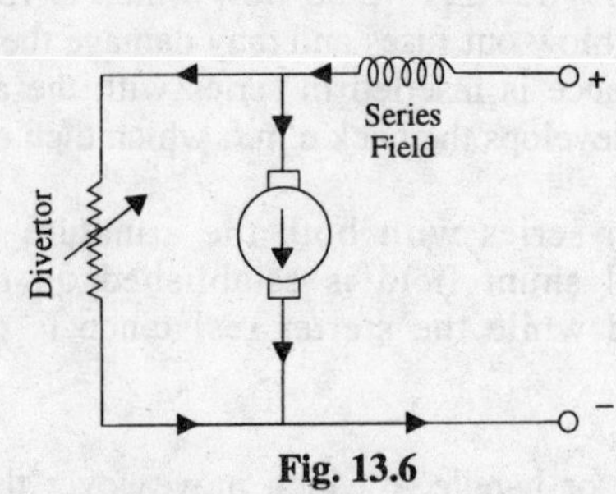

Fig. 13.6

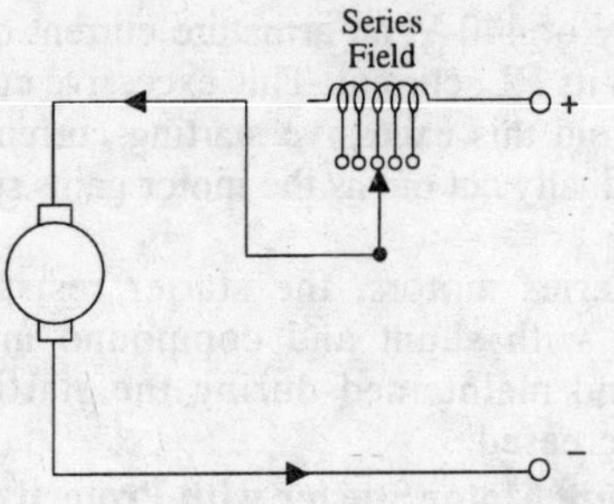

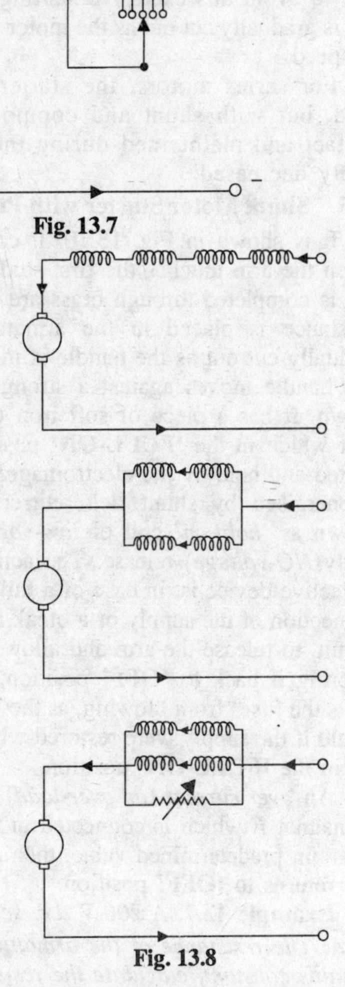

Fig. 13.7

Fig. 13.8

can be controlled by varying the divertor resistance.

(c) Tapped Field Control

This method is often used in electric traction and is shown in Fig. 13.7.

The number of series field turns in the circuit can be changed at will as shown. With full field, the motor runs at its minimum speed which can be raised in steps by cutting out some of the series turns.

(d) Paralleling Field Coils

In this method, used for fan motors, several speeds can be obtained by regrouping the field coils as shown in Fig. 13.8. It is seen that for a 4-pole motor, three fixed speeds can be obtained easily.

(*ii*) **Variable Resistance in Series with Motor Armature**

By increasing the resistance in series with the armature (Fig. 13.9), voltage applied across the armature terminals can be decreased.

With reduced voltage across the armature, the speed is reduced. However, it will be noted that since full motor current passes through this resistance, there is a considerable loss of power in it.

13.4. Motor Starters-their Necessity

When a motor is at rest, there is, as yet, no back e.m.f and if full supply voltage is applied, then the starting current is very high because armature resistance is very small. Suppose, a 200-kW motor has a cold armature resistance of 0.28 Ω and a F.L. current of 50 A. If this motor is directly switched on

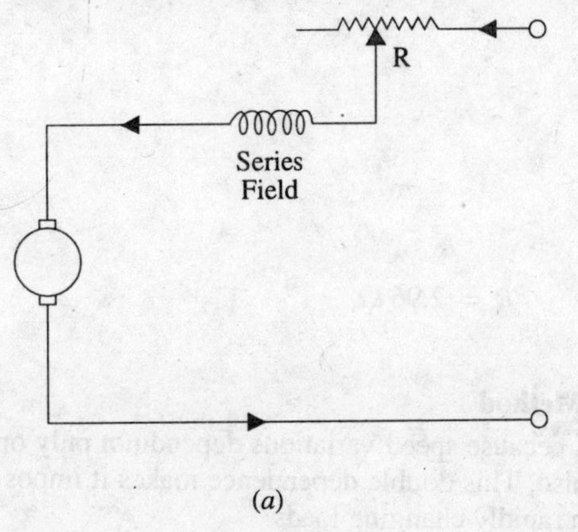

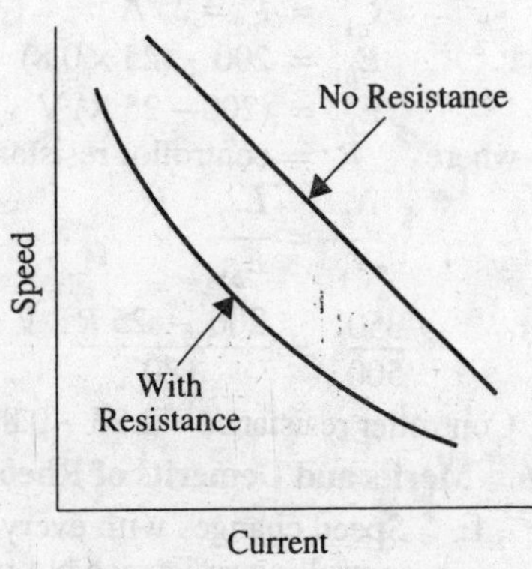

Fig. 13.9

to a supply of 440-V, an armature current of 440/0.28 = 1572 A would flow which is 1572/50 = 31.4 times its F.L. current. This excessive current will blow out fuses and may damage the brushes etc. To avoid this excessive starting current, a resistance is inserted in series with the armature and is gradually cut out as the motor gains speed and develops the back e.m.f. which then regulates its speed.

For series motors, the starter resistance is in series with both the armature and the field, but with shunt and compound motors, full shunt field is established on the first contact and maintained during the starting period while the starter resistance is progressively decreased.

13.5 Shunt Motor Starter with Protective Devices

It is shown in Fig. 13.10. It consists of an arm or handle *A* which moves over the studs. When the arm touches the first stud, field circuit is completed through brass are *B* and full resistance is placed in the armature but is gradually cut out as the handle is moved over. The handle moves against a strong spring as shown. It has a piece of soft iron *C* attached to it which in the **'FULL-ON'** position is attracted and held by the electromagnet *E* which is energised by shunt field current. This is known as *'hold-on'* coil or *low-voltage'* (formerly *NO-voltage*) release. The action of this protective device is, in case of a failure or disconnection of the supply or a break in the field circuit, to release the arm and allow the spring to bring it back to **'OFF'** position. This prevents the fuses from blowing, as they otherwise would if the supply were restored with the handle in the **'FULL-ON'** position.

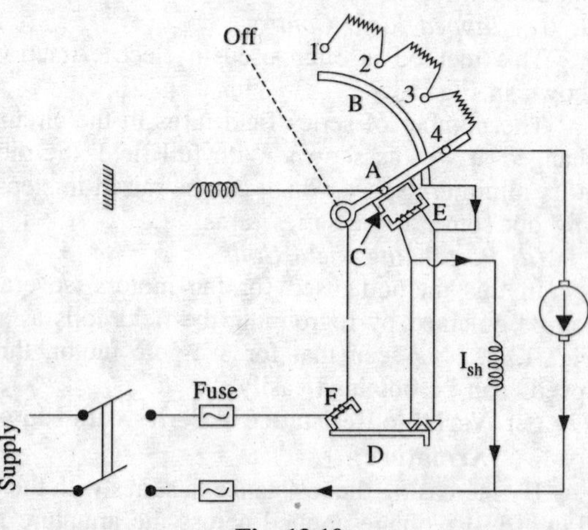

Fig. 13.10

An *over-current (or over-load)* release is also fitted in the starter. This consists of an electromagnet *F* which is connected in the supply line. If the machine becomes over-loaded beyond a certain predetermined value, then *D* is lifted and short-circuits *E*. Hence, the handle is released and returns to **'OFF'** position.

Example 13.7. *A 200-V d.c. series motor runs at 500 r.p.m. when taking a line current of 25 A. The resistance of the armature is 0.3 Ω and that of the field 0.5 Ω. If the current taken remains constant, calculate the resistance necessary to reduce the speed to 350 r.p.m.*

Solution. Total armature circuit resistance is

$$= 0.3 + 0.5 = 0.8\ \Omega$$

$$I_{a1} = I_{a2} = 25\text{ A}$$

$$\therefore \quad E_{b1} = 200 - (25 \times 0.8) = 180\text{ V}$$

$$E_{b2} = (200 - 25\text{ R})\text{ V}$$

where R = controllor resistance + $R_a + R_{se}$.

$$\therefore \quad \frac{N_2}{N_1} = \frac{E_{b2}}{E_{b1}}$$

$$\therefore \quad \frac{350}{500} = \frac{200 - 25R}{180} \qquad \therefore \quad R = 2.96\ \Omega$$

Controller resistance = 2.96 – 0.8 = **2.16 Ω**

13.6. Merits and Demerits of Rheostatic Control Method

1. Speed changes with every change in load, because speed variations depend not only on controlling resistance but on load current also. This double dependence makes it impossible to keep the speed sensibly constant on rapidly changing loads.

2. A large amount of power is wasted in the controller resistance. Loss of power is directly proportional to the reduction in speed. Hence, efficiency is decreased.
3. Maximum output power developed is diminished in the same ratio as speed.
4. It needs expensive arrangement for dissipation of heat produced in the controller resistance.
5. It gives speeds *below* the normal speed, not above it because armature voltage can be decreased (not increased) by the controller resistance.

This method is, therefore, employed when low speeds are required for a short period only and that too occasionally as in printing machines and for cranes and hoists where the motor is continually started and stopped.

13.7. Advantages of Field Control Method

This method is economical, more efficient and convenient though it can give speeds *above* (not below) the normal. The only limitation of this method is that commutation becomes unsatisfactory, because the effect of armature reaction is greater on a weaker field.

It should, however, be noted that by combining the two methods, speeds above and below the normal may be obtained.

Example 13.8. *A series motor, with an unsaturated field and negligible resistance when running at a certain speed on a given load takes 50 A at 460–V. If the load torque varies as the cube of speed, calculate the resistance required to reduce the speed by 25%.*

Solution. As series field is unsaturated, $\therefore \quad \Phi \propto I_a$

Now $\quad T_a \propto \Phi I_a$

or $\quad T_a \propto I_a^2 \quad$ ...(*i*)

Also $\quad T_a \propto N^3 \quad$ ... given $\quad$ ...(*ii*)

From (*i*) and (*ii*) $\quad I_a^2 \propto N^3$

In the first case, $\quad I_{a1}^2 \propto N_1^3$

In the second case, $\quad I_{a2}^2 \propto N_2^3$

$$\therefore \quad \left(\frac{I_{a2}}{I_{a1}}\right)^2 = \left(\frac{N_2}{N_1}\right)^3 = \left(\frac{3}{4}\right)^3 = \frac{27}{64} \qquad \left(\because \; N_2 = \frac{3}{4}N_1\right)$$

$$I_{a2} = I_{a1}\sqrt{\frac{27}{64}} = 50 \times \sqrt{\frac{27}{64}} = 32.45 \text{ A}$$

Hence $\quad E_{b1} = 460 - I_{a1} \times 0 = 460$ V

$\quad E_{b2} = (460 - 32.45\, R_t)$ V

where R_t is controller resistance in series with the armature.

$$\text{Now} \quad \frac{N_2}{N_1} = \frac{E_{b2}}{E_{b1}} \times \frac{I_{a1}}{I_{a2}} \qquad (\because \; \Phi \propto I_a)$$

$$\therefore \quad \frac{3}{4} = \frac{460 - 32.45\, R_t}{460} \times \frac{50}{32.45}$$

$$R_t = \mathbf{7.28\ \Omega\ (approx)}$$

Tutorial Problems No. 13.2

1. The full-load efficiency of a 250-V, 7.46 kW, 1000 r.p.m., shunt motor is 80%. The field and armature resistance are 130 Ω and 0.4 Ω respectively. Calculate the value of resistance to be inserted in series with the armature circuit to reduce the speed to 800 r.p.m. with full-load torque being developed. **[1.34 Ω]**
2. A d.c. 4-pole shunt motor has a wave-wound armature with 738 conductors, flux per pole 0.02 Wb, armature circuit resistance 0.2 Ω and total contact drop of 2.5 V. When connected to a 500-V supply, the armature current is 20 A.
Calculate the value of the resistance which must be connected in series with the armature so that the motor speed is reduced to 700 r.p.m. when the armature current is 20 A. **[7.53 Ω]**

3. A 220-V d.c. shunt motor has an armature resistance of 0.5 Ω and runs at 850 r.p.m. when taking a full-load current of 32 A. The shunt field resistance is 110 Ω. Calculate the speed at which the motor-will run
(*a*) if a 1.5 Ω resistor were connected in series with the armature,
(*b*) if a 30 Ω resistor were connected in series with the field winding, the load torque remaining the same throughout. Assume that the field flux is proportional to the field current.
[(*i*) 664 r.p.m. (*ii*) 1062 r.p.m.]

4. A 500-V d.c. series motor has a total resistance of 0.2 Ω and when taking a current of 60 A its speed is 500 r.p.m. When taking 30 A, the flux is 65% of that at 60 A. Calculate under the new conditions (i) the speed, (*ii*) the new torque expressed as a percentage of the original torque.
[(*i*) 778 r.p.m. (*ii*) 32.5%]

5. A 200-V shunt motor running at 1000 r.p.m takes an armature current of 17-5 A. It is required to reduce the speed to 500 r.p.m. by the addition of a resistance in the armature circuit, the armature current being unaltered. What must be the magnitude of this resistance if the armature resistance is 0.4 Ω?
[5.51 Ω]

6. A 220-V d.c. shunt motor has a speed of 1200 r.p.m., an armature resistance of 0.2 Ω and negligible voltage drop at the brushes. The motor draws an armature current of 20 A when connected to a rated supply voltage for a given load. As the mechanical load is increased, the field flux is also increased by 15% and the armature current is found to rise to 45 A. Find (i) the back e.m.f. at 45 A load, (*ii*) the speed at 45 A load, (*iii*) the internal power developed at 20 A and 45 A loads respectively.
[(*i*) 211 V (*ii*) 1019 r.p.m. (*iii*) 4.32 kW; 9.495 kW] *(A.M.I.E. Summer 1984)*

HIGHLIGHTS

1. Two factors controlling the speed of a given motor are
   (*i*) $N \propto E_b$ and (*ii*) $N \propto 1/\Phi$
2. Based on these two factors, there are two main methods of speed control
   (*i*) Flux or field control method
   (*a*) For shunt motors, this method consists of putting a variable field rheostat in the shunt winding for changing the shunt current and hence the flux d.c. shunt motor.
   (*b*) For series motors, flux is changed by using a divertor across the series field windings.
3. In the armature or rheostatic control method, a variable controlling resistance is used in series with the armature circuit.
4. Flux control method is more economical and convenient whereas rheostatic control method is expensive and less efficient.
5. Flux control method gives speeds above the normal whereas rheostatic control method gives speeds below the normal.
6. Motor starters are essential for starting the motors in order to avoid excessive starting current.

OBJECTIVE TESTS—13

A. Fill in the following blanks

1. The speed of a d.c. shunt motor may be increased by .... its shunt field flux.
2. Flux control method gives maximum speed change for d.c. motors fitted with................ .
3. As controller resistance of a d.c. motor is decreased, its speed is ....
4. Rheostatic speed control method is unsuitable for .... changing loads.
5. The flux control method of varying the speed of a d.c. series motor requires a field....................... .
6. Armature divertor method of controlling the speed of a d.c. series motor gives speeds .... than the normal.
7. For a 4-pole d.c. series motor, the paral-

leling field coils method gives .... fixed speeds.

8. The flux control method of controlling the speed of a d.c. motor gives speeds .... the normal speed.

B. Answer True or False

1. Too much weakening of flux adversely affects the commutation of a d.c. motor not fitted with interpoles.
2. As compared to flux control method, the rheostatic speed control method is more efficient.
3. Armature control method gives speeds, higher than the normal speed of a d.c. motor.
4. Field divertors are used to control the speed of a d.c. series motor.
5. The tapped field control method is often used for series motors in electric traction.
6. Armature voltage control method for speed control of a d.c. motor involves considerable loss of power.
7. A motor starter is essential for starting a motor from rest irrespective of its rating.
8. Rheostatic speed control method is employed only when low motor speeds are required for short periods only.

C. Multiple ChoiceQuestions :

1. The only disadvantage of field control method for controlling the speed of a d.c. shunt motor is that it
   (*a*) gives speeds lower than the normal speed
   (*b*) adversely affects commutation
   (*c*) is wasteful
   (*d*) needs a large field rheostat.
2. The rheostatic speed control method is very
   (*a*) economical
   (*b*) efficient
   (*c*) unsuitable for rapidly changing loads
   (*d*) suitable for getting speeds above the normal.
3. The speed of d.c. series motors used for electric traction is usually changed by employing
   (*a*) tapped field control method
   (*b*) armature divertor
   (*c*) field divertors
   (*d*) variable resistance in series with armature.
4. Motor starters are essential for
   (*a*) accelerating the motor
   (*b*) starting the motor
   (*c*) avoiding excessive starting current
   (*d*) preventing fuse blowing.
5. The flux control method using paralleling of field coils when applied to a 4-pole d.c. series motor can give .... speeds.
   (*a*) 4 (*b*) 2
   (*c*) 8 (*d*) 3
6. The speed of a d.c. motor can be controlled by varying
   (*a*) its flux
   (*b*) armature circuit resistance
   (*c*) applied voltage
   (*d*) all of the above.

ANSWERS

A. 1. decreasing 2. interpoles 3. increased 4. rapidly 5. divertor 6. lower 7. three 8. above

B. 1. T 2. F 3. F 4. T 5. T 6. T 7. F 8. T

C. 1. *b* 2. *c* 3. *a* 4. *c* 5. *d* 6. *d*

14 CHEMICAL EFFECTS OF CURRENT

14.1. Types of Electric Conductors

In a very broad sense, conductors of electricity may be divided into two general classes as under :

(*i*) *Those conductors whose composition is not affected due to the passage of current through them.*

This class includes metals, alloys and liquids like mercury. Consider the case of copper wire. When current passes through it, there is no change in its physical and chemical states, except for a rise in temperature. This is so, because current flow in such metallic conductors is due to the movement of free electrons present in them and because the number of electrons entering at one end of the conductor is the same as that leaving it from the other end.

(*ii*) The second type of conductors are those *which undergo decomposition when electric current is passed through them.*

Such substances, which conduct an electric current and are, at the same time, decomposed by it, are known as *electrolytes.*

To this class belong acids (like H_2SO_4, HNO_3 and HCl etc.), bases (like NaOH, KOH etc.), salt solutions (like NaCl, $CuSO_4$ and $AgNO_3$ etc.) and fused salts. This phenomenon of conduction through fused salt is the basis of some important electrochemical industries, like the production of aluminium and caustic soda etc.

It has been found that the conductivity of electrolytes is due to the presence in them of small electrified particles, to which, Faraday gave the name of '*ions*' The process of ionization is discussed below.

14.2. Ionization or Dissociation

It is found that when a substance is dissolved in water, some of its molecules break up spontaneously into two groups (of atoms) *carrying equal but opposite charges.* These two groups of charged atoms are known as *ions.*

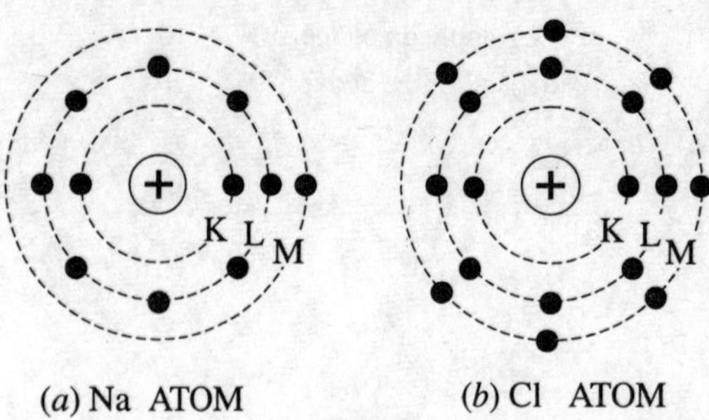

(*a*) Na ATOM (*b*) Cl ATOM

Fig. 14.1

Ions are composed of atoms or groups of atoms, but they differ from them in that they carry electric charge, positive or negative. This process of splitting up of molecules into negative and positive ions is known as *dissociation* or *ionization,* although it is referred to as *ionic* dissociation.

To understand the process of ionization, let us first consider the formation of a NaCl (sodium chloride) molecule. The atomic number of Na is 11 and that of Cl is 17. Hence, the number of electrons is 11 in Sodium and 17 in Chlorine and these electrons are arranged in their orbits, as shown in Fig. 14.I. The various orbits are numbered as *K, L, M* etc., starting from the orbit nearest to the nucleus. The maximum number of electrons that can occupy an orbit is given by $2n^2$ where n is the serial number of the orbit. Thus, the *K* orbit (n = 1) can hold a maximum number of two electrons, the *L* orbit (n = 2) 8 electrons and M orbit (n = 3) 18 electrons, and so on.

It is seen that the *valency* orbit of Na atom contains only *one* electron and that of Cl contains *seven* electrons. The orbits of both the atoms are incomplete, hence the atoms are chemically active. When these atoms are brought together, Na atom loses its outermost electron, thereby acquiring a positive charge, whereas Cl atom gains one electrons thereby acquiring a negative charge. This exchange of electrons results in the completion of the outermost orbits of both the atoms, hence they no longer show any chemical activity. But the electrostatic force existing between a positive sodium atom and a negative chlorine atom brings the two atoms together to form a sodium chloride (NaCl) molecule, as shown in Fig. 14.2. The formation of HCl may likewise be explained by the association of a hydrogen atom (which becomes positive due to loss of one electron) and a chlorine atom (which becomes negative due to the gain of one electron) because of electrostatic force existing between them.

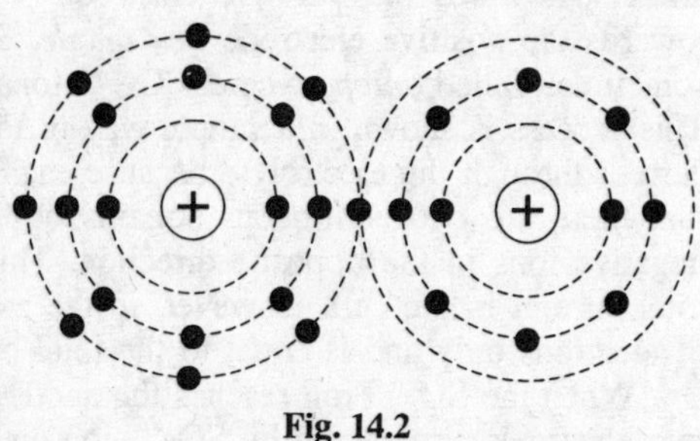

Fig. 14.2

When a NaCl molecule is dissolved in water, then the interatomic electrostatic force $\left[F = \frac{1}{4\pi \in_0 \in_r} \frac{Q_1 Q_2}{d^2}\right]$ is reduced to 1/80 of its value in air (or vacuum) because the relative permittivity of water $\in_r = 80$. This reduced force is too small to hold the two atoms together, hence the molecule breaks up under the action of normal thermal agitation (which is ever present) and now exists in solution as positively-charged sodium ion (Na^+) and negatively-charged chlorine ion (Cl^-) as in Fig. 14.3 (*b*). Similarly as shown, HCl is decomposed or dissociated into H^+ ions and Cl^- ions. *The electric charge associated with an ion is decided by its valency. Monovalent* ions are those which have gained or lost one electron each *i.e.*, sodium or chlorine or hydrogen ions. In the case of dissociation of a $CuSO_4$ (copper sulphate) molecule, Cu^{++} ions and SO_4^{--} ions are produced. Both these ions are *divalent*, because Cu atom loses two electrons and SO_4 gains two electrons.

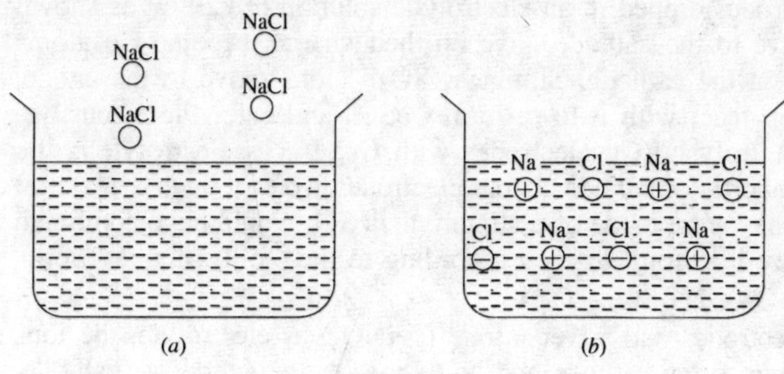

Fig. 14.3

Acids always dissociate into positive hydrogen H^+ ions and negative ions of acid radical.

Bases always dissociate into positive metal ions and negative hydroxyl ions (OH).

In the case of salts, the metallic elements like Na, Cu, Ag develop into positive ions, whereas non-metallic elements like Cl, SO_4 NO_3 etc. give negative ions.

14.3. Electrolysis

Electrolysis (electric analysis) was the name given by Faraday to the chemical decomposition which occurs in electrolytes when traversed by an electric current. Con-

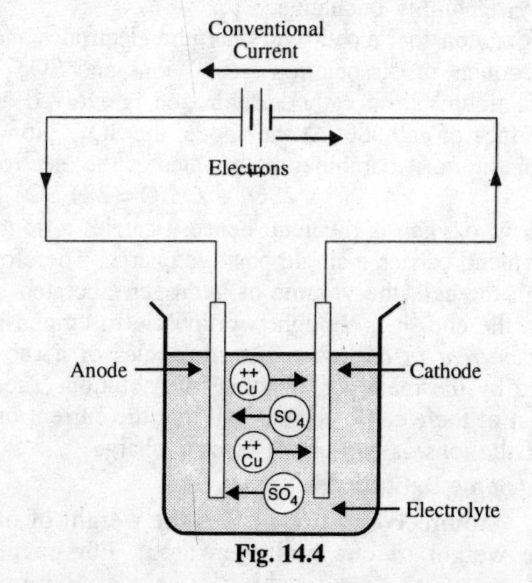

Fig. 14.4

sider a solution of $CuSO_4$ in water. As discussed above, the moment $CuSO_4$ is dissolved in water, its molecules split up into Cu^{++} ions and SO_4^{--} ions. If, now two Cu conductors or Cu electrodes are dipped in $CuSO_4$ solution and a p.d. is applied, then the positive copper ions (Cu^{++}) are attracted towards the negative electrode or *cathode* and the negative SO_4^{--} ions are attracted towards the positive electrode *i.e., anode.* Since SO_4^{--} ions move towards the anode, they are sometimes called *anions* whereas Cu^{++} ions which move towards the cathode are called *cations.* This process is shown in a simple way in Fig. 14.4. The movement of ions constitutes a flow of current through the electrolyte because each ion is a carrier of electricity. But it should be noted that there are two continuous streams of ions, positive ions moving in one direction and the negative ions in the opposite direction. This oppositely-directed movement of ions give us the total current in the cell. However, in the external circuit, the current is due to the axial motion of electrons only and is equal to the total current flowing in the electrolyte.

When the SO_4^{--} ion reaches the anode, it '*gives up its charge*' and ceases to be an ion. The two electrons surrendered by SO_4^{--} ion enter the anode and become part of the electron stream in the external circuit. Similarly, when Cu^{++} ion reaches the cathode, it makes up its deficiency of two electrons from the cathode.

14.4. Electrode Reactions

Now, let us see what happens to the ions when they give up their charges at their respective electrodes. It is found that if the material of the electrode is suitable, they enter into combination with it and if it is not, then they are removed by some reaction with the electrolyte or pass off as a gas or else in the case of metal ions, they are deposited in the metallic state on the electrode.

Take the case of two copper electrodes dipped in an electrolytic solution of $CuSO_4$ as shown in Fig. 14.4 above. The Cu^{++}ions move to the cathode, give up their charge, become Cu atoms and are deposited as metallic copper on the cathode. Similarly, SO_4^{--} ions move to the anode, give up their negative charge and then react with it to re-form $CuSO_4$ and enter the electrolye. Hence, the final result of such an electrolysis (Cu electrodes with $CuSO_4$, as electrolyte)is the transference of metallic copper from anode to cathode. If the electrodes were of such a material that SO_4 could not react with it *i.e.* if they were made of platinum, then SO_4 would have combined with the hydrogen of the water thereby liberating oxygen according to the following reaction

$$2SO_4 + 2H_2O \rightarrow 2H_2SO_4 + O_2$$

Similarly, in a cell with silver electrodes and silver nitrate ($AgNO_3$) as electrolyte, the ions are Ag^+ and NO_3^-, the Ag^+ ions moving to the cathode and NO_3^- ions to the anode. At cathode, Ag^+ ions leave the ionic state and are deposited as metallic silver on the surface of the cathode. At the anode, NO_3^- ions leave their ionic state and combine with the anode to form $AgNO_3$. So the cathode gains in weight exactly by the same amount what the anode loses, the electrolyte thereby remaining unchanged.

Next, consider a cell with platinum electrodes and a weak solution of H_2SO_4 as an electrolyte. The products of dissociation are H^+ ions and SO_4^{--} ions. These ions can form no compounds with platinum, so at cathode hydrogen is evolved as a gas and collects in the form of bubbles on the face of cathode. At the anode, the SO_4^{--} ion gives up its charge and since it cannot react with platinum, it combines with water of the electrolytic solution and liberates oxygen thus :

$$2SO_4 + 2\,H_2O = 2\,H_2SO_4 + O_2$$

Now, oxygen is bivalent, hence it carries a double negative charge whereas hydrogen, being monovalent, carries a single positive charge. Therefore, for a given quantity of electricity passing through the cell, the volume of hydrogen liberated is twice that of oxygen.

In the end, it is thought worthwhile to emphasize that *ions are not produced by the action of the current in any way.* The molecules of a substance in solution are already dissociated or ionized by the mere act of going into solution. The ionization is due to the action of water and not that of the electric current. All that the current or applied potential difference does is to make use of the ions as carricrs of electric charge.

14.5. Some Definitions

1. **Atomic Weight (A) :** It is the weight of one atom of the given element as compared to the weight of one hydrogen atom. For example, when we say that atomic weight of

sodium is 23, it simply means that one sodium atom is as heavy as 23 hydrogen atoms put together.

2. **Valency (v).** It measures the combining capacity of an atom. The valency of an atom is equal to the number of hydrogen atoms which the given atom can combine with or replace in any chemical reaction.

For example, in HCl, one atom of chlorine combines with one hydrogen atom. Hence, valency of chlorine is one.

Similarly, when aluminium (Al) reacts with HCl, it forms $AlCl_3$. In other words, one Al atom replaces *three* hydrogen atoms in HCl to from one molecule of $AlCl_3$. Hence, valency of Al is three.

3. **Chemical Equivalent Weight (E).** It is given by the ratio of the above two quantities.

$$\therefore \qquad E = \frac{A}{v}$$

4. **Electrochemical Equivalent Weight (Z).** It is equal to the mass of ions of a substance liberated by the passage of one coloumb charge through its electrolytic solution. Its unit is kg/C.

Also, $$Z = \frac{E}{F} = \left(\frac{1}{F} \cdot \frac{A}{v}\right)$$

5. **Faraday's Constant (F).** It is defined as the charge required to liberate electrolytically one gram-equivalent of any substance.

It is also given by the ratio of chemical equivalent weight (E) and the electro-chemical equivalent weight (Z).

$$\therefore \qquad F = \frac{E}{Z}$$

It is a universal constant and its value is 96,500 C/gram-equivalent (or 96.5×10^6 C/kg-equivalent)

6. **Gram-equivalent Weight.** It is the unit of mass. It is equal to the equivalent weight of a substance expressed in grams. Sliver has a chemical equivalent weight of 118. Hence, for silver one gram-equivalent means an amount of 118 gram. Also, one kg-equivalent would mean an amount of 118 kg.

14.6. Faraday's Laws of Electrolysis

From his experiments, Faraday deduced two fundamental laws which govern the phenomenon of electrolysis. These are :

1. **First Law.** The mass of an ion liberated at an electrode is directly proportional to the quantity of electricity which passes through the electrolyte.
2. **Second Law.** The masses of ions of different substances liberated by the same quantity of electricity are proportional to their chemical equivalent weights.

Explanation of the First Law

If $\quad m$ = mass of the ions liberated

Q = quantity of electricity passed

$= I \times t$ where I is the current and t is the time, then according to the first law

$$m \propto Q \quad \text{or} \quad m = ZQ \quad \text{or} \quad m = Z\,It$$

where Z is constant and is known as the electro-chemical equivalent (E.C.E.) of the substance.

If $Q = 1$ coulomb *i.e.* I = 1 amp and $t = 1$ second, then $m = Z$

Hence, E.C.E. of a substance is *equal to the mass of its ions liberated by the passage of one ampere current for one second through its electrolytic solution or by the passage of a charge of one coulomb.*

Faraday's First Law can also be expressed as given below :

$$m = \left(\frac{1}{F} \cdot \frac{A}{v}\right) It = \left(\frac{1}{F} \cdot \frac{A}{v}\right) Q = \frac{E}{F} \cdot Q$$

where E is the chemical equivalent weight ($= A/v$)

Explanation of the Second Law

Suppose an electric current is passed for the same time through acidulated water, solution of $CuSO_4$ and $AgNO_3$, then for every 1.0078 (or 1.008) gram of hydrogen evolved, 107.88 gram of silver and 31.54 gram of Cu are liberated. The values 107.88 and 31.54 the equivalent weights* of silver and copper respectively *i.e.* their atomic weights (as referred to hydrogen) divided by their respective valencies.

Example 14.1. *Calculate the quantity of electricity and the steady current required to deposit 5 gram of copper from copper sulphate solution in one hour. Electrochemical equivalent of copper is 0.3294 mg/C.*

Solution. $m = ZQ$

$5 = 0.3294 \times 10^{-3} \times Q$

$\therefore$ $Q = 5 \times 10^3/0.3294 =$ **15,180 coulomb**

Now $I = Q/t = 15{,}180/3600 =$ **4.27 A**

Example 14.2. *A steady current of 20 A is passed through a solution of copper sulphate for 30 minutes. Find the weight of copper deposited. The chemical equivalent for copper is 31.8 and E.C.E of hydrogen is 0.01044 miligram per coulomb.*

Solution. Let us first find out E.C.E. of copper

$$\frac{\text{E.C.E. of Cu}}{\text{E.C.E. of H}_2} = \frac{\text{Chemical Equivalent of Cu}}{\text{Chemical Equivalent of H}_2}$$

$$\therefore \text{ E.C.E. of Cu} = \text{E.C.E. of hydrogen} \times \frac{\text{C.E. of copper}}{\text{C.E. of hydrogen}}$$

$$= 0.01044 \times 10^{-3} \times \frac{31.8}{1} = 0.000332 \text{ g/C}$$

Now $m = ZIt = 0.000332 \times 20 \times (30 \times 60) =$ **11.95 g**

Example 14.3. *In a copper refinery, copper is deposited on the cathodes with a current density of 0.02 A/cm². Find approximately the time needed to deposit a layer of copper on the cathode 1 cm thick. The density of copper may be taken as 8.9 g/cm³ and the E.C.E. as 0.000328 g/C. If 0.3 V is needed to send the current through the cell, find the number of kWh needed to deposit one kg of copper.*

Solution. Let A cm² be the cross-section of the cathode.

Then $I = 0.02\,A$ amperes

Vol. of Cu $= A \times 1 = A\ cm^3$

Mass of Cu $= 8.9\,A$ gram

Using $m = ZIt$, we get

$t = m/ZI = 8.9\,A/0.000328 \times 0.02\,A = 1.357 \times 10^6$ seconds $=$ **377 hr.**

We know that a current of 0.02 A amperes is needed to deposit $8.9\,A \times 10^{-3}$ kg of Cu, therefore current required for depositing 1 kg is given by

$0.02\,A/8.9\,A \times 10^{-3} = 2.247$ A

$\therefore$ electric energy used $= 2.247 \times 0.3 \times 377/1000 =$ **0.254 kWh**

Example 14.4. *A coating of nickel 1 mm thick is to be built on a cylinder 20 cm in diameter and 30 cm in length in 2 hours. Calculate the electrical energy used in the process if the voltage is 10 V. Electro-chemical equivalent of nickel is 0.000304 g/C. Specific gravity of nickel is 8.9.*

Solution. Area of the curved surface to be coated

---

*The electro-chemical equivalents and chemical equivalents of different substances are inter-related thus:

$$\frac{\text{E.C.E of } A}{\text{E.C.E. of } B} = \frac{\text{chemical equivalent of } A}{\text{chemical equivalent of } B}$$

Further, it m_1 and m_2 are masses of ions deposited at or liberated from an electrode, E_1 and E_2 their chemical equivalents and Z_1 and Z_2 their electro-chemical equivalent weights, then

$$m_1/m_2 = E_1/E_2 = Z_1/Z_2$$

$= \pi D \times L = \pi \times 20 \times 30 = 1885 \text{ cm}^2$

Volume of nickel deposited $= 1885 \times 0.1 = 188.5 \text{ cm}^3$

Weight of nickel deposited $= 188.5 \times 8.9 = 1677.7$ gram

Now, $m = ZIt$; $\therefore$ $1677.7 = 0.000304 \times I \times 2 \times 3600$

$I = 1677.7/2 \times 3600 \times 0.000304 = 766.5$ A

Watt-hours reqd. $= VIt$ ... t in hours $= 10 \times 766.f5 \times 2 = 15{,}330$

$\therefore$ energy reqd. $= 15{,}330/1000 =$ **15.33 kWh**

Example 14.5. *A current of 0.5 A is passed through a dilute solution of sulphuric acid for 15 minutes. Given that E.C.E. of hydrogen is 10.44×10^{-6} g/C and that 1 gram of hydrogen occupies 11.2 litres at N.T.P., calculate the volume of hydrogen liberated at an atmospheric pressure of 750 mm of Hg at 20°C.* **(A.M.I.E. Winter 1989)**

Solution. $m = ZIt = 10.44 \times 10^{-6} \times 0.5 \times (15 \times 60) = 0.0047$ g

Volume at N.T.P. $= 0.0047 \times 11.2 = 0.0526$ litre $= 52.6 \text{ cm}^3$

This volume can be converted to the value for the given condition by using general gas equation.

$$\frac{P_1 V_1}{T_1} = \frac{P_2 V_2}{T_2}$$

$$\therefore \quad \frac{76 \times 52.6}{273} = \frac{75 \times V_2}{293}; \quad V_2 = \mathbf{57.23 \text{ cm}^3}$$

Tutorial Problems No. 14.1

1. A metal plate having a surface of 100 cm^2 is plated with silver 0.0019 mm thick. To deposit this silver, a current of 2.5 amperes is used for 12 minutes. Calculate the electrochemical equivalent of silver. Density of silver is 10.6 g/cm^3. **[0.001119 g/C]**
2. The atomic weights of sliver and copper are 108 and 63 respectively and E.C.E. of silver is 0.001118, calculate the E.C.E. of copper in cupric condition (valency = 2). Also, calculate the quantity of silver deposited in an hour by a current of 3 amperes. **[31.5×10^{-5} g/C ; 11.99 g]**
3. A metal plate having a surface of 120 cm^2 is to be silver-plated. What thickness of silver will be deposited if a current of 1.5 A is used for 2 hours. Density of silver is 10.6 and its E.C.E. = 0.001118 g/C. **[0.095 mm]**
4. An ammeter connected in series with a silver electrolyte cell reads 0.5 ampere and it is found that 0.322 gram of the metal deposits in a run lasting 10 minutes. What is the error of the ammeter at this point of its range (E.C.E. of silver = 1.118 mg/C). **[0.02 A]**
5. Electric energy is supplied to an electroplating company at 200 V, it electroplates 1013 gram of nickel per day by working 5 hours daily. If the E.C.E. of nickel be 20.26×10^{-5} g/C, calculate the power that the company is using. **[55.5 kW]**
6. A steady current was passed for 10 minutes through an ammeter in series with a silver voltameter and 3.489 gram of silver were deposited. The reading of the ammeter was 5 A. Calculate the percentage error. Electro-chemical equivalent of silver is 1.1183 milligram per coulomb. **[3.85%]**

14.7. Polarisation or Back E.M.F.

Let us consider the case of two platinum electrodes dipped in dilute sulphuric acid solution. When a small potential difference is applied across the electrodes, no current is found to flow. When, however, the applied voltage is increased, a time comes when a temporary flow of current takes place. The H^+ ions move towards the cathode and O^{--} ions move towards the anode and are absorbed there. These absorbed ions have a tendency to go back into the electrolytic solution, thereby leaving them as oppositely-charged electrodes. This tendency produces an e.m.f. which is in opposition to the applied voltage which is consequently reduced.

This opposing e.m.f. which is produced in an electrolyte due to the absorption of gaseous ions by the electrolyte from the two electrodes is known as the back e.m.f. of electrolysis or polarisation.

The value of this back e.m.f. is different for different electrolytes. The minimum voltage required to decompose an electrolyte is called the *decomposition* voltage for that electrolyte.

14.8. Value of Back E.M.F.

For producing electrolysis, it is necessary that the applied voltage must be greater than the back e.m.f. of electrolysis for that electrolyte. The value of this back e.m.f. of electrolysis can be found thus :

Let us, for example, find the decomposition voltage of water. We will assume that the energy required to separate water into its constituents (*i.e.* oxygen and hydrogen) is equal to the energy liberated when hydrogen and oxygen combine to form water. Let H be the amount of heat energy absorbed when 9 gram of water are decomposed into 1 gram of hydrogen and 8 gram of oxygen. If the electrochemical equivalent of hydrogen is Z gram/coulomb, then a passage of Q coulomb liberates ZQ gram of hydrogen. Now, H is the heat energy required to release 1 gram of hydtogen. hence for releasing ZQ gram of hydrogen, heat energy required is HZQ calories or $JHZQ$ joules. If E is the decomposition voltage, then energy spent in circulating Q coulombs of charge is EQ joules. Equating the two amounts of energies, we have

$$EQ = JHZQ \quad \text{or} \quad E = JHZ$$

where J is 4.2 joules/cal.

The e.m.f. of a cell can be calculated by determining the two electrode potentials. The electrode potential is calculated on the assumption that the electrical energy comes entirely from the heat of the reactions of constituents. Let us take a zinc electrode. Suppose, it is given that 1 gram of zinc when dissolved liberates 540 calories of heat and that the electro-chemical equivalent of zinc is 0.000338 gram/coulomb. As calculated above,

$$E = JHZ = 4.2 \times 540 \times 0.000338 = 0.76 \text{ volt}$$

The electrode potentials are usually referred to in terms of the potential of a standard hydrogen electrode *i.e.* an electrode of hydrogen gas at normal atmospheric pressure and in contact with a normal acid solution. In Table 14.1 are given the electrode potentials of various elements as referred to the standard hydrogen electrode. The elements are assumed to be in normal solution and at atmospheric pressure.

In the case of Daniel cell having copper and zinc electrodes, copper electrode potential with respect to hydrogen ion is + 0.345 V and that of the zinc electrode is – 0.758 V. Hence, the cell e.m.f. is = 0.345 – (– 0.758) = 1.103 volt. The e.m.f. of other primary cells can be found in a similar way.

Table No. 14.1

Electrodes	*Potential (volts)*
Cadmium	– 0.398
Copper	+ 0.345
Hydrogen	0
Iron	– 0.441
Lead	– 0.122
Mercury	+ 0.799
Nickel	– 0.231
Potassium	– 2.922
Silver	+ 0.80
Zinc	– 0.758

Example 14.6. *One gram of hydrogen on burning to form water yields 34,500 calories and E.C.E. of hydrogen is* 1.05×10^{-5} *gram per coulomb. What is the minimum e.m.f. necessary to decompose water?* $J = 4.2$ *joules/calorie.* **(A.M.I.E. Summer 1993)**

Solution. The decomposition voltage is given by

$$E = JHZ = 4.2 \times 34{,}500 \times 1.05 \times 10^{-5} = \mathbf{1.49\ V}$$

Example 14.7. *Calculate the weight of zinc and* MnO_2 *required to produce 1 ampere–hour in a Leclanche cell.*

Atomic weight : Mn, 55 ; O, 16 ; Zn, 65. ;   E.C.E. of hydrogen = 0.0000104 g/C

Solution. 1 ampere–hour = 3600 A–s = 3600 coulombs

Wt. of hydrogen liberated = $ZQ = 0.0000104 \times 3600 = 0.03744$ g

Now, the chemical reactions in the cell are,

$$Zn + 2\,NH_4Cl = ZnCl_2 + 2NH_3 + H_2$$

It is seen that 1 *atom* of zinc is used up in liberating *two atoms* of hydrogen. In other words, to produce 2 gram of hydrogen, 65 gram of zinc will have to go into chemical combination.

∴ zinc required to produce 0.03744 gram of hydrogen

$$= 0.03744 \times 65/2 = 1.217 \text{ gram}$$

The hydrogen so liberated combines with manganese dioxide as under :

$$2MnO_2 + H_2 = H_2O + Mn_2O_3$$

Atomic weight of MnO_2 = $2(55 + 16 \times 2) = 174$

It is seen that 174 gram of MnO_2 combine with 2 gram of hydrogen, hence weight of MnO_2 needed to combine with 0.03744 gram of hydrogen

$$= 0.03744 \times 174/2 = 3.258 \text{ gram}$$

Hence, for 1 ampere–hour, **1.217** gram of zinc and **3.258** gram of MnO_2 are needed.

14.9. Storage Cells–Definition

The function of 'storage' cells is to convert electrical energy into chemical energy during the process known as 'charging' and the reverse of it when 'discharging'.

Diring charging of the cell, when current is passed through it, certain chemical changes take place in the active materials of the cell. Such chemical changes absorb energy during their formation. When these chemical reactions are completed and the electric current produces no further chemical changes, the cell is said to be fully charged.

When the cell is next connected to an external circuit, the active materials of the cell revert to their original condition, thereby reversing the changes which occurred during charging. In this process of undoing the chemical changes, absorbed energy is released in the form of electric current, the process being known as discharging.

It should be noted that the cell does not 'store' electricity as such but absorbs electric energy in the form of chemical energy, the whole process being reversible.

We will discuss two types of storage cells or accumulators or secondary cells *i.e.* Lead–acid cell and Edison alkali cell.

14.10. Materials of a Lead–acid Cell

Those substances of the cell which take active part in chemical combination and hence absorb or produce electrictity during charging or discharging, are known as the **active materials** of the cell.

The active materials of a lead acid cell are :

1. *Lead Peroxide* (PbO_2) for +ve plate
2. *Sponge Lead* (Pb) for –ve plate
3. *Dilute Sulphuric Acid* (H_2SO_4) as electrolyte

(i) Lead Peroxide

It is a combination of lead and oxygen, is dark chocolate brown in colour and is quite hard but brittle substance. It is made up of one atom of lead (Pb) and two atoms of oxygen (O_2) and its chemical formula is PbO_2. As said earlier, it forms the positive active material.

(ii) Sponge Lead

It is pure lead in soft spongy or porous condition. Its chemical formula is Pb and forms the negative active material.

(iii) Dilute Sulphuric Acid

It is approximately 3 parts water and one part sulphuric acid. The chemical formula of the acid is H_2SO_4. The positive and negative plates are immersed in this solution which is known as 'electrolyte'. It is this medium through which the current produces chemical changes.

Hence, the lead–acid cell depends for its action on the presence of two plates of PbO_2 and Pb in a solution of dilute H_2SO_4 of specific gravity 1.21 or so.

Lead in the form of PbO_2 or sponge Pb has very little mechanical strength, hence it is supported by plates of pure lead. Those plates covered with or otherwise supporting PbO_2, are known as +ve plates and those supporting spongy lead are called −ve plates. The +ve and −ve plates are arranged alternately and are connected to two common +ve and −ve terminals. These plates are assembled in a suitable jar or container to make a complete cell.

14.11. Chemical Changes

Following chemical changes take place during the charging and discharging of a lead–acid cell.

Discharging (Fig. 14.5).

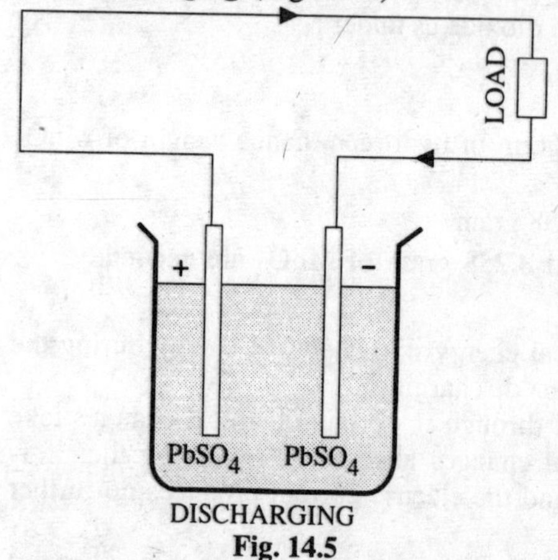

DISCHARGING
Fig. 14.5

When the cell is fully charged, its positive plate or anode is PbO_2 (dark chocolate brown) and the negative plate or cathode is Pb (slate grey). When the cell discharges *i.e.* it sends current through the external load, then H_2SO_4 is dissociated into positive H_2 ions and negative SO_4 ions. As the current within the cell is flowing from cathode to anode, H_2 ions move to anode and SO_4 ions move to the cathode.

At anode (PbO_2), H_2 combines with the oxygen of PbO_2 and H_2SO_4 attacks lead to form $PbSO_4$

$$PbO_2 + H_2 + H_2SO_4 \rightarrow PbSO_4 + 2H_2O$$

At the cathode (Pb), SO_4 combines with it to form $PbSO_4$

$$Pb + SO_4 \rightarrow PbSO_4$$

It will be noted that during discharging,

1. *Both anode and cathode become $PbSO_4$ which is somewhat whitish in colour*
2. *Due to formation of water, specific gravity of the acid decreases.*
3. *Voltage of the cell decreases.*
4. *The cell gives out energy.*

Charging (Fig. 14.6)

When the cell is re-charged, H_2 ions move to cathode and SO_4 ions go to anode and the following changes take place.

At cathode, $PbSO_4 + H_2 \rightarrow Pb + H_2SO_4$

At anode, $PbSO_4 + SO_4 + 2H_2O \rightarrow PbO_2 + 2H_2SO_4$

Hence, the anode and cathode again become PbO_2 and Pb respectively.

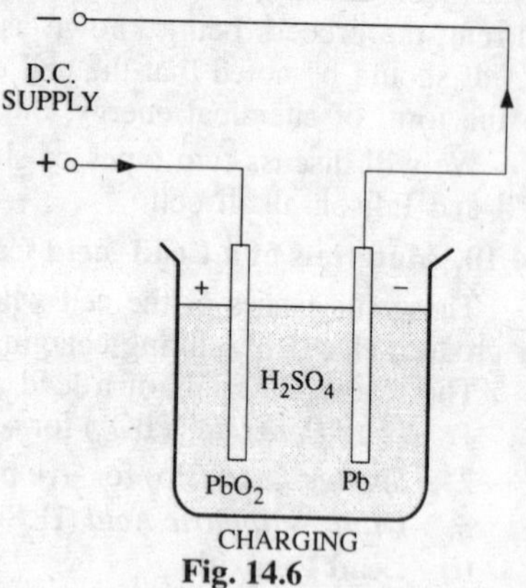

CHARGING
Fig. 14.6

It will be noted that during charging :–

1. *The anode becomes dark chocolate brown in colour (PbO_2) and cathode becomes grey metallic lead (Pb).*
2. *Due to consumption of water, specific gravity of H_2SO_4 is ncreased.*
3. *There is a rise in voltage.*
4. *Energy is absorbed by the cell.*

The charging and discharging of the cell can be represented by a single reversible equation given below :

$$\underset{\text{Pos. Plate}}{PbO_2} + 2H_2SO_4 + \underset{\text{Neg. Plate}}{Pb} \underset{\text{Charge}}{\overset{\text{Discharge}}{\rightleftharpoons}} \underset{\text{Pos. Plate}}{PbSO_4} + 2H_2O + \underset{\text{Neg. Plate}}{PbSO_4}$$

For discharge, the equation should be read from left to right and for charge from right to left.

14.12. Formation of Plates

There are, in general, two methods of producing the active materials of the cell and attaching them to the lead plates. These are known after the names of their inventors (*i*) **Plante** plates or formed plates and (*ii*) **Faure** plates or pasted plates.

14.13. Plante Process

In this process, two sheets of lead are taken and immersed in dilute H_2SO_4. When a current is passed into this lead–acid cell from a dynamo or some other external source of supply, then due to electrolysis, hydrogen and oxygen are evolved. At anode, oxygen attacks lead converting it into PbO_2 whereas cathode is unaffected because hydrogen can form no compound with Pb.

If the cell is now discharged (or current is reversed through it) then peroxide coated plate becomes cathode, so hydrogen forms on it and combines with the oxygen of PbO_2 to form water thus

$$PbO_2 + 2H_2 \rightarrow Pb + 2H_2O$$

At the same time, oxygen goes to anode (the plate previously unattacked)which is lead and reacts to form PbO_2. Hence, the anode becomes covered with a thin film of PbO_2.

By continuous reversal of the current or by charging and discharging the above electrolytic cell, the thin film of PbO_2 will become thicker and thicker and the polarity of the cell will take increasingly longer time to reverse. Two lead plates after being subjected to hundreds of reversals will acquire a skin of PbO_2 thick enough to possess sufficiently high capacity. This process of making positive plates is known as *formation.* The negative plates are also made by the same process. They are turned from positive to negative plates by reversing the current through them until whole PbO_2 is converted into spongy lead. Although Plante positives are very commonly used for stationary work, Plante negatives have been completely replaced by the Faure or pasted type plates. However, owing to the length of time required and enormous expenditure of electrical energy, this process is commercially impracticable. The process of formation can be accelerated by forming agents such as acetic, nitric or hydrochloric acid or their salts but still this method is expensive and slow and plates are heavy.

14.14. Internal Resistance and Capacity of a Cell

The secondary cell possesses internal resistance due to which some voltage is lost in the form of potential drop across it when current is flowing. Hence, the internal resistance of the cell has to be kept to the minimum.

One obvious way to lessen internal resistance is to increase the size of the plates. However, there is a limit to this because the cell will become too big to handle. hence, in practice, it is usual to multiply the number of plates inside the cell and to join all the negatives together and all the positives together.

The effect is equivalent to joining many cells in parallel. At the same time, the length of the electrolyte between the electrodes is decreased with a consequent reduction in the internal resistance.

The 'capacity' of a cell is given by the product of current in amperes and the time in hours during which the cell can supply current until its e.m.f. falls to 1.8 volts. It is expressed in ampere–hours (Ah).

The interlacing of plates not only decreases the internal resistance but, additionally, increases the capacity of the cell also. There is always one more negative plate than the positive plates *i.e.* there is a negative plate at both ends. This gives not only more mechanical strength but assures that both sides of a positive plates are used.

Since in this arrangement, the plates are quite close to each other, something must be done to make sure that a positive plate does not touch the negative plate otherwise an internal short–circuit will take place. This separation between the two plates is achieved by using separators which, in the case of small cells, are made of wood, ebonite or hard rubber and in the case of large stationary cells, are in the form of glass rods.

14.15. Two Efficiencies of the Cell

The efficiency of a cell can be considered in two ways :

1. **The quantity or ampere-hour (Ah) efficiency**
2. **The energy or watt-hour (Wh) efficiency**

The Ah efficiency does not take into account the varying voltages of charge and discharge. The Wh efficiency does so and is always less than Ah efficiency because average p.d. during discharging is less than that during charging. Usually, during discharge, the e.m.f. falls from about 2.1 V to 1.8 V whereas during charge, it rises from 1.8 volts to about 2.6 V.

$$\therefore \quad \text{Ah. eff.} = \frac{\text{ampere–hours on discharge}}{\text{ampere–hours on charge}}$$

The Ah efficiency of a lead–acid cell is normally between 90 to 95% meaning that about 100 Ah must be put back into the cell for every 90 – 95 taken out of it.

If Ah efficiency is given, Wh efficiency can be found from the following relation :

$$\text{Wh efficiency} = \text{Ah efficiency} \times \frac{\text{average volts on discharge}}{\text{average volts on charge}}$$

The Wh efficiency varies between 72–80%.

From the above, it is clear that anything that increases the charge volts or reduces the discharge volts will decrease Wh efficiency. Because high charge and discharge rates will do this, it is advisable to avoid these.

14.16. Electrical Characteristics of the Lead-acid Cell

The three important features of an accumulator, of interest to an engineer, are (*i*) voltage, (*ii*) capacity and (*iii*) efficiency.

1. Voltage

The open-circuit voltage of a fully-charged cell is approximately 2.1 volts. This value is not fixed but depends on (*a*) length of time since it was last charged (*b*) specific gravity—voltage increasing with increase in specific gravity and *vice–versa*. If specific gravity comes near to density of water *i.e.* 1.00, then voltage of the cell will disappear altogether (*c*) temperature-voltage increases (through not much) with increase in temperature.

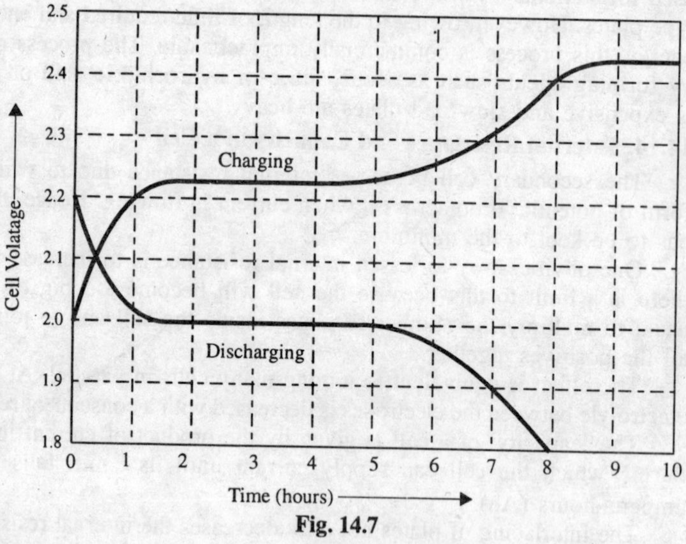

Fig. 14.7

CHARGE AND DISCHARGE VOLTAGE CURVES

The variation in the terminal p.d. of a cell on charge and discharge are shown in Fig. 14.7. The voltage fall depends on the rate of discharge. Rates of discharge are generally specified by the number of hours during which the cell will sustain the rate in question before falling to 1.8 V. The voltage falls rapidly in the beginning (rate of fall depending on the rate of discharge), then very slowly up to 1.85 V and again suddenly to 1.8 V. The voltage should not be allowed to fall to lower than 1.8 V, otherwise hard insoluble lead sulphate is formed on the plates which increases the internal resistance of the cell.

The general from of the voltage-time curves corresponding to 1–, 3–, 5– and 10 – hour rates of discharge, are shown in Fig. 14.8 *i.e.* corresponding to the steady currents which would discharge the cell in the above mentioned times (in hours). It will be seen that both the terminal voltage and the rate at which the voltage falls depends on the rate of discharge. The more rapid fall in voltage at higher rates of discharge is due to rapid increase in the internal resistance of the cell.

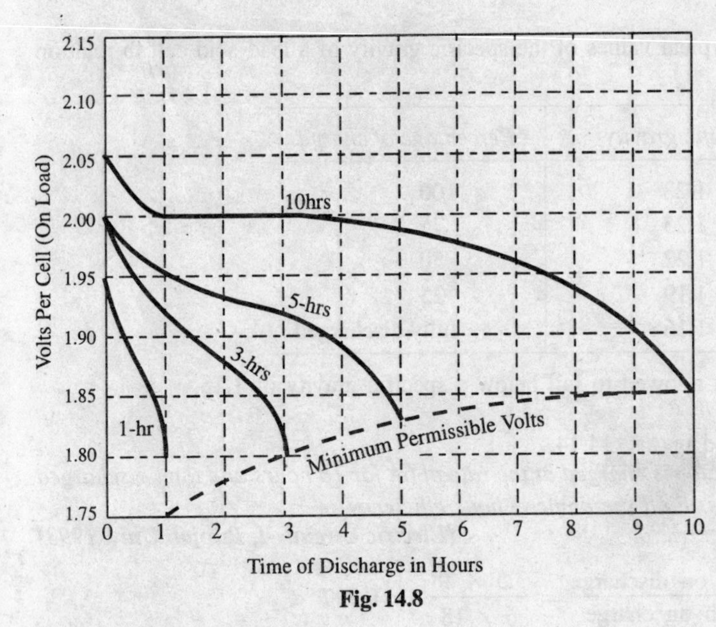

Fig. 14.8

During charging, the p.d. increases (Fig. 14.7). The curve is similar to the discharge curve reversed but is everywhere higher due to the increased density of H_2SO_4 in the pores of the positive plates.

2. Capacity

It is measured in ampere-hours (Ah). One ampere-hour (Ah) is the amount of electricity conveyed by one ampere in one hour.

The capacity is always given at a specified rate of discharge (10-hour discharge in U.K., **8**-hour discharge in U.S.A.). The capacity of a cell depends on the amount of the active material on its plates. In other words, it depends on the size and thickness of the plates However, for a given battery, the capacity is affected by the following factors:

(*i*) *Rate of Discharge.* The capacity of a cell, as measured in Ah, depends on the discharge rate. It decreases with increased rate of discharge. Rapid rate of discharge means greater fall in p.d. of the cell due to internal resistance of the cell. Moreover, with rapid discharge, the weakening of the acid in the pores of the plates is also greater. Hence, the chemical change produced at the plates by 1 ampere for 10 hours is not the same as produced by 2 amperes for 5 hours or 4 amperes for 2.5 hours. It is found that a cell having a 100 Ah capacity at 10-hour discharge rate, has its capacity reduced to 82.5 Ah at 5-hour rate and 50 Ah at 1-hour rate. The variation of capacity with discharge rate is shown in Fig. 14.9.

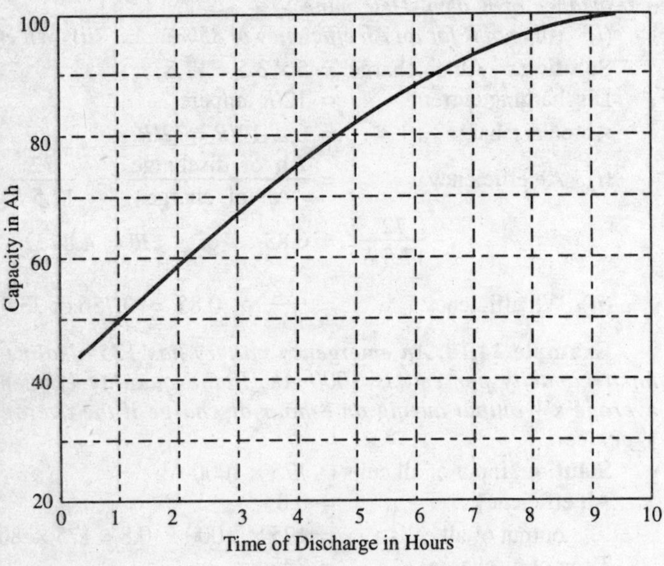

Fig. 14.9

(*ii*) *Temperature.* Capacity increases with increase in temperature, the increase in capacity being more marked at higher rates of discharge. This is due to the fact that at higher temperatures (*a*) chemical action is more vigorous (*b*) the resistance of the acid decreases and (*c*) there is better diffusion of the electrolyte. With decrease in temperature, available voltage and capacity decrease until at freezing point, the capacity is zero even when the cell is fully charged.

(*iii*) *Density of electrolyte.* As the density of electrolyte affects the internal resistance and the vigour of chemical reaction, it has an important effect on the capacity. Capacity increases with the density.

The following table gives typical values of the specific gravity of a lead-acid cell in relation to charge.

Specific gravity	*Percentage of charge*
1.28	100
1.25	75
1.22	50
1.19	25
1.16	fully discharged

In practice, the cells are not allowed to fall below a specific gravity of 1.18.

3. Efficiency

It has already been discussed in Art. 14.14.

Example 14.8. *An accumulator is charged at the rate of 6A for 18 hours and then discharged at the rate of 3.5 A for 28 hours. Find the ampere–hour efficiency.*

(Electric Circuits-I, Punjab Univ. 1993)

Solution. $\text{Ah } \eta = \dfrac{\text{Ah on discharge}}{\text{Ah on charge}} = \dfrac{3.5 \times 28}{6 \times 18} = \mathbf{0.907}$

Example 14.9. *A discharged battery is put on charge at 5 A for $3\frac{1}{2}$ hours at a mean charging voltage of 13.5 V. It is then discharged in 6 hours at a constant terminal voltage of 12 V through a resistance of R ohm. Determine :*

(i) value of R for an Ah efficiency of 85%. *(ii) Wh efficiency of the battery.*

Solution. Ah on charge $= 5 \times 3.5 = 17.5$

Discharging current $= 12/R$ ampere

Ah on discharge $= 6 \times 12/R = 72/R$

(*i*) Ah efficiency $= \dfrac{\text{Ah of discharge}}{\text{Ah of charge}} = \dfrac{72}{17.5\,R}$

$\therefore \quad \dfrac{72}{17.5\,R} = 0.85 \quad \text{or} \quad R = \mathbf{4.84\ \Omega}$

(*ii*) Wh efficiency $= \dfrac{12}{13.5} \times 0.85 = 0.756$ or **75.6%**

Example 14.10. *An emergency battery has 175 alkaline cells connected in series and the input capacity per cell is 1000 Ah. If the quantity efficiency is 80 per cent, calculate the average kW output during an 8-hour discharge if the average voltage during discharge is 1.2 V per cell.*

Solution. Input of all cells $= 175 \times 1000$ Ah

Ah efficiency $= 0.8$

$\therefore$ output of all cells $= 175 \times 1000 \times 0.8 = 175 \times 800$ Ah

Time of discharge $= 8$ hours

Out put current $= 175 \times 800/8 = 17{,}500$ A

Average discharge voltage $= 1.2$ V

$\therefore$ power output $= 17{,}500 \times 1.2$ watt $=$ **21 kW**

14.17 Indications of a Fully-charged Cell

The indications of a fully–carged cell are :

1. *gassing* 2. *voltage* 3. *specific gravity* 4. *colour of plates.*

1. Gassing

When the cell is fully charged, it freely gives off hydrogen at cathode and oxygen at the anode, the process being known as 'gassing'. Gassing at *both* plates indicates that the current is no longer doing any useful work and hence should be stopped.

2. Voltage

The voltage ceases to rise when the cell becomes fully charged. The value of the voltage of a fully-charged cell is a variable quantity being affected by the rate of charging, the temperature and specific gravity of the electrolyte etc. The approximate value of the e.m.f. is 2.1 V or so.

3. Specific Gravity of the Electrolyte

A third indication of the state of charge of a battery is given by the specific gravity of the electrolyte. We have seen from the chemical equations of Art. 14.11, that during discharging, the density of electrolyte decreases due to the production of water, whereas it increases during charging due to the absorption of water. The value of density when the cell is fully charged is 1.21 and 1.18 when discharged up to 1.8 V. Specific gravity can be measured with a suitable hydrometer.

4. Colour

The colour of plates, on full charge, is deep chocolate brown for positive plates and slate grey for negative plates and the cell looks quite brisk and alive.

14.18. Applications of Lead-acid Batteries

Storage batteries are, these days, used for a great variety and range of purposes, some of which are summarised below:

1. In central stations for supplying the whole load during light load periods, also to assist the generating plant during peak load periods for providing reserve emergency supply during periods of plant breakdown and finally, to store energy at times when load is light for use at times when load is at its peak value.
2. In private generating plants both for industrial and domestic use, for much the same purpose as in Central Stations.
3. In sub-stations, they assist in maintaining the declared voltage by meeting a part of the demand and so reducing the load on and the voltage drop in the feeder during peak-load periods.
4. As a power source for industrial and mining battery locomotives and for road vehicles like cars and trucks.
5. As a power source for submarines when submerged.
6. For petrol motor-car starting and ignition etc.
7. As a low-voltage supply for operating purposes in many different wahys such as high–tension switchgear, automatic telephone exchange and repeater stations, broadcasting stations and for wireless receiving sets.
8. Semi-sealed portable lead-acid batteries find many applications such as in electronic cash registers, alarm systems, cordless TV sets, mini-computers and terminals, electronically-controlled petrol pumps, portable instruments and tools etc.

14.19. Charging Systems

In various installations, batteries are kept 'floating' on the line and are so connected that they are being charged when load demands are light and automatically discharged during peak periods when load demands are heavy or when the usual power supply fails or is disconnected. In some other installations, the battery is connected to the feeder circuit as and when desired, allowed to discharge to a certain point, then removed and re–charged for further requirements.

For batteries other than the 'floating' and 'system–governed' type, the following two general methods (though there are some variations of these) are employed :

1. *The constant-current system ;* 2. *The constant-voltage system*

14.20. Constant Current System

In this method (Fig. 14.10), the charging current is kept constant by varying the supply voltage to overcome the increased back e.m.f. of cells. If a charging booster (which is just a shunt dynamo directly driven by a motor) is used, then current supplied by it can be kept constant by adjusting its excitation. If charged on a d.c. supply, then current is controlled by varying the rheostat connected in the circuit. The value of charging current should be so chosen that there

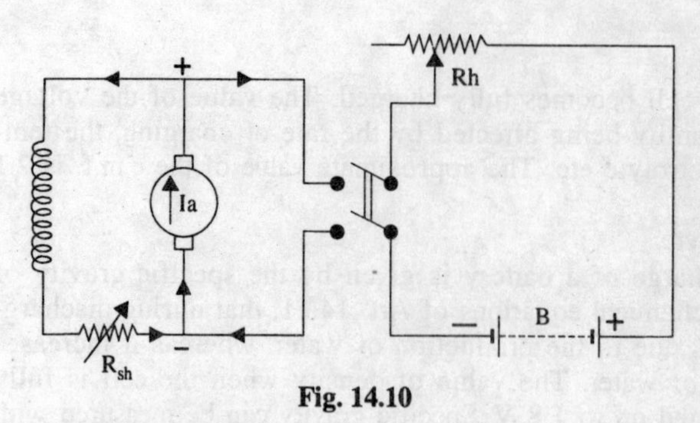

Fig. 14.10

would be no excessive gassing during final stages of charging and also the cell temperature does not exceed 45°C. The method takes a comparatively longer time.

14.21. Constant Voltage System

In this method (Fig. 14.11), the voltage is kept constant but it results in very large charging current in the beginning when the back e.m.f. of the cells is low and a small current when their back e.m.f. increases on being charged. This automatic decrease in charging current as cells near their full-charge condition, is the most satisfactory feature of this system.

With this method, time of charging is almost reduced to half. It increases the capacity by approximately 20% but reduces the efficiency by 10% or so.

Calculations

When a secondary cell or a battery of such cells is being charged, the e.m.f. of the cells acts in opposition to the applied voltage. If V is the supply voltage which sends a charging current of I against the back e.m.f. E_b, then input is VI but the work done in overcoming the opposition (Fig 14.11) is $E_b I$. This work $E_b I$ minus that lost as heat due to circuit resistance is converted into the chemical energy which is stored in the cell. The charging current can be found from the following equation.

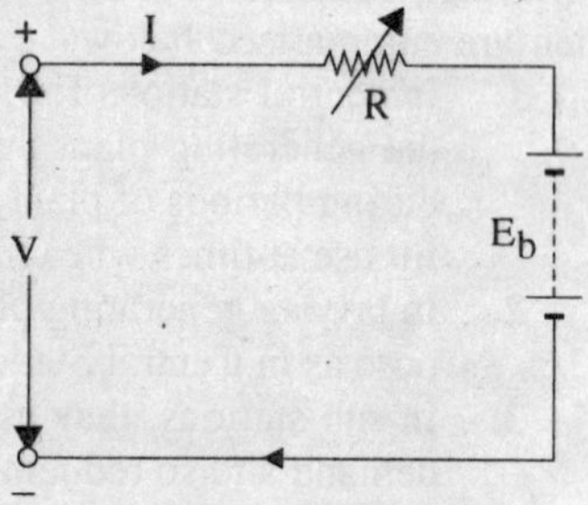

Fig. 14.11

$$I = \frac{V - E_b}{R}$$

where V = supply voltage (assumed constant)
E_b = back e.m.f. of the battery
R = total circuit resistance including internal resistance of the battery
I = charging current

By varying R, the charging current can be kept constant throughout.

14.22 Battery Charging from AC Source

Cells and batteries are charged by connecting them to a controlled dc source. This source may be obtained in many different ways : (*i*) motor generator set (*ii*) full–wave rectified ac (*iii*) rotary convertor and (*iv*) dc mains supply. The most commonly used method is full-wave rectified ac. which is used for both constant-voltage charging and constant-current charging.

1. Constant Voltage Charging

The circuit is shown in Fig. 14.12. It consists of a two-winding transformer and a full-wave bridge rectifier consisting of four diodes. The battery to be charged is connected across the rectifier circuit, as shown.

The charging voltage is kept constant with the help of variable resistance R. This is the most popular method for everyday use.

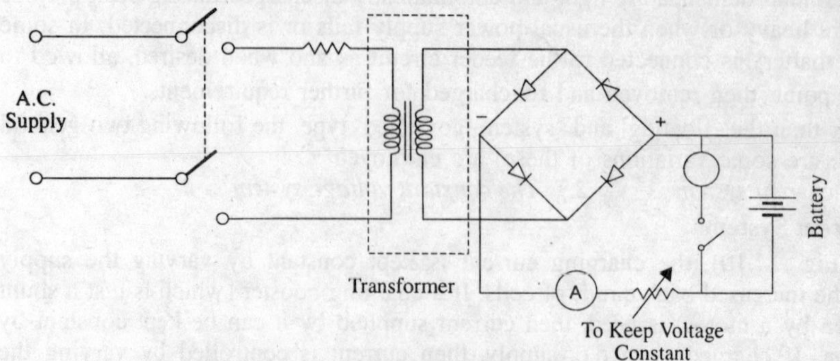

Fig. 14.12

2. **Constant Current Charging**

The rectifier circuit is the same as that used for constant voltage charging. However, in this case the resistance R is varied to keep the charging current constant even as the battery emf increases. As shown in Fig. 14.13, the battery is connected across similar terminals of the rectifier output circuit.

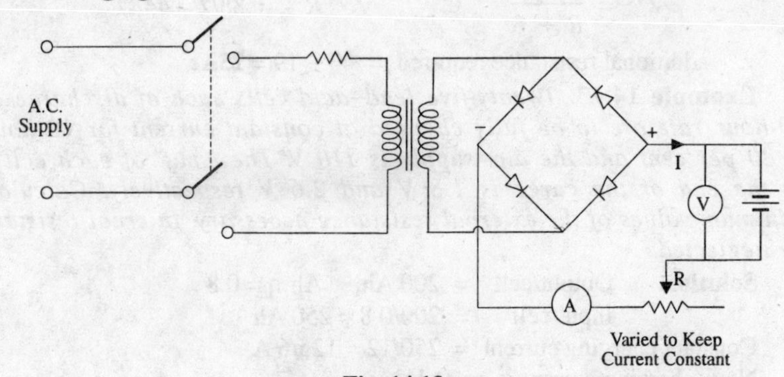

Fig. 14.13

Cells, batteries and their associated charging equipment are frequently used in installation work such as indicator and call systems in hospitals and hotels, fire alarm and burglar alarm systems and emergency lighting installation.

Example 14.11. *A battery of accumulators of e.m.f. 50 volts and internal resistance 2 Ω is charged on 100-volt direct mains. What series resistance will be required to give a charging current of 2 A ?*

If the price of energy is 50 paise per kWh, what will it cost to charge the battery for 8 hours and what percentage of energy supplied will be used in the form of heat?

Solution. Applied voltage $= 100$ V; $E_b = 50$ V

Net charging voltage $= 100 - 50 = 50$ V

Let R be the required resistance, then,

$$2 = \frac{50}{R+2} \qquad \therefore \quad R = \frac{46}{2} = 23\ \Omega$$

Input for eight hours $= 100 \times 2 \times 8 = 1600$ Wh $= 1.6$ kWh

Cost $= 1.6 \times 50 =$ **80 paise**

Power wasted on total resistance $= 25 \times 2^2 = 100$ W

Total input $= 100 \times 2 = 200$ W

Percentage waste $= 100 \times 100/200 =$ **50%**

Example 14.12. *A battery of 60 cells is charged from a supply of 250 V. Each cell has an e.m.f. of 2 volts at the start of charge and 2.5 V at the end. If internal resistance of each cell is 0.1 Ω and if there is an external resistance of 1.9 Ω in the circuit, calculate (a) the initial charging current (b) the final charging current and (c) the additional resistance which must be added to give a finishing charge at 2 A rate.*

Solution. Supply voltage

$V = 250$ V

Back e.m.f. E_b of the battery

at start $= 60 \times 2 = 120$ V and

at the end $= 60 \times 2.5 = 150$ V

Internal resistance of the battery $= 60 \times 0.1 = 6\ \Omega$

Total circuit resistance $= 19 + 6 = 25\ \Omega$

(*a*) Net charging voltage at start $= 250 - 120 = 130$ V

$\therefore$ initial charging current $= 130/25 =$ **5.2 A**

(*b*) Final charging current $= 100/25 =$ **4 A**

(*c*) Let R be the external resistance, then

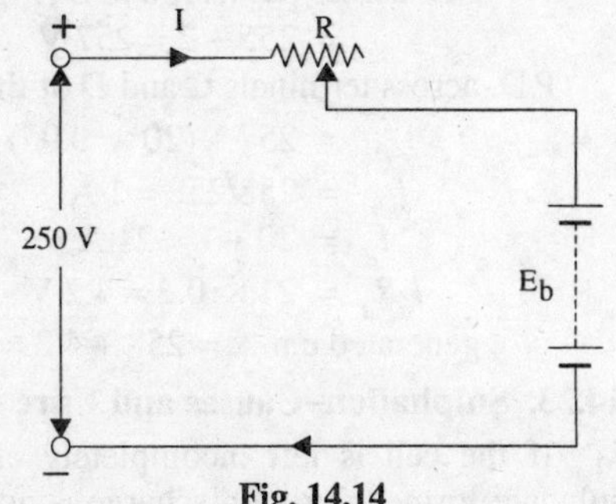

Fig. 14.14

$$2 = \frac{100}{R + 6} \qquad \therefore \quad R = 88/2 = 44\ \Omega$$

$\therefore$ additional resistance required = 44 – 19 = **25 Ω**

Example 14.13. *Twenty-five lead–acid cells each of discharge capacity of 200 Ah at 10-hour rate are to be fully charged at constant current for 12 hours. The Ah efficiency is 80 per cent and the d.c. supply is 110 V. The e.m.f. of each cell at the beginning and at the end of the carge is 1.8 V and 2.6 V respectively. Calculate the maximum and minimum values of the external resistance necessary. Internal resistance of the cells should be neglected.*

Solution. Output/cell = 200 Ah, Ah $\eta = 0.8$

Input/cell = 200/0.8 = 250 Ah

Constant charging current = 250/12 = 125/6 A

Now, charging current is given by

$$I = \frac{V - E_b}{R} \qquad \text{— Art 14.20}$$

At the beginning of charge

$$E_b = 25 \times 1.8 = 45 \text{ V}$$

$$\therefore \quad \frac{125}{6} = \frac{110 - 45}{R} \qquad \text{or} \qquad R = \mathbf{3.12\ \Omega}$$

At the end of the charge

$$E_b = 25 \times 2.6 = 65 \text{ V}$$

$$\therefore \quad \frac{125}{6} = \frac{110 - 65}{R} \qquad \therefore \quad R = \mathbf{2.16\ \Omega}$$

Example 14.14. *A charging booster (shunt generator) is to charge a storage battery of 100 cells each of internal resistance 0.001 Ω. Terminal p.d. of each cell at completion of charge is 2.55 V. Calculate the e.m.f. which the booster must generate to give a charging current of 20 A at the end of charge. The armature and shunt field resistances of the generator are 0.2 Ω and 258 Ω respectively and the resistance of the cable connectors is 0.05 Ω.*

Solution. Terminal p.d./cell = 2.55 volt

The charging voltage across the battery must be capable of overcoming the back e.m.f. and also to supply the voltage drop across the internal resistance of the battery.

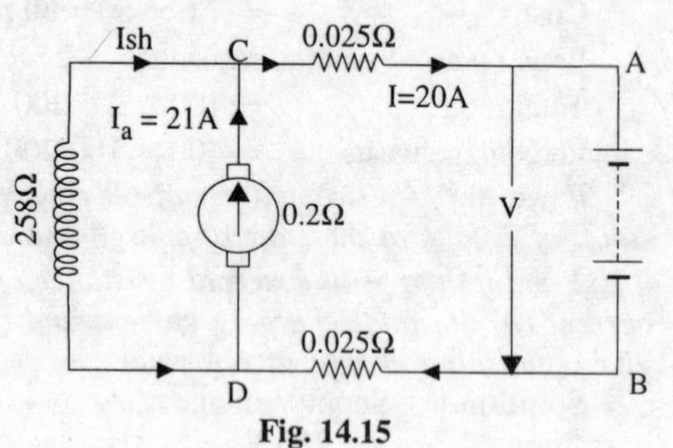

Fig. 14.15

Back e.m.f. = 100 × 2.55 = 255 V

Voltage drop on internal resistance

= 100 × 0.001 × 20 = 2 V

$\therefore$ P.D. across points A and B (Fig. 14.15)

= 255 + 2 = 257 V

P.D. across terminals C and D of the generator

= 257 + (20 × 0.05) = 258 V

$\therefore \quad I_{sh}$ = 258/258 = 1 A

I_a = 20 + 1 = 21 A

I_aR_a = 21 × 0.2 = 4.2 V

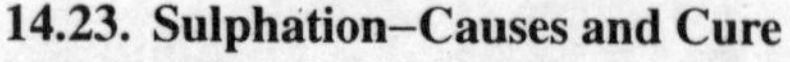

$\therefore$ generated e.m.f. = 258 + 4.2 = **262.2 V**

14.23. Sulphation–Causes and Cure

If the cell is left incompletely charged or is not fully charged periodically, then the lead sulphate formed during discharge is not converted back into PbO_2 and Pb. Some of the unreduced $PbSO_4$ which is left, gets deposited on the plates which are then said to be sulphated. $PbSO_4$ is in the form of minute crystals which gradually increase in size if not reduced by thoroughly charging the cell. It increases the internal resistance of the cell thereby reducing its efficiency and capacity. Sulphation also sets in if the battery is overcharged or left discharged for a long time.

Sulphated cells can be cured by giving them successive overcharges, for which purpose they are cut out of the battery during discharge so that they can get two charges with no intervening discharge. The other method in which sulphated cells need not be cut out of the battery is to continue charging them with a **'milking booster'** even after the battery as a whole has been charged. A milking booster is a motor-driven low-voltage dynamo which can be connected directly across the terminals of the sulphated cells.

14.24. Maintenance of Lead-acid Cells

The following important points should be kept in mind for keeping the battery in good condition:

1. Discharging should not be prolonged after the minimum value of the voltage for the particular rate of discharge is reached.
2. It should not be left in discharged condition for long.
3. The level of the electrolyte should always be above the top of the plates which must not be left exposed to air. Evaporration of electrolyte should be made up by adding distilled water occasionally.

14.25. Alkaline Accumulators

Alkaline batteries are ideally suited for portable work. There are two types of alkaline batteries which are in general use (*i*) the nickel-iron type (or Edison type) and (*ii*) nickel-cadmium type (or Junger type) also commercially known as Nife battery. Another alkaline battery which differs from the above only in the mechanical details of its plates, is known as Alkum accumulator which uses nickel hydroxide and graphite in the positive plates and a powered alloy of iron and chromium in the negative plates.

14.26. Edison Alkali Cell

The active materials in the cell are nickel oxide for the positive plate and iron oxide for the negative plate. The electrolyte is 21 per cent solution of caustic potash KOH (potassium hydrate) to which is added a small quantity of lithium hydrate LiOH for increasing the capacity of the cell.

14.27. Construction

The positive plates are built up from a number of tubes made of nickel-plated perforated steel ribbon wound spirally. The tubes are usually 10 cm long and 6 mm in diameter and are held in shape by steel rings. The active material, consisting of apple–green oxide of nickel NiO_2 or $Ni(OH)_4$ and flakes of metallic nickel (added to increase conductivity of the electrolyte) is packed into the steel tubes in alternate layers. After being filled up, the tubes are then clamped in a nickel-plated steel frame.

The negative plate consists of a number of oblong or rectangular pockets stamped from a finely-perforated nickelled-steel ribbon or strip, the pockets being filled with powdered iron oxide. A little bit of mercury is added to improve its conductivity.

The plates, in the form of tubes, are held in a nickel-plated steel container with welded seams and are kept insulated from each other by hard rubber strip separators.

14.28. Chemical Changes

The exact nature of the chemical changes taking place in such a cell is not clearly understood because the exact formula for the nickel oxide is not yet well established but the action of the cell can be understood by assuming the peroxide NiO_2 or its hydrated form $Ni(OH)_4$.

First, let us assume that at positive plate, nickel oxide is in its hydrated form $Ni(OH)_4$. During discharge, electrolyte KOH splits up into K ions and OH ions. The K ions go to anode and reduce $Ni(OH)_4$ to $Ni(OH)_2$. The OH ions travel towards the cathode and oxidise iron. During charging, just the opposite reactions take place *i.e.* K ions go to cathode and OH ions go to anode. The chemical reactions can be written thus:

$$KOH \rightarrow K + OH$$

During Discharge

Positive Plate : $Ni(OH)_4 + 2K \rightarrow Ni(OH)_2 + 2KOH$

Negative Plate : $Fe + 2OH \rightarrow Fe(OH)_2$

During Charging

Positive Plate : $Ni(OH)_2 + 2OH \rightarrow Ni(OH)_4$

Negative Plate : $Fe(OH)_2 + 2K \rightarrow Fe + 2KOH$

The charge and discharge can be represented by a single reversible equation thus

$$\underset{+\text{ Plate}}{Ni(OH)_4} + KOH + \underset{-\text{ve Plate}}{Fe} \underset{\text{Charge}}{\overset{\text{Discharge}}{\rightleftharpoons}} \underset{+\text{ Plate}}{Ni\,(OH)_2} + KOH + \underset{-\text{ve Plate}}{Fe(OH)_2}$$

It will be observed from the above equation that as no water is formed, there is no overall change in the strength of the elctrolyte. Its function is merely to serve as a conductor or as a vehicle for the transfer of OH ions from one plate to another. Hence, the specific gravity of the electrolyte remains practically constant, both during charging and discharging. That is why only a small amount of electrolyte is required which fact enables the cells to be small in bulk.

Note. If, however, we assume the nickel oxide to be in the form NiO_2, then the above reactions can be represented by the following reversible equation.

$$\underset{+\text{ Plate}}{6NiO_2} + 8KOH + \underset{-\text{ve Plate}}{3Fe} \underset{\text{Charge}}{\overset{\text{Discharge}}{\rightleftharpoons}} \underset{+\text{ Plate}}{2Ni_3O_4} + 8KOH + \underset{-\text{ve Plate}}{Fe_3O_4}$$

14.29. Electrical Characteristics

The e.m.f. of an Edison cell, when fully charged, is nearly 1.4 V which decreases rapidly to 1.3 V and then very slowly to 1.1 or 1.0 V on discharge. The average discharge voltage for a 5-hour discharge rate is 1.2 V. Hence, for the same average value of the voltage, an alkali accumulator will consist of 1.6 to 1.7 times as many cells as in a lead acid battery. The internal resistance of an alkali cell is nearly five times that of the lead-acid cell, hence there is a relatively greater difference between its terminal voltages when charging and discharging.

The average charging voltage for an alkali cell is about 1.7 V. The general shapes of the charge and discharge curves for such cells are, however, similar to those for lead-acid cells. The rated capacity of nickel accumulators usually refers to 5-hours discharge rate unless stated otherwise.

The plates of such cells have greater mechanical strength because of all-steel construction. They are comparatively lighter because (*i*) their plates are lighter and (*ii*) they require less quantity of electrolyte. They can withstand heavy charge and discharge currents and do not deteriorate even if left discharged for long periods.

Due to its relatively higher internal resistance, the efficiencies of an Edison cell are lower than those of the lead acid cell. On the average, its Ah efficiency is about 80% and Wh efficiency 60 or 50%.

With increase in temperature, e.m.f. is increased slightly but capacity increases by an appreciable amount. With decrease in temperature, the capacity decreases becoming practically zero at 4°C even though the cell is fully charged. This is a serious drawback in the case of electrically-driven vehicles in cold weather and precautions have to be taken to heat up the battery before starting, though in practice, the I^2R loss in the internal resistance of the battery is sufficient to keep the battery cells warm when running.

The principal disadvantage of the Edison battery or nickel-iron battery is its high initial cost (which will probably be sufficiently reduced when patents expire). At present, an Edison battery costs approximately twice as much as a lead-acid battery designed for similar service. But since the alkaline battery outlasts an indeterminate number of lead–acid batteries, it is cheaper in the end.

Because of their lightness, compact construction, increased mechanical strength, ability to withstand rapid charging and discharging without injury and freedom from corrosive liquids and fumes, alkali batteries are idealy suited for traction work such as propulsion of electric factory trucks, mine locomotives, miner's lamps, lighting and starting of public service vehicles and other services involving rough usage etc.

14.30. Nickel-cadmium Cell

The positive plate of such a cell is similar to that of an Edison cell with the same active materials and the same electrolyte but its negative plate is of cadmium. The use of cadmium results in reduced internal resistance of the cell. Such batteries are more suitable than Edison batteries for floating duties in conjunction with a charging dynamo because, in their case, the difference between charging and discharging e.m.fs. is not as great as in nickel-iron cell batteries.

Nickel-cadmium (Nicad) battery is extremely reliable, rechargeable, maintenance-free and has a life expectancy of about 20 years. They have excellent low-temperature performance, light mass and high energy density. They are used in many cordless electrical devices in consumer, industrial, commercial and scientific applications.

14.31. Comparison : Lead-acid Cell and Edison Cell

The relative strong and weak points of the cells are summarized as under :

Particulars	*Lead-acid cell*	*Edison cell*
1. Positive plate	PbO_2, lead-peroxide	Nickel hydroxide $Ni(OH)_4$ or NiO_2
2. Negative plate	Spongy iead	Iron oxide
3. Electrolyte	diluted H_2So_4	KOH
4. Average e.m.f.	2.0 V/cell	1.2 V/cell
5. Internal resistance	Comparatively low	Comparatively higher
6. Efficiency : amp-hour watt-hour	90 – 95% 72 – 80%	nearly 80% about 60%
7. Cost	comparatively less than alkali cell	almost twice that of Pb-acid cell. Easy maintenance.
8. Life	gives nearly 1250 charges and discharges	Five years at least
9. Strength	needs much care and maintenance. Sulphation occurs often due to incomplete charge or discharge.	due to all-steel construction, they are roubust mechanically strong, can withstand vibration, are light, unlimited rates of charge and discharge. Can be left discharged. Free from corrosive liquids and fumes.

Tutorial Problem No. 14.7

1. An alkaline cell is discharged at a steady current of 4 A for 12 hours, the average terminal voltage being 1.2 V. To restore it to its original state of charge, a steady current of 3 A for 20 hours is required, the average terminal voltage being 1.44 V. Calculate the ampere-hour and watt-hours efficiencies. **[80% ; 66.7%]**
2. Twenty alkaline secondary cells, each of discharge capacity 60 Ah at the 10-hr rate, are to be charged with constant current for 9 hours.
Calculate the maximum and minimum values of the charging resistance required if the direct supply is 120 V and the Ah efficiency of the cells is 80%. The e.m.f. of each cell at commencement and end of charge respectively is 1.3 V and 1.7 V **[11.28 Ω ; 10.32Ω]**
3. A battery of 10 cells is to be charged at constant current of 8 A from a generator of e.m.f. 35 V and internal resistance 0.2 Ω. The internal resistance of each cell is 0.08 Ω and its em.f. ranges from

1.85 Vdischarged to 2.15 V charged. Determinme the maximum and minimum values of series resistance which must be included in the circuit. **[1.6025 Ω ; 0.6875 Ω]**

4. A separately-excited d.c. generator has an armature resistance of 0.2 Ω. It is used to change a battery having an open-circuit e.m.f. of 105 V and an internal resistance of 0.1 Ω. The copper conductors between generator and battery have a total length of 100 m and are 2.5 mm in diameter. What will be the value of the generated e.m.f. of the generator when the battery is being charged at 10 A?
How much power will be wasted in the conductors between generator and battery?
The resistance of a copper wire 1 metre long and 1 mm^2 cross–section is 1/58 Ω
[111.5 V ; 35.1 W]

HIGHLIGHTS

1. Faraday's Laws of electrolysis are :

(*i*) First Law : The mass of an ion liberated at an electrode is directly proportional to the quantity of electricity that passes through the electrolyte. Mathematically,

$$M = Zlt$$

$$= \left(\frac{1}{F} \times \frac{A}{\nu}\right) \quad \text{It} = \frac{E}{F}\text{ It} = \frac{E}{F} \cdot \text{Q}$$

where Z = *E.C.E.* of the element

E = chemical equivalent weight = atomic wt./valency.

(*ii*) Second Law : The masses of ions of different substances liberated by the same quantity of electricity are proportional to their electro–chemical equivalent weights. In other words,

$$m_1 : m_2 : M_3 :: Z_1 : Z_2 : Z_3$$

2. Electrochemical equivalent weight of a substance is defined as the mass of its ions liberated by the passage of a charge of one coulomb through the electrolytic solution.

3. Faraday's constant is defined as the charge required to liberate one gram-equivalent of a substance.

For all substances,

$$\frac{\text{chemical equivalent}}{\text{electrochemical equivalent}} = \text{Faraday's constant} = 96{,}500 \text{ C}$$

4. The E.C.E.s and chemical equivalents of different substances are inter–related thus

$$\frac{\text{E.C.E. of } A}{\text{E.C.E. of } B} = \frac{\text{chemical equivalent of } A}{\text{chemical equivalent of } B} \quad \text{or} \quad \frac{E_1}{E_2} = \frac{Z_1}{Z_2}$$

5. The value of back e.m.f. of electrolysis is $E = JHZ$

6. The two efficiencies of a cell are :

(*i*) The quantity or ampere–hour efficiency

(*ii*) The energy or watt–hour efficiency

$$\text{Ah } \eta = \frac{\text{ampere–hours on discharge}}{\text{ampere-hours on charge}}$$

$$\text{Wh } \eta = \frac{\text{Watt–hours on discharge}}{\text{Watt-hours on chart}} = \text{Ah } \eta \times \frac{\text{average volts on discharge}}{\text{average volts on charge}}$$

OBJECTIVE TESTS—14

A. Fill in the following blanks :

1. Those conductors which undergo decomposition when electric current is passed through them are called ....

2. Chemical equivalent of a substance is given by the ratio of its atomic weight and ....

3. Electrochemical equivalent weight is given by the mass of ions of a substance liberated by the passage of one .... charge through its electrolytic solution.

4. The active material at the positive electrode of a lead acid cell is ....

5. During charging of a lead-acid cell, the specific gravity of the electrolyte is ....
6. The two efficiencies of a lead-acid cell are known as Ah efficiency and .... efficiency.
7. Capacity of a cell is measured in ....
8. In an Edison cell, specific gravity of the electrolyte remains practically .... both during charging and discharging.

B. Answer True or False :

1. Conductivity of electrolytes is due to the presence of ions in them.
2. Valency measures the combining capacity of an atom.
3. Faraday's constant is given by the ratio of electrochemical equivalent weight and chemical equivalent weight of a substance.
4. During discharging of a storage cell, its specific gravity increases due to consumption of water.
5. The watt-hour efficiency of a cell is always less than its ampere-hour efficiency.
6. The capacity of a cell is measured in watt-hours.
7. As compared to constant-current method of charging a cell, the constant-voltage charging method takes much longer time.
8. The internal resistance of an Edison cell is comparatively higher than that of lead-acid cell.

C. Multiple Choice Questions.

1. The electrochemical equivalent weight of a substance depends on
   (*a*) Faraday's constant
   (*b*) its chemical equivalent weight
   (*c*) its valency
   (*d*) all of the above.
2. The e.m.f. of a fully-charged lead-acid cell is about .... volt.
   (*a*) 1.21 (*b*) 2.1
   (*c*) 1.8 (*d*) 1.4
3. The capacity of a storage cell is measured in ....
   (*a*) Joule (*b*) watt-hour
   (*c*) ampere-hour (*d*) coulomb
4. The indications of a fully-charged cell are given by
   (*a*) voltage
   (*b*) gassing
   (*c*) specific gravity of the electrolyte
   (*d*) all of the above.
5. As compared to the Edison alkali cell, a lead-acid has
   (*a*) higher efficiency
   (*b*) greater internal resistance
   (*c*) lower voltage
   (*d*) higher cost.
6. Sulphation in a lead-acid battery occures due to
   (*a*) trickle charging
   (*b*) incomplete charging
   (*c*) heavy discharging
   (*d*) fast charging.

ANSWERS

A. 1. electrolyte 2. valency 3. coulomb 4. lead peroxide 5. increased 6. Wh 7. ampere-hour 8. constant
B. 1. T 2. T 3. F 4. F 5. T 6. F 7. F 8. T
C. 1. *d* 2. *b* 3. *c* 4. *d* 5. *a* 6. *b*

15 ELECTRICAL INSTRUMENTS AND MEASUREMENTS

15.1. Absolute and Secondary Instruments

The various electrical instruments may, in a very broad sense, be divided into (*i*) *absolute* instruments and (*ii*) *secondary* instruments. Absolute instruments *are those which give the value of the quantity to be measured in terms of the constant of the instruments and their deflection only.* No previous calibration or comparison is necessary in their case. The example of such an instrument is the tangent galvanometer, which gives the value of current in terms of the tangent of deflection produced by the current and of the radius and number of turns of wire used and the horizontal component of earth's field.

Secondary instruments are those in which the value of electrical quantity to be measured can be determined from the deflection of the instruments only when they have been pre-calibrated by comparison with an absolute instrument. Without calibration, the defelction of such instruments is meaningless.

It is the secondary instruments which are most generally used in everyday work, the use of the absolute instruments being merely confined within laboratories as standardizing instruments.

15.2. Electrical Principles of Operation

All electrical measuring instruments depend for their action on one of the many physical effects of an electric current or potential and are generally classified according to which of these effects is utilized in their operation. The effects generally utilized are :

1. Magnetic Effect .... for ammeters, voltmeters usually.
2. Electrodynamic Effect ....for ammeters, voltmeters but particularly for wattmeters.
3. Electromagnetic effect .... for ammeters, voltmeters, wattmeters and watt-hour meters.
4. Thermal Effect .... for ammeters and voltmeters.
5. Chemical Effect .... for d.c. ampere-hour meters.
6. Electrostatic Effect .... for voltmeters only.

Another way to classify secondary instruments is to divide them into (*i*) *indicating instruments, (ii) recording instruments* and (*iii*) *integrating instruments.*

Indicating instruments are those which indicate the instantaneous value of the electrical quantity being measured *at the time* at which it is being measured. Their indications, are given by pointers moving over calibrated dials. Ordinary ammeters, voltmeters and wattmeters belong to this class.

Recording instruments are those which, instead of indicating by means of a pointer and a scale the instantaneous value of an electrical quantity, give a *continuous record* of the variations of such a quantity over a selected period of time. The moving system of the instrument carries an inked-pen which rests lightly on a chart or graph that is moved at a uniform and low speed, in a direction perpendicular to that of the deflection of the pen. The path traced out by the pen presents a continuous record of the variations in the deflection of the instruments.

Integrating instruments are those which measure and register by a set of dials and pointers either the *total* quantity of electricity (in ampere-hours) or the *total* amount of electrical energy (in watt-hours or kWh) supplied to a circuit in a given time. Their summation gives the product of time and the electrical quantity but gives no direct indication as to the *rate* at which the quantity or energy is being supplied because their registrations are independent of this rate provided the current flowing through the instrument is sufficient to operate it.

Ampere-hour and watt-hour meters belong to this class.

15.3. Essentials of Indicating Instruments

As defined above, indicating instruments are those which indicate the value of the quantity that is being measured at the time at which it is measured. Such instruments consist essentially of a pointer which moves over a calibrated scale and which is attached to a moving system pivoted in jewelled bearings. The moving system is subjected to the following three torques :

1. A deflecting (or operating) torque.
2. A controlling (or restoring) torque.
3. A damping torque.

15.4. Deflecting Torque

The deflecting or operating torque (T_d) is produced by utilizing one or other effects mentioned in Art. 15.2 *i.e.* magnetic, electrostatic, electrodynamic, thermal or chemical etc. The actual method of torque production depends on the type of instrument and will be discussed in the succeeding paragraphs. This deflecting torque causes the moving system (and hence the pointer attached to it) to move from its *'zero'* position *i.e.* its position when the instrument is disconnected from the supply.

15.5. Controlling Torque

The deflection of the moving system would be indefinite, if there were no controlling or restoring torque. This torque opposes the deflecting torque and increases with the deflection of the moving system. The pointer is brought to rest at a position where the two opposing torques are numerically equal. The deflecting torque ensures that currents of different magnitudes shall produce deflections of the moving system in proportion to their magnitudes. Without such a torque, the pointer would swing over to the maximum deflection position, irrespective of the magnitude of the current to be measured. Moreover, in the absence of a restoring torque, the pointer, once deflected, would not return to its zero position on removing the current. The controlling or restoring or balancing torque in indicating instruments is either obtained by a spring or by gravity, as described below.

(*a*) Spring Control

A hair-spring, usually of phosphor-broze, is attached to the moving system of the instrument, as shown in Fig. 15.1 (*a*).

With the deflection of the pointer, the spring is twisted in the opposite direction. This twist in the spring produces restoring torque, which is directly proportional to the angle of deflection of the moving system. The pointer comes to a position of rest (or equilibrium) when the deflecting torque (T_d) and controlling torque (T_c) are equal. For eample, in permanent-magnet moving coil type of instruments, the deflecting torque is proportional to the current I passing through them.

$\therefore\ T_d \propto I$ and for spring control $T_c \propto \theta$

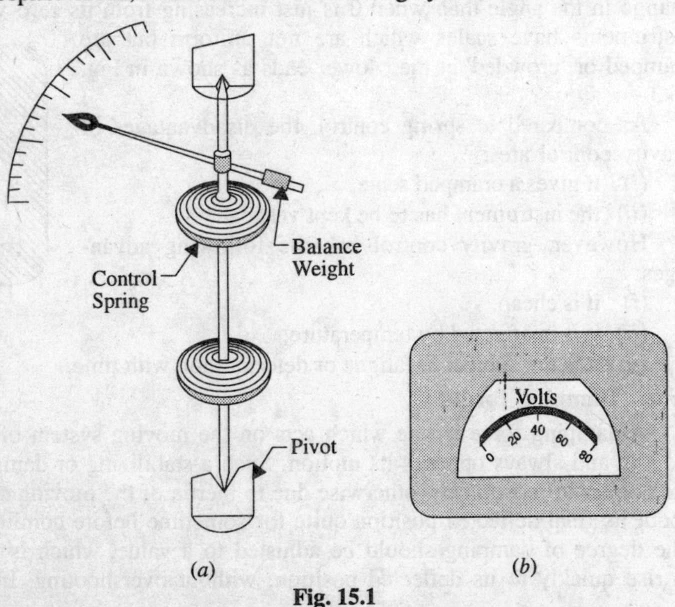

Fig. 15.1

As $T_c = T_d$ $\therefore \theta \propto I$

Since deflection θ is directly proportional to current I, the spring-controlled instruments have a uniform or equally-spaced scales over the whole of their range as shown in Fig. 15.1 (*b*).

To ensure that controlling torque is proportional to the angle of deflection, the spring should have a fairly large number of turns so that angular deformation per unit length, on full-scale defelction, is small. Moreover, the stress in the spring should be restricted to such a value that it does not produce a permanent set in it.

Springs are made of such materials which

1. are non-magnetic,
2. are not subject to much fatigue,
3. have low specific resistance-especially in cases where they are used for leading the current in or out of the instrument,
4. have low temperature-resistance coefficient.

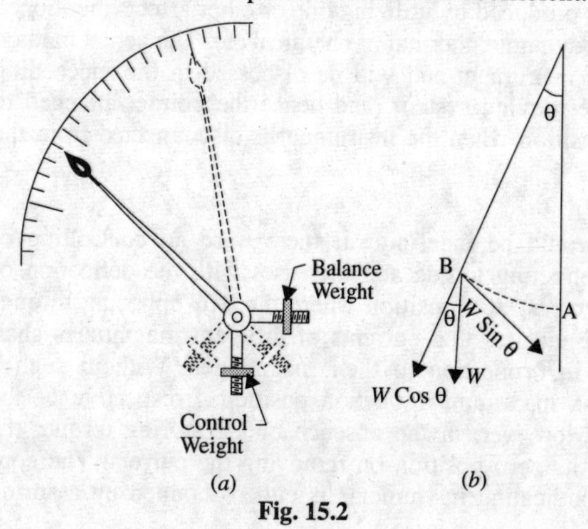

Fig. 15.2

(*b*) Gravity Control

Gravity control is obtained by attaching a small adjustable weight to some part of the moving system, such that the two exert torques in the opposite directions. The usual arrangement is shown in Fig. 15.2 (*a*).

It is seen from Fig. 15.2(*b*) that the controlling or restoring torque is proportional to the sin of the angle of deflection *i.e.* $T_c \propto \sin\theta$. The degree of control is adjusted by screwing the weight up or down the carrying system.

If $T_d \propto I$, then for position of rest

$T_d \propto T_c$ or $I \propto \sin\theta$ (*not* θ)

It will be seen from Fig. 15.2 (*b*) that as θ approaches 90°, the distance *AB* increases by a relatively smaller amount for a given change in the angle than when θ is just increasing from its zero value. Hence, gravity-controlled instruments have scales which are not uniform but are cramped or 'crowded' at their lower ends as shown in Fig. 15.3.

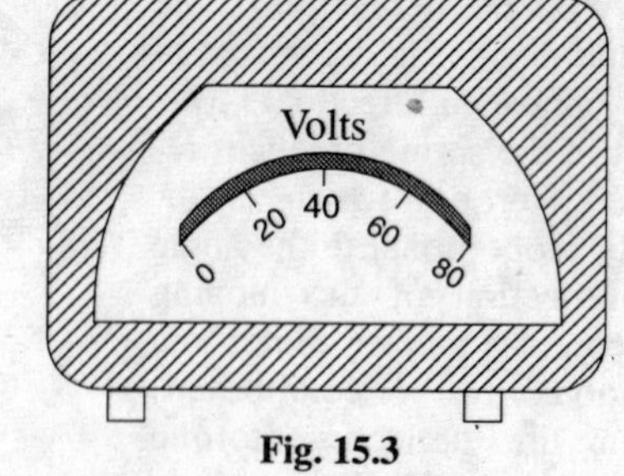

Fig. 15.3

As compared to spring control, the disadvantages of gravity control are :

(*i*) it gives a cramped scale,
(*ii*) the instrument has to be kept vertical.

However, gravity control has the following advantages:

(*i*) it is cheap,
(*ii*) it is unaffected by temperature,
(*iii*) it is not subject to fatigue or deterioration with time.

15.6. Damping Torque

A damping force is one which acts on the moving system of the instrument *only when it is moving* and always opposes its motion. Such a stabilizing or damping force is necessary to bring the pointer to rest *quickly,* otherwise due to inertia of the moving system, the pointer will oscillate about its final deflected position quite for sometime before coming to rest in the steady position. The degree of damping should be adjusted to a value, which is sufficient to enable the pointer to rise quickly to its deflected position, without overshooting. In that case, the instrument is said to be *dead-beat.* Any increase of damping above this limit *i.e.* over-damping, will make the instrument slow and lethargic. In Fig. 15.4 is shown the effect of damping on the variation of position, with time, of the moving system of an instrument.

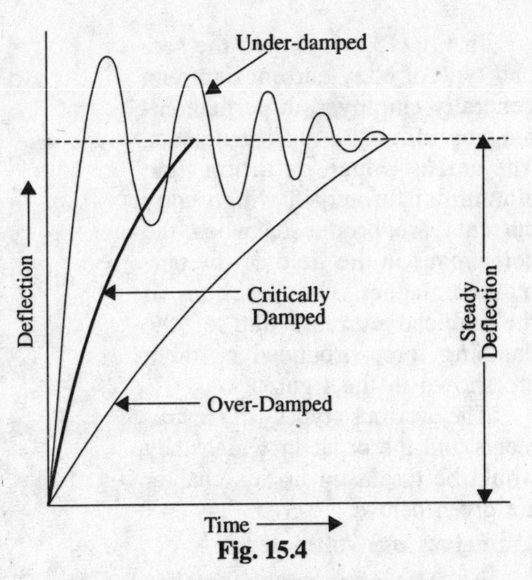

Fig. 15.4

The damping force can be provided by (*i*) *air friction* (*ii*) *eddy current* and (*iii*) *fluid-friction* (used occasionally).

Two methods of air-friction damping are shown in Fig. 15.5 (*a*) and (*b*). In Fig. 15.5 (*a*), the light aluminium piston, attached to the moving system of the instrument, is arranged to travel with a very small clearance, in fixed air chamber closed at one end. The cross-section of the chamber is either circular or rectangular. Damping of the oscillations is affected by the compression and suction actions of the piston on the air enclosed in the chamber. Such a system of damping is not much favoured these days, those shown in Fig. 15.5 (*b*) and (*c*) being preferred. In the latter method, one or two light aluminium vanes are mounted on the spindle of the moving system, which moves in a closed sector-shaped box, as shown.

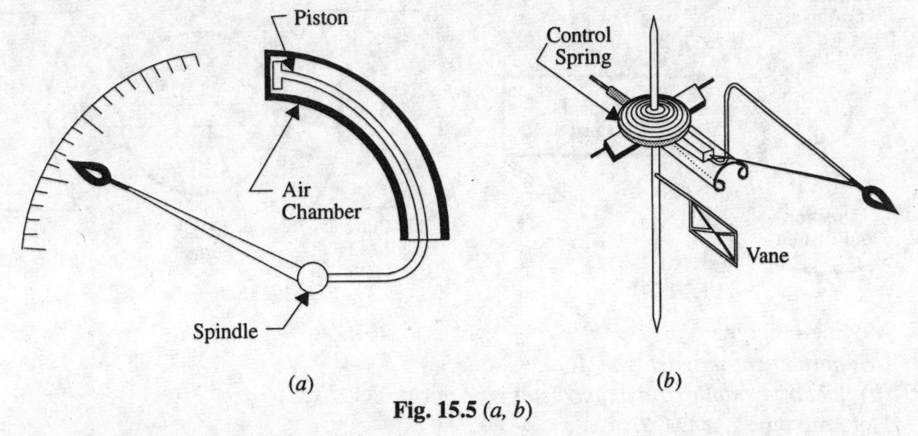

Fig. 15.5 (*a, b*)

Fluid friction is similar in action to the air friction. Due to greater viscosity of oil, the damping is more effective. However, oil damping is not much used because of several disadvantages such as objectionable creeping of oil, the necessity of using the instrument always in the vertical position and its obvious unsuitablity for use in protable instruments.

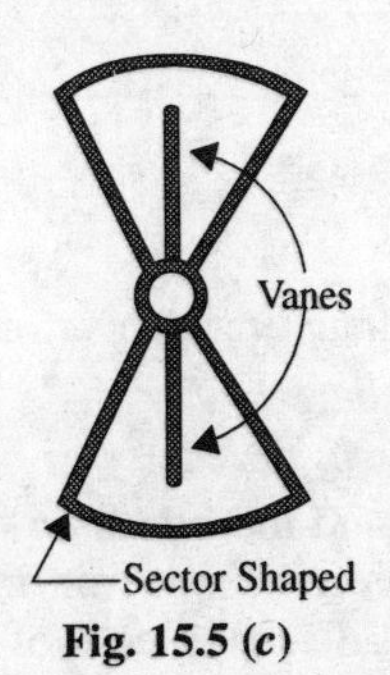

Fig. 15.5 (*c*)

The eddy-current form of damping is the most efficient of the three. The two forms of such a damping are shown 15.6 and 15.7. In Fig. 15.6 is shown a thin disc of a conducting but *non magnetic* material like copper or aluminium mounted on the spindle which carries the moving system and the pointer of the instrument. The disc is so positioned that its edge, when in rotation, cuts the magnetic flux between the poles of a permanent magnet. Hence eddy currents are produced in the disc which flow and so produce a damping force in such a direction as to oppose the very cause producing them (Lengz's Law Art. 8-6). Since the cause producing them is the rotation of the disc, these eddy currents retard the motion of the disc and the moving system as a whole.

In Fig. 15.8 is shown the second type of eddy-current damping generally employed in permanent-magnet moving-coil instruments. The coil is wound on a thin light aluminium former, in which eddy currents are produced, when the coil moves in the field of the permanent magnet. The directions of the induced currents and of the damping force produced by them are shown in the figure.

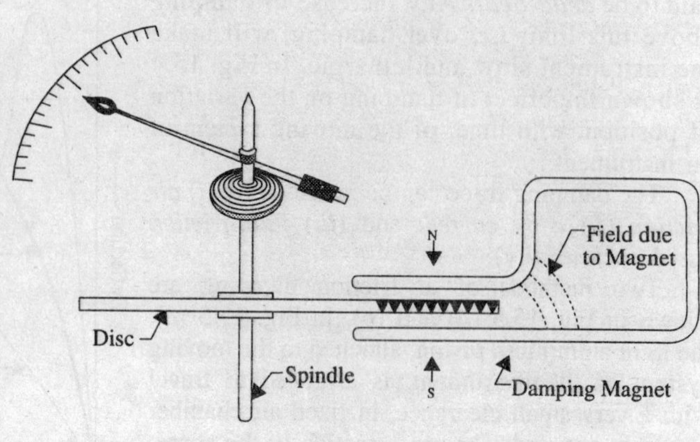

Fig. 15.6

The various types of instruments and the order in which they would be discussed in this chapter are given below:

Ammeters and Voltmeters

1. Moving-iron type (both for D.C./A.C.)
   (*a*) *the attraction type* (*b*) *the repulsion type*

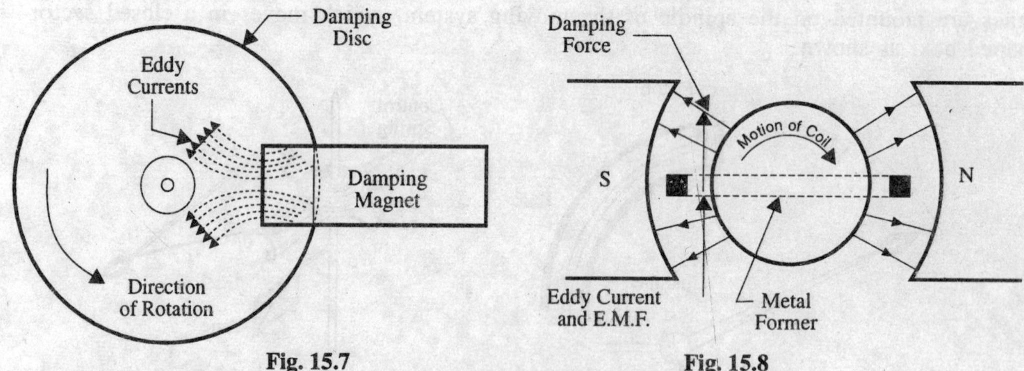

Fig. 15.7 **Fig. 15.8**

2. Moving-coil type
   (*a*) *permanent-magnet type* (for D.C. only)
   (*b*) *electrodynamic or dynamometer type of* (for D.C./A.C)
3. Hot-wire type (for D.C./A.C.)
4. Induction type (for A.C. only)
   (*a*) *Split-phase winding type* (*b*) *Shaded-pole type.*
5. Electrostatic type ....for voltmeters only (for D.C./A.C.)

Wattmeters

6. Dynamometer type (for D.C./A.C.)
7. Induction type (for A.C. only)

Energy Meters

8. Electrolytic type (for D.C. only)
9. Motor Meters
   (*i*) *Mercury Motor Meter.* For D.C. work only. Can be used as ampere-hour or watt-hour meter.
   (*ii*) *Induction type.* (For A.C. only)

15.7. Moving-iron Ammeters and Voltmeters

There are two basic forms of these instruments *i.e.* the *attraction* type and the *repulsion* type. the operation of the attraction type depends on the attraction of a single piece of soft iron into a magnetic field and that of repulsion type depends on the repulsion of two adjacent pieces of iron magnetised by the same magnetic field. For both types of these instruments, the necessary magnetic field is produced by the ampere-turns of a current-carrying coil. In case the instrument is to be used as an ammeter, the coil has comparatively few turns of thick wire so that the ammeter has low resistance because it is connected in series with the circuit. In case it is to be used as a voltmeter, the coil has high impedance so as to draw as small a current as possible since it is connected in parallel with the circuit. As the current through the coil is small, it has large number of turns in order to produce sufficient ampere-turns.

15.8. Attraction Type M.I. Instruments

The basic working principle of an attraction type moving-coil instrument is illustrated in Fig. 15.9. It is well-known that if a piece of unmagnetised soft iron is brought up near either of the two ends of a current-carrying coil, it would be attracted into the coil in the same way as it would be attracted by the pole of a bar magnet. Hence, if we pivot an oval-shaped disc of soft iron on a spindle between bearings and near the coil, the iron disc will swing into the coil when the latter has an electric current passing through it. As the field strength would be strongest at the centre of the coil, the oval-shaped iron disc is pivoted in such a way that the greatest bulk of iron moves into the centre of the coil. If a pointer is fixed to the spindle carrying the disc, then passage of current through the coil will cause the pointer to deflect. The amount of deflection produced would be greater when the current producing the magnetic field is greater. Another point worth noting is that *whatever the direction of current through the coil, the iron disc would always be magnetised in such a way that it is pulled inwards.* Hence, such instruments can be used both for direct as well as for alternating currents.

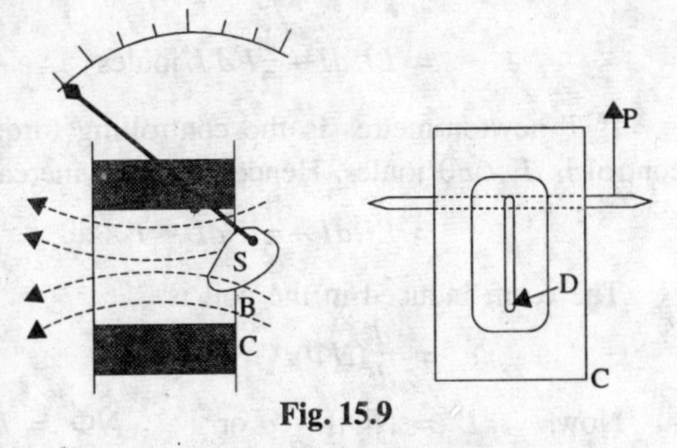

Fig. 15.9

When the current to be measured is passed through the coil *C*, the magnetic field is produced which attracts the eccentrically mounted disc inwords, thereby deflecting the points which moves over a calibrated scale that is made as uniform as possible by shaping the disc suitably.

Deflecting Torque

The force (*F*) pulling the iron disc inwards depends on (*i*) the strength of the field *H* produced by the coil and (*ii*) the pole strength (*m*) developed by the disc which again depends on *H*. In other words,

$$F \propto mH \propto H^2$$

Hence, the deflecting torque,

$$T_d \propto F \propto H^2$$

If the relative permeability of iron is assumed constant, then $H \propto$ current *I*.

$$\therefore \quad T_d \propto H^2 \propto I^2 \quad ...(i)$$

If spring control is used, then controlling torque is given by

$$T_c \propto \text{deflection of the disc, } \theta \quad \text{or} \quad T_c \propto \theta \quad ...(ii)$$

Hence, from Eq. (*i*) and (*ii*), we get $\theta \propto I^2$

In the steady position of defelction, $T_d = T_c$

If alternating current is used, then $\theta \propto I^2$ r.m.s.

Obviously, in both cases, the scale is uneven, although, as stated earlier, the disc is specially shaped to give a scale as nearly uniform as possible.

Alternate Treatment

The value of the deflecting torque can also be found in terms of the inductance *L* of the instrument. The inductance of the coil of the instrument changes with the deflection θ in both the attraction type and repulsion type moving-iron instruments. In the attraction type, inductance of the instrument increases with increase in the deflection of the iron disc. In the case of repulsion type, the increase in instrument inductance is due to the increase of flux produced by the decrease in demagnetising effect of one iron rod on the other as the two move farther apart. Hence, there is relationship between the deflecting torque and inductance of the instrument as shown below.

Suppose that when a direct current of I passes through the instrument, its deflection is θ and the inductance is L. Also, suppose that when current changes from I to $(I + dI)$, deflection changes from θ to $(\theta + d\theta)$ and L changes to $(L + dL)$. Then, the increase in the energy stored in the magnetic field is given by

$$dE = d\left(\frac{1}{2}\,LI\right) = \frac{1}{2}\,L.2I.dI + \frac{1}{2}\,I^2.dL$$

$$= LI.dI + \frac{1}{2}I^2 dL \text{ joules}$$

If T newton-metres is the controlling torque for deflection θ, then extra energy stored in control is $T \times d\theta$ joules. Hence, the total increase in energy of the system is

$$= LI.dI + \frac{1}{2}I^2 dL + T \times d\theta \qquad ...(i)$$

The e.mf. induced in the coil is

$$e = \frac{d}{dt}(N\Phi)$$

Now, $L = N\Phi/I$ or $N\Phi = LI$

$$\therefore \quad e = \frac{d}{dt}(LI)$$

The energy drawn from the supply to overcome this e.m.f.

$$= eI.dt = \frac{d}{dt}(LI).I.dt = I.d(LI) \qquad ...(ii)$$

$$= I(L.dI + I.dL) = LI.dI + I^2 dL \text{ joules}$$

Equating Eq. (*i*) and (*ii*), we get

$$LI.dI + \frac{1}{2}I^2 dL + T\,d\theta = LI.dI + I^2.dL \quad \therefore \quad T = \frac{1}{2}\,I^2\,\frac{dL}{d\theta} \text{ N–m}$$

Damping

As shown, air-friction damping is provided, the actual arrangement being a light piston moving in an air-chamber.

15.9. Repulsion Type M.I. Instruments

A simple sketch of such an instrument is shown in Fig. 15.10. The arrangement consists of a fixed coil C inside which are placed two soft-iron rods (or bars) A and B parallel to one another and along the axis of the coil C. One of them *i.e.* A is fixed and the other, B, which is movable, carries a pointer that moves over a calibrated scale. When the current to be measured is passed through the fixed coil, it sets up its own magnetic field which magnetises the two rods similarly *i.e.* the adjacent points on the lengths of the rods will have the same magnetic polarity. Hence, they repel each other with the result that the pointer is deflected against the controlling torque of a spring or gravity. The force of repulsion is approximately proportional to the square of the current passing through the coil. Moreover, whatever may be the direction of the current through the coil, the two rods will be magnetised similarly and hence will repel each other.

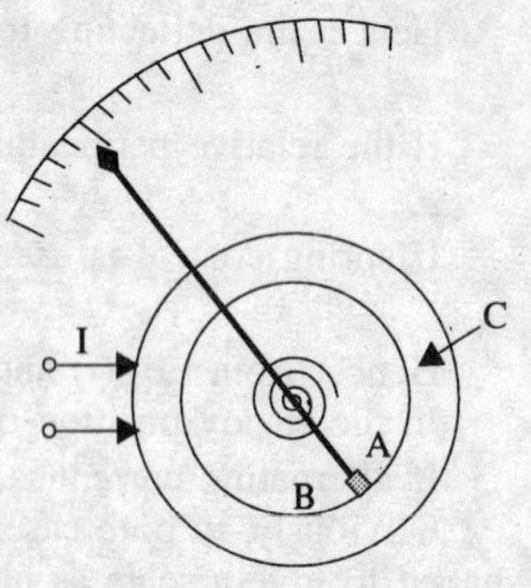

Fig. 15.10

In order to achieve uniformity of scale, two tongue-shaped strips of iron are used instead of two rods. As shown in Fig. 15.11 (*a*), the fixed iron consists of a tongue-shaped sheet iron

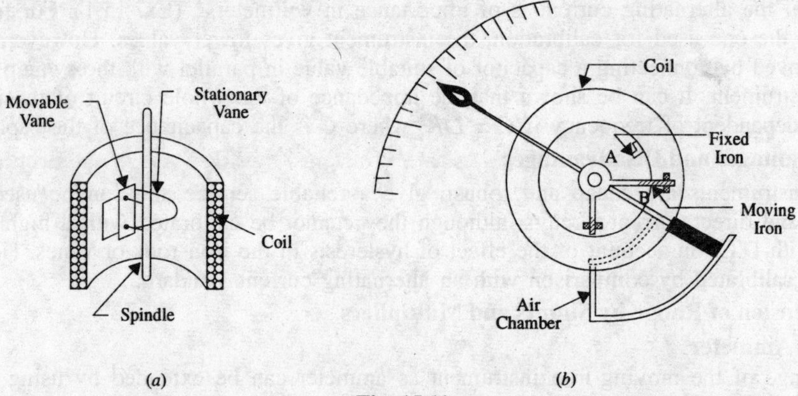

Fig. 15.11

bent into a cylindrical form, the moving iron also consists of another sheet of iron and is so mounted as to move parallel to the fixed iron and towards its narrower end.

Deflecting torque

The deflecting torque is due to the repulsive force between the two similarly-magnetised iron rods or sheets.

Instantaneous torque $\propto$ repulsive force

$\propto m_1 m_2$ ...product of pole strength

Since pole strengths are proportional to the magnetising field H of the coil.

$\therefore$ instantaneous torque $\propto H^2$

Since H itself is proportional to current (assuming *constant* permeability) passing through the coil,

$\therefore$ instantaneous torque $\propto i^2$

Hence, the deflecting torque, which is proportional to the mean torque is, in effect, proportional to the mean value of i^2. Therefore, when used on alternating current, the instrument reads the r.m.s. value of the current.

Scales of such instruments are uneven if rods are used and uniform if suitably-shaped pieces of iron sheet are used.

The instrument is spring-controlled.

Damping is pneumatic, eddy current damping cannot be employed because the presence of a permanent magnet required for such a purpose would affect the deflection and hence the reading of the instrument.

Since the polarity of both iron rods reverses simultaneously, the instrument can be used both for alternating and direct currents *i.e.* instrument belongs to the unpolarised class.

15.10. Sources of Error

There are two types of possible errors in such instruments, firstly, those which occur both in alternating current and direct current work and secondly, those which occur in alternating current work alone.

(*a*) **Errors with both D.C. and A.C.**

(*i*) *Error due to hysteresis.* Because of hysteresis in the iron parts of the moving system, the readings are higher for descending values but lower for ascending values.
The hysteresis error is almost completely eliminated by using Mumetal or Permalloy which have negligible hysteresis loss.

(*ii*) *Error due to stray fields.* Unless shielded effectively from the effect of stray external fields, it will give wrong reading. Magnetic shielding of the working parts is obtained by using a covering case of cast-iron.

(*b*) **Errors with A.C. only**

Changes of frequency produce (*i*) change in the impedance of the coil and (*ii*) change in the magnitude of the eddy currents. The increase in the impedance of the coil with increase in the frequency of the alternating current is of importance in voltmeters* (Ex. 15.1). For frequencies higher than the one used for calibration, the instrument gives lower values. However, this error can be removed by connecting a capacitor of suitable value in parallel with the swamp resistance R of the instrument. It can be shown that the impedance of the whole circuit of the instrument becomes independent of frequency if $C = L/R^2$ where C is the capacitance of the capacitor.**

15.11. Advantages and Disadvantages

Such instruments are cheap and robust, give a reliable service and can be used both on alternating and direct current circuits, although they cannot be calibrated with a high degree of precision with D.C. on account of the effect of hysteresis in the iron rods or vanes. Hence, they are usually calibrated by comparison with an alternating current standard.

15.12. Extension of Range by Shunts and Multipliers

(*i*) As Ammeter

The range of the moving-iron instrument as ammeter can be extended by using a suitable shunt across its terminals. So far as the operation with direct currents is concerned, there is no trouble but with alternating currents, the division of current between the instrument and the shunt changes with the change in the applied frequency. For a.c. work, both the inductance and the resistance of the instrument coil and shunt have to be considered. Obviously,

$$\frac{\text{Current through instrument, } i}{\text{Current through shunt, } I_s} = \frac{\sqrt{R_s^2 + \omega^2 L_s^2}}{\sqrt{R^2 + \omega^2 L^2}} = \frac{Z_s}{Z}$$

where R and L are the resistance and inductance of the instrument coil and R_s and L_s are the corresponding values for the shunt. It can be shown that the above ratio *i.e.* division of current between the instrument and the shunt would be independent of frequency if the time constant of the instrument coil and shunt are the same *i.e.* if L/R equals L_s/R_s. the multiplying power (N) of the shunt is given by

$$N = \frac{I}{i} = 1 + \frac{R}{R_s}$$

where I = line current

i = full-scale deflection current of the instrument (without shunt)

(*ii*) As Voltmeter

The range of this instrument when used as a voltmeter can be extended or multiplied by using a high non-inductive series resistance R connected in series with it as shown in Fig. 15.12 (*a*). That is why these series resistances are known as 'multipliers'. When used for d.c. work, suppose, its range is to be extended from v to V. Then, obviously, the excess voltage of $(V - v)$ is to be dropped across R. If i is the full-scale deflection current of the instrument, then

$$iR = V - v$$

$$\therefore \quad R = \frac{V - v}{i} = \frac{V - ir}{i} = \frac{V}{i} - r$$

Now, voltage multiplication = V/v

Since, $iR = V - v \qquad \therefore \quad \frac{iR}{v} = \frac{V}{v} - 1$

or, $\frac{iR}{ir} = \frac{V}{v} - 1 \qquad \therefore \quad \frac{V}{v} = \left(1 + \frac{R}{r}\right)$

---

*At higher frequencies, an error is introduced due to the capacitance existing between different turns of the coil of the instrument when used as an ammeter. This capacitance error can be compensated by an additional circuit consisting of resistance and capacitance and series for connected across the instrument terminals.

**This is the value of capacitance required when resistance of the instrument coil r is not negligible as compared to high series resistance R. However, if r is negligible as compared to R, then $C = 0.41\ L/R^2$.

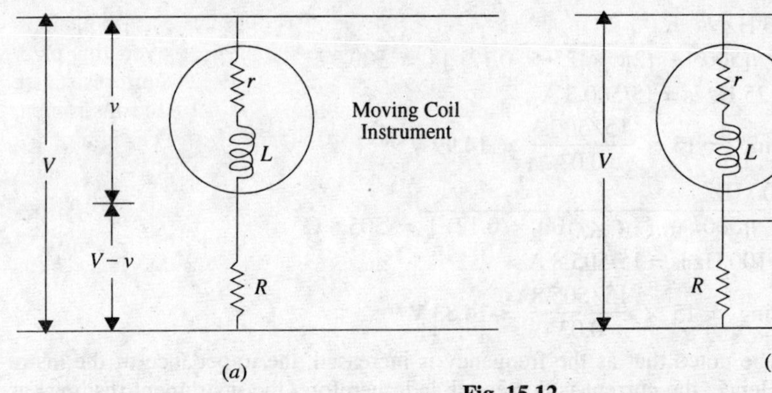

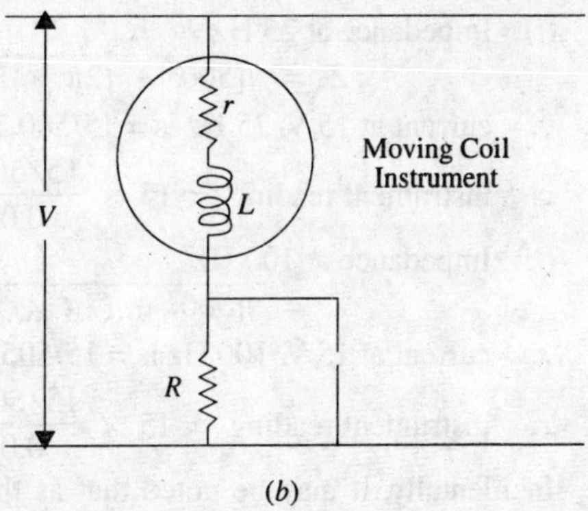

Fig. 15.12

Hence, greater the value of R, greater is the extension in the voltage range of the instrument. For d.c. work, the principal requirement of R is that its value should remain constant *i.e.* it should have a low temperature coefficient. For a.c. work, it is essential that total impedance of the voltmeter and the series resistance R should remain as nearly constant as possible at different frequencies. That is why R is made as non-inductive as possible in order to keep the inductance of the whole circuit to the minimum. The frequency error introducted by the inductance of the instrument coil can be compensated for by shunting R by a capacitor C as shown in Fig. 15.12 (*b*).

Example 15.1. *A moving-iron instrument require 400 ampere-turns to give full-scale deflection. Calculate (a) the number of turns required if the instrument is to be an ammeter reading up to 50 A, (b) the number of turns and the total resistance if the instrument is to be used for a voltage reading up to 30 V with current 20 mA.* ***(Electrical Engg. A.M.Ae.S.I. Dec. 1994)***

Solution. (*a*) Since the full-scale reading of the ammeter is 50 A and 400 AT are required for full-scale deflection, the number of turns required for 50 A are given by

$$50 \times N = 400 \text{ or } N = 400/50 = \mathbf{8}$$

(*b*) It is given that with 30 V applied across the voltmeter, the current is 20 mA. Hence, total resistance of the voltmeter circuit = $30/20 \times 10^{-3} = 1{,}500\ \Omega$

Also, $N \times 20 \times 10^{-3} = 400$ or $N = \mathbf{20{,}000}$

Example 15.2. *The coil of a 250-V moving-iron voltmeter has a resistance of 500 Ω and an inductance of 1 henry. The current taken by the instrument when placed on 250-V, d.c. supply is 0.05 A. Determine the percentage error when the instrument is placed on 250 V, a.c. supply at 100 Hz.* ***(A.M.I.E. Summer 1993)***

Solution. Here, original calibration of the instrument is with direct current.

Total ohmic resistance = $250/0.05 = 5000\ \Omega$

Reactance of the coil = $2\pi \times 100 \times 1 = 628\ \Omega$

Coil impedance = $\sqrt{5000^2 + 628^2} = 5040\ \Omega$

$\therefore$ voltage reading on a.c. = $250 \times 5000/5040 = 248$ V

$\therefore$ error = -2 V

percentage error = $2 \times 100/250 = \mathbf{0.806\%}$

Example 15.3. *A 15.V moving-iron voltmeter has a resistance of 500 Ω and an inductance of 0.12 H. Assuming that this instrument reads correctly on D.C., what will ve its reading on A.C. at 15 volts when the frequency is (i) 25 Hz and (ii) 100 Hz.*

Solution. When used for d.c. work, only ohmic resistance is involved. Hence, full-scale deflection of the instrument on d.c. is

$$= 15/500 = 0.03 \text{ A}$$

When used for a.c. work, it is the impedance of the instrument which has to be taken into account.

(*i*) Impedance at 25 Hz

$$Z = \sqrt{[500^2 + (2\pi \times 25 \times 0.12)^2]} = 500.3\ \Omega$$

$\therefore$ current at 15 V, 25 Hz is = 15/500.3 A

$$\therefore \text{ instrument reading } = 15 \times \frac{15/500.3}{0.03} = \mathbf{14.99\ V^*}$$

(*ii*) Impedance at 100 Hz

$$Z = \sqrt{[500^2 + (2\pi \times 100 \times 0.12)^2]} = 505.8\ \Omega$$

$\therefore$ current at 15 V, 100 Hz is = 15/505.8 A

$$\therefore \text{ instrument reading } = 15 \times \frac{15/505.8}{0.03} = \mathbf{14.83\ V^{**}}$$

Incidentally, it may be noted that as the frequency is increased, the impedance of the instrument is also increased. Hence, the current is decreased and, therefore, the instrument readings are lower.

Example 15.4. *A moving-iron instrument gives full-scale deflection with 100 V. It has a coil of 10,000 turns and a resistance of 2,000 Ω. If the instrument is to be used as an ammeter to give full-scale deflection at 20 A, calculate the necessary number of turns in the coil.*

Solution. It should be kept in mind that in such instruments, the strength of the magnetic field (and hence the deflecting torque) depends on the number of ampere-turns.

Full-scale deflection current = 100/2,000 = 1/20 A

Full-scale deflection ampere-turns = 10,000 × 1/20 = 500 AT

When current is 20 A, the number of turns required is = 500/20 = **25**

Example 15.5. *The inductance of a certain moving-iron ammeter is* $(8 + 4\theta - {}^1/_2\,\theta^2)\ \mu H$ *where θ is the deflection in radians from the zero position. The control spring torque is* 12×10^{-6} *N-m/rad. Calculate the scale position in radians for a current of 2 A.*

Solution. The torque is given by

$$T = \frac{1}{2} I^2 \frac{dL}{d\theta}$$

Now, $$L = \left(8 + 4\theta - \frac{1}{2}\theta^2\right) \times 10^{-6}\ \text{H}$$

$$\therefore \quad \frac{dL}{d\theta} = \left(0 + 4 - 2.\frac{1}{2}\theta\right) \times 10^{-6}\ \text{H/rad} = (4 - \theta) \times 10^{-6}\ \text{H/rad}$$

Let the deflection be θ radians for a current of 2A, then

$$12 \times 10^{-6} \times \theta = \frac{1}{2} \times 2^2 \times (4 - \theta) \times 10^{-6}$$

$$12\,\theta = 8 - 2\theta \qquad \therefore \theta = 8/14 = \mathbf{0.571\ radian}$$

Tutorial Problems No. 15.1

1. The total resistance of a moving-iron voltmeter is 1000 Ω and the coil has an inductance of 0.765 H. The instrument is calibrated with a full-scale deflection on 50 V d.c. Calculate the percentage error when the instrument is used on (*a*) 25 Hz supply (*b*) 250 Hz supply, the applied voltage being 50 V in each case. **[(*a*) 0.72% (*b*)36%]**
2. A moving-iron voltmeter has full-scale deflection of 75 V and was calibrated to read correctly at 50 Hz when the current taken for full-scale deflection was 50 mA. When a capacitor of 6.36 μF was connected in series with the meter, the current taken was again 50 mA with 75 V applied. Calculate the percentage error in the reading if the meter were used on a 75 V d.c. supply. **[–1.4%]**

---

*Or reading = 15 × 500/500.3 = 14.95 V

**Similarly, reading = 15 × 500/505.8 = 14.83 V

15.13. Moving-coil Instruments

There are two types of such instruments (*i*) *permanent-magnet type* which can be used for d.c. only and (*ii*) the *dynamometer type* which can be used both for A.C. and D.C.

15.14. Permanent-magnet Moving-coil (PMMC) Type Instruments

The operation of a moving-coil permanent-magnet type instrument is based upon the principle that when a current-carrying conductor is placed in a magnetic field, it is acted upon by a force which tends to move it to one side and out of the field.

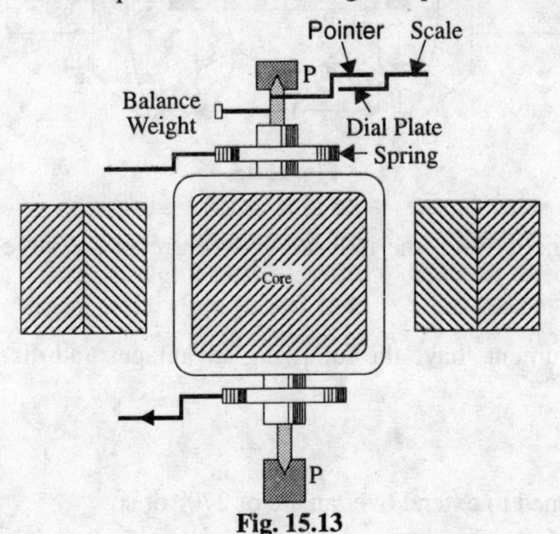

Fig. 15.13

Construction

As the name indicates, the instrument consists of a permanent magnet and a rectangular coil of many turns wound on light aluminium or copper former inside which is an iron core as shown in Fig. 15.13. the powerful U-shaped permanent magnet is made of Alnico and has solft-iron pole pieces which are bored out cylindrically. Between the magnetic poles is fixed a soft-iron cylinder whose function is (*i*) to make the field radial and uniform and (*ii*) to decrease the reluctance of the air path between the poles and hence increase the magnetic flux. Surrounding the core is a rectangular coil of many turns wound on a light aluminium frame which is supported by delicate bearings and to which is attached a light pointer. The aluminium frame not only provides support for the coil but also provides damping by eddy currents induced in it. The sides of the coil are free to move in the two air-gaps between the poles and core as shown in Fig. 15.13 and 15.14. Control of the coil movement is affected by two phosphor-bronze hair springs, one above and one below, which additionally serve the purpose of leading the current in and out of the coil. The two springs are spiralled in opposite directions for neutralizing the effects of temperature changes.

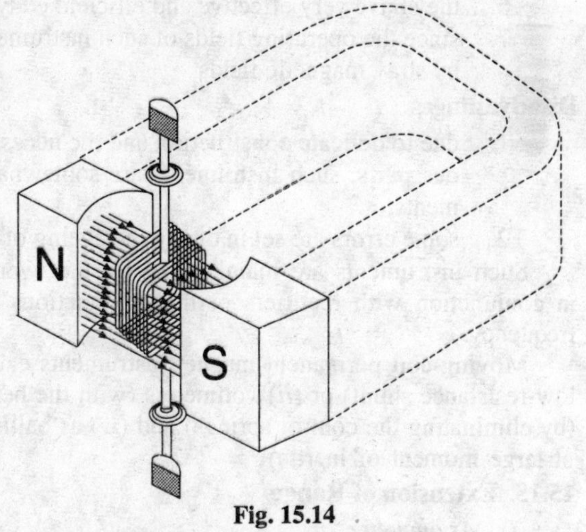

Fig. 15.14

Deflecting Torque

When current is passed through the coil, forces are set up on its both sides which produce a deflection torque as shown in Fig. 15.15. Let

B = flux density in Wb/m^2

l = length or depth of the coil in metres

b = breadth of coil in metres

N = number of turns in the coil

Then, if I amperes is the current passing through the coil, the magnitude of the force experienced by each of its sides is

$= BIl$ newton

For N turns, the force on each side of the coil is

$= NBIl$ newton

$\therefore$ deflecting torque T_d = force × perpendicular distance

$= NBIl \times b = NBI\,(l \times b) = NBIA$ N-m

where A is the face area of the coil*.

It is seen that if B is constant, then, T_d is proportional to the current passing through the coil.

Since such instruments are variably spring-controlled $T_c \propto$ deflection θ.

Since in the final deflected position $T_d = T_c$

$\therefore \quad \theta \propto I$

Hence, such instruments have uniform scales. Damping is electromagnetic *i.e.* by eddy currents induced in the metal frame over which the coil is wound. Since the frame moves in an intense magnetic field, the induced eddy currents are large and damping is very effective.

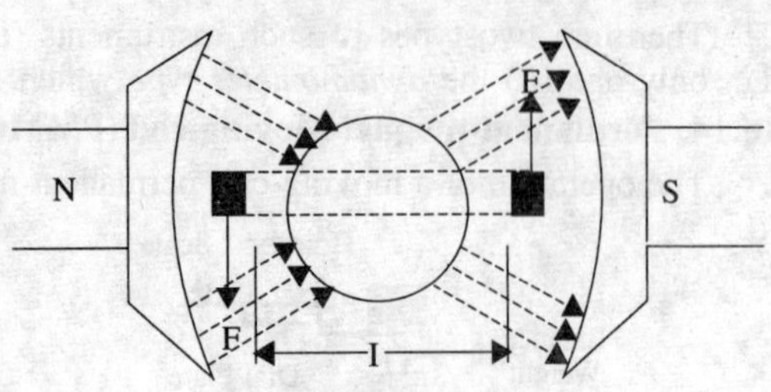

Fig. 15.15

Advantages and Disadvantages

The moving-coil permanent-magnet type instruments have the following advantages and disadvantages :

Advantages

1. they have low power consumption;
2. their scales are uniform and can be designed to extend over an arc of 270° or so;
3. they possess high (torque/weight) ratio;
4. they can be modified with the help of shunts and resistances to cover a wide range of currents and voltages;
5. they have no hysteresis loss;
6. they have very effective and efficient eddy-current damping.;
7. since the operating fields of such instruments are very strong, they are not much affected by stray magnetic fields

Disadvantages

1. due to delicate construction and the necessary accurate machining and assembly of various parts, such instruments are somewhat costlier as compared to moving-iron instruments;
2. some errors are set in due to the ageing of control springs and the permanent magnets.

Such instruments are mainly used for d.c. work alone, but they have been sometimes used in conjunction with rectifiers or thermo-junctions for a.c. measurements over a wide range of frequency.

Moving-coil permanent-magnet instruments can be used as (*i*) ammeters (with the help of a low-resistance shunt) or (*ii*) voltmeters (with the help of a high series resistance) (*iii*) flux-meters (by eliminating the control springs) and (*iv*) as ballistic galvanometers (by making control springs of large moment of inertia).

15.15. Extension of Range

(*i*) *As ammeter*

When such an instrument is used as an ammeter, its range can be extended with the help of a low-resistance shunt usually made of manganin and nichrome.

---

*Now $T_d = NBIl$ N-m. Suppose the deflection of the coil is θ and it changes by $d\theta$, then the change in the flux linked with the coil is $d\Phi = BA.d\theta$.

$$\frac{d\Phi}{d\theta} = BA \qquad \text{Hence,} \quad T_d = NIBA = NI\frac{d\Phi}{d\theta}$$

It is another useful expression for the deflecting torque.

This has been fully discussed in Art. 2-16 Fig. 15.16 (*a*) shows the actual arrangement whereas Fig. 15.16 (*b*) shows the diagrammatic representation.

(*ii*) *As voltmeter*

The range of this instrument, when used as a voltmeter can be increased by using a high resistance in series with it (Fig. 15.17).

Let I = full-scale deflection current

R_g = its resistance

ν = IR_g = full-scale p.d. across it

V = voltage to be measured

R = series resistance required.

Then, it is seen that the voltage drop across R is = $V - \nu$

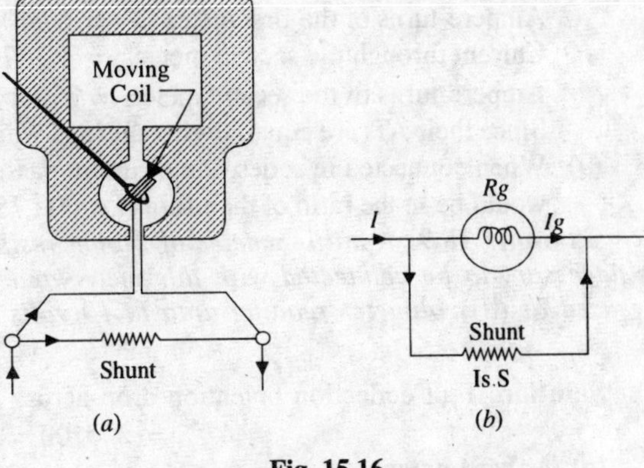

Fig. 15.16

$$\therefore \quad R = \frac{V-\nu}{I} \quad \text{or} \quad IR = V - \nu$$

Dividing both sides by υ, we get

$$\frac{IR}{\nu} = \frac{V}{\nu} - 1 \quad \text{or} \quad \frac{IR}{I.R_g} = \frac{V}{\nu} - 1$$

$$\therefore \quad \frac{V}{\nu} = \left(1 + \frac{R}{R_g}\right)$$

$$\therefore \quad \text{voltage multiplication} = \left(1 + \frac{R}{R_g}\right)$$

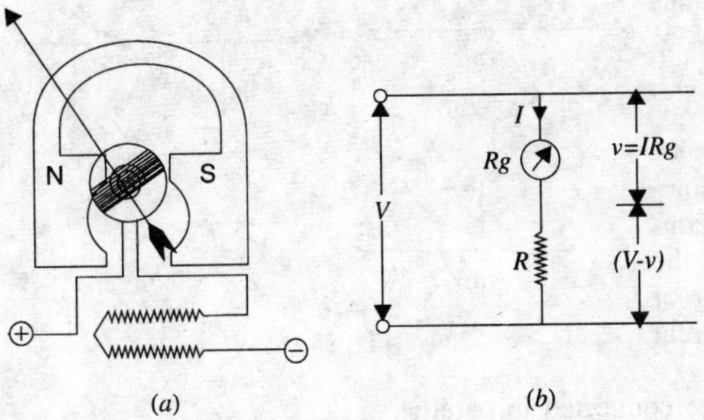

Fig. 15.17

Example 15.6. *A moving-coil voltmeter with a resistance of 10 ohms gives a full-scale deflection with a potential difference of 45 mV. The coil has 100 turns, an effective depth of 3 cm and a width of 2.5 cm. The controlling torque exerted by the spring is 49 × 10⁻⁶ Nm for full-scale deflection. Calculate the flux density in the gap.* **(A.M.I.E. Summer 1989)**

Solution. Full-scale deflection current, $I = 45/10 = 4.5$ mA

$$T_d = NBIlb = 100\,B \times 4.5 \times 10^{-2} \times 0.03 \times 0.025 = 0.3375 \times 10^{-3}\,B \text{ Nm}$$

Now $T_d = T_c$ —at full-scale deflection

$$\therefore \quad 0.3375 \times 10^{-3}\,B = 49 \times 10^{-6}$$

$$\therefore \quad B = \mathbf{0.145\ Wb/m^2}$$

Example 15.7. *A moving-coil ammeter is wound with 40 turns and gives full-scale deflection with 5 A. How many turns would be required on the same bobbin to give full-scale deflection with 20 A?*

Solution. Since full-scale deflection ampere-turns have to be the same in both cases,

$$\therefore \quad \text{No. of turns} = \frac{5}{20} \times 40 = \mathbf{10}$$

Example 15.8. *Two moving-coil meters differ only in that the moving-coil in one has 60 turns and a resistance of 10 Ω, while in the other it has 750 turns and 700 Ω resistance. Find the ratios of their deflections when*

(*a*) *each is connected in turn across a battery of 1.5 V e.m.f. and 50 Ω internal resistance,*

(*b*) *the two meters are connected in series across this battery.*

Solution. The readings of the two meters depend on their deflection torque which itself depends on the ampere-turns of the meter coil.

(*a*) Current through the first meter $= 1.5/(10 + 50) = 1/40$ A
Ampere-turns of the first meter $= 60/40 = 1.5$ AT
Current through the second meter $= 1.5/(700 + 50) = 1/500$ A
Ampere-turns of the second meter $= 750 \times 1/500 = 1.5$ AT
Since their ATs are equal, **their readings would also be equal.**

(*b*) When connected in series, same current passes through the meters. Hence, their deflection would be in the ratio of their turns *i.e.* 60 : 750 or **1 : 12.5.**

Example 15.9. *A milliammeter of 2 ohm resistance reads upto 100 mA. What resistance is necessary to be connected with this meter and how (in series or parallel) to enable it to be used as (a) voltmeter reading upto to 10 volts (b) ammeter reading upto 10 amperes.*

(Elements of Elect. Engg., Punjab Univ. 1993)

Solution. Full-deflection potention drop across the milliammeter is

$$= 2 \times 100 = 200 \text{ mV} = 0.2 \text{ V}$$

(*a*) As voltmeter

As shown in Fig. 15.18 (*a*), a resistance of *R* has been connected in series with the milliammeter and the two are connected across the supply lines. As seen, with full-deflection the voltage to be dropped across *R* is $= (10 - 0.2) = 9.8$ V. Hence, $R = 9.8/0.1 = 98\Omega$.

(*b*) As ammeter

As shown in Fig. 15.18 (*b*), a low-resistance shunt S has been connected across the milliameter and the shunted milliameter has been connected in one of the supply lines. Since, the milliammeter can carry a maximum current of 0.1 A, the shunt carries the rest of the line current *i.e.* $= (10 - 0.1) = 9.9$ A.

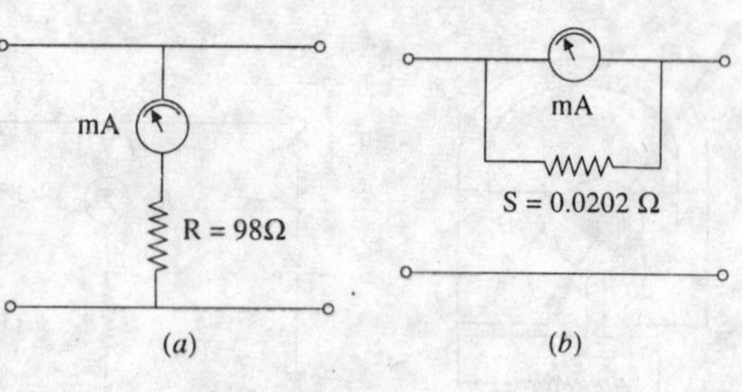

Fig. 15.18

Since the milliammeter and the shunt are connected in parallel.

$\therefore \quad 0.1 \times 2 = S \times 9.9, \; S = 0.2/9.9 = \mathbf{0.0202\ \Omega}$

Example 15.10. *A moving-coil instrument has a resistance of 5Ω between terminals and full-scale deflection is obtained with a current of 0.015 A. This instrument is to be used with a manganin shunt to measure 100 A full scale. Calculate the error caused by a 10°C rise in temperature.*

(*a*) *when the internal resistance of 5Ω is due to copper only,*

(*b*) *when a 4-W manganin resistor is used in series with a copper resistor of 1 Ω.*

The temperature-resistance coefficients of copper and manganin are 0.4% per °C and 0.015% per °C respectively.

Solution. Current through instrument $= 0.015$ A
Current through shunt $= 100 - 0.015 = 99.985$ A
Voltage across shunt $= 5 \times 0.015 = 0.075$ V
$\therefore$ shunt resistance $= 0.075/99.985 = 0.00075\ \Omega$
Shunt resistance after a rise of 10°C

$$= 0.00075\,(1 + 10 \times 0.00015) = 0.000751\ \Omega$$

In case (*a*), the instrument resistance (which is wholly copper) after a rise of 10°C

$$= 5(1 + 10 \times 0.004) = 5.2\ \Omega$$

Hence, current through the instrument corresponding to 100 A

$$= \frac{0.000751}{5.20075} \times 100 = 0.0144 \text{ A}$$

Reading of the instrument

$$= 0.01444 \times 100/0.015 = 96.3 \text{ A}$$

$\therefore$ percentage error $= 100 - 96.3 = 3.7$

In case (*b*), the instrument resistance after a rise of 10°C

$= 1(1 + 10 \times 0.004) + 4(1 + 10 \times 0.00015) = 5.046\ \Omega$

Instrument current with a line current of 100A

$= \dfrac{0.000751}{5.046751} \times 100 = 0.01488\ \text{A}$

$\therefore$ instrument reading $= 0.01488 \times 100/0.015 = 99.2\ \text{A}$

Percentage error $= 100 - 99.2 = \mathbf{0.8}$

Tutorial Problems No. 15.2

1. The flux density in the gap of a 1 mA (full scale) moving-coil ammeter is 0.1 Wb/m^2. The rectangular moving coil is 8 mm wide by 1 cm deep and is wound with 50 turns. Calculate the full-scale torque which must be provided by the springs. **[4×10^{-7} N–m]**

2. A moving-coil instrument has 100 turns of wire with a resistance of 10Ω, an active length in the gap of 3 cm and a width of 2 cm. A p.d. of 45 mV produces full-scale deflection. The control spring exerts a torque of 490.5 × 10^{-7} N-m at full-scale deflection. Calculate the flux density in the gap . **[0.1817 Wb/m^2**

3. A moving-coil insturment has a resistance of 75 Ω and gives a full-scale deflection of 100 scale divisions for a current of 1 mA. The instrument is connected in parallel with a shunt of resistance 25 Ω and the combination is then connected in series with a load and a supply. What is the current in the load when the instrument gives an indication of 80 scale divisions? **[3.2 A]**

4. The coil of a moving-coil meter has a resistance of 15 Ω and gives a full-scale deflection with 5 mA. Show how could this meter be adopted to measure 2A? If full-scale deflection were 90° of arc, what current would be required to give a deflection of 60°? **[375.9 mΩ ; 1.33 A]**

5. A moving-coil instrument which has a resistance of 12.5 Ω gives a full-scale deflection with 8 mA. It is required to adopt this instrument to read 25 A. Calculate the value of the resistor required.
The resistor is to be made of three manganin strips, connected in parallel, each strip being 0.5 mm thick and 1 cm wide. Calculate the length of each strip. The resistivity of manganin is 50 μΩ-cm. **[4.001 mΩ ; 12.003]**

6. A moving-coil voltmeter of range 0 – 0.5 V has the following details:
Moving-coil resistance of copper = 5Ω
Series resistor of manganing = 128.3 Ω
The meter was calibrated correctly at 18.5°C. What would be the error in the full-scale deflection (expressed as a percentage of full-scale deflection) if used at a temperature of 65.5°?
What would have been the error in the full-scale deflection if both resistors had been made of copper (resistance tempt. coeff. for copper 1/234.5 per °C and for manganing assume zero) **[–0.71% ; –15.8%]**

7. A moving-coil instrument, which gives full-scale deflection with 0.015 A, has a copper coil having a resistance of 1.5 Ω at 15°C and a temperature coefficient of 1/234.5 at 0°C in series with a swamp resistance of 3.5 Ω having a negligible temperature coefficient.
Determine (*a*) the resistance of shunt required for a full-scale deflection of 20 A and (*b*) the resistance required for a full-scale deflection of 250 V.
If the instrument reads correctly at 15°C, determine the percentage error in each case when the temperature is 25°C. **[(*a*) 0.00376 ; 1.3% (*b*) 16,662 Ω, negligible]**

8. The following data refer to a moving-coil voltmeter ; resistance = 10,000 Ω, dimensions of coil = 3 × 3 cm, number of turns of coil = 100, flux density in air-gap = 0.08 Wb/m, stiffiness of springs = 3 × 10^{-6} N-m per degree. Find the deflection produced by 110 V. **[48.2°]**

9. A moving-coil instrument has a resistance of 10 ohms and gives a full-scale deflection when carrying 50 mA. Show how it can be adopted to measure voltage upto 500 volts and currents upto 50 amperes. **[0.01 Ω ; 999 Ω]** ***(A.M.I.E. Winter 1983)***

15.16. Electrodynamic or Dynamometer Type Instruments

An electrodynamic instrument is a moving-coil instrument in which the operating field is produced, not by a permanent magnet, but by another fixed coil. This instrument can be used either as an ammeter or as voltmeter but is generally used as a wattmeter.

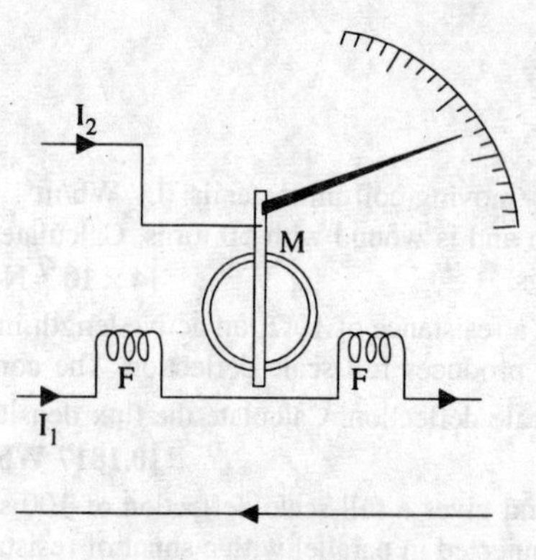

Fig. 15.19

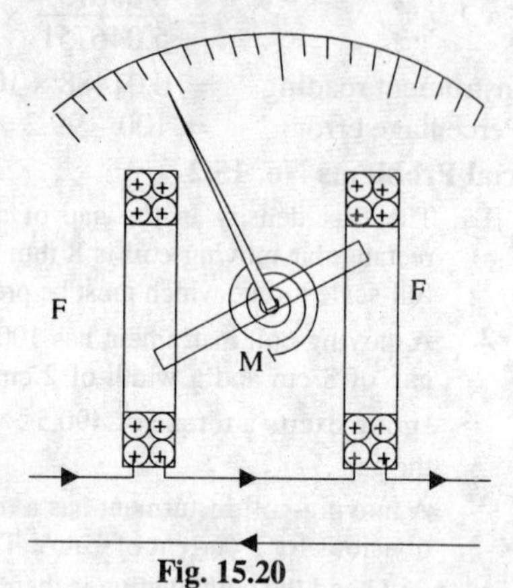

Fig. 15.20

As shown in Figs. 15.19 and 15.20, the fixed coil is usually arranged in two equal sections *F* and *F* placed close together and parallel to each other. The two fixed coils are air-cored to avoid hysteresis effects when used on a.c. circuits. This has the effect of making the magnetic field in which moves the moving coil *M*, more uniform, The moving coil is spring-controlled and has a pointer attached to it as shown.

Deflection Torque

The production of the deflecting torque can be understood from Fig. 15.21. Let the current passing through the fixed coil be I_1 and that through the moving coil be I_2. Since there is no iron, the field strength and so the flux density is proportional to I_1.

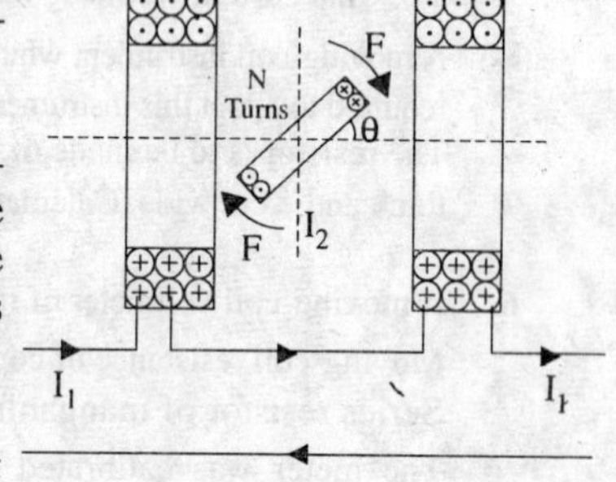

Fig. 15.21

$\therefore \quad B = KI_1$ where K is a constant.

Let us assume, for simplicity, that the moving coil is rectangular (it can be circular also) and of dimensions $l \times b$. Then, force on each side of the coil having N turns is = (NBI_2l) newton.

The turning moment or deflecting torque on the coil is given by

$$T_d = (NBI_2lb) = (NKI_1I_2lb) \text{ N-m}$$

Now putting $NKlb = K_1$ we have

$$T_d = K_1I_1I_2, \text{ where } K_1 \text{ is another constant*.}$$

---

*It has been shown on page 277 that the value of torque for an instrument of inductance L and for a current of I is $\quad T = \frac{1}{2}I^2\frac{dL}{d\theta}$

The equivalent inductance of the fixed and moving coils when joined in series is

$$L = L_1 + L_2 + 2M$$

where M is the mutual inductance between them and L_1 and L_2 are their individual self-inductances

Since L_1 and L_2 are fixed

$$\therefore \quad \frac{dL}{d\theta} = 2\frac{dM}{d\theta} \qquad \therefore \quad T = \frac{1}{2}I^2 \times 2\frac{dM}{d\theta} = I^2\frac{dM}{d\theta}$$

If the current in the fixed and moving coils are different, say, I_1 and I_2, then $T = I_1 I_2 \frac{dM}{d\theta}$

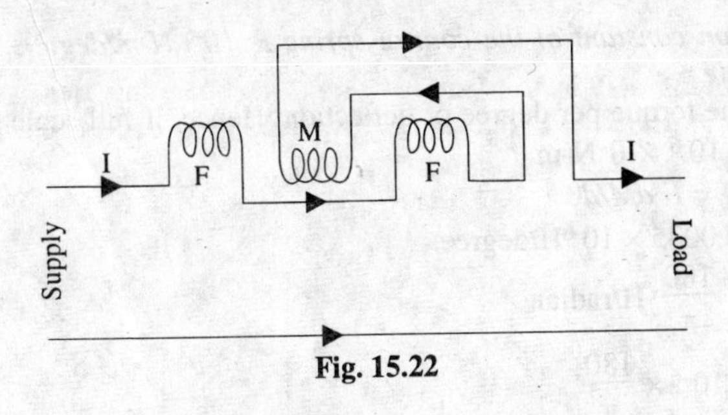

Fig. 15.22

It shows that the deflecting torque is proportional to the product of the currents flowing in the fixed coils and the moving coil. Since the instrument is spring controlled, the restoring or controlling torque is proportional to the angular deflection θ.

$\therefore \quad T_c \propto \theta \qquad$ or $\qquad T_c = K_2\theta$

In the final deflected position, the two torques are equal and opposite

$\therefore \quad T_d = T_c \quad$ or $\quad K_1 I_1 I_2 = K_2\theta$

$\therefore \quad \theta \propto I_1 I_2$

(*i*) When the instrument is used as an ammeter, the same current passes through both the fixed and the moving coils as shown in Fig. 15.22

In that case $I_1 = I_2 = I$, hence $\theta \propto I^2$ or $I \propto \sqrt{\theta}$. The connections of Fig. 15.22 are used when small currents are to be measured. In the case of heavy currents, a shunt *S* is used to limit current through the moving coil as shown in Fig. 15.13.

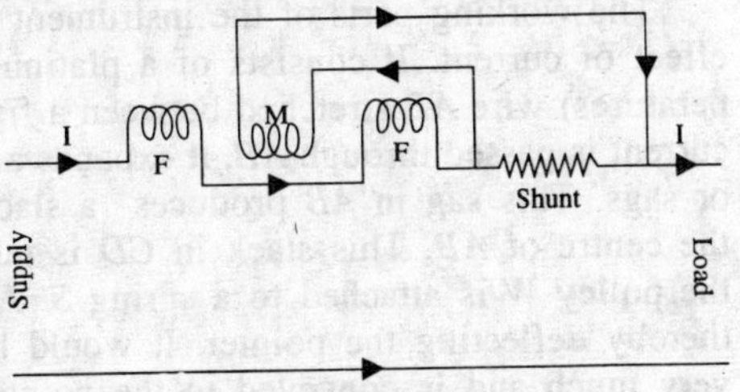

Fig. 15.23

(*ii*) When used as a voltmeter, the fixed and moving coils are joined in series along with a high resistance and connected as shown in Fig. 15.24. Hence, again $I_1 = I_2 = I$ where $I = V/R$ in d.c. circuits and $I = V/Z$ in a.c. circuits.

$\therefore \quad \theta \propto V \times V$ or $\theta \propto V^2 \qquad$ or $\qquad V \propto \sqrt{\theta}$

Hence, it is found that whether the instrument is used as an ammeter or voltmeter, its scale is uneven throughout the whole of its range and is particularly cramped or crowded near the zero.

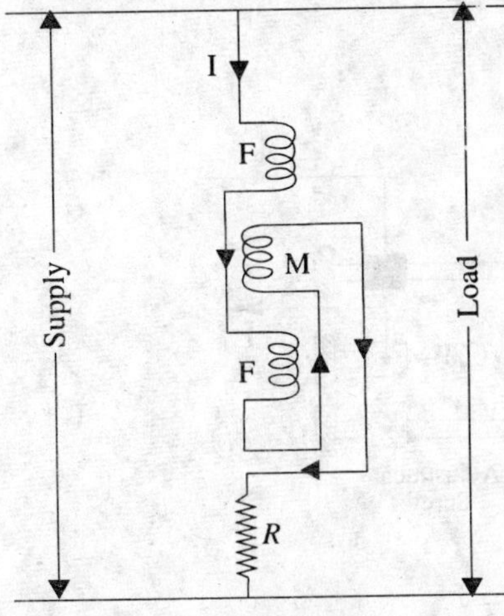

Fig. 15.24

Damping is pneumatic since, owing to weak operating field, eddy current damping is inadmissible. Such instruments can be used for a.c. and d.c. measurements. But it is more expensive and inferior to a moving-coil instrument for d.c. measurements.

As mentioned earlier, the most important application of electrodynamic principle is for wattmeters and is discussed in details elsewhere.

Errors

Since the coils are air-cored, the operating field produced is small. For producing an appreciable deflecting torque, a large number of turns is necessary for the moving coil. The magnitude of the current is also limited because the two control springs are used both for leading in and for leading out the current. Both these factors lead to a heavy moving system resulting in frictional losses which are somewhat larger than in other types and so frictional errors tend to be relatively higher. The current in the field coils is limited for the fear of heating the coils which results in the increase of their resistance. A good amount of screening is necessary to avoid the influence of stray fields.

Advantages and Disadvantages

1. Such instruments are free from hysteresis and eddy-current errors.
2. Since (torque/weight) ratio is small, such instruments have low sensitivity.

Example 15.11. *The mutual inductance of a 25-A electro-dynamic ammeter changes uniformly* *at the rate of 0.0035 μ H /degree. The torsion constant of the control spring is 10^{-6} N-m/degree. Determine the angular deflection at full scale.*

Solution. By torsion constant is meant the torque per degree of deflection. Hence, if full-scale deflection is θ°, then torque on full scale = $10^{-6} \times \theta$ N-m

Now, for an electrodynamic instrument, $T = I^2\, dM/d\theta$

Here $I = 25\text{A},\ dM/d\theta = 0.0035 \times 10^{-6}\text{H/degree}$

$$= 0.0035 \times 10^{-6} \times \frac{180}{\pi}\ \text{H/radian}$$

$$\therefore \quad 10^{-6} \times \theta^\circ = 25^2 \times 0.0035 \times 10^{-6} \times \frac{180}{\pi}$$

$$\therefore \quad \theta = 25^2 \times 0.0035 \times \frac{180}{\pi}\ \text{degrees} = \mathbf{125.4^\circ}$$

15.17. Hot-wire Instruments

The working parts of the instrument are shown in Fig. 15.25. It is based on the heating effect of current. It consists of a platinum-iridium (it can withstand oxidation at high temperatures) wire *AB* stretched between a fixed end *B* and a tension adjusting screw at *A*. When current is passed through *AB*, it expands according to I^2R formula. Due to heating, *AB* expands or sags. This sag in *AB* produces a slack in phosphor-bronze tension wire *CD* attached to the centre of *AB*. This slack in *CD* is taken up by the silk fibre which after passing round the pulley *W* is attached to a spring *S*. As the silk thread is pulled by *S*, the pulley rotates, thereby deflecting the pointer. It would be noted that even a small sag in *AB* is magnified very much and is conveyed to the pointer. Expansion of *AB* is magnified by *CD* which is further magnified by the silk thread.

It will be seen that the deflection of the pointer is proportional to the extension of *AB* which is itself proportional to I^2. Hence, deflection is $\propto I^2$. If spring control is used, then $T_c \propto \theta$.

Hence, $\theta \propto I^2$

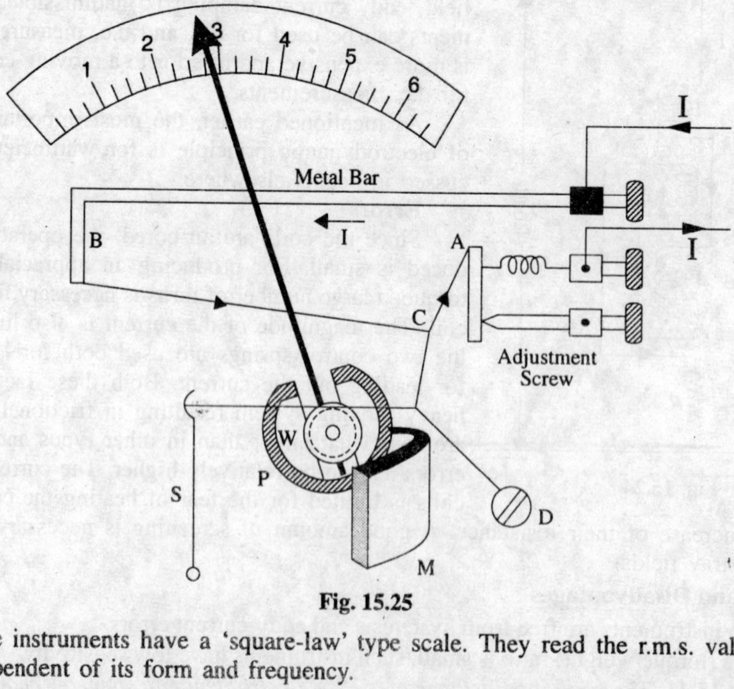

Fig. 15.25

So, these instruments have a 'square-law' type scale. They read the r.m.s. value of current and are independent of its form and frequency.

Damping

A thin light aluminium disc P is attached to the pulley such that its edge moves between the poles of a permanent magnet M. Eddy currents produced in this disc give the necessary damping.

These instrument are primarily meant for being used as ammeters but can be adopted as voltmeters by connecting a high resistance in series with the insturment. These instruments are suited both for alternating and direct currents.

Advantages of Hot-wire Instruments

1. As their deflection depends on the r.m.s. value of the alternating current, they can be used on direct current also.
2. Their readings are independent of waveform and frequency.
3. The calibration is the same both for a.c. as well d.c. measurements.
4. They are unaffected by stray fields.

Disadvantages

1. They are unable to withstand overload without burning out.
2. They have a cramped scale.
3. They are sluggish owing to the time taken up for the wire to heat up.
4. They have a high power consumption as compared to moving coil instruments. Current consumption is about 200 mA at full load.
5. Their zero position needs frequent adjustment.
6. They are fragile.

15.18. Range Extension

The current-carrying capacity of such instruments is limited owing to the fineness of the wire. Normally, they can carry a current of 5 A without shunt. But for higher currents, a shunt is essential. Its range as a voltmeter can be extended by connecting a high resistance in series with the hot wire.

15.19. Induction Type Instruments

Induction type instruments are used only for a.c. measurement and can be used either as ammeter, voltmeter or wattmeter. However, the induction principle finds its widest application as a watt-hour or energy meter. In such instruments, the deflecting torque is produced due to the reaction between the flux of an a.c. magnet and the eddy currents induced by this flux. Before discussing the two types of most commonly-used induction instruments, we will first discuss the underlying principle of their operation.

Principle

The operation of all induction instruments depends on the production of torque due to the reaction between a flux Φ_1 (whose magnitude depends on the current or voltage to be measured) and eddy currents induced in a metal disc or drum by another flux Φ_2 (whose magnitude also depends on the current or voltage to be measured). Since the magnitude of eddy currents also depends on the flux producing them, the *instantaneous* value of deflecting torque is proportional to the square of the current or voltage under measurement and the value of *mean* deflecting torque is proportional to the mean square of the current or voltage.

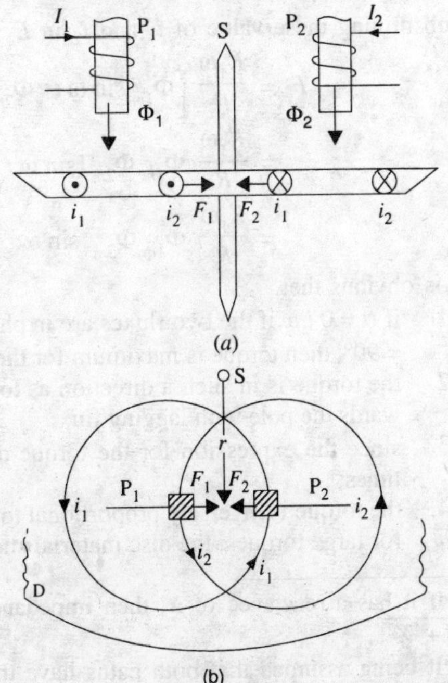

Fig. 15.26

Consider a thin aluminium or Cu disc D free to rotate about an axis passing through its centre, as shown in Fig. 15.26. Two a.c. magnetic poles P_1 and P_2 produce alternating fluxes Φ_1 and Φ_2 which cut this disc. Consider any annular portion of the disc around P_1 with centre on the axis of P_1. This protion will be linked by Φ_1 and so an alternating e.m.f. e_1 will circulate the eddy current i_1 which, as shown in Fig. 15.26 (*b*) will pass under P_2. Similarly, Φ_2 will induce an e.m.f. e_2 which will further induce the eddy current i_2 in the annular portion of the disc around P_2. This eddy current i_2 flows under pole P_1

Let us take the downward directions of fluxes as positive and further assume that at the instant under consideration, both Φ_1 and Φ_2 and carrying current i_1 and i_2 can be found and are as indicated in Fig. 15.26 (*b*).

The portion of the disc which lies in flux Φ_1 and carries current i_2 experiences a force F_1 along the direction as indicated ($\because\ F = Bll$). The force $F_1 \propto \Phi_1\, i_2$. Similarly, the portion of the disc lying in flux Φ_2 and carrying current i_1 experiences a force $F_2 \propto \Phi_2 i_1$

$$\therefore \quad F_1 \propto \Phi_1 i_2 = K\Phi_1 i_2 \quad \text{and} \quad F_2 \propto \Phi_2 i_2 = K\Phi_2 i_1$$

It is assumed that constant K is the same in both cases due to the symmetrical positions of P_1 and P_2 with respect to the disc.

If r is the effective radius at which these forces act, then the instantaneous torque T acting on the disc which is the *difference of* the two torque, is given by

$$T = r(K\Phi_1 i_2 - K\Phi_2 i_1) = K_1(\Phi_1 i_2 - \Phi_2 i_1) \quad ...(i)$$

Let the alternating flux Φ_1 be given by $\Phi_1 = \Phi_{1m}$ sin ωt. The flux Φ_2 which is assumed to lag Φ_1 by angle α radian, is given by $\Phi_2 = \Phi_{2m}$ sin $(\omega t - \alpha)$.

$$\text{Induced e.m.f.,} \quad e_1 = \frac{d\Phi_1}{dt} = \frac{d}{dt}\left(\Phi_{1m} \sin \omega t\right) = \omega\,\Phi_{1m} \cos \omega t$$

Assuming the eddy current path to be purely resistive and of value R*, the value of eddy current is

$$i_1 = \frac{e_1}{R} = \frac{\omega\,\Phi_{1m}}{R} \cos \omega t$$

$$\text{Similarly,} \quad e_2 = \omega\,\Phi_{2m} \cos(\omega t - \alpha) \quad \text{and} \quad i_2 = \frac{\omega\,\Phi_{2m}}{R} \cos(\omega t - \alpha)^{**}$$

Substituting these value of i_1 and i_2 in E_q. (*i*), above, we get

$$T = \frac{K_1\omega}{R}\left[\Phi_{1m} \sin \omega t \,.\, \Phi_{2m} \cos(\omega t - \alpha) - \Phi_{2m} \sin(\omega t - \alpha)\, \Phi_{1m} \cos \omega t\right]$$

$$= \frac{K_1\omega}{R}\,\Phi_{1m}\,\Phi_{2m}\,[\sin \omega t \cos(\omega t - \alpha) - \cos \omega t \sin(\omega t - \alpha)]$$

$$= \frac{K_1\omega}{R}\,\Phi_{1m}\,\Phi_{2m} \sin \alpha = K_2\omega\,\Phi_{1m}\,\Phi_{2m} \sin \alpha \qquad (\text{put } K_1/R = K_2)$$

It is obvious that

1. if $\alpha = 0$ *i.e.* if the two fluxes are in phase, then net torque is zero. If, on the other hand, $\alpha = 90°$, then torque is maximum for the given values of Φ_{1m} and Φ_{2m}.;
2. the torque is in such a direction as to rotate the disc from the pole with leading flux towards the pole with lagging flux;
3. since the expression for the torque does not involve 'ωt', it has the same value at all times;
4. the torque is inversely proportional to R—the resistance of the eddy current path. Hence, for large torques, the disc material should have low resistivity. Usually, it is made of Cu

---

* If it has a reactance of X, then impedance Z should be taken whose value is given by $Z = \sqrt{R^2 + X^2}$.

**It being assumed that both paths have the same resistance.

or, more often, of aluminium.

15.20. Induction Ammeters

It has been shown above that

$$T = K_2 \omega \Phi_{1m} \Phi_{2m} \sin \alpha$$

Obviously, if both fluxes are produced by the alternating current (of maximum value I_m) to be measured, then

$$T = K_3 \omega Im^2 \sin \alpha$$

Henre, for a given frequency and angle α, the torque is proportional to the square of the current. If the disc has spring control, it well take up a steady deflected position where controlling torque becomes equal to the deflecting torque. By attaching a suitable pointer to the disc, the device can be used as an ammeter.

Two possible arrangements, by which the operational requirements of induction/ammetre can be net are discussed below.

15.21. Disc Ammeter with Split-phase Windings

In this arrangement, the windings on the two laminated a.c. magnets P_1 and p_2 are connected in series. But the winding of P_2 is shunted by a resistance R with the result that the current. In this way necessary phase anbgle α is produced between two fluxes, Φ_1 and Φ_2 produced by P_1 and P_2 respectively. This angle is of the order of 60°. If the hysteresis effects etc are neglected, then each flux will be proportional to the current to be measured *i.e.* line current I.

$\therefore \quad T_d \propto \Phi_{1m} \Phi_{2m} \sin \alpha \quad$ or $\quad T_d \propto i^2 \quad$ —where I is the r.m.s. value

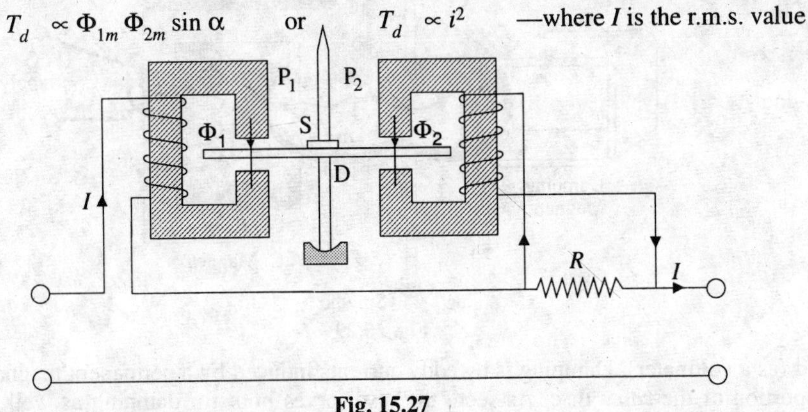

Fig. 15.27

If spring control is used, then $T_c \propto \theta$

In the final deflected position, $T_d = T_c \quad \therefore \quad \theta \propto I^2$

Eddy current damping is employed in this instrument. When the disc rotates, it cuts the flux in the air-gap of the magner and has eddy currents induced in it which provided efficient damping.

15.22. Shaded-pole Induction Ammeters

In the shaded-pole disc type induction ammeter (Fig. 15.28) only one single flux-producing winding is use. The flux Φ produced by this winding is split up into two fluxes Φ_1 and Φ_2 which are made to have the necessary phase difference of α by the device shown in Fig. 15.28. The portions of the upper and lower poles near the disc D are divided by a slot into two halves, one of which carries a closed 'shading' winding or ring. This shading winding or ring acts as a shortcircuited secondary and the main winding as the primary. The current induced in the ring by

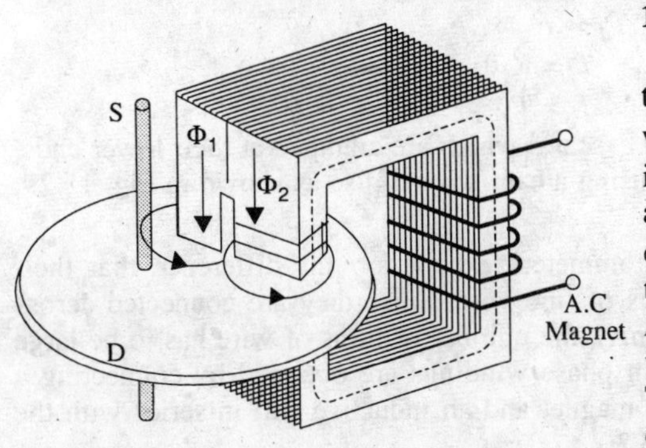

Fig. 15.28

transformer action retards the phase of flux Φ_2 with respect to that of Φ_1 by about 50° or so. The fluxes Φ_1 and Φ_2 passing through unshaded and the shaded part, react with eddy currents i_2 and i_1 respectively and so produce the driving torque whose value is $T_d \propto \Phi_{1m}\Phi_{2m} \sin \alpha$

Assuming that both Φ_1 and Φ_2 are proportional to the current I we have $T_d \propto I^2$

This torque is balanced by the controlling torque provided by the springs.

The actual shaded-pole type induction instrument is shown in Fig. 15.29. It consists of a suitably-shaped aluminium or copper disc mounted on a spindle which is supported by jewelled bearings. The spindle carries a pointer and has a control spring attached to it. The edge or periphery of the disc moves in the air-gap of a laminated a.c. electromagnet which is energised either by the current to be measured (as ammeter) or by the current proportional to the voltage to be

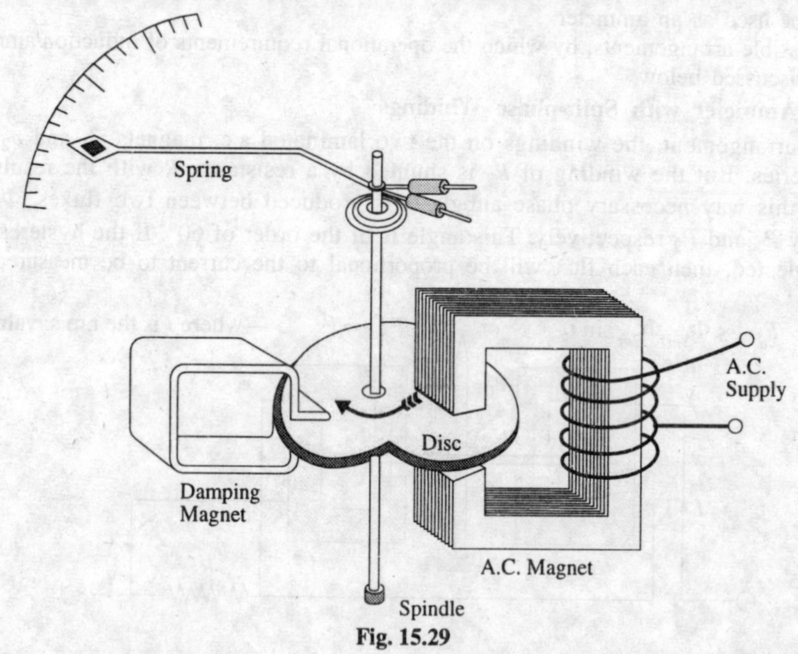

Fig. 15.29

measured (as a voltmeter). Damping is by eddy currents induced by a permanent magnet embracing another portion of the *same* disc. As seen, the disc serves both for damping as well as operating purposes. The main flux is split into two component fluxes by shading one-half of each pole. These two fluxes have a phase difference of 40° to 50° between them and they induce two eddy currents in the disc. The fluxes and eddy currents produce a resultant torque which *deflects* the disc—continuous rotation being prevented by the control spring and the deflection produced is proportional to the square of the current or voltage being measured.

As seen from above, for a given frequency $T_d \propto I^2 = KI^2$

For spring control $T_c \propto \theta$ or $T_c = K_1\theta$

For steady deflection, we have $T_c = T_d \quad \therefore \quad \theta \propto I^2$

Hence, such instruments have uneven scales *i.e.* scales which are cramped at their lower ends. A more even scale can, however, be obtained by using a cam-shaped disc as shown in Fig. 15.29.

15.23. Induction Voltmeters

Their construction is similar to that of the ammeters except for this difference that their windings are wound with a large number of turns of fine wire. Since they are connected across the lines, they carry very small currents (5 – 10 mA), the number of turns of wire has to be large for producing an adequate amount of m.m.f. Split phase windings are obtained by connecting a high resistance in series with the winding of one magnet and an inductive coil in series with the winding of the other magnet is shown in Fig. 15.30.

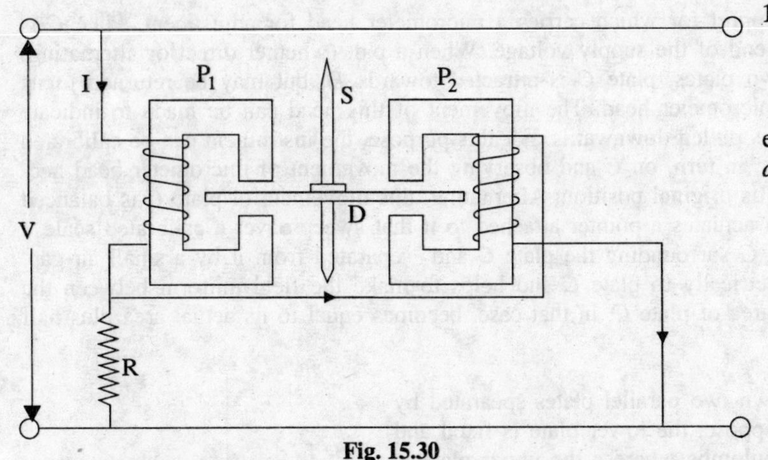

Fig. 15.30

15.24. Errors in Induction Ammeters and Voltmeters

There are two types of errors (*i*) *frequency error and (ii) temperature* error.

1. Since deflecting torque depends on frequency, hence unless the alternating current to be measured has the same frequency with which the instrument was calibrated, there will be large error in its readings. Frequency error can be compensated for by the use of a non-inductive shunt in the case of ammeters. In voltmeters, such errors are not large and, to a great extent, are self-compensating.
2. Serious errors may occur due to the variations of temperature, because the resistances of eddy current paths depend on the temperature. Such errors can, however, be compensated for by shunting in the case of ammeters and by a combination of shunt and swamping resistance in the case of voltmeters.

15.25. Advantages and Disadvantages

1. A full-scale deflection of over 200° can be obtained with such instruments. Hence, they have long open scales.
2. Damping is very efficient.
3. They are not much affected by external stray fields.
4. Their power consumption is fairly large and cost relatively high.
5. They can be used for a.c. measurements only.
6. Unless compensated for frequency and temperature variations, serious errors may be introduced.

15.26. Electrostatic Voltmeters

Electrostatic instruments are almost always used as voltmeters and that too more as a laboratory rather than as industrial instruments. The underlying principle of their operation is the force of attraction between electric charges on neighbouring plates between which a p.d. is maintained. This force gives rise to a deflecting torque. Unless the p.d. is sufficiently large, the force is small. Hence, such instruments are used for the measurement of very high voltages.

There are two general types of such instruments:

(*i*) *the quadrant type*-used up to 20 kV,

(*ii*) the attracted-disc type - used up to 500 kV

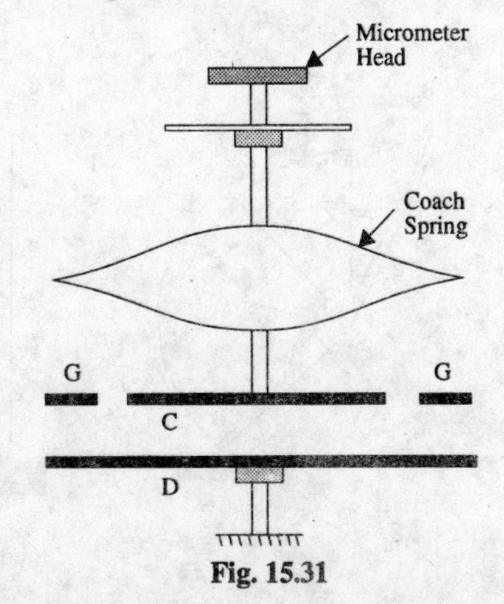

Fig. 15.31

15.27. Attracted-disc Type Voltmeter

As shown in Fig. 15.31, it consists of two discs or plates C and D mounted parallel to each other. Plate D is fixed and is earthed while C is suspended by a coach spring, the support for which carries a micrometer head for adjustment. Plate C is connected to the positive end of the supply voltage. When a p.d. (whether direct or alternating) is applied between the two plates, plate C is attracted towards D, but may be returned to its original position by the micrometer head. The movement of this head can be made to indicate the force F with which C is pulled downwards. For this purpose, the instrument can be calibrated by placing known weights, in turn, on C and observing the movement of micrometer head necessary to bring C back to tis original position. Alternating, this movement of plate C is balanced by a control device which actuates a pointer attached to it that sweeps over a calibrated scale.

There is a guard ring G surrounding the plate C and separated from it by a small air-gap. The ring is connected electrically to plate C and helps to make the field uniform between the two plates. The effective area of plate C, in that case, becomes equal to its actual area plus half the area of the air-gap.

Theory

In Fig. 15.32 are shown two parallel plates spearated by a distance of x metres. Suppose, the lower plate is fixed and carries a charge of $-Q$ coulombs whereas the upper plate is movable and carries a charge of $+Q$ coulombs. Let the mutual force of attraction between the two plates be F newtons. Suppose, the upper plate is moved apart by a distance dx. The, mechanical work done during this movement is $F \times dx$ joules. Since charge on the plate is constant, no electrical energy can move into the system from outside. This work is done at the cost of the energy stored in the parallel-plate capacitor formed by the two plates.

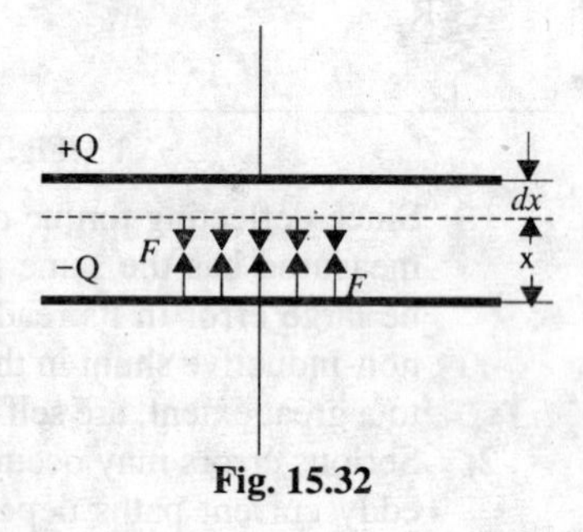

Fig. 15.32

$$\therefore \quad F \times dx = d\left(\frac{1}{2}\ CV^2\right)$$

Now $\quad C = \in A / x$ where x is the plate separation

$$\therefore \quad F \times dx = d\left(\frac{1}{2}\ \frac{\in A}{x}\ .\ V^2\right) = \frac{\in A V^2}{2}\ .\ d\left(\frac{1}{x}\right) = -\frac{1}{2}\ \frac{\in A V^2}{x^2}\ .\ dx \qquad \therefore\ F = -\frac{1}{2}\ V^2\ \frac{\in A}{x^2}$$

Hence, we find that force is directly proportional to the square of the voltage to be measured. The negative sign merely shows that it is a force of attraction.

15.28. Quadrant Type Voltmeters

The working principle and basic construction of such instruments can be understood from Fig. 15.33. A light aluminium vane C is mounted on a spindle S and is situated partially within

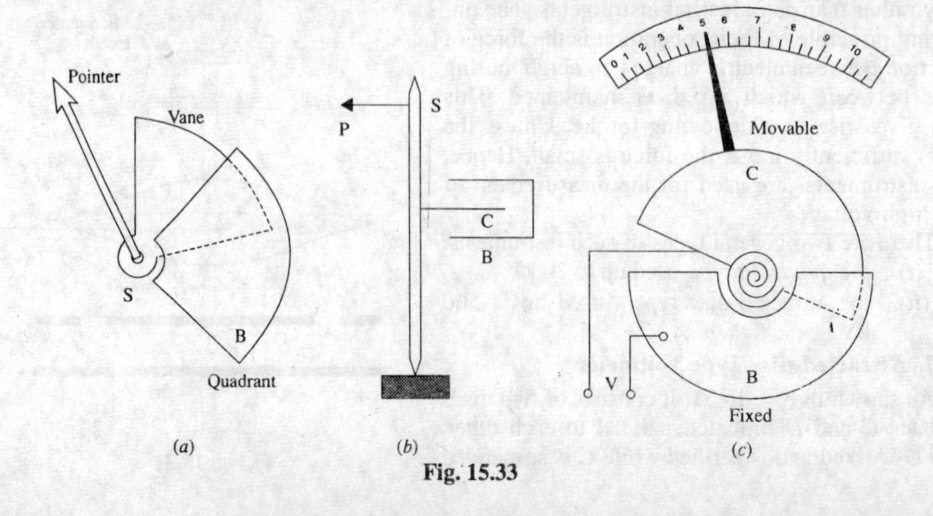

Fig. 15.33

a hollow metal quadrant *B*. Alternatively, the vane may be suspended in the quadrant. When the vane and the quadrant are oppositely-charged by the voltage under measurement, then vane is further attracted inwards into the quadrant thereby causing the spindle and hence the pointer to rotate. The amount of rotation and hence the deflecting torque is found proportional to V^2. The deflecting torque in the case of arrangement shown in Fig. 15.33 is very small unless *V* is extremely large. The force on the vane may be increased by using a greater number of quadrants and a double-ended vane. Such voltmeters have an uneven scale [Fig 15.33 (*c*)]. Controlling torque is produced by torsion of the suspension strip or by the spring (used in pivoted type voltmeters). Damping is by a disc or vane immersed in oil in the case of suspended type or by air friction in the case of pivoted type instruments.

Theory

With reference to Fig. 15.33, suppose the quadrant and vane are connected across a source of *V* volts and let the resultant deflection be θ. If *C* is the capacitance between the quadrant and vane in the deflected position, then the charge on the instrument will be *CV* coulombs. Suppose the voltage is changed from *V* to $(V + dV)$, then as a result, let θ, *C* and *Q* change to $(\theta + d\theta)$, $(C + dC)$ and $(Q + dQ)$ respectively. Then, the energy stored in the electrostatic field is increased by $dE = d\left(\frac{1}{2}CV^2\right) = \frac{1}{2}V^2 . dC + CV . dV$ joules.

If *T* newton-metre is the value of controlling torque corresponding to a deflection of θ, then the additional energy stored in the control system will be $T \times d\theta$ joules.

Total increase in stored energy

$$= T \times d\theta + \frac{1}{2}V^2\, dC + C\,V\,dV \text{ joules}$$

It is seen that during this change, the source supplies a charge dQ at potential *V*. Hence, the value of energy supplied is

$$= V \times dQ = V \times d(CV)\ V^2 \times dC + CV . dV$$

Since the energy supplied by the source must be equal to the extra energy stored in the field and the control,

$$\therefore\ T \times d\theta + \frac{1}{2}V^2\ dC + CV . dV = V^2 . dC + CV . dV$$

$$T \times d\theta = \frac{1}{2}V^2 . dC \quad \therefore\ T = \frac{1}{2}V^2\frac{dC}{d\theta} N\text{–}m$$

The torque is found to be proportional to the square of the voltage to be measured whether that voltage is alternating or direct. However, on alternating circuits the scale will read r.m.s. value.

15.29. Kelvin's Multicellular Voltmeter

As shown in Fig. 15.34, it is essentially a quadrant type instrument as described above but with the difference that instead of four quadrants and one vane it has a large number of fixed quadrants and vanes mounted on the same spindle. In this way, the deflecting torque for a given voltage is increased many times. Such voltmeters can be used to measure voltages as low as 30 V. As said above, this reduction in the minimum limit of voltage is due to the increase in the operating force in proportion to the number of component units. Such an instrument has a torsion head for zero adjustment and a coach spring for protection against accidental fracture of suspension due to vibration etc.

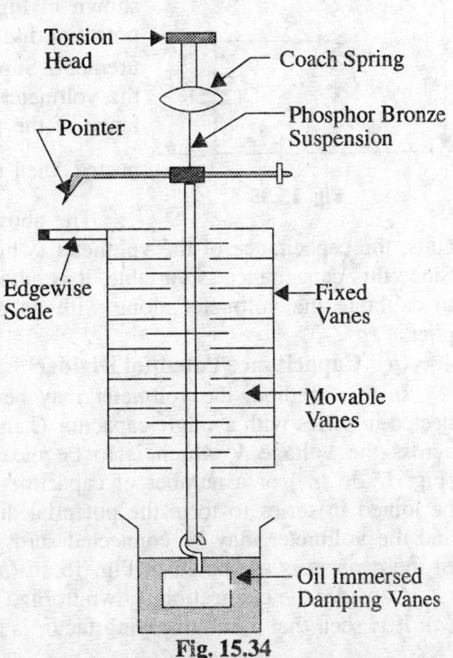

Fig. 15.34

There is a pointer and scale of edgewise pattern and damping is by a vane immersed in an oil dashpot.

15.30. Advantages and Limitations of Electrostatic Voltmeters.

Some of the main advantages and uses of electrostatic voltmeters are as follows--

1. They can be manufactured with first grade accuracy.
2. They give correct reading on both d.c. and a.c. circuits. On a.c. circuits, the scale will, however, read r.m.s. values whatever the wave-form.
3. Since no iron is used in their construction, such instruments are free from hysteresis and eddy current losses and temperature errors.
4. They do not draw any continuous current on d.c. circuits and that drawn on a.c. cirlcuits (due to the capacitance of the instrument) is extremely small. Hence, such voltmeters do not cause any disturbance to the circuits to which they are connected.
5. Their power loss is negligibly small.
6. They are unaffected by stray magnetic fields although they have to be guarded against any stray electrostatic field.
7. They can be used up to 1000 kHz without any serious loss of accuracy.

However, their main limitations are :

1. Low-voltage voltmeters (like Kelvin's Multicellular voltmeter) are liable to friction errors.
2. Since torque is proportional to the square of the voltage, their scales are not uniform although some uniformity can be obtained by suitably shaping the quadrants of the voltmeters.
3. They are expensive and cannot be made robust.

15.31. Range Extension of Electrostatic Voltmeters

The range of such voltmeters can be extended by the use of multipliers which are in the form of a resistance potential divider or capacitance potential divider. The former method can be used both for direct and alternating voltage.

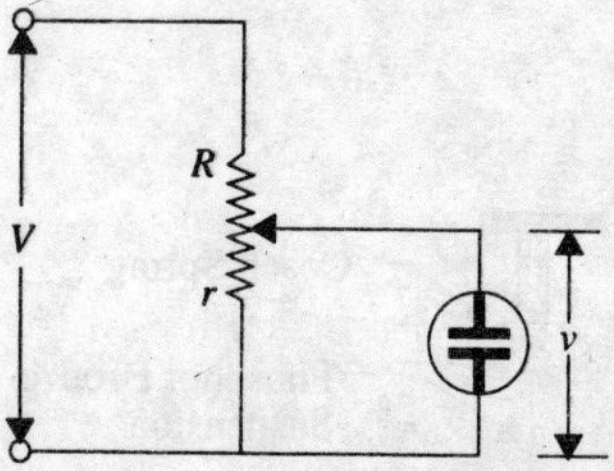

(*i*) Resistance Potential Divider

This divider consists of a high non-inductive resistance across a small portion of which is attached the electrostatic voltmeter as shown in Fig. 15.35. let R be the resistance of the whole of the potential divider across which is applied the voltage V under measurement. Suppose υ is the maximum value of the voltage which the voltmeter can measure without the multiplier. If r is the resistance of the portion of the divider across which voltmeter is connected, then the multiplying factor is given by $\frac{V}{\upsilon} = \frac{R}{r}$.

Fig. 15.35

The above expression is true for d.c. circuits, but for a.c. circuits, the capacitance of the voltmeter (which is in parallel with r) has to be taken into account. Since this capacitance is variable, it is advisable to calibrate the voltmeter along with its multiplier.

(*ii*) Capacitance Potential Divider

In this method, the voltmeter may be connected in series with a single capacitor C and put across the voltage V which is to be measured [Fig. 15.36 (*a*)] or a number of capacitors may be joined in series to form the potential divider and the voltmeter may be connected across one of the capacitors as shown in Fig. 15.36 (*b*)

Consider the connection shown in Fig. 15.36 (*a*). It is seen that the multiplying factor is given

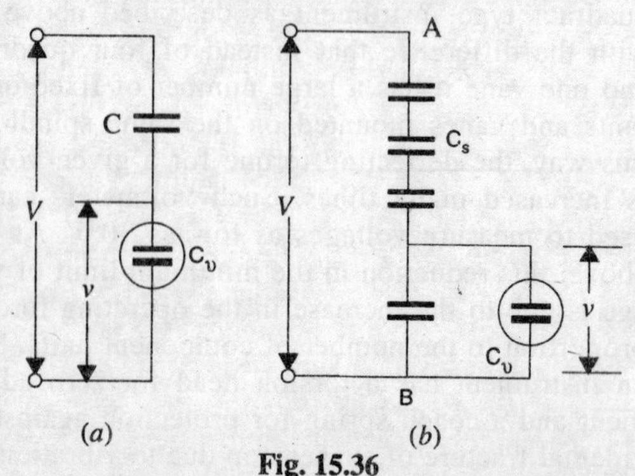

Fig. 15.36

by

$$\frac{V}{v} = \frac{\text{reactance of total circuit}}{\text{reactance of voltmeter}}$$

Now, capacitance of the total circuit is $\dfrac{C_v C}{C + C_v}$ and its reactance is

$$= \frac{1}{\omega \times \text{capacitance}} = \frac{C + C_v}{\omega C_v}$$

Reactance of the voltmeter $= \dfrac{1}{\omega C_v}$

$$\therefore \quad \frac{V}{v} = \frac{(C + C_v)/\omega CC_v}{1/\omega C_v} = \frac{C + C_v}{C}$$

$$\therefore \quad \text{Multiplying factor*} = \frac{C + C_v}{C} = 1 + \frac{C_v}{C}.$$

Example 15.12. *The reading '100' of a 120-V electrostatic voltmeter is to represent 10,000 volts when its range is extended by the use of a capacitor in series. If the capacitance of the voltmeter at the above reading is 70 μμF, find the capacitance of the capacitor multiplier required.*

Solution. Multiplying factor $= \dfrac{V}{v} = 1 + \dfrac{C_v}{C}$

Here, V = 10,000 Volt ; v = 100 volt

C_v = capacitance of the voltmeter = 70 μμF

C = capacitance of the multiplier

$$\therefore \quad \frac{10{,}000}{100} = 1 + \frac{70}{C} \quad \therefore \quad \frac{70}{C} = 99 \quad \therefore \quad \frac{70}{99} = \mathbf{0.7\mu\mu F\ approx.}$$

15.32. Wattmeters

We will discuss the two main types of wattmeters in general use, that is, (*i*) the dynamometer or electrodynamic type and (*ii*) the induction type.

15.33. Dynamometer Wattmeter

The basic principle of dynamometer instrument has already been explained in detail in Art 15.16. The connections of a dynamometer type wattmeter are shown in Fig. 15.37. The fixed circular coil, which carries the main circuit current I_1, is wound in two halves positioned parallel to each other. The distance between the two halves can be adjusted to give a uniform magnetic field. The moving coil, which is pivoted centrally, carries a current I_2 which is proportional to the voltage V. Current I_2 is led into the moving coil by two springs which also supply the necessary controlling torque.

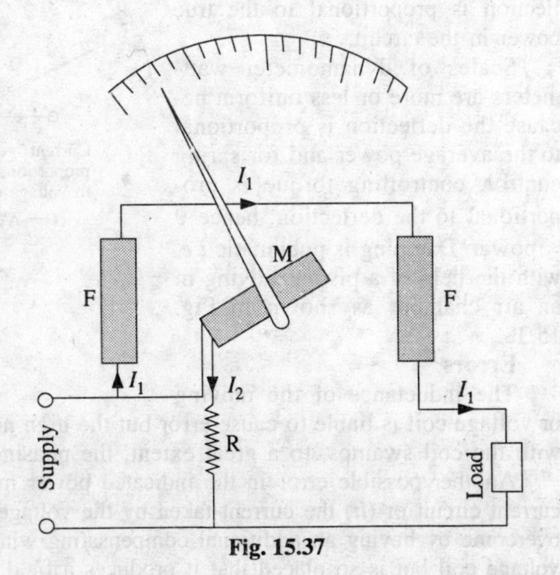

Fig. 15.37

Deflecting torque

Since coils are air-cored, the flux density produced is directly proportional to the current I_1.

*It is helpful to compare it with a similar expression in Art. 15.15 for permanent magnet moving-coil instruments.

ls are air-cored, the flux density produced is directly proportional to the current I_1.

$\therefore \quad B \propto I_1 \quad$ or $\quad B = K_1 I_1$; $\quad$ Current $\quad I_2 \propto V \quad$ or $\quad I_2 = K_2 V$

Now $\quad T_d \propto BI_2 \propto I_1 V \quad \therefore \quad T_d = KVI_1 = K \times$ power

In d.c.circuits, power is given by the product of voltage and current in amperes, hence the torque is directly proportional to power.

Let us see how this instrument indicates true power on a.c. circuits.

For alternating currents, the value of instantaneous torque is given by

$$T_{inst} \propto vi = Kvi$$

where $\quad v$ = instantaneous value of voltage across moving coil

i = instantaneous value of current through *fixed* coils.

However, owing to the large inertia of the moving system, the instrument indicates the mean or average power.

$\therefore \quad$ mean deflecting torque $T_{mean} \propto$ average value of vi

Let $\quad v = V_{max} \sin\theta$

$i = I_{max} \sin(\theta - \phi)$

$$\therefore \quad T_{mean} \propto \frac{1}{2\pi}\int_0^{2\pi} V_{max} \sin\theta \times I_{max} \sin(\theta - \phi)\, d\theta$$

$$\propto \frac{V_{max} I_{max}}{2\pi}\int_0^{2\pi} \sin\theta \sin(\theta - \phi)\, d\theta$$

$$\propto \frac{V_{max} I_{max}}{2\pi}\int_0^{2\pi} \frac{\cos\phi - \cos(2\theta - \phi)}{2}\, d\theta$$

$$\propto \frac{V_{max} I_{max}}{4\pi}\left[\theta \cos\phi - \frac{\sin(2\theta - \phi)}{2}\right]_0^{2\pi} \propto \frac{V_{max}}{\sqrt{2}} \cdot \frac{I_{max}}{\sqrt{2}} \cos\phi \propto VI\cos\phi$$

where V and I are the r.m.s. values.

$\therefore \quad T_{mean} \propto VI\cos\phi \propto$ true power.

Hence, we find that in the case of an alternating current also, the deflection is proportional to the true power in the circuit.

Scales of dynamometer wattmeters are more or less uniform because the deflection is proportional to the average power and for spring control, controlling torque is proportional to the deflection, hence $\theta \propto$ power. Damping is pneumatic *i.e.* with the help of a piston moving in an air chamber as shown in Fig. 15.38.

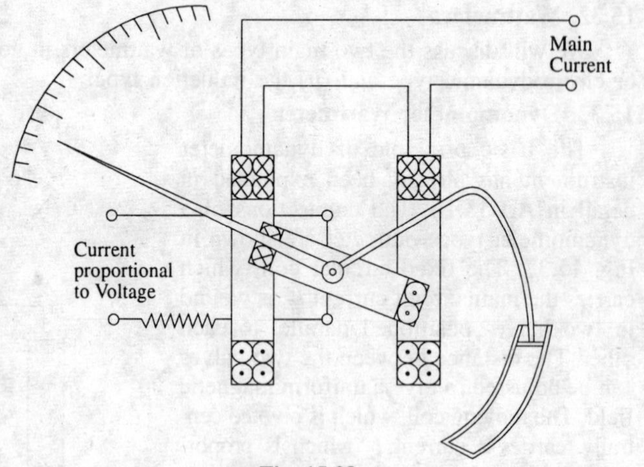

Fig. 15.38

Errors

The inductance of the moving or voltage coil is liable to cause error but the high non-inductive resistance connected in series with the coil swamps, to a great extent, the phasing effect of the voltage-coil inductance.

Another possible error in the indicated power may be due to (*i*) some voltage drop in the current circuit or (*ii*) the current taken by the voltage coil. In standard wattmeters, this defect is overcome by having an additional compensating winding which is connected in series with the voltage coil but is so placed that it produces a field in a direction opposite to that of the fixed or current coils.

Advantages and Disadvantages

By careful design, such instruments can be built to give a very high degree of accuracy. hence, they are used as standard for calibration purposes. They are equally accurate on d.c. as well as a.c. circuits.

However, at low power factors, the inductance of the voltage coil causes serious error unless special precautions are taken to reduce this effect.

Example 15.13. *The current coil of a dynamometer wattmeter has a resistance of 1.5 Ω and its pressure coil circuit has a resistance of 10,000 Ω. It is connected into a.d.c. circuit to measure the power taken by a load from a 200-V supply. The pressure coil terminals of the wattmeter are connected directly across the supply and the wattmeter reading is 600 W. Calculate the true power taken by the load.*

Solution. The circuit connections are shown in Fig. 15.39.

Load current = 600/200 = 3 A.

Power consumed by the current coil of the wattmeter is

$$= I^2R = 3^2 \times 1.5 = 13.5 \text{ W}$$

Total wattmeter reading = 600 W

This includes power consumed by the load and the current coil and not by the pressure coil because being connected at point *a*, its current does not flow through the instrument. If, however, its one end were connected to point *b*, then wattmeter reading would include its power consumption as well.

∴ load power consumption is = 600 – 13.5 = **586.5 W**

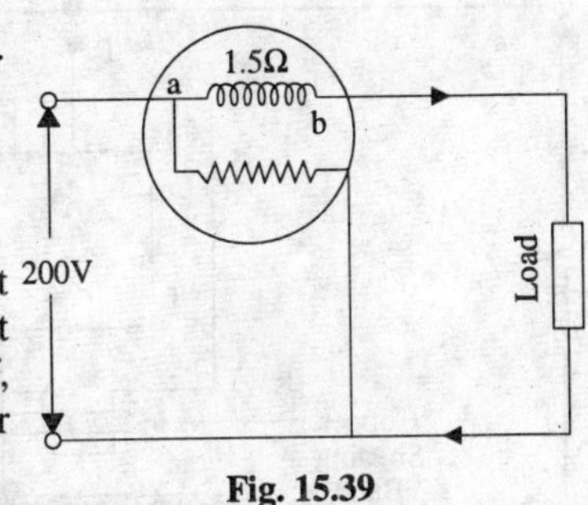

Fig. 15.39

15.34. Induction Wattmeters

Principle of induction wattmeters is the same as that of induction ammeters and voltmeters. They can be used on a.c. circuits only in contrast with dynamometer wattmeters which can be used both on d.c. and a.c. circuits. Induction wattmeters are useful only when the frequency and supply voltage are constant.

Since both a current and a pressure element are required in such instruments, it is not essential to use the shaded-pole principle. Instead of this, two separate a.c. magnets are used which produce two fluxes which have the required phase difference.

Construction

The wattmeter has two laminated electromagnets one of which is excited by the current in the main circuit—its exciting winding being joined in series with the circuit, hence it is also called a *series* magnet. The other is excited by current which is proportional to the voltage of the circuit. Its exciting coil is joined in parallel with the circuit, hence this magnet is sometimes referred to as *shunt* magnet.

A thin aluminium disc is so mounted that it cuts the fluxes of both magnets. Hence, two eddy currents are produced in the disc. The deflection torque is produced due to the interaction of these eddy currents and the inducing fluxes. Two or three copper rings are the fitted on the central limb of the shunt magnet and can be so

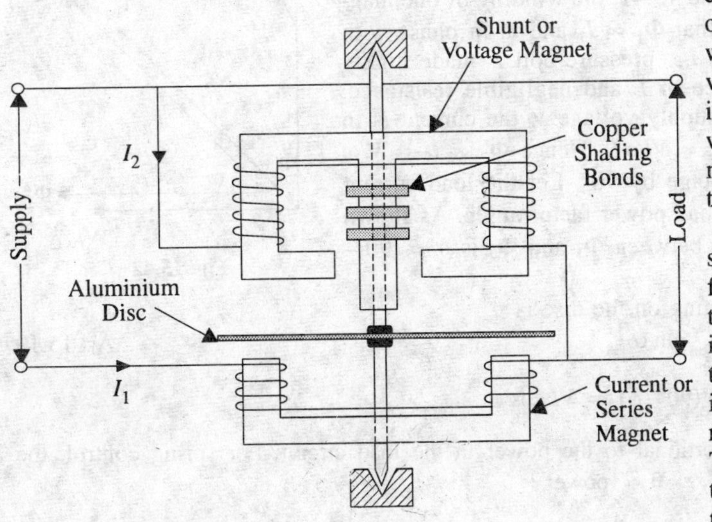

Fig. 15.40

adjusted as to make the resultant flux in the shunt magnet lag behind the applied voltage by a suitable angle.

Two most common forms of the electromagnets are shown in Fig. 15.40 and 15.41. It is seen that in both cases one magnet is placed above and the other below the disc. The magnets are so positioned and shaped that their fluxes are cut by the disc.

In Fig. 15.40 the two pressure coils are joined in series and are so wound that both send the flux through the central limb in the same direction. The series magnet carries two coils joined in series and so wound that they magnetise their respective cores in the same direction. Correct phase displacement between the shunt and series magnet fluxes can be obtained by adjusting the position of the coper shading bands as shown.

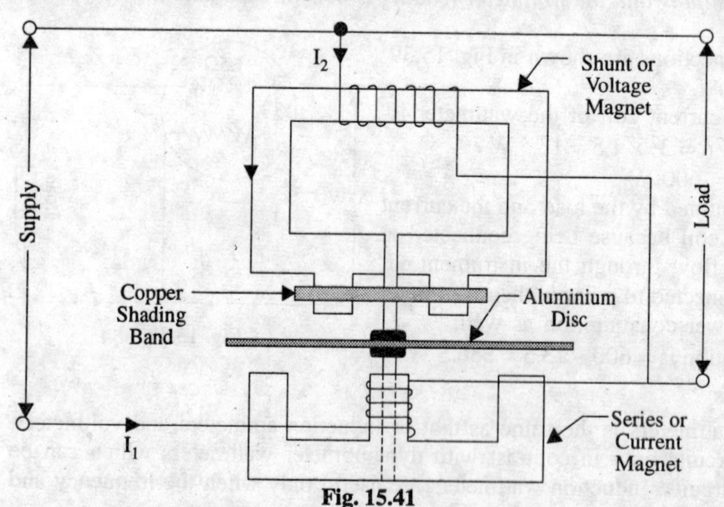

Fig. 15.41

In the type of instrument shown in Fig. 15.41, there is only one pressure winding and one current winding. The two projecting poles of the shunt magnet are surrounded by a copper shading band whose position can be adjusted for correcting the phase of the flux of this magnet with respect to the voltage.

Both types of induction wattmeters shown above are spring-controlled, the springs being fitted to the spindle of the moving system which also carries the pointer. The scale is unifromly even and extends over 300°.

Currents up to 100 A can be handled by such wattmeters directly, but for currents greater than this value, they are used in conjunction with current transformers. The pressure coil is purposely made as much inductive as possible in order that the flux through it may lag behind the voltage by a suitable value.

Theory

As seen from Fig. 15.40 and 15.41, the winding of one magnet carries line current I_1 so that $\Phi_1 \propto I_1$ and is in phase with I_1 (Fig. 15.42). The other coil *i.e.* pressure coil is made highly inductive having an inductance of L and negligible resistance. This is connected across the supply voltage V. the current I_2 in the pressure coil is, therefore, $\propto V/\omega L$. Hence $\Phi_2 \propto I_2 \propto V/\omega L$ and Φ_2 lags behind the voltage by 90°. Let the load current I_1 lag behind V by ϕ *i.e.* the load power factor angle. As shown in Fig. 15.42, the phase angle between Φ_1 and Φ_2 is $\alpha = (90 - f)$.

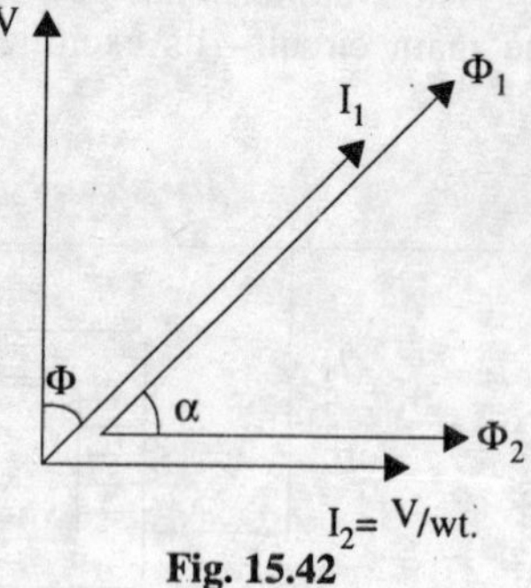

Fig. 15.42

The value of the torque acting on the disc is

$$T = K\omega\,\Phi_{1m}\,\Phi_{2m}\sin\alpha \qquad \text{...Art 15.19}$$

or $$T \propto \omega.\,\text{I}.\frac{V}{\omega L}\sin(90-\phi) \propto VI\cos\phi$$

Hence, the torque is proportional to the power in the load circuit. For spring control, the controlling torque $T_c \propto \theta \quad \therefore \quad \theta \propto$ power

Hence, the scale is even.

15.35. Advantages and Limitation of Induction Wattmeters

These wattmeters possess the advantage of fairly long scale (extending over 300°), are free from the effects of stray fields and have good damping. They are practically free from frequency errors. However, they are subject to (sometime) serious temperature errors because the main effect of temperature is on the resistance of the eddy current paths (Art 15.24).

15.36. Energy Meters

Energy meters are integrating instruments used to measure the quantity of electric energy supplied to a circuit in a given time. They give no direct indication of power *i.e.* as to the rate at which energy is being supplied because their registrations are independent of the rate at which a given quantity of electric energy is being consumed. We will consider the following energy meters.

(*i*) *Electrolytic meters*—their operation depends on electrolytic action and (*ii*) *Motor meters*—they are really small electric motors.

15.37. Electrolytic Meter

It is used on d.c. circuits* only and is essentially an ampere-hour meter and not a true watt-hour meter. However, its registrations are converted to watt-hour by multiplying them by the voltage (assumed constant) of the circuit in which it is used. Such instruments are usually calibrated to read kWh directly at the declared voltage. Their readings would, obviously, be incorrect when used on any other voltage. *Because of the question of power factor, such instruments cannot be used on a.c. circuits.*

The advantages of simplicity, cheapness and of low power consumption of ampere-hour meters are to a large extent, discounted by the fact that variations in supply voltage are not taken into account by them. As an example, suppose that the voltage of a supply whose nominal value is 220 V, has an average value of 216 volts in one hour during which a consumer draws a current of 100 A. Quantity of electricity as indicated by the instrument which is calibrated on 220 V, is $220 \times 100/1000 = 22$ kWh. Actually, the energy consumed by the customer is only $216 \times 100/1000 = 21.6$ kWh. Obviously, the customer is being overcharged to the extent of the cost of $22–21.6 = 0.4$ kWh or energy per hour. A true wattmeter hour would have taken into account the decrease in the supply voltage and would have, therefore, resulted in a saving to the customer. If the supply voltage would have been higher by that amount, then the supply company would have been the loser (see Ex. 15.14).

In this instrument, the operating current is passed through a suitable electrolyte contained in a voltmeter. Due to electrolysis, a deposit of mercury is given or a gas is librated (depending on the type of meter) in proportion to the quantity of electrocity passed (Faraday's Law of Electrolysis). The quantity of electricity passed is indicated by the level of mercury in a graduated tube. Hence, such instruments are calibrated in ampere-hours or if constancy of supply voltage is assumed, are calibrated in watt-hours or kWh.

Such instruments are cheap, simple and are accurate even at very small loads. They are not affected by stray magnetic fields and due to the absence of any moving parts, are free from friction errors.

15.38. Motor Meters

Most commonly-used instruments of this type are :

(*i*) *Mercury motor* meter and (*ii*) *induction motor* meter.

Of these, mercury motor meter is normally used on d.c. circuits whereas the induction type instrument is used only on a.c. circuits.

Instruments used for d.c. work can be either in the form of amp-hour meters or watt-hour meters. In both cases, the moving system is allowed to revolve continuously instead of being merely allowed to deflect or rotate through a fraction of a revolution as in indicating instruments. The speed of rotation is directly proportional to the current in the case of amp-hour meter and

---

*Recently such instruments have been marketed for measurement of kilovolt -ampere-hours on a.c. circuits using a small rectifier unit which consists of a current transformer and full-wave copper oxide rectifier.

to power in the case of watt-hour meter. Hence, the number of revolutions made in a given time is proportional, in the case of an amp-hour meter, to the quantity of electricity ($Q = I \times t$) and in the case of Wh meter to the quantity of energy supplied to the circuit. The number of revolutions made are registered by a counting mechanism consisting of a train of gear wheels and dials.

The control of speed of the rotating system is brought about by a permanent magnet (known as braking magnet) which is so placed as to set up eddy currents in some part of the rotating system. These eddy currents produce a retarding torque which is proportional to their magnitude—their magnitude itself depending on the speed of rotation of the rotating system. The rotating system attains a steady speed when the braking torque exactly balances the driving torque which is produced either by the current or power in the circuit.

The essential parts of motor meters are :

1. An operating system which produces an operating torque prportional to the current or power in the circuit and which causes the rotation of the rotating system.
2. A retarding or braking device usually a permanent magnet, which produces a braking torque in proportion to the speed of rotation. Steady speed of rotation is achieved when braking torque becomes equal to the operating torque.
3. A registering mechanism for the revolutions of the rotating system. Usually, it consists of a train of wheels driven by the spindle of the rotating system. A worm which is cut on the spindle engages a pinion and so drives a wheel-train.

15.39. Errors in Motor Meters

The two main errors in such instruments are : (*i*) friction error and (*ii*) braking error. Friction error is of much more importance in their case than the corresponding error in indicating instruments becuse (*a*) it operates continuously and (*b*) it affects the speed of the rotor. The braking action in such meters corresponds to damping in indicating instruments. The braking torque directly affects the speed for a given driving torque, and also the number of revolutions made in a given time.

Friction torque can be compensated for by providing a small constant driving torque to be applied to the moving system independent of the load.

As said earlier, steady speed of such instruments is reached when driving torque is equal to the braking torque. The braking torque is proportional to the flux of the braking magnet and the eddy current induced in the moving system due to its rotation in the field of brake magnet.

$$\therefore \quad T_B \propto \Phi i$$

where Φ is the flux of the braking magnet and i the induced current. Now $i = e/r$ where e is the induced e.m.f. and r the resistance of theeddy current path. Also, $e \propto \Phi n$ where n is the speed of the moving part of the instrument

$$\therefore \quad T_B \propto \Phi \times \frac{\Phi n}{r} \propto \frac{\Phi^2 n}{r}$$

The torque T_B' at the steady speed of N is given by $T_B' \propto \Phi^2 N/r$

Now $\quad T_B' = T_D \quad$ ...the driving torque

$\therefore \quad T_D \propto \Phi^2 N/r \quad$ or $\quad N \propto T_D\, r/\Phi^2$

Hence, for a given driving torque, the steady speed is directly proportional to the resistance of the eddy current path and inversely to the square of the flux.

Obviously, it is very important that the strength of the field of the brake magnet should be constant throughout the time the meter is in service. The constancy of field strength can be assured by careful design and treatment during the manufacture of the brake magnet. Variations in temperture will affect the braking torque, since the resistance of the eddy current path will change. This error is difficult to fully compensate for.

15.40. Quantity or Ampere-hour (Ah) Meters

The use of such meters is mostly confined to d.c. circuits. Their operation depends on the production of two torques (*i*) a driving torque which is proportional to the current I in the circuit and (*ii*) a braking torque which is proportional to the speed N of the spindle. This speed attains a steady value when these two torques become numerically equal. In that case spead becomes proportional to current *i.e.* $N \propto I$. Over a certain period of time, the total number of revolutions $\int Ndt$ will be proportional to the quantity of electricity $\int Idt$ passing through the meter. A worm cut in the spindle at its top, engages gear wheels or the recording mechanism which has suitably-marked dials reading directly in ampere-hours. Since electric supply charges are based on watt-hours, rather than on ampere-hours, the dials of ampere-hour meters are frequently marked in corresponding watt-hours at the normal supply voltage. Hence, their indications of watt-hours are correct only when the supply voltage remains constant, otherwise readings will be wrong.

15.41. Ampere-hour Mercury Motor Meter

It is one of the best and most popular form of mercury Ah meter used for d.c. work.

Construction

It consists of a thin Cu disc *D*, mounted at the base of a spindle working in jewelled cup bearing and revolving between a pair of *permanent* magnets M_1 and M_2. One of the two magnets *i.e.* M_2 is used for driving purposes whereas M_1 is used for braking. In between the poles of M_1 and M_2 is a hollow circular box *B* in which rotates the Cu disc and the rest of the space is filled up with mercury which exerts a considerable upward thrust on the disc, thereby reducing the pressure on the bearings.

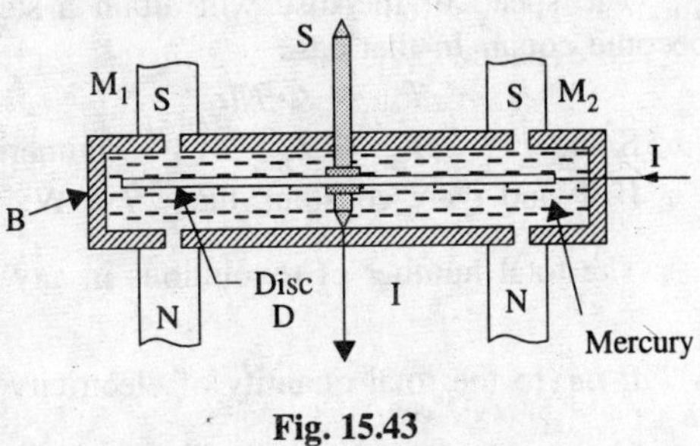

Fig. 15.43

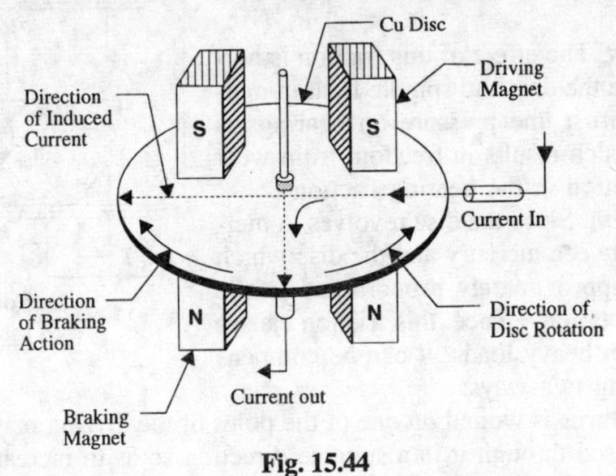

Fig. 15.44

Principle of Action

Its principle of action can be understood from Fig. 15.44 which shows a separate line drawing of the motor element.

The current to be measured is led in the disc through the mercury, at a point at its circumference on the right-hand side. As shown by arrows, it flows radially to the centre of the disc where it passes out to the external circuit through the spindle and its bearings. It is worth noting *that current flow takes place only under the right-hand side magnet* M_2 and not under the left-hand side magnet M_1. The field of M_2 will, therefore, exert a force on the right-side portion of the disc which carries the current (motor action). The direction of the force, as found by Fleming's Left-hand rule, is as shown by the arrow. The magnitude of the force depends on the flux density and current ($\because$ $F = BIl$). The driving or motoring torque T_d so produced is given by the product of the force and the distance from the spindle at which this force acts. When the disc rotates under the influence of this torque, it cuts through the field of left-hand side magnet M_1 and hence eddy currents are produced in it which results in the production of braking torque. The magnitude of the retarding or braking torque is proportional to the speed of rotation of the disc.

Theory

Driving torque $\quad T_d \propto$ force on the disc $\propto BI$

If the flux density of M_2 remains constant, then $T_d \propto I$

The braking torque T_B is proportional to the flux Φ of braking magnet M_1 and the eddy current i induced in the disc due to its rotation in the field of M_1.

$\therefore \quad T_B \propto \Phi i$

Now $\quad i = e/r$ where e is the induced e.m.f. and r the resistance of eddy-current path.

Also $\quad e \propto \Phi n$ where n is the speed of the disc

$$\therefore \quad T_B = \Phi \times \frac{\Phi n}{r} \propto \frac{\Phi^2 n}{r}$$

The speed of the disc will attain a steady value N when the driving and braking torques become equal. In that case

$T_B \propto \Phi^2 N/r$

Since $\quad T_d = T_B$ —numerically ; $\quad \therefore \quad I \propto \Phi^2 N/r$

If Φ and r are constant, then $I \propto N$

The total number of revolutions in any given time t *i.e.* $\int_0^t N.dt$ will become proportional to $\int_0^t I.dt$ *i.e.* to the total quantity of electricity passed through the meter.

15.42. Friction Compensation

There are two types of friction in this ampere-hour meter.

(i) Bearing Friction. The effect of this friction is normally negligible because the disc and spindle float in mercury. Due to upward thrust, the pressure on bearings is considerably reduced which results in freedom from wear as well as a great reduction in the bearing friction.

(ii) Mercury Friction. Since the disc revolves in mercury, there is friction between mercury and the disc which gives rise to a torque approximately proportional to the square of the speed of rotation. Hence, this friction causes the meter to run slow on heavy loads. It can be compensated for in the following two ways:

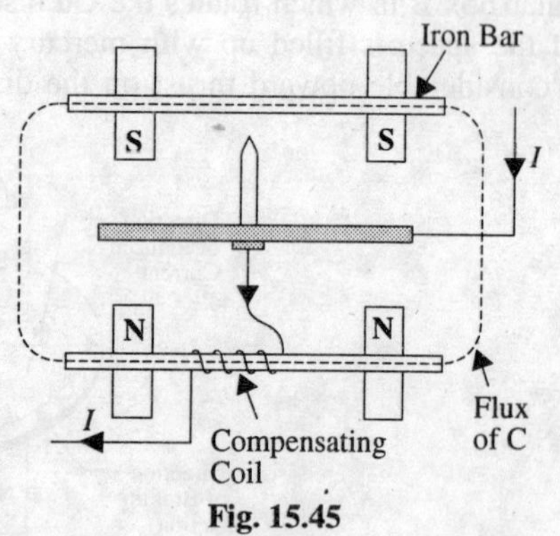

Fig. 15.45

(*a*) a coil of few turns is wound on one of the poles of the driving magnet M_2 and the meter current is passed through it in a suitable direction so as to increase the strength of M_2. The additional driving torque so produced can be made just sufficient to compensate for the mercury friction.

(*b*) In the other method, two iron bars are placed across the two permanent magnets, one above and one below the mercury chamber as shown in Fig. 15.45. The lower bar carries a small compensating coil through which is passed the load current. The local magnetic field set up by this coil strengthens the field of driving magnet M_2 and weakens that of the braking magnet M_1 thereby compensating for mercury friction.

15.43. Mercury Meter Modified as Watt-hour Meter

If the permanent magnet M_2 of the amp-hour meter, used for producing the driving torque, is replaced by a wound electromagnet, connected across the supply, the result is a watt-hour meter. The exciting current of this electromagnet is proportional to the voltage of the supply. The driving torque is exerted on the aluminium disc immersed in the mercury chamber below which is placed this electromagnet. The aluminium disc has radial slots cut in it for ensuring the radial flow of current through it—the current being led into and out of this disc through mercury contacts situated at diametrically opposite points. These radial slots, moreover, prevent the same disc being used for braking purposes. Braking is by a separate aluminium disc mounted on the same spindle and revolving in the air-gap of a separate braking magnet.

15.44. Induction Type Single-phase Watthour Meter

Induction type meters are, by far, the most common form of a.c. meters met with in every-day domestic and industrial installations. These meters measure electric energy in kilowatthours. The principle of these metres is practically the same as that of the induction wattmeters. Constructionally, the two are similar except that the control spring and the pointer of the watt-meter are replaced, in the case of watthour meter, by a brake magnet and by a spindle of the meter.

The brake magnet induces eddy currents in the disc which revolves continuously instead of rotating through only a fraction of a revolution as in the case of wattmeters.

Construction

The meter consists of two a.c. magnets of laminated construction as shown in Fig 15.46 (*a*), one of which *i.e.* M_1 is excited by line current and is known as 'series' or 'current' magnet. The alternating flux Φ_1 produced by it is directly proportional to and in phase with current provided the effects of hysteresis and saturation are neglected. The winding of the other magnet M_2 is connected across the supply lines and so it carries a current I_2 which is proportional to the supply voltage V. Hence, it is called 'voltage' or 'shunt' magnet. The flux Φ_2 produced by magnet M_2 lags behind the voltage V by 90° (Fig. 15.42). This phase displacement of exact 90° is achieved by the adjustment of the copper shading band C (also known as power factor compensator) on the shunt magnet M_2. Major protion of Φ_2 crosses the narrow gap between the centre and side limbs of M_2 but a small amount, which is the useful flux, passes through the disc D. The two fluxes Φ_1 and Φ_2 induce e.m.fs in the disc which further produce the circulatory eddy currents. The reaction between these fluxes and eddy currents produces the driving torque on the disc in a manner similar to that explained in Art 15.34. The braking torque is produced by a pair of magnets [Fig. 15.46 (*b*)] which are mounted diametrically opposite to the magnets M_1 and M_2. This arrangement minimizes the interaction between the fluxes of M_1 and M_2 and that of the braking magnet. When the peripheral portion of the rotating disc passes through the air gap of the braking magnet, the eddy currents are induced in it which give rise to the necessary torque. The braking torque $T_B \propto N/r$ where Φ is the flux of braking magnet, N the *steady* speed of the rotating disc and r the resistance of the eddy current path. If Φ and r are constant, then $T_B \propto N$.

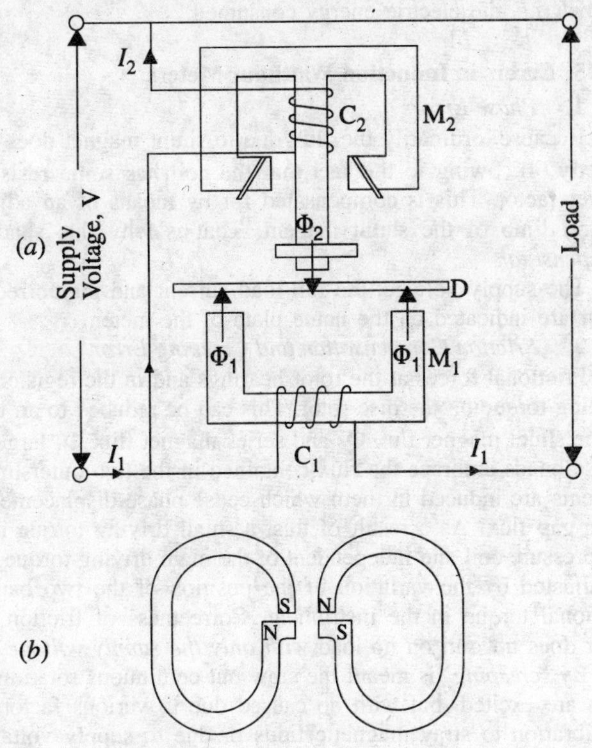

Fig. 15.46

The register mechanism is either of pointer type or cyclometer type. In the former type, the pinion on the rotor shaft drives, with the help of a suitable train of reduction gears, a series of five or six pointers rotating on dials marked in kWh.

Theory

As shown in Art. 15.34 and with reference to Fig. 15.42 the driving torque is given by

$$T_d \propto \omega \Phi_{1m} \Phi_{2m} \sin \alpha$$

where Φ_{1m} and Φ_{2m} are the maximum fluxes produced by magnets M_1 and M_2 and α the angle between these fluxes. Assuming that fluxes are proportional to the currents, we have

Current I_2 through the windings of $M_2 = \dfrac{V}{\omega L}$

Current through the winding of $M_1 = I_1$ ...the line current

where ϕ is the load p.f. angle. $\alpha = 90° - \phi$

$$T_i \propto \omega . I . \frac{V}{\omega L} \sin(90° - \phi) \propto VI \cos\phi \propto \text{power} \; ; \quad \text{Also,} \; T_B \propto N$$

The disc achieves a steady speed when the two torque are equal *i.e.* when

$$T_d = T_B \quad \text{or} \quad N \propto \text{power } W$$

Hence, in a given period of time, the total number of revolutions $\int_0^t N\,dt$ is proportional to $\int_0^t W\,dt$ *i.e.* the electric energy consumed.

15.45. Errors in Induction Watthour Meters

1. *Phase Errors*

Because, ordinarily the flux due to shunt magnet does not lag behind the supply voltage by exactly 90° owing to the fact that the coil has some resistance, the torque is not zero at zero power factor. This is compensated for by means of an adjustable shading ring placed over the central limb of the shunt magent. That is why this shading ring is known as *power factor compensator.*

The supply voltage, the full-load current and the correct number of revolutions per kilowatthour are indicated on the name plate of the meter.

2. *Friction Compensation and Creeping Error*

Frictional forces at the rotor bearings and in the register mechanism give rise to an unwanted braking torque on the disc rotor. This can be reduced to an unimportant level by making the ratio of the shunt magnet flux Φ_2 and series magnet flux Φ_1 large with the help of two shading bands. These bands embrace the flux contained in the two outer limbs of the shunt magnet and so eddy currents are induced in them which cause phase displacement between the enclosed flux and the main-gap flux. As a result of this, a small driving torque is exerted on the disc rotor solely by the pressure coil and independent of the main driving torque. The amount of this corrective torque is adjusted by the variation of the position of the two bands, so as to exactly compensate for frictional torque in the instrument. Correctness of friction compensation is achieved when the rotor does not run on no-load *with only the supply voltage connected.*

By *'creeping'* is meant the slow but continuous rotation of the rotor, when only the pressure coils are excited, but with no caused due to various factors like incorrect friction composition, to vibration to stray magnetic fidds or due to supply voltage beity in excess of the normal. In order to preused creapany, on no-load, two holes are drilled in the disc, on a diameter i.e. a the opposite sides of the spindle.

Thus causes sufficient distinction of the field to prevent rotates, whereone of the holes comes and are pole of the shunt magnet.

3. *Error due to Temperator Variations*

The error due to temperature variations of the instruments are usually small because the various effects produced tend to neutralise one another.

Example 15.14. *The name-plate of a meter reads '1 kWh = 15,000 revolutions'. In a check-up, the meter completed 150 revolutions during 50 seconds. Calculate the power in the circuit.*

Solution. Power metered in 150 revolutions

$= 1 \times 150/15{,}000 = 1/100$ kWh

If P kilowatt is the power in the circuit, then energy consumed in 50 seconds is

$= P \times 50/3600$ kWh.

Equating the two amounts of energy, we have

$$P \times 50/3600 = 1/100 \; ; \; P = 0.72 \text{ kW} = \mathbf{720\ W}$$

Example 15.15. *When a current of 25 A is passed through an Ah metre for 10 minutes from*

220-V mains, the disc makes 600 revolutions. Calculate the testing constant.

Solution. Energy consumed in 10 minutes

$$= (220 \times 25) \times \frac{10}{60} \times \frac{1}{1000} = 0.9167 \text{ kWh}$$

$\therefore$ testing constant = 0.9167/600 = **0.00153 kWh/rev**

Example 15.16. *An ampere-hour meter is calibrated to give reading of energy in kilowatt-hours on 220-volt circuit. When used on 250-V circuit, it records 845 kWh in a certain time. Calculate the energy actually supplied.*

Solution. As explained Art. 15.38, ampere-hour meters are calibrated to read direct in kWh *at the declared voltage.* Obviously, their readings would be incorrect when used on any other voltage.

Reading on the assumed 220 V = 845 kWh

Actual reading on 250-V circuit = 845 ×250/220 = **960**

Example 15.17. *A 5-ampere, 230-V meter on full-load unity power factor test makes 60 revolutions in 360 seconds. If the normal disc speed is 520 revolutions per kWh, what is the percentage error?*

Solution. Testing time = 360 second = 1/10 hour

Energy consumed in 1/10 hour = 230 × 5 × 1/10 = 115 Wh = 115/1000 = 0.115 kWh

If the meter were running correctly *i.e.* at 520 rev/kWh, then revolutions for 0.115 kWh would have been = 0.115 × 520 = 59.8

Actual revolutions made during this test time are 60.

$\therefore$ error = 60 – 59.8 = 0.2

Percentage error = 0.2 × 100/60 = **0.33%**

Example 15.18. *In a test run of 20 minutes' duration, an Ah meter was found to register 0.753 kWh when a constant current of 10 A was passed. If the meter is to be used on a 220-V supply line, find the error and state if it is running fast or slow.*

Solution. Energy consumed $= (220 \times 10) \times \frac{20}{60} \times \frac{1}{1000} = 0.733$ kWh

Energy recorded = 0.753 kWh

$$\text{percentage error} = \frac{(0.753 - 0.733)}{0.733} \times 100 = \mathbf{2.7\%}$$

Since the error is positive, the meter runs **fast.**

Example 15.19. *A 230 - V ampere-hour type meter is connected to a 230-V d.c. supply. If the meter completes 225 revolutions in 10 minutes when carrying 14 A, calculate :*

(a) the kWh registered by the meter and

(b) the percentage error of the meter above or below the original calibration.

The timing constant of the meter is 40 A-s/revolution.

Solution. The declared constant of the meter is 40 A-s/rev. During 225 revolutions, it would register 40 × 225 A-s or coulomb. Since time taken is 10 minutes or 600 seconds, it corresponds to a current of 40 × 225/600 = 15 A

(*a*) $\therefore$ energy as registered by the meter is $= \frac{230 \times 15}{1000} \times \frac{1}{6} = \mathbf{0.575\ kWh}$

(*b*) Actual energy consumed $= \frac{230 \times 14}{1000 \times 6} = \mathbf{0.5367\ kWh}$

$$\text{percentage error} = \frac{0.575 - 0.5367}{0.575} \times 100 = \mathbf{6.66\%}$$

Alternatively, we may find the speed thus :

Ampere-seconds passed in 10 minutes = 14 × (10 × 60)

No. of revolutions that should have been made at the rate of 40 A-s/rev is

= 14 × 600/40 = 210

Revolutions actually made during that period are 225.

$$\therefore \quad \text{error} = \frac{225 - 210}{225} \times 100 = \mathbf{6.66\%}$$

Obviously, the meter reads high.

Example 15.20. *The meter constant of a 250-V, 10-A Ah meter is 5 ampere-second per coulomb. What is full-load disc speed in r.p.m. and the number of revolutions per kWh.*

The meter was subjected to a test run at half-load and was found to take 98 seconds to complete 100 revolutions. Calculate the percentage error of the meter.

Solution. Energy consumption per revolution

$$= 250 \times 5 \text{ W–s} = \frac{250 \times 5}{1000} \times \frac{1}{3600} = 0.347 \times 10^{-3} \text{ kWh}$$

$$\text{Energy consumed at full-load in one minute} = \frac{250 \times 10}{1000} \times \frac{1}{60} = 0.04167 \text{ kWh}$$

Disc speed $= 0.04167/0.347 \times 10^{-3} = 120$ r.p.m.

No. of rev./kWh $= 1/0.347 \times 10^{-3} = \mathbf{2880}$

$$\text{Energy consumed at half-load in 98 seconds} = \frac{250 \times (10/2) \times 98}{1000 \times 3600} = 0.034 \text{ kWh}$$

Since it made 100 revolutions in the test run, energy actually consumed by the meter is

$$= 0.347 \times 10^{-3} \times 100 = 0.0347 \text{ kWh}$$

Since energy indicated is more than actually consumed, the meter is running fast.

$$\text{percentage error} = \frac{0.0347 - 0.034}{0.034} \times 100 = \mathbf{2.06\%}$$

Example 15.21. *A wattmeter and watt-hour meter are connected in series with a loading resistance across a 200-V d.c. supply. The reading of the watthour meter $1\frac{1}{2}$ hours after the supply was switched on was 4,320 Wh, the initial reading being zero. Find*

(a) the wattmeter reading (b) the load resistance

If the supply voltage is increased to 300 V, what will be the readings on the instruments after a further hour? (Neglect any instrument errors)

Solution. (*a*) Wattmeter reading = 4,320 Wh/1.5 h = **2880 W**

(*b*) $V^2/R = W$ or $R = V^2/W = 200^2/2880 = = \mathbf{13.9\ \Omega}$

When voltage is increased, wattmeter reading would increase in the ratio of $(300/200)^2$ because $W = V^2/R$

$\therefore$ wattmeter reading with 300 V $= 2{,}880 \times (300/200)^2 = \mathbf{6{,}480\ W}$

Similarly, watthour reading would increase by an amount

$$= 2880 \times (300/200)^2 \text{ Wh} = 6{,}480 \text{ kWh}$$

Being an integrating instrument, the watthour meter readings go on adding over the length of time.

Total reading of the watthour meter is

= previous reading + reading in the next hour

$= 4{,}320 + 6{,}480 = 10{,}800$ Wh = **10.8 kWh**

15.46. Megger

It is a portable insturment used for testing the insulation resistance of a circuit and for measuring resistances of the order of megohms which are connected across the outside terminals *XY* in Fig. 15.47 (*b*).

Working Principle

The working principle of a 'cross-coil' type megger may be understood from Fig. 15.47 (*a*) which shows two coil *A* and *B* mounted rigidly at right angles to each other on a common axis and free to rotate in a magnetic field. When currents are passed through them, the two coils are acted upon by torques which are in opposite directions. The torque of coil *A* is proportional to $I_1 \cos \theta$ and that of *B* is proportional to $I_2 \cos (90° - \theta)$ or $I_2 \sin \theta$. The two coils come to a position of equilibrium where the two torques are equal and opposite *i.e.*

$$I_1 \cos\theta = I_2 \sin\theta \quad \text{or} \quad \tan\theta = I_1/I_2$$

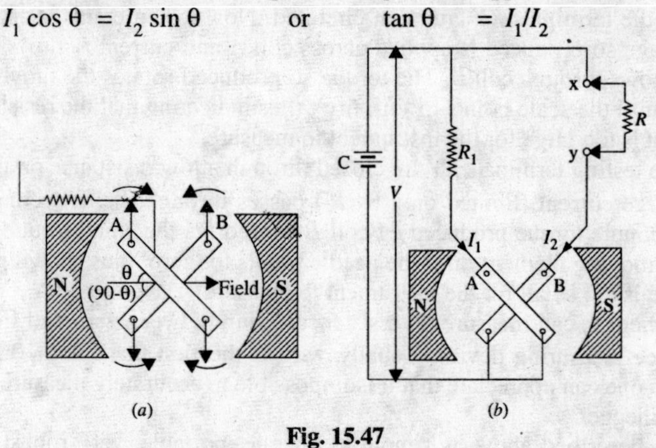

Fig. 15.47

In practice, however, by modifying shape of pole faces and the angle between the two coils, the ratio I_1/I_2 is made proportional to θ instead of tan θ in order to achieve a linear scale.

Suppose the two coils are connected across a common source of voltage *i.e.* battery *C* as shown in Fig. 15.47 (*b*). Coil *A* which is connected direclty across *V* is called the voltage (or control) coil. Its current $I_1 = V/R_1$. The coil *B* called current or deflecting coil carries the current $I_2 = V/R$ where *R* is the external resistance to be measured. This resistance may vary from infinity (for good insulation or open circuit) to zero (for poor insulation or a short-circuit). The two coils are free to rotate in the field of a permanent magnet. The deflection θ of the instrument is proportional to I_1/I_2 which is equal to R/R_1. If R_1 is fixed, then the scale can be calibrated to read *R* directly (in practice, a current-limiting resistance is connected in the circuit of coil *B*, but the presence of this resistance can be allowed for in scaling). The value of *V* is immaterial so long as it remains constant and is large enough to give suitable currents with the high resistances to be measured.

Construction

The essential parts of the megger are shown in Fig. 15.48. Instead of battery *C* of Fig. 15.47, there is a hand-driven d.c. generator. The crank turns the generator armature through a clutch mechanism which is designed to slip at a pre-determined speed. In this way, the generator speed and voltage are kept constant and at their correct values when testing.

The generator voltage is applied across the voltage coil *A* through a fixed resistance R_1 and across deflecting coil *B* through a current-limiting resistance *R′* and the external resistance connected across the testing terminals *XY*. The two coils, in fact, constitute moving-coil voltmeter

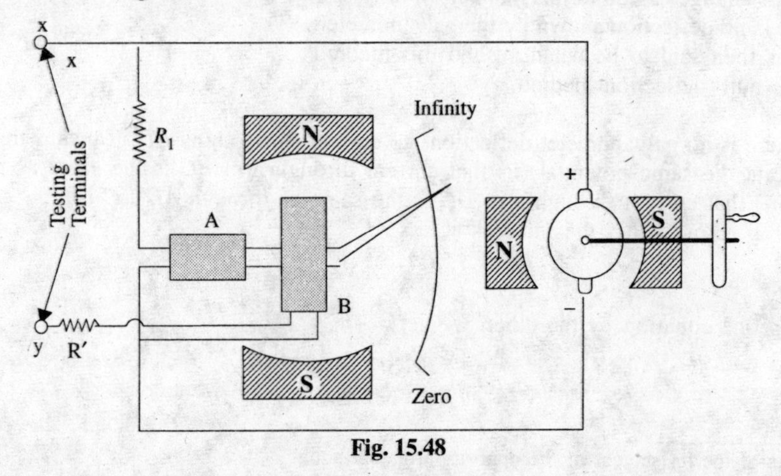

Fig. 15.48

and an ammeter combined into one instrument.

(*i*) Suppose the terminals *XY* are open-circuited. Now, when crank is operated, the generator voltage so produced is applied across coil *A* and current I_1 flows through it but no current flows through coil *B*. The torque so produced rotates the moving element of the megger until the scale points to 'infinity', thus indicating that the resistance of the external circuit is too large for the instrument to measure.

(*ii*) When the testing terminals *XY* are closed through a low resistance or are short-circuited, then a large current (limited only by *R′*) passes through the deflecting coil *B*. The deflecting couple torque produced eg coil *B* overcorres the small torque of coil *A* and rotates the moving element until the needle points to 'zero' thus showing that the external resistance is too small for the instrument to measure.

Although, the megger can measure all resistances lying between zero and infinity, essentially it is a high-resistance measuring device. Usually, zero is the first mark and 10 kΩ is the second mark on its scale, so one can appreciate that it is impossible to accurately measure small resistances with the help of a megger.

The instrument described above is simple to operate, portable, very robust and independent of the external supplies.

15.47. Wheatstone Bridge

It is a four-arm bridge and is extensively used for the measurement of medium-range resistances (1 to 100,000Ω). It also forms the basis from which many other 'bridge' circuits have been developed both for direct and alternating currents. This method is, however, a comparative one, because the value of the unknown resistance is obtained in terms of a known resistance.

Layout

At shown in Fig. 15.49, four resistances are connected to form a square *ABCD*. A sensitive detector or galvanometer is connected across *BD*. The arms *P* and *Q* are known as ratio arms and are generally adjustable to one of the four values 1,10, 100 and 1000 Ω in steps of 1.0 Ω.

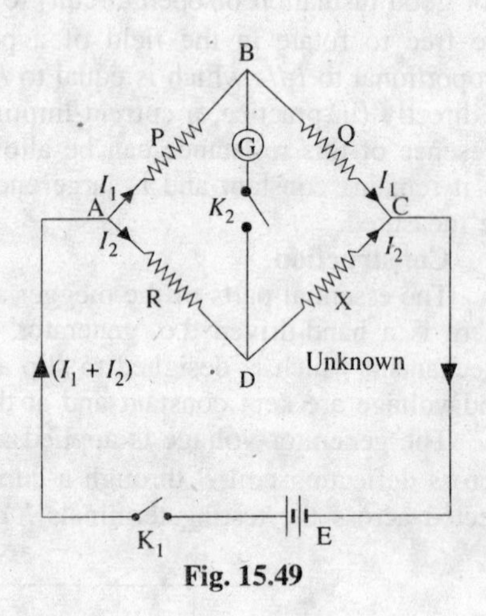

Fig. 15.49

First, the values of *P* and *Q* are set at 10 Ω each. The unknown resistance *X* is connected in the arm *CD*. Next, the battery key K_1 is closed, causing current to flow through the two sets of bridge arms. Now, when galvanometer key K_2 (spring type) is closed, there will be some deflection shown by the galvanometer. The resistance *R* is changed suitably and K_2 is pressed again. If still the galvanometer shows some deflection, then K_2 is opened. *R* is changed again in this manner till on pressing K_2, there is no deflection shown by the galvanometer. The bridge is then said to be balanced and this method is known as 'null' deflection method.

Theory

Since there is no galvanometer deflection, no current flows along *BD*. It means that (*i*) points *B* and *D* are at the same potential (*ii*) that current through arm *BC* is the same as through *AB* *i.e.* I_1 and (*iii*) that current through arm *DC* is the same as through *AD* *i.e.* I_2

Since drop across *AB* is the same as across *AD*

$\therefore$ $$I_1 P = I_2 R \qquad \text{...(i)}$$

Similarly $$I_1 Q = I_2 X \qquad \text{...(ii)}$$

Dividing one equation by the other, we get

$$\frac{P}{Q} = \frac{R}{X} \qquad \text{or} \qquad PX = QR^*$$

*Products of the resistances of the opposite arms are equal.

$\therefore$ unknown resistance $$X = R \times \frac{Q}{P}$$

Hence, if P, Q and R are known, X can be easily found. The experiment is again repeated by keeping Q at 10 Ω as before and changing P to 100 Ω and then 1000 Ω. Again, balance is obtained for each ratio and X found in each case by using the above relation. Mean of the three readings gives the value of X.

It may be noted that

(*i*) position of the battery and galvanometer could be interchanged in their connection to the square *ABCD*. However, with the connections as shown in Fig. 15.49, the bridge has greater sensitivity.

(*ii*) Since galvanometer is merely an indicator of zero current, it need not be calibrated in any particular units.

The Wheatstone bridge principle has been used in the construction of Post-office box, slide-wire bridge and many loop-tests like Murray's and Varley's which are used for finding the position of earth faults on telegraph and telephone lines.

Example 15.22. *In a Wheatstone bridge experiment for determining the resistance of a wire, a balance point was obtained when* R = 30 Ω,*P* = *Q* = 10 Ω. *Find the resistance of the wire. Next, the length of wire was found to be 110 cm and its thickness as measured by a screw gauge 0.014 cm. Calculate the specific resistance of the material of the wire.*

Solution. If X is the unknown resistance, then

$$\frac{P}{Q} = \frac{R}{X} \quad \text{or} \quad X = \frac{R \times Q}{P} = \frac{30 \times 10}{10} = \mathbf{30\ \Omega}$$

Now, $$R = \rho\frac{l}{A} \qquad \therefore \quad \rho = A\,R/l$$

$$\rho = \frac{\pi(0.014)^2 \times 30}{4 \times 110} = \mathbf{42 \times 10^{-6}\ ohm\text{-}cm}$$

15.48. D.C. Potentiometer

A potentiometer is used for measuring and comparing the e.m.fs. of different cells and for calibrating and standardizing voltmeters, ammeters and wattmeters etc. In its simplest form, it consists of a German silver or manganin wire usually one meter long and stretched between two terminals as shown in Fig. 15.50.

This wire is connected in series with a suitable rheostat *Rh* and battery *B* which sends a steady current throught the resistance wire *AC*. As the wire is of uniform cross-section throughout, the fall in potential across it is uniform and the drop between any two points is proportional to the distance between them. As seen, the battery voltage is spread over the rheostat and the resistance wire *AC*. As we go along *AC*, there is a progressive fall of potential. If ρ is the resistance/cm of this wire, l its length, then for a current of I amperes, the fall of potential over the whole length of wire is $\rho l I$ volts.

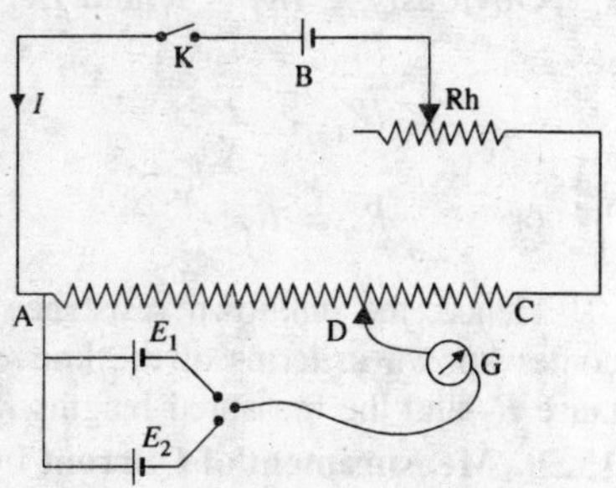

Fig. 15.50

The two cells whose e.m.fs. are to be compared are joined as shown in Fig. 15.50, always remembering that *positive terminals of the cells and the battery must be joined together.* The cells can be joined with the galvanometer in turn through a two-way key. The other end of the galvanometer is connected to a movable contact on *AC*. By this movable end, a point like *D* is found where there is no current in and hence no deflection of *G*. Then, it means that the e.m.f. of the cell just balances the potential fall on *AD* due to the battery current passing through it.

Suppose that the balance or null point for first cell of e.m.f. E_1 occurs at length l_1 as measured from point *A*. Then

$$E_1 = \rho\, l_1 I$$

Similarly, if the balance point is at l_2 for other cell, then

$$E_2 = \rho l_2 I$$

Dividing one equation by the other, we have

$$\frac{E_1}{E_2} = \frac{\rho l_1 I}{\rho l_2 I} = \frac{l_1}{l_2}$$

If one of the cells is a standard cell, the e.m.f. of the other cell can be found.

15.49. Measurement of Low Resistance by Potentiometer

In this method, instead of comparing two e.m.fs., potential drops across an unknown low resistance R_2 and a standard known resistance R_1 are compared. In the circuit of Fig. 15.51, AC is the slide wire and D is the sliding contact. Battery B_1 sends current through wire AC thereby producing a progressive fall in potential from A to C. When S_2 is closed, battery B_2 sends current I through R_1 and R_2 thereby producing voltage drops of IR_1 and IR_2 respectively.

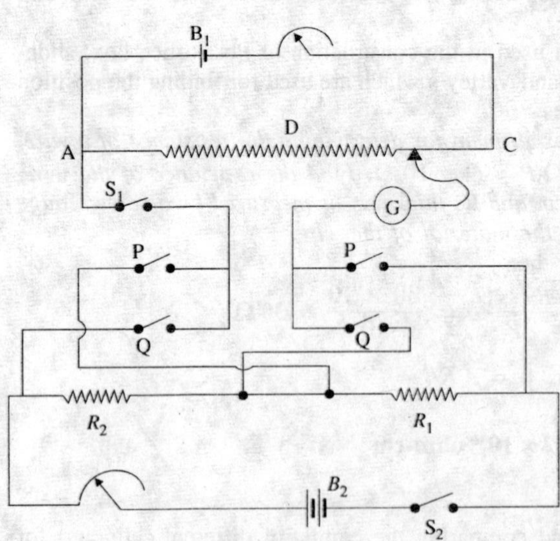

Fig. 15.51

When switches QQ are closed, drop across R_2 is impressed on the galvanometer circuit. By pressing S_1, this drop is applied across the slide wire and is compared with the drop produced by B_1. The sliding contact D is so adjusted that galvanometer G reads zero. This will happen only when drop over AD produced by B_1 is equal to the drop over R_2. Let l_2 be the length of the slide wire as measured from point A over which the balance is obtained.

Next, switches QQ are opened and PP are closed. Now, the drop across R_1 is similarly balanced against the drop produced on the slide wire by B_1 Suppose that the balance is obtained over length l_1.

Obviously, $IR_1 \propto l_1$ and $IR_2 \propto l_2$

$$\therefore \quad \frac{IR_2}{IR_1} = \frac{l_2}{l_1}$$

or

$$R_2 = R_1 \cdot \frac{l_1}{l_2}$$

Hence, the unknown resistance R_2 becomes known in terms of the known resistance R_1 and the measured lengths l_1 and l_2

15.50. Measurement of Current by Potentiometer

In this method, the current to be measured is passed through a known standard resistance R and the potential drop across it is compared with the e.m.f. of a standard cell. The value of the standard resistance is so chosen that potential drop across it is very nearly equal to the e.m.f. of the cell.

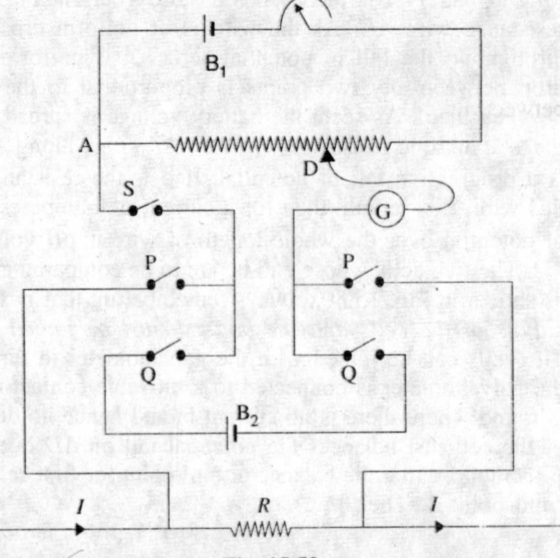

Fig. 15.52

In Fig. 15.52, B_2 is the standard cell and R is the standard resistance which carries the current under measurement. When switches *QQ* and *S* are closed, the e.m.f. of B_2 is applied across the slide wire via the galvanometer *G*. Sliding contact *D* is adjusted for zero galvanometer deflection. Suppose drop over the length l_1 of the slide wire is balanced against the e.m.f. *E* of B_2. Next, switches *QQ* are opened and *PP* closed. The drop over *R i.e. IR* is now impressed on the galvanometer circuit and is again balanced over the slide wire by adjusting the movable contact *D*. Let the new length of *AD* for zero galvanometer deflection be l_2

$$E \propto l_1 = Kl_1 \text{ and } IR \propto l_2 = kl_2$$

$$\therefore \quad \frac{IR}{E} = \frac{Kl_2}{Kl_1} = \frac{l_2}{l_1} \quad \text{or} \quad I = \frac{E}{R} \times \frac{l_2}{l_1}$$

15.51. Direct-reading Potentiometer

The simple potentiometer described in Art. 15.48 is used for educational purposes only. But in its commercial form, it is so calibrated that the readings of the potentiometer give the voltage directly, thereby eliminating tedious arithmetical calculations and hence saving appreciable time.

Such a direct-reading potentiometer is shown in Fig. 15.53. These resistance *R* consists of 14 equal resistances joined in series, the resistance of each unit being equal to that of the whole slide wire *S* (which is divided into 100 equal parts). The battery current is controlled by slide-wire resistance *W*.

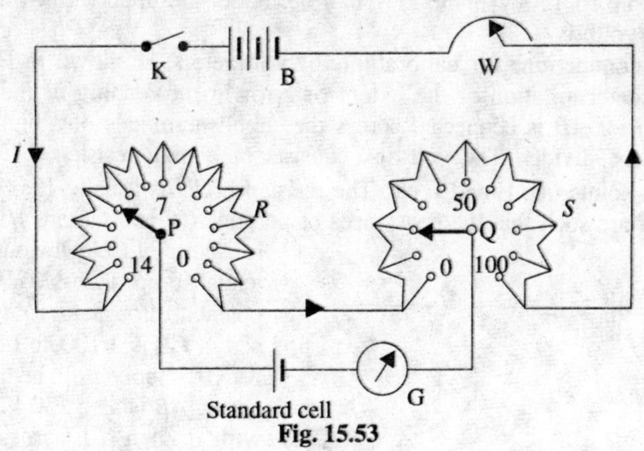

Fig. 15.53

15.52. Standardizing the Potentiometer

A standard cell *i.e.* Weston Cadmium cell of e.m.f. 1.0183 V is connected to sliding contacts *P* and *Q* through a sensitive galvanometer *G*. First, *P* is put on stud No. 10 and *Q* on 18.3 division on *S* and then *W* is adjusted for zero deflection on *G*. In that case, potential difference between *P* and *Q* is equal to cell voltage *i.e.* 1.0183 V so that ptential drop on each resistance of *R* is 1/10 = 0.1 V and every division of *S* represents 0.1/100 = 0.001 V. After standardizing this way, *the position of W is not to be changed in any case* otherwise the whole adjustment would go wrong. After this, the instrument becomes direct reading. Suppose in a subsequent experiment for obtaining balance, *P* is moved to stud No. 7 and *Q* to 84 division, then voltage would be (7 × 0.10 + (84 × 0.001) = 0.784 V.

It should be noted that since most potentiometers have fourteen steps on *R* it is usually not possible to measure p.ds. exceeding 1.5 V. Four measuring higher voltages, it is necessary to use a volt box.

15.53. Calibration of Ammeters

The ammeter to be calibrated is connected in series with a variable resistance and standard resistance *F*, say, of 0.1 Ω across a battery B_1 of ample current capacity as shown in Fig. 15.54. Obviously, the resistance of *F* should be such that with maximum current flowing through the ammeter *A* the potential drop across it should not exceed 1.5 V. Some convenient current, say, 6 amperes (as indicated by *A*) is passed through the circuit by adjusting the rheostat *RH*.

The potential drop across *F* is applied between *P* and *Q* as shown. Next, the sliding contact

P and Q are adjusted for zero deflection on G. Suppose P reads 5 and Q reads 86.7. Then, it means that p.d. across F is 0.5867 V and since F is of 0.1 Ω, hence, *true* value of current through F is 0.5867/0.1 = 5.867 ampere. Hence, the ammeter reads high by 6 – 5.867 = 0.133 A. The test is repeated for various values of the current over the entire range of the ammeter*.

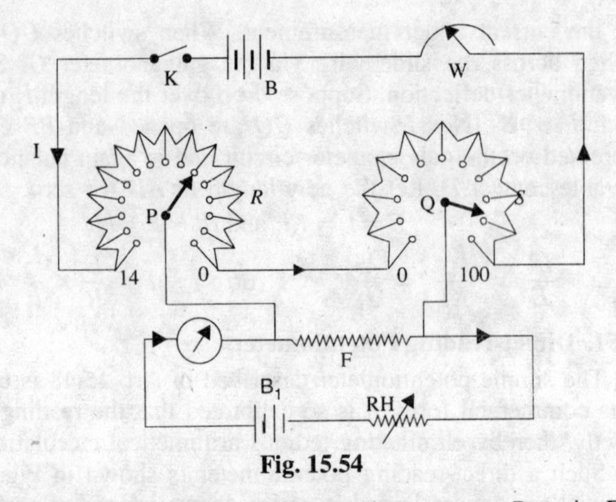

Fig. 15.54

15.54. Calibration of Voltmeter

As pointed out in Art. 15.52, a voltage higher than 1.5 V cannot be measured by the potentiometer directly, the limit being set by the standard cell and the type of the potentiometer (as it has only 14 resistances on R as in Fig. 15.53). However, with the help of a volt-box which is nothing else but a voltage reducer, measurements of voltages up to 150 V or 3000 V can be made, the upper limit of voltage depending on the design of the voltbox.

The diagram of connections for calibrations of voltmeters, is shown in Fig. 15.55. By calibration is meant the determination of the extent of error in the reading of the voltmeter through its range. A high resistor AB is connected across the supply terminals of a high-voltage battery B_1 and it acts as a voltage divider. The volt-box consists of a high resistance CD with tappings at accurately-determined points like E and F etc. The resistance CD is usually 15,000 to 30,000 Ω. The two tappings F and E are such that the resistances of portions CF and CE are 1/10th and 1/100th the resistance of CD. Obviously, whatever the potential drop across CD, the corresponding potential drop across CF is 1/10th and that across CE is 1/1000th of that across CD.

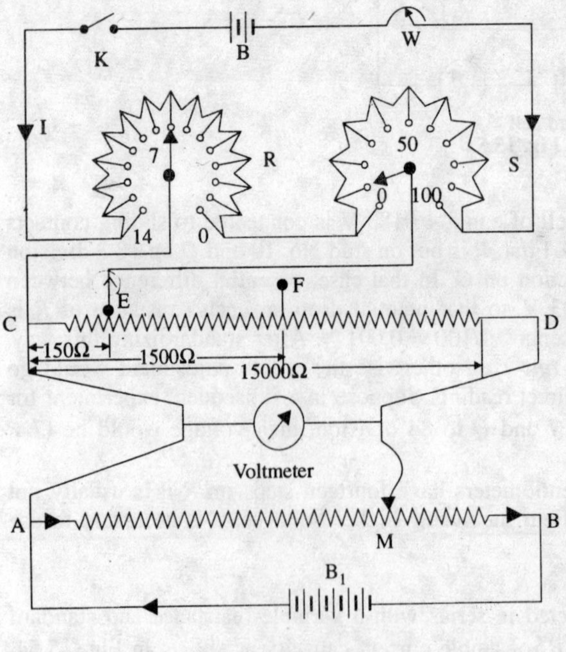

If supply voltage is 150 V, then p.d. across AB is also 150 V, and if M coincides with B, then p.d. across CD is also 150 V, so across CF is 15 volts and across CE is 1.5 V. The p.d. across CE can be balanced over the potentiometer as shown in Fig. 15.55. Various voltages can be applied across the voltmeter by moving the contact point M on the resistance AB.

Suppose that M is so placed that voltmeter V reads 70 V and p.d. across CE is balanced by adjusting P and Q. If the readings on P and Q, to give balance, are 7 and 8.4 respectively, then p.d. across CE is 0.7084 V.

Hence, the true p.d. across AM or CD or voltmeter is 0.7084 × 100 = 70.84 V (because resistance of CD is 100 times greater than that of CE). In other words, the reading of the voltmeter is low by 0.84 V.

Fig. 15.55

*This method may also be used for finding the value of unknown resistance. In that case, p.d. across the resistance is measured and then divided by current to give the value of the unknown resistance (Art. 15.59).

By shifting the position of M and then balancing the p.d. across CE on the potentiometer, the voltmeter can be calibrated throughout its range. By plotting the errors on a graph, a calibration curve of the insturment can also be drawn.

Example 15.23. *A standard cell of 1.0185 V, when used in a one meter long slide-wire potentiometer, balances at 60 cm. Calculate (i) the percentage error in a voltmeter which balances at 65 cm when reading 1.1 V and (ii) the percentage error in an ammeter that reads 0.345 A when balance is obtained at 40 cm with the potential drop across a 2-ohm resistance in the ammeter circuit.*

Solution. It is obvious that 1.0185 V drops over 60 cm length of the slide wire. Hence, potential drop across one cm length of the potentiometer wire is = 1.0185/60 V/cm.

(*i*) Voltage drop across 65 cm of slide-wire

$$= 65 \times 1.0185/60 = 1.1035 \text{ V}$$

Reading indicated by voltmeter = 1.1 V

Obviously, the voltmeter reads low by 0.0035 V.

$$\% \text{ error} = \frac{0.0035}{1.1035} \times 100 = \mathbf{0.31}$$

(*ii*) Voltage drop across 40 cm of wire

$$= 40 \times 1.0185/60 = 0.679 \text{ V}$$

Obviously, this is the potential drop across the 2-ohm resistance.

Current through 2 ohm resistance

$$= 0.679/2 = 0.3395 \text{ A}$$

Ammeter reading = 0.345 A

Absolute error = 0.345 – 0.3395 = 0.0055 A

Percentage error = 0.0055 × 100/0.3395 = **1.62**

It is obvious that the ammeter reads a bit high.

Example 15.24. *A potentiometer consists of 14 equal decade resistors and slide wire with 100 scale divisions, the total resistance of the slide wire being equal to that of one of the decade resistors.*

The potentiometer is calibrated to read voltages up to 1.5 V and is connected in turn across a standard 5 Ω resistor and an unknown resistor which are both in series with an ammeter carrying a current. The readings of the potentiometer at balance and the ammeter are as follows:

Across standard resistor : decade 12, slide wire 73

Across unknown resistor : decade 5, slide wire 56.

Ammeter reading ; 0.25 A

Calculate;

(*a*) *the value of the unknown resistor ;* (*b*) *the error in the ammeter reading.*

Solution. As explained in Art. 15.52, drop across each decade resistor is 0.1 V and each division of the slide wire represents 0.1/100 = 0.001 V.

Voltage drop across 5 Ω standard resistor is

$$= (12 \times 0.1) + (73 \times 0.001) = 1.273 \text{ V}$$

Correct value of circuit current = 1.273/5 = 0.2546 A

It is obvious that the ammeter is reading less.

(*b*) ammeter reading = 0.25 A

error = 0.25 - 0.2546 = – 0.0046 A

$$\text{percentage error} = \frac{-0.0046}{0.25} \times 100 = \mathbf{-1.84\%}$$

(*a*) p.d. across unknown resistance

$$= (5 \times 0.1) + (0.001 \times 56) = 0.556 \text{ V}$$

actual value of circuit current

$$= 0.2546 \text{ A}$$

unknown resistance = 0.556/0.2546 = **2.184 Ω**

Tutorial Problems No. 15.3

1. A 230-V, 1-ϕ energy meter has a constant load current of 10 A at unity p.f. If the meter disc makes 1150 revolutions during 2 hous, calculate the meter constant. If the p.f. were 0.8, what would be the number of revolutions made by the disc in that time?
[250 rev/kWh ; 920] *(A.M.I.E. Summer 1984)*
2. During a test, a 5-A Ah meter registered 0.48 kWh in 30 minutes. The testing voltage was 200 V. Calculate its percentage error and state whether the meter ran fast or slow. **[– 4% ; slow]**
3. An Ah meter is calibrated to measure the consumption of energy in a 200-V d.c. line. When a steady current of 10 A is passed in 5 hours, the meter reads 9.75 kWh. Calculate the percentage error. **[– 2.5% ; 0.0012 kWh/rev.]**

OBJECTIVE TESTS—15

A. Fill in the following blanks.

1. The moving system of an indicating instrument is subjected to deflecting torque, controlling torque and .... torque.
2. Moving iron instruments are either of attraction type or .... type.
3. Permanent-magnet moving-coil instruments are used for measuring .... current only.
4. The range of an ammeter can be extended with the help of a low-resistance ....
5. Electrodynamic instruments are almost always used as ....
6. Electrostatic instruments are almost always used as ....
7. The speed of an Ah meter becomes steady when its driving torque equals the .... torque.
8. Megger is used for measuring resistances of the order of..................... .

B. Answer True or False:

1. The pointer of an indicating instrument comes to rest when its deflecting torque equals the damping torque.
2. The moving-iron instruments can be used for both d.c. and a.c. circuits.
3. Permanent-magnet moving-coil instruments have uniform scales.
4. Induction type instruments are used only for a.c. measurement.
5. In dynamometer wattmeters, damping is by eddy currents.
6. The disc of an induction type watt-hour meter becomes stationary when its driving torque becomes equal to its braking torque.
7. Wheatstone bridge is particularly useful for measuring very high resistances.
8. A. d.c. potentiometer can be used for measuring current.

C. Multiple Choice Questions.

1. Which is NOT essential for the working of an indicating instrument?
(*a*) deflecting torque
(*b*) braking torque
(*c*) damping torque
(*d*) controlling torque.
2. The main function of a damping torque in an indicating electrical instrument is to
(*a*) bring the pointer to rest quickly
(*b*) prevent sudden movement of the pointer
(*c*) make pointer deflection gradual
(*d*) provide friction.
3. If an ammeter is used as a voltmeter, in all probability it will
(*a*) indicate much higher reading
(*b*) give extremely low reading
(*c*) indicate no reading at all
(*d*) burn out .
4. In electrodynamic instruments, the damping is invariably
(*a*) pneumatic
(*b*) electromagnetic
(*c*) by fluid friction
(*d*) by braking magnet.
5. In all induction instruments, deflection torque is produced due to the reaction between
(*a*) two eddy currents
(*b*) two alternating fluxes
(*c*) flux and eddy current
(*d*) voltage and current.
6. In an induction type Wh meter

(*a*) there is no braking magnet

(*b*) two d.c. magnets are used

(*c*) no shading bands are used

(*d*) disc revolves continuously.

7. In a d.c. Wheatstone bridge, the current through the galvanometer at balance condition is

(*a*) 1.08 a (*b*) 0.0 A

(*c*) 1.414 A (*d*) 1.732 A.

(Elect. Engg. A.M.Ae. S.I.Dec 1994)

8. A moving coil instrument has a resistance of 0.5 ohm and a full-scale deflecting of 0.1 A. To convert it into an ammeter of 0– 10 A, the shunt resistance should be .... ohm

(*a*) 0.004 (*b*) 0.005

(*c*) 0.050 (*d*) 0.1

(Elect. Engg. A.M.Ae.S.I.Dec. 1993)

9. A metre with a resistance of 100 ohm and a full-scale deflection of 1 mA is to be converted into a voltmeter of 0.5 V range. The multiplier resistance should be

(*a*) 490 ohm (*b*) 600 ohm

(*c*) 4900 ohm (*d*) none of these

(Elect. Engg. A.M.Ae. S.I. June 1994)

10. The reading of a moving-coil ammeter connected in series with a diode across a 110 V ac supply (the diode provides 50 ohms to current in one direction and an infinite resistance in the reverse direction) is given by

(*a*) 1.98 A (*b*) 0.99 A

(*c*) 1.02 A (*d*) 50 A

(Elect, Engg. A.M.Ae. S.I. Dec. 1994)

ANSWERS

A. 1. damping 2. repulsion 3. direct 4. shunt 5. wattmeter 6. voltmeters 7. braking 8. megohms

B. 1. F 2. T 3. T 4. T 5. F 6. F 7. F 8. T

C. 1. *b* 2. *a* 3. *d* 4. *a* 5. *c* 6. *d* 7. *b* 8. *b* 9. *c* 10. *b*

16 A.C. FUNDAMENTALS

16.1. Generation of Alternating Voltages and Alternating Currents

Referring to Art. 10.2 and Fig. 10.1 and 10.2, it is clear that alternating voltage may be generated by rotating a coil in a magnetic field, as shown again in Fig. 16.1 *(a)* or by rotating a magnetic field within a stationary coil, as shown in Fig. 16.1 (*b*).

The value of the voltage generated depends, in each case, upon the number of turns in the coil, strength of the field and the speed at which the coil or magnetic field rotates. Alternating voltage may be generated in either of the two ways shown above, although rotating field method is the one which is mostly used in practice.

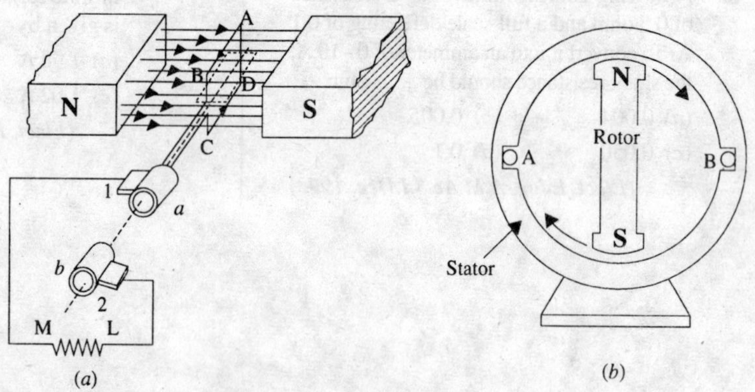

Fig. 16.1

16.2. Equations of the Alternating Voltages and Currents

Consider a rectangular coil having N turns rotating in a uniform magnetic field with an angular velocity of ω radian/second as shown in Fig. 16.2. Let time be measured from the X-axis. Maximum flux Φ_m is linked with the coil when its plane coincides with the X-axis. In time t seconds, this coil rotates through an angle $\theta = \omega t$. In this deflected position, the component of the flux which is perpendicular to the plane of the coil is $\Phi = \Phi_m \cos \omega t$. Hence, 'flux-linkages' of the coil in this deflected position are

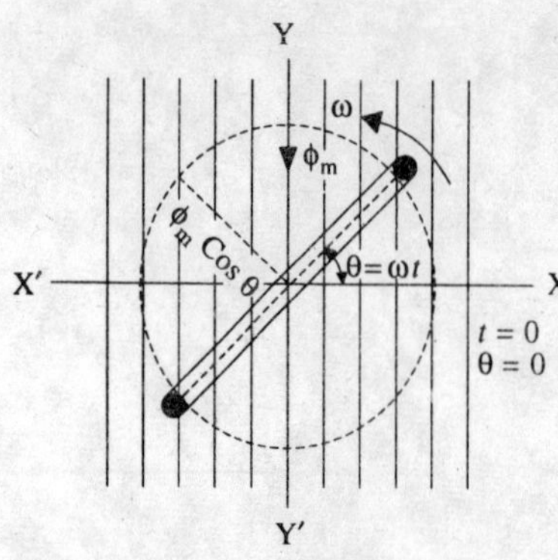

Fig. 16.2

$$N\Phi = N\Phi_m \cos \omega t$$

According to Faraday's Laws of Electromagnetic Induction, the e.m.f. induced in the coil is given by the rate of change of flux-linkages of the coil. Hence, the value of the induced e.m.f. at this instant (*i.e.* when $\theta = \omega t$) or the instantaneous value of the induced e.m.f. is

$$e = -\frac{d}{dt}(N\Phi) \text{ volt}$$

$$= -N \cdot \frac{d}{dt}(\Phi_m \cos \omega t) \text{ volt}$$

$$= -\omega N \Phi_m(-\sin \omega t) \text{ volt}$$

$$= \omega N \Phi_m \sin\theta \text{ volt} \qquad ...(i)$$

When the coil has turned through 90° *i.e.* when $\theta = 90°$, then $\sin \theta = 1$, hence e has maximum value, say E_m. Therefore, from E_q (*i*) we get

$$E_m = \omega N \Phi_m \text{ volt} = \omega N B_m A \text{ volt} = 2\pi f N B_m A \text{ volt} \qquad ..(ii)$$

where B_m = maximum flux density in Wb/m^2
A = area of the coil in m^2
f = frequency of rotation of the coil in rev/s

Substituting this value of E_m in Eq. (*i*) we get

$$e = E_m \sin \pi = E_m \sin \omega t \quad ...(iii)$$

Similarly, the equation of the induced alternating current is

$$i = I_m \sin \omega t \quad ...(iv)$$

provided the coil circuit has been closed through a resistive load.

Since $\omega = 2\pi f$ is the frequency of rotation of the coil, the above equations of the voltage and current can be written as

$$e = E_m \sin 2\pi ft \quad = E_m \sin\left(\frac{2\pi}{T}\right)t;$$

$$i = I_m \sin 2\pi ft = I_m \sin\left(\frac{2\pi}{T}\right)t$$

where T = time period of the alternating voltage or current
$= 1/f$

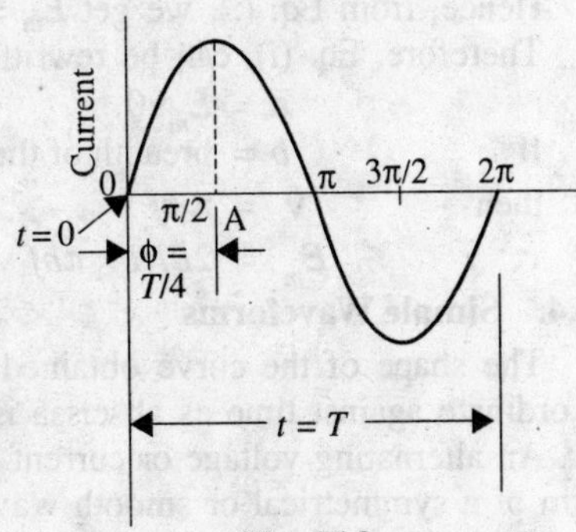

Fig. 16.3

It is seen that the induced e.m.f. varies as sine function of the time angle ωt and when e.m.f. is plotted against time, a curve similar to one shown in Fig. 16.3 is obtained. This curve is known as sine curve and the e.m.f. which varies iin this manner is known as *sinusoidal* e.m.f. Such a sine curve can be conveniently drawn as shown in Fig. 16.4. A vector equal in length to E_m is drawn. It rotates in the counter clockwise direction with the velocity of ω radian/second making one revolution while the generated e.m.f. makes two loops or one cycle. The projection of this vector on *Y*-axis gives the instantaneous value of the induced e.m.f. *i.e.* $E_m \sin \omega t$.

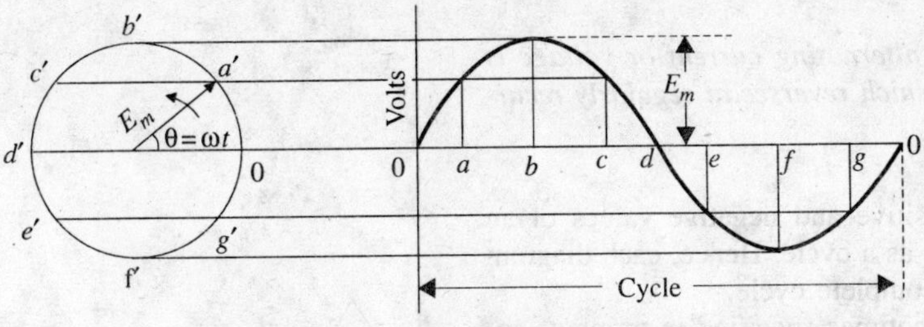

Fig. 16.4

To construct the curve, lay off along *X*-axis equal angular distances *oa, ab, bc, cd*, etc. corresponding to suitable angular displacements of the rotating vector. Now, erect ordinates at the points *a,b,c* and *d* etc. (Fig. 16.4) and then project the free ends of the vector E_m at the corresponding positions *a',b',c'*, etc. to meet these ordinate. Next, draw a curve passing through these intersecting points. The curve so obtained is the graphic representation of equation (*iii*).

16.3. Alternative Method for the Equations of Alternating Voltages and Currents

In fig. 16.5 is shown a rectangular coil *AC* having *N* turns and rotating in the magnetic field of flux density *B* Wb/m^2. Let the length of each of its sides *A* and *C* l metres and their peripheral velocity ν be *l* metres/second. Let the angle be measured from the horizontal position *i.e.* from the *X*-axis. When in horizontal position, the two sides *A* and *C* move parallel to the lines of magnetic flux. Hence, no flux is cut and so no e.m.f. generated in the coil.

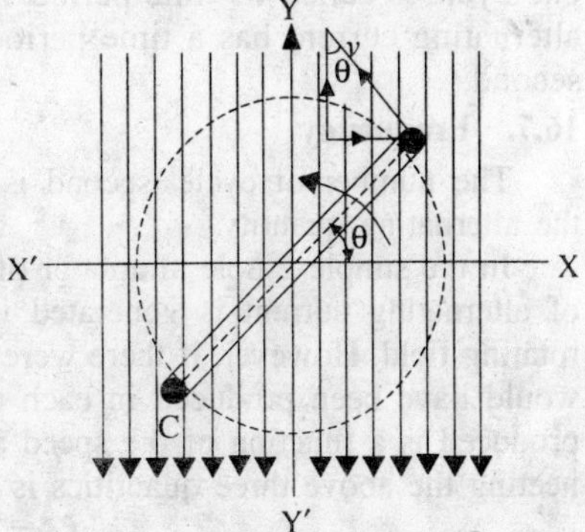

Fig. 16.5

When the coil has turned through angle θ, its velocity can be resolved into two mutually perpendicular components (*i*) $v \cos\theta$ component parallel to the direction of the magnetic flux and (*ii*) $v \sin\theta$ component perpendicular to the direction of the magnetic flux.

The e.m.f. generated is due entirely to the perpendicular component *i.e.* $v\sin\theta$.

Hence, the e.m.f. generated in one side of the coil which contains N conductors as seen from Art. 8-8 is given by

$$e = N \times Blv \sin\theta$$

Total e.m.f. generated in both sides of the coil is

$$e = 2BNl\, v\sin\theta \text{ volt} \qquad \text{...}(i)$$

Now, e has maximum value of E_m (say), when $\theta = 90°$

Hence, from Eq. (*i*), we get $E_m = 2BNlv$ volt

Therefore. Eq. (*i*), can be rewritten as

$$e = E_m v\theta \qquad \text{...as before}$$

If b = breadth of the coil in meters ; f = frequency of rotation of coil in rev/s

then $v = \pi bf$

$$\therefore \quad E_m = 2BNl \times \pi bf \text{ volts} = 2\pi f NBA \text{ volts} \qquad \text{....as before}$$

16.4. Simple Waveforms

The shape of the curve obtained by plotting the instantaneous values of voltage or current as ordinate against time as abscissa is called its *waveform* or *wave-shape.*

An alternating voltage or current may not always take the form of a symmetrical or smooth wave such as that shown in Fig. 16.3. Thus, Fig. 16.6 also represents alternating waves. But while it is scarcely possible for the manufacturers to produce sine-wave generators or alternators, yet sine wave is the ideal form sought by the designer and is the accepted standard. The waves deviating from the standard sine wave are termed as distorted waves.

In general, however, *an alternating current or voltage is one, the circuit direction of which reverses at regularly recurring intervals.*

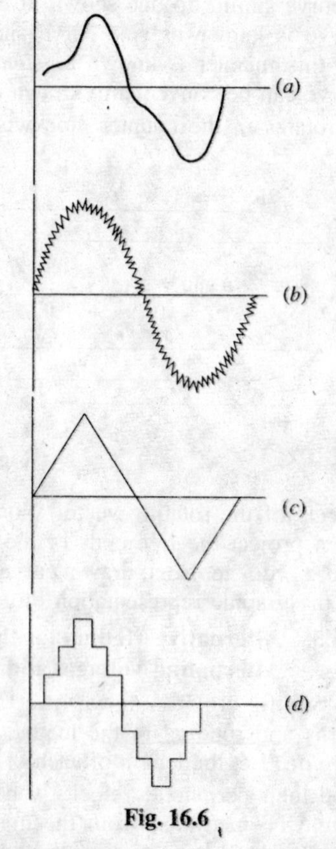

Fig. 16.6

16.5. Cycle

One complete set of positive and negative values of an alternating quantity is known as a cycle. Hence, each diagram of Fig. 16.6 represents one complete cycle.

A cycle may also be sometimes specified in terms of angular measure. In that case, one complete cycle is said to spread over 360° or 2π radians.

16.6. Time Period

The time taken by an alternating quantity to complete one cycle is called its time period T. For example, a 50 Hz alternating current has a time period of 1/150 second=0.02 second.

16.7. Frequency

The number of cycles/second is called the frequency of the alternating quantity.

In the simple 2-pole alternator of Fig. 16.1 (*b*), one cycle of alternating current is generated in one revolution of the rotating field. However, if there were 4-poles, then two cycles would have been produced in each revolution. In fact, the frequency of the alternating voltage produced is a function of the speed and the number of poles of the alternator. The relation connecting the above three quantities is given by

$$f = PN/120$$

where N = revolutions in r.p.m. ; P = number of poles.

For example, an alternator having 20 poles and running at 300 r.p.m. will generate alternating voltage and current whose frequency is $20 \times 300/120 = 50$ Hz

It may be noted that the frequency is given by the reciprocal of the time period of the alternatingg quantity.

$\therefore \quad f = 1/T \quad$ or $\quad T = 1/f$

16.8. Amplitude

The maximum value, positive or negative, of an alternating quantity is known an its amplitude.

16.9. Different Forms of E.M.F. Equation

The standard form of an alternating voltage, as already given in Art. 16.2, is

$$e = E_m \sin\theta = E_m \sin\omega t = E_m \sin 2\pi f t = E_m \sin\frac{2\pi}{T} t$$

By closely looking at the above equation, we find that (*i*) the maximum value or peak value or amplitude of an *alternating voltage is given by the coefficient of the sine function.*

(ii) the frequency f is given by dividing the coefficient of time in the angle by 2 p.

For example if the equation of an alternating voltage is given by $e = 50 \sin 314\, t$ then its maximum value is 50 V and its frequency is $314/2\pi = 50$ Hz

Similarly, if the equation is of the form $e = I_m\sqrt{(R^2 + 4\omega^2L^2)} \sin 2\omega t$ then its maximum value is $E_m = I_m\sqrt{(R^2 + 4\omega^2L^2)}$ and the frequency is $2\omega/2\pi$ or ω/π.

Example 16.1. *(a) What is the equation of a 25-Hz current sine wave having an r.m.s. value of 30 A?*

(b) A 60-Hz engine-driven alternator has a speed of 1200 r.p.m. How many poles has it?

Solution. (*a*) $I_{r.m.s.} = 30$ A

$I_{max} = 30 \times \sqrt{2} = 42.42$ A ...Art. 16.14

Hence, the equation of the alternating current is

i = **42.42 sin 157 t**

(*b*) frequency, $f = PN/120$ Hz

$\therefore \quad 60 = 1200\,P/120 \qquad \therefore \; P = \mathbf{6}$

The alternator has 6 poles.

Example 16.2. *A square coil of 10 cm side and with 100 turns is rotated at the uniform speed of 500 r.p.m. about an axis at right angles to a uniform field of 0.5 Wb/m². Calculate the instantaneous value of the induced e.m.f. when the plane of the coil is (a) at right angles to the plane of the field (b) ar 30° to the plane of the field and (c) in the plane of the field.*

(Electrical Engg. A.M.Ae.S.I. Dec. 1991)

Solution. The instantaneous value of the e.m.f. is

$$e = 2\pi f NBA \sin\theta \text{ volt}$$

where θ is measured from the position of the coil *when its plane is at right angles to the flux* (Fig. 16.2).

Here $f = 500/60 = 25/3$ r.p.s.

$N = 100;\; B = 0.5$ Wb/m²

$A = 10 \times 10 = 100\text{ cm}^2 = 10^{-2}\text{ m}^2$

(*a*) $\theta = 0, \sin\theta = 0 \qquad \therefore \; e = 0$

(*b*) $\theta = 60^0, \sin 60^0 = 0.866$

$e = 2\pi \times (25/3) \times 100 \times 10^{-2} \times 0.5 \times 0.866 =$ **22.6V**

(*c*) $\theta = 90°, \sin 90° = 1$

$\therefore \quad e = 2\pi \times (25/3) \times 100 \times 0.5 \times 10^{-2} =$ **26.18V**

Example 16.3. *An alternating current of frequency 60 Hz has a maximum value of 120 A. Write down the equation for its instantaneous value. Reckoning time from the instant the current* *is zero and is becoming positive, find (a) the instantaneous value after 1/360 second and (b) the time taken to reach 96 A for the first time.*

Solution. The instantaneous equation is

$i = 120 \sin 2\pi ft = 120 \sin 120\pi t$

Now, then $t = 1/360$ second, then

(*a*) $i = 120 \sin (120 \times \pi \times 1/360)$ ...angle in redians

$= 120 \sin (120 \times 180 \times 1/360)$ ...angle in degrees

$= 120 \sin 60° =$ **103.9A**

(*b*) $96 = 120 \times \sin 360 \times 60 \times t$ ...angle in degrees

or $\sin (360 \times 60 \times t) = 4/5 = 0.8$

$\therefore \quad 360 \times 60 \times t = \sin^{-1} 0.8 = 53°$ (approx)

$\therefore \quad t = 53/360 \times 60 =$ **0.00245 second**

Example 16.4. *The e.m.f. produced by an alternator is given by*

$e = E_m \sin (314t + \pi/2)$

If the rotor runs at 75 r.p.m., find the number of poles of the alternator. Also, find the angle through which rotor turns in T/4 second.

Solution. As seen from its equation, the e.m.f. has a frequency of $314/2\pi = 50$ Hz.

Now, $f = PN/120$

$\therefore \quad 50 = P \times 75/120 \qquad \therefore$ **P = 80**

Obviously, the number of pair of poles is $= 80/2 = 40$. With one pair of poles, rotor turns through 90° (electrical) in one quarter time period *i.e.* in $T/4$ second. With 40 pairs of poles, angle turned would be 40 times smaller *i.e.* it would be $= 90/40 =$ **2.25°**.

16.10. Phase

By phase of an alternating current is meant the fraction of the time period of that A.C. that has elapsed since it last passed through the zero position of reference. For example, the phase of current at point *A* is $T/4$ second where *T* is time period or expressed in terms of angle, it is $\pi/2$ radian (Fig. 16.7). Similarly, the phase of the rotating coil at the instant shown in Fig. 16.2 is ωt which is, therefore, called its phase angle.

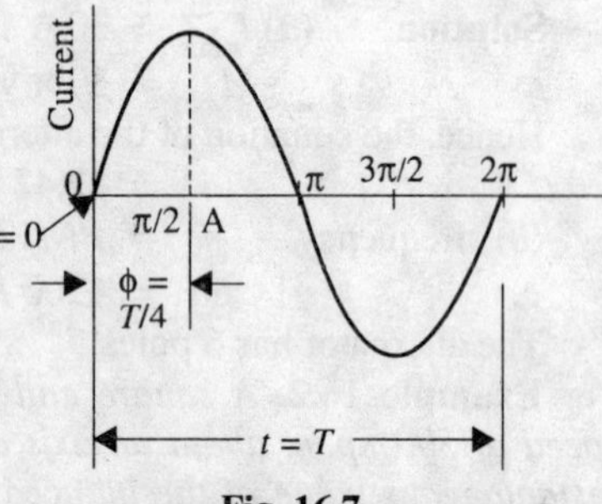

Fig. 16.7

In electrical engineering, we are, however, more concerned with relative phases *i.e.* phase differences between different alternating quantities rather than with their absolute phases. Consider two single-turn coils of different sizes [Fig. 16.8] (*a*)] arranged radially in the same plane and rotating with the same angular velocity, in a common magnetic field of uniform intensity. The e.m.f.s. induced in both coils will be of the same frequency and of sinusoidal shape, although the values of instantaneous e.m.fs. induced would be different. However, the two alternating e.m.fs. would reach their maximum and zero values at the same time as shown in Fig.16.8(*b*). Such alternating voltages (or currents) are said to be in phase with each other. The two voltages will have the equations $e = E_{m1} \sin \omega t$ and $e_2 = E_{m2} \sin \omega t$.

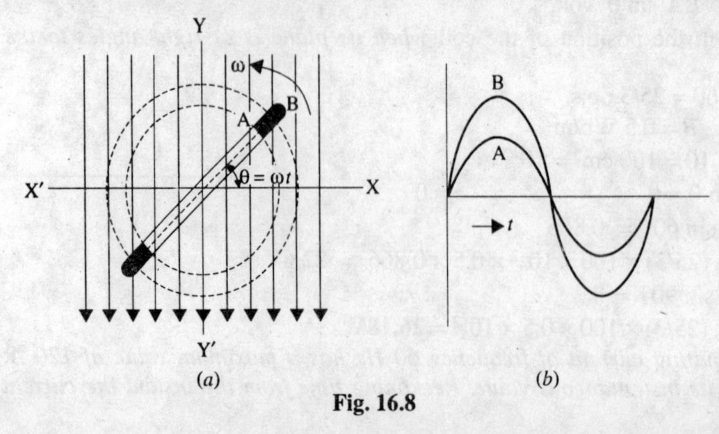

Fig. 16.8

16.11. Phase Difference

Now, consider three similar single-turn coils displaced from each other by angles α and β and rotating in a uniform field with the same angular velocity [Fig. 16.9 (*a*)]

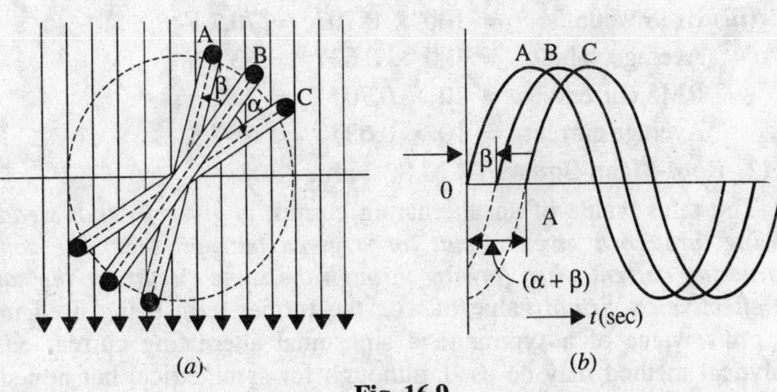

Fig. 16.9

In this case, the values of induced e.m.f.s. in the three coils, are the same, but there is one important difference. The e.m.fs. in these coils do not reach their maximum or zero values simultaneously, but one after another. The three sinusoidal waves are shown in Fig. 16.9 (*b*). It is seen that curves *B* and *C* are displaced from curve *A* by angles β and (α + β) respectively. Hence, it means that phase difference between *A* and *B* is β and between *B* and *C* is α but between *A* and *C* is (α + β). This statement, however, does not give indication as to which e.m.f. reaches its maximum value first. This deficiency is supplied by using term *'lag'* or *'lead'*.

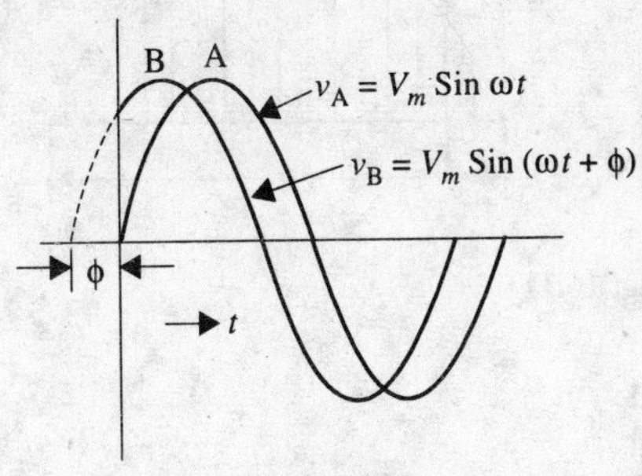

Fig. 16.10

A leading alternating quantity is one which reaches its maximum (or zero) value earlier as compared with the other quantity.

Similarly, a lagging alternating quantity is one which reaches its maximum or zero value later than the other quantity. For example, in Fig. 16.9(*b*), *B* lags behind *A* by β and *C* lags behind *A* by (α + β).

The three equations for the instantaneous induced e.m.fs. are:

$$e_A = E_m \sin \omega t$$
$$e_B = E_m \sin (\omega t - \beta)$$
$$e_c = E_m \sin [\omega t - (\alpha + \beta)]$$

In Fig. 16.10, quantity *B leads A by* an angle φ. Hence, their equations are

$$v_A = V_m \sin \omega t \,; \qquad v_B = V_m \sin (\omega t + \phi)$$

A plus (+) sign when used in connection with phase difference denotes 'lead' whereas a minus (–) sign denotes 'lag'.

Example 16.5. *An alternating current is given by the expression*

$$i(t) = 300 \sin (157t + \pi/3)$$

Find (i) maximum value of current, (ii) frequency and (iii) periodic time.

(Elect. Engg.-I, Punjab Univ. 1993.)

Solution. Comparing the given current equation with the standard equation for a sinusoidal alternating current, we get the following :

(*i*) I_m = **300 A**

(*ii*) ω = 157 or 2 π*f* = 157 or *f* = **25 Hz**

(*iii*) Time period, *T* = 1/*f* = 1/25 = **0.04 second**

Example 16.6. *An a.c. voltage of 50 Hz frequency has a peak value of 100 V. (i) write an equation to calculate the instantaneous value of the voltage, (ii) write an equation for a current having a maximum value of 10 A and lagging the voltage wave by 45°, (iii) find average and effective values of the voltage and the current.* ***(Elect. Engg.-I, Punjab Univ. 1994.)***

Solution. Taking the standard equation into account, we have the following:

(*i*) v = 100 sin 100 π t = **100 sin 314t**

(*ii*) *i* = **10 sin (100 π t – π/4)**

(*iii*) RMS value, V = 100 × 0.707 = **70.7 V**
Average value = 100 × 0.637 = **63.7 V**
RMS current = 10 × 0.707 = **7.07 A**
Average current = 10 × 0.637 = **6.37 A**

16.12. Root-Mean-Square (R.M.S.) Value

The r.m.s. value of an alternating current is *given by that steady current (d.c.) which when flowing through a given circuit for a given time produces the same* **heat** *as produced by the alternating current when flowing through the same circuit for the same time.* It is also known as the *effective* or *virtual* value of a.c., the former term being used more extensively. For finding the r.m.s. value of a symmetrical sinusoidal alternating current either mid-ordinate method or analytical method may be used, although for symmetrical but non-sinusoidal waves, the mid-ordinate method would be found more convenient.

16.13. Mid-ordinate Method

In Fig. 16.11 are shown the positive half-cycles for both symmetrical sinusoidal and a non-sinusoidal alternating current. Divide the time base '*t*' into *n* equal intervals of time each of duration *t/n* seconds. Let the mean instantaneous values of currents during these intervals be respectively $i_1, i_2, i_3 .....i_n$ *i.e.* mid-ordinates in Fig. 16.11. Suppose this alternating current is passed through a circuit of resistance *R* ohm. Then,

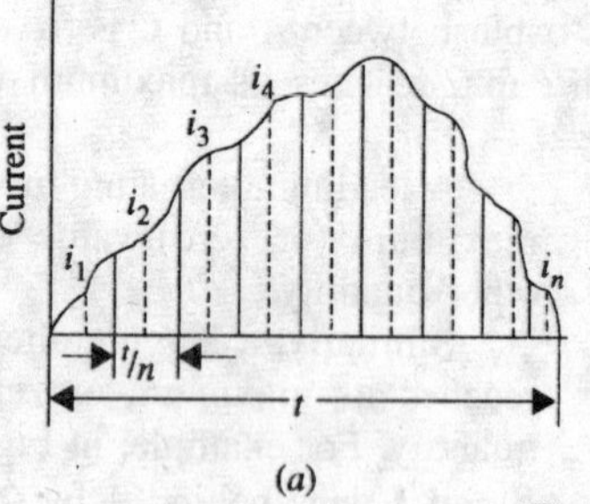

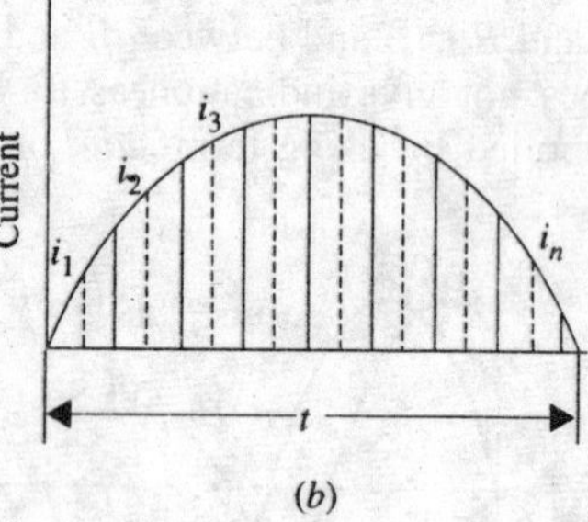

Fig. 16.11

Heat produced in 1st inverval = $0.24\, i_1^2 R\, t/n$ cal
" " " 2nd " = $0.24\, i_2^2 R\, t/n$ cal
: : : : :
Heat produced in *n*th interval = $0.24\, i_n^2 R\, t/n$ cal
Total heat produced in *t* seconds is

$$= 0.24\, R\, t \left(\frac{i_1^2 + i_2^2 + \ldots + i_n^2}{n} \right)$$

Now, suppose that a direct current of value *I* produces the same heat through the same resistance during the same time *t*. Heat produced by it is = $0.24\, I^2 R\, t$ cal. By definition, the two amounts of heat produced should be equal.

$$\therefore \quad 0.24\, I^2 Rt = 0.24\, Rt \left(\frac{i_1^2 + i_2^2 + \ldots + i_n^2}{n} \right) \qquad \therefore \quad I^2 = \frac{i_1^2 + i_2^2 + \ldots + i_n^2}{n}$$

$$\therefore \quad I = \sqrt{\left(\frac{i_1^2 + i_2^2 + \ldots i_n^2}{n} \right)}$$

= square root of the mean of squares of the instantaneous currents.

Similarly, the r.m.s. value of the alternating voltage is given by the expression

$$V = \sqrt{\left(\frac{v_1^{\,2} + v_2^{\,2} + \ldots\ldots + v_n^{\,2}}{n}\right)}$$

16.14. Analytical Method

The standard form of a sinusoidal alternating current is

$$i = I_m \sin \omega t = I_m \sin \theta$$

The mean of the squares of the instantaneous values of current over one complete cycle is (even the value over half a cycle will do),

$$= \int_0^{2\pi} \frac{i^2 d\theta}{(2\pi - 0)}$$

The square root of this value is

$$= \sqrt{\left(\int_0^{2\pi} \frac{i^2 d\theta}{2\pi}\right)}$$

Hence, the r.m.s. value of the alternating current is

$$I = \sqrt{\left(\int_0^{2\pi} \frac{i^2\, d\theta}{2\pi}\right)} = \sqrt{\left(\frac{Im^2}{2\pi}\int_0^{2\pi} \sin^2\theta\, d\theta\right)} \qquad \text{(Putting } i = I_m \sin\theta\text{)}$$

Now, $\cos 2\theta = 1 - 2\sin^2\theta$ or $\sin^2\theta = \dfrac{1 - \cos 2\theta}{2}$

$$\therefore \quad I = \sqrt{\left(\frac{Im^2}{4\pi}\int_0^{2\pi} (1 - \cos 2\theta)\, d\theta\right)}$$

$$= \sqrt{\left(\frac{Im^2}{4\pi}\left|\theta - \frac{\sin 2\theta}{2}\right|_0^{2\pi}\right)} = \sqrt{\left(\frac{Im^2}{4\pi} \times 2\pi\right)} = \frac{Im^2}{2}$$

$$\therefore \quad I = I_m/\sqrt{2} \text{ or } \quad I = 0.707\, I_m$$

Hence, we find that for a symmetrical sinusoidal current,

r.m.s. value of current = 0.707 × Max. value of current

The r.m.s. value of an alternating current is of considerable importance in practice, because the ammeters and voltmeters record the r.m.s. value of alternating current and voltage respectively. In electrical engineering work, *unless indicated otherwise, the values of the given current and voltage are always taken as the r.m.s. values.*

It should be noted that the average heating effect produced during one cycle is

$$= I^2R = \left(\frac{I_m}{\sqrt{2}}\right)^2 R = \frac{1}{4} I_m^2 R$$

Example 16.7. *An alternating current, when passed through a resistance immersed in water for 5 miniutes just raised the temperature of the water to boiling point. When a direct current of 4 A was passed through the same resistance under identical conditions, it took 8 minutes to boil the water. Find the r.m.s. value of the alternating current. Neglect factors other than heat given to the water.*

Solution. Let I be the r.m.s. value of the alternating current and R the value of the resistance.

Heat produced in the first case $= I^2Rt/4.2 = 0.24\, I^2 R\, (5 \times 60)$ cal

Heat produced in the second case $= 4^2 \times R \times (8 \times 60)/4.2$ cal

By definition of the r.m.s. value, the two quantities of heat should be equal.

$\therefore \quad 0.24\, I^2 R \times 300 = 0.24 \times 16 R \times 480 \qquad \therefore$ **$I = 5.06$ A**

16.15. Average Value

The average value I_{av} of an alternating current is expressed *by that steady current which transfers across any circuit the same charge as is transferred by that alternating current.* In the case of a symmetrical alternating current (*i.e.* one whose two half-cycles are exactly similar, whether sinusoidal or non-sinusoidal), the average value over a complete cycle is zero. Hence, in their case, the average value is obtained by adding or integrating the instantaneous values of current over half-cycle only. *But in the case of an unsymmetrical alternating current (like half-wave rectified current) the average value must always be taken over the whole cycle.*

(*i*) Mid-ordinate Method

With reference to Fig.16.11.

$$I_{av} = \frac{i_1 + i_2 + \ldots + i_n}{n}$$

This method may be used both for sinusoidal and non-sinusoidal waves, although it is specially convenient for the latter.

(*ii*) Analytical Method

The standard equation of an alternating current is

$$i = I_m \sin\theta$$

$$\therefore \quad I_{av} = \int_0^{\pi} \frac{i\,d\theta}{(\pi - 0)} = \frac{I_m}{\pi}\int_0^{\pi} \sin\theta\, d\theta \qquad \text{(put value of } i\text{)}$$

$$= \frac{I_m}{\pi}\left|-\cos\theta\right|_0^{\pi} = \frac{I_m}{\pi}\left|+1-(-1)\right| = \frac{I_m}{\pi/2} = \frac{2I_m}{\pi}$$

$$\therefore \quad I_{av} = \frac{\text{twice the maximum current}}{\pi} \quad \text{or} \quad I_{av} = I_m/\tfrac{1}{2}\pi = 0.637\, I_m$$

$\therefore$ *average value of current* = 0.637 × *maximum value*

16.16. Form Factor

It is defined as the ratio,

$$K_f = \frac{\text{r.m.s. value}}{\text{average value}} = \frac{0.707\, I_m}{0.637\, I_m} = 1.11 \qquad \text{(for sinusoidal A.C. only)}$$

In the case of sinusoidal alternating voltage also

$$K_f = \frac{0.707\, E_m}{0.637\, E_m} = 1.11$$

Obviously, the knowledge of form factor will enable the r.m.s. value to be found from the arithmetic mean value and *vice versa.*

16.17. Crest or Peak or Amplitude Factor

It is defined as the ratio

$$K_a = \frac{\text{maximum vlaue}}{\text{r.m.s. value}} = \frac{I_m}{I_m/\sqrt{2}}$$

$$\therefore \quad K_a = \sqrt{2} = 1.414 \qquad \text{(for sinusoidal A.C. only)}$$

For sinusoidal alternating voltage also

$$K_a = \frac{E_m}{E_m/\sqrt{2}} = 1.414$$

Knowledge of this factor is of importance in dielectric insulation testing because the dielectric stress to which the insulation is subjected, is proportional to the maximum or peak value of the appiied voltage. This knowledge is also necessary when measuring iron losses because the iron loss depends on the value of maximum flux.

Example 16.8. *The graph in Fig. 16.12 shows the variation of voltage with time. Use this* *graph to calculate the r.m.s. value of the voltage. What is the frequency of the voltage? Calculate the r.m.s. current which woluld flow through a 5 Ω resistor if this voltage were connected across it.*

What would be the r.m.s. voltage of a sinusodial wave if it had the same peak value as the above voltage?

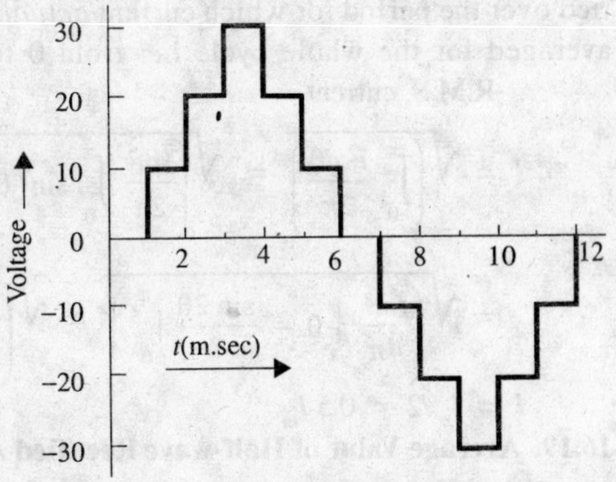

Solution. Since the graph (Fig. 16.12) is symmetrical about the time axis, we may consider only the positive half-cycle.

Fig. 16.12

Mean value of v^2 is

$$= \frac{0^2 + 10^2 + 20^2 + 30^2 + 20^2 + 10^2}{6} = \frac{950}{3}$$

∴ R.M.S. value, $V = \sqrt{950/3} = \mathbf{17.8\ V}$

Time period $T = 12$ millisecond $= 12 \times 10^{-3}$ second

Frequency, $f = 1/T = 1/12 \times 10^{-3} = \mathbf{83.3\ Hz}$

R.M.S.current $= 17.8/5 = 3.56$ A

R.M.S. value of a sinusoidal voltage of peak value 30 V is $= 30/\sqrt{2} = \mathbf{21.2V}$

Example 16.9. *A sinusoidal alternating voltage of 50 Hz has an r.m.s. value of 200 V. Write down the equation for the instantaneous value and find this value 0.0125 second after passing through a positive maximum value.*

At what time measured from a positive maximum value will the instantaneous voltage be 141.4 V?

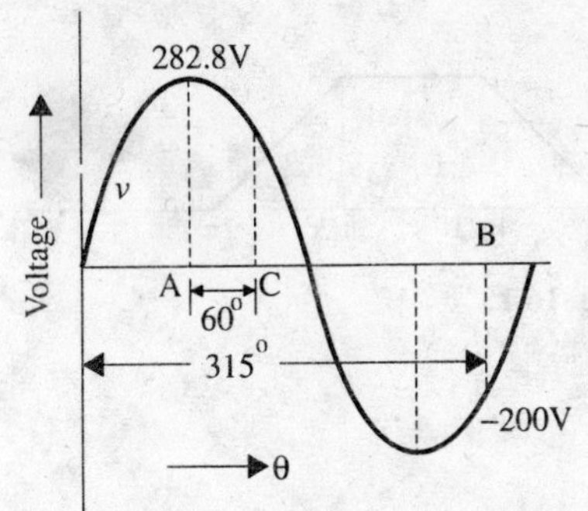

Fig. 16.13

Solution. $V_m = 200\sqrt{2} = 282.8$ V

$\omega = 2\pi \times 50 = 314$ rad/second

The instantaneous equation of the voltage wave with reference to 0 point in Fig. 16.13 is

$$v = \mathbf{282.8 \sin 100\,\pi\, t}$$

Since time values are given from point A where voltage has positive and maximum value, the equation may itself be referred to point A. In that case, the equation becomes

$v = 282.8 \cos 100\pi t$

Now, $t = 0.0125$ second —given

∴ $v = 282.8 \cos 100\pi \times 0.0125$ —angle in radians

$= 282.8 \cos 100 \times 180 \times 0.0125$ —angle in degrees

$= 282.8 \cos 225° = 282.8 \times (-1/\sqrt{2})$

$= \mathbf{-200V}$ **...point B**

Here $v = +141.4$ V

∴ $141.4 = 282.8 \cos 100 \times 180\, t$

∴ $\cos 100 \times 180t = 0.5$

or $100 \times 180t = \cos^{-1}(0.5) = 60^0$

$t = \mathbf{1/300\ second}$ **...point C**

16.18. R.M.S. Value of Half-wave Rectified A.C.

Half-wave (H.W.) rectified alternating current is one whose one-half cycle has been suppressed *i.e.* one which flows for half the time during one cycle. It is shown in Fig. 16.14 where suppressed half-cycle is shown dotted.

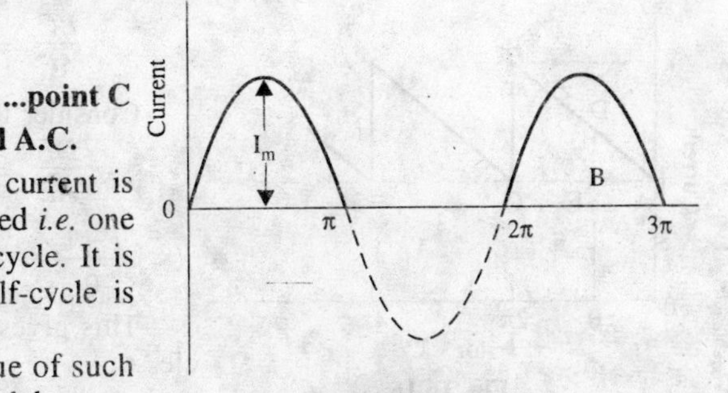

Fig. 16.14

As said earlier, for finding r.m.s. value of such an alternating current, summation should be carried over the period for which current *actually* flows *i.e.*, from 0 to π, though it would be averaged for the whole cycle *i.e.* from 0 to 2π.

$\therefore$ R.M.S. current

$$I = \sqrt{\left(\int_0^{\pi} \frac{i^2\, d\theta}{2\pi}\right)} = \sqrt{\left(\frac{Im^2}{2\pi}\int_0^{\pi} \sin^2\theta\, d\theta\right)} = \sqrt{\left(\frac{Im^2}{4\pi}\int_0^{\pi}(1 - \cos 2\theta)\, d\theta\right)}$$

$$= \sqrt{\left(\frac{Im^2}{4\pi}\left|\theta - \frac{\sin 2\theta}{2}\right|_0^{\pi}\right)} = \sqrt{\left(\frac{Im^2}{4\pi} \times \pi\right)} = \sqrt{\left(\frac{Im^2}{4}\right)}$$

$$I = I_m/2 = 0.5\, I_m$$

16.19. Average Value of Half-wave Rectified A.C.

For the same reasons as given in Art. 16.18, integration would be carried over from 0 to π.

$$\therefore \quad I_{av} = \int_0^{\pi} \frac{i\, d\theta}{2\pi} = \frac{I_m}{2\pi}\int_0^{\pi} \sin\theta\, d\theta \qquad (\because\ i = I_m \sin\theta)$$

$$= \frac{I_m}{2\pi}\left| -\cos\theta\right|_0^{\pi} = \frac{I_m}{2\pi} \times 2 = \frac{I_m}{\pi}$$

16.20. Form Factor of Half-wave Rectified A.C.

$$\text{From factor} = \frac{\text{r.m.s. value}}{\text{average value}} = \frac{I_m/2}{I_m/\pi} = \frac{\pi}{2} = 1.57$$

Example 16.10. *Find the average value of the periodic function shown Fig. 16.15.*

(A.M.I.E. Summer 1992)

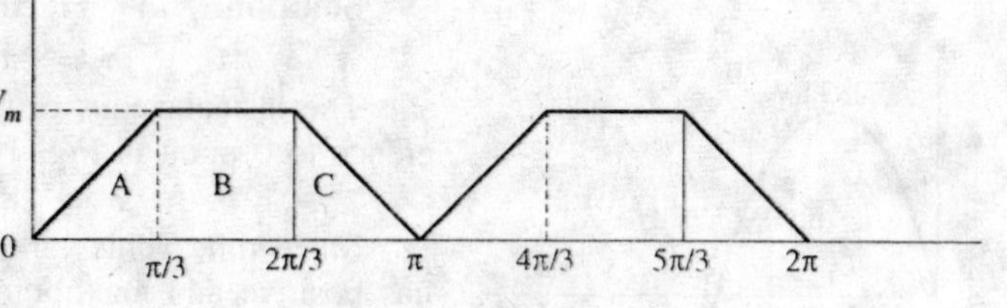

Fig. 16.15

Solution. Since in parts *A* and *C*, function changes linearly, average value in *A* as well as in $C = V_m/2$. Since in Part *B* function remains constant, average value = V_m

Average value of the three parts or for one half cycle is

$$= \frac{V_m/2 + V_m + V_m/2}{3} = \frac{2V_m}{3}$$

Same will be the average for the whole cycle.

Example 16.11. (*a*) *Calculate the r.m.s. and average value of the current 'i' represented by Fig. 16.16.*

(*b*) *Calculate the r.m.s. value of current given by $i = 10 + 5\cos(628\,t + 30^0)$*

Solution. (*a*) The slope of the curve *AB* of Fig. 16.16 is

$$\frac{BC}{AC} = \frac{1}{2\pi}$$

Consider the current *i* at any angle θ. It is seen that

$$\frac{DE}{AE} = \frac{BC}{AC} = \frac{1}{2\pi} \quad \text{or} \quad \frac{DE}{AE} = \frac{1}{2\pi}$$

$$\therefore \quad \frac{i-1}{\theta} = \frac{1}{2\pi} \quad \text{or} \quad i = 1 + \frac{\theta}{2\pi}$$

This gives us the equation of the current for one cycle.

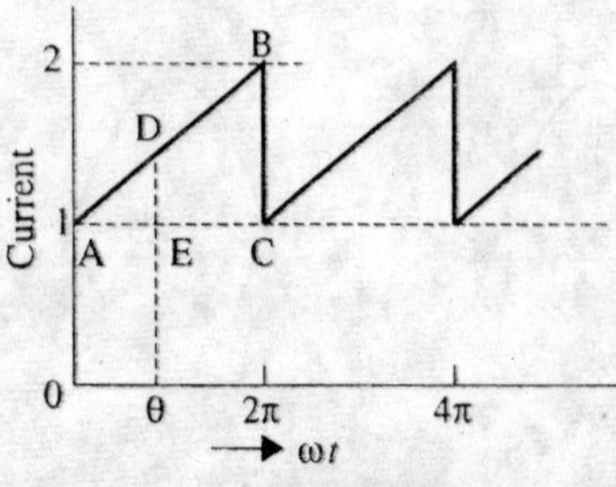

Fig. 16.16

$$\text{Average current} = \frac{1}{2\pi}\int_0^{2\pi} i\,d\theta = \frac{1}{2\pi}\int_0^{2\pi}\left(1+\frac{\theta}{2\pi}\right)d\theta$$

$$= \frac{1}{2\pi}\int_0^{2\pi}\left(d\theta + \frac{\theta\,.\,d\theta}{2\pi}\right) = \frac{1}{2\pi}\left|\theta + \frac{1}{4\pi}\theta^2\right|_0^{2\pi}$$

$$= \frac{1}{2\pi}(2\pi+\pi) = \frac{3\pi}{2\pi} = \frac{3}{2} = \mathbf{1.5^*}$$

$$\text{Mean square value} = \frac{1}{2\pi}\int_0^{2\pi} i^2\,d\theta = \frac{1}{2\pi}\int_0^{2\pi}\left(1+\frac{\theta}{2\pi}\right)^2 d\theta = \frac{1}{2\pi}\int_0^{2\pi}\left(1+\frac{\theta^2}{4\pi^2}+\frac{\theta}{\pi}\right)d\theta$$

$$= \frac{1}{2\pi}\left|\theta + \frac{\theta^3}{12\pi^2} + \frac{\theta^2}{2\pi}\right|_0^{2\pi} = \frac{7}{3}$$

$\therefore$ R.M.S. value $= \sqrt{7/3}$

(*b*)The given current is a mixture of a d.c. component of 10 A and an alternating current of maximum value of 5 A.

$$\text{R.M.S. value} = \sqrt{(\text{d.c.})^2 + (\text{r.m.s. value of a.c.})^2}$$

$$= \sqrt{(10)^2 + (5/\sqrt{2})^2} = \sqrt{225/2} = \mathbf{15/\sqrt{2}\ A^{**}}$$

Example 16.12..*Four alternating currents of peak value 200 A have the following waveforms:*
(a) sinusoidal (b) full-wave rectified sinusoidal (c) rectangular (d) triangular.
If these currents are passed, in turn, through
(i) a moving-coil ammeter, (ii) a moving-iron ammeter connected in series,
find the readings of the instruments in each case.

Solution. It should be kept in mind that a moving-coil ammeter reads the average value of the current over the whole cycle. However, moving-iron ammeter reads r.m.s. value of the current. Various wave-forms are shown in Fig. 16.17.

(*a*) M.C. ammeter reading = 0

M.I. ammeter reading = $200/\sqrt{2}$ = **141.4A**

(*b*) Average value of current over one cycle is [Fig. 16.17 (*b*)]

$$= \frac{2\int_0^{\pi} I_m \sin\theta\ d\theta}{2\pi} = \frac{I_m}{\pi}\int_0^{\pi}\sin\theta\ d\theta = \frac{2I_m}{\pi} = \frac{2\times 200}{\pi} = \mathbf{127.4\ A}$$

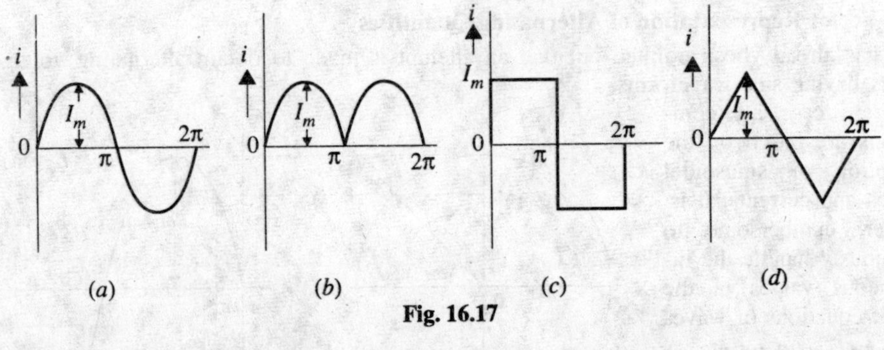

(*a*) (*b*) (*c*) (*d*)

Fig. 16.17

*The average value could be found by taking the mean of the initial and final values. Mean value = $\frac{1+2}{2}$ = 1.5 A.

**The phase difference of 30° and the fact that it is a cosine function makes no difference to the r.m.s. or

$\therefore$ M.C. ammeter reads **127.4 A**

R.M.S. value of full-wave rectified current is

$$= 2\int_0^{\pi} \frac{i^2\, d\theta}{(2\pi - 0)} = \frac{1}{\pi}\int_0^{\pi} Im^2 \sin^2\theta \, d\theta = \frac{Im^2}{2\pi}\int_0^{\pi} (1 - \cos 2\theta)\, d\theta = \frac{Im^2}{2}$$

$\therefore$ $I = I_m/\sqrt{2} = 200/\sqrt{2} = 141.4$ A

M.I. ammeter reading = **141.4 A**

(*c*) Average value of current over one cycle is zero.

$\therefore$ M.C. ammeter reading = 0

R.M.S. value = 200A

$\therefore$ M.I. ammeter readings = **200A**

(*d*) For a triangular wave-form

Average value over one cycle = 0

R.M.S. value $I = I_m/\sqrt{3}$

$\therefore$ M.C. ammeter reading = **0**

M.I. ammeter reading = $200/\sqrt{3}$ = **115.4 A**

Tutorial Problems No. 16.1

1. Calculate the maximum value of the e.m.f. generated in a coil which is rotating at 50 rev/second in a uniform magnetic field of 0.3 Wb/m^2. The coil is wound on a square former having sides 5 cm in length and is wound with 300 turns. **[188.5 V]**

2. (*a*) What is the peak value of a sinusoidal alternating current of 4.78 r.m.s. amperes?
   (*b*) What is the r.m.s. value of a rectangular voltage wave with an amplitude of 9.87 V?
   (*c*) What is the average value of sinusoidal alternating current of 31 A maximum value?
   (*d*) An alternating current has a periodic time of 0.03 second. What is its frequency?
   (*e*) An alternating current is represented by $i = 70.7 \sin 520\, t$. Determine (*i*) the frequency (*ii*) the current 0.0015 second after passing through zero and increasing positively.
   [6.76 A; 9.87 V; 19.75 A; 33.3 Hz; 82.8 Hz; 49.7 A]

3. A sinusoidal alternating voltage has an r.m.s. value of 200 V and a frequency of 50 Hz. It crosses the zero axis in a positive direction when $t = 0$. Determine (*i*) the time when the voltage first reaches the instantaneous value of 200 V and (*ii*) the time when voltage after passing through its maximum positive value reaches the value of 14.4 V. **[(*i*) 0.0025 second (*ii*) 1/300 second]**

4. A circuit offers a resistance of 10 Ω in one direction and 50 Ω in the opposite direction to the flow of current through it. A sinusoidal voltage of maximum value 100 V is applied to the above circuit in series with (*i*) an M.C. ammeter (*ii*) a moving-iron ammeter (*iii*) a moving-coil ammeter with a full-wave rectifier. Find the reading of each instrument. **[(*i*) 5.1 A (*ii*) 2.55 A (*iii*) 3.82 A]**

16.21. Vector Representation of Alternating Quantities

It has already been pointed out that an attempt is made to obtain alternating voltages and currents having sine waveforms.

In any case, a.c. computations are based on the assumption of sinusoidal voltages and currents. It is, however, cumbersome to continuously handle the instantaneous values in the form of equations of waves like $e = E_m \sin \omega t$ etc. A conventional method is to employ vector method of representing these sine waves. These vectors may then be manipulated in-

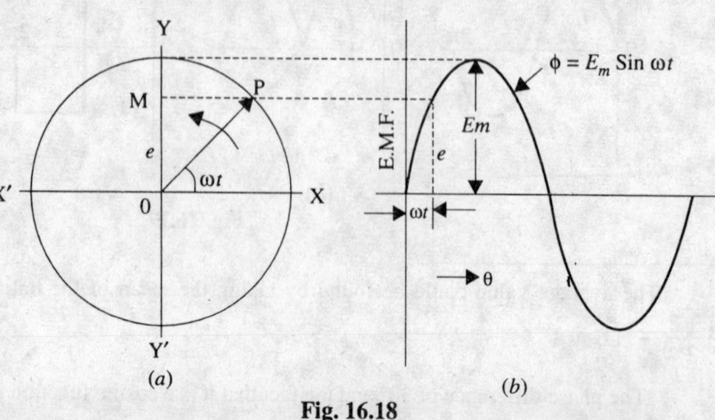

Fig. 16.18

stead of the sine functions to achieve the desired result. *In fact, vectors are a short-hand for the representation of alternating voltages and currents and their use greatly simplifies the problems in a.c. work.*

A vector quantity is a physical quantity which has both magnitude and direction. Such vector quantities are completely known only when particulars of their magnitude, direction and the sense in which they act, are given. They are graphically represented by straight lines called vectors. The length of the line represents the magnitude of the alternating quantity, the inclination of the line with respect to some axis of reference gives the direction of that quantity and an arrow-head placed at one end indicates the direction in which that quantity acts.

The alternating voltages and currents are represented by such vectors rotating counter-clockwise with the same frequency as that of the alternating quantity. In Fig. 16.18 (*a*), *OP* is such a vector which represents the maximum value of the A.C. and its angle with *X-axis* gives its phase. Let the alternating current be represented by the equation $e = E_m \sin \omega t$. It will be seen that the projection of *OP* on *Y*-axis at any instant gives the instantaneous value of the altrenating current.

$$OM = OP \sin \omega t \quad \text{or} \quad e = OP \sin \omega t = E_m \sin \omega t$$

It should be noted that a line like *OP* can be made to represent an alternating voltage or current only if it satisfies the following conditions:

1. Its length should be equal to the peak or maximum value of the sinusoidal alternating quantity to a suitable scale.
2. It should be in the horizontal position at the same instant as the alternating quantity is zero and increasing positively.
3. Its angular velocity should be such that it completes one revolution in the same time as taken by the alternating quantity to complete one cycle.

16.22. Vector Diagrams Using R.M.S. Values

Instead of using maximum values as above, it is more common practice to draw vector diagrams using r.m.s. values of alternating quantities. But it should be understood that in that case, the projection of the rotating vector on the *Y*-axis does not give the instantaneous value of that alternating quantity.

16.23. Vector Diagrams of Sine Waves of Same Frequency

Two or more sine waves of the same frequency can be shown on the same vector diagram because the various vectors representing different waves all rotate counter-clockwise at the same frequency and so maintain a fixed position relative to each other. This is illustrated in Fig.16.19 where a voltage *e* and current *i* of the same frequency are shown. The current wave is supposed to pass upwards through zero at the instant when $t = 0$ while at the same time the voltage wave has already advanced by an angle α from its zero value. Hence, their equations can be written as

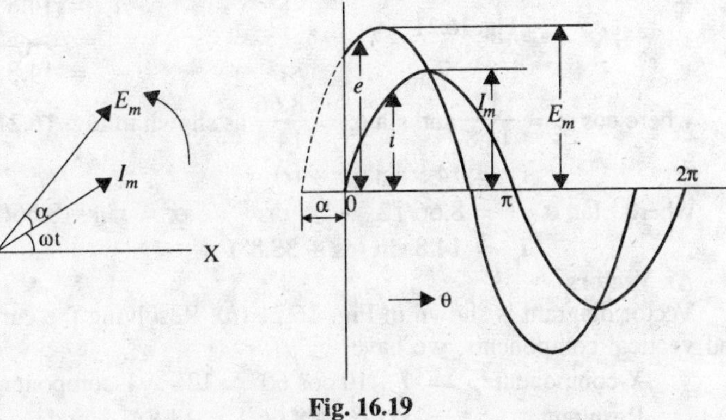

Fig. 16.19

$$i = I_m \sin \omega t \quad \text{and} \quad e = E_m \sin (\omega t + \alpha)$$

Sine waves of different frequencies cannot be represented on the same vector diagram in a still picture because due to difference in speed of different vectors, the phase angles between them will be continuously changing.

16.24. Addition of Two Alternating Quantities

In Fig. 16.20 (*a*) are shown two rotating vectors representing the maximum values of two sinusoidal voltage waves represented by $e_1 = E_{m1} \sin \omega t$ and $e_2 = E_{m2} \sin(\omega t - \phi)$. It is seen that the sum of the two sine waves of the same frequency is another sine wave of the same frequency but of a different maximum value and phase. The value of the instantaneous resultant voltage e_r at any instant is obtained by algebraically adding the projections of the two vectors on the Y-axis. If these projections are e_1 and e_2, then $e_r = e_1 + e_2$ at that time. The resultant curve has been drawn in this way by adding the ordinates. It is found that the resultant wave is a sine wave of the same frequency as the component waves but lagging behind E_{m1} by angle α. The vector diagram of Fig. 16.20 (*a*) can be very easily drawn. Lay off E_{m2} lagging ϕ behind E_{m1} and then complete the parallelogram. So E_r is obtained.

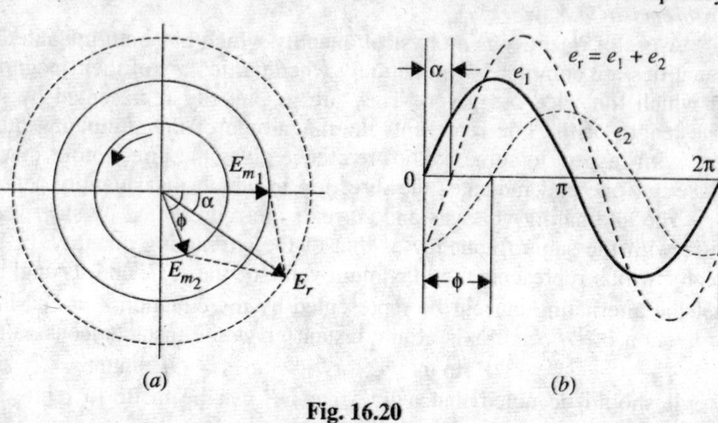

Fig. 16.20

Example 16.12. *Add the following currents as waves and as vectors.*

$$i_1 = 7 \sin \omega t \text{ and } i_2 = 10 \sin (\omega t + \pi/3)$$

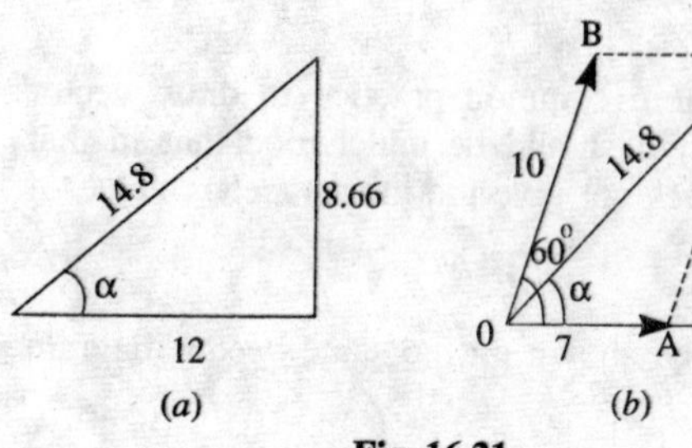

Fig. 16.21

Solution. As waves

$$i_r = i_1 + i_2 = 7 \sin \omega t + 10 \sin (\omega t + 60°)$$

$$= 7 \sin \omega t + 10 \sin \omega t \cos 60° + 10 \cos \omega t \sin 60°$$

$$= 12 \sin \omega t + 8.66 \cos \omega t$$

Dividing and multiplying both sides by $\sqrt{(12^2 + 8.66^2)} = 14.8$, we get

$$i_r = 14.8 \left(\frac{12}{14.8} \sin \omega t + \frac{8.66}{14.8} \cos \omega t \right)$$

$$= 14.8 (\cos \alpha \sin \omega t + \sin \alpha \cos \omega t)$$

where $\cos \alpha = \frac{12}{14.8}$ and $\sin \alpha = \frac{8.66}{14.8}$ as shown in Fig. 16.21 (*a*)

$$i_r = 14.8 \sin (\omega t + \alpha)$$

where $\tan \alpha = 8.66/12$ or $\alpha = \tan^{-1} (8.66/12) = 35.8°$

$$i_r = \mathbf{14.8 \sin (\omega t + 35.8°)}$$

As Vectors

Vector diagram is shown in Fig. 16.21 (*b*). Resolving the current vectors into their horizontal and vectical components, we have

X-component $= 7 + 10 \cos 60° = 12$; *Y*-component $= 0 + 10 \sin 60° = 8.66$

Resultant $= \sqrt{(12^2 + 8.66^2)} = 14.8$ A and $\alpha = \tan^{-1} (8.66/12) = 35.8°$

Hence, the resultant equation can be written as

$$i_r = \mathbf{14.8 \sin (\omega t + 35.8°)}$$

16.25. Addition and Subtraction of Vectors

(*i*) **Addition.** In a.c. circuit problems, we may be concerned with a number of alternating voltages or currents of the same frequency but of different phases and it may be required to obtain the resultant voltage or current. As explained earlier (Art. 16.21) if the quantities are sinusoidal, they may be represented by a number of rotating vectors having common axis of rotation and displaced from one another by fixed angles which are equal to the phase differences between the respective alternating quantities. The instantaneous value of the resultant voltage is given by the algebraic sum of the projections of the different vectors on Y-axis. The maximum value (or r.m.s. value if vectors represent that value) is obtained by compounding the several vectors by using the parallelogram and polygon laws of vector addition.

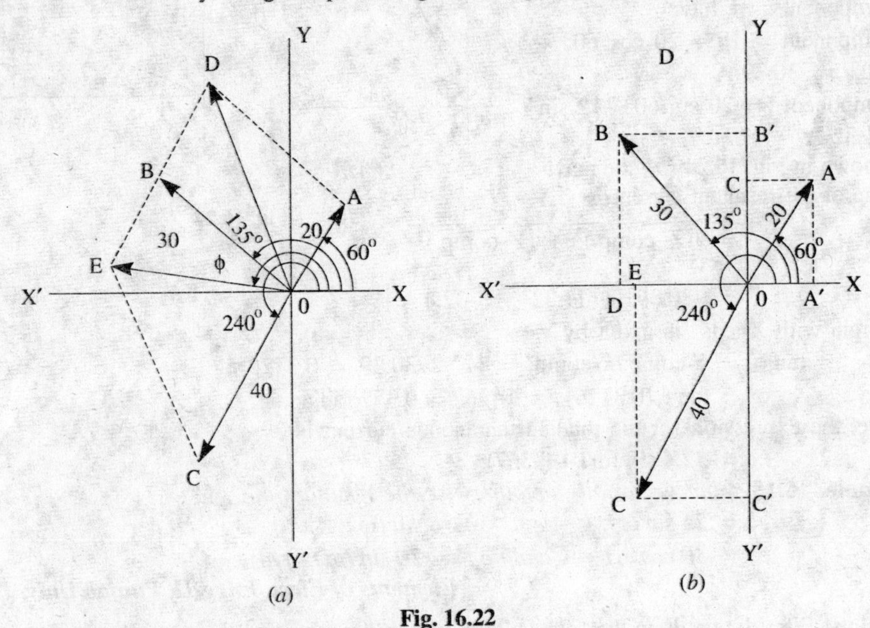

Fig. 16.22

However, another easier method is to resolve the various vectors into their X- and Y-components and then to add them up as shown in Example 16.12.

Suppose, we are given the following three alternating e.m.fs. and it is required to find the equation of the resultant e.m.f.

$$e_1 = 20 \sin (\omega t + \pi/3)$$
$$e_2 = 30 \sin (\omega t + 3\pi/4)$$
$$e_3 = 40 \sin (\omega t + 4\pi/3)$$

Then, the vector diagrams can be drawn as explained before and solved in any of the following three ways:

1. By compounding acording to parallelogram law as in Fig. 16.22 (*a*).
2. By resolving the various vectors into their X- and Y-components as in Fig. 16.22 (*b*).
3. By laying the various vectors end-on-end at their proper phase angles and then measuring the closing vector as shown in Fig. 16.23.

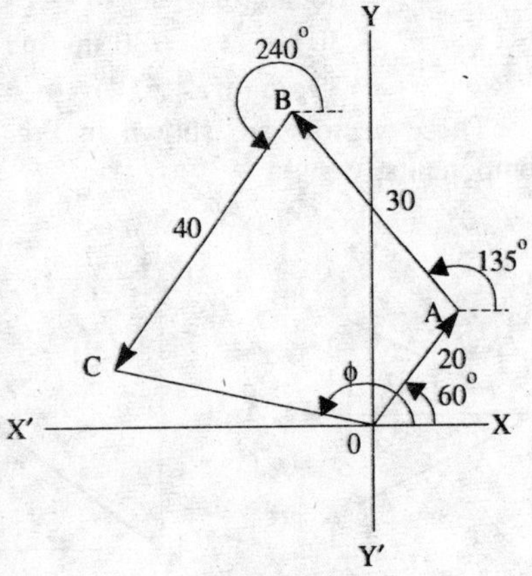

Fig. 16.23

Knowing the magnitude of the resultant vector and its inclination ϕ with X-axis, the equation of the resultant e.m.f. can be written as

$$e = E_m \sin (\omega t \pm \phi)$$

Example 16.14. *In a parallel circuit with three branches, the instantaneous branch currents are represented by*

$$i_1 = 10 \sin \omega t; \; i_2 = 20 \sin (\omega t + \pi/3); \; i_3 = 12 \sin (\omega t - \pi/6)$$

Write down the expression for the total instantaneous current in the form of

$$i = I_m \sin (\omega t + \phi)$$

Solution. The vectors representing the maximum values of the three currents in their proper phases are shown in Fig. 16.24 (*a*). Resolving them into their *X*- and *Y*-components, we have

X-component = 10 + 20 cos 60° + 12 cos 30° = 30.39 A

Y-component = 20 sin 60° −12 sin 30° = 11.32 A

As shown in Fig. 16.24 (*b*), the maximum value of the resultant current is

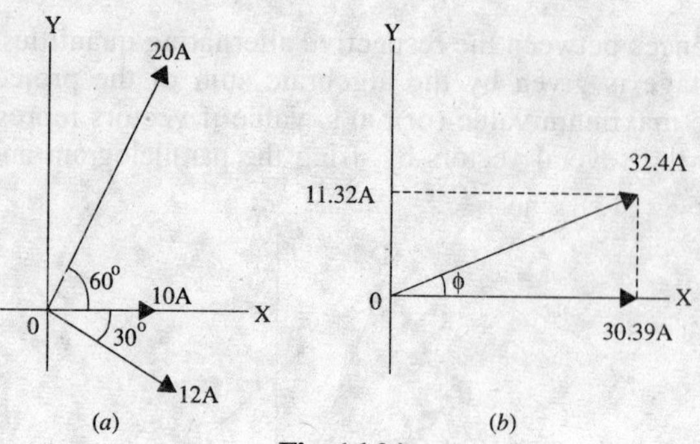

Fig. 16.24

$$I_m = \sqrt{(X\text{–comp.})^2 + (Y\text{–comp.})^2}$$

$$= \sqrt{30.39^2 + 11.32^2} = 32.4 \text{ A}$$

Its angle with *X*-axis is given by

$$\tan\phi = Y\text{-comp.}/X\text{-comp.} = 11.32/30.39 = 0.3726$$

$$\phi = \tan^{-1}(0.3726) = 20°26' = 0.357 \text{ radian}$$

Hence, the expression for the total instantaneous current is

$$\boldsymbol{i = 32.4 \sin(\omega t + 0.357)}\text{*}$$

Example. 16.15. *Find vectorially or otherwise the resultant of*

$e_1 = 25 \sin \omega t,$ $\quad e_2 = \sin(\omega t + \pi/6)$

$e_3 = 30 \cos \omega t,$ *and* $e_4 = 20 \sin(\omega t - \pi/4)$

(Elements of Elect. Engg-II, Punjab Univ. 1993.)

Solution. First, let us draw the four voltage vectors representing the maximum values of the given alternation voltages.

$e_1 = 25 \sin \omega t$ — here, phase angle with *X*-axis is zero, hence its vector will be drawn parallel to the X-axis.

$e_2 = 30 \sin(\omega t + \pi/6)$ — its vector will be above *X*-axis by 30°

$e_3 = 30 \cos \omega t = 30 \sin(\omega t + 90°)$ — its vector will be 90° above *X*-axis.

$e_4 = 20 \sin(\omega t - \pi/4)$ — its vector will be 45° below *X*-axis.

These vectors are shown in Fig. 16.25 (*a*). Resolving these vectors into their *X*- and *Y*-components, we get

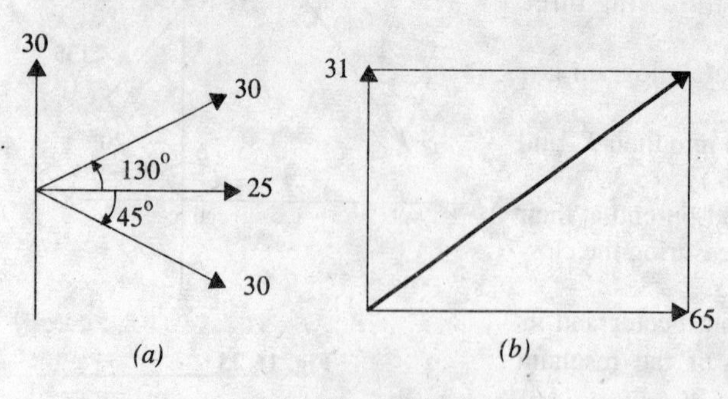

Fig. 16.25

X-component = 25 + 30 cos 30° + 20 cos 45° = 25 + 26 + 14 = 65 V

Y-component = 30 + 30 sin 30° − 20 sin 45° = 30 + 15 − 14 = 31 V

As seen from Fig. 16.25 (*b*), the maximum value of resultant voltage is

$$OA = \sqrt{65^2 + 31^2 + 72 \text{ V}}$$

The phase angle of resultant voltage is given by

$$\tan\phi = \frac{Y\text{–component}}{X\text{–component}} = \frac{31}{65} = 2.097$$

*The angle could also be given in degrees instead of radians,

$\therefore \quad \phi = \tan^{-1}(2.097) = 64.5° = 1.126$ rad.

Hence, the equation of the resultant voltage is given by

$$e = 72 \sin(\omega t + 64.5°) \quad \text{or} \quad e = \mathbf{72 \sin(\omega t + 1.126)}$$

***(ii)* Subtraction of Vectors**

If difference of two vectors is required, then one of the vectors is reversed and this reversed vector is then compounded with the other as usual.

Suppose it is required to subtract vector *OB* from vector *OA*. Then, *OB* is reversed as shown in Fig. 16.26 *(a)* and compounded with *OA* according to the parallelogram law. The vector difference (A – B) is given by vector *OC*.

Similarly, the vector *OC* in Fig. 16.26 *(b)* represents (B – A). *i.e.,* the subtraction of *OA* from *OB*.

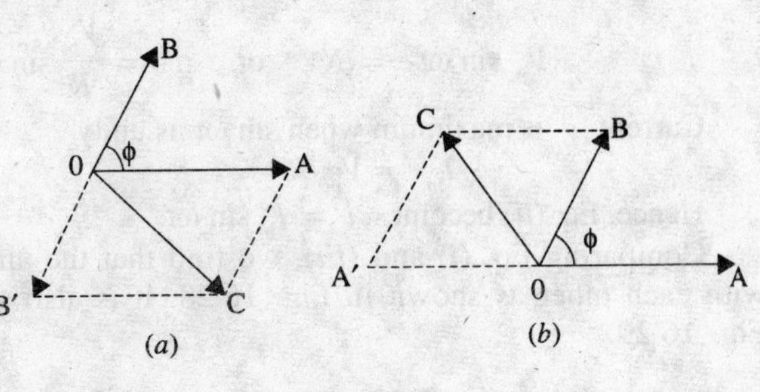

Fig. 16.26

Tutorial Problems No. 16.2

1. The currents taken by two parallel circuits are 12 A in phase with the applied voltage and 20 A lagging 30° behind the applied voltage respectively. Determine the current taken by the combined circuit and its phase with respect to the applied voltage. **[31 A; 18.8°]**
2. Three circuits, *A, B* and *C* are connected in series across a 200 V supply. The voltage across circuit *A* is 50 V lagging the supply voltage by 45° and the voltage across circuit *C* is 100 V leading the supply voltage by 30°. Determine graphically or by calculation, the voltage across circuit *B* and its phase displacement from its supply voltage. **[79.4 V; 10° 38' lagging]**
3. Three alternating currents are given by

   $i_1 = 141 \sin(\omega t + \pi/4); \quad i_2 = 30 \sin(\omega t + \pi/2) \quad \text{and} \quad i_3 = 20 \sin(\omega t - \pi/6)$

   and are fed into a common conductor. Find graphically or otherwise the equation of the resultant current and its r.m.s. value. **[i = 167.4 sin (ωt + 0.797); $I_{r.m.s.}$ = 118.4 A]**
4. Four e.m.fs. $e_1 = 100 \sin \omega t$, $e_2 = 80 \sin(\omega t - \pi/6)$, $e_3 = 120 \sin(\omega t + \pi/4)$ and $e_4 = 100 \sin(\omega t - 2\pi/3)$ are induced in four coils connected in series so that the vector sum of four e.m.fs. is obtained. Find graphically or by calculation, the resultant e.m.f. and its phase difference with *(a)* e_1 and *(b)* e_2.

   If the connections to the coil in which the e.m.f. e_3 is induced are reversed, find the new resultant e.m.f. **[208 sin (ωt – 0.202); *(a)* 11°34′ lag *(b)* 18° 26′ lead; 76 sin (ωt + 0.528)]**
5. Draw to scale a vector diagram showing the following voltages:

   $v_1 = 100 \sin 500\, t \qquad v_3 = -50 \cos 500\, t$

   $v_2 = 200 \sin(500\, t + \pi/3) \qquad v_4 = 150 \sin(500\, t - \pi/4)$

   Obtain graphically or otherwise, their vector sum and express this in the form $V_m \sin(500\, t + \theta)$, using v_1 as the reference vector. Give the r.m.s. value and frequency of the resultant voltage. **[306.5 sin (500t + 0.056); 217 V; 79.6 Hz]**

16.26. A.C. Through Resistance, Inductance and Capacitance

We will now consider the phase angle introduced between an alternating voltage and current when the circuit contains resistance only, inductance only and capacitance only. *In each case, we will assume that we are given the alternating voltage of equation* $v = V_m \sin \omega t$ and will proceed to find the equation and phase of the alternating current produced in each case.

16.27. A.C. Through Pure Ohmic Resistance Only

The circuit is shown in Fig. 16.27. Let the applied voltage be given by the equation

$$v = V_m \sin\theta = V_m \sin \omega t \qquad ...(i)$$

Let R = ohmic resistance

i = instantaneous current

Obviously, the applied voltage has to overcome ohmic voltage drop only. Hence, for equilibrium

$$v = iR$$

Putting the value of 'v' from above, we get

$$V_m \sin \omega t = iR \quad \text{or} \quad i = \frac{V_m}{R} \sin \omega t \qquad ...(ii)$$

Current 'i' is maximum when $\sin \omega t$ is unity

$$I_m = V_m / R$$

Hence, Eq. (*ii*) becomes, $i = I_m \sin \omega t$. ...(*iii*)

Comparing Eq. (*i*) and (*iii*), we find that the alternating voltage and current are in phase with each other as shown in Fig. 16.28. It is also shown vectorially by vectors V_R and I in Fig. 16.29.

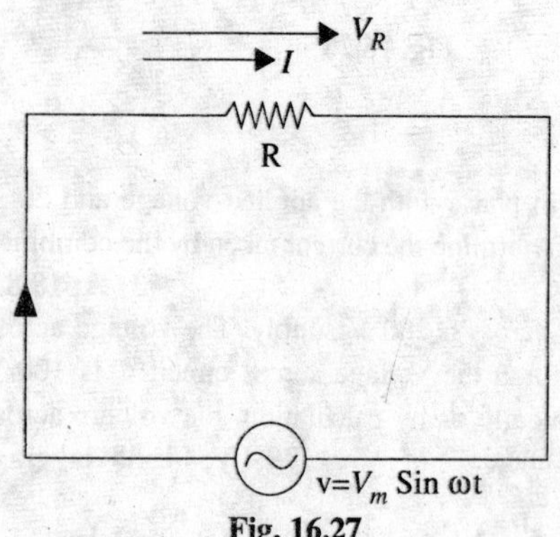

Fig. 16.27

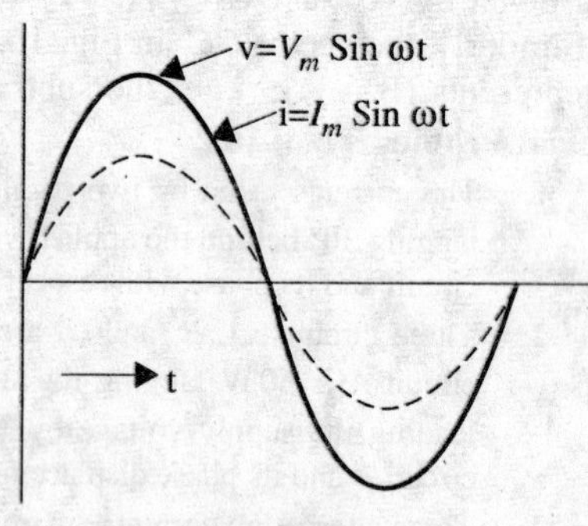

Fig. 16.28

Power

Instantaneous power $p = vi$ ...(Fig. 16.29)

$$= V_m I_m \sin^2 \omega t = \frac{V_m I_m}{2}(1 - \cos 2\omega t) = \frac{V_m I_m}{2} - \frac{V_m I_m}{2} \cos \omega t$$

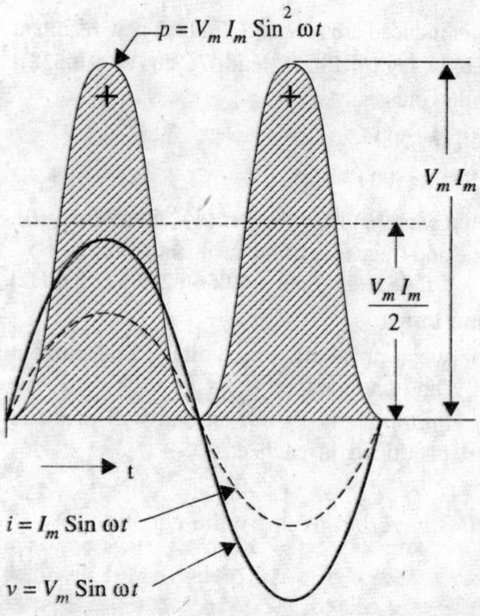

Power consists of a constant part $\frac{V_m I_m}{2}$ and fluctuating part $\frac{V_m I_m}{2} \cos 2\omega t$ of frequency double that of voltage and current waves. For a complete cycle, the average value of $\frac{V_m I_m}{2} \cos 2\omega t$ is zero.

Hence, power for whole cycle is

$$P = \frac{V_m I_m}{2} = \frac{V_m}{\sqrt{2}} \times \frac{I_m}{\sqrt{2}}$$

or $P = V \times I$ watts

where V = r.m.s. value of applied voltage

I = r.m.s. value of the current

It is seen from Fig. 16.29 that no part of the power cycle becomes negative at any time. In other words, in a purely resistive circuit, power is never zero. This is so because the instantaneous values of voltage and current are always either both positive or negative and hence the product is always positive.

Fig. 16.29

16.28. A.C. Through Inductance Only

Whenever an alternating voltage is applied to a pure inductive coil*, a back e.m.f. is produced due to the self-inductance of the coil. This back e.m.f. at every step, opposes the rise or fall of current through the coil. As there is no ohmic voltage drop, the applied voltage has to overcome this self-induced e.m.f. only. So, at every step

$$v = L\frac{di}{dt}$$

Now, $$v = V_m \sin \omega t$$

$$\therefore \quad V_m \sin \omega t = \frac{di}{dt}$$

$$\therefore \quad di = \frac{V_m}{L} \sin \omega t \, dt$$

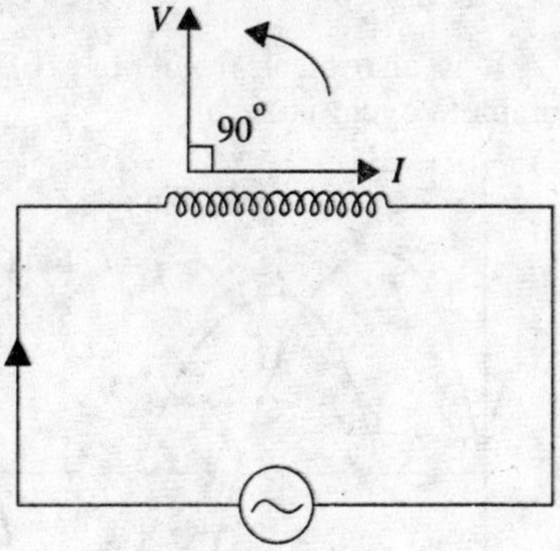

Integrating both sides, we get

$$i = \frac{V_m}{L} \int \sin \omega t \; dt$$

$$= \frac{V_m}{\omega L}(-\cos \omega t)$$

... (constant of integration = 0)

$$= -\frac{V_m}{\omega L} \cos \omega t$$

$$\therefore \quad i = \frac{V_m}{\omega L} \sin\left(\omega t - \frac{\pi}{2}\right)$$

Fig. 16.30

Maximum value of i is $I_m = \frac{V_m}{\omega L}$ when $\sin\left(\omega t - \frac{\pi}{2}\right)$ is unity. Hence, equation of the current becomes

$$i = I_m \sin(\omega t - \pi/2)$$

So, we find that if applied voltage is represented by

$$v = V_m \sin \omega t$$

then current, flowing in *purely* inductive circuit, is given by

$$i = I_m \sin\left(\omega t - \frac{\pi}{2}\right)$$

Clearly, the current lags behind the applied voltage by a quarter cycle (Fig. 16.31) or the phase difference between the two is $\pi/2$ with voltage leading. Vectors are shown in Fig. 16.30.

We have seen that $$I_m = \frac{V_m}{\omega L}$$

Here, 'ωL', plays the part of resistance. It is called the (inductive) *reactance* of the coil and is given in ohms if L is in henrys and ω in radian/second. It is denoted by X_L.

Power

Instantaneous power $p = vi$

---

*By pure inductive coil is meant one that has no ohmic resistance and hence no I^2R loss. Pure inductance is actually not attainable though it is very nearly approached by a coil wound with such thick wire that its resistance is negligible. If it has some actual resistance, then it is represented by a separate equivalent resistance joined in series with it.

$$= V_m I_m \sin \omega t \sin\left(\omega t - \frac{\pi}{2}\right) = -V_m I_m \sin \omega t \cos \omega t = -\frac{V_m I_m}{2} \sin 2\omega t$$

Power for the whole cycle is

$$P = \frac{V_m I_m}{2} \int_0^{2\pi} \sin 2\omega t \, dt = 0$$

It is also clear from Fig. 16.32 that the average demand of power from the supply for a complete cycle is zero.

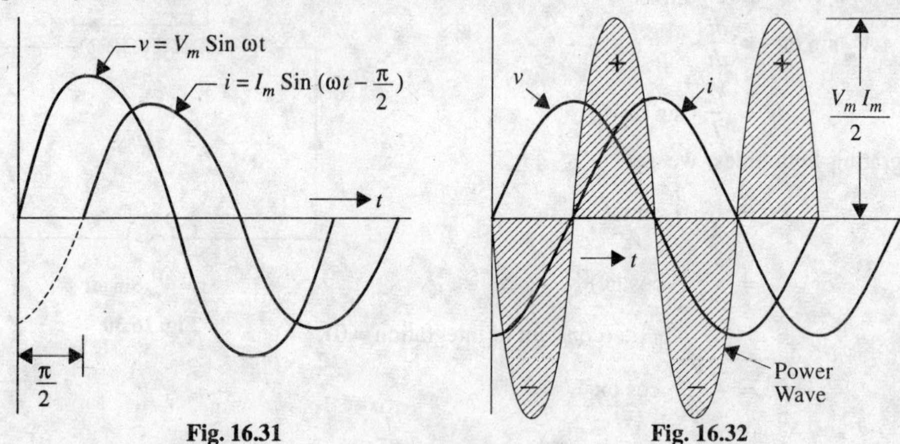

Fig. 16.31 **Fig. 16.32**

Here, again it is seen that power wave is a sine wave of frequency double that of the voltage and current waves. The maximum value of the instantaneous power is $\frac{V_m I_m}{2}$.

16.29. A.C. Through Capacitance Alone

When an alternating voltage is applied to the plates of a capacitor, it is charged first in one direction and then in the opposite direction. With reference to Fig. 16.33,

let

υ = p.d. between plates at any instant

q = charge on plates at that instant.

Then,

$q = C_\upsilon$ ...*C* is the capacitance

$= C V_m \sin \omega t$ ...(putting value of v)

Now, current i is given by the rate of flow of charge

$$\therefore \quad i = \frac{dq}{dt} = \frac{d}{dt}(C V_m \sin \omega t) = \omega C V_m \cos \omega t$$

$$\text{or} \quad i = \frac{V_m}{1/\omega C} \cos \omega t = \frac{V_m}{1/\omega C} \sin\left(\omega t + \frac{\pi}{2}\right)$$

$$\text{Obviously,} \quad I_m = \frac{V_m}{1/\omega C} = \omega C V_m \qquad \ldots(i)$$

$$\therefore \quad i = I_m \sin\left(\omega t + \frac{\pi}{2}\right)$$

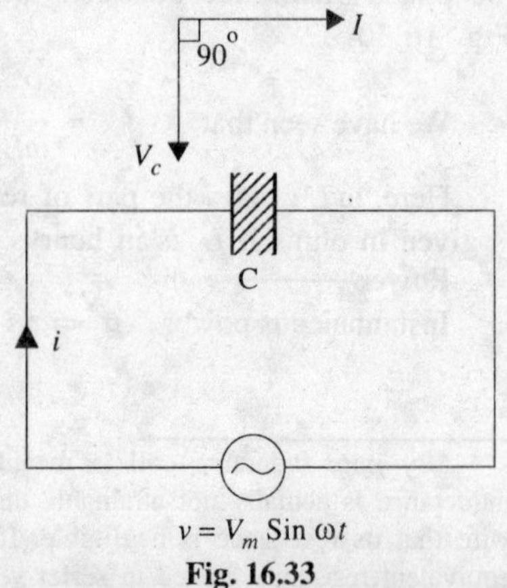

Fig. 16.33

The denominator '$1/\omega C$' is known as capacitive *reactance* and is in ohms if C is in farads and ω in radian/second. It is denoted by X_c.

It is seen that if the applied voltage is given by

$$\upsilon = V_m \sin \omega t$$

the current is given by $\quad i = I_m \sin (\omega t + \pi/2)$

Hence, we find that the current in a pure capacitor leads its voltage by a quarter cycle as shown in Fig. 16.35 or phase difference between its voltage and current is $\pi/2$ with the current leading. Vector representation is given in Fig. 16.33. Note that V_c is along the *negative* direction of *Y*-axis.

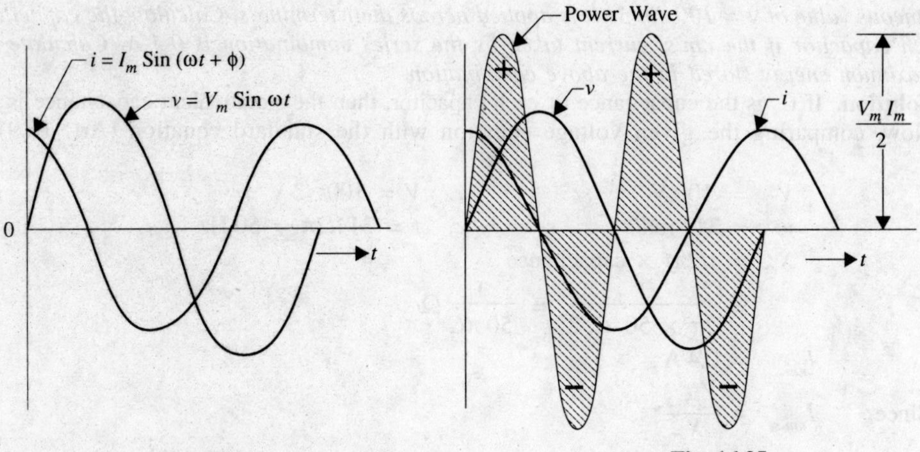

Fig. 16.34 **Fig. 16.35**

Power

Instantaneous power $p = vi = V_m \sin \omega t \,.\, I_m \sin (\omega t + 90°)$

$$= V_m I_m \sin \omega t \cos \omega t = \frac{1}{2} V_m I_m \sin 2\omega t$$

Power for the whole cycle

$$P = V_m I_m \int_0^{2\pi} \sin 2\omega t \; dt = 0$$

This fact is graphically illustrated in Fig. 16.35. We find that in a purely capacitive circuit*, the average demand of power from supply is zero (as in a purely inductive circuit). Again, it is seen that power wave is a sine wave of frequency double that of the voltage and current waves. The maximum value of the instantaneous power is $V_m I_m /2$.

Example 16.16. *Two capacitors of 80 μF and 50 μ F respectively are connected in series. Find (i) the current and (ii) the maximum energy stored in the circuit when 200-V at 50 Hz are applied across the series circuit.*

Solution. The combined or equivalent capacitance of the two series-connected capacitors is

$$C = \frac{C_1 \times C_2}{C_1 + C_2} = \frac{80 \times 50}{80 + 50} = \frac{400}{13} \mu \text{F}$$

Equivalent reactance is

$$X_C = 1/\omega C = 1/2\pi fC = \frac{10^6}{2\pi \times 50 \times 400/13} = 103.4 \; \Omega$$

(*i*) r.m.s. value of current is

$$I = V/X_C = 200/103.4 = \mathbf{1.93 \; A}$$

(*ii*) Maximum energy $E = \frac{1}{2} CV_m^2$

*By pure capacitor is meant one that has neither resistance nor dielectric loss. If there is loss. If there is loss in a capacitor, then it may be represented by loss in (*a*) a comparatively high resistance joined in parallel with the pure capacitor or (*b*) by a comparatively low resistance in series with the capacitor. But out of the two alternatives, usually (*a*) is chosen.

Now $V_m = 200 \times \sqrt{2} = 283$ V

$$\therefore \quad E = \frac{1}{2} \times \frac{400}{13} \times 10^{-6} \times 283^2 = \mathbf{1.23\ J}$$

Example 16.17. *Two similar capacitors are connected in series and a voltage with an instantaneous value of v = 100 sin 314t is applied acroos their terminals. Calculate the capacitance of each capacitor if the r.m.s. current taken by the series combination is 0.4 A. Calculate also the maximum energy stored in the above combination.*

Solution. If C is the capacitance of each capacitor, then their equivalent capacitance is $C/2$.

Now, comparing the given voltage equation with the standard equation (Art. 16.9) we have

$$V_m = 100 \text{ V}; \qquad \therefore \quad V = 100/\sqrt{2} \text{ V}$$

$$\omega = 314 \text{ rad/s} \qquad \therefore \quad f = 314/2\pi = 50 \text{ Hz}$$

Now $X_C = 1/2\pi f \times$ capacitance

$$= \frac{1}{2\pi \times 50\ C/2} = \frac{1}{50\ \pi C}\ \Omega$$

$$I_{r.m.s.} = 0.4 \text{ A}$$

Since, $I_{r.m.s.} = \dfrac{V_{r.m.s.}}{X_C}$

$$\therefore \quad 0.4 = \frac{100/\sqrt{2}}{1/50\ \pi C} \qquad \therefore \quad C = \mathbf{36\ \mu F}$$

$$\text{Maximum energy} = \frac{1}{2}\left(\frac{C}{2}\right)V_m^{\ 2} = \frac{1}{2} \times \left(\frac{36}{2} \times 10^{-6}\right) \times 100^2 = \mathbf{0.09\ J}$$

HIGHLIGHTS

1. The equation of the e.m.f. induced in a rectangular coil of N turns and area A m² rotating in a field of flux density B Wb/m² with a frequency of f revolutions/second is,
$$e = 2\pi f NBA \sin\theta = E_m \sin\theta = E_m \sin\omega t = E_m \sin 2\pi f t$$
2. Simiiarly, the equation of the induced current is
$$i = I_m \sin\theta = I_m \sin\omega t = I_m \sin 2\pi f t$$
3. Those waveforms which depart from the ideal sine wave aer known as *complex* waveforms. They are made up of pure sine wave (fundamental) and its harmonics.
4. The general equation of an alternating voltage and current is
$$v = V_m \sin(\omega t \pm \phi)\ ; \quad i = I_m \sin(\omega t \pm \phi)$$
where ϕ is the phase difference (in radians or degrees)
5. R.M.S. value of an alternating current is measured in terms of the direct current which when flowing through a given circuit for a given time produces the same heat as produced by the alternating current when flowing through the same circuit for the same time.
$$I_{r.m.s.} = \sqrt{\frac{i_1^{\ 2} + i_2^{\ 2} + ...+ i_n^{\ 2}}{n}}$$
$$= I_m/\sqrt{2} \qquad \text{... for sinusoidal A.C.}$$
6. The average value of an alternating current is expressed by that steady current (direct current) which transfers aross any circuit the same charge as is transferred by that alternating current.
7. Form Factor $= \dfrac{\text{r.m.s. vlaue}}{\text{average value}} = \dfrac{I_m/\sqrt{2}}{2I_m/\pi} = 1.11$ ... sinusoidal A.c.

8. Peak Factor $= \dfrac{\text{maximum vlaue}}{\text{r.m.s. value}} = \dfrac{I_m}{I_m/\sqrt{2}} = 1.414$ ... sinusoidal A.C.

9. When an alternating voltage given by $\upsilon = V_m \sin \omega t$ is applied across a pure resistive circuit, the alternating current is given by

$i = I_m \sin \omega t$ where $I_m = V_m/R$

power $= V \times I$ watt $=$ r.m.s. volts $\times$ r.m.s. amperes

10. In a purely inductive circuit, if the applied voltage is

$\nu = V_m \sin \omega t$

then $i = I_m \sin (\omega t - \pi/2)$

where $I_m = V_m / \omega L$ and ωL = reactance offered by the coil

Power absorbed $= 0$

11. In a purely capacitive circuit, if

$\nu = V_m \sin \omega t$

then $i = I_m \sin (\omega t + \pi/2)$

where $I_m = \dfrac{V_m}{1/\omega C} = \omega C V_m$

and $\dfrac{1}{\omega C}$ = reactance offered $= X_C$

Power absorbed $= 0$

OBJECTIVE TESTS—16

A. Fill in the following blanks.

1. A sinusoidal ac voltage which undergoes 100 reversals of polarity per second has a frequency of .... Hz.

2. An ac current given by $i = 141.4 \sin (\omega t + \pi/6)$ has an r.m.s. value of .... amperes.

3. The time period of a sine wave of frequency 50 Hz is .... second.

4. The r.m.s. value of a sinusoidal alternating current is .... times its maximum value.

5. The average value of a symmetrical sinusoidal alternating current is ....

6. Net power consumed by a pure inductive coil or pure capacitor is ....

7. Higher the frequency of an ac current, —the reactance offered by a capacitor.

8. Alternating voltages and currents can be represented by .... rotating counter clockwise.

B. Answer True or False.

1. A leading alternating quantity is one which achieves its maximum or zero value first.

2. The alternating voltage $\nu = 100 \cos (\omega t - \pi/3)$ has a leading phase of 60°.

3. In electrical engineering, phase differences of alternating quantities are more important than their absolute phases.

4. Average value of a sinusoidal ac is slightly greater than its r.m.s. value.

5. The crest factor of a sine wave is greater than its form factor.

6. R.M.S. value of full-wave rectified ac is twice that of the half-wave rectified ac.

7. A moving-iron ammeter reads average value of the ac current over the whole cycle.

8. If frequency of the applied alternating voltage is doubled, the inductive reactance offered by a pure coil is also doubled.

C. Multiple Choice Questions.

1. The direction of an ac current

(*a*) keeps changing

(*b*) cannot be found

(*c*) keeps reversing

(*d*) is fixed.

2. The r.m.s. value of a sinusoidal ac current is equal to its value at an angle of—degrees.

(*a*) 60 (*b*) 45

(*c*) 30 (*d*) 90

3. Two sinusoidal currents are given by the equations; $i_1 = 10 \sin(\omega t + \pi/3)$ and $i_2 = 15 \sin(\omega t - \pi/4)$. The phase difference between them is— degrees.

(*a*) 105 (*b*) 75

(*c*) 15 (*d*) 60

4. A sine-wave has a frequency of 50 Hz. Its angular velocity is — radian/second.

(*a*) $50/\pi$ (*b*) $50/2\pi$

(*c*) 50π (*d*) 100π

5. An ac current is given by $i = 100 \sin 100 \pi t$. It will achieve a value of 50 A after —second.

(*a*) 1/600 (*b*) 1/300

(*c*) 1/1800 (*d*) 1/900

6. A coil has $X_L = 1000\ \Omega$. If both its inductance and frequency are doubled, ts reactance will become — ohm.

(*a*) 2000 (*b*) 500

(*c*) 250 (*d*) 4000

7. If a moving coil ammeter is used to measure the value of an alternating sinusoidal current having a peak value of 100 A, it will read — amperes

(*a*) 50 (*b*) 15.7

(*c*) 63.7 (*d*) 70.7

8. An alternator having 20 poles and running at 300 r.p.m. will produce an alternating voltage having a frequency of — Hz

(*a*) 50 (*b*) 60

(*c*) 300 (*d*) 20

9. A capacitor used on a 230 V ac supply should have a peak voltage rating of

(*a*) 230 V (*b*) 115 V

(*c*) 325 V (*d*) none of these

(Elect. Engg. A.M.Ae. S.I. June 1994.)

ANSWERS

A. 1. 50 2. 100 3. 0.02 4. 0.707 5. 0 6. 0 7. lower 8. vector

B. 1. T 2. F 3. T 4. F 5. T 6. F 7. F 8. T

C. 1. *c* 2. *b* 3. *a* 4. *d* 5. *a* 6. *d* 7. *c* 8. *a* 9. *c*

17 SERIES A.C. CIRCUITS

17.1. A.C. Through Resistance and Inductance in Series

A pure resistance of R ohms and a pure inductive coil of inductance L henrys are shown connected in series in Fig.17.1.

Let $\quad V$ = r.m.s. value of applied voltage

$\quad I$ = r.m.s. value of resultant current

$\quad V_R = IR$ = voltage drop across R (in phase with I)

$\quad V_L = IX_L$ = voltage drop over coil (at right angles to I)

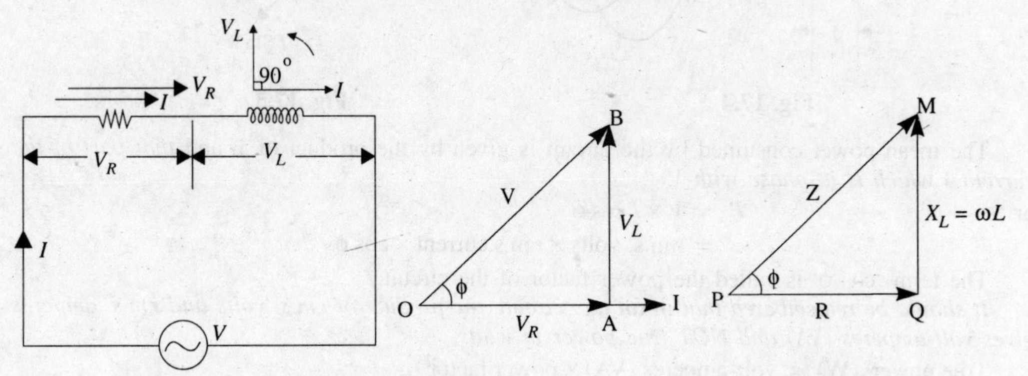

Fig. 17.1 **Fig. 17.2** **Fig. 17.3**

These voltage drops are shown in the voltage triangle *OAB* in Fig.17.2. Vector *OA* represents ohmic drop V_R and *AB* represents inductive drop V_L. The applied voltage V is the *vector* sum of the two *i.e.* it equals *OB*.

$$\therefore \quad V = \sqrt{V_R^2 + V_L^2}$$

$$= \sqrt{[(IR)^2 + (IX_L)^2]} = 1\sqrt{(R^2 + X_L^2)}$$

$$= \frac{V}{\sqrt{(R^2 + X_L{}^2)}}$$

The quantity $\sqrt{(R^2 + X_L^2)}$ is known as the impedance (Z) of the circuit. As seen from the impedance triangle PQM (Fig.17.3)

$$Z^2 = R^2 + X_L^2$$

i.e. $\quad$ *(impedance)*^{2} = *(resistance)*^{2} + *(reactance)*^{2}

From Fig.17.2, it is clear that the current I lags behind the applied voltage by an angle ϕ such that

$$\tan\phi = \frac{X_L}{R} = \frac{\omega L}{R} = \frac{\text{reactance}}{\text{resistance}}$$

$$\therefore \quad \phi = \tan^{-1}(X_L / R)$$

The same fact is illustrated graphically in Fig. 17.4.

Hence, if applied voltage is given by $\upsilon = V_m \sin \omega t$, then the current equation is,

$$i = I_m \sin(\omega t - \phi)$$

where $I_m = V_m / Z$.

Power

In Fig. 17.5, I has been resolved into two mutually perpendicular components, $I \cos \phi$ along the applied voltage V and $I \sin \phi$ in quadrature **with it** (*i.e.* perpendicular to V).

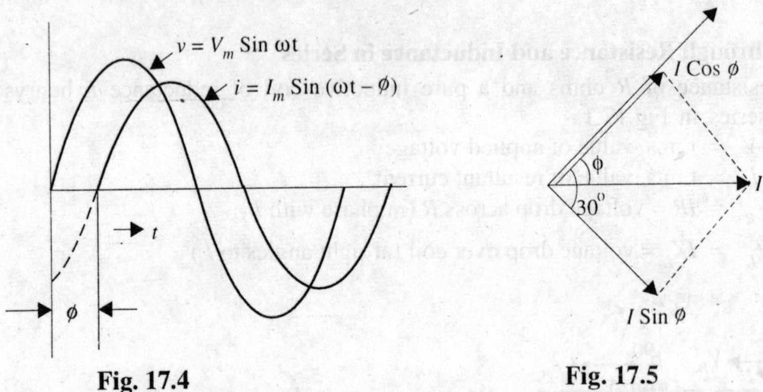

Fig. 17.4 **Fig. 17.5**

The mean power consumed by the circuit is given by the product of V and *that part of the current I which is in phase with V.*

or $P = V \times I \cos \phi$

$= \text{r.m.s. volts} \times \text{r.m.s.current} \times \cos \phi$

The term 'cos ϕ' is called the power factor of the circuit.

It should be remembered that in an a.c. circuit, the product of r.m.s. volts and r.m.s. amperes gives volt-amperes (VA) and NOT true power in watts.

True power **(W)** = volt-amperes (VA) × power factor

or **Watts = VA × cos ϕ**

Also **kW** = kVA × cos ϕ

It should be noted that power consumed is due to ohmic resistance only because pure inductance does not consume any power (Art.16-28).

$$P = VI \cos \phi = VI \times (R/Z) \qquad \ldots(\cos \phi = R/Z)$$

$$= (V/Z) \times I \times R = I \times I \times R$$

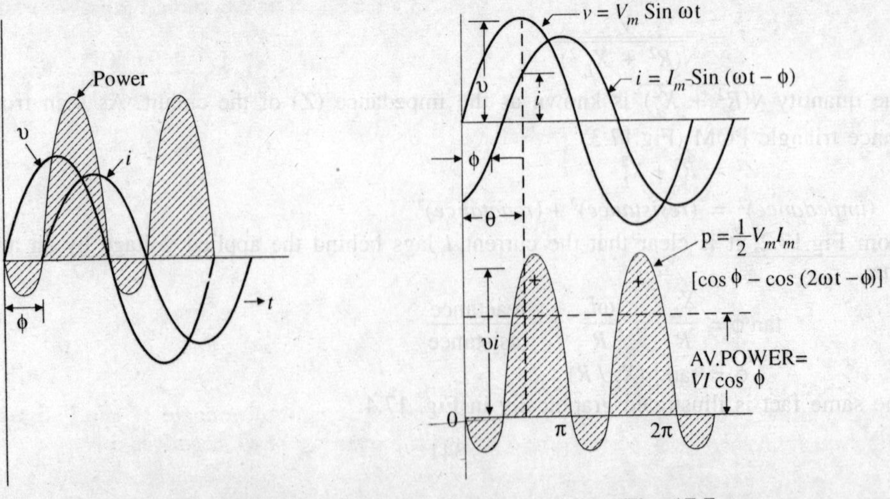

Fig. 17.6 **Fig. 17.7**

$\therefore \quad P = I^2 R$ watt

Graphical representation of the power consumed is shown in Fig.17.6.

Let us calculate power in terms of instantaneous values.

Instantaneous power is

$$
\begin{aligned}
p &= vi \\
&= V_m \sin \omega t \times I_m \sin (\omega t - \phi) \\
&= V_m I_m \sin \omega t \sin (\omega t - \phi) \\
&= \frac{1}{2} V_m I_m [\cos \phi - \cos (2\omega t - \phi)]
\end{aligned}
$$

Obviously, this power consists of two parts (Fig.17.7)

(*i*) a constant part $\frac{1}{2} V_m I_m \cos (2\,\omega t - \phi)$ which contributes to real power,

(*ii*) a pulsating component $\frac{1}{2} V_m I_m \cos (2\omega t - \phi)$ which has a frequency twice that of the voltage and current. It does not contribute to actual power since its average value over a complete cycle is zero.

Hence, average power consumed

$$
\begin{aligned}
&= \frac{1}{2} V_m I_m \cos \phi \\
&= \frac{V_m}{\sqrt{2}} \cdot \frac{I_m}{\sqrt{2}} \cdot \cos \phi = VI \cos \phi
\end{aligned}
$$

where V and I represent the r.m.s. values.

17.2. Power Factor

It may be defined as

(*i*) cosine of the angle of lead or lag

(*ii*) the ratio $\frac{R}{Z} = \frac{\text{resistance}}{\text{impedance}}$ ...(Fig.17.3)

(*iii*) the ratio $\frac{\text{watts}}{\text{volt–amperes}} = \frac{W}{VA}$

17.3. Active and Reactive Components of Current

(*i*) Active component is that which is in phase with the applied voltage *i.e.* $I \cos \phi$. It is also known as 'wattful' component.

(*ii*) Reactive component is that which is in quadrature with V *i.e.,* $I \sin \phi$. It is also known as 'wattless' or 'idle' component.

It should be noted that the product of volts and amperes in an a.c. circuit gives volt-amperes (VA). Out of this, the actual power is **VA** cos ϕ = W and reactive power is VA sin ϕ. Expressing the values in kVA, we find that it has two rectangular components,

(*i*) active component which is obtained by multiplying kVA by cos ϕ. It gives power in kW.

(*ii*) The other is known as reactive component and is obtained by multiplying kVA by sin ϕ. It is written as kVAR (kilovar).

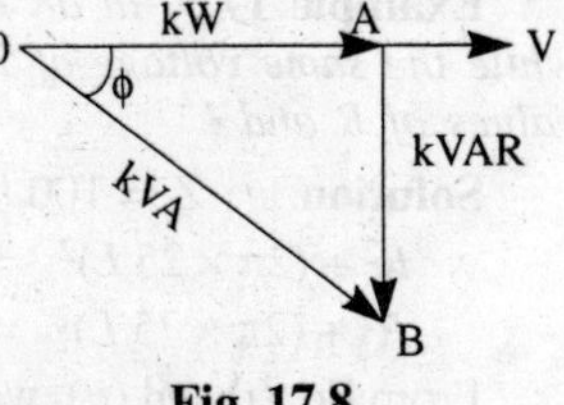

Fig. 17.8

The following relations can be easily deduced

$$
\begin{aligned}
\text{kVA} &= \sqrt{\text{kW}^2 + \text{kVAR}^2} \\
\text{kW} &= \text{kVA} \cos \phi \\
\text{kVAR} &= \text{kVA} \sin \phi
\end{aligned}
$$

These relations can be easily understood by referring to **kVA** triangle of Fig.17.8 where *it should be specifically noted that lagging kVARs have been taken as negative.*

For example, suppose a circuit draws a current of 10 A at a voltage of 20,000 V and its power factor is 0.8. Then

$$\text{kVA} = \frac{10 \times 20{,}000}{1000} = 200$$

$$\cos\phi = 0.8; \quad \sin\phi = 0.6$$

Hence, $\text{kW} = 200 \times 0.8 = 160$

$\text{kVAR} = 200 \times 0.6 = 120$

Obviously, $\sqrt{160^2 + 120^2} = 200$

i.e. $\text{kVA} = \sqrt{\text{kW}^2 + \text{kVAR}^2}$

Example 17.1. *A coil having a resistance of 6 Ω and an inductance of 0.03 H is connected across a 50-V, 60-Hz supply. Calculate (i) the current, (ii) the phase angle between the current and the applied voltage, (iii) the power factor, (iv) the volt-amperes and (v) the power.* **(A.M.I.E., Winter 1990)**

Solution. $X_L = 2\pi \times 60 \times 0.03 = 11.3\ \Omega$

$Z = \sqrt{R^2 + X_L^2} = \sqrt{6^2 + 11.3^2} = 12.8\ \Omega$

(*i*) $I = V/Z = 50/12.8 = \mathbf{3.9\ A}$

(*ii*) $\phi = \cos^{-1}(0.468) = \mathbf{27.9^0}$

(*iii*) $\cos\phi = R/Z = 6/12.8 = \mathbf{0.468}$

(*iv*) Volt-ampere $= 50 \times 3.9 = \mathbf{185\ VA}$

(*v*) $P = VI\cos\phi = 50 \times 3.9 \times 0.468 = \mathbf{91.3\ W.}$

Example 17.2. *A single-phase motor operating from 400-V, 50-Hz supply is developing 7.46 kW output with an efficiency of 84% and a p.f. of 0.7 lagging. Calculate (i) the input kVA (ii) the active and reactive components of current and (iii) the reactive kilovoltamperes.* **(A.M.I.E., Summer 1991)**

Solution. Motor output = 7.46 kW = 7,460 W

Motor input = 7,460 / 0.84 = 8880 W

Now, $VI\cos\phi$ = input

Motor input in volt-amperes = 8,880 / 0.7 = 12,685 VA

(*i*) kVA drawn by motor = 12,685 / 1000 = 12.685 kVA

(*ii*) Current drawn by motor I = 12,685 / 400 = 31.7 A

Active component $= I\cos\phi = 31.7 \times 0.7 = 22.2\text{A}$

Reactive component $= I\sin\phi = 31.7 \times 0.7 = 22.2\text{A}$

(*iii*) reactive kilovolt-amperes $= \text{kVA} \times \sin\phi$

$= 12.685 \times 0.7 = \mathbf{8.88\ kVAR}$

Example 17.3. *In an R-L series circuit, a voltage of 100 V at 25 Hz produces one ampere while the same voltage of 75 Hz produces half ampere. Draw the circuit diagram and insert the values of R and L.* **(A.M.I.E., Summer 87)**

Solution. $Z_1 = 100/1 = 100\ \Omega$; $Z_2 = 100/0.5 = 200\ \Omega$

$\therefore\ R^2 + (2\pi \times 25\ L)^2 = 100^2$ ...(*i*)

$R^2 + (2\pi \times 75\ L)^2 = 200^2$ ...(*ii*)

From Eq. (*i*) and (*ii*), we get $R = \mathbf{79}\ \Omega$ and $L = \mathbf{0.39\ H}$

Example 17.4. *An alternating voltage of* $v = 100 \sin 376.8\, t$ *is applied to a circuit consisting of a coil having a resistance of 6 Ω and an inductance of 21.22 mH.*

(*a*) *Express the current flowing through the circuit in the form*

$$i = I_m \sin(376.8\, t \pm \phi)$$

(*b*) *If a moving-iron voltmeter, a wattmeter and a frequency meter are connected in the circuit, what would be the respective readings on the instruments.*

Solution. $X_L = \omega L = 376.8 \times 21.22 \times 10^{-3} = 8\ \Omega$

$$Z = \sqrt{6^2 + 8^2} = 10\ \Omega$$

$$I_m = V_m / Z = 100 / 10 = 10\ \text{A}$$

$$\phi = \tan^{-1}(8/6) = 53.1° \text{ (lag)}$$

(*a*) The circuit current equation is

$$i = \mathbf{10 \sin (376.8t - 53.1°)}$$

(*b*) A moving-iron voltmeter measures the r.m.s. voltage.

Hence, its reading would be $= 100 / \sqrt{2} = \mathbf{70.7\ V}$

Wattmeter reading $= I^2 R = \frac{1}{2} I_m^2 R = \frac{1}{2} \times 10^2 \times 6 = \mathbf{300\ W}$

Frequency-meter reading $= 376.8 / 2\pi = \mathbf{60\ Hz}$

Example 17.5. *A sinusoidal alternating supply has an r.m.s. value of 100 volt and a frequency of 50 hertz and is connected to a series circuit having both resistance and inductance. The current taken from supply has an r.m.s. value of 5 A and power taken is 250 watts. Calculate:*

(i) The resistance of the circuit (ii) The impedance of the circuit (iii) The power factor of the circuit (iv) The inductance (v) The peak value of current (vi) The time taken for one complete cycle of supply and (vii) The apparent power taken by the circuit.

Assuming supply voltage as reference phasor, write expression for the instantaneous current as a function of time. **(Electrical Science, AMIE Winter 1994)**

Solution. (*i*) Since power is consumed by R only,

$\therefore$ $I^2 R = 250$ or $5^2 \times R = 250$, $R = \mathbf{10\ \Omega}$

(*ii*) $Z = V / I = 100 / 5 = \mathbf{20\ \Omega}$

(*iii*) p.f. $= R / Z = 10 / 20 = \mathbf{0.5}$

(*iv*) $X_L = \sqrt{Z^2 - R^2} = \sqrt{20^2 - 10^2} = 17.32\ \Omega$

$X_L = 2\pi f L$ or $17.32 = 2\pi \times 50 \times L$,

$\therefore$ $\Psi = \mathbf{0.055\ H}$

(*v*) Peak current $= \sqrt{2} \times$ r.m.s. current

$= \sqrt{2} \times 5 = \mathbf{7.07\ A}$

(*vi*) Time period, $T = 1 / f = 1 / 50 = \mathbf{0.028}$

(*vii*) Apparent power $= VI = 100 \times 5 = \mathbf{500\ VA}$

Current lags the voltage by $\phi = \cos^{-1} 0.5 = 60° = \pi/3$ radian. Hence, equation for the instantaneous value of the current is given by

$$\mathbf{i = 7.07 \sin (\omega t - \pi / 3)}$$

Example 17.6. *A 100-V, 60-W lamp is to be operated off 220-V, 50-Hz mains. What* (*i*) *pure resistance and* (*ii*) *pure inductance, placed in series with the lamp, will enable it to be used without being over-run? Which method would be more economical?* **(A.M.I.E., Winter 1987)**

Solution. Rated current of the bulb $= 60 / 100 = 0.6$ A

(*i*) A resistor R has been shown connected in series with the lamp in Fig.17.9.

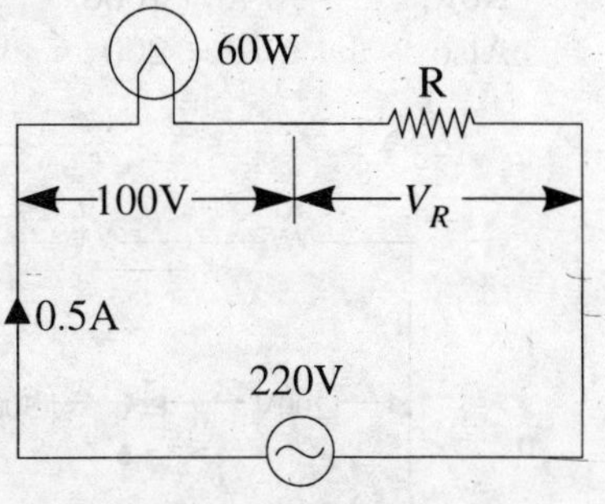

p.d. across R is

$V_R = 220 - 100 = 120$ V

It is in phase with the applied voltage

$\therefore$ $R = 120 / 0.6 = \mathbf{200\ \Omega}$

(*ii*) p.d. across bulb = 100 V

p.d. across pure inductance

$V_L = \sqrt{220^2 - 100^2} = 196$ V

(Remember that V_L is in quadrature with V_R 2 – Fig.17.2)

Now, $V_L = IX_L = 0.6 \times X_L$

or $196 = 0.6 \times 2\pi \times 50 \times L$ $\therefore$ $L = \mathbf{1.04H}$

60W
R
100V
V_R
0.5A
220V

Fig. 17.9

Method (*ii*) is preferrable to (*i*) because in method (*ii*) there is no loss of power.

Ohmic resistance of 200 Ω dissipates large power = 120 × 0.6 = 72 W — more than the bulb itself.

Example 17.7. *An arc lamp (which may be regarded as being non-inductive) takes 10 A at 50 V. Calculate the impedance of a choke of 1 ohm resistance to be placed in series with it in order that it may be worked off a 200-V, 50-Hz supply. Find also the total power used and the power factor.* **(A.M.I.E. Summer, 1988)**

Solution. Resistance of the arc lamp is 50 / 10 = 5 Ω. as seen from Fig. 17.10,

Z = 200 / 10 = 20 Ω

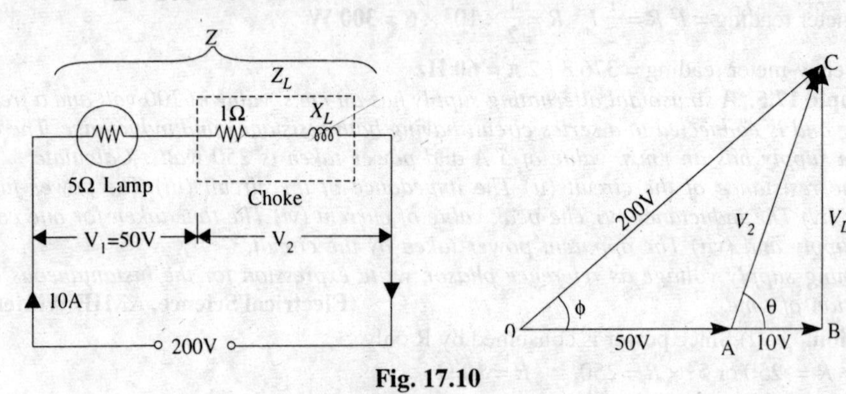

Fig. 17.10

$$\therefore \quad 6^2 + X_L^2 = 20^2$$

or $$X_L = \sqrt{364} = 19.08\ \Omega$$

$$Z_L = \sqrt{1^2 + 19.08^2} = \mathbf{19.1\ \Omega}$$

Total power used $= 10^2 \times (5+1) = \mathbf{600\ W}$

Power factor $= \cos\phi = OB / OC = 60 / 200 = \mathbf{0.3\ (lag)}$

Alternative Solution

$$BC = \sqrt{OC^2 - OB^2} = \sqrt{200^2 - 60^2} = 190.8\text{ V}$$

$\therefore \quad V_L = 190.8$ V or $IX_L = 190.8$ or $X_L = 190.8/10 = 19.08\Omega$

Example 17.8. *A circuit consists of a pure resistance and a coil connected in series. Powers dissipated in the resistance and in the coil are 1000 W and 250 W respectively. Voltage drops across the resistance and the coil are 200 V and 300 V respectively. Determine the reactance of the coil and the supply voltage.*

Solution. The ohmic resistance *R* and coil have been shown connected in series in Fig.17.11(*a*). The voltage phasor diagram is shown in Fig.17.11 (*b*).

Now, $V^2 / R = 1000$; $R = 40\ \Omega$

Also $IR = 200$ $\therefore$ $I = 200 / 40 = 5$ A

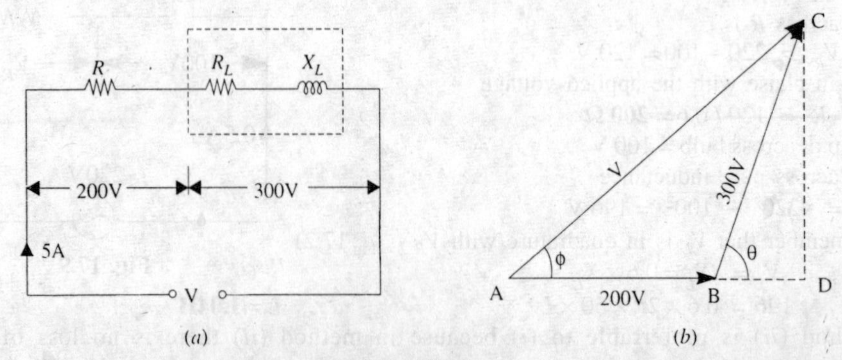

Fig. 17.11

Now, $I^2 R_L = 250$; $R_L = 250 / 5^2 = 10\ \Omega$

Coil impedance $Z = 300 / 5 = 60\ \Omega$, $X_L = \sqrt{60^2 - 10^2} = \mathbf{59.2\ \Omega}$

Combined resistance $= R + R_L = 40 + 10 = 50\ \Omega$

Combined impedance $= \sqrt{50^2 + 59.2^2} = 77.5\ \Omega$

Supply voltage $= IZ = 5 \times 77.5 = \mathbf{387.5\ V}$

Example 17.9. *An inductive load is connected in series with a non-inductive resistance of 8 ohms. The combination is connected across an a.c. supply of 100 volts, 50 Hz. A voltmeter connected across the non-inductive resistor and then across the inductive load gives the reading of 64 volts and 48 volts respectively. Calculate the following:*

(i) impedance of the load (ii) impedance of the combination (iii) power absorbed by the load (iv) power absorbed by the resistor (v) total power taken from the supply (vi) power factor of load (vii) power factor of the whole circuit. **(A.M.I.E., Summer 1989)**

Solution. (*i*) $I = 64 / 8 = 8\text{A}$, $Z = 48 / 8 = \mathbf{6\ \Omega}$

(*ii*) Combination $Z = 100 / 8 = \mathbf{12.5\ \Omega}$

From $\Delta\, BCD$ of Fig. 17.12 (*b*), $BC^2 + CD^2 = 48^2$ ...(*i*)

From $\Delta\, ACD$, $(64 + BC)^2 + CD^2 = 100^2$ ...(*ii*)

$\therefore\ BC = 28.1$ V, $CD = \sqrt{48^2 - 28.1^2} = \mathbf{38.9\ V}$

$R_L = 28.1 / 8\ 3.5\ \Omega$; $X_L = 38.9 / 8 = \mathbf{4.87\ \Omega}$

(*iii*) $P_{coil} = I^2 R_L = 8^2 \times 3.5 = \mathbf{224\ W}$

(*iv*) $P = I^2 R = 8^2 \times 8 = \mathbf{512\ W}$

(*v*) Total $P = 224 \times 512 = \mathbf{736\ W}$

(*vi*) $\cos\theta = 28.1 / 48 = \mathbf{0.585}$

(*vii*) $\cos\phi = AC / AD = 92.1 / 100 = \mathbf{0.921}$

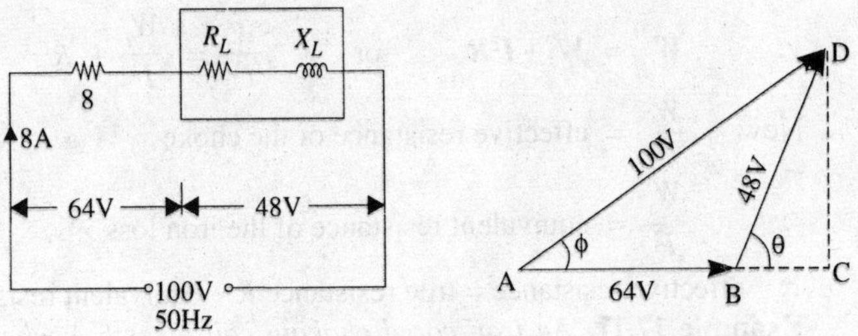

Fig. 17.12

Example 17.10. *Two coils A and B are connected in series across a 240-V, 50-Hz supply. The resistance of A is 5 Ω and the inductance of B is 0.015 H. If the input from the supply is 3 kW and 2 kVAR, find the inductance of A and the resistance of B. Calculate the voltage across each coil.*

Solution. The kVA triangle is shown in Fig.17.13 (*b*) and the circuit in Fig.17.13(*a*). The circuit kVA is given by

$$\text{kVA} = \sqrt{3^2 + 2^2} = 3.606 \text{ or VA} = 3606 \text{ voltamperes}$$

Circuit current $= 3606 / 240 = 15.03$ A

$\therefore\ 15.03^2 (R_A + R_B) = 3000$

$\therefore\ R_A + R_B = 3000 / 15.03^2 = 13.3\ \Omega$

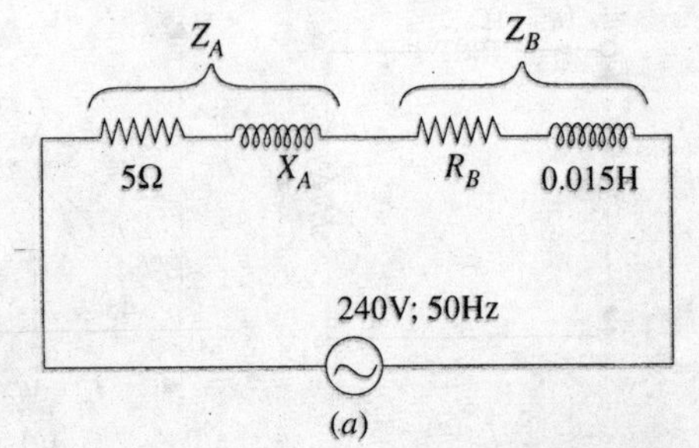

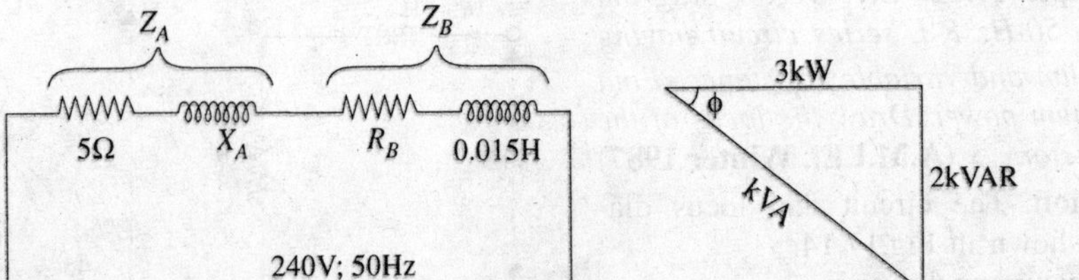

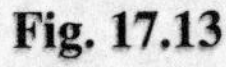

Fig. 17.13

$\therefore \quad R_B = 13.3 - 5 = \mathbf{8.3\ \Omega}$

Now, impedance of the whole circuit is given by

$$Z = 240 / 15.03 = 15.97\ \Omega$$

$$\therefore \quad X_A + X_B = \sqrt{Z^2 - (R_A + R_B)^2}$$

$$= \sqrt{15.97^2 - 13.3^2} = 8.843\ \Omega$$

Now, $X_B = 2\pi \times 50 \times 0.015 = 4.713\ \Omega$

$\therefore \quad X_A = 8.843 - 4.713 = 4.13\ \Omega$

$\therefore \quad 2\pi \times 50 \times L_A = 4.13 \quad \therefore \quad L_A = \mathbf{0.0132\ H\ (approx)}$

Now $Z_A = \sqrt{R_A^2 + X_A^2} = \sqrt{5^2 + 4.13^2} = 6.485\ \Omega$

p.d across coil $A = IZ_A = 15.03 \times 6.485 = \mathbf{97.5\ V}$

$$Z_B = \sqrt{8.3^2 + 4.713^2} = 9.545\ \Omega$$

$\therefore$ p.d. across coil $B = IZ_B = 15.03 \times 9.545 = \mathbf{143.5\ V}$

17.4. Power in an Iron-cored Choking Coil

Total power W taken by an iron-cored choking coil is used to supply,

(*i*) power loss in ohmic resistance *i.e.* I^2/R

(*ii*) iron-loss in core, W_i

$$\therefore \quad \boldsymbol{W = W_i + I^2R} \quad \text{or} \quad \frac{W}{I^2} = \frac{W_i}{I^2} + R$$

Now $\dfrac{W}{I^2}$ = effective resistance of the choke

$\dfrac{W_i}{I^2}$ = equivalent resistance of the iron loss

$\therefore$ effective resistance = true resistance R + equivalent resistance W_i / I^2.

Example 17.11. *An iron-cored choking coil takes 5 A when connected to a 20-V d.c. supply and takes 5 A at 100-V a.c. and consumes 250 W. Determine* (*a*) *impedance* (*b*) *the power factor* (*c*) *the iron loss* (*d*) *the inductive reactance of the coil.*

Solution. (*a*) $Z = 100 / 5 = \mathbf{20\ \Omega}$

(*b*) $W = VI \cos\phi$ or $250 = 100 \times 5 \times \cos\phi$

$\cos\phi = 250 / 500 = \mathbf{0.5}$

(*c*) Total loss = loss in resistance + iron loss

$\therefore \quad 250 = 20 \times 5 + W_i$

$\therefore \quad W_i = 250 - 100 = \mathbf{150\ W}$

(*d*) Effective resistance of the choke is

$$\frac{W}{I^2} = \frac{250}{25} = 10\ \Omega$$

$$X_L = \sqrt{(Z^2 - R^2)} = \sqrt{(400 - 100)} = \mathbf{17.32\ \Omega}$$

Example 17.12. *Give vector diagram for 100-V, 50-Hz R-L series circuit having ωL = 1 ohm and variable resistance. Find the maximum power. Draw the locus of the current vector.* **(A.M.I.E., Winter 1987)**

Solution. The circuit and locus diagram are shown in Fig.17.14.

The maximum power is

$$P_{max} = \frac{V^2}{2X_L} = \frac{100^2}{2 \times 1}$$

$$= \mathbf{5000\ W}$$

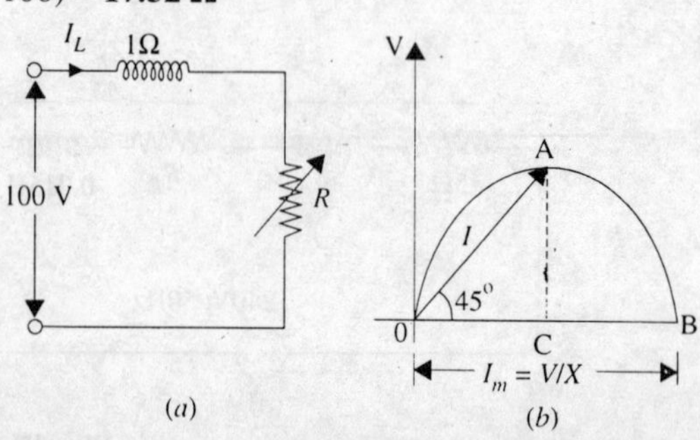

Fig. 17.14

Tutorial Problems No. 17.1

1. When a coil is connected to a 200-V, 50-Hz supply, it takes a current of 1.5 A and the power consumption is 70 W. Calculate the resistance and inductance of the coil. **[31.15 Ω; 0.413 H]**
2. A voltage of 260 *V* at 50 Hz applied to a coil produced a current of 2 A. The application of 2 V (d.c.) was necessary to produce 2 A of direct current through the coil. Calculate the inductance of the coil. **[382.5 mH]**
3. An inductor having a resistance of 10 Ω is connected to a 240-V, 50-Hz alternating current supply. The current flowing through the coil is found to be 12 A.
   (*a*) Calculate the impedance, inductive reactance and inductance of the inductor
   (*b*) Determine the phase angle between the current and the applied voltage
   (*c*) Draw a vector diagram showing the relation between the current and various voltages across the circuit components. **[(*a*) 20Ω; 17.32Ω; 0.055H (*b*) 60°]**
4. In a series circuit, a voltage of 10 V at 25 Hz produces 100 mA while the same voltage at 75 Hz produces 60 mA. Draw the circuit diagram and insert the values of the constants. ***(A.M.I.E. Winter 1986)***
5. An inductive coil takes a current of 2 A and consumes 160 W when connected to a 240-V a.c. supply. A second coil when connected across the same supply takes 3 A and 500 W. Find the current and total power when the two coils are connected in series to this supply. **[1.23 A, 151 W]**
6. A 1-henry inductor having a resistance of 50 Ω is connected in series with a 250 Ω resistor. A 20-V, 50-Hz supply is connected across this series circuit. Determine the voltage across the 250 Ω resistor. **[11.5 V]**
7. A certain iron-cored coil when connected to a 400-V, 50-Hz supply takes a current of 10 A and power of 3 kW while when it is connected to a 100 V d.c. supply, the current is 5 A. Calculate for operation on the a.c. supply:
   (*a*) the power factor, (*b*) the iron loss and (*c*) coil resistance. **[(*a*) 0.75 lag (*b*) 1000 W (*c*) 26.46 Ω]**
8. A 100 ohm resistor is connected in series with an inductor across a sinusoidal, 400-V, 50-Hz supply. If the p.ds. across the resistor and inductor are respectively 200 V and 300 V, calculate the resistance, inductance and p.f of the inductor. **[37.5 Ω; 0.462 H; 0.25 lag]**
9. A 1-phase load takes 8 A at 50 Hz when the supply is 220 V. The power factor of the load is then 0.9. Calculate the effective resistance and reactance of the circuit and the active and reactive components of the current. **[10.98 Ω; 53 Ω, 16.2 A, 7.85 A]**
10. Coil *A* draws 5 A and absorbs 300 W when connected to a 110-V a.c. supply. Similar figures for coil *B* are 8 A and 400 W. Determine the current, power and reactive volt-amperes when ACB are joined in series to a 220 V supply at the same frequency. **[6.16A; 6.95 kW; 1,170 kVAR]**
11. An iron-cored choking coil of resistance 6 Ω takes a current of 8 A when connected to a 200-V, 50-Hz supply and the power dissipated is 750 W. Assuming the coil to be equivalent to a series impedance, calculate (*i*) the iron loss (*ii*) the inductance at the given value of the current and (*iii*) the power factor. **[(*i*) 366 W (*ii*) 70.2 mH (*iii*) 0.469]**

17.5. A.C. Through Resistance and Capacitance in Series

The circuit is shown in Fig.17.15.

Here, $V_R = IR$ = drop across R (in phase with I)

$V_C = IX_C$ = drop across capacitor (lagging I by $\pi / 2$)

As capacitive reactance (X_C) is taken negative, V_C is shown along negative direction of Y-axis as shown in voltage triangle (Fig.17.16).

Now $$V = \sqrt{V_R^2 + V_C^2} = \sqrt{(IR)^2 + (-IX_C)^2}$$

$$= I\sqrt{R^2 + X_C^2}$$

or $$I = \frac{V}{\sqrt{R^2 + X_C^2}} = \frac{V}{Z}$$

The denominator is called the *impedance* of the circuit. Impedance triangle is shown in Fig. 17.17.

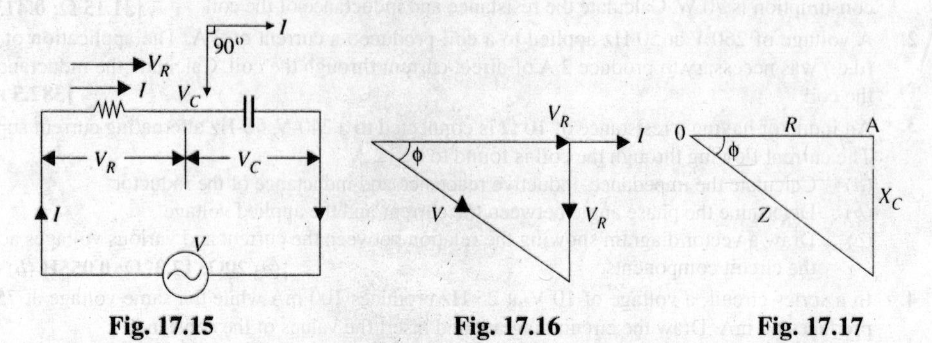

Fig. 17.15 **Fig. 17.16** **Fig. 17.17**

From Fig. 17.16 it is found that I leads V by angle ϕ such that

$$\tan \phi = X_C / R$$

Hence, it means that if the equation of the applied alternating voltage is $\upsilon = V_m \sin \omega t$, the equation of the resultant current in an R-C circuit is

$$i = I_m \sin (\omega t + \phi)$$

so that current leads the applied voltage by an angle ϕ. This fact is shown graphically in Fig. 17.18.

Power consumed is $VI \cos \phi$.

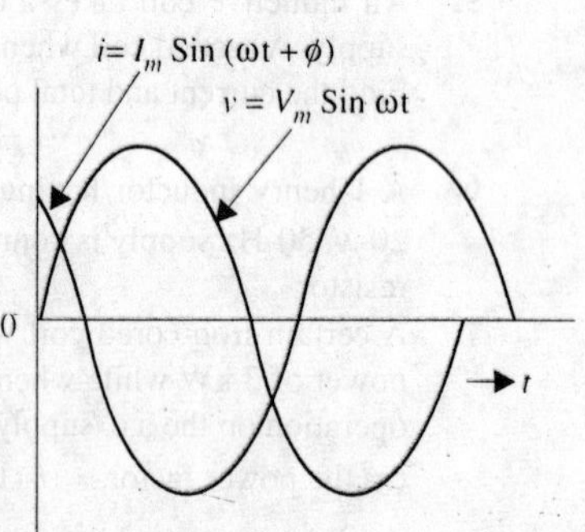

Fig. 17.18

Example 17.13. *A capacitor having a capacitance of 10 μF is connected in series with a non-inductive resistance of 120 Ω across a 100-V, 50-Hz supply. Calculate (a) the current, (b) the phase difference between the current and the supply voltage and (c) the power.* **(Elect. Engg.-I, Punjab Univ. 1993)**

Solution. $R = 120\ \Omega$

$$X_C = \frac{10^6}{2\pi \times 50 \times 10} = 318.3\ \Omega$$

$$Z = \sqrt{120^2 + 318.3^2} = 340\ \Omega$$

(*a*) $I = 100 / 340 = \mathbf{0.294\ A}$

(*b*) $\tan \phi = 318.3 / 120$

$\therefore$ $\phi = \tan^{-1} (318.3 / 120) = \mathbf{69°20'}$

(*c*) $W = I^2 R = 0.294^2 \times 120 = \mathbf{10.4\ W}$

Example 17.14. *A capacitor having a capacitance of 10 μF is connected in series with a non-inductive resistance of 120 Ω across a 100-V, 50-Hz supply. Calculate (i) current (ii) phase difference between the currrent and the supply voltage and (iii) power.*

(Elect.Engg.-I, Punjab Univ. 1993.)

120 Ω R 10 μF C 100V, 50 Hz

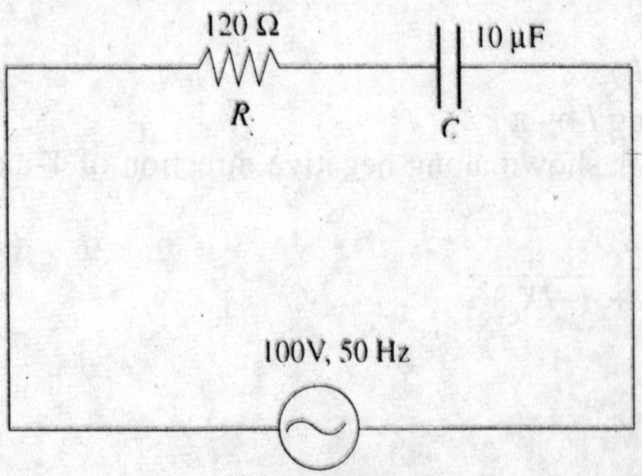

Fig. 17.19

Solution. $X_C = 10^6 / 2\pi \times 50 \times 10 = 31.8\ \Omega$

$$Z = \sqrt{120^2 + 31.8^2} = 124\ \Omega$$

(*i*) $I = 100 / 124 = \mathbf{0.806\ A}$

(*ii*) $\cos \phi = 120 / 124 = 0.968$, or ϕ **14.6°**

(*iii*) $P = VI \cos \phi = 100 \times 0.806 \times 0.968 = \mathbf{78\ W}$

Alternatively, $P = I^2 R = 0.806^2 \times 120 = \mathbf{78\ W}$

The circuit is shown in Fig. 17.19.

Example 17.15. *A resistance R and inductance L = 0.01 H and a capacitance C are connected in series. When a voltage V = 400 (3000 t – 10°) volts is applied to the series combination, the current flowing is $10\sqrt{2}$ cos (3000 t – 55°) ampere. Find R and C.*

(Electric Circuits-I, Punjab Univ.1994.)

Solution. As seen from the given equations,

$V_m = 400$ V, $I_m = 10\sqrt{2}$ A,

$\therefore \quad Z = 400 / 10\sqrt{2} = 28.3\ \Omega$

The given voltage lags behind the reference quantity by 10° whereas current lags behind by 55°. The phase difference between the two = 55° – 10° = 45°.

$R = Z\cos\phi = 28.3 \times \cos 45° = \mathbf{20\ \Omega}$

$Z^2 = R^2 + X^2$ or $X = \sqrt{Z^2 - R^2}$, where X is the net reactance of the circuit *i.e.* $(X_L - X_C)$

$\therefore \quad X = \sqrt{28.3^2 - 20^2} = 20\ \Omega$

Now, $X_L = \omega L = 3000 \times 0.01 = 30\ \Omega$

$\therefore \quad 20 = (30 - X_C)$ or $X_C = 10\ \Omega$

Now, $X_C = 1/\omega C$,

$\therefore \quad 10 = 10^6 / 3000 \times C$ or $C = \mathbf{33.3\ \mu F}$

Example 17.16. *A voltage of 125 V at 60 Hz is applied across a non-inductive resistor connected in series with a capacitor. The current in the circuit is 2.2 A. The power loss in the resistor is 96.8 watts and that in the capacitor is negligible. Calculate the resistance and the capacitance. Draw the vector diagram.*

Solution. I^2R = Power loss $\quad \therefore \quad 2.2^2 \times R = 96.8$

$\therefore \quad R = 96.8 / 2.2^2 = \mathbf{20\ \Omega}$

Now $\quad Z = 125 / 2.2 = 56.82\ \Omega$

$\therefore \quad X_C = \sqrt{Z^2 - R^2} = \sqrt{56.82^2 - 20^2} = \mathbf{53.2\ \Omega}$; $X_C = 1/\omega C = 53.2\ \Omega$

$C = 1/2\pi \times 60 \times 53.2 = 0.00005\ \text{F} = \mathbf{50\ \mu F}$

Example 17.17. *A resistor R in series with a capacitor C is connected to a 50-Hz, 240-V supply. Find the value of C so that R. absorbs 300 W at 100 V. Find also the maximum charge and maximum energy stored in C.* **(A.M.I.E., Summer 1986)**

Solution. $100^2 / R = 300$; $R = 100/3\ \Omega$

$I^2 R = 300$; $I = 3$ A, $Z = 240/3 = 80\ \Omega$

$X_C = \sqrt{Z^2 - R^2} = \sqrt{80^2 - (100/3)^2} = 72.7\ \Omega$

$X_C = 1/2\pi fC$; $C = 1/2\pi f X_C = 1/2\pi \times 50 \times 72.7 = \mathbf{43.7\ \mu F}$

$Q_{max} = C\,V_{max} = 43.7 \times 10^{-6} \times 240 \times \sqrt{2} = \mathbf{0.0148\ J}$

$E_{max} = \frac{1}{2} C\,V^2_{max} = \frac{1}{2} \times 43.7 \times 10^{-6} \times (240\sqrt{2})^2 = \mathbf{2.5\ J}$

Example 17.18. *A metal filament lamp, rated at 750-W, 100-V is to be connected in series with a capacitance across a 230-V, 60-hertz supply. Calculate (i) the capacitance required and (ii) the phase angles between the current and the supply voltage when the lamp obtains its rated voltage.*

Solution. (*i*) The circuit diagram is shown in Fig.17.20 (*a*).

$V_C = \sqrt{230^2 - 100^2} = 207.1$ V

Rated lamp current $= 750/100 = 7.5$ A

$$I = \frac{V_C}{X_C} = \frac{V_C}{1/\omega C} = \omega C V_C$$

$\therefore \quad 7.5 = 2\pi \times 60 \times C \times 207.1 \quad \therefore \quad C = \mathbf{96\ \mu F}$

(*ii*) As seen from Fig. 17.20 (*b*), $\tan\phi = 207.1/100 = 2.071$; $\phi = \tan^{-1}(2.071) = \mathbf{64.2°}$.

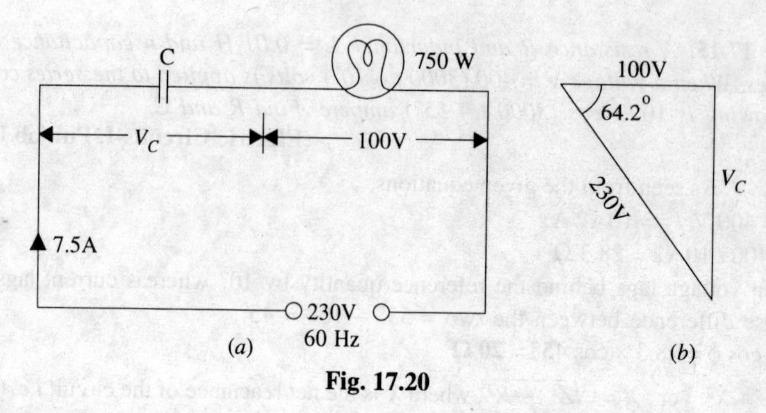

Fig. 17.20

Example 17.19. *A voltage* υ = *100 sin 314t is applied to a circuit consisting of a 25 Ω resistor and an 80 μ F capacitor in series. Determine:*

(*a*) *an expression for the value of the current flowing at any instant*

(*b*) *the power consumed*

(*c*) *the p.d. across the capacitor at the instant when the current is one half of its maximum value.*

Solution. $X_C = 10^6 / 314 \times 80 = 39.8\ \Omega$

$\phi = \tan^{-1}(39.8/25) = 57°52' = 1.01$ radian (lead)

$Z = \sqrt{25^2 + 39.8^2} = 47\ \Omega$

$I_m = V_m / Z = 100/47 = 2.13$ A

(*a*) $i = I_m \sin(\omega t + \phi) = \mathbf{2.13 \sin (314\, t + 1.01)}$

(*b*) $P = I^2 R = (2.13/\sqrt{2})^2 \times 25 = \mathbf{56.7\ W}$

(*c*) Now, maximum voltage across the capacitor is

$= I_m \cdot X_C = 2.13 \times 39.8 = \mathbf{84.8\ V}$

Since capacitor voltage lags behind the circuit *current* by π / 2 radian, the instantaneous voltage across the capacitor is

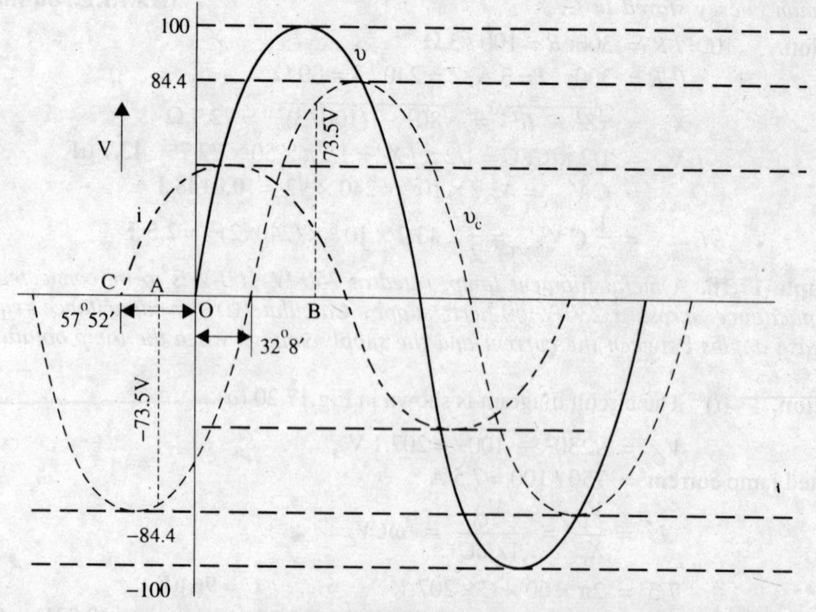

Fig. 17.21

$$\upsilon_C = 84.4 \sin(314\,t + 1.01 - \pi/2)$$

As shown in Fig. 17.21, the applied voltage has been taken as the reference voltage. The current wave *leads* the applied voltage wave by 57°52′. The capacitor voltage wave lags behind the *current* wave by 90° but lags behind voltage wave by (90° − 57°52′) = 32°8′).

When current is half its maximum value, we have

$$0.5 \times 2.13 = 2.13 \sin(314\,t + 1.01)$$

$$\therefore \quad 314\,t + 1.01 = \sin^{-1}(0.5) = \frac{\pi}{2} \text{ radian or } \frac{5\pi}{6} \text{ rad.}$$

It means that current is half its maximum value at angles of 30° or 150° as measured from *point C*. It means at points A and B respectively.*

At point A

$$\upsilon_C = 84.4 \sin(\pi/6 - \pi/2)$$
$$= 84.4 \sin(-\pi/3) = \mathbf{-73.5\ V}$$

At point B

$$\upsilon_C = 84.4 \sin(5\pi/6 - \pi/2) = 84.4 \sin(\pi/3) = \mathbf{73.5\ V}$$

Hence, p.d. across the capacitor when current is half its maximum value is 73.5 V.

Example 17.20. *Give vector diagram for 100-V, 50-Hz, R-C series circuit having ωC = 1 mho and variable resistance. Find the maximum power. Draw the locus of the current vector.*

(A.M.I.E., Winter 1987)

Solution. The circuit and locus diagram are shown in Fig.17.22.

The maximum power is given by

$$P_{max} = \frac{V^2}{2X_C} = \frac{V^2\,\omega C}{2} = \frac{100^2 \times 1}{2}$$

$$= \mathbf{5000\ W}$$

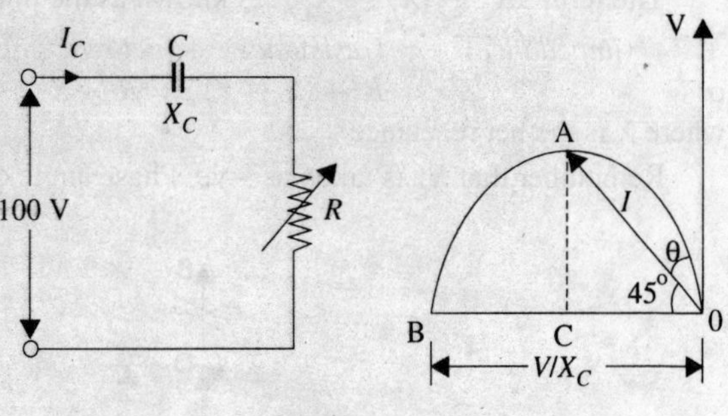

Fig. 17.22

Tutorial Problems No.17.2

1. A 110-V, 40-W lamp is connected to a 240-V, 50-Hz supply with a capacitor in series with the lamp so that it operates at its normal voltage. Determine the capacitance of the capacitor used and p.f. of the circuit. **[5.42 μF; 0.458 lead]**
2. An 80-W, 100-V lamp is to be used on a 240-V, 50-Hz supply. Calculate the value of
   (*a*) the inductance of an iron-cored choke of effective resistance of 12.5Ω
   (*b*) the capacitance of a capacitor
   either of which could be connected in series with the lamp so that it operates at its correct voltage. **[0.85 H; 11.7 μF]**
3. A voltage represented by 230 sin 314 *t* is applied to a series circuit, the resulting current is represented by the equation $i = 10 \sin(314t + \pi/6)$. Sketch the current and voltage waveforms over a complete cycle. What is the frequency, resistance and reactance of the circuit? **[50Hz; 19.92 Ω; 11.5 Ω]**

17.6. Resistance, Inductance and Capacitance in Series

The three are shown in Fig. 17.23 joined in series across an a.c. supply of r.m.s. voltage *V*.

Let $V_R = IR$ = voltage drop across *R* (in phase with *I*)

$V_L = I.\,X_L$ = voltage drop across *L* (leading *I* by π/2)

*Or at angles of (30° − 57°52′) = -27°52′ (point *A*) and (150° − 7°52′) = 92°8′ (point *B*) as measured from reference point *O*.

$V_C = I.X_C$ = voltage drop across C (lagging behind I by $\pi/2$)

In voltage triangle of Fig.17.24, OA represents V_R, AB and AC represents the inductive and capacitive drops respectively. It will be seen that V_L and V_C are 180° out of phase with each other *i.e.* they are in direct opposition to each other. Moreover, it has been assumed that V_L is greater than V_C in magnitude.

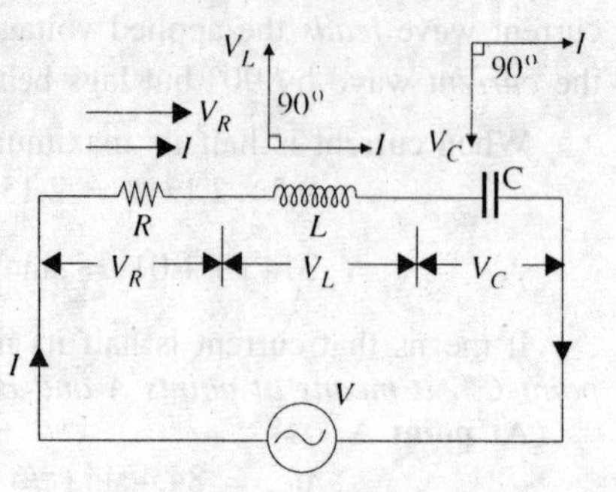

Fig. 17.23

Subtracting BD (=AC) from AB, we get the net reactive drop

$$AD = (V_L - V_C) = I\,(X_L - X_C)$$

The applied voltage V is represented by OD and is the vector sum of OA and AD.

$$OD = \sqrt{(OA^2 + AD^2)}$$

or $$V = \sqrt{[(IR)^2 + (IX_L - IX_C)^2]}$$

$$= I\sqrt{[R^2 + (X_L - X_C)^2]}$$

or $$I = \frac{V}{\sqrt{R^2 + (X_L - X_C)^2}} = \frac{V}{Z}$$

The term $\sqrt{R^2 + (X_L - X_C)^2}$ is known as the impedance of the circuit. Obviously,

$$(impedance)^2 = (resistance)^2 + (net\ reactance)^2$$

or $$Z^2 = R^2 + (X_L - X_C)^2 = R^2 + X^2$$

where X is the net reactance ...(Fig.17.25)

Remember that X_C is taken as –ve. Phase angle ϕ is given by

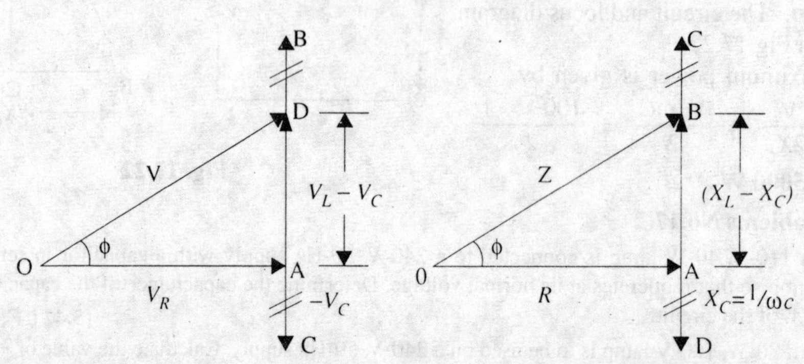

Fig. 17.24 Fig. 17.25

$$\tan\phi = (X_L - X_C)/R = X/R$$

= net reactance / resistance

$$\text{Power factor } \cos\phi = \frac{R}{Z} = \frac{R}{\sqrt{R^2 + (X_L - X_C)^2}} = \frac{R}{\sqrt{R^2 + X^2}}$$

Hence, it is seen that if the equation of the applied voltage is

$$\upsilon = V_m \sin\omega t$$

then equation of the resulting currrent in an R-L.C circuit is given by

$$i = I_m \sin(\omega t \pm \phi)$$

The +ve sign is to be used when current leads *i.e.* when $X_C > X_L$. The -ve sign is used when current lags *i.e.* when $X_L > X_C$. In general, the current lags or leads the supply voltage by an angle ϕ such that tan ϕ X / R.

Summary of Results of Series Circuits

Type of Impedance	*Value of Impedance*	*Phase angle for current*	*Power factor*
Resistance only	R	0°	1
Inductance only	ωL	90° lag	0
Capacitance only	1/ωC	90° lead	0
Resistance and Inductance	$\sqrt{[R^2 + (\omega L)^2]}$	$0 < \phi < 90°$ lag	1 > p.f. > 0 lag
Resistance and Capacitance	$\sqrt{[R^2 + (-1/\omega C)^2]}$	$0 < \phi < 90°$ lead	1 > p.f. > 0 lead
R – L – C	$\sqrt{[R^2 + (\omega L - 1/\omega C)^2]}$	between 0° and 90° lag or lead	between 0 and unity lag or lead

Example 17.21. *A resistance of 5 ohm, an inductance of 10 mH, and a capacitor of 200 μF are connected in series and the combination is connected across a 230-V, 50-Hz supply. Calculate the current flowing through the circuit and the power factor.*

(Elements of Elect.Engg.-II, Punjab Univ.1994.)

Solution. $X_L = 2\pi \times 50 \times 10 \times 10^{-3} = 3.14\ \Omega$

$X_C = 10^6 / 2\pi \times 50 \times 200 = 15.9\ \Omega$

$X = X_L - X_C = 3.14 - 15.9 = -12.76\ \Omega$ (capacitive)

$Z = \sqrt{5^2 + (-12.76)^2} = 13.7\ \Omega$

$I = 230 / 13.7 =$ **16.8 A**

p.f. $= R/Z = 5/13.7 =$ **0.365 (leading)**

Example 17.22. *A 230-V, 50-Hz ac supply is applied to a coil of 0.06 H inductance and 2.5Ω resistance connected in series with a 6.8 μF capacitor. Calculate (i) impedance, (ii) current, (iii) phase angle between current and voltage, (iv) power factor and (v) power consumed.*

(Electrical Engg., A.M.Ae.S.I., June 1992)

Solution. $X_L = 2\pi \times 50 \times 0.06 = 18.85\ \Omega$

$X_C = 1/\omega C = 10^6 / 2\pi \times 50 \times 6.8 = 468\ \Omega$

Net $X = X_L - X_C = 18.85 - 468 = -449\ \Omega$ (capacitive)

(*i*) $Z = \sqrt{25^2 + 449^2} = 450\ \Omega$ (*ii*) $I = 230/450 =$ **0.51 A**

(*iii*) p.f. $= R/Z = 25/450 =$ **0.055 (lead)**

(*iv*) $\phi = \cos^{-1}(0.055) =$ **86.8°**

(*v*) Power $= VI\cos\phi = 230 \times 0.51 \times 0.055 =$ **6.45 W**

Example 17.23. *A coil of power factor 0.6 is in series with a 100 μF capacitor. When connected to 50-Hz supply, the p.d across the coil is equal to p.d. across the capacitor. Find the resistance and inductance of the coil.* **(A.M.I.E., Winter 1988)**

Solution. $X_C = 1/\omega C = 1/2\pi f C$

$= 10^6 / 2\pi \times 50 \times 100 = 31.8\ \Omega$

Since for a supply of 50-Hz, drop across coil (IZ) equals the drop across the capacitor ($I \cdot X_C$), it means that $Z = X_C = 31.8\ \Omega$

∴ coil $Z = 31.8\ \Omega$

Also, coil p.f. = $\cos\phi = 0.6, \quad \sin\phi = 0.8$

$\therefore$ coil $R = Z\cos\phi = 31.8 \times 0.6 = \mathbf{19.1\ \Omega}$

Coil $X_L = Z\sin\phi = 31.8 \times 0.8 = \mathbf{25.4\ \Omega}$.

Example 17.24. *A coil of resistance 10 ohms and inductance 0.1 H is connected in series with a capacitor of capacitance 150 μF across a 200 V, 50 Hz supply. Determine the following:*

(i) impedance, (ii) current, (iii) power factor, (iv) voltage across the coil and (v) voltage across the capacitor. **(A.M.I.E., Winter 1989)**

Solution. $X_L = 2\pi \times 50 \times 0.1 = 31.2\ \Omega$

$X_C = 1/2\,\pi \times 50 \times 150 \times 10^{-6} = 21.2\ \Omega$

$X = X_L - X_C = 31.2 - 21.2 = 20\ \Omega$

(*i*) $Z = \sqrt{10^2 + 20^2} = \mathbf{22.4\ \Omega}$ (*ii*) $I = 200/224 = \mathbf{8.93\ A}$

(*iii*) p.f. = $R/Z = 10/22.4 = \mathbf{0.446\ (lag)}$

(*iv*) $Z_{coil} = \sqrt{10^2 + 31.2^2} = 32.8\ \Omega$; $V_L = 8.93 \times 32.8 = \mathbf{293\ V}$

(*v*) $V_C = 8.93 \times 21.2 = \mathbf{189\ V}$

Example 17.25. *A coil of insulated wire of resistance 8 Ω and inductance 0.03 H is connected to an a.c. supply at 240-V, 50-Hz. Calculate:*

(*a*) *the current p.f and the power,*

(*b*) *the value of capacitance which when connected in series with the above coil, causes no change in the values of the current and power taken from the supply.*

Solution. (*a*) $X_L = 314 \times 0.03 = 9.42\ \Omega$

$Z = \sqrt{8^2 + 9.42^2} = 12.36\ \Omega$

$I = 240 / 12.36 = \mathbf{19.4\ A}$

$W = I^2 R = 19.4^2 \times 8 = \mathbf{3011\ W}$

p.f. = $R/Z = 8/12.36 = \mathbf{0.65\ (lag)}$

(*b*) If the circuit is to draw the same current and at the same power factor, then the total reactance of the *R-L-C* circuit must be 9.42 Ω. This can be achieved by selecting a capacitor which should not only neutralize the inductive reactance of 9.42 Ω of the coil but must add a further capacitive reactance of 9.42 Ω. In other words, capacitor must have a reactance of 2 × 9.42 = 18.84 Ω

$\therefore$ $1/\omega C = 18.84$ or $C = 10^6 / 314 \times 18.84 = \mathbf{169\ \mu F}$

Example 17.26. *A 75 ohm resistor is connected in series with 30 μF capacitor and 2 H inductor across an a.c. variable frequency source. At what frequency will the phase angle of circuit will be 45° (i) lagging (ii) leading the e.m.f.* **(Electrical Circuits-I, Punjab Univ. 1993.)**

Solution. As explained in Art.17.12, the phase angle at half-power points is 45° or power factor is 0.707. The power factor is leading at the lower half-power point of frequency f_1 and is lagging at the upper half-power point of frequency f_2.

Let us first find the resonant frequency f_0 of the circuit.

$$f_0 = \frac{1}{2\pi\sqrt{LC}} = \frac{1}{2\pi\sqrt{2 \times 30 \times 10^{-6}}} = 20.5\ \text{Hz}$$

(*i*) $f_2 = f_0 + R/4\pi L = 20.5 + 75/4\,\pi \times 2 = \mathbf{23.5\ Hz}$

(*ii*) $f_1 = f_0 - R/4\pi L = 20.5 - 3 = \mathbf{17.5\ Hz}$

Tutorial Problems No. 17.3.

1. A circuit consists of a resistance of 20Ω in series with an inductance of 95.6 mH and a capacitor of 318 μF. It is connected to a 500-V, 25-Hz supply. Calculate the current in the circuit and the power factor. **[24.3A; 0.97 lead]**

2. A circuit is made up of 10 Ω resistance, 12 mH inductance and 281.5 μF capacitance in series. The supply voltage is 10 V (constant). Calculate the value of the current when the supply frequency is (*a*) 50 Hz and (*b*) 150 Hz. **[(*a*) 8 A leading (*b*) 8 A lagging]**

3. A circuit takes a current of 3 A at a power factor of 0.6 lagging when connected to a 115-V, 50-Hz supply. Another circuit takes a current of 5 A at a power factor of 0.707 leading when connected to the same supply. If the two circuits are connected in series across a 230-V, 50-Hz supply calculate: (*a*) the current, (*b*) the power consumed and (*c*) the power factor.

[(*a*) 5.5 A (*b*) 1,188 W (*c*) 0.939 lag]

4. A series circuit consists of a choke coil, having a resistance of 10 Ω and a capacitor. When connected to a 240-V, 50-Hz supply, the current was 6 A leading the applied voltage. When connected to a 240-V, 100-Hz supply, the current was 6 A lagging the applied voltage. Calculate the inductance of the coil and the capacitance of the capacitor **[124 mH; 41 μF]**

5. The applied e.m.f. of frequency 45 Hz has a peak value of 100 volts. The resistance, inductance and capacitance in the circuit are 75 Ω, 0.2 H and 50 μF respectively. Find the maximum current and the average rate of consumption of energy. **[1.3 A; 169W]** ***(A.M.I.E. Summer 1983)***

17.7. Resonance in R-L-C Circuit

A series circuit is said to be in electrical resonance when its net reactance is zero. The frequency f_o at which this happens is known as 'resonant frequency'.

As seen in Art.17.6, net reactance of an *R-L-C* circuit is

$$X = X_L - X_C \text{ and } Z = \sqrt{R^2 + X^2}$$

At resonance, $X = 0$ or $X_L - X_C = 0$

$\therefore$ $X_L = X_C$

or $\omega L = 1/\omega C$ or $\omega^2 = 1/LC$

$\therefore$ $(2\pi f_0)^2 = \dfrac{1}{LC}$ or $f_0 = \dfrac{1}{2\pi\sqrt{LC}}$

If L is in henry and C in farad, then f_o is in Hz. Under resonant conditions, $X = 0$, hence $Z=R$. This is the minimum possible value of impedance. Hence, circuit current is maximum and is given by

$$I_m = V/Z = V/R$$

The circuit behaves like a pure resistive circuit and current is in phase with the applied voltage. Obviously, the power factor of the circuit is unity under resonant conditions.

As current is maximum, it produces large voltage drops across L and C. But these drops being equal and opposite, cancel each other out. Taken together, L and C form part of a circuit across which no voltage is developed, however, large the current flowing. If it were not for the presence of R, such a circuit would act like a short-circuit to currents of the frequency to which it resonates. Hence, it is sometimes called *acceptor circuit* and the series resonance is often referred to as *voltage resonance.*

17.8. Graphic Representation of Series Resonance

Suppose an alternating voltage of constant magnitude but of varying frequency is applied to an *R-L-C* circuit. The variations of resistance, inductive reactance X_L and capacitive reactance X_C with frequency are as shown in Fig. 17.26.

(*i*) R : It is independent of f, hence it is represented by a straight horizontal line.

(*ii*) $X_L = \omega L = 2\pi f L$

Inductive reactance X_L is directly proportional to f, hence its graph is a straight line through the origin. So, X_L increases linearly with f.

(*iii*) $X_C = 1/\omega C = 1/2\pi f C$

Capacitive reactance X_C is inversely proportional to f. Its graph is a rectangular hyperbola which is drawn in fourth quadrant because X_C is regarded negative.

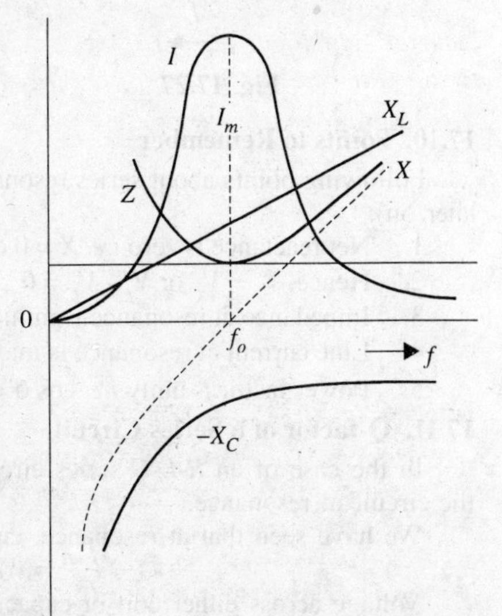

Fig. 17.26

(*iv*) *Net reactance*

$$X = X_L - X_C$$

Its graph is a hyperbola (not rectangular) and crosses the X-axis at point A. The value of frequency at A is called resonant frequency f_o.

(*v*) $Z = \sqrt{[R^2 + (X_L - X_C)^2]}$

At low frequencies, Z is large (Fig.17.26). But as $X_C > X_L$, net impedance is capacitive and the p.f is leading. At high frequencies Z is again large, but is inductive because $X_L > X_C$ and the p.f. is lagging. The minimum value of Z is R when $X_L = X_C$.

(*vi*) *Current I*

It has low value on both sides of resonant frequency (because Z is large) but has maximum value of $I_m = V/R$ at resonance as shown by the peaked curve (Fig.17.26).

17.9. Resonance Curve

The curve between current and frequency is known as resonance curve. The shape of such a curve for various values of R is shown in Fig. 17.27. For smaller values of R, the current-frequency curve is sharply-peaked, but for larger values of R, the curve is flat. The variations of Z, power factor and I are separately shown in Fig. 17.28.

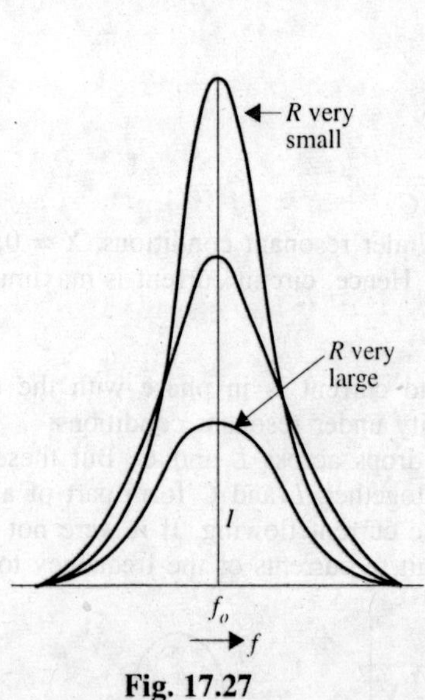

Fig. 17.27

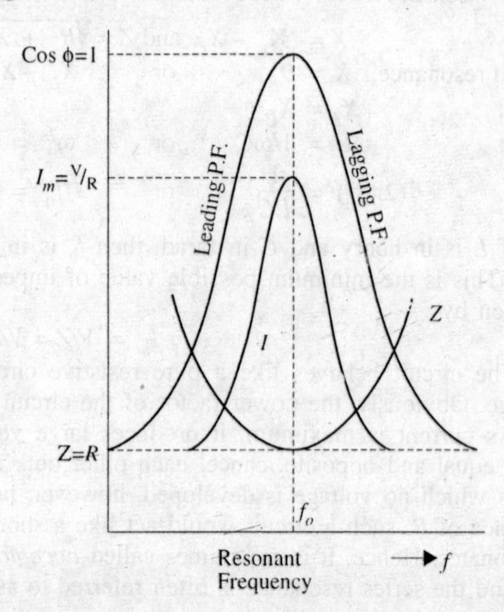

Fig. 17.28

17.10. Points to Remember

Following points about series resonance must be noted (and compared with parallel resonance later on):

1. Net reactance is zero *i.e.* $X = 0$ or $X_L = X_C$
2. Hence, $V_L = V_C$ or $V_L - V_C = 0$
3. Impedance at resonance is minimum and = R ohm
4. Line current at resonance is maximum and is equal to V/R and is in phase with V.
5. Power factor is unity *i.e.* $\cos\phi = 1$.

17.11. Q-factor of a Series Circuit

In the case of an R-L-C series circuit, It is defined as equal to the voltage magnification in the circuit at resonance.

We have seen that at resonance, current is maximum *i.e.*

$$I_m = V/R$$

Voltage across either coil or capacitor = $I_m X_L$

Supply voltage $V = I_m R$

$\therefore$ voltage magnification $= \dfrac{I_m X_L}{I_m R} = \dfrac{X_L}{R} = \dfrac{\omega L}{R}$

$$\therefore \quad Q\text{-factor} = \frac{\omega L}{R} = \frac{2\pi f_0 L}{R} \quad \text{...(i)}$$

We have seen that resonant frequency is

$$f_0 = \frac{1}{2\pi\sqrt{LC}} \quad \text{or} \quad 2\pi f_0 = \frac{1}{\sqrt{LC}}$$

Putting this value in (*i*) above, we get

$$Q\text{-factor} = \frac{1}{R}\sqrt{\left(\frac{L}{C}\right)}$$

In the case of series resonance, higher Q-factor means not only higher voltage magnification, but also a higher selectivity of the tuning coil.

Obviously, Q-factor can be increased by having a coil of large inductance but of small ohmic resistance.

17.12. Bandwidth of a Series Circuit

Bandwidth of a series circuit is given by the band of frequencies which lie between two points on either side of its resonance curve where current is $1/\sqrt{2}$ or 0.707 of its maximum value at resonance. The p.f. of the circuit at the two frequencies is $1/\sqrt{2}$ or $\phi = 45°$. The p.f. is leading at point A but lagging at point B. As shown in Fig. 17.29, band width or passband is given by

$$BW = \Delta f = f_2 - f_1$$

It can be shown that

$$BW = f_2 - f_1 = \frac{R}{2\pi L} \quad \text{or} \quad BW = \frac{f_0}{Q} \quad \text{...from (i) above.}$$

It can also be proved that edge frequencies are given by

$$f_1 = f_0 - \frac{\Delta f}{2} = f_0 - \frac{BW}{2} = f_0 - \frac{R}{4\pi L}$$

Similarly, $f_2 = f_0 + \dfrac{\Delta f}{2} = f_0 + \dfrac{BW}{2} = f_0 + \dfrac{R}{4\pi L}$

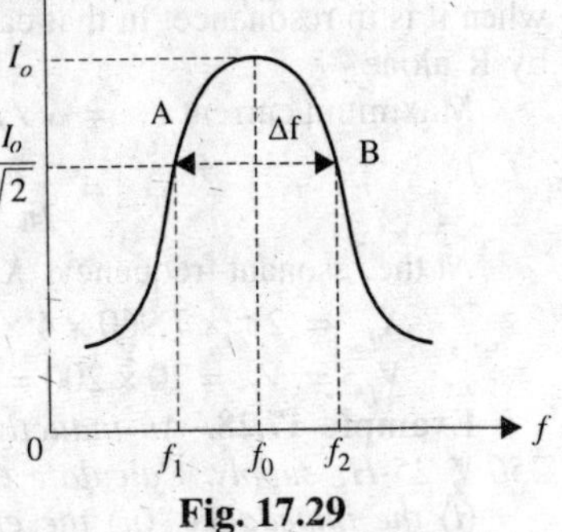

Fig. 17.29

It may be noted that points A and B in Fig. 17.29 are also called half-power points because power developed at each of these points is half the power developed at resonance. Hence, edge frequencies f_1 and f_2 are known as lower half-power frequency and upper half-power frequency respectively. All the frequencies lying between f_1 and f_2 are called useful frequencies because they produce currents which when passed through headphones produce a sound that is not much weaker than that produced by maximum current I_m at resonance. Hence, bandwidth of a resonant circuit, in fact, represents the range of its useful frequencies.

17.13. Sharpness of Resonance

It is defined as the ratio of the bandwidth of the circuit to its resonance frequency.

$$\text{Sharpness of resonance} = \frac{f_2 - f_1}{f_0} = \frac{\Delta f}{f_0}$$

$$= \frac{BW}{f_0} = \frac{1}{Q_0}$$

It shows that as value of Q_o increases, bandwidth decreases. As bandwidth decreases, selectivity of the circuit increases because selectivity is inversely proportional to bandwidth.

Fig.17.30 shows the resonance curve of a circuit having comparatively large value of R. The passband of the circuit is from 980 kHz to 1020 kHz. Its bandwidth is 40 kHz so that its sharpness of resonance is 40/1000 = 0.04. Similarly, its value for the circuit of Fig. 17.31 is 20 / 1000 = 0.02. This circuit has a bandwidth of 20 kHz as compared to 40 kHz of the first circuit. The circuit shown in Fig.17.30 has poorer selectivity because it does not reject frequencies close to its resonance frequency.

It can be proved that at half-power points A and B, the net reactance of the circuit equals its resistance. Hence, the power factor of the circuit is $1/\sqrt{2}$ = 0.707. In other words, the phase angle of the circuit is 45°. Since at lower half-power points, $X_C > X_L$, the power factor is leading. On the other hand, at the upper half-power point, $X_C < X_L$, the power factor is lagging.

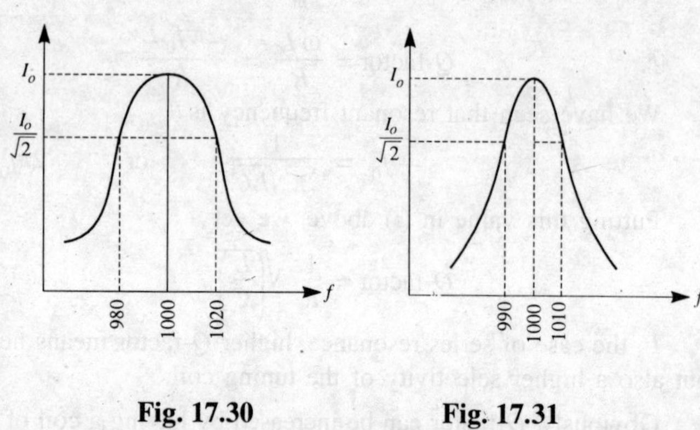

Fig. 17.30 **Fig. 17.31**

Example. 17.27. *(a) A generator supplies a variable frequency voltage of constant amplitude, 100 V (r.m.s.) to a series RLC circuit where R = 5 Ω, L = 4 mH and C = 0.1 μF. The frequency is to be varied until a maximum current flows. Predict the maximum current, the frequency at which it occurs and the resulting voltages across the inductance and the capacitance.*

(Electrical Science, AMIE Summer 1994.)

Solution. As explained in Art.17.7, maximum current flows through a series RLC circuit when it is in resonance. In that case, X_L cancels X_C and only opposition to current flow is provided by R alone.

Maximum current $= V/R = 100/5 = \mathbf{20\ A}$

$$f_0 = \frac{1}{2\pi\sqrt{LC}} = \frac{1}{2\pi\sqrt{4 \times 10^{-3} \times 0.1 \times 10^{-6}}} = 7{,}960 \text{ Hz}$$

At the resonant frequency, $X_L = X_C$ so that $V_L = V_C$

$X_L = 2\pi \times 7{,}960 \times 4 \times 10^{-3} = 200\ \Omega$

$\therefore\ V_L = V_C = 20 \times 200 = \mathbf{4000\ V}$

Example 17.28. *An inductive coil takes 10 A and dissipates 1000 W when connected to a 250-V, 25-Hz supply. Calculate the following.*

(i) the impedance, (ii) the effective resistance, (iii) the reactance, (iv) the power factor, (v) the value of the capacitance required to be connected in series with the coil to make the power factor of the circuit unity, (vi) what is now the current taken by the coil? **(A.M.I.E., Summer 1990)**

Solution. *(i)* $Z = 250/10 = \mathbf{25\ \Omega}$

(ii) $10^2 \times R = 1000$, $R = \mathbf{10\ \Omega}$

(iii) $X_C = \sqrt{25^2 - 10^2} = \mathbf{22.9\ \Omega}$

(iv) p.f. $= R/Z = 10/25 = \mathbf{0.4}$

(v) $X_C = X_L = 22.9\ \Omega$

or $1/2\pi \times 25 \times C = 22.9$; $C = \mathbf{278\ \mu F}$

(vi) $I = 250/10 = \mathbf{25\ A}$

Example 17.29. *A circuit having a resistance of 4 Ω, and inductance of 0.5 H and a variable capacitance in series is connected across 100-V, 50-Hz supply. Calculate (i) the capacitance to give resonance and (ii) the voltage across inductance and the capacitance.*

(Elect.Engg.-I, Punjab Univ. 1994.)

Solution. *(i)* $50 = \frac{1}{2\pi\sqrt{0.5 \times C}}$, $C = \mathbf{20\ \mu F}$

(*ii*) Under resonance, $I = V/R = 100/4 = 25$ A

$X_L = 2\pi \times 50 \times 0.5 = 157\ \Omega$

$\therefore\ V_L = IX_L = 20 \times 157 =$ **3927 V**

Same would be the voltage drop across the capacitance.

Example 17.30. *A series RLC circuit has the following parameter values : R = 10 ohm, L = 0.01 H, C = 100 μF voltage-source, e(t) = 10 sin 1000 t.*

Find (i) impedance of the circuit, (ii) power dissipated in the circuit, (iii) resonance frequency of the circuit, (iv) bandwidth and (v) quality factor. **(Electric Circuits-I, Punjab Univ.1993)**

Solution. As seen from the given voltage equation,

$\omega = 1000$ rad./s

$X_L = \omega L = 1000 \times 0.01 = 10\ \Omega$

$X_C = 1/\omega C = 10^6 / 1000 \times 100 = 10\ \Omega$

Net $X = X_L - X_C = 10 - 10 = 0$

It is obvious that circuit resonates at the angular frequency of 1000 rad./s

(*i*) $Z = R + jX = 10 + j0 =$ **10 Ω**

(*ii*) Since under resonant conditions, the entire applied voltage drops across R, power dissipated in it is

$P = V^2 / R$ where V is the rms value of the applied voltage.

Since maximum value of the applied voltage is 10 V, its rms value is $10 / \sqrt{2}$ V.

$\therefore\ P = (10/\sqrt{2})^2 / 10 =$ **5 W**

(*iii*) $2\pi f_0 = 1000,\ f_0 = 1000 / 2\pi =$ **159 Hz**

(*iv*) Bandwidth $= R / 2\pi L = 10 / 2\pi \times 0.01 =$ **159 Hz**

(*v*) Quality factors $= f_0 /$ BW $= 159 / 159 = 1$

Example 17.31. *An R.L.C. series circuit consists of a resistance of 1000 Ω, an inductance of 100 mH and a capacitance of 10 μμ F. If a voltage of 100 V is applied across the combination, find (i) the resonance frequency, (ii) Q-factor and (iii) half-power points.*

(Electric Circuits-I, Punjab Univ. 1994.)

Solution. (*i*) $f_0 = \dfrac{1}{2\pi\sqrt{100 \times 10^{-3} \times 10 \times 10^{-12}}}$

$= 159{,}155$ Hz $=$ **159.155 kHz**

(*iii*) The frequencies of the half-power points are given as under:

$f_1 = f_0 - R / 4\pi L = 159{,}155 - 1000 / 4\pi \times 100 \times 10^{-3}$

$= 159155 - 796 =$ **158,359 Hz**

$f_2 = f_0 + R / 4\pi L = 159155 + 1000 / 4\pi \times 100 \times 10^{-3}$

$=$ **159,951 Hz**

These frequencies are shown in Fig. 17.32.

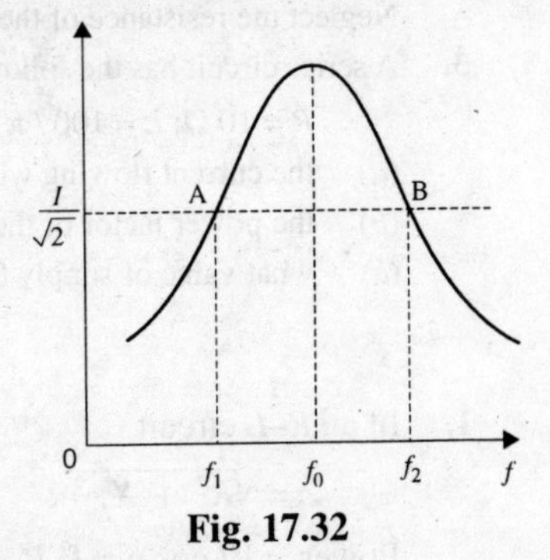

Fig. 17.32

(*ii*) The bandwidth of the circuit is given by

$BW = f_2 - f_1 = 2 \times 796 = 1{,}592$ Hz

$Q = f_0 / BW = 159{,}155 / 1592 =$ **100**

Example 17.32. *A practical resonant circuit has an inductor of 0.24 H and a capacitor of 3 μF, the capacitor can be assumed to be lossless. The resistance of the inductor is 150 Ω, find the resonance frequency, Q-factor and bandwidth.* **(Electrical Engg. A.M.Ae.S.I., Dec 1993)**

Solution. $f_0 = \dfrac{1}{2\pi\sqrt{LC}} = \dfrac{1}{2\pi\sqrt{0.24 \times 3 \times 10^{-6}}} =$ **188 Hz**

Q - factor $= \dfrac{1}{R}\sqrt{\left(\dfrac{L}{C}\right)} = \dfrac{1}{150}\sqrt{\left(\dfrac{0.24}{3 \times 10^{-6}}\right)} =$ **1.88**

Bandwidth $= f_0 / Q = 188 / 1.88 =$ **10**

Example 17.33. *A constant voltage at a frequency of 1 MHz is applied to a choke coil in series with a variable capacitor. When the capacitor is set at 500 pF, the current is maximum. When the capacitance is 600 pF, the current is half the maximum value. Find resistance, inductance and Q-factor of the choke coil.* **(Electrical Engg. A.M.Ae.S.I. June, 1994.)**

Solution. When an R.L.C. circuit is in resonance,

$X_L = X_C$ or $2\pi f L = 1/2\pi f C$ or $4\pi^2 f^2 LC = 1$

$\therefore$ $4\pi^2 (1 \times 10^6)^2 \times 500 \times 10^{-12} \times L = 1$ or $L =$ **0.05 mH**

At resonance, the current drawn is maximum and is given by $I_m = V/R$

When Capacitance is 600 pF

$X_C = 1/2\pi \times 10^6 \times 600 \times 10^{-12} = 265\ \Omega$

$X_L = 2\pi \times 10^6 \times 0.05 \times 10^{-3} = 314\ \Omega$

Net reactance, $X = X_L - X_C = 314 - 265 = 49\ \Omega$

$\therefore$ $Z = i = V/Z$

Now, $I = I_m / 2$ or $V/Z = V/2R$ or $Z = 2R$

$\therefore$ $\sqrt{R^2 + 49^2} = 2R$, $\therefore R = 28.3\ \Omega$

Q-factor of the coil $= 2\pi fL/R$

$= 277 \times 10^6 \times 0.05 \times 10^{-3} / 28.3 =$ **11.1**

Tutorial Problems No. 17.4

1. An a.c. series circuit has a resistance of 10 Ω, an inductance of 0.2 H and a capacitance of 60 μF. Calculate:
(*a*) the resonant frequency, (*b*) the current and (*c*) the power at resonance.
Given that applied voltage is 200 V. **[(*a*) 46 Hz (*b*) 20 A (*c*) 4kW]**

2. A resistor and a capacitor are connected in series with a variable inductor. When the circuit is connected to a 240-V, 50-Hz supply, the maximum current given by varying the inductance is 0.5 A. At this current, the voltage across the capacitor is 250 V. Calculate the value of:
(*a*) the resistance, (*b*) the capacitance and (*c*) the inductance.
Neglect the resistance of the inductor. **[(*a*) 480 Ω (*b*) 6.36 μF (*c*) 1.59 H]**

3. A series circuit has the following characteristics:
$R = 10\ \Omega$; $L = 100/\pi$ mH; $C = 500/\pi$ μF. Find:
(*a*) the current flowing when the applied voltage is 100 V at 50 Hz
(*b*) the power factor of the circuit
(*c*) what value of supply frequency would produce series resonance?
[(*a*) 7.07 A (*b*) 0.707 lead (*c*) 70.71 Hz]

HIGHLIGHTS

1. In an *R–L* circuit

$Z = \sqrt{R^2 + X_L^2}$; $I = V/Z$ and $\tan\phi = X_L / R$

Power $= VI\cos\phi = I^2 R$

2. kVA input of a circuit has two rectangular components
(*i*) active kVA or kilowatt (kW)
(*ii*) reactive kVA or kVAR
$\text{kVA} = \sqrt{\text{kW}^2 + \text{kVAR}^2}$

3. In an *R-C* circuit,

$Z = \sqrt{R^2 + X_C^2}$; $I = V/R$ and $\tan\phi = -X_C / R$

Power $= VI\cos\phi = I^2 R$

4. In an *R-L-C* circuit,

$Z = \sqrt{R^2 + (X_L \sim X_C)^2} = \sqrt{R^2 + X^2}$

$I = V/Z$, $\tan\phi = X/R$

Power $= VI\cos\phi = I^2 R$

5. An *R-L-C* circuit is said to be in electrical resonance when $X_L = X_C$. Under resonant conditions:

(*i*) net reactance $X = X_L \sim X_C = 0$

(*ii*) resonant frequency $f_o = \dfrac{1}{2\pi\sqrt{LC}}$ Hz

(*iii*) circuit current is maximum and equals *V/R* amperes.

(*iv*) voltages across the inductance and capacitance are equal in magnitude although opposite in phase *i.e.* $V_L = V_C$.

(*v*) impedance $Z = R$.

(*vi*) power factor of the resonant circuit is unity.

6. The *Q*-factor of a series circuit is given by the voltage magnification produced in the circuit under resonant conditions.

$$Q\text{-factor} = \frac{1}{R}\frac{\sqrt{L}}{C}$$

OBJECTIVE TESTS—17

A. Fill in the following blanks.

1. In a series *R-L* circuit, current always .... behind the applied voltage.
2. Impedance of an *R-L* circuit is given by the................ sum of its resistance and reactance.
3. Power factor is given by the ratio of circuit resistance and ....
4. In a series *R-L-C* circuit, voltage drops across capacitor and inductor have a phase difference of .... degrees.
5. A series *R-L-C* circuit becomes resonant when algebraic sum of X_L and X_C equals ....
6. Resonance curve shows variation of circuit current with ....
7. Higher the *Q*-factor of a circuit, ... its bandwidth.
8. At half-power frequencies, the current in a series *R-L-C* resonant circuit is—times the maximum value of the current.

B. Answer True or False.

1. In a.c. circuits, product of r.m.s. volts and r.m.s. amperes gives power in watts.
2. The power factor of an a.c. circuit depends only on its resistance.
3. The kVA of an a.c. circuit is given by the vector sum of kW and kVAR.
4. The power factor of an *R-L-C* circuit varies between zero and unity lag or lead.
5. An *R-L-C* circuit has maximum impedance under resonant conditions.
6. The *Q*-factor of an *R-L-C* circuit depends only on circuit resistance and inductance and not on capacitance.
7. At half-power points of an *R-L-C* circuit, current is half its maximum value.
8. The power factor of an a.c. circuit can lie between 0 and ± 1.

C. Multiple Choice Questions.

1. The power in an a.c. circuit is given by

   (*a*) $VI\sin\phi$ (*b*) $I^2 X_L$
   (*c*) $I^2 R$ (*d*) $I^2 Z$.

2. The *kVA* of an a.c. circuit having kW = 80 and kVAR = 60 is

   (*a*) 100 (*b*) 140
   (*c*) 20 (*d*) 53.

3. In a series circuit with $R = 10\ \Omega$, $X_L = 25\ \Omega$ and $X_C = 35\ \Omega$ and carrying an effective current of 5 A, the power dissipated is watt.

   (*a*) $250\sqrt{2}$ (*b*) 250
   (*c*) 500 (*d*) 50

4. Higher the *Q* of an a.c. series circuit,

   (*a*) greater its bandwidth
   (*b*) sharper its resonance curve
   (*c*) broader its resonance curve
   (*d*) narrower its passband.

5. In a series resonant circuit,

   (*a*) $X_L\ X_C$ (*b*) $X_L\ X_C$
   (*c*) $X_L = X_C$ (*d*) $X_L = 1/X_C$.

6. The resonant frequency of a series *R-L-S* circuit depends on

(*a*) *R* (*b*) *L*
(*c*) *C* (*d*) both (*a*) and (*b*)
(*e*) (*a*), (*b*) and (*c*).

7. A series R-L-C circuit has a resonance frequency of 10 Hz and a bandwidth of 5 Hz, the Q factor of the circuit is
(*a*) 0.5 (*b*) 2.5
(*c*) 50 Hz (*d*) none of these

(Elect. Engg. A.M.Ae.S.I., Dec. 1993)

ANSWERS

A. 1. lags 2. algebraic 3. impedance 4. 90 5. zero 6. current 7. narrower 8. 0.707

B. 1. F 2. F 3. T 4. T 5. F 6. F 7. F 8. T

C. 1. *c* 2. *a* 3. *b* 4. *d* 5. *c* 6. *e* 7. *d*

18 PARALLEL A.C. CIRCUITS

18.1. Solving Parallel Circuits

When impedances are joined in parallel, then we have three methods to solve such circuits:

(*a*) *Vector Method*

(*b*) *Admittance Method*

(*c*) *Complex Algebra (Chapter 19)*

18.2. Vector Method

Consider the circuit shown in Fig. 18.1. Here, two reactors *A* and *B* have been joined in parallel across an r.m.s. supply of *V* volt. The voltage across *A* and *B* is the same but currents through them are different.

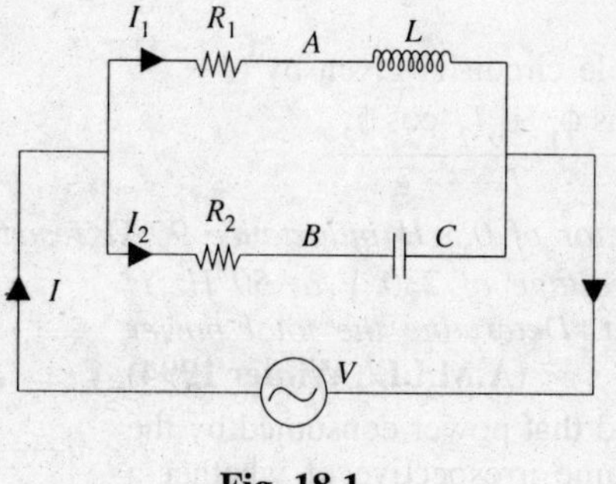

Fig. 18.1

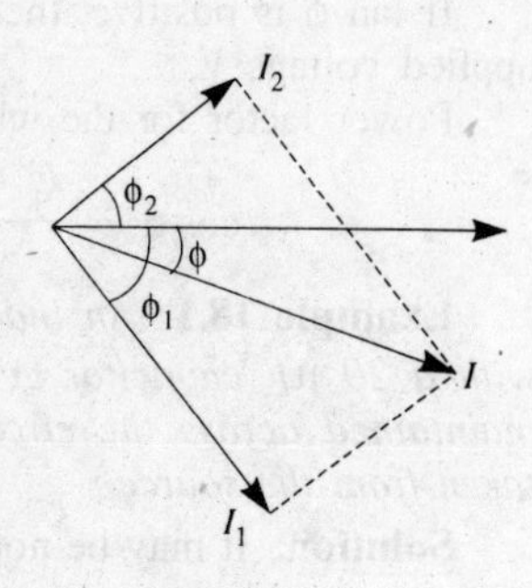

Fig. 18.2

For Branch A

$$Z_1 = \sqrt{(R_1^2 + X_L^2)};$$

$$I_1 = V/Z_1$$

$$\cos \phi_1 = R_1/Z_1$$

or $\quad \phi_1 = \cos^{-1} (R_1/Z_1)$

Current I_1 lags behind the applied voltage by ϕ_1 (Fig. 18.2).

For Branch B

$$Z_2 = \phi (R_2^2 + X_C^2); \qquad I_2 = V/Z_2$$

$$\cos \phi_2 = R_2/Z_2, \qquad \phi_2 = \cos^{-1}(R_2/Z_2)$$

Current I_2 leads *V* by ϕ_2 (Fig. 18.2).

Resultant Current I

The resultant circuit current *I* is the vector sum of the branch currents I_1 and I_2 and can be found (*i*) by using parallelogram law of vectors as shown in Fig. 18.2 or (*ii*) by resolving I_1 and I_2 into their *X*- and *Y*-components (or active and reactive components) respectively and then by combining these components as shown in Fig. 18.3. Method (*ii*) is preferable as it is quick and convenient.

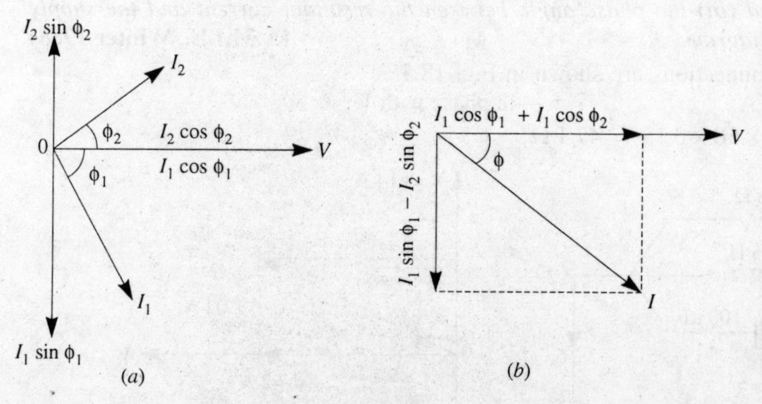

Fig. 18.3

With reference to Fig. 18.3 (*a*), we have

Sum of the active components of I_1 and I_2

$$= I_1 \cos\phi_1 + I_2 \cos\phi_2$$

Sum of reactive components of I_1 and I_2

$$= I_2 \sin\phi_2 - I_1 \sin\phi_1$$

If I is the resultant current and ϕ is its phase angle, then its active and reactive components must be equal to these [Fig. 18.3(*b*)]

$$\therefore \quad I\cos\phi = I_1 \cos\phi_1 + I_2 \cos\phi_2$$

$$\text{and} \quad I\sin\phi = I_2 \sin\phi_2 - I_1 \sin\phi_1$$

$$\therefore \quad I = \sqrt{[(I_1 \cos\phi_1 + I_2 \cos\phi_2)^2 + (I_2 \sin\phi_2 - I_1 \sin\phi_1)^2]}$$

$$\text{and} \quad \tan\phi = \frac{I_2 \sin\phi_2 - I_1 \sin\phi_1}{I_1 \cos\phi_1 + I_2 \cos\phi_2}$$

If $\tan\phi$ is positive, then current leads and if $\tan\phi$ is negative, then current lags behind the applied voltage V.

Power factor for the whole circuit is given by

$$\cos\phi = \frac{I_1 \cos\phi_1 + I_2 \cos\phi_2}{I}$$

Example 18.1. *An inductor of 0.5 H inductance 90 Ω resistance is connected in parallel with a 20 μF capacitor. A voltage of 230 V at 50 Hz is maintained across the circuit. Determine the total power taken from the source.* **(A.M.I.E., Winter 1994)**

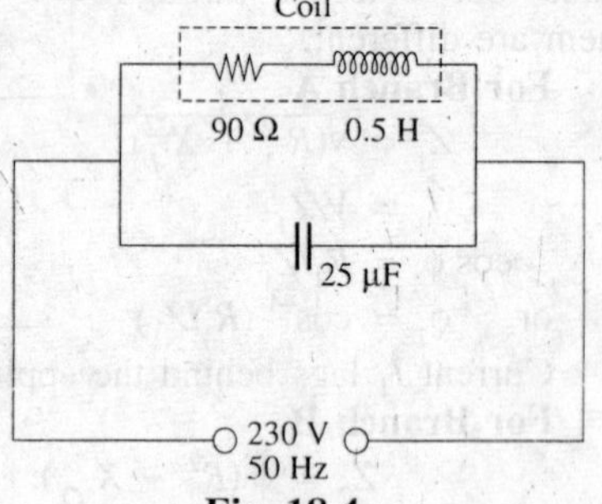

Fig. 18.4

Solution. It may be noted that power consumed by the inductor would remain the same irrespective of whether a capacitor is connected in parallel with it or not. It is so because a pure capacitor does not consume any power.

$$X_L = 2\pi \times 50 \times 0.5 = 157\Omega$$

$$Z = \sqrt{90^2 + 157^2} = 181\Omega$$

$$I = 230/181 = 1.27 \text{ A}$$

power taken from supply

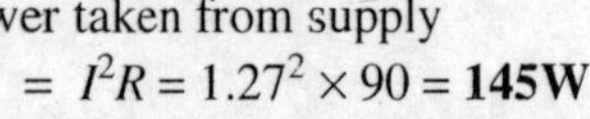

$$= I^2R = 1.27^2 \times 90 = \mathbf{145W}$$

Example 18.2. *A resistance of 50Ω, an inductance of 0.15 H and a capacitance of 100 μF are connected in parallel across a 100-V, 30-Hz supply. Calculate (i) the current in each circuit (ii) the resultant current and (iii) the phase angle between the resultant current and the supply voltage. Draw the phasor diagram.* **(A.M.I.E., Winter 1988)**

Solution. The circuit connections are shown in Fig. 18.5.

(*i*) $I_1 = 100/50 = 2$ A —in phase with V

$X_L = 2\pi f L = 2\pi \times 50 \times 0.15 = 47.1\ \Omega$

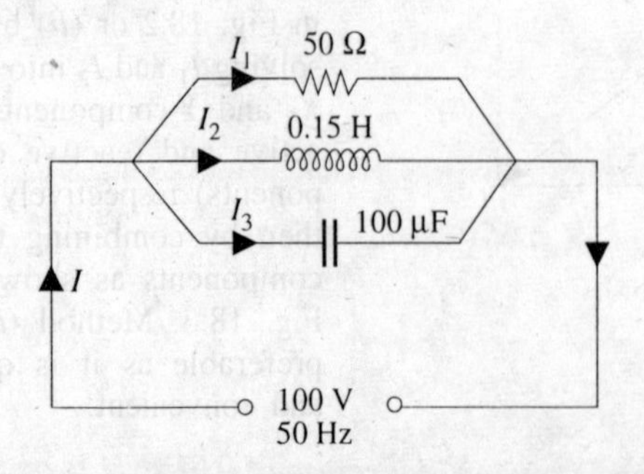

Fig. 18.5

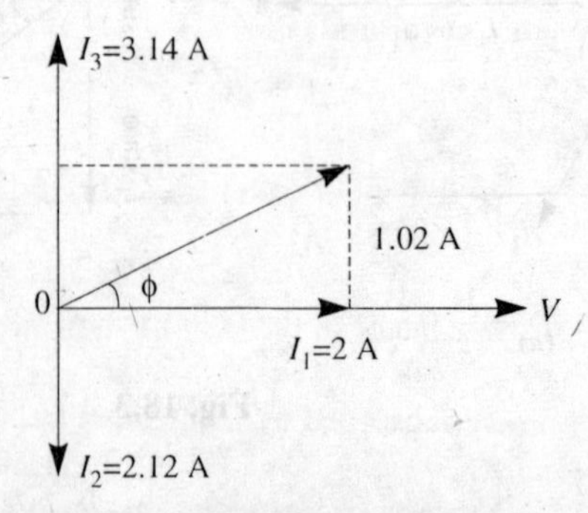

Fig. 18.6

$I_2 = V/X_L = 100/47.1 = \mathbf{2.12\ A}$ —lagging V by 90°

$X_C = 1/2\pi f C = 10^6/2\pi \times 50 \times 100 = 31.8\ \Omega$

$I_2 = V/X_C = 100/31.8 = \mathbf{3.14\ A}$ —lagging V by 90°

(*ii*) As seen from the phasor diagram of Fig. 18.5 (*b*).

$$I = \sqrt{2^2 + 1.02^2} = \mathbf{2.245\ A}$$

(*iii*) $\tan\phi = 1.02/2 = 0.51,\ \phi = \mathbf{27^0}$

Hence, resultant circuit current *leads* the supply voltage by 28°.

Example 18.3. *An inductive coil is connected in parallel across a non-inductive resistance of 30 Ω and this parallel circuit is connected across I-phase, 50 Hz main. The total current taken from the mains is 8 A when the current in the non-inductive resistance is 4 A and that in the inductive coil is 6 A. Calculate (a) the inductance of the coil and (b) power absorbed by the parallel circuit.*

Draw a complete vector diagram for the circuit showing the voltage and currents.

Solution. The circuit is shown in Fig. 18.7 and its vector diagram in Fig. 18.8 where *AB* represents current drawn by 30 Ω non-inductive resistance in phase with the applied voltage and *BD* represents current drawn by the coil lagging behind the supply voltage by ϕ. The supply voltage is obviously = 4 × 30=120 V.

It is seen that current *BD* drawn by the coil has two components:

(*i*) in-phase component *BC* and

(*ii*) quadrature component *CD*.

$$BC^2 + CD^2 = 6^2 \quad ...(i)$$

$$(4 + BC)^2 + CD^2 = 8^2 \quad ...(ii)$$

Substracting (*i*) from (*ii*), we get

$$(4 + BC)^2 - BC^2 = 8^2 - 6^2 = 28$$

$$16 + 8BC = 28 \qquad \therefore \quad BC = 1.5\ \text{A}$$

$$CD = \sqrt{6^2 - 1.5^2} = 5.81\ \text{A}$$

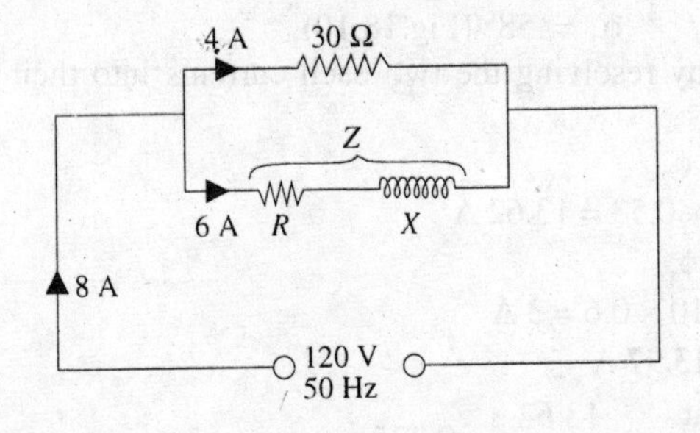

Fig. 18.7

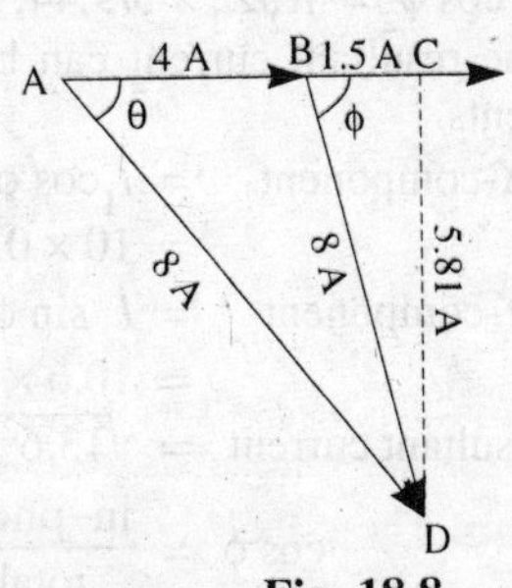

Fig. 18.8

P.f. angle of the coils is ϕ.

Now $\cos\phi = 1.5/6 = 0.25$ and $\sin\phi = 5.81/6$

Impedance of the coil $Z = 120/6 = 20\ \Omega$

Now $\sin\phi = X/Z \qquad \therefore \quad X = Z\sin\phi = 20 \times 5.81/6\ \Omega$

(*a*) $\therefore \quad 2\pi \times 50 \times L = 20 \times 5.81/6$

$$L = 20 \times 5.81/6 \times 314 = \mathbf{0.0617\ H}$$

(*b*) Power factor of the combined parallel circuit is

$$\cos\theta = 5.5/8$$

Power absorbed by the parallel circuit

$$= 120 \times 8 \times 5.5/8 = \mathbf{660\ W}$$

Example 18.4. *A coil having a resistance of 8 Ω and an inductance of 0.0191 H is connected* *in parallel with a capacitor having a cpacitance of 398 μF and resistance of 5Ω. If 100 V at 50 Hz are applied across the terminals of the above parallel circuit, calculate (a) the total current from the supply and (b) its phase angle with respect to the supply voltage. Draw a complete vector diagram for the circuit showing the currents and the supply voltage.*

Solution. The circuit is shown in Fig. 18.9.

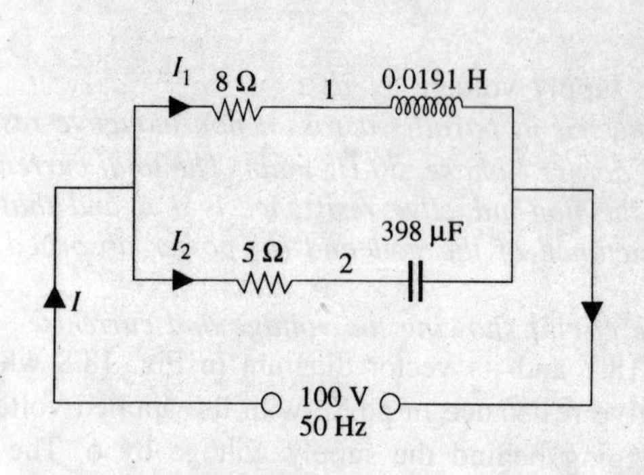

Fig. 18.9

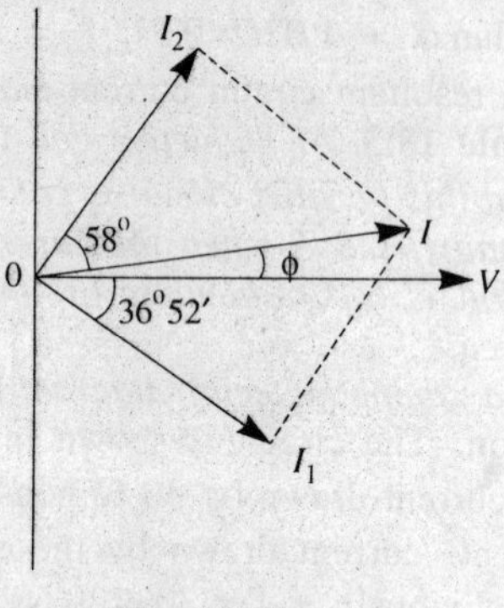

Fig. 18.10

$$X_L = 2\pi f L = 2\pi \times 50 \times 0.0191 = 6\ \Omega$$

$$Z_1 = \sqrt{6^2 + 8^2} = 10\ \Omega; \quad I_1 = V/Z_1 = 100/10 = 10\ \text{A}$$

This current lags behind V by an angle ϕ_1 such that

$$\cos\phi_1 = R_1/Z_1 = 8/10 = 0.8; \qquad \phi_1 = 36°\ 52'$$

$$X_C = 10^6/2\pi \times 50 \times 398 = 8\ \Omega; \qquad R_2 = 5\ \Omega$$

$$Z_2 = \sqrt{5^2 + 8^2} = 9.44\ \Omega \qquad I_2 = 100/9.44 = 10.6\ \text{A}$$

Its angle of lead is given by

$$\cos\phi_2 = R_2/Z_2 \times 5/9.44; \qquad \phi_2 = 58° \text{ (Fig.18.10)}$$

(*a*) The resultant current can be found by resolving the two each currents into their *X*- and *Y*-components.

$$\text{Total } X\text{-component} = I_1 \cos\phi_1 + I_2 \cos\phi_2$$
$$= 10 \times 0.8 + 10.6 \times 0.53 = 13.62\ \text{A}$$

$$\text{Total } Y\text{-component} = I_2 \sin\phi_2 - I_1 \sin\phi_1$$
$$= 10.6 \times 8/9.44 - 10 \times 0.6 = 3\ \text{A}$$

$$\therefore \quad \text{resultant current} = \sqrt{13.6^2 + 3^2} = \mathbf{13.97}\ \text{A}$$

$$\cos\phi = \frac{\text{in–phase current}}{\text{total current}} = \frac{13.62}{13.97} = 0.975$$

$$\therefore \quad \phi = \cos^{-1}(0.975) = 12°\ 48' = \mathbf{12.8°}$$

18.3. Admittance Method

Admittance of a circuit is defined as the reciprocal of its impedance. Its symbol is *Y*.

$$Y = \frac{I}{Z} \qquad \text{or} \qquad Y = \frac{\text{r.m.s. amperes}}{\text{r.m.s. volts}}$$

Its unit is siemens and its symbol is Ψ.

One siemens is defined as the admittance of a circuit having an impedance of 1ohm.

As impedance *Z* has two rectangular components *R* and *X* [Fig. 18.11 (*a*)] similarly, admittance *Y* also has two components as shown in Fig. 18.11 (*b*).

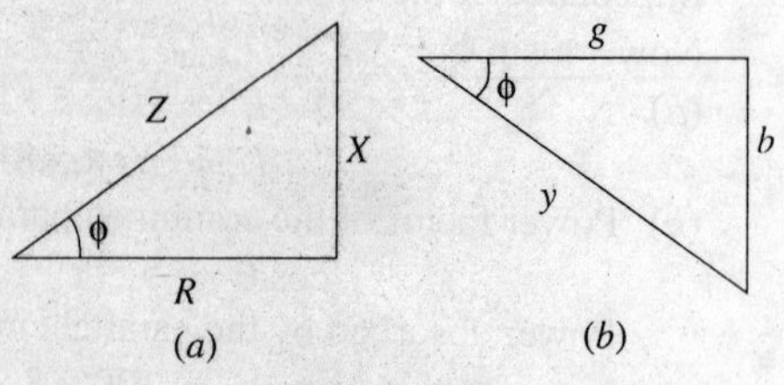

Fig. 18.11

The *X*-component is called *conductance (g) and the Y*-component is known as *suceptance (b)*

Obviously, conductance, $g = Y \cos \phi$ ...Fig. 18.11 (*b*)

$$= \frac{1}{Z} \cdot \frac{R}{Z} = \frac{R}{Z^2} = \frac{R}{R^2 + X^2} \text{ siemens}$$

Similarly, susceptance

$$b = Y \sin \phi = \frac{1}{Z} \cdot \frac{X}{Z}$$

$$= \frac{X}{Z^2} = \frac{X}{R^2 + X^2} \text{ siemens}$$

Admittance $Y = \sqrt{g^2 + b^2}$ as $Z = \sqrt{R^2 + X^2}$

The unit of *g*, *b* and *Y* is siemens. We will regard *capacitive susceptance as positive and inductive susceptance as negative.*

18.4. Application of Admittance Method

Consider the 3-branched circuit of Fig. 18.12. Total conductance is found by merely adding the conductances of the branches. But total susceptance is found by *algebraically* adding the individual susceptance of each branch.

Total conductance, $G = g_1 + g_2 + g_3$

Total susceptance, $B = -b_1 - b_2 + b_3$

Total admittance, $Y = \sqrt{G^2 + B^2}$

Circuit current $I = VY$

Power factor, $\cos \phi = G/Y$

Note. The branch currents are

$I_1 = V_{y1}$

where $y_1 = \sqrt{g_1^{\,2} + (-b_1)^2}$

$I_2 = V_{y2}$

where $y_2 = \sqrt{g_2^{\,2} + (-b_2)^2}$

$I_3 = V_{y3}$

where $y_3 = \sqrt{g_3^{\,2} + b_3^{\,2}}$

Similarly, $\cos \phi_1 = g_1/y_1$; $\cos \phi_2 = g_2/y_2$ and $\cos \phi_3 = g_3/y_3$.

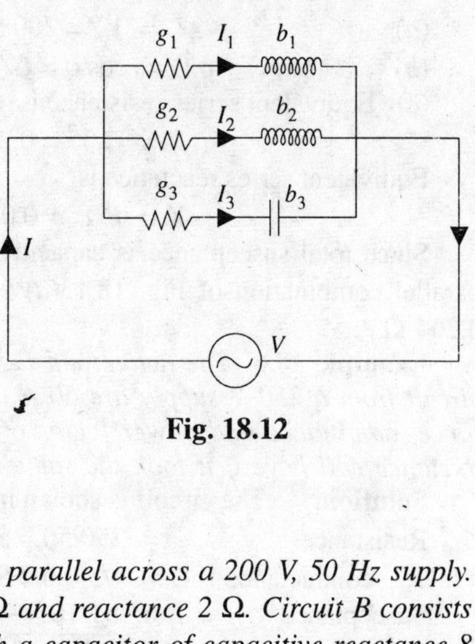

Fig. 18.12

Example 18.5. *A and B are two circuits connected in parallel across a 200 V, 50 Hz supply. Circuit A consists of a choking coil whose resistance is 5 Ω and reactance 2 Ω. Circuit B consists of a non-inductive resistor of 6 Ω connected in series with a capacitor of capacitive reactance 8 Ω. Calculate:*

(*a*) *total current,*
(*b*) *power factor of the combined circuit and*
(*c*) *the resistance and reactance of a series circuit which will take the same current at the same p.f. as the parallel combination.*

(A.M.I.E. , Winter 1994)

Solution. $g_1 = \frac{R_1}{R_1^2 + X_1^2} = \frac{5}{5^2 + 2^2} = 0.1724 \text{ S}$

$$b_1 = -\frac{X_1}{R_1^2 + X_1^2} = -\frac{2}{5^2 + 2^2} = -0.06897 \text{ S}$$

Similarly, $g_2 = \frac{6}{6^2 + 8^2} = 0.06 \text{ S}$

$$b_2 = \frac{8}{6^2 + 8^2} = 0.08 \text{ S}$$

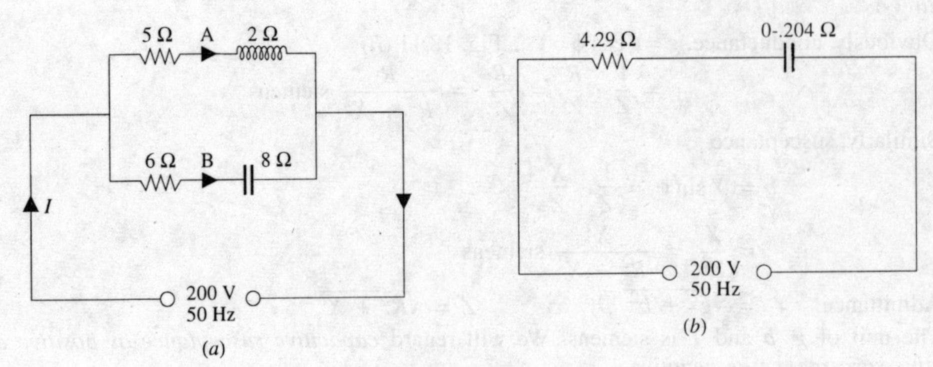

Fig. 18.13

$$G = g_1 + g_2 = 0.1724 + 0.06 = 0.2324 \text{ siemens}$$
$$B = -b_1 + b_2 = -0.06897 + 0.08 = 0.01103 \text{ S (capacitive)}$$
$$Y = \sqrt{G^2 + B^2} = \sqrt{0.2334^2 + 0.01103^2} = 0.2327 \text{ siemens}$$

(*a*) $I = VY = 200 \times 0.2327 = \mathbf{46.5\ A}$

(*b*) p.f. $= \cos\phi = G/Y = 0.2324/0.2327 = \mathbf{0.9987\ lead}$

(*c*) Equivalent series resistance is

$$R = G/Y^2 = 0.2324/0.2327^2 = 4.29\ \Omega$$

Equivalent series reactance is

$$X = B/Y^2 = 0.01103/0.2327^2 = 0.204\ \Omega$$

Since total susceptance is capacitive, this series reactance is capacitive in nature. Hence, the parallel combination of Fig. 18.13(*a*) is equivalent to an *R-C* circuit with R = **4.29** Ω and X_C = **0.204** Ω.

Example 18.6. *The active and lagging reactive components of the current taken by an a.c. circuit from a 250 V supply are 50 A and 25 A respectively. Calculate the conductance, susceptance, admittance and power factor of the circuit. What resistance and reactance would an inductance coil have if it took the same current from mains at the same power factor?*

Solution. The circuit is shown in Fig. 18.14 (*a*)

Resistance $= 250/50 = 5\Omega$; Reactance $= 250/25 = 10\Omega$

∴ conductance $G = 1/5 = \mathbf{0.2\ S}$

Susceptance $B = -1/10 = -\mathbf{0.1\ S}$

Admittance $Y = \sqrt{G^2 + B^2} = \sqrt{0.2^2 + (-0.1)^2}$

$= \sqrt{0.05} = \mathbf{0.224\ S}$

$\cos\phi = G/Y = 0.2/0.224 = \mathbf{0.894\ (lag)}$

Equivalent resistance of series circuit is [Fig.18.14 (*b*)]

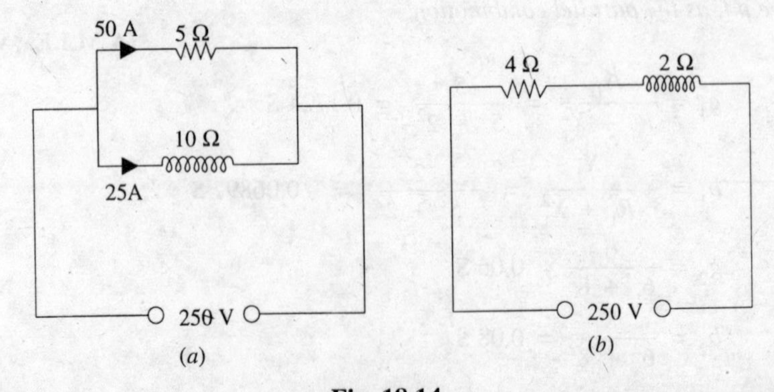

Fig. 18.14

$$R = Z\cos\phi = \frac{1}{Y}\cdot\frac{G}{Y} = \frac{G}{Y^2}$$

$= 0.2/0.224^2 = \mathbf{4\ \Omega}$

$X_L = B/Y^2 = 0.1/0.224^2 = \mathbf{2\ \Omega}$

Example 18.7. *(a) Define conductance, susceptance and admittance with reference to alternating current circuits.*

(b) Calculate the values of conductance and susceptance which when connected in parallel to each other will be equivalent to a circuit comprising of a resistance of 20 Ω in series with a reactance of 10 Ω.

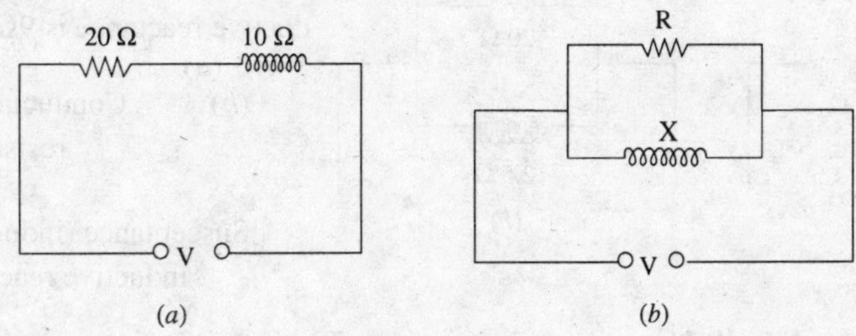

Fig. 18.15

Solution. (*a*) Please refer to Art. 18.3.

(*b*) Conductance of series circuit is

$G = R/Z^2 = 20/(20^2 + 10^2) = 0.04$ siemens

Susceptance of series circuit is

$B = -X/Z^2 = -10/(20^2 + 10^2) = -0.02$ siemens

Since parallel circuit has to have the same conductance and susceptance, the elements of the parallel circuit are:

$R = 1/G = 1/0.04 = \mathbf{25\ \Omega}$; $X = 1/B = 1/0.02 = \mathbf{50\ S}$

Hence, the series circuit of Fig. 18.15 (*a*) is equivalent to the parallel circuit of Fig. 18.15 (*b*).

Example 18.8. *A circuit having a resistance of 6 Ω and an inductive reactance of 8Ω is connected across 200 V, 50 Hz mains in parallel with another circuit having a resistance of 8Ω and a capacitive reactance of 6Ω. Calculate (a) the admittance, conductance and susceptance of the combind circuit and (b) the total current taken from the mains and its power factor. Find also (c) the value of the reactance that must be connected in parallel with this combination to raise the resultant p.f. to unity.*

Solution. **Branch A:**

$R = 6\ \Omega$, $X_L = 8\ \Omega$ $\therefore$ $Z_1^2 = 8^2 + 6^2 = 100$

$g_1 = 6/100 = 0.06$ S ; $b_1 = -8/100 = -0.08$ S

Branch B:

$R_2 = 8\ \Omega$, $X_C = 6\ \Omega$; $Z_2^2 = 8^2 + 6^2 = 100$

$g_2 = 8/100 = 0.08$ S; $b_2 = 6/100 = 0.06$ S

(*a*) $G = g_1 + g_2 = 0.06 + 0.08 = \mathbf{0.14\ S}$

$B = -b_1 + b_2 = -0.08 + 0.06 = \mathbf{-0.02\ S}$ (inductive)

$Y = \sqrt{[0.14^2 + (-0.02)^2]} = \mathbf{0.1414\ S}$

(*b*) $I = VY = 200 \times 0.1414 = \mathbf{28.28\ A}$

p.f. $= \cos\phi = G/Y = 0.14/0.1414 = \mathbf{0.99\ lag}$

(*c*) The circuit possesses a (negative) inductive susceptance of 0.02 S. Hence, to neutralize this we need a (positive) capacitive susceptance of an equal value. So the value of the capacitive reactance required in parallel is

$1/X_C = 0.02$ or $X_C = 1/0.02 = \mathbf{50\ \Omega}$

Example 18.9. *The admittance of a circuit is (0.05-siemens/j 0.08) Find the values of the resistance and inductive reactance of the circuit if they are in (a) in parallel, (b) in series. Draw the circuit diagrams for the two cases.* **(Electrical Engg., A.M.Ae.S.I., Dec.1994.)**

Solution. (*a*) $Y = (0.05 - j\,0.08)$ S. Since susceptance is negative, it is due to inductive reactance.

$$Z = \frac{1}{(0.05 - j\,0.08)} = \frac{0.05 + j\,0.08}{0.05^2 + j\,0.08^2} = \mathbf{5.6 + j\,9}$$

Hence, if the circuit consits of a resistance and an inductive reactance in series, then resistance is 5.6 Ω and inductive reactance is 9Ω as shown in Fig. 18.16 (*a*).

(*b*) Conductance $= 0.05$ S

∴ resistance $= 1/0.05$ $= \mathbf{20\ \Omega}$

Susceptance (inductive) $= 0.08$ S

∴ inductive reactance $= 1/0.08$ $= \mathbf{12.5\ \Omega}$

The equivalent parallel circuit is shown in Fig.18.16 (*b*).

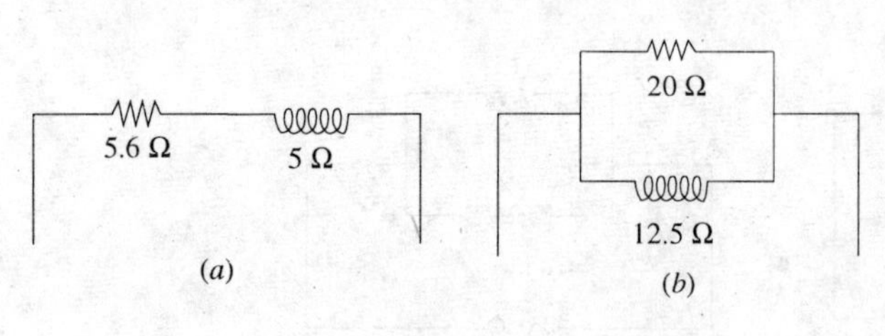

Fig. 18.16

Example 18.10. *Applying admittance method to the series-parallel circuit shown in Fig. 18.17, find*

(*i*) *total impedance in ohms,*
(*ii*) *total supply current and*
(*iii*) *power factor of the entire circuit.*

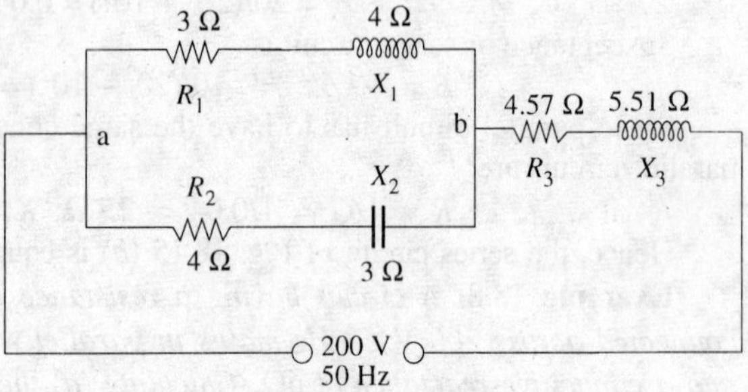

3 Ω 4 Ω
R_1 X_1
a b
4.57 Ω 5.51 Ω
R_3 X_3
R_2 X_2
4 Ω 3 Ω
200 V
50 Hz

Fig. 18.17

Solution.
$Z_1^2 = 3^2 + 4^2 = 25$

$g_1 = 3/25 = 0.12$ S; $b_1 = -4/25 = -0.16$ S

$Z_2^2 = 4^2 + 3^2 = 25$

$g_2 = 4/25 = 0.16$ S; $b_2 = 3/25 = 0.12$ S

$G_{ab} = 0.12 + 0.16 = 0.28$ S

$B_{ab} = -0.16 + 0.12 = -0.04$ S (inductive)

$Y_{ab} = \sqrt{0.28^2 + (-0.04)^2} = \sqrt{0.06} = 0.283$ S

The parallel circuit may now be reduced to an equivalent series circuit as shown in Fig.18.18.

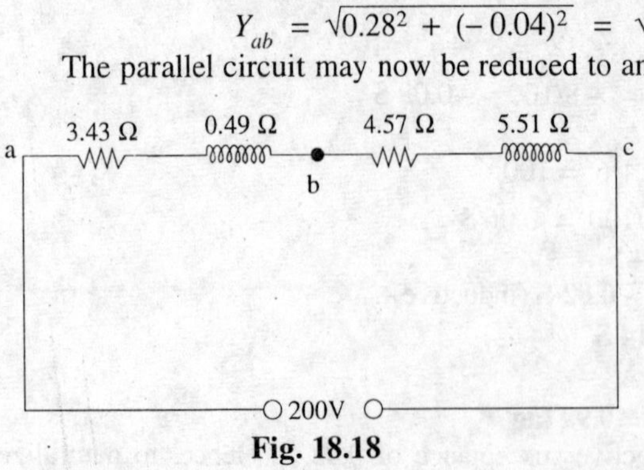

$Z_{ab} = 1/Y_{ab} = 1/0.283 = 3.5\ \Omega$

$R_{ab} = G_{ab}.Z_{ab2} = 0.283 \times 3.52 = 3.43\ \Omega$

$X_{ab} = B_{ab}.Z_{ab2} = 0.04 \times 3.52 = 0.49\ \Omega$

3.43 Ω 0.49 Ω 4.57 Ω 5.51 Ω
a b c
200V

Fig. 18.18

As seen from Fig.18.18

$R_{ac} = 3.43 + 4.57 = 8\ \Omega$

$X_{ac} = 0.49 + 5.51 = 6\ \Omega$

(*i*) $Z_{ac} = \sqrt{82 + 6^2} = \mathbf{10\ \Omega}$

(*ii*) $I_{ac} = 200/10 = \mathbf{20\ A}$

(*iii*) p.f. $= \cos\phi = R_{ac}/Z_{ac} = 8/10 = $ **0.8 (lag)**

Example 18.11. (*a*) *A coil which has 6 Ω resistance and 25.5 mH inductance is energised from a 440-V, 50-Hz supply. Calculate the current.*

(*b*) *A capacitor is then connected in parallel with the coil so that the overall p.f. is unity. Calculate the capacitance of the capacitor.*

(c) What current will be flowing in the main supply cable when the capacitor is connected in the circuit and what is capacitor current?

Solution. The circuit is shown in Fig. 18.19.

(*a*) $X_L = 2\pi \times 50 \times 25.5 \times 10^{-3} = 8\ \Omega$

$Z = \sqrt{6^2 + 8^2} = 10\ \Omega$

$I_L = 400/10 = \mathbf{40\ A}$; $\phi_L = \cos^{-1}(0/10) = 53.1°$

As shown in phasor diagram of Fig. 18.19, the coil current lags the voltage by 53.1°.

(*b*) If the combined p.f. is to be unity, then current I_C drawn by the capacitor should be such that when vectorially combined with I_L, it should give the resultant current I in phase with V. A' seen from Fig. 18.20, in that case

$$I_C = I_L \sin\phi_L = 40 \times \sin 53.1° = 40 \times 0.8 = 32\ \text{A}$$

Now $$I_C = \frac{V}{X_C} = \frac{V}{1/\omega C} = V\omega C$$

$\therefore$ $32 = 400 \times 2\pi \times 50 \times C$

$\therefore$ $C = \mathbf{255\ \mu F}$

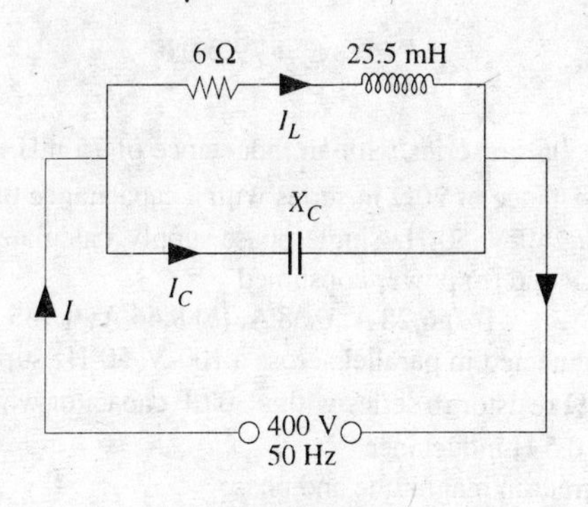

Fig. 18.19

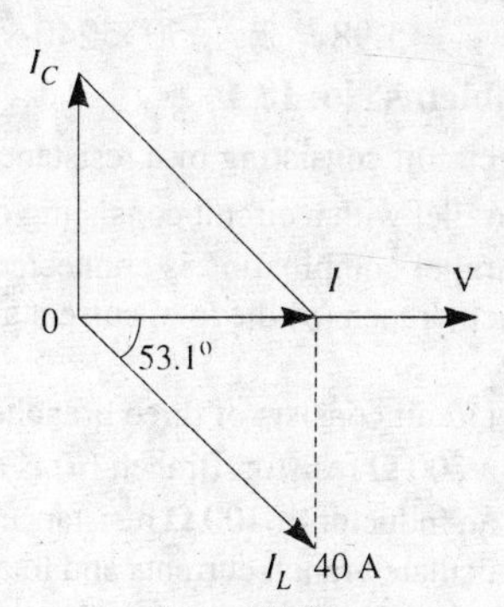

Fig. 18.20

(*c*) When the capacitor is connected to the circuit, the current in the main supply cable is

$$I = I_L \cos\phi_L = 40 \times 0.6 = \mathbf{24\ A}$$

As found in (*b*) above, $I_C = \mathbf{32\ A}$

Note: If required, the kVA rating of the capacitor may be found. It is = 32 × 400 = 12,800 VA = 12.8 kVA. This kVA is reactive, since capacitor does not absorb any power.

Example 18.12. *With the use of a capacitor in parallel with an induction motor, it is desired to raise the power factor to 0.9. The induction motor is a 240-V single-phase motor and takes 20 A at 0.75 lagging power factor. Calculate the capacitance of the capacitor to be used.*

Solution. The current taken by the motor is I_m = 20 A and lags behind the applied voltage (as shown in Fig. 18.22) by an angle $\phi_m = \cos^{-1}(0.75) = 41°\ 24'$.

When a capacitor is connected in parallel with the motor as shown, then the current I_C drawn by it should be such that when combined with the motor current I_m it should produce a resultant current I which lags behind the applied voltage by an angle.

$$\phi = \cos^{-1}(0.9) = 25°\ 48'$$

It is obvious from the vector diagram of Fig. 18.22 that

$$I \cos\phi = I_m \cos\phi_m \qquad \text{(each} = AB)$$

or $I \times 0.9 = 20 \times 0.75$ $\therefore$ $I = 16.7$ A

Also $I_m \sin\phi_m - I \sin\phi = I_C$ $(\because ID = BD - BI)$

$$20 \sin 41°\ 24' - 16.7 \sin 25°\ 48' = I_C$$

$$I_C = 13.25 - 7.27 = 5.98\ \text{A}$$

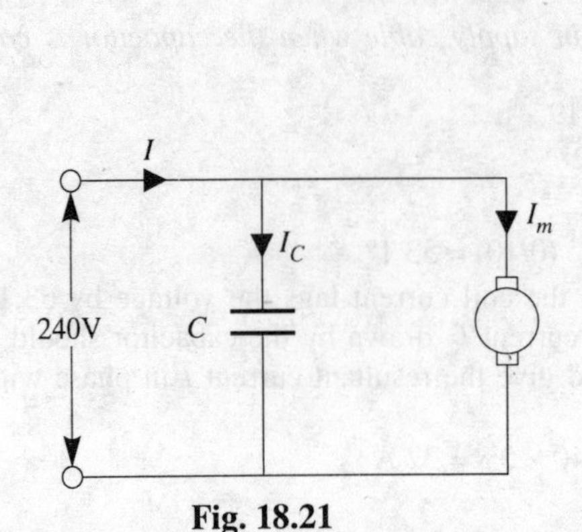

Fig. 18.21

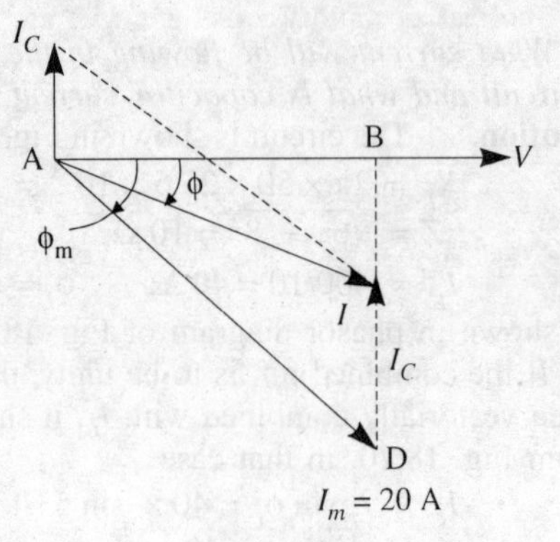

Fig. 18.22

Now $\quad I_C = \dfrac{V}{1/\omega C} = V\omega C = 2\pi f CV$

Assuming a frequency of 50 Hz, we have

$5.98 = 2\pi \times 50 \times 240 \times C \qquad \therefore \quad C = \mathbf{79.3\ \mu F}$

Tutorial Problems No. 18.1

1. A circuit consisting of a resistance of 30Ω in series with an inductance of 75 mH is connected in parallel with a circuit consisting of resistance of 20Ω in series with a capacitance of 100μF. If the parallel combination is connected to a 240-V, 50-Hz single-phase supply, calculate (*a*) current in each branch (*b*) the total current and p.f. and (*c*) power consumed.
[(*a*) 6.28 A; 6.38 A, (*b*) 8.46 A; 0.985 and (*c*) 2 kW]

2. A circuit consists of three branches connected in parallel across a 100-V, 50-Hz supply. Branch (*i*) is a 200 Ω resistor. Branch (*ii*) is a 30 Ω resistor in series with a 30 μF capacitor while Branch (*iii*) is an inductor of 100 Ω resistance and 0.5 H inductance.
Calculate branch currents and total current in magnitude and phase
[0.5 A; 0.855 A; 0.535 A; 1.21 A; 13.5° leading]

3. During a test on a circuit possessing resistance and inductance in series, the following figures were obtained:
Power, 1.5 kW; voltage 200 V; frequency 50 Hz; current 10 A. Find the resistance and inductance of the circuit.
If a cpacitor of 10 5μF capacitance is connected in parallel with this circuit, what is the p.f. of the new load? **[15Ω; 42 mH; unity]**

4. An inductive circuit in parallel with a non-inductive resistance of 20Ω is connected across a 50-Hz supply. The currents through the inductive circuit and the non-inductive resistance are respectively 4.3 A and 2.7 A and the current taken from the supply is 5.8 A. Find:
(*a*) the power absorbed by the inductive circuit,
(*b*) its inductance and
(*c*) the p.f. of the combined circuit. **[(*a*) 72.8 W (*b*) 0.038 H (*c*) 0.72]**

5. .A circuit consists of two branches, *A* and *B*, connected in parallel. Branch *A* is a coil having a resistance of 20Ω and an inductance of 0.12 H. Branch *B* consists of a non-inductive resistor of 50Ω. Determine the kW, kVA and kVAR taken by this circuit from a 230-V, 50-Hz supply.
[1.64 kW, 1.97 kVA, 1.09 kVAR]

6. A 230-V single-phase circuit takes 20 A at 0.7 power factor lagging. Determine the capacitance of a capacitor to be connected in parallel with the circuit in order to raise the overall p.f. to 0.95 lagging. **[134 μF]**

7. A mercury-vapour lamp unit is connected to a 240-V, 50-Hz, single-phase supply. The unit consists of the lamp in series with an inductor of negligible resistance and having a capacitor connected across this series combination for the purpose of power factor correction. If the lamp itself is purely resistive and under normal operating conditions has a voltage across it of 175 V and a current of 2 A flowing through it, calculate:

(*a*) the inductance of the inductor

(*b*) capacitor value needed to make the supply power factor unity

(*c*) the power taken from the supply.

[(*a*) 0.262 (*b*) 18.3 μF (*c*) 350 W]

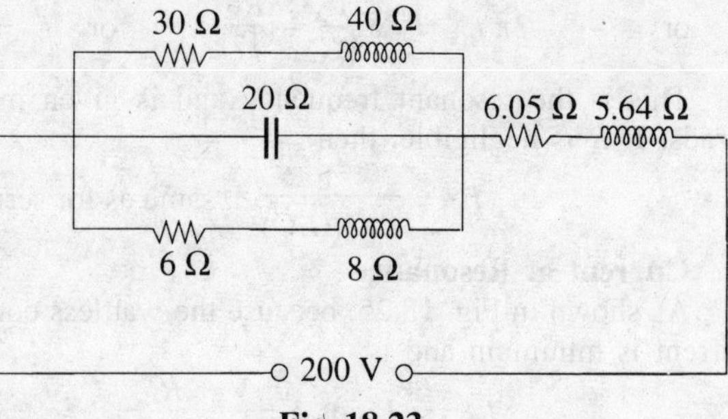

Fig. 18.23

8. Applying admittance method to the circuit shown in Fig.18.23, find

(*i*) total circuit impedance

(*ii*) total current taken from the mains

(*iii*) p.f. of the combined circuit.

[(*i*) 20 Ω(*ii*) 10 A (*iii*) 0.8(lag)]

18.5. Resonance in Parallel Circuits

We will consider the practical case of a coil in parallel with a capacitor as shown in Fig.18.24. Such a circuit is said to be in electrical resonance when the reactive (or wattless) component of line current becomes zero. The frequency at which this happens is known as resonant frequency.

The vector diagram for the circuit is shown in Fig. 18.25.

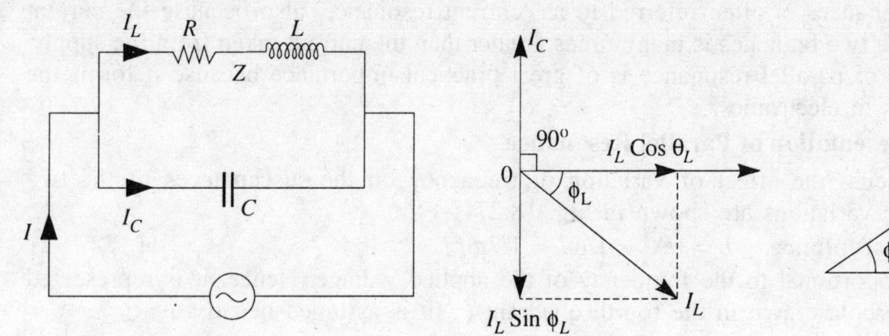

Fig. 18.24 **Fig. 18.25** **Fig. 18.26**

Net reactive or wattless component

$$= I_C - I_L \sin \phi$$

As at resonance its value is zero

$\therefore \quad I_C - I_L \sin \phi = 0 \quad$ or $\quad I_L \sin \phi = I_c$

Now $\quad I_L = V/Z; \sin \phi = X_L/Z \qquad I_C = V/X_C$

Hence, condition for resonance becomes

$$\frac{V}{Z} \times \frac{X_L}{Z} = \frac{V}{X_C} \quad \text{or} \quad X_L \times X_C = Z^2$$

Now, $\quad X_L = \omega L, \qquad X_C = \dfrac{1}{\omega C}$

$$\therefore \quad \frac{\omega L}{\omega C} = Z^2 \quad \text{or} \quad \frac{L}{C} = Z^2 \qquad \ldots(i)$$

$$\text{or} \quad \frac{L}{C} = R^2 + X_{L2}$$

$$\text{or} \quad \frac{L}{C} = R^2 + (2\pi f_0 L)^2 \quad \text{or} \quad (2\pi f_0 L)^2 = \frac{L}{C} - R^2$$

or $\quad 2\pi f_0 = \sqrt{\dfrac{1}{LC} - \dfrac{R^2}{L^2}}$ $\quad$ or $\quad f_0 = \dfrac{1}{2\pi}\sqrt{\dfrac{1}{LC} - \dfrac{R^2}{L^2}}$

This is the resonant frequency and is given in Hz if R is in ohms, L in henrys and C in farads. If R is negligible, then

$$f_0 = \frac{1}{2\pi\sqrt{(LC)}} \text{ ... same as for series resonance}$$

Current at Resonance

As shown in Fig. 18.25, because the wattless component of the line current is zero, the circuit current is minimum and is

$$I_{min} = I_L \cos\phi = \frac{V}{Z} \cdot \frac{R}{Z}$$

or $\quad I_{min} = \dfrac{VR}{Z^2}$

Putting the value of $Z^2 = L/C$ from (i) above, we get

$$I_{min} = \frac{VR}{L/C} = \frac{V}{L/CR}$$

The denominator 'L/CR' is known as the *equivalent or dynamic impedance* of the parallel circuit at resonance. It should be noted that this impedance is resistive only. Since current is minimum at resonance, L/CR must then represent the maximum impedance of the circuit.

Current at resonance is minimum, hence such a circuit (when used in radio work) is sometimes known as *rejector* circuit because it rejects (or takes minimum current of) that frequency to which it resonates. This resonances is often referred to as 'current resonance' also because the current circulating *between* the two branches is many times greater than the current taken from the supply.

The phenomenon of parallel resonance is of great practical importance because it forms the basis of tuned circuits in electronics.

18.6. Graphic Representation of Parallel Resonance

We will now discuss the effect of variation of frequency on the susceptances of the two parallel branches. The variations are shown in Fig. 18.27.

1. Inductive Susceptance: $\quad b = 1/X_L = 1/\omega L = 1/2\pi f L$

It is inversely proportional to the frequency of the applied voltage. Hence, it is represented by a rectangular hyperbola drawn in the fourth quadrant ($\because$ it is assumed negative).

2. Capacitive Susceptance $\quad b = 1/X_C = \omega C = 2\pi f C$

It increases with the increase in frequency of the applied voltage. Hence it is represented by a straight line drawn in the first quadrant ($\because$ It is assumed positive).

3. Net Susceptance B

It is the difference of the two susceptances and is represented by the dotted hyperbola. At point A, net susceptance is zero.

4. Admittance

Variations of admittance (Y) are as shown in Fig. 18.27. It has large values for frequencies lower as well as higher than the resonant frequency f_0. For $f < f_0$, Y is inductive whereas for $f > f_0$, it is capacitive in nature. It has minimum value of $Y = G$ at $f = f_0$. Its value would have been zero if R had been zero (because in that case $G = 0$).

5. Current

It varies in exactly the same manner as Y. It has minimum value at point A (where $f = f_0$) and equals $VY = VG$. Its value would have been zero if R had been equal to zero.

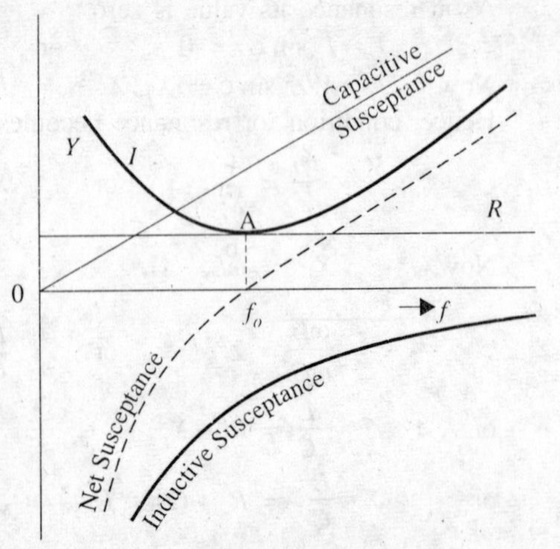

Fig. 18.27

Obviously, below resonant frequency (corresponding to point *A*), inductive susceptance predominates, hence line current lags behind the applied voltage. But for frequencies above the resonant frequency, capacitive susceptance predominates, hence line current leads.

18.7. Points to Remember

Following points about parallel resonance should be noted and compared with those about series resonance. For parallel resonance:

1. net susceptance is zero *i.e.* $1/X_C = X_L/Z^2$ or $X_L \times X_C = Z^2$ or $Z^2 = L/C$.
2. reactive or wattless component of line current is zero.
3. dynamic impedance = L/CR ohm
4. line current at resonance is minimum and equals $\dfrac{V}{L/CR}$ and is in phase with the applied voltage.
5. power factor of the circuit is unity.

Comparison of Series and Parallel Resonance Circuits

Item	*Series circuit*	*Parallel circuit*
Impedance at resonance	Minimum	Maximum
Current at resonance	Minimum $= V/R$	Maximum $= V/(L/CR)$
Effective impedance	R	L/CR
Power factor at resonance	Unity	Unity
Resonant frequency	$1/2\pi\sqrt{(LC)}$	$\dfrac{1}{2\pi}\sqrt{\left(\dfrac{1}{LC} - \dfrac{R^2}{L^2}\right)}$
It magnifies	Voltage	Current
Magnification is	$\omega L/R$	$\omega L/R$

18.8. Q-factor of a Parallel Circuit

It is defined as *'the ratio' of the branch circulating current to the line current'* or as *'the current magnification'*.

Let the circulating current between capacitor and coil be I_C.

Then $\quad Q\text{-factor} = I_C/I$

Now $\quad I_C = V/X_C = V/(1/\omega C) = V\omega C$

$\quad I = V/(L/CR)$

$$Q\text{-factor} = V\omega C \div \frac{V}{L/CR} = \frac{\omega L}{R} = \frac{2\pi f_0 L}{R}$$

(same as for series circuit)

Now, resonant frequency when R is negligible is, $f_0 = \dfrac{1}{2\pi\sqrt{(LC)}}$

Putting this value above, we get

$$Q\text{-factor} = \frac{2\pi L}{R} \cdot \frac{1}{2\pi\sqrt{(LC)}} = \frac{1}{R}\sqrt{\left(\frac{L}{C}\right)}$$

It should be noted that, in series circuits, Q-factor gives the voltage magnification whereas in parallel circuits, it gives current magnification.

Example 18.13. *Explain the meaning of resonance in a parallel circuit. Show that in a parallel circuit, the resistance of the inductor affects the frequency of resonance. Is this true in a series resonant circuit?*

A resonant circuit consists of a 4 μF capacitor in parallel with an inductor of 0.25 H having a resistance of 50 Ω. Calculate the frequency of resonance.

Solution. The resonance frequency is

$$f_0 = \frac{1}{2\pi}\sqrt{\frac{1}{LC} - \frac{R^2}{L^2}}$$

It is seen that f_o is dependent on R. When R is large, f_o is reduced. In a series circuit $f_o = 1/2\pi\sqrt{LC}$. Obviously, f_o does not depend on R.

$$f_0 = \frac{1}{2\pi}\sqrt{\frac{10^6}{4 \times 0.25} - \frac{50^2}{0.25^2}} = \mathbf{156\ Hz}$$

Example 18.14. *An impedance coil having an inductance of 0.04 H and a resistance of 5Ω is placed in parallel with a variable condenser. The circuit is connected to a 100 V, 50 Hz supply. For what value of capacitor will the line current be minimum and what are the values of two branch currents, the total power and power factor of the circuit?*

(Electrical Engg., A.M.Ae.S.I. , June.1993)

Solution. As is well-known, a two-branched circuit draws minimum line current under resonant condition which occurs when $C = L/Z^2$

Now $X_L = 2\pi f L = 2\pi \times 50 \times 0.04 = 12.57\ \Omega$

$Z = \sqrt{R^2 + X_L^{\ 2}} = \sqrt{5^2 + 12.57^2} = 13.53\ \Omega$

$\therefore \quad C = L/Z^2 = 0.04/13.53^2 = \mathbf{218 \times 10^{-6}\ F}$

Power factor of the circuit under resonant conditions is unity.

$I_{min} = VCR/L = 100 \times 218 \times 10^{-6} \times 5/0.04 = 2.725$ A

Total power drawn $= VI_{min} = 100 \times 2.725 = \mathbf{272.5\ W}$

$I_L = V/Z = 100/\sqrt{13.53} = \mathbf{7.4\ A}$

$$I_C = \frac{V}{X_C} = \frac{V}{1/\omega C} = V\omega C = 2\pi f VC = 2\pi \times 50 \times 100 \times 218 \times 10^{-6} = \mathbf{6.85\ A}$$

Example 18.15. *A series combination of a capacitance C and resistance R is shunted by an inductive coil having resistance R and inductance L and the combination is connected to an a.c. source. Show that the impedance of the circuit will be independent of frequency if* $R = \sqrt{L/C}$.

(A.M.I.E., Winter 1990)

Solution. The impedance of the circuit of Fig.18.28 is

$$\mathbf{Z} = \frac{\mathbf{Z}_1\mathbf{Z}_2}{\mathbf{Z}_1 + \mathbf{Z}_2} = \frac{(R + j\omega L)(R - j/\omega C)}{(R + j\omega L) + (R - j/\omega C)}$$

$$= \frac{R^2 + (L/C) - jR(\omega L - 1/\omega C)}{2R + j(\omega L - 1/\omega C)}$$

Putting $R = \sqrt{L/C}$

or $R^2 = L/C$, we get

$$Z = \frac{R^2 + R^2 + jR(\omega L - 1/\omega C)}{2R + j(\omega L - 1/\omega C)}$$

$$= R\left[\frac{2R + j(\omega L - 1/\omega C)}{2R + j(\omega L - 1/\omega C)}\right]$$

or $Z = R$

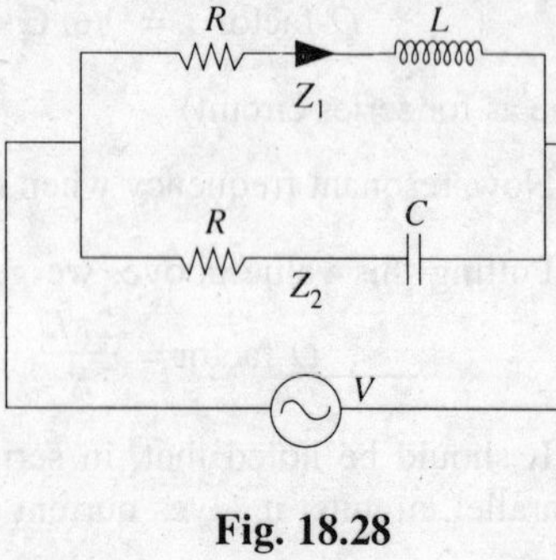

Fig. 18.28

Tutorial Problems No. 18.2

1. A circuit consisting of a coil of inductance 10 mH and resistance 300Ω connected in parallel with a certain capacitor resonates at 48 kHz. Determine the size of this capacitor and the dynamic impedance of the circuit.

[1090 μμ F; 30.6 k Ω]

2. A coil of inductance 31.8 mH and resistance of 10Ω is connected in parallel with a capacitor

across a 250-V, 50-Hz supply. Determine the value of the capacitance if no reactive current is taken from the supply. **[159 μF]**

3. A series circuit, consisting of a coil ($R = 30\ \Omega$, $L = 0.5\ H$) and a capacitor resonates at a frequency of 48 Hz. Calculate the capacitance of a capacitor which, when connected in parallel with this circuit, will increase the resonant frequency to 60 Hz. Calculate in the latter case the total current and the current in each branch of the circuit if the line voltage is 100 V. **[32.8 μF; 0.546 A; 135 A; 1.235 A]**

4. A 100 V, 80 W lamp is to be operated on 230 V, 50 Hz a.c. supply. Calculate the inductance of the choke required to be connected in series with the lamp for this operation. The lamp can be taken as equivalent to a non inductive resistance. If the p.f. the lamp circuit is to be improved to unity, calculate the value of the capacitor which is to be connected across the circuit. **[L = 0.823 H; 10 μF]**

5. A coil of resistance 10Ω and inductance 0.1 H is connected in parallel with a capacitor of 200 μF. Calculate the frequency at which the circuit behaves as non-inductive resistance and the current taken if supply voltage is 250 V. **[31.83 Hz; 5A]**

HIGHLIGHTS

1. Admittance of a circuit is defined as the reciprocal of its impedance. Its unit is siemens (S).

$$Y = \sqrt{G^2 + B^2}$$

2. Admittance has two rectangular components:
(*i*) conductance (G) and (*ii*) susceptance (B)

3. $G = R/Z^2$; $B = X/Z^2$

4. Equivalent resistance and reactance of a series circuit are given by

$$R = G/Y^2 \quad \text{and} \quad X = B/Y^2$$

5. A parallel a.c. circuit is said to be in resonance when it does not draw any reactive current. Resonant frequency is

$$f_0 = \frac{1}{2\pi}\sqrt{\frac{1}{LC} - \frac{R^2}{L^2}}$$

6. Condition for parallel resonance is $Z^2 = L/C$

OBJECTIVE TESTS—18

A. Fill in the following blanks

1. Total current drawn by a parallel a.c. circuit is the.............. sum of branch currents.
2. Admittance of an a.c. circuit is given by the............. of its impedance.
3. Unit of admittance is ....
4. Like impedance, admittance also has .... rectangular components.
5. Capacitive susceptance is taken as ....
6. In a parallel a.c. circuit, total susceptance is the .... sum of branch susceptances.

B. Answer True or False.

1. The unit of admittance, conductance and susceptance is siemens.
2. The admittance of a parallel a.c. circuit is the vector sum of its conductance and susceptance.
3. Current drawn by a parallel a.c. circuit is given by the product of the component voltage and its admittance.
4. A parallel a.c. circuit is said to be in electrical resonance when active component of the line current drawn by it becomes zero.
5. A parallel a.c. circuit draws maximum current when in resonance.
6. When in resonance, the power factor of a parallel a.c. circuit is unity.

C. Multiple Choice Questions.

1. The power factor of a parallel a.c. circuit is given by
(*a*) G/B (*b*) G/Y
(*c*) G/Y^2 (*d*) Y/G
2. A parallel a.c. circuit has a conductance of 0.6 S and a susceptance of 0.8 S. Its admittance is — siemens.
(*a*) 0.14 (*b*) 0.75

(*c*) 1.0 (*d*) 1.33

3. The equivalent series reactance of a parallel a.c. circuit is given by

(*a*) $X = G/Y^2$ (*b*) $X = G/Y$

(*c*) $X = B/Y$ (*d*) $X = B/Y^2$

4. A parallel a.c. circuit possesses a net susceptance of-0.02 siemens. The capacitive reactance required to be connected in parallel for raising the circuit power factor to unity is — ohm.

(*a*) 50 (*b*) 0.02

(*c*) 2500 (*d*) 25

5. When a parallel a.c. circuit is in resonance, it

(*a*) draws maximum current

(*b*) offers minimum impedance

(*c*) is called a rejector circuit

(*d*) has no branch currents.

6. The line current drawn by a parallel a.c. circuit when in resonance is

(*a*) wholly wattless

(*b*) zero

(*c*) much greater than circulatory branch current

(*d*) much less than circulatory branch current

7. The impedance offered by a parallel L.C. circuit is given by

(*a*) L/CR (*b*) LC

(*c*) LCR (*d*) LC/R

where R is the resistance and L inductance of the inductive coil, and C is the capacitor

(Elect.Engg.A.M.Ae.S.I. Dec.1994.)

8. Two voltage sources $5\angle 20°$ V and $5\angle 200°$ V are connected in parallel and feed a resistance of 10 ohm. The currents through the resitance is

(*a*) 0.5 A (*b*) 1.0 A

(*c*) 0 A (*d*) none of these

(Elect.Engg., A.M.Ae.S.I., June 1994)

ANSWERS

A. 1. vector 2. reciprocal 3. siemens 4. two 5. positive 6. algebraic

B. 1. T 2. T 3. T 4. F 5. F 6. T

C. 1. *b* 2. *c* 3. *d* 4. *a* 5. *c* 6. *d* 7. *a* 8. *c*

19 COMPLEX ALGEBRA AND A.C. CIRCUITS

19.1. Mathematical Representation of Vectors

There are various forms or methods of representing vector quantities all of which enable those operations which are carried on graphically in a vector diagram, to be performed analytically. The various methods are :

1. *Symbolic Notation*—according to which a vector quantity is expressed algebraically in terms of its rectangular components. Hence, this form of representation is also known as rectangular or Cartesian form of notation or representation.
2. *Trigonometrical Form*
3. *Exponential Form*
4. *Polar Form*

19.2. Symbolic Notation

A vector can be specified in terms of its *X*-component and *Y*-component. For example, the vector OV_1 (Fig.19.1) may be completely described by stating that its horizontal component is a_1 and vertical component is b_1. But, instead of stating this verbally, we may express symbolically as

$$V_1 = a_1 + jb$$

where symbol *j*, known as an operator, indicates that component b_1 is perpendicular to component a_1 and that the two terms are not to be treated like terms in any algebraic expression. The vector written in this way is said to be written in '*complex form*'. In mathematics, a_1 is known as real component and b_1 as imaginary component but in electrical engineering these are known as *in-phase* (or active) and *quadrature* (or reactive) components respectively.

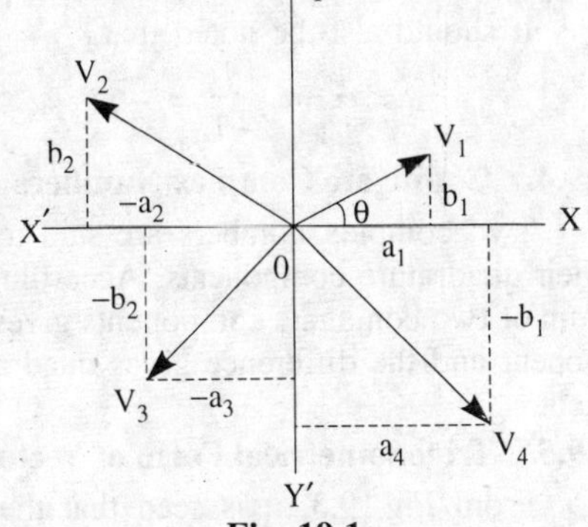

Fig. 19.1

The other vectors OV_2, OV_3, and OV_4 can similarly be expressed in the form:

$V_2 = -a_2 + jb_2$; $V_3 = -a_3 - jb_3$; $V_4 = +a_4 - jb_4$

It should be noted that in this book, a vector quantity would be represented by letters in heavy type and its numerical or scalar value by the same letter in ordinary type*. Other methods adopted for indicating a vector quantity are to put a dot above or below the letter such as V or V

The numerical value of vector V_1 is $\sqrt{a_1^2 + b_1^2}$

Its angle with *X*-axis is given by $\phi = \tan^{-1}(b_1/a_1)$

19.3. Significance of Operator j

The letter *j* used in the above expression is a symbol of an operation. Just as symbol ×, +, √, ∫ etc. are used with numbers for indicating certain operations to be performed on those numbers, similarly symbol *j* is used to indicate the counter-clockwise rotation of a vector through 90°. It

*The magnitude of a vector is sometimes called 'modulus' and is represented by | V | or *V*.

is assigned a value of $\sqrt{(-1)}$*. The double operation of j on a vector rotates it counter-clockwise through 180° and hence reverses its sense because

$$j = j^2 = [\sqrt{(-1)}]^2 = -1$$

When operator j is operated on vector V, we get the new vector j**V** which is displaced by 90° in counter-clockwise direction from **V** (Fig. 19.2). Further application of j will give j^2 **V** = −**V** as shown.

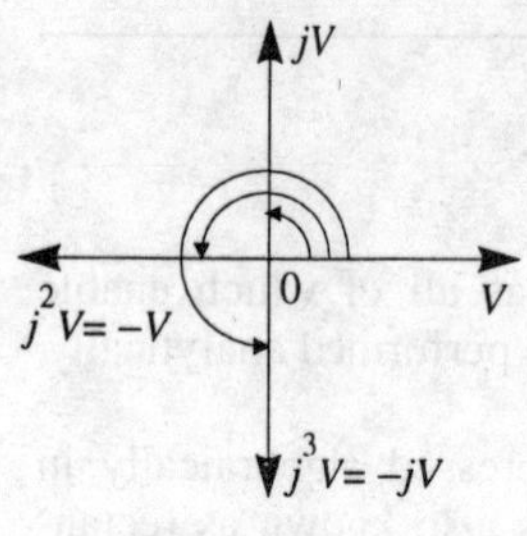

Fig. 19.2

If the operator j is applied to the vector j^2 **V**, the result is j^3 **V** = − j**V**. The vector j^3**V** is 270° couter-clockwise from the reference axis and is directly opposite to j**V**. If the vector j^2**V** is, in turn, operated on by j, the result will be

$$j^4\mathbf{V} = [\sqrt{(-1)}]^4\ \mathbf{V} = \mathbf{V}$$

Hence, it is seen that successive applications of the operator j to the vector **V** produce successive 90° steps of rotation of the vector in the counter-clockwise direction without in any way affecting the magnitude of the vector.

It will also be seen from Fig. 19.2 that the application of $-j$ to **V** yields $-j$**V** which is a vector of identical magnitude but rotated 90° *clockwise* from **V**.

Summarising the above, we have

$$j = 90°\text{ CCW rotation} = \sqrt{(-1)}$$
$$j^2 = 180°\text{ CCW rotation} = [\sqrt{(-1)}]^2 = -1$$
$$j^3 = 270°\text{ CCW rotation} = [\sqrt{(-1)}]^3 = -\sqrt{(-1)} = -j$$
$$j^4 = 360°\text{ CCW rotation} = [\sqrt{(-1)}]^4 = +1$$
$$j^5 = 450°\text{ CCW rotation} = [\sqrt{(-1)}]^5 = \sqrt{(-1)} = j$$

It should also be noted that

$$\frac{1}{j} = \frac{1}{j^2} = \frac{j}{-1} = -j$$

19.4. Conjugate Complex Numbers

Two complex numbers are said to be conjugate if they differ only in the algebraic sign of their quadrature components. Accordingly, the numbers $(a + jb)$ and $(a - jb)$ are conjugate. The sum of two conjugate components gives in phase (or active) component and the difference gives quadrature (or reactive) component.

19.5. Trigonometrical Form of Vector Representation

From Fig.19.3, it is seen that the X-component of **V** is V cos θ and Y- component is V sin θ. Hence, we can represent the vector **V** in the form

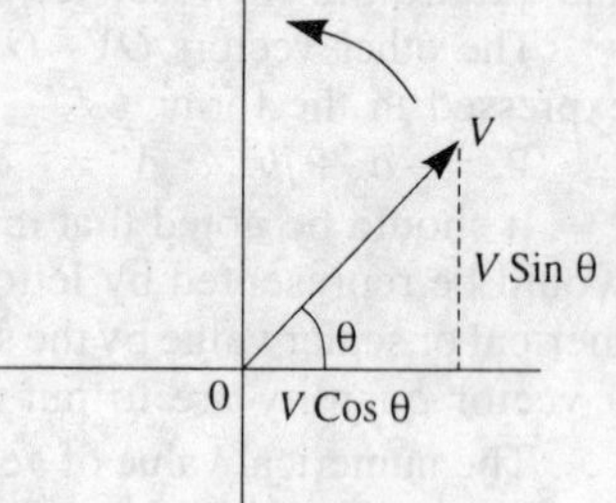

Fig. 19.3

$$\mathbf{V} = V\cos\theta + jV\sin\theta$$
$$= V(\cos\theta + j\sin\theta)$$

This is equivalent to the rectangular form **V** = $a + jb$ because

$a = V\cos\theta$ and $b = V\sin\theta$

In general

$$\mathbf{V} = V(\cos\theta \pm j\sin\theta)$$

19.6. Exponential Form of Vector Representation

It can be proved that

$$e^{\pm j\theta} = (\cos\theta \pm j\sin\theta)$$

This equation is known as Euler's equation after the famous mathematician of 18th century, Leonard Euler.

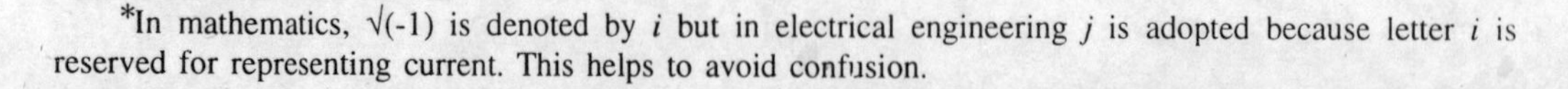
*In mathematics, $\sqrt{(-1)}$ is denoted by i but in electrical engineering j is adopted because letter i is reserved for representing current. This helps to avoid confusion.

This equation follows directly from an inspection of Maclaurin's* series expansionsof sin θ, cos θ and $e^{j\theta}$

When expanded into series form

$$\cos\theta = 1 - \frac{\theta^2}{\angle 2} + \frac{\theta^4}{\angle 4} - \frac{\theta^6}{\angle 6} + \ldots$$

$$\sin\theta = \theta - \frac{\theta^3}{\angle 3} + \frac{\theta^5}{\angle 5} - \frac{\theta^7}{\angle 7} + \ldots$$

$$e^{j\theta} = 1 + j\theta + \frac{(j\theta)^2}{\angle 2} + \frac{(j\theta)^3}{\angle 3} + \frac{(j\theta)^4}{\angle 4} + \frac{(j\theta)^5}{\angle 5} + \frac{(j\theta)^6}{\angle 6} + \ldots$$

Keeping in mind that

$$j^2 = -1, j^3 = -j, j^4 = 1, j^5 = j, j^6 = -1$$

we get

$$e^{j\theta} = \left(1 - \frac{\theta^2}{\angle 2} + \frac{\theta^4}{\angle 4} - \frac{\theta^6}{\angle 6} + \ldots\right) + j\left(\theta - \frac{\theta^3}{\angle 3} + \frac{\theta^5}{\angle 5} - \frac{\theta^7}{\angle 7} + \ldots\right)$$

$$\therefore \quad e^{j\theta} = \cos\theta + j\sin\theta$$

Similarly, it can be shown that

$$e^{-j\theta} = \cos\theta - j\sin\theta$$

Hence, $\mathbf{V} = V(\cos\theta \pm j\sin\theta)$ or $\mathbf{V} = Ve^{\pm j\theta}$

This is known as exponential form of representing vector quantities. It represents a vector of numerical value V and having a phase angle of ±θ with the reference axis.

19.7. Polar Form of Representation

The expression $V(\cos\theta + j\sin\theta)$ is written in the simplified form of $V\ \theta$. In this expression, V represents the magnitude of the vector and θ its inclination (in *CCW* direction) with the *X*-axis. For angles in *CW* direction, the expression becomes $V\ -\theta$. In general, the expression is written as $V \angle \pm\theta$. *It may be pointed out here that* $V \pm \theta$ *is simply a short-hand or symbolic style of writing* $Ve^{\pm j\theta}$. Also, this form is purely conventional and does not possess the mathematical elegance of the other various forms of vector representation given above.

Summarizing, we have the following alternative ways of representing vector quantities:

1. Rectangular Form (or complex form) $\mathbf{V} = a \pm jb$
2. Trigonometrical Form $\mathbf{V} = V(\cos\theta \pm j\sin\theta)$
3. Exponential Form $\mathbf{V} = Ve^{\pm j\theta}$
4. Polar Form (conventional) $\mathbf{V} = V\angle\theta \pm \theta$

Example 19.1. *Write the equivalent exponential and polar forms of vector 3 + j4. How will you illustrate the vector by means of a vector diagram?*

Solution. With reference to Fig.19.4, magnitude of the vector is

$$= \sqrt{3^2 + 4^2} = 5$$

$$\tan\theta = 4/3; \quad \theta = \tan^{-1}(4/3) = 53.1°$$

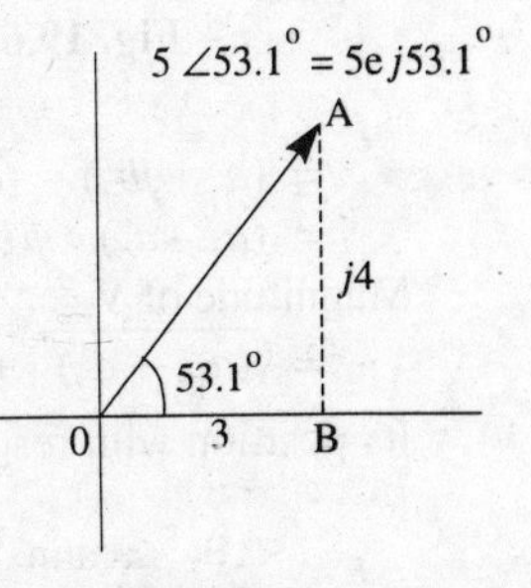

Fig. 19.4

*Functions like cos θ, sin θ and $e^{j\theta}$ etc. can be expanded into series form with the help of Maclaurin's Theorem. The theorem states :

$$f(\theta) = f(0) + \frac{f'(0)\theta}{1} + \frac{f''(0)\,\theta^2}{\angle 2} + \frac{f'''(0)\,\theta^3}{\angle 3} + \ldots$$

where $f(\theta)$ is the function of θ which is to be expanded, $f(0)$ is the value of the function when θ = 0, $f'(0)$— is the value of first derivative of $f(\theta)$ when θ = 0; $f''(0)$ is the value of second derivative of the function $f(\theta)$ when θ = 0 etc.

∴ Exponential form

$$= 5\, e^{j53.1°}$$

The angle may also be expressed in radians.

Polar form $= 5\angle 53.1°$

Example 19.2. *A vector is represented by 20 $e^{-j2\pi/3}$. Write the various equivalent forms of the vector and illustrate by means of a vector diagram, the magnitude and position of the above vector.*

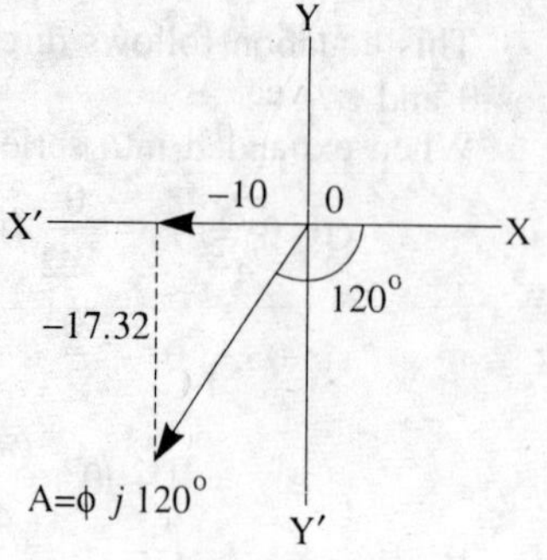

Fig. 19.5

Solution. The vector is drawn in a direction making an angle of $2\pi/3 = 120°$ in the clockwise direction (Fig. 19.5). The clockwise direction is taken because the angle is negative.

(*i*) **Rectangular form**

$$a = 20 \cos(-120°) = -10$$

$$b = 20 \sin(-120°) = -17.32$$

expression is $= -10 - j\,17.32$

(*ii*) **Polar form** is $20\angle 120°$

19.8. Addition and Subtraction of Complex Quantities

Rectangular form is best suited for addition and subtraction of vector quantities.

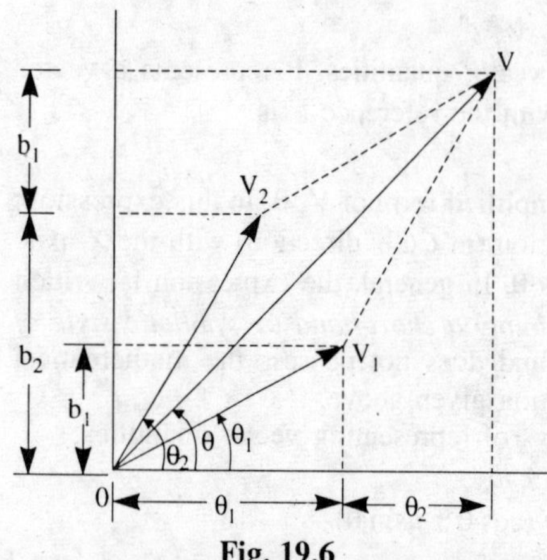

Fig. 19.6

Suppose we are given two vector quantities

$$\mathbf{V}_1 = a_1 + jb_1 \text{ and } \mathbf{V}_2 = a_2 + jb_2$$

and it is required to find their sum and difference.

Addition

$$V = \mathbf{V_1} + \mathbf{V_2} = a_1 + jb_1 + a_2 + jb_2$$

$$= (a_1 + a_2) + j(b_1 + b_2)$$

The magnitude of the resultant vector **V** is

$$= \sqrt{(a_1 + a_2)^2 + (b_1 + b_2)^2}$$

The position of **V** with respect to *X*-axis is

$$\theta = \tan^{-1}\left(\frac{b_1 + b_2}{a_1 + a_2}\right)$$

A graphic representation of the addition process is shown in Fig. 19.6.

Subtraction

$$V = V_1 - V_2$$

$$= (a_1 + jb_1) - (a_2 + jb_2)$$

$$= (a_1 - a_2) + j(b_1 - b_2)$$

Magnitude of **V**

$$= \sqrt{(a_1 - a_2)^2 + (b_1 - b_2)^2}$$

Its position with respect to *X*-axis is given by the angle

$$\theta = \tan^{-1}\left(\frac{b_1 - b_2}{a_1 - a_2}\right)$$

The graphic representation of the process of subtraction is shown in Fig. 19.7.

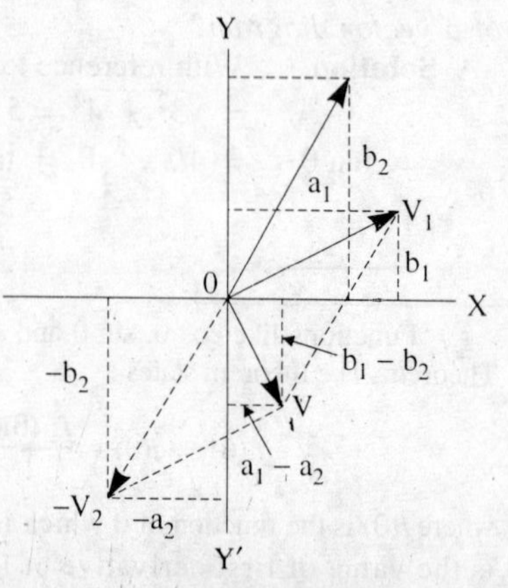

Fig. 19.7

19.9. Multiplication and Division of Complex Quantities

Multiplication and division of vector becomes very simple and easy *if they are represented in the polar or exponential form.* As will be shown below, the rectangular form of representation is not well suited for this process.

1. **Multiplication**—*Rectangular Form*

Let the two vectors be given by

$$\mathbf{A} = a_1 + jb_1 \quad \text{and} \quad \mathbf{B} = a_2 + jb_2$$

$$\begin{aligned}\mathbf{A} \times \mathbf{B} &= \mathbf{C} = (a_1 + jb_1)(a_2 + jb_2)\\ &= a_1a_2 + j^2b_1b_2 + j(a_1b_2 + b_1a_2)\\ &= (a_1a_2 - b_1b_2) + j(a_1b_2 + b_1a_2) \qquad (\because\ j^2 = -1)\end{aligned}$$

The magnitude of $\mathbf{C} = \sqrt{[(a_1a_2 - b_1b_2)^2 + (a_1b_2 + b_1a_2)^2]}$

Its angle with respect to X-axis is given by

$$\theta = \tan^{-1}\left(\frac{a_1 b_2 + b_1 a_2}{a_1 a_2 - b_1 b_2}\right)$$

2. **Division**—*Rectangular Form*

$$\frac{\mathrm{A}}{\mathrm{B}} = \frac{a_1 + jb_1}{a_2 + jb_2} = \frac{(a_1 + jb_1)(a_2 - jb_2)}{(a_2 + jb_2)(a_2 - jb_2)}$$

Both the numerator and denominator have been multiplied by the conjugate of $(a_2 + jb_2)$ *i.e.* by $(a_2 - jb_2)$

$$\therefore \quad \frac{\mathrm{A}}{\mathrm{B}} = \frac{(a_1 a_2 + b_1 b_2) + j(b_1 a_2 - a_1 b_2)}{a_2^2 + b_2^2}$$

$$= \frac{a_1 a_2 + b_1 b_2}{a_2^2 + b_2^2} + j\,\frac{b_1 a_2 - a_1 b_2}{a_2^2 + b_2^2}$$

The magnitude and the angle with respect to X-axis can be found in the same way as given above.

As will be noted, both results are somewhat awkward, but unfortunately, there is no easier way to perform multiplication in rectangular form.

3. **Multiplication**—*Polar Form*

Let $\mathbf{A} = a_1 + jb_1 = A\angle\alpha = Ae^{j\alpha}$ where $\alpha = \tan^{-1}(b_1/a_1)$

$\mathbf{B} = a_2 + jb_2 = B\angle\beta = Be^{j\beta}$ where $\beta = \tan^{-1}(b_2/a_2)$

$\therefore \quad \mathbf{AB} = A\angle\alpha \times B\angle\beta = AB(\alpha + \beta)$*

or $\quad \mathbf{AB} = Ae^{j\alpha} \times B\,e^{j\beta} = AB\,e^{j(\alpha + \beta)}$

Hence, product of any two vectors A and B is given by another vector equal in length to $A \times B$ and having a phase angle equal to the sum of angles of A and B.

4. **Division**—*Polar Form*

$$\frac{\mathrm{A}}{\mathrm{B}} = \frac{A\angle\alpha}{B\angle\beta} = \frac{A}{B}\angle(\alpha - \beta)$$

Hence, the quotient $\mathbf{A} \div \mathbf{B}$ is another vector having a magnitude of $A \div B$ and having phase angle equal to the angle of **A** minus the angle of **B.**

$$\text{Also} \quad \frac{\mathrm{A}}{\mathrm{B}} = \frac{Ae^{j\alpha}}{Be^{j\beta}} = \frac{A}{B}\,e^{j(\alpha - \beta)}$$

As seen, the division and multiplication become extremely simple if vectors are represented in their polar or exponential form.

Example 19.3. *Add the following vectors given in rectangular form and illustrate the process graphically.*

---

* $\mathbf{A} = A(\cos\alpha + j\sin\alpha)$ and $\mathbf{B} = B(\cos\beta + j\sin\beta)$

$$\begin{aligned}AB &= AB(\cos\alpha\cos\beta + j\sin\alpha\cos\beta + j\cos\alpha\sin\beta + j^2\sin\alpha\sin\beta)\\ &= AB[(\cos\alpha\cos\beta - \sin\alpha\sin\beta) + j(\sin\alpha\cos\beta + \cos\alpha\sin\beta)]\\ &= AB[\cos(\alpha + \mathrm{B}) + j\sin(\alpha + \beta)] = \mathbf{AB}\ \angle(\alpha + \beta)\end{aligned}$$

$A = 16 + j12, \quad B = -6 + j10.4$

Solution. $\mathbf{A} + \mathbf{B} = \mathbf{C} = (16 + j12) + (-6 + j\,10.4)$

$= 10 + j22.4$

$\therefore$ Magnitude of $\mathbf{C} = \sqrt{(10^2 + 22.4)^2}$

$= 24.5$ units

Slope of $\mathbf{C} = \theta = \tan^{-1}\sqrt{(22.4/10)}$

$= 65.95°$

The vector addition is shown in Fig. 19.8

$\alpha = \tan^{-1}(12/16) = 36.9°$

$\beta = \tan^{-1}(-10.4/6) = -240°$ or $120°$

The resultant vector is found by using parallelogram law of vectors (Fig. 19.8)

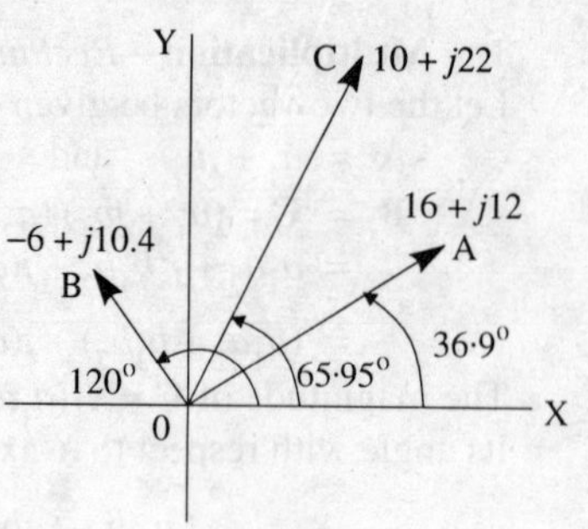

Fig. 19.8

Example 19.4. *Subtract the following given vectors from one another.*

$A = 30 + j\,52 \text{ and } B = -39.5 - j\,14.36$

Solution. $\mathbf{A} - \mathbf{B} = \mathbf{C} = (30 + j\,52) - (39.5 - j\,14.36)$

$= 69.5 + j66.36$

$\therefore$ magnitude of $\mathbf{C} = \sqrt{(69.5^2 + 66.36^2)} = 96$

Slope of $\mathbf{C} = \tan^{-1}(66.36/69.5) = 43.6°$; $\mathbf{C} = 43.6°$

Similarly $\mathbf{B} - \mathbf{A} = -69.5 - j\,66.36$

$= 96\text{A}$ or $96 \angle -136.4°$

Example 19.5. *Given the following two vectors:*

$A = 20 \angle 60°, B = 5 \angle 30°$

Perform the following indicated operations and illustrate graphically. (*i*) *A* × *B and* (*ii*) *A/B*.

Solution. $\text{A} \times \text{B} = \text{C} = 20 \angle 60° \times 5 \angle 30° = 100 \angle 90°$

Vectors are shown in Fig. 19.9.

(*ii*) $\dfrac{\mathbf{A}}{\mathbf{B}} = \dfrac{20 \angle 60°}{5 \angle 30°} = 4 \angle 30°$ (Fig. 19.10)

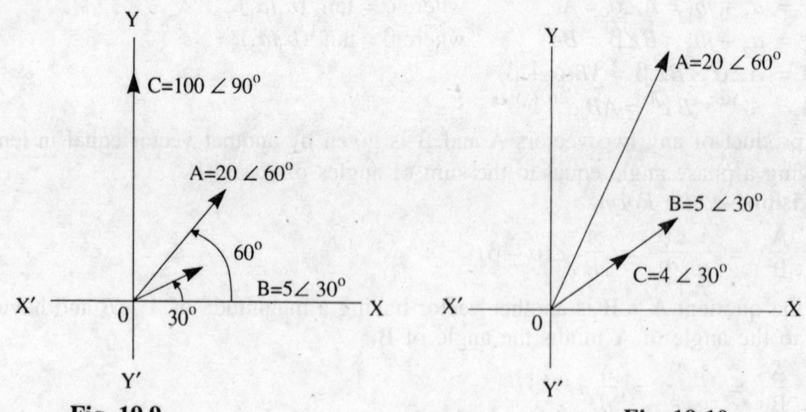

Fig. 19.9

Fig. 19.10

Example 19.6. *The following three vectors are given:*

$A = 20 + j\,20, \quad B = 30\angle 120° \text{ and } C = 10 + j0$

Perform the following indicated operations:

(*i*) $\dfrac{AB}{C}$ *and* (*ii*) $\dfrac{BC}{A}$

Solution. Rearranging all three vectors in polar form, we get

$\mathbf{A} = 28.3\angle 45°, \quad \mathbf{B} = 30 -\angle\, 120°, \quad \mathbf{C} = 10\angle 0°$

(*i*) $\frac{\mathbf{AB}}{\mathbf{C}} = \frac{28.3 \angle 45° \times 20 \angle -120°}{10\angle 0°} = 84.9 \angle -75°$

(*ii*) $\frac{\mathbf{BC}}{\mathbf{A}} = \frac{30 \angle -120° \times 10\angle 45°}{28.3 \angle 45°} = 10.6 \angle -165°$

19.10. Powers and Roots of Vectors

(*a*) **Powers**

Suppose it is required to find the cube of the vector $3\angle 15°$. For this purpose, the vector has to be multiplied by itself three times

$\therefore \quad (3 \angle 15)^3 = 3 \times 3 \times 3 \angle (15° + 15° + 15°) = 27\angle 45°$

In general

$$A^n = A^n \angle n\alpha$$

Hence, *nth* power of vector **A** is a vector whose magnitude is A^n and whose phase angle with respect to *X*-axis is $n\alpha$.

It is also clear that

$$\mathbf{A}^n\mathbf{B}^n = A^nB^n \angle(n\alpha + n\beta)$$

(*b*) **Roots**

It is clear that $\sqrt[3]{(8 \angle 45°)} = 2 \angle 15°$

It general, $\sqrt[n]{A} = \sqrt[n]{A} \angle \alpha/n$

Hence, *nth* root of a vector **A** is a vector whose magnitude is $\sqrt[n]{A}$ and whose phase angle with respect to *X*-axis is α/n

19.11. Complex Algebra Applied to Series Circuit

Consider the series circuit of Fig. 19.11 (*a*). The impedance of the circuit may be written as

$$Z = R + j X_L$$

The numerical value of impedance is $|\mathbf{Z}| = \sqrt{R^2 + X_L^2}$

Its phase angle with respect to the reference axis is

$$\phi_L = \tan^{-1} (X_L / R)$$

It may be expressed in polar form as $Z = Z \angle\phi$

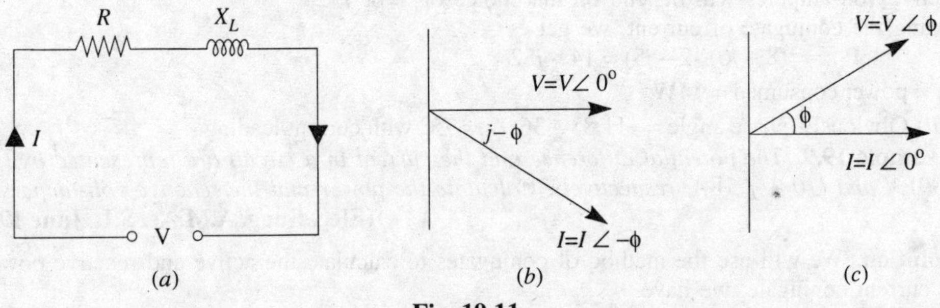

Fig. 19.11

(*i*) If we take voltage vector along the reference axis [Fig.19.11(*b*)] then it may be expressed as

$$\mathbf{V} = V \angle 0°$$

$$\therefore \quad = \frac{\mathbf{V}}{\mathbf{Z}} = \frac{V \angle 0°}{Z \angle \phi°} = \frac{V}{Z} \angle -\phi°$$

It shows that current vector lags behind the voltage vector by 0°. The numerical value of current is $|\mathbf{I}| = V/Z$.

(*ii*) However, if current vector, is taken along reference *X*-axis [Fig.19.11(*c*)], then it may be written as

$\mathbf{I} = I \angle 0°$

Then $\mathbf{V} = \mathbf{IZ} = I \angle 0° \times Z \angle \phi° = IZ \angle \phi°$

It shows that voltage vector is $\phi°$ ahead of the current vector in the *CCW* direction as shown.

Example 19.7. *In a given R-L circuit, R = 35 Ω and L = IH. And the current flowing through the series circuit if a 50-Hz voltage V = 110 ∠30° volts is applied across the circuit.*

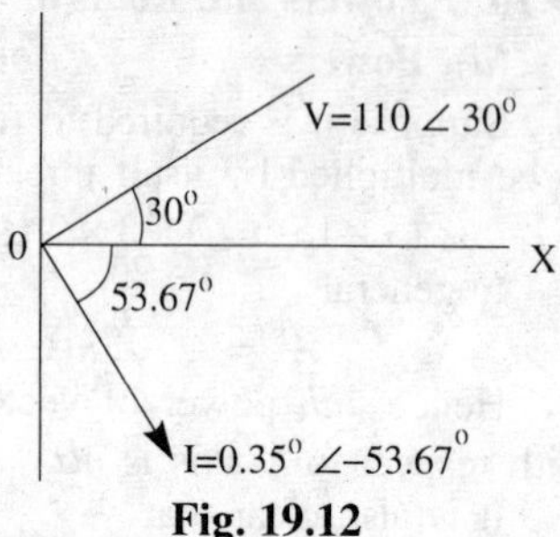

Fig. 19.12

Solution. $R = 35\Omega, \quad X_L = \alpha L$

$X_L = 1 \times 314 = 314\Omega$

$Z = 35 + j\,314$

$= \sqrt{(35^2 + 314^2)}\angle \tan^{-1}(314/35)$

$= 315.9\angle 83.67°$

$\therefore \quad \mathbf{I} = \dfrac{\mathbf{V}}{\mathbf{Z}} = \dfrac{110 \angle 30°}{315.9/83.67°}$

$= \mathbf{0.35\angle{-53.67°}}$

Example 19.8. *An alternating voltage (8 + j 6) V is applied to a series A.C. circuit and the current flowing is (–2 + j 5) A. Find (i) the impedance, (ii) the power consumed and (iii) the phase angle.*

Solution. $\mathbf{V} = 8 + j\,6 = 10 \angle 36.9°$

$\mathbf{I} = -2 + j\,5 = 5.39 \angle \tan^{-1} - 5/2$

$= 5.39 \angle (180° - 68.2°) = 5.39 \angle 111.8°$

(i) $\therefore$ impedance $= \mathbf{Z} = \dfrac{\mathbf{V}}{\mathbf{I}} = \dfrac{10 \angle 36.9°}{5.39 \angle 111.8°} = 1.86 \angle -75°$

$\therefore$ impedance $= \mathbf{1.86\ \Omega}$

(*ii*) The method of conjugate will be used to determine the real power and reactive volt-amperes. This is a convenient way of calculating these quantities when both voltage and current are expressed in Cartesian form. If the *conjugate* of current is multiplied by the voltage in Cartesian form, the result is a complex quantity, the real part of which gives the real power and *j* part of which is the reactive volt-amperes (vars). It should, however, be noted the real power as obtained by this method of conjugates is the same regardless of whether **V** or **I** is reversed although sign of reactive volt-amperes will depend on the choice of **V** or **I***.

Using the conjugate of current, we get

$\mathbf{P}_{VA} = (8 + j6)(-2 - j5) = 14 - j52$

$\therefore$ power consumed = **14W**

(*iii*) Obviously, phase angle = 111.8° – 36.9° = 75° with current leading.

Example 19.9. *The potential difference and the current in a circuit are represented by (100 + j 200) V and (10 + j 5) A, respectively. Calculate the power and the reactive volt-ampers.*

(Elec. Engg. A.M.Ae. S.I. June 1995)

Solution. We will use the method of conjugates to calculate the active and reactive powers. Using current conjugate, we have

$P_{VA} = (100 + j\,200)(10 - j\,5)$

$= 100 \times 10 - j\,5 \times 100 + j\,200 \times 10 + 200$

$= 2000 + j\,1500$

Active power $= 2000\text{ W} = \mathbf{2\ kW}$

Reactive power $= \mathbf{1.5\ kVAR}$

Example 19.10. *If the potential difference across a circuit is represented by (40 + j 25) volts and the circuit consists of a resistance of 20 Ω in series with an inductance of 0.06 henry and the frequency is 79.5 Hz, find the complex number representing the current in amperes.*

---

*If voltage conjugate is used, then capacitive VARs are positive and inductive VARs negative. If current conjugate is used, then capacitive VARs are negative and inductive VARs are positive.

Solution. $X_L = \omega L = 0.06 \times 2\pi \times 79.5 = 29.97\Omega$

$$\mathbf{Z} = 20 + j\,29.27; \qquad \mathbf{V} = 40 + j\,25$$

$$\mathbf{I} = \frac{\mathbf{V}}{\mathbf{Z}} = \frac{40 + j\,25}{20 + j\,29.97}$$

$$= \frac{(40\ j\,25)\,(20 - j\,29.97)}{(20 + j\,29.97)\,(20 - j\,29.97)}$$

$$\mathbf{I} = (1.19 - j\,0.54)\text{ A}$$

Example 19.11. *A 400-Hz generator has an induced e.m.f. of 100 V and an internal impedance of (5 + j 0) ohm. If it supplies an impedance consisting of a 40 ohm capacitive reactance in series with a 10 ohm resistance, what is the magnitude of the current passing through? Determine also the voltage at the terminal of the generator.* **(A.M.I.E., Summer 1985)**

Solution. The circuit diagram is shown in Fig. 19.13.

Here, $\mathbf{Z}_1 = 5 + j\,0 = 5\ \angle 0°$

$$\mathbf{Z}_2 = 10 - j\,40 = 41.2\ \angle{-76°}$$

$$\mathbf{Z} = (5 + j\,0) + (10 - j\,40)$$

$$= 15 - j\,40$$

$$= \mathbf{42.7\ \angle{-69.4°}}$$

$$\mathbf{I} = \mathbf{V/Z} = 100\ \angle 0°/42.7\angle{-69.4°} = 2.34\ \angle 69.4°$$

Drop $= \mathbf{I} \times$ internal impedance of the generator

$$= 2.34\ \angle 69.4° \times 5\ \angle 0° = 11.7\ \angle 69.4°$$

$$= 4.1 + j\,11$$

Terminal voltage $= 100\ \angle 0°$ —internal drop

$$= (100 + j\,0) - (4.1 + j\,11) = 96.5\ \angle{-6.5°}$$

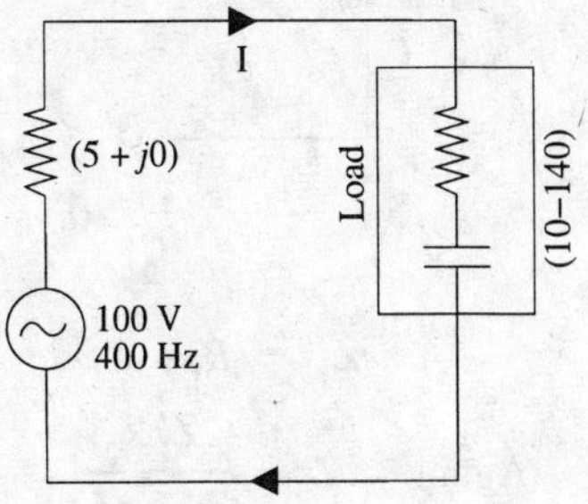

Fig. 19.13

Example 19.12. *The current in a circuit is given by (4.5 + j 12) when the applied voltage is (100 + j 150). Determine (a) the complex expression for the impedance stating whether it is inductive or capacitive, (b) the power and (c) the phase angle between the current and voltage.*

Solution. $\mathbf{I} = 4.5 + j\,12; \qquad \mathbf{V} = 100 + j\,150$

(*a*) $$\mathbf{Z} = \frac{\mathbf{V}}{\mathbf{I}} = \frac{100 + j\,150}{4.5 + j\,12} = \frac{(100 + j\,150)\,(4.5 - j\,12)}{(4.5 + j\,12)\,(4.5 - j\,12)}$$

$$= 13.7 - j\,3.2$$

Because the *j*-term is negative, hence impedance is capacitive.

(*b*) Using current conjugate

$$\mathbf{P}_{VA} = (100 + j\,150)(4.5 - j\,12) = 2250 - j525$$

$\therefore$ real power = **2250 W**

(*c*) $$\mathbf{V} = \sqrt{(100^2 + 150^2)}\ \angle \tan^{-1} 150/100 = 180.2\ \angle 56°\ 26'$$

$$\mathbf{I} = \sqrt{(4.5^2 + 12^{2)}}\ \angle \tan^{-1}\ 12/4.5 = 12.8\ \angle 69°\ 26'$$

$\therefore$ Phase difference $= 69°\ 26' - 56° - 26' = 13°$.

19.12. Complex Algebra Applied to Parallel Circuits

First, consider the parallel circuit of Fig. 19.14 (*a*). The two impedances $\mathbf{Z}_1$ and $\mathbf{Z}_2$ have the same p.d. across them.

$$\therefore \qquad \mathbf{I}_1 = \frac{\mathbf{V}}{\mathbf{Z}_1} = \frac{V}{R_1 + jX_1}\ \angle \tan^{-1} R_1/X_1$$

$$\mathbf{I}_2 = \frac{\mathbf{V}}{\mathbf{Z}_2} = \frac{V}{R_2 + jX_2}\ \angle \tan^{-1} R_2/X_2$$

Total current $\mathbf{I} = \mathbf{V/Z}$ where $\mathbf{Z}$ is the combined impedance of the parallel circuit which may be found by using law of parallel impedances.

$$\frac{1}{\mathbf{Z}} = \frac{1}{\mathbf{Z}_1} + \frac{1}{\mathbf{Z}_2} \quad \text{or} \quad \mathbf{Z} = \frac{\mathbf{Z}_1 \mathbf{Z}_2}{\mathbf{Z}_1 + \mathbf{Z}_2}$$

Now, consider the circuit of Fig. 19.14 (*b*)

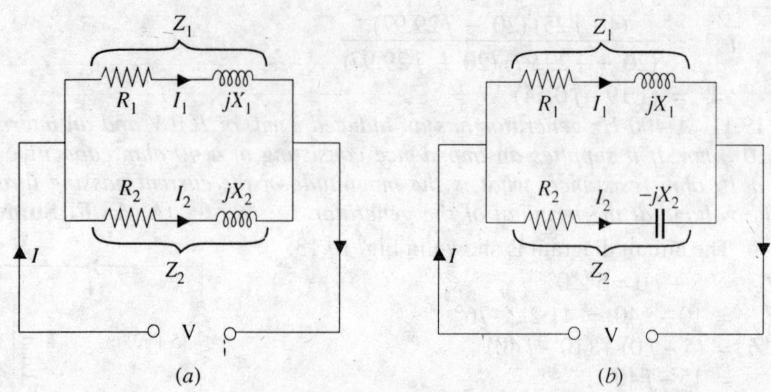

Fig. 19.14

$$\mathbf{Z}_1 = R_1 + jX_1 \quad \text{and} \quad \mathbf{Z}_2 = R_2 - jX_2$$

Again $\mathbf{Z} = \dfrac{\mathbf{Z}_1 \mathbf{Z}_2}{\mathbf{Z}_1 + \mathbf{Z}_2}$ and $\mathbf{I} = \dfrac{\mathbf{V}}{\mathbf{Z}}$

Example 19.13. *Two impedance given by $Z_1 = (10 + j\,5)$ and $Z_2 = (8 + j\,6)$ are joined in parallel and adross a voltage of $V = (200 + j\,0)$ volts. Calculate the circuit current, its phase and the branch currents. Draw the vector diagram.*

(Electric Circuits-I, Punjab Univ.1994.)

Fig. 19.15

Solution. We are given that $V = 200\angle 0°$

$\mathbf{Z}_1 = 10 + j\,5 = 11.2\angle 26.56°$

$\mathbf{Z}_2 = 8 + j\,6 = 10\angle 36.87°$

$\mathbf{I}_1 = \mathbf{V}/\mathbf{Z}_1$

$= 200\angle 0° / 11.2\angle 26.56° = 17.86\angle -26.56°$

$\mathbf{I}_2 = \mathbf{V}/\mathbf{Z}_2 = 200\angle 0°/10\angle 36.87° = \mathbf{20\angle -36.78°}$

As expected, both currents lag behind the applied voltage.

$$\mathbf{Z} = \frac{\mathbf{Z}_1 \mathbf{Z}_2}{\mathbf{Z}_1 + \mathbf{Z}_2} = \frac{11.2\angle 26.56° \times 10\angle 36.87°}{18 + j\,11} = 5.3\angle 32°$$

$$\mathbf{I} = \frac{200\angle 0°}{5.3\angle 32°} = \mathbf{37.7\angle -32°}$$

Power factor = cos (–32°) = **0.848 lag**

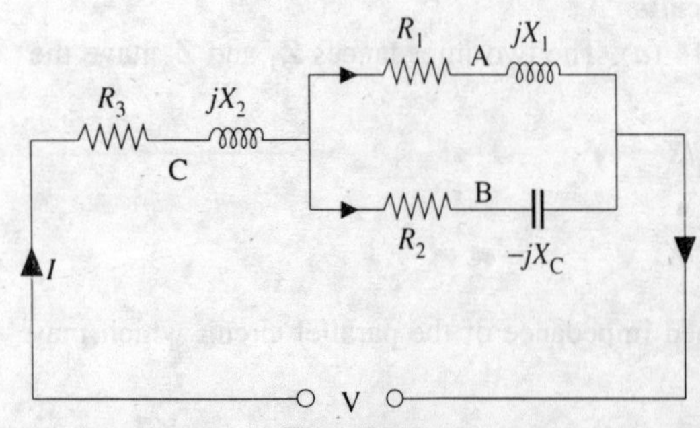

Fig. 19.16

19.13. Series-parallel Circuits

Consider the circuit of Fig. 19.16. First, equivalent impedance of parallel branches is calculated and it is then added to the series impedance to get the total circuit impedance. Then,circuit current can be easily found.

$$\mathbf{Y}_A = \frac{1}{R_1 + jX_1}; \quad \mathbf{Y}_B = \frac{1}{R_2 - jX_C}$$

$$\therefore \quad \mathbf{Y} = \frac{1}{R_1 + jX_1} + \frac{1}{R_2 - jX_C}$$

$$\therefore \quad \mathbf{Z}_P = \frac{1}{\mathbf{Y}}; \ \mathbf{Z}_C = R_3 + jX_3$$

$$\therefore \quad \mathbf{Z} = \mathbf{Z}_P + \mathbf{Z}_C \text{ and } \mathbf{I} = \mathbf{V}/\mathbf{Z}$$

Example 19.14. *Two circuits A and B are connected in parallel to a 200 V supply. Circuit A consists of a resistance of 10 Ω in series with an inductive reactance of 10 Ω and circuit B consists of a resistance of 20 Ω in series with a capacitive reactances of 10 Ω. Find the current taken by each circuit and the total current. Find also the p.f. of the circuit.*

Solution. Take voltage vector as the reference vector as shown in Fig. 19.17, so that $\mathbf{V} = 200° \angle 0°$

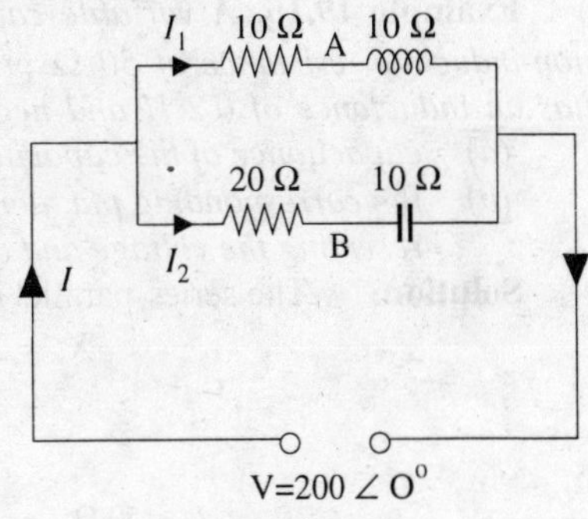

Fig. 19.17

$$\mathbf{Z}_1 = 10 + j\,10 = 14.14 \angle 45°$$

$$\mathbf{I}_1 = \frac{200 \angle 0°}{14.14 \angle 45°} = 14.14 \angle -45°$$

Obviously, current lags behind the voltage.

$$\mathbf{Z}_2 = 20 - j\,10 = 22.35 \angle -26.6°$$

$$\mathbf{I}_2 = \frac{200 \angle 0°}{22.35 \angle -26.6°}$$

$$= 8.95 \angle 26.6°$$

The current leads the applied voltage.

$$\mathbf{Z} = \frac{\mathbf{Z}_1 \mathbf{Z}_2}{\mathbf{Z}_1 + \mathbf{Z}_2}$$

$$= \frac{(10 + j\,10) \times (20 - j\,10)}{(10 + j\,10) + (20 - j\,10)} = \frac{(300 + j\,100)}{30}$$

$$= 10 - j\,3.33 = 10.54 \angle 18° \, 26'$$

$$\mathbf{I} = \frac{200 \angle 0°}{10.54 \angle 18° \, 26'} = 18.98 \angle -18° \, 26'$$

$$\text{p.f.} = \cos 18° \, 26' = \mathbf{0.9487 \text{ (lag)}}$$

Example 19.15. *The following three impedances are expressed at a frequency of 25 Hz; $Z_1 = 5 + j\,4$; $Z_2 = 2 + j\,8$; $Z_3 = 3 - j\,8$. Write down the corresponding expressions at a frequency of 50 Hz.*

If the three impedances were connected in series, calculate the effective impedance of the combination. Hence, find

(a) the current taken by this combination from a 250-V, 50-Hz supply
(b) the power factor of the circuit
(c) the inductance or capacitance, as the case may be, of the parallel-connected component which would give unity power factor.

Solution. Since $\mathbf{Z}_1$ and $\mathbf{Z}_2$ have positive *j*-terms, they are inductive impedances. As inductive reactance is directly proportional to frequency, hence at 50 Hz

$$\mathbf{Z}_1 = 5 + j\,8 \quad \text{and} \quad \mathbf{Z}_2 = 2 + j\,16$$

Now, negative *j*-term in $\mathbf{Z}_3$ represents capacitive reactance which is inversely proportional to frequency. Hence, at 50-Hz

$$\mathbf{Z}_3 = 3 - j\,4$$

Total impedance, $\mathbf{Z} = \mathbf{Z}_1 + \mathbf{Z}_2 + \mathbf{Z}_3$

$$= 5 + j\,8 + 2 + j16 + 3 - j\,4 = 10 + j\,20$$

(*a*) $\mathbf{I} = \frac{\mathbf{V}}{\mathbf{Z}} = \frac{250 + j0}{10 + j20} = \frac{250(10 - j20)}{(10 + j20)(10 - j20)}$

$= (5 - j10)$ A

$I = \sqrt{5^2 + 10^2} = \mathbf{11.2\ A}$

(*b*) $\text{p.f.} = \cos\phi = \frac{\text{'real' component of current}}{\text{total current}}$

$= 5/11.2 = \mathbf{0.447\ lag}$

(*c*) For raising the p.f. to unity, the lagging reactive component–j 10 must be neutralized by an equal but leading current. For this purpose, a capacitor should be connected in parallel with the series circuit such that it should draw a current of j 10 A

Now, $I = V\omega C$ $\quad\therefore\quad C = I/\omega V$

$\therefore\quad C = 10/2\pi \times 50 \times 250 = 127 \times 10^{-6}$ F $= \mathbf{127\ \mu F}$

Example 19.16. *A variable capacitor is connected in series with a circuit consisting of a non-inductive resistance of 50 Ω in parallel with a coil across a 200-V, 50-Hz supply. The coil has an inductance of 0.2 H and negligible resistance. Calculate:*

(*a*) *capacitance of the capacitor when the p.f. of the circuit is unity,*

(*b*) *the corresponding p.d. across the capacitor. Draw the vector diagram, not to scale, representing the voltage and currents.*

Solution. The series-parallel circuit is shown in Fig. 19.18 (*a*).

$X_L = 2\pi \times 50 \times 0.2 = 62.8\ \Omega$

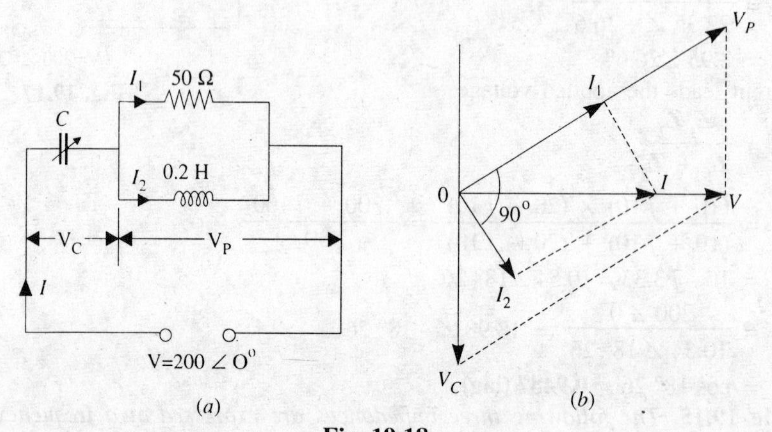

Fig. 19.18

The admittance of the parallel circuit is

$$\mathbf{Y} = \mathbf{Y}_1 + \mathbf{Y}_2 = \frac{1}{50 + j0} + \frac{1}{0 + j62.8}$$

$$= \frac{1}{50} - \frac{j}{62.8} = 0.02 - j0.0159 \text{ mho}$$

Impedance of the parallel circuit is

$$\mathbf{Z} = \frac{1}{\mathbf{Y}} = \frac{1}{0.02 - j0.0159}$$

$$= 30.6 + j24.35\ \Omega$$

The impedance of the capacitor required for making the power factor of the whole circuit unity should be equal in magnitude to the reactive component of the impedance of the parallel circuit.

Impedance of the capacitor $= -j\,24.35\ \Omega$

$$\therefore \qquad \frac{1}{\omega C} = 24.35$$

or $$C = 1/2\pi \times 50 \times 24.35 = \mathbf{130.9\ \mu F}$$

Circuit current $$I = \frac{(220 + j\,0)}{30.6} = \frac{200}{30.6}\angle 0°$$

$$\mathbf{V}_C = -j\,24.35 \times \frac{200}{30.6}\angle 0° = -j\,\mathbf{159\ V}$$

The vector diagram, not to scale, is shown in Fig. 19.17(*b*) where I_1 represents current through the resistor I_2 through the coil and I through the capacitor.

Example 19.17. *A transmission line having a resistance R of 1.4 ohm and an inductive reactance of 1.6 ohm at 60 Hz, supplies power to a load connected at the terminal at 460 volts (r.m.s.) and takes a current of 27 A (r.m.s.) at a p.f. of 0.8 lagging. Find the voltage at the generator terminal and its power factor.*

Solution. In vector diagram of Fig. 19.19, the receiving end voltage V_R has been taken as the reference vector.

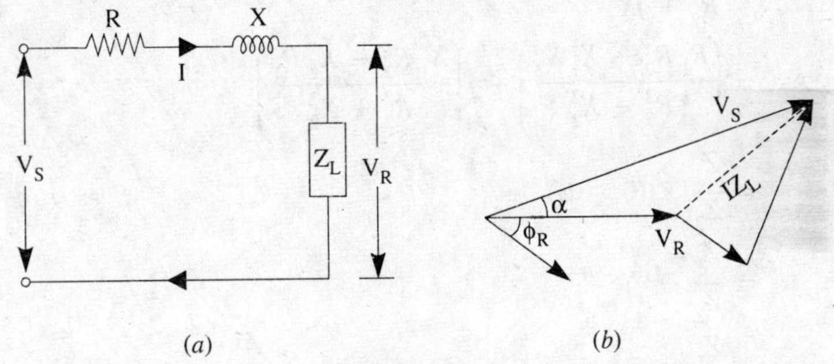

Fig. 19.19

$$\mathbf{V_R} = 460 \angle 0° = (460 + j\,0)$$
$$\mathbf{Z_L} = 1.4 + j\,1.6 = 2.12 \angle 48.8°$$
$$\cos\phi_R = 0.8\,; \qquad \phi_R = \cos^{-1}(0.8) = 36°\ 52'$$
$$\mathbf{I} = 27 \angle -36°\ 52'$$
Line drop $$= \mathbf{IZ} = 27 \angle -36°\ 52' \times 2.12 \angle 48°\ 48'$$
$$\mathbf{I} = 57.24 \angle 11°\ 56' = 57.24\,(\cos 11°\ 56' + j \sin 11°\ 56')$$
$$= 56.2 + j11.9$$
$$\mathbf{V_S} = \mathbf{V_R} + \mathbf{IZ}$$
$$= (460 + j\,0) + (56.2 + j\,11.9)$$
$$= 516.2 + j\,11.9 = 516.3 \angle 1.3°$$
$$\phi_S = \phi_R + \alpha = 36°\ 52' + 1°\ 18' = 38°\ 10'$$
$$\cos\phi_S = \cos 38°\ 10' = \mathbf{0.783}$$

HIGHLIGHTS

Summary of Complex Algebra

1. Rectangular form of an impedance is
$$\mathbf{Z} = R \pm jX$$
2. Polar form is
$$\mathbf{Z} = \mathbf{Z} \pm \theta \text{ where } \theta = \tan^{-1}(X/R) \text{ and } \mathbf{Z} = \sqrt{(R^2 + X^2)}$$
3. Trigonometrical form is
$$\mathbf{Z} = \mathbf{Z}\,(\cos\theta \pm j\sin\theta)$$
4. Exponential form is
$$\mathbf{Z} = \mathbf{Z}_e^{\pm j\theta}$$

5. Addition

Let $\mathbf{Z}_1 = R_1 + jX_1$ and $\mathbf{Z}_2 = R_2 + jX_2$

$$\mathbf{Z}_1 + \mathbf{Z}_2 = (R_1 + R_2) + j(X_1 + X_2)$$

6. Subtraction

$$\mathbf{Z}_1 - \mathbf{Z}_2 = (R_1 - R_2) + j(X_1 - X_2)$$

7. Multiplication

$$\begin{aligned}\mathbf{Z}_1\mathbf{Z}_2 &= (R_1 + jX_1)(R_2 + jX_2)\\ &= (R_1R_2 - X_1X_2) + j(X_1R_2 + X_2R_1)\\ &= \mathbf{Z}_1\mathbf{Z}_2 \angle(\theta_1 + \theta_2)\\ &= \mathbf{Z}_1\mathbf{Z}_2\, e^{j(\theta_1 \theta_2)}\\ &= \mathbf{Z}_1\mathbf{Z}_2 [\cos(\theta_1 + \theta_2) + j\sin(\theta_1 + \theta_2]\end{aligned}$$

8. Division

$$\begin{aligned}\frac{\mathbf{Z}_1}{\mathbf{Z}_2} &= \frac{R_1 + jX_1}{R_2 + jX_2}\\ &= \left(\frac{R_1R_2 + X_1X_2}{R_2^2 + X_2^2}\right) + j\left(\frac{X_1R_2 - X_2R_1}{R_2^2 + X_2^2}\right)\\ &= \frac{Z_1}{Z_2}\angle(\theta_1 - \theta_2)\\ &= \frac{Z_1}{Z_2}\, e^{j(\theta_1 - \theta_2)}\\ &= \frac{Z_1}{Z_2}[\cos(\theta_1 - \theta_2) + j\sin(\theta_1 - \theta_2)]\end{aligned}$$

20 THREE PHASE CIRCUITS

20.1. Generation of Three-phase Voltages

The kind of alternating currents and voltages discussed in chapters 16 to 18 are known as single-phase voltages and currents because they consist of a single alternating current and voltage wave. A single-phase alternator was diagrammatically depicted in Fig. 16.1 (*b*) and it was shown to have one armature winding only. But if the number of armature windings is increased to three, then it becomes three-phase alternator and it produces as many independent voltage waves as the number of windings or phases. These windings are displaced from one another by equal angles, the value of these angles being determined by the number of phases or windings.

In a two-phase alternator, the armature windings are displaced 90 electrical degrees apart. A 3-phase alternator, as the name shows, has three independent armature windings which are 120 electrical degrees apart. Hence, the voltages induced in the three windings are 120° apart in time phase. With the exception of two-phase windings, it can be stated that, in general, the electrical displacement between different phases is $360/n$ where n is the number of phases or windings.

Three-phase systems are the most common although, for certain special jobs, greater number of phases is also used. For example, almost all mercury-arc rectifiers for power purposes are either six-phase or twelve-phase and most of the rotary convertors in use are six phase. All modern generators are practically three-phase. For transmitting large amounts of power, three-phase is invariably used. The reasons for the immense popularity of three-phase apparatus are that (*i*) it is more efficient (*ii*) it uses less material for a given capacity and (*iii*) it costs less than single-phase apparatus etc.

In Fig. 20.1 is shown a two-pole stationary-armature, rotating-field type three-phase alternator. It has three armature coils *aa*′, *bb*′, *cc*′ displaced 120° apart from one another. With the position and clockwise rotation of the poles shown in Fig. 20.1, it is found that the e.m.f. induced in conductor '*a*' of coil *aa*′ is maximum and its direction* is *away* from the reader. The e.m.f. in conductor '*b*' of coil *bb*′ would be maximum and away from the reader when the *N*-pole has turned through 120° *i.e.,* when *N-S* axis lies along *bb*′. It is clear that the induced e.m.f. in conductor '*b*' reaches its maximum value 120° later than the maximum value in conductor '*a*'. In the like manner, the maximum e.m.f. induced (in the direction away from the reader) in conductor '*c*' would occur 120° later than '*b*' or 240° later than that in '*a*'.

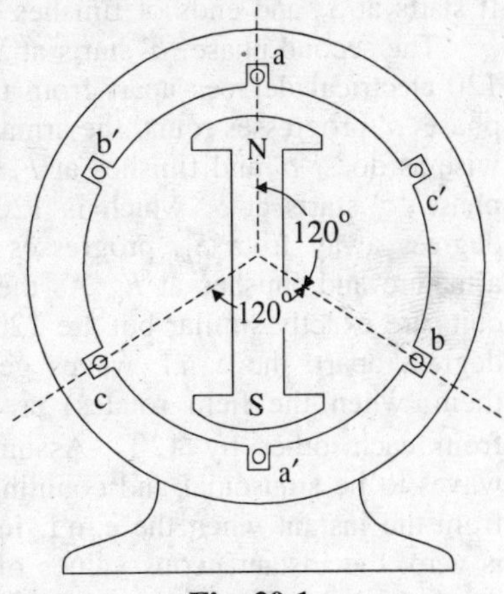

Fig. 20.1

Thus, the three coils have three e.m.fs. induced in them which are similar in all respects except that they are 120° out of time phase with one another as pictured in Fig. 20.2. Each voltage

*The direction is found with the help of Fleming's Right-hand rule. But while applying the rule, it should be remembered that the relative motion of the conductor with respect to the field is anti-clockwise although the motion of the field with respect to the conductor is clockwise as shown. Hence, thumb should point to the left.

wave is assumed to be sinusoidal and having maximum value of E_m.

In practice, the space on the armature is completely covered and there are many slots/phase/pole.

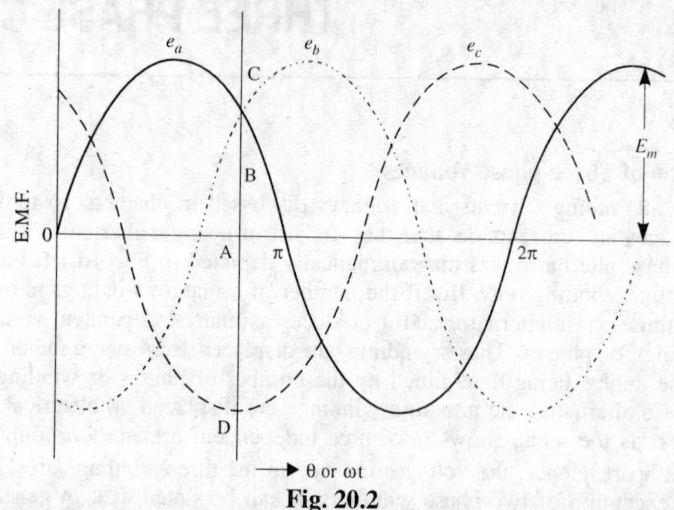

Fig. 20.2

Fig. 20.3 illustrates relative positions of the three phases of a 4-pole alternator and Fig. 20.4 shows the developed diagram of its armature windings. Assuming full-pitched winding and the direction of rotation as shown, phase '*a*' occupies the position under the centres of *N* and *S*-poles. It starts at S_a and ends of finishes at F_a.

The second phase '*b*' starts at S_b which is 120 electrical degrees apart from the start of phase '*a*' progresses round the armature clockwise as does '*a*' and finishes at F_b. Similarly, phase '*c*' starts at S_c which is 120 electrical degrees away from S_b, progresses round the armature and finishes at F_c. As the three circuits are exactly similar but are 120 electrical degrees apart, the e.m.f. waves generated in them (when the field rotates) are displaced from each other by 120°. Assuming these waves to be sinusoidal and counting the time from the instant when the e.m.f. in phase '*a*' is zero, the instantaneous values of the three e.m.fs, will be given by curves of Fig. 20.2. The equations are:

$$e_a = E_m \sin \omega t \quad ...(i)$$

$$e_b = E_m \sin (\omega t - 120°) \quad ...(ii)$$

$$e_c = E_m \sin (\omega t - 240°) \quad ...(iii)$$

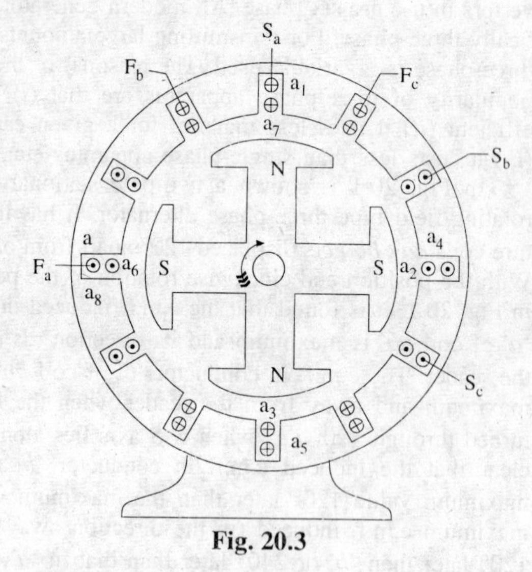

Fig. 20.3

As discussed in Chapter 16, alternating voltages may be represented by revolving vectors which indicate their maximum values (or r.m.s. values if desired). The actual values of these voltages vary from peak positive to zero and to peak negative values in one revolution of the vectors. In Fig. 20.5 are shown the three vectors representing the r.m.s. voltages of the three phases: E_a, E_b and E_c [in the present case, $E_a = E_b = E_c = E$ (say)].

It can be shown that the sum of the three phase e.m.s. is zero in the following three ways:

(*i*) The sum of the above three equations (*i*), (*ii*) and (*iii*) is zero as shown below:

Resultant instantaneous e.m.f. is

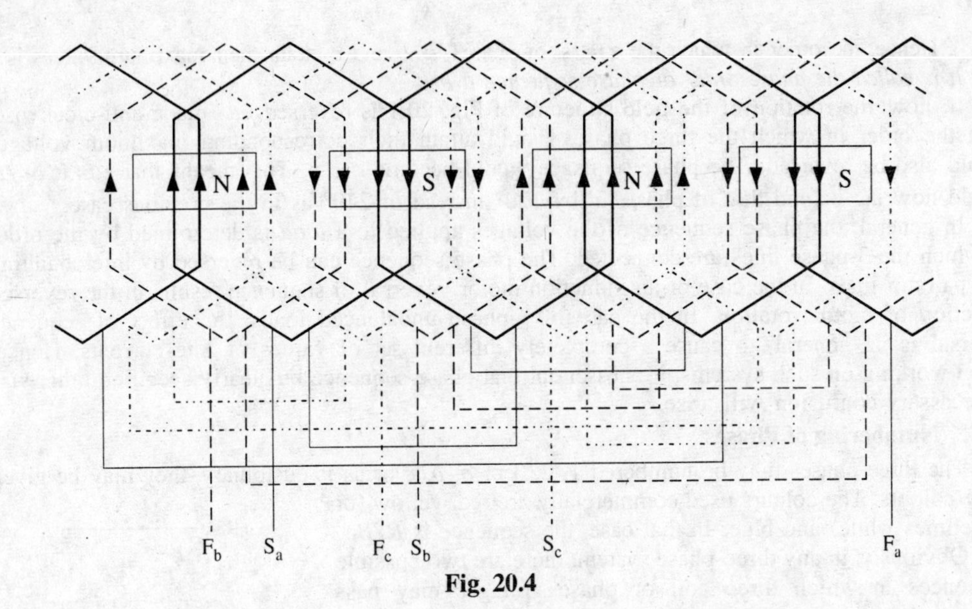

Fig. 20.4

$$= e_a + e_b + e_c$$
$$= E_m \sin \omega t + E_m \sin (\omega t - 120°) + E_m \sin (\omega t - 240°)$$
$$= E_m [\sin \omega t + 2 \sin (\omega t - 180°) \cos 60°]$$
$$= E_m [\sin \omega t - 2 \sin \omega t \cos 60°] = 0$$

(*ii*) The sum of ordinates of the three e.m.f. curves of Fig. 20.2 is zero. For example, taking ordinates AB and AC as positive and AD as negative, it can be shown that

$$AB + AC + (-AD) = 0$$

(*iii*) If we add the three vectors of Fig. 20.5 either vectorially or by calculation, the result is zero.

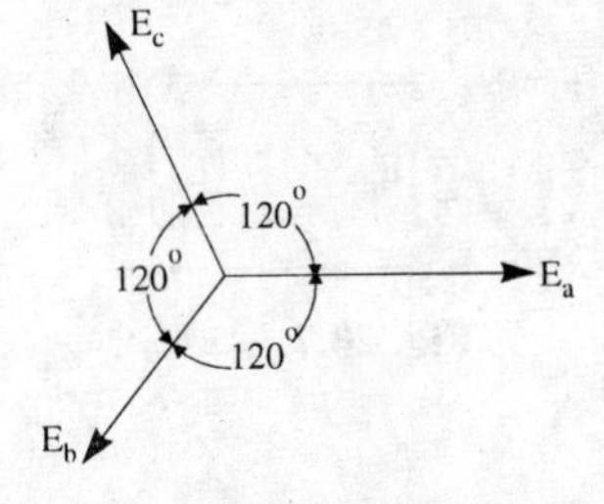

Fig. 20.5

Vector Addition

As shown in Fig. 20.6, the resultant of E_a and E_b is E_r and its magnitude is $2E \cos 60° = E$, where

$$E_a = E_b = E_c = E$$

This resultant is equal and opposite to E_c. Hence, their resultant is zero.

By Calculation

Let us take $\mathbf{E_a}$ is reference voltage and assuming clockwise phase sequence,

$$\mathbf{E}_a = E \angle 0° = E + j0$$
$$\mathbf{E}_b = E \angle -120°$$
$$= E [\cos (-120°) + j \sin (-120°)]$$
$$= E(-0.5 - j0.866)$$
$$\mathbf{E}_c = E \angle -240° = E \angle 120° = E(-0.5 + j0.866)$$
$$\therefore \quad \mathbf{E}_a + \mathbf{E}_b + \mathbf{E}_c = (E + j0) + E(-0.5 - j0.866) + E(-0.5 + j0.866) = 0$$

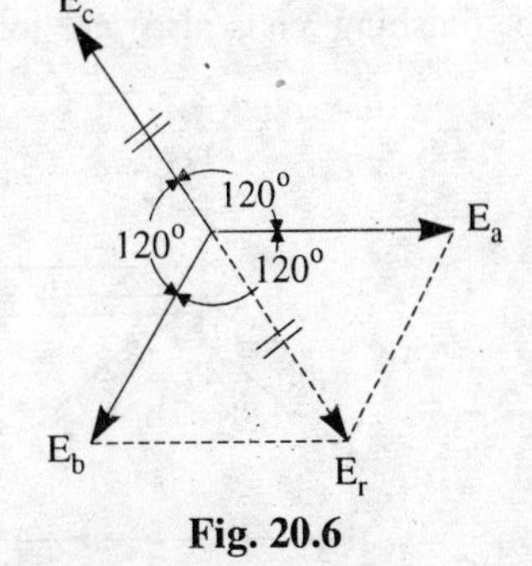

Fig. 20.6

20.2. Phase Sequence

By phase sequence is meant the order in which the three phases attain their peak or maximum values. In the development of the three-phase e.m.fs. in Fig. 20.2, clockwise rotation of the field system in Fig. 20.1 was assumed. This assumption made the e.m.f. of phase 'b' lag behind that of 'a' by 120° and in a similar way, made that of 'c' lag behind that of 'b' by 120° (or that of 'a' by

240°). Hence, the order in which the e.m.fs. of phases, *a, b* and *c* attain their maximum values is *a b c. It is called the phase order or phase sequence a b c.*

If, now, the rotation of the field structure of Fig. 20.1 is reversed *i.e.,* made anti-clockwise, then the order in which the three phases would attain their corresponding maximum voltages would also be reversed. The phase sequence would become *a c b.* This means that e.m.f. of '*c*' would now lag behind that of phase '*a*' by 120° instead of 240° as in the previous case.

In general, the phase sequence of the voltages applied to a *load* is determined by the order in which the 3-phase lines are connected. The phase sequence can be reversed by interchanging any pairs of lines. In the case of an induction motor, reversal of sequence results in the reversed direction of motor rotation. In the case of 3-phase unbalanced loads, the effect of sequence reversal is, in general, to cause a completely different set of values of line currents. Hence, when working on such systems, it is essential that phase sequence be clearly specified otherwise unnecessary confusion will arise.

20.3. Numbering of Phases

The three phases may be numbered 1, 2, 3 or *a, b, c* or as is customary, they may be given three colours. The colours used commercially are red, yellow (or sometimes white) and blue. In that case, the sequence is *RYB.*

Obviously, in any three-phase system, there are two possible sequences in which three coil or phase voltages may pass through their maximum value *i.e.,* red → yellow → blue *(RYB)* or red → blue → yellow *(RBY).* By convention, *RYB* is regarded as positive sequence and *RBY* as negative sequence.

20.4. Inter-connection of Three Phases

If the three armature coils of the 3-phase alternator (Fig. 20.3) are not inter-connected but are kept separate as shown in Fig. 20.7, then each phase or circuit would need two conductors, the total number of conductors, in that case, being six. It means that each transmission cable would contain six conductors which will make the whole system complicated and expensive. Hence, the three phases are, generally, inter-connected which results in substantial saving of copper. The general methods of inter-connection are:

(*a*) Star or Wye (*Y*) connection and

(*b*) Mesh or Delta (Δ) connection.

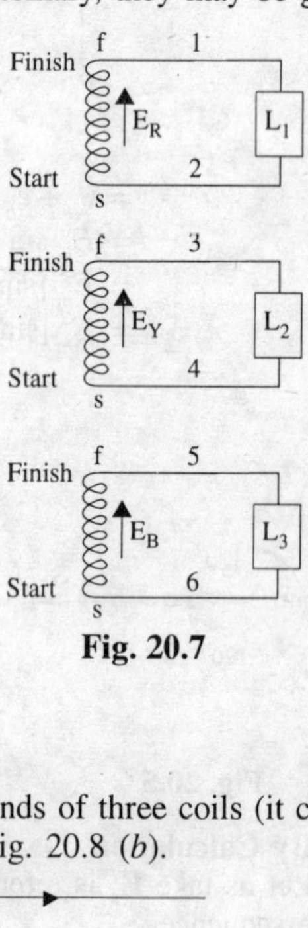

Fig. 20.7

20.5. Star or Wye (Y) Connection

In this method of inter-connection, the *similar* ends*, say, 'start' ends of three coils (it could be 'finishing' ends also) are joined together at point *N* as shown in Fig. 20.8 (*b*).

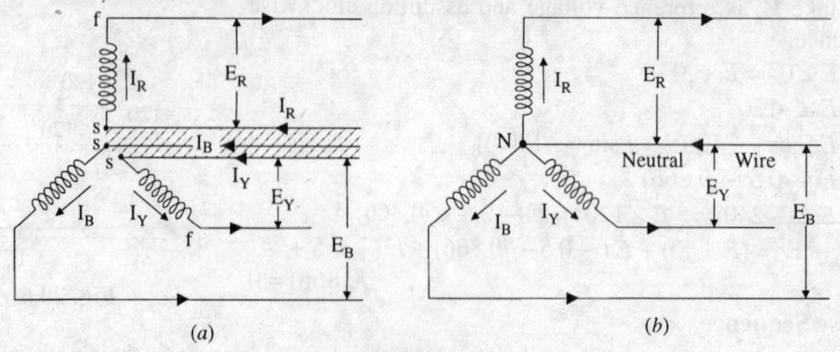

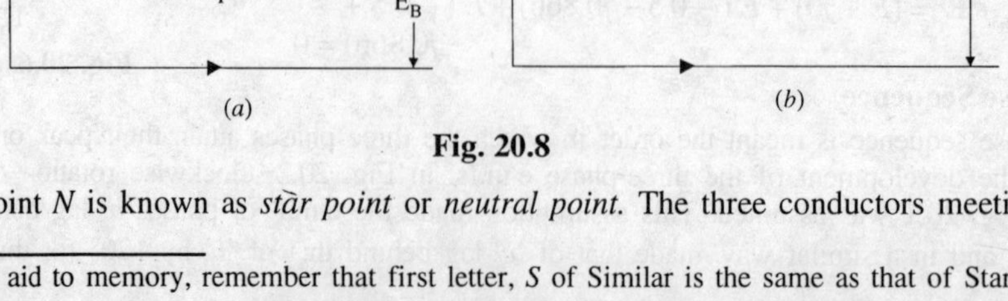

Fig. 20.8

The point *N* is known as *star point* or *neutral point.* The three conductors meeting at point

*As an aid to memory, remember that first letter, *S* of Similar is the same as that of Star.

N are replaced by a single conductor known as *neutral conductor* as shown in Fig. 20.8 (*b*). Such an inter-connected system is known as 3-phase four-wire system and is diagrammatically shown in Fig. 20.8 (*b*). If this three-phase voltage is applied across a balanced symmetrical load, the neutral wire will be carrying three currents which are exactly equal in magnitude but are 120° out of phase with each other. Hence, their *vector* sum is zero.

i.e., $I_R + I_Y + I_B = 0$ ...vectorially

The neutral wire, in that case, may be omitted although its retention is useful for supplying loads at low voltages (Ex. 20.13). The p.d. between any terminal (or line) and neutral (or star) point gives the *phase* or *star* voltage. But the p.d. between any two lines gives the line voltage. This connection is also referred to as 3-phase, 4-wire system.

Note. When considering the distribution of current in a 3-phase system, it is extremely important to bear in mind that:

(*i*) the arrows placed alongside the currents I_R, I_Y and I_B flowing in three phases [Fig. 20.8 (*b*)] indicate the directions of currents when they are assumed to be *positive* and not the directions at a particular instant. It should be clearly understood *that at no instant will all the three currents flow in the same direction either outwards or inwards.* The three arrows indicate that first the current flows outwards in phase R, then after a phase-time of 120°, it will flow outwards from phase Y and after a further 120°, outwards from phase B.

(*ii*) the current flowing outwards in one or two conductors is always equal to that flowing inwards in the remaining conductor or conductors. In other words, *each conductor, in turn, provides a return path for the currents of the other conductors.*

In Fig. 20.9 are shown the three phase currents having the same peak value of 20 A but displaced from each other by 120°. At instant '*a*', the currents in phase R and B are each + 10 A (*i.e.,* flowing outwards) whereas the current in phase Y is –20 A (*i.e.,* flowing inwards). In other words, at the instant '*a*', phase Y is acting as return path for the currents in phases R and B. At instant *b*, I_R = + 15 A and I_Y = + 5 A but I_B = – 20 A which means that now phase B is providing the return path.

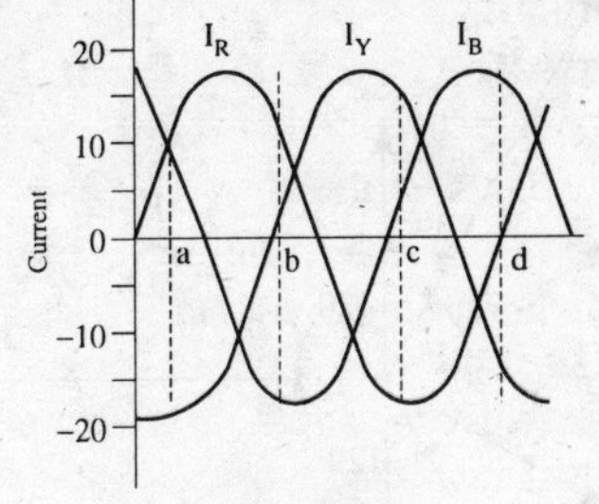

Fig. 20.9

At instant *c*, I_Y = + 15 A and I_B = + 5 A and I_R = – 20A.

Hence, now phase R carries current inwards whereas Y and B carry currents outwards. Similarly, at point *d*, I_R = 0, I_B = 17.3 A and I_Y = – 17.3 A. In other words, current is flowing outwards from phase B and returning *via* phase Y.

In addition, it may be noted that although the distribution of current between the three lines is continuously changing, yet at any instant, the algebraic sum of the *instantaneous* values of the three currents is zero *i.e.,*

$$i_R + i_Y + i_B = 0 \quad \text{—algebraically}$$

20.6. Voltages and Currents in Y-Connection

The voltage induced in each winding is called the 'phase' voltage and current in each winding is likewise known as 'phase' current. However, the voltage available between any pair of terminals (or outers) is called *line* voltage (V_L) and the current flowing in each *line* is called line current (I_L).

As seen from Fig. 20.10 (*a*), in this form of inter-connection, there are two phase windings between each pair of terminals but since their *similar* ends have been joined together, they are in *opposition.* Obviously, the instantaneous value of p.d. between any two terminals is the *difference* of the two phase e.m.fs. concerned. However, the r.m.s. value of this p.d. is given by the *vector difference* of two phase e.m.fs.

The vector diagram for phase voltages and currents in a star connection is shown in Fig. 20.10 (*b*) where a balanced system has been assumed*. It means that $E_R = E_Y = E_B = E_{ph}$ (phase e.m.f.).

*A balanced system is one in which (*i*) the voltages in all phases are equal in magnitude and differ in phase from one another by equal angles, in this case, the angle = 360/3 = 120° (*ii*) the currents in the three phases are equal in magnitude and also differ in phase from one another by equal angles.

A 3-phase balanced load is that in which the loads connected across three phases are identical in magnitude and phase.

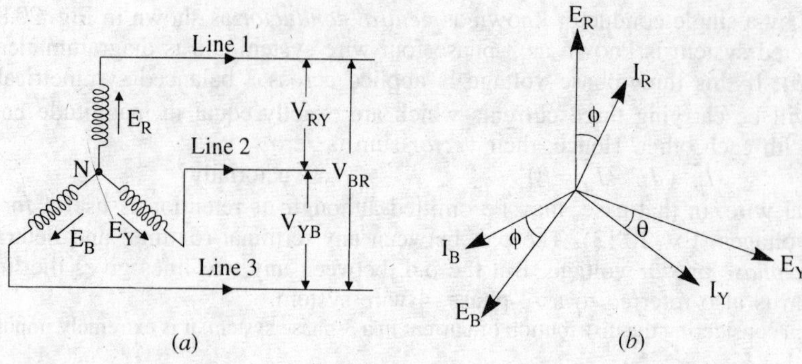

Fig. 20.10

Line voltage V_{RY} between line 1 and line 2 is the vector difference of E_R and E_Y.
Line voltage V_{YB} between line 2 and line 3 is the vector difference of E_Y and E_B.
Line voltage V_{BR} between line 3 and line 1 is the vector difference of E_B and E_R.

(i) Line Voltages and Phase Voltages

The p.d. between lines 1 and 2 is

$$V_{RY} = E_R - E_Y \quad \text{...vector difference}$$

Hence, V_{RY} is found by compounding E_R and E_Y reversed and its value is given by the diagonal of the parallelogram of Fig. 20.11. Obviously, the angle between E_R and E_Y reversed is 60°. Hence, if $E_R = E_Y = E_B$ = say, E_{ph} – the phase e.m.f., then

$$V_{RY} = 2 \times E_{ph} \times \cos\ (60°/2)$$
$$= 2 \times E_{ph} \times \cos\ 30°$$
$$= 2 \times E_{ph} \times \frac{\sqrt{3}}{2} = \sqrt{3}\, E_{ph}$$

Similarly,

$$V_{YB} = E_Y - E_B \quad \text{(vector difference)}$$
$$= \sqrt{3}.E_{ph}$$

and

$$V_{BR} = E_B - E_R \quad \text{(vector difference)}$$
$$= \sqrt{3}.E_{ph}$$

Now $V_{RY} = V_{YB} = V_{BR}$ = line voltage, say, V_L

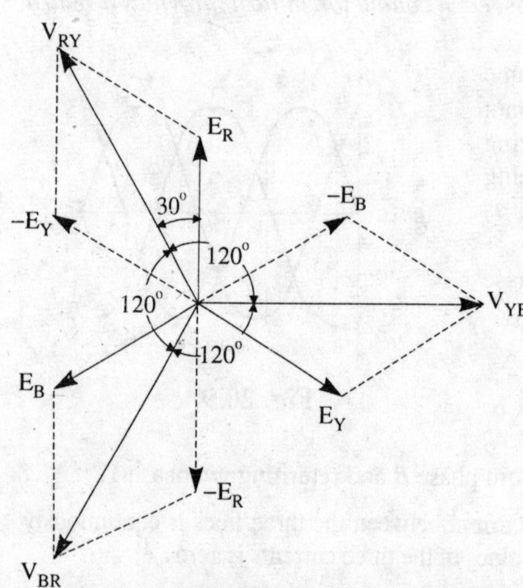

Fig. 20.11

Hence, in star connection

$$V_L = \sqrt{3}.E_{ph}$$

It will be noted from Fig. 20.11 that

(a) Line voltages are 120° apart.
(b) Line voltages are 30° ahead of their respective phase voltages.
(c) The angle between the line currents and the corresponding line voltages is $(30 + \phi)$ with current lagging.

(ii) Line Currents and Phase Currents

It is seen from Fig. 20.10 (*a*) that each line is in series with its individual phase winding, hence the line current in each line is the same as the current in the phase winding to which it is connected.

Current in line 1 = I_R
Current in line 2 = I_Y
Current in line 3 = I_B

Since $I_R = I_Y = I_B =$ say, I_{ph} – the phase current

$\therefore$ line current $I_L = I_{ph}$.

(*iii*) **Power**

The total power in the circuit is the sum of the three phase powers. Hence

Total power = 3 × phase power

or $$P = 3 \times E_{ph} I_{ph} \cos \phi$$

Now $E_{ph} = V_L/\sqrt{3}$ and $I_{ph} = I_L$

Hence, in terms of line values, the above expression becomes,

$$P = 3 \times \frac{V_L}{\sqrt{3}} \times I_L \times \cos \phi$$

or $$P = \sqrt{3}\, V_L I_L \times \cos \phi$$

It should be noted that ϕ is the angle between *phase* voltage and *phase* current and not between the line voltage and line current.

Example 20.1. *400 V (line to line) is applied to three star-connected identical impedances each consisting of a 4 Ω resistance in series with 3 Ω inductive reactance. Find (a) line current and (b) total power supplied.*

Solution. The circuit is shown in Fig. 20.12.

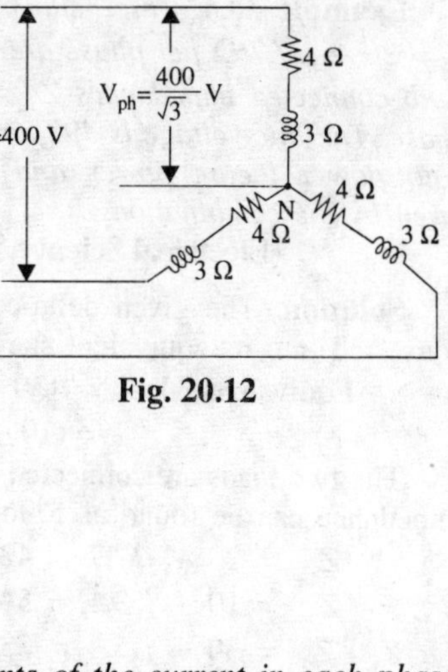

Here $R = 4\,\Omega;\; X_L = 3\,\Omega$

$\therefore\quad Z_{ph} = \sqrt{4^2 + 3^2} = 5\,\Omega$

p.f. $= \cos \phi = R/Z = 4/5 = 0.8$

$V_L = 400$ V

$V_{ph} = 400/\sqrt{3}$ V

$I_{ph} = V_{ph}/Z_{ph} = 400/\sqrt{3} \times 5 = 80/\sqrt{3}$ A

(*a*) $\therefore\quad I_L = I_{ph} = \mathbf{80/\sqrt{3}\ A}$

Fig. 20.12

(*b*) Total power

$$P = \sqrt{3}\, V_L I_L \cos \phi$$

$$\therefore\quad P = \sqrt{3} \times 400 \times \frac{80}{\sqrt{3}} \times 0.8 = \mathbf{25{,}600\ W}$$

[**Note.** $P = 3 \times I_{ph}^2\, R_{ph} = 3 \times \left(\frac{80}{\sqrt{3}}\right)^2 \times 4 = 25{,}600$ W]

Example 20.2. *Calculate the active and reactive components of the current in each phase of a star-connected, 5000 V, 3-phase alternator supplying 3,000 kW at a power factor 0.8.*

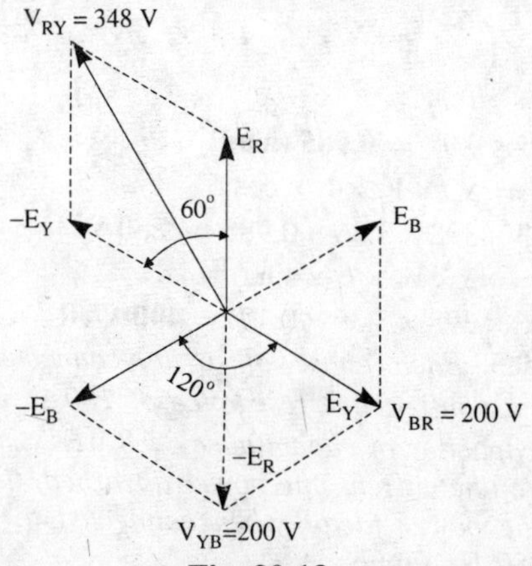

Fig. 20.13

Solution. $P = \sqrt{3}\, V_L I_L \cos \phi$

or $3000 \times 1000 = \sqrt{3} \times 5000 \times I_L \times 0.8$

$\therefore\quad I_L = 433$ A $\qquad \therefore\quad I_{ph} = 433$ A

Active component $= I_{ph} \times \cos \phi$
$= 433 \times 0.8 = \mathbf{346.2\ A}$

Reactive component $= I_{ph} \times \sin \phi = 433 \times 0.6$
$= \mathbf{260\ A}$

Example 20.3. *Deduce the relationship between the phase and line voltages of a 3-phases star-connected alternator. If the phase voltage of a 3-phase star-connected alternator be 200 V, what will be the line voltages (a) when the phases are correctly connected and (b) when the connections to one of the phases are reversed.*

Solution. (*a*) When phases are correctly connected, the vector diagram is as shown in Fig. 20.11. As proved in Art. 20.6 (*i*).

$$V_{RY} = V_{YB} = V_{BR} = \sqrt{3}.E_{ph}$$

Each line voltage = $\sqrt{3} \times 200 =$ **346 V**

(*b*) Suppose connections to *B*-phase have been reversed. Then voltage vector diagram for such a case is shown Fig. 20.13. It should be noted that E_B has been drawn in the reversed direction, so that angles between the three phase voltages are 60°.

$$V_{RY} = E_R - E_Y \qquad \text{...vector difference}$$
$$= 2.\,E_{ph} \times \cos 30°$$
$$= \sqrt{3} \times 200 = \mathbf{346\ V}$$

$$V_{YB} = E_Y - E_B \qquad \text{...vector difference}$$
$$= 2 \times E_{ph} \times \cos 60°$$
$$= 2 \times 200 \times \frac{1}{2} = \mathbf{200\ V}$$

$$V_{BR} = E_B - E_R \qquad \text{...vector difference}$$
$$= 2 \times E_{ph} \times \cos 60° = 2 \times 200 \times \frac{1}{2} = \mathbf{200\ V}$$

Example 20.4. *Three star-connected impedances $Z_1 = 20 + j37.7\ \Omega$ per phase are in parallel with three delta-connected impedances $Z_2 = 30 - j159.3\ \Omega$ per phase. The line voltage is 398 volts. Find the line current, power factor, power and reactive volt-ampere taken by the combination.*

(Electrical Science, AMIE Winter 1993.)

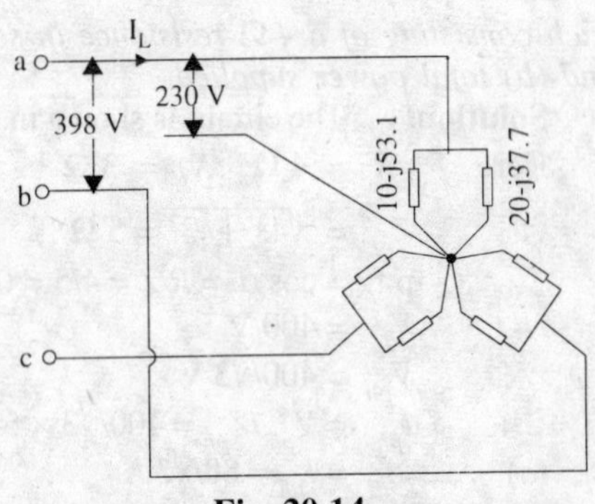

Fig. 20.14

Solution. The given delta-connected load can be converted into its equivalent star-connected load.

$$\text{Equivalent } Y\text{-load} = (30 - j159.3)/3$$
$$= (10 - j53.1)\ \Omega$$

The two loads are connected in parallel in each phase as shown in Fig. 20.14. Their combined impedance can be found as follows:

$$Z_1 = 20 + j37.7 = 42.7 \angle 62°$$
$$Z_2 = 10 - j53.1 = 54 \angle -79.3°$$
$$Z_1 + Z_2 = 30 - j15.4 = 33.7 \angle -27.2°$$
$$Z = \frac{Z_1 Z_2}{Z_1 + Z_2} = \frac{42.7 \angle 62° \times 54 \angle -79.3°}{33.7 \angle -27.2°}$$
$$= 68.4 \angle 9.9°$$
$$V_{ph} = 398/\sqrt{3} = 230\ \text{V}$$
$$I_{ph} = I_L = 230/68.4 \angle 9.9° = \mathbf{3.36 \angle -9.9°\ A}$$

Power factor = cos 9.9° = **0.985 (lead)**

$$\text{Power consumed} = \sqrt{3} \times V_L \times I_L \times \cos\phi$$
$$= \sqrt{3} \times 398 \times 3.36 \times 0.985 = \mathbf{2280\ W}$$

$$\text{Reactive power} = \sqrt{3} \times V_L \times I_L \times \sin\phi$$
$$= \sqrt{3} \times 398 \times 3.36 \times 0.172 = \mathbf{400\ VAR}$$

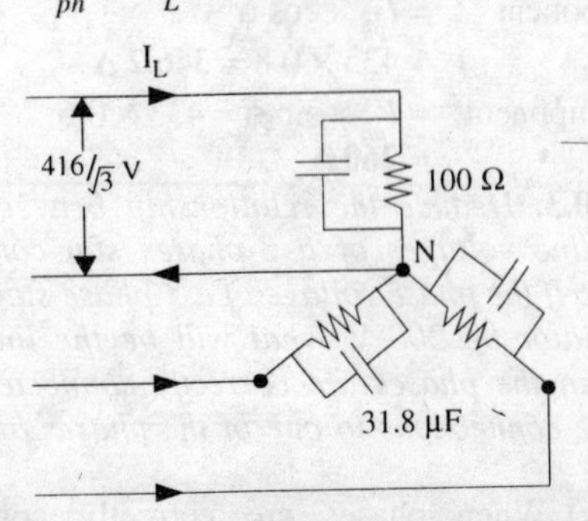

Fig. 20.15

Example 20.5. *Each phase of a star-connected load consists of a non-reactive resistance of 100 Ω in parallel with a capacitor of capacitance 31.8 μF.*

Calculate the line current, the power absorbed, the total kVA and the power factor when connected to a 416V, 3-phase, 50 Hz supply.

Solution. The circuit is shown Fig. 20.15.

$$V_{ph} = (416/\sqrt{3}) \angle 0° = 240 \angle 0° = (240 + j0)$$

Admittance of each phase is

$$Y_{ph} = \frac{1}{R} + j\omega C = \frac{1}{100} + j2\pi \times 50 \times 31.8 \times 10^{-6} = 0.01 + j0.01$$

$$\therefore \quad \mathbf{I}_{ph} = \mathbf{V}_{ph}.\mathbf{Y}_{ph} = 240\,(0.01 + j0.01) = 2.4 + j2.4 = 3.39 \angle 45°$$

Since $I_{ph} = I_L$ —for a star-connection

$\therefore \quad I_L$ = **3.39 A**

Power factor = cos 45° = **0.707 (leading)**

Now $\quad \mathbf{V}_{ph} = (240 + j0)$ and $\mathbf{I}_{ph} = (2.4 + j2.4)$

$$\mathbf{P}_{VA} = (240 + j0)(2.4 - j2.4)$$
$$= 240 \times 2.4 - j2.4 \times 240 = 576 - j576$$
$$= 814.4 \angle -45° \quad \text{...per phase}$$

Now total power $= 3 \times 576 = 1728$ W = **1.728 kW**

Total volt-amperes $= 814.4 \times 3 = 2443$ VA

Kilovolt–amperes = **2.443 kVA**

20.7. Neutral Current in Unbalanced Star-Connection

With the help of a vector diagram, it can be proved that the neutral current I_N in a balanced star-connection is zero. Fig. 20.16 (*a*) shows the vectors for the three equal phase currents. I_R, I_Y and I_B drawn at a mutual phase difference of 120°. The vector combination I_R and I_Y is I_{RY}. This vector is equal in magnitude to I_B but is in phase opposition to it. Hence, their vector sum is zero which means that $I_N = 0$.

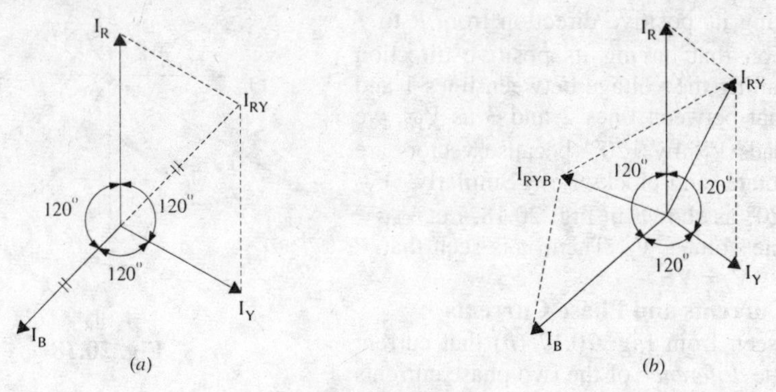

Fig. 20.16

Fig. 20.16 (*b*) represents current vectors for an unbalanced star-connection. As seen, the three currents are unequal in magnitude but are at 120° to each other. Here, vector combination of I_R and I_Y gives I_{RY} whereas vector addition of I_{RY} and I_B gives I_{RYB} which represents I_N. As seen, in this case I_N is not zero.

20.8. Delta (Δ) or Mesh Connection

In this form of inter-connection, the *dissimilar* ends of the three phase windings are joined together i.e., the 'starting' end of one phase is joined to the 'finishing' end of the other phase and so on as shown in Fig. 20.17* (*a*). In other words, the three windings are joined in series to form a closed mesh as shown in Fig. 20.17 (*b*).

Three leads are taken out from the three junctions as shown and outward directions are taken as positive.

It might look as if this sort of inter-connection results in short-circuiting the three windings. However, if the system is balanced, then sum of three voltages round the closed mesh is zero,

*As an aid to memory, remember that first letter D of Dissimilar is the same as that of Delta.

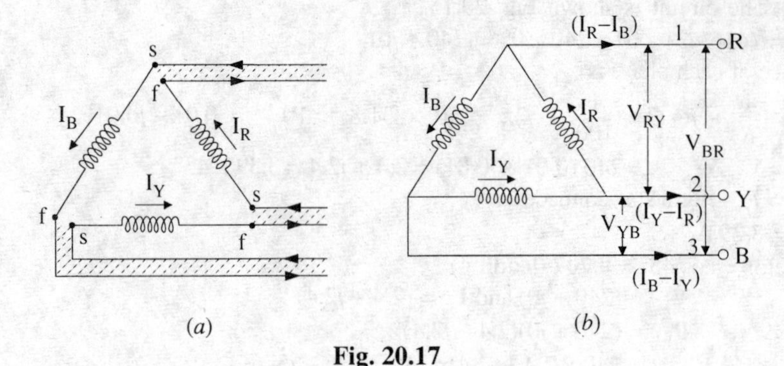

Fig. 20.17

hence no current of fundamental frequency can flow around the mesh when the terminals are open. It should be clearly understood that at any instant, the e.m.f. in one phase is equal and opposite to the resultant of those in the other two phases.

This type of connection is also referred to as 3-phase, 3-wire-system.

(*i*) **Line Voltages and Phase Voltages**

It is seen from Fig. 20.17 (*b*) that there is only one phase winding completely included between any pair of terminals. Hence, in Δ-connection, the voltage between any pair of lines is equal to the phase voltage of the phase winding connected between the two lines considered. Since phase sequence is *R, Y, B,* hence the voltage having its positive direction from *R* to *Y* leads by 120° on that having its positive direction from *Y* to *B*. Calling the voltage between lines 1 and 2 as V_{RY} and that between lines 2 and 3 as V_{YB}, we find that V_{RY} leads V_{YB} by 120° (because vectors are supposed to rotate anti-clockwise). Similarly, V_{YB} leads V_{BR} by 120° as shown in Fig. 20.18. Let V_{RY} = V_{YB} = V_{BR} = line voltage V_L. Then, it is seen that

$$V_L = V_{ph}$$

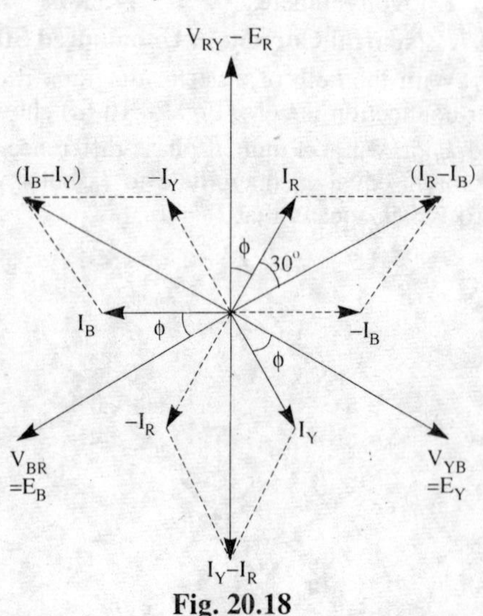

Fig. 20.18

(*ii*) **Line Currents and Phase Currents**

It will be seen from Fig. 20.17 (*b*) that current in each line is the *difference* of the two phase currents flowing through that line.

For example

Current in line 1 is $I_1 = I_R - I_B$
Current in line 2 is $I_2 = I_Y - I_R$
Current in line 3 is $I_3 = I_B - I_Y$
} vector difference

Current in line No. 1 is found by compounding I_R with I_B reversed and its value is given by the diagonal of the parallelogram of Fig. 20.18. The angle between I_R and I_B reversed (*i.e.*, $-I_B$) is 60°. If $I_B = I_R = I_Y$ = phase current I_{ph} (say), then

Current in line No. 1 is

$$I_1 = 2 \times I_{ph} \times \cos(60°/2)$$
$$= 2 \times I_{ph} \times \sqrt{3}/2$$
$$= \sqrt{3}\, I_{ph}$$

Similarly, current in line No. 2 is

$$I_2 = I_Y - I_R \qquad \text{...vector difference}$$
$$= \sqrt{3}\, I_{ph}$$

and current in line No. 3 is

$$I_3 = I_B - I_Y \quad \text{...vector difference}$$

$$= \sqrt{3}\, I_{ph}$$

Since all line currents are equal in magnitude *i.e.*,

$$I_1 = I_2 = I_3 = I_L$$

$$\therefore \quad I_L = \sqrt{3}\, I_{ph}$$

With reference to Fig. 20.18 it should be noted that

(*a*) line currents are 120° apart

(*b*) line currents are 30° behind the respective phase currents

(*c*) the angle between the line current and the corresponding line voltage is $(30 + \phi)$ with the current lagging.

(*iii*) **Power**

Power/phase = $V_{ph}\, I_{ph} \cos \phi$

Total power $= 3 \times V_{ph}\, I_{ph} \cos \phi$

Now $V_{ph} = V_L$, but $I_{ph} = I_L/\sqrt{3}$

Hence, in terms of line values, the above expression for power becomes

$$P = 3 \times V_L \times \frac{I_L}{\sqrt{3}} \times \cos \phi = \sqrt{3}\, V_L I_L \cos \phi$$

where ϕ is the phase power factor angle.

20.9. Balanced *Y*/Δ and Δ/*Y* Conversions

In view of the above relationship between line and phase currents and voltages, any balanced *Y*-connected system may be completely replaced by an equivalent Δ-connected system. For example, a 3-phase, *Y*-connected system having the voltage of V_L and line current I_L may be replaced by a Δ-connected system in which phase voltage is V_L and phase currrent is $I_L/\sqrt{3}$.

Similarly, a balanced *Y*-connected load having equal branch impedances each of ZL ϕ may be replaced by an equivalent Δ-connected load whose each phase impedance in 3Z ϕ. This equivalence is shown in Fig. 20.19 (*a, b*).

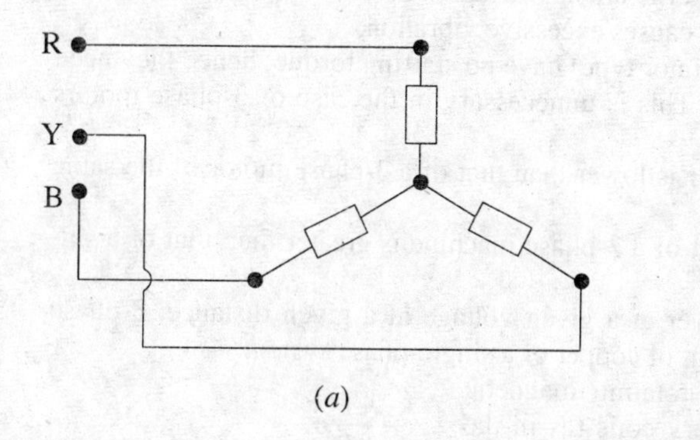

(*a*)

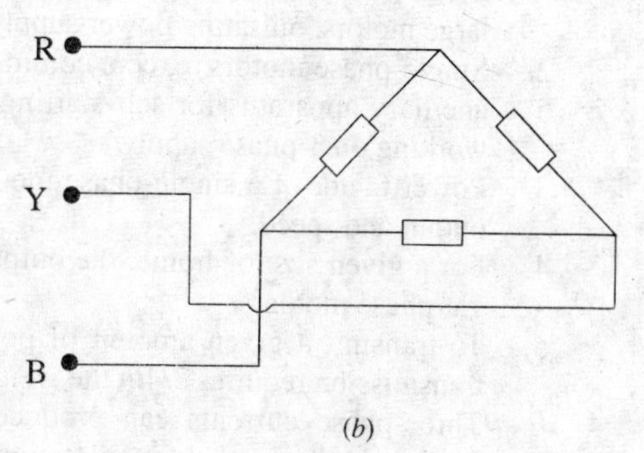

(*b*)

Fig. 20.19

For a balanced star-connected load, let

V_L = line voltage; $\quad I_L$ = line current

Z_Y = impedance/phase; $\quad \therefore \; V_{ph} = V_L/\sqrt{3}$,

$I_{ph} = I_L$; $\quad Z_Y = V_L/(\sqrt{3}\, I_L)$

Now, in the equivalent Δ-connected system, the line voltages and currents must have the same values as in the *Y*-connected system, hence we must have

$$V_{ph} = V_L, \quad I_{ph} = I_L/\sqrt{3}$$

$\therefore \quad Z_\Delta = V_L/(I_L/\sqrt{3}) = \sqrt{3}\, V_L/I_L = 3Z_Y$ $\quad (\because V_L/I_L = \sqrt{3}\, Z_Y)$

$\therefore \quad Z_\Delta \angle \phi = 3Z_Y \angle \phi$

or $\quad Z_\Delta = 3Z_Y$ or $Z_Y = Z_\Delta / 3$

20.10. Comparison: Star and Delta Connections

(a) Star Connection

1. Alternators are usually star-wound for the reason that phase voltage has to be only $1/\sqrt{3}$ of the line voltage, whereas for a delta-winding, the phase voltage has to be equal to the line voltage. Now, the number of conductors per phase in the winding of an alternator or motor, for a given frequency and flux, is directly proportional to the phase voltage. Hence, for a given line voltage, fewer turns/phase are required with *Y*-connection than with a delta-connection.
2. With star connection, the system of distribution mains can be arranged to suit both lighting and power circuits without using transformers. Heating and lighting circuits (up to 260 V) are put across neutral and any line wire (Fig. 20.26) whereas the 3-phase motors, running at higher voltages can be joined across the lines directly (Fig. 20.26).
3. Another advantage of *Y*-connection is that the neutral point of the alternators can be (and usually is) earthed. In that case, the p.d. between each line and earth is equal to the phase voltage *i.e.*, $1/\sqrt{3}$ of the line voltage. Hence, if through fault, line conductor is earthed, the insulators will have to bear $1/\sqrt{3}$ of the line voltage (*i.e.*, 57.7%) only. But in the case of a delta connection, if any line conductor is earthed, the insulators will have to bear full line voltage. Hence, there would be produced a correspondingly higher stress in the insulators with greater liability to breakdown.

(b) Delta Connection

1. The advantage of this connection is that transformers, in general, work more satisfactorily.
2. It is the only connection suitable for such machines as rotary converters.
3. This connection is much used for comparatively small low-voltage three-phase motors.

20.11. Comparison Between Single- and 3-phase Supply Systems

1. Power in a single-phase system is pulsating (at twice the frequency of voltage). This is not objectionable for lighting (if flicker is not appreciable) or for small motors. But with large motors, pulsating power supply causes excessive vibration.
2. Single-phase motors (except commutator type) have no starting torque, hence they need ancillary apparatus for self-starting. This is unnecessary in the case of 3-phase motors working on 3-phase supply.
3. Power factor of a single-phase motor is lower than that of a 3-phase motor of the same output and speed.
4. For a given size of frame, the output of a 3-phase machine is greater than that of a single-phase motor.
5. To transmit a given amount of power at a given voltage at a given distance, 3-phase transmission requires 3/4*th* the weight of copper of a single-phase system.
6. Three-phase currents can produce rotating magnetic fields when passed through stationary coils (as in the case of induction and synchronous motors).

Example. 20.6. *A 3-phase, 3-line, 400 V a.c. source is connected to energise a balanced star-connected load drawing 5 A at 0.5 p.f. (lagging). Calculate:*

(*i*) *impedance of the one leg of the star-connected load;*

(*ii*) *impedance of one leg of the equivalent delta-connected load;*

(*iii*) *Draw the phasor diagram for each of the above load-showing phase voltage, line voltage, phase current, line current.* **(Electric Circuits-I, Punjab Univ. 1993)**

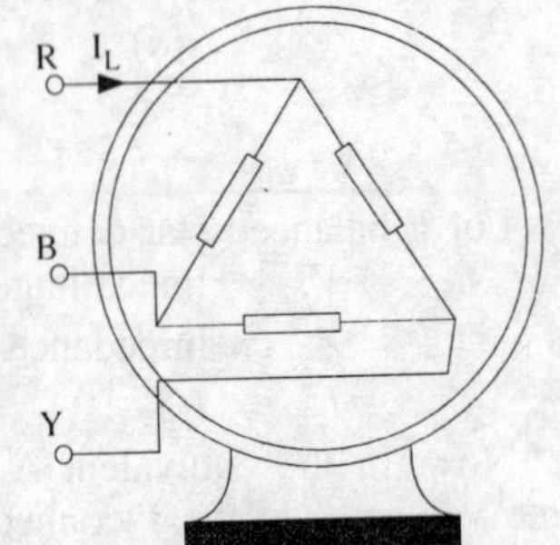

Fig. 20.20

Solution. $V_{ph} = 400/\sqrt{3} = 331$ V, $I_{ph} = I_L = 5$A

(*i*) $Z_{ph} = 231/5 = \mathbf{46.2\ \Omega}$

(*ii*) $Z_\Delta = 3Z_Y = 3 \times 46.2 = \mathbf{138.6\ \Omega}$

The phasor diagram is shown in Fig. 20.20.

Example. 20.7. *Describe the basic features of a balanced 3-phase system.*

A 3-phase motor operating off a 400 V balanced system develops 18.65 kW at an efficiency of 0.87 per unit and a power factor of 0.82. Calculate the line current and the phase current if the windings are delta-connected. **(Electrical Science, AMIE Summer 1993)**

Solution. The delta-connected induction motor is shown in Fig. 20.20. Since the motor output is 18.65 kW and its efficiency is 0.87.

Motor input $= 18.65/0.87 = 21.437$ kW

$\therefore \quad 21437 = \sqrt{3} \times 400 \times I \times 0.82$

$\therefore \quad I_L = \mathbf{37.7\ A}, I_{ph} = 37.7/\sqrt{3} = \mathbf{21.77\ A}$

Example. 20.8. *Three impedances each of (15 – j20) Ω are connected in mesh across a 3-phase 400 V a.c. supply. Determine the phase current, line current, active power and reactive power drawn from the supply.* **(Electric Circuits-I, Punjab Univ. 1994)**

Solution. The mesh-connected impedances are shown in Fig. 20.21.

$V_{ph} = V_L = 400$ V

$I_{ph} = V_{ph}/Z_{ph}, \quad Z_{ph} = 15 - j20 = 25\angle{-53.1°}$

$\therefore \quad I_{ph} = 400/25\angle{-53.1°} = 16 - \angle{-53.1°}$ A

$I_L = 16\sqrt{3} = \mathbf{27.7\ A}$, $\cos\phi = \cos 53.1° = 0.6$

Active power $= \sqrt{3} \times V_L \times I_L \times \cos\phi$

$= \sqrt{3} \times 400 \times 27.7 \times 0.6 = \mathbf{11{,}520\ \ W}$

Reactive power $= \sqrt{3} \times V_L \times I_L \times \sin\phi$

$= \sqrt{3} \times 400 \times 27.7 \times 0.8 = \mathbf{15{,}360\ \ VAR}$

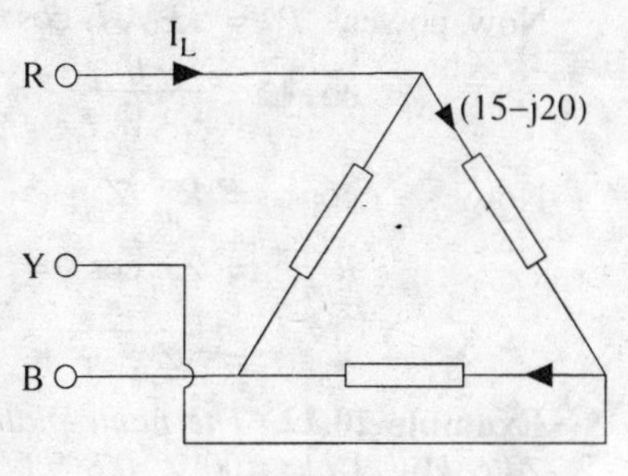

Fig. 20.21

Example 20.9. *Three similar coils, each of resistance 20 Ω and inductance 0.5 H are connected (a) in star (b) in delta to a 3-phase, 50-Hz, 400 V (between lines) supply. Calculate the line current and the total power absorbed.*

Solution. $R_{ph} = 20\ \Omega, X_L = 2\pi f L = 2\pi \times 0.5 \times 50 = 157\ \Omega$

$Z_{ph} = \sqrt{20^2 + 157^2} = 158.2\ \Omega$

$\cos\phi = R_{ph}/Z_{ph} = 20/158.2 = 0.1264$

(*a*) Star Connection

$V_{ph} = 400/\sqrt{3} = 231$ V; $\quad Z_{ph} = 158.2\ \Omega$

$I_{ph} = 231/158.2 = 1.46$ A; $I_L = \mathbf{1.46\ A}$

Power $= \sqrt{3}\ V_L I_L \cos\phi = \sqrt{3} \times 400 \times 1.46 \times 0.1264 = \mathbf{127.8\ W}$

or $\quad P = 3 I_{ph}^2 R_{ph} = 3 \times 1.46^2 \times 20 = \mathbf{127.8\ W}$

(*b*) Delta Connection

$V_{ph} = V_L = 400$ V; $\quad Z_{ph} = 158.2\ \Omega$

$I_{ph} = 400/158.2$ A

$I_L = \sqrt{3} \times I_{ph} = \sqrt{3} \times 400/158.2 = \mathbf{4.38\ A}$

(It may be noted that for the same V_L and Z_{ph}, line current for delta connection is three times the line current for star connection)

Power $= \sqrt{3} \times 400 \times 4.38 \times 0.1264 = \mathbf{383.4\ W}$

(**Note**: Power is also three times)

Example 20.10. *Three similar resistors are connected in star across 400-V, 3-phase lines.*

The line current is 5 A. Calculate the value of each resistor. To what value should the line voltage be changed to obtain the same line current with the resistors delta-connected?

Solution. **Star Connection**

$I_L = I_{ph} = 5 \text{ A}$

$V_{ph} = 400/\sqrt{3} = 231 \text{ V}$

$R_{ph} = 231/5 = \mathbf{46.2\ \Omega}$

Delta Connection

$I_L = 5 \text{ A}$ ...(given)

$I_{ph} = 5/\sqrt{3} \text{ A}$

$R_{ph} = 46.2\ \Omega$ ...found above

$V_{ph} = I_{ph} \times R_{ph} = 5 \times 46.2/\sqrt{3} = \mathbf{133.3\ V}$

(**Note**: Voltage needed is 1/3rd the star value)

Example 20.11. *Three identical impedances are connected in delta to a 3-phase supply of 400-V. The line current is 34.65 A and the total power taken from the supply is 14.4 kW. Calculate the resistance and reactance values of each impedance.*

Solution. $V_{ph} = V_L = 400 \text{ V}; \quad I_L = \sqrt{3}\, I_{ph}$

$I_{ph} = I_L/\sqrt{3} = 34.65/\sqrt{3} \text{ A}$

$Z_{ph} = V_{ph}/I_{ph} = 400 \times \sqrt{3}/34.65 = 20\ \Omega$

Now, power, $P = \sqrt{3}\, V_L I_L \cos\phi$

$$\therefore \quad \cos\phi = \frac{P}{\sqrt{3}\, V_L I_L} = \frac{14{,}400}{\sqrt{3} \times 400 \times 34.65}$$

Now $\cos\phi = R_{ph}/Z_{ph}$

$$\therefore \quad R_{ph} = Z_{ph}\cos\phi = 20 \times \frac{14{,}400}{\sqrt{3} \times 400 \times 34.65} = \mathbf{12\ \Omega}$$

$$X_{ph} = \sqrt{Z_{ph}^2 - R_{ph}^2} = \sqrt{20^2 - 12^2} = \mathbf{16\ \Omega}$$

Example 20.12. *The name-plate of an a.c. motor reads as follows:*

74.6 kW; 3-ph; 400-V; 0.8 p.f. delta-connected. If the efficiency of the motor at the rated output is given as 88%, calculate the current drawn from the supply and the phase currents in the motor windings.

Solution.

Motor output $= 746 \times 1000 \text{ W}$

Motor input $= 74{,}600/0.88 \text{ W}$

$V_L = 400 \text{ V}, \quad \cos\phi = 0.8$

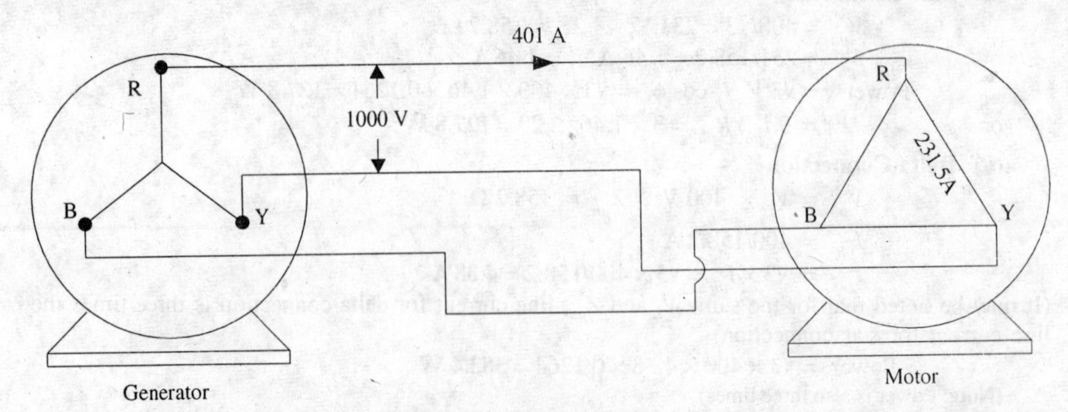

Fig. 20.22

$\therefore$ motor input $= \sqrt{3}\, V_L I_L \cos\phi$

or $74{,}600/0.88 = \sqrt{3} \times 400 \times I_L \times 0.8$

I_L = **153 A**; $I_{ph} = 153/\sqrt{3}$ = **88.3 A**

Example 20.13. *A 3-phase star-connected 1000 volt alternator supplies power to a 500 kW delta-connected induction motor. If the motor power factor is 0.8 lagging and its efficiency 0.9, find the current in each alternator and motor phase.* **(Electrical Science, AMIE Winter 1994)**

Solution. The alternator and induction motor circuit connections are shown in Fig. 20.22

Motor input power $= 500 \times 10^3/0.9\ \Omega$

$\therefore \quad 500 \times 10^3/0.9 = \sqrt{3} \times 1000 \times I_L \times 0.8$

$\therefore \quad I_L = 401$ A, $I_{ph} = 401\sqrt{3} = 231.5$ A

Hence, current in each motor phase is **231.5 A** and the current in each alternator phase is **401 A.**

Example 20.14. *Three 40 Ω non-inductive resistances are connected (i) in star (ii) in mesh across 200 V, 3-phase lines. Calculate the power taken from the mains in each case. If one of the two resistors is disconnected, what would be the power taken from the mains in each case?* **(Electrical Engg. A.M.Ae.S.I. June 1992)**

Solution. *(i)* **Star Connection**

$V_{ph} = 200/\sqrt{3}$ V

$I_{ph} = 200/\sqrt{3} \times 40 = 5/\sqrt{3}$ A

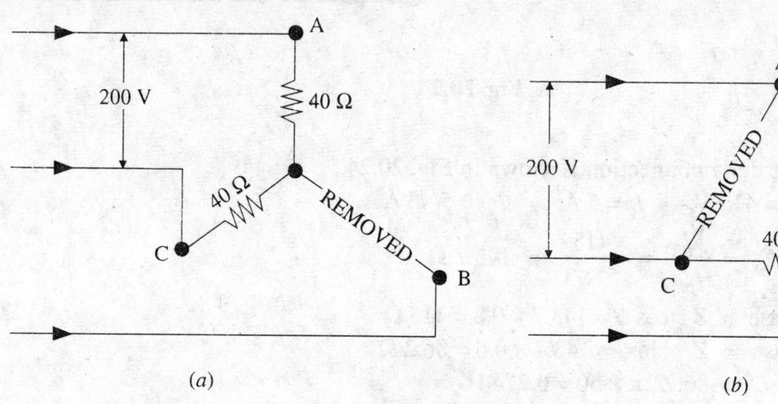

Fig. 20.23

$$P = \sqrt{3}\, V_L I_L \cos\phi$$

$$= \sqrt{3} \times 200 \times \frac{5}{\sqrt{3}} \times 1 = \mathbf{1000\ W}$$

(ii) **Delta Connection**

$V_{ph} = 200$ V; $\quad R_{ph} = 40\ \Omega$

$I_{ph} = 200/40 = 5$ A; $\quad I_L = 5 \times \sqrt{3}$ A

$P = \sqrt{3} \times 200 \times 5 \times \sqrt{3} \times 1 =$ **3,000 W**

When one of the resistors is disconnected

(i) **Star Connection** [Fig. 20.23 *(a)*]

The circuit no longer remains a 3-phase circuit but consists of two 40 Ω resistors in series across a 200 V supply. Current in lines *A* and *C* is = 200/80 = 2.5 A.

Power absorbed in both = 200 × 2.5 = **500 W**

Hence, by disconnecting one resistor, the power consumption is reduced by half.

(ii) **Delta Connection** [Fig. 20.23 *(b)*]

In this case, currents in A and C lines remain as usual 120° out of phase with each other.

Current in each phase = 200/40 = 5 A

Power consumption in both = $2 \times 5^2 \times 40$ = **2,000 W**

(or $P = 2 \times 5 \times 200 = 2000$ W)

In this case, when one resistor is disconnected, the power consumption is reduced by one-third.

Example 20.15. *Three coils are connected in delta to a 3-phase, 3-wire, 415 V, 50 Hz supply and take a line current of 5 A at 0.8 power factor lagging. Calculate the resistance and inductance of the coils.*

If the coils are star-connected to the same supply, calculate the line current and the total power.

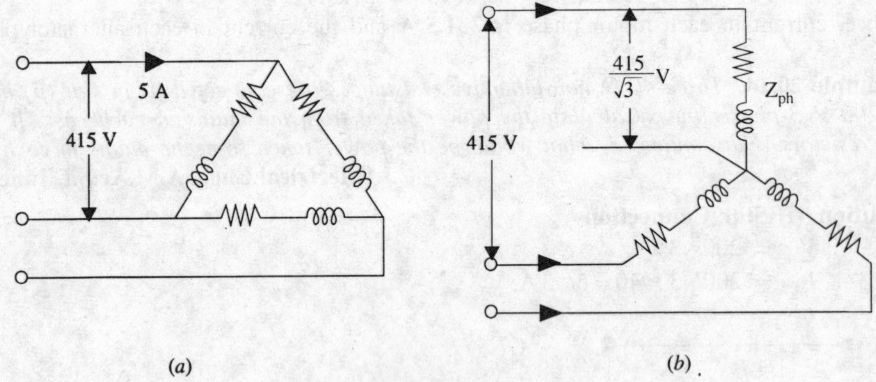

Fig. 20.24

Solution. The delta connection is shown in Fig. 20.24.

$V_{ph} = V_L = 415$ V; $\quad I_L = 5$ A; $\quad I_{ph} = 5/\sqrt{3}$ A

$$\therefore \quad Z_{ph} = \frac{V_{ph}}{I_{ph}} = \frac{415}{5/\sqrt{3}} = 143.7\ \Omega$$

Resistance/phase = $Z_{ph} \cos \phi = 143.7 \times 0.8 =$ **115 Ω**

Reactance/phase = $Z_{ph} \sin \phi = 143.7 \times 0.6 = 86.2\ \Omega$

$\therefore \quad L = 86.2/2\pi \times 50 =$ **0.274 H**

The star connection is shown in Fig. 20.24 (*b*).

$V_{ph} = 415/\sqrt{3}$ V; $\quad Z_{ph} = 143.7\ \Omega$

$I_{ph} = 415/143.7 \times \sqrt{3} = 2.89/\sqrt{3}$ A

$I_L = 2.89/\sqrt{3}$ A

Power consumed $= \sqrt{3}\, V_L I_L \cos \phi$

$= \sqrt{3} \times 415 \times (2.89/\sqrt{3}) \times 0.8 =$ **960 W**

Example 20.16. *The load connected to a 3-phase supply comprises three similar coils connected in star. The line currents are 25 A and the kVA and kW inputs are 20 and 11 respectively. Find the line and phase voltages, the kVAR input and the resistance and reactance of each coil.*

If the coils are now connected in delta to the same three-phase supply, calculate the line currents and the power taken.

Solution. **Star Connection**

$\cos \phi$ = kW/kVA = 11/20; $\quad I_L = 25$ A

P = 11 kW = 11,000 W

Now $\quad P = \sqrt{3} V_L I_L \cos \phi$

$\therefore \quad 11{,}000 = \sqrt{3} \times V_L \times 25 \times 11/20 \qquad \therefore \quad$ **V_L = 462 V**

$$V_{ph} = 462/\sqrt{3} = \mathbf{267\ V}$$
$$kVAR = \sqrt{kVA^2 - kW^2} = \sqrt{20^2 - 11^2} = \mathbf{16.7}$$
$$Z_{ph} = 267/25 = 10.68\ \Omega$$
$$\therefore \quad R_{ph} = Z_{ph} \times \cos\phi = 10.68 \times 11/20 = \mathbf{5.87\ \Omega}$$
$$X_{ph} = Z_{ph} \times \sin\phi = 10.68 \times 0.835 = \mathbf{8.92\ \Omega}$$

Delta Connection

$$V_{ph} = V_L = 462\ V; \qquad Z_{ph} = 10.68\ \Omega$$
$$\therefore \quad I_{ph} = 462/10.68\ A; \qquad I_L = \sqrt{3} \times 462/10.68 = \mathbf{75\ A}$$
$$P = \sqrt{3} \times 462 \times 75 \times 11/20 = \mathbf{33{,}000\ W}$$

Example 20.17. *A 3-phase, star-connected system with 230-V between each phase and neutral has resistances of 4, 5 and 6 Ω respectively in the three phases. Estimate the current flowing in each phase and the neutral current. Find the total power absorbed.*

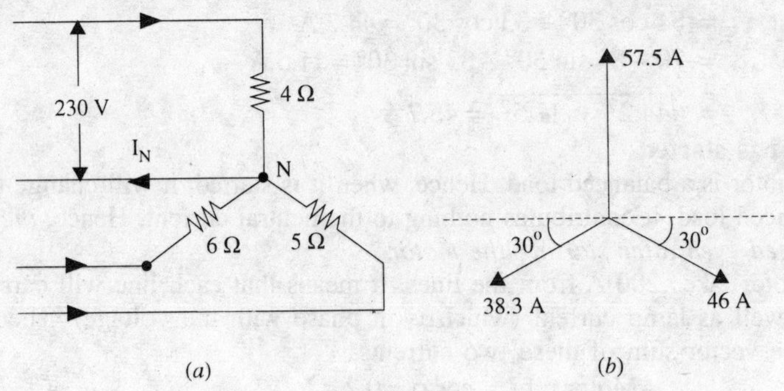

Fig. 20.25

Solution. Here V_{ph} = 230 V [Fig. 20.25 (*a*)]

Current in 4 Ω resistor = 230/4 = 57.5 A

Current in 5 Ω ,, = 230/5 = 46 A

Current in 6 Ω ,, = 230/6 = 38.3 A

These currents are mutually displaced by 120°. The neutral current I_N is the vector sum* of these three currents. I_N can be obtained by splitting up these three phase currents into their *X*-components and *Y*-components and then by combining them together.

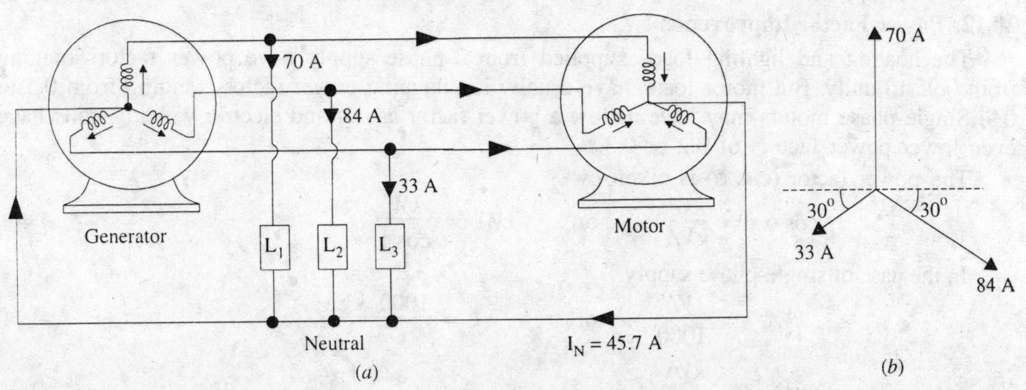

Fig. 20.26

*Some writers disagree with this on the ground that according to Kirchhoff's First Law, at any junction,

$$I_N + I_R + I_Y + I_B = 0 \qquad \therefore \quad I_N = -(I_R + I_Y + I_B)$$

Hence, according to them, numerical value of I_N is the same but its phase is changed by 180°.

X-component $= 46 \cos 30° - 38.3 \cos 30° = 6.64$ A

Y-component $= 57.5 - 46 \sin 30° - 38.3 \sin 30° = 15.3$ A

$\therefore \quad I_N = \sqrt{6.64^2 + 15.3^2} = $ **16.71 A**

Total power absorbed = 230 (57.5 + 46 + 38.3) = **32,610 W**

Example 20.18. *A 3-phase, 4-wire system supplies power at 400 V and lighting at 230 V. If the lamps in use require 70, 84 and 33 A in each of the lines, what should be the current in the neutral wire? If a 3-phase motor is now started, taking 200 A from the lines at a power factor of 0.2, what should be the current in each line and the neutral? Find also the total power supplied to the lamps and the motor.*

Solution. The lamp and motor connections are shown in Fig. 20.26.

When motor is not started

The neutral current is the vector sum of lamp currents. Again, splitting up the currents into their X- and Y-components, we get

X-component $= 84 \cos 30° - 33 \cos 30° = 44.2$ A

Y-component $= 70 - 84 \sin 30° - 33 \sin 30° = 11.5$ A

$I_N = \sqrt{44.2^2 + 11.5^2} = $ **45.7 A**

When motor is started

A 3-phase motor is a balanced load. Hence, when it is started, it will change the line currents but being a balanced load, it contributes nothing to the neutral current. Hence, *the neutral current remains unchanged even after starting the motor.*

Now, the motor takes 200 A from the lines. It means that each line will carry motor current (which lags) as well as lamp current (which is in phase with the voltage). The current in each line would be the vector sum of these two currents.

Motor p.f. $= \cos \phi = 0.2$

$\sin \phi = 0.9799$ ...from tables

Active component of motor current $= 200 \times 0.2 = 40$ A

Reactive component of motor current $= 200 \times 0.9799 = 196$ A

(*i*) Current in first line $= \sqrt{(40 + 70)^2 + 196^2} = $ **224.8 A**

(*ii*) Current in second line $= \sqrt{(40 + 84)^2 + 196^2} = $ **232 A**

(*iii*) Current in third line $= \sqrt{(40 + 33)^2 + 196^2} = $ **210.6 A**

Power supplied to lamps $= 230(33 + 84 + 70) = $ **43,000 W**

Power supplied to motor $= \sqrt{3} \times 200 \times 400 \times 0.2 = $ **27,700 W**

20.12. Power Factor Improvement

The heating and lighting loads supplied from 3-phase supply have power factors ranging from 0.95 to unity. But motor loads have usually low lagging power factors ranging from 0.5 to 0.9. Single-phase motors may have as low a power factor as 0.4 and electric welding units have even lower power factors of 0.2 or 0.3.

The power factor ($\cos \phi$) is given by

$$\cos \phi = \frac{kW}{kVA} \quad \text{or} \quad kVA = \frac{kW}{\cos \phi}$$

In the case of single-phase supply,

$$kVA = \frac{VI}{1000} \quad \text{or} \quad I = \frac{1000\ kVA}{V}$$

$$\therefore \quad I \propto kVA$$

In the case of 3-phase supply

$$kVA = \frac{\sqrt{3}\, V_L I_L}{1000} \quad \text{or} \quad I_L = \frac{1000\ kVA}{\sqrt{3} \times V_L}$$

$\therefore \quad I_L = kVA$

In each case, the *kVA* is directly proportional to current. The chief disadvantage of a low p.f. is that the current required for a given power's is very high. This fact leads to the following undesirable results:

(*i*) Large kVA for a given amount of power

All electrical machinery like alternators, transformers, switch-gears and cables are limited in current carrying capacity by the permissible temperature rise which is proportional to I^2. Hence, they may all be fully loaded with respect to their rated *kVA* without delivering their full power. Obviously, it is possible for an existing plant of a given *kVA* rating to increase its earning capacity (which is proportional to the power supplied in kW) if the overall power factor is improved *i.e.*, raised.

(*ii*) Poor voltage regulation

When a load having a low lagging power factor is switched on, there is a large voltage drop in the supply voltage because of the increased voltage drop in the supply lines and transformers. This drop in voltage adversely affects the starting torques of motors and necessitates expensive voltage stabilizing equipment for keeping the consumer's voltage fluctuations within the statutory limits. Hence, all supply undertakings try to encourage consumers to have a high power factor.

Example 20.19. *A 50 MVA, 11 kV, 3-ϕ alternator supplies full load at a lagging power factor of 0.7. What would be the percentage increase in earning capacity if the power factor is increased to 0.95?*

Solution. The earning capacity is proportional to the power (in MW or kW) supplied by the alternator.

MW supplied at 0.7 lagging p.f. $= 50 \times 0.7 = 35$

MW supplied at 0.95 lagging p.f. $= 50 \times 0.95 = 47.5$

Increase in MW $= 12.5$

The increase in earning capacity is proportional to 12.5

$\therefore$ percentage increase in earning capacity $= (12.5/35) \times 100 = \mathbf{35.7}$

20.13. Power Factor Correction Equipment

The following equipment is generally used for improving or correcting the power factor.

1. Synchronous Motors (or capacitors)

These machines draw leading *kVAR* when they are over-excited and, especially, when they are running idle. They are employed for correcting the power factor in bulk and have the special advantage that the amount of correction can be varied by changing their excitation.

2. Static Capacitors

They are installed to improve the power factor of a group of a.c. motors and are practically loss free (*i.e.*, they draw a current leading in phase by 90°). Since their capacitances are not variable, they tend to over-compensate on light loads unless arrangements for automatic switching off the capacitor bank are made.

3. Phase Advancers

They are fitted with individual machines.

However, it may be noted that the economic degree of correction to be applied in each case, depends upon the tariff arrangement between the consumers and the supply authorities.

Example 20.20. *A 3-phase, 37.3 kW, 440 V, 50 Hz induction motor operates on full-load with an efficiency of 89% and at a power factor of 0.85 lagging. Calculate the total kVA rating of capacitors required to raise the full-load power factor to 0.95 lagging. What will be the capacitance per phase if the capacitors are delta-connected?*

Solution. Motor input $= \dfrac{\text{output}}{\text{efficiency}} = \dfrac{37.3}{0.89} = 41.91$ kW

Power factor 0.85 (lag)

$\cos\phi_1 = 0.85; \; \sin\phi_1 = 0.527$ ...from tables

$\therefore$ motor $kVA_1 = kW/\cos\phi_1 = 41.91/0.85 = 49.3$

Motor $kVAR_1 = kVA_1 \times \sin\phi_1 = 49.3 \times 0.527 = 25.98$ (lag)

Power factor 0.95 (lag)

The power supplied [= *AB* in Fig. 20.27(*b*)] has to remain the same but the *kVAR* is to be reduced from *BD* to *BC* such that *AC* subtends the new power factor angle ϕ_2 with *AB*.

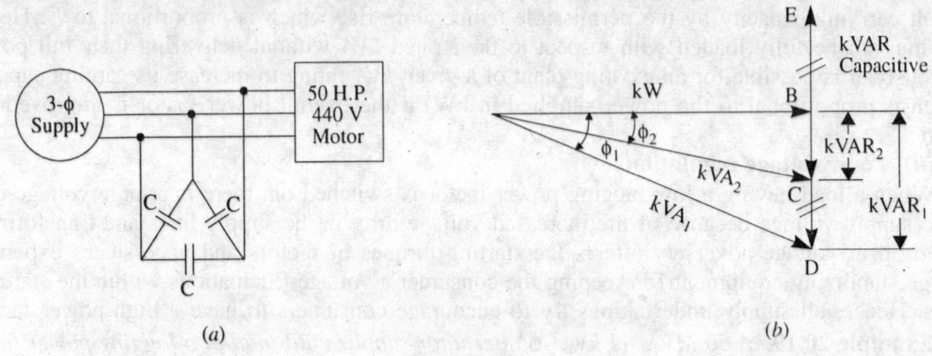

Fig. 20.27

$\cos\phi_2 = 0.95; \qquad \tan\phi_2 = 0.329$

Now, $BC/AB = \tan\phi_2$

$\therefore \quad BC = 41.91 \times 0.329 = 13.79 \qquad \therefore \quad kVAR_2 = 13.79$

Hence, *leading kVAR* supplied by capacitors

$= kVAR_1 - kVAR_2 = 25.98 - 13.79 = \mathbf{12.19}$

This is represented by *BE* in the *kVA* diagram of Fig. 20.27 (*b*) and equals *DC*. This is the *kVAR* (or *kVA* because capacitors are assumed loss-free) of the capacitors required.

Capacitor *VAR* (or *VA*) = 12,190 $\qquad \therefore \quad 12{,}190 = \sqrt{3}\, V_L\, I_L$

$\therefore$ capacitor line current, $I_L = \dfrac{12{,}190}{\sqrt{3} \times 440} = 15.99$ A

Since capacitors are delta-connected,

$\therefore$ current drawn by each capacitor = $15.99/\sqrt{3} = 9.23$ A

$\therefore$ capacitive reactance/phase = $440/9.23 = 47.65\ \Omega$

Now $\quad X_C = \dfrac{1}{2\,\pi f C} \quad$ or $\quad 47.65 = \dfrac{1}{2\pi \times 50 \times C}$

$\therefore \quad C = 1/314 \times 47.65 = 66.79 \times 10^{-6}$ F = $\mathbf{66.8\ \mu F}$

Example 20.21. *Three impedance coils, each having a resistance of 20 Ω and a reactance of 15 Ω, are connected in star to a 400-V, 3-φ, 50-Hz supply. Calculate (i) the line current (ii) power supplied and (iii) the power factor.*

If three capacitors, each of the same capacitance, are connected in delta to the same supply so as to form parallel circuit with the above impedance coils, calculate the capacitance of each capacitor to obtain a resultant power factor of 0.95 lagging.

Solution. $V_{ph} = 400/\sqrt{3}$ V, $Z_{ph} = \sqrt{20^2 + 15^2} = 25\ \Omega$

$\cos\phi_c = R_{ph}/Z_{ph} = 20/25 = 0.8$ (lag); $\sin\phi_c = 0.6$ (lag)

where ϕ_c is the power factor angle of the coils.

When capacitors are not connected

(*i*) $I_{ph} = 400/25 \times \sqrt{3} = 9.24$ A $\qquad \therefore \quad I_L = \mathbf{9.24\ A}$

(*ii*) $P = \sqrt{3}\, V_L\, I_L \cos\phi_c$

$= \sqrt{3} \times 400 \times 9.24 \times 0.8 = \mathbf{5{,}120\ W}$

(*iii*) Power factor = **0.8 (lag)**

When capacitors are connected (Fig. 20.28)

As stated earlier, by using a capacitor bank, the power factor can be raised. But it should be remembered that capacitors themselves do not consume any power. Their function is merely to neutralize the *lagging kVAR* of the load by their leading *kVAR*.

Let ϕ be the phase angle of the combined load such that $\cos \phi = 0.95$.

Total *VA* of the load $= \sqrt{3} \times 400 \times 9.24 = 6400$

$\therefore \quad kVA_1 = 6.4 \text{ kVA}$

$kW = kVA_1 \times \cos \phi_c = 6.4 \times 0.8$

$= 5.12$ (= *AB* in Fig. 20.29)

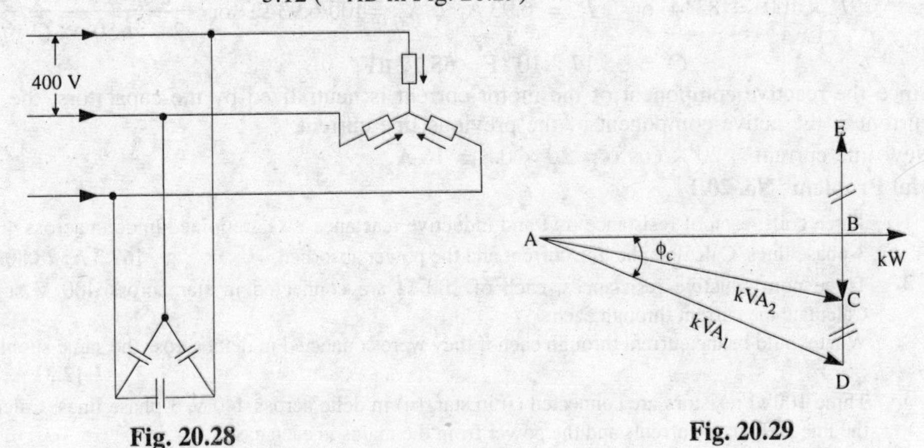

Fig. 20.28 **Fig. 20.29**

$kVAR_1 = kVA_1 \times \sin \phi_c = 6.4 \times 0.6$

$= 3.84$ (= *BD* in Fig. 20.29)

Since the capacitors themselves do not absorb any power, the power input of the combined load remains unaltered *i.e.*, 5.12 kW.

Now, $\quad kVA \text{ input} = 5.12/\cos \phi = 5.12/0.95 = 5.39$

$kVA_2 = 5.39$ (= *AC* in Fig. 20.29)

$kVAR_2 = 5.39 \times \sin \phi = 5.39 \times 0.3123$

$= 1.6884$ (= *BC* in Fig. 20.29)

It means that before joining the capacitors, the $kVAR_1 = 3.84$. Now it is 1.684. Hence, the leading *kVAR* contributed by the capacitor bank

$= BD - BC = kVAR_1 - kVAR_2$

$= 3.84 - 1.684 = 2.156$ (leading)

The leading *kVAR* of the capacitor has been shown separately by the vector *BE*.

If I_L is the line current, then

$2156 = \sqrt{3}\, V_L I_L = \sqrt{3} \times 400 \times I_L$

$\therefore \quad I_L = 2156/\sqrt{3} \times 400 = 3.112 \text{ A}$

Since the capacitors are delta-connected, current drawn by each capacitor

$= 3.112/\sqrt{3} = 1.797 \text{ A}$

$\therefore$ reactance/phase $= 400/1.797 = 222.6\ \Omega$

$$X_C = \frac{1}{2\pi f C} \quad \therefore \quad 222.6 = \frac{1}{2\pi \times 50 \times C}$$

$\therefore \quad C = 14.31 \times 10^{-6} = \mathbf{14.31\ \mu F}$

Example 20.22. *A three-phase, star-connected motor, connected across a 3-phase, star-connected supply of 400 V, 50 Hz takes a current of 20 A at 0.8 power factor lagging. Determine the capacitance of the capacitors per phase that are to be connected in delta across the terminals of the motor to raise the power factor to unity and also the new value of the supply line current with the capacitors connected.*

Solution. Motor input $kVA = \sqrt{3}.\ V_L I_L/1000 = \dfrac{\sqrt{3} \times 400 \times 20}{1000} = 13.856$

$kVAR = kVA \sin\phi = 13.856 \times 0.6 = 8.314$ (lagging)

$kW = kVA \cos\phi = 13.856 \times 0.8 = 11.085$

Since the resultant power factor is unity, it means that all the capacitors together should contribute as much *leading kVAR* as the lagging *kVAR* of the motor.

If I_C is the current drawn by each capacitor, then total volt-amperes taken by the three delta-connected capacitors $= 3I_C \times 400$

$\therefore \quad 3I_C \times 400 = 8314$ or $I_C = 6.93$ A; $X_C = 400/6.93\ \Omega$ or $\dfrac{1}{2\pi \times 50 \times C} = \dfrac{400}{6.93}$,

$\therefore \quad C = 55.17 \times 10^{-6}$ F $=$ **55.17 μF**

Since the reactive component of the motor current is neutralized by the capacitors, the new line current is the active component of the previous line current.

New line current $= 20 \times \cos\phi = 20 \times 0.8 =$ **16 A.**

Tutorial Problems No. 20.1

1. Three coils, each of resistance 6 Ω and inductive reactance 8 Ω are joined in delta across 400 V, 3-phase lines. Calculate the line current and the power absorbed. **[69.3 A; 28,800 W]**
2. Three non-inductive resistances, each of 100 Ω are connected in star across 400 V supply. Calculate the current through each.
   What would be the current through each if they were connected in delta across the same supply? **[2.31 A; 4 A]**
3. Three 100 Ω resistors are connected (*i*) in star, (*ii*) in delta across 440 V, 3-phase lines. Calculate the line and phase currents and the power from the mains in each case. **[(*i*) 2.54 A; 1936 W (*ii*) 17.62 A; 5808 W]**
4. Three similar coils, each having a resistance of 5 Ω and an inductance of 0.021 H, are connected (*i*) in star (*ii*) in delta to a 440 V, 3 phase, 50 Hz supply. Calculate the line current and the total power absorbed in each case. **[(*i*) 30.8 A; 14.2 kW (*ii*) 92.5 A, 42.6 kW]**
5. A star-connected 3-phase load consists of three similar impedances. When the load is connected to a 3-phase, 500 V, 50 Hz supply, the line current is 28.85 A and the p.f. is 0.8 lagging.
   (*a*) Calculate:
   (*i*) the total power taken by the load ;
   (*ii*) the resistance of each phase of the load
   (*b*) If the load were re-connected in delta and supplied from the same 3-phase system, calculate the current flowing in each line. **[(*a*) 20 kW; 8 W (*b*) 86.55 A]**
6. Three identical coils connected in delta to a 400 V, 3-phase supply take a total power of 50 kW and a line current of 90 A. Find (*a*) the phase currents (*b*) the power factor (*c*) the total apparent power taken by the coils. **[(*a*) 52 A (*b*) 0.73 (*c*) 68.5 kVA]**
7. Three similar choke coils are connected in star to a 3-phase supply. If the line currents are 15A, the total power consumed is 11 kW and volt-ampere input is 15 kVA, find the line and phase voltages, the kVAR input and the resistance of reactance of each coil.
   [577.3 V; 333.3 V; 10.2 kVAR; 16.3 Ω ; 15.1 Ω]
8. A balanced load takes 20 kVA at a p.f. of 0.8 lagging from a 250 V, 50 Hz, 3-phase supply. Calculate the values of the components in the circuit if the load is (*i*) delta-connected (*ii*) star-connected.
   What would be the values of the three line currents if one of the fuses 'blows' whilst in star-connection. **[(*i*) 7.5 Ω; 17.9 mH (*ii*) 2.5 Ω; 5.96 mH; 40 A; 40 A; 0]**
9. Three coils each having a resistance of 40 Ω and reactance 30 Ω are connected in star across a 415 V line, 3-phase, 50 Hz supply. Calculate:
   (*a*) the line current (*b*) the total kVA (*c*) the p.f. (*d*) the total power (*e*) value of each of the three capacitors which must be connected in star to the supply in order to bring the overall p.f. to unity.
   [(*a*) 4.8 A (*b*) 3.45 kVA (*c*) 0.8 lag (*d*) 2.76 kW (*e*) 38.2 μF]

20.14. Power Measurement in 3-phase Circuits

Following methods are available for measuring power in a 3-phase load.

(*a*) Three Wattmeter Method

In this method, three wattmeters are inserted in each phase and the algebraic sum of their readings gives the total power consumed by the 3-phase load.

(*b*) Two Wattmeter Method

(*i*) This method gives true power in a 3-phase circuit without regard to balance or wave-form provided in the case of *Y*-connected load, the neutral of the source of power. Or if there is a neutral connection, the neutral wire should not carry any current. This is only possible if the load is perfectly balanced and there are no harmonics present of triple frequency or any other multiples of that frequency.

(*ii*) This method can also be used for 3-phase, 4-wire system in which the neutral wire carries the neutral current. In this method, the current coils of the wattmeters are supplied from current transformers inserted in the principal line wires in order to get the correct magnitude and phase differences of the currents in the current coils of the wattmeters, because in the 3-phase, 4-wire system, the sum of the instantaneous currents in the principal line wires is not necessarily equal to zero as in 3-phase 3-wire system.

(*c*) One Wattmeter Method

In this method, a single wattmeter is used to obtain the two readings which are obtained by two wattmeters in the two-wattmeter method. This method can, however, be used when the load is balanced.

20.15. Three Wattmeter Method

A wattmeter consists of (*i*) a low-resistance current coil which is inserted in series with the line carrying the current and (*ii*) a high-resistance pressure coil which is connected across the two points whose potential difference is to be measured.

A wattmeter shows a reading which is proportional to the product of the current through its current coil, the p.d. across its potential or pressure coil and cosine of the angle between this voltage and current.

As shown in Fig. 20.30, in this method, three wattmeters are inserted in each of the three phases of the load whether Δ-connected or *Y*-connected. The current coil of each wattmeter carries

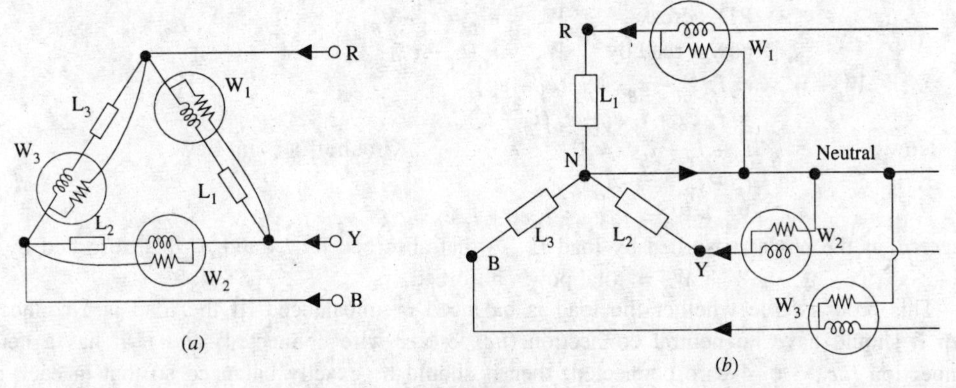

Fig. 20.30

the current of one phase only and the pressure coil measures the phase voltage of that phase. Hence, each wattmeter measures the power in a single phase. The algebraic sum of the readings of the three wattmeters must give the total power in the load.

The difficulty with this method is that under ordinary conditions, it is not generally feasible to break into the phases of a delta-connected nor is it always possible, in the case of a *Y*-connected load, to get at the neutral point which is required for connection as shown in Fig. 20.30 (*b*). However, it is not necessary to use three wattmeters to measure power, two wattmeters can be used for the purpose as shown below.

20.16. Two Wattmeter Method—*Balanced or Unbalanced load*

As shown in Fig. 20.31, the current coils of the two wattmeters are inserted in *any* two lines and their potential coils joined to the third line. It can be proved that the sum of the instantaneous

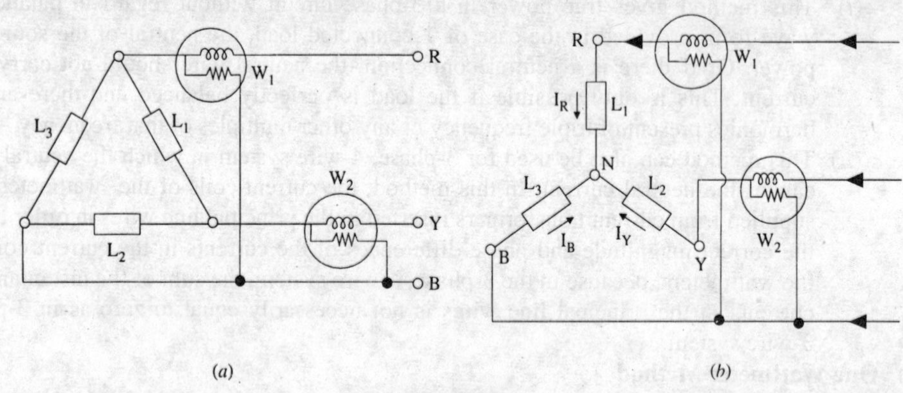

Fig. 20.31

powers indicated by W_1 and W_2 gives the instantaneous power absorbed by the three loads L_1, L_2 and L_3. The star-connected load is considered in the following discussion, although it can be equally applied to a Δ-connected load because a Δ-connected load can always be replaced by an equivalent *Y*-connected load.

Now, before we consider the currents through and p.d. across each wattmeter, it may be pointed out that *it is important to take the direction of the voltage through the circuit the same as that taken for the current when establishing the readings of the two wattmeters.*

Instantaneous current through $W_1 = i_R$

" " P.D. across $W_1 = e_{RB} = e_R - e_B$

" " power read by $W_1 = i_R(e_R - e_B)$

Instantaneous current through $W_2 = i_Y$

" " P.D. across $W_2 = e_{YB} = e_Y - e_B$

" " power read by $W_2 = i_Y(e_Y - e_B)$

$$\therefore \quad W_1 + W_2 = i_R(e_R - e_B) + i_Y(e_Y - e_B)$$

$$= i_R e_R + i_Y e_Y - e_B(i_R + i_Y)$$

Now $i_R + i_Y + i_B = 0$ ...Kirchhoff's Point Law

$$\therefore \quad i_R + i_Y = -i_B$$

$$\therefore \quad W_1 + W_2 = i_R e_R + i_Y e_Y + i_B e_B = p_1 + p_2 + p_3$$

where p_1 is the power absorbed by load L_1, p_2 that absorbed by L_2 and p_3 that absorbed by L_3.

$$\therefore \quad W_1 + W_2 = \text{total power absorbed.}$$

This proof is true whether the load is balanced or unbalanced. If the load is *Y*-connected, then it should have no neutral connection (*i.e.*, 3-ϕ, 3-wire connected) and if it has a neutral connection (*i.e.*, 3-ϕ, 4-wire connected) then it should be exactly balanced so that in each case there is no neutral current i_N, otherwise Kirchhoff's Point Law will give $i_N + i_R + i_Y + i_B = 0$.

We have considered *instantaneous readings* but, in fact, the moving system of the wattmeter, due to its inertia, cannot quickly follow the variations taking place in a cycle, hence it indicates the *average* power.

$$\therefore \quad W_1 + W_2 = \frac{1}{T}\int_0^T i_R\, e_{RB}\; dt + \frac{1}{T}\int_0^T i_Y\, e_{YB}\; dt$$

20.17. Two Wattmeter Method—*Balanced load*

If the load is balanced, then power factor of the load can also be found from the two wattmeter readings. The *Y*-connected load in Fig. 20.31 (*b*) will be assumed inductive. The phasor diagram for such a balanced *Y*-connected load is shown in Fig. 20.32. We shall now consider the problem in terms of r.m.s. values instead of instantaneous values.

Let V_R, V_Y and V_B be the r.m.s. values of the three-phase voltages and I_R, I_Y and I_B be the r.m.s. values of the currents. Since these voltages and currents are assumed sinusoidal, they can be represented by vectors, the currents lagging behind their phase voltages by ϕ.

Current through wattmeter W_1 [Fig. 20.31 (*b*)] = I_R, p.d. across voltage coil of W_1 is $V_{RB} = V_R - V_B$ (vectorially).

Thus, V_{RB} is found by compounding V_R and V_B reversed as shown in Fig. 20.32. It is seen that phase difference between V_{RB} and

$$I_R = (30 - \phi).$$

$\therefore$ reading of $W_1 = V_{RB} I_R \cos(30° - \phi)$

Similarly, as seen from Fig. 20.25 (*b*).

Current through $W_2 = I_Y$

p.d. across $W_2 = V_{YB} = V_Y - V_B$ ...vectorially

Again V_{YB} is found by compounding V_Y with V_B reversed as shown in Fig. 20.32. The angle between I_Y and V_{YB} is $(30° + \phi)$.

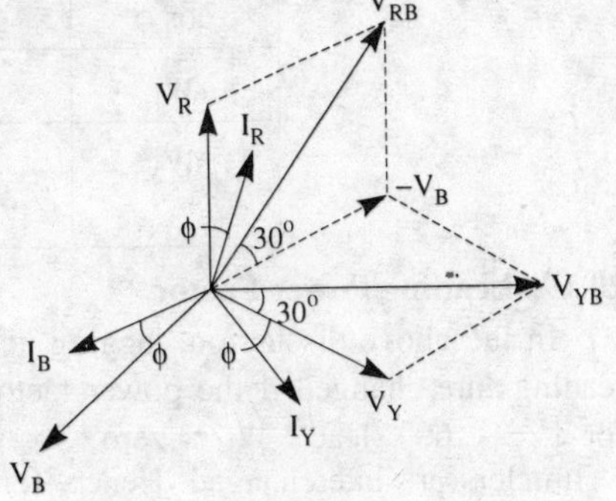

Fig. 20.32

Reading of $W_2 = V_{YB} I_Y \cos(30° + \phi)$

Since the load is balanced,

$$V_{RB} = V_{YB} = \text{line voltage, } V_L$$
$$I_R = I_Y = \text{line current, } I_L$$

$$\therefore \quad W_1 = V_L I_L \cos(30 - \phi)$$
$$W_2 = V_L I_L \cos(30 + \phi)$$

$$\therefore \quad W_1 + W_2 = V_L I_L \cos(30 - \phi) + V_L I_L \cos(30 + \phi)$$
$$= V_L I_L [\cos 30° \cos\phi + \sin 30° \sin\phi + \cos 30° \cos\phi - \sin 30° \sin\phi]$$
$$= V_L I_L (2 \cos 30° \cos\phi) = \sqrt{3}\, V_L I_L \cos\phi$$
$$= \text{total power in the 3-phase load}$$

Hence, the sum of the two wattmeter readings gives the total power consumption in the 3-phase load.

It should be noted that phase sequence *RYB* has been assumed above. Reversal of phase sequence will interchange the readings of the two wattmeters.

20.18. Variations in Wattmeter Readings

It has been shown above that for a *lagging power factor*

$$W_1 = V_L I_L \cos(30 - \phi)$$
$$W_2 = V_L I_L \cos(30 + \phi)$$

From this it is clear that individual readings of the wattmeters not only depend on *the load but upon its power factor also*. We will consider the following cases:

(*a*) When $\phi = 0$ *i.e.*, power factor is unity (*i.e.*, resistive load) then

$$W_1 = W_2 = V_L I_L \cos 30°$$

Both wattmeters indicate equal and positive *i.e.*, up-scale readings.

(*b*) When $\phi = 60°$ *i.e.*, power factor = 0.5 (lagging)

Then, $W_2 = V_L I_L \cos(30° + 60°) = 0$. Hence, the power is measured by W_1 alone.

(*c*) When $90° > 60°$ *i.e.*, 0.5 > p.f. > 0, then W_1 is still positive but reading of W_2 is reversed because the phase angle between the current and voltage is more than 90°. For getting the total power, the reading of W_2 is to be subtracted from that of W_1.

Under this condition, W_2 will read 'down-scale' *i.e.*, backwards. Hence, to obtain a reading on W_2, it is necessary to reverse either its pressure coil or current coil, usually the former. *All readings taken after reversal of pressure coil are to be taken as negative.*

(*d*) When $\phi = 90°$ (*i.e.*, p.f. = 0 as for pure inductive or capacitive load), then

$$W_1 = V_L I_L \cos(30° - 90°) = V_L I_L \sin 30°$$
$$W_2 = V_L I_L \cos(30° + 90°) = -V_L I_L \sin 30°$$

Then, the two readings are equal but of opposite sign.

$$\therefore \quad W_1 + W_2 = 0$$

Summarizing the above, we have for a *lagging* power factor,

ϕ	0°	60°	90°
cos ϕ	1	0.5	0
W_1	+ ve	+ ve	+ ve
W_2	+ ve $W_1 = W_2$	0	– ve $W_1 = W_2$

20.19. Leading Power Factor

In the above discussion, lagging angles are taken positive. Now, we will see how wattmeter readings are changed if the power factor becomes leading. For $\phi = +60°$ (lag), W_2 is zero. But for $\phi = -60°$ (lead), W_1 is zero. So, we find that for angles of lead, the readings of the two wattmeters are interchanged. Hence, for a *leading* power factor

$$W_1 = V_L I_L \cos(30° + \phi)$$
$$W_2 = V_L I_L \cos(30° - \phi)$$

20.20. Power Factor—*Balanced Load*

In case the load is balanced (and currents and voltages are sinusoidal) and for a *lagging* power factor

$$W_1 + W_2 = V_L I_L \cos(30° - \phi) + V_L I_L \cos(30° + \phi)$$
$$= \sqrt{3}\, V_L I_L \cos\phi \quad \text{...(i)}$$

Similarly

$$W_1 - W_2 = V_L I_L \cos(30 - \phi) - V_L I_L \cos(30 + \phi)$$
$$= V_L I_L (2 \times \sin\phi \times 1/2) = V_L I_L \sin\phi \quad \text{...(ii)}$$

Dividing (*ii*) by (*i*), we have

$$\tan\phi = \sqrt{3}\,\frac{(W_1 - W_2)^*}{(W_1 + W_2)} \quad \text{...(iii)}$$

Knowing tan ϕ and hence ϕ, the value of power factor cos ϕ can be found by consulting the trigonometrical tables. It should, however, be kept in mind that if W_2 reading has been taken after reversing the pressure coil *i.e.*, if W_2 is negative, then the above relation becomes

$$\tan\phi = \sqrt{3}\,\frac{W_1 - (-W_2)}{W_1 + (-W_2)} = \sqrt{3}\,\frac{W_1 + W_2}{W_1 - W_2}$$

We may express power factor in terms of the ratio of the two wattmeter readings as under.

Let $\dfrac{\text{smaller reading}}{\text{larger reading}} = \dfrac{W_2}{W_1} = r$

*For a leading power factor, this expression becomes

$$\tan\phi = \frac{\sqrt{3}\,(W_1 - W_2)}{W_1 + W_2}.$$

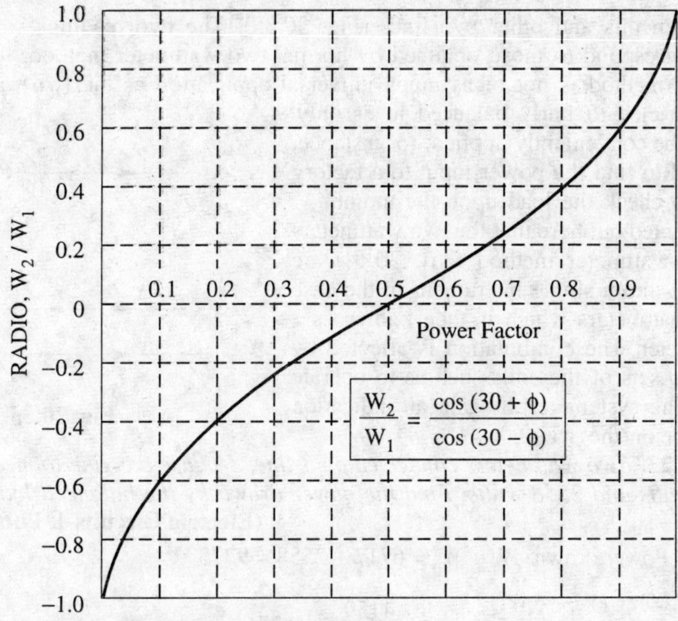

Fig. 20.33

Then, from equation (*iii*) above,

$$\tan\phi = \frac{\sqrt{3}\,[1-(W_2/W_1)]}{1+(W_2/W_1)} = \frac{\sqrt{3}\,(1-r)}{1+r}$$

Now $\qquad \sec^2\phi = 1+\tan^2\phi \qquad$ or $\qquad \frac{1}{\cos^2\phi} = 1+\tan^2\phi$

$$\cos\phi = \frac{1}{\sqrt{1+\tan^2\phi}} = \frac{1}{\sqrt{1+3\left(\frac{1-r}{1+r}\right)^2}} = \frac{1+r}{2\sqrt{1-r+r^2}}$$

If r is plotted against $\cos\phi$, then a curve called watt-ratio curve is obtained as shown in Fig. 20.33.

20.21. Reactive Power

We have seen that $\tan\phi = \sqrt{3}\,\frac{(W_1 - W_2)}{(W_1 + W_2)}$

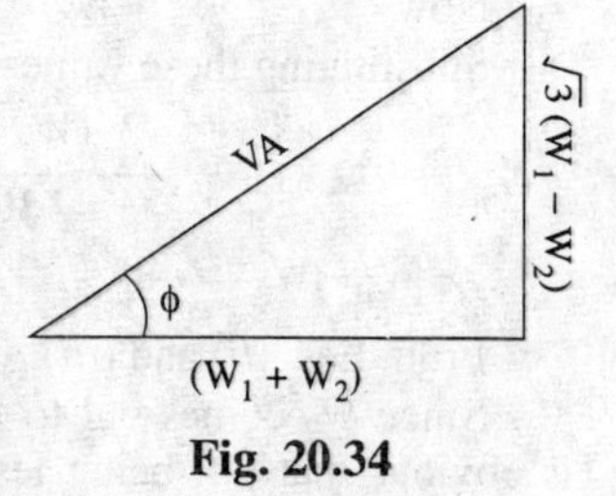

Fig. 20.34

Since the tangent of the angle of lag between phase current and phase voltage of a circuit is always equal to the ratio of the reactive power to the true power (in watts), it is clear that $\sqrt{3}\,(W_1 - W_2)$ represents the reactive power (Fig. 20.34). Hence, for a balanced load, the reactive power is given by $\sqrt{3}$ times the difference of the readings of the two wattmeters used to measure the power for a 3-phase circuit by the two wattmeter method. It may also be proved mathematically as follows:

$$\sqrt{3}(W_1 - W_2) = \sqrt{3}[V_L I_L \cos(30° - \phi) - V_L I_L \cos(30° + \phi)]$$
$$= \sqrt{3}\, V_L I_L (\cos 30° \cos\phi + \sin 30° \sin\phi - \cos 30° \cos\phi + \sin 30° \sin\phi)$$
$$= \sqrt{}\, V_L I_L \sin\phi$$

20.22. One Wattmeter Method

In this case, it is possible to apply two-wattmeter method by means of one wattmeter without breaking the circuit. The current coil is connected in any one line and the pressure coil is connected alternately between this and other two lines (Fig. 20.35). The two readings so obtained, for a balanced load, correspond to those obtained by normal two wattmeter method. It should be kept in mind that this method is not of as much universal application as the two-wattmeter method because it is restricted to fairly balanced loads only. However, it may be conveniently applied, for instance, when it is desired to find the power input to a factory motor in order to check the load upon the motor.

It may be pointed out here that the two wattmeters used in the two-wattmeter method (Art. 20.14) are usually combined into a single instrument in the case of switchboard wattmeters which is then known as a polyphase wattmeter. The combination is affected by arranging the two sets of the coils such as to operate on a single moving system resulting in an indication of the total power on the scale.

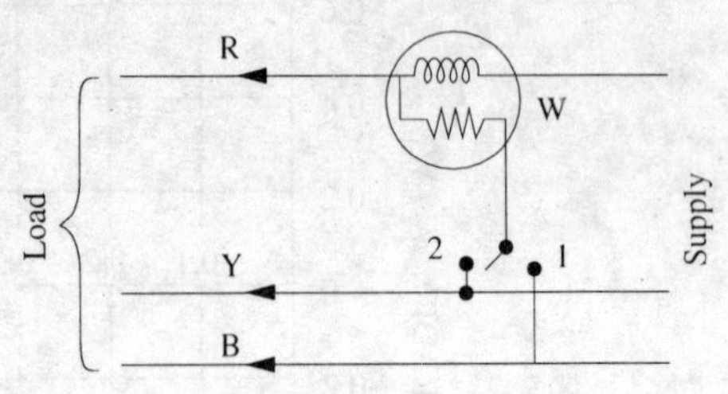

Fig. 20.35

Example 22.23. *Two wattmeters connected on 3-line, 3-phase a.c. line to measure a.c. power and read 6717 watts and 2558 watts. Find the power drawn by the balanced load and its power factor.* **(Electric Circuits-I, Punjab Univ. 1993)**

Solution. Power drawn, $W_1 + W_2 = 6717 + 2558 =$ **9275 W**

$$\tan\phi = \sqrt{3}\frac{(W_1 - W_2)}{(W_1 + W_2)} = \sqrt{3}\,\frac{4159}{9275} = 0.7766$$

$\therefore$ $\phi = 37.83°$, p.f. $= \cos 37.83° =$ **0.79 lag**

Example 20.24. *A 3-phase, 500 V motor load has a power factor of 0.4. Two wattmeters connected to measure the power show the input to be 30 kW. Find the readings on each instrument.* **(Electrical Engg. A.M.Ae.S.I. June 1994)**

Solution. As seen from Art. 17.20

$$\tan\phi = \frac{\sqrt{3}\,(W_1 - W_2)}{W_1 + W_2} \qquad ...(i)$$

Now, $\cos\phi = 0.4$

$\therefore$ $\phi = \cos^{-1}(0.4) = 66.6°$; $\tan 66.6° = 2.311$

Now, $W_1 + W_2 = 30$ ...(*ii*)

Substituting these values in equation (*i*) above, we get

$$2.311 = \frac{\sqrt{3}\,(W_1 - W_2)}{30}$$

$\therefore\ W_1 - W_2 = 40$ ...(*iii*)

From Eqs. (*ii*) and (*iii*), we have $W_1 =$ **35 kW** and $W_2 =$ **–5 kW**

Since W_2 comes out to be negative, second wattmeter reads 'down scale'. Even otherwise it is obvious that p.f. being less than 0.5, W_2 must be negative (Art. 17.18).

Example 20.25. *A 500 V, 3-phase motor has an output of 37.3 kW and operates at a p.f. of 0.85 lagging with an efficiency of 90%. Calculate the reading on each wattmeter connected to measure the input.*

Solution. Motor output $= 37.3 \times 1000 = 37{,}300$ W

Motor input $= 37{,}300/0.9 = 41{,}440$ W $= 41.44$ kW

$W_1 + W_2 = 41.44$ kW ...(*i*)

Now $\cos\phi = 0.85$, $\phi = 31°48'$, $\tan 31°48' = 0.62$

$$\therefore \quad 0.62 = \sqrt{3}\,\frac{W_1 - W_2}{W_1 + W_2} = \sqrt{3}\,\frac{W_1 - W_2}{41.44}$$

or $\quad W_1 - W_2 = 0.62 \times 41.44/\sqrt{3} = 14.86$ kW ...(ii)

From (i) and (ii), we get

$$W_1 = \mathbf{28.15\ kW}, \quad W_2 = \mathbf{13.29\ kW}$$

Example 20.26. *Two wattmeters connected to measure the input to a balanced 3-phase circuit indicate 2000 W and 500 W respectively. Find the power factor of the circuit (a) when both readings are positive (b) when latter reading is obtained after reversing the connection to the current coil of one instrument.*

Solution. (a) $W_1 + W_2 = 2500$ W; $\quad W_1 - W_2 = 1500$ W

$$\tan\phi = \sqrt{3} \times \frac{1500}{2500} = 1.0392$$

$\therefore \quad \phi = 46°6'$ and $\cos 46°6' = \mathbf{0.6934}$

(b)
$$\tan\phi = \sqrt{3}\frac{(W_1 + W_2)}{(W_1 - W_2)} = \sqrt{3} \times \frac{2500}{1500} = 2.887$$

$\therefore \quad \phi = 70°54' \quad \therefore \cos 70°54' = \mathbf{0.3272}$

Example 20.27. *Three identical coils, each having reactance of 20 Ω and a resistance of 20 Ω are connected in (a) star (b) in delta across 440 V, 3-phase line. Calculate for each method of connection, the line current and readings on each of the two wattmeters connected to measure the power.* **(Electrical Engg. Nagpur Univ. 1993)**

Solution. (a) **Star Connection**

Impedance/phase $= \sqrt{(20^2 + 20^2)} = 20\sqrt{2}\ \Omega$

$$V_{ph} = 440/\sqrt{3}\ \text{V}$$

$$I_{ph} = I_L = 440/\sqrt{3} \div 20\sqrt{2} = \mathbf{8.98\ A}$$

$$\tan\phi = 20/20 = 1 \quad \therefore \phi = 45°$$

$\therefore$ power factor $= \cos\phi = \cos 45° = 1/\sqrt{2}$

Total power taken $= \sqrt{3}\, V_L I_L \cos\phi = \sqrt{3} \times 440 \times 8.98 \times 1/\sqrt{2} = \mathbf{4{,}840\ W}$

If W_1 and W_2 are the wattmeter readings, then

$$W_1 + W_2 = 4{,}840 \quad \text{...(i)}$$

As
$$\tan 45° = 1 = \sqrt{3}\frac{(W_1 - W_2)}{W_1 + W_2}$$

$$W_1 - W_2 = 4840/\sqrt{3} = 2{,}795\ \text{W} \quad \text{...(ii)}$$

From equations (i) and (ii), we find

$$W_1 = \mathbf{3817.5\ W}; \quad \mathbf{W_2 = 1022.5\ W}$$

(b) **Delta Connection**

$$Z_{ph} = 20\sqrt{2}; \quad V_{ph} = 440\ \text{V}$$

$$I_{ph} = 440/20\sqrt{2} = 15.56\ \text{A}$$

$$I_L = \sqrt{3}\, I_{ph} = \sqrt{3} \times 15.56 = \mathbf{26.95\ A}$$

$$\cos\phi = \cos 45° = 1/\sqrt{2}$$

Total power $= \sqrt{3}\, V_L I_L \cos\phi$

$= \sqrt{3} \times 440 \times 26.95 \times 1/\sqrt{2} = 14{,}520$ W

(**Note.** This power is 3 times the *Y*-power)

$$\therefore \quad W_1 + W_2 = 14{,}520 \quad \text{...(iii)}$$

Again,
$$\tan 45° = 1 = \sqrt{3}\frac{(W_1 - W_2)}{W_1 + W_2}$$

$$\therefore \quad W_1 - W_2 = 14{,}520/\sqrt{3} = 8{,}385\ \text{W} \quad \text{...(iv)}$$

From equations (iii) and (iv), we find

$$W_1 = \mathbf{11{,}452.5\ W} \text{ and } W_2 = \mathbf{3067.5\ W}$$

Example 20.28. *Three identical coils are connected in star to a 200 V, three-phase supply and each takes 500 W. The power factor is 0.8 lagging. What will be the current and the total power if the same coils are connected in delta to the same supply? If the power is measured by two wattmeters, what will be their readings? Prove any formulae used.* **(Electrical Engg. A.M.Ae.S.I. Dec. 1991)**

Solution. **Star Connection**

Total power taken by the three identical star-connected coils $= 3 \times 500 = 1500$ W

$\therefore \quad 1500 = \sqrt{3} \times 200 \times I_L \times 0.8, \qquad \therefore \quad I_L = 5.4$ A

Delta Connection

When the same coils are connected in delta, the line current drawn as well as the power taken will be three times the star values.

$\therefore \quad P = 3 \times 1500 = \mathbf{4500\ W}, \ I_L = 3 \times 5.4 = \mathbf{16.2\ A}$

Now, $\cos\phi = 0.8$, $\phi = 36.87°$, $\tan\phi = 0.75$

If W_1 and W_2 are the wattmeters readings, then

$W_1 + W_2 = 4500$

$$\tan\phi = \sqrt{3}\left(\frac{W_1 - W_2}{W_1 + W_2}\right) \quad \text{or} \quad 0.75 = \sqrt{3}\left(\frac{W_1 - W_2}{4500}\right)$$

$W_1 - W_2 = 4500 \times 0.75/\sqrt{3} = 1949$ W ....(*ii*)

From (*i*) and (*ii*), we get

$W_1 =$ **3222 W,** $\quad W_2 =$ **1278 W**

Tutorial Problems No. 20.2

1. In a 3-phase, star-connected balanced circuit, the line voltage is 400 V, the line current 25 A and the p.f. is lagging. What will be the readings of the two wattmeters connected to measure the power in the circuit? **[3,928 W; 9,929 W]**
2. The power taken by a 3-phase, 400 V motor is measured by the two-wattmeter method and the readings of the two wattmeters are 460 and 780 W of respectively. Estimate the p.f. of the motor and the line current. **[0.9129; 1.96 A]**
3. Each of the two wattmeters connected to measure the input to a 3-phase circuit reads 10 kW on a balanced load when the p.f. is unity. What does each instrument read when the power factor falls to (*a*) 0.866 lagging (*b*) 0.5 lagging, the total 3-phase power remaining unchanged.
[(*a*) W1 = 13.33 kW; W2 = 6.66 kW (*b*) W1 = 20 kW, W2 = 0]

HIGHLIGHTS

1. For a *Y*-connection:
$V_{ph} = V_L/\sqrt{3}; \quad I_{ph} = I_L; \quad P = \sqrt{3}\, V_L I_L \cos\phi$
2. For a Δ-connection :
$V_{ph} = V_L; \quad I_{ph} = I_L/\sqrt{3}; \quad P = \sqrt{3}\, V_L I_L \cos\phi$
3. Both in 1-phase and 3-phase systems, the *kVA* is directly proportional to current *I*. The disadvantage of a low power factor is that current required for a given power is very high which fact leads to many undesirable results. The power factor can be improved by the following:
(*a*) Synchronous capacitors—used for correcting p.f. in bulk to any desired amount.
(*b*) Static capacitors—used for a group of motors but gives a fixed amount of correction.
(*c*) Phase advancers—are fitted to individual motors.
4. Three-phase power is generally measured by the two-wattmeter method. In this method, the current coils of the two wattmeters are connected in *any* two lines and their potential coils to the remaining third line. The sum of the two wattmeter readings gives the total power in the circuit. If the load is balanced, then its power factor can also be calculated from these two readings. The readings of the two wattmeters are:

$$\left.\begin{array}{ll} (i) & W_1 = V_L I_L \cos(30° - \phi) \\ & W_2 = V_L I_L \cos(30° + \phi) \\ (ii) & \tan\phi = \sqrt{3}\dfrac{(W_1 - W_2)}{W_1 + W_2} \end{array}\right\} \text{lagging p.f.}$$

$$\left.\begin{array}{ll} (iii) & W_1 = V_L I_L \cos(30 + \phi) \\ & W_2 = V_L I_L \cos(30 - \phi) \\ & \tan\phi = -\sqrt{3}\dfrac{(W_1 - W_2)}{W_1 + W_2} \end{array}\right\} \text{leading p.f.}$$

OBJECTIVE TESTS—20

A. Fill in the following blanks.

1. In a balanced 3-ϕ system, the line voltages are .... electrical degrees apart.
2. In a 3-ϕ, 4-wire system supplying a balanced load, current in the neutral is ....
3. In a 3-ϕ, *Y*-connection, line voltages are the vector .... of respective phase voltages.
4. If connections to one of the phases of a 3-ϕ, *Y*-connected alternator with phase voltage of 200 V are reversed, only one line voltage would be .... volt.
5. In a 3-ϕ delta connection, line voltage equals .... voltage.
6. When one of the three equal *Y*-connected load resistors fed by a balanced 3-phase supply is disconnected, power consumption is reduced by ....
7. Two-wattmeter method can be used to find the .... factor of a balanced 3-phase load.
8. Sum of the two wattmeter readings gives the .... power consumption in the load.

B. Answer True or False.

1. In a 3-ϕ, *Y*-connected system, line current equals phase current.
2. In a 3-phase, 3-wire system, the vector sum of voltages round the closed mesh is zero even when the system is unbalanced.
3. Everything else being the same, 3-phase transmission requires less copper than a single-phase system.
4. Three similar coils consume same power whether connected in delta or star to the same three-phase supply.
5. Two-wattmeter method gives true power in a 3-phase system only if it is balanced.
6. In the two-wattmeter method of power measurement, readings of the two wattmeters depend only on the load.
7. Two-wattmeter method can be used for measuring reactive power.
8. In the two-wattmeter method of 3-phase power measurement, one wattmeter reads zero when load p.f. lies between zero and 0.5.

C. Multiple Choice Questions

1. In a 3-phase, *Y*-connected system
   (*a*) line current equals phase current
   (*b*) line voltage equals phase voltage
   (*c*) line voltage and line current have a phase difference of 30°
   (*d*) line voltages and phase voltages are (30° + ϕ) apart.
2. Alternators are usually star-wound because
   (*a*) both lighting and power circuits can be supplied without using transformers
   (*b*) neutral wire can be used
   (*c*) fewer turns/phase are required
   (*d*) lighter insulation is needed
3. Three equal impedances are first connected in star across a balanced 3-phase supply. If they are now connected in delta across the same supply
   (*a*) phase current will be trippled
   (*b*) phase current will be doubled
   (*c*) line current will become one-third
   (*d*) power factor will be improved

4. A 3-phase, 4-wire, 230/440 V system is supplying lamp load at 230 V. If a 3-phase motor is now switched on across the same supply, then

(*a*) neutral current will increase

(*b*) all line currents will decrease

(*c*) neutral current will remain unchanged

(*d*) power factor will be improved

5. In the two-wattmeter method of measuring 3-phase power, the two wattmeters indicate equal and opposite readings when load power factor angle is —— degrees lagging.

(*a*) 60 (*b*) 0

(*c*) 30 (*d*) 90

6. When phase sequence at the 3-phase load is reversed

(*a*) phase powers are changed

(*b*) phase currents are changed

(*c*) phase currents change in angle but not in magnitude

(*d*) total power consumed to changed

7. In a three-phase measurement by two-wattmeter method both the wattmeters had identical readings. The power factor of the load would be

(*a*) 0 (*b*) 0.5

(*c*) 0.8 (*d*) unity

(Elect. Engg. A.M.Ae.S.I. Dec. 1993)

ANSWERS

A. 1. 120 2. zero 3. difference 4. 346 5. phase 6. half 7. power 8. total

B. 1. T 2. F 3. T 4. F 5. F 6. F 7. T 8. T

C. 1. *a* 2. *c* 3. *d* 4. *c* 5. *d* 6. *c* 7. *d*

21 TRANSFORMER

21.1. Working Principle of a Transformer

A transformer is a static (or stationary) piece of apparatus by means of which electric power in one circuit is transformed to electric power of the same frequency in another circuit. It can raise or lower the voltage in a circuit but with a corresponding decrease or increase in current. The physical basis of a transformer is *mutual induction* between two circuits linked by a common magnetic flux. In its simplest form, it consists of two inductive coils which are electrically separate but magnetically linked through a path of low reluctance as shown in Fig. 21.1. The two coils possess high mutual inductance. If one coil is connected to a source of alternating voltage, an alternating flux is set up in the laminated core, most of which is linked with the other coil in which it produces mutually-induced e.m.f. (according to Faraday's Laws of Electromagnetic Induction *viz.*, $e = M dI/dt$). If the second coil circuit is closed, a current flows in it and so electric energy is transferred (entirely magnetically) from the first coil to the second coil. The first coil, in which electric energy is fed from the a.c. supply mains, is called *primary winding* and the other, from which energy is drawn out, is called *secondary winding*. In brief, a transformer is a device that

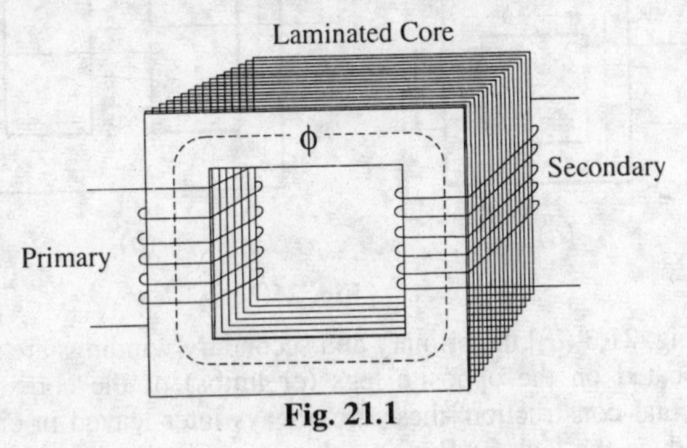

Fig. 21.1

1. transfers electric power from one circuit to another;
2. does so without change of frequency;
3. accomplishes this by electromagnetic induction; and
4. where the two electric circuits are linked by mutual induction.

21.2. Transformer Construction

The simple elements of a transformer consist of two coils having mutual inductance and a laminated steel core. The two coils are insulated from each other and from the steel core. Other necessary parts are: some suitable container for the assembled core and windings; a suitable medium for insulating the core and its windings from its container; suitable bushings (either of porcelain, oil-filled or capacitor-type) for insulating and bringing out the terminals of the winding from the tank.

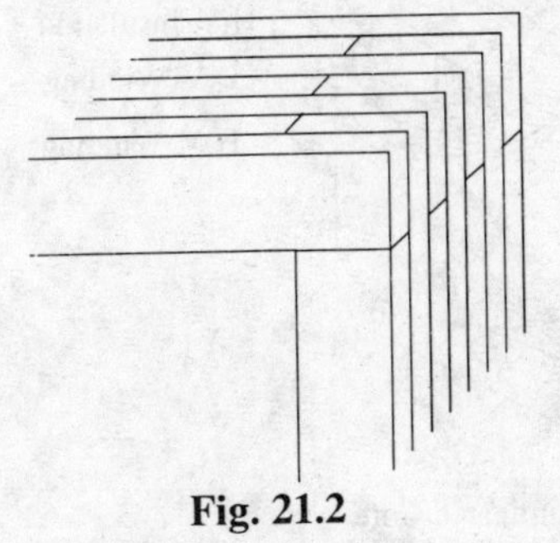
Fig. 21.2

In all types of transformers, the core is constructed of transformer sheet steel laminations assembled to provide a continuous magnetic path with minimum of air-gap included. The steel used is of high silicon content sometimes heat treated to produce a high permeability and a low hysteresis loss at the usual operating flux densities. The eddy current loss is minimised by laminating the core, the laminations being insulated from each other by a light coat of coreplate varnish or by an oxide layer on the surface. The thickness of laminations varies from 0.35 mm for a frequency of

50 Hz to 0.5 mm for a frequency of 25 Hz. The core laminations (in the form of strips) are joined as shown in Fig. 21.2. It is seen that the joints in the alternate layers are staggered in order to avoid the presence of narrow gaps right through the cross-section of the core. Such staggered joints are said to be 'imbricated'.

Constructionally, the transformers are of two general types, distinguished from each other merely by the manner in which the primary and secondary coils are placed around the laminated steel core. The two types are known as (*i*) **core-type** and (*ii*) **shell-type.** Another recent development is *spiral core* or *wound-core type,* the trade name being *spirakore* transformer.

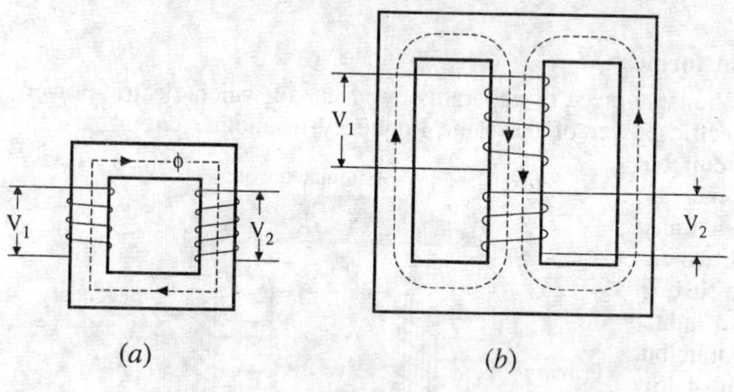

Fig. 21.3

In the so-called core-type transformers, the *windings surround a considerable part of the core* whereas in shell-type transformers, the *core surrounds a considerable portion of the windings* as shown schematically in Fig. 21.3 (*a*) and (*b*) respectively.

In the simplified diagram for the core-type transformers [Fig. 21.3 (*a*)] the primary and secondary windings are shown located on the opposite legs (or limbs) of the core, but in actual construction, these are always inter-leaved in order to reduce the leakage flux. As shown in Fig. 21.4, half the primary and half the secondary winding have been placed side by side or concentrically on each limb, not primary on one limb (or 'leg') and the secondary on the other.

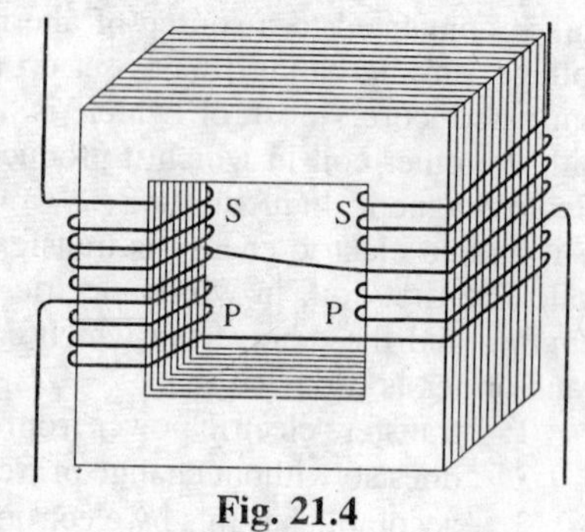

Fig. 21.4

21.3. Core-type Transformers

The coils used are form-wound and are of the cylindrical type. The general form of these coils may be circular or oval rectangular. In small size core-type transformers, simple rectangular core is used with cylindrical coils which are either circular or rectangular in form. But for large size core-type transformers, round or circular cylindrical coils are used which are so wound as to fit over a cruciform core section as shown in Fig. 21.5 (*a*). The circular cylindrical coils are used in most of the core-type transformers because of their mechanical strength. Such cylindrical coils are wound in helical layers with the different layers insulated from each other by paper, cloth, micarta board or cooling ducts. Fig. 21.5 (*c*) shows the general arrangement of these coils with respect to the core. Insulating cylinders of fuller board are used to separate the cylindrical windings from the core and from each other. Since the low-voltage (*LV*) winding is easiest to insulate, it is placed nearest to the core.

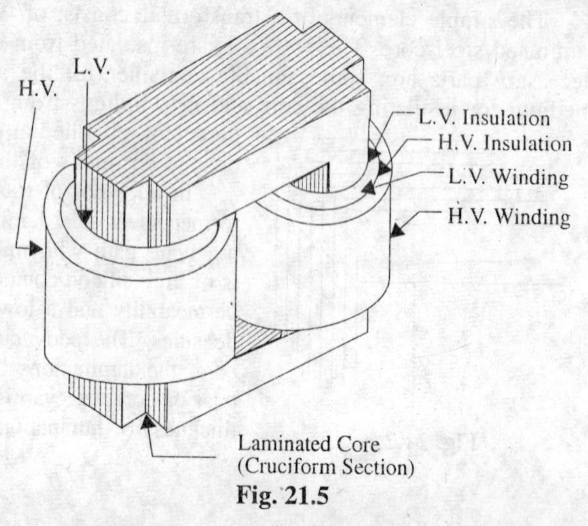

Fig. 21.5

Because of laminations and insulation, the net or effective core area is reduced, due allowance for which has to be made. It is found that, in general, the reduction

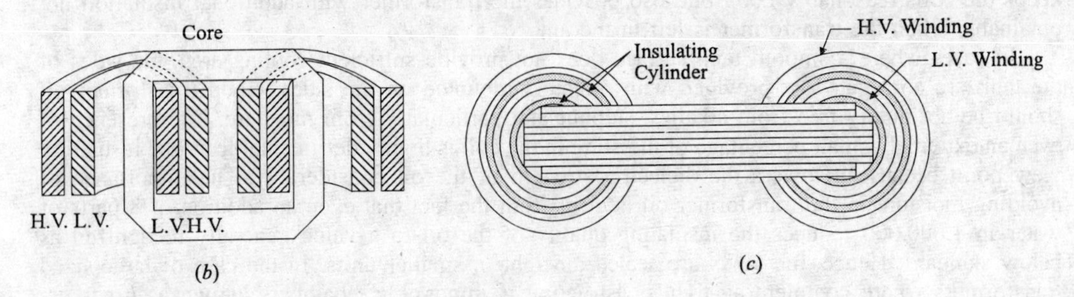

(*b*) (*c*)

Fig. 21.5

in core sectional area due to the presence of paper, surface oxide etc., is of the order of 10% approximately.

21.4. Shell-type Transformers

In their case also, the coils are form-wound but are multi-layer disc type usually wound in the form of pancakes. The different layers of such multi-layer discs are insulated from each other by paper. The complete winding consists of stacked discs with insulation spaces between the coils—the spaces forming horizontal cooling and insulating ducts. A shell type transformer may have a simple rectangular form as shown in Fig. 21.6 (*a*) or it may have a distributed form as shown in Fig. 21.7.

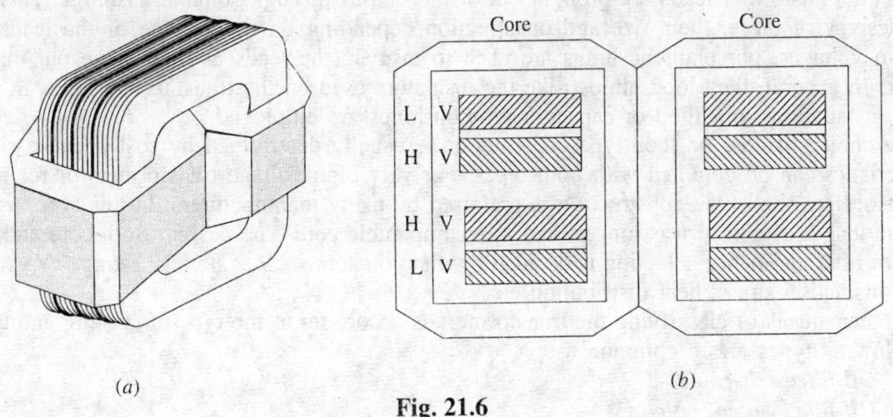

(*a*) (*b*)

Fig. 21.6

It may be pointed out that cores and coils of transformers must be provided with rigid mechanical bracing in order to prevent movement and possible insulation damage. Good bracing reduces vibration and the objectionable noise—a humming sound—during operation.

Transformers are generally housed in tightly-fitted sheet-metal tanks filled with special insulating oil. This oil has been highly developed and its function is two-fold. By circulation, it not only

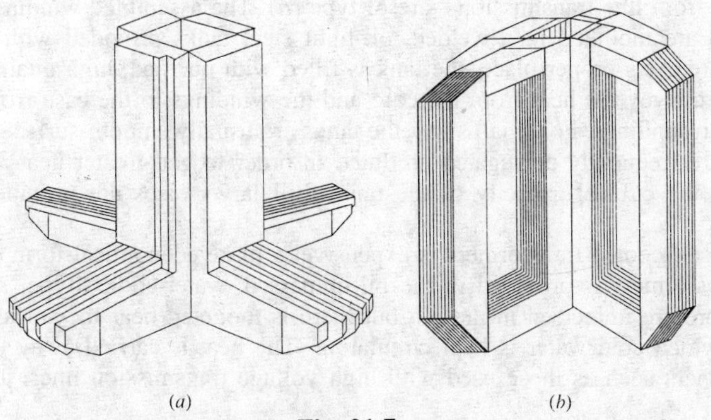

(*a*) (*b*)

Fig. 21.7

keeps the coils reasonably cool, but also provides the transformer with additional insulation not obtainable when the transformer is left in the air.

In cases where a smooth tank surface does not provide sufficient cooling area, the sides of the tank are corrugated or provided with radiators mounted on the sides. Good transformer oil should be absolutely free from alkalies, sulphur and particularly from moisture. The presence of even an extremely small percentage of moisture in the coil is highly detrimental from the insulation view point because it lowers the dielectric strength of the oil considerably. The importance of avoiding moisture in the transformer oil is clear from the fact that even an addition of 8 parts of water in 1,000,000 reduces the insulating quality of the oil to a value generally recognized as below standard. Hence, the tanks are sealed air-tight in smaller units. In the case of large-sized transformers where complete air-tight construction is impossible, chambers known as *breathers* are provided to permit the oil inside the tank to expand and contract as its temperature increases and decreases. The atmospheric moisture is entrapped in these breathers and is not allowed to pass on to the oil. Another thing to avoid in the oil is sledging which is simply the decomposition of oil with long and continued use. Sledging is caused principally by exposure to oxygen during heating and results in the formation of large deposits of dark and heavy matter that eventually clogs the cooling ducts in the transformer.

No other fauture in the construction of a transformer is given more attention and care than the insulating materials, because the life of the unit almost solely depends on the quality, durability and handling of these insulating materials. All the insulating materials are selected on the basis of their high quality and their ability to preserve this high quality even after many years of normal use.

All the transformer leads are brought out of their cases through suitable bushings. There are many designs of these, their size and construction depending on the voltage of the leads. For moderate voltages, porcelain bushings are used to insulate the leads as they come out through the tank. In general, they look almost like the insulators used on the transmission lines. In high-voltage installations, oil-filled or capacitor-type bushings are employed.

The choice of core or shell-type construction is usually determined by cost because similar characteristics can be obtained with both types. For very high voltage transformers or for multi-winding design, shell-type construction is preferred by many manufacturers. In this type, usually the mean length of coil turn is longer than in a comparable core-type design. Both core and shell forms are used and hence selection is decided by many factors such as voltage rating, kVA rating, weight, insulation stress, heat distribution etc.

Another means of classifying the transformers is according to the type of cooling employed. The following types are in common use:

(*a*) *Oil-filled self-cooled;*

(*b*) *Oil-filled water-cooled;*

(*c*) *Air-blast type.*

Small and medium-sized distribution transformers—so called because of their use on distribution systems as distinguished from line transmission—are of type *(a)*. The assembled windings and cores of such transformers are mounted in a welded, oil-tight steel tanks provided with a steel cover. After putting the core at its proper place, the tank is filled with purified, high quality insulating oil. The oil serves to convey the heat from the core and the windings to the case from where it is radiated out to the surroundings. For small sizes, the tanks are usually smooth-surfaced, but for larger sizes, the cases are frequently corrugated or fluted in order to get greater heat-radiation area without increasing the cubical capacity of the tank. Still larger sizes are provided with radiators or pipes.

Construction of very large self-cooled transformers is expensive, a more economical form of construction for such large transformers is provided in the oil-immersed, water-cooled type. As before, the windings and the core are immersed in the oil, but there is mounted near the surface of oil, a cooling coil through which cold water is kept circulating. The heat is carried away by this water. The largest transformers such as those used with high voltage transmission lines, are constructed in this manner.

Oil-filled transformers are built for outdoor duty and as these require no housing other than their own, a great saving is thereby effected. These transformers require only periodic inspection.

For voltages below 25,000 V, transformers can be built for cooling by means of an air-blast. The transformer is not immersed in oil but is housed in a thin sheet-metal box open at both ends through which air is blown from the bottom to the top by means of a fan or blower.

21.5. Elementary Theory of an Ideal Transformer

An ideal transformer is one which has no losses *i.e.,* whose windings have no ohmic resistance, so that there is no I^2R loss and no core loss and in which there is no magnetic leakage. In other words, an ideal transformer consists of two purely inductive coils wound on a loss-free core. It may, however, be noted *that it is impossible to realize such a transformer in practice, yet for convenience we will start with such a transformer and step-by-step approach an actual transformer.*

Consider an ideal transformer [Fig. 21.8 (*a*)] whose secondary is open and whose primary is connected to a sinusoidal alternating voltage V_1. This potential difference causes the flow of an alternating current in primary winding. Since the primary coil is purely inductive and there is

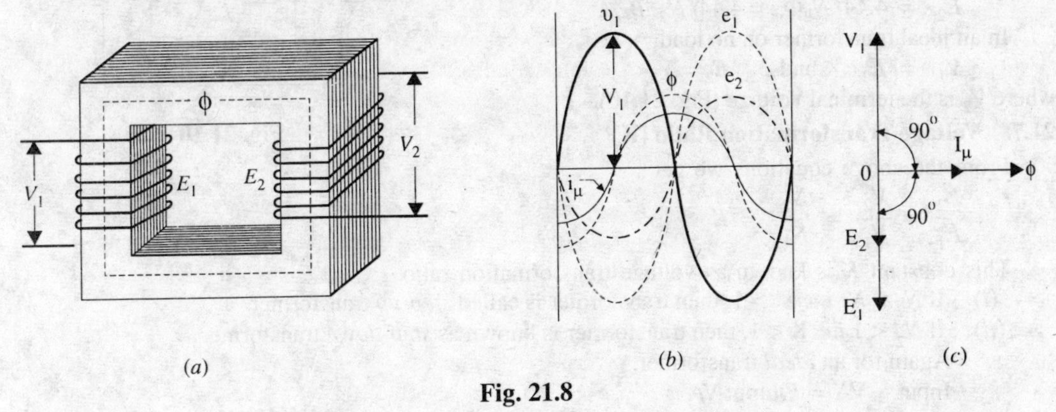

Fig. 21.8

no output (secondary being open), the primary draws the magnetising current I_μ only. The function of this current is merely to magnetise the core, it is small in magnitude and lags V_1 by 90°. This alternating current I_μ produces an alternating flux Φ which is, at all times, proportional to the current (assuming permeability of the magnetic circuit to be constant) and hence is in phase with it. This changing flux is linked both with the primary and the secondary windings. Therefore, it produces self-induced e.m.f. in the primary. This *self-induced* e.m.f. E_1 is, at every time, equal to and in opposition to V_1. It is also known as *counter* e.m.f. or *back* e.m.f. of the primary.

Similarly, there is produced in the secondary an induced e.m.f. E_2 which is known as *mutually*-induced e.m.f. This e.m.f. is anti-phase with V_1 and its magnitude is proportional to the rate of change of flux and the number of secondary turns.

The instantaneous values of applied voltage, induced e.m.fs., flux and magnetising current are shown by sinusoidal waves in Fig. 21.8 (*b*). Fig. 21.8 (*c*) shows the vectorial representation of the effective values of the above quantities.

21.6. E.M.F. Equation of a Transformer

Let N_1 = No. of turns in primary

N_2 = No. of turns in secondary

Φ_m = maximum flux in the core in webers = $B_m \times A$

f = frequency of a.c. input in hertz (Hz)

As shown in Fig. 21.9, the core flux increases from its zero value to maximum value Φ_m in one quarter of the cycle *i.e.,* in 1/4 *f* second.

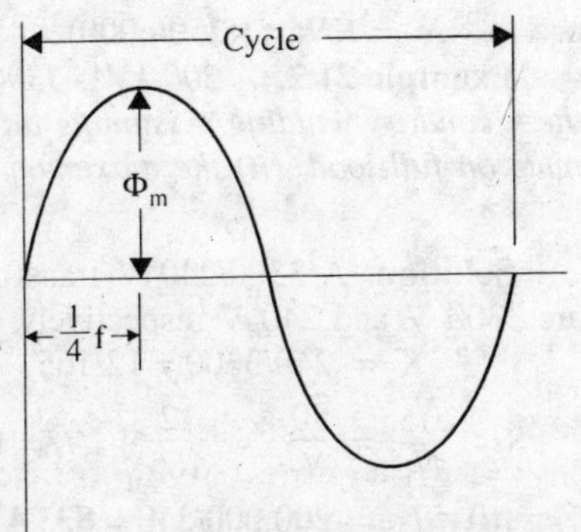

Fig. 21.9

$\therefore$ average rate of change of flux $= \dfrac{\Phi_m}{1/4f} = 4f\,\Phi_m$ Wb/s

Now, rate of change of flux per turn means induced e.m.f. in volts.

$\therefore$ average e.m.f induced/turn $= 4f\,\Phi_m$ volt

If flux Φ varies *sinusoidally,* then r.m.s. value of induced e.m.f. is obtained by multiplying the average value with form factor.

$$\text{Form factor} = \frac{\text{r.m.s. value}}{\text{average value}} = 1.11$$

$\therefore$ r.m.s. value of e.m.f./turn $= 1.11 \times 4f\,\Phi_m = 4.44f\,\Phi_m$ volt

Now, r.m.s. value of induced e.m.f. in the whole of primary winding

$$= (\text{induced e.m.f./turn}) \times \text{No. of primary turns}$$

$$\therefore \quad E_1 = 4.44 f N_1 \Phi_m = 4.44 f N_1 B_m A$$

Similarly, r.m.s value of induced e.m.f. in secondary is

$$E_2 = 4.44 f N_2 \Phi_m = 4.44 f N_2\ B_m A$$

In an ideal transformer on no load,

$$V_1 = E_1 \quad \text{and} \quad E_2 = V_2$$

where V_2 is the terminal voltage (Fig. 21.10).

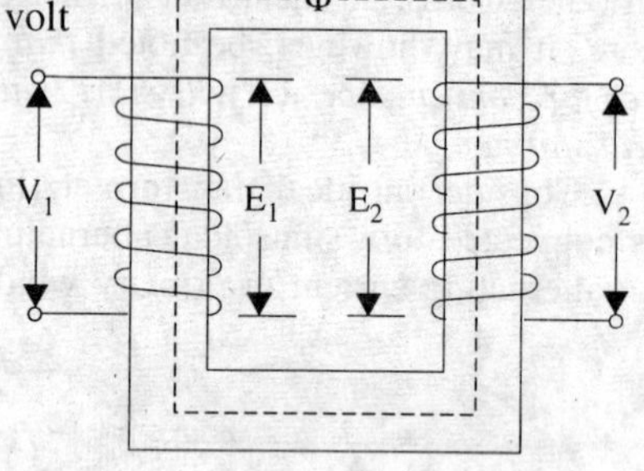

Fig. 21.10

21.7. Voltage Transformation Ratio (K)

From the above equations we get

$$\frac{E_2}{E_1} = \frac{V_2}{V_1} = \frac{N_2}{N_1} = K$$

This constant K is known as voltage transformation ratio.

(*i*) If $N_2 > N_1$ *i.e.* $K > 1$, then transformer is called *step-up* transformer.

(*ii*) If $N_2 < 1$ *i.e.* $K < 1$, then transformer is known as *step-down* transformer.

Again for an *ideal* transformer,

$$\text{Input VA} = \text{Output VA}$$

$$V_1 I_1 = V_2 I_2$$

or
$$\frac{I_2}{I_1} = \frac{V_1}{V_2} = \frac{1}{K}$$

Hence, currents are in the inverse ratio of the (voltage) transformation ratio.

Example 21.1. *The no-load ratio of a 50 Hz, single-phase transformer is 6,000/250 V. Estimate the number of turns in each winding if the maximum flux is 0.06 Wb in the core.*

Solution. Using the transformer e.m.f. equation, we get

$$E_1 = 4.44 f N_1\,\Phi_m$$

$$\therefore \quad 6{,}000 = 4.44 \times 50 \times N_1 \times 0.06 \qquad \therefore\ N_1 = 450$$

Similarly, $\quad 250 = 4.44 \times 50 \times N_2 \times 0.06 \qquad \therefore\ N_2 = 19$

or $\quad N_2 = KN_1 = [(250/6000) \times 450 = 19]$

Example 21.2. *A 200 kVA, 3300/240 volts, 50 Hz single-phase transformer has 80 turns on the secondary winding. Assuming an ideal transformer, calculate (i) primary and secondary, currents on full-load, (ii) the maximum value of flux and (iii) the number of primary turns.*

[A.M.I.E., Summer 1990]

Solution. A 3300/240 V transformer is one whose normal primary and secondary voltages are 3300 V and 240 V respectively.

$$K = 240/3300 = 12/165$$

(*i*) $\dfrac{N_2}{N_1} = \dfrac{80}{N_1} = \dfrac{12}{165}$, $N_1 = \mathbf{1100}$

(*ii*) $I_2 = 200{,}000/240 = \mathbf{833\ A}$

$$I_1 = KI_2 = (12/165)833 = \mathbf{60.6\ A}$$

(*iii*) $3300 = 4.44 \times 50 \times 110.0 \times \Phi_m$;

$$\Phi_m = \mathbf{13.5\ mWb}$$

Example. 21.3. *A 200 kVA, 6600/400 V, 50 Hz single-phase transformer has 80 turns on the secondary. Calculate (i) the approximate values of the primary and secondary current, (ii) the approximate number of primary turn and (iii) the maximum value of flux.*

(Elect. Engg.-I, Punjab Univ. 1994)

Solution. (*i*) Secondary current, $I_2 = 200{,}000/400 = \mathbf{500\ A}$

Assuming input kVA to be equal to the output kVA, we have

$$I_1 = 200{,}000/6{,}600 = \mathbf{30.3\ A}$$

Alternatively, $I_1 = KI_2$. Now, $K = 400/6{,}600 = 2/33$

$\therefore \quad I_1 = (2/33) \times 500 = 30.3$ A

(*ii*) $$K = \frac{N_2}{N_1} = \frac{E_2}{E_1}$$

$\therefore \quad \dfrac{80}{N_1} = \dfrac{400}{6600}, \quad N_1 = \mathbf{1320}$

Alternatively, voltage drop per turn in secondary = 400/80 = 5. It has the same value for primary as well. Hence, the number of turns in primary = 6600/5 = 1320

(*iii*) Using voltage equation of the transformer, we have

$$E_1 = 4.44\, fN_1\Phi_m \quad \text{or} \quad 6600 = 4.44 \times 50 \times 1320 \times \Phi_m$$

or $\Phi_m = 0.0225$ Wb or **22.5 mWb**

Example 21.4. *A single-phase, 50 Hz core-type transformer has square cores of 20 cm side. The permissible flux density in the core is 1.0 Wb/m². Calculate the number of turns per limb on the high and low voltage sides for a 3000/220 V ratio. To allow for insulation of stampings, assume the net iron length to be 0.9 × gross iron length.* **[A.M.I.E., Nov.1992]**

Solution. Iron-length on one side of the core (Fig. 21.11)

$= 0.9 \times 20 = 18$ cm

Since, the other side of the square is unaffected by insulation, core area is

$= 18 \times 20 = 360\ \text{cm}^2$

$E_1 = 4.44 f N_1 B_m A$

$3000 = 4.44 \times 50 \times N_1 \times 1 \times (360 \times 10^{-4})$

$N_1 = 375$ ... h.v. side

$N_2 = kN_1 = 375\ (220/3000)$

$= 28$ ...l.v. side

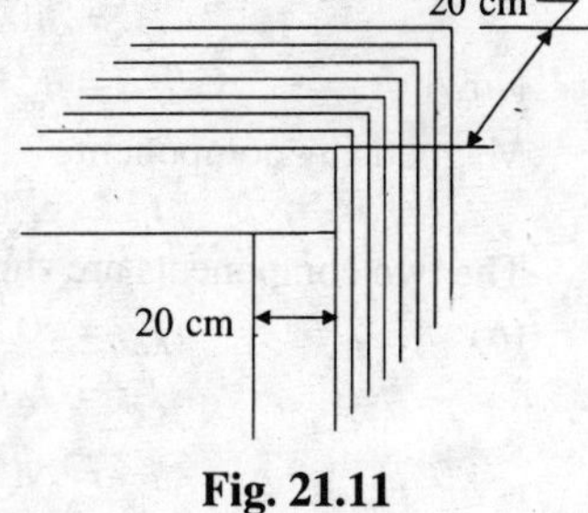

Fig. 21.11

21.8. Transformer with Losses but no Magnetic Leakage

We will consider two cases (*i*) when such a transformer is on no-load and (*ii*) when it is loaded.

(*i*) Transformer on No-load

In the above discussion, we assumed an ideal transformer *i.e.,* one in which there were no core losses and copper losses. But practical conditions require that certain modifications be made in the foregoing theory. When an *actual* transformer is on load, there is iron loss in the core and copper loss in the windings (both primary and secondary) and these losses are not entirely negligible.

Hence, even when the transformer is on no-load, the primary input current is not wholly reactive. The primary input current under no-load conditions has to supply (*i*) iron losses in the core *i.e.,* hysteresis loss and eddy current loss and (*ii*) a very small amount of copper loss in primary (there being no Cu loss in secondary as it is open). Hence, the no-load primary input current I_0 is not 90° behind V_1 but lags it by an angle $\phi_0 < 90°$. No-load primary input power

$$W_0 = V_1 I_0 \cos \phi_0$$

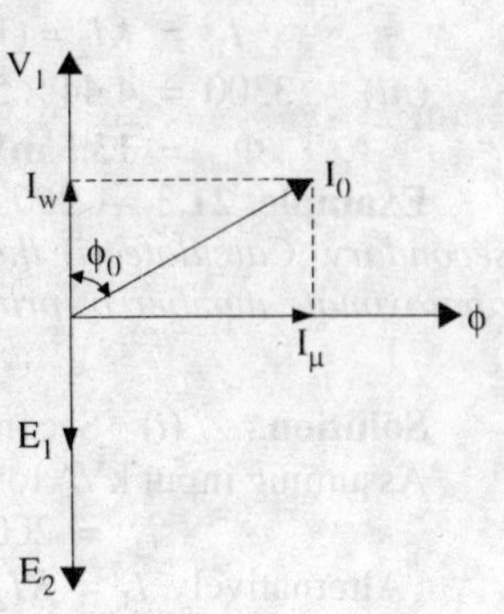

Fig. 21.12

where $\cos \phi_0$ is primary power factor under no-load conditions. No-load condition of an actual transformer is shown vectorially in Fig. 21.12.

As seen from Fig. 21.12, primary current I_0 has two components:

(*i*) one in phase with V_1. It is known as *active* or *working* or *iron* loss component I_w, because it supplies iron loss plus a small quantity of primary Cu loss.

$$I_w = I_o \cos \phi_0$$

Obviously, I_o is the vector sum of I_w and I_μ hence

$$I_o = \sqrt{(I_\mu^2 + I_w^2)}$$

The following points should be noted carefully:

1. The no-load primary current I_0 is very small as compared to the full-load primary current.
2. As I_0 is very small, no-load primary Cu loss is negligibly small which means *that no-load primary input is practically equal to the iron-loss in a transformer.*
3. As it is principally the core-loss which is responsible for shift in the current vector, angle ϕ_0 is known as *hysteresis angle of advance.*

Example 21.5. *(a) A 2,200/200 V transformer draws a no-load primary current of 0.6 A and absorbs 400 watts. Find the magnetising and iron-loss currents.*

(b) A 2,200/250 V transformer takes 0.5 A at a p.f. of 0.3 on open circuit. Find magnetising and working components of no-load primary current.

Solution. Fig. 21.12 may please be referred to.

(*a*) Iron-loss current $= \dfrac{\text{no-load input in watts}}{\text{primary voltage}}$

$= 400/2200 = \mathbf{0.182\ A}$

Now $\quad I_0^2 = I_w^2 + I_\mu^2$

Magnetising component

$$I_\mu = \sqrt{(0.6^2 - 0.182^2)} = \mathbf{0.572\ A}$$

The two components are shown in Fig. 21.12.

(*b*) $\quad I_o = 0.5\text{ A}; \quad \cos \phi_0 = 0.3$

$$I_w = I_0 \cos \phi_0 = 0.5 \times 0.3 = 0.15\text{ A}$$

$$I_\mu = \sqrt{(0.5^2 - 0.15^2)} = \mathbf{0.477\ A}$$

Example 21.6. *The no-load current of a transformer is 5 A at 0.25 power factor when supplied at 235 V, 50 Hz. The number of turns on the primary winding is 200. Calculate:*

(a) the maximum value of flux in the core,

(b) the core loss,

(c) the magnetising component. **[Elect. Machines-I, Nagpur Univ. 1993]**

Solution. (*a*) $E_1 = 4.44 f N_1 \Phi_m$

$$235 = 4.44 \times 50 \times 200\ \Phi_m$$

$$\Phi_m = \mathbf{5.29\ mWb}$$

(*b*) Primary no-load input represents the core loss.

$$\therefore \text{core loss} = V_1 I_o \cos \phi_0 = 235 \times 5 \times 0.25 = \mathbf{294\ W}$$

(*c*) Magnetising component is

$$I_\mu = I_0 \sin \phi_0$$

Now $\quad \sin \phi_0 = \sqrt{1 - \cos \phi_0} = \sqrt{15/16} = 0.9682$

$\therefore \quad I_\mu = 5 \times 0.9682 = \mathbf{4.84\ A}$

Tutorial Problems No. 21.1

1. A 3000/200 V, 50-Hz single-phase transformer is built on a core having an effective cross-sectional area of 150 cm^2 and has 80 turns in the low-voltage winding. Calculate:
   (*a*) the value of maximum flux density in the core
   (*b*) the number of turns in the high-voltage winding **[(*a*) 0.75 Wb/m^2 (*b*) 1200]**
2. A 40 kVA, 3,300/240 V, 50 Hz, single-phase transformer has 660 turns on the primary. Determine:
   (*a*) the number of turns on the secondary
   (*b*) the maximum value of the flux in the core
   (*c*) the approximate values of primary and secondary full-load currents. Internal drops in the windings are to be ignored. **[(*a*) 48 (*b*) 22.5 mWb (*c*)12.1 A; 166.7 A]**
3. A double-wound, single-phase transformer is required to step down from 1900 V to 240 V, 50 Hz. It is to have 1.5 V per turn. Calculate the required number of turns on the primary and secondary windings respectively. The peak value of the flux density is required to be not more than 1.2 Wb/m^2. Calculate the required cross-sectional area of the steel core.
   If the output is 10 kVA, calculate the secondary current. **[1,267; 160; 56.4 cm^2; 41.75 A]**
4. The number of turns in the low-voltage winding of a 200 kVA, 50 Hz, 2,000/250 V single-phase transformer is 25. Calculate:
   (*a*) peak value of the magnetic flux in the core
   (*b*) the full-load current in the low-voltage winding
   (*c*) the number of turns in the high-voltage winding. **[(*a*) 45 mWb (*b*) 800 A (*c*) 200]**
5. The no-load current of a transformer is 15 A at a p.f. of 0.2 when connected to a 460 V, 50 Hz supply. If the primary windings has 550 turns, calculate
   (*a*) the magnetising component of the no-load current
   (*b*) the iron loss
   (*c*) the maximum value of flux in the core. **[(*a*) 14.7 A (*b*) 1,380 W (*c*) 3.77 mWb]**
6. The design requirements of a 6600 V/440 V, 50 Hz, single-phase, core-type transformer are: approximate e.m.f./turn = 15 V; maximum flux density = 1.5 tesla (Wb/m^2). Find a suitable number of primary and secondary turns and the net cross-sectional area of the core. **[462, 28, 428 cm^2]**
7. The primary winding of a single-phase transformer is connected to a 230 V, 50 Hz supply. The secondary winding has 1500 turns. If the maximum value of the core flux is 2.07 mWb, determine:
   (*a*) the number of turns on the primary winding (*b*) the secondary induced voltage
   (*c*) the net cross-sectional core area if the flux density has a maximum value of 0.465 tesla (Wb/m^2). **[(*a*) 500 (*b*) 690 V (*c*) 44.5 cm^2]**
8. A 6600 V/250 V, 50 Hz, single-phase core-type transformer has a cross-section 25 cm × 25 cm. Allowing for a space factor of 0.9, find a suitable number of primary and secondary turns if the flux density is not to exceed 1.2 tesla (Wb/m^2). **[475, 18]**
9. The no-load current of transformer is 5.0 A at 0.3 power factor when supplied at 230 V, 50 Hz. The number of turns on the primary winding is 200. Calculate:
   (*a*) the maximum value of the flux in the core (*b*) the core loss (*c*) the magnetizing current. **[(*a*) 5.18mWb (*b*) 345W (*c*) 4.77 A (r.m.s.)]**

(*ii*) Transformer on Load

When the secondary is loaded, secondary current I_2 is set up. The magnitude and phase of I_2 with respect to V_2 are determined by the characteristics of the load. Current I_2 is in phase with V_2 if load is resistive it lags if load is inductive and it leads if load is capacitive.

The secondary current sets up its own m.m.f. (= $N_2 I_2$) and hence its own flux Φ_2 which is in opposition to the main no-load primary flux Φ_o which is due to I_0. The secondary ampere-turns $N_2 I_2$ are known as *demagnetising* ampere-turns. The opposing secondary flux Φ_2 weakens the primary flux Φ_0 momentarily, hence primary back e.m.f. E_1 tends to be reduced. For a moment, V_1 gains the upper hand over E_1 and hence causes more current to flow in the primary.

Let the additional primary current be I_2'. It is known as *load component of primary current.* This current is anti-phase with I_2. The additional primary m.m.f. $N_2 I_2'$ sets up its own flux Φ_2' which is in opposition to Φ_2 (but in the same direction as Φ_o) and is equal to it in magnitude. Hence, Φ_2 and Φ_2' cancel each other out, leaving the only flux Φ_0 in the core. So, we find that the magnetic effects of secondary current I_2 are immediately neutralised by the additional primary current I_2' which is brought into existence exactly at the same instant as I_2. The whole process is illustrated in Fig. 21.13.

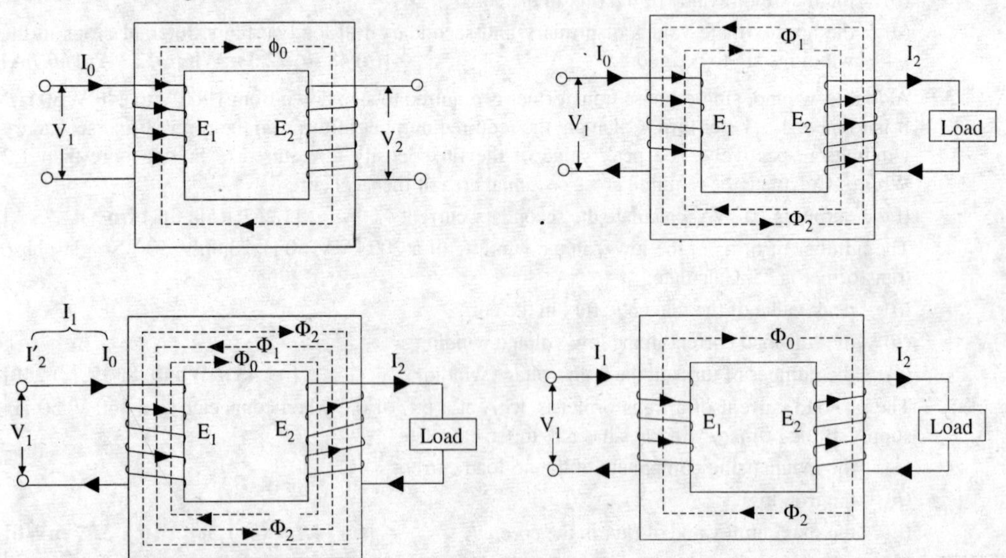

Fig. 21.13

Hence, whatever the load condition, *the net flux passing through the core is approximately the same as at no-load.* An important deduction is that due to the constancy of core flux at all loads, *the core-loss* is also practically the *same under all load conditions* (Art. 21.8).

Since $\quad \Phi_2 = \Phi_2' \quad \therefore \quad N_2 I_2 = N_1 I_2'$

$$\therefore \quad I_2' = \frac{N_2}{N_1} \times I_2 = KI_2$$

Hence, when transformer is on load, the primary winding has two currents in it; one is I_0 and the other is I_2' which is anti-phase with I_2 and K times in magnitude. *The total primary current is the vector sum of I_0 and I_2'.*

In Fig. 21.14 are shown the current vector diagrams for a transformer when load is non-inductive and when it is inductive (a similar diagram could be drawn for the capacitive load). Voltage transformation ratio of unity is assumed so that primary vectors are equal to the secondary vectors. With reference to Fig. 21.14 (*a*), I_2 is secondary current in phase with E_2 (strictly speaking it should be V_2). It causes primary current I_2' which is anti-phase with it and equal to its magnitude ($\because K = 1$). Total primary current I_1 is the vector sum of I_0 and I_2' and lags behind V_1 by an angle ϕ_1.

In Fig. 21.14 (*b*), vectors are drawn for an inductive load. Here, I_2 lags E_2 (actually V_2) by ϕ_2. Current I_2' is again anti-phase with I_2 and equal to it in magnitude. As before, I_1 is the vector sum of I_2' and I_0 and lags behind V_1 by ϕ_1.

It will be observed that ϕ_1 is slightly greater than ϕ_2. But if we neglect I_0 as compared to I_2' as in Fig. 21.14 (*c*), then $\phi_1 = \phi_2$. Moreover, under this assumption

$$N_1 I_2' = N_1 I_1 = N_2 I_2$$

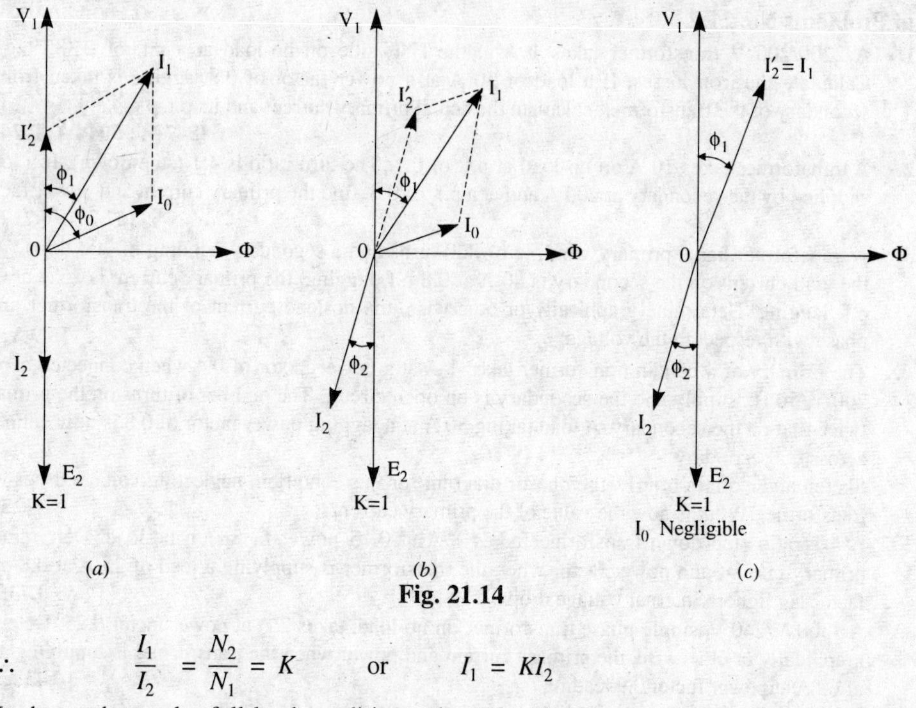

Fig. 21.14

$$\therefore \quad \frac{I_1}{I_2} = \frac{N_2}{N_1} = K \quad \text{or} \quad I_1 = KI_2$$

It shows that under full-load conditions, the ratio of primary and secondary currents is constant. This important relationship is made the basis of current transformer—a transformer which is used with a low-range ammeter for measuring currents in circuits where the direct connection of the ammeter is impracticable.

The components of the primary current are as shown below:

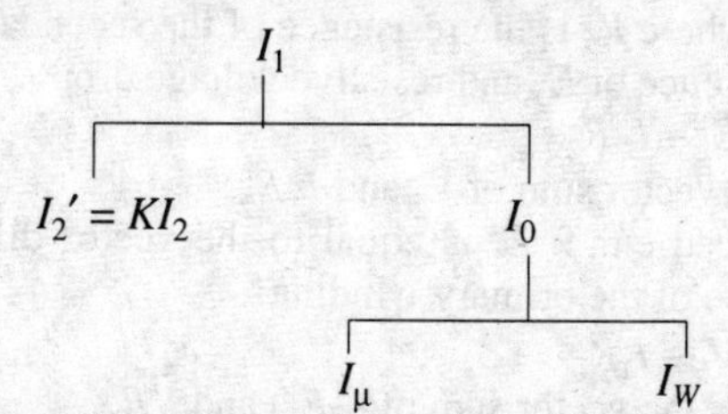

Example 21.7. *A single-phase transformer with a ratio of 440/110 V takes a no-load current of 5 A at 0.2 power factor lagging. If the secondary supplies a current of 120 A at a p.f. of 0.8 lagging, estimate the current taken by the primary.* **[Elect. Machines-I, Bangalore Univ. 1992)**

Solution. The vector diagram is shown in Fig. 21.15

$\cos \phi_2 = 0.8, \quad \phi_2 = \cos^{-1}(0.8) = \mathbf{36°54'}$

$\cos \phi_0 = 0.2, \quad \phi_0 = \cos^{-1}(0.2) = \mathbf{78°30'}$

Now, $K = V_2 / V_1 = 110/440 = 1/4$

$I_2' = KI_2 = 120 \times 1/4 = 30 \text{ A}$

$I_0 = 5 \text{ A}$

Angle between I_0 and $I_2' = 78°30' - 36°54' = 41°36'$

Using parallelogram law of vectors (Fig. 21.15), we get

$$I_1 = \sqrt{(5^2 + 30^2 + 2 \times 5 \times 30 \times \cos 41°36')} = \mathbf{33.9 \text{ A}}$$

The resultant current could also have been found by resolving I_2' and I_0 into their *X*- and *Y*-components.

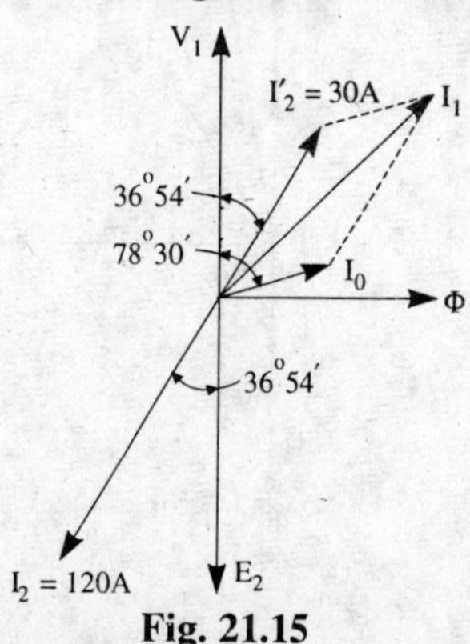

Fig. 21.15

Tutorial Problems No. 21.2

1. A 2200/200 V transformer takes 1 A at the H.T. side on no-load at a p.f. of 0.385 lagging. Calculate the iron losses. If a load of 50 A at a power factor of 0.8 lagging is taken from the secondary of the transformer, calculate the actual primary current and its p.f. **[847 W; 5044 A; 0.74 lag]**
2. A transformer takes 10 A on no-load at p.f. of 0.1. The turn ratio is 4:1 (step-down). If a load is supplied by the secondary at 200 A and at a p.f. of 0.8, find the primary current and power factor. **[57.2 A; 0.717 lag]**
3. A transformer has a primary winding of 800 turns and a secondary winding of 200 turns. When the load current on the secondary is 80 A at 0.8 p.f. lagging, the primary current is 25 A at 0.707 p.f. lagging. Determine graphically or otherwise, the no-load current of the transformer and its phase with respect to the voltage. **[5.9 A; 74°]**
4. The primary of a certain transformer takes 1 A at a power factor of 0.4 when connected across a 200 V, 50 Hz supply and the secondary is on open circuit. The number of turns on the primary is twice that on the secondary. A load taking 50 A at a lagging power factor of 0.8 is now connected across the secondary.
   Sketch and explain briefly, the phasor diagram for this condition, neglecting voltage drops in the transformer. What is now the value of the primary current? **[25.9 A]**
5. A 4:1 ratio step-down transformer takes 1 A at 0.15 power factor on no-load. Determine the primary current and power factor when the transformer is supplying a load of 25 A at 0.8 power factor lag. Ignore internal voltage drops. **[7 A; 0.735 lag]**
6. A 3300 V/240 V, single-phase transformer on no load, takes 2 A at power factor 0.25. Determine graphically or otherwise, the primary current and power when the transformer is supplying a load of 60 A at power factor 0.9 leading. **[4.43 A; 1.0]**

21.9. Transformer with Winding Resistance but no Magnetic Leakage

An ideal transformer was supposed to possess no resistance but in any actual transformer, there is always present some resistance of the primary and secondary windings. Due to this resistance, there is some voltage drop in the two windings. The result is that:

(*i*) the secondary terminal voltage V_2 is vectorially less than the secondary induced e.m.f. E_2 by an amount $I_2 R_2$ where R_2 is the resistance of the secondary winding. Hence, V_2 is equal to the vector difference of E_2 and resistive voltage drop $I_2 R_2$.

$$V_2 = E_2 - I_2 R_2 \qquad \text{...vector difference}$$

In other words, E_2 is the vector sum of V_2 and $I_2 R_2$.

(*ii*) similarly, primary induced e.m.f. E_1 is equal to the vector difference of V_1 and $I_1 R_1$ where R_1 is the resistance of the primary winding.

$$\therefore \quad E_1 = V_1 - I_1 R_1 \qquad \text{...vector difference}$$

In other words, V_1 equals the vector sum of $(-E_1)$ and $I_1 R_1$

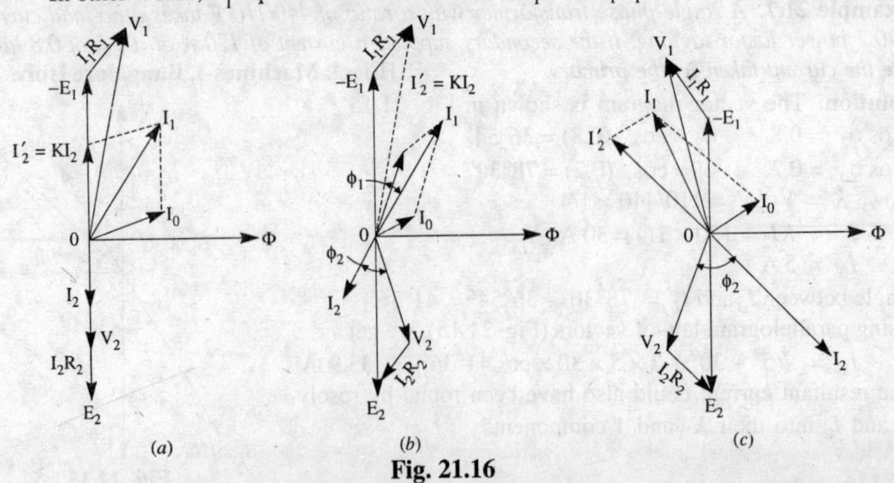

Fig. 21.16

The voltage vector diagrams for non-inductive, inductive and capacitive loads are shown in Fig. 21.16 (*a*), (*b*) and (*c*) respectively. It should be noted that vector for E_1 has been reversed so that it is written as $(-E_1)$.

It would be interesting to note that a transformer can be looked upon as a dual machine. Its primary behaves electrically as a motor and its secondary as a generator. The voltage equation of the motor applies to the primary (Art. 12.4) whereas generator equation applies to the secondary.

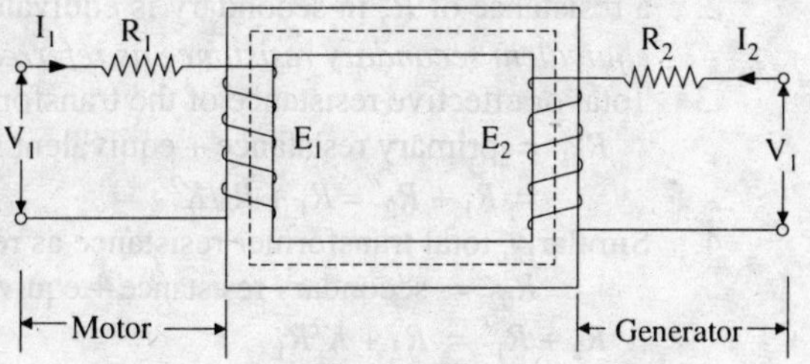

Fig. 21.17

21.10. Equivalent or Referred Resistances

In Fig. 21.18 is shown a transformer whose primary and secondary windings have resistances of R_1 and R_2 respectively. The resistances have been shown external to the windings.

It would now be shown that the resistances of the two windings can be transferred to any one of the two windings. The advantage of concentrating both the resistances in one winding is that it makes calculations very simple and easy because one has then to work in one winding only. It will be proved that a resistance of R_2 in secondary is equivalent to R_2/K^2 in primary. The value of R_2/K^2 will be denoted by R_2' — *the equivalent secondary resistance as referred to primary.*

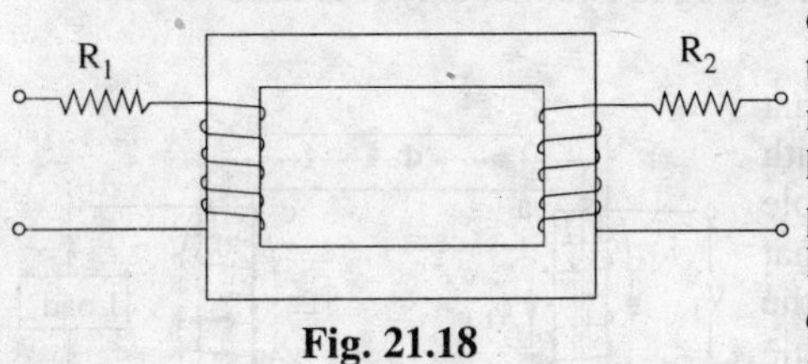

Fig. 21.18

The copper loss in secondary is $I_2^2 R_2$. This loss is supplied by primary which takes a current of I_1. Hence, if R_2' is the *equivalent resistance in primary which would have caused the same loss as* R_2 in secondary, then

$$I_1^2 R_2' = I_2^2 R_2 \quad \text{or} \quad R_2' = (I_2/I_1)^2 R_2$$

Now, if *we neglect no-load current* I_0, then $I_2 / I_1 = 1/K$*

$$\therefore \quad R_2' = R_2 / K^2$$

Similarly, equivalent primary resistance as referred to secondary is

$$R_1' = K^2 R_1$$

In Fig. 21.19, secondary resistance has been transferred to primary side leaving secondary circuit resistanceless. The resistance $R_1 + R_2' = R_1 + R_2 / K^2$ is known as the *equivalent* or *effective resistance of the transformer as referred to primary* and may be designated as R_{01}.

$$\therefore \quad R_{01} = R_1 + R_2' = R_1 + R_2 / K^2$$

Similarly, the *equivalent resistance of the transformer as referred to secondary is*

$$R_{02} = R_2 + K^2 R_1$$

This fact is shown in Fig. 21.20 where all the resistance of the transformer has been concentrated in the secondary winding.

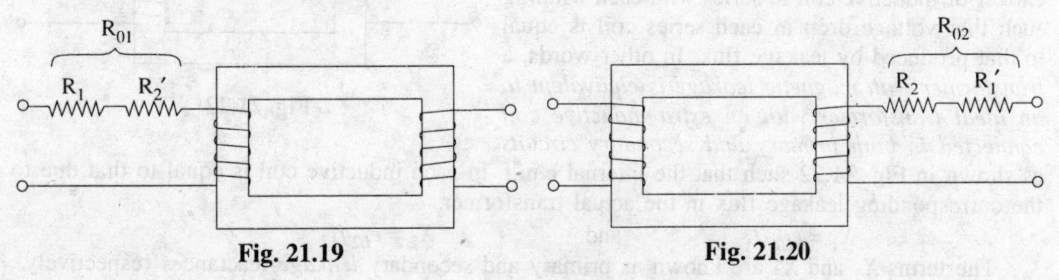

Fig. 21.19 **Fig. 21.20**

It will be noted that

*Actually $I_2 / I_2' = 1/K$ and not I_2 / I_1. However, if I_0 is neglected, then $I_2' = I_1$.

1. a resistance of R_1 in primary is equivalent to K^2R_1 in the secondary. Hence, it is called the *equivalent primary resistance as referred to secondary i.e.,* R_1'.
2. a resistance of R_2 in secondary is equivalent to R_2/K^2 in primary. Hence, it is called the *equivalent secondary resistance as referred to primary i.e.,* R_2'.
3. Total or effective resistance of the transformer as referred to primary is

$$R_{01} = \text{primary resistance + equivalent secondary resistance referred to primary}$$
$$= R_1 + R_2' = R_1 + R_2/K^2$$

4. Similarly, total transformer resistance as referred to as secondary is

$$R_{02} = \text{secondary resistance + equivalent primary resistance as referred to secondary}$$
$$R_2 + R_1' = R_2 + K^2R_1$$

It is important to remember that

(*a*) When shifting primary resistance to the secondary, *multiply* it by K^2 *i.e.,* (transformation ratio)^2.

(*b*) When shifting secondary resistance to the primary, *divide* it by K^2.

(*c*) However, when shifting voltage from one winding to another, only K is used.

21.11. Magnetic Leakage

In the preceding discussion, it was assumed that all the flux linked with primary winding also links with the secondary winding. But, in practice, it is impossible to realize this condition. It is found, in practice, that all the flux linked with primary does not link with the secondary but part of it *i.e.,* Φ_{L1} completes its magnetic circuit by passing through air rather than around the core, as shown in Fig. 21.21. This leakage flux is produced when the m.m.f. due to primary ampere-turns existing between points *a* and *b*, acts along the leakage paths. Hence, this flux is known as *primary leakage flux* and is proportional to the primary ampere-turns alone, because the secondary turns do not link the magnetic circuit of Φ_{L1}. The flux Φ_{L1} is in time phase with I_1. It induces an e.m.f. e_{L1} in primary but none in secondary.

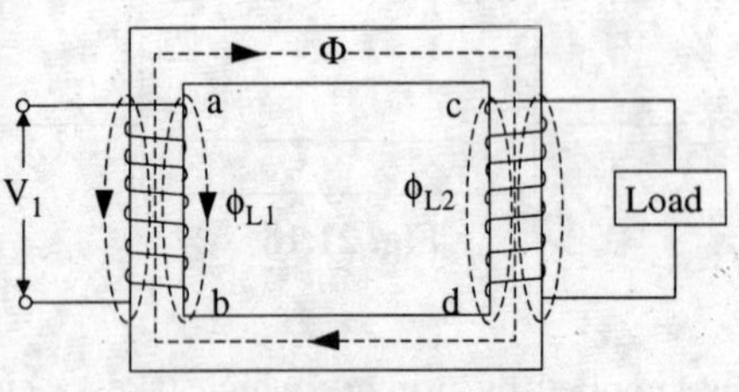

Fig. 21.21

Similarly, secondary ampere-turns (or m.m.f.) acting across points *c* and *d* set up leakage flux Φ_{L2} which is linked with secondary winding alone (and not with primary turns). This flux Φ_{L2} is in time phase with I_2 and produces a self-induced e.m.f. e_{L2} in secondary (and not in primary).

At no-load and light-loads, the primary and secondary ampere-turns are small, hence leakage fluxes are negligible. But when load is increased, both primary and secondary windings carry huge currents. Hence, large m.m.fs. are set up which, while acting on leakage paths, increase the leakage flux.

As said earlier, the leakage flux linking with each winding, produces a self-induced e.m.f. in that winding. Hence, in effect, it is equivalent to a small choker or inductive coil in series with each winding such that voltage drop in each series coil is equal to that produced by leakage flux. In other words, a *transformer with magnetic leakage is equivalent to an ideal transformer with an extra inductive coil connected in both primary and secondary circuits* as shown in Fig. 21.22 such that the internal e.m.f. in each inductive coil is equal to that due to the corresponding leakage flux in the actual transformer.

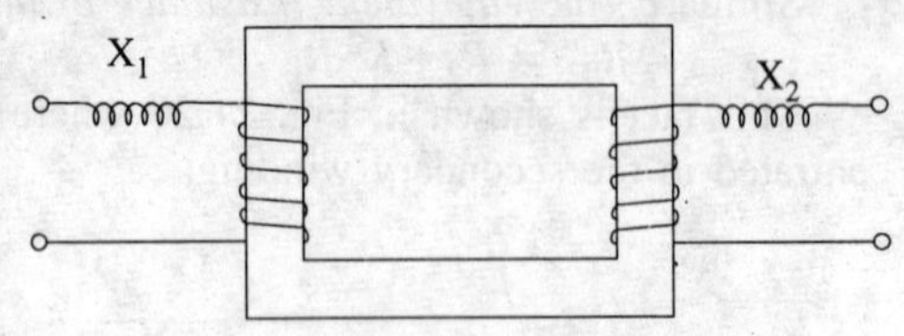

Fig. 21.22

$$\therefore \quad X_1 = e_{L1}/I_1 \quad \text{and} \quad X_2 = e_{L2}/I_2$$

The terms X_1 and X_2 are known as primary and secondary *leakage* reactances respectively.

Following few points should be kept in mind :

1. The leakage flux links one or the other windings but *not both,* hence it in no way contributes to the transfer of energy from the primary to the secondary winding.

2. The primary voltage V_1 will have to supply reactive drop I_1X_1 in addition to I_1R_1. Similarly, E_2 will have to supply I_2R_2 and I_2X_2
3. In an actual transformer, the primary and secondary windings are not placed on separate legs or limbs as shown in Fig.21.33 because due to their being widely-separated, large primary and secondary leakage fluxes would result. These leakage fluxes are minimised by sectionalizing and interleaving the primary and secondary windings.

21.12. Transformer with Resistance and Leakage Reactance

In Fig. 21.23 are shown the primary and secondary windings of a transformer with resistances and leakage reactances taken out of the windings. The primary impedance is given by

$$Z_1 = \sqrt{R_1^2 + X_1^2}$$

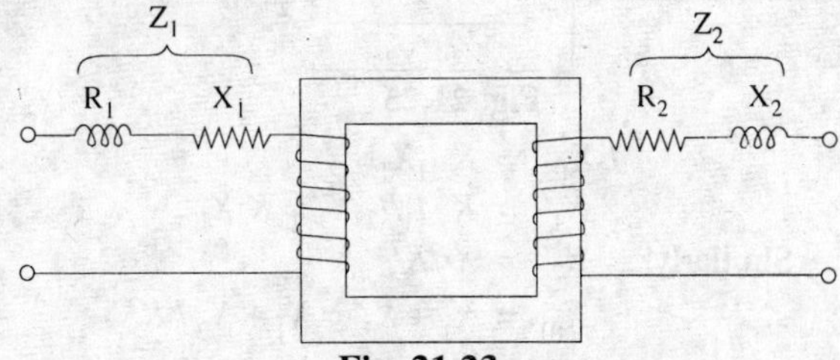

Fig. 21.23

Similarly, secondary impedance is given by

$$Z_2 = \sqrt{R_2^2 + X_2^2}$$

The resistance and leakage reactance of each winding is responsible for some voltage drop in each winding. In primary, the leakage reactance drop is I_1X_1 (usually 1 or 2% of V_1). Hence,

$$\mathbf{V_1} = \mathbf{E_1} + \mathbf{I_1}(R_1 + jX_1) = \mathbf{E_1} + \mathbf{I_1 Z_1}$$

Similarly, there are I_2R_2 and I_2X_2 drops in secondary which combine with V_2 to give E_2.

$$\mathbf{E_2} = \mathbf{V_2} + \mathbf{I_2}(R_2 + jX_2) = \mathbf{V_2} + \mathbf{I_2 Z_2}$$

The voltage vector diagrams for such a transformer for different kinds of loads are shown in Fig. 21.24. In these diagrams, vectors for resistive drops are drawn parallel to current vector I_1 whereas reactive drops are perpendicular to the current. The angle ϕ_1 between V_1 and I_1 gives the power factor angle of the transformer.

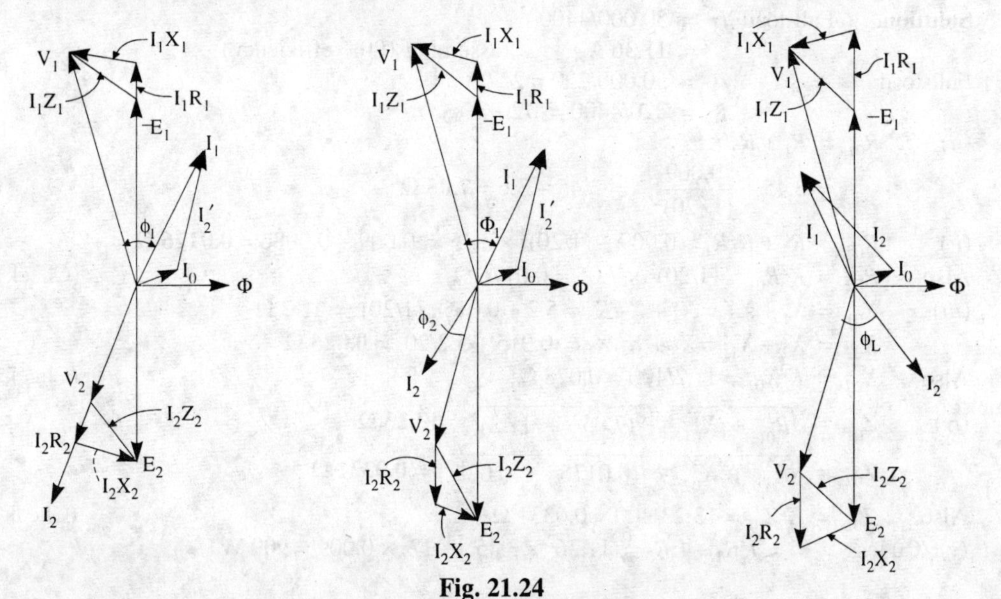

Fig. 21.24

It may be noted that leakage reactance can also be transferred from one winding to the other in the same way as resistance.

Let X_1' be the primary reactance as referred to secondary. Voltage drop across it when in secondary = I_2X_1'. But voltage drop when it is in primary = I_1X_1. Remembering that voltages between the two windings are related through K, we get

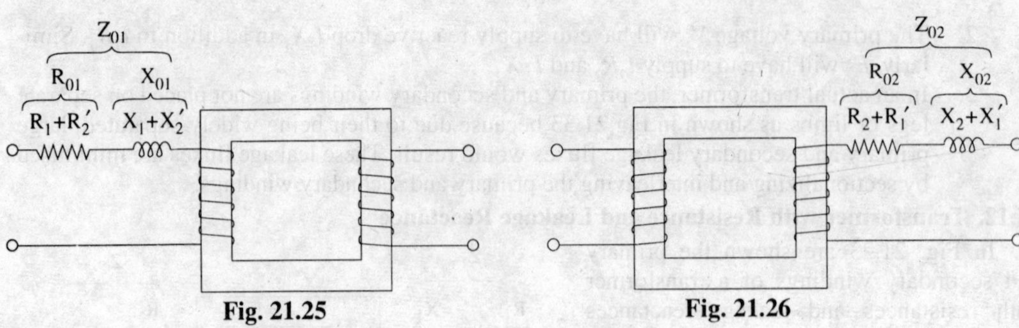

Fig. 21.25 **Fig. 21.26**

$$I_2X_1' = K\,I_1X_1$$

$\therefore$ $X_1' = K.(I_1/I_2)X_1 = K^2X_1$

Similarly $X_2' = X_2/K^2$

$\therefore$ $X_{01} = X_1 + X_2' = X_1 + X_2 / K^2$

$$X_{02} = X_2 + X_1' = X_2 + K^2X_1$$

It is obvious that total impedance of the transformer as referred to primary is given by

$$Z_{01} = \sqrt{R_{01}^2 + X_{01}^2} \qquad \text{...Fig 21.25}$$

and $Z_{02} = \sqrt{R_{02}^2 + X_{02}^2}$ ...Fig.21.26

Example 21.8. *A 50-kVA, 4400/220 V transformer has $R_1 = 3.45\ \Omega$. $R_2 = 0.009\ \Omega$. The values of reactances are $X_1 = 5.2\ \Omega$ and $X_2 = 0.015\ \Omega$. Calculate for the transformer (i) equivalent resistance as referred to primary (ii) equivalent resistance as referred to secondary (iii) equivalent reactance as referred to both primary and secondary (iv) equivalent impedance as referred to both primary and secondary (v) total Cu loss, first using individual resistances of the two windings and secondly, using equivalent resistances as referred to each side.*

Solution. Full-load $I_1 = 50{,}000/4400$

$= 11.36$ A (assuming 100% efficiency)

Full-load $I_2 = 50{,}000/220 = 227$ A

$K = 220/4400 = 1/20$

(*ii*) $R_{01} = R_1 + R_2/K^2$

$$= 3.45 + \frac{0.009}{(1/20)^2} = 3.45 + 3.6 = \mathbf{7.05\ \Omega}$$

(*ii*) $R_{02} = R_2 + K^2R_1 = 0.009 + (1/20)^2 \times 3.45 = 0.009 + 0.0086 = \mathbf{0.0176\ \Omega}$

Also $R_{02} = K^2R_{01} = (1/20)^2 \times 7.05 = 0.0176\ \Omega$ (Check)

(*iii*) $X_{01} = X_1 + X_2' = X_1 + X_2/K^2 = 5.2 + 0.015 / (1/20)^2 = \mathbf{11.2\ \Omega}$

$X_{02} = X_2 + X_1' = X_2 + K^2 X_1 = 0.015 + 5.2/20^2 = \mathbf{0.028\ \Omega}$

Also $X_{02} = K^2X_{01} = 11.2/400 = 0.028\ \Omega$ (Check)

(*iv*) $Z_{01} = \sqrt{R_{01}^2 + X_{01}^2} = \sqrt{(7.05^2 + 11.2)^2} = \mathbf{13.23\ \Omega}$

$Z_{02} = \sqrt{R_{02}^2 + X_{02}^2} = \sqrt{(0.0176^2 + 0.028^2)} = \mathbf{0.0331\ \Omega}$

Also $Z_{02} = K^2Z_{01} = 13.23/400 = 0.0331\ \Omega$ (Check)

(*v*) Cu loss $= I_1^2R_1 + I_2^2R_2 = 11.36^2 \times 3.45 + 227^2 \times 0.009 = \mathbf{909\ W}$

Also, Cu loss $= I_1^2R_{01} = 11.36^2 \times 7.05 = \mathbf{910\ W}$

$= I_2^2R_{02} = 227^2 \times 0.0176 = \mathbf{910\ W}$

21.13. Total Approximate Voltage Drop in a Transformer

As seen from Fig. 21.27 and discussed in Art. 21.12, the primary applied voltage V_1 has to

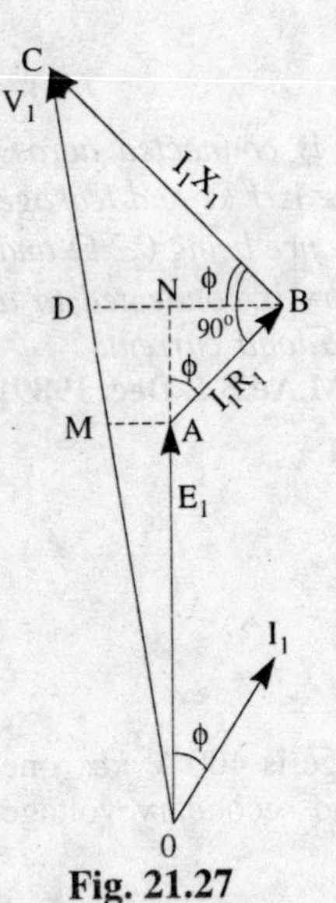

Fig. 21.27

supply (*i*) resistive drop I_1R_1 (*ii*) leakage reactance drop I_1X_1 and (*iii*) has to overcome the back e.m.f. E_1 in the primary winding. In other words, V_1 is the vector sum of E_1, I_1R_1 and I_1X_1. If Z_1 is the primary impedance, then the voltage drop in the primary winding is I_1Z_1. Hence, V_1 is the vector sum of E_1 and I_1Z_1 as shown in Fig. 21.27.

Drop in primary $= I_1Z_1 = AC = MC$ (approx.)

$= MD + DC$

From Δs *ANB* and *DBC*, we get the approximate primary voltage drop as

$= AB \cos\phi + BC \sin\phi$

$= I_1R_1 \cos\phi + I_1X_1 \sin\phi$

Similarly, as seen from Fig. 21.28, the voltage drop in the secondary winding is

$= I_2 R_2 \cos\phi + I_2X_2 \sin\phi$*

The total drop in the transformer is the sum of the two. This total drop can be referred either to primary or secondary. Let us refer it to secondary. The primary drop as referred to secondary is

$= K \times$ primary drop

$= K(I_1R_1 \cos\phi + I_1X_1 \sin\phi)$

Hence, the total drop as referred to secondary is

$=$ secondary drop $+ K \times$ primary drop

$= (I_2 R_2 \cos\phi + I_2 X_2 \sin\phi) + K(I_1R_1 \cos\phi + I_1 X_1 \sin\phi)$

If we neglect the no-load primary current I_0, then

$I_1 = KI_2$

Eliminating I_1 from the above equation, we get

Total drop $= (I_2 R_2 \cos\phi + I_2 X_2 \sin\phi) + K(KI_2 R_1 \cos\phi + KI_2 X_2 \sin\phi)$

$= I_2 [(R_2 + K^2R_1) \cos\phi + (X_2 + K^2X_1) \sin\phi]$

$= I_2 (R_{02} \cos\phi + X_{02} \sin\phi)$

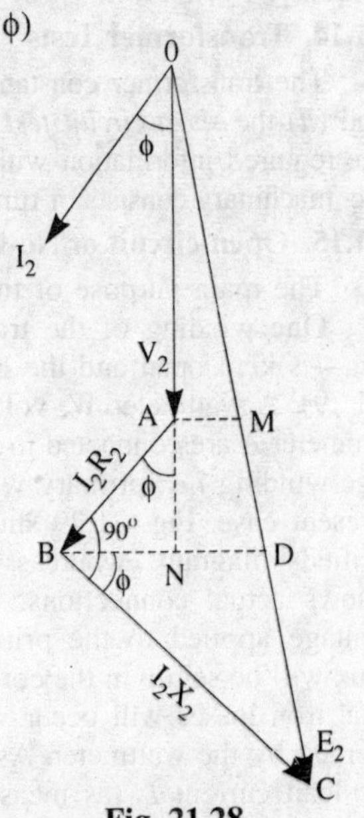

Fig. 21.28

Hence, the total voltage drop of the transformer as referred to secondary is

$= I_2(R_{02} \cos\phi + X_{02} \sin\phi)$

This is the drop for a *lagging* power factor. However, if the power factor is leading, then the sine term will become negative. Hence, for a *leading* power factor the expression is

$= I_2(R_{02} \cos\phi - X_{02} \sin\phi)$

Hence, if the primary voltage V_1 is kept constant, the secondary terminal voltage will decrease by this amount from no-load to full-load.

Example 21.9. *A 230/460 V transformer has a primary resistance of 0.2 Ω and a reactance of 0.5 Ω and the corresponding values for the secondary are 0.75 Ω and 1.8 Ω respectively. Find the secondary terminal voltage when supplying (a) 10 A at 0.8 power factor lagging (b) 10 A at p.f. 0.8 leading.* **[Elect. Machines-I, Allahabad Univ. 1993]**

Solution. $K = 2$, $R_{02} = 0.75 + 4(4 \times 0.2) = 1.55\ \Omega$

Similarly, $X_{02} = 1.8 + (4 \times 0.5) = 3.8\ \Omega$

$I_2 = 10$ A; $\cos\phi = 0.8$, $\sin\phi = 0.6$

(*a*) Voltage drop $= I_2 (R_{02} \cos\phi + X_{02} \sin\phi)$

$= 10(1.55 \times 0.8 + 3.8 \times 0.6) = 35.2$ V

$V_2 = 460 - 35.2 =$ **424.8 V**

*It being assumed that power factor angles of both primary and secondary are equal.

(c) Voltage drop $= 10\,(1.55 \times 0.8 - 3.8 \times 0.6) = -10.4$ V

$V_2 = 460 - (-10.4) = \mathbf{470.4\ V}$

Example 21.10. *A single-phase transformer with turns ratio 400:200 is connected across 400 V, 50 Hz a.c. supply across the 400 turns. The primary winding resistance is 1 Ω and leakage reactance is 3 Ω, the corresponding secondary-side resistance and leakage reactance being 0.2 Ω and 0.8 Ω respectively. Determine approximately the load voltage when secondary is connected to a load which draws a current of 20 A at 0.9 lagging power factor. Neglect no-load current.*

[Electrical Engg. A.M.Ae.S.I. Dec. 1989]

Solution. $K = N_2/N_1 = 200/400 = 1/2$

$R_{02} = 0.2 + (1/2)^2 \times 1 = 0.45\ \Omega$

Similarly, $X_{02} = 0.8 + (1/2)^2 \times 3 = 1.55\ \Omega$

Voltage drop $= I_2\,(R_{02} \cos\phi + X_{02} \sin\phi)$

$= 20\,(0.45 \times 0.9 + 1.55 \times 0.436) = 21.6$ V

∴ no-load, $V_2 = 200 - 21.6 = \mathbf{178.4\ V}$

It may be noted that with 400 primary winding turns, the primary voltage is 400 V *i.e.,* one volt per primary turn. Hence, for a 200 turn secondary winding, the no-load secondary voltage is 200 × 1 = 200 V.

21.14. Transformer Tests

The transformer constants or parameters can be easily determined by two tests (*i*) *open-circuit* and (*ii*) the *short-circuit test.* These tests are very economical and convenient, because they furnish the required information without actually loading the transformer. In fact, the testing of every large a.c. machinary consists of running two tests similar to the open and short-circuit tests of a transformer.

21.15. Open-circuit or No-load Test

The main purpose of this test is to determine no-load losses.

One winding of the transformer—whichever is convenient but usually high voltage winding—is kept open and the other is connected to a supply of normal voltage and frequency (Fig. 21.29). A wattmeter *W*, voltmeter *V* and ammeter *A* are connected in the low-voltage winding *i.e.*, primary winding in the present case. Fig. 21.29 shows the simplified diagram whereas Fig. 21.30 shows actual connections. With normal voltage applied to the primary, normal flux will be set up in the core, hence normal iron losses will occur which are recorded by the wattmeter. As the primary no-load current I_0 (as measured by ammeter) is small (usually 2 to 10% of rated load current), Cu loss is negligibly small in primary and nil in secondary (being open). Hence, the *wattmeter reading represents practically the core-loss under no-load conditions (and which is the same for all loads as pointed out earlier in Art. 21.8)*

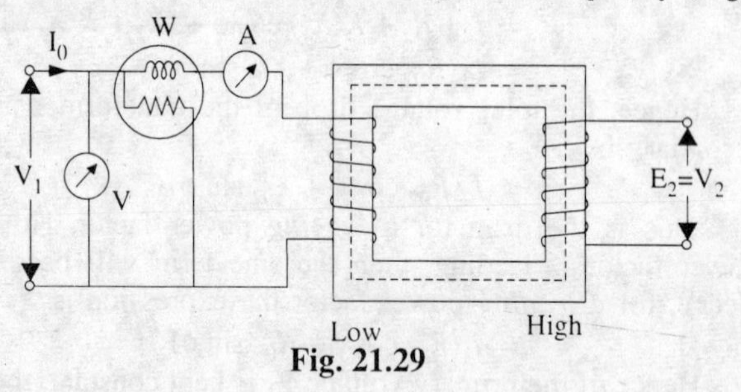

Fig. 21.29

Sometimes a high resistance voltmeter is connected across the secondary. The reading of the voltmeter gives the induced e.m.f. in the secondary winding. This helps in finding the transformation ratio *K*.

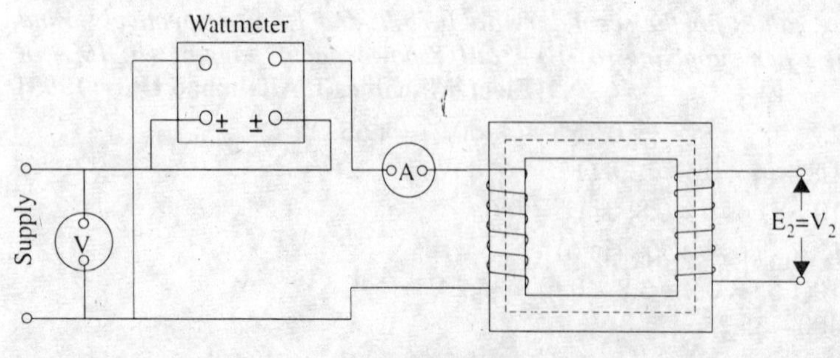

Fig. 21.30

21.16. Short-circuit or Impedance Test

It is an economical method for determining the following:

1. Equivalent impedance (Z_{01} or Z_{02}), leakage reactance (X_{01} or X_{02}) and resistance (R_{01} or R_{02}) of the transformer as referred to the winding in which the measuring instruments are placed.
2. Cu loss at full-load (and at any desired load). This loss is used in calculating the efficiency of the transformer.
3. Knowing Z_{01} or Z_{02}, the total voltage drop in the transformer as referred to primary or secondary can be calculated and hence regulation of the transformer determined.

In this test, one winding—usually the low-voltage winding is solidly short-circuited by a thick conductor (or through an ammeter which may serve the additional purpose of indicating rated load current) as shown in Fig. 21.31.

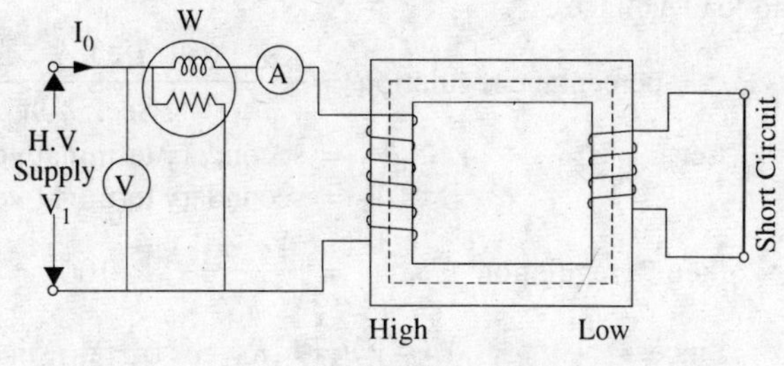

Fig. 21.31

A low voltage (usually 5 to 10% of normal primary voltage) at correct frequency (though for Cu losses it is not essential) is applied to the primary and is cautiously increased till full-load currents are flowing both in primary and secondary (as indicated by the respective ammeters).

Since, in this test, the applied voltage is a small percentage of the normal voltage, the mutual flux Φ produced in the core is also a small percentage of its normal value (because flux is proportional to the voltage as shown in the e.m.f. equation of the transformer in Art. 21.6). Hence, core losses are very small with the result that the wattmeter reading represents the full-load Cu loss or $I^2 R$ loss for the whole transformer *i.e.*, both primary Cu loss and secondary Cu loss. If V_{SC} is the voltage required to calculate rated load current on short circuit, then

$$Z_{01} = V_{SC} / I_1$$

Also $$W = I_1^2 R_{01} \quad \therefore \quad R_{01} = W / R_1^2$$

$$\therefore \quad X_{01} = \sqrt{(Z_{01}^2 - R_{01}^2)}$$

Example 21.11. *The ratio of turns of a single-phase transformer is 8, the resistances of the primary and secondary windings are 0.85 Ω and 0.012 Ω respectively and the leakage reactances of these windings are 4.8 Ω and 0.07 Ω respectively. Determine the voltage to be applied to the primary to obtain a current of 150 A in the secondary when the secondary terminals are short-circuited. Ignore the magnetising current.*

Solution. Turn ratio $= N_1 / N_2 = 8 \qquad \therefore \; K = 1/8$

Now
$$R_{01} = R_1 + R_2 / K^2 = 0.85 + 0.012 \times 64 = 1.618\ \Omega$$
$$X_{01} = X_1 + X_2 / K^2 = 4.8 + 64 \times 0.07 = 9.28\ \Omega$$
$$Z_{01} = \sqrt{(1.618^2 + 9.28^2)} = 9.419\ \Omega$$
$$I_1 = KI_2 = 150/8\ \text{A}$$

Now, $$V_{SC} = I_1 Z_{01} = \frac{150}{8} \times 9.419 = \mathbf{176.6\ V}$$

Example 21.12. *The low-voltage winding of a 300 kVA, 11,000/2,200 volt, 50 Hz transformer has 190 turns and a resistance of 0.06 ohm. The high voltage winding has 910 turns and a resistance of 1.6 ohm. When the low voltage winding is short-circuited, the full-load current is obtained with 550 volts applied to h.v. winding. Calculate the equivalent resistance and leakage reactance referred to high-voltage side. Assume full-load efficiency of 98.5%.*

[A.M.I.E. Summer 1987]

Solution. Full-load primary current

$= 300{,}000 / 0.985 \times 11{,}000 = 27.7$ A

$Z_{01} = 550 / 27.7 = 19.8\ \Omega; \quad R_2' = R_2 / K^2 = 0.06 / (910/190)^2 = 1.38\ \Omega$

$R_{01} = R_1 + R_2' = 1.6 + 1.38 = \mathbf{2.98\ \Omega}$

$X_{01} = \sqrt{(Z_{01}^2 - R_{01}^2)} = \sqrt{(19.8^2 - 2.98^2)} = \mathbf{19.5\ \Omega}$

21.17. Voltage Regulation of a Transformer

It has been discussed in Art. 21.13 that when a transformer is loaded, its secondary terminal voltage falls (for a lagging p.f.) *provided the applied primary voltage V_1 is held constant.* This variation in secondary terminal voltage from no-load (N.L.) to full-load (F.L.) is called the (voltage) regulation of the transformer and is usually expressed as a percentage of the secondary no-load voltage. **[A.M.I.E. Winter 1988]**

$$\therefore \text{ percentage regulation} = \frac{\text{Sec. Volt (N.L.)} - \text{Sec. Volt (F.L.)}}{\text{Sec. Volt (N.L.)}} \times 100$$

Let $_0V_2$ = secondary terminal voltage on no-load

V_2 = secondary terminal voltage on load

then % regulation $= \dfrac{_0V_2 - V_2}{_0V_2} \times 100$

Since $_0V_2 - V_2 = I_2(R_{02} \cos\phi \pm X_{02} \sin\phi)$

$$\therefore \quad \text{\% age regn.} = \frac{I_2 (R_{02} \cos\phi \pm X_{02} \sin\phi)}{_0V_2}$$

21.18. Losses in a Transformer

In a static transformer, there are no friction or windage losses. Hence, the only losses occurring are:

(*i*) **Core or Iron Loss.** It includes both hysteresis loss and eddy current loss. Because the core flux in a transformer remains practically constant for all loads (its variation being 1 to 3% from no-load to full-load), the core loss is practically the same at all loads.

Hysteresis loss $W_h = \eta B_{max}^{1.6} f V$ watt

Eddy current loss $W_e = P B_{max}^2 f^2 t^2$ watt

These losses are minimized by using steel of high silicon content for the core and by using very thin laminations. Iron or core-loss is found from the O.C. test. *The input of the transformer when on no-load measures the core loss.* This loss is found from the no-load test (Art. 21.15).

(*ii*) **Copper Loss.** This loss is due to the ohmic resistance of the transformer windings. Total Cu loss $= I_1^2 R_1 + I_2^2 R_2 = I_1^2 R_{01} = I_2^2 R_{02}$. It is clear that Cu loss is proportional to (current)^2 or (kVA)^2. In other words, Cu loss at half the full-load is one-fourth of that at full-load. The value of Cu loss is found from the short-circuit test (Art. 21.16).

Example. 21.13. *The flux in a magnetic core is alternating sinusoidally at a frequency of 600 Hz. The maximum flux density is 2T and the eddy current loss is 15 W. Find the eddy current loss in the core if the frequency is raised to 800 Hz and the maximum flux density is reduced to 1.5T.* **[Elements of Elect.Engg., Punjab Univ. 1994.]**

Solution. The eddy-current loss, $W_e \propto B_{max} f^2$

$\therefore \quad 15 \propto 2^2 \times 600^2, \quad W_e \propto 1.5^2 \times 800^2$

$$\therefore \quad \frac{W_e}{15} = \left(\frac{1.5}{2}\right)^2 \times \left(\frac{800}{600}\right)^2 \qquad \therefore \quad W_e = \mathbf{35\ W}$$

Example 21.14. *The total loss in a sample of sheet steel weighing 13 kg is 20 W at 50 Hz and 35 W at 75 Hz, both measured at the same peak flux density. Separate the loss at 50 Hz into hysteresis and eddy current components.* **[Electrical Engg. A.M.Ae.S.I. Dec. 1991]**

Solution. Since peak flux density is the same at both frequencies, the eddy current loss is proportional to f^2 and hysteresis loss to f.

Hysteresis loss $\propto f = Af$, eddy current loss $\propto f^2 = Bf^2$
where A and B are constants.

Total iron loss, $W_i = Af + Bf^2$

At 50 Hz, $20 = 50\text{ A} + 2500\text{ B}$ ...(*i*)

At 75 Hz, $35 = 75\text{ A} + 5625\text{ B}$ ...(*ii*)

From (*i*) and (*ii*) we get

$A = 0.2667$ and $B = 2.666 \times 10^{-3}$

Hence at 50 Hz, hysteresis loss $= Af = 0.2667 \times 50 =$ **13.3 W**

Eddy current loss $= Bf^2 = 2.666 \times 10^{-3} \times 50^2 =$ **6.7 W**

Example. 21.15. *In a 440 V, 50 Hz transformer the total iron loss is 2500 W, when applied voltage is 220 V at 25 Hz the corresponding loss is 850 W. Calculate the eddy current loss at normal frequency and potential difference.* **[Electrical Engg. A.M.Ae.S.I. Dec. 1992]**

Solution. The flux density in both cases is the same because in second case voltage as well as frequency are halved. Flux density remaining the same, the eddy current loss is proportional to f^2 and hysteresis loss $\propto f$.

Hysteresis loss $\propto f = Af$ and eddy current loss $\propto f^2 = Bf^2$ where A and B are constants.

Total iron loss $W_i = Af + Bf^2$

$\therefore \quad W_i / f = A + Bf$ ...(*i*)

Now, when $f = 50$ Hz; $W_i = 2500$ W

when $f = 25$ Hz; $W_i = 850$ W

Using these values in (*i*) above, we get

$2500 / 50 = A + 50\,B$ and $850 / 25 = A + 25\,B \quad \therefore \quad B = 16 / 25 = 0.64$

Hence, at normal p.d. and frequency

Eddy current loss $= Bf^2 = 0.64 \times 50^2 =$ **1600 W**

Hysteresis loss $= 2500 - 1600 = 900$ W

21.19. Efficiency of a Transformer

As is the case with other types of electrical machines, the efficiency of a transformer at a particular load and power factor is defined as the output divided by the input—the two being measured in the same units (either watts or kilowatts).

$$\therefore \quad \text{efficiency} = \frac{\text{output}}{\text{input}}$$

But a transformer being a highly efficient piece of equipment, has very small losses, hence, it is impractical to try to measure transformer efficiency by measuring its input and output. These quantities are nearly of the same size. A better method is to determine the losses and then to calculate the efficiency from

$$\text{Efficiency} = \frac{\text{output}}{\text{output + losses}} = \frac{\text{output}}{\text{output + Cu loss + Iron loss}}$$

or

$$\eta = \frac{\text{input} - \text{losses}}{\text{input}} = 1 - \frac{\text{losses}}{\text{input}}$$

It may be noted here that efficiency is based on power output in watts and not on volt-amperes, although losses are proportional to VA. Hence, at any volt-ampere load, the efficiency depends on power factor, being maximum at a power factor of unity.

Efficiency can be computed by determining core loss from no-load or open-circuit test and Cu loss from the short-circuit test.

21.20. Condition for Maximum Efficiency

Cu loss $= I_1^2 R_{01}$ or $I_2^2 R_{02} = W_{cu}$

Iron loss = hysteresis loss + eddy current loss $= W_h + W_e = W_i$

Considering primary side,

Primary input $= V_1 I_1 \cos \phi_1$

$$\therefore \quad \eta = \frac{V_1 I_1 \cos\phi_1 - \text{losses}}{V_1 I_1 \cos\phi_1} = \frac{V_1 I_1 \cos\phi_1 - I_1^2 R_{01} - W_i}{V_1 I_1 \cos\phi_1}$$

$$= 1 - \frac{I_1 R_{01}}{V_1 \cos\phi_1} - \frac{W_i}{V_1 I_1 \cos\phi_1}$$

Differentiating both sides with respect to I_1, we get

$$\frac{d\eta}{dI_1} = 0 - \frac{R_{01}}{V_1 \cos\phi_1} + \frac{W_i}{V_1 I_1^2 \cos\phi_1}$$

For η to be maximum, $\frac{d\eta}{dI_1} = 0$. Hence, the above equation becomes

$$\frac{R_{01}}{V_1 \cos\phi_1} = \frac{W_i}{V_1 I_1^2 \cos\phi_1} \quad \text{or} \quad W_i = I_1^2 R_{01} \quad \text{or} \quad = I_2^2 R_{02}$$

or **Cu loss = Iron loss**

The output current corresponding to maximum efficiency is

$$I_2 = \sqrt{\frac{W_i}{R_{02}}}$$

It is this value of the output current which will make the Cu loss equal to the iron loss. By proper design, it is possible to make the maximum efficiency occur at any desired load.

In Fig. 21.32, Cu and iron losses are plotted as a percentage of power input and the efficiency curve as deduced from these is also shown. It is obvious that the point of inter-section of the Cu and iron loss curves gives the point of maximum efficiency. It would be seen that the efficiency is high and is practically constant from 25% load to 25% overload.

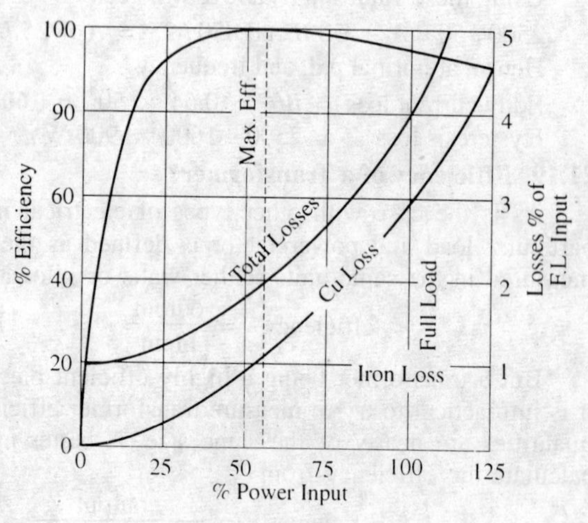

Fig. 21.32

21.21. Load Corresponding to Maximum Efficiency

If full-load Cu loss and iron loss of a transformer are known, the load at which it will operate at maximum efficiency can be easily found.

Let W_i be the iron loss and W_{cu} the full-load Cu loss. Efficiency will be maximum at that load for which Cu loss becomes equal to W_i. If x is that load, then Cu loss when the transformer delivers this output will be W_i *i.e.*, equal to iron loss. So, W_i becomes the Cu loss for x kVA output. Remembering the Cu loss is proportional to kVA^2, we have

$$W_{cu} \propto (\text{full-load kVA})^2$$

$$W_i \propto x^2$$

$$\therefore \quad \left(\frac{x}{\text{F.L. kVA}}\right)^2 = \frac{W_i}{W_{cu}}$$

or

$$x = \text{F.L. kVA} \times \sqrt{\frac{W_i}{W_{cu}}}$$

$$\therefore \quad x = \text{F.L. kVA} \times \sqrt{\frac{\text{iron loss}}{\text{full–load Cu loss}}}$$

Example 21.16. *A 200/400 V, 10 kVA, 50-Hz single-phase transformer has at full-load, a copper loss of 120 W. If it has an efficiency of 98% at full-load unity power factor, determine the iron losses.*

What would be the efficiency of the transformer at half load 0.8 power factor lagging?

[A.M.I.E. Summer 1985]

Solution. Output at full-load u.p.f. = 10 kW

Input = 10 / 0.98 = 10.204 kW

Total full-load losses = 10.204 – 10 = 0.204 kW = 204 W

Iron losses = 204 – 120 = 84 W

Half-load 0.8 p.f.

$$\text{Cu loss} = \left(\frac{1}{2}\right)^2 \times 120 = 30 \text{ W, Iron loss} = 84 \text{ W}$$

$$\text{Total loss} = 30 + 84 = 114 \text{ W} = 0.114 \text{ kW}$$

$$\text{Output} = (10/2) \times 0.8 = \mathbf{4\ kW}$$

$$\eta = \frac{\text{output}}{\text{output + losses}} = \frac{4}{4.114} = 0.97 \text{ or } 97\%$$

Example 21.17. *A single-phase transformer working at unity power factor has an efficiency of 92% at both half-load and at full-load of 500 kW. Determine the efficiency at 80% of full-load.*

[Electrical Engg. A.M.Ae.S.I. June 1995]

Solution. The fact that efficiency is the same *i.e.*, 92% at both full-load and half-load will help us to find the iron and copper lossess.

At Full-load

Output = 500 kW, input = 500 / 0.92 = 543.5 kW

$\therefore$ total loss = 543.5 – 500 = 43.5 kW

Let x = iron loss ...it remains constant at all loads

y = F.L. Cu loss ...It is $\propto (\text{kVA})^2$

$\therefore x + y = 43.5$ ...(*i*)

At Half-load

Output = 500 / 2 = 250 kW, input = 250 / 0.92 = 271.7 kW

$\therefore$ total loss = 271.7 – 250 = 21.7 kW

Since Cu loss becomes $(1/2)^2$ *i.e.*, one-fourth of its F.L. value, hence

$\therefore x + y/4 = 21.7$ ...(*ii*)

Solving for x and y we get, x = 14.4 kW and y = 29.1 kW

At 80% Full-load

Cu. loss = $0.8^2 \times 29.1$ = 18.6 kW

$\therefore$ total loss = 14.4 + 18.6 = 33 kW

Output = 0.8 × 500 = 400 kW, input = 400 + 33 = 433 kW

$\therefore \eta$ = 400 / 433 = 0.924 or **92.4 %**

Example 21.18. *A 600 kVA, single-phase transformer when working at u.p.f. has an efficiency of 92% at full load and also at half-load. Determine the efficiency when it operates at unity p.f. and 60% of full load.* **[Electrical Engg. A.M.Ae.S.I. June 1993]**

Solution. The fact that efficiency is the same *i.e.*, 92% at both full-load and half-load will help us to find the iron and copper losses.

At full-load

Output = 600 kW; input = 600 / 0.92 = 652.2 kW; Total loss = 652.2 – 600 = 52.2 kW

Let x = iron loss ... it remains constant at all loads

y = F.L. Cu loss ... it is $\propto (\text{kVA})^2$

$\therefore x + y = 52.2$ ...(*i*)

At half-load

Output = 300 kW; input = 300 / 0.92; losses = (300 / 0.92 – 300) = 26.1 kW

Since Cu loss becomes one-fourth of its F.L. value, hence

$x + y/4 = 26.1$ ...(*ii*)

Solving for x and y, we get $x = 17.4$ kW; $y = 34.8$ kW

At 60% full-load

Cu loss $= 0.6^2 \times 34.8 = 12.53$ kW; Total loss $= 17.4 + 12.53 = 29.93$ kW

Output $= (600 \times 0.6) \times 1 = 360$ kW $\therefore \eta = 360 / 389.93 = 0.965$ or **96.5%**

Example 21.19. *A 1000 kVA transformer has 94 per cent efficiency at full-load and at 50 per cent of full-load. The power factor is unity in both cases.*

(i) Segregate the losses.

(ii) Determine the efficiency of the transformer for unity p.f. and 75 per cent of full load.

[Electrical Science, AMIE Winter 1994]

Solution. The fact that the efficiency is the same at both full-load and half-load helps us to find the transformer losses.

At Full load

Output $= 1000 \times 1 = 1000$ kW, input $= 1000 / 0.94 = 1063.8$ kW

Total losses $= 1063.8 - 1000 = 63.8$ kW

Let, x = iron loss ... it remains constant at all loads

y = F.L. Cu loss ... it is $\propto$ (kVA)^2

$x + y = 63.8$ kW ...(*i*)

At Half Load

Output $= 500 \times 1 = 500$ kW, input $= 500 / 0.94 = 531.9$ kW.

Total losses $= 531.9 - 500 = 31.9$ kW

Since, Cu loss becomes $(500 / 1000)^2 \times$ F.L. Cu loss hence, Cu loss at half-load is $y / 4$

$\therefore \quad x + y/4 = 31.9$ ...(*ii*)

From (*i*) and (*ii*) above we get

$x =$ **21.3 kW** and $y =$ **42.5 kW**

At 75% Full Load

Cu loss $= 42.5 \times (750 / 1000)^2 = 23.9$ kW

Total loss $= 23.9 + 21.3 = 45.2$ kW

Output $= 750 \times 1 = 750$ kW

$= 750 / (750 + 45.2) = 0.943$ or **94.3%**

Example 21.20. *The primary and secondary windings of a 500 kVA transformer have resistances of 0.42 Ω and 0.0011 Ω respectively. The primary and the secondary voltages are 6600 V and 400 V respectively and the iron loss is 2.9 kW. Calculate the efficiency at full-load, assuming the power factor of the load to be 0.8.* **[Electrical Science, AMIE Summer 1993]**

Solution. For calculating copper losses, let us transfer primary resistances to the secondary winding.

$\therefore \quad R_{02} = R_2 + K^2 R_1$

Now, $K = 400 / 6600 = 2 / 33$, $R_1 = 0.42\ \Omega$ and $R_2 = 0.0011\ \Omega$

$\therefore \quad R_{02} = 0.0011 + (2/33)^2 \times 0.42 = 0.0026\ \Omega$

Full-load $I_2 = 500 \times 10^3 / 400 = 1250$ A

F.L. Cu loss $= I_2^2 R_{02} = 1250^2 \times 0.0026 = 4060$ W $= 4.06$ kW

Total loss $= 4.06 + 2.9 = 6.96$ kW

F.L. output $= 500 \times 0.8 = 400$ kW

F.L. $\eta = 400 / (400 + 6.96) = 0.983$ or **98.3%**

Example 21.21. *In a 25 kVA, 3300 / 230 V, 1-phase transformer, the iron and full-load copper losses are respectively 350 and 400 W. Calculate (i) the efficiency at half-load 0.8 p.f. and (ii) the laod at which the efficiency is maximum.* **(A.M.I.E. Winter 1988)**

Solution. (*i*) **Half-load 0.8 p.f.**

Cu loss at half-load $= (1/2)^2 \times$ F.L. loss

$= 400/4 = 100$ W

Output $= (25/2) \times 0.8 = 10$ kW

Input $= 10 + 0.1 = 10.1$ kW

$$\therefore \quad \eta = \frac{10 \times 100}{10.1} = \mathbf{99.02\%}$$

(*ii*) Load for maximum efficiency

$$= \text{Full-load kVA} \times \sqrt{\frac{\text{iron loss}}{\text{F.L. Cu loss}}}$$

$$= 25 \times \sqrt{350/400} = \mathbf{23.4\ kVA}$$

Tutorial Problems No. 21.3

1. Iron loss of a 100-kVA single-phase transformer is 1.5 kW and full-load copper loss in 1 kW. Calculate transformer efficiency at:
   (*a*) full-load unity p.f. and
   (*b*) half full-load 0.8 p.f. **[97.56%; 95.81%]**
2. The primary resistance of 440 / 110V transformer is 0.5 Ω and the secondary resistance is 0.04 Ω. When 440 V is applied to the primary and the secondary is left open-circuited, 200 W is drawn from the supply. Find the secondary current which will give maximum efficiency and calculate this efficiency for a load having unity p.f. **[53 A; 93.58%]**
3. A 200 kVA transformer has an efficiency of 98% at full-load. If the maximum efficiency occurs at three-quarters of full-load, calculate (*a*) iron loss (*b*) Cu loss at F.L. (*c*) efficiency at half-load. Ignore magnetising current and assume a p.f. of 0.8 at all loads. **[(*a*) 1.777 kW (*b*) 2.09 kW (*c*) 97.92%]**
4. At full-load, the Cu and iron losses in a 100 kVA transformer are each equal to 2.5 kW. Find the efficiency at a load of 65 kVA power factor 0.8. **[93.58%]**
5. A 150 kVA, single-phase transformer has a core loss of 1.5 kW and a full-load Cu loss of 2 kW. Calculate the efficiency of the transformer (*a*) at full load 0.8 p.f. lagging (*b*) at one-half full-load unity p.f. Determine also the secondary current at which the efficiency is maximum if the secondary voltage is maintained at its rated value of 240 V. **[(*a*) 97.71% (*b*) 97.4%; 541 A]**
6. A 100 kVA, single-phase transformer has an iron loss of 600 W and a copper loss of 1.5 kW at full-load current. Calculate the efficiency at (*a*) 100 kVA output at 0.8 p.f. lagging (*b*) 50 kVA output at unity power factor. **[(*a*) 97.44% (*b*) 98.09%]**
7. A 500 kVA, single-phase transformer has an iron loss of 2.5 kW and the maximum efficiency at 0.8 p.f. occurs when the load is 268 kW.
   Calculate (*a*) the maximum efficiency at unity power factor and (*b*) the efficiency on full-load at 0.71 power factor. **[(*a*) 98.53% at 335 kW (*b*) 97.78%]**
8. A 500 / 250 V, 1-phase transformer has constant losses of 150 W. When delivering an output of 8 kW to a load, the transformer takes 21 A at a p.f. of 0.8 from 500 V supply mains. Calculate the variable losses of the transformer at this load. **[250 W]**
9. Calculate the efficiency at full-load of 500 kVA transformer from the following results when the p.f. of the load is 0.8 (lag).
   Primary volts = 4000
   Secondary volts = 500
   Primary resistance = 0.2 Ω
   Secondary resistance = 0.002 Ω
   Iron-loss (by no-load test) = 2,500 W **[98.2%]**
10. Find the efficiency of a 100 kVA transformer at (*i*) half-load (*ii*) full-load both at 0.8 p.f. lagging if the copper loss at full-load is 2 kW and iron loss is 1.2 kW. **[95.94%; 96.16%]**
11. A 400 / 230 V, 150 kVA, 1-phase transformer has a core loss of 1 kW and full-load copper loss of

2 kW. Calculate (*a*) the kVA load for maximum efficiency and (*b*) the efficiency at 50 per cent full-load at 0.8 p.f. lagging. **[106.05 kVA; 97.57%]**

21.22. All-day Efficiency

The ordinary or commercial efficiency of a transformer is given by the ratio

$$\eta = \frac{\text{output in watts}}{\text{input in watts}}$$

But there are certain types of transformers whose performance cannot be judged by this efficiency. Transformers used for supplying lighting and general network *i.e.,* distribution transformers have their primaries energised all the twenty-four hours, although their secondaries supply little or no-load much of the time during the day except during the house-lighting period. It means that whereas core loss occurs throughout the day, the Cu loss occurs only when the transformer is loaded. Hence, it is considered a good practice to design such transformers so that core losses are very low. The Cu losses are relatively less important because they depend on the load. The performance of such a transformer should be judged by all-day efficiency also known as operational efficiency which is computed on the basis of energy consumed during certain time period, usually a day of 24 hours.

$$\therefore \quad \eta_{all\text{-}day} = \frac{\text{output in kWh}}{\text{input in kWh}} \quad \text{(for 24 hours)}$$

To find this all-day efficiency or (as it is also called) energy efficiency, we have to know the load cycle on the transformer *i.e.,* how much and how long the transformer is loaded during 24 hours. Practical calculations are facilitated by making use of a *load factor*.

Example 21.22. *A 100 kVA lighting transformer has a full-load loss of 3 kW, the losses being equally divided between iron and copper. During a day, the transformer operates on full-load for 3 hours, one half-load for 4 hours, the output being negligible for the remainder of the day. Calculate the all-day efficiency.* **(Elect. Technology, Allahabad Univ. 1991)**

Solution. It is usually assumed that lighting transformers have a load p.f. of unity.

Iron loss for 24 hours	$= 1.5 \times 24$	$= 36$ kWh
Full-load Cu loss	$= 1.5$ kW	
Cu loss for 3 hours on full-load	$= 3 \times 1.5$	$= 4.5$ kWh
Cu loss at half full-load	$= 1.5 / 4$ kW	
Cu loss for 4 hours at half F.L.	$= 4 \times (1.5 / 4)$	$= 1.5$ kWh
Total losses in 24 hours	$= 36 + 4.5 + 1.5$	$= 42$ kWh
Total output in 24 hours	$= (100 \times 3) + (50 \times 4)$	$= 500$ kW

$$\therefore \quad \eta_{all\text{-}day} = \frac{500 \times 100}{(500 + 42)} = \mathbf{92.25\%}$$

Note. It may be noted that the ordinary or commercial or power efficiency of this transformer is = 100 / 103 = 97.1%.

Example 21.23. *A 200 kVA, single-phase transformer is in circuit continuously. For 8 hours in a day, the load is 160 kW at 0.8 p.f. For 6 hours, the load is 80 kW at unity p.f. and for the ramaining period of 24 hours, it runs on no-load. Full-load copper losses are 3.02 kW and iron losses are 1.6 kW. Find all-day efficiency.* **(Electrical Engg. A.M.Ae.S.I. June 1994)**

Solution. It should be noted that the transformer kVA = kW / cos ϕ. As given, the transformer runs at a kVA or 160/ 0.8 = 200 kVA for 8 hours. In the 6 hours, the transformer kVA = 80/1= 80

Obviously, F.L. Cu loss takes place for 8 hours whereas during 6 hours, the Cu loss = 3.02 $(80/200)^2$ = 1.91 kW.

F.L. Cu loss for 8 hours	$= 8 \times 3.02 = 24.16$ kW
Cu loss for 6 hours	$= 6 \times 1.91 = 11.46$ kWh
Total Cu loss in 24 hours	$= 24.16 + 11.46 = 35.62$ kWh
Iron loss for 24 hours	$= 38.4$ kWh
Total transformer loss in 24 hours	$= 35.62 + 38.4 = 74.02$ kWh

Transformer output for 24 hours $= 8 \times 160 + 6 \times 80 = 1760$ kWh

$\therefore \quad \eta_{all\text{-}day} = 1760 \times 100 / (1760 + 74.02) = \mathbf{96\%}$

Example 21.24. *A 20 kW distribution transformer has an efficiency of 0.95 at full-load. The copper losses at full-load equal the iron losses. The transformer is kept connected to the mains all the time but is loaded with full-load for 6 hours a day the load during the remaining time being zero. Find the all-day efficiency.* **(Electrical Engg. A.M.Ae.S.I. Dec. 1993)**

Solution. Input = output / efficiency = 20 / 0.95 = 21.052

full-load loss = 21.052 – 20 = 1.052 kW

full-load Cu loss = 1.052 / 2 = 0.526 kW, iron loss = 0.526 kW

The Cu loss takes place only when the transformer is fully or partially loaded whereas iron loss takes place all the time the transformer is energised.

F.L. Cu loss per day for 6 hours $= 0.526 \times 6 = 3.156$ kWh

Iron loss per day $= 24 \times 0.526 = 12.624$ kWh

Total loss in 24 hours $= 3.156 + 12.624 = 15.780$ kWh

Output in 24 hours $= 20 \times 6 = 120$ kWh

All-day $\eta = 120 \times 100 / (120 + 15.78) = \mathbf{88.4\%}$

21.23. Three-phase Transformers

Large scale generation of electric power is usually 3-phase at generated voltages of 132 kV or somewhat higher. Transmission is generally accomplished at higher voltages of 66, 110, 132, 220 and 275 kV for which purpose 3-phase transformers are necessary to step up the generated voltage to that of the transmission line. Next at load centres, transmission voltages are reduced to distribution voltages or 6600, 4600 and 2300 volts. Further at most of the consumers, the distribution voltages are further reduced to utilization voltages of 440, 220 or 110 volts. Years ago, it was a common practice to use suitable inter-connected three single-phase transformers instead of a single 3-phase transformer. But these days, the latter is gaining popularity partly because of improvement in design and manufacture but principally because of better acquaintance of operating men with the three-phase type. As compared to a bank of 3 single-phase transformers, the main advantages of a single 3-phase transformer are that it occupies less floor space for equal rating, weighs less, costs about 15% less and further, that only one unit is to be handled and connected.

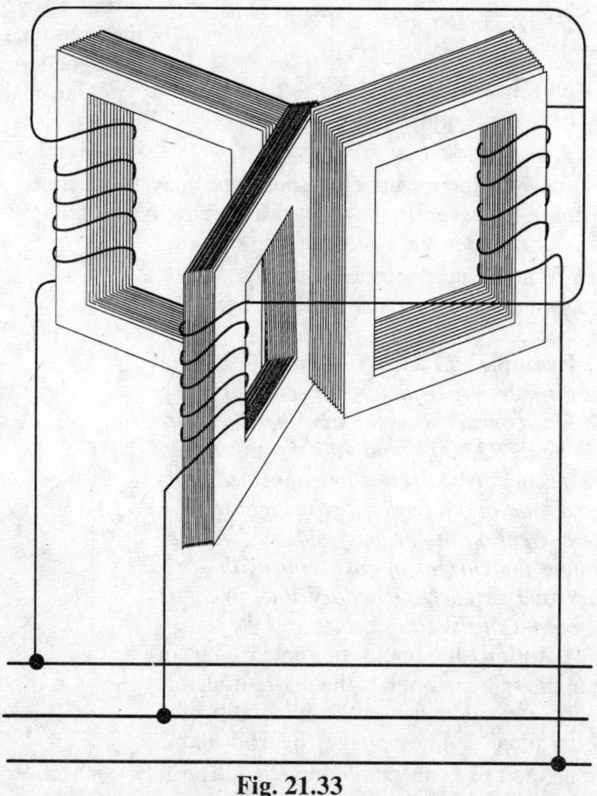

Fig. 21.33

Like single-phase transformers, the three-phase transformers are also of the core type or shell type. The basic principle of a 3-phase transformer is illustrated in Fig. 21.33 in which only primary windings have been shown inter-connected in star and put across 3-phase supply. The three cores are 120° apart and their empty legs are shown in contact with each other. The centre-leg, formed by these three, carries the flux produced by the three-phase currents I_R, I_Y and I_B. As, at any instant, $I_R + I_Y + I_B = 0$, hence the sum of three fluxes is also zero. Hence, it will make no difference if the common leg is removed. In that case, any two legs will act as the return for the third just as in a 3-phase system any two conductors act as the return for the current in the third conductor. This improved design is shown in Fig. 21.34 where dotted rectangles indicate the three windings and numbers in the cores and yokes represent the directions and magnitudes of fluxes at a particular instant. It will be seen that at any instant, the amount of 'up' flux in any leg is equal to the sum of 'down' fluxes in the other two legs. The core-type transformers are usually wound with circular cylindrical coils.

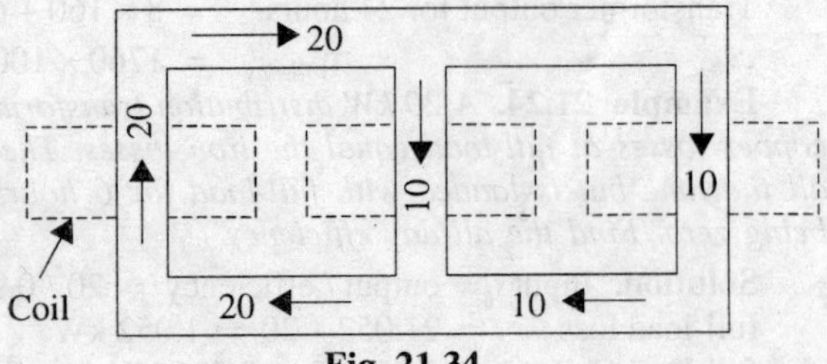

Fig. 21.34

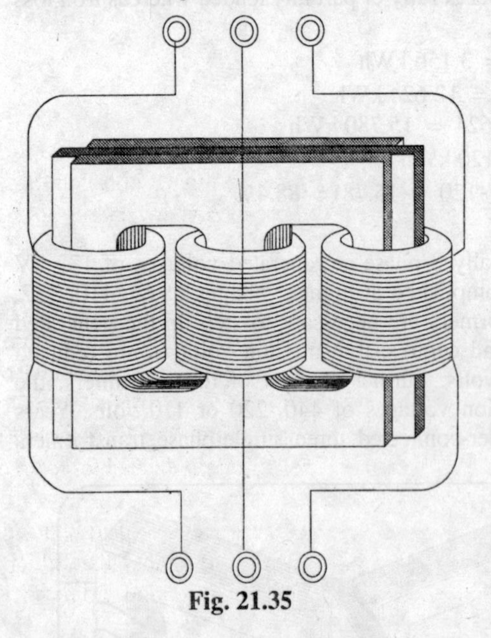

Fig. 21.35

In a similar way, three single-phase shell-type transformers can be combined together to form a 3-phase shell-type unit. But some saving in iron can be achieved in constructing a single 3-phase transformer as shown in Fig. 21.36. It does not differ from three single phase transformers put side by side. Saving in iron is due to the joint use of the magnetic paths between the coils. The three phases in this case are more independent than they are in the core-type transformer, because each phase has a magnetic circuit independent of the other.

One main drawback in a 3-phase transformer is that if any one-phase becomes disabled, then the whole transformer has to be ordinarily removed from service for repairs (the shell-type may be operated open-Δ or Vee but this is not always feasible). However, in the case of a 3-phase bank of single-phase transformers, if one transformer goes out of order, the system can still be run open-Δ at reduced capacity or the faulty transformer can be rapidly replaced by a single spare.

Example 21.25. *A 3-phase, 50 Hz transformer has a delta-connected primary and star-connected secondary, the line voltages being 22,000 V and 400 V respectively. The secondary has a star-connected balanced load at 0.8 power factor lagging. The line current on the primary side is 5 A. Determine the current in each coil of the primary and in each secondary line. What is the output of the transformer in kW?*

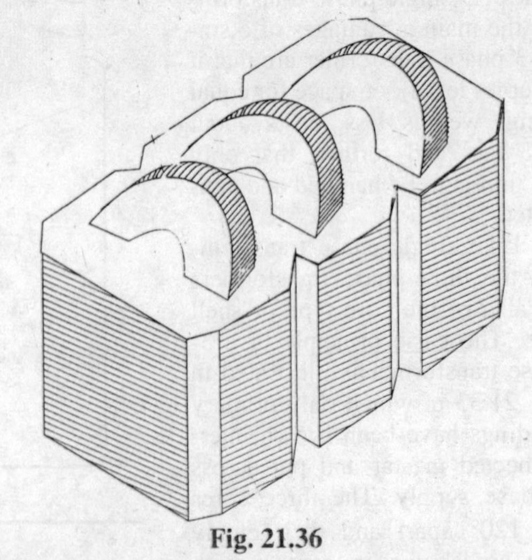

Fig. 21.36

Solution. It should be noted that in three-phase transformers, the *phase* transformation ratio is equal to the turn ratio but the terminal or line voltages depend upon the method of connection employed. The Δ/*Y*-connection is shown in Fig. 21.37.

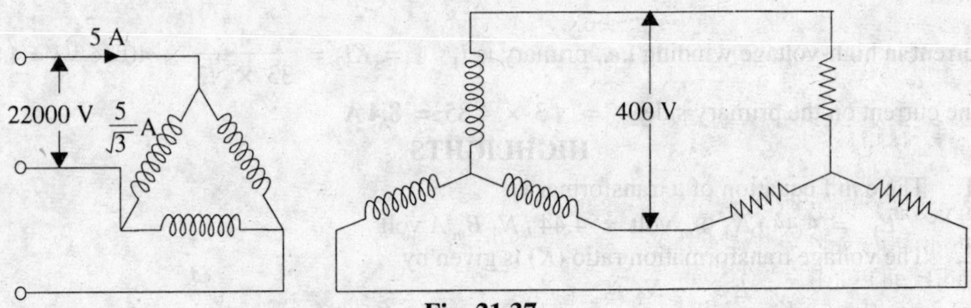

Fig. 21.37

Phase voltage on primary side	= 22,000 V
Phase voltage on secondary side	= $400 / \sqrt{3}$
K	= $400 / 22,000 \times \sqrt{3} = 1 / 55\sqrt{3}$
Primary phase current	= $5 / \sqrt{3}$ A
Secondary phase current	= $\frac{5}{\sqrt{3}} \div \frac{1}{55\sqrt{3}} = 275$ A
Secondary line current	= **275 A**
Output	= $\sqrt{3}\, V_L\, I_L \cos\phi$
	= $\sqrt{3} \times 400 \times 275 \times 0.8$ = **152.42 kW**

Example 21.26. *A 3-phase, 3,300 / 400V transformer has its high-voltage winding connected in delta and the low-voltage winding connected in star. If a load consisting of three impedances each of 6 + j8 ohm are jointed in delta across the low-voltage side. calculate (a) the kW delivered to the load (b) currents in the low and high-voltage windings and the current drawn by the transformer from line. Neglect losses and no-load current of the transformer.*

Solution. The transformer connection diagram is shown in Fig. 21.38.

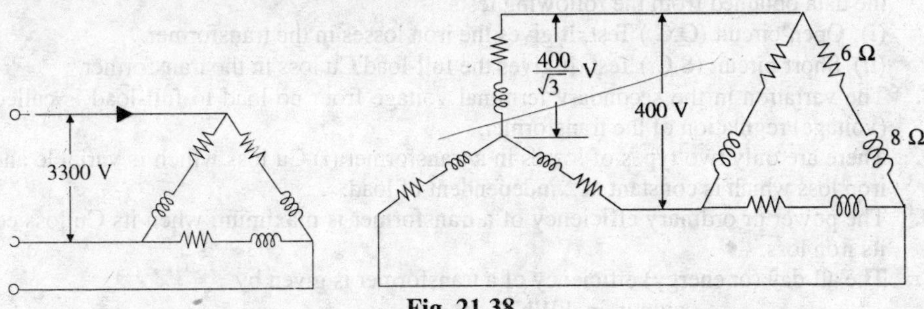

Fig. 21.38

(*a*) Power delivered to the load = $\sqrt{3}\, V_L\, I_L \cos\phi$

Now, consider the Δ-connected load.

$V_{ph} = V_L = 400$ V $\qquad Z_{ph} = \sqrt{6^2 + 8^2} = 10\,\Omega$

$I_{ph} = 400/10 = 40$ A $\qquad I_L = \sqrt{3} \times 40 = 69.3$ A

$\cos\phi = 6/10 = 0.6$

$P = \sqrt{3} \times 400 \times 69.3 \times 0.6$ = **28,800 W**

(*b*) Primary phasse voltage = 3,300 V

Secondary phase voltage = $400 / \sqrt{3}$

$\therefore \qquad K = \frac{400/\sqrt{3}}{3300} = \frac{4}{33 \times \sqrt{3}}$

Current in low-voltage winding *i.e.*, secondary is $I_2 = 40 \times \sqrt{3}$ = **69.3 A**

Current in high-voltage winding *i.e.*, primary is I_1* $= KI_2 = \dfrac{4}{33 \times \sqrt{3}} \times 40 \times \sqrt{3} = \mathbf{4.85\ A}$

Line current on the primary side $= \sqrt{3} \times 4.85 = \mathbf{8.4\ A}$

HIGHLIGHTS

1. The e.m.f equation of a transformer is

$$E_1 = 4.44 f N_1 \Phi_m \text{ volt} = 4.44 f N_1 B_m A \text{ volt}$$

2. The voltage transformation ratio (K) is given by

$$K = \frac{E_2}{E_1} = \frac{N_2}{N_1}$$

3. The no-load primary current (I_o) of a transformer has two rectangular components:

(*i*) working component, $I_w = I_0 \cos \phi_0$

(*ii*) magnetising component $I_\mu = I_0 \sin \phi_0$

4. The full-load primary current (I_1) also has the following two non-rectangular components:

(*i*) no-load component I_0 which is very small but constant

(*ii*) load component I_2' which is large but variable

$$I_2' = KI_2$$

5. The total resistance and leakage reactance of the transformer as referred to secondary are given by

$$R_{02} = R_2 + K^2 R_1 \text{ and } X_{02} = X_2 + K^2 X_1$$

6. The total voltage drop of the transformer as referred to secondary is

$$= I_2 (R_{02} \cos \phi \pm X_{02} \sin \phi)$$

The + ve sign is for lagging power factor and the –ve sign for a leading power factor.

7. The efficiency and voltage regulation of a transformer can be calculated with the help of the data obtained from the following tests:

(*i*) Open-circuit (O.C.) Test. It gives the iron losses in the transformer.

(*ii*) Short-circuit (S.C.) Test. It gives the full-load Cu loss in the transformer.

8. The variation in the secondary terminal voltage from no-load to full-load is called the (voltage) regulation of the transformer.

9. There are only two types of losses in a transformer (*i*) Cu loss which is variable and (*ii*) iron loss which is constant *i.e.*, independent of load.

10. The power or ordinary efficiency of a transformer is maximum when its Cu loss equals its iron loss.

11. The all-day (or energy) efficiency of a transformer is given by

$$\eta_{all\text{-}day} = \frac{\text{output in kWh}}{\text{input in kWh}} \quad \text{— for 24 hours}$$

OBJECTIVE TESTS—21

A. Fill in the following blanks.

1. A transformer works on the principle of .... induction.

2. In a shell-type transformer, the............. surrounds a considerable portion of the windings.

3. Whatever the load condition, the net flux passing through the core of a transformer is nearly the .... as at no-load.

4. With a load of leading power factor, the full-load secondary voltage of a transformer is...............than its no-load voltage.

5. Short-circuit test of a transformer helps in finding its .... loss at full-load and at any desired load.

6. A transformer has maximum efficiency when its copper loss equals its . ... loss.

---

*Because, as given, the no-load current I_0 is neglected.

7. The power efficiency of a lighting transformer is always .... than its all-day efficiency.
8. The performance of a .... transformer is best judged by the all-day efficiency.

B. Answer True or False.

1. A transformer is often used for multiplying electric power.
2. Constructionally, the transformers are of two general types: core type and shell-type.
3. The no-load primary input is practically equal to the iron-loss in a transformer.
4. The open-circuit test of a transformer is used for finding its copper loss.
5. The commercial efficiency of a transformer is about 95%.
6. The all-day efficiency of a transformer is also called its energy efficiency.
7. For large power transmission, three single-phase inter-connected transformers are better than a single 3-phase transformer.
8. In 3-phase transformers, voltage transformation is ratio given by the ratio of secondary to primary line voltages.

C. Multiple Choice Questions

1. The steel used for transformer cores has
   (*a*) high silicon content
   (*b*) high permeability
   (*c*) low hysteresis loss
   (*d*) all of the above
2. In a transformer, leakage flux
   (*a*) helps in transfer of energy
   (*b*) is negligible at full-load
   (*c*) is minimised by inter-leaving the primary and secondary windings
   (*d*) produces mutually induced e.m.f.
3. Short-circuit test of a transformer helps us to find its
   (*a*) iron loss
   (*b*) full-load Cu loss
   (*c*) Cu loss at no-load
   (*d*) Cu loss at any desired load
4. A two-winding transformer operates at maximum efficiency when its
   (*a*) hysteresis loss equals eddy-current loss
   (*b*) Cu loss equals iron loss
   (*c*) primary resistance equals secondary resistance
   (*d*) voltage regulation is minimum
5. All-day efficiency is meant to judge the performance of a transformer
   (*a*) distribution (*b*) auto
   (*c*) power (*d*) two-winding
6. A 100 kVA transformer has full-load Cu loss of 1600 W and iron loss of 900 W. It will have maximum efficiency for a load of —kVA.
   (*a*) 100 (*b*) 56.2
   (*c*) 75 (*d*) 133.3
7. In a 10 kVA, 230 / 1000 V, single-phase transformer, the no-load current will be about
   (*a*) 0.5 A (*b*) 3 A
   (*c*) 8 A (*d*) 10 A
   [Elect. Engg. A.M.Ae.S.I. June 1994]
8. Open-circuit and short-circuit tests on a transformer give
   (*a*) windage losses
   (*b*) friction losses
   (*c*) iron and copper losses respectively
   (*d*) copper and iron losses respectively

ANSWERS

A. 1. mutual 2. core 3. same 4. greater 5. copper 6. iron 7. higher 8. distribution.
B. 1. F 2. T 3. T 4. F 5. T 6. T 7. F 8. F
C. 1. *d* 2. *c* 3. *d* 4. *b* 5. *a* 6. *c* 7. *a* 8. *c*

22 THREE PHASE INDUCTION MOTOR

22.1. Induction Motor: General Principle

Of all the a.c. motors, the 3-phase induction motor is the one which is extensively used for various kinds of industrial drives. It has the following main advantages as well as some disadvantages.

Advantages

1. It has very simple and extremely rugged, almost unbreakable construction (especially squirrel-cage type).
2. Its cost is low and it is very reliable.
3. It has sufficiently high efficiency. It has a reasonably good power factor.
4. It requires minimum of maintenance.
5. It starts up from rest and needs no extra starting motor and has not to be synchronized. Its starting arrangement is simple, especially for squirrel-cage rotor type motor.

Disadvantages

1. Its speed can't be varied without sacrificing some of its efficiency.
2. Just like a d.c. shunt motor, its speed decreases somewhat with increase in load.
3. Its starting torque is somewhat inferior to that of a d.c. shunt motor.

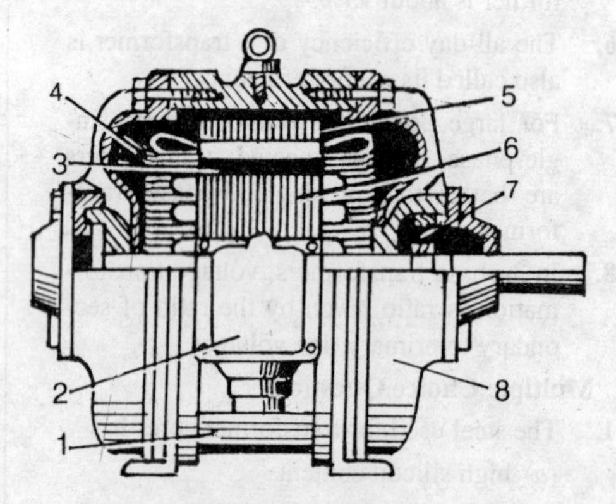

Fig. 22.1 : Three-phase squirrel-cage Induction Motor
1. frame 2. terminal 3. air gap 4. stator winding 5. stator laminations 6. rotor laminations 7. bearing 8. end shield.

22.2. Construction

An induction motor consists of two main parts:

(*a*) a stator and (*b*) a rotor.

(*a*) **Stator**

The stator of an induction motor is, in principle, the same as that of an alternator. It is made up of a number of stampings which are slotted to receive the windings (Fig. 22.2). The stator carries a 3-phase winding and is fed from a 3-phase supply (Fig. 22.3). It is wound for a definite number of poles, the exact number of poles being determined by the requirements of speed. Greater the number of poles, lesser the speed and *vice versa*. It will be seen in Art. 22.3 that the stator windings, when supplied with 3-phase current, produce a magnetic field or flux which is of constant value but which revolves or rotates at synchronous speed (given by $N_S = 120f / P$). This revolving magnetic flux induces an e.m.f. in the rotor by mutual induction.

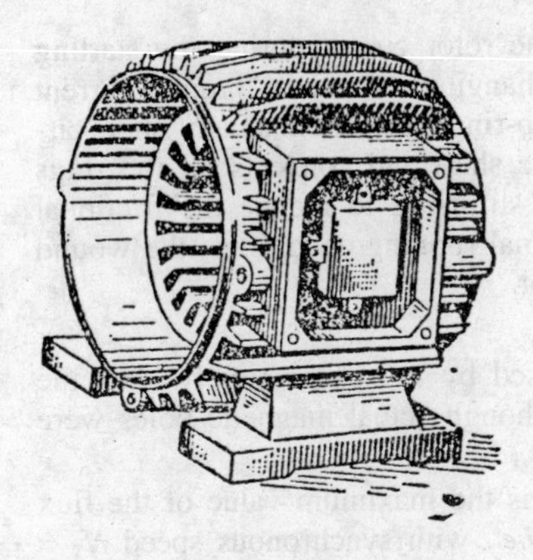

Fig. 22.2 : Induction motor frame with unwound stator.

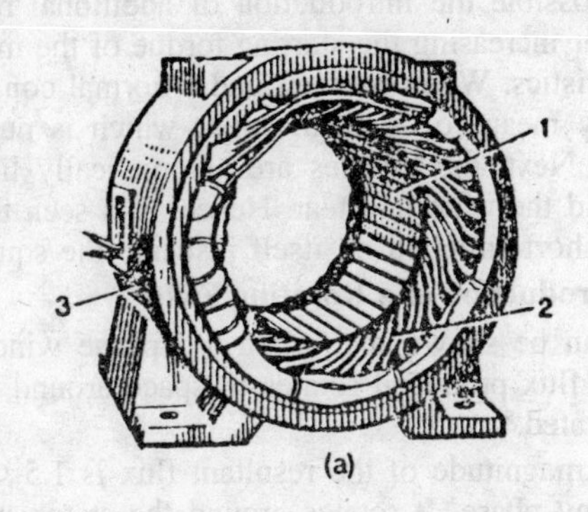

Fig. 22.3 : Wound stator. 1 core 2. winding 3. frame.

(*b*) **Rotor**

(*i*) *Squirrel-cage rotor.* Motors employing this type of rotor are known as squirrel-cage induction motors. [Fig. 22.4 (*a*)].

(*ii*) *Phase-wound or Wound rotor.* Motors employing this type of rotor are variously known as 'phase-wound' motors or 'wound' motors or as 'slip-ring' motors. [Fig. 22.5 (*b*)].

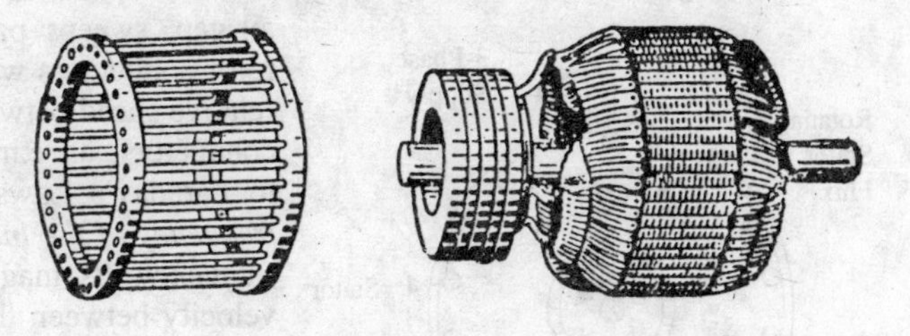

Fig. 22.4

1. Squirrel-cage Rotor

Almost 90 per cent of induction motors are squirrel-cage type, because this type of rotor has the simplest and most rugged construction imaginable and is almost indestructible. The rotor consists of a cylindrical laminated core with parallel slots for carrying the rotor conductors which, it should be noted clearly, are not wires but consist of heavy bars of copper, aluminium or alloys. One bar is placed in each slot, rather the bars are inserted from the end when semi-closed slots are used. The rotor bars are brazed or electrically welded to two heavy and slout short-circuiting end-rings, thus giving us, what is so picturesquely called a squirrel-cage construction (Fig. 22.4)*.

It should be noted that the rotor bars are permanently short-circuited on themselves, hence it is not possible to add any external resistance in series with the rotor circuit for starting purposes. There are no brushes in this motor which is one of the reasons for its ruggedness.

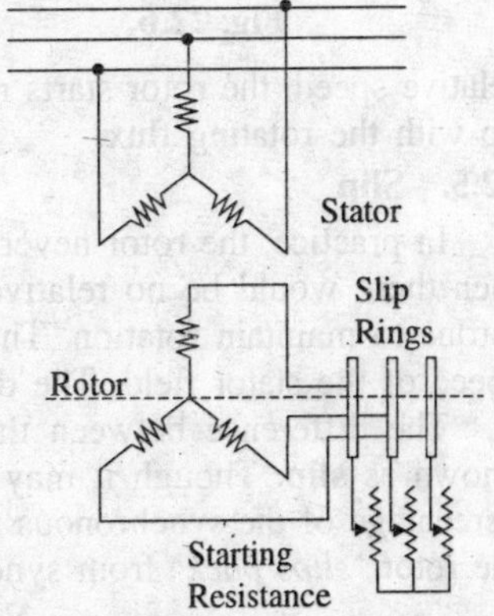

Fig. 22.5

2. Phase-wound Rotor

This type of rotor (Fig. 22.4 (*b*)] is provided with 3-phase, double-layer, distributed winding consisting of coils as used in alternators. The rotor is wound for as many poles as the number of stator poles and is always wound 3-phase even *when the stator is wound two-phase.*

The three phases are started internally. The other three winding terminals are brought out and connected to three insulated slip-rings mounted on the shaft with brushes resting on them. These three brushes are further externally connected to a 3-phase star-connected rheostat (Fig. 22.5). This

*Such cylindrical cages were originally made for exercising pet squirrel.

makes possible the introduction of additional resistance in the rotor circuit during the starting period for increasing the starting torque of the motor and for changing its speed / torque / current characteristics. When running under normal conditions, the slip-rings are automatically short-circuited by means of a metal collar which is pushed along the shaft and connects all the rings together. Next, the brushes are automatically lifted from the slip-rings to reduce the frictional losses and the wear and tear. Hence, it is seen that under normal running conditions, the wound rotor is short-circuited on itself just like the squirrel-cage rotor.

22.3. Production of a Rotating Field

It can be shown that when a 3-phase winding is energised by a 3-phase supply, then the resultant flux produced rotates in space around the stator as though actual magnetic poles were being rotated.*

The magnitude of the resultant flux is 1.5 Φ_m where Φ_m is the maximum value of the flux due to any phase. It rotates around the stator synchronously *i.e.,* with synchronous speed $N_S = 120\, f / P$ where P is the number of motor poles and f is the frequency of the stator a.c. supply.

22.4. Principle of Operation

The reason why the rotor of an induction motor is set into rotation is as follows: when the 3-phase stator windings are fed by a 3-phase supply, then, as said above, a magnetic flux of constant magnitude but rotating at synchronous speed, is set up. This flux passes through the air-gap, sweeps past the rotor surface and so cuts the rotor conductors which, as yet, are stationary. Due to the relative speed between the rotating flux and the stationary conductors, an e.m.f. is induced in the latter according to Faraday's Laws of Electro-magnetic Induction. *The frequency of the induced e.m.f. is the same as the supply frequency.* Its magnitude is proportional to the relative velocity between the flux and the conductors and its direction is as given by Fleming's Right-hand rule. Since the rotor bars or conductors form a closed circuit, rotor current is produced whose direction, as given by Lenz's law, is such as to oppose the very cause producing it. In this case, the cause which produces the rotor current is the relative speed between the rotating flux of the stator and the stationary rotor conductors. Hence, to reduce the relative speed, the rotor starts running in the *same* direction as that of the flux and tries to catch up with the rotating flux.

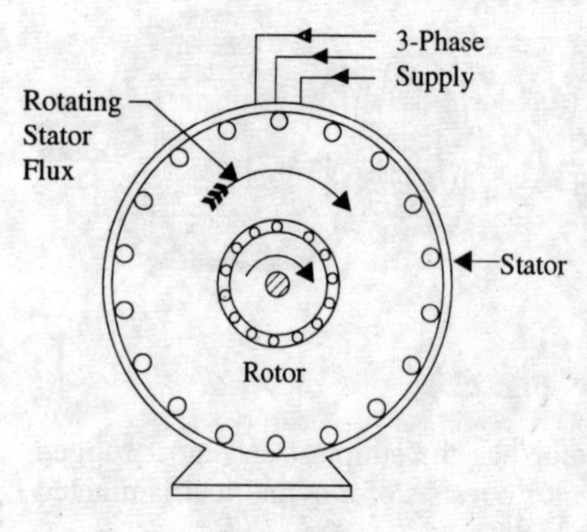

Fig. 22.6

22.5. Slip

In practice, the rotor never succeeds in "*catching up*" with the stator flux. If it really did so, then there would be no relative speed between the two, hence no e.m.f., no current and so, no torque to maintain rotation. That is why the rotor runs at a speed which is always less than the speed of the stator field. The difference in speed depends upon the load on the motor.

The difference between the synchronous speed N_S and the actual speed N of the rotor is known as **slip.** Though it may be expressed in so many rev/s, yet it is usual to express it as a percentage of the synchronous speed. Actually, the term '*slip*' is descriptive of the way in which the rotor '*slips back*' from synchronism.

$$\therefore \quad \% \text{ slip } s = \frac{N_S - N}{N_S} \times 100$$

Sometimes, $N_S - N$ is called the *slip speed.*

22.6. Frequency of Rotor Current

When the motor is stationary, the frequency of rotor current *is the same as the supply frequency.* But when the rotor starts revolving, then the frequency depends upon the relative speed

*Strictly speaking, it is the seat of the resultant flux which keeps on shifting around synchronously.

or slip-speed. Let at any slip-speed the frequency of the rotor current be f'. Then

$$N_S - N = \frac{120 f'}{P}$$

Also $$N_S = \frac{120\, f}{P}$$

Dividing one by the other, we get

$$\frac{f'}{f} = \frac{N_S - N}{N_S} = s \qquad \therefore\ f' = sf$$

22.7 Speed of Rotor Field

As seen rotor currents have a frequency of $f_r = f' = sf$ and when flowing through the individual phases of rotor winding give rise to rotor magnetic fields. These individual rotor magnetic fields produce a combined rotating magnetic field whose speed relative to rotor is

$$= \frac{120\, f'}{P} = \frac{120\, sf}{P} = sN_S$$

However, the rotor itself is running at speed N with respect to space. Hence

speed of rotor field in space = speed of field relative to rotor + speed of rotor relative to space

$$= sN_s + N = sN_s + N_s(1 - s) = N_s$$

It means that no matter what the value of slip, rotor currents and stator currents each produce a sinusoidally distributed magnetic field of constant magnitude and constant space speed of N_S. In other words, both rotor and stator fields rotate synchronously which means that they are stationary with respect to each other. These two synchronously rotating magnetic fields, in fact, super-impose on each other and give rise to the actually existing rotating field which corresponds to the magnetising current of the stator winding.

Example 22.1. *A 12-pole, 3-phase alternator driven at a speed of 500 r.p.m. supplies power to an 8-pole, 3-phase induction motor. If the slip of the motor at full-load is 3%, calculate the full-load speed of the motor.* **(Elect. Machines, Nagpur Univ. 1993)**

Solution. Let N = actual motor speed

Supply frequency $f = 12 \times 500 / 120 = 50$ Hz.

Synchronous speed $N_s = 120 \times 50 / 8 = 750$ r.p.m.

$$\%\text{ slip } s = \frac{N_S - N}{N_S} \times 100$$

$$3 = \frac{750 - N}{750} \times 100 \qquad \therefore\ N = \mathbf{727.5\ r.p.m.}$$

Example 22.2. *If a 6-pole motor running from a 50 Hz supply has an e.m.f. in the rotor of frequency 2.5 Hz, determine (a) the slip (b) the speed of the motor.*

Solution. We have seen that $f' = sf$

(*a*) $\therefore\ 2.5 = s \times 50 \qquad \therefore\ s = 2.5 / 50 = 0.05$ or **5%**

$N_s = 120 \times 50 / 6 = 1000$ r.p.m.

(*b*) $\therefore\ 5 = \frac{1000 - N}{1000} \times 100\ ; \qquad N = \mathbf{950\ r.p.m.}$

Example 22.3. *A three-phase induction motor is wound for 4 poles and is supplied from a 50 Hz system. Calculate (i) the synchronous speed (ii) the speed of the rotor when the slip is 4 per cent (iii) the rotor frequency when speed of the rotor is 600 r.p.m.*

(Elect. Engg.-I, Punjab Univ. 1994)

Solution. (*i*) $N_S = 120 f / P = 120 \times 50 / 4 =$ **1500 r.p.m.**

(*ii*) $s = \frac{N_S - N}{N_S}$ or $0.04 = \frac{1500 - N}{1500}, \qquad \therefore\ N = \mathbf{1440\ r.p.m.}$

(*iii*) $s = \dfrac{1500 - 600}{1500} = 0.6$

$\therefore \quad f_r = sf = 0.6 \times 50 =$ **30 Hz**

Example 22.4. *An 8-pole alternator runs at 750 r.p.m. and supplies power to a 6-pole induction motor which has at full-load a slip of 3%. Find the full-load speed of the induction motor and the frequency of its rotor e.m.f.*

Solution. The supply frequency is found from the speed of the alternator and its number of poles.

Supply frequency $f = PN/120 = 8 \times 750/120 = 50$ Hz

Synchronous speed of the motor

$$N_S = 120 \times 50/6 = 1000 \text{ r.p.m.}$$

$$3 = \frac{1000 - N}{1000} \times 100$$

$\therefore$ actual full-load motor speed $N =$ **970 r.p.m.**

$$f' = s \times f = 3 \times 50/100 = \mathbf{1.5\ Hz}$$

22.8. Relation Between Torque and Rotor P.F.

In Art.12.7 it has been shown that in the case of a d.c. motor, the torque T_a is proportional to the product of armature current and flux per pole *i.e.*, $T_a \propto \Phi I_a$. Similarly, in the case of an induction motor, the torque is also proportional to the product of flux per stator pole and the rotor current. However, there is another factor to be taken into account *i.e.*, the power factor of the rotor.

$\therefore \quad T \propto \Phi I_2 \cos \phi_2 \quad$ or $\quad T = k\Phi I_2 \cos \phi_2$

where I_2 = rotor current at *standstill*

ϕ_2 = angle between rotor e.m.f. and rotor current

k = a constant

Denoting rotor e.m.f. at *standstill* by E_2, we have that $E_2 \propto \Phi$

$$T \propto E_2 I_2 \cos \phi_2$$

or $\quad T = k_1 E_2 I_2 \cos \phi_2 \quad$ where k_1 is another constant

Example 22.5. *A 3-phase induction motor runs at almost 1000 r.p.m. at no-load and 950 r.p.m. at full-load when supplied with power from a 50 Hz, 3-phase line.*

(*i*) *How many poles has the motor?*

(*ii*) *What is the percentage slip at full load?*

(*iii*) *What is the corresponding frequency of the rotor voltages?*

(*iv*) *What is the corresponding speed of the rotor field with respect to rotor?*

(*v*) *What is the corresponding speed of the rotor with respect to the stator?*

(*vi*) *What is the corresponding speed of the rotor field with respect to the stator field?*

(*vii*) *What is the rotor frequency at the slip of 10 per cent?*

(Electrical Science, AMIE Winter 1994)

Solution. (*i*) $f = PN/120$ or $50 = P \times 1000/120 \quad \therefore \quad P =$ **6**

(*ii*) $s = (50/1000) \times 100 =$ **5%**

(*iii*) $f_r = sf = 0.05 \times 50 =$ **2.5 Hz**

(*iv*) As explained in Art. 22.7 speed of the rotor field with respect to rotor is

$sN_s = 0.05 \times 1000 =$ **50 r.p.m.**

(*v*) Speed of the rotor with respect to stator is

$N = N_s(1 - s) = 1000(1 - 0.05) =$ **950 r.p.m.**

(*vi*) As explained in Art. 22.6, the rotor and stator fields rotate synchronously which means that they are stationary with respect to each other. Hence, speed of rotor field with respect to stator field is **zero.**

(*vii*) $f_r = 0.1 \times 50 =$ **5 Hz.**

22.9. Starting Torque

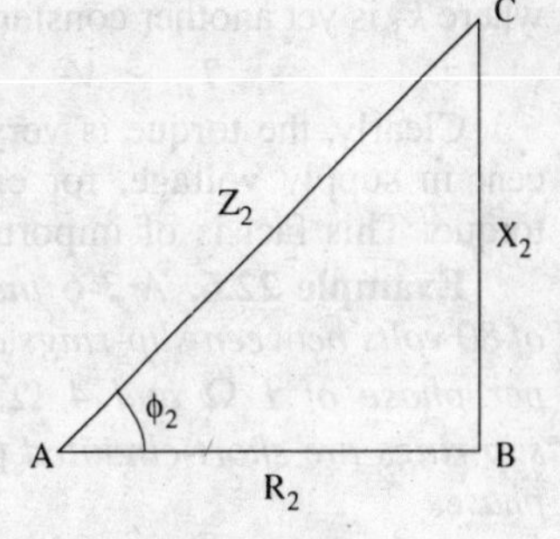

Fig. 22.7

E_2 = rotor e.m.f. per phase *at standstill*

R_2 = rotor resistance / phase

X_2 = rotor reactance / phase *at standstill*

$\therefore \quad Z_2 = \sqrt{(R_2^2 + X_2^2)}$

= rotor impedance / phase *at standstill*

then $\quad I_2 = E_2 / Z_2 = E_2 / \sqrt{(R_2^2 + X_2^2)}$

$$\cos \phi_2 = \frac{R_2}{\sqrt{(R_2^2 + X_2^2)}} \quad \text{(Fig. 22.7)}$$

Standstill or starting torque $T_{st} = k_1 E_2 I_2 \cos \phi_2$

$$= k_1 E_2 \frac{E_2}{\sqrt{(R_2^2 + X_2^2)}} \times \frac{R_2}{\sqrt{(R_2^2 + X_2^2)}} = \frac{k_1 E_2^2 R_2}{R_2^2 + X_2^2}$$

If supply voltage ,' is constant, then flux Φ and hence E_2 both are constant.

$$\therefore \quad T_{st} = k_2 \frac{R_2}{R_2^2 + X_2^2} = k_2 \frac{R_2}{Z_2^2} \quad \text{...(i)}$$

where k_2 is some other constant.

22.10. Starting Torque of a Squirrel-cage Motor

The resistance of a squirrel-cage rotor is fixed and small as compared to its reactance which is very large especially at the start because at standstill, the frequency of the rotor current equals the supply frequency. Hence, the starting current I_2 of the rotor, though very large in magnitude, lags by a very large angle behind E_2 with the result that the starting torque per ampere is very poor. It is roughly 1.5 times the full-load torque although the starting current is 5 to 7 times the full-load current. Hence, such motors are not useful where the motor has to start against heavy loads.

22.11. Starting Torque of a Slip-ring Motor

The starting torque of such a motor is increased by improving its power factor by adding external resistance in the rotor circuit from the star-connected rheostat, the rheostat resistance being progressively cut out as the motor catches speed. Addition of external resistance, however, increases the rotor impedance and so reduces the rotor current. At first, the effect of improved power factor predominates the current-decreasing effect of impedance. Hence, starting torque is increased. But after a certain point, the effect of increased impedance predominates the effect of improved power factor and so the torque starts decreasing.

22.12. Condition for Maximum Starting Torque

It can be proved that starting torque is maximum when rotor resistance equals rotor reactance.

$$\text{Now} \quad T_{st} = \frac{k_2 R_2}{R_2^2 + X_2^2}$$

$$\therefore \quad \frac{dT_{st}}{dR_2} = k_2 \left[\frac{1}{R_2^2 + X_2^2} - \frac{R_2 (2R_2)}{(R_2^2 + X_2^2)^2} \right] = 0$$

or $\quad R_2^2 + X_2^2 = 2R_2^2 \quad$ or $\quad R_2 = X_2$

22.13. Effect of Change in Supply Voltage

We have seen that

$$T_{st} = \frac{k_1 E_2^2 R_2}{R_2^2 + X_2^2} \quad \text{... from Art. 22.9}$$

Now, $E_2 \propto$ supply voltage V

$$\therefore \quad T_{st} = \frac{k_3 V^2 R_2}{R_2^2 + X_2^2}$$

where k_3 is yet another constant.

$\therefore \quad T_{st} \propto V^2$

Clearly, the torque is very sensitive to any changes in the supply voltage. A change of 5 per cent in supply voltage, for example, will produce a change of approximately 10% in the rotor torque. This fact is of importance in star-delta and auto-transformer starters.

Example 22.6. *A 3-φ induction motor having a star-connected rotor has an induced e.m.f. of 80 volts between slip-rings at standstill on open-circuit. The rotor has a resistance and reactance per phase of 1 Ω and 4 Ω respectively. Calculate current/phase and power factor when (a) slip-rings are short-circuited (b) slip-rings are connected to a star-connected rheostat of 3 Ω per phase.* **(Induction Machines, Madras Univ. 1993)**

Solution. Standstill e.m.f. / rotor phase = $80/\sqrt{3} = 46.2$ V

(*a*) Rotor impedance / phase $= \sqrt{(1^2 + 4^2)} = 4.12\ \Omega$

Rotor current / phase $= 46.2/4.12 = \mathbf{11.2\ A}$

Power factor $= \cos\phi = 1/4.12 = \mathbf{0.243}$

As p.f. is low, the starting torque is also small.

(*b*) Rotor resistance / phase $= 3 + 1 = 4\ \Omega$

Rotor impedance / phase $= \sqrt{(4^2 + 4^2)} = 5.66\ \Omega$

$\therefore$ rotor current / phase $= 46.2/5.66 = \mathbf{8.16\ A}$

$\cos\phi = 4/5.66 = \mathbf{0.707}$

Hence, the starting torque is increased due to improvement in the power factor. It will also be noted that improvement in p.f. is much more than the decrease in current due to increased impedance.

Tutorial Problems No. 22.1

1. A 4-pole, 50 Hz induction motor on full-load runs at 1,440 r.p.m. Calculate
(*i*) the synchronous speed
(*ii*) the speed of slip
(*iii*) the frequency of the rotor currents. **[(*i*) 1500 r.p.m. (*ii*) 60 r.p.m. (*iii*) 2 Hz.]**

2. Calculate the number of poles in the cases of the undermentioned 3-phase, 50 Hz induction motors, of which the full-load shaft speeds are given:
(*i*) A 482 r.p.m. winder motor
(*ii*) A 725 r.p.m. coal-cutter motor
(*iii*) A 1,445 r.p.m. centrifugal pump motor. **[(*i*) 12 (*ii*) 8 (*iii*) 4]**

3. A 3-phase, 50 Hz, 4-pole, wound-rotor induction motor has a rotor standstill e.m.f. per phase of 115 V. If the motor is running at 1,440 r.p.m., calculate for this speed (*a*) the slip (*b*) the frequency of the rotor induced e.m.f. (*c*) the value of the rotor induced e.m.f. per phase. **[(*a*) 4% (*b*) 2 Hz (*c*) 4.6 V]**

4. A 3-phase, 50-Hz, 4-pole, wound-rotor induction motor has a rotor standstill e.m.f. per phase of 115 V. If the motor is running at 1,440 r.p.m., calculate for this speed (*a*) the slip (*b*) the frequency of the rotor induced e.m.f. (*c*) the value of the rotor induced e.m.f. per phase. **[(*a*) 4% (*b*) 2 Hz (*c*) 4.6 V]**

5. Calculate the number of poles in the cases of the undermentioned 3-phase, 50-Hz induction motors, of which the full-load shaft speeds are given (*i*) a 482 r.p.m. winder motor (*ii*) a 725 r.p.m. coal-cutter motor (*iii*) a 1,445 r.p.m. centrifugal pump motor. **[12, 8, 4]**

6. A 3-phase, 50 Hz, 8-pole induction motor drives a triple-ram pump having a speed of 60 r.p.m. If the motor slip is 4%, what is the total gear ratio between the motor and the pump? **[1 : 12]**

7. A 2-pole, 3-phase, 50 Hz induction motor is running on load with a slip of 4 per cent. Calculate the actual speed and the synchronous speed of the machine. **[2880 r.p.m.; 3000 r.p.m.]**

8. An induction motor has four poles and is energized from a 50 Hz supply. If the machine runs on full load at 2 per cent slip, determine the running speed and the frequency of the rotor currents. **[1470 r.p.m.; 1 Hz]**

9. A 14-pole, 50 Hz induction motor runs at 415 r.p.m. Deduce the frequency of the currents in the rotor winding and the slip. **[1.585 Hz; 3.17%]**

10. A centre-zero d.c. galvanometer, suitably shunted, is connected in one lead of the rotor of a 3-phase 6-pole, 50 Hz slip-ring induction motor and the pointer makes 85 complete oscillations per minute. What is the rotor speed? **[971.7 r.p.m.]**

22.14. Rotor E.M.F. and Reactance Under Running Conditions

Let E_2 = *standstill* rotor induced e.m.f. / phase

X_2 = " " reactance / phase

f_2 = rotor current frequency *at standstill*

When rotor is stationary *i.e.*, $s = 1$, then frequency of rotor e.m.f. is the same as that of the stator supply frequency. The value of e.m.f. induced in the rotor at standstill is maximum because the relative speed between the rotor and the revolving stator flux is maximum. In fact, the motor is equivalent to a 3-phase transformer with a short-circuited rotating secondary.

When rotor starts running, the relative speed between it and the rotating stator flux is decreased. Hence, the rotor induced e.m.f. which is directly proportional to this relative speed, is also decreased (and may disappear altogether if rotor speed were to become equal to the speed of stator flux). Hence, for a slip s, the rotor induced e.m.f. will be s times the induced e.m.f. at standstill.

Therefore, under *running* conditions

$$E_r = sE_2$$

The frequency of the induced e.m.f. will likewise become

$$f_r = sf_2$$

Due to decrease in frequency of the rotor e.m.f.: the rotor reactance will also decrease,

$$X_r = sX_2$$

where E_r and X_r are rotor e.m.f. and reactance under *running* conditions.

22.15. Torque Under Running Conditions

$$T \propto E_r I_r \cos\phi_2$$

or $$T \propto \Phi I_r \cos\phi_2 \qquad (\because E_r \propto \Phi)$$

where E_r = rotor e.m.f. / phase under *running conditions*

I_r = current / phase " "

Now $E_r = sE_2$

$$\therefore \quad I_r = \frac{E_r}{Z_r} = \frac{sE_2}{\sqrt{[R_2^2 + (sX_2)^2]}}$$

$$\cos\phi_2 = \frac{R_2}{\sqrt{[R_2^2 + (sX_2)^2]}} \qquad \text{(Fig. 2.8)}$$

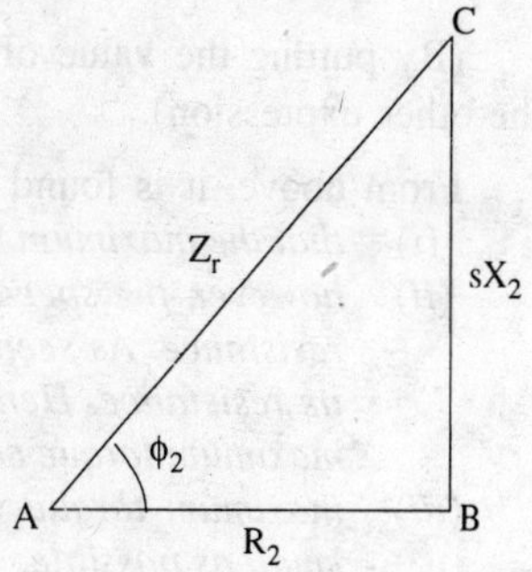

Fig. 22.8

$$\therefore \quad T \propto \frac{\Phi\, s\, E_2 R_2}{R_2^2 + (sX_2)^2} = \frac{k\,\Phi\,.\,s\,.\,E_2 R_2}{R_2^2 + (sX_2)^2}$$

Also $$T = \frac{k_1\, s\, E_2^2 R_2}{R_2^2 + (sX_2)^2} \qquad (\because \; E_2 \propto \Phi)$$

where k_1 is another constant.

It can be proved that for a 3-phase induction motor, $k_1 = 3 / 2\,\pi\, N_S$

$$\therefore \quad T = \frac{3}{2\pi N_S} \cdot \frac{s\, E_2^2 R_2}{R_2^2 + (sX_2)^2}$$

At standstill when $s = 1$, obviously

$$T = \frac{k_1 E_2^2 R_2}{R_2^2 + X_2^2} \qquad \text{... the same as in Art. 22.9.}$$

22.16. Condition for Maximum Torque

The torque of a rotor under running conditions is

$$T = \frac{k\,\Phi\, s\, E_2\, R_2}{R_2^2 + (sX_2)^2}$$

The condition for maximum torque may be obtained by differentiating the above expression with respect to slip s and then by putting it equal to zero. However, it is simpler to put $Y = \frac{1}{T}$ and then differentiate it.

$$\therefore \quad Y = \frac{R_2^2 + (sX_2)^2}{k\,\Phi\, s\, E_2\, R_2} = \frac{R_2}{k\,\Phi\, s\, E_2} + \frac{s\, X_2^2}{k\,\Phi\, E_2\, R_2}$$

$$\frac{dY}{ds} = \frac{-R_2}{k\,\Phi\, E_2\, s^2} + \frac{X_2^2}{k\,\Phi\, E_2\, R_2} = 0$$

$$\therefore \quad \frac{R_2}{k\,\Phi\, E_2\, s^2} = \frac{X_2^2}{k\,\Phi\, E_2\, R_2}$$

or $\quad R_2^2 = s^2\, X_2^2 \quad$ or $\quad R_2 = s\, X_2$

Hence, torque *under running conditions* is maximum at that value of slip s which makes rotor reactance/phase equal to rotor resistance/phase.

Slip corresponding to maximum torque is

$$s = R_2 / X_2$$

Putting $R_2 = sX_2$ in the above equation for the torque, we get

$$T_{max} = \frac{k\,\Phi\, s^2\, E_2\, X_2}{2s^2\, X_2^2} \quad \left(\text{or} \quad \frac{k\,\Phi\, s\, E_2\, R_2}{2R_2^2}\right)$$

$$= \frac{k\,\Phi\, E_2}{2X_2} \quad \left(\text{or} \quad \frac{k\,\Phi\, sE_2}{2R_2}\right)$$

(By putting the value of s in the expression in bracket it can be seen that it is the same as the other expression).

From above, it is found

(*i*) *that the maximum torque is independent of rotor resistance as such;*

(*ii*) *however, the speed or slip at which maximum torque occurs is determined by the rotor resistance. As seen from above, torque becomes maximum when rotor reactance equals its resistance. Hence, by varying rotor resistance (possible only with slip-ring motors) maximum torque can be made to occur at any desired slip (or motor speed);*

(*iii*) *maximum torque varies inversely as standstill reactance. Hence, it should be kept as small as possible.*

22.17. Relation Between Torque and Slip

A family of torque/slip curves is shown in Fig. 22.9 for a range of $s = 0$ to $s = 1$ with R_2 as the parameter. We have seen above that

$$T = \frac{k\,\Phi\, s\, E_2\, R_2}{R_2^2 + (sX_2)^2}$$

It is clear that when $s = 0$, $T = 0$, hence the curve starts from point O.

At normal speeds, close to synchronism, the term (sX_2) is small and hence negligible with respect to R_2

$$\therefore \quad T \propto \frac{s}{R_2}$$

or $\quad T \propto s \quad$ if R_2 is constant

Hence, for low values of slip, the torque/slip curve is approximately a straight line. As slip increases (for increasing load on the motor) the torque also increases and becomes maximum when $s = R_2/X_2$. This torque is known as 'pull out' or 'break down' torque. As the slip further increases (*i.e.*, motor speed falls) with further increase in motor load, R_2 becomes negligible as compared to (sX_2). Therefore,

$$T \propto \frac{s}{(sX_2)} \propto \frac{1}{s}$$

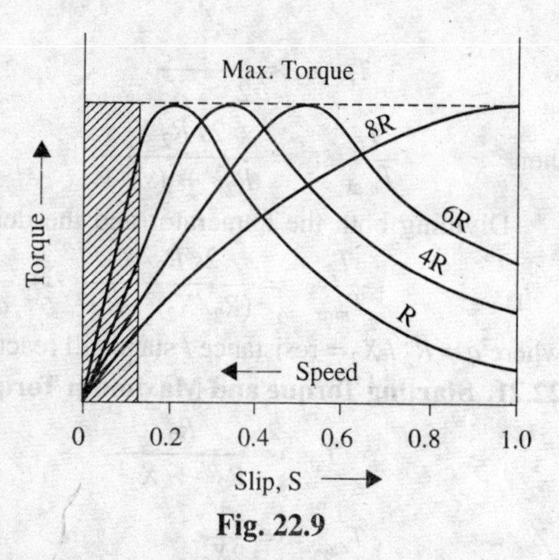

Fig. 22.9

Hence, the torque / slip curve is a rectangular hyperbola. So we see that beyond the point of maximum torque, any further increase in motor load results in decrease of torque developed by the motor. The result is that the motor slows down and eventually stops. The circuit-breakers will be tripped open if the circuit has been so protected. In fact the stable operation of the motor lies between the values of $s = 0$ and that corresponding to maximum torque. The operating range is shown shaded in Fig. 22.9.

It is seen that although maximum torque does not depend on R_2, *yet the exact location of T_{max} is dependent on it.* Greater the R_2, greater is the value of slip at which the maximum torque occurs.

22.18. Speed Regulation of an Induction Motor

By speed regulation of a motor is meant the *natural* change in its speed from no-load to full-load without using any speed controlling apparatus. Suppose that the no-load speed of a motor is 980 r.p.m. and its full-load speed is 940 r.p.m. Its speed regulation is (980 – 940) = 40 r.p.m. or percentage speed regulation is 40 × 100/940 = 4.25%.

As we know, in an induction motor, slip is necessary in order to induce the rotor currents required for developing motor torque. At no-load, only a small torque is required for overcoming only mechanical losses, hence motor slip is very small *i.e.,* about 2 per cent or so. But when motor load is increased, the rotor slip has to increase in order to increase induced rotor currents. The increased rotor currents will, in turn, produce higher torque required by the increase in motor load.

Since the short-circuited rotor bars of a squirrel-cage motor have low resistance and their reactance under running conditions is also low, the rotor impedance is also relatively low. Consequently, a small increase in the rotor induced voltage produces a relatively large increase in the rotor current. Therefore, as the motor is loaded from no-load to full-load, only a small decrease in speed is required for producing a relatively large increase in rotor currents. That is why the speed regulation of a squirrel-cage induction motor is very small and hence it is often classified as a constant-speed motor.

22.19. Effect of Change in Supply Voltage on Torque and Speed

$$T = \frac{k\,\Phi\, s\, E_2\, R_2}{R_2^2 + (sX_2)^2}$$

As $E_2 \propto \Phi \propto V$ where V is supply voltage

$\therefore$ $T \propto V^2$

Change in supply voltage not only affects the starting torque T_{st} but torque under running conditions also. If V decreases, then T also decreases. Hence, for maintaining the same torque, slip increases *i.e.,* speed falls.

Let V change to V', s to s' and T to T'

then $$\frac{T}{T'} = \frac{sV^2}{s'V'^2}$$

22.20. Full-load Torque and Maximum Torque

Let s be the slip corresponding to full-load torque, then

$$T_f \propto \frac{sR_2}{R_2^2 + (sX_2)^2}$$

$$T_{max} \propto \frac{1}{2 \times X_2} \qquad \text{... Art. 22.16}$$

then
$$\frac{T_f}{T_{max}} = \frac{2s\,R_2\,X_2}{R_2^2 + (sX_2)^2}$$

Dividing both the numerator and the denominator by X_2^2, we get

$$\frac{T_f}{T_{max}} = \frac{2s\,R_2/X_2}{(R_2/X_2)^2 + s^2} = \frac{2\,a\,s}{a^2 + s^2}$$

where $a = R_2 / X_2$ = resistance / standstill reactance per phase.

22.21. Starting Torque and Maximum Torque

$$T_{st} \propto \frac{R_2}{R_2^2 + X_2^2} \qquad \text{... Art. 22.9}$$

$$T_{max} \propto \frac{1}{2X_2} \qquad \text{... Art. 22.16}$$

$$\therefore \quad \frac{T_{st}}{T_{max}} = \frac{2R_2\,X_2}{R_2^2 + X_2^2} = \frac{2R_2/X_2}{1 + (R_2/X_2)^2} = \frac{2a}{1 + a^2}$$

where
$$a = \frac{R_2}{X_2} = \frac{\text{rotor resistance}}{\text{standstill reactance}} - \text{per phase.}$$

22.22. Induction Motor Power Factor

The presence of air-gap between the stator and rotor of an induction motor increases the reluctance of the stator magnetic circuit. At start-up, the squirrel cage rotor has large reactance because the frequency of its induced currents is the highest (equal to stator supply frequency). Due to these two factors, the starting current drawn by the motor has a very large magnetising component but a very small in-phase component. Hence, motor current lags behind the supply voltage by a very large angle. Consequently, at the instant of start-up, the squirrel cage induction motor has a very low lagging power factor of the order of 0.05.

When running at no-load, the motor draws in-phase current just to meet no-load losses. This component is still relatively small as compared with the magnetising current (which would, of course, be much less than its value at the instant of starting). Hence, at no-load also, the motor will have a low lagging power factor of about 0.1.

However, as load on the motor is increased, the in-phase component of the motor is increased but its magnetising component remains about the same. Hence, the power factor of a squirrel-cage induction motor improves as its load is increased. In fact, it may become as high as 0.9 lagging at rated load.

Example 22.7. *A 3300 V, 24-pole, 50 Hz, 3-ϕ star-connected induction motor has a slipping motor resistance of 0.016 Ω and standstill reactance of 0.265 Ω per phase. Calculate (a) the speed at maximum torque (b) ratio of ϕ full-load torque to maximum torque if full-load torque is obtained at 2.47 r.p.m.* **(Elect. Machines-II, Anna Univ. 1992)**

Solution. We have seen that for maximum torque

$$R_2 = sX_2$$

or $s = R_2 / X_2 = 0.016 / 0.265 = 0.0604$

or $s = 6.04\%$

Rotor speed $N = (1 - s)\,N_S$

Now $N_S = 120 \times 50 / 24 = 250$ r.p.m.

(*a*) ∴ $N = (1 - 0.0604) \times 250 =$ **235 r.p.m.**

(*b*) Full-load slip $= \dfrac{250 - 247}{250} = \dfrac{3}{250} = 0.012$ or 1.2%

Now
$$\frac{T_f}{T_{max}} = \frac{2\,a\,s}{a^2 + s^2}$$

where $a = X_2 / R_2 = 0.0604$

$$\therefore \quad \frac{T_f}{T_{max}} = \frac{2 \times 0.0604 \times 0.012}{0.012^2 + 0.0604^2} = \mathbf{0.382}$$

22.23. Power Stages in an Induction Motor

Stator iron loss (consisting of eddy current and hysteresis losses) depends on the supply frequency and the flux density in the iron core. It is practically constant. The iron loss of the rotor is, however, negligible because frequency of rotor currents under normal running conditions is always small.

Total rotor Cu loss $= 3 I_2^2 R_2$.

A better visual for power flow within an induction motor is given in Fig. 22.10.

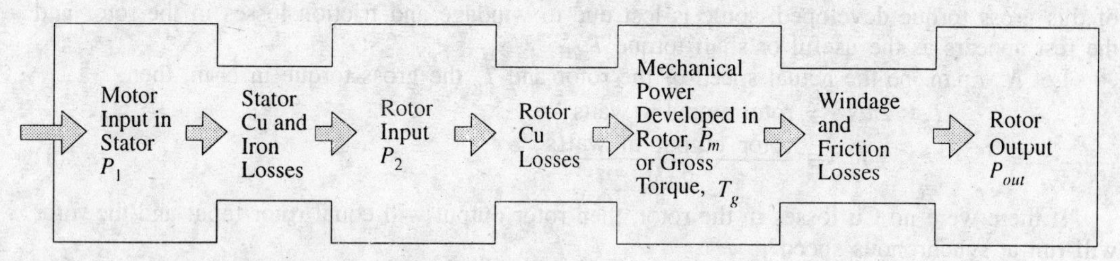

22.24. Torque Developed by an Induction Motor

An induction motor develops gross torque T_g due to gross rotor output P_m. Its value can be

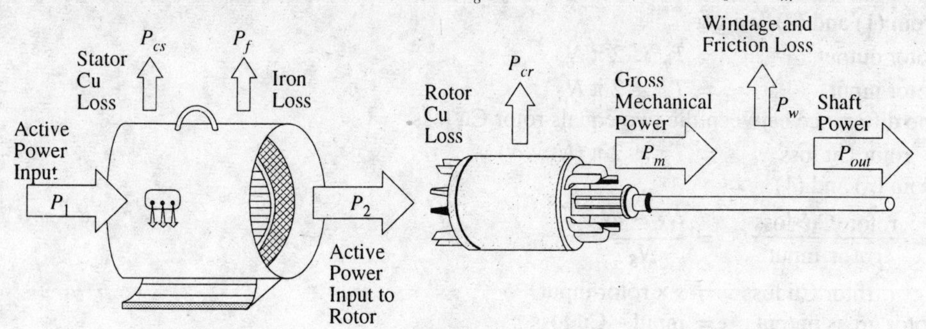

Fig. 22.10

expressed either in terms of rotor input P_2 or rotor gross output P_m as given below:

$$T_g = \frac{P_2}{\omega_s} = \frac{P_2}{2\pi N_S} \qquad \text{— in terms of rotor input}$$

$$= \frac{P_m}{\omega} = \frac{P_m}{2\pi N} \qquad \text{— in terms of rotor output}$$

The shaft torque T_{sh} is due to output power P_{out} which is less than P_m because of rotor friction and windage losses.

$$T_{sh} = \frac{P_{out}}{\omega} = \frac{P_{out}}{2\pi N}$$

The difference between T_g and T_{sh} equals the torque lost due to friction and windage loss in the motor.

In the above expression, N and N_S are in r.p.s. However, if they are in r.p.m., the above expressions for motor torque become

$$T_g = \frac{P_2}{2\pi N_S/60} = \frac{60}{2\pi} \cdot \frac{P_2}{N_S} = 9.55 \frac{P_2}{N_S} \text{ N-m}$$

$$= \frac{P_m}{2\pi N/60} = \frac{60}{2\pi} \cdot \frac{P_m}{N} = 9.55 \frac{P_m}{N} \text{ N-m}$$

$$T_{sh} = \frac{P_{out}}{2\pi N/60} = \frac{60}{2\pi} \cdot \frac{P_{out}}{N} = 9.55 \frac{P_{out}}{N} \text{ N-m}$$

22.25. Rotor Output

Stator input = stator output + stator losses

The stator output is transferred entirely inductively to the rotor circuit.

Obviously, rotor input = stator output

Rotor ouput = rotor input – rotor Cu losses

This rotor output is converted into mechanical energy and gives rise to gross torque T_g.. Out of this gross torque developed, some is lost due to windage and friction losses in the rotor and the rest appears as the useful or shaft torque T_{sh}.

Let N r.p.m. be the actual speed of the rotor and T_g the gross torque in N-m, then

$$T_g \times 2\pi N = \text{rotor output in watts}$$

$$\therefore \quad T_g = \frac{\text{rotor output in watts}}{2\pi N} \qquad \text{...(1)}$$

If there were no Cu losses in the rotor, then rotor output will equal rotor input and the rotor will run at synchronous speed.

$$\therefore \quad T_g = \frac{\text{rotor input}}{2\pi N_S} \qquad \text{...(2)}$$

From (1) and (2), we get

$$\text{Rotor output} = T_g \times 2\pi N$$

$$\text{Rotor input} = T_g \times 2\pi N_S \qquad \text{...(3)}$$

The difference between the two equals rotor Cu loss.

$$\therefore \text{ rotor Cu loss} = T_g \times 2\pi (N_S - N) \qquad \text{...(4)}$$

From (3) and (4)

$$\frac{\text{rotor Cu loss}}{\text{rotor input}} = \frac{N_S - N}{N_S} = s$$

$$\therefore \quad \text{rotor Cu loss} = s \times \text{rotor input} \qquad \text{...(5)}$$

$$\text{Rotor gross output} = \text{input} - \text{Cu loss}$$

$$= \text{input} - s \times \text{rotor input} = (1 - s) \text{ input}$$

$$\therefore \text{ rotor gross output} = (1 - s) \text{ rotor input} \qquad \text{... (6)}$$

$$\therefore \quad \frac{\text{rotor gross output}}{\text{rotor input}} = 1 - s = 1 - \frac{N_S - N}{N_S} = \frac{N}{N_S}$$

$$= \frac{\text{actual speed of rotor}}{\text{synchronous speed}}$$

$$\therefore \text{ rotor efficiency} = N/N_S$$

$$\text{Also} \quad \frac{\text{rotor Cu loss}}{\text{rotor gross output}} = \frac{s}{1 - s}$$

Important Conclusion

If some power P_2 is delivered to a rotor, then a part sP_2 is lost in the rotor itself as copper loss (and appears as heat) and the remaining $(1–s)$ P_2 appears as gross machanical power P_m (including friction and windage losses).

$$\therefore \quad P_2 : P_m : I^2R :: 1 : 1–s : s$$

The rotor input power will always divide itself in this ratio, hence it is advantageous to run the motor with as small a slip as possible.

Example 22.8. *A 7.46 kW, 230 V, 3-phase 50 Hz, 6-pole squirrel-cage induction motor operates at a full-load slip of 4% when rated voltage frequency are applied. Determine*

(i) full-load speed (ii) full-load torque in newton-metres (iii) frequency of rotor current under this condition (iv) speed of rotation of the stator m.m.f. **(A.M.I.E. Winter 1991)**

Solution. *(iv)* Stator m.m.f. revolves with synchronous speed given by $N_S = 120\ f/P = 120 \times 50\ /\ 6 =$ **1000 r.p.m.**

(i) $N = N_S\ (1 - s) = 1000\ (1 - 0.04) =$ **960 r.p.m.**

(ii) If we neglect the windage and friction losses of the rotor, then 7.46 kW also represents the gross mechanical power developed in rotor (Art. 22.23)

(iii) $f' = sf = 0.04 \times 50 =$ **2 Hz**

$$\therefore \quad T_g = 9.55\ \frac{P_{out}}{N} = 9.55 \times \frac{7.46 \times 10^3}{960} = \mathbf{74.2\ N\text{-}m}$$

Example 22.9. *A 3-phase, 6-pole induction motor delivers 3.73 kW at 950 r.p.m. What is the stator input if the stator loss is 250 W?*

Solution. Rotor output = 3,730 W

$$N_S = 120 \times 50\ /\ 6 = 1000 \text{ r.p.m.}$$

$$N = 950 \text{ r.p.m.}$$

Now, $\dfrac{\text{rotor output}}{\text{rotor input}} = \dfrac{N}{N_S}$

$$\therefore \quad \text{rotor input (or stator output)} = \frac{3{,}730 \times 1000}{950} = 3{,}930 \text{ W}$$

Stator input = stator output + stator losses

= 3930 + 250 = **4180 W**

Example 22.10. *A 6-pole, 3-phase induction motor runs at a speed of 960 r.p.m. and the shaft torque is 135.7 N-m. Calculate the rotor Cu loss if the friction and windage losses amount to 150 watts. The frequency of supply is 50 Hz.*

Solution. Output $= 2\,\pi \times N \times T_{sh}$

$= 2\,\pi \times (960\ /\ 60) \times 135.7 = 13{,}642$ W

Gross rotor output = 13,642 + 150 = 13,792 W

$$N_S = 120 \times 50\ /\ 6 = 1000 \text{ r.p.m.}$$

$$s = (1000 - 960)\ /\ 1000 = 0.04$$

Now $\dfrac{\text{rotor Cu loss}}{\text{rotor gross output}} = \dfrac{s}{1 - s}$

$$\therefore \quad \text{rotor Cu loss} = \frac{13{,}792 \times 0.04}{1 - 0.04} = \mathbf{575\ W}$$

Example 22.11. *A 6-pole, 3-phase induction motor develops 30 horse-power (22.38 kW) including mechanical losses totalling 2 horse-power (1.492 kW) at a speed of 950 r.p.m. on 550 volts, 50 Hz mains. The p.f. is 0.88. Neglect core losses. Calculate for this load the following: (i) the slip (ii) the rotor copper loss (iii) the total input if the stator losses are 2000 watts (iv) the line current.* **(Electrical Science, AMIE Winter 1993)**

Solution. Synchronous speed $= N_s = 120 \times 50\ /\ 6 = 1000$ r.p.m.

(i) $s = (1000 - 950)\ /\ 1000 = 0.05$ or **5%**

(ii) Since 22.38 kW includes total mechanical losses, it represents the mechanical power developed in the rotor *i.e.*, gross mechanical power developed.

$$\text{Rotor Cu loss} = \frac{s}{1 - s} \times \text{gross mech. power}$$

$$= \frac{0.05}{1 - 0.05} \times 22{,}380 = \mathbf{1178\ W}$$

(iii) $\text{Rotor power input} = \dfrac{\text{motor Cu. loss}}{s} = 1178/0.05 = 23{,}560 \text{ W}$

Stator Cu loss $= 2000$ W ... (given)

Total stator input $= 23{,}560 + 2000 =$ **25,560 W**

(*iv*) $I_L = 25{,}560 / \sqrt{3} \times 550 \times 0.88 = 30.5$ A

Example 22.12. *The power supplied to a 3-phase induction motor is 40 kW and the corresponding stator losses are 1.5 kW. Calculate (i) the total mechanical power developed and the rotor $I^2 - R$ loss when the slip is 0.04 per unit and (ii) the efficiency of the motor. (Neglect the rotor iron-loss).* **(Electrical Science, AMIE Summer 1993)**

Solution. Rotor power input $= 40 - 1.5 = 38.5$ kW

(*i*) Rotor Cu loss $= 0.04 \times 38.5 =$ **1.54 kW**

Gross mechanical power developed $= 38.5 - 1.54 =$ **36.96 kW**

(*ii*) Since the rotor iron losses have been neglected, 36.96 kW represents the output power of the motor.

$\eta = 36.96 / 40 = 0.924$ or **92.4%**

22.26. Starting Methods for Cage Motors

The following three methods are commonly used for starting squirrel-cage induction motors :

(*a*) Direct-on-line (DOL) Starting

It is also called across-the-line starting and is used for small motors below about 5 kW. It is customary to employ a straight single-throw switch or a contactor as shown in Fig. 22.11. The starting current is high *i.e.*, about 4 to 7 times the full-load current, the actual value depending on the size and design of the motor. Such a high starting current causes a relatively large voltage drop in the cables and thereby affects the supplies to other consumers on the same system. If the supply is isolated as in factories having their own generators, there is no such limit and motors of several hundred kW rating are started by direct-on-line (DOL) method.

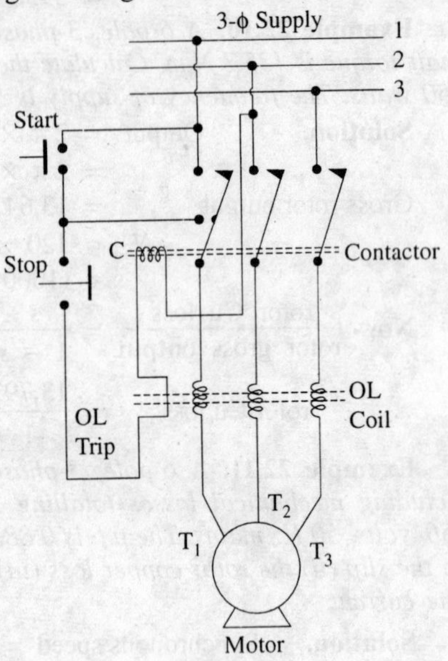

Fig. 22.11

Large motors fed from a common supply are started with reduced voltage by using any of the following two methods.

(*b*) Auto-transformer Starter

Fig. 22.12 shows a 3-phase star-connected auto-transformer used for applying reduced voltage across motor stator during starting period. In this way, starting current drawn by the motor can be reduced to any desired value. The auto-transformer usually has tappings which reduce the line voltage to 50%, 65% and 80% of the normal value. Since torque developed by the motor varies as the square of the applied voltage, the starting torque is considerably reduced. If, for example, 50% tap is used, the starting torque developed by the motor would be $(0.5)^2$ or 1/4 or 25% of the full-load torque. For 65% and 80% taps, the starting torques would be $(0.65)^2 = 42.2\%$ and $(0.8)^2 = 64\%$ of the full-load torque respectively.

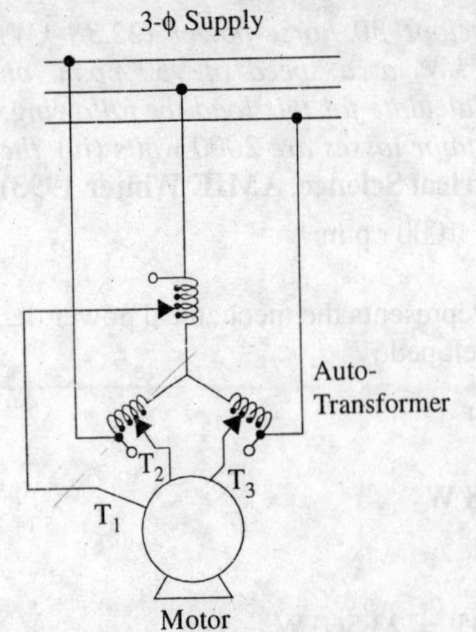

Fig. 22.12

Generally, a double-throw switch is used for connecting

the auto-transformer in the circuit for starting purposes. After the motor has run up to speed, the switch is moved into the RUN position which connects the motor directly to the supply.

(*c*) **Star-delta Starter**

In this method, the stator winding of the motor is connected in star for starting and in delta for normal running (Fig. 22.13). The voltage of each phase at starting is reduced to $1/\sqrt{3}$ of the line voltage. The current in each *phase* is also reduced by the same factor so that line current at starting becomes $1/\sqrt{3} \times 1/\sqrt{3} = 1/3$ of the current which the motor would have taken if connected directly across the supply.

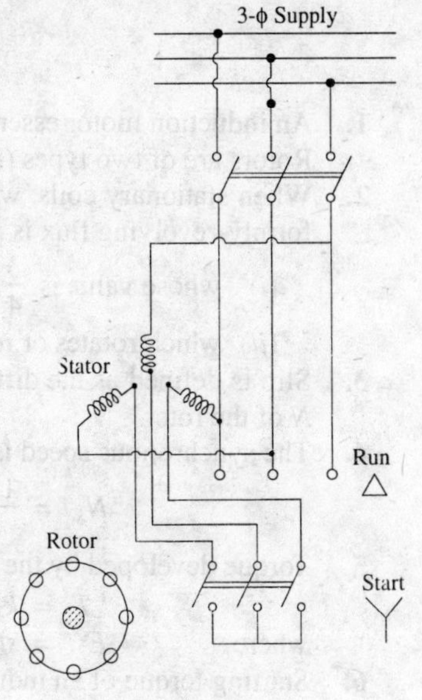

Fig. 22.13

Since $T \propto V^2$, the starting torque is reduced to $(1/\sqrt{3})^2 = 1/3 = 13.3\%$ of the normal value.

The change-over from star-starting to delta-running is made by a double-throw switch with interlocks to prevent motor starting with the switch in the RUN position.

22.27. Starting of Slip-ring Motors

Fig. 22.14 shows the schematic diagram of the circuit used for starting wound-rotor or slip-ring induction motor. As seen, variable resistance can be added into the rotor circuit for reducing the starting current and improving the starting torque. The rotor windings are connected to the *Y*-connected external resistors through slip-rings and brushes. The starter resistance is usually of the face-plate type, the starter arms forming the star point. The motor is started with all the starter resistance circuit in thus giving maximum starting torque. As motor speed increases, the starter resistance is gradually decreased. When motor comes upto its full-load speed, the starting resistance is reduced to zero and the slip-rings are short-circuited.

Tutorial Problems No. 22.2

1. A three-phase, 50 Hz induction motor has 4 poles and runs at a speed of 1440 r.p.m. when the total torque developed by the rotor is 70 N-m. Calculate (*a*) the total input to the rotor (*b*) the rotor copper loss in watts. **[(*a*) 11 kW (*b*) 440 W]**

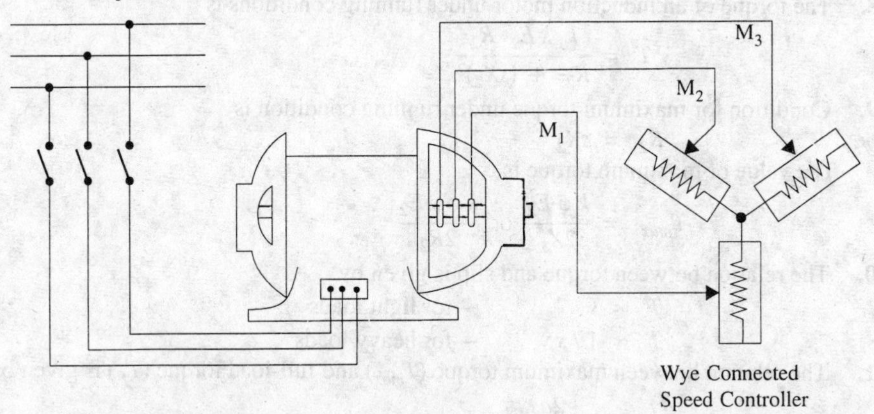

Fig. 22.14

2. Estimate the copper loss in the rotor circuit of an induction motor running at 50 per cent of synchronous speed with a useful output of 41.03 kW and mechanical losses totalling 1.492 kW. If the stator losses total 3.5 kW, at what efficiency is the motor operating?
3. A 50 Hz, 8-pole induction motor has a full-load slip of 4 per cent. The rotor resistance is 0.001 Ω

per phase and the standstill reactance is 0.005 Ω per phase. Find the ratio of the maximum to the full-load torque and the speed at which the maximum torque occurs. **[2.6 ; 600 r.p.m.]**

4. A six-pole, 415 V, three-phase, 50 Hz, mesh-connected induction motor develops 7.46 kW with a 5% slip, efficiency of 80% and power factor of 0.75. Find its speed, torque and line current.
[950 r.p.m. 75 N-m, 17.3 A] (*City and Guilds, London*)

HIGHLIGHTS

1. An induction motor essentially consists of (*a*) a stator and (*b*) a rotor.
   Rotors are of two types (*i*) squirrel-cage rotor and (*ii*) phase-wound or wound rotor.
2. When stationary coils, wound for three phase, are fed by three-phase supply, then a uniformly-revolving flux is produced.
   (*i*) whose value is $\frac{3}{4}\Phi_m = 1.5$ times the maximum flux due to either phase; and
   (*ii*) which rotates or revolves at synchronous speed.
3. Slip is defined as the difference between the synchronous speed N_S and the actual speed N of the rotor.
4. The synchronous speed is given by
   $$N_S = \frac{120f}{P}$$
5. Torque developed by the rotor is
   $$T = k_1 E_2 I_2 \cos\phi_2$$
   where E_2 = *standstill* e.m.f. induced in the rotor/ phase
6. Starting torque of an induction motor is
   $$T_{st} = \frac{k_1 E_2^2 R_2}{R_2^2 + X_2^2} = k_2 \cdot \frac{R_2^2}{R_2^2 + X_2^2}$$
   where R_2 = rotor resistance / phase
   X_2 = rotor reactance / phase *at standstill.*
7. Condition for maximum starting torque is
   $$R_2 = X_2$$
8. The torque of an induction motor under running conditions is
   $$T = \frac{k_1 s E_2^2 R_2}{R_2^2 + (sX_2)^2}$$
9. Condition for maximum torque under running condition is
   $$R_2 = sX_2$$
   The value of maximum torque is
   $$T_{max} = \frac{k\phi E_2}{2X_2} \text{ or } \frac{k\phi s E_2}{2R_2}$$
10. The relation between torque and slip is given by
    $T \propto s$ —for light loads
    $T \propto 1/s$ —for heavy loads
11. The relation between maximum torque (T_{max}) and full-load torque (T_f) is given by
    $$\frac{T_f}{T_{max}} = \frac{2as}{a^2 + s^2}$$
    where $a = R_2/X_2$
    = resistance/standstill reactance per phase
12. The relation between standstill (or starting) torque (T_{st}) and maximum torque is given by
    $$\frac{T_{st}}{T_{max}} = \frac{2a}{1 + a^2}$$

13. The gross torque developed by a motor is

$$T_{sh} = 9.55 \frac{P_2}{N_S} \text{ newton-metre}$$

$$= 9.55 \frac{P_m}{N} \text{ newton-metre}$$

14. The shaft torque developed by an induction motor is

$$T_{sh} = 9.55 \frac{P_{out}}{N} \text{ newton-metre}$$

15. (*i*) Gross torque $T_g = \dfrac{\text{rotor output in watts}}{2\pi \times N}$

(*ii*) rotor Cu loss $= T_g \times 2\pi (N_S - N)$

(*iii*) $\dfrac{\text{rotor Cu loss}}{\text{rotor input}} = s$

(*iv*) rotor Cu loss $= s \times$ rotor input

(*v*) $\dfrac{\text{rotor gross output}}{\text{rotor input}} = 1 - s = \dfrac{N}{N_S}$

(*vi*) rotor efficiency $= N/N_S$

(*vii*) $\dfrac{\text{rotor Cu loss}}{\text{rotor gross output}} = \dfrac{s}{1 - s}$

OBJECTIVE TESTS — 22

A. Fill in the following blanks.

1. The two main parts of an induction motor are stator and ....
2. Slip-ring motors have phase- .... rotors.
3. In an induction motor, the resulting flux revolves synchronously around the inside surface of the.................. core.
4. The rotor of an induction motor always revolves in the ....direction as the stator flux.
5. At the instant of start-up, a squirrel-cage induction motor has very .... power factor.
6. Under running conditions, the torque of an induction motor becomes maximum at that value of slip which makes rotor .... per phase equal to rotor resistance / phase.
7. For low values of slip, the torque / slip curve of an induction motor is approximately a ....line.
8. The slip of induction motor also represents the ratio of rotor Cu loss to ....input.

B. Answer True or False.

1. Squirrel-cage motor and slip-ring motor have similar stators.
2. There are no brushes in a squirrel-cage induction motor.
3. In the rotor of a 3-phase induction motor, the voltage is induced by transformer action.
4. The rotor bars of a squirrel-cage induction motor are insulated from the rotor core.
5. The slip of an induction motor varies from zero to hundred per cent depending on the load.
6. The torque of an induction motor varies as the square of the supply voltage.
7. An induction motor has a lagging power factor both at start-up and during operation.
8. The efficiency and power factor of an induction motor increase with load upto its rated load.

C. Multiple Choice Questions.

1. An induction motor is so-called because its operation depends on the phenomenon of

   (*a*) self-induction

   (*b*) mutual induction

   (*c*) eddy currents

(*d*) hysteresis

2. The no-load speed of an induction motor depends on

(*a*) the supply frequency

(*b*) the number of its poles

(*c*) the maximum flux / phase

(*d*) only (*a*) and (*b*) above

3. The frequency of rotor current in a 6-pole, 50 Hz, 3-ϕ induction motor running at 950 r.p.m. is .... Hz.

(*a*) 2.5 (*b*) 1.5

(*c*) 5.0 (*d*) 0.05

4. The torque of an induction motor is proportional to

(*a*) $1 / V$ (*b*) $\sqrt{V}$

(*c*) V^2 (*d*) V^3

5. In an induction motor, the ratio of rotor Cu loss and input is given by

(*a*) $1 / s$ (*b*) s

(*c*) $(1-s)$ (*d*) $s / (1-s)$

6. The speed regulation of a squirrel-cage induction motor is very good primarily because of its relatively low............ under normal operating conditions.

(*a*) rotor impedance

(*b*) losses

(*c*) power factor

(*d*) rotor frequency

ANSWERS

A. 1. rotor 2. wound 3. stator 4. same 5. low 6. reactance 7. straight 8. rotor

B. 1. T 2. T 3. T 4. F 5. F 6. T 7. T 8. F

C. 1. *b* 2. *d* 3. *a* 4. *c* 5. *b* 6. *a*

23 SINGLE-PHASE MOTORS

23.1. Types of Single-phase Motors

As the name shows, such motors are designed to operate from a single-phase supply and are manufactured in a large number of types to perform a wide variety of useful services in home, offices, factories, workshops and in business establishments. These motors are classified according to their construction and method of starting:

1. Induction motors. These may be further sub-divided into split-phase, capacitor and shaded-pole type.
2. Repulsion type Motors or Single-Phase Wound-rotor Motors.
3. A.C. Series Motors.
4. Universal Motors.
5. Unexcited Synchronous Motors (like Reluctance Motor and Hysteresis Motor).

23.2. Single-phase Induction Motor

Like 3-phase motor, a single-phase motor also consists of

(*i*) a stator which carries single-phase winding, and

(*ii*) a squirrel-cage type of rotor.

However, there is one very fundamental difference between the two: whereas a 3-phase motor is self-starting, a single-phase motor is non-self-starting. This is due to the fact that whereas the stator winding of a 3-phase motor produces a rotating (or revolving) flux, the single-phase winding produces merely *a pulsating* or *alternating flux.* A synchronously rotating flux can be produced only by either a 2-phase or 3-phase stator winding when energised from a 2-phase or 3-phase supply respectively. An alternating flux acting on a *stationary* squirrel-cage rotor cannot produce rotation (only a rotating flux can).

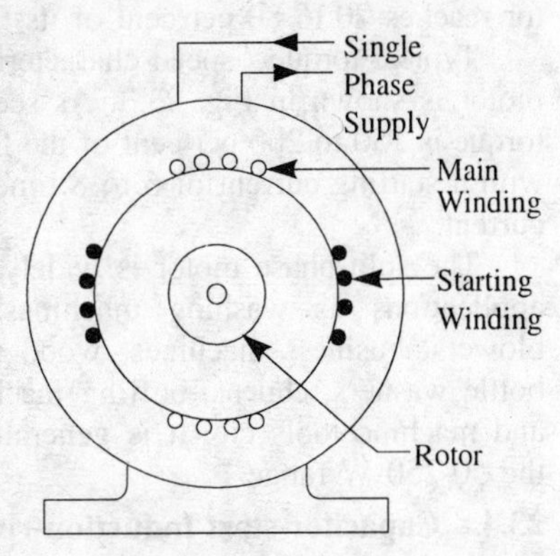

Fig. 23.1

However, if the rotor of a single-phase motor is given an initial start by hand or otherwise in *either* direction, then immediately a torque arises and the motor accelerates to its final speed provided it is not heavily loaded.

There are many methods to make a single-phase motor self-starting and single-phase induction motors are classified and named according to the method used as detailed below:

(*a*) *Split-phase motors:* These motors are started by two-phase motor action with the help of an additional winding known as starting winding.

(*b*) *Capacitor motors:* These are also started by the two-phase motor action by using a capacitor.

(*c*) *Shaded-pole motors:* These are started by using shaded-pole principle.

23.3. Split-phase Induction Motor

As shown in Fig. 23.2 (*a*), a single-phase motor is temporarily converted into a 2-phase motor by providing an extra winding on the stator in addition to the main or running winding.

The circuit connections are shown in Fig. 23.2 (*b*). By making starting winding highly resistive and main winding highly reactive, the phase difference between the currents drawn by them can be made sufficiently large (the ideal value being 90°). The motor behaves like a two-phase motor

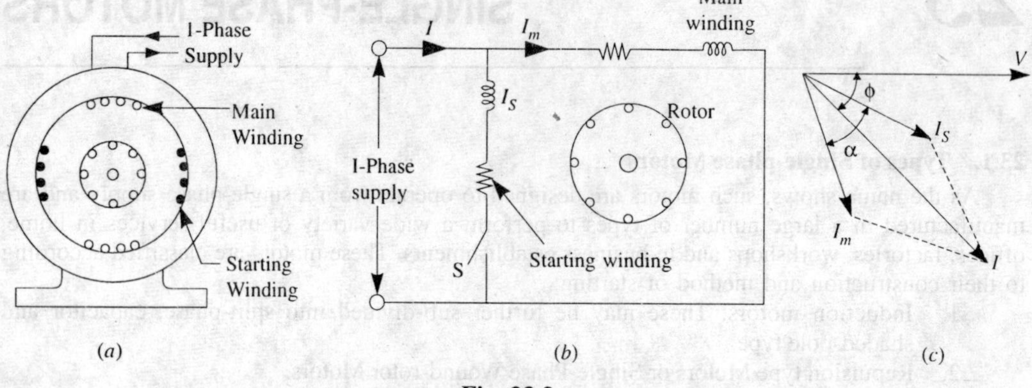

Fig. 23.2

for starting purposes [Fig. 23.2 (*c*)]. The two currents produce a revolving flux and hence make the motor self-starting. The starting torque $T_{st} = k\, I_s\, I_m \sin \alpha$ where k is a constant governed by motor design parameters.

A centrifugal switch S is connected in series with the starting winding and is located outside the motor. Its function is to automatically disconnect the starting winding from the supply when the motor reaches 70 to 80 per cent of its full-load speed.

Typical torque / speed characteristic for such a motor is shown in Fig. 23.3. As seen, the starting torque is 150 to 200 per cent of the full-load torque with a starting current of 6 to 8 times the full-load current.

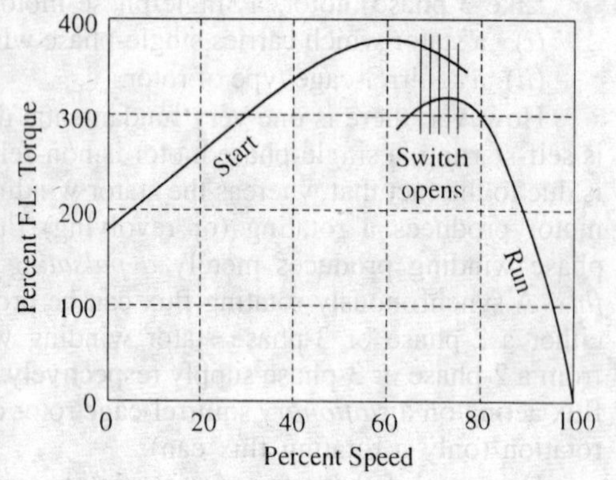

Fig. 23.3

The split-phase motor is widely used for such applications as washing machines, oil burners, blowers, business machines, wood working tools, bottle washers, churns, buffing machines, grinders and machine tools etc. It is generally available in the 50-250 W range.

23.4. Capacitor-start Induction-run Motors

These motors have a higher starting torque because in their case angle α between currents I_s and I_m is large. The angle α is increased by connecting a capacitor in series with the starting winding as shown in Fig. 23.4 (*a*). Usually, the capacitor is mounted on top of the motor frame and is generally an electrolytic capacitor. As before, the centrifugal switch cuts off both the starting winding and the capacitor when motor runs up to nearly 75 per cent of its full-load speed. As seen from Fig. 23.4 (*b*), the current I_s drawn by starting winding leads the voltage whereas the current I_m in the main winding, as before, lags V. In this way, value of α is increased to about 80° which increases the starting torque to twice that developed by a standard split-phase induction motor. Such motors have starting torques as high as 450 per cent of the full-load value. Typical performance curve of such a motor is shown in Fig. 23.4 (*c*). They are usually manufactured in the 100-500 W range.

The capacitor-start motors are very popular for heavy-duty general-purpose applications requiring high starting torques. These are generally used for such applications as compressors, jet pumps, farm and home-workshop tools, swimming pool pumps and conveyors etc.

23.5. Capacitor-start-and-run Motors

These are similar to the capacitor-start induction-run motors except that the starting winding and capacitor are connected in the circuit at *all times.* In other words, the capacitor remains in the circuit permanently as shown in Fig. 23.5. That is why such motors are also known as permanent-split capacitor motors. Obviously, there is no need for centrifugal switch. The advantages of leaving the capacitor permanently in circuit are (*i*) improvement of overload capacity of the motor (*ii*) a higher p.f. (*iii*) higher efficiency and (*iv*) quieter running of the motor.

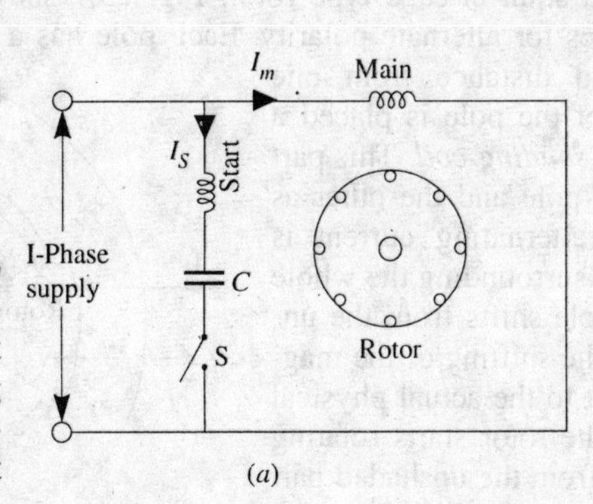

(*a*)

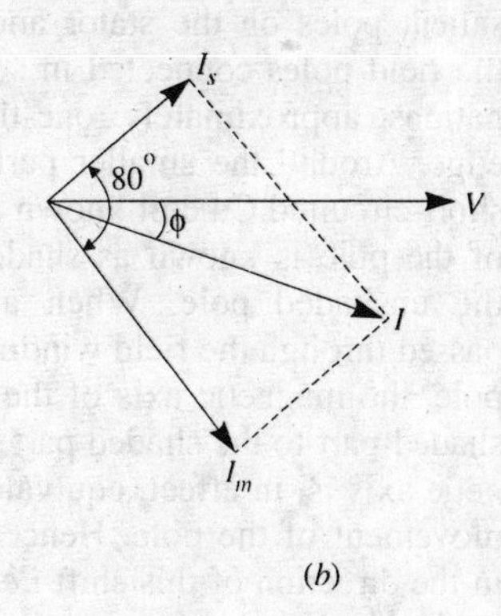

(*b*)

Fig. 23.4

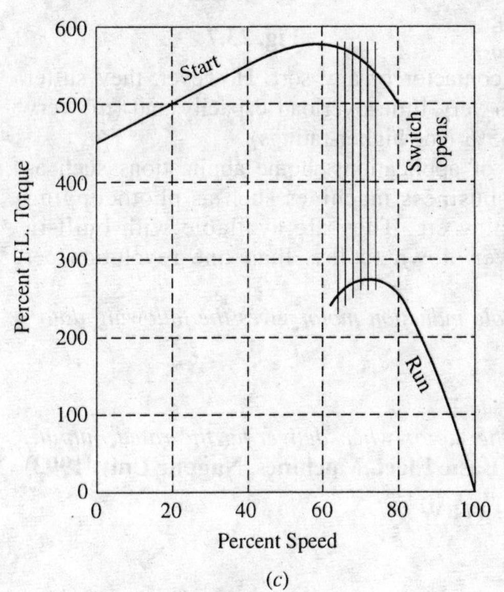

(*c*)

Fig. 23.4

Such motors have a comparatively low starting torque which is about 50 to 100 per cent of the rated torque.

Such motors which start and run with one value of capacitance in the circuit are called single-value capacitor-run motors. Others which start with a high value of capacitance but run with a low value of capacitance are known as two-value capacitor-run motors (Fig. 23.6).

Such motors find applications in refrigerators, compressors, stokers, ceiling fans and blowers etc. In the smaller sizes, they often compete with shaded-pole motors. Generally, they are not suitable for belted applications or for any other continuous-duty application requiring large locked-rotor torque.

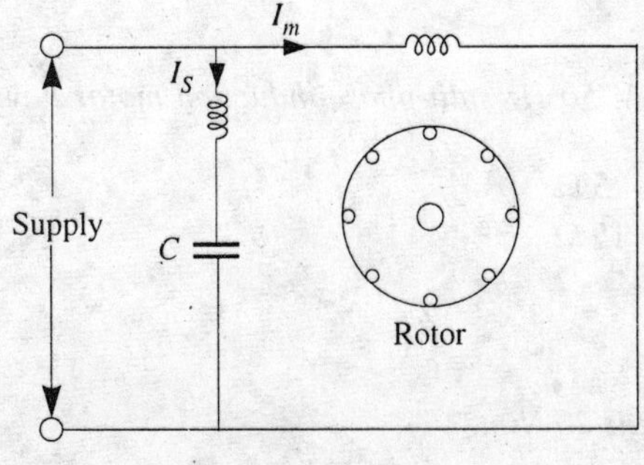

Fig. 23.5

Fig. 23.6

23.6. Shaded-pole Motors

In such motors, the necessary phase-splitting is produced by induction. These motors have salient poles on the stator and a squirrel-cage type rotor. Fig. 23.7 shows a 4-pole motor with the field poles connected in series for alternate polarity. Each pole has a slot cut across its laminations approximately one-third distance from one edge. Around the smaller part of the pole is placed a short-circuited Cu coil known as *shading coil.* This part of the pole is known as shaded pole and the other as the unshaded pole. When an alternating current is passed through the field winding surrounding the whole pole, the magnetic axis of the pole shifts from the unshaded part to the shaded part. The shifting of the magnetic axis is, in effect, equivalent to the actual physical movement of the pole. Hence, the rotor starts rotating in the direction of this shift *i.e.*, from the unshaded part to the shaded part.

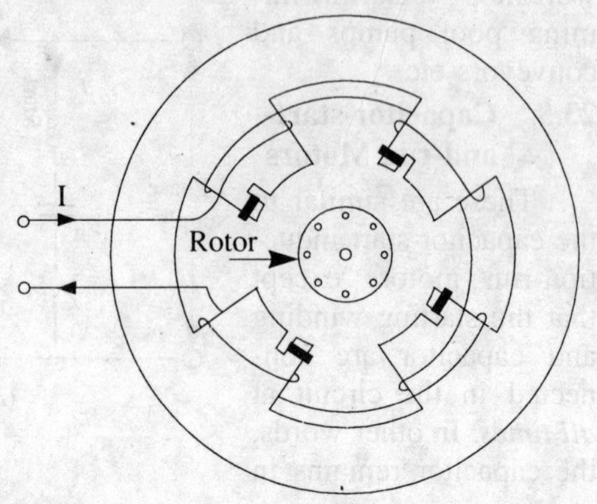

Fig. 23.7

Such motors are built in very small sizes (5-50 W) but are simple in construction and are extremely rugged, reliable and cheap. They do not need any commutator, switch, brushes, collector rings, governors or contactor of any sort. However, they suffer from the disadvantages of (*i*) low starting torque (*ii*) very little overload capacity and (*iii*) very low efficiency ranging from 5% (for tiny sizes) to 35% (for higher-ratings).

Shaded-pole motors are used in a wide-variety of applications: home applications such as fans of all kinds, slide projectors, humidifiers, small business machines such as photocopying machines, vending machines and advertising display etc. They are available with built-in gear reducers for obtaining almost any speed even down to less than one revolution per month!

Example 23.1. *The name plate of a 1-phase, 4-pole induction motor gives the following data:*
Output 373 W; 230 V
Frequency 50 Hz; input current 2.9 A
Power factor 0.71; speed 1410 r.p.m.
Calculate (a) the efficiency and (b) the slip of the motor when delivering the rated output.

(Basic Elect. Machines, Nagpur Univ. 1993)

Solution. Input $= VI\cos\phi = 230 \times 2.9 \times 0.71 = 474$ W

Output $= 373$ W

(*a*) $\eta = 373/474 = 0.785$ or **78.5%**

(*b*) $\text{Slip} = \dfrac{N_S - N}{N_S} \times 100$

Now $N_S = 1200 \times 50/4 = 1500$ r.p.m.

$\therefore \quad s = \dfrac{1500 - 1410}{1500} \times 100 = \mathbf{6\%}$

Example 23.2. *At starting, the windings of a 220 V, 50 Hz split-phase induction motor have the following values:*

Main winding:	$R = 10\ \Omega$,	$X_L = 25\ \Omega$
Starting winding:	$R = 25\ \Omega$,	$X_L = 15\ \Omega$

Find (a) current in the main winding and ϕ_m;
(b) current in the starting winding and ϕ_s;
(c) phase displacements between I_m *and* I_s;
(d) value of constant k if initial starting torque is 20 N-m;
(e) line current.

Solution. (*a*) $I_m = \dfrac{V}{Z_m} = \dfrac{220}{\sqrt{10^2 + 25^2}} = \dfrac{220}{26.9} = \mathbf{8.17\ A}$

$$\cos \phi_m = R/Z_m = 10/26.9 = 0.37\ ; \quad \phi_m = \mathbf{68.2° \ (lag)}$$

(*b*) $$I_s = \frac{V}{Z_s} = \frac{220}{\sqrt{15^2 + 25^2}} = \frac{220}{29.1} = \mathbf{7.56\ A}$$

$$\cos \phi_s = R/Z_s = 25/29.1 = 0.86, \quad \phi_s = \mathbf{30.8°}$$

(*c*) $$\alpha = \phi_m - \phi_s = 68.2° - 30.8° = \mathbf{37.4°}$$

(*d*) $$T_{st} = k\, I_m I_s \sin \alpha$$

$$\therefore \quad k = \frac{T_{st}}{I_m I_s \sin \alpha}$$

$$= \frac{20}{8.17 \times 7.56 \times \sin 37.4°} = \mathbf{0.53}$$

(*e*) As shown in Fig. 23.8, line current I is equal to the vector sum of I_m and I_s.

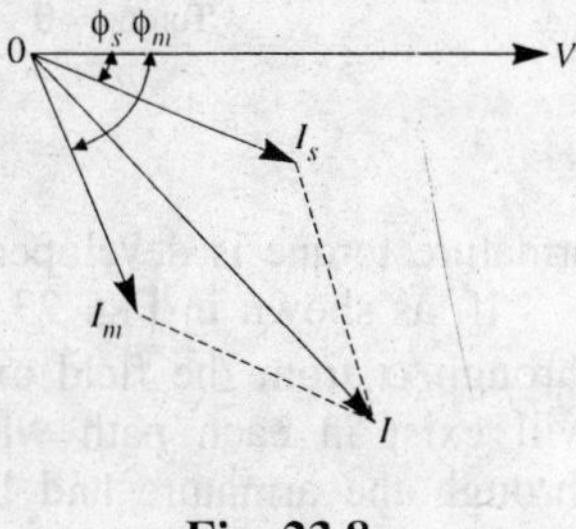

Fig. 23.8

Total X-component $= I_m \cos \phi_m + I_s \cos \phi_s$
$= 8.17 \cos 68.2° + 7.56 \cos 30.8°$
$= 9.55$ A

Total Y-component $= I_m \sin \phi_m + I_s \sin \phi_s$
$= 8.17 \sin 68.2° + 7.56 \sin 30.8°$
$= 11.42$ A

$$\therefore \quad I = \sqrt{9.55^2 + 11.42^2} = \mathbf{14.9\ A}$$

Example 23.3. *If a 160 μF capacitor is connected in series with the starting winding of the motor discussed in Example 23.2, what will be the value of its starting torque?*

Solution. $$X_C = \frac{1}{2\pi fC} = \frac{10^6}{2\pi \times 50 \times 160} = 19.9\ \Omega$$

$$I_s = \sqrt{25^2 + (15 - 19.9)^2} = 25.5\ \Omega$$

$$I_s = 220 / 25.5 = 8.62 \text{ A}$$

$$\cos \phi_s = R / Z_s = 25 / 25.5 = 0.98 \text{ (lead)}$$

$$\phi_s = 11°$$

$$\alpha = \phi_m + \phi_s$$

$$= 68.2° + 11° = 79.2°$$

The phasor diagram is shown in Fig. 23.9.

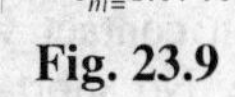

Fig. 23.9

$$T_{st} = kI_m I_s \sin \alpha$$

$$= 0.53 \times 8.17 \times 8.62 \times \sin 79.2° = \mathbf{36.7\ N\text{-}m}$$

The increase in the value of starting torque is considerable.

23.7. Repulsion Principle

Repulsion motors work on the repulsion principle which can be explained with the help of a 2-pole motor having a wound armature with commutator and brushes. The brushes are not connected to the supply but are short-circuited. The current through the stator winding is alternating and would reverse direction every half-cycle. In Fig. 23.10 (*a*), suppose that the a.c. supply current in the stator winding is increasing and is in the direction as indicated. The stator flux so produced will induce voltages in the armature conductors in the directions indicated by small arrows. These voltages are additive on both sides of the brushes and so send a large current through the armature and the short-circuited brushes. It is seen that half of the conductors under each pole carry current in one direction and the other half carries current in the opposite direction. Hence, no torque is developed by the armature.

In Fig. 23.10 (*b*), brushes have been shifted through 90° *i.e.,* now brush axis lies along the neutral axis*. In this position, induced voltages in each path get neutralised. As a result, no voltage exists at the brushes and hence, no current flows through the armature. Consequently, again no

*This axis lies at right angles to the stator field axis.

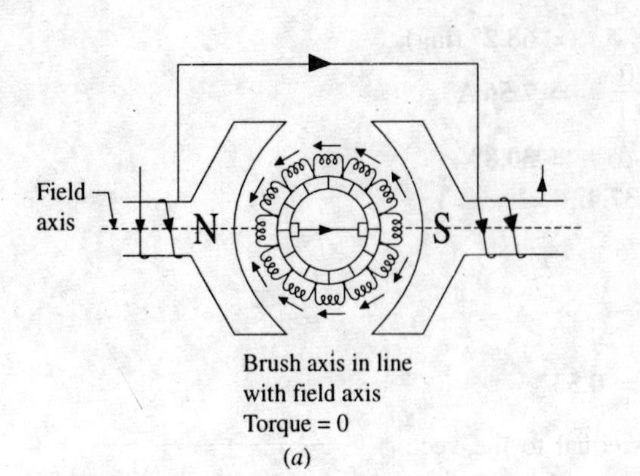

Brush axis in line
with field axis
Torque = 0
(*a*)

Neutral axis
N
S

Brush axis on Neutral axis,
Torque = 0
(*b*)

Fig. 23.10

armature torque is developed.

If, as shown in Fig. 23.11, brushes are shifted through α from the field axis, a resultant voltage will exist in each path which will send current through the armature and brushes. Hence, torque would be developed which will rotate the armature, in the present case, in the clockwise direction. If brush axis is shifted to the opposite side of the field axis, the torque so developed will rotate the armature in the counter-clockwise direction. In fact, a repulsion motor develops torque *in the direction in which the brushes are shifted from the field axis.*

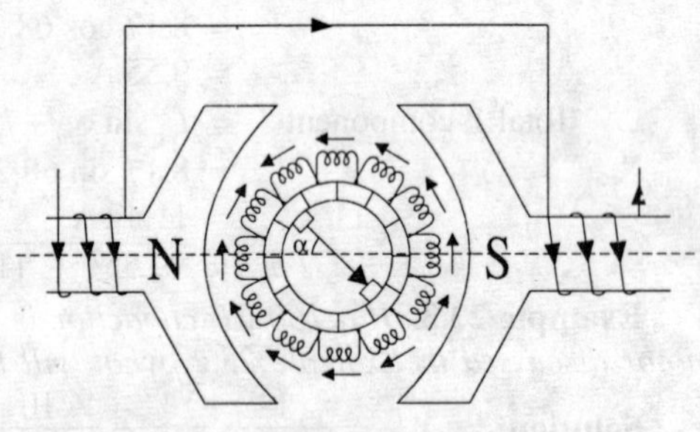

Brush axis at angle α with field axis
torque developed

Fig. 23.11

23.8. Repulsion Type Motors

These run on the repulsion principle explained in Art. 23.7. Such motors can be further sub-divided into following three types:

1. Repulsion Motor

It is a single-phase motor which consists of (*a*) one stator winding (*b*) one rotor which is wound like a d.c. armature (*c*) commutator and (*d*) a set of brushes which are short-circuited and remain in contact with the commutator at all times. It operates continuously on the 'repulsion' principle. No short-circuiting mechanism is required for this type.

The torque developed by such a motor depends on the product of the field flux and armature current which itself depends on the field flux. Therefore, torque depends on the square of the field flux or field or a.c. supply current as in a d.c. series motor. Hence, such motors have a high starting torque and poor speed regulation. The magnitude of the starting torque as well as speed can be varied by varying the angle of shift of the brush axis from the field axis. The direction of rotation of the motor can be reversed during rotation.

2. Repulsion-start Induction-run Motor

This motor starts as a repulsion motor but normally runs as an induction motor with constant speed characteristics. It consists of (*a*) one stator winding (*b*) one rotor which is similar to the wire-wound d.c. armature (*c*) a commutator and (*d*) a centrifugal mechanism which short-circuits the commutator bars all the way round (with the help of a short-circuiting necklace) when the motor has reached nearly 75 per cent of its full-load speed. From now onwards, the motor runs as an induction motor.

At one time, such motors were almost exclusively used where high starting torques were required such as in commercial refrigerators, compressors and pumps etc., but have been now replaced in nearly all cases by capacitor motors because they

(*i*) require more maintenance because of commutator, brushes and the centrifugal switch;

(*ii*) are relatively more expensive;

(*iii*) cause radio interference during starting;
(*iv*) are noisier.

However, despite all these disadvantages, these motors are still used in integral horse-power sizes because they

(*i*) develop more locked-rotor torque per line ampere;
(*ii*) can withstand longer starting periods than capacitor type motors.

3. Repulsion-induction Motor

It works on the combined principle of repulsion and induction. It consists of (*a*) stator winding (*b*) two independent rotor windings: one squirrel-cage and the other usual d.c. winding connected to the commutator and (*c*) a short-circuit the commutator. Fig. 23.12 shows the connections of a 4-pole repulsion-induction motor for 230 V operation. In the field of repulsion motors, this type is becoming very popular because of its good all-round characteristics which are comparable to those of a d.c. compound motor. It has fairly constant speed owing to squirrel-cage winding and also develops large torque to carry suddenly-applied loads. It has high starting torque and uses no centrifugal short-circuiting mechanism.

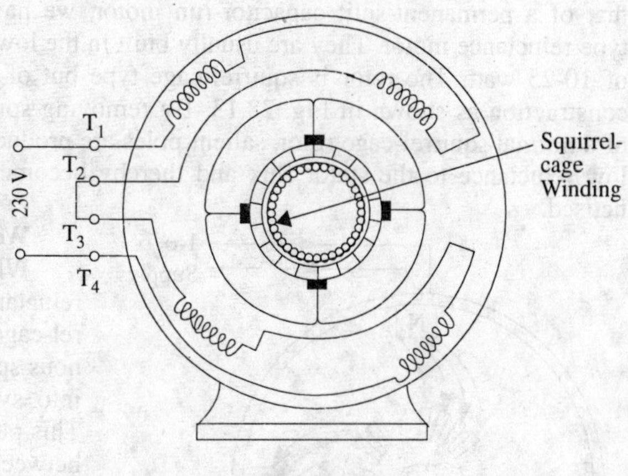

Fig. 23.12

23.9. Universal Motor

It is defined as a motor which can be operated either on d.c. supply or a.c. supply at approximately the same speed and output. It is a smaller version of a d.c. series motor suitably modified for operation on a.c. supply. Their power ratings vary from 5 W to 200 W and full speed varies from 6000 to 8000 r.p.m. Usually, universal motors are custom-built for a specific application and are often sold as parts rather than as complete motors.

Construction

Basically, the construction of a universal motor is similar to that of a d.c. series motor. However, the adverse effects caused by a high field impedance, iron losses, eddy currents and hysteresis are minimized by using few series field turns and a low reluctance magnetic path. Moreover, the armature and the field are laminated and low flux densities are used. The number of armature conductors and commutator segments is also increased.

Working

Since field windings and the armature are connected in series, the same current passes through both when motor is connected to either d.c. or a.c. supply. The magnetic fluxes of the series field and armature produced by this current react with each other and hence produce rotation. The direction of rotation on a.c. supply is the same as on a d.c. supply. However, direction of rotation of a series motor can be reversed by reversing the current flow through either the armature or the field (but not through both). However, they are usually wound for operation in only one direction. Since universal motors are series-wound, they have high starting torque and a variable speed characteristic. They run at dangerously high speed at no-load and because of this, they are built into the device they drive.

Applications

Such motors are commonly used for the following devices:

1. sewing machines
2. vacuum cleaners
3. food mixers and blenders
4. hair driers
5. electric shavers
6. saws
7. projectors
8. portable power tools like drills etc.

23.10. Reluctance Synchronous Motor

It is a single-phase non-excited synchronous motor which does not require any d.c. power for exciting its rotor.

Construction

It consists of a stator which carries both the main and starting windings for producing a synchronously-rotating magnetic flux. If the stator has a centrifugal switch for cutting out the starting winding, we have split-phase type reluctance motor. If the stator is similar to that of a permanent-split capacitor-run motor, we have a capacitor-type reluctance motor. They are usually built in the low wattage range of 10-25 watt. The rotor is squirrel-cage type but of unsymmetrical construction as shown in Fig. 23.13. By removing some of the teeth of a normal squirrel-cage rotor, salient poles are produced which offer low reluctance to the stator flux and thereby become strongly-magnetised.

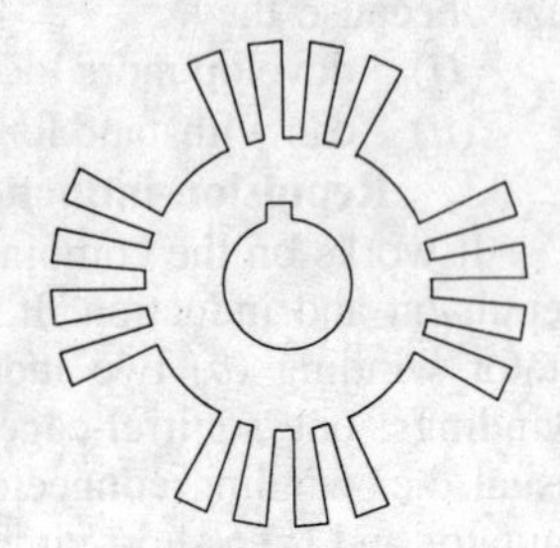

Fig. 23.13

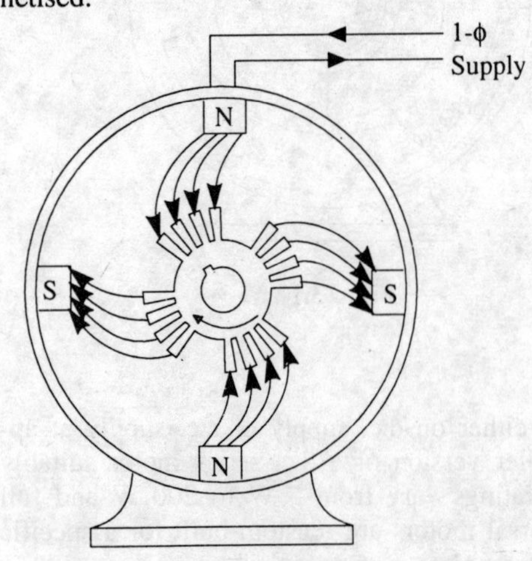

Fig. 23.14

Working

When single-phase supply is switched on, the reluctance motor starts and accelerates like a squirrel-cage motor. But as it approaches the synchronous speed, a point is reached when the rotor snaps into synchronism with the revolving stator flux. This pull-in action takes place due to the attraction between the revolving stator poles and the magnetised low-reluctance rotor poles (Fig. 23.14). This motor adjusts its torque angle for a change in load in a similar way to that described for a 3-phase synchronous motor (Art. 25.2). If the load is too high at the starting, the motor may not pull into synchronism and if already running, it may pull out of synchronism.

Applications

As compared to an equivalent induction motor, such motors have poorer torque, power factor and efficiency. However, despite these drawbacks, such motors are widely used for constant-speed applications, such as in:

1. recording instruments
2. timing devices
3. control apparatus
4. signalling devices
5. phonograph turntables
6. control apparatus etc.

23.11. Hysteresis Synchronous Motor

It is a single-phase non-excited synchronous motor having a non-salient pole rotor which has no d.c. excitation. Usually, they are made in the 2–4 watt range extending, in certain cases, upto 16 W.

Construction

It consists of a stator and a rotor.

(*a*) **Stator.** For working on the split-phase principle, the stator has two windings which remain connected to the single-phase supply continuously both at starting as well as during the running of the motor. Usually, shaded-pole principle is applied and the stator winding similar to that used in capacitor-type motor is used giving us capacitor shaded-pole hysteresis motor.

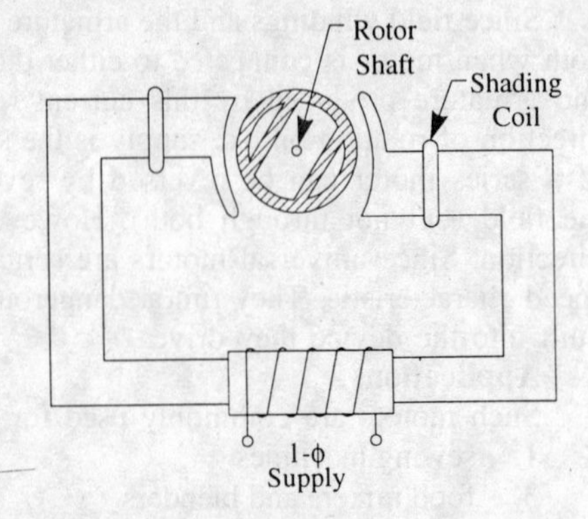

Fig. 23.15

(*b*) **Rotor.** As shown in Fig. 23.15, the rotor consists of two or more outer rings and cross-bars of a specially selected hard steel which has high retentivity and permeability *i.e.*, large hysteresis loop. As seen, it has no salient or projecting poles.

Another type of rotor design used is a hardened cast cobalt steel cylinder firmly mounted to the shaft with a non-magnetic support.

Working

When the stator is energised by the single-phase a.c. supply, a revolving flux is produced which moves past the rotor and induces current in the cross-bars which magnetises them permanently. Hence, motor starts as an induction motor. As the motor develops speed, hysteresis loss occurs in the rotor rings which produces large driving torque. The motor, therefore, quickly accelerates to its synchronous speed.

Applications

Since the rotor has no salient poles or windings of any sort, the motor is extremely quiet in operation and free from mechanical and magnetic vibrations. Hence, a hysteresis synchronous motor is ideally suited for driving

1. high-fidelity tape decks
2. phono turn-tables
3. precision audio equipment
4. electric clocks
5. various indicating devices

23.12. Motor Troubles

Common single-phase motor troubles as diagnosed by different human senses are summarized in Table 23.1 given below:

Table 23.1 : Motor Troubles

Sense	*Symptom*	*Possible Causes*	
		Mechanical	*Electrical*
Sight	motor will not start smoking	seized bearing, over-heated bearing	supply faulty, burnt out winding
Hearing	bearing rumble vibration	faulty bearing, installation fault or out-of-balance rotor	—
Touch	overheating	insufficiently or incorrectly lubricated bearing	continued overload or faulty winding
Smell	burning smell	over-heated bearing	burning winding

HIGHLIGHTS

1. A single-phase split-phase induction motor has a squirrel-cage rotor and a stator which has two windings. One is running winding and the other is starting winding which is disconnected from the supply by a centrifugal switch when motor reaches 70 to 80 per cent of its full-load speed.

2. Single-phase split-phase capacitor motors are of three types :

(*a*) **Capacitor-start Induction-run Motor**

It has a rotor and a stator with running and starting windings. However, starting winding has a capacitor in series with it and both are disconnected from the supply by a centrifugal switch when motor reaches about 75 per cent of its full-load speed. It has higher efficiency and power factor than split-phase induction motor.

(*b*) **Permanent-split Capacitor Motor**

It is similar to the above motor except that starting winding and capacitor are connected in the circuit *at all times.* It does not use any centrifugal switch. This motor is also called single-value capacitor-run motor.

(*c*) **Two-value Capacitor Motor**

It is similar to the above motor except that it uses two capacitors: one for starting and the other for running condition. It is also called capacitor-start-and-run motor.

3. Shaded-pole Motor

It works on the split-phase principle and consists of a squirrel-cage rotor and a stator with projecting poles. Here, shading coil acts as the starting coil.

4. Repulsion Motor

It uses the principle of repulsion for starting purposes *i.e.*, repulsion between like stator and rotor poles. It consists of (*a*) one stator winding connected to a single-phase a.c. supply (*b*) a rotor which is wound like a d.c. armature (*c*) commutator and (*d*) a set of brushes which are short-circuited.

5. Universal Motor

It is a series-wound motor which can be operated on either direct current or 1-phase a.c. supply at nearly the same speed. Constructionally, it is almost similar to a d.c. motor.

6. Reluctance Synchronous Motor

It is a single-phase non-excited synchronous motor which needs no rotor excitation. It consists of (*a*) a stator with starting and running windings and (*b*) a squirrel-cage rotor of unsymmetrical construction for varying the reluctance.

7. Hysteresis Motor

It is a single-phase non-excited synchronous motor having a stator and a rotor. The stator has shaded poles and the rotor is a smooth cobalt steel cylinder or has rings and cross-bars made of hard steel having large hysteresis loop.

OBJECTIVE TESTS—23

A. Fill in the following blanks.

1. A split-phase induction motor has two stator windings: main winding and .... winding.
2. Capacitor-start-and-run motors are also called ....-split motors.
3. The rotor of a shaded-pole motor rotates from the unshaded part of the pole to its .... part.
4. When brush axis of a repulsion motor lies along its pole axis, it develops .... torque.
5. Repulsion-induction motor has characteristics comparable to those of a d.c. .... motor.
6. A universal motor can be run on either d.c. or .... supply.
7. Vacuum cleaners usually use .... motors.
8. The rotor of a hysteresis synchronous motor has no .... poles.

B. Answer True or False.

1. Single-phase winding when fed from a 1-phase supply produces a pulsating flux.
2. Capacitor-start induction-run motor has a much higher starting torque than a split phase induction motor.
3. Shaded-pole motors have non-salient poles on the stator.
4. Repulsion motors have commutators and brushes which are connected to the a.c. supply.
5. Repulsion-start induction-run motor is the same as repulsion-induction motor.
6. A universal motor is nothing else than a smaller and modified version of a d.c. series motor.
7. Reluctance motor is a non-synchronous motor.
8. Hysteresis motor is extremely quiet in operation.

C. Multiple Choice Questions.

1. In a split-phase induction motor, the two stator windings
   (*a*) have equal R / X ratio
   (*b*) are mutually displaced by 90° electrical
   (*c*) draw only in-phase currents
   (*d*) draw equal currents
2. As compared to capacitor-start induction-run motors or split-phase motors, a permanent-split capacitor motor has lower
   (*a*) efficiency

(*b*) cost

(*c*) noise

(*d*) power factor

3. A universal motor has variable speed characteristics because it is

(*a*) series-wound

(*b*) compound-wound

(*c*) operated on both d.c. and a.c. supplies

(*d*) wholly laminated

4. The rotor of a hysteresis synchronons motor

(*a*) has salient poles

(*b*) has low permeability

(*c*) slips behind the stator flux

(*d*) is made of high-retentivity hard steel

5. A universal motors is idealy suited for driving

(*a*) food mixers

(*b*) ceiling fans

(*c*) electric clocks

(*d*) photocopying machines

6. Vacuum cleaners generally use   motor.

(*a*) reluctance

(*b*) repulsion

(*c*) shaded-pole

(*d*) universal

7. Which of the following single-phase motor is available with a speed as low as one revolution per month?

(*a*) shaded-pole

(*b*) reluctance

(*c*) hysteresis

(*d*) universal

8. Hysteresis synchronous motor is used for high fidelity tape decks because it

(*a*) is extremely quiet in operation

(*b*) is free from mechanical and magnetic vibrations

(*c*) has no revolving magnetic flux

(*d*) both (*a*) and (*b*)

(*e*) both (*b*) and (*c*)

ANSWERS

A. 1. starting/auxiliary 2. permanent 3. shaded 4. no 5. compound 6. a.c. 7. universal 8. salient / projecting

B. 1. T 2. T 3. F 4. F 5. F 6. T 7. F 8. T

C. 1. *b* 2. *c* 3. *a* 4. *d* 5. *c* 6. *d* 7. *a* 8. *d*

24 ALTERNATORS

24.1. Basic Principle and Construction

A.C. generators or alternators (as they are usually called) operate on the same fundamental principle of electromagnetic induction as d.c. generators. They also consist of an armature winding and a magnetic field. But there is one important difference between the two. Whereas in d.c. generators, the *armature rotates* and field system is stationary, the arrangement in alternators is just the reverse. In their case, standard construction consists of armature windings mounted on a stationary element called '*stator*' and field windings on a rotating element called '*rotor*'. The details of construction are shown in Fig. 24.1.

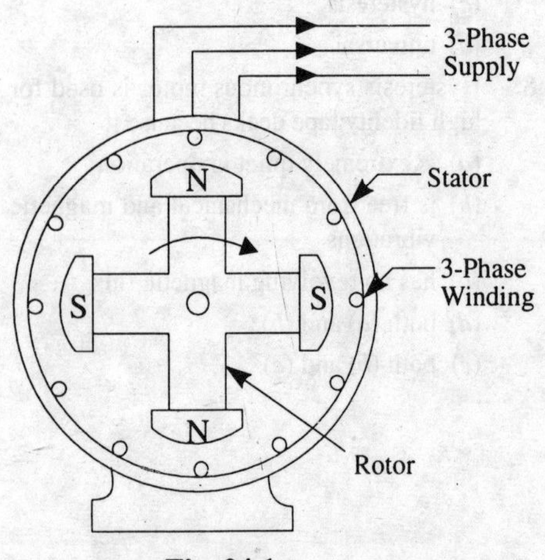

Fig. 24.1

(*a*) **Stator**

It consists of a cast-iron frame which supports the laminated armature core having slots on its inner-periphery for housing the 3-phase winding.

(*b*) **Rotor**

These are of two types:

(*i*) *Salient (or projecting) pole type*

It is like a flywheel which has a large number of alternate North and South poles bolted to it as shown in Fig. 24.1 or in more details in Fig. 24.2. The magnetic wheel is made of cast iron or steel of good magnetic quality. The magnetic poles are excited by a small d.c. generator mounted on the shaft of the alternator itself.

Such rotors are used in low and medium-speed alternators which are characterised by their large diameters and short axial lengths. Alternator driven by diesel or gas engines and gas turbines have salient pole rotors.

(*ii*) *Smooth Cylindrical Type*

It consist of a smooth solid forged-steel cylinder having a number of slots milled out at intervals along the outer periphery for accommodating field coils as shown in Fig. 24.3. Two or four regions corresponding to the central polar areas are left unslotted. The central polar areas are surrounded by the field windings placed in slots. Obviously, in this case, the poles are non-salient *i.e.*, they do not project out from the surface of the rotor.

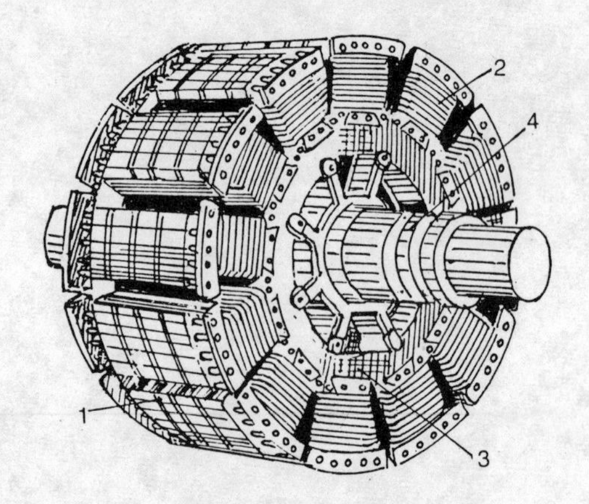

Fig. 24.2 : Different parts of the rotor are:
1. poles 2. pole coils 3. rotor core
4. slip-rings

Such rotors are used in steam turbine-driven alternators *i.e.*, turbo-alternators or turbo-generators which run at very high speeds and are characterised by their small diameters and very long axial lengths. Since steam turbines run at very high speeds, nearly all turbo-alternators are 2-pole machines.

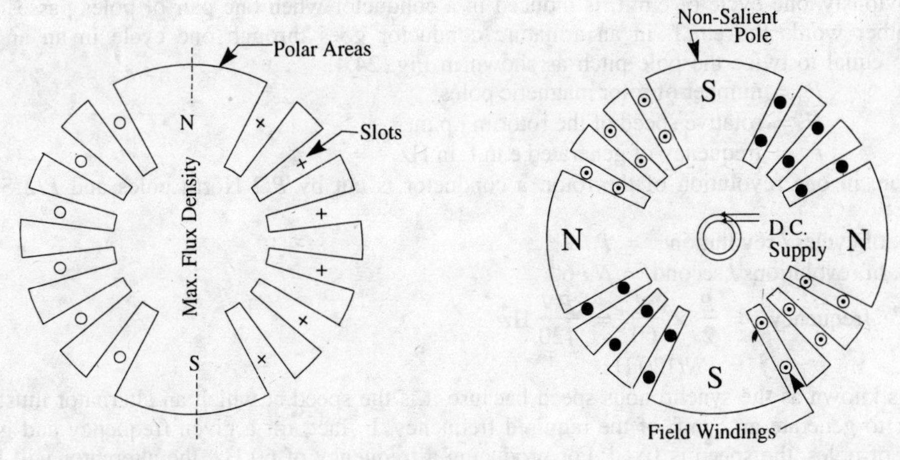

Fig. 24.3

24.2. Principle of Operation

When the rotor is rotated by the prime-mover, the stator winding or conductors are cut by the magnetic flux of the rotor poles. Hence, an e.m.f. is induced in the stator conductors. Because the rotor poles are alternately *N* and *S*, they induce an alternating e.m.f. in the stator conductors. The frequency of this induced e.m.f. is given by $f = PN/120$ (Art. 24.3) and its direction can be found by applying Fleming's Right-hand rule.

The e.m.f. generated in the stator conductors is taken out from the three leads connected to the stator winding as shown in Fig. 24.1.

24.3. Speed and Frequency

In an alternator, there exists a definite relationship between the rotational speed (*N*) of the rotor, the frequency (*f*) of the generated e.m.f. and the number of poles, *P*.

Consider an armature conductor marked *X* in Fig. 24.4 situated at the centre of a *N*-pole rotating in clockwise direction. Since the conductor is situated at the place of maximum flux density, it will have highest rate of flux cutting and hence will have maximum e.m.f. induced in it.

The direction of the induced e.m.f. is given by the Fleming's Right-hand rule. But while applying this rule, one should be careful to note that the thumb indicates the direction of the motion of the *conductor* relative to the *field*. To an observer stationed on the clockwise revolving poles, the conductor would seem to be rotating anti-clockwise. Hence, thumb should point to the left. The direction of the induced e.m.f. is downwards in a direction at right angles to the plane of the paper.

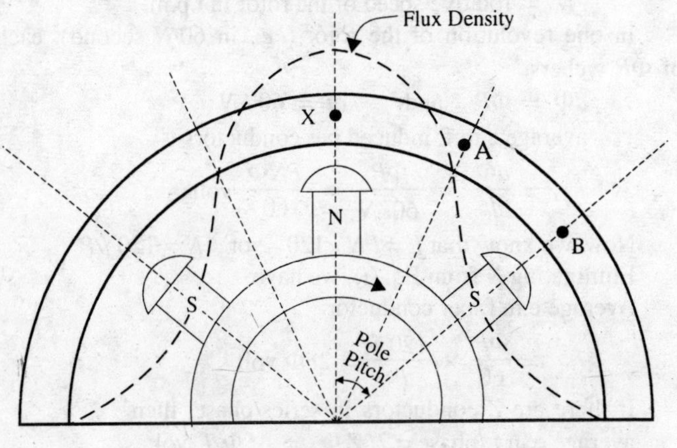

Fig. 24.4

When the conductor is in inter-polar gap as at *A* in Fig. 24.4, it has minimum e.m.f. induced in it because flux cutting rate is minimum there. Again, when it is at the centre of a *S*-pole, it has maximum e.m.f. induced in it because flux density at *B* is maximum. But the direction of the e.m.f. when conductor is over a *N*-pole is opposite to that when it is over a *S*-pole.

Obviously, one cycle of e.m.f. is induced in a conductor when one pair of poles passes over it. In other words the e.m.f. in an armature conductor goes through one cycle in an angular distance equal to twice the pole pitch as shown in Fig. 24.4.

Let P = number of rotor magnetic poles
N = rotative speed of the rotor in r.p.m.
f = frequency of generated e.m.f. in Hz

Then, in one revolution of the rotor, a conductor is cut by $P/2$ North poles and $P/2$ South poles.

No. of cycles / revolution $= P/2$
No. of revolutions / second $= N/60$

$$\therefore \quad \text{frequency} = \frac{P}{2} \times \frac{N}{60} = \frac{PN}{120} \text{ Hz}$$

or $$f = PN/120 \text{ Hz}$$

N is known as the synchronous speed because it is the speed at which an alternator must run in order to generate an e.m.f. of the required frequency. In fact, for a given frequency and given number of poles, the speed is fixed. For producing a frequency of 60 Hz, the alternator will have to run at the following speeds:

No. of poles	2	4	6	12	24	36
Speed (r.p.m.)	3600	1800	1200	600	300	200

Referring to the above equation, we get $P = 120\, f/N$.

It is clear from the above that because of slow rotative speeds of engine-driven alternators, their number of poles is much greater as compared to that of the turbo-generators which run at very high speeds.

24.4. Equation of Induced E.M.F.

Let Z = No. of conductors or coil sides in series / phase
= $2T$ where T is the No. of coils or turns per phase
(remember one turn or coil has two sides)
P = No. of rotor poles
f = frequency of induced e.m.f. in Hz
Φ = flux / pole in webers
N = rotative speed of the rotor in r.p.m.

In one revolution of the rotor (*i.e.,* in $60/N$ second), each stator conductor is cut by a flux of ΦP webers.

$$\therefore \quad d\Phi = \Phi P \quad \text{and} \quad dt = 60/N$$

$\therefore$ average e.m.f. induced per conductor

$$= \frac{d\Phi}{dt} = \frac{\Phi P}{60/N} = \frac{PN\Phi}{60} \text{ volt} \qquad \ldots(i)$$

Now, we know that $f = PN/120$ or $N = 120 f/P$

Eliminating N from Eq. (*i*), we have

Average e.m.f. per conductor

$$= \frac{\Phi P}{60} \times \frac{120 f}{P} = 2f\Phi \text{ volt}$$

If there are Z conductors in series/phase, then

average e.m.f./phase $= 2fZ\Phi = 4f\Phi T$ volt

R.M.S. value of e.m.f. / phase $= 1.11 \times 2f\Phi Z = 2.22 f Z \Phi$ volt

$= 4.44 f \Phi T$ volt*

If the alternator is star-connected (as is usually the case), then the line voltage is $\sqrt{3}$ times the phase voltage (as found from the above formula).

Example 24.1. *Calculate the synchronous speed of a four-pole 50 Hz alternator.*

Solution. $f = \dfrac{PN}{120}$ or $N = \dfrac{120 f}{P}$

$\therefore \quad N = 120 \times 50 / 4 = \mathbf{1500\ r.p.m.}$

Example 24.2. *What is the frequency of voltage generated by an alternator having 10- poles and rotating at 720 r.p.m.?*

Solution. $f = PN / 120 = 10 \times 720 / 120 = \mathbf{60\ Hz.}$

Example 24.3. *A 3-phase, 16-pole alternator has a star-connected winding with 144 slots and 10 conductors per slot. The flux per pole is 30 mWb sinusoidally distributed and the speed is 375 r.p.m. Find the frequency, the phase and the line e.m.f.*

Solution. Formula used

$E = 2.22 f Z \Phi$ volt–per phase

$f = PN / 120 = 16 \times 375 / 120 = \mathbf{50\ Hz}$

No. of slots per phase $= 144 / 3 = 48$

No. of conductors / slot $= 10$

No. of conductors / phase, $z = 48 \times 10 = 480$

e.m.f. per phase $= 2.22 \times 50 \times 30 \times 10^{-3} \times 480 = \mathbf{1{,}600\ V}$

For a star-connection, $V_L = \sqrt{3}\, V_{ph}$

Line voltage $= \sqrt{3} \times 1600 = \mathbf{2770\ V}$

Example 24.4. *A 3-phase water-wheel generator is rated at 100 MVA. unity p.f., 11 kV, star-connected, 50 Hz, 120 r.p.m. Determine:*

(i) the number of poles (ii) the kW ratings (iii) the current rating (iv) the input at rated kW load if the efficiency is 97 per cent (excluding the field loss) (v) prime-mover torque applied to the generator shaft. **(Electrical Science, AMIE Winter 1994)**

Solution. *(i)* $f = PN / 120$ or $P = 120 \times 50 / 120 = \mathbf{50\ Hz}$

(ii) Since the power factor is unity, the kW rating is

$= 100 \times 1.0 = 100\text{ MW} = \mathbf{100{,}000\ kW}$

(iii) Current rating is given by the relation

$= 100 \times 10^6 = \sqrt{3} \times 11{,}000 \times I_L, \quad \therefore \quad I_L = \mathbf{5{,}250\ A}$

(iv) Input $= 100 / 0.97 = \mathbf{103.1\ MW}$

It also represents the output of the prime mover

(v) $T_{sh} = 9.55 \times 103.1 \times 10^6 / 120 = \mathbf{10^6\ N}$ (Art. 22.24)

Tutorial Problems No. 24.1

1. A 3-phase, star-connected, 2-pole alternator runs at 3600 r.p.m. If there are 500 conductors per phase in series on the armature winding and the sinusoidal flux per pole is 0.1 Wb, calculate the magnitude and frequency of the generated voltage. **[60 Hz; 11.5 kV]**

2. A 4-pole, 3-phase, star-connected alternator has 24 slots with 12 conductors per slot and the flux per pole is 0.1 Wb. Calculate the line e.m.f. generated when the alternator is driven at 1,500 r.p.m. **[1,785 V]**

24.5. Alternator on Load

As load on an alternator is varied, its terminal voltage V is also found to vary as in d.c. generators. This variation in V is due to the following reasons:

1. voltage drop due to armature resistance R_a ;

*It is exactly the same as e.m.f. equation of a transformer (Art. 21.6).

2. voltage drop due to armature leakage reactance X_L;
3. voltage drop due to armature reaction.

The drop due to armature reaction is theoretically accounted for by assuming a fictitious reactance X_a in the armature winding. The vector sum of X_L and X_a gives *synchronous reactance* X_S.

Hence, it can now be said that an alternator possesses (*i*) resistance R_a and (*ii*) synchronous reactance, X_S. Their *vector* sum gives synchronous impedance Z_S.

$$\therefore \qquad Z_S = \sqrt{R_a^2 + X_S^2}$$

When an alternator is on load, there is voltage drop due to R_a and X_S or due to Z_S alone. If V is the terminal voltage/phase on load and E the generated e.m.f./phase on no-load, then

$$\mathbf{E} = \mathbf{V} + \mathbf{I}_a (R_a + jX_S) = \mathbf{V} + \mathbf{I}_a \mathbf{Z}_S$$

24.6. Phasor Diagram of a Loaded Alternator

Let
E = no-load e.m.f. / phase
V = terminal voltage / phase
I_a = armature current / phase
ϕ = load p.f. angle
Z_S = armature synchronous impedance / phase

(*a*) **Unity Load p.f.**

In Fig. 24.5, V is taken as the reference vector. Current vector I_a is in phase with V. The voltage drop $I_a R_a$ is in phase with I_a whereas drop $I_a X_S$ is at right angles to it, their vector sum giving $I_a Z_S$. When $I_a Z_S$ is combined with V, we get E.

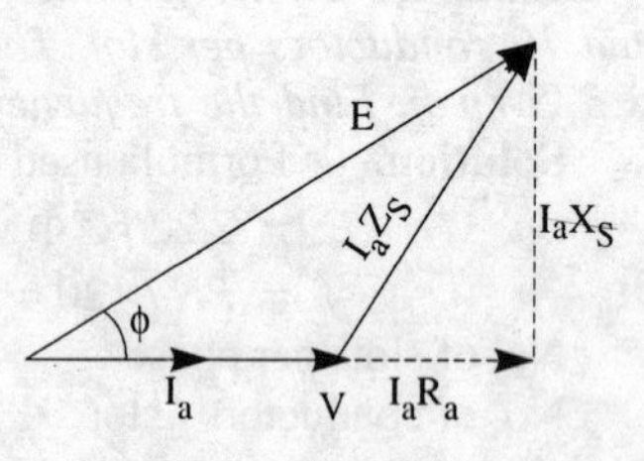

Fig. 24.5

(*b*) **Lagging Load p.f.**

In this case, I_a lags behind V by an angle ϕ. As usual, $I_a R_a$ is in phase with I_a whereas $I_a X_S$ is at right angles to it. When $I_a Z_S$ is combined with V, we get E as shown in Fig. 24.6 (*a*). Fig. 24.6 (*b*) shows the diagram when instead of V, I_a is taken as the reference vector.

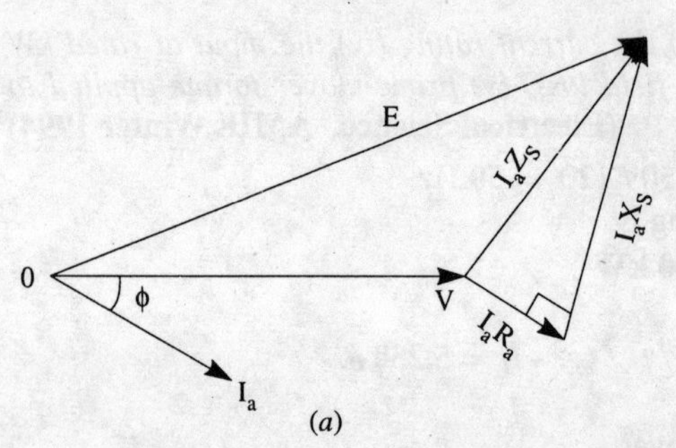

Fig. 24.6

(*c*) **Leading Load p.f.**

Such a case is shown in Fig. 24.7. Here, I_a leads V by ϕ. As usual, vector sum of V and $I_a Z_S$ gives E.

24.7. Voltage Regulation

The terminal voltage V of an alternator is found to vary with load and its power factor. Voltage regulation of an alternator is defined as the *rise* in voltage when full-load is removed divided by the rated terminal voltage

$$\% \text{ regulation 'up'} = \frac{E - V}{V} \times 100$$

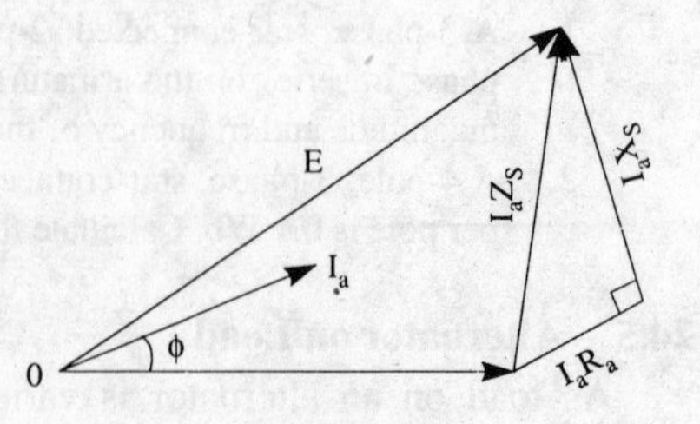

Fig. 24.7

Example 24.5. *A 500 VA, 1100 V, 50 Hz, Y-connected, 3-ϕ alternator has armature resistance/phase of 1.0 Ω and synchronous reactance / phase of 1.5 Ω. Find its voltage regulation for (a) unity p.f. (b) 0.9* *lagging p.f. and (c) 0.8 leading p.f. Also, calculate voltage regulation in each case.*

(Elect. Machines, Allahabad Univ. 1993)

Solution. $\sqrt{3}\, V_L I_L = 500{,}000$

$$I_L = 500{,}000 / \sqrt{3} \times 1100 = 262 \text{ A}$$

armature current, $I_a = 262$ A

$$I_a R_a = 262 \times 0.1 = 26.2 \text{ V}$$

$$I_a X_S = 262 \times 1.5 = \mathbf{393\ V}$$

$$V = V_L / \sqrt{3} = 1100 / \sqrt{3} = \mathbf{635\ V}$$

(*a*) **Unity p.f.**

As shown in Fig. 24.8 (*a*)

$$E = \sqrt{(V + I_a R_a)^2 + (I_a X_S)^2}$$

$$= \sqrt{(635 + 26.2)^2 + (393)^2} = \mathbf{769.2\ V}$$

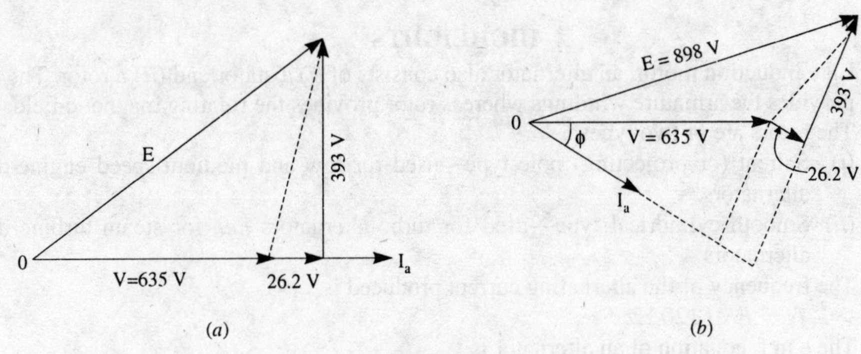

Fig. 24.8

$$\% \text{ regn.} = \frac{769.2 - 635}{635} \times 100 = \mathbf{21.1\%}$$

(*b*) **Lagging p.f.**

This case is shown in Fig. 24.8(*b*)

$$\cos\phi = 0.9\ \phi = 25.83°;\ \sin\phi = \sin 25.83° = 0.436$$

As seen from Fig. 24.6 (*b*)

$$E = \sqrt{(V\cos\phi + I_a R_a)^2 + (V\sin\phi + I_a X_S)^2}$$

$$= \sqrt{(635 \times 0.9 + 26.2)^2 + (635 \times 0.436 + 393)^2} = \mathbf{898\ V}$$

$$\% \text{ regn.} = \frac{898 - 635}{635} \times 100 = \mathbf{41.4\%}$$

(*c*) **Leading p.f.**

As seen from Fig. 24.9

$$E = \sqrt{V\cos\phi + I_a R_a)^2 + (V\sin\phi - I_a X_S)^2}$$

$$= \sqrt{(635 \times 0.8 + 26.2)^2 + (635 \times 0.6 - 393)^2}$$

$$= \mathbf{534\ V}$$

$$\% \text{ regn.} = \frac{534 - 635}{635} \times 100 = \mathbf{-15.9\%}$$

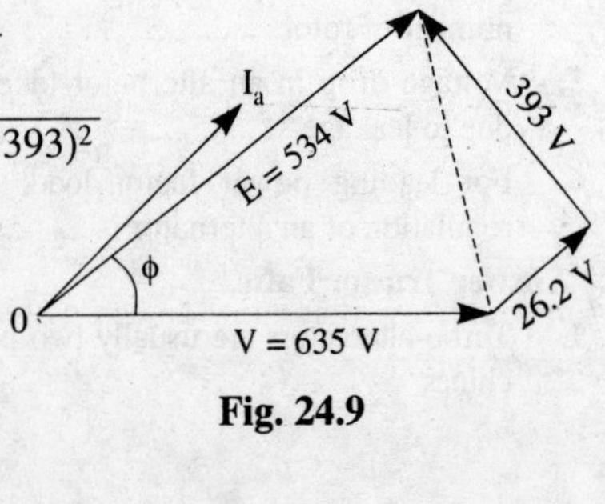

Fig. 24.9

Example 24.6. *A 3-phase star-connected alternator is rated at 1600 kVA, 13.5 kV. The armature effective resistance* *and synchronous reactance are 1.5 Ω and 30 Ω respectively per phase. Calculate the percentage regulation for a load of 1280 kW at p.f. of 0.8 leading.*

(Electrical Science, AMIE Winter 1993)

Solution. Let us first find the phase current of the star-connected alternator for a load of 1280 kW at p.f. of 0.8 leading.

$$1280 \times 10^3 = \sqrt{3} \times 13{,}500 \times I_L \times 0.8, \quad \therefore I_L = 68.4 \text{ A}$$

Since the alternator is star-connected, it also represents the phase current. Hence, armature phase current $I_a = 68.4$ A

$$I_a R_a = 68.4 \times 1.5 = 103 \text{ V}, \quad I_a X_S = 68.4 \times 30 = 2050 \text{ V},$$

$$V = 13500 / \sqrt{3} = 7795 \text{ V}.$$

As seen from Fig. 24.10.

$$E = \sqrt{(7795 \times 0.8 + 103)^2 + (7795 \times 0.6 - 2050)^2} = 6910 \text{ V}$$

$$\therefore \quad \% \text{ regn.} = \frac{6910 - 7795}{7795} = -0.1135 \text{ or } \mathbf{-11.35\%}.$$

HIGHLIGHTS

1. Like induction motor, an alternator also consists of (*i*) a stator and (*ii*) a rotor. The stator provides the armature windings whereas rotor provides the rotating magnetic field.
2. The rotors are of two types:
   (*i*) Salient (or projecting) pole type—used for low and medium-speed engine-driven alternators.
   (*ii*) Smooth cylindrical type—used for turbo-alternators *i.e.*, for steam-turbine driven alternators.
3. The frequency of the alternating current produced is
   $$f = PN / 120 \text{ Hz}$$
4. The e.m.f. equation of an alternator is
   $$E = 2.22 f Z \Phi \text{ volt / phase}$$
   $$= 4.44 f \Phi T \text{ volt / phase}$$
5. The voltage equation of the alternator is
   $$\mathbf{E} = \mathbf{V} + \mathbf{I}_a \mathbf{Z}_S$$

OBJECTIVE TESTS—24

A. Fill in the following blanks.

1. In an alternator, the field system rotates whereas armature is ....
2. Smooth cylindrical rotor is used in ....
3. Diesel-driven alternators have salient pole ....
4. The frequency of the e.m.f. generated by an alternator depends on its r.p.m. and number of rotors ....
5. Voltage drop in an alternator takes place due to leakage ....
6. For leading power factor load, voltage regulation of an alternator is ....

B. Answer True or False.

1. Turbo-alternators are usually two-pole machines.
2. A 50 Hz alternator will run at highest speed if wound for two poles.
3. Turbo-alternators are generally driven by gas turbines.
4. Terminal voltage of an alternator always decreases with increase in load irrespective of its power factor.
5. Load power factor affects the voltage regulation of an alternator.
6. Greater the number of poles, greater the e.m.f. generated in an alternator.

C. Multiple Choice Questions.

1. The frequency of a 2-pole alternator running at 3600 r.p.m. is .... Hz.
   (*a*) 50 (*b*) 60
   (*c*) 7200 (*d*) 120
2. When speed of an alternator is reduced by

half, the genera l e.m.f. will become

(*a*) one-half (*b*) twice

(*c*) four times (*d*) one-fourth

3. The magnitude of the e.m.f generated by an alternator depends on

(*a*) number of its poles

(*b*) rotor speed

(*c*) flux per pole

(*d*) all of the above

4. Voltage drop in an alternator when under load is due to

(*a*) armature resistance

(*b*) armature reactance

(*c*) armature reaction

(*d*) all of the above

5. For producing an e.m.f. of a given frequency, an alternator will run at greatest speed if wound for .... poles.

(*a*) 4 (*b*) 8

(*c*) 2 (*d*) 12

6. An alternator supplying a load with leading power factor always has .... voltage regulation.

(*a*) positive (*b*) negative

(*c*) unity (*d*) zero

— ANSWERS —

A. 1. stationary 2. turbo-alternators 3. rotors 4. poles 5. reactance 6. negative

B. 1. T 2. T 3. F 4. F 5. T 6. T

C. 1. *b* 2. *a* 3. *d* 4. *d* 5. *c* 6. *b*

25 SYNCHRONOUS MOTOR

25.1. Synchronous Motor–Construction

A synchronous motor is electrically identical with an alternator. A synchronous machine may be used, at least theoretically, as an alternator when driven mechanically or as a motor when driven electrically, just as in the case of d.c. machines (Art. 12.2).

As shown in Fig. 25.1, essential mechanical elements of such a motor are the same as those of an alternator *viz.*

(*i*) a stator–which houses 3-phase windings in the slots of the stator core.

(*ii*) a rotor–which has a number of alternately North and South poles.

The rotor can be either salient pole type (usually) or smooth cylindrical type (Art. 24.1). The rotor poles are excited by a small d.c. shunt generator mounted on the rotor shaft.

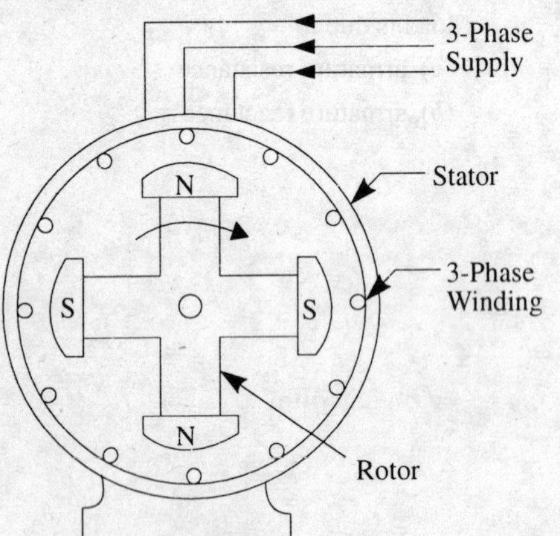

Fig. 25.1

25.2. Principle of Operation

One of the basic limitations of a synchronous motor is that it is not inherently self-starting as may be understood from the discussion to follow.

It was shown in Art. 22.3 that when a 3-phase winding is fed from a 3-phase supply, a resultant magnetic flux is produced whose magnitude is constant but which revolves synchronously. For all practical purposes it can be assumed as if two actual magnetic poles rotate around the stator. Fig. 25.2 shows two *stator* poles marked N_S and S_S revolving synchronously in clockwise direction. With the rotor position as shown in the figure, suppose the stator poles are at that instant at points *A* and *B*. The two similar poles N_R (of rotor) and N_S (of stator) poles as well as S_R and S_S will repel each other, with the result that the rotor tends to rotate in the anti-clockwise direction [Fig. 25.2 (*a*)]. But half a period later [Fig. 25.2 (*b*)], stator poles, having travelled half-way around, interchange their positions *i.e.,* N_S goes to point *B* and S_S to point *A*. During this period, the rotor has hardly moved its original position. Under these conditions, N_S attracts S_R and S_S attracts N_R. Hence, rotor tends to rotate clockwise (which is just the

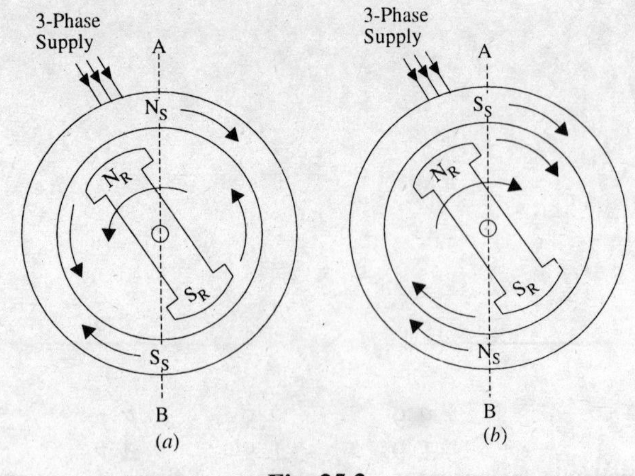

Fig. 25.2

opposite of the condition half a period earlier). So, we find that due to continuous and rapid rotation of stator poles, rotor is subjected to a rapidly reversing torque. In other words, rotor is subjected to a torque which tends to move it first in one direction and then in the opposite. Owing to its large moment of inertia, rotor cannot respond to such quickly-reversing torques with the result that it remains stationary. Now, consider the condition shown in Fig. 25.3 (*a*). The stator and rotor poles attract each other. Within half a period, stator poles will inter-change their position as shown in Fig. 25.3 (*b*). Suppose that, by some means, the rotor poles are also made to rotate through half a revolution during the same period as shown in Fig. 25.3 (*b*). Then, again the stator and rotor poles will attract each other. Obviously, if the rotor poles also shift their positions along with the stator poles, they will continuously experience a undirectional torque.

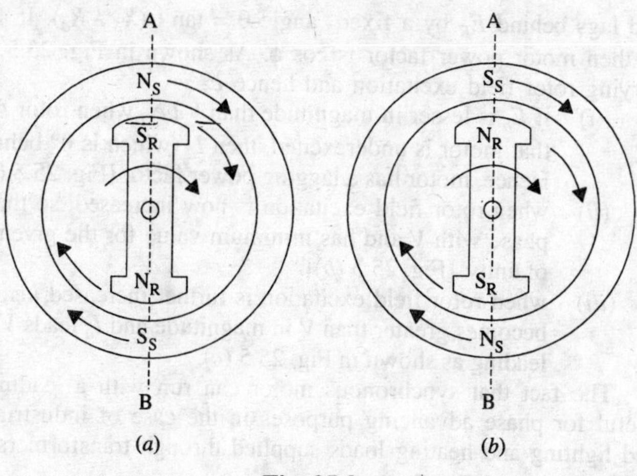

Fig. 25.3

One method of starting the synchronous motor is as under:

The rotor is speeded up by another motor (called pony motor) to its synchronous or near synchronous speed. Then, the exciting current of the rotor poles is switched on. The moment rotor poles are excited, they get magnetically locked into position with the stator poles of opposite polarity *i.e.*, the rotor poles are engaged with the stator poles and both run synchronously in the same direction. It is because of this inter-locking of rotor and stator poles that the motor has either to run at synchronous speed or not at all.

25.3. Making Synchronous Motor Self-starting

The motor can be made self-starting by providing squirrel-cage winding (also called damper winding) consisting of Cu bars embeded in the *rotor pole-shoes* and short-circuited at both ends (Fig. 25.1). Obviously, such a motor will start readily, acting as a 3-phase induction motor during the starting period.

The supply voltage is switched on to the stator winding while *the rotor pole field circuits is left open.* The synchronously rotating stator flux will produce e.m.f. in the squirrel-cage rotor bars as discussed in Art. 22.4. The motor starts up as an induction motor but when it reaches nearly 90 per cent of its synchronous speed, the rotor field circuit is closed. The rotor poles are magnetised and get immediately interlocked with the stator poles thus pulling the motor into synchronism (Fig. 25.4).

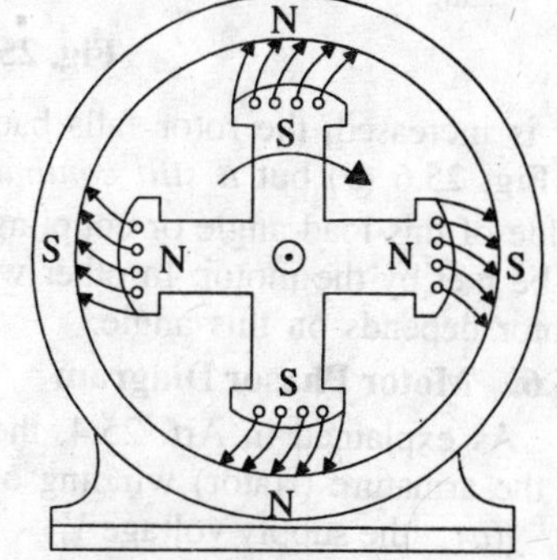

Fig. 25.4

25.4. Characteristics of a Synchronous Motor

One of the most important property of this motor is that it can be made to operate at a power factor *varying from lagging to a leading one.* This peculiar behaviour of the motor can be understood by noting that, as in the armature of a d.c. motor, there are two opposing voltages in the stator (or armature) winding of a synchronous motor. The two voltages are:

(*a*) the supply voltage V;

(*b*) the back e.m.f. E_b induced in the stator winding by the rotor pole flux.

The net voltage across the stator winding is the vector difference of V and E_b. If $V - E_b = E_R$ and Z_S is the impedance of the stator winding, then stator winding current is $I_a = E_R / Z_S$

and lags behind E_R by a fixed angle $\theta = \tan^{-1}(X_S / R_a)$. If ϕ is the phase angle between V and I_a, then motor power factor is $\cos\phi$. As shown in Fig. 25.5, the power factor can be varied by varying rotor field excitation and hence E_b.

(*i*) if E_b is lesser in magnitude than V *i.e.*, when rotor field excitation is weak which means that motor is underexcited, then I_a (which is θ° behind E_R) lags behind V by an angle ϕ. Hence, motor has a lagging power factor [Fig. 25.5 (*a*)].

(*ii*) when rotor field excitation is now increased, so that E_b equals V in magnitude, I_a is in phase with V and has minimum value for the given load. The motor has a power factor of unity. [Fig. 25.5 (*b*)].

(*iii*) when rotor field excitation is further increased *i.e.*, when motor is overexcited, then E_b becomes greater than V in magnitude and I_a leads V so that motor power factor becomes leading as shown in Fig. 25.5 (*c*).

The fact that synchronous motor can run with a leading power factor makes it extremely useful for phase advancing purposes in the case of industrial loads driven by induction motors and lighting and heating loads supplied through transformers (Art. 20.11).

25.5. Motor on Load

As explained in Art. 25.2, the stator and rotor poles of a running synchronous motor are magnetically inter-locked and hence both run with the same synchronous speed. But the two poles are not in phase with each other even when motor runs on no-load. There is a small phase difference between the two poles, with rotor pole trailing behind stator pole *even though running synchronously* as shown in Fig. 25.6 (*a*). This phase angle is essential for developing motor torque sufficient enough to meet no-load losses.

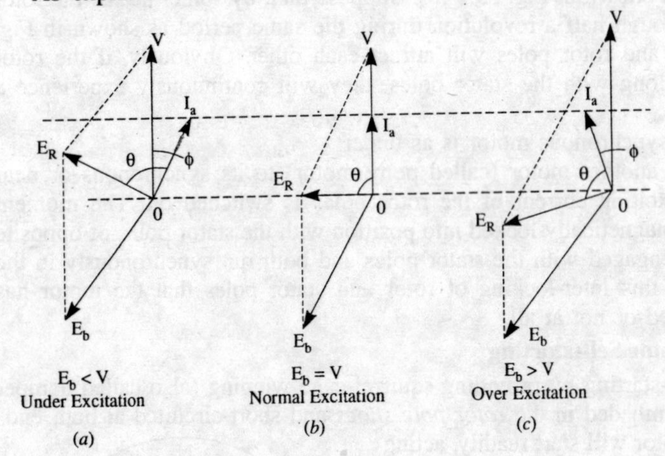

Fig. 25.5

As the load on the motor is increased, the rotor falls back in *phase* (but not in speed) by a greater angle α as shown in Fig. 25.6 (*b*) but *it still continues to run synchronously.* The value of this load angle or coupling angle α depends on the load to be met by the motor. In other words, torque developed by the motor depends on this angle.

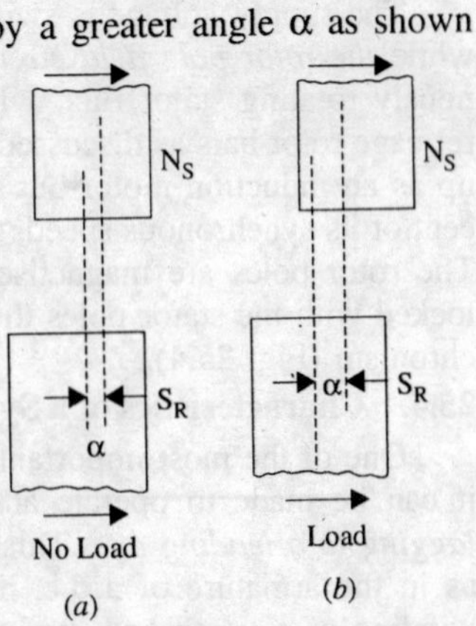

25.6. Motor Phasor Diagram

As explained in Art. 25.4, there are *two* opposing voltages in the armature (stator) winding of a synchronous motor:

(*a*) the supply voltage V;

(*b*) back e.m.f. E_b induced in the stator winding by the synchronously-revolving pole flux of the rotor.

N_S N_S S_R α S_R No Load (*a*) Load (*b*)

Fig. 25.6

This E_b depends on the rotor speed and flux. Since rotor runs at a constant speed, its value depends on flux produced by rotor poles *i.e.*, on the value of rotor excitation.

The net voltage / phase in stator windings is

$$E_R = V - E_b \quad \text{—vector difference}$$

If R_a is stator resistance / phase, X_S stator synchronous reactance / phase, Z_S, stator synchronous impedance/phase, then

$$Z_S = \sqrt{R_a^{\,2} + X_a^{\,2}}$$

Stator or armature current/phase is $I_a = E_R / Z_S$.

This current I_a lags behind E_R by a fixed angle $\theta = \tan^{-1}(X_S/R_a)$ in the phasor diagram of Fig. 25.7 (*a*) which is similar to Fig. 25.5 (*a*) tought turned 90° clockwise. If $R_a = 0$, then $\tan\theta = X_S/R_a = X_S / 0 = \infty$, hence $\theta = 90°$.

In practice, however, $R_a \neq 0$ though negligible as compared to X_S and so $\theta \cong 85°$ or so.

Fig. 25.7(*a*), in fact, represents the no-load condition of an under-excited motor* *i.e.*, one in which $E_b \angle V$. If ϕ is the angle between V and I_a, motor power factor is $\cos\phi$. Obviously, motor power input per phase = $V I_a \cos\phi$.

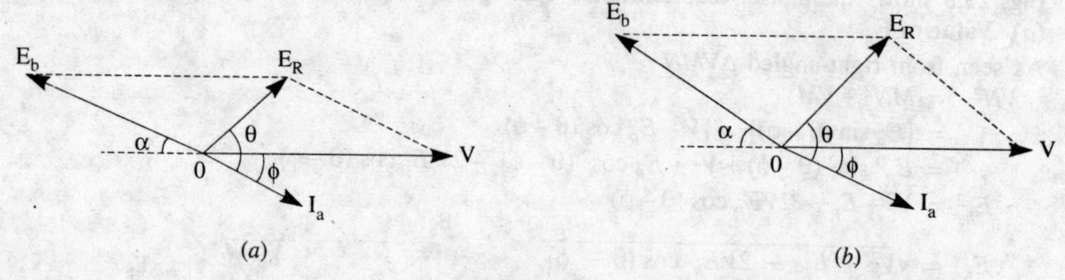

Fig. 25.7

When load is applied to the motor, its rotor further falls behind the stator poles in phase as shown in Fig. 25.7 (*b*). Consequently, vector for E_b further falls behind vector for V (you should imagine the vectors in Fig. 25.7 to be revolving in anti-clockwise direction) though its magnitude remains unaffected. The angle α in Fig. 25.7 (*b*) is greater than its value in Fig. 25.7(*a*). It is seen that magnitude of E_R and hence of I_a is increased with the result that now motor draws more power in order to supply the extra load put on it.

The mechanical power developed by the motor is

P_m = back e.m.f. × stator current × cosine of the angle between the two

$= E_b I_a \cos(\alpha - \phi)$ per phase.

25.7. Power Stages in a Synchronous Motor

We will consider the phase values.

Power input / phase = $V I_a \cos\phi$

Out of this, some power goes to meet armature Cu loss $I_a^2 R_a$ and the rest is converted into gross mechanical power in armature *i.e.*, $P_m = E_b I_a \cos(\alpha - \phi)$. Out of it, some power is used for meeting iron, friction and excitation losses and the balance is available as output power as shown to below:

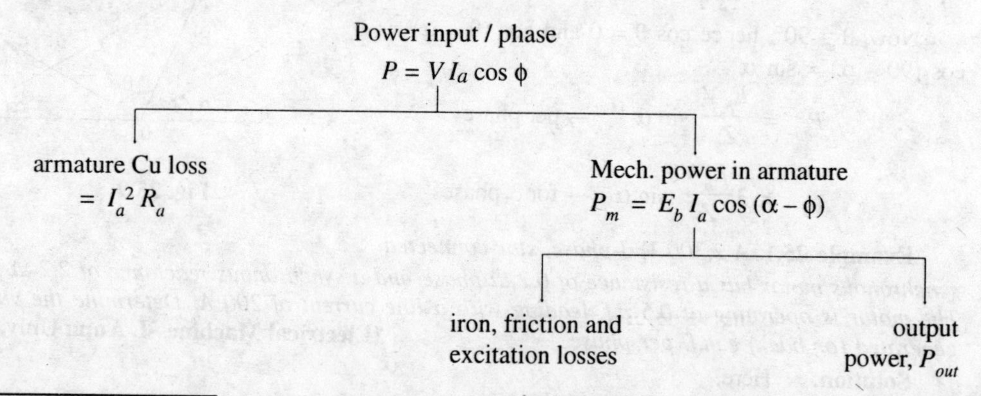

*If $E_b > V$, then motor is said to be overexcited and draws a *leading* current. If $E_b = V$, motor draws in-phase current.

Different power stages in a synchronous motor are as under:

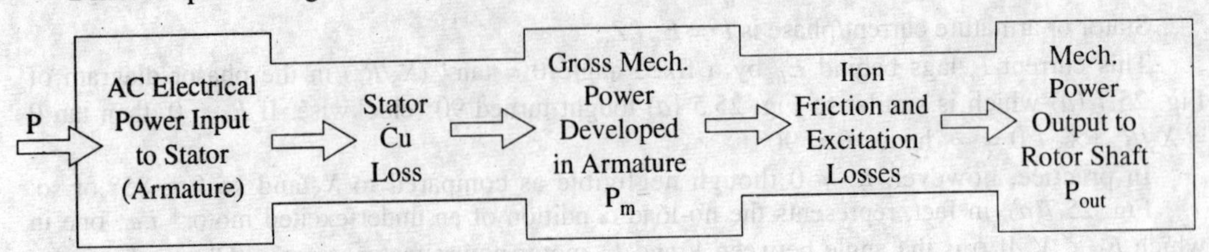

25.8. Values of E_b and E_R

Fig. 25.8 shows the phasor diagram of an under-excited motor running loaded.

(*a*) **Value of E_b**

As seen from right-angled ΔVMN

$$VN^2 = MN^2 + VM$$
$$= [E_R \sin(\theta - \phi)]^2 + [V - E_R \cos(\theta - \phi)$$
$$= E_R^2 \sin^2(\theta - \phi) + V^2 + E_R \cos^2(\theta - \phi) - 2V E_R \cos(\theta - \phi)$$
$$E_b^2 = V^2 + E_R^2 - 2VE_R \cos(\theta - \phi)$$

or
$$E_b = \sqrt{V^2 + E_{R^2} - 2VE_R \cos(\theta - \phi)}$$

(*b*) **Value of E_R**

Again, with reference to Fig. 25.8,

$$E_R = \sqrt{OM^2 + MN^2}$$
$$= \sqrt{(V - E_b \cos\alpha)^2 + (E_b \sin\alpha)^2}$$
$$= \sqrt{V^2 + E_b^2 - 2V E_b \cos\alpha}$$

Fig. 25.8

The angle between V and E_R is $NOM = (\theta - \phi)$

$$\tan(\theta - \phi) = \tan\delta = NM/OM$$

$$(\theta - \phi) = \tan^{-1}\left(\frac{E_b \sin\alpha}{V - E_b \cos\alpha}\right)$$

... Fig. 25.8

25.9. Mechanical Power Developed by Motor

With reference to Fig. 25.9, it can be proved that mechanical power / phase developed by a synchronous motor is given by

$$P_m = \frac{E_b V}{Z_S} \cos(\theta - \alpha) - \frac{E_b^2}{Z_S} \cos\theta$$

Now, $\theta \cong 90°$, hence $\cos\theta = 0$ and $\cos(\theta - \alpha) =$ $\cos(90 - \alpha) = \sin\alpha$

$$\therefore \quad P_m = \frac{E_b V}{Z_S} \sin\alpha \quad \text{— per phase}$$

$$= 3\frac{E_b V}{Z_S} \sin\alpha \quad \text{— for 3-phases}$$

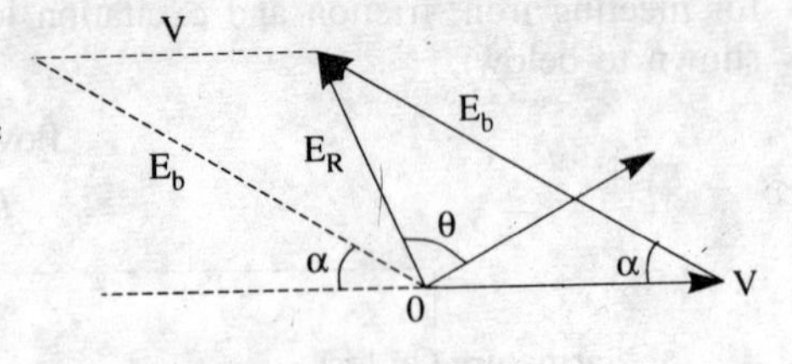

Fig. 25.9

Example 25.1. *A 2,300 V, 3-phase, star-connected synchronous motor has a resistance of 0.2 Ω/phase and a synchronous reactance of 2.2 Ω/phase. The motor is operating at 0.5 p.f. leading with a line current of 200 A. Determine the value of generated (or back) e.m.f. per phase.* **(Electrical Machines-I, Anna Univ. 1991)**

Solution. Here,

$$\phi = \cos^{-1}(0.5) = 60° \text{ (lead)}$$
$$\tan\theta = X_S / R_a = 2.2 / 0.2 = 11$$

$\therefore \quad \theta = \tan^{-1}(11) = 84.8°$

$\therefore \quad \theta + \phi = 84.8° + 60° = 144.8°$

$\cos 144.8° = -\cos 35.2°$

$V / \text{phase} = 2300 / \sqrt{3} = 1328 \text{ V}$

$Z_S = \sqrt{R_a^2 + X_S^2} = \sqrt{0.2 + 2.2^2} = 2.21\ \Omega$

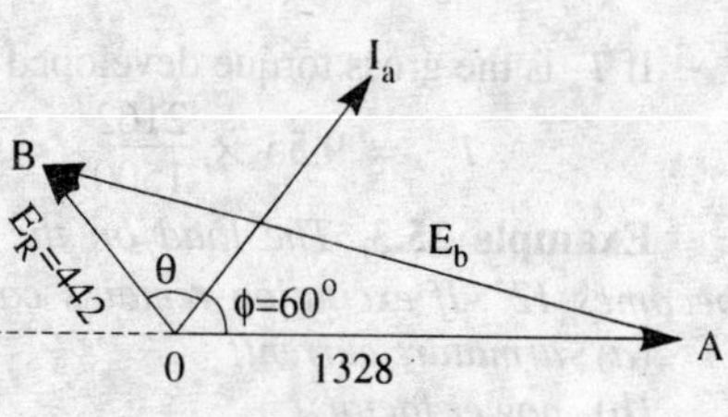

Fig. 25.10

Now, for *Y*-connection, phase current equals line current

$E_R = I_a Z_S = 200 \times 2.21 = 442 \text{ V}$

As seen from vector (or phasor) diagram of Fig. 25.10

$E_b^2 = V^2 + E_R^2 - 2V E_b \cos(\theta + \phi)$

$= 1328^2 + 442^2 - 2 \times 1328 \times 442 \times \cos 144.8°$

$E_b = \sqrt{1328^2 + 442^2 - 2 \times 1328 \times 442 \times (-\cos 35.2°)}$

$=$ **1708 volts / phase**

Example 25.2. *A 18.65 kW, 220 V, 50 Hz, 4-pole, Y-connected synchronous motor is running with a light load. The load angle is 4° and back e.m.f. generated per phase is 110 V. If armature resistance/phase is 0.1 Ω and synchronous reactance/phase is 1.5 Ω, find*

(*a*) *resultant armature voltage / phase;*
(*b*) *armature current / phase;*
(*c*) *internal angle θ;*
(*d*) *p.f. angle φ and p.f.;*
(*e*) *power input to motor ;*
(*f*) *gross torque developed by motor.*

Solution. Here, $\alpha = 4°$, $V = 220 / \sqrt{3} = 127$ V, $E_b = 110$ V

(*a*) $E_R = \sqrt{V^2 + E_b^2 - 2V E_b \cos\alpha}$

$= \sqrt{127^2 + 110^2 - 2 \times 127 \times 110 \times \cos 4°}$

$=$ **18.9 V per phase**

(*b*) $Z_S = \sqrt{R_a^2 + X_S^2} = \sqrt{0.1^2 + 1.5^2} \cong 1.5\ \Omega$

$I_a = E_R / Z_S = 18.9 / 1.5 =$ **12.6 A**

(*c*) $\tan\theta = X_S / R_a$, $\theta = \tan^{-1}(1.5 / 0.1) =$ **86.2°**

The phasor diagram is shown in Fig. 25.11.

(*d*) As seen from Art. 25.8,

$$(\theta - \phi) = \tan^{-1}\left(\frac{E_b \sin\alpha}{V - E_b \cos\alpha}\right)$$

$$= \tan^{-1}\left(\frac{110 \times \sin 4°}{127 - 110 \cos 4°}\right)$$

$= \tan^{-1}(0.4435) = 23.9°$

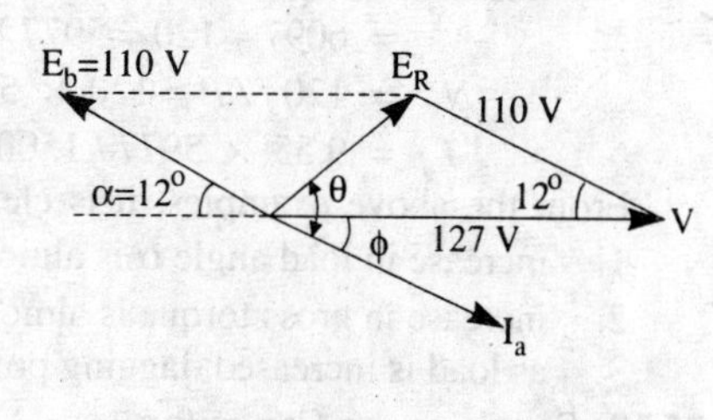

Fig. 25.11

$\therefore \quad \phi = \theta - 23.9° = 86.2° - 23.9° =$ **62.3°**

power factor $= \cos 62.3° =$ **0.46 (lag)**

(*e*) $P_{in} = \sqrt{3}\, V_L I_L \cos\phi$

$= \sqrt{3} \times 220 \times 12.6 \times \cos 62.3° =$ **2232 W**

(*f*) Total armature Cu loss

$= 3 I_a^2 R_a = 3 \times 12.6^2 \times 0.1 = 48 \text{ W}$

Internal power developed

$= 2232 - 48 = 2184 \text{ W}$

$N_S = 120f / P = 120 \times 50 / 4 = 1500 \text{ r.p.m.}$

If T_g is the gross torque developed in newton-metre (N-m), then

$$T_g = 9.55 \times \frac{2162}{1500} = \mathbf{13.9\ N\text{-}m.}$$

Example 25.3. *The load on the motor of example 25.2 is increased so that torque angle becomes 12°. If excitation remains constant, calculate*

(*a*) *armature current;*
(*b*) *power factor;*
(*c*) *total power input to the motor;*
(*d*) *gross torque developed by the motor.*

Solution. Now, $\alpha = 12°$, $\cos\alpha = 0.978$, $\sin\alpha = 0.208$

From example 25.2, we have

$$Z_S = 1.5\ \Omega;\ \theta = 86.2°$$
$$V = 127\ \text{V / phase and } E_b = 110\ \text{V / phase}$$

(*a*) $E_R = \sqrt{V^2 + E_b^{\,2} - 2VE_b\cos\alpha}$

$= \sqrt{127^2 + 110^2 - 2 \times 127 \times 110 \times 0.987} = 30\ \text{V}$

$I_a = E_R / Z_S = 30 / 1.5 = \mathbf{20\ A}$

(*b*) $\tan(\theta - \phi) = \dfrac{E_b \sin\alpha}{V - E_b\cos\alpha}$

$= \dfrac{110 \times 0.208}{127 - 110 \times 0.978} = 1.167$

$(\theta - \phi) = \tan^{-1}(1.167) = 49.4°$

$\therefore \quad \phi = 86.2° - 49.4° = 36.8°$

power factor $= \cos\phi = \cos 36.8° = \mathbf{0.8\ (lag)}$

Phasor diagram is shown in Fig. 25.12

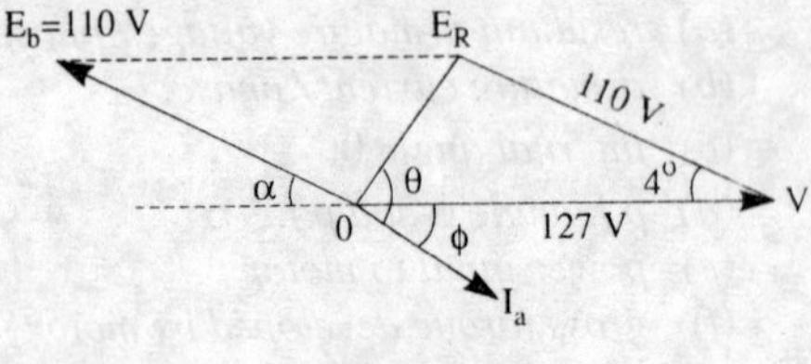

Fig. 25.12

(*c*) $P_{in} = \sqrt{3}\, V_L I_L \cos\phi$

$= \sqrt{3} \times 220 \times 20 \times 0.8 = \mathbf{6097\ W}$

(*d*) Total armature Cu loss

$= 3 I_a^{\,2} R_a = 3 \times 20^2 \times 0.1 = 120\ \text{W}$

$\therefore$ internal power developed

$= 6097 - 120 = 5977\ \text{W}$

$N_S = 120 f / P = 120 \times 50 / 4 = 1500\ \text{r.p.m.}$

$T_g = 9.55 \times 5977 / 1500 = \mathbf{38\ N\text{-}m.}$

From the above examples, it is clear that

1. increase in load angle α is almost directly proportional to increase in motor input;
2. increase in gross torque is almost directly proportional to increase in load angle α;
3. as load is increased, lagging power factor tends to approach unity.

25.10. Synchronous Capacitor

When a synchronous motor is over-excited, its generated or back e.m.f. E_b is much greater than supply voltage V. When such a motor operates without load, α is small and E_b is almost 180° ahead of V. As shown in Fig. 25.13, I_a lags E_R by nearly 90° and hence *leads* V by about 90°.

Due to this leading power factor, a synchronous motor acts as a capacitor and can be used to correct the p.f. of a lagging load. Such a motor, designed to carry no load, is called a *synchronous capacitor*, is used where load is so large that construction of a static capacitor is impractical.

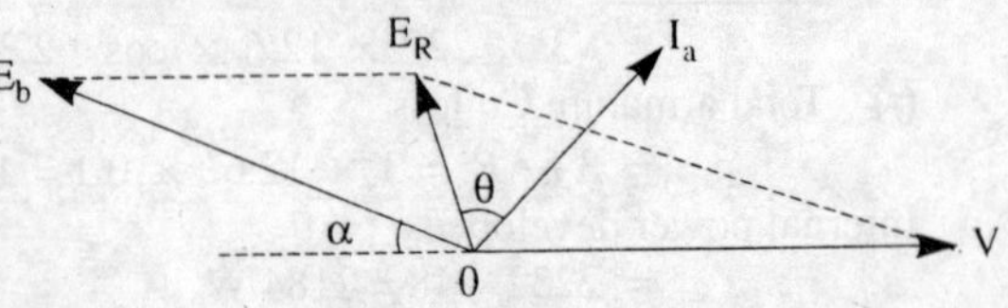

Fig. 25.13

In factories which use induction motors, the p.f. rarely exceeds 85 per cent. Taking into account the fact that starting power factor of such motors is poor and also that not all the motors operate at full-load, the overall p.f. is as low as 60 per cent.

As shown in Fig. 25.14, installation of a synchronous capacitor in parallel with the induction motors improves the p.f. of the line and hence of supply generators and transformers. It is obvious that in order to raise the p.f. to unity, current taken by the capacitor must equal the total reactive current drawn by the induction motors.

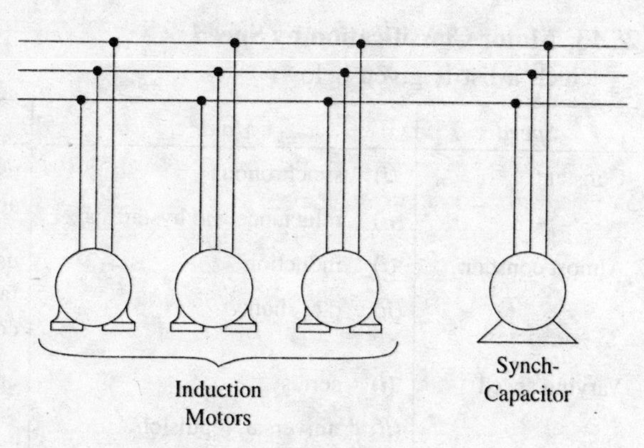

Fig. 25.14

Example 25.4. *A load connected to a 3-ϕ, 2300 V, supply line draws 1500 kW at a lagging p.f. of 0.6. Find (a) kVA rating of a synchronous capacitor required to be installed in parallel with the load in order to raise the line p.f. to unity and (b) kVA rating of an alternator required to supply the new load.*

Solution. (*a*) Load kW = 1500

$$\cos\phi = 0.6, \quad \sin\phi = 0.8$$

$$\text{kVA} = \text{kW} / \cos\phi = 1500 / 0.6 = 2500$$

$$\text{kVAR} = \text{kVA} \sin\phi = 2500 \times 0.8 = 2000 \text{ (lag)}$$

For bringing the power factor to unity, the synchronous capacitor must supply an equal number of lagging kVAR. Hence, kVAR supplied by synchronous motor is **2000 (leading).**

(*b*) The alternator now has to supply only 1500 kW as shown in Fig. 25.15 which represents the alternator rating *i.e.*, **1500 kVA.**

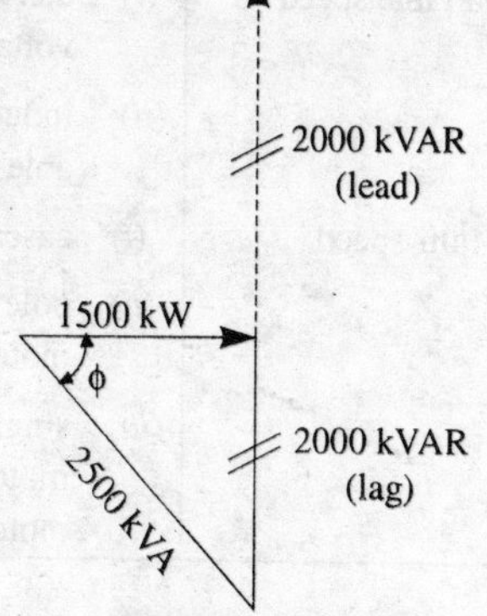

Fig. 25.15

25.11. Applications of Synchronous Motors

1. In view of their high efficiency and constant speed, these motors can be advantageously employed for loads where constant speeds are desirable as in ammonia and air compressors, blowers, motor-generator sets, continuous rolling mills, paper and cement industries.
2. Since these motors have high efficiency and can be built in low speeds, they are well-suited for direct connection to reciprocating compressors.
3. Over-excited synchronous motors are most commonly used for power factor improvement of lagging industrial loads. When employed in such a role, they are referred to as synchronous capacitors.
4. They are generally used for improving voltage regulation of long transmission lines.

25.12. Comparison between Synchronous and Induction Motors

1. Whereas synchronous motor runs only at synchronous speed, an induction motor never runs with synchronous speed.
2. Synchronous motor can be operated under a wide range of power factors both lagging and leading. But induction motors always run with lagging power factor.
3. A synchronous motor is not inherently self-starting whereas an induction motor is.
4. The torque of a synchronous motor is much less affected by changes in applied voltage than that of an induction motor.
5. A d.c. excitation is required by a synchronous motor but not by an induction motor.
6. Synchronous motors are usually more costly and complicated than induction motors.

25.13. Motor Classification by Speed

Such a list is given below:

Speed	*Motor*	*Applications*
Constant	(*i*) synchronous (*ii*) reluctance and hysteresis	useful for p.f. correction, small power, applications, clocks, turntables, conveyors
Almost constant	(*i*) induction (*ii*) d.c. shunt	used when small speed change can be tolerated, fans, blowers, machine tools, pumps, compressors, crushers
Varying speed	(*i*) series (*ii*) universal, repulsion	cranes, hoists, traction
Marginal speed control	(*i*) wound rotor induction (*ii*) d.c. compound with field control	cranes, hoists
Variable speed	(*i*) d.c. motor with armature voltage control (*ii*) induction motor with variable frequency supply	used for any drive requiring a wide speed variation–expensive
Multi-speed	(*i*) cascaded induction motor (*ii*) pole changing induction motors	used when two discrete speeds are required. Two speed fans
	(*iii*) single-phase induction motor, FHP universal motors	all domestic applications: cleaners, washers, tumble dryers, central heating, pumps

HIGHLIGHTS

1. Synchronous motor is nothing else but an alternator operated in the reverse direction.
2. A synchronous motor runs at one speed only *i.e.*, synchronous speed and can work at both lagging and leading power factors.
3. Mechanical power developed by a synchronous motor is

$$P_m = E_b I_a \cos(\alpha - \phi) \cong \frac{E_b V}{Z_S} \sin\alpha \quad \text{— per phase}$$

4. Back e.m.f. generated in stator / phase is

$$E_b^2 = V^2 + E_R^2 - 2VE_R \cos(\theta - \phi)$$

5. The resultant voltage in stator / phase is

$$E_R^2 = V^2 + E_b^2 - 2VE_b \cos\alpha$$

OBJECTIVE TEST—25

A. Fill in the following blanks.

1. A synchronous motor is electrically identical with an ....
2. A synchronous motor starts as an .... motor.
3. At the instant of start-up of a synchronous motor, its rotor field circuit is left ....
4. A synchronous motor can operate at a power factor varying from lagging to a .... one.
5. There are .... opposing voltages in the stator winding of a synchronous motor.
6. The stator and rotor poles of a synchronous motor run with the same .... speed.

7. As the load on a synchronous motor is increased, its rotor falls back in ....

8. An over-excited synchronous motor acts as a ....

B. Answer True or False.

1. A three-phase synchronous motor is made self-starting by providing squirrel-cage winding in its stator poles.

2. The p.f. of a synchronous motor can be varied by varying its rotor field excitation.

3. Even when fully-loaded, synchronous motor runs with synchronous speed.

4. The armature current drawn by a synchronous motor lags behind the net stator voltage by a fixed angle.

5. Mechanical power developed by a synchronous motor depends on the phase angle between the rotor and stator poles.

6. Under-excited synchronous motors are commonly used to raise the p.f. of the line supplying induction motors.

7. The armature current drawn by a synchronous motor is minimum at unity power factor.

8. A synchronous motor is ideally suited for applications where there are sudden changes in load.

C. Multiple Choice Questions.

1. In a synchronous motor, squirrel-cage winding is provided for making the motor

(*a*) noise-free (*b*) self-starting
(*c*) cheap (*d*) quick-start

2. An electric motor in which stator and rotor poles run with exactly the same speed is called a / an .... motor.

(*a*) induction (*b*) universal
(*c*) shaded-pole (*d*) synchronous

3. A three-phase synchronous motor is made self-starting by providing

(*a*) squirrel-cage winding
(*b*) salient-pole rotor
(*c*) revolving flux
(*d*) rotor excitation

4. A 3-phase synchronous motor is well-suited for driving

(*a*) fluctuating loads
(*b*) printing presses
(*c*) motor-generator sets
(*d*) vacuum cleaners

5. When load on a three-phase synchronous is increased, its

(*a*) rotor falls in phase
(*b*) rotor slows down
(*c*) armature current decreases
(*d*) power factor increases

6. An over-excited synchronous motor running in parallel with induction motors can be used to improve the power factor of

(*a*) supply line
(*b*) transformers
(*c*) generators
(*d*) all of the above

ANSWERS

A. 1. alternator 2. induction motor 3. open 4. leading 5. two 6. synchronous 7. phase 8. capacitor

B. 1. F 2. T 3. T 4. T 5. T 6. F 7. T 8. F

C. 1. *b* 2. *d* 3. *a* 4. *c* 5. *a* 6. *d*

26 Q AND A ON ELECTRIC MACHINERY

A. Direct Current Generators

26.1. *What type of voltage is generated in a standard d.c. generator?*

Ans. It is an alternating *i.e.*, a.c. voltage.

26.2. *How can the output voltage of a d.c. generator be regulated?*

Ans. Either by changing excitation with the help of field rheostat or by controlling generator speed.

26.3. *Which parts, if any, of a d.c. generator need to be laminated? And why?*

Ans. Armature has to be laminated in order to reduce eddy current loss. Field poles need not be laminated because they have constant field and hence there is no eddy current loss.

26.4. *What is meant by the term voltage 'build up' of a d.c. generator?*

Ans. It means the gradual increase in the generator voltage to its maximum value after the generator is started from rest

26.5. *How should a generator be started?*

Ans. It is usually brought up to speed with the help of the driving engine called prime mover.

26.6. *How should a shunt or compound generator be started?*

Ans. Such machines excite best when all switches controlling the external circuits are open.

26.7. *How about a series generator?*

Ans. In this case, the external circuit must be closed otherwise the generator will not build up.

26.8. *What is the procedure for shutting down a generator?*

Ans. First, the load should be gradually reduced, if possible by easing down the driving engine, then when the generator is supplying little or no current, the main switch should be opened. When the voltmeter reads almost zero, the brushes should be raised from the commutator.

26.9. *What are the indications and causes of an overloaded generator?*

Ans. A generator is said to be overloaded if a greater output is taken from it than it can safely carry. Overloading is indicated by (*i*) excessive sparking at brushes and (*ii*) overheating of the armature and other parts of the generator.

Most likely causes of overloading are:

(*a*) excessive voltage–as indicated by the voltmeter or the increased brilliance of the pilot lamp. This could be due to over-excitation of field magnets or too high speed of the driving engine.

(*b*) excessive current–which could be due to bad feeding of the load.

(*c*) reversal of polarity–this happens occasionally when the series or compound-wound generators are running in parallel. Polarity reversal occurs during stopping by the current from the machines at work.

(*d*) short-circuit or ground in the generator itself or in the external circuit.

26.10. *Mention and explain the various causes for the failure of the generator to build up.*

Ans. Principal causes due to which a generator may fail to excite are:

1. *brushes not properly adjusted*–if brushes are not in their proper positions, then whole of the armature voltage will not be utilized and would be insufficient to excite the machine.

2. *defective contacts*–unclean contacts may interpose large resistance in the path of the exciting current and reduce it to such a small value that it fails to excite the machine.
3. *incorrect adjustment of regulators*–in the case of shunt and compound generators, it is possible that the resistance of field regulator may be too great to permit the passage of sufficient current through the field windings.
4. *speed too low*–in the case of shunt and compound-wound generators, there is a certain critical armature speed below which they will not excite (Art.11.7).
5. *open-circuit*–in the case of series machines.
6. *short-circuit*–in the generator or external circuit.
7. *wrong connections*–particularly when connecting the field coils to the armature.
8. *reversed field polarity*–usually caused by the reversed connections of the field coils.
9. *insufficient residual magnetism*–this trouble normally occurs when the generator is new. It can be remedied by passing a strong direct current through the field coils.

26.11. *Is there any major difference between a d.c. motor and a d.c. generator?*

Ans. Fundamentally, there is none. A d.c. motor can run as a d.c. generator when driven by a prime mover. Similarly, a d.c. generator will run as a d.c. motor when connected across suitable d.c. supply.

B. Direct Current Motors

26.12. *Which end of the motor is called 'front end'?*

Ans. The end opposite the shaft.

26.13. *What is the standard direction of rotation of a d.c. motor?*

Ans. Counter-clockwise when looking at the front of the motor.

26.14. *How may the direction of rotation of a d.c. motor be reversed?*

Ans. By reversing either the field current or current through the armature. Usually, reversal of current through the armature is adopted.

26.15. *What will happen if both currents are reversed?*

Ans. The motor will run in the original direction.

26.16. *What will happen if the direction of current at the terminals of a series motor is reversed?*

Ans. It will not reverse the direction of rotation of the motor.

26.17. *What will happen if the field of a d.c. motor is opened?*

Ans. The motor will achieve dangerously high speed and may destroy itself.

26.18. *Explain what happens when a d.c. motor is connected across an a.c. supply?*

Ans.
(*i*) since on a.c. supply, reactance will come into the picture, the a.c. supply will be offered impedance (not resistance) by the armature winding. Consequently, with a.c. supply, current will be much less. The motor will run but it would not carry the same load as it would on d.c. supply.
(*ii*) there would be more sparking at the brushes.
(*iii*) though armature is laminated, the field poles are not. Consequently, the eddy currents will cause the motor to heat up and eventually burn on a.c. supply.

26.19. *What will happen if a shunt motor is directly connected to the supply line* ?

Ans. Small motors up to 1 kW rating may be line-started without any adverse results being produced. High rating motors must be started through a suitable starter in order to avoid the huge starting current which will

(*i*) damage the motor itself; and
(*ii*) badly affect the voltage of the supply line.

26.20. *What is the function of interpoles and how are the interpole windings connected?*

Ans. Interpoles are small poles placed in between the main poles. Their function is to assist commutation by producing the auxiliary or commutating flux. Consequently, brush sparking is practically eliminated. Interpole windings are connected in series with the armature winding.

26.21. *How can we control the speed of a d.c. motor?*

Ans. By changing either the pole flux or the voltage across armature.

C. Three-phase Induction Motors

26.22. *What is, in brief, the basis of operation of a 3-phase induction motor?*

Ans. The revolving magnetic field which is produced when the 3-phase stator winding is fed from a 3-phase supply.

26.23. *What factors determine the direction of rotation of the motor?*

Ans. The phase sequence of the supply lines and the order in which these lines are connected to the stator winding.

26.24. *How can the direction of rotation of the motor be reversed?*

Ans. By transposing or changing over any two line leads as shown in Fig. 26.1.

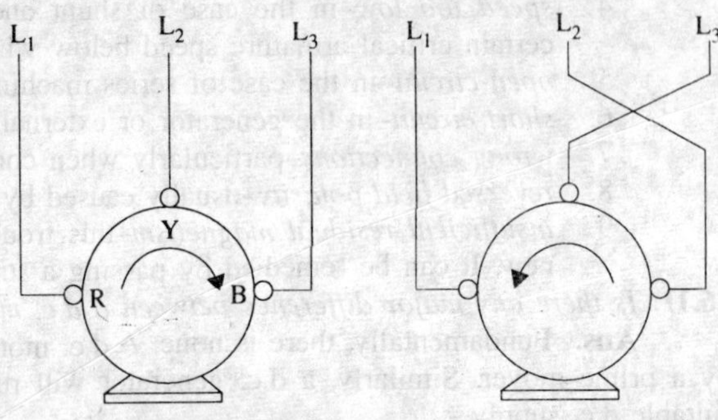

Fig. 26.1

26.25. *Why are induction motors called 'asynchronous'?*

Ans. Because their rotors can never run with the synchronous speed.

26.26. *How does the slip vary with load?*

Ans. The greater the load, greater is the slip or slower is the rotor speed.

26.27. *What modifications would be necessary if a motor is required to operate on a voltage different from that for which it was designed?*

Ans. The number of conductors per slot will have to be changed in the same ratio as the change in voltage. If the voltage is doubled, the number of conductors per slot will have to be doubled.

26.28. *Enumerate the possible reasons if a 3-phase motor fails to start.*

Ans. Any one of the following reasons could be responsible:

1. one or more fuses may be blown;
2. voltage may be too low;
3. the starting load may be too heavy;
4. worn bearings due to which the armature may be touching field laminae thus introducing excessive friction.

26.29. *A motor stops after starting i.e., it fails to call load. What could be the causes?*

Ans. Any one of the following:

1. hot bearings which increase the load by excessive friction;
2. excessive tension on belt which causes the bearings to heat;
3. failure of short cut-out switch;
4. single-phasing on the running position of the starter.

26.30. *Which is the usual cause of blow outs in induction motors?*

Ans. The commonest cause is single-phasing.

26.31. *What is meant by 'single-phasing' and what are its causes?*

Ans. By single-phasing is meant the opening of one wire (or leg) of a three-phase circuit whereupon the remaining leg at once becomes single-phase. When a three-phase circuit functions normally, there are three distinct currents flowing in the circuit. As is known, any two of these currents use the third wire as the return path *i.e.*, one of the three phases acts as a return path for the other two. Obviously, an open circuit in one leg kills two of the phases and there will be only one current or phase working even though two wires are left intact. The remaining phase attempts to carry all the load.

The usual cause of single-phasing is, what is generally referred to as, *running fuse* which is a fuse whose current-carrying capacity is equal to the full-load current of the motor connected in the circuit. This fuse will blow out whenever there is overload (either momentary or sustained) on the motor.

26.32. *What happens if single-phasing occurs when the motor is running? And when it is stationary?*

Ans. (*i*) If already running and carrying half load or less, the motor will continue running as a single-phase motor on the remaining single-phase supply without damage because half loads do not blow normal fuses.

(*ii*) If it be running on full load or overload when single-phasing occurs, it will try to keep on running throwing all the load on the intact winding which sooner or later burns out.

(*iii*) If motor is very heavily loaded, then it stops under single-phasing and since it can neither restart nor blow out the remaining fuses, the burn out is very prompt.

A stationary motor will not start with one line broken. In fact, due to heavy standstill current, it is likely to burn out quickly unless immediately disconnected.

26.33. *Which phase is likely to burn out in a single-phasing delta-connected motor.*

Ans. The *Y*-phase connected across the live or operative lines carries nearly three times its normal current and is the one most likely to burn out (Fig. 26.2).

The other two phases *R* and *B* which are in series across L_2 and L_3 carry more than their full-load current.

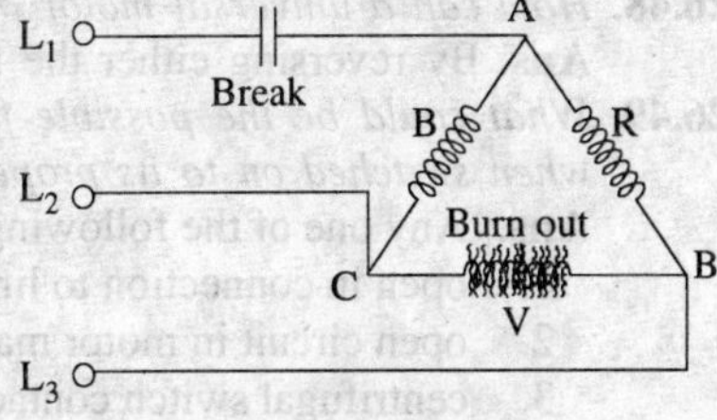

Fig. 26.2

26.34. *What currents flow in single-phasing star-connected motor?*

Ans. With L_1 disabled, the currents flowing in L_2 and L_3 and through phases *Y* and *B* in series will be of the order of 250 per cent of the normal full-load current, 160 per cent on 3/4 load and 100 per cent on 1/2 load.

26.35. *How can the motors be protected against single phasing?*

Ans. (*i*) By incorporating a combined overload and single-phasing relay in the control gear.

(*ii*) By incorporating a phase-failure relay in the control gear.

The relay may be either voltage or current-operated.

26.36. *Can a 3-phase motor be run on a single-phase line?*

Ans. Yes, it can be. But a phase-splitter is essential.

26.37. *What is meant by a phase-splitter?*

Ans. It is a device consisting of a number of capacitors so connected in the motor circuit that it produces, from a single input wave, three output waves which differ in phase from each other.

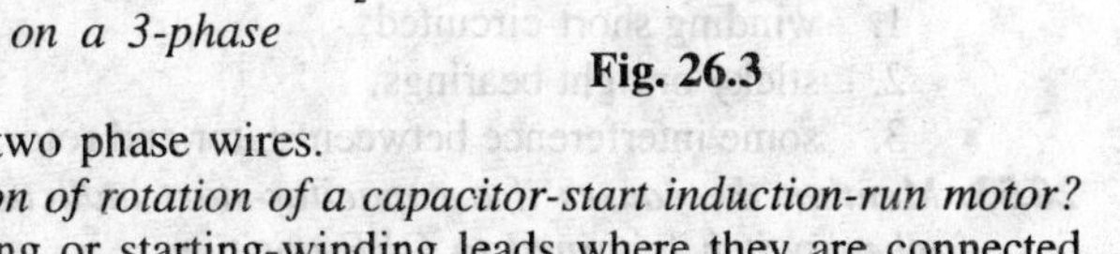

Fig. 26.3

D. Single-phase Motors

26.38. *Can a single-phase motor be run on a 3-phase supply?*

Ans. Yes, by connecting it across two phase wires.

26.39. *How would you reverse the direction of rotation of a capacitor-start induction-run motor?*

Ans. By reversing either the running or starting-winding leads where they are connected to the lines. Both must not be reversed.

26.40. *Does a resistance-start split-phase motor have a high starting torque?*

Ans. No. It has low starting torque which is 150-200 per cent of full-load torque.

26.41. *How about the starting torque of a capacitor-start induction-run motor?*

Ans. It has a starting torque as high as 350 to 450 per cent of its full-load torque.

26.42. *Are capacitor-start-and-run motors single-value or two-value capacitor motors?*

Ans. Two-value capacitor motors.

26.43. *What are the advantages of a capacitor-start-capacitor-run motor over the capacitor-start induction-run motors?*

Ans. (*i*) smoother running;

(*ii*) higher p.f.;

(*iii*) less input current.

26.44. *In which direction does a shaded-pole motor turn?*
Ans. It turns from the unshaded to the shaded pole.
26.45. *Do shaded-pole motors have much starting torque?*
Ans. No. That is why they are built in very small sizes.
26.46. *Can such a motor be reversed?*
Ans. Normally, such motors are not reversible because that would involve mechanical dismantling and re-assembly. However, special motors are made having two rotors on a common shaft, each having one stator assembled for rotation in opposite direction.
26.47. *What is a universal motor?*
Ans. It is built like a series d.c. motor with the difference that both its stator and armature are laminated. They can be used either on d.c. or a.c. supply although the speed and power are greater on direct current. They cannot be satisfactorily made to run at less than about 2000 r.p.m.
26.48. *How can a universal motor be reversed?*
Ans. By reversing either the field leads or armature leads but not both.
26.49. *What could be the possible faults if a capacitor-start induction-run motor fails to start when switched on to its proper supply?*
Ans. Any one of the following could be responsible:
1. open in connection to line;
2. open circuit in motor main or running winding;
3. centrifugal switch contacts open;
4. defective capacitor;
5. open in starting winding.

26.50. *A permanent-split capacitor motor does not always start even on no-load but runs in either direction when started manually. What could be the reasons?*
Ans. Any one of the following:
(*i*) defective capacitor;
(*ii*) one or both windings open.
26.51. *A capacitor-start-and-run motor does not start when switched on but runs in either direction when started manually, though it overheats. List the causes responsible for this behaviour.*
Ans. The possible causes are as under:
1. defective capacitor;
2. either winding short-circuited or grounded;
3. one or more winding open.

26.52. *A shaded-pole motor is found to develop less power but gets too hot. Reason?*
Ans. Any of the following:
1. winding short-circuited;
2. sticky or tight bearings;
3. some interference between stator and rotor.

26.53. *Mention the causes if a capacitor-start induction-run motor blows fuse or does not stop when switch is turned to OFF position.*
Ans. (*i*) winding short-circuited or grounded;
(*ii*) ground near switch end of the winding.

E. Alternators

26.54. *Basical, what does an alternator consist of?*
Ans. (*a*) stator-which has 3-ϕ armature winding;
(*b*) rotor—which has salient or smooth magnetic poles.
26.55. *How is the field of an alternator usually excited?*
Ans. From a small d.c. generator mounted on the shaft of the alternator itself.
26.56. *What kind of rotor is most suitable for turbo-alternators which run at very high speed?*
Ans. Non-salient pole type *i.e.*, smooth cylindrical type.
26.57. *What is the formula for the frequency of the a.c. voltage produced by an alternator?*

Ans. $f = PN / 120$ Hz
P = No. of magnetic poles on the rotor
N = r.p.m. of the rotor

26.58. *When will a 50Hz alternator run at the greatest possible speed?*
Ans. When its rotor has two poles.

26.59. *How will the generated e.m.f. / phase of an alternator change when its speed is doubled?*
Ans. It will be doubled.

26.60. *How can the generated voltage of an alternator be controlled?*
Ans. By varying field excitation.

26.61. *Which factors cause variations in the terminal voltage of an alternator when loaded?*
Ans. (*a*) drop across armature resistance *i.e.*, $I_a R_a$;
(*b*) drop due to armature leakage reactance *i.e.*, $I_a X_L$;
(*c*) armature reaction.

26.62. *Is voltage regulation of an alternator always negative?*
Ans. No. It depends on the load power factor. It is positive for lagging p.f. and negative for leading p.f.

26.63. *What is the unit for expressing rating of an alternator?*
Ans. kVA.

F. Synchronous Motors

26.64. *What happens when the field of a synchronous motor is (i) under-excited (ii) over-excited?*
Ans. (*i*) It has a lagging p.f. and (*ii*) it has a leading p.f.

26.65. *When is a synchronous motor said to be over-excited?*
Ans. When its excitation is such that generated or back e.m.f. / phase E_b becomes greater than applied voltage/phase, V.

26.66. *What is a synchronous capacitor?*
Ans. An overexcited unloaded synchronous motor is called synchronous capacitor because, like a capacitor, it takes a leading current.

26.67. *What happens when load on a synchronous motor increases?*
Ans. (*a*) first, torque angle α increases as rotor falls back in phase;
(*b*) then, E_R increases;
(*c*) next, I_a increases;
(*d*) hence, ultimately motor input increases in order to meet the extra load.

26.68. *What are the causes of faulty starting of a synchronous motor?*
Ans. It could be due to the following causes:
1. voltage may be too low—at least half voltage is required for starting;
2. there may be open—circuit in any phase—due to which motor may heat up;
3. static friction may be large—either due to great belt tension or too light bearings;
4. stator windings may be incorrectly connected;
5. field excitation may be too much.

27 SEMI-CONDUCTOR PHYSICS

27.1. Bohr's Atomic Model and Electron Orbits

According to the model of an atom proposed by Nobel Prize-winner physicist Niels Bohr, an atom is composed of negatively-charged electrons moving in fixed circular or elliptical orbits around a relatively heavy nucleus made up of two fundamental particles known as protons and neutrons (Fig. 27.1). The proton is a relatively heavy particle (1840 times heavier than an electron) while the neutron has no charge *i.e.*, it is electrically neutral. The positive charge of the proton is equal in magnitude to the negative charge of the electron and has a value of 1.6×10^{-19} coulomb. The diameter of an atom is about 10^{-19} m and that of the nucleus is approximately 10^{-15} m*. By applying Quantum Theory, Bohr was able to determine the radii of the electron orbits as well as the energy associated with them.

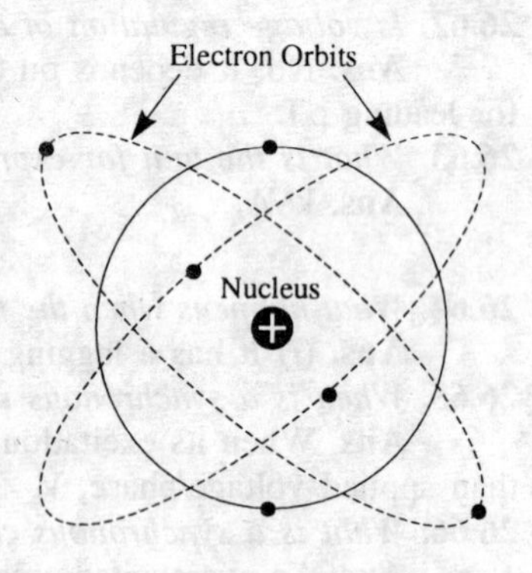

Fig. 27.1

The electron orbit nearest to the nucleus is designated by the letter *K* and is given the principal quantum number $n = 1$. The succeeding orbits are referred to alphabetically as *L, M, N*...etc., and have principal quantum numbers of $n = 2, 3, 4$ etc. The maximum number of electrons that can be accommodated in any orbit $= 2n^2$ where n is the principal quantum number of the electron orbit or shell.

Fig. 27.2 shows the schematic distribution of electrons in different orbits for hydrogen, boron, phosphorus and germanium.

The outermost electrons *i.e.*, those farthest from the nucleus are known as *valence* electrons. These electrons have the highest energy** and may fully or partially fill the valence orbit. For example, boron has 3 valence electrons whereas it could accommodate 8 electrons in *L* shell.

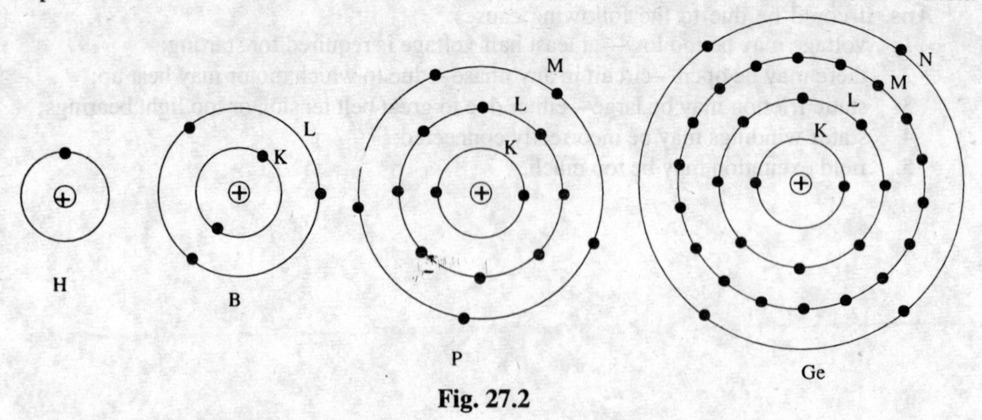

Fig. 27.2

*The relative distance of the electron from the nucleus is about the same as the relative distance of the earth from Pluto.

**Algebraically but not in magnitude.

The radii of the various permitted electron orbits (assumed circular) are given by

$$r_n = \frac{\epsilon_0 n^2 h^2}{\pi m Z e^2}$$

where
n = principal quantum number
h = Planck's constant = 6.62×10^{-34} J-s
m = electron mass = 9.1×10^{-31} kg
e = electron charge = 1.6×10^{-19} C
ϵ_0 = absolute permittivity of free space
= 8.854×10^{-12} F / m
Z = atomic number

Substituting values of different constants, we get

$$r_n = 5.3 \times 10^{-11} n^2 / Z \text{ metre} \qquad ...(i)$$

When an electron moves in any permitted orbit, it has a definite energy which is the sum of

(*i*) kinetic energy due to motion, and
(*ii*) potential energy due to the attraction of the nucleus.

Total energy

$$E_n = -\frac{m \epsilon^4 Z^2}{8 \epsilon_0{}^2 n^2 h^2} \text{ joule*}$$

Again, substituting different values, we have

$$E_n = -21.76 \times 10^{-19} Z^2 / n^2 \text{ joule}$$

Now, $1 \text{ eV} = 1.6 \times 10^{-19}$ joule

Hence when expressed in electron-volts, the energy becomes

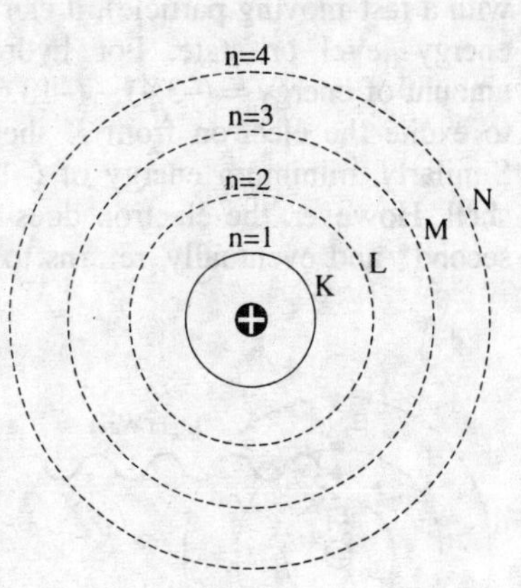

Fig. 27.3

$$E_n = \left(\frac{-21.76 \times 10^{-19}}{1.6 \times 10^{-19}}\right)\frac{Z^2}{n^2} = -13.6\frac{Z^2}{n^2} \text{ eV} \qquad ...(ii)$$

27.2. Energy Levels in a Single Atom

Let us find out the value of energy associated with different electron orbits for a *single isolated* hydrogen atom which has one proton in the nucleus surrounded by one orbital electron. Obviously, for hydrogen $Z = 1$. Substituting this value in Eq. (*ii*) of Art. 27.1, we have

$E_n = -13.6 / n^2$ eV

Energies of various electron orbits are:

$E_1 = -13.6$ eV
$E_2 = -13.6 / 2^2 = -3.39$ eV
$E_3 = -13.6 / 3^2 = -1.51$ eV
$E_4 = -13.6 / 4^2 = -0.85$ eV

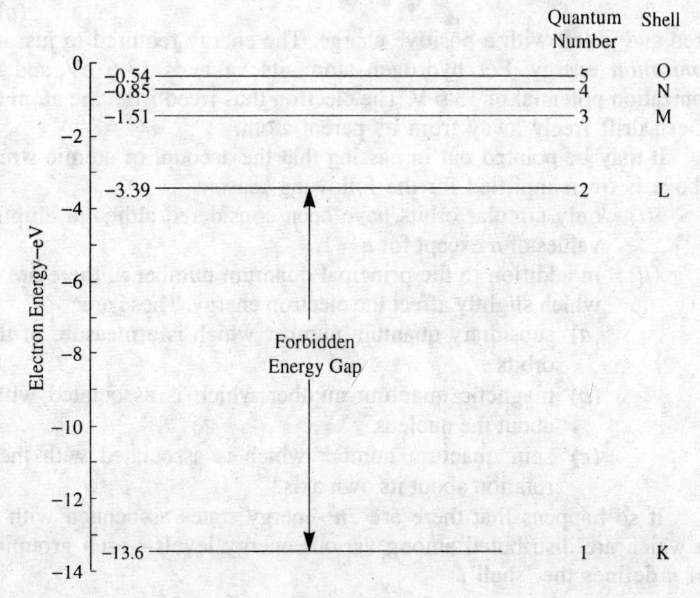

Fig. 27.4

*The negative sign appears because the energy of the electron when at infinity is taken as zero.

Different permitted (or possible) energy levels are shown in Fig. 27.4. Each level is represented by a horizontal line of the same length although the length of the lines has no significance. Moreover, there is no indication as to the actual occupancy of these levels. These levels are separated by forbidden energy gaps. Which orbit an electron occupies at a given time depends on the total energy possessed by it. At 0°K or 273°C, the electron in a hydrogen atom occupies the first orbit or first quantum level in the atom (which is associated with the least energy). If the electron acquires correct amount of additional energy from an external source (as by heating or by impact with a fast-moving particle), it can be excited to a higher energy level or state. For hydrogen atom, an extra amount of energy = (–3.4) – (–13.6) = 10.2 eV is needed to excite the electron from *K* shell to the *L* shell. In that case, the atom is said to be *excited.* Similarly, minimum energy of (–1.51) – (–13.6)=12.5 eV is required to raise the electron to *M* shell. However, the electron does not remain in the excited state or level longer than about 10^{-8} second* and eventually, returns to the ground state (or unexcited or stable state) in one or more 'jumps'. During each jump, a photon of energy or light waves are emitted (or radiated out) in accordance with the relation [Fig. 27.6 (*a*)].

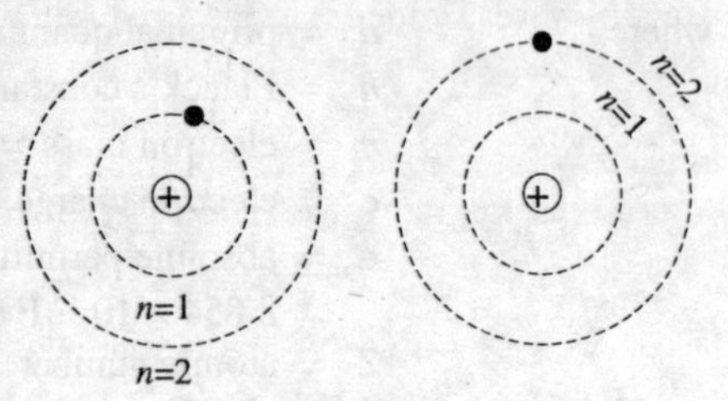

Fig. 27.5

$$E_2 - E_1 = hf$$

where *f* is the frequency of the light wave emitted. It is worth noting that when revolving in its normal orbit, the electron does not radiate out any energy.

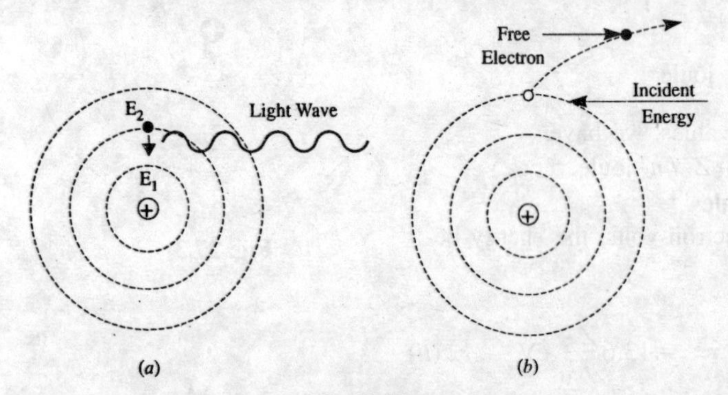

Fig. 27.6

When an electron is removed altogether [Fig. 27.6 (*b*)], the atom is said to be ionized and is left with a positive charge. The energy required to just ionize a normal atom is called ***ionization*** energy. For hydrogen atom, its value is 13.6 eV and hydrogen is said to have an ionization potential of 13.6 V. The electron thus freed from the atom becomes free electron because it can drift freely away from its parent atom.

It may be pointed out in passing that the account of atomic structure and energy levels given above is oversimplified for the following reasons:

(*i*) only circular orbits have been considered although elliptical orbits are possible for all values of *n* except for *n* = 1.

(*ii*) in addition to the principal quantum number *n*, there are three other quantum numbers which slightly affect the electron energy. These are:

(*a*) subsidiary quantum number which is a measure of the eccentricity of the electron orbits.

(*b*) magnetic quantum number which is associated with the rotation of the electron about the nucleus.

(*c*) spin quantum number which is associated with the spin of the electorn *i.e.*, its rotation about its own axis.

It so happens that there are $2n^2$ energy states associated with a particular quantum number ***n*** which are distributed among various energy levels—each grouping of states for a given value of ***n*** defines the 'shell'.

---

*Some atoms like those of mercury have excitation levels which are occupied for quite long time of 10^{-2} second. Such atoms are said to be in a metastable state.

27.3. Energy Bands in Solids

So far we have considered electron energy levels in a *single isolated* atom. But there are significant changes in the energy levels when atoms are brought close together as in solids. However, in gases, under normal conditions of pressure and temperature, atomic spacing is so great that the changes in energy levels are almost negligible.

As is well-known, a solid consists of an ordered array of many closely-packed atoms which is usually referred to as crystal lattice. Since atoms are brought very close to each other, there is some intermingling of electrons from adjacent atoms. The effect of inter-mixing is greatest with

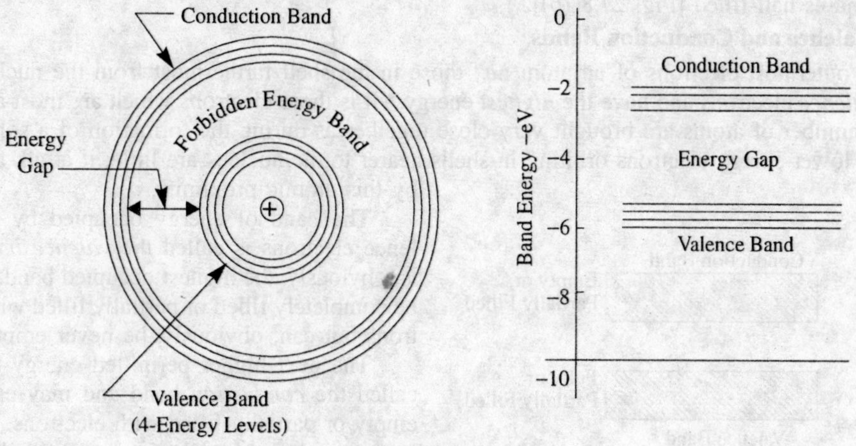

Fig. 27.7

the outermost or valence electrons which have the *highest* energy. This fact results in the modification of energy levels in a solid. The number of permissible orbits is vastly increased which in other words, means that the number of permitted energy levels is increased. Consequently, the single energy level of an *isolated* atom becomes a band of energies in a solid (Fig. 27.7) The individual energies within the band are so close together that the energy band may be considered to be continuous for many purposes.

For example, in an assembly of N electrons, the number of possible energy states is N. Since only two electrons of opposite spin can occupy the same state*, maximum number of electrons which these N states can accommodate is $2N$.

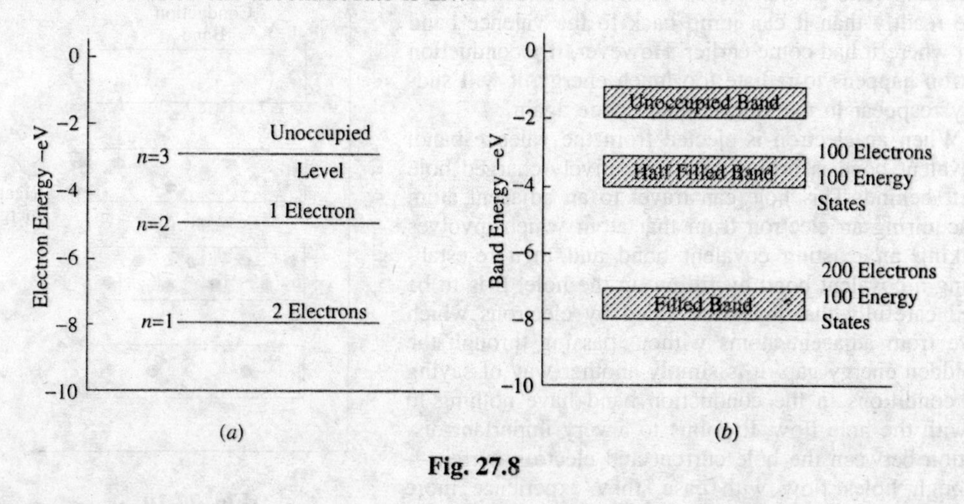

Fig. 27.8

*According to Pauli's Exclusion Principle which states that no two electrons may have the same four quantum numbers described in Art. 27.2.

Consider the case of lithium metal—the simplest atom which forms a solid at ordinary temperature. Its atom has three electrons, two of which have the same energy and the third one has higher value of energy. In an isolated single atom, two electrons move round the electron orbit with $n = 1$ whereas the third occupies the orbit with $n = 2$ as shown in Fig. 27.8 (*a*). Now, consider a piece of lithium metal containing 100 atoms. It will be found that the lower *level* (with $n = 1$) forms a *band* of 200 electrons occupying 100 different energy states. The higher level (with $n = 2$) forms a wider band also of 100 energy states which could, as before accommodate 200 electrons. But as there are only 100 electrons available (one from each atom) this energy band remains half-filled [Fig. 27.8 (*b*)].

27.4. Valence and Conduction Bands

The outermost electrons of an atom *i.e.,* those in the shell farthermost from the nucleus are called *valence* electrons and have the *highest* energy.* It is these electrons which are most affected when a number of atoms are brought very close together as during the formation of a solid. The states of lower-energy electrons orbiting in shells nearer to the nucleus are little, if at all, affected by this atomic proximity.

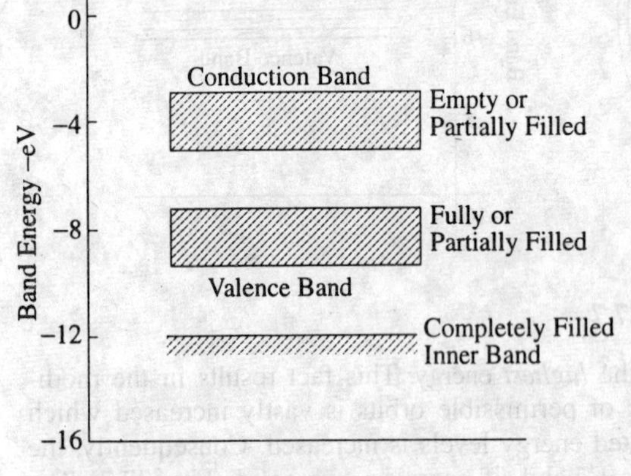

Fig. 27.9

The band of energy occupied by the valence electrons is called the *valence band* and is, obviously, the highest occupied band. It may be completely filled or partially filled with electrons but can, obviously, be never empty.

The next higher permitted energy band is called the *conduction* band and may either be empty or partially-filled with electrons. In fact, it may be defined as the lowest unfilled energy band.

In conduction band, electrons can move freely and hence are known as *conduction* electrons. As shown in Fig. 27.7, the gap between these two bands is known as the forbidden energy gap.

It may be noted that the covalent forces of the crystal lattice have their source in the valence band. If a valence electron happens to absorb enough energy, it jumps across the forbidden energy gap and enters the conduction band. An electron in the conduction band can jump to an adjacent conduction band more readily than it can jump back to the valence band from where it had come earlier. However, if a conduction electron happens to radiate too much energy, it will suddenly reappear in the valence band once again.

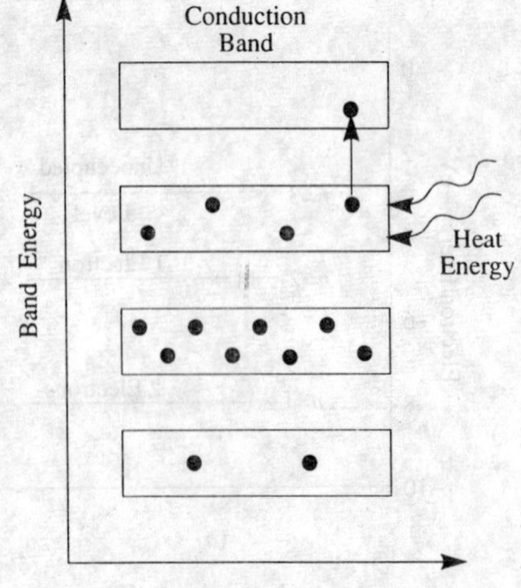

Fig. 27.10

When an electron is ejected from the valence band, a covalent bond is broken and a positively-charged hole is left behind. This hole can travel to an adjacent atom by acquiring an electron from that atom which involves breaking an existing covalent bond and then re-establishing a covalent bond by filling up the hole. It is to be noted carefully that holes are filled by electrons which move from adjacent atoms without passing through the forbidden energy gap. It is simply another way of saying that conditions in the conduction band have nothing to do with the hole flow. It points to a very important distinction between the hole current and electron current—although holes flow with ease, they experience more

*Algebraically, not numerically.

opposition than electron flow in the conduction band.

To summarize the above, it may be repeated that:

1. conduction electrons are found in and freely flow in the conduction band.
2. holes exist in and flow in the valence band.
3. conduction electrons move almost twice as fast as the holes.

Fig. 27.10 shows the energy-band diagram of a silicon atom ($Z = 14$) with its electronic distribution. When a silicon crystal is given some heat energy from outside, some electrons gain sufficient energy to jump across the gap from the valence band to the conduction band as shown in Fig. 27.10. In this way, they become free electrons. For every electron that jumps into the conduction band, a hole is created in the valence band. In this way, an electron-hole pair is created.

27.5. Conductors, Semi-conductors and Insulators

The electrical conduction properties of different elements and compounds can be explained in terms of the electrons having energies in the valence and conduction bands. The electrons lying in the lower energy bands, which are normally filled, play no part in the conduction process.

(*i*) **Insulators.** Stated simply, insulators are those materials in which valence electrons are bound very tightly to their parent atoms thus requiring very large electric field to remove them from the attraction of the nuclei.

In terms of energy bands, it means that insulators

1. have a full valence band;
2. have an empty conduction band; and
3. have a large energy gap (of several eV) between them.

This is shown in Fig. 27.11 (*a*). For conduction to take place, electrons must be given sufficient energy to jump from the valence band to the conduction band. Increase in temperature enables some electrons to go to the conduction band which fact accounts for the negative resistance temperature coefficient of insulators.

(*ii*) **Conductors.** Conducting materials are those in which plenty of free electrons are available for electric conduction.

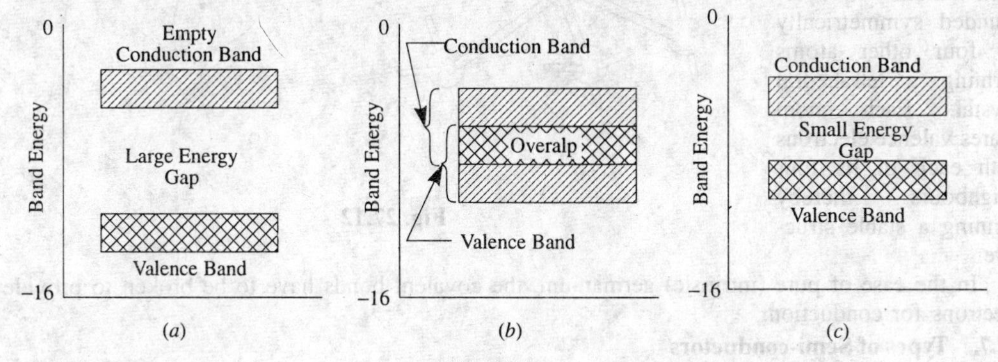

Fig. 27.11

In terms of energy bands, it means that electrical conductors are those which have overlapping valence and conduction bands as shown in Fig. 27.11 (*b*).

In fact, there is no physical distinction between the two bands. Hence, the availability of a large number of conduction electrons.

Another point worth noting is that in the absence of forbidden energy band in good conductors, there is no structure to establish holes. The total current in such conductors is simply a flow of electrons. It is exactly for this reason that the existence of holes was not discovered until semi-conductors were studied thoroughly.

(*iii*) **Semi-conductors.** A semi-conductor material is one whose electrical properties lie in between those of insulators and good conductors. Examples are germanium and silicon.

In terms of energy bands, semi-conductors can be defined as those materials which have

In terms of energy bands, semi-conductors can be defined as those materials which have almost an empty conduction band and an almost filled valence band with a very narrow energy gap (of the order of 1 eV) separating the two as shown in Fig. 27.11 (*c*).

At 0°K, there are no electrons in the conduction band and the valence band is completely filled. However, with increase in temperature, width of the forbidden energy band is decreased so that some of the electrons are liberated into the conduction band. In other words, conductivity of semi-conductors increases with rise in temperature.

Semi-conductors are of two types:

(*i*) pure or intrinsic semi-conductors; and

(*ii*) impure or extrinsic semi-conductors.

These are further discussed in Arts. 27.8 and 27.9.

27.6. Atomic Binding in Semi-conductors

Semi-conductors like germanium and silicon have crystalline structure. Their atoms are arranged in an ordered array known as crystal lattice. Both these materials are tetravalent *i.e.*, each has four valence electrons in its outermost shell.* The neighbouring atoms form covalent bonds by sharing four electrons with each other so as to achieve inert gas structure (*i.e.*, 8 electrons in the outermost orbit). A two-dimensional view of the germanium crystal lattice is shown in Fig. 27.12 (*b*) in which circles represent atom cores consisting of the nucleus and inner 28 electrons. Each pair of lines represents a covalent bond. The dots represent the valence electrons. It is seen that each atom has 8 electrons under its influence.

A 3-dimensional view of germanium crystal lattice is shown in Fig. 27.12 (*a*) where each atom is surrounded symmetrically by four other atoms forming a tetrahedral crystal. Each atom shares valence electrons with each of its four neighbours thereby forming a stable structure.

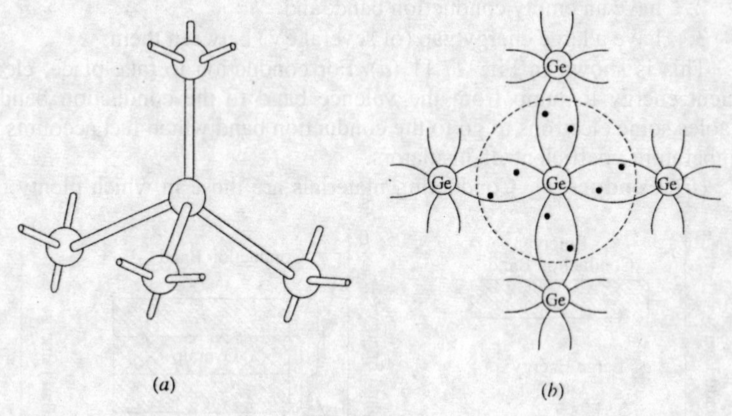

Fig. 27.12

In the case of pure (intrinsic) germanium, the covalent bonds have to be broken to provide electrons for conduction.

27.7. Types of Semi-conductors

Semi-conductors are classified as shown below:

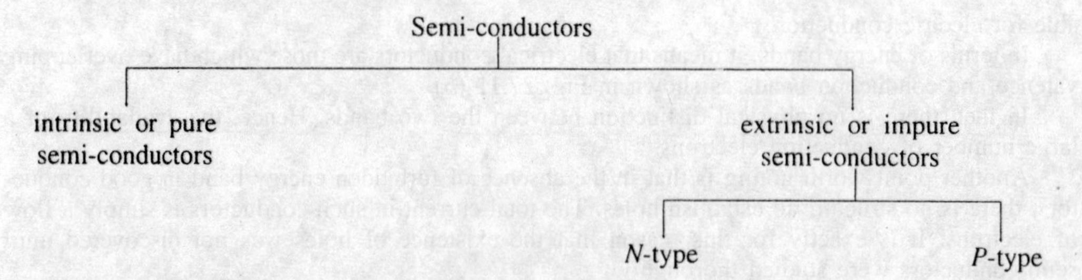

*A single germanium atom has 32 electrons out of which only four electrons take part in the electrical properties of germanium, the remaining 28 electrons being tightly bound to the nucleus. The four electrons revolve in the outermost shell and are called valence electrons.

27.8. Intrinsic Semi-conductors

An intrinsic semi-conductor is one which is made of the semi-conductor material in its extremely pure form.

Common examples of such semi-conductors are pure germanium and silicon which have forbidden energy gap of 0.72 eV and 1.1 eV respectively. The energy gap is so small that even at ordinary room temperature, there are many electrons which possess sufficient energy to jump across the small energy gap from the valence to the conduction band. However, it is worth noting that for each electron liberated into conduction band, a positively-charged hole is created in the valence band. When an electric field is applied to an intrinsic semi-conductor at a temperature greater than 0°K, conduction electrons move to the anode and the holes in the valence band move to the cathode. Hence, semi-conductor current consists of movement of electrons and holes in opposite directions.

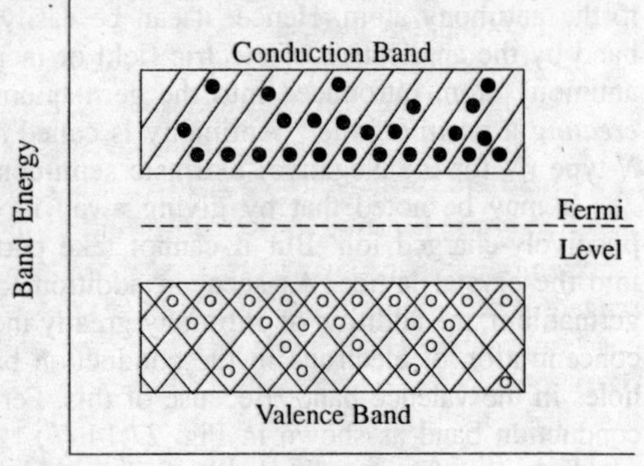

Fig. 27.13

Alternatively, an intrinsic semi-conductor may be defined as one in which the number of conduction electrons is equal to the number of holes.

Schematic energy band diagram of an intrinsic semi-conductor at room temperature is shown in Fig. 27.13. Here, Fermi level* lies in the middle of the forbidden energy gap.

27.9. Extrinsic Semi-conductors

Those intrinsic semi-conductors to which some suitable impurity or doping agent has been added in extremely small amount (about 1 part in 10^8) are called *extrinsic* or *impurity* semi-conductors.

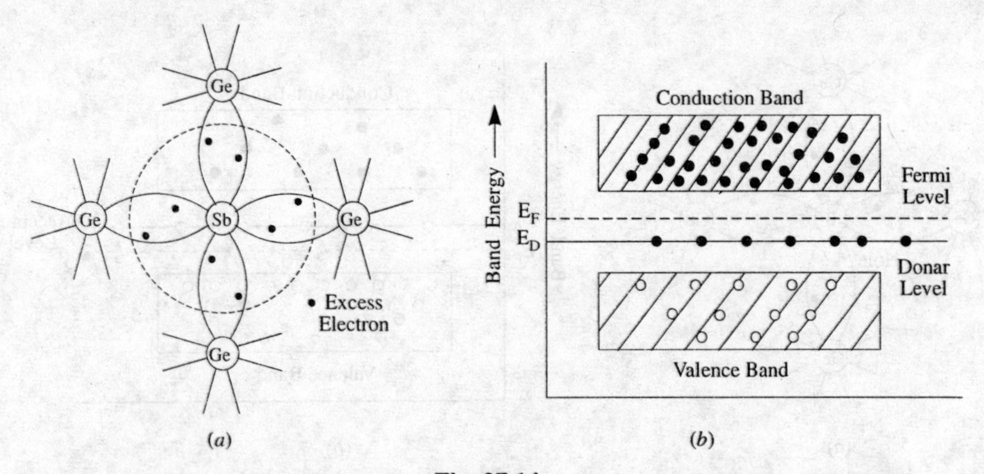

Fig. 27.14

Usually, the doping agents are pentavalent atoms having five valence electrons (arsenic, antimony, phosphorus) or trivalent atoms having three valence electrons (gallium, indium, aluminium, boron). Pentavalent doping atom is known as *donor* atom because it donates or contributes one electron to the conduction band of pure germanium. The trivalent atom, on the other hand, is called *acceptor* atom because it accepts one valence electron from the germanium atom.

---

*For the present discussion, Fermi level may be defined as the energy which corresponds to the centre of gravity of the conduction electrons and holes weighted according to their energies.

Depending on the type of doping material used, extrinsic semi-conductors can be further subdivided into two classes (*i*) *N*-type semi-conductors and (*ii*) *P*-type semi-conductors.

(*i*) **N-type Extrinsic Semi-conductor.** This type of semi-conductor is obtained when a pentavalent material like antimony (Sb) is added to a pure germanium crystal. As shown in Fig. 27.14 (*a*), each antimony atom forms covalent bonds with the surrounding four germanium atoms with the help of four of its five electrons. The fifth electron is superfluous and is loosely bound to the antimony atom. Hence, it can be easily excited from the valence band to the conduction band by the application of electric field or increase in its thermal energy. Thus, practically every antimony atom introduced into the germanium lattice contributes a conduction electron *without creating* a positive hole.* Antimony is called *donor impurity* and makes the pure germanium an *N*-type (*N* for the Negative) extrinsic semi-conductor.

It may be noted that by giving away its one valence electron, the donor atom becomes a positively-charged ion. But it cannot take part in conduction because it is firmly fixed or tied into the crystal lattice. As seen, in addition to the electrons and holes intrinsically available in germanium, the addition of antimony greatly increases the number of conduction electrons. Hence, concentration of electrons in the conduction band is increased and exceeds the concentration of holes in the valence band. Because of this, Fermi level shifts upwards towards the bottom of the conduction band as shown in Fig. 27.14 (*b*).**

In terms of energy levels, the fifth antimony electron has an energy level (called donor level) just below the conduction band. Usually, the donor level is 0.01 eV below conduction band for germanium and 0.054 eV for silicon.

It is seen from the above description that in an *N*-type semi-conductor, electrons are the majority carriers while holes constitute the minority carriers.

(*ii*) **P-type Extrinsic Semi-conductor.** This type of semi-conductor is obtained when traces of a trivalent impurity like boron (B) are added to a pure germanium crystal.

In this case, the three valence electrons of boron atom form covalent bonds with four surrounding germanium atoms but one bond is left incomplete and gives rise to a hole as shown in Fig. 27.15 (*a*).

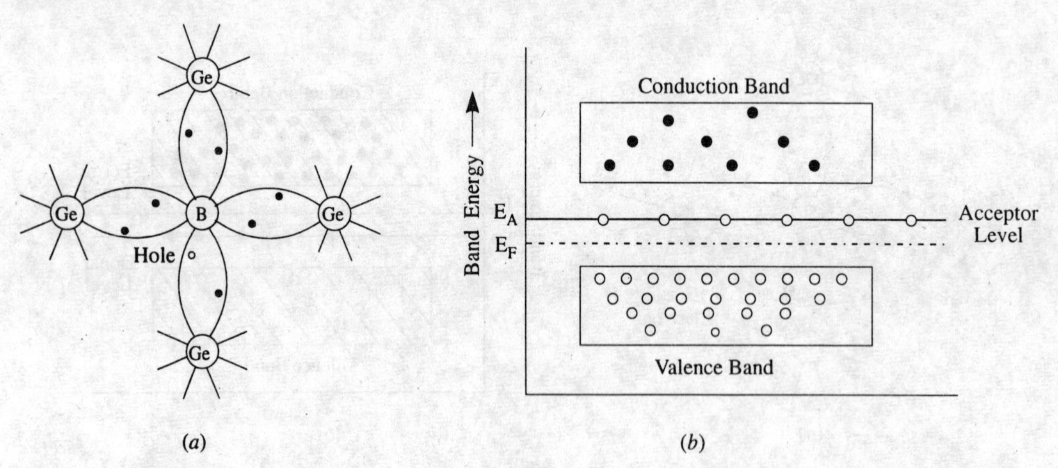

Fig. 27.15

Thus, boron which is called an acceptor impurity causes as many positive holes in a germanium crystal as there are boron atoms thereby producing a *P*-type (*P* for the Positive) extrinsic semi-conductor.

---

*However, *N*-type extrinsic semi-conductor does not have negative charge. It is electrically neutral because antimony atoms add as much positive charge as the negative charge.

**Since the number of electrons as compared to the number of holes increases with temperature, the position of Fermi level also changes considerably with temperature.

In this type of semi-conductor, conduction is by means of holes in the valence band. Accordingly, holes form the majority carriers whereas electrons constitute minority carriers.

Since concentration of holes in the valence band is more than the concentration of electrons in the conduction band, Fermi level shifts nearer to the valence band [Fig. 27.15 (*b*)] The acceptor level lies immediately above the Fermi level. Conduction is by means of hole movement at the top of valence band, the acceptor level readily accepting electrons from the valence band.

27.10. Majority and Minority Charge Carriers

In a piece of pure germanium or silicon, no free charge carriers are available at 0°K. However, as its temperature is raised to room temperature, some of the covalent bonds are broken by heat energy and as a result, electron-hole pairs are produced. These are called thermally-generated charge carriers. They are also known as intrinsically-available charge carriers. Ordinarily, their number is quite small.

An intrinsic or pure germanium can be converted into a *P*-type semi-conductor by the addition of an acceptor impurity which adds a large number of holes to it. Hence, a *P*-type material contains following charge carriers:

(*a*) large number of positive holes—most of them being the added impurity holes with only a very small number of thermally-generated ones.

(*b*) a very small number of thermally-generated electrons (the companions of the thermally-generated holes mentioned above).

Obviously, in a *P*-type material, the number of holes (both added and thermally-generated) is much more than that of electrons. Hence, in such a material, holes constitute majority carriers and electrons form minority carriers as shown in Fig. 27.16 (*a*).

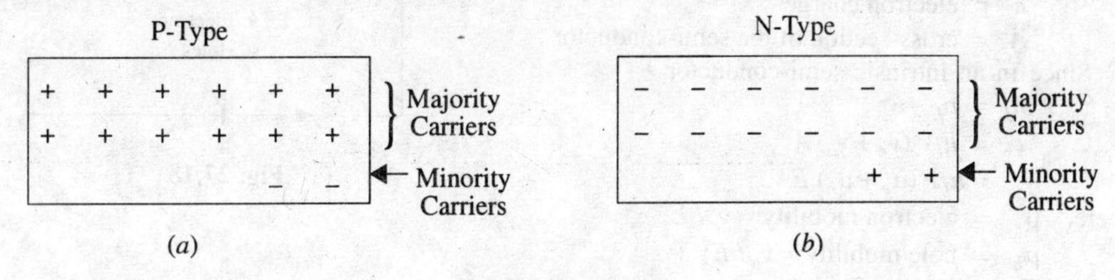

Fig. 27.16

Similarly, in an *N*-type material, the number of electrons (both added and thermally-generated) is much larger than the number of thermally-generated holes. Hence, in such a material, electrons are majority carriers whereas holes are minority carriers as shown in Fig. 27.16 (*b*).

27.11. Mobile Charge Carriers and Immobile Ions

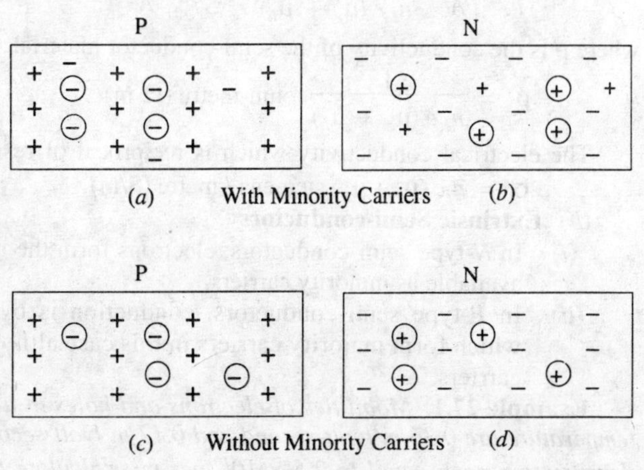

Fig. 27.17

As discussed in Art. 27.9, *P*-type material is formed by the addition of acceptor impurity atoms like boron to the pure Ge or Si crystals. The number of holes added is equal to the number of boron atoms because each such atom contributes one hole. Now, when a hole moves *away* from its parent atom, the remaining atom becomes a negative ion. Unlike the mobile and free-moving hole, this ion cannot take part in conduction because it is fixed in the crystal lattice. In Fig. 27.17 (*a*) these immobile ions are shown by circled minus signs whereas free and mobile holes are shown by uncircled plus signs. Thermally-generated electrons (which form minority carriers) are shown by uncircled minus signs.

Similarly, addition of pentavalent atoms like antimony to pure Ge or Si crystal produces an *N*-type material. The number of free and mobile electrons which are added equals the number of donor Sb atoms. Again, when an electron moves *away* from its parent atom, it leaves behind a positive ion. This ion, being fixed in crystal structure, cannot take part in conduction. As shown in Fig. 27.17 (*b*), these immobile ions are represented by *circled* plus signs whereas free and mobile electrons are represented by *uncircled* minus signs. The thermally-generated holes (which form minority carriers in his case) are shown by *uncircled* plus signs. In Fig. 27.17 (*c*) and (*d*), minority carriers of both types have been neglected. Hence, the figure does not show the small number of free electrons in the *P*-type material or the small number of holes in the *N*-type material.

27.12. Current Carriers in Semi-conductors

We will consider the following cases separately:

(*a*) **Intrinsic Semi-conductors.** In their case, current flow is due to the movement of electrons and holes in opposite directions (Fig. 27.18). Even though the number of electrons equals the number of holes, hole mobility μ_h is practically half of electron mobility.

The total current flow which is due to the flow of electrons and holes is given by

$$I = I_e + I_h = n_I\, e\, v_e A + p_I\, e\, v_h A$$

where v_e = drift velocity of electrons
v_h = drift velocity of holes
n_I = density of free electrons in an intrinsic semi-conductor
p_I = density of holes in an intrinsic semi-conductor
e = electron charge
A = cross-section of the semi-conductor

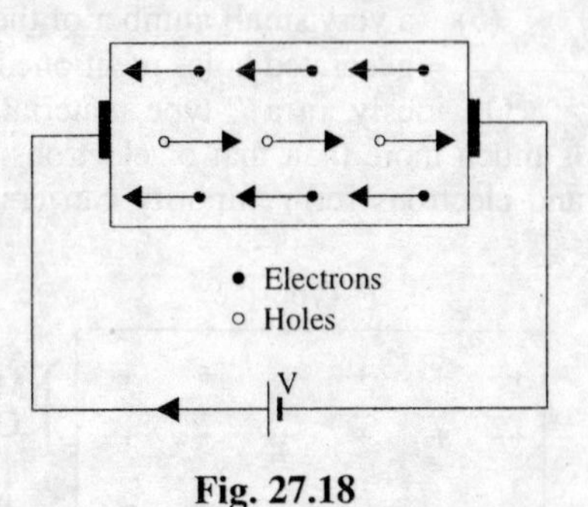

Fig. 27.18

Since in an intrinsic semi-conductor

$$n_I = p_I$$

$$\therefore \quad I = n_I\, e\,(v_e + v_h)\, A$$

$$= n_I\, e\,(\mu_e + \mu_h)\, EA$$

where μ_e = electron mobility $= v_e / E$
μ_h = hole mobility $= v_h / E$
E = electric field

Now, $E = V/l$ where l is the length of the intrinsic semi-conductor.

$$\therefore \quad I = n_I\, e\,(\mu_e + \mu_h)\, AV/l$$

$$\therefore \quad \frac{V}{I} = \frac{l}{A} \cdot \frac{1}{n_I\, e\,(\mu_e + \mu_h)} = \rho \cdot \frac{l}{A}$$

where ρ is the conductivity of the semi-conductor material. It is given by

$$\rho = \frac{1}{n_I\, e\,(\mu_e + \mu_h)} \text{ ohm-metre } (\Omega\text{-m})$$

The electrical conductivity which is reciprocal of resistivity is given by

$$\sigma = n_I e\,(\mu_e + \mu_h) \text{ siemens / metre (S/m)}$$

(*b*) **Extrinsic Semi-conductors**

(*i*) In *N*-type semi-conductors, electrons form the majority carriers although holes are also available as minority carriers.

(*ii*) In *P*-type semi-conductors, conduction is by means of holes in the valence band which form majority carriers in this case although electrons are available as minority carriers.

Example 27.1. *Mobilities of electrons and holes in a sample of intrinsic germanium at room temperature are 0.36 m²/volt-second and 0.17 m²/volt-second respectively. If the electron and hole densities are each equal to 2.5 × 10¹⁹ per m², calculate the conductivity.*

Solution. As shown in Art. 27.12, the conductivity of an intrinsic conductor is

$\sigma = n_I e (\mu_e + \mu_h)$ S/m

$n_I = 2.5 \times 10^{19} / m^3$; $e = 1.6 \times 10^{-19}$ C, $\mu_e = 0.36 \, m^2/V\text{-s}$

$\mu_h = 0.17 \, m^2/V\text{-s}$

$\therefore \quad \sigma = 2.5 \times 10^{19} \times 1.6 \times 10^{-19} (0.36 + 0.17) =$ **2.12 S/m**

27.13. The P-N Junction

It is possible to manufacture a single piece of a semi-conductor material (either Ge or Si) one half of which is doped by *P*-type impurity and the other half by *N*-type impurity as shown in Fig. 27.19. The plane dividing the two halves or zones is called a *P-N junction.*

During the formation of junction, following two phenomena take place:

1. a thin depletion layer or region is set up on both sides of the junction and is so-called because it is depleted or devoid of free charge carriers. Its width is about 1 μm (10^{-6} m).
2. a junction or barrier potential V_B is developed across the junction whose value is about 0.3 V for Ge and 0.7 V for Si.

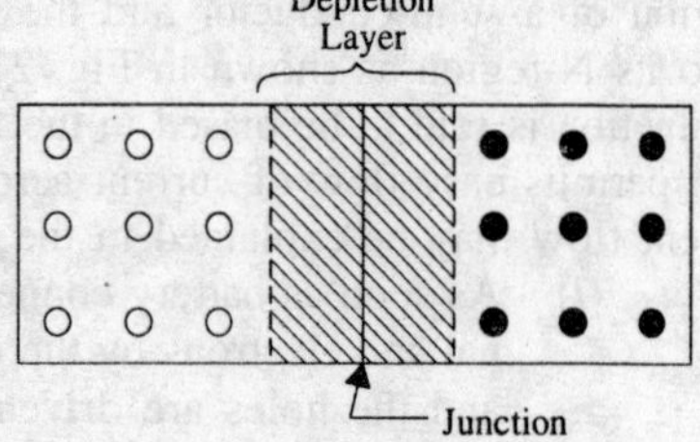

Fig. 27.19

When a *P-N* junction is packed as a semi-conductor device, it is called a *P-N* junction diode.

27.14. Formation of Depletion Layer

Suppose that a *P-N* junction has just been formed. At that instant, holes are still in the P-region and electrons in the N-region. However, there is greater concentration of holes in P-region (where they form majority carriers) than in N-region (where they form minority carriers). Similarly, concentration of electrons is greater in N-region then in P-region (where they exist as minority carriers). The difference in concentration establishes a density gradient across the junction resulting in carrier diffusion. Holes diffuse from P-to-N-region and electrons from N-to-P region and terminate their existance by recombination as shown in Fig. 27.20.

This recombination of free and mobile holes and electrons produces the narrow region at the junction called *depletion layer.* It is so-called because this region is *devoid or depleted of free and mobile charge carriers* though it does contain fixed or immobile positive and negative ions.

This production of ions is due to the fact that the impurity atoms which provide migratory electrons and holes are themselves left behind in an ionised state bearing a charge which is opposite to that of the departed carrier. For example, when an electron migrates across the junction from N-region to P-region, it leaves behind an atom that is one electron short of its normal quota. This atom is now *ionised* and has a *positive* charge. These impurity ions so produced are fixed in their positions in the crystal lattice in the P and N-regions. Hence, as shown in Fig. 27.20 (*b*), they form parallel rows or *plates* of opposite charges facing each other across the depletion layer. As seen, depletion layer contains *no free and mobile charge* carriers but only fixed and immobile ions. Hence, this layer or region behaves like an insulator (and due to the presence of rows of fixed charges, it possesses capacitance as explained in Art. 27.20).

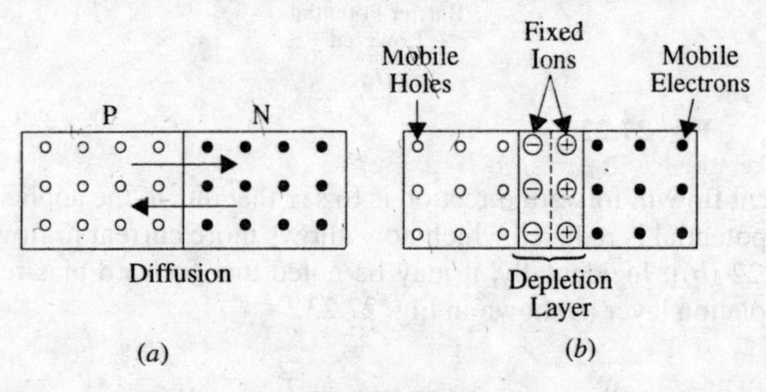

Fig. 27.20

27.15. Junction or Barrier Voltage (V_B)

As explained in Art. 27.14, depletion layer of a P-N junction diode has no free charge carriers but only fixed rows of oppositely-charged ions on its two sides. Because of this charge separation, an electric potential V_B is established across the junction even when *the junction is not connected to any external source of e.m.f.* (Fig. 27.21). It is known as *junction* or barrier potential. It stops further flow of carriers across the junction unless supplied by energy from an external source. At room temperature of 300° K, V_B is about 0.3 V for Ge and 0.7 V for Si.

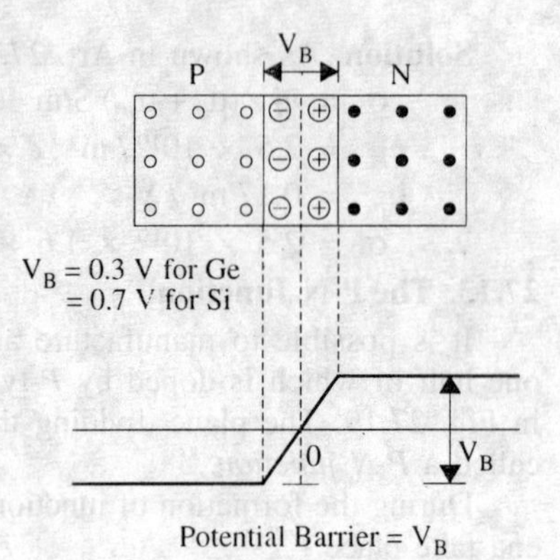

Fig. 27.21

27.16. Forward Biased P-N Junction

Suppose, positive battery terminal is connected to P-region of a semi-conductor and the negative battery terminal to its N-region as shown in Fig. 27.22 (*a*). In that case, the junction is said to be biased in the forward direction because it permits easy flow of current across the junction. The current flow may be explained in the following two ways:

(*i*) As soon as battery connection is made, holes are *repelled* by the positive battery terminal and electrons by the negative battery terminal with the result that both the electrons and the holes are driven *towards* the junction where they recombine. This *en masse* movement of electrons to the left and that of holes to the right of the junction constitutes a large current flow through the semi-conductor. Obviously, the crystal offers *low resistance in the forward direction.*

Incidentally, it may be noted that though there is movement of both electrons and holes *inside* the crystal, only free electrons move in the external circuit *i.e.,* in the battery-connected wires.

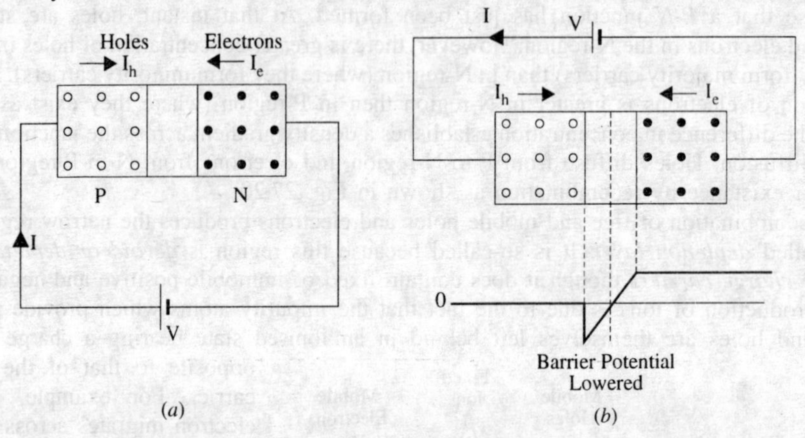

Fig. 27.22

(*ii*) Another way to explain current flow in forward direction is to say that due to the applied external voltage, the barrier potential is reduced which now allows more current to flow across the junction [Fig. 27.22 (*b*)]. Incidentally, it may be noted that forward bias reduces the thickness of the depletion layer as shown in Fig. 27.23.

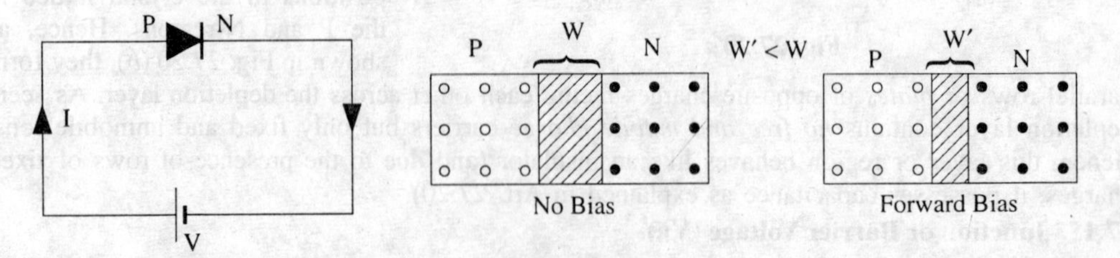

Fig. 27.23

A typical *V/I* characteristic for a forward-biased P-N junction is shown in Fig. 27.24. It is seen that forward current rises exponentially with the applied forward voltage. However, at ordinary room temperature, a p.d. of about 0.3 V is required before a reasonable amount of forward current starts flowing in a germanium junction. This voltage is known as *threshold* voltage (V_{th}) or *cut-in* voltage. It is the same as barrier voltage V_B. Its value for silicon junction is about 0.7 volt. For $V < V_{th}$, current flow is negligible. But as applied voltage increases beyond the threshold value, the forward current increases sharply. If forward voltage is increased beyond a certain safe value, it will produce an extremely large current which may destroy the junction due to overheating. Ge devices can stand junction temperatures around 100°C whereas Si units can function upto 175°C.

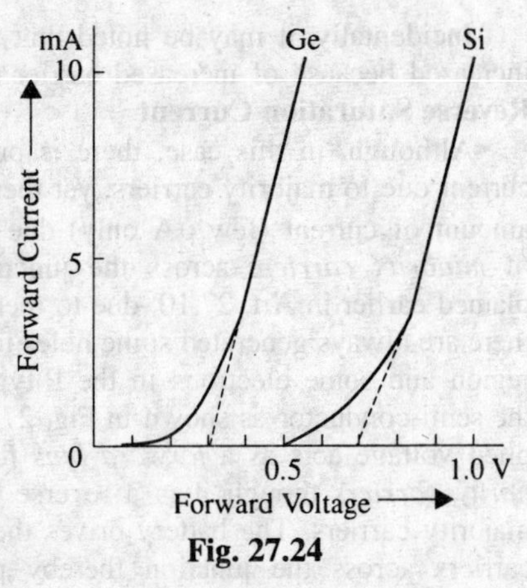

Fig. 27.24

27.17. Reverse Biased P-N Junction

When battery connections to the semi-conductor are made as shown in Fig. 27.25 (*a*), the junction is side to be reverse-biased. In this case, holes are attracted by the negative battery terminal and electrons by the positive terminal so that both holes and electrons move *away* from the junction and away from each other. Since there is no electron-hole combination, no current flows and the junction offers high resistance.

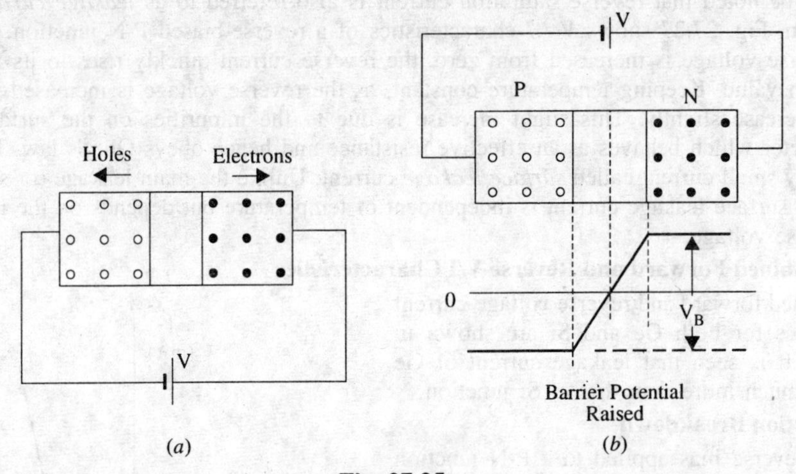

Fig. 27.25

Another way of looking at the above process is that, in this case, the applied voltage increases the barrier potential thereby blocking the flow of *majority* carriers as shown in Fig. 27.25 (*b*).

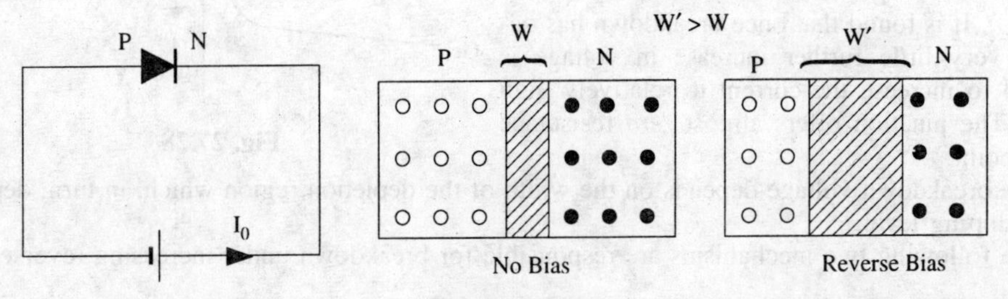

Fig. 27.26

Incidentally, it may be noted that under reverse bias conditions, width of depletion layer is increased because of increased barrier potential (Fig. 27.26)

Reverse Saturation Current

Although, in this case, there is practically no current due to majority carriers, yet there is a small amount of current (few μA only) due to the flow of *minority carriers* across the junction. As explained earlier in Art. 27.10, due to thermal energy, there are always generated some holes in the N-type region and some electrons in the P-type region of the semi-conductor as shown in Fig. 27.16. The applied voltage acts as a *forward bias for these minority carriers* though it is a reverse bias for the majority carriers. The battery drives these minority carriers across the junction thereby producing a small current called *reverse current* or *reverse saturation current* I_s or I_o. Since minority carriers are thermally-generated, I_o is extremely temperature-dependent. For the same reason, forward current is also temperature-dependent but to a much less degree because minority current forms a very small percentage of the majority current.

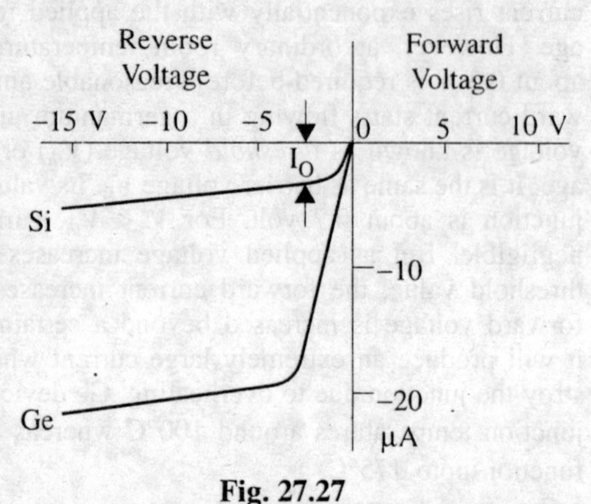

Fig. 27.27

I_o is found to double for every 10°C rise for Ge and for every 6°C rise in the case of Si. Usually, it is of the order of μA for Ge and nA for Si.

It may be noted that reverse saturation current is also referred to as *leakage current* of the P-N junction. Fig. 27.27 shows *V / I* characteristics of a reverse-biased P-N junction. It is seen that as reverse voltage is increased from zero, the reverse current quickly rises to its maximum or saturation value. Keeping temperature constant, as the reverse voltage is increased, I_o is also found to increase slightly. This slight increase is due to the impurities on the surface of the semi-conductor which behaves as an effective resistance and hence obeys Ohm's law. This gives rise to a very small current called *surface leakage* current. Unlike the main leakage (or saturation) current, this surface leakage current is independent of temperature but depends on the magnitude of the reverse voltage.

27.18. Combined Forward and Reverse V/I Characteristics

Combined forward and reverse voltage-current characteristics for both Ge and Si are shown in Fig. 27.28. It is seen that leakage current of Ge junction is much more than that of Si junction.

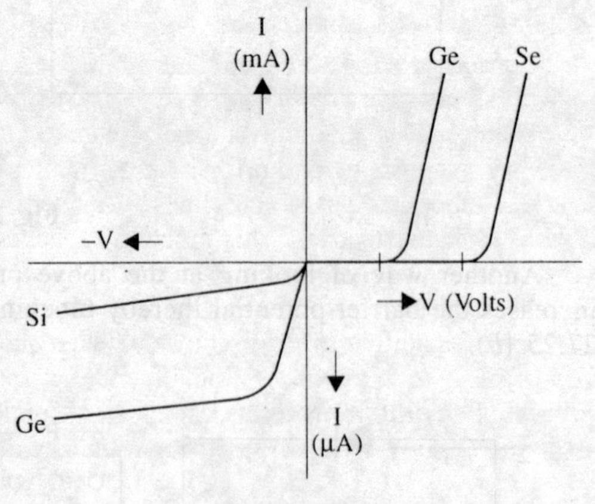

Fig. 27.28

27.19. Junction Breakdown

If the reverse bias applied to a P-N junction is increased, a point is reached when the junction breaks down and reverse current rises sharply to a value limited only by the resistance connected in series with the junction (Fig. 27.29). This critical value of the voltage is known as *breakdown* voltage (V_{BR}). It is found that once breakdown has occurred, very little further increase in voltage is required to increase the current to relatively high values. The junction offers almost *zero* resistance at this point.

The breakdown voltage depends on the width of the depletion region which, in turn, depends on the doping level.

The following two mechanisms are responsible for breakdown under increasing reverse voltage:

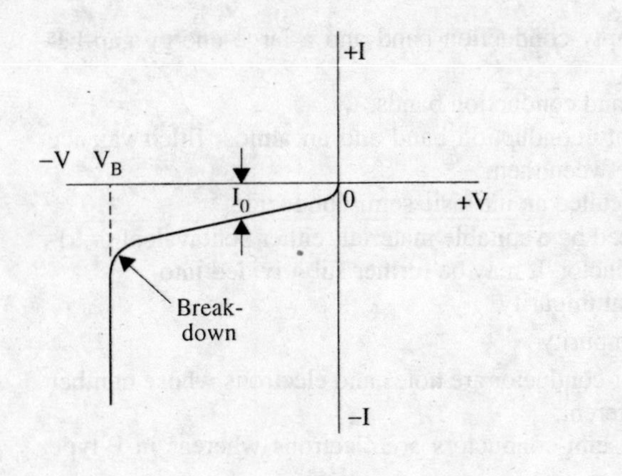

Fig. 27.29

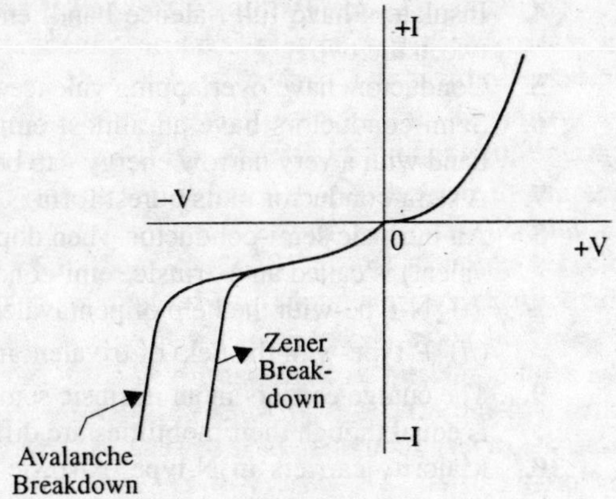

+I
−V
0
+V
Zener
Break-
down
Avalanche
Breakdown
−I

Fig. 27.30

(*a*) **Zener Breakdown**

This form of breakdown occurs in junctions which, being heavily doped, have *narrow* depletion layers. The breakdown voltage sets up a very strong electric field (about 10^8 V/m) across this narrow layer. This field is strong enough to break or *rupture* the covalent bonds thereby generating electron-hole pairs. Even a small further increase in reverse voltage is capable of producing large number of current carriers. That is why junction has very low resistance in the breakdown region.

(*b*) **Avalanche Breakdown**

This form of breakdown occurs in junctions which, being lightly-doped, have *wide* depletion layers where the electric field is not strong enough to produce Zener breakdown. Instead, the minority carriers (accelerated by this field) collide with the semi-conductor atoms in the depletion region. Upon collision, covalent bonds are broken and electron-hole pairs are generated. These newly-generated charge carriers are also accelerated by the electric field resulting in more collisions and hence further production of charge carriers. This leads to an avalanche of charge carriers and, consequently, to a very low reverse resistance. The two breakdown phenomena are shown in Fig. 27.30.

27.20. Junction Capacitance

Capacitive effects are exhibited by P-N junctions when they are both forward and reverse-biased.

(*a*) *Transition Capacitance* (C_T) *or Space-charge Capacitance*

When a P-N junction is *reverse-biased,* the depletion ragion acts like an insulator or as a dielectric material essential for making a capacitor. The P and N-type regions on either side have low resistance and act as the plates. We, therefore, have all the components necessary for making a parallel-plate capacitor. This junction capacitance is called *transition* or *space charge* capacitance (C_T). It is voltage-dependent.

(*b*) *Diffusion or Storage Capacitance* (C_D)

This capacitive effect is present when the junction is *forward-biased.* It is called *diffusion capacitance* to account for the time delay in moving charge across the junction by diffusion process. Due to this fact, this capacitance cannot be identified in terms of a dielectric and plates. It varies directly with the magnitude of forward current.

HIGHLIGHTS

1. The electrons in the atoms of a solid occupy specific energy bands.
2. The outermost electrons of an atom are called valence electrons. They occupy valence band which is the highest occupied band of an atom.
3. The higher band next to the valence band is called conduction band which may be empty or partially filled by electrons.

4. Insulators have full valence band, empty conduction band and a large energy gap between the two.
5. Conductors have overlapping valence and conduction bands.
6. Semi-conductors have an almost empty conduction band and an almost filled valence band with a very narrow energy gap between them.
7. A semi-conductor in its purest form is called an intrinsic semi-conductor.
8. An intrinsic semi-conductor when doped by a suitable material (either pentavalent or trivalent) is called an extrinsic semi-conductor. It may be further sub-divided into
(*i*) N-type-with the help of pentavalent impurity
(*ii*) P-type-with the help of trivalent impurity.
9. The charge carriers in an intrinsic semi-conductor are holes and electrons whose number is equal though their mobilities are different.
10. Majority carriers in N-type extrinsic semi-conductors are electrons whereas in P-type holes are the majority carriers.
11. A P-N junction has a depletion region and a barrier voltage.
12. A forward-biased P-N junction conducts current whereas a reverse-biased junction does not except for an extremely small leakage current.
13. A P-N junction suffers Zener breakdown when heavily-doped and avalanche breakdown when lightly-doped.
14. A P-N junction possesses diffusion capacitance when forward-biased and transition capacitance when reverse-biased.

OBJECTIVE TESTS—27

A. Fill in the following blanks.

1. The outermost electrons of an atom are called .... electrons.
2. The single energy level of an isolated atom becomes a............ of energies in a solid.
3. The conduction band of a solid may be either .... or partially-filled with electrons.
4. Insulators are those materials which have full .... band.
5. The atoms of semi-conductor materials form .... bands.
6. When a pentavalent material is added to a pure Ge crystal, we get............. type extrinsic semi-conductor.
7. In an intrinsic semi-conductor, number of electrons .... the number of holes.
8. In silicon, barrier potential is nearly ....volt.

B. Answer True or False.

1. The outermost electrons of an atom have the highest binding energy.
2. Conductors have overlapping valence and conduction bands.
3. The trivalent doping atom is called donor atom.
4. A P-type extrinsic semi-conductor is electrically positive.
5. Electrons form majority carriers in N-type semi-conductors.
6. The depletion region around a P-N junction is free of charge carriers.
7. Zener breakdown occurs when semi-conductor material is heavily-doped.
8. A forward-biased P-N junction possesses diffusion capacitance.

C. Multiple Choice Questions.

1. An atom is said to be ionised when any one of its orbiting electron
(*a*) jumps from one orbit to another
(*b*) is raised to a higher orbit
(*c*) comes to the ground state
(*d*) is completely removed

2. The outermost electrons of an atom are called .... electrons.
(*a*) free (*b*) valence
(*c*) conduction (*d*) bound

3. The conduction band of a semi-conductor material may be ....
(*a*) completely-filled
(*b*) empty
(*c*) partially-filled

(*d*) either (*b*) or (*c*)

4. .... is a donor material

(*a*) gallium (*b*) antimony

(*c*) boron (*d*) indium

5. The depletion region around a P-N junction

(*a*) is quite wide

(*b*) contains mobile ions

(*c*) has no free charge carriers

(*d*) is of constant width

6. When a P-N junction is reverse-biased,

(*a*) it offers high resistance

(*b*) its depletion layer becomes narrow

(*c*) its barrier potential decreases

(*d*) it breaks down

ANSWERS

A. 1. valence 2. band 3. empty 4. valence 5. covalent 6. P 7. equals 8. 0.7

B. 1. F 2. T 3. F 4. F 5. T 6. T 7. T 8. T

C. 1. *d* 2. *b* 3. *d* 4. *b* 5. *c* 6. *a*

28 SEMI-CONDUTOR DIODES

28.1. P-N Junction Diode

(a) Construction

It is a two-terminal device consisting of a P-N junction formed either in Ge or Si crystal. Its circuit symbol is shown in Fig. 28.1 (*a*). The P- and N-type regions are referred to as *anode* and *cathode* respectively. In Fig. 28.1 (*b*), arrowhead indicates the *conventional* direction of current flow when forward-biased. It is the same direction in which hole flow takes place.

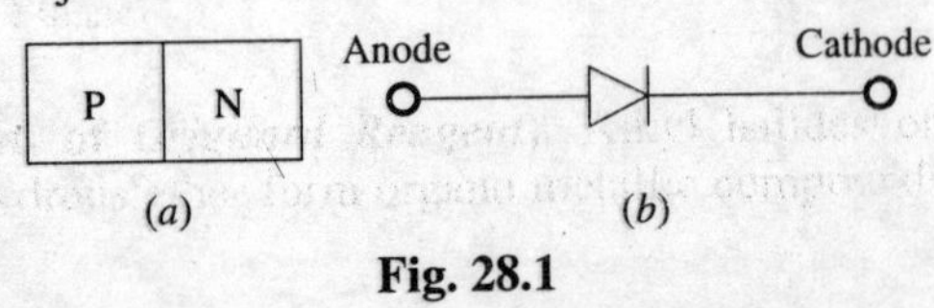

Fig. 28.1

Commercially available diodes usually have some means to indicate which leads is P and which lead is N. Standard notation consists of type numbers preceded by 'IN' such as IN 240 and IN 1250. Here, 240 and 1250 correspond to colour bands.

Fig. 28.2 (*a*) shows typical diodes having a variety of physical structures whereas Fig. 28.2 (*b*) shows terminal identifications.

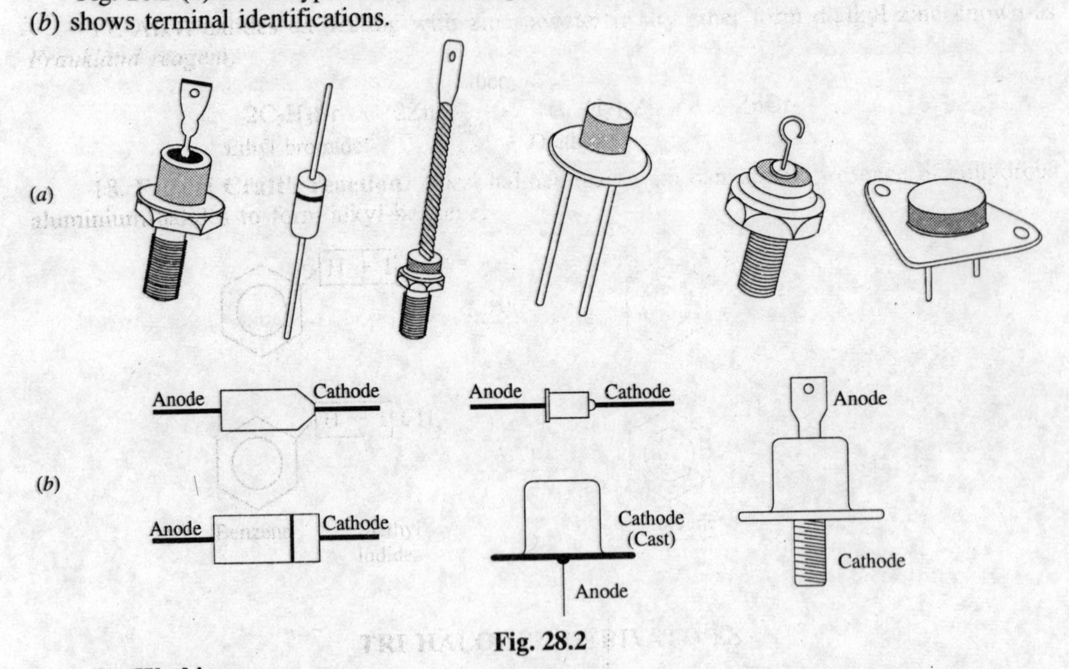

Fig. 28.2

(b) Working

A P-N junction diode is a one-way device offering low resistance when forward-biased and behaving almost as an insulator when reverse-biased. Hence, such diodes are mostly used as rectifiers *i.e.*, for converting alternating current into direct current.

(c) V/I Characteristic

Fig. 28.3 shows the static voltage-current characteristics for a low-power P-N junction diode.

(*i*) **Forward Characteristic**

When the diode is forward-biased and the applied voltage is increased from zero, hardly any current flows through the device in the beginning. It is so because the external voltage is being opposed by the internal barrier voltage V_B whose value is 0.7 V for Si and 0.3 V for Ge. As soon as V_B is neutralised, current through the diode increases rapidly with increasing battery voltage. It is found that as little a voltage as 1.0 V produces forward current of about 50 mA.

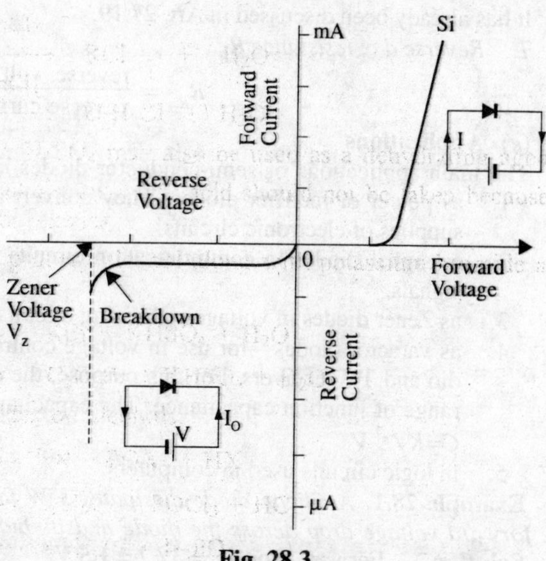

Fig. 28.3

(*ii*) **Reverse Characteristic**

When the diode is reverse-biased, majority carriers are blocked and only a small current (due to minority carriers) flows through the diode. As the reverse voltage is increased from zero, the reverse current very quickly reaches its maximum or saturation value I_0 which is also known as *leakage current.* It is of the order of nanoamperes (nA) for Si and microamperes (μA) for Ge.

As seen from Fig. 28.3, when reverse voltage exceeds a certain value called *Zener voltage* V_z, the leakage current suddenly and sharply increases, the curve indicating zero resistance at this point. When P-N junction diodes are employed primarily because of this property as voltage regulators, they are called *Zener diodes* (Art. 28.7).

(*d*) **Diode Parameters**

The diode parameters of greatest interest are:

1. *Bulk resistance* (r_B)

It is the sum of the resistance values of the P- and N-type semi-conductor materials of which the diode is made of. Usually, it is very small.

2. *Junction resistance* (r_j)

Its value for forward-biased junction depends on the magnitude of forward d.c. current.

$$r_j \cong \frac{25 \text{ mV}}{I_F} \quad \text{... for Ge}$$

$$\cong \frac{50 \text{ mV}}{I_F} \quad \text{... for Si}$$

where I_F is the forward current in mA.

3. *Dynamic or a.c. resistance*

r_{ac} or $r_d = r_B + r_j$

As seen from forward characteristic of Fig. 28.4.

$$r_{ac} = \frac{\Delta V_F}{\Delta I_F}$$

... at a given d.c. forward current.

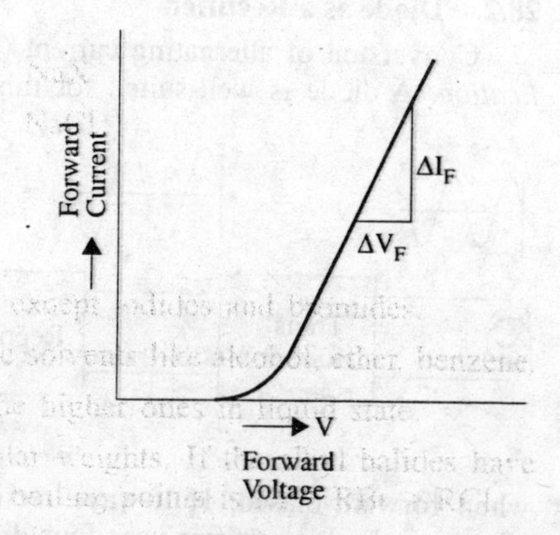

Fig. 28.4

4. *Forward voltage drop*

It is given by the relation: forward voltage drop

$$= \frac{\text{power dissipated}}{\text{forward d.c. current}}$$

5. *Reverse saturation current* (I_0)

It has already been discussed in Art. 27.17.

6. *Reverse breakdown voltage* (V_{BR})

It has already been discussed in Art. 27.19.

7. *Reverse d.c. resistance* R_R

$$R_R = \frac{\text{reverse voltage}}{\text{reverse current}}$$

(*e*) **Applications**

The main applications of semi-conductor diodes in modern electronic circuitry are as under:

1. as power or rectifier diodes. They convert a.c. current into d.c. current for d.c. power supplies of electronic circuits.
2. as signal diodes in communication circuits for modulation and demodulation of small signals.
3. as Zener diodes in voltage stabilizing circuits.
4. as varactor diodes—for use in voltage controlled tuning circuits as may be found in radio and TV receivers. For this purpose, the diode is deliberately made to have a certain range of junction capacitance. The capacitance of the reversed-biased diode is given by $C = K/\sqrt{V}$.
5. in logic circuits used in computers.

Example 28.1. *A silicon diode dissipates 3 W for a forward d.c. current of 2 A. Calculate the forward voltage drop across the diode and its bulk resistance.*

Solution. Forward drop $V_F = 3/2 = \mathbf{1.5\ V}$

Now, $V_F = V_j + r_B I_F$

or $1.5 = 0.7 + r_B \times 2 \qquad \therefore \quad r_B = \mathbf{0.4\ \Omega}$

Example 28.2. *A silicon diode has a forward voltage drop of 1.2 V for a forward d.c. current of 100 mA. It has a reverse current of* 1 μA *for a reverse voltage of 10 V. Calculate*

(*a*) *bulk and reverse resistances of the diode;*

(*b*) *a.c. resistance at forward d.c. current of (i) 2.5 mA and (ii) 25 mA.*

Solution. (*a*)
$$r_B = \frac{V_F - V_j}{I_F} = \frac{1.2\ \text{V} - 0.7\ \text{V}}{100\ \text{mA}} = \mathbf{5\ \Omega}$$

$$R_R = \frac{V_R}{I_R} = \frac{10\ \text{V}}{1\ \mu\text{A}} = \mathbf{10\ M\Omega}$$

(*b*) (*i*)
$$r_j = \frac{50\ \text{mA}}{2.5\ \text{mA}} = \mathbf{20\ \Omega}$$

$$r_{ac} = r_B + r_j = 5 + 20 = \mathbf{25\ \Omega}$$

(*ii*)
$$r_j = \frac{50\ \text{mV}}{25\ \text{mA}} = \mathbf{2\ \Omega}$$

$$r_{ac} = 5 + 2 = \mathbf{7\ \Omega}$$

28.2. Diode as a Rectifier

Conversion of alternating current (or voltage) into direct current (or voltage) is called *rectification.* A diode is well-suited for this job because it conducts only in one direction *i.e.*, only

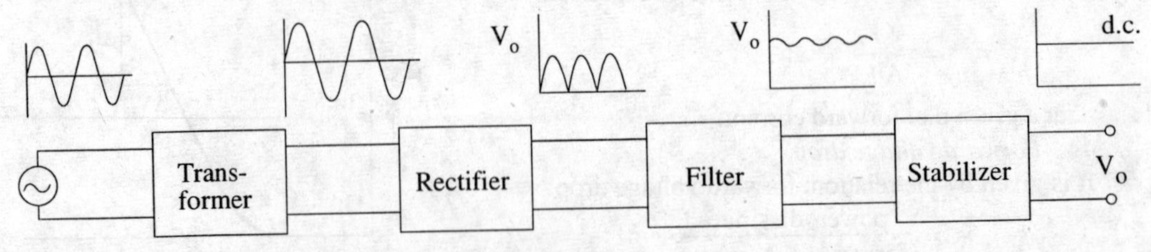

Fig. 28.5

when forward-biased. Half-wave rectifier uses one diode, full-wave rectifier uses two diodes whereas a bridge rectifier uses four diodes.

In India, electric energy is available at an alternating voltage of 220 V (r.m.s. value) at 50 Hz. However, for the operation of most of the electronic equipment, d.c. voltage is needed. Hence, practically, every electronic equipment includes a circuit which provides it with a stabilized d.c. output voltage from the input a.c. voltage. In fact, this circuit forms the power supply of the equipment and consists of:

1. a transformer—for either stepping up or down the a.c. supply voltage.
2. rectifier—for converting this stepped up (or down) a.c. voltage into d.c. voltage.
3. smoothing filter—for removing any variations or ripples in the d.c. output voltage.
4. stabilizer—for keeping output d.c. voltage constant even when input voltage or load current changes.

A block diagram of such a power supply is shown in Fig. 28.5. Obviously, rectifier is the heart of a power supply.

Incidentally, in Fig. 28.5, full-wave rectifier has been assumed.

28.3. Half-wave Rectifier

Though such a circuit is not much used, yet we will discuss it here for the sake of explaining the basic principle involved in the working of a rectifier. The simple circuit of a half-wave rectifier is shown in Fig. 28.6 along with the input and output voltage waveforms. An a.c. voltage is applied to a single diode connected in series with a load resistor R_L.

(*a*) **Working**

During the positive half-cycle of the input a.c. voltage *i.e.*, when point *M* is positive, the diode *D* is forward-biased (ON) and conducts. While conducting, the diode acts as a short-circuit so that circuit current flows and hence, positive half-cycle of the input a.c. voltage is dropped across R_L. It constitutes the output voltage υ_0 as shown in Fig. 28.6 (*b*). Waveform of the diode current (which equals load current) is similar to the voltage waveform.

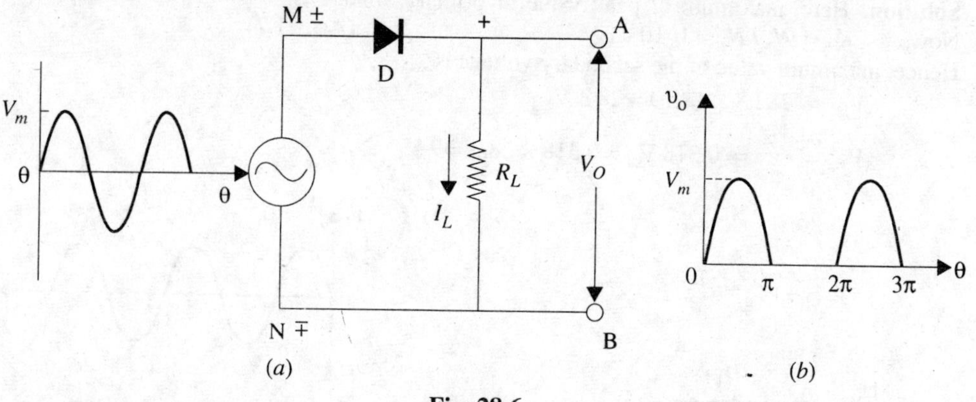

Fig. 28.6

During the negative input half-cycle *i.e.*, when point *M* becomes negative, the *diode* is reverse-biased (OFF) and, so, does not conduct *i.e.*, there is no current flow. Hence, there is no voltage drop across R_L. In other words $i_L = 0$ and $\upsilon_L = \upsilon_0 = 0$. Obviously, the negative input half-cycle is suppressed *i.e*, it is not utilized for delivering power to the load. As seen, the output is not a steady d.c. but only a *pulsating d.c. wave* having a ripple frequency equal to that of the in-

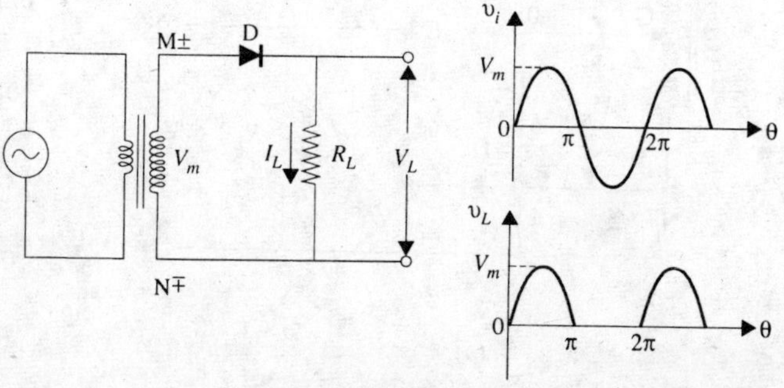

Fig. 28.7

put voltage frequency. This wave can be observed by an oscilloscope connected across R_L. When measured by a d.c. meter, it will show some average positive value both for voltage and current. Since only one half-cycle of the input wave is used, it is called a *half-wave recitifier*. It should be noted that forward voltage drop across the diode has been neglected in the above discussion. We have, in fact, assumed an ideal diode. If it is required to step up or step down the input voltage, we will have to use a power transformer as shown in Fig. 28.7.

Another advantage of using this transformer is that it isolates the circuit from the a.c. mains thereby reducing the risk of electrical shock.

(*b*) **Average Values**

The output voltage of a half-wave rectifier is found to consist of positive half-cycles only. Their average value (Fig. 28.8) is

$$V_{av} = V_{d.c.} = \frac{V_m}{\pi} = 0.318\ V_m$$

Similarly,

$$I_{av} = I_{d.c.} = \frac{I_m}{\pi} = 0.318\ I_m$$

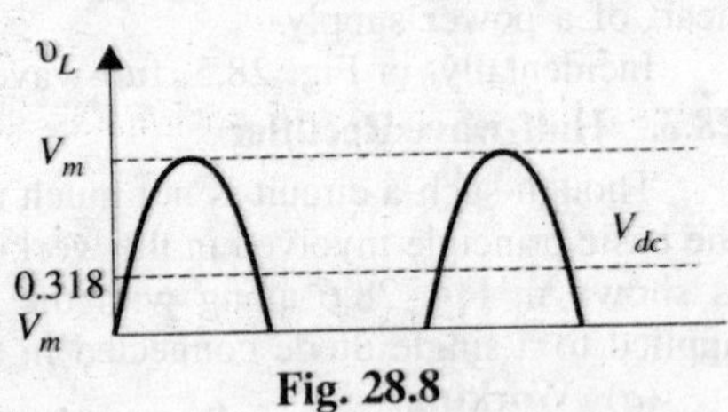

Fig. 28.8

$V_{d.c.}$ is the d.c. component of the load voltage V_L and, in a similar way, $I_{d.c.}$ is the d.c. component of I_L. Both V_L and I_L have a.c. components.

Example 28.3. *A single-phase half-wave rectifier supplies power to a 1k Ω load. The sinusoidal a.c. supply has an r.m.s. value of 200 V. The step-down transformer has a turn ratio* $N_1 / N_2 = 10$. *Neglecting forward resistance of the diode, calculate the d.c. voltage across the load.*

Solution. Here, maximum or peak value of primary voltage is $= 200 \times \sqrt{2} = 282.8$ V (Art. 21.7)

Now, $K = N_2 / N_1 = 1/10$

Hence, maximum value of the secondary voltage is

$$V_m = 282.8 \times 1/10 = 28.3 \text{ V}$$

$$\therefore \quad V_{d.c.} = \frac{V_m}{\pi} = 0.318\ V_m = 0.318 \times 28.3 = \mathbf{9\ V}$$

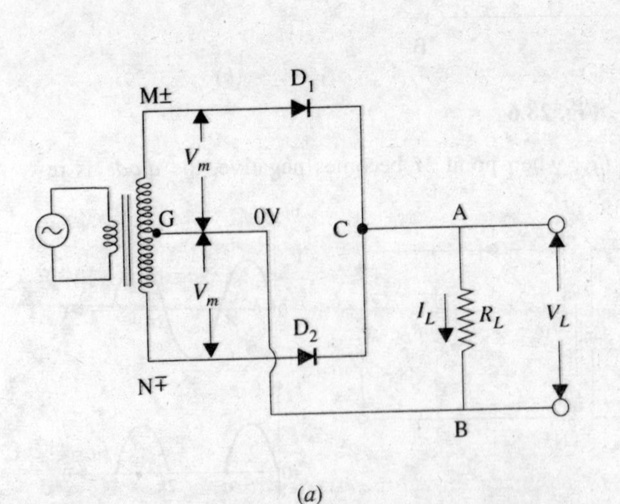

(*a*)

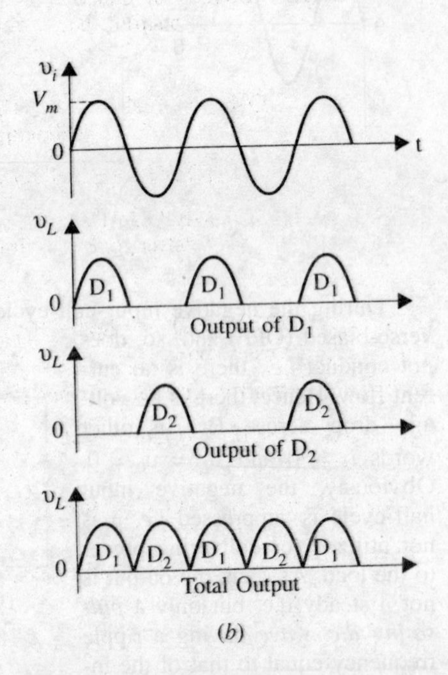

(*b*)

Fig. 28.9

28.4. Full-Wave Rectifier

In this case, *both* half-cycles of the input are utilized with the help of two diodes working alternately. For full-wave rectification, use of a centre-tap transformer is essential (though it is optional for half-wave rectification).

The full-wave rectifier circuit using two diodes and a centre-tapped transformer is shown in Fig. 28.9. The centre-tap is usually taken as the ground or zero voltage reference point.

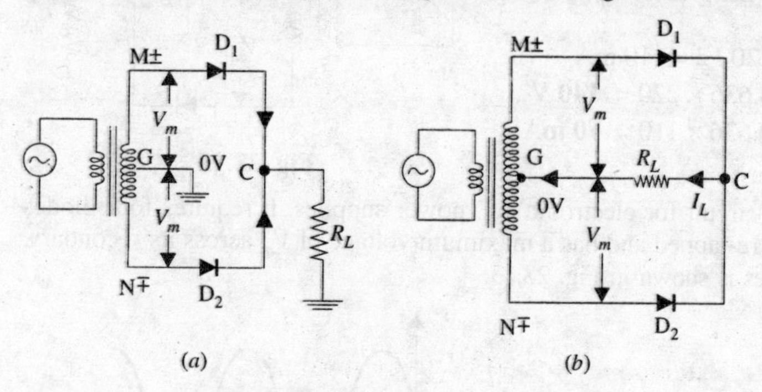

Fig. 28.10

Fig. 28.10 shows two different ways of drawing the circuit. In Fig. 28.10 (*a*), R_L becomes connected to point *G via* the earth whereas in Fig. 28.10 (*b*), it is connected directly to *G*.

(*a*) Working

When input a.c. supply is switched on, the ends *M* and *N* of the transformer secondary become positive and negative alternately. During the positive half-cycle of the a.c. input, terminal *M* is positive, *G* is at zero potential and *N* is at negative potential. Hence, being forward-biased, diode D_1 conducts (but not D_2 which is reverse-biased) and current flows along MD_1CABG (Fig. 28.9). As a result positive half-cycle of the voltage appears across R_L.

During the negative half-cycle, when terminal *N* becomes positive then D_2 conducts (but not D_1) and current flows along ND_2CABG. So, we find that current keeps on flowing through R_L in the same direction (*i.e.*, from *A* to *B*) in both half-cycles of the a.c. input. It means that both half-cycles of the input a.c. supply are utilized as shown in Fig. 28.9 (*b*). Also, the frequency of the rectified output voltage is twice the supply frequency.

(*b*) Average Values

As seen from Fig. 28.11, the output voltage of a full-wave rectifier consists of both positive and negative half-cycles. The d.c. value of load voltage V_L is given by

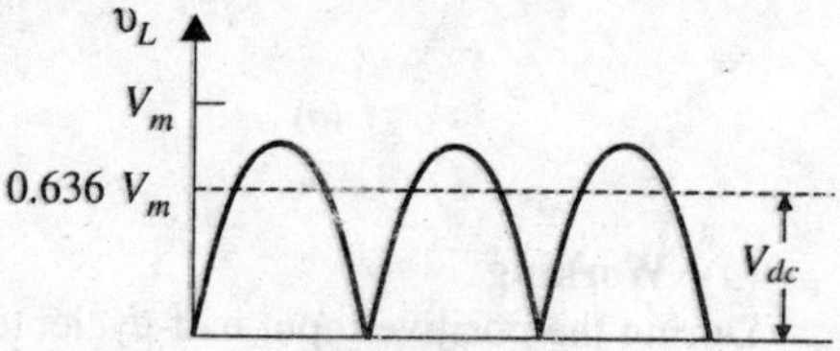

$$V_{dc} = \frac{2V_m}{\pi} = 0.636\,V_m$$

= twice the HW rectified value

where V_m is the maximum voltage across *each half* of the transformer secondary.

Fig. 28.11

Similarly, the d.c. component of load current* is

$$I_{dc} = \frac{2I_m}{\pi} = 0.636\quad I_m = 0.636\frac{V_m}{R_L}$$

Example 28.4. *A centre-tap full-wave rectifier supplies a load of 1k Ω. The a.c. voltage across the secondary is 200-0-200 V (r.m.s.). If diode resistance is neglected, find the d.c. voltage across the load and d.c. load current.* **(Applied Electronics, Maharashtra 1993)**

Solution. $V_m = \sqrt{2}\,V_{rms} = \sqrt{2} \times 200 = 282.8$ V

$\therefore$ $V_{dc} = 0.636 \times 282.8 = \mathbf{180}$

Now, $I_m = V_m / R_L = 282.8\text{ V} / 1\text{ k}\Omega = 282.8$ mA

$\therefore$ $I_{dc} = 0.636 \times 282.8 = 180\text{ mA} = \mathbf{0.18\ A}$

Example 28.5. *In a centre-tap full-wave rectifier, load resistance is 2 K. The a.c. supply across the primary winding is 220 sin 314 t. Taking transformer turn ratio $N_1/N_2 = 1/2$ and*

*It also has an a.c. component.

neglecting diode and secondary winding resistance, compute the d.c. load voltage and current.

Solution. The peak secondary voltage across the whole secondary is = 220 (2/1) = 440 V.

Hence, maximum value of the voltage across *each half* of the secondary winding is = V_m = 440/2 = 220 V.

$$I_m = V_m / R_L = 220/2 = 110 \text{ mA}$$
$$V_{dc} = 0.636\ V_m = 0.636 \times 220 = \mathbf{140\ V}$$
$$I_{dc} = 0.636\ I_m = 0.636 \times 110 = \mathbf{70\ mA}$$

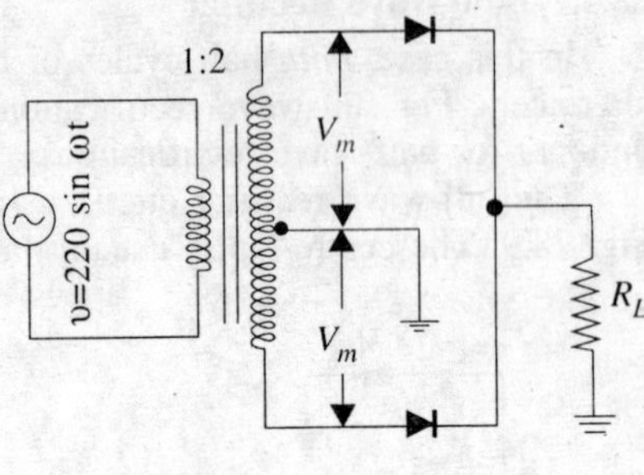

Fig. 28.12

28.5. Full-Wave Bridge Rectifier

It is the most frequently-used circuit for electronic d.c. power supplies. It requires four diodes but the transformer used is not centre-tapped and has a maximum voltage of V_m across its secondary. The circuit using four discrete diodes is shown in Fig. 28.13.

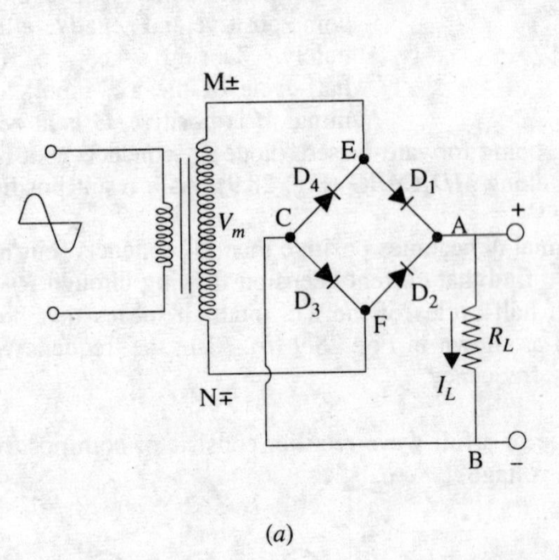

Fig. 28.13

(a) Working

During the positive input half-cycle, terminal *M* of the secondary is positive and *N* is negative as shown separately in Fig. 28.14 (*a*). Diodes D_1 and D_2 become forward-biased (ON) whereas

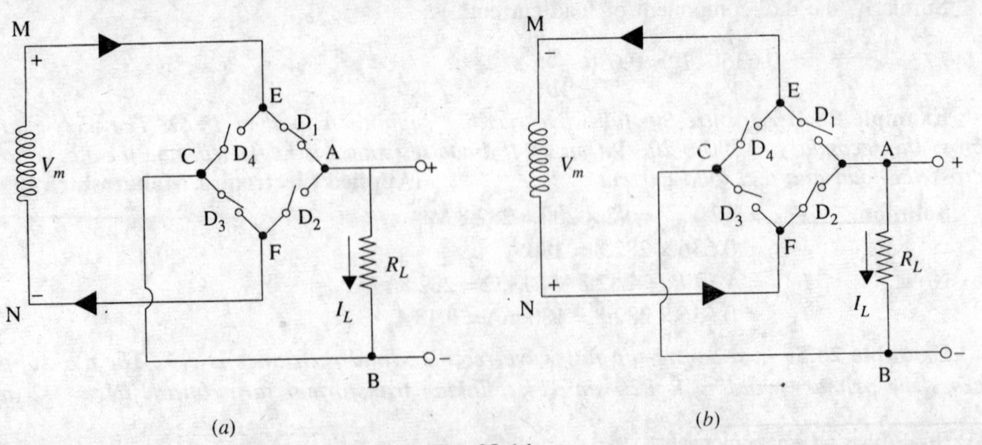

Fig. 28.14

D_2 and D_4 are reverse-biased (OFF). Hence, current flows along the path *MEABCFN* producing a drop across R_L.

During the negative input half-cycle, secondary terminal *N* becomes positive and *M* negative. Now, D_2 and D_4 are forward-biased. Circuit current flows along *NFABCEM* as shown in Fig. 28.14 (*b*). Hence, we find that current keeps flowing through load resistance R_L in *the same direction AB* during both half-cycles of the a.c. input supply. Consequently, point *A* of the bridge rectifier always acts as an anode and point *B* as cathode.

The output voltage across R_L is as shown in Fig. 28.13 (*b*). Its frequency is *twice* that of the supply frequency. Moreover, for a given power transformer, a bridge rectifier produces an output voltage nearly *twice* as large as produced by an ordinary full-wave rectifier. It is due to the fact that the *entire* secondary voltage of the transformer is applied across each of the diode pairs.

(*c*) **Average values**

$$V_{dc} = 0.636\, V_m \text{ and } I_{dc} = 0.636\, I_m \text{ where } I_m = V_m / R_L$$

(*d*) **Advantages**

After the advent of low-cost, highly-reliable and small-sized silicon diodes, bridge circuit has become much more popular than the centre-tapped transformer FW rectifier. The main reason for this is that for a bridge rectifier, a much smaller transformer is required for the same output because it utilizes the transformer secondary continuously unlike the 2-diode FW rectifier which uses the two halves of the secondary alternately.

So, the advantages of the bridge rectifier are:

1. no centre-tap is required on transformer;
2. much smaller transformers are required;
3. it is suitable for high-voltage applications;
4. it has less PIV rating per diode.

The obvious disadvantage is the need for twice as many diodes as for the centre-tapped transformer version. But ready availability of low-cost silicon diodes has made it more economical despite its requirement of four diodes.

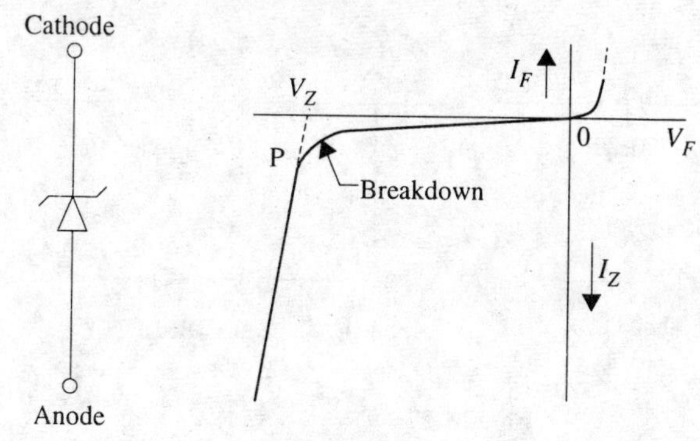

Fig. 28.15

28.6. Zener Diode

It is a reverse-biased heavily-doped silicon or germanium P-N junction diode which is always reverse-biased and operates in the breakdown region where current is limited only by both external resistance and the power dissipation of the diode. Conventional diodes and rectifiers never operate in the breakdown region but the Zener diode makes a virtue of it and operates at this very point like *P* shown in Fig. 28.15. The Zener breakdown occurs due to breaking of covalent bonds by the strong electric field set up in the depletion region by the reverse voltage.

Fig. 28.15 (*a*) shows schematic symbol of a Zener diode and is similar to that of a normal diode except that the line representing the cathode is bent at both ends. With a little mental effort, the cathode symbol can be imagined to look like the letter *Z* for Zener.

The typical *V/I* characteristic is shown in Fig. 28.15 (*b*).

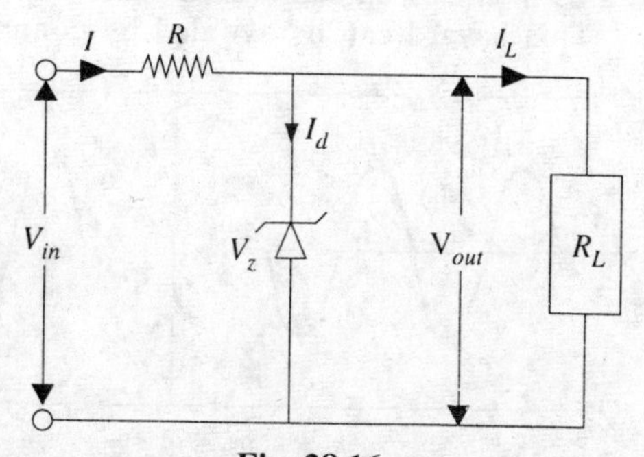

Fig. 28.16

28.7. Zener Diode as Voltage Regulator

Voltage regulation is a measure of a circuit's ability to maintain a constant output voltage even when either input voltage or load current varies. A Zener diode when working in the break-down region can serve as a voltage regulator. In Fig. 28.16, V_{in} is the input d.c. voltage whose variations are to be regulated. The Zener diode is reverse-connected across V_{in}. When p.d. across the diode is greater than V_Z, it conducts and draws relatively large current through the series resistance R. The load resistance R_L across which a constant voltage V_{out} is required, is connected in parallel with the diode. The total current I passing through R equals the sum of diode current and load current *i.e.*, $I = I_d + I_L$.

It will be seen that under all conditions, $V_{out} = V_Z$.

Hence, $\quad V_{in} = IR + V_{out} = IR + V_Z$.

Case 1. Suppose R is kept fixed but supply voltage V_{in} is increased slightly. It will increase I. This increase in I will be absorbed by the Zener diode without affecting I_L. The increase in V_{in} will be dropped across R thereby keeping V_{out} constant. Conversely, if supply voltage V_{in} falls, the diode takes a smaller current and voltage drop across R is reduced thus again keeping V_{out} constant. Hence, when V_{in} changes, I and IR drop change in such a way as to keep V_{out} ($=V_Z$) constant.

Case 2. In this case V_{in} is fixed but I_L is changed. When I_L increases, diode current I_d decreases thereby keeping I and IR drop constant. In this way, V_{out} remains unaffected.

Should I_L decreases I_d would increase in order to keep I and hence IR drop constant. Again, V_{out} would remain unchanged because

$$V_{out} = V_{in} - IR = V_{in} - (I_d + I_L)\,R$$

Incidentally, it may be noted that

$$R = \frac{V_{in} - V_{out}}{I_d + I_L}$$

It may also be noted that when diode current reaches its maximum value, I_L becomes zero. In that case

$$R = \frac{V_{in} - V_{out}}{I_{d(max)}}$$

28.8. Zener Diode for Meter Protection

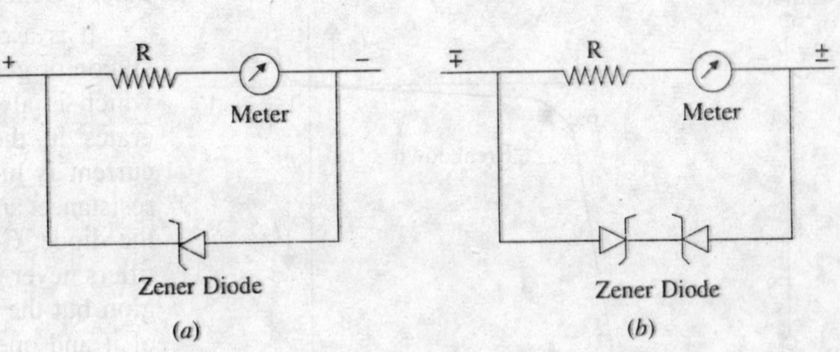

Fig. 28.17

Zener diodes are frequently used in volt-ohm-milliammeters for protecting meter movement against burn-out from accidental overloads. If VOM is set to its 2.5 V range and the test leads are accidentally connected to a 25 V circuit, an unprotected meter will be burned out or at least get severely damaged.

This hazard can be avoided by connecting a Zener diode in parallel with the meter as shown in Fig. 28.17(*a*). In the event of an accidental overload, most of the current will pass through the diode. Two Zener diodes connected as shown in Fig. 28.17 (*b*) can provide overload protection regardless of the applied polarity.

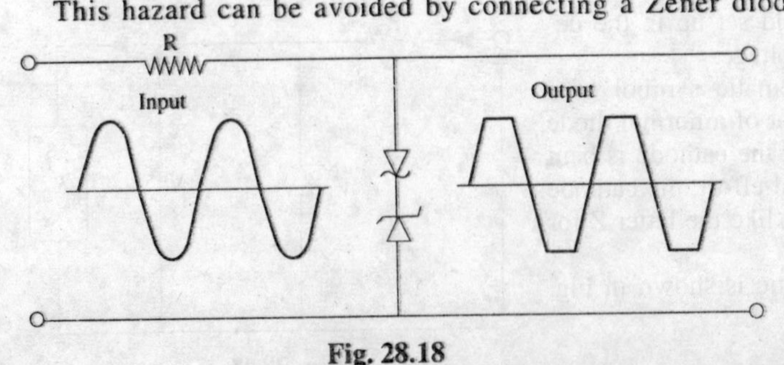

Fig. 28.18

28.9. Zener Diode as Peak Clipper

Use of Zener diodes in wave-shaping circuits is illustrated in Fig. 28.18. Here, a semi-square wave is produced by clipping a sine wave. Two Zener diodes are shunted across a sine wave source. The shunt resistance remains very high until the positive and negative voltage across the diodes exceeds the Zener value. Thereafter, the diodes become low-resistance shunt circuits and increasing voltage values are dropped across the series resistance *R*. Consequently, the output wave is clipped on both peaks.

Example 28.6. *A 24-V, 600 mW Zener diode is to be used for providing a 24 V stabilized supply to a variable load. If input voltage is 32 V, calculate*

(i) series resistance R required; and

(ii) diode current when load resistance is 1200 Ω.

Solution. (*i*) maximum value of I_d *i.e.*, $I_{d(max)} = 600 / 24 = 25$ mA

$$\therefore \quad R = \frac{V_{in} - V_{out}}{I_{d(max)}} = \frac{32 - 24}{25 \times 10^{-3}} = \mathbf{320\ \Omega}$$

(*ii*) When, $R_L = 1200\ \Omega,\ I_L = 24 / 1200 = 20$ mA

$$\therefore \quad I_d = I - I_L = 25 - 20 = \mathbf{5\ mA}$$

28.10. Diode Clipper Circuits

Such circuits are used to clip off some portions of signal voltages above or below certain levels. They are also known as limiters or slicers and are used primarily to prevent signals from exceeding some preset amplitude limit. Applications include the prevention of damage to sensitive microcircuits by overload conditions appearing on input leads, reduction of noise in receivers and amplifiers and in limiting frequency modulated signals to remove amplitude modulation before demodulation.

A diode clipper circuit consists of three elements: (*i*) a current-limiting resistor (*ii*) a diode and (*iii*) a reference voltage source, usually a d.c. battery. There are many ways of combining these elements to obtain clipping action. Four of the commonly used combinations are discussed below.

28.11. Shunt Positive Clipper

It is shown in Fig. 28.19 where a sinusoidal input voltage has been taken. It has a resistor connected in series and the diode connecting the output to the d.c. battery. It is called *shunt positive clipper* because the diode shunts the output signal and also all signals above the threshold level are removed.

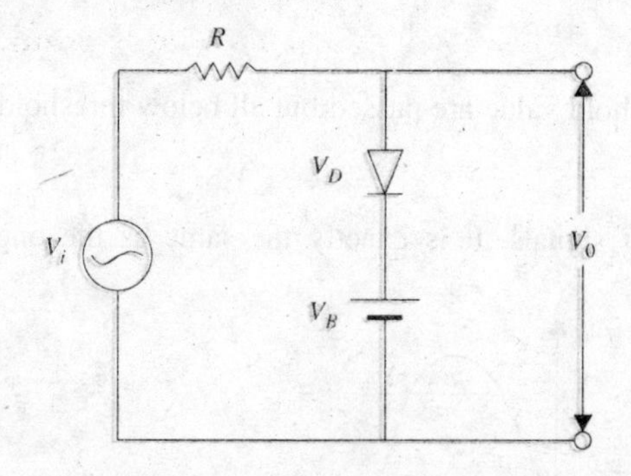

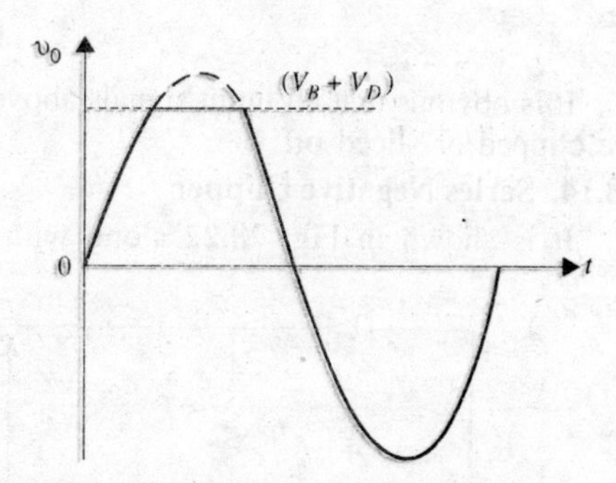

Fig. 28.19

In this circuit the threshold voltage is the sum of the battery voltage and the diode voltage *i.e.*, $(V_B + V_D)$. When the input signals is more positive than the threshold voltage, the diode becomes forward biased and holds the output voltage at the threshold voltage level. When the input voltage is more negative than the threshold voltage, the diode becomes reverse-biased and the output follows the input voltage directly.

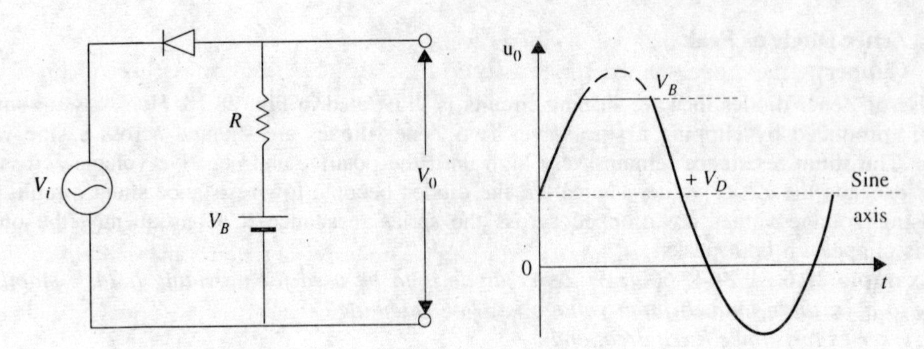

Fig. 28.20

28.12. Series Positive Clipper

It is shown in Fig. 28.20 along with its output voltage for sinusoidal input. As seen, the diode is connected in series but the resistor is connected to the battery across the output terminals. When the input voltage is less than $(V_B - V_D)$, the diode conducts and the output voltage follows the input voltage. However, when the input voltage is greater than $(V_B - V_D)$, the diode becomes reverse-biased and the battery voltage V_B is transmitted to the output.

It may be noted that this circuit shifts the signals upwards by the voltage drop on the diode as well as clipping it.

28.13. Shunt Negative Clipper

It is shown in Fig. 28.21. The only difference between this circuit and that of Fig. 28.19 is that the diode is reversed.

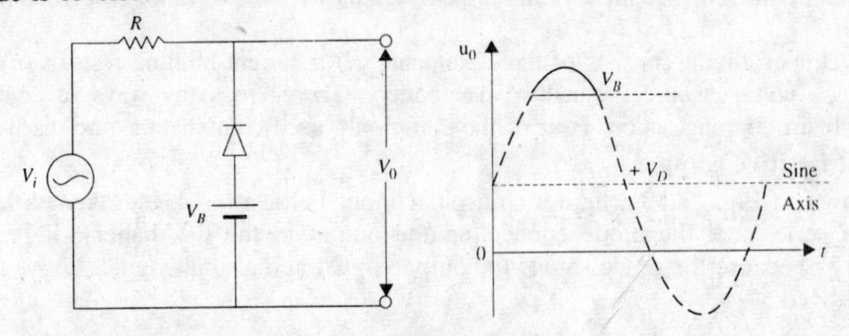

Fig. 28.21

It is obvious that all input signals above the threshold value are passed but all below threshold are clipped or sliced off.

28.14. Series Negative Clipper

It is shown in Fig. 28.22 along with its output signals. It is exactly the same as the one

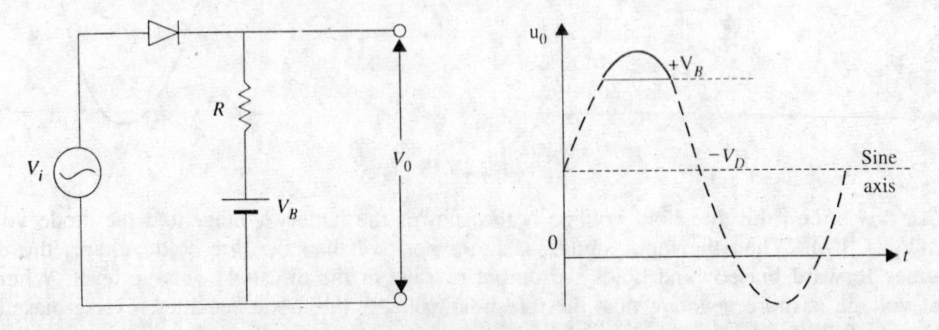

Fig. 28.22

shown in Fig. 28.20 except that the diode is reversed.

By reversing the diode, the direction of shift of the signal waveform has been changed. The waveform is shifted negative by V_D. All the signals above the value of V_B are passed while all below V_D are sliced off.

28.15. Clamper Circuits or Clampers

Such circuits are also known as d.c. restoration circuits or d.c. restorers. They are often used in television receivers for restoring the d.c. reference level. In a television, the incoming composite video signal is usually processed through capacitively-coupled amplifier which eliminate the dc component thereby loosing the black and white reference levels and the blanking level. Before being applied to the picture tube, these reference levels must be restored with the help of a clamper.

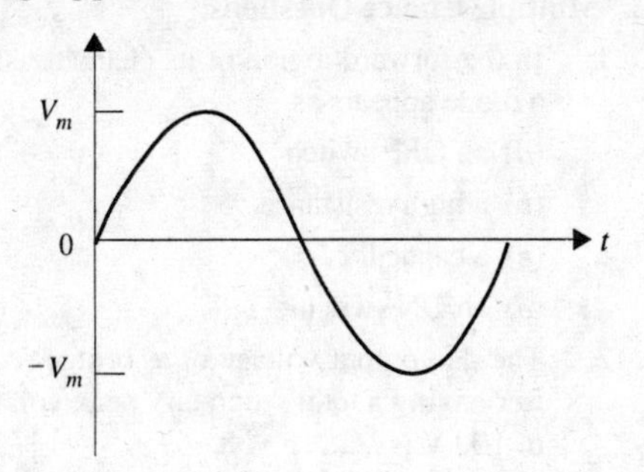

Fig. 28.23

The minimum requirement of a clamping circuit is: a diode, a resistor and a capacitor, usually, such circuits also have a d.c. battery. The magnitudes of R and C must be such that their time constant = RC is large enough to ensure that the voltage across the capacitor does not change much during the interval of time.

A relatively simple clamping circuit is shown in Fig. 28.23 in which a sinusoidal voltage of maximum value $\pm V_m$ is applied across the input terminals.

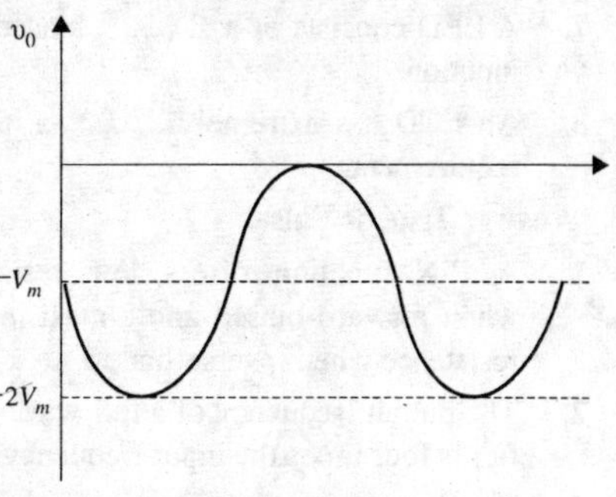

Fig. 28.24

The output of the circuit is shown in Fig. 28.24. It is seen that the output is an undistorted replica of the input except that its d.c. level is at $-V_m$ volt and not zero volt.

HIGHLIGHTS

1. A P-N junction consists of P- and N-regions called anode and cathode respectively. It is a one-way device because it conducts only in one direction. Hence, it is commonly used as a rectifier.
2. Zener diode is a reverse-biased heavily-doped two-terminal P-N junction diode which operates in the breakdown region. It is generally used for (*i*) voltage regulation (*ii*) fixed reference voltage (*iii*) peak clipping (*iv*) meter protection and (*v*) reshaping a waveform etc.
3. A tunnel diode is a high-conductivity P-N junction diode which conducts large current even when forward bias is as low as 0.05 V. It is used (*i*) as an ultra-high speed switch (*ii*) as logic storage memory device (*iii*) as a microwave oscillator and (*iv*) in relaxation oscillator circuits.
4. A PIN diode is a two-junction high-resistance diode having three sections. There are P- and N-regions separated by a layer of an intrinsic material. It is used as (*i*) a switching diode for signal frequencies up to GHZ range and (*ii*) as AM modulator of VH frequencies.
5. A light-emitting diode (LED) is a P-N junction diode which emits light when forward-biased. It is commonly used (*i*) in burglar-alarm circuits (*ii*) for video displays (*iii*) for

optical communication and (*iv*) for numeric displays in hand-held calculators.

6. Liquid Crystal Display (LCD) is a liquid crystal cell which consists of a thin layer of a liquid crystal sandwiched between two glass sheets with electrodes deposited on their inside faces. As compared to a LED, it has extremely low power requirement. It is generally used in (*i*) watches and portable instruments and (*ii*) screens of B and W pocket TV receivers.

OBJECTIVE TESTS—28

A. Fill in the following blanks.

1. A P-N junction diode offers .... resistance when forward-biased.
2. A P-N junction diode is often used as a .... *i.e.* for converting alternating current into direct current.
3. A full-wave rectifier uses a centre-tap ....and two diodes.
4. A full-wave bridge rectifier uses .... diodes.
5. A Zener diode operates in the .... region.
6. A tunnel diode is a high .... two-terminal P-N junction diode.
7. A LED consists of a .... biased P-N junction.
8. An LCD has extremely .... power requirement.

B. Answer True or False.

1. A P-N junction offers low resistance when forward-biased and almost infinite resistance when reverse-biased.
2. The output frequency of a full-wave rectifier is four times the input frequency.
3. The d.c. output of a bridge rectifier equals the total secondary voltage.
4. A Zener diode maintains an essentially constant voltage across its terminals over a specified range of reverse current values.
5. A tunnel diode consists of a very heavily-doped P-N junction diode.
6. A tunnel diode is so-called because a small tunnel exists between its two regions for electrons to flow through.
7. A PIN diode acts like a variable resistance when forward-biased.
8. Field effect LCD's are normally used in watches and portable instruments.

C. Multiple Choice Questions.

1. In the forward region of its characteristic, a diode appears as
   (*a*) an OFF switch
   (*b*) a high resistance
   (*c*) a capacitor
   (*d*) an ON switch.
2. The d.c. output voltage of a bridge rectifier having a total secondary peak voltage of 100 V is .... volt.
   (*a*) 63.6 (*b*) 31.8
   (*c*) 90 (*d*) 70.7
3. When used in circuit, the Zener diode is always
   (*a*) forward-biased
   (*b*) connected in series
   (*c*) troubled by overheating
   (*d*) reverse-biased
4. The main reason why electrons can tunnel through a P-N junction is that
   (*a*) they have high energy
   (*b*) barrier potential is very low
   (*c*) depletion layer is extremely thin
   (*d*) impurity level is low
5. A LED is made up of a .... junction.
   (*a*) PNP (*b*) NPN
   (*c*) PIN (*d*) PN
6. As compared to a LED, an LCD has the distinct advantage of
   (*a*) extremely low power requirement
   (*b*) providing a silver display
   (*c*) being extremely thin
   (*d*) giving two types of displays

ANSWERS

A. 1. low 2. rectifier 3. transformer 4. four 5. breakdown 6. conductivity 7. forward 8. low

B. 1. T 2. F 3. F 4. T 5. T 6. F 7. T 8. T

C. 1. *d* 2. *a* 3. *d* 4. *c* 5. *d* 6. *a*

29 OPTOELECTRONIC DEVICES

29.1. Introduction

Optoelectronic devices are of two categories (*i*) light emitters and (*ii*) photodetectors. Light emitters convert electrical energy into light energy. Examples are : junction lasers and light-emitting diodes (LEDs).

Photodetectors do just the opposite *i.e.*, they convert light energy into electrical energy. In their case, the light energy carried by the incident photons is absorbed and transferred to electrons in the photodetector material.

Photodetectors may be further sub-divided into the following sub-groups:

(*a*) Bulk photoconductive devices where conductivity of the material changes with the amount of incident light.

(*b*) Photo junction devices which contain a P-N junction which is exposed to light energy. Examples are : photodiodes, phototransistors, solar cells and photo-FETs etc.

29.2. Photoconductive Cell

It is a two-terminal conductor device whose resistance varies with the intensity of the incident light. Such devices are also known as photoresistive cells or simply photoresistors. As seen from the Fig. 29.1, when there is no light, the cell resistance is maximum and is called the *dark resistance*. As the light intensity increases, the cell resistance decreases significantly. A typical photoconductive cell and its schematic symbol along with its characteristics is shown in Fig. 29.1.

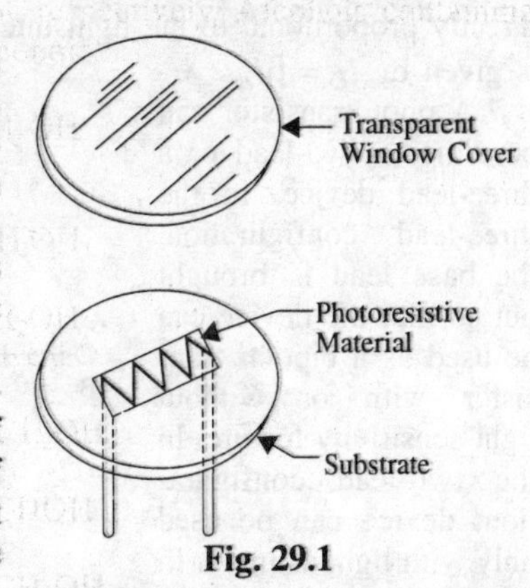

Fig. 29.1

Photoconductive cells are made from either cadmium selenide (CdSe) or cadmium sulphide (CdS). A thin layer of the semi-conductor material is deposited on a substrate of glass or ceremic and connections are made to lead pins at each end. A glass cover is used to allow light to fall on the semi-conductor.

Photoconductive cells are used in many applications that require the control of a certain function or event according to the absence, presence, colour or intensity of light. They are widely used in door openers, burglar alarms, flame detectors, smoke detectors and light control for street lamps in the residential and industrial areas. Often, these are used in conjunction with relays.

29.3. Photodiode

It is a P-N junction device which is designed to operate with reverse bias applied to its junction. The reverse current increases almost linearly with the increase in the incident light as shown in Fig. 29.2 (*c*). The basic biasing arrangement, construction and symbol of the device are shown in Fig. 29.2 (*a*) and (*b*) respectively. When there is no incident light, the reverse current I_λ is almost nil and is called the *dark current*. As the amount of light energy increases, there is increase in I_λ.

The advantages of photodiodes over photoconductive cells are their fast response time and wider bandpass. Their disadvantages are that their current is very low even at high light intensities. Hence, photodiodes are not used in relay circuits but their fast response time makes them useful in high-speed switching and logic circuits.

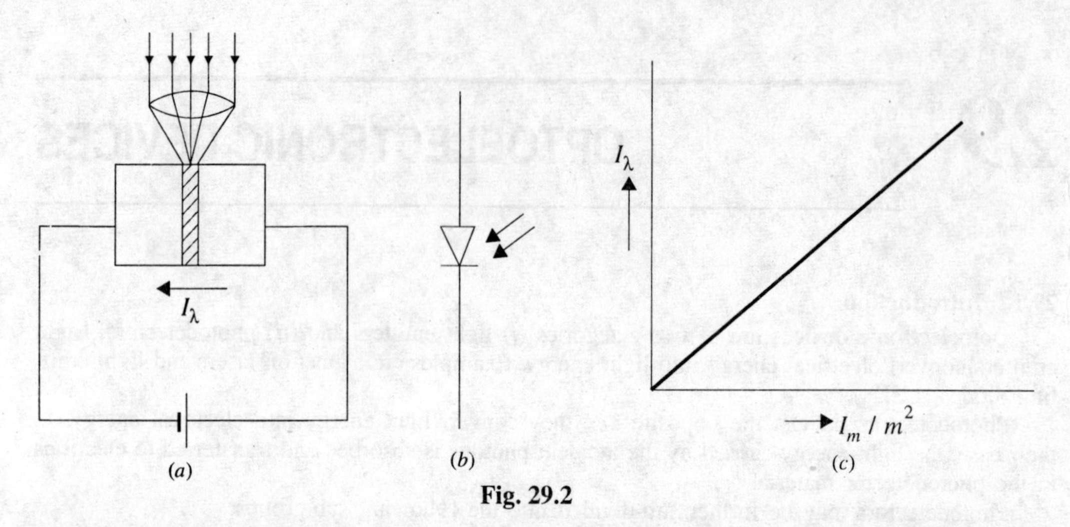

Fig. 29.2

29.4. Phototransistor

It has a light-sensitive collector-base P-N junction which is exposed to incident light through a lens. When there is no incident light, there is an extremely small leakage current called *dark current* of a few nA. When light strikes the P-N junction, a base current I_λ is produced which is directly proportional to the light intensity. This action produces collector current I_C whose value is given by $I_C = \beta I_\lambda$.

A phototransistor can be either a two-lead or a three-lead device. In the three-lead configuration, the base lead is brought out so that the device can be used as a bipolar transistor with or without light-sensitivity feature. In the two-lead configuration, device can be used only with light as input. In many applications, the phototransistor is used in the two-lead version.

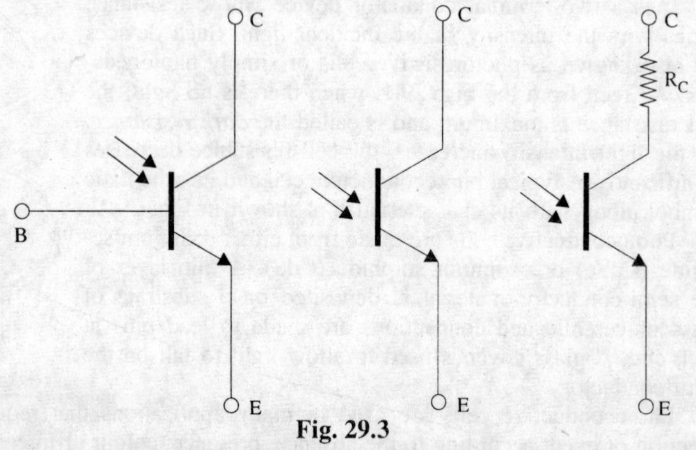

Fig. 29.3

Fig. 29.3 (*a*) shows the symbol for a three-lead phototransistor whereas Fig. 29.3 (*b*) shows the two-lead version. Fig. 29.3 (*c*) shows the phototransistor biasing method.

Phototransistors are used in a wide variety of applications particularly in light-operated relay circuits used for automatic door activators, process counters and various alarm systems.

29.5. Solar Cells

Their operation is based on the principle of photovoltaic cells. That is why they are also called *photovoltaic cells.* They are two-terminal devices which produce voltage between the terminals when exposed to light. The voltage produced is directly proportional to the intensity of light.

A basic solar cell consists of an N-type and P-type semiconductor materials forming a P-N junction. Fig. 29.4 shows the construction of a silicon solar cell comprising a P-N junc-

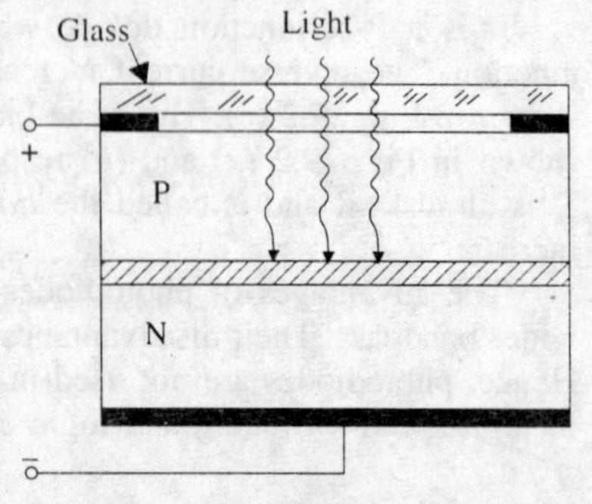

Fig. 29.4

tion with two metal electrodes with leads attached. The top electrode is formed in a ring shape so that a window can be placed in the centre for light to enter the silicon material.

A photon after penetrating through the P-type material hits an atom in the depletion region causing an electron to break out of its covalent band. This causes a free electron-hole pair. The electron moves into the N-type material thereby adding a negative charge whereas hole drifts towards the P-type material thereby adding positive charge. The process is repeated millions of times and causes a potential difference of about 0.5 V between the P-type and N-type materials.

If an external load is applied across a solar cell, a small current is set up. The current capability of the solar cell is directly proportional to the area of the cell exposed to the sun or other source of radiation. On a clear day, when the sun is directly overhead, a silicon solar cell with an area of 0.5 cm^2 can deliver 25 mA at 0.4 V and one with an area of 75 cm^2 can deliver 2 A at 0.4 V. To get higher power levels, it will be necessary to connect solar cells in series and parallel. When cells are connected in series, the output voltage is increased and when connected in parallel, the output current is increased. A typical series-parallel solar array is shown in Fig. 29.5.

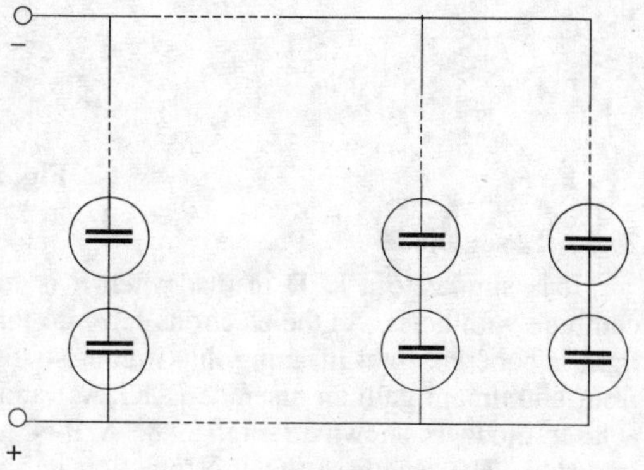

Fig. 29.5

Silicon solar cells are the most popular photovoltaic cells because they deliver the greatest output power for a given exposed area. Silicon cells respond best to infrared light whereas selenium cells have a spectral response similar to the human eye. Solar cells are extensively used in space and missile systems. In space vehicles, these cells are used as secondary power source which can supply electrical power when the satellite is illuminated by sunlight. When the satellite is in the dark, the solar cells do not operate and storage battery provides the vehicle power. The cells serve to recharge the storage batteries during the light hours.

29.6. Light Emitting Diode (LED)

Its a special P-N junction diode that emits light when forward-biased. When a normal diode is forward-biased, current flows which causes a voltage drop of approximately 0.7 volt in a silicon diode. The voltage drop times the magnitude of current flow results in power dissipation at the junction. In a normal diode, this power is dissipated in the form of heat energy. However, in the case of a light-emitting diode, this power is dissipated in the form of light.

There are two main differences between normal diodes and LED. One difference is that LEDs are made from gallium arsenide (Ga/As), gallium arsenide phosphide (GaAsP) or gallium phosphide (GaP). Diodes manufactured from these elements radiate visible light energy when forward-biased. The second difference is that LED are so designed as to have a window over the junction so that light energy can be seen.

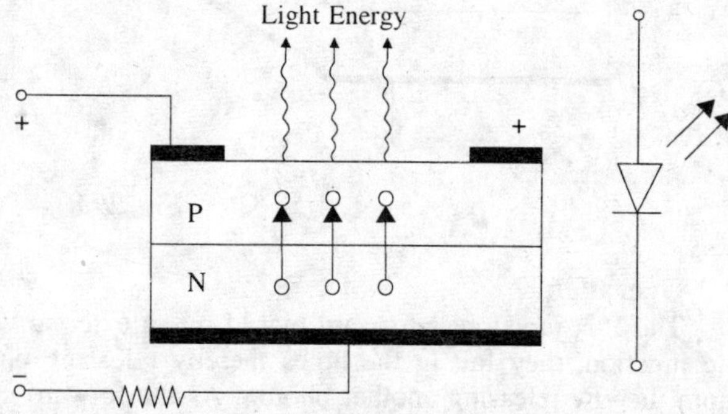

Fig. 29.6

The different wavelengths of light emitted by the above LED are as under :

(*i*) GaAs — 8800 Å infrared radiation
(*ii*) GaP — 5500 Å (green) or 7000 A (red)
(*iii*) GaAsP — 5800 Å (amber) or 6600 A (red)

Fig. 29.6 shows a LED in diagramatic form along with its symbol.

LED are commonly used for indicator lamps and read out displays in many instruments varying from consumer appliances to scientific apparatus. A very common type of display device using LED is the seven segment display shown in Fig. 29.7. Combination of these segments can form the ten decimal digits. Infrared emitting LED are used in optical coupling applications often in conjuction with fibre optics.

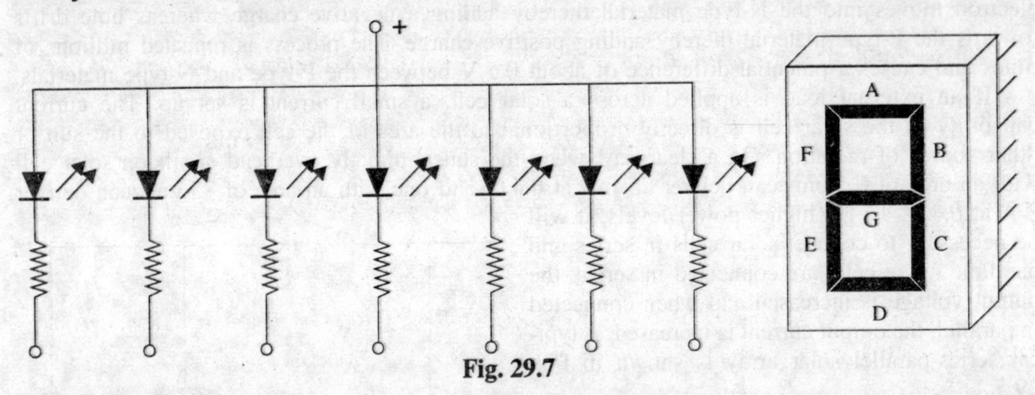

Fig. 29.7

29.7. Laser Diode

It is similar to a LED in that when it is forward-biased, current flows causing electrons to combine with holes. As the electrons fall into the holes, they release photons or light energy. This light is coherent light meaning that it consists of a single wavelength. The laser diodes are made from aluminium gallium arsenide (AlGaAs) and gallium arsenide (GaAs). The basic structure of a laser diode is shown in Fig. 29.8. A P-N junction is formed by the above two mentioned materials. The length of the P-N junction has an exact relationship with the wavelength of the light to be emitted. There is a highly reflective surface at one end of the junction and a partially reflective surface at the other end. External leads provide the anode and cathode connections.

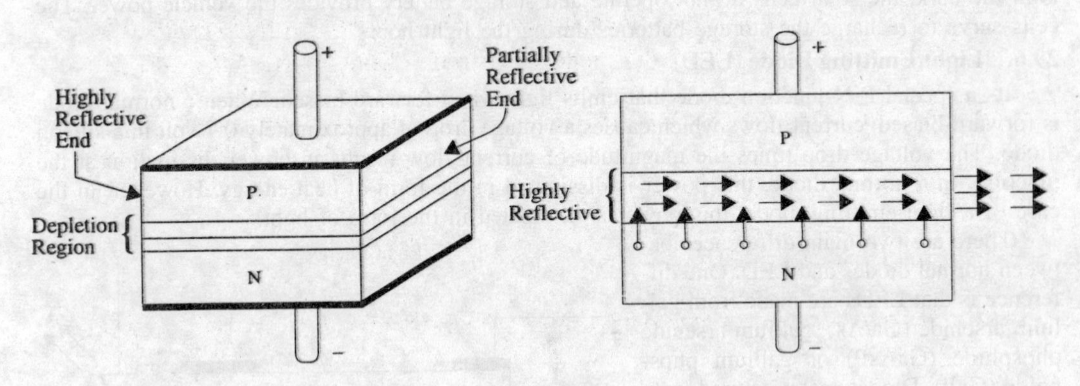

Fig. 29.8

Operation

The P-N junction is forward-biased by an external voltage source. As electrons move across the junction, they fall in the holes thereby releasing photons. A released photon can strike an atom thereby releasing another photon. As the forward current is increased, more electrons enter the depletion region and cause more photons to be released. Some of these photons strike the reflective surface perpendicular. These reflected photons move along within the depletion region, striking atoms and releasing additional photons. This back-and-forth movement of photons increases as the generation of photons 'snowballs' until very strong beam of laser light is formed by the photons that pass through the partially reflective surface at the end of the P-N junction as shown in Fig. 29.8 (*b*).

29.8. Fibre Optics

Modern man has often used light beams for communication purposes employing atmosphere as a medium for this purpose. However, atmosphere has two shortcomings. First, the atmosphere changes constantly due to rain, snow and fog etc., these changes affect light transmission. Second, light beams provide line-of-sight communication which is limited to two points that are in-sight of each other.

Recently, extremely thin fibres of glass or plastic (about 0.1 mm in diameter) have been used to transfer information with the help of light beams. Even though RF transmission and electrical signals are being sent over copper conductors, fibre optic communication possess the following two advantages:

1. Light signals have a much wider bandwidth *i.e.*, they can transmit much more information.
2. Light signals are not affected by electromagnetic interference.

29.9. Light Transmission through Optic Fibre

The basic idea behind using glass or plastic fibres for light transmission is that the light entering from one end of the fibre must travel down to the other end of fibre. For achieving this purpose, the light must be contained in the fibre and not allowed to escape from its sides. As shown in Fig. 29.9 the input light ray is reflected from side to side as it travels down the thin glass fibre. The light ray is incident at point *A* on the glass-air boundary at an angle θ_1 which is more than the critical angle for the two media involved. Hence, it suffers total internal reflection *i.e.*, instead of getting refracted into air it is reflected back into the glass fibre. The light is again internally reflected from point *B* because its angle of incidence θ_2 is greater than the critical angle. After suffering further reflections from point *C*, *D* and *E*, the light ray finally comes out from the other end of the glass fibre.

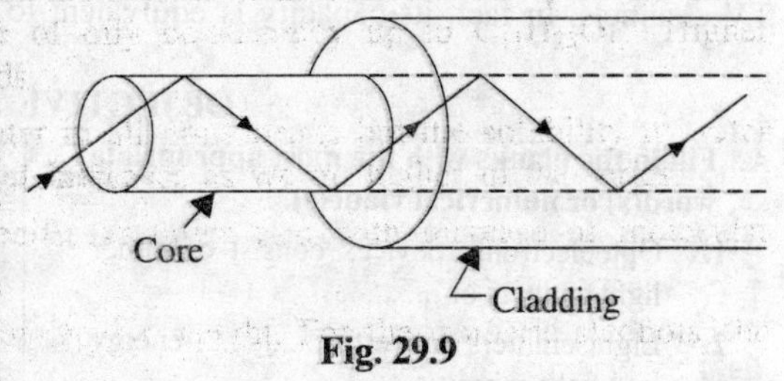

Fig. 29.9

For total internal reflection to take place, the following two conditions must be fulfilled:

1. The light must travel from an optically denser medium to an optically rarer medium.
2. The light must be incident at the boundary at an angle greater than the critical angle.

The value of the critical angle θ_c for two media is given by

$$\sin \theta_c = \mu_1 / \mu_2$$

where μ_1 is the refractive index of the more dense medium and μ_2 is the refractive index of the less dense medium.

As shown in Fig. 29.10, when light ray is incident at the boundary at an angle just equal to critical angle θ_c, it is refracted through 90° and hence grazes along the boundary.

If it is incident at an angle $\theta > \theta_c$, it is reflected back into the denser medium according to the ordinary laws of reflection. The light ray is said to suffer total internal reflection because whole of it is reflected inwards and no part of it is refracted outside into the rarer medium.

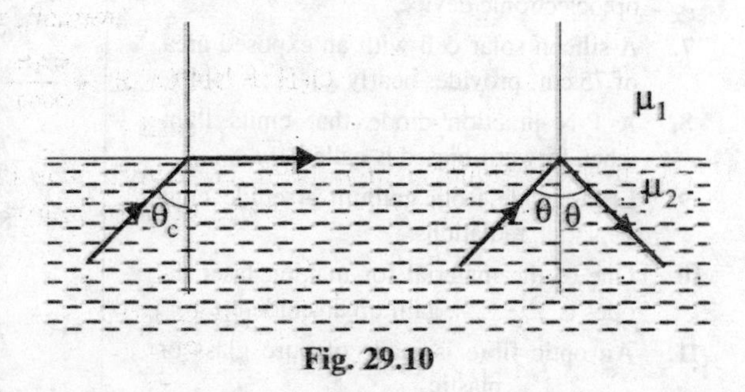

Fig. 29.10

29.10. Construction of Optic Fibre Cables

The optic fibre is made of either glass or transparent plastic. Glass is more expensive and is harder to work with but it transmits light with less loss per metre than plastic. While the glass or plastic fibre could be uncovered and use air as the second material surrounding it to determine the critical angle, most often covering called *cladding* is used as shown in Fig. 29-11. The cladding material has lower refractor index than that of glass (or plastic) fibre core. In this way, incident light is trapped in the core by total internal reflection at the core-cladding boundary. Incidentally, the glass in optical fibres is so pure that a 2 km length absorbs less light than a sheet of window glass.

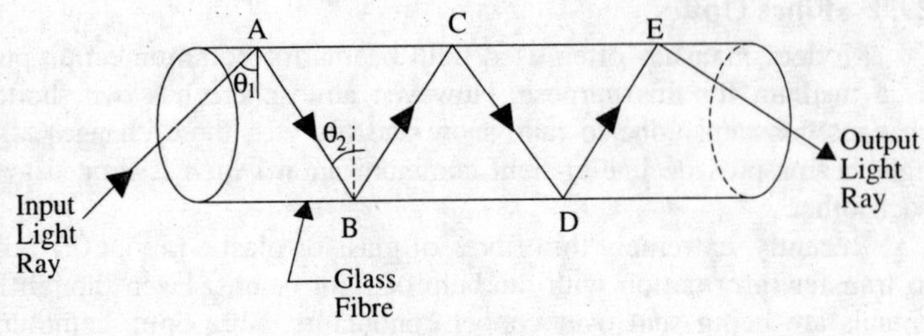

Fig. 29.11

29.11 Capacity of Optical Fibre Cables

The information carrying capacity of an optical fibre system is typically 140 MB/s. One MB/s = one million bits of information *i.e.*, 1 s and 0 s per second. A 140 *MB/s* can carry 1,920 telephone channels or the same number of A4 pages of electronic mail per second or 4 colour TV channels. In fact, its capacity is equivalent to delivering 8 average size books every second!

OBJECTIVE TESTS—29

A. Fill in the blanks with the most appropriate word(s) or numerical vlaue(s).

1. Optoelectronic devices consist of either light emitters or ....
2. Light emitters convert .... energy into light energy.
3. A photodetector converts .... energy into electrical energy.
4. Photoconductive cells are made from either cadmium selenide or cadmium ....
5. Photodiode is a P-N junction device with .... bias applied to its junction.
6. A solar cell is a .... terminal optoelectronic device.
7. A silicon solar cell with an exposed area of 75 cm^2 provides nearly .... volt.
8. A P-N junction diode that emits light when forward-biased is called ....
9. LEDs made from gallium arsenide emit .... radiations.
10. One of the material for making laser diodes is .... gallium arsenide.
11. An optic fibre is made of pure glass or .... plastic.
12. The basic material used for making glass fibres is .... dioxide.
13. The thickness of an optical fibre is a fraction of a ....
14. Transmission of light through an optical fibre is based on .... internal reflection.

B. Answer True or False.

1. All optoelectronic devices depend on the conversion of light energy into electrical energy.
2. The resistance of a photoconductive cell is maximum, when it is not exposed to light.
3. When no light is incident on a photodiode, its reverse current is negligibly small.
4. As compared to a photoconductive cell, a photodiode has much faster response time.
5. In a photodiode, current is very low even at high light intensities.
6. A phototransistor has a light-sensitive emitter-base P-N junction which is exposed to light.
7. Solar cells are also called photovoltaic cells.
8. When a number of solar cells are connected in series, their output current is increased.
9. Selenium solar cells supply much higher output power than silicon cells.
10. In LEDs, power dissipated at the junctions is emitted in the form of light.
11. LEDs are commonly used for read out displays in many instruments.
12. The coherent light emitted by a laser diode depends on the length of its P-N junction.

13. Unlike LEDs, the P-N junction of a laser diode is reversed-biased.

14. The light travelling in an optical fibre is trapped with the help of total internal reflection.

15. The cladding material of an optical fibre has higher refractive index than the material of the fibre.

16. Fibre optic communication systems are immune to RF interference.

C. Multiple Choice Questions

1. Optoelectronic devices convert
(*a*) light energy into electrical energy
(*b*) electrical energy into light energy
(*c*) chemical energy into light energy
(*d*) both (*a*) and (*b*)

2. A photoconductive cell is also known as
(*a*) photoresistor
(*b*) photoconductor
(*c*) photodiode
(*d*) phototransistor

3. The advantage of a photodiode over a photoconductive cells are its
(*a*) fast responds time
(*b*) wider bandpass
(*c*) very low current
(*d*) both (*a*) and (*b*).

4. A phototransistor has a light-sensitive .... P-N junction which is exposed to incident light through a lens.
(*a*) emitter-base
(*b*) collector-base
(*c*) collector-emitter
(*d*) base-emitter

5. Working of solar cells is based on .... principle
(*a*) photodetector
(*b*) photodiode
(*c*) phototransistor
(*d*) photovoltaic

6. A series-parallel solar array provides higher output ....
(*a*) voltage (*b*) current
(*c*) power (*d*) all of the above

7. A LED made of gallium phosphide semiconductor material emits .... radiations.
(*a*) green (*b*) red
(*c*) amber (*d*) either (*a*) or (*b*)

8. All laser diodes provide a beam of .... light.
(*a*) coherent (*b*) red
(*c*) green (*d*) blue

9. The critical angle in an optical fibre depends on the refractive index of the
(*a*) fibre material
(*b*) cladding material
(*c*) light wavelength
(*d*) both (*a*) and (*b*)

10. Since optical fibres do not radiate the energy within them, they provide a high degree of
(*a*) security
(*b*) privacy
(*c*) approachability
(*d*) both (*a*) and (*b*)

11. Optical fibres reject radio-frequency and electromagnetic interference because they are made of .... material.
(*a*) thin (*b*) transparent
(*c*) insulator (*d*) conducting

12. The approximate thickness of fibres used in optical fibre cables is .... millimetre
(*a*) 2.5 (*b*) 1.5
(*c*) 1.0 (*d*) 0.1

ANSWERS

A. 1. photodetectors 2. electrical 3. light 4. sulphide 5. reverse 6. two
7. 0.4 8. LED 9. infrared 10. aluminium 11. transparent 12. silicon
13. millimetre 14. total

B. 1. F 2. T 3. T 4. T 5. T 6. F 7. T 8. F 9. F 10. T 11. T
12. T 13. F 14. T 15. F 16. T

C. 1. *a* 2. *a* 3. *d* 4. *b* 5. *d* 6. *d* 7. *d* 8. *a* 9. *d* 10. *d* 11. *c* 12. *d*

30 BIPOLAR JUNCTION TRANSISTORS

30.1. The Bipolar Junction Transistor*

Basically, it consists of two back-to-back P-N junctions manufactured in a single piece of a semi-conductor crystal. These two junctions give rise to three regions called *emitter, base* and *collector*. As shown in Fig. 30.1, a junction transistor is simply a sandwich of one type of semi-conductor material between two layers of the other type. Fig. 30.1 (*a*) shows a layer of N-type material sandwiched between two layers of P-type material. It is described as PNP transistor. Fig. 30.1 (*b*) shows an NPN transistor consisting of a layer of P-type material sandwiched between two layers of N-type material.

The emitter, base and collector are provided with terminals which are labelled as *E, B* and *C*. The two junctions are : emitter-base (*E/B*) junction and collector-base (*C/B*) junction.

The symbols employed for PNP and NPN transistors are also shown in Fig. 30.1. The arrow-head is *always* at the emitter (not at the collector) and in each case, its direction indicates the *conventional* direction of current flow. For a PNP transistor, arrowhead points from emitter to base meaning that emitter is positive with respect to base (and also with respect to collector)**. For an NPN transistor, it points from base to emitter meaning that base (and collector as well)** is positive with respect to emitter.

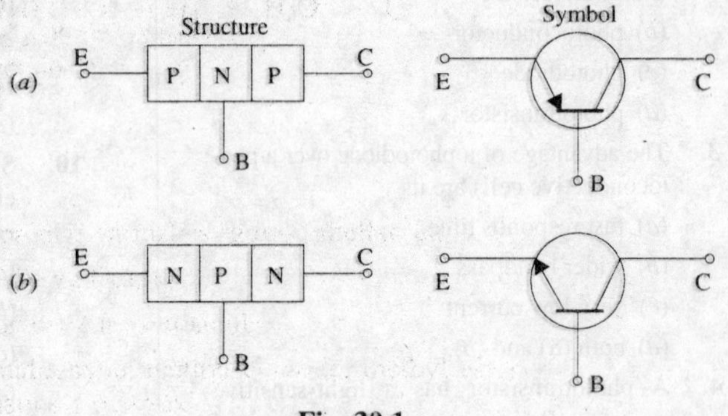

Fig. 30.1

1. Emitter

It forms the left-hand section or region of the transistor as shown in Fig. 30.1. It is more *heavily doped* than any of the other regions because its main function is to supply majority charge carriers (either electrons or holes) to the base.

2. Base

It forms the middle section of the transistor. It is very thin (10^{-6} m) as compared to either the emitter or collector and is very *lightly doped.*

3. Collector

It forms the right-hand side section or region of the transistor as shown in Fig. 30.1 and its main function is (as indicated by its name) to collect majority charge carriers coming from the emitter and passing through the base.

In most transistors, collector region is made physically larger than the emitter region because it has to dissipate much greater power.

---

*It is a contraction of *transfer resistor*. It is so because a transistor is basically a resistor that amplifies electrical impulses as they are transferred through it from its input to output terminals.

**In a transistor, for normal operation, collector and base have *same* polarity with respect to emitter (Art. 30.3).

It may be noted, in passing, that transistors are made by growing, alloying or diffusing processes.

30.2. Transistor Biasing

For proper working of a transistor, it is essential to apply voltages of correct polarity across its two junctions. It is worthwhile to remember that for normal operation

1. emitter-base junction is *always* forward-biased ; and
2. collector-base junction is *always* reverse-biased.

This type of biasing is known as FR biasing.

In Fig. 30.2, two batteries respectively provide the d.c. emitter supply voltage V_{EE} and collector supply voltage V_{CC} for properly biasing the two junctions of the transistor. In Fig. 30.2 (*a*), Positive terminal of V_{EE} is connected to P-type emitter in order to *repel* or *push* holes into the base.

The negative-terminal of V_{CC} is connected to the collector so that it may attract or pull holes through the base. Similar considerations apply to the NPN transistor of Fig. 30.2 (*b*). It must be remembered that a transistor will never conduct any current *if its emitter-base junction is not forward-biased.**

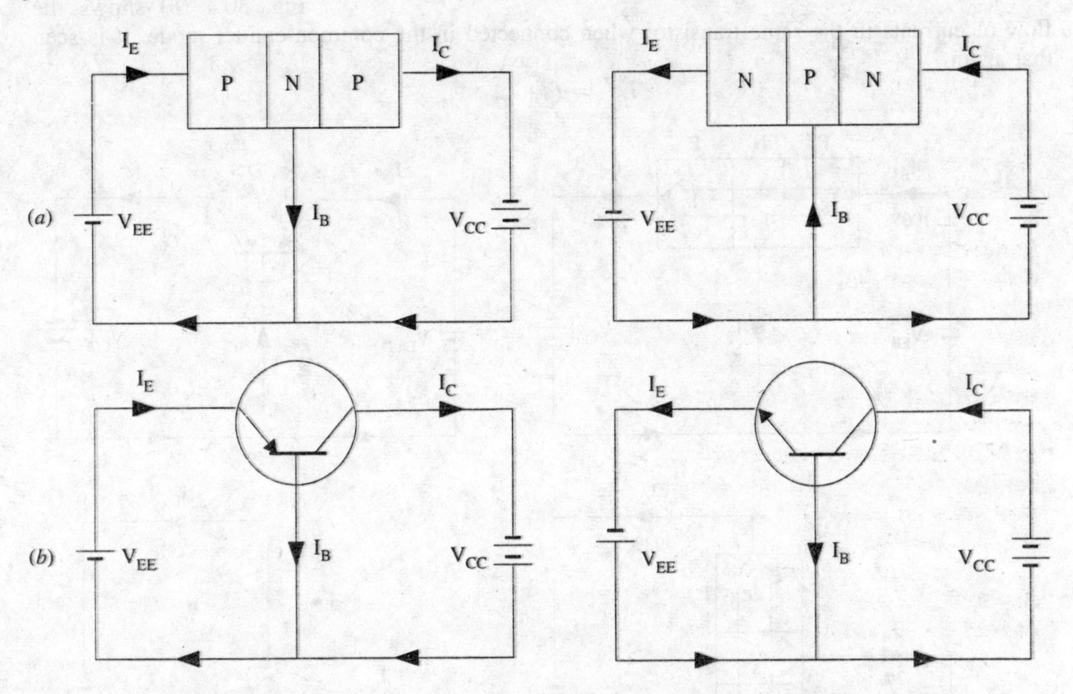

Fig. 30.2

30.3. Important Biasing Rule

For a PNP transistor, both collector and base are *negative* with respect to the emitter (the letter N of *Negative* being the same as the middle letter of PNP). Of course, collector is *more* negative than base [Fig. 30.3 (*a*)]. Similarly, for an NPN transistor, both collector and base are *positive* with respect to the emitter (the letter P of *Positive* being the same as the middle letter of NPN). Again, collector is *more* positive than the base as shown in Fig. 30.3 (*b*).

30.4. Transistor Currents

The three primary currents which flow in a properly-biased transistor are I_E, I_B and I_C. In

*There would be no current due to majority charge carriers. However, there would be an extremely small current due to minority charge carriers which is called leakage current of the transistor (Art. 30.12).

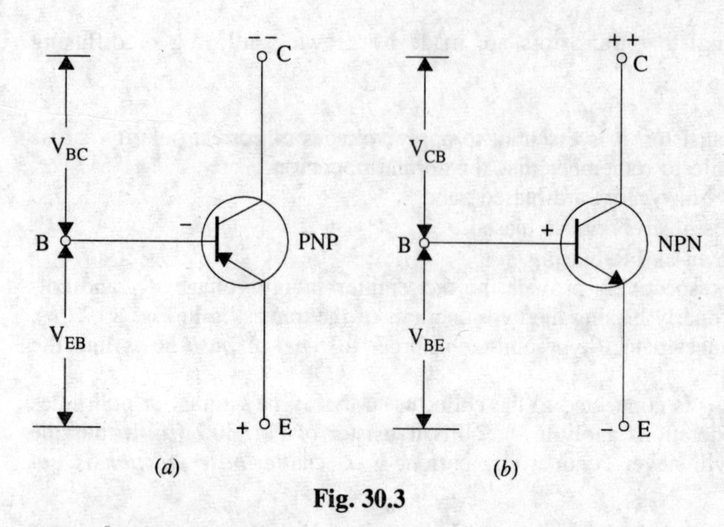

Fig. 30.3

Fig. 30.4 (*a*) are shown the directions of flow as well as relative magnitudes of these currents for a PNP transistor connected in the common-base mode. It is seen that

$$I_E = I_B + I_C$$

It is seen that a small part (about 1–2%) of emitter current goes to supply base current and the remaining major part (99–98%) goes to supply collector current.

Moreover, I_E flows *into* the transistor whereas both I_B and I_C flow out of it.

Fig. 30.4 (*b*) shows the flow of currents in the same transistor when connected in the common-emitter mode. It is seen that again

$$I_E = I_B + I_C$$

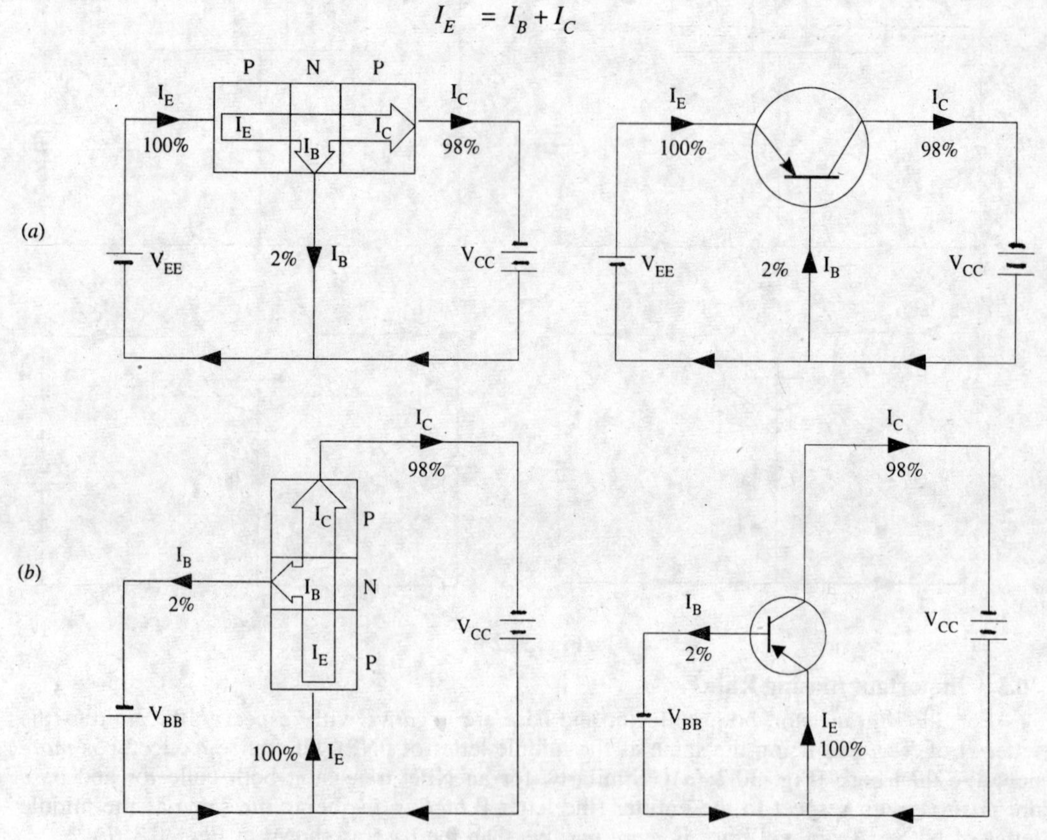

Fig. 30.4

By normal convention, currents flowing *into* a transistor are taken as positive whereas those flowing *out* of it are taken as *negative.* Hence, I_E is positive whereas both I_B and I_C are negative. Applying Kirchhoff's Current Law, we have

$$I_E + (-I_B) + (-I_C) = 0$$

or $$I_E - I_B - I_C = 0$$

$$\therefore \quad I_E = I_B + I_C$$

This statement is true *regardless of transistor type or transistor configuration.*

Note. For the time being, we have not taken into account the leakage currents which exist in a transistor.

30.5. Summing Up

The four basic guideposts about all transistor circuits are:

1. conventional current flows along the arrow whereas electrons flow against it.
2. *E/B* junction is always forward-biased.
3. *C/B* junction is always reverse-biased.
4. $I_E = I_B + I_C$.

30.6. Transistor Circuit Configurations

Basically, there are three types of circuit connections (called configurations) for operating a transistor

1. common-base (CB);
2. common-emitter (CE);
3. common-collector (CC).

The term 'common' is used to denote the electrode that is common to the input and output circuits. Because the common electrode is generally grounded, these modes of operation are frequently referred to as grounded-base, grounded-emitter and grounded-collector configurations as shown in Figs. 30.5, 30.7 and 30.8.

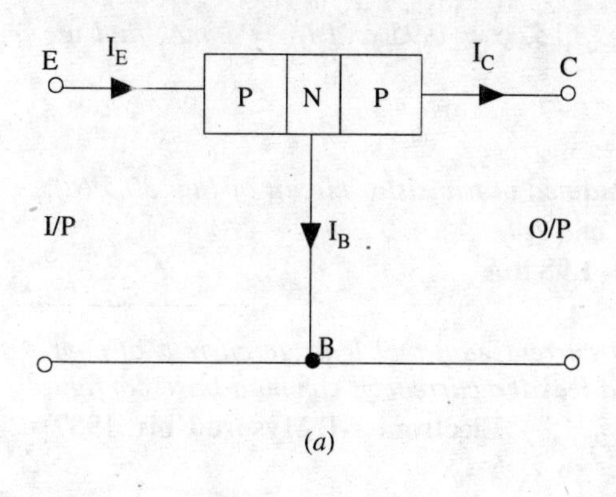

(*a*)

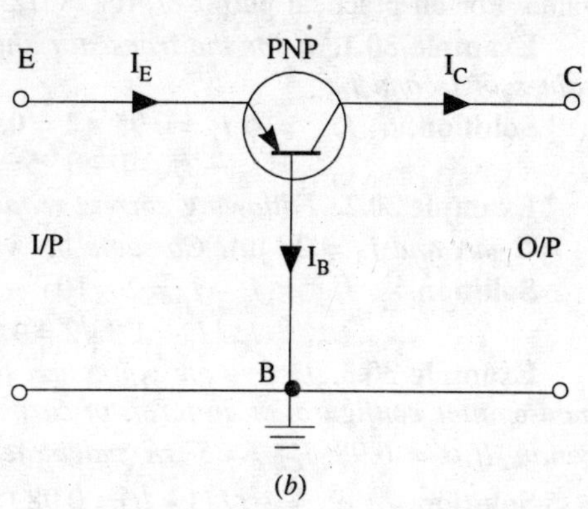

(*b*)

Fig. 30.5

30.7. CB Configuration

In this configuration (Fig. 30.5), emitter current I_E is the input current and collector current I_C is the output current. The input signal is applied between the emitter and base whereas output is taken out from the collector and base.

The ratio of the collector current to the emitter current is called d.c. alpha (α_{dc}) of a transistor.

$$\therefore \quad \alpha_{dc} = \frac{-I_C{}^*}{I_E} \quad \text{or} \quad I_C = \alpha_{dc} I_E{}^{**}$$

If we write α_{dc} simply as α, then

$$\alpha = -I_C / I_E \quad \text{or} \quad I_C = \alpha I_E$$

---

*More accurately, $\alpha_{dc} = -\dfrac{I_C - I_{CBO}}{I_E}$. —Art. 30.12

**Negative sign has been omitted since we are here concerned with only magnitudes of the currents involved.

It is also called *forward current* transfer ratio ($-h_{FB}$). In h_{FB}, subscript F stands for forward and B for common base. The negative sign merely indicates that emitter and collector currents flow in *opposite* directions. The subscript *d.c.* on α signifies that this ratio is defined from d.c. values of I_C and I_E .

The α of a transistor is a measure of the quality of a transistor: higher the value of α, better the transistor in the sense that collector current more closely equals the emitter current. Its value ranges from 0.95 to 0.999. Obviously, it applies *only to CB configuration of a transistor.* As seen from above.

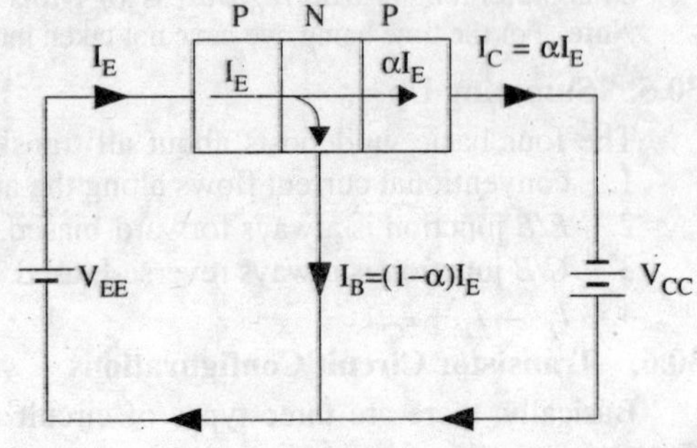

Fig. 30.6

$$I_C = \alpha I_E$$

Now, $$I_B = I_E - I_C = I_E - \alpha I_E$$

$\therefore$ $$I_B = (1-\alpha) I_E$$

Incidentally, there is also an *a.c.* α for a transistor. It refers to *change* in collector current to *change* in emitter current.

$\therefore$ $$\alpha_{ac} = \frac{\Delta I_C}{\Delta I_E}$$

It is also known as *short-circuit gain* of a transistor and is written as $-h_{fb}$. It may be noted that upper case subscripts *FB* indicate *d.c.* value whereas lower case subscripts '*f b*' indicate *a.c.* value. For all practical purposes, $\alpha_{dc} = \alpha_{ac}$.

Example 30.1. *If for the transistor shown in Fig. 30.5, $\alpha = 0.95$ and $I_E = 1$ mA, find the values of I_C and I_B.*

Solution. $I_C = \alpha I_E = 0.95 \times 1 = \mathbf{0.95\ mA}$

$I_B = I_E - I_C = 1 - 0.95 = \mathbf{0.05\ mA}$

Example 30.2. *Following current readings are obtained in transistor circuit of Fig. 30.5 (a): $I_E = 2$ mA and $I_B = 20$ μA. Compute the values of α and I_C.*

Solution. $I_C = I_E - I_B = 2 \times 10^{-3} - 20 \times 10^{-6} = \mathbf{1.98\ mA}$

$\alpha = I_C / I_E = 1.98/2 = \mathbf{0.99}$

Example 30.3. *Derive an expression for forward current gain and leakage current of common-emitter configuration in terms of current gain and leakage current of common-base configuration. If $\alpha = 0.98$, $I_{CBO} = 5$ μA, calculate β and I_{CEO}.* **(Electronics-I, Mysore Univ. 1987)**

Solution. $\beta = \alpha / (1-\alpha) = 0.98 / (1 - 0.98) = \mathbf{49}$

$I_{CEO} = (1 + \beta) I_{CO} = (1 + 49) \times 5 = 250\ \mu A = \mathbf{0.25\ mA}$

30.8. CE Configuration

Here, input signal is applied between the base and emitter and output signal is taken out from collector and emitter circuits. As seen from Fig. 30.7 (*b*), I_B is the input current and I_C is the output current.

The ratio of d.c. collector current to d.c. base current is called d.c. beta (β_{dc}) or just β of the transistor.

$$\beta = \frac{I_C}{I_B} \quad \text{or} \quad I_C = \beta I_B$$

It is also called *common-emitter* d.c. forward transfer ratio and is written as h_{FE}. It is possible for β to have as high a value as 500.

While analysing a.c. operation of a transistor, we use a.c. β which is given by

$$\beta_{ac} = \frac{\Delta I_C}{\Delta I_B}$$

It is also written as h_{fe}. Since I_C and I_B are both outgoing currents, they have the *same* sign.

Hence, β is positive unlike α which is negative.

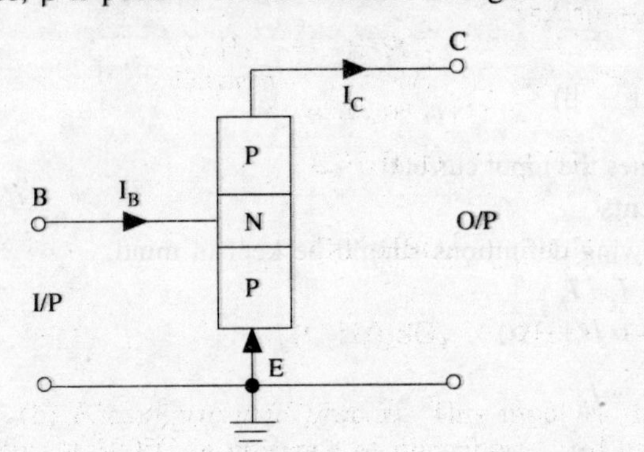

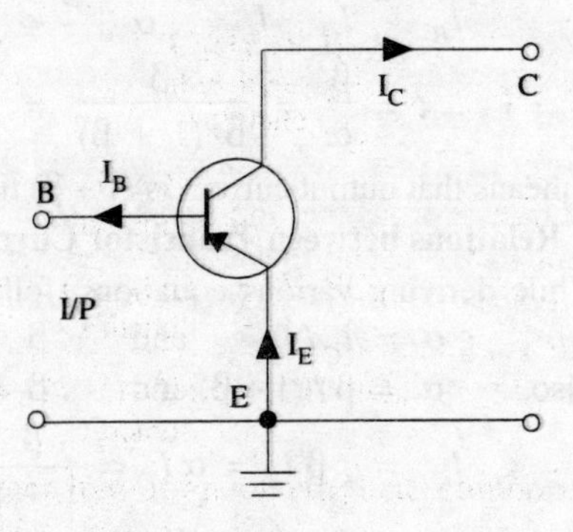

Fig. 30.7

30.9. Relation between α and β

$$\beta = \frac{I_C}{I_B} \quad \text{and} \quad \alpha = \frac{I_C}{I_E} \qquad \therefore \quad \frac{\beta}{\alpha} = \frac{I_E}{I_B}$$

Now, $\quad I_B = I_E - I_C$

$$\therefore \quad \beta = \frac{I_C}{I_E - I_C} = \frac{I_C / I_E}{I_E / I_E - I_C / I_E}$$

$$\therefore \quad \beta = \frac{\alpha}{1 - \alpha} \qquad \text{...(i)}$$

Cross-multiplying the above equation and simplifying it, we get

$$\beta(1 - \alpha) = \alpha \quad \text{or} \quad \beta = \alpha(1 + \beta)$$

$$\alpha = \frac{\beta}{1 + \beta} \qquad \text{..(ii)}$$

It is seen from the above two equations that

$$1 - \alpha = \frac{1}{1 + \beta}$$

30.10. CC Configuration

In this case (Fig. 30.8), input signal is applied between base and collector and output signal is taken out from emitter-collector circuit. Obviously, conventionally speaking, here I_B is the input current and I_E is the output current as shown in Fig. 30.8. The current gain of the circuit is

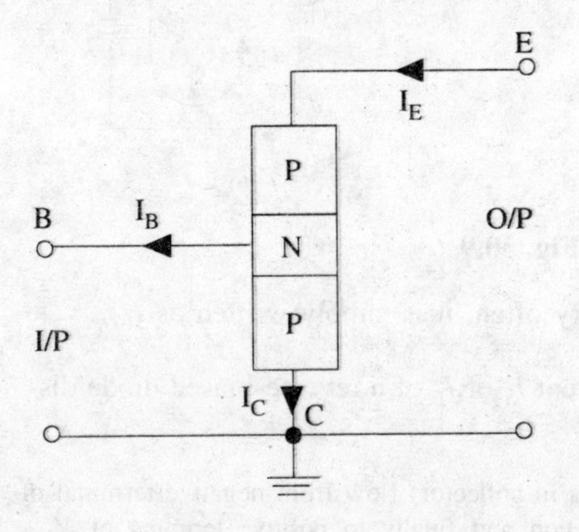

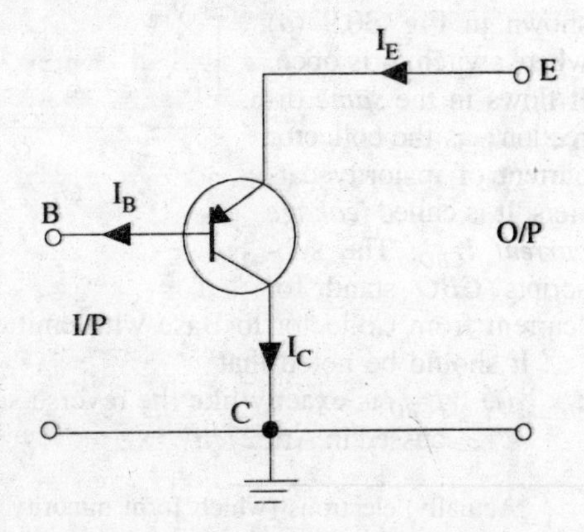

Fig. 30.8

$$\frac{I_E}{I_B} = \frac{I_E}{I_C} \cdot \frac{I_C}{I_B} = \frac{1}{\alpha} \cdot \beta$$

$$= \frac{\beta}{\alpha} = \frac{\beta}{\beta/(1+\beta)} = (1+\beta)$$

It means that output current is $(1 + \beta)$ times the input current.

30.11. Relations between Transistor Currents

While deriving various equations, following definitions should be kept in mind,

$$\alpha = I_C / I_E \quad \text{and} \quad \beta = I_C / I_B$$

Also, $\alpha = \beta / (1 + \beta)$ and $\beta = \alpha / (1 - \alpha)$

1. $I_C = \beta I_B = \alpha I_E = \dfrac{\beta}{1+\beta} \cdot I_E$

2. $I_B = \dfrac{I_C}{\beta} = \dfrac{I_E}{(1+\beta)} = (1-\alpha) I_E$

3. $I_E = \dfrac{I_C}{\alpha} = \dfrac{(1+\beta)}{\beta} I_C$

$$= (1+\beta) I_B = \frac{I_B}{(1-\alpha)}$$

4. The three transistor d.c. currents always bear the following ratio:

$$I_E : I_B : I_C :: 1 : (1 - \alpha) : \alpha$$

Incidentally, it may be noted that for *a.c.* currents, small letters i_e, i_b and i_c are used.

30.12. Leakage Currents in a Transistor

(*a*) CB Circuit

Consider the CB transistor circuit shown in Fig. 30.9. The emitter current (due to majority carriers) initiated by the forward-biased emitter-base junction is split into two parts:

(*i*) $(1 - \alpha) I_E$ which becomes base current I_B in the external circuit; and

(*ii*) αI_E which becomes collector current I_C in the external circuit.

As mentioned earlier (Art. 30.2) though *C/B* junction is reverse-biased for majority charge carriers (*i.e.*, holes in this case), it is forward-biased so far as thermally-generated minority charge carriers (*i.e.*, electrons) are concerned. This current flows *even when emitter is disconnected from its d.c. supply* as shown in Fig. 30.9 (*a*) where switch S_1 is open. It flows in the *same* direction* as the collector current of majority carriers. It is called *leakage current* I_{CBO}. The subscripts *CBO* stand for 'current from Collector to Base with emitter Open'. Very often, it is simply written as I_{CO}.

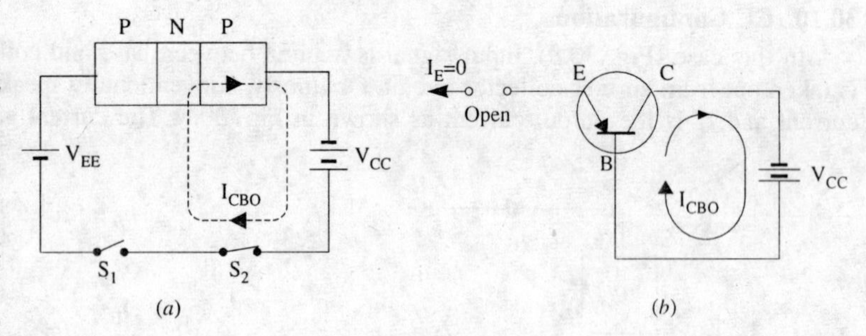

Fig. 30.9

It should be noted that

(*i*) I_{CBO} is exactly like the reverse saturation current I_S or I_0 of a reverse-biased diode discussed in Art. 27.17.

*Actually, electrons (which form minority charge carriers in collector) flow from negative terminal of collector battery, to collector, then to base through *C/B* junction and finally to positive terminal of V_{CC}. However, conventional current flows in the opposite direction as shown by the dotted line in Fig. 30.9 (*a*).

(*ii*) I_{CBO} is extremely temperature-dependent because it is made up of thermally-generated minority carriers. As mentioned earlier, I_{CBO} doubles for every 10°C rise in temperature for Ge and 6°C for Si (Art. 27.17).

If we take into account the leakage current, the current distribution in a *CB* transistor circuit becomes as shown in Fig. 30.10 both for PNP and NPN type transistors.

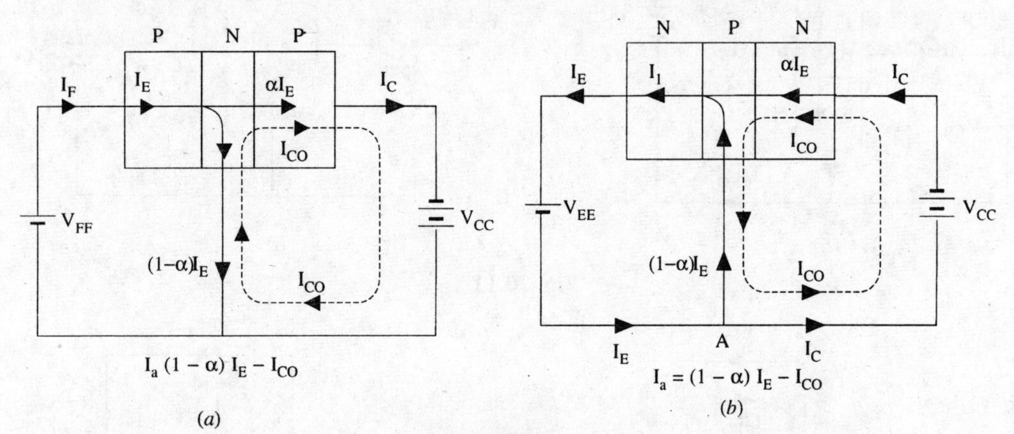

Fig. 30.10

It is seen that total collector current is actually the sum of two components

(*i*) current produced by normal transistor action *i.e.*, component controlled by emitter current. Its value is αI_E and is due to majority charge carriers.

(*ii*) temperature-dependent leakage current I_{CO} due to minority carriers

$$\therefore \quad I_C = \alpha I_E + I_{CO}$$

$$= \underset{\text{majority}}{\alpha I_E} + \underset{\text{minority}}{I_{CO}}$$

It can also be proved that

$$I_C = \frac{I_{CO}}{1 - \alpha} + \frac{\alpha I_B}{1 - \alpha}$$

In view of the above, it would be appreciated that

$$\alpha = -\frac{I_C - I_{CO}}{I_E} \cong -\frac{I_C}{I_E}$$

It can also be proved that

$$I_B = (1 - \alpha) I_E - I_{CO}$$

(*b*) **CE Circuit**

In Fig. 30.11 (*a*) is shown a grounded-emitter circuit of an NPN transistor whose base lead is open. It is found that despite $I_B = 0$, there is a leakage current from collector to emitter. It is called I_{CEO}, the subscripts *CEO* standing for 'Collector to Emitter with base Open.'

Taking this leakage current into account, the current distribution through a *CE* circuit becomes as shown in Fig. 30.12.

$$I_C = \beta I_B + I_{CEO} = \beta I_B + (1 + \beta) I_{CO}$$

$$= \beta I_B + I_{CO} / (1 - \alpha) = \alpha I_E + (1 - \alpha) I_{CEO}$$

$$= \frac{\alpha I_B}{1 - \alpha} + \frac{I_{CO}}{1 - \alpha}$$

Now, $\beta I_B = \alpha I_E$. — Substituting this value above, we get

$$I_C = \alpha I_E + I_{CEO}$$

Also, $I_B = I_E - I_C$

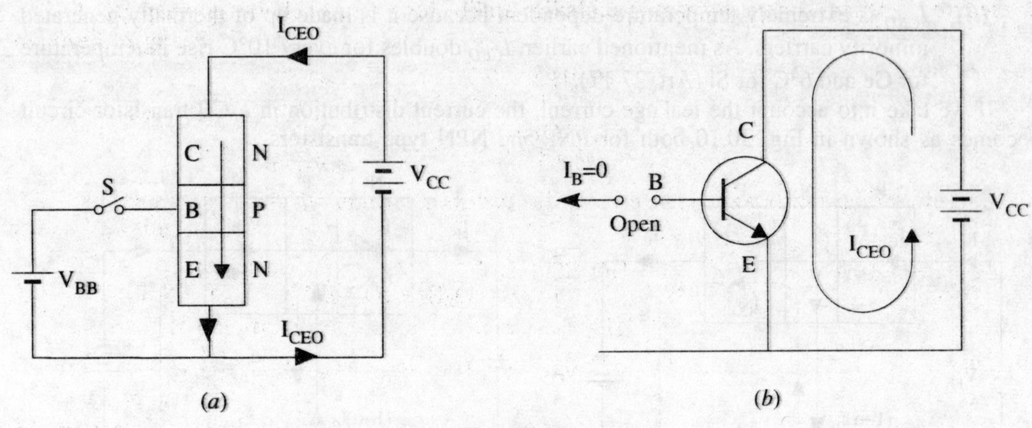

(a) (b)

Fig. 30.11

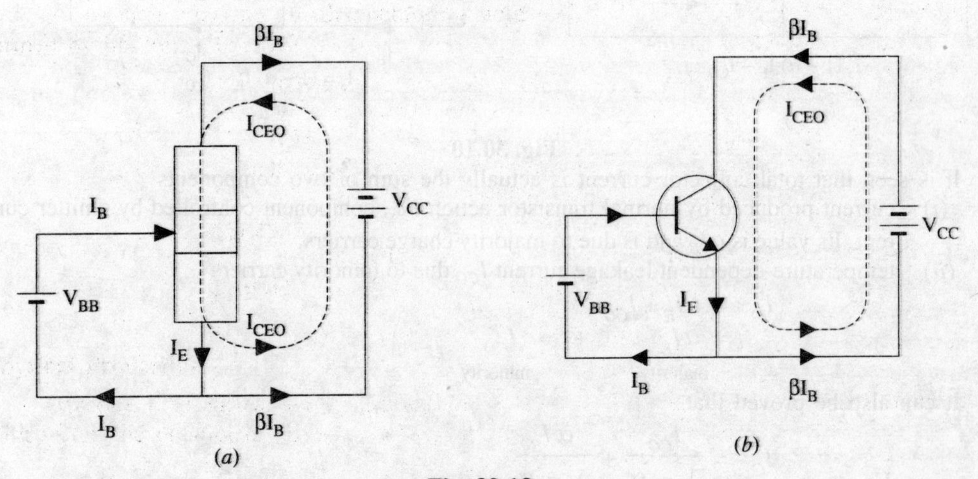

(a) (b)

Fig. 30.12

Substituting the value of I_C from above, we have

$$I_B = I_E - \alpha I_E - I_{CEO}$$
$$= (1-\alpha)I_E - I_{CEO}$$

Example 30.4. *A transistor has* $\alpha = 0.98$, $I_B = 100\ \mu A$ *and* $I_{CO} = 6\ \mu A$. *Calculate* I_C *and* I_E.

Solution.
$$I_C = \frac{\alpha I_B}{1-\alpha} + \frac{I_{CO}}{1-\alpha}$$
$$= \frac{0.98 \times 100}{1 - 0.98} + \frac{6}{1 - 0.98}$$
$$= 4900 + 300 = 5200\ \mu A = \mathbf{5.2\ mA}$$
$$I_E = I_C + I_B = 5200 + 100 = 5300\ \mu A = \mathbf{5.3\ mA}$$

30.13. Transistor Static Characteristics

These are the curves which represent relationships between different d.c. currents and voltages of a transistor. These are helpful in studying the operation of a transistor when connected in a circuit. The three important characteristics of a transistor are

1. *input characteristic ;*
2. *output characteristic ;*
3. *constant-current transfer characteristic.*

30.14. Common Base Test Circuit

The static characteristics of an NPN transistor connected in common-base configuration can be determined by the use of the test circuit shown in Fig. 30.13. Milliammeters are included in

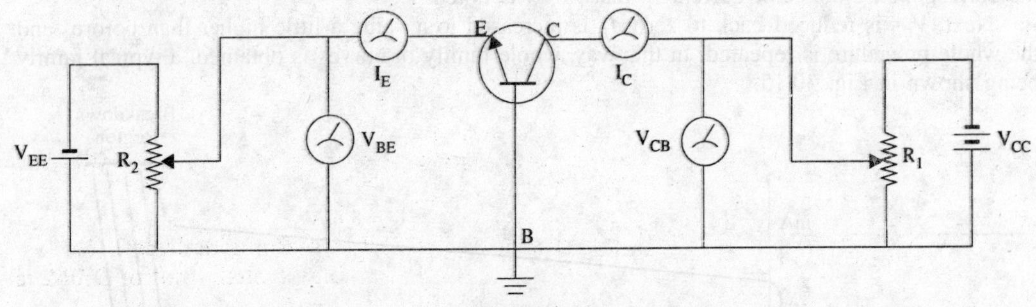

Fig. 30.13

series with the emitter and collector circuits for measuring I_E and I_C. Similarly, voltmeters are connected across E and B to measure voltage V_{BE}* and across C and B to measure V_{CB}.* The two potentiometer resistors R_1 and R_2 supply variable voltages from the collector and emitter d.c. supplies.

30.15. Common Base Static Characteristics

(a) Input characteristic

It shows low I_E varies with V_{BE} when voltage V_{CB} is held constant. The method of determining the characteristic is as follows:

First, voltage V_{CB} is adjusted to a suitable value with the help of R_1 (Fig. 30.13). Next, voltage V_{BE} is increased in a number of discrete steps and corresponding values of I_E are noted from the milliammeter connected for the purpose. When plotted, we get the input characteristics shown in Fig. 30.14, one for Ge and the other for Si. Both curves are exactly similar to the forward characteristic of a P-N diode which, in essence, is what the emitter-base junction is.

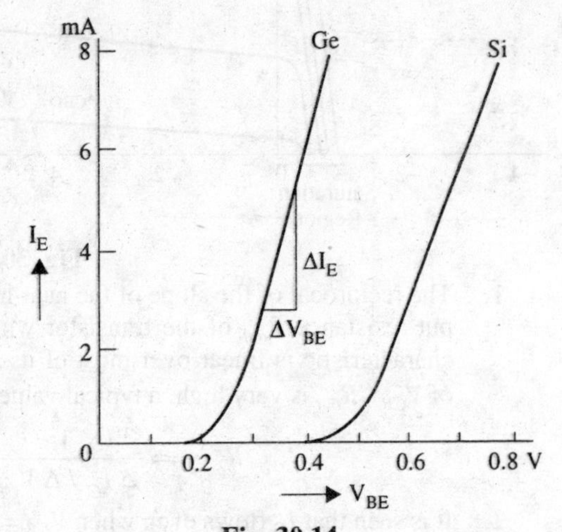

Fig. 30.14

This characteristic may be used to find the input resistance of the transistor. Its value is given *by the reciprocal of its slope.*

$$R_{in} = \frac{\Delta V_{BE}}{\Delta I_E} \quad — V_{CB} \text{ constant}$$

Since the characteristic is initially non-linear, R_{in} will vary with the point of measurement. Its value over linear part of the characteristic is about 50 Ω but for low values of V_{BE} it is considerably greater. This change in R_{in} with change in V_{BE} gives rise to distortion of signals handled by the transistor.

This characteristic is hardly affected by changes either in V_{CB} or temperature.

Note. Strictly speaking both I_E and V_{BE} should be shown negative for an NPN transistor.

(b) Output Characteristic

It shows the way I_C varies with V_{CB} when I_E is held constant.

*In these subscripts, first letter indicates that electrode which is at higher potential as compared to the other electorde.

The method of obtaining the characteristic is as follows:

First, movable contact on R_2 (Fig. 30.13) is changed to get a suitable value of V_{BE} and hence I_E. While keeping I_E constant at this value, V_{CB} is increased from zero in a number of steps and the corresponding collector current I_C that flows is noted.

Next, V_{CB} is reduced back to zero, I_E is increased to a value a little higher than before and the whole procedure is repeated. In this way, whole family of curves is obtained, a typical family being shown in Fig. 30.15.

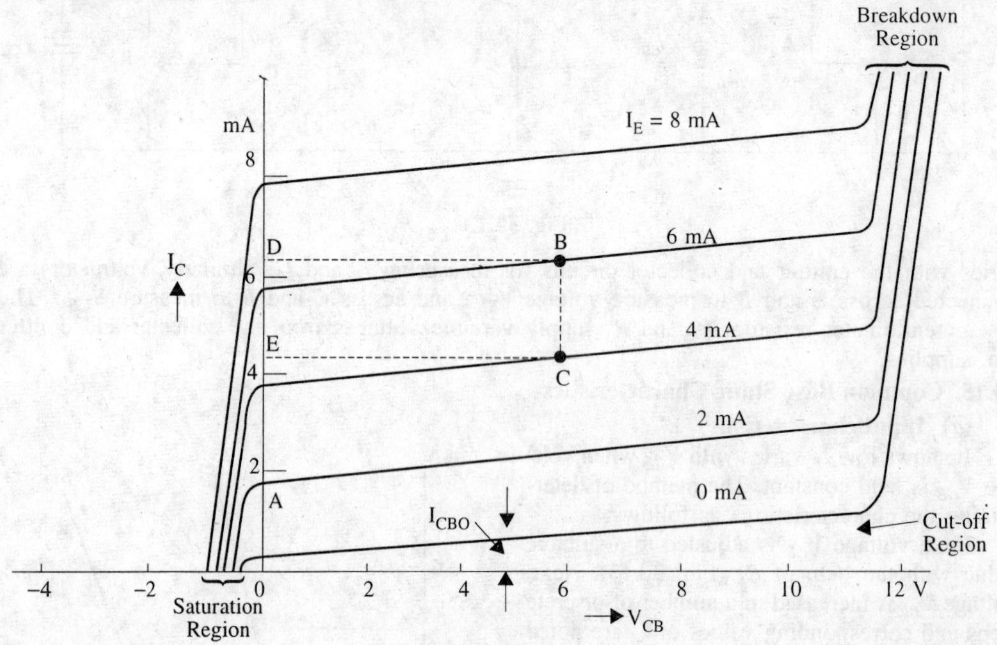

Fig. 30.15

1. The reciprocal of the slope of the near horizontal part of the characteristic gives the output resistance R_{out} of the transistor which it would offer to an input signal. Since the characteristic is linear over most of its length (meaning that I_C is virtually independent of V_{CB}), R_{out} is very high, a typical value being 500 kΩ.

$$R_{out} = \frac{1}{\Delta I_C / \Delta V_{CB}} = \frac{\Delta V_{CB}}{\Delta I_C}$$

2. It is seen that I_C flows even when $V_{CB} = 0$. For example, it has a value = OA corresponding to $I_E = 2$ mA as shown in Fig. 30.15. It is due to the fact that electrons are being injected into the base under the action of forward-biased *E/B* junction and are being collected by the collector due to the action of the internal junction voltage at the *C/B* junction (Art. 30.2). For reducing I_C to zero, it is essential to neutralize this potential barrier by applying a small forward bias across *C/B* junction.
3. Another important feature of the characteristic is that a small amount of collector current flows even when emitter current $I_E = 0$. As we know (Art. 30.12), it is collector leakage current I_{CBO}.
4. This characteristic may be used to find α_{ac} of the transistor as shown in Fig. 30.15.

$$\alpha_{ac} = \frac{\Delta I_C}{\Delta I_E} = \frac{DE}{BC} = \frac{6.2 - 4.3}{2} = 0.95$$

5. Another point worth noting is that although I_C is practically independent of V_{CB} over the working range of the transistor, yet if V_{CB} is permitted to increase, I_C eventually increases rapidly due to avalanche breakdown as shown.

(*c*) **Current Transfer Characteristic**

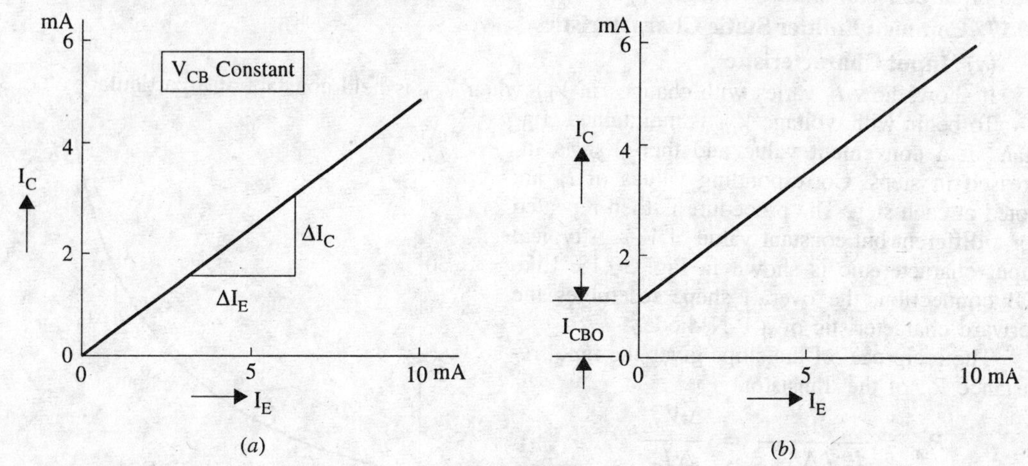

Fig. 30.16

It shows how I_C varies with changes in I_E when V_{CB} is held constant. For drawing this characteristic, first V_{CB} is set to a convenient value and then I_E is increased in steps and corresponding values of I_C noted. A typical transfer characteristic is shown in Fig. 30.16 (*a*). Fig. 30.16 (*b*) shows a more detailed view of the portion, near the origin.

As seen, α_{ac} may be found from the equation

$$\alpha_{ac} = \frac{\Delta I_C}{\Delta I_E}$$

Usually, α_{ac} is found from output characteristic rather than from this characteristic.

It may be noted in the end that CB connection is rarely employed for audio-frequency circuits because firstly, its current gain is less than unity and secondly, its input and output resistances are so different.

Note. Strictly speaking, for an NPN transistor, I_C should be positive but I_E should be negative.

30.16. Common Emitter Test Circuit

The static characteristics of a transistor connected in CE configuration may be determined by the use of circuit diagram shown in Fig. 30.17. A milliammeter (or a microammeter in the case of a low-power transistor) is connected in series with the base to measure I_B. Similarly, an ammeter is included in the collector circuit to measure I_C. A voltmeter with a typical range of 0–1 V is connected across base and emitter terminals for measuring V_{BE} .

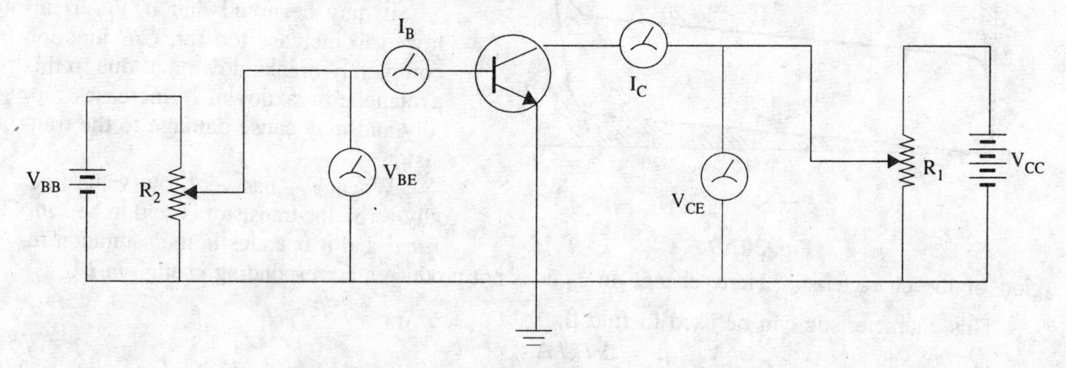

Fig. 30.17

Potentiometer R_2 connected across d.c. supply V_{BB} is used to vary I_B and V_{BE}. A second voltmeter with a typical range of 0–20 V is connected across collector-emitter terminals to measure the output collector-emitter voltage V_{CE}.

30.17. Common Emitter Static Characteristics

(*a*) Input Characteristic

It shows how I_B varies with changes in V_{BE} when V_{CE} is held constant at a particular value.

To begin with, voltage V_{CE} is maintained constant at a convenient value and then V_{BE} is increased in steps. Corresponding values of I_B are noted at each step. The procedure is then repeated for a different but constant value of V_{CE}. A typical input characteristic is shown in Fig. 30.18. Like CB connection, the overall shape resembles the forward characteristic of a P-N diode.

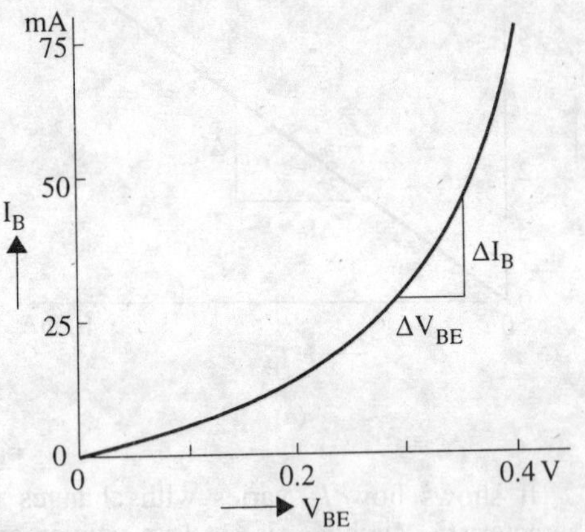

The reciprocal of the slope gives the input resistance R_{in} of the transistor.

$$R_{in} = \frac{1}{I_B / \Delta V_{BE}} = \frac{\Delta V_{BE}}{\Delta I_B}$$

Fig. 30.18

Due to initial non-linearity of the curve, R_{in} varies considerably from a value of 4 kΩ near the origin to a value of 600 Ω over the more linear part of the curve.

(*b*) Output or Collector Characteristic

It indicates the way in which I_C varies with changes in V_{CE} when I_B is held constant.

For obtaining this characteristic, first I_B is set to a convenient value and maintained constant and then V_{CE} is increased from zero in steps, I_C being noted at each step. Next, V_{CE} is reduced to zero and I_B increased to another convenient value and the whole procedure repeated. In this way, a family of curves (Fig. 30.19) is obtained.

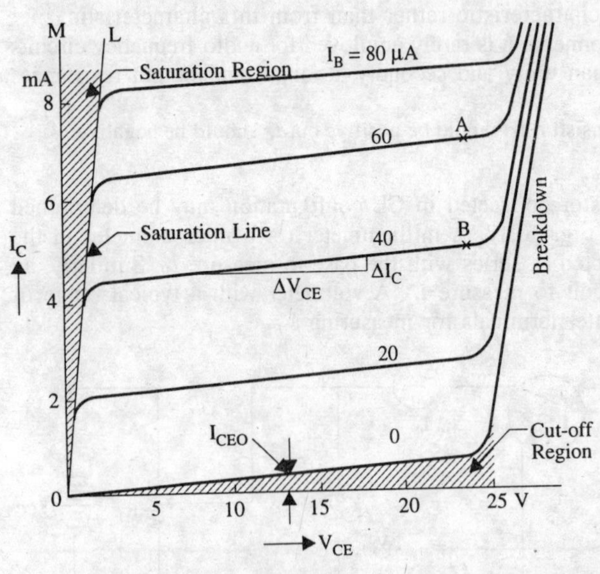

Fig. 30.19

It is seen that as V_{CE} increases from zero, I_C rapidly increases to a near saturation level for a fixed value of I_B. As shown, a small amount of collector current flows even when $I_B = 0$. It is called I_{CEO} (Art. 30.12). Since main collector current is zero, the transistor is said to be *cut-off*.

It may be noted that if V_{CE} is allowed to increase too far, *C/B* junction completely breaks down and due to this avalanche breakdown, I_C increases rapidly and may cause damage to the transistor.

When V_{CE} has very low value (ideally zero), the transistor is said to be *saturated* and it operates in the saturation region of the characteristic. Here, change in I_B does not produce a corresponding change in I_C.

This characteristic can be used to find β.

$$\beta = \Delta I_C / \Delta I_B$$

We may select any two points *A* and *B* on the I_B = 60 μA and 40 μA lines respectively and measure corresponding values of I_C from the diagram for finding ΔI_C. Since ΔI_B = (60 – 40) = 20 μA, β can be found.

The value of output resistance $R_{out} = \Delta V_{CE} / \Delta I_C$ over the near horizontal part of the characteristic varies from 10 kΩ to 50 kΩ.

(c) Current Transfer Characteristic

It indicates how I_C varies with changes in I_B when C_{CE} is held constant at a given value.

Such a typical characteristic is shown in Fig. 30.20 (*a*). Its slope gives $\beta = \Delta I_C / \Delta I_B$.

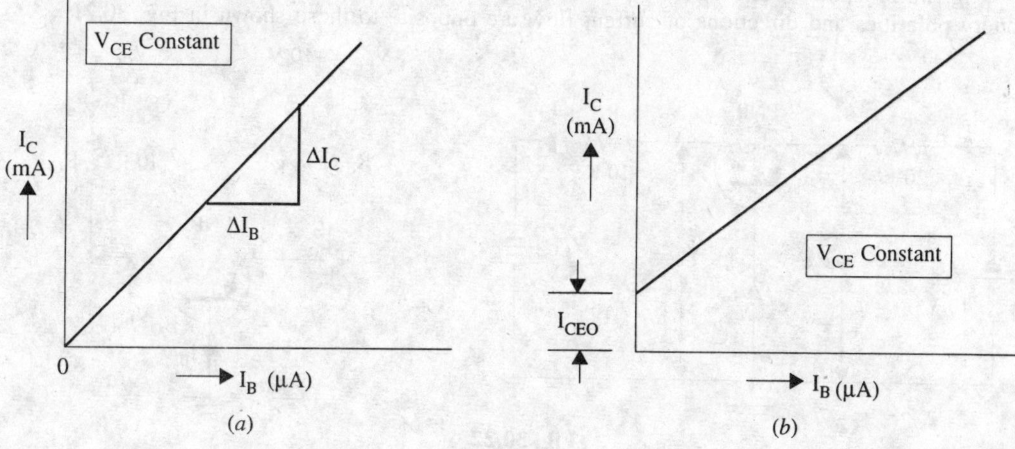

Fig. 30.20

From Fig. 30.20 (*b*), it is seen that a small collector current flows even when $I_B = 0$. It is the common-emitter leakage current $I_{CEO} = (1 + \beta) I_{CO}$. Like I_{CO}, it is also due to the flow of minority carriers across the reverse-biased *C/B* junction.

30.18. Different ways of Drawing Schematic Transistor Circuits

In Fig. 30.21 (*a*) is shown a CB transistor circuit which derives its voltage and current requirements from two independent power sources *i.e.*, two different batteries. Correct battery

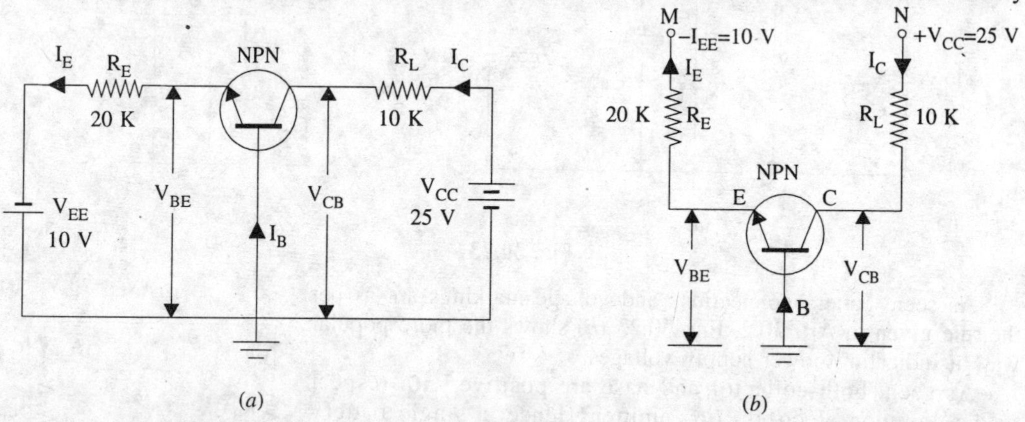

Fig. 30.21

connections can be done by remembering the transistor polarity rule (Art. 30.2) that in an NPN transistor, both collector and base have to be POSITIVE with respect to the emitter. Of course, collector is a little more positive than base which means that between themselves, collector is at a slightly higher positive potential with respect to base. Conversely, base is at a little lower potential with respect to the collector.

Putting it in a slightly different way, we can say that collector is positive w.r.t. base and conversely, base is negative w.r.t. collector. That is why, potential difference between collector and base is written as V_{CB} (and not V_{BC}) because terminal at higher potential is mentioned first (footnote Art. 30.14). Same reasoning applies to V_{BE}. Fig. 30.21 (*b*) shows another and more popular way of indicating power supply voltage. Only one terminal of the battery is shown, the other terminal of each power supply is understood to be grounded so as to provide a complete path for the current. For example, negative terminal of V_{CC} and positive terminal of V_{EE} are supposed to be grounded (as is the base) even though not shown in the diagram.

Fig. 30.22 shows same circuit drawn for a PNP transistor. As expected, in this case, the battery polarities and directions of current flow are opposite to those shown in Fig. 30.21.

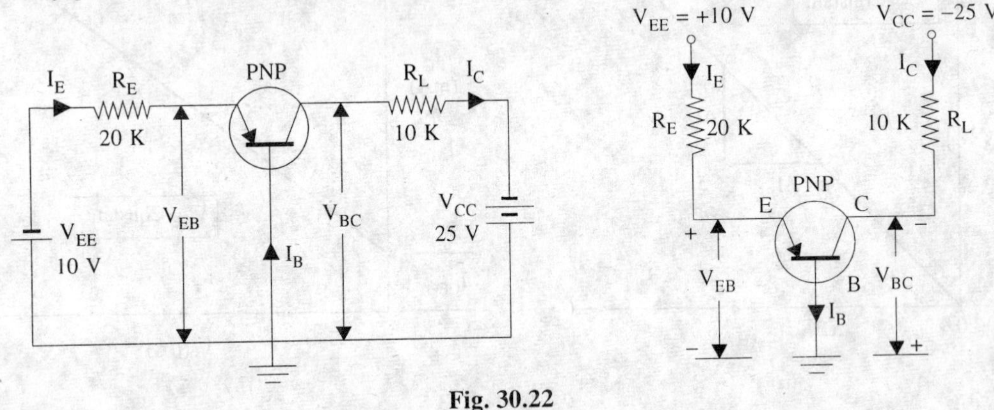

Fig. 30.22

Fig. 30.23 (*a*) shows an NPN transistor connected in CE configuration with voltages and currents drawn from two independent power sources.

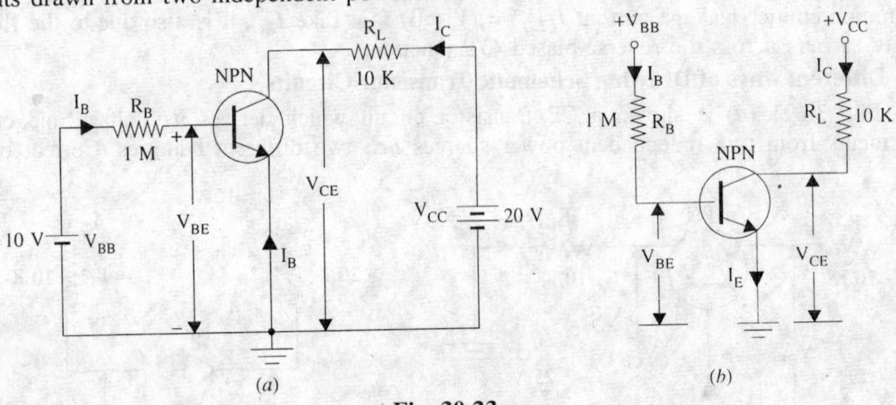

Fig. 30.23

As seen, battery connections and voltage markings are as per the rule given in Art. 30.2. Fig. 30.23 (*b*) shows the more popular way of indicating power supply voltages.

As seen, both collector and base are positive with respect to the common electrode *i.e.*, emitter. Hence, a single battery can be used to get proper voltages across the two as shown in Fig. 30.24.

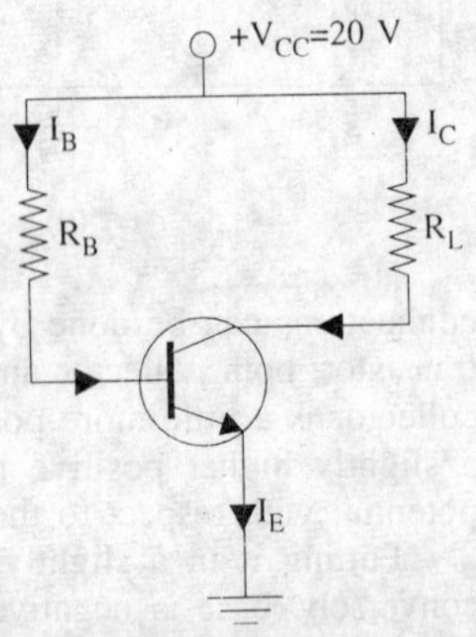

Fig. 30.24

Fig. 30.25 (*a*) shows the CC configuration of an NPN transistor and Fig. 30.25 (*b*) shows the same circuit drawn differently.

30.19. Common Base Formulas

Let us find the values of different voltages and currents for the circuit shown in Fig. 30.21 (*b*). Consider circuit *MEBM*. Starting from *M*, used applying Kirchhoff's current law, we get

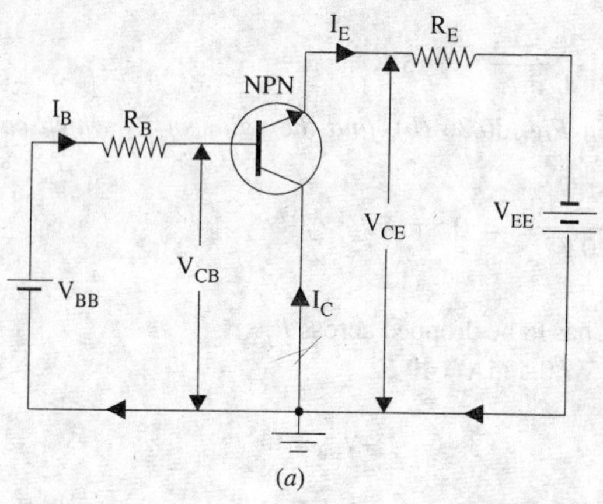

(*a*)

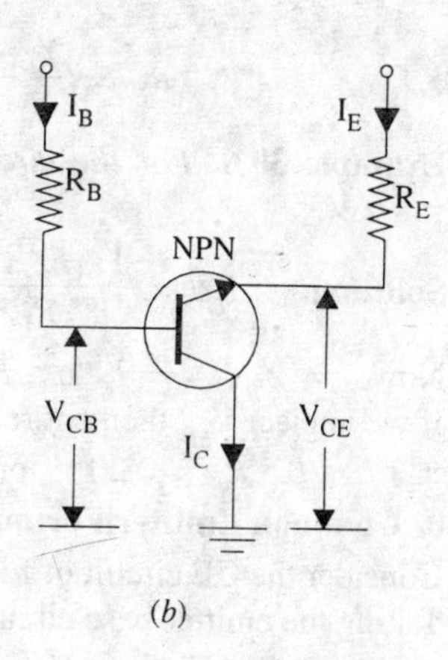

(*b*)

Fig. 30.25

$$+I_E R_E + V_{BE} - V_{EE} = 0$$

(*a*) or $$I_E = \frac{V_{EE} - V_{BE}}{R_E}$$

where $V_{BE} = 0.3$ V — for Ge

$= 0.7$ V — for Si

Since generally, $V_{EE} \gg V_{BE}$, we can simplify the above to

$$I_E \cong V_{EE} / R_E$$

$$\cong 10 \text{ V}/20 \text{ K} = 0.5 \text{ mA}$$

Taking V_{BE} into account and assuming silicon transistor,

$$I_E = (10 - 0.7) \text{ V}/20 \text{ K} = 0.465 \text{ mA}$$

(*b*) $I_C = \alpha I_E \cong I_E$ —neglecting leakage current I_{CO} (Art. 30.12)

(*c*) From circuit *NCBN*, we get

$$V_{CB} = V_{CC} - I_C R_L$$

$$\cong V_{CC} - I_E R_L \qquad \because I_C \cong I_E$$

$$V_{CB} = 25 - 0.5 \times 10 = 20 \text{ V}$$

Example 30.5. *In the circuit of Fig. 30.26 (a), what value of R_L causes $V_{CB} = 5$ V?*

Solution. $I_E \cong V_{EE} / R_E = 10 \text{ V} / 10 \text{ K} = 1 \text{ mA}$

$I_C = \alpha I_E \cong I_E = 1$ mA

Now, $V_{CC} = I_C R_L + V_{CB}$

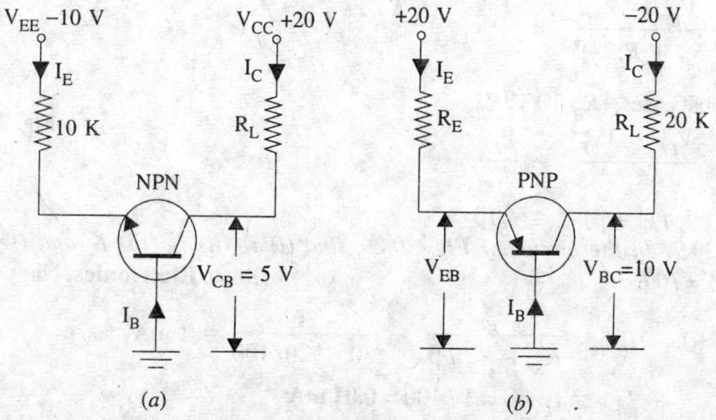

(*a*) (*b*)

Fig. 30.26

$$\therefore \quad R_L = \frac{V_{CC} - V_{CB}}{I_C} = \frac{20 - 5}{1 \text{ mA}} = \mathbf{15\ K}$$

Example 30.6. *For the circuit shown in Fig. 30.26 (b), find the value of R_E which causes $V_{BC} = 10$ V.*

Solution. $$I_C = \frac{V_{CC} - V_{BC}}{R_L} = \frac{20 - 10}{20 \text{ K}} = 0.5 \text{ mA}$$

Now, $I_E = I_C/\alpha \cong I_C = 0.5$ mA

If we neglect $V_{EB,}$ then entire $V_{EE} = 20$ V has to be dropped across R_E.

$\therefore \quad 0.5\, R_E = 20 \quad$ or $\quad R_E = 20 \text{ V} / 0.5 \text{ mA} = \mathbf{40\ K}$

30.20. Common Emitter Formulas

Consider the CE circuit of Fig. 30.27.

Taking the emitter-base circuit, we have

$$I_B = \frac{V_{BB} - V_{BE}}{R_B} \cong \frac{V_{BB}}{R_B}$$

$I_C = \beta I_B$ – neglecting leakage current I_{CEO} (Art. 30.12)

$V_{CE} = V_{CC} - I_C R_L$

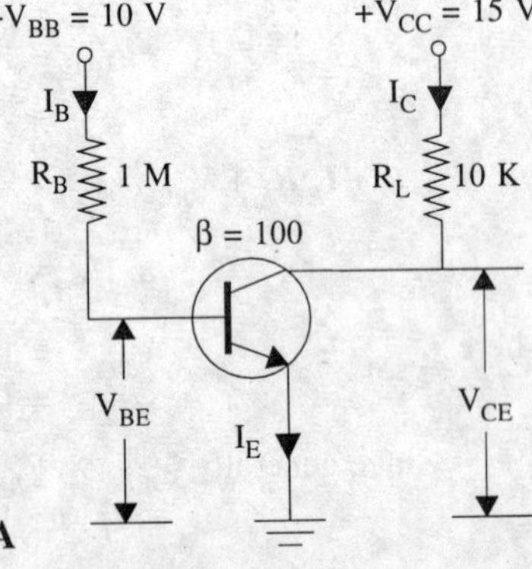

Fig. 30.27

Example 30.7. *From the circuit of Fig. 30.27, find (i) I_B (ii) I_C (iii) I_E and V_{CE}. Neglect V_{BE}.*

Solution. (i) $I_B = V_{BB}/R_B = 10/1 \text{ M} = \mathbf{10\ \mu A}$

(ii) $I_C = \beta I_B = 100 \times 10\ \mu\text{A} = \mathbf{1\ mA}$

(iii) $I_E = I_B + I_C = 1 \text{ mA} + 10\ \mu\text{A} = \mathbf{1.01\ mA}$

(iv) $V_{CE} = V_{CC} - I_C R_L = 15 - 1 \times 10 = \mathbf{5\ V}$

Example 30.8. *Find the exact value of the emitter current I_E in the two-supply emitter-bias circuit of Fig. 30.28.* **(Electronics-1, Bangalore Univ. 1988)**

Solution. Let us apply Kirchhoff's voltage law to the loop containing R_B, R_E and V_{EE}. Starting from emitter and going clockwise, we get

$$-I_E R_E + V_{EE} - I_B R_B - V_{BE} = 0$$

or $$I_E R_E + I_B R_B = V_{EE} - V_{BE}$$

Now, $$\beta = \frac{I_C}{I_B} \cong \frac{I_E}{I_B} \quad \therefore \quad I_B \cong \frac{I_E}{\beta}$$

Substituting this value in Eq. (*i*) above, we get

$$I_E R_E + \frac{I_E R_B}{\beta} = V_{EE} - V_{BE}$$

$$\therefore \quad I_E = \frac{V_{EE} - V_{BE}}{R_E + R_B/\beta}$$

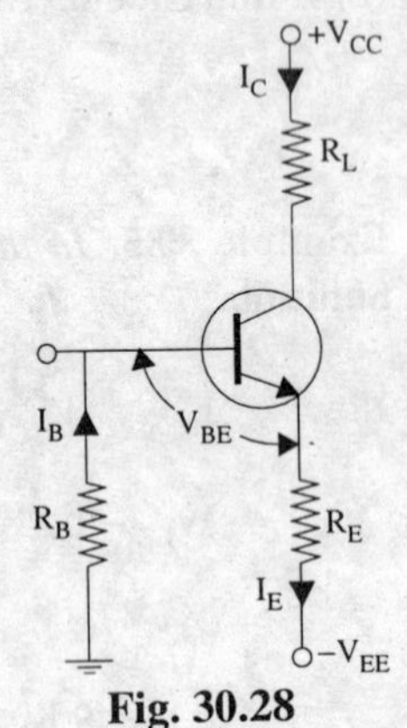

Fig. 30.28

Since, in most cases, $(R_B/\beta) \ll R_E$,

$$\therefore \quad I_E \cong \frac{V_{EE} - V_{BE}}{R_E} \cong \frac{V_{EE}}{R_E}$$

Also, $I_B = I_E/(1 + \beta) \cong I_E/\beta$

Example 30.9. *In the circuit of Fig. 30.29, find (i) I_E (ii) I_B (iii) I_C and (iv) V_{CE}. Neglect V_{BE} and take $\beta = 100$.* **(Basic Electronics, Nagpur Univ. 1992)**

Solution. (*i*) $$I_E = \frac{V_{EE}}{R_E + R_B/\beta} = \frac{30}{50 + 20/100} \cong \mathbf{1\ mA}$$

(*ii*) $$I_B \cong I_E/\beta = 1/100 = \mathbf{0.01\ mA}$$

(*iii*) $I_C = I_E - I_B = \mathbf{0.99\ mA}$

(*iv*) $V_{CE} = V_{CC} - I_C R_L$

$= 30 - 10 \times 0.99 \cong \mathbf{20\ V}$

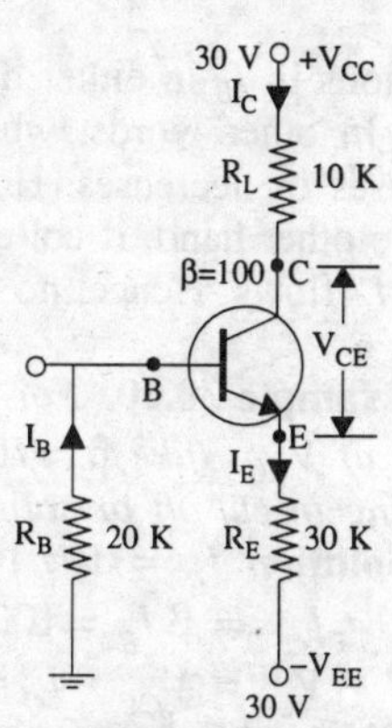

Fig. 30.29

30.21. Cut-off and Saturation Points

Consider the circuit of Fig. 30.30 (*a*). As seen from Art. 30.20

$V_{CE} = V_{CC} - I_C R_L$

Since $I_B = 0$, $I_C = 0$. Hence

$V_{CE} = V_{CC}$

Under these conditions, the transistor is said to *cut-off* for the simple reason that transistor does not conduct any current.* This value of V_{CE} is written as $V_{CE(cut\text{-}off)}$. Moreover, in cut-off, both the base-emitter and the base-collector junctions are reverse-biased.

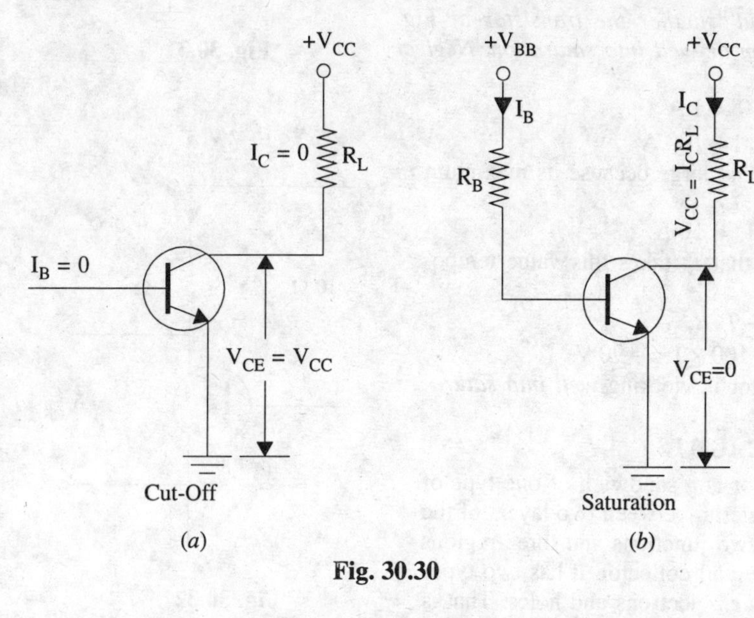

Fig. 30.30

If in Fig. 30.30 (*b*), I_B is increased by increasing V_{BB}, then I_C is increased because $I_C = \beta I_B$. This increases drop across R_L as a result of which V_{CE} is decreased because $V_{CE} = V_{CC} - I_C R_L$. A certain value of I_C is reached when $I_C - R_L$ becomes equal to V_{CC} itself.

In that case

$V_{CE} = V_{CC} - I_C R_L$

$= V_{CC} - V_{CC} = 0$

When $V_{CE} = 0$, the transistor is said to be in saturation because it then carries the maximum collector current called $I_{C(sat)}$. Even if I_B is increased now, I_C does not increase beyond its saturation value because under saturation condition, the relation $I_C = \beta I_B$ does not hold good. To summarize the above, we have that when a transistor is in saturation:

1. whole of V_{CC} drops across R_L.
2. I_C has maximum value of $I_{C(sat)} = V_{CC} / R_L$.

Normal operation of a transistor lies between the above two extreme conditions of out-off and saturation.

30.22. Importance of V_{CE}

The voltage V_{CE} is very important in checking whether the transistor is

(*a*) defective;

(*b*) working in cut-off;

(*c*) in saturation or well into saturation.

When $V_{CE} = V_{CC}$, the transistor is in cut-off *i.e.,* it is turned OFF. When $V_{CE} = 0$, the transistor is in saturation *i.e.*, it is turned full ON. When V_{CE} is less than zero *i.e.*, negative, the transistor is said to be *well into saturation.* In practice, both these conditions are avoided. For amplifier operation, $V_{CE} = \frac{1}{2} V_{CC}$ *i.e.,* transistor is operated at approximately $\frac{1}{2}$ ON. In this way,

*Except leakage current I_{CEO}.

variations in I_B in either direction will control I_C in both directions. In other words, when I_B increases or decreases, I_C also increases or decreases. However, if I_B is OFF, I_C is also OFF. On the other hand, if collector has been turned fully ON, maximum I_C flows. Hence, no further increase in I_E can be reflected in I_C.

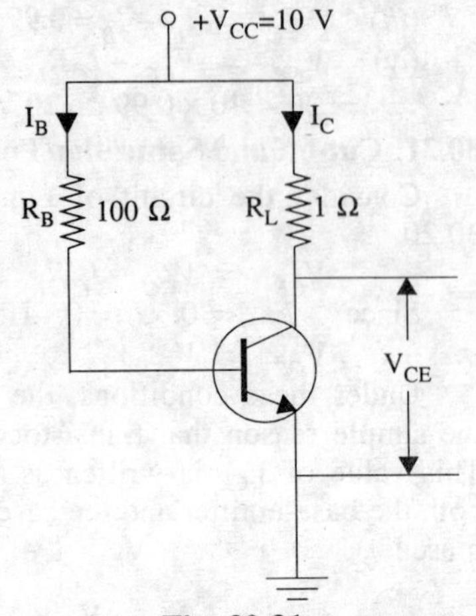

Fig. 30.31

Example 30.10. *For the CE circuit of Fig. 30.31, find the value of V_{CE}. Take β =100 and neglect V_{BE}. Is the transistor working in cut-off or saturation?*

Solution. $I_B = 10/100 = 0.1$ A

$I_C = \beta I_B = 100 \times 0.1 = 10$ A

$V_{CE} = V_{CC} - I_C R_L = 10 - 10 \times 1 = 0$

Obviously, the transistor is operating at saturation and not well into saturation.

Example 30.11. *Find out whether the transistor of Fig. 30.32 is working in saturation or well into saturation. Neglect V_{BE}.*

Solution. $I_B = 10/10 = 1$ A

$I_C = 100 \times 1 = 100$ A

Obviously, I_C cannot be that large because its maximum value is given by

$V_{CC}/R_L = 10/1 = 10$ A.

However, let us assume that I_C takes this value temporarily, then,

$$V_{CE} = V_{CC} - I_C R_L$$
$$= 10 - 100 \times 1 = -90 \text{ V}$$

It means that the transistor is working *well into saturation.*

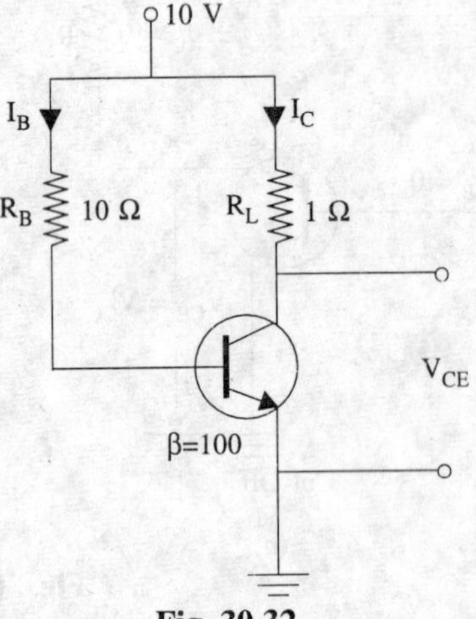

Fig. 30.32

HIGHLIGHTS

1. A junction transistor is a sandwich of one type of semi-conductor material between two layers of the other type. It has two junctions and three regions called emitter, base and collector. It has two types of charge carriers *i.e.*, electrons and holes. That is why it is called a bipolar junction transistor (BJT).
2. In a BJT, both collectors and base have same polarity with respect to the base. In the case of NPN transistors, this polarity is positive whereas for PNP transistors, it is negative.
3. In a properly-biased transistor,
   $$I_E = I_B + I_C$$
4. The d.c. alpha of a transistor is: $\alpha = -I_C / I_E$. It is always less than unity.
5. The d.c. beta of a transistor is: $\beta = I_C / I_B$. It is always greater than unity and positive.
6. The ralation between β and α is
   $$\beta = \frac{\alpha}{1-\alpha}; \quad \alpha = \frac{\beta}{1+\beta} \quad \text{and} \quad 1-\alpha = \frac{1}{1+\beta}$$
7. A transistor is said to be cut-off when it does not conduct current. In that condition, $I_B = 0, I_C = 0$ and $V_{CE} = V_{CC}$.
8. A transistor is said to be in saturation when it carries maximum current $I_{C(sat)} = V_{CC}/R_L$. In saturation, $V_{CE} = 0$.

OBJECTIVE TESTS—30

A. Fill in the following blanks.

1. In a transistor, collector-base junction is always .... biased.
2. In a NPN transistor, both collector and base are .... with respect to the emitter.
3. The ratio of I_C and I_E gives the .... of a transistor.
4. The alpha of a transistor is always less than ....
5. In leakage current I_{CEO}, the subscript CEO stands for Collector to Emitter with .... open.
6. In CC configuration of a transistor, the ratio of I_E and I_B equals (1 + –).
7. If, in a circuit, $V_{CE} = V_{CC}$, the transistor is said to be ....
8. A transistor is said to be in saturation when V_{CE} equals ....

B. Answer True or False.

1. A bipolar transistor consists of three equally-doped regions.
2. In a normally-biased NPN transistor, collector is at higher positive potential than base.
3. Amongst the three currents of a transistor, I_E is the largest.
4. The current I_C in a transistor is about 98 per cent of I_E.
5. Theoretically, the CE configuration of a transistor yields the highest current gain.
6. Temperature-dependent leakage current I_{CO} of a transistor is due to majority carriers.
7. When in a circuit $I_C = 0$, the transistor is said to be cut-off.
8. A transistor is said to be saturated when $V_{CE} = 0$.

C. Multiple Choice Questions.

1. In a properly-biased NPN transistor, most of the electrons from the emitter
   (*a*) recombine with holes in the base
   (*b*) recombine in the emitter itself
   (*c*) pass through the base to the collector
   (*d*) are stopped by the junction barrier
2. The current amplification factor alpha d.c. (α_{dc}) is given by
   (*a*) I_C / I_E (*b*) I_C / I_B
   (*c*) I_B / I_E (*d*) I_B / I_C
3. The common-emitter forward amplification factor β_{dc} is given by
   (*a*) I_C / I_E (*b*) I_C / I_B
   (*c*) I_E / I_B (*d*) I_B / I_E
4. The following relationships between α and β are correct EXCEPT
   (*a*) $\beta = \dfrac{\alpha}{1 - \alpha}$
   (*b*) $\alpha = \dfrac{\beta}{1 - \beta}$
   (*c*) $\alpha = \dfrac{\beta}{1 + \beta}$
   (*d*) $1 - \alpha = \dfrac{1}{1 + \beta}$
5. In the case of a junction bipolar transistor, α is
   (*a*) positive and > 1
   (*b*) positive and < 1
   (*c*) negative and > 1
   (*d*) negative and < 1
6. When the *E/B* junction of a transistor is reverse-biased, collector current
   (*a*) is reversed (*b*) increases
   (*c*) decreases (*d*) stops

ANSWERS

A. 1. reverse 2. positive 3. alpha 4. unity 5. base 6. β 7. cut-off 8. V_{CC}
B. 1. F 2. T 3. T 4. T 5. F 6. F 7. T 8. F
C. 1. *c* 2. *a* 3. *b* 4. *b* 5. *b* 6. *d*

31 LOAD LINE AND BIASING CIRCUITS

31.1 D.C. Load Line and Active Region

For drawing the d.c. load line of a transistor, one needs to know only its cut-off and saturation points (Art. 30.21). It is a straight line joining these two points. For the CE circuit of Fig. 30.27, the load line is drawn in Fig. 31.1. *A* is the *cut-off* point and *B* is the *saturation* point.

The voltage equation of the collector-emitter circuit is

$$V_{CC} = I_C R_L + V_{CE}$$

$$\therefore \quad I_C = \frac{V_{CC}}{R_L} - \frac{V_{CE}}{R_L}$$

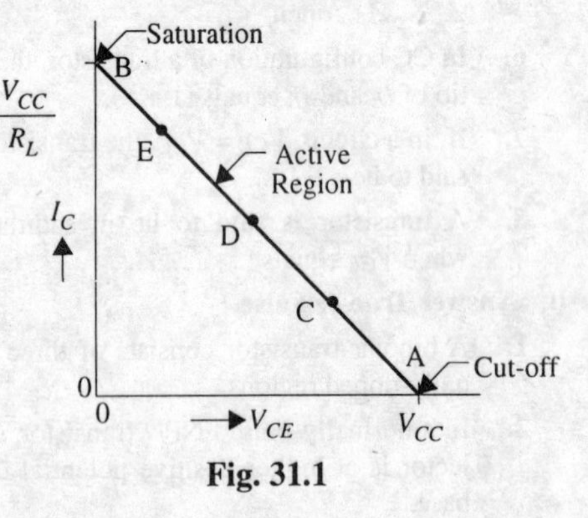

Fig. 31.1

Consider the following two particular cases:

(*i*) When $I_C = 0$;
$V_{CE} = V_{CC}$ —cut-off point *A*

(*ii*) When $V_{CE} = 0$;
$I_C = V_{CC} / R_L$ —saturation point *B*

Obviously, load line can be drawn if only V_{CC} and R_L are known.

Incidentally, slope of load line $AB = -1/R_L$.

Note. The above-given equation can be written as

$$I_C = -\frac{V_{CE}}{R_L} + \frac{V_{CC}}{R_L}$$

It is a linear equation similar to

$$y = mx + C$$

The graph of this equation is a straight line whose slope is $m = -1 / R_L$.

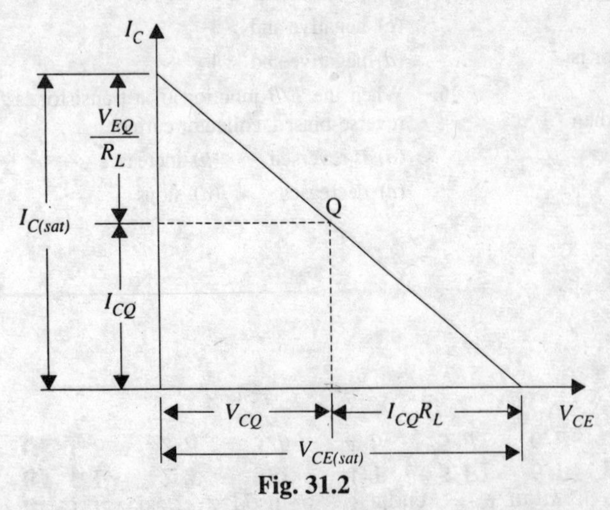

Fig. 31.2

Active Region

All operating points (like *C, D, E*) etc., in Fig. 31.1 lying between cut-off and saturation points form the active region of the transistor. In this region, *E/B* junction is forward-biased and *C/B* junction is reverse-biased-conditions necessary for the proper operation of a transistor.

Quiescent Point

It is a point on the d.c. load line which represents values of I_C and V_{CE} that exist in a transistor circuit when no input signal is applied.

It is also known as the d.c. *operating* point or *working* point. The best positions for this point is midway between cut-off and saturation points where $V_{CE} = \frac{1}{2} V_{CC}$ (like point D in Fig. 31.1).

The general location of Q-point is as shown in Fig. 31.2. The different components of I_C and V_{CE} are also shown. It will be seen that

$$I_{C(sat)} = \frac{V_{CC}}{R_L} = I_{CQ} + \frac{V_{CEQ}}{R_L}$$

$$V_{CE(sat)} = V_{CC} = V_{CEQ} + I_{CQ} R_L$$

Example 31.1. *For the circuit shown in Fig. 31.3 (a), draw the d.c. load line and locate its quiescent or d.c. working point.*

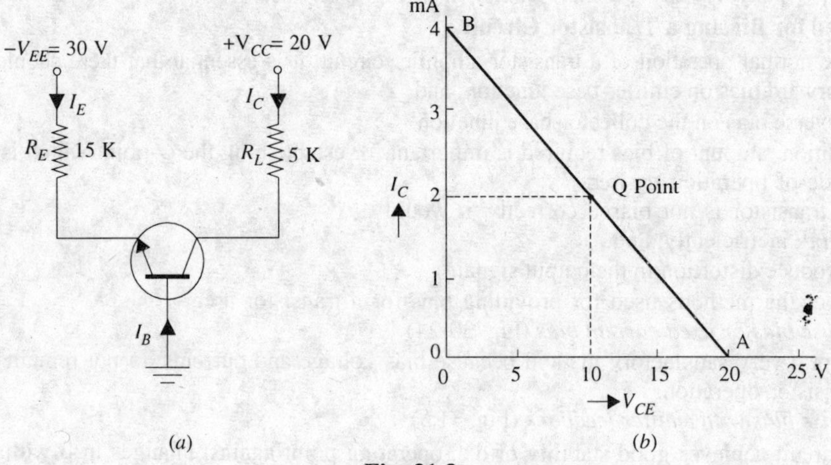

Fig. 31.3

Solution. The cut-off point is easily found because it lies along X-axis where $V_{CE} = V_{CC}$ = 20 *i.e.*, point A in Fig. 31.3 (*b*). At saturation point B, saturation value of the collector current is

$$I_{C(sat)} = V_{CC} / R_L = 20 / 25 \text{ K} = 4 \text{ mA} \qquad \text{— point } B$$

The line AB represents the load line for the given circuit.

We will now find the actual operating point.

$$I_E = V_{EE} / R_E = 30 / 15 \text{ K} = 2 \text{ mA} \qquad \text{— neglecting } V_{BE} \text{ assuming } \alpha = 1$$

$$I_C = \alpha I_E = I_E \cong 2 \text{ mA}$$

$$\therefore \quad V_{CE} = V_{CC} - I_C R_L = 20 - 2 \times 5 = 10 \text{ V}$$

Hence, Q-point is located at (10 V, 2 mA) as shown in Fig. 31.3 (*b*).

Example 31.2. *For the CE circuit shown in Fig. 31.4 (a), draw the d.c. load line and mark the d.c. working point on it. Assume* β = *100 and neglect* V_{BE}.

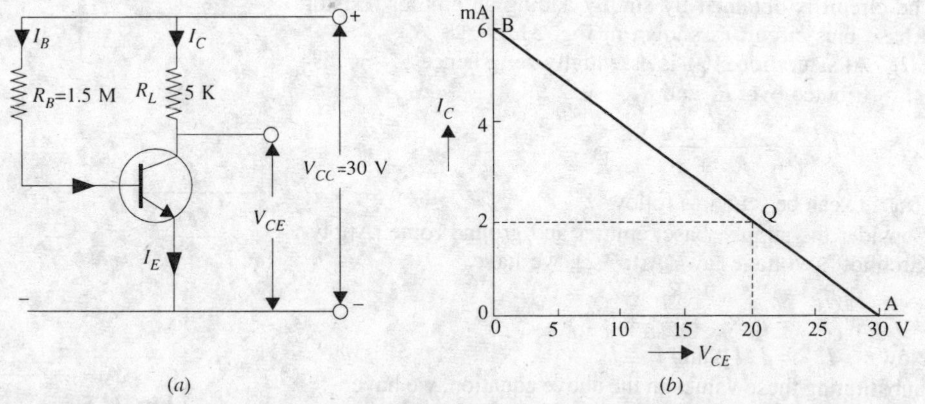

Fig. 31.4

Solution. Cut-off point A is located where

$$I_C = 0 \quad \text{and} \quad V_{CE} = V_{CC} = 30 \text{ V}$$

Saturation point B is given where

$$V_{CE} = 0 \quad \text{and} \quad I_{C(sat)} = 30 / 5 \text{ K} = 6 \text{ mA}$$

Line AB represents the load line in Fig. 31.4 (*b*).

Let us find the d.c. working point from the given values of resistance and supply voltage.

$$I_B = 30 / 1.5 \text{ M} = 20 \ \mu\text{A}$$

$$I_C = \beta I_B = 100 \times 20 = 2000 \ \mu\text{A} = 2 \text{ mA}$$

$$V_{CE} = V_{CC} - I_C R_L = 30 - 2 \times 5 = 20 \text{ V}$$

Hence, Q-point is (20 V, 2 mA) as shown in Fig. 31.4 (*b*).

31.2. Need for Biasing a Transistor Circuit

For the normal operation of a transistor amplifier circuit, it is essential that there should be a

(*a*) forward bias on emitter-base junction; and

(*b*) reverse bias on the collector-base junction.

In addition, amount of bias required is important for establishing the Q-point which is dictated by the mode of operation desired.

If the transistor is not biased correctly, it would

(*i*) work inefficiently; and

(*ii*) produce distortion in the output signal.

Some of the methods used for providing bias for a transistor are:

1. *base bias or fixed current bias* (Fig. 30-24)

It is not a very satisfactory method because bias voltage and currents do not remain constant during transistor operation.

2. *base bias with emitter feedback* (Fig. 31.5)

This circuit achieves good stability of d.c. operating point against changes in β with the help of emitter resistor which causes degeneration to take place.

3. *base bias with collector feedback* (Fig. 31.7)

It is also known as collector-to-base bias or collector feedback-bias. It provides better bias stability.

4. *base bias with collector and emitter feedback*

Here, feedback is provided with the help of R_B and R_E (Fig. 31.8).

5. *voltage divider bias*

It is most widely used in linear discrete circuits because it provides good bias stability. It is also called universal bias circuit or base bias with one supply.

Each of the above circuits will now be discussed separately.

31.3. Base Bias

It has already been discussed in Art. 30.28 and is shown in Fig. 30.24.

31.4. Base Bias with Emitter Feedback

The circuit is obtained by simply adding an emitter resistor to the base bias circuit as shown in Fig. 31.5.

(*i*) At saturation, V_{CE} is essentially zero, hence V_{CC} is distributed over R_L and R_E.

$$\therefore \quad I_{C(sat)} = \frac{V_{CC}}{R_E + R_L}$$

(*ii*) I_C can be found as follows :

Consider the supply, base, emitter and ground route. Applying Kirchhoff's voltage law (Art. 3.2), we have

$$-I_B R_B - V_{BE} - I_E R_E + V_{CC} = 0$$

or $$V_{CC} = I_B R_B + V_{BE} + I_E R_E \quad \text{...}(i)$$

Now, $I_B = I_C / \beta$ and $I_E \cong I_C$

Substituting these values in the above equation, we have

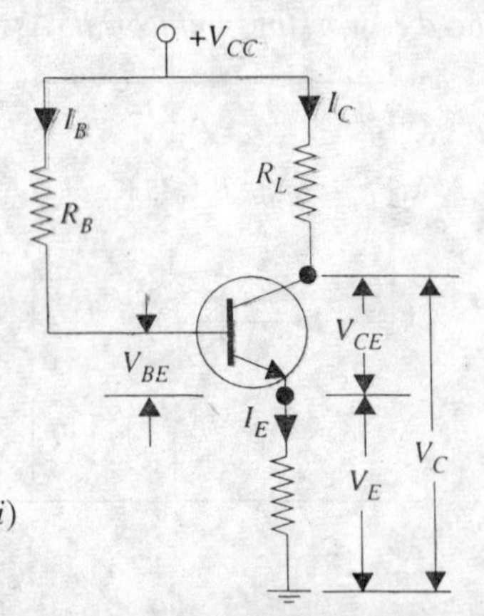

Fig. 31.5

$$V_{CC} \cong \frac{I_C R_B}{\beta} + V_{BE} + I_C R_E$$

$$\therefore \quad I_C \cong \frac{V_{CC} - V_{BE}}{R_E + R_B / \beta} \cong \frac{V_{CC}}{R_E + R_B / \beta} \quad \text{—neglecting } V_{BE}$$

(*iii*) collector-to-ground voltage $V_C = V_{CC} - I_C R_L$

(*iv*) emitter-to-ground voltage, $V_E = I_E R_E \cong I_C R_E$

$$\therefore \quad V_{CE} = V_C - V_E$$

Example 31.3. *For the circuit shown in Fig. 31.6, find (i) $I_{C(sat)}$ (ii) I_C (iii) V_C (iv) V_E and (v) V_{CE}.*

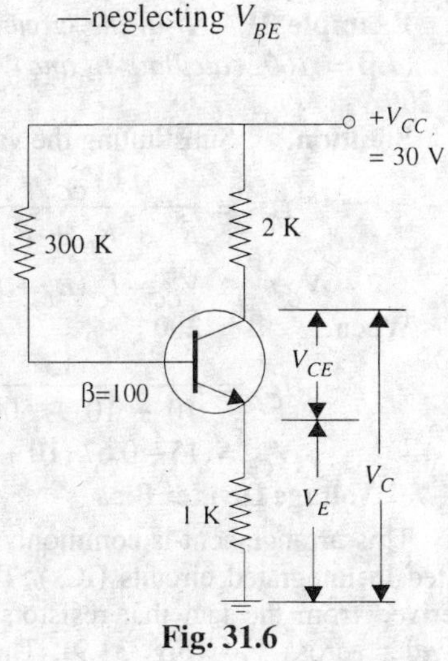

Fig. 31.6

Solution. (*i*) $I_{C(sat)} = \frac{V_{CC}}{R_E + R_L} = \frac{30}{1+2} = \mathbf{10\ mA}$

(*ii*) actual $I_C \cong \frac{V_{CC}}{R_E + R_E / \beta} = \frac{30}{1 + 300/100}$

$= \mathbf{7.5\ mA}$

(*iii*) $V_C = V_{CC} - I_C R_L$
$= 30 - 7.5 \times 2 = \mathbf{15\ V}$

(*iv*) $V_E = I_E R_E \cong I_C R_E$
$= 7.5 \times 1 = \mathbf{7.5\ V}$

(*v*) $V_{CE} = V_C - V_E = 15 - 7.5 = \mathbf{7.5\ V}$

31.5. Base Bias with Collector Feedback

This circuit (Fig. 31.7) is like the base bias circuit except that base resistor is returned to collector rather than to the V_{CC} supply. It derives its name from the fact that since voltage for R_B is derived from collector, there exists a negative-feedback effect which tends to stabilise I_C against changes in β. To understand this action, suppose that β increases. It will increase I_C as well as $I_C R_L$ but decrease V_C which is applied across R_B. Consequently, I_E will be decreased which will partially compensate for the original increase in β.

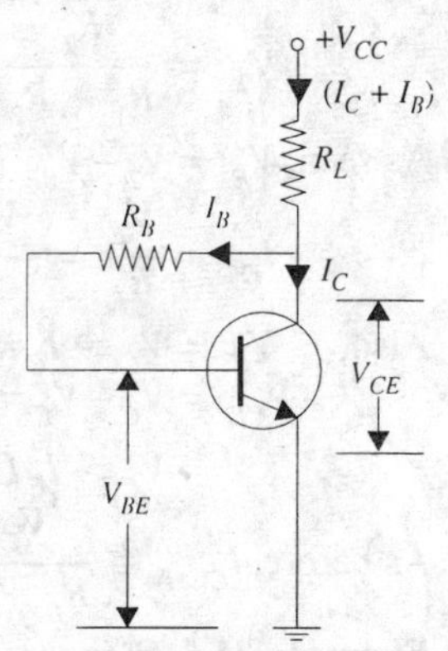

Fig. 31.7

(*i*) $I_{C(sat)} = V_{CC} / R_L$ —since $V_{CE} = 0$

(*ii*) $V_C = V_{CC} - (I_B + I_C) R_L \cong V_{CC} - I_C R_L$

Also, $V_C = I_B R_B + V_{BE}$ — via base-emitter route.

Equating the two expressions for V_C, we have

$I_B R_B + V_{BE} \cong V_{CC} - I_C R_L$

Since $I_B = I_C / \beta$, we get

$$\frac{I_C}{\beta} \cdot R_B + V_{BE} \cong V_C - I_C R_L$$

$$\therefore \quad I_C = \frac{V_{CC} - V_{BE}}{R_L + R_B / \beta} \cong \frac{V_{CC}}{R_L + R_B / \beta}$$

31.6. Base Bias with Collector and Emitter Feedback

As shown in Fig. 31.8, both collector and emitter feedbacks have been used with the help of R_B and R_E respectively. This makes the circuit less sensitive to changes in β. Assuming I_B to be negligible as compared to I_C so that $I_C \cong I_E$ and applying KVL (Art. 31.2) to the V_{CC}, base and R_E circuit, we have

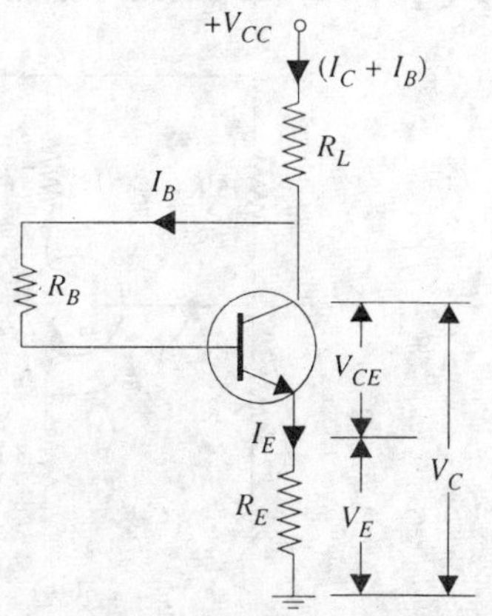

Fig. 31.8

$$-(I_C + I_B) R_L - V_{BE} - I_C (R_E + R_B / \beta) + V_{CC} = 0$$

$$\therefore \quad I_C = \frac{V_{CC} - V_{BE}}{R_E + R_L + R_B / \beta} \cong \frac{V_{CC}}{R_E + R_L + R_B / \beta}$$

Example 31.4. *If in the circuit of Fig. 31.8, V_{CC} = 15 V, R_L = 10 K, R_B = 500 K, R_E = 10 K and β = 100, calculate I_C and V_{CE}. Neglect V_{BE}. How will these values change if β increases to 200?*

Solution. Substituting the values in the equation of Art. 31.6, we have

$$I_C = \frac{V_{CC}}{R_E + R_L + R_B/\beta} = \frac{15}{10 + 10 + 500/100} = \mathbf{0.6\ mA}$$

$$V_{CE} \cong V_{CC} - I_C (R_L + R_E) \cong 15 - 0.6\,(10 + 10) = \mathbf{3\ V}$$

When β = **200**

$$I_C = \frac{15}{10 + 10 + 500/200} = \mathbf{0.67\ mA}$$

$$V_{CE} = 15 - 0.67\,(10 + 10) = \mathbf{1.7\ V}$$

31.7. Voltage Divider Bias

This arrangement is commonly used for transistors incorporated in integrated circuits (ICs). The name 'voltage divider' is derived from the fact that resistors R_1 and R_2 form a potential divider across V_{CC} (Fig. 31.9). The voltage drop V_2 across R_2 forward-biases the emitter whereas V_{CC} supply reverse-biases the collector. As per voltage divider theorem,

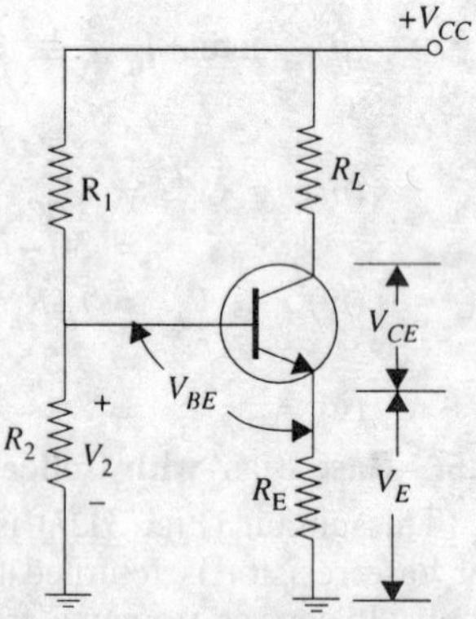

$$V_2 = \frac{R_2}{R_1 + R_2} \cdot V_{CC}$$

As seen, $V_E = V_2 - V_{BE}$

$$\therefore \quad I_E = \frac{V_E}{R_E} = \frac{V_2 - V_{BE}}{R_E} \cong \frac{V_2}{R_E}$$

Also, $V_C = V_{CC} - I_C R_L$

$$V_{CE} = V_C - V_E = V_{CC} - I_C R_L - I_E R_E$$

$$= V_{CC} - I_C (R_L + R_E) \qquad \because I_C \cong I_E$$

As before, $I_{C(sat)} \cong \dfrac{V_{CC}}{R_L + R_E} \qquad \because V_{CE} \cong 0$

Fig. 31.9

Example 31.5. *For the circuit shown in Fig. 31.10 (a), draw the d.c. load line and mark in the Q-point of the circuit. Assume silicon transistor material.*

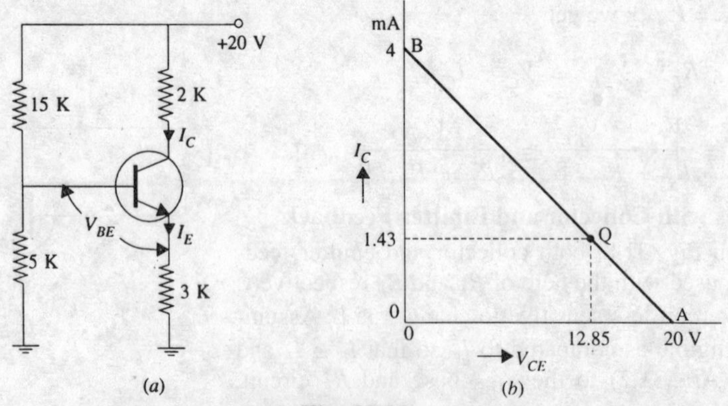

Fig. 31.10

Solution. Under cut-off condition, $V_{CE} = V_{CC}$.

$\therefore\ V_{CE(cut\text{-}off)} = V_{CC} = 20$ V — point *A* in Fig. 31.10 (*b*)

$$I_{C(sat)} = \frac{V_{CC}}{R_L + R_E} = \frac{20}{2+3} = 4 \text{ mA}$$ — point *B* in Fig. 31.10 (*b*)

$$V_2 = \frac{R_2}{R_1 + R_2} \cdot V_{CC} = \frac{5}{15+5} \times 20 = 5 \text{ V}$$

$$I_E = \frac{V_2 - V_{BE}}{R_E} = \frac{5 - 0.7}{3} = 1.43 \text{ mA}$$

Now, $I_C \cong I_E = 1.43$ mA

$V_{CE} = V_{CC} - I_C (R_L + R_E) = 20 - 1.43\,(2 + 3) = \mathbf{12.85\ V}$

The *Q*-point is shown in Fig.31.10 (*b*).

31.8. Load Line and Output Characteristics

In order to study the effect of bias conditions on the performance of a CE circuit, it is necessary to superimpose the d.c. line on the transistor output (V_{CE}/I_C) characteristics. Consider a silicon NPN transistor which is connected in CE configuration (Fig. 31.11) and whose output characteristics are given in Fig. 31.12. Let its β = 100.

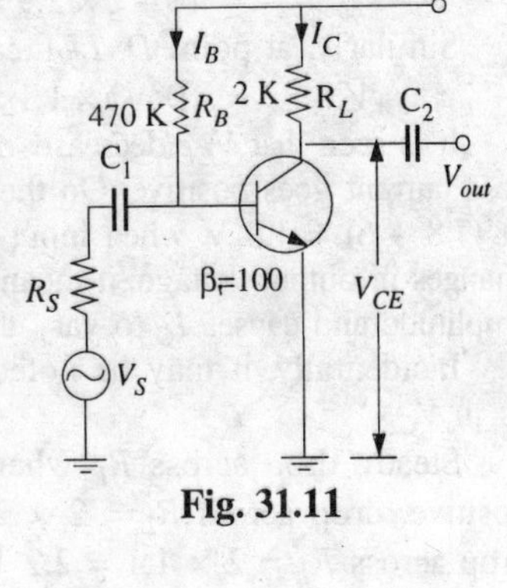

Fig. 31.11

First, let us find the cut-off and saturation points for drawing the d.c. load line and then mark in the *Q*-point

$I_{C(sat)} = 10/2 = 5$ mA — point *B*

$V_{CE(cut\text{-}off)} = V_{CC} = 10$ V — point *A*

The load line is drawn in Fig. 31.12 below.

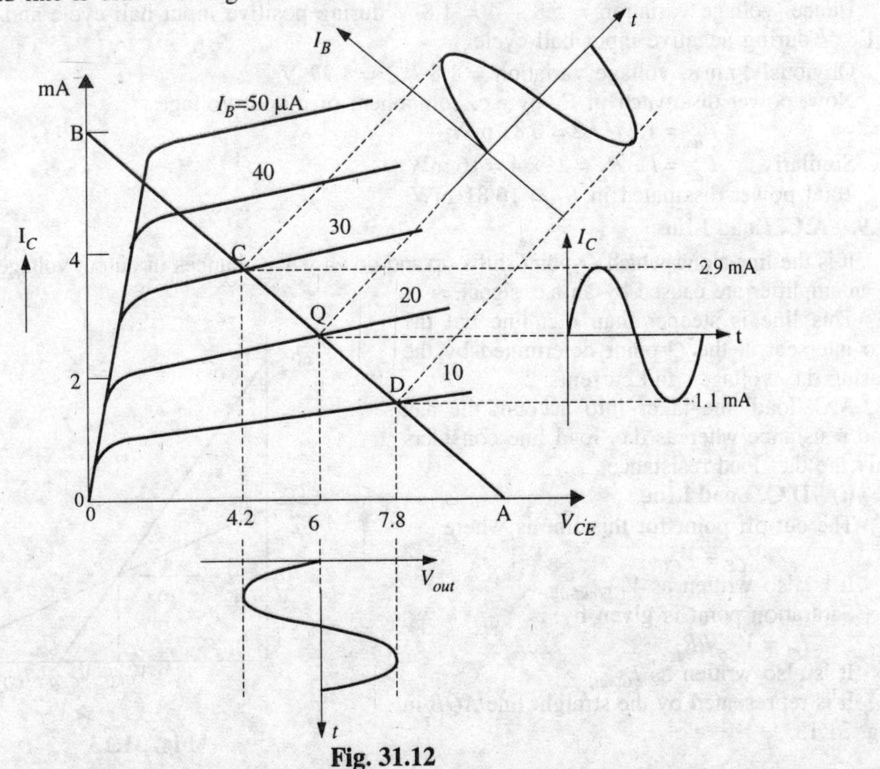

Fig. 31.12

Actual $I_B = \dfrac{V_{CC} - V_{BE}}{R_B} = \dfrac{10 - 0.7}{470} = 20\ \mu A$

$I_C = \beta I_B = 100 \times 20 = 2000\ \mu A = 2\ mA$

$V_{CE} = V_{CC} - I_C R_L = 10 - 2 \times 2 = 6\ V$

This locates the *Q*-point in Fig. 31.12.

Suppose an a.c. input signal voltage injects a sinusoidal base current of peak value of 10 μA into the circuit of Fig. 31.12. Obviously, it will swing the operating of *Q*-point up and down along the load line.

When positive half-cycle of I_B is applied, the *Q*-point shifts to point *C* which lies on the (20 + 10) = 30 μA line.

Similarly, during negative half-cycle of the input base current, *Q*-point shifts to point *D* which lies on the (20 – 10) = 10 μA line.

By measurement at point *C*, I_C = 2.9 mA, Hence,

$V_{CE} = 10 - 2 \times 2.9 = 4.2\ V$

Similarly, at point *D*, I_C measures 1.1 mA.

$\therefore\ V_{CE} = 10 - 2 \times 1.1 = 7.8\ V$

It is seen that V_{CE} decreases from 6 V to 4.2 V *i.e.*, by a peak value of (6 – 4.2) = 1.8 V when base current goes positive. On the other hand, V_{CE} increases from 6 V to 7.8 V *i.e.*, by a peak value of (7.8 – 6) = 1.8 V when input base current signal goes negative. Since changes in V_{CE} represent changes in output voltage, it means that when input signal is applied, I_B varies according to the signal amplitude and causes I_C to vary thereby producing output voltage variations.

Incidentally, it may be noted that variations in voltage drop across R_L are exactly the same as in V_{CE}.

Steady drop across R_L when no signal is applied = 2 × 2 = 4 V. When base signal goes positive, drop across R_L = 2 × 2.9 = 5.8 V. When base signal goes negative, I_C = 1.1 mA and drop across R_L = 2 × 1.1 = 2.2 V.

Hence, voltage variation = 5.8 – 4 = 1.8 V during positive input half-cycle and = (4 – 2.2) = 1.8 V during negative input half-cycle.

Obviously, r.m.s. voltage variation = 1.8 /√2 = 1.27 V.

Now, power dissipated in R_L by a.c. component of output voltage

$P_{ac} = 1.27^2 / 2 = 0.81\ mW$

Similarly, $P_{dc} = I_C^2 R_L = 2^2 \times 4 = 16\ mW$

Total power dissipated in R_L = 16.81 mW.

31.9. A.C. Load Line

It is the line along which *Q*-point shifts up and down when changes in output voltage and current of an amplifier are caused by an a.c. signal.

This line is steeper than d.c. line but the two intersect at the *Q*-point determined by the biasing d.c. voltages and currents.

A.C. load line takes into account the a.c. load resistance whereas d.c. load line considers only the d.c. load resistance.

(*i*) D.C. Load Line

The cut-off point for this line is where

$V_{CE} = V_{CC}$

It is also written as $V_{CE(cut\text{-}off)}$.

Saturation point is given by

$I_C = V_{CC} / R_L$

It is also written as $I_{C(sat)}$.

It is represented by the straight line *AQB* in Fig. 31.13.

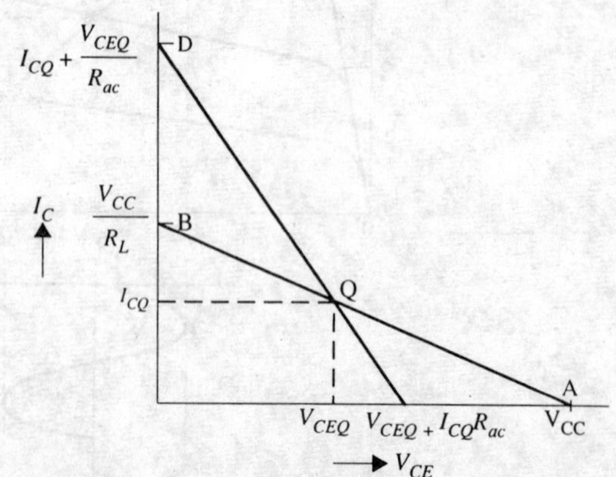

Fig. 31.13

(*ii*) A.C. Load Line

The cut-off point is given by

$V_{CE(cut\text{-}off)} = V_{CEQ} + I_{CQ} R_{ac}$

where R_{ac} is the *a.c. load resistance.*

Saturation point is given by

$I_{C(sat)} = I_{CQ} + V_{CEQ} / R_{ac}$ —as shown in Fig. 31.13.

It is represented by the straight line *CQD* in Fig. 31.13.

Example 31.6. *Draw the d.c. and a.c. load lines for the CE circuit shown in Fig.31.14 (a).*

Solution. D.C. Load Line [Fig. 31.14 (*b*)]

$V_{CE(cut\text{-}off)} = V_{CC} = 20$ V —point *A*

$I_{C(sat)} = V_{CC} / (R_L + R_E) = 20 / 5 = 4$ mA —point *B*

Hence, *AB* represents d.c. load line for the given circuit.

Approximate bias conditions can be quickly found by assuming that I_B is too small to affect the base bias in Fig. 31.14 (*a*).

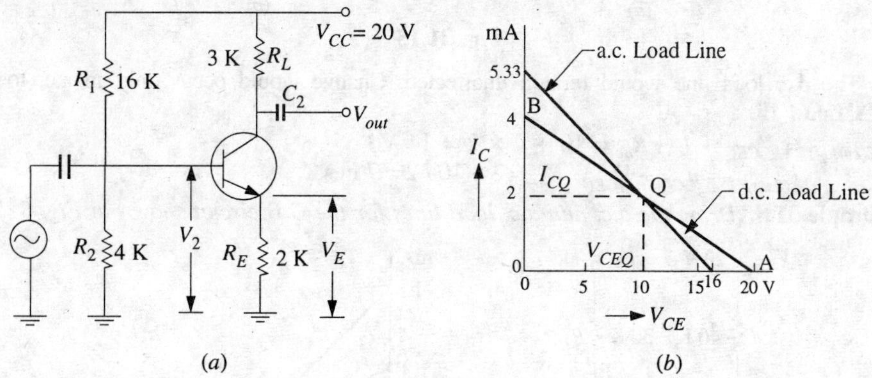

Fig. 31.14

$$V_2 = \frac{4}{4+16} \times 20 = \mathbf{4\ V}$$

If we neglect V_{BE}, $V_2 = V_E$

∴ $I_E = V_E / R_E = V_2 / R_E = 4 / 2 = 2$ mA

Also, $I_C \cong I_E = 2$ mA

Hence, $I_{CQ} = 2$ mA

The corresponding value of V_{CEQ} can be found by drawing dotted line in Fig. 31.14 (*b*) or may be calculated as under.

$$V_{CEQ} = V_{CC} - I_{CQ} (R_L + R_E)$$
$$= 20 - 2 (3 + 2) = 10 \text{ V}$$

A.C. Load Line

Cut-off point, $V_{CE(cut\text{-}off)} = V_{CEQ} + I_{CQ} R_{ac}$

Now, for the given circuit, a.c. load resistance is

$$R_{ac} = R_L = \mathbf{3\ K}$$

∴ cut-off point $= 10 + 2 \times 3 = 16$ V —Fig. 31.14 (*b*)

Saturation point, $I_{C(sat)} = I_{CQ} + V_{CEQ} / R_{ac}$

$= 2 + 10 / 3 = 5.33$ mA

Hence, the line joining 16 V point and 5.13 mA points gives a.c. load line as shown in Fig. 31.14 (*b*).

As expected, this line passes through the *Q*-point.

Example 31.7. *Find the d.c. and a.c. load lines for the CE circuit shown in Fig. 31.15 (a).*

Solution. The given circuit is identical to that shown in Fig. 31.14 (*a*) except for the addition of 6 K resistor. This makes R_{ac} = 3 K ‖ 6 K = 2 K because collector *feeds* these two resistors in

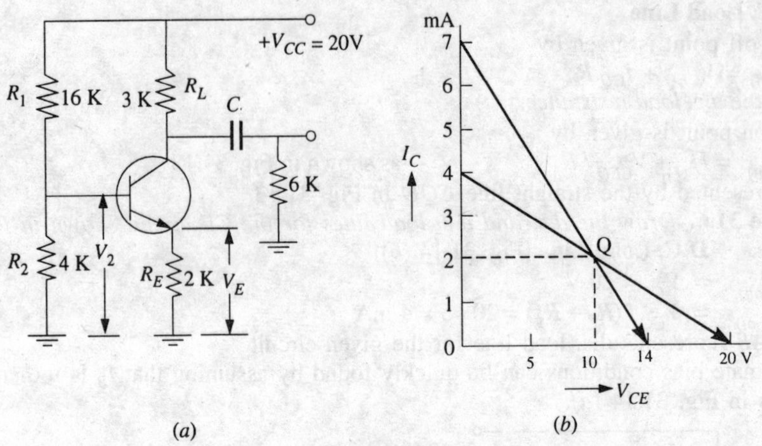

(a) (b)

Fig. 31.15

parallel. The d.c. load line would remain unaffected. Change would occur only in a.c. load line.

A.C. Load Line

$$V_{CE(cut\text{-}off)} = V_{CEQ} + I_{CQ}\, R_{ac} = 10 + 2 \times 2 = 14 \text{ V}$$

$$I_{C(sat)} = I_{CQ} + V_{CEQ} / R_{ac} = 2 + 10/2 = 7 \text{ mA}$$

Example 31.8. *Draw the d.c. and a.c. load lines for the C B circuit shown in Fig. 31.16 (a).*

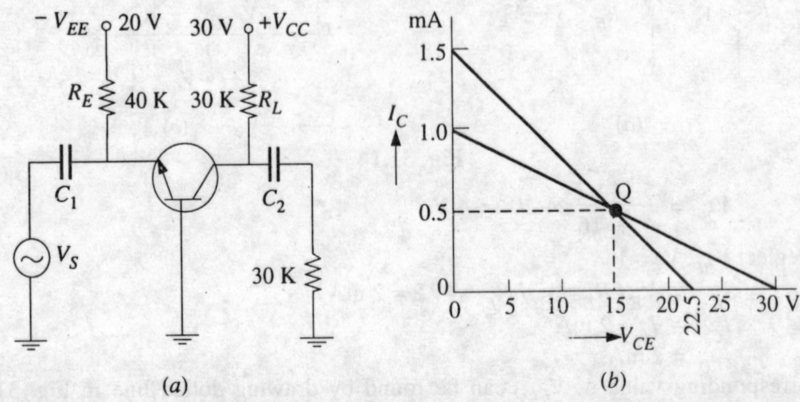

(a) (b)

Fig. 31.16

Solution. The d.c. load line passes through cut-off point of 30 V and saturation point of $V_{CC} / R_L = 1$ mA

Now, $I_E \cong 20/40 = 0.5$ mA $\quad \therefore \quad I_C \cong I_E = 0.5$ mA

$\therefore \quad I_{CQ} = 0.5$ mA

$V_{CE} = V_{CC} - I_C R_L = 30 - 0.5 \times 30 = 15$ V

$\therefore \quad V_{CEQ} = 15$ V

Hence, d.c. operating point or Q-point is (15 V, 0.5 mA) as shown in Fig. 31.16 (*b*).

A.C. Load Line

The cut-off point for a.c. load line is

$$= V_{CEQ} + I_{CQ}\, R_{ac}$$

Since collector *sees* an a.c. load of two 30 K resistors in parallel

$\therefore \quad R_{ac} =$ **30 K || 30 K = 15 K**

$\therefore \quad V_{CEQ} + I_{CQ}\, R_{ac} = 15 + 0.5 \times 15 = 22.5$ V

Saturation current for a.c. line is

$$= I_{CQ} + V_{CEQ} / R_{ac} = 0.5 + 15/15 = 1.5 \text{ mA}$$

The line joining these two points (and also passing through Q) gives the a.c. load line as shown in Fig. 31.16 (*b*).

Note. Knowing a.c. cut-off point and Q-point, we can draw the a.c. load line. Hence, we need not find the value of saturation current for this purpose.

HIGHLIGHTS

1. D.C. load line of a transistor is a straight line which joins its cut-off and saturation points.
2. Quiescent point is that point on the d.c. load line which represents the values of I_C and V_{CE} that exist when no signal is applied.
3. For proper operation of a transistor, it is essential that its *E/B* junction should be given forward bias and *C/B* junction reverse bias. Different methods are used to get bias stability.
4. A.C. load line takes a.c. load into consideration and is steeper than the d.c. load line.

OBJECTIVE TESTS —31

A. Fill in the following blanks.

1. The straight line joining the cut-off and .... points gives the d.c. load line of a transistor.
2. The best position for Q-point is .... between cut-off and saturation points.
3. The saturation point for a transistor lies along............... axis.
4. Improperly-biased transistor produces .... in the output signal.
5. A.C. load line of transistor is .... than its d.c. load line.
6. Voltage divider bias is commonly used for transistors incorporated in .... circuits.

B. Answer True or False.

1. D.C. load line of transistor can be drawn if supply voltage and load resistor are known.
2. Load line of a transistor has negative-slope.
3. The Q-point of a transistor keeps shifting its position even when there is no input signal.
4. D.C. and a.c. load lines of a transistor intersect at the Q-point.
5. A.C. load line takes into consideration the a.c. load resistance of the circuit.
6. The a.c. load resistance of a transistor is always higher than its d.c. load resistance.

C. Multiple Choice Questions.

1. The d.c. load line of a transistor circuit
(*a*) has a negative slope
(*b*) is a curved line
(*c*) gives graphic relation between I_C and I_B
(*d*) does not contain the Q-point
2. The maximum peak-to-peak output voltage swing is obtained when the Q-point of a circuit is located
(a) near saturation point
(*b*) near cut-off point
(*c*) at the centre of the load line
(*d*) at least on the load line
3. Improper biasing of a transistor circuit leads to
(*a*) excessive heat production in collector
(*b*) distortion in output signal
(*c*) faulty location of load line
(*d*) heavy loading of emitter terminal
4. The a.c. load line of a transistor circuit is steeper than its d.c. load line because
(*a*) a.c. signal sees less load resistance
(*b*) it has greater slope
(*c*) I_C is higher
(*d*) input signal varies in magnitude
5. The universal bias stabilization circuits is most popular because
(*a*) I_C does not depend on transistor characteristics
(*b*) its sensitivity is high
(*c*) voltage divider is heavily loaded by transistor base
(*d*) I_C equals I_E
6. The d.c. and a.c. load lines of a transistor
(*a*) have equal slopes
(*b*) have positive slopes
(*c*) are curved lines
(*d*) intersect each other

ANSWERS

A. 1. saturation 2. midway 3. Y 4. distortion 5. steeper 6. integrated
B. 1. T 2. T 3. F 4. T 5. T 6. F
C. 1. *a* 2. *c* 3. *b* 4. *a* 5. *a* 6. *d*

32 TRANSISTOR EQUIVALENT CIRCUITS AND MODELS

32.1 General

We will begin by idealizing a transistor with the help of simple approximations that will retain its essential features while discarding its less important qualities. These approximations will help us to analyse transistor *circuits easily and rapidly.*

We will discuss only the *small signal* equivalent circuits in this chapter. Small signal operation is that in which the a.c. input signal voltages and currents are in the order of ±10 per cent of *Q*-point voltages and currents.

There are two prominent schools of thought to-day regarding the equivalent circuits to be substituted for the transistor. The two approaches make use of

(*a*) four *h*-parameters of the transistor and the values of circuit components;

(*b*) the beta (β) of the transistor and the values of the circuit components.

Since long, industrial and educational institutions have heavily relied on the hybrid parameters because they produce more accurate results in the analysis of amplifier circuits. In fact, hybrid-parameter equivalent circuit continues to be popular even to-day. But their use is beset with the following difficulties:

1. the values of *h*-parameters are not so readily or easily available;
2. their values vary considerably with individual transistors even of the same type and number;
3. their values are limited to a particular set of operating conditions for reasonably accurate results.

The second method which employs transistor beta and resistance values is gaining more popularity of late. It has the following advantages:

1. the required values are easily available;
2. the procedure followed is simple and easy to understand;
3. the results obtained are quite accurate for the study of amplifier circuit characteristics.

To begin with, we will consider the second method first.

32.2. The Beta Rule

According to this rule, resistance from one part of a transistor circuit can be referred to another of its parts (as we do with the primary and secondary winding impedances of a transformer). For example, resistance R_L in the collector circuit can be referred to the base circuit and *vice versa*. Similarly, R_E can be referred to the base circuit and reciprocally, R_B can be referred to the emitter circuit. Since current through R_L is I_C (= βI_B), hence β-factor comes into the picture. Similarly, current through R_E is I_E which is (1 + β) times I_B, hence (1+β) or approximately β comes into the picture again. Use of 'β-rule' makes transistor circuit calculations quite quick and easy. It makes the calculation of I_B quite simple.

The 'β-rule' may be stated as under:

1. When referring R_L (or R_C) to the base circuit **multiply** it by β. When referring R_B to the collector circuit, **divide** it by β.
2. When referring R_E to base circuit, **multiply** it by (1+β) or just β (as a close approximation).

3. Similarly, when referring R_B to emitter circuit, **divide** it by $(1 + \beta)$ or β.

Before you apply this rule to any circuit, you must remember one very important point otherwise you are likely to get wrong answers. The point is that *only those resistances are transferred which lie in the path of the current being calculate.* Not otherwise. The utility of this rule will be demonstrated by solving the following problems.

Example 32.1. *Calculate the value of V_{CE} in the collector stabilisation circuit of Fig. 32.1.*

(Electronics-I, Madras Univ. 1993)

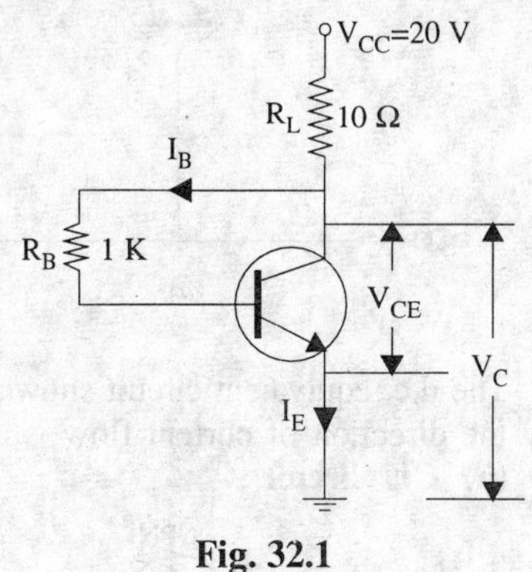

Fig. 32.1

Solution. We will use the β-rule to find I_C in the following two ways :

(i) **First Method**

Here, we will transfer R_L to the base circuit.

$$I_B = \frac{V_{CC}}{R_B + \beta R_L} = \frac{20}{1000 + 100(10)} = 10 \text{ mA}$$

$$I_C = \beta I_B = 100 \times 10 = 1000 \text{ mA} = 1 \text{ A}$$

$$V_{CE} \cong V_{CC} - I_C R_L = 20 - 1 \times 10 = \mathbf{10\ V}$$

(ii) **Second Method**

Now, we will refer R_B to collector circuit

$$I_C \cong \frac{V_{CC}}{R_L + R_B/\beta} = \frac{20}{10 + (1000/100)} = 1 \text{ A}$$

$$V_{CE} = V_{CC} - I_C R_L = 10 \text{ V} \qquad \text{—as above}$$

It was a simple circuit because $R_E = 0$ and R_B was connected to V_{CC} through R_L and not directly (in which case, R_L would not lie in the path of I_B). Now, we will consider the case when R_E is present and R_L does not lie in the path of I_B.

Example 32.2. *Calculate the three transistor currents in the circuit of Fig. 32.2.*

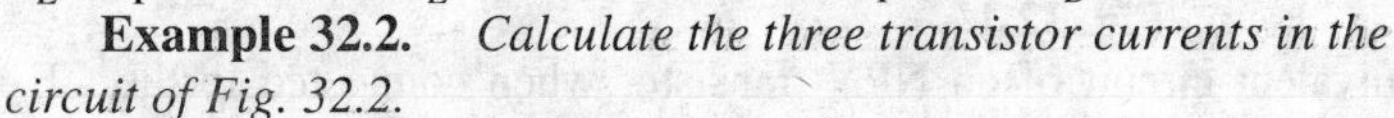

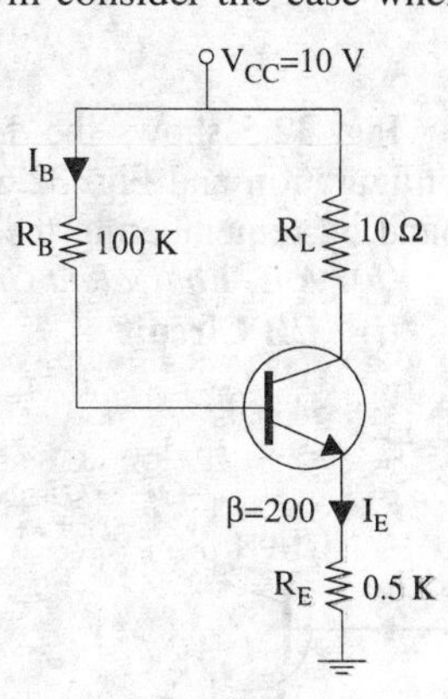

Fig. 32.2

Solution. *(i)* **First Method**

Since R_E lies in the path of I_B

$$\therefore \quad I_B = \frac{V_{CC}}{R_B + \beta R_E} \qquad \text{—neglecting } V_{BE}$$

$$= \frac{10}{100 + 200 \times 0.5} = \mathbf{0.05\ mA}$$

$$I_C = \beta I_B = 200 \times 0.05 = \mathbf{10\ mA}$$

$$I_E = I_B + I_C = \mathbf{10.05\ mA}$$

(ii) **Second Method**

Now, we will transfer R_B to emitter circuit and find I_E directly.

$$I_E = \frac{V_{CC}}{R_E + R_B/\beta} = \frac{10}{0.5 + 100/200} = 10 \text{ mA} \qquad \text{— as before}$$

$$I_B = I_C / \beta \cong 10/200 = \mathbf{0.05\ mA}$$

32.3. Ideal Transistor Equivalent Circuits

(a) D.C. Equivalent Circuit

(i) **CB Circuit**

In an ideal transistor, $\alpha = 1$ which means that $I_C = I_E$.

The emitter diode acts like any forward-biased ideal diode. However, due to transistor action, collector diode acts as a current source. In other words, for the purpose of drawing d.c. equivalent circuit, we can view an ideal transistor as nothing more than a rectifier diode in emitter and a current source in collector. In the d.c. equivalent circuit of Fig. 32.3 (*b*), current arrow always points in the direction of conventional current.

As per the polarities of transistor terminals (Art. 30.2) shown in Fig. 32.3 (*a*), emitter current flows from *E* to *B* and collector current from *B* to *C*.

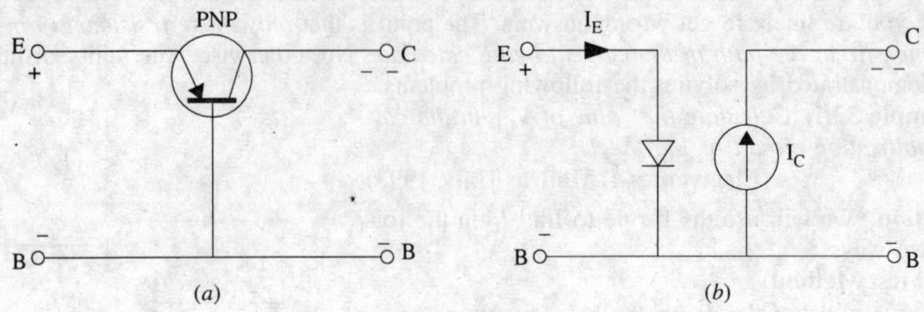

Fig. 32.3

The d.c. equivalent circuit shown in Fig. 32.4 for an NPN transistor is exactly similar except that the direction of current flow is opposite.

(*ii*) CE Circuit

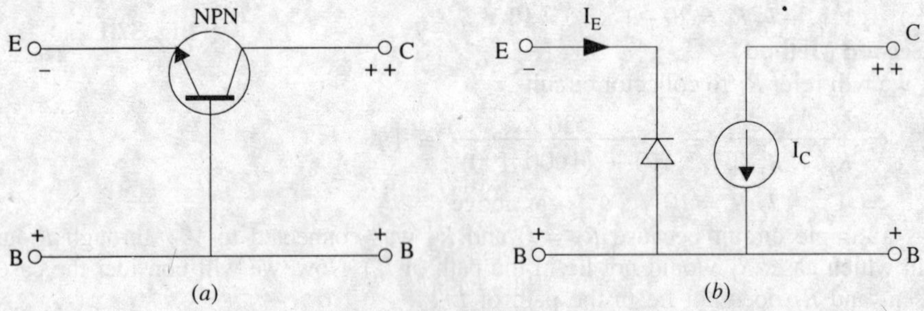

Fig. 32.4

Fig. 32.5 shows the d.c. equivalent circuit of an NPN transistor when connected in the CE configuration and Fig. 32.6 shows that of PNP transistor. Direction of current flow can be easily found by remembering the transistor polarity rule given in Art. 30.2.

(*b*) *A.C. Equivalent Circuit*

(*i*) CB Circuit

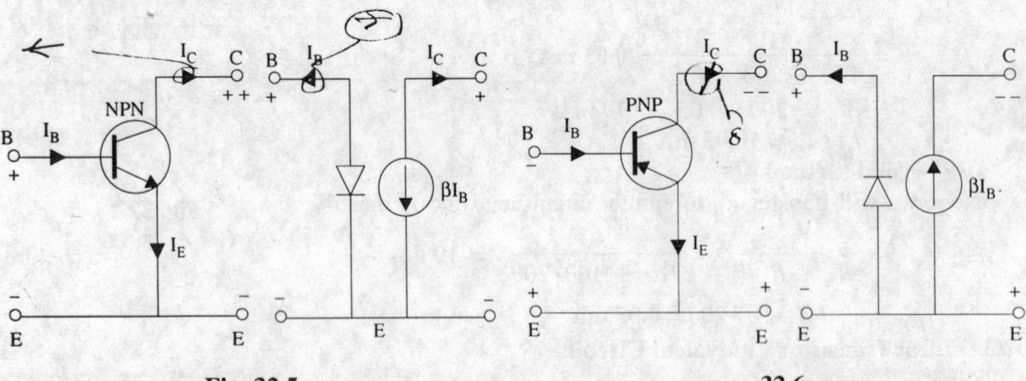

Fig. 32.5 **32.6**

In the case of *small* input a.c. signals, the emitter *diode does not rectify,* instead it offers resistance called *a.c. resistance.* As usual, collector diode acts as a current source.

Fig. 32.7 (*b*) shows the a.c. equivalent circuit of a transistor connected in the CE configuration.* Here, a.c. resistance offered by the emitter diode is

*This circuit is valid both for PNP as well as NPN transistors because difference in the direction of a.c. currents does not matter.

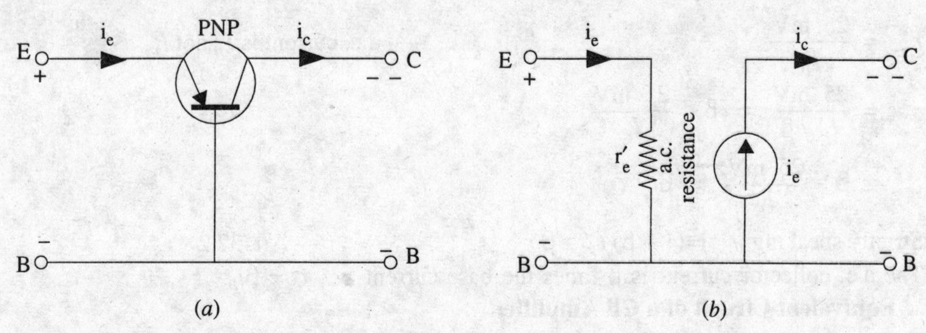

Fig. 32.7

$$r_{ac} = \text{junction resistance } (r_j) + \text{ bulk resistance } (r_B)$$
$$= r_j \qquad (\because\ r_B \text{ is negligible})$$
$$= \frac{25 \text{ mV}}{I_E} \qquad \text{—where } I_E \text{ is d.c. emitter current}$$

It is written as r_e' signifying junction resistance of the *emitter* diode *i.e.*, a.c. resistance *looking into the emitter.*

$$\therefore \qquad r_e' = \frac{25 \text{ mV}}{I_E}$$

Hence, the a.c. equivalent circuit of a CE circuit becomes as shown in Fig. 32.8 (*b*). Since changes in collector current are almost equal to changes in emitter current, $\Delta\ i_c \cong \Delta\ i_e$.

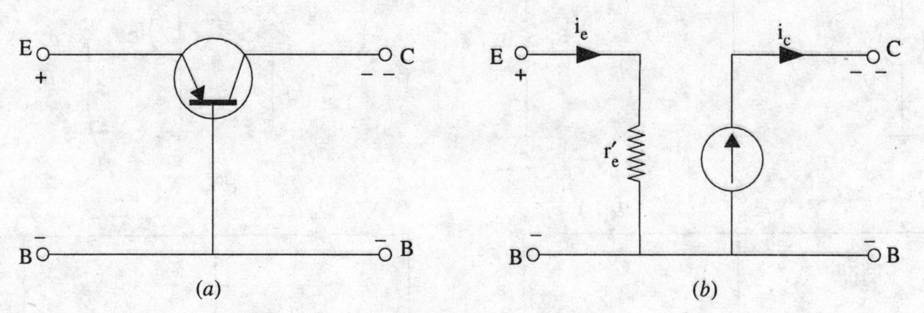

Fig. 32.8

(*ii*) CE Circuit

Fig. 32.9 (*b*) shows the equivalent circuit when transistor is connected in the CE configuration. The a.c. resistance *looking into the base* is

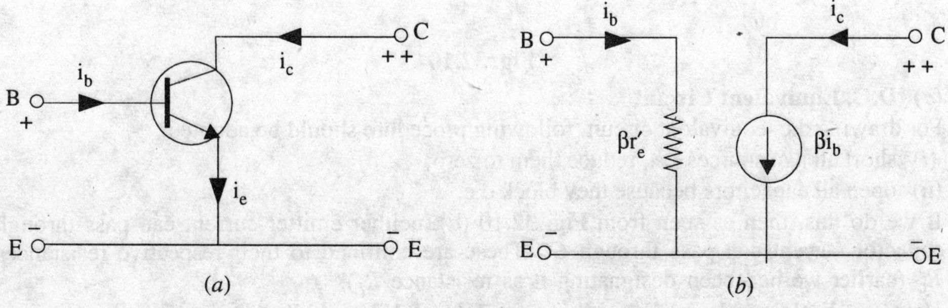

Fig. 32.9

$$r_{ac} = \frac{25 \text{ mV}}{I_B} \qquad \text{— d.c. current is } I_B \text{ not } I_E.$$

$$= \frac{25 \text{ mV}}{I_C / \beta} = \beta \cdot \frac{25 \text{ mV}}{I_C}$$

$$\cong \beta \cdot \frac{25 \text{ mV}}{I_E} = \beta r_e'$$

Strictly speaking, $r_{ac} = (1 + \beta)\, r_e' \cong \beta\, r_e'$ — Art. 32.2.

The a.c. collector current is β times the base current *i.e.*, $i_c = \beta i_b$.

32.4. Equivalent Circuit of a CB Amplifier

In Fig. 32.10 (*a*) is shown the circuit of a common-base amplifier. As seen, emitter is forward-biased by V_{EE} and collector is reverse-biased by V_{CC}. The a.c. signal voltage v_{in} drives the emitter. It produces small fluctuations in transistor voltages and currents in the output circuit. It will be seen that a.c. output voltage is amplified because it is more than v_{in}.

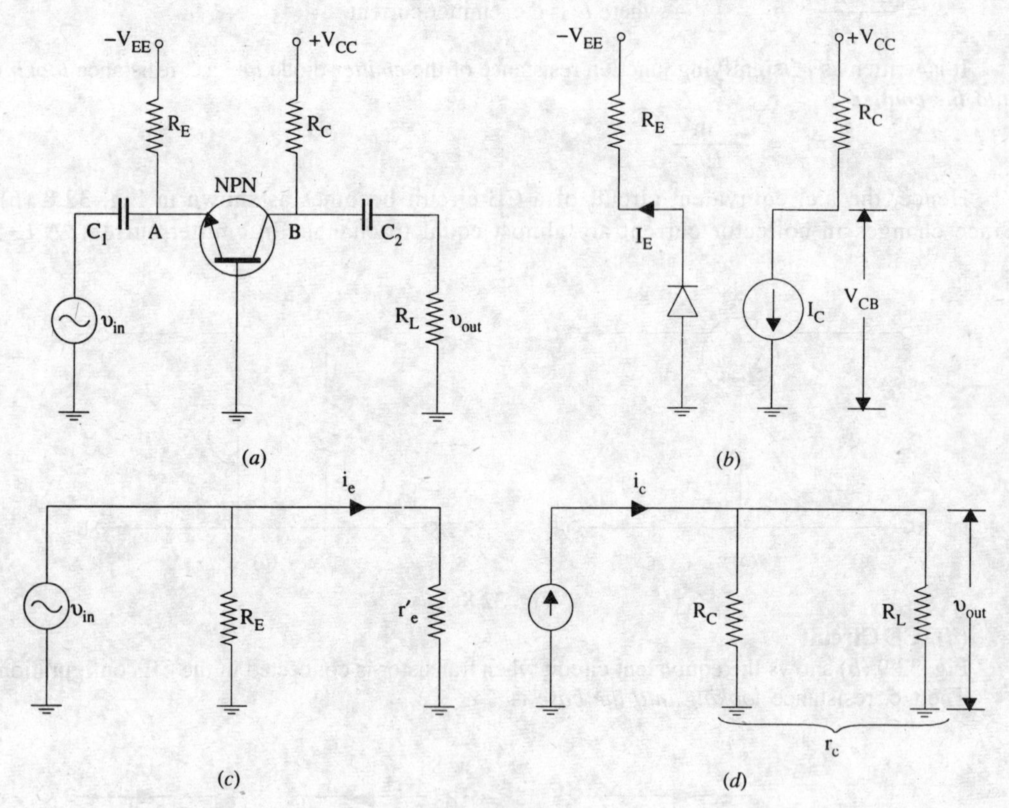

Fig. 32.10

(*a*) D.C. Equivalent Circuit

For drawing d.c. equivalent circuit, following procedure should be adopted:

(*i*) short all a.c. sources *i.e.*, reduce them to zero;

(*ii*) open all capacitors because they block d.c.

If we do this, then as seen from Fig. 32.10 (*a*), neither emitter current can pass through C_1 nor collector current can pass through C_2. These are confined to their respective resistances R_E and R_C (earlier we had been designating it as resistance R_L).

Here, $I_C \cong I_E$ and $V_{CB} = V_{CC} - I_C R_C$

Hence, the d.c. equivalent circuit becomes as shown in Fig. 32.10 (*b*).

(*b*) A.C. Equivalent Circuit

For drawing a.c. equivalent circuit, following procedure is adopted:

(*i*) all d.c. sources are shorted *i.e.*, they are treated as a.c. ground.

(*ii*) all coupling capacitors like C_1 and C_2 in Fig. 32.10 (*a*) are shorted; and

(*iii*) emitter diode is replaced by its a.c. resistance* $r_{ac} = r_e'$

$$\therefore \quad r_e' = \frac{25 \text{ mV}}{I_E}$$

where I_E is d.c. emitter current.

As seen by the input a.c. signal, it has to feed R_E and r_e' in parallel [Fig. 32.10 (*c*)]. As seen from point *A*, R_E is grounded through V_{EE} which has been shorted and r_e' is grounded via the base. Hence, $r_{in} = r_e' \parallel R_E \cong r_e'$ since $R_E \gg r_e'$ in practice.

Similarly, collector has to feed R_C and R_L which are connected in parallel across it *i.e.*, at point *B*. The collector a.c. load resistance is

$$r_e = R_C \parallel R_L$$

Hence, a.c. equivalent circuit is as shown in Fig. 32.10 (*d*). Here, collector diode itself has been shown by a current source.

Following two points are worth noting:

(*i*) changes in collector current are very nearly equal to changes in emitter current. Hence, $\Delta i_c \cong \Delta i_e$.

(*ii*) directions of a.c. currents shown in the circuit diagram are those which correspond to *positive* half-cycle of the a.c. input voltage. That is why i_e is shown flowing upwards in Fig. 32.10 (*d*).

(*c*) Voltage Gain (A_v)

Since a.c. input voltage drives the emitter diode,

$$\therefore \quad i_e = v_{in} / r_e' \quad \text{or} \quad v_{in} = i_e r_e'$$

The collector current drives a load** $r_e = R_C \parallel R_L$.

$$\therefore \quad v_{out} = i_c r_e = i_c (R_C \parallel R_L).$$

$$\text{Voltage gain, } A_v = \frac{v_{out}}{v_{in}} = \frac{i_c r_e}{i_e r_e'} \cong \frac{r_e}{r_e'} \qquad (\because \; i_e \cong i_c)$$

Example 32.3. *For the CB amplifier circuit shown in Fig. 32.11 find*

(i) emitter a.c. resistance r_e' *(ii)* a.c. output voltage v_{out} *and*

(iii) voltage gain A_v *Take* $V_{BE} = 0.7$ *V*

Solution. Here, $I_E = (20 - 0.7)/30 = 0.64$ mA

(*i*) $r_e' = 25 \text{ mV} / I_E = 25 \text{ mV}/0.64 \text{ mA} = \mathbf{39\ \Omega}$

(*ii*) $i_e = v_{in} / r_e' = 1.6/39 = 0.04$ mA

$i_c \cong i_e = 0.04$ mA

$r_c = R_C \parallel R_L = 15 \text{ K} \parallel 30 \text{ K} = 10 \text{ K}$

$\therefore \quad v_{out} = i_c r_e = 0.04 \times 10 = \mathbf{0.4\ V}$

(*iii*) $A_v = \dfrac{v_{out}}{v_{in}} = \dfrac{0.4}{1.6 \times 10^{-3}} = \mathbf{250}$

---

*Emitter diode a.c. resistance is the sum of junction resistance (r_j) and bulk resistance (r_B). Since r_B is negligible,

$$r_{ac} = r_j = 25 \text{ mV}/I_E$$

**On the emitter side, we are only concerned with emitter input a.c. current i_e and not with that going via R_E. Total a.c. current supplied by a.c. voltage would be definitely more than i_e. On the collector side, i_C has to pass through both R_L and R_C. Hence, these two parallel resistances constitute collector load resistance r_C.

or $A_v = \frac{r_c}{r_e'} = \frac{10\ K}{39\ \Omega} = \mathbf{250}$

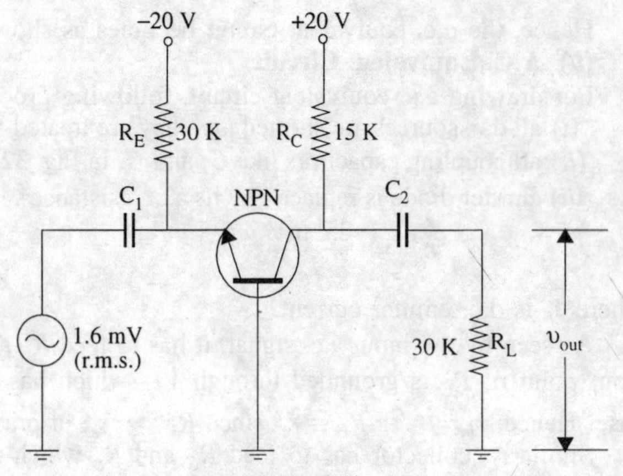

Fig. 32.11

Example 32.4. *For the circuit shown in Fig. 32.12, find*

(i) voltage gain A_v and

(ii) output voltage v_{out}

Solution. This circuit is exactly similar to the one shown in Fig. 32.11 except for the addition of a.c. voltage source resistance R_S.

Since R_S and $r_e' \parallel R_E \cong r_e'$ are in series so far as input a.c. signal is concerned,

$\therefore$ input resistance $= R_S + r_e'$

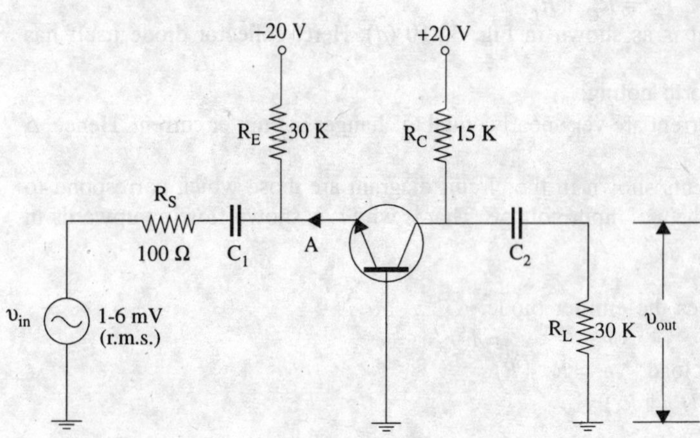

Fig. 32.12

(i) $\therefore A_v = \frac{r_c}{R_S + r_e'} = \frac{10\ K}{(100 + 39)\ \Omega} = \mathbf{72}$

It is seen that overall voltage gain is drastically reduced *i.e.* from 250 to 72.

(ii) $v_{out} = v_{in} \times A_v = 1.6 \times 72 = 115$ mV $= \mathbf{0.115\ V}$

32.5. Equivalent Circuit of a CE Amplifier

Consider the simple CE circuit of Fig. 32.13 (*a*) in which base bias has been employed.

Remembering the rules given in Art. 32.3 for drawing d.c. and a.c. equivalent circuits, we get Fig. 32.13 (*b*) and Fig. 32.14.

In Fig. 32.13 (*b*), $I_B \cong V_{CC} / R_B$ and $I_C = \beta I_B$.

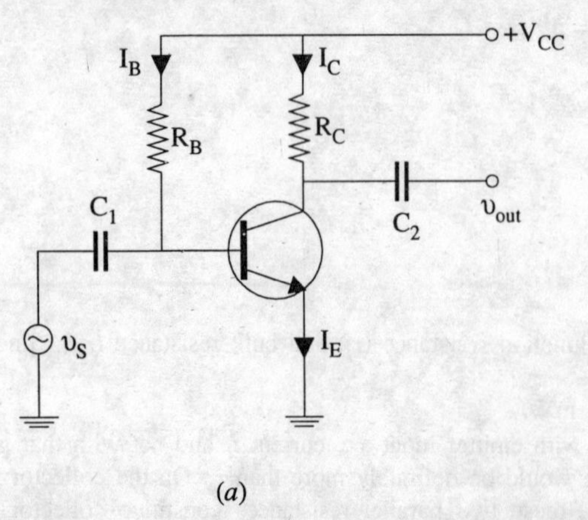

I_B I_C R_B R_C

(*b*)

Fig. 32.13

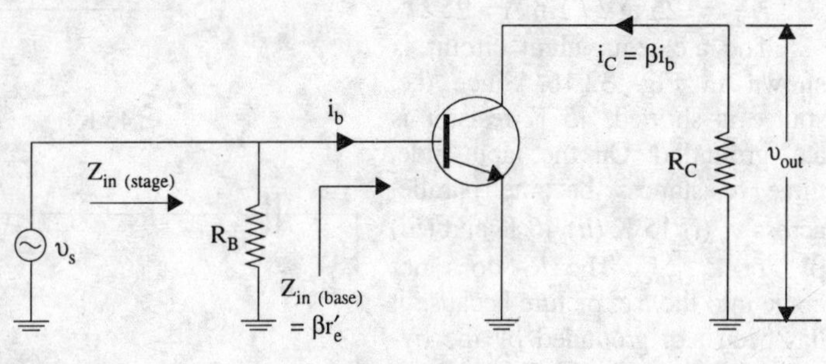

Fig. 32.14

Hence, $I_E \simeq I_C = \beta I_B$

Let us now analyse the a.c. equivalent circuit of Fig. 32.14.

Looking into the base, the a.c. input signal sees an a.c. resistance which is called $Z_{in(base)}$

$Z_{in(base)} = \beta r_e' = \beta \times 25\text{ mV} / I_E$

The a.c. input signal voltage of r.m.s. value v_s acts across R_B as well as $Z_{in(base)} = \beta r_e'$ because the two are connected in parallel across it.

∴ a.c. input base current is

$i_b = v_s / Z_{in(base)} = v_s / \beta r_e'$ (or $v_s = \beta r_e' . i_b$)

a.c. collector current $i_c = \beta i_b$.

a.c. output voltage, $\upsilon_{out} = i_c R_C = \beta i_b R_C$

$$\therefore \quad A_v = \frac{v_{out}}{v_s}^* = \frac{v_{out}}{v_{in}} = \frac{\beta i_b R_C}{\beta r_e' i_b} = \frac{R_C}{r_e'}$$

Note. In case, some load resistance R_L is connected across collector (Fig. 32.15) then

$\upsilon_{out} = i_c r_c$ — where $r_c = R_C \parallel R_L$

$\therefore \quad A_v = \dfrac{r_c}{r_e'}$ — as usual

Example 32.5. *If in the CE circuit of Fig. 32.13 (a), V_{CC} = 20 V, R_C = 10 K, R_B = 1 M, v_s = 2 mV and β = 50, find*

(i) $Z_{in(base)}$ *(ii)* i_b *(iii)* i_c *(iv)* v_{out} and *(v) voltage gain A_v. Neglect V_{BE}.*

Solution. $I_B \cong V_{CC} / R_B = 20 / 1\text{ M} = 20\ \mu\text{A}$

$I_C = \beta I_B = 50 \times 20\ \mu\text{A} = 1\text{ mA}$

$r_e' = 25\text{ mV} / 1\text{ mA} = \mathbf{25\ \Omega}$

(i) $Z_{in(base)} = \beta r_e' = 50 \times 25 = \mathbf{1250\ \Omega}$

(ii) a.c. base current is

$i_b = \upsilon_s / Z_{in(base)} = 2 / 1250 = \mathbf{1.6\ \mu A}$

(iii) a.c. collector current $= \beta i_b = 50 \times 1.6 =$ **80 μA**

(iv) $v_{out} = i_c R_C = 80\ \mu\text{A} \times 10\text{ K} = \mathbf{0.8\ V}$

(v) $A_v = v_{out} / v_s = 0.8\text{ V} / 0.2\text{ mV} = \mathbf{400}$

or $A_v = R_C / r_e' = 10\text{ K} / 25\ \Omega = 400$ (Here, $r_C = R_C$)

Example 32.6. *For the circuit shown in Fig. 32.15, compute*

(i) $Z_{in(base)}$ *(ii)* υ_{out} *(iii)* A_υ *(iv)* Z_{in}

Neglect V_{BE} and take β = 200.

Solution. It may be noted that voltage divider bias has been used in the circuit.

$$V_2 = \frac{15}{15 + 45} \times 30 = 7.5\text{ V}$$

$V_E = V_2 - V_{BE} \cong V_2 = 7.5\text{ V}$

$I_E = 7.5 / 7.5\text{ K} = 1\text{ mA}$

*Here $v_s = v_{in}$ because there is no source rsistance R_S. If R_S is there, then $v_{in} < v_s$ and is given by $v_{in} = v_2$ − a.c. drop across R_S.

$r_e' = 25 \text{ mV} / 1 \text{ mA} = 25\ \Omega$

The a.c. equivalent circuit is shown in Fig. 32.16. Since d.c. source is shorted, 45 K resistor is a.c. grounded. On the input side, three resistances become parallel across v_s (*i*) 15 K (*ii*) 45 K and (*iii*) $\beta r_e'$ or $Z_{in(base)}$. The R_E does not come into the a.c. picture because it has been a.c. grounded by the by-pass capacitor C_2. If C_2 were removed, then $Z_{in(base)} = (\beta r_e' + R_E)$.

Capacitor C_2 grounds the emitter, so does C_3 to 5 K and V_{CC} to 10 K.

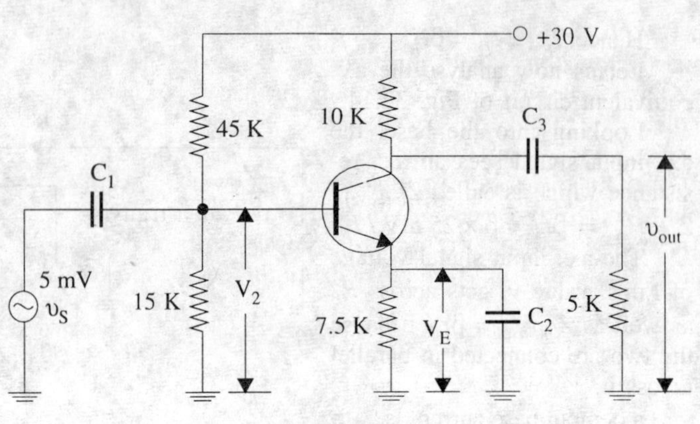

Fig. 32.15

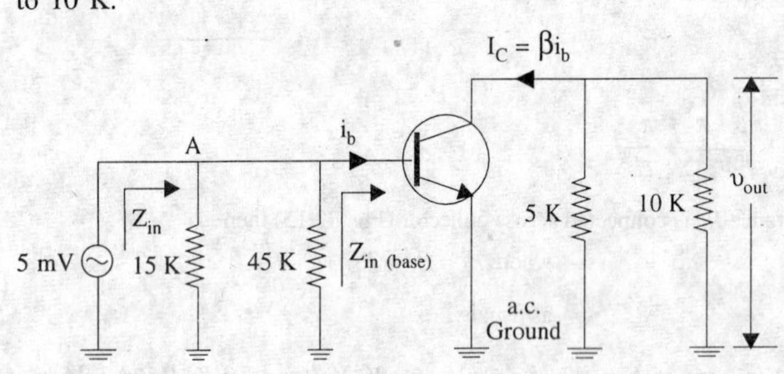

Fig. 32.16

(*i*) $Z_{in(base)} = \beta r_e'$
$= 200 \times 25 = \mathbf{5\ K}$

(*ii*) $i_b = 5 \text{ mV} / 5 \text{ K}$
$= \mathbf{1\ \mu A}$

(Obviously, i_b is not the current which leaves the source but that part of the source current i_s which enters the base).

$i_c = \beta i_b = 200 \times 1$
$= 200\ \mu\text{A} = 0.2 \text{ mA}$

Now, collector load resistance is

$r_c = 10 \text{ K} \parallel 5 \text{ K} = 10/3 \text{ K} = 3.33 \text{ K}$

$v_{out} = i_c r_c = 0.2 \times 3.33 = \mathbf{0.667\ V}$

(*iii*) $A_v = v_{out} / v_{in} = 0.667 / 5 \text{ mV} = \mathbf{133}$

or $A_v = r_c / r_e' = 3.33 \text{ K} / 25\ \Omega = 133$

(*iv*) Z_{in} means the input a.c. resistance as seen from the source *i.e.*, from point *A* in Fig. 32.16. In fact, it is $Z_{in(stage)}$. It is different from $Z_{in(base)}$. Obviously, Z_{in} is equal to the equivalent resistance of three resistances connected in parallel.

$\therefore \quad Z_{in} = Z_{in(stage)} = 15 \text{ K} \parallel 45 \text{ K} \parallel 5 \text{ K} = \mathbf{3.46\ K}$

32.6. Transistor Models

Although many transistor model circuits have been suggested and widely used, the equivalent T-model is the easiest to understand because, in this representation, component parts retain their identity in all configurations leading to rapid appreciation of a given network.

32.7. T-model

(*a*) CB Circuit

In Fig. 32.17 is shown the low-frequency T-equivalent circuit of a transistor connected in CB configuration. Here, r_e represents the a.c. resistance of the forward-biased emitter-base junction. Its value is

$$r_e = \frac{25 \text{ mV}}{I_E} \quad \text{— for Ge}$$

$$= \frac{50 \text{ mV}}{I_E} \quad \text{— for Si}$$

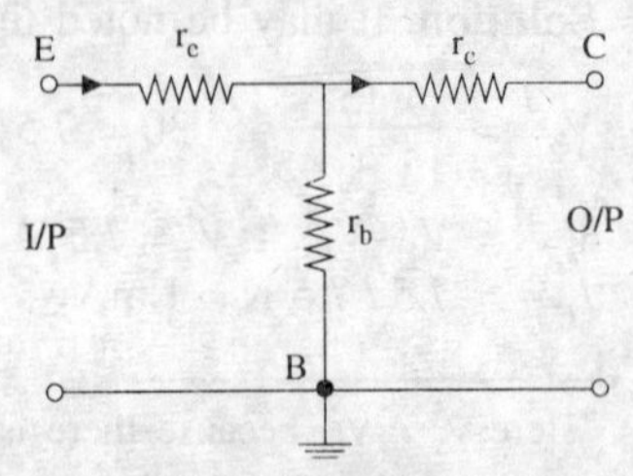

Fig. 32.17

This resistance is fairly small and depends on I_E. Also, r_c represents the a.c. resistance of the reverse-biased *C/B* junction. It is of the order of a few MΩ. Finally, r_b represents the resistance of the base region which is common to both junctions. Its value depends on the degree of doping. Usually, r_b is larger than r_e but much smaller than r_c.

However, circuit shown in Fig. 32.17 is not complete because it does not illustrate the forward current transfer ratio. Since current in the output of a transistor depends on the current at the input, a current-generator in parallel with r_c must be included as shown in Fig. 32.18.

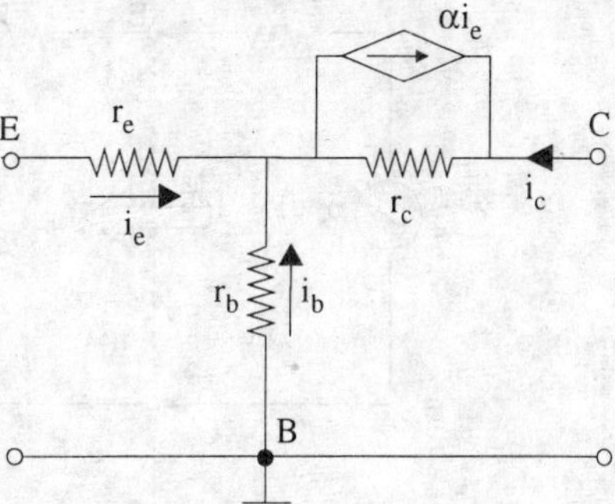

Fig. 32.18

As per convention, *all currents are shown flowing towards the transistor.*

The current generator may be replaced by a voltage generator with the help of Thevenin's theorem as shown in Fig. 32.19 (*a*). In that case, the T-equivalent circuit becomes as shown in Fig. 32.19 (*b*). The generator has a voltage of $\alpha \, i_e \, r_c = r_m \, i_e$ where $r_m = \alpha \, i_c$.

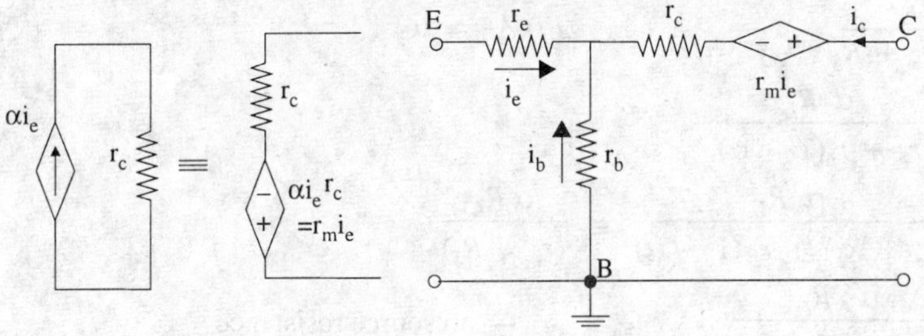

(*a*) (*b*)

Fig. 32.19

(*b*) CE Circuit

The T-equivalent circuit for such a configuration is shown in Fig. 32.20. Whereas the circuit shown in Fig. 32.20 contains a parallel current-generator that shown in Fig. 32.21 contains a series-voltage generator.

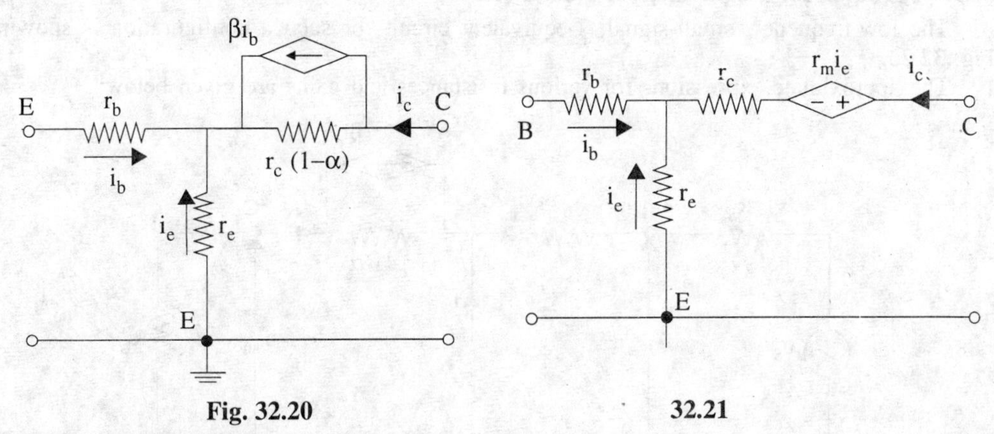

Fig. 32.20 **32.21**

32.8. Formulas for T-equivalent of a CB Circuit

In Fig. 32.22 is shown a small-signal, low-frequency, T-equivalent circuit for CB configuration. The a.c. input signal source has a resistance of R_S and voltage of v_s.

As seen, d.c. biasing circuit has been omitted and only a.c. equivalent circuits is shown. The approximate expressions for input and output resistances and voltages and current gains are given below without derivation :

(*i*) $$r_{in} = r_e + r_b (1 - \alpha) = r_e + \frac{r_b}{(1 + \beta)}$$

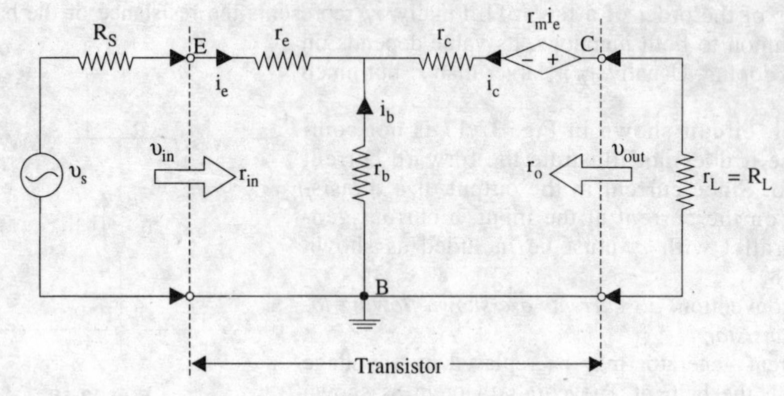

Fig. 32.22

(*ii*) $r_{out} = r_c - \dfrac{\alpha\ r_b\ r_c}{r_e + r_b + R_S}$ (*iii*) $A_i = \alpha$

(*iv*) $A_v = \dfrac{v_{out}}{v_{in}} = \dfrac{\alpha\ R_L}{r_e + r_b(1 - \alpha)}$

$= \dfrac{v_{out}}{v_s} = \dfrac{\alpha\ R_L}{(r_e + R_S) + r_b(1 - \alpha)} \cong \dfrac{\alpha\ R_L}{(r_e + R_S)}$

(*v*) $A_p = A_i A_v = \dfrac{\alpha^2\ R_L}{r_e + r_b(1 - \alpha)}$ — no source resistance

$= \dfrac{\alpha^2\ R_L}{(r_e + R_S) + r_b(1 - \alpha)}$ — with source resistance

32.9. Formulas for T-equivalent of a CE Circuit

The low-frequency, small-signal, T-equivalent circuit for such a configuration is shown in Fig. 32.23.

The approximate expressions for various resistances and gains are given below:

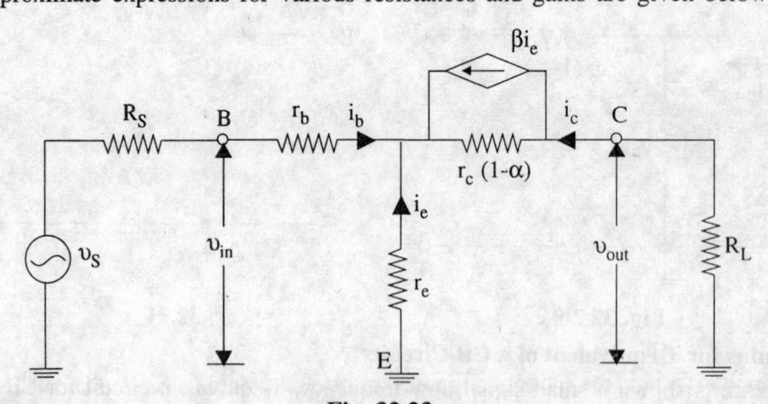

Fig. 32.23

(*i*) $r_{in} = r_b + \dfrac{r_e}{(1 - \alpha)} = r_b + (1 + \beta)\ r_e$

(*ii*) $r_{out} = r_c(1 - \alpha) + \dfrac{\alpha\ r_c\ r_e}{r_b + r_c + R_S}$

(*iii*) $A_i = \beta = \alpha/(1 - \alpha)$

(*iv*) $A_v = \frac{v_{out}}{v_{in}} = \frac{\alpha R_L}{r_e + r_b(1-\alpha)} = \frac{v_{out}}{v_s} = \frac{\alpha R_L}{r_e + (r_b + R_S)(1-\alpha)}$

(*v*) $A_p = A_p A_i = \frac{\alpha^2 R_L}{(1-\alpha)[r_e + r_b(1-\alpha)]}$ or $\frac{\alpha^2 R_L}{(1-\alpha)[r_e + (r_b + R_S)(1-\alpha)]}$

Example 32.7. *A junction transistor has $r_e = 50\ \Omega$, $r_b = 1$ K, $r_c = 1$ M and $\alpha = 0.98$. It is used in common-base circuit with a load resistance of 10 K. Calculate the current, voltage and power gains and the input resistance.* **(A.M.I.E. Summer 1988)**

Solution. (*i*) $A_i = \alpha = \mathbf{0.98}$

(*ii*) $A_v = \frac{\alpha R_L}{r_e + r_b(1-\alpha)} = \frac{0.98 \times 10{,}000}{50 + 1000\ (1 - 0.98)} = \mathbf{140}$

(*iii*) $A_p = A_v A_i = 140 \times 0.98 = \mathbf{137}$

(*iv*) $r_{in} = r_e + r_b(1-\alpha) = 50 + 1000\ (1 - 0.98) = \mathbf{70\ \Omega}$

Example 32.8. *Calculate the input and output resistances, overall current, voltage and power gains for a CE connected transistor having following r-parameters:*

$r_b = 30\ \Omega$, $r_e = 400\ \Omega$, $r_c = 0.75$ M, $\alpha = 0.95$

$R_L = 10$ K and $R_S = 400\ \Omega$

Also, calculate the power gain in decibels. **(Electronics, Bangalore Univ. 1990)**

Solution. (*i*) $r_{in} = r_b + \frac{r_c}{1-\alpha} = 30 + 400/(1 - 0.95) = \mathbf{8030\ \Omega}$

(*ii*) $r_{out} = r_c(1-\alpha) + \frac{\alpha\, r_c r_e}{r_b + r_c + R_S}$

$= 750{,}000\ (1 - 0.95) + \frac{0.95 \times 750{,}000 \times 400}{(30 + 400 + 400)} = 380{,}870 = \mathbf{0.38\ M}$

(*iii*) $A_i = \beta = \alpha/(1-\alpha) = 0.95/(1 - 0.95) = \mathbf{19}$

(*iv*) $A_v = \frac{\alpha R_L}{r_e + (r_b + R_S)(1-\alpha)} = \frac{0.95 \times 10 \times 10^3}{400\ (30 + 400) \times 0.05} = \mathbf{22.5}$

(*v*) $A_p = A_v A_i = 22.5 \times 19 = \mathbf{427.5}$

Power gain in decibels

$G_p = 10 \log_{10}^{427.5} = \mathbf{26.3\ dB}$

32.10. The h-parameters of a Transistor

Every linear circuit has a set of parameters associated with it which completely describe its behaviour. When *small* a.c. signals are involved, a transistor behaves like a linear device because its output a.c. signal varies directly as the input signal. Hence, for small a.c. signals, each transistor has its own characteristic set of *h*-parameters or constants. The letter 'h' has come from the word 'hybrid' which means mixture of distinctly different items. These constants are hybrid because they have different units.

The four *h*-parameters of a transistor are as under:-

h_i = Input impedance with **output shorted**

h_f = Forward current gain with **output shorted**

h_r = Reverse voltage ratio with **input open**

h_o = Output admittance with **input open.**

Consider a transistor connected in CE configuration as shown in Fig. 32.24 (*a*). In this case, output circuit has been shorted. The input impedance under this condition is given by

$$h_i = \frac{v_b}{i_b}$$

The forward current gain is given by

$$h_f = \frac{i_c}{i_b}$$

In Fig. 32.24 (*b*), the output collector circuit is driven with an a.c. voltage of v_c which produces an a.c. current of i_c. However, input circuit has been left open. Though no current can flow in the open input circuit, a voltage v_b does appear at its terminals. The reverse voltage ratio is given by

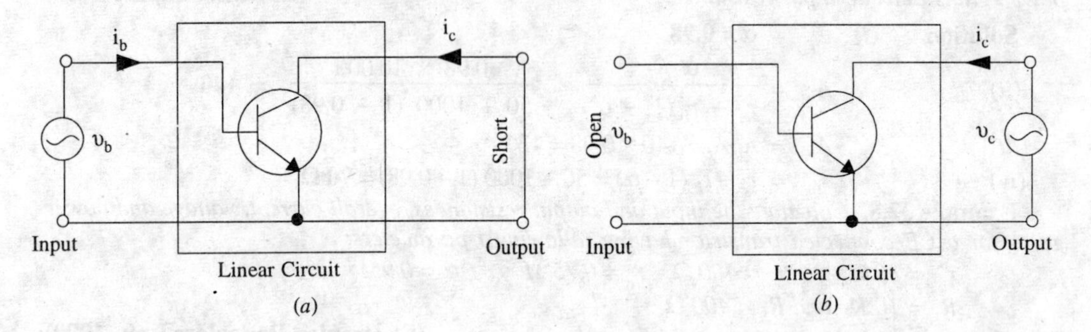

Fig. 32.24

$$h_r = \frac{v_b}{v_c}$$

Also, under this condition, admittance of the output circuit is given by

$$h_o = \frac{i_c}{v_c}$$

It may be noted that a second subscript is added to the above parameters to indicate the particular circuit configuration used.

For example, for CE connection, the four parameters are written as h_{ie}, h_{fe}, h_{re} and h_{oe}.

Similarly, for CB connection, these are written as h_{ib}, h_{fb}, h_{rb} and h_{ob} and for CC connection as h_{ic}, h_{fc}, h_{rc} and h_{oc}.

32.11. Hybrid Equivalent Circuits

(*a*) CB Circuit

In Fig. 32.25 (*a*) is shown an NPN transistor connected in CB configuration. Its equivalent circuit employing *h*-parameters is shown in Fig. 32.25 (*b*). It should be noted that no external biasing resistor or any signal source (with its internal resistance R_S) has been shown connected to the transistor.

However, if we connect such an a.c. signal source across input terminals and also add a load

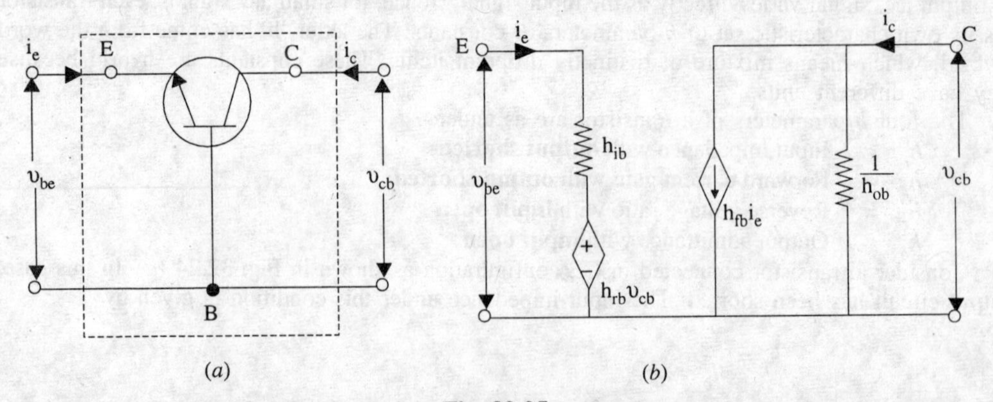

Fig. 32.25

resistance R_L across output terminals, then amplifier circuit becomes as shown in Fig. 32.26. Here R_{in} represents the input resistance of the *transistor only.* Similarly, R_{out} represents output resistance of the *transistor itself* whereas R_{out}' takes into account load resistance R_L. Hence,

$$R_{out}' = R_{out} \parallel R_L$$

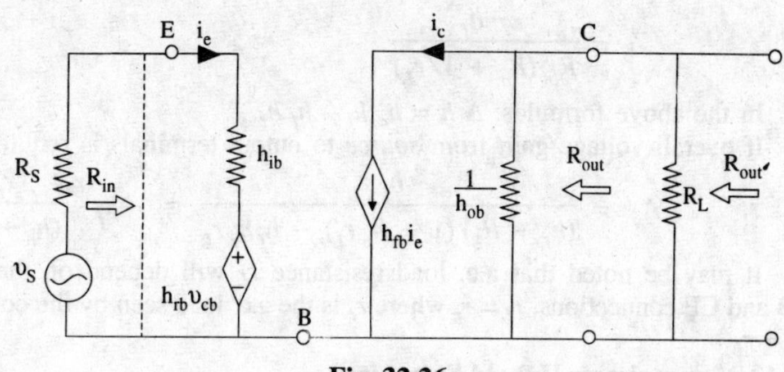

Fig. 32.26

(*b*) CE Circuit

The hybrid equivalent circuit of the transistor *alone* when connected in *CE* configuration is shown in Fig. 32.27 (*b*). As before, we may connect input signal source across the input terminals and load resistance across the output terminals.

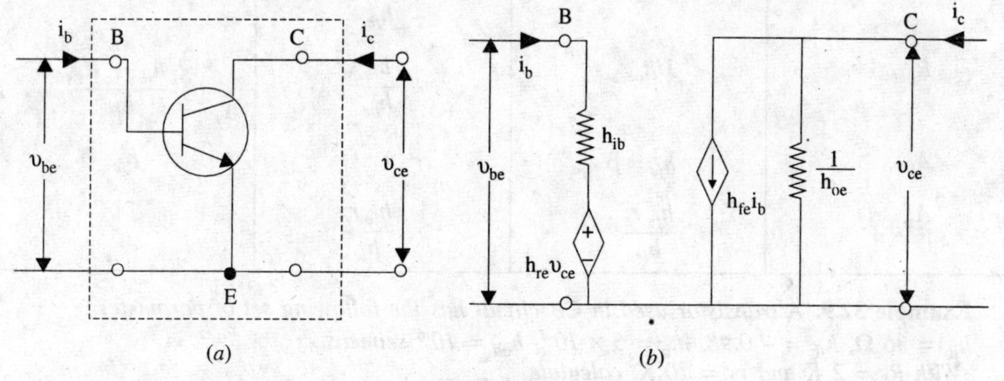

Fig. 32.27

32.12. Hybrid Formulas

Sufficiently accurate formulas for the various gains and resistances of a transistor amplifier circuit in terms of *h*-parameters are given below. These are applicable to all transistor configurations

(*i*) R_{in} = input resistance at transistor terminals

$$= h_i - \frac{h_r h_f}{h_o + 1/r_L} \quad \text{— where } r_L \text{ is a.c. load resistance}$$

(*ii*) R_{out} = output resistance at transist or terminals

$$= \frac{h_i + R_S}{(h_i + R_S)\, h_o - h_f h_r} \cong \frac{h_i + R_S}{h_o R_S + \Delta h}$$

It is worth noting that *input resistance depends on a.c. load resistance r_L and output resistance depends on input signal source resistance R_S.*

(*iii*) A_i = transistor current gain from input terminals *to load* $= \dfrac{h_f}{1 + h_o r_L}$

where h_f is the short-circuit current gain of the transistor *alone i.e., from input terminals to output terminals only.*

(*iv*) A_v = voltage gain from input terminals to output terminals

$$= \frac{-h_f r_L}{h_i(1 + h_o r_L) - h_f h_r r_L} = \frac{-h_f r_L}{h_i + \Delta h \cdot r_L} \cong \frac{-h_f r_L}{h_i}$$

$$= \frac{-h_f}{R_{in}(h_o + 1/r_L)}$$

In the above formulas, $\Delta h = h_o h_i - h_f h_r$.

If overall voltage gain from *source* to output terminals is required, then add R_S to h_i.

$$\therefore \quad A_v' = \frac{-h_f r_L}{(h_i + R_S)(1 + h_o r_L) - h_f h_r r_L} \cong -h_f \cdot \frac{r_L}{(h_i + R_S)}$$

It may be noted that a.c. load resistance r_L will depend on the transistor connections. For CB and CE connections, $r_L = r_c$ where r_c is the a.c. load seen by the collector. For a CC connection, $r_L = r_E$.

32.13. Approximate Hybrid Formulas

The *approximate* hybrid formulas for the three connections are listed below :

Item	*CE*	*CB*	*CC*
R_{in}	h_{ie}	$\frac{h_{ie}}{h_{fe}}$	$h_{fe} \cdot r_E$
R_{out}	$1/h_{oe}$	$\frac{h_{fe}}{h_{oe}}$	$\frac{h_{ie} + R_s}{h_{fe}}$
A_i	$h_{fe} = \beta$	1	h_{fe}
A_v	$\frac{h_{fe} r_c}{h_{ie}}$	$\frac{h_{fe} r_c}{h_{ie}}$	1

Example 32.9. *A transistor used in CB circuit has the following set of parameters.*
$h_{ib} = 36\ \Omega$, $h_{fb} = -0.98$, $h_{rb} = 5 \times 10^{-4}$, $h_{ob} = 10^{-6}$ seimens.
With $R_S = 2$ K and $r_L = 10$ K, calculate
(i) R_{in} (ii) R_{out} (iii) A_i and (iv) A_v **(Applied Electronics-I, Punjab Univ.1984)**

Solution. (*i*) $R_{in} = h_{ib} - \frac{h_{rb} h_{fb}}{h_{ob} + 1/r_L} = 36 - \frac{-0.98 \times 5 \times 10^{-4}}{10^{-6} + 1/10 \times 10^3} = 36 + 4.9 = \mathbf{40.9\ \Omega}$

(*ii*) $$R_{out} = \frac{h_{ib} + R_S}{h_o(h_{ib} + R_S) - h_{fb} h_{rb}}$$

$$= \frac{36 + 2000}{10^{-6}(36 + 2000) - (-0.98) \times 5 \times 10^{-4}} = \mathbf{0.8\ M\Omega}$$

(*iii*) $$A_i = \frac{h_{fb}}{1 + h_{ob} r_L} = \frac{-0.98}{1 + 10^{-6} \times 10^4} = \mathbf{-0.97}$$

(*iv*) $$A_v' = \frac{-h_{fb} r_L}{(h_{ib} + R_S)(1 + h_{ob} r_L) - h_{fb} \cdot h_{rb} \cdot r_L}$$

$$= \frac{-(-0.98) \times 10^4}{(36 + 2000)(1 + 10^{-6} \times 10^4) - (-0.98) \times 5 \times 10^{-4} \times 10^4}$$

Note. If we consider A_v for transistor *alone*, then putting $R_S = 0$, we get

$$A_v = \frac{-(-0.98) \times 10^4}{36(1 + 10^{-6} \times 10^4) - (-0.98) \times 5 \times 10^{-4} \times 10^4}$$

Example 32.10. *A transistor used in CE connection (Fig. 32.28) has the following set of h-parameters:*

$h_{ie} = 1$ *K*, $h_{fe} = 100$, $h_{re} = 5 \times 10^{-4}$ *and* $h_{oe} = 2 \times 10^{-5}$ *seimens. With* $R_S = 2$ *K*, $r_L = 5$ *K, determine*

(i) R_{in} (ii) R_{out}
(iii) A_i *and* (iv) A_v.

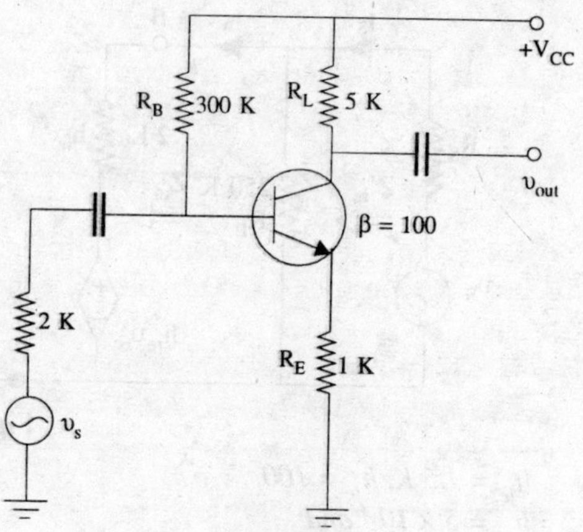

Fig. 32.28

Solution. (i) $R_{in} = h_{ie} - \dfrac{h_{re}\, h_{fe}}{h_o + 1/r_L}$

$$= 1000 - \frac{5 \times 10^{-4} \times 100}{2 \times 10^{-5} + 1/5 \times 10^3}$$

$= \mathbf{773\ \Omega}$

(ii) $R_{out} = \dfrac{h_{ie} + R_S}{(h_{ie} + R_S)\, h_{oe} - h_{fe}\, h_{re}}$

$$= \frac{1000 + 2000}{(1000 + 2000) \times 2 \times 10^{-5} - 100 \times 5 \times 10^{-4}} = 300{,}000\ \Omega = \mathbf{300\ k\Omega}$$

(iii) $$A_i = \frac{h_{fe}}{1 + h_{oe}\, r_L} = \frac{100}{1 + 2 \times 10^{-5} \times 5 \times 10^3} = \mathbf{91}$$

(iv) $$A_v = \frac{-h_{fe}\, r_L}{(h_{ie} + R_S)(1 + h_{oe}\, r_L) - h_{fe}\, h_{re}\, r_L}$$

$$= \frac{-100 \times 5 \times 10^3}{(1000 + 2000)(1 + 2 \times 10^{-5} \times 5 \times 10^3) - 100 \times 5 \times 10^{-4}\ 5 \times 10^3}$$

$= \mathbf{-164}$

The negative sign indicates that there is 180° phase shift between the input and output a.c. signals.

Example 32.11. *In the CE circuit shown in Fig. 32.29, the transistor parameters are:* $h_{ie} = 2$ *K*, $h_{fe} = 100$, $h_{re} = 5 \times 10^{-4}$, $h_{oe} = 2 \times 10^{-5}$ *seimens*

Calculate (i) $Z_{in(base)}$ *(ii)* $Z_{in(stage)}$ *(iii)* Z_{out} *(iv)* Z_{out}' (v) A_i *and (vi)* A_v.

(Electronic Engg. M.S. Univ. Vadodara, 1991)

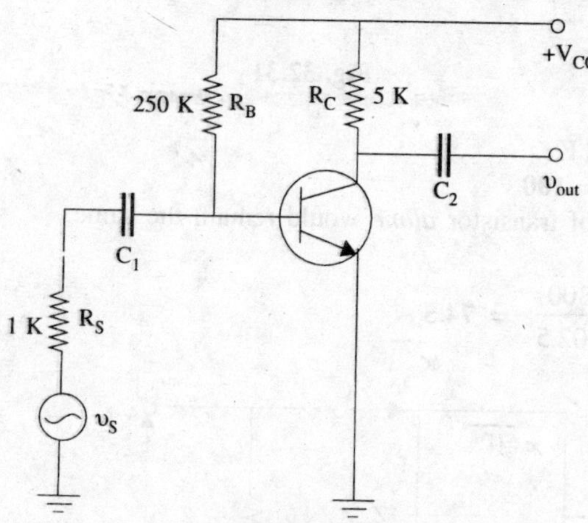

Fig. 32.29

Solution. The hybrid equivalent circuit is shown in Fig. 32.30. We will use approximate formulas given in Art. 31.13.

(i) $Z_{in(base)} = Z_{in} \cong h_{ie} = \mathbf{2\ K}$

(ii) $Z_{in(stage)} = Z_{in}' = 2\ \text{K} \parallel 250\ \text{K} = \mathbf{1.98\ K}$

(iii) $Z_{out} \cong 1/h_{oe} = 1/2 \times 10^{-5} = \mathbf{50\ K}$

It is the output impedance of the transistor *only*.

(iv) $Z_{out}' = 50\ \text{K} \parallel 5\ \text{K} = \mathbf{4.54\ K}$

The impedance takes into account collector load.

(v) $A_i \cong h_{fe} = \mathbf{100}$

(vi) $$A_v = \frac{-h_{fe}\, r_L}{h_{ie}} = \frac{-100 \times 5}{2} = \mathbf{-250}$$

Example 32.12. *In the CE amplifier circuit shown in Fig. 32.31, the transistor parameters are:*

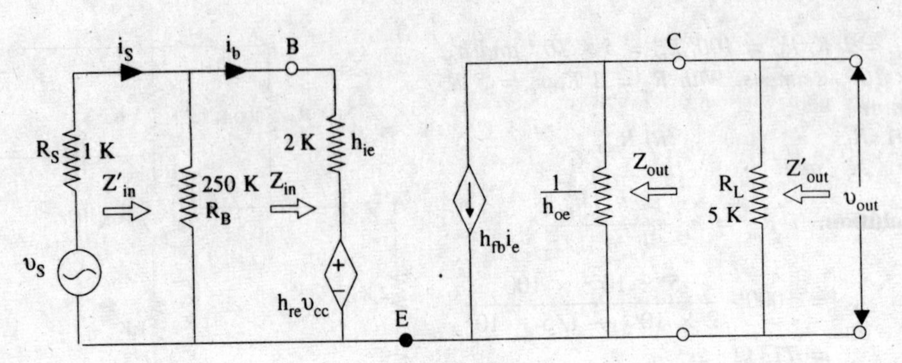

Fig. 32.30

$h_{ie} = 1.5$ K, $h_{fe} = 100$,
$h_{re} = 5 \times 10^{-4}$ and
$h_{oe} = 2 \times 10^{-5}$ seimens
***Calculate the approximate** values of*

(i) $Z_{in(base)}$	*(ii)* $Z_{in(stage)}$
(iii) Z_{out}	*(iv)* Z_{out}'
(v) A_i	*(vi)* A_v

(Electronic Technology, Bangalore Univ. 1990)

Fig. 32.31

Solution. *(i)* $Z_{in(base)} = Z_{in}$
$= h_{ie} + (1+\beta)\, R_E$ *
$= h_{ie} + (1+h_{fe})\, R_E$
$= 1.5 + 101 \times 1 = \mathbf{102.5\ K}$

(ii) $Z_{in(stage)} = Z_{in}' = Z_{in(base)} \parallel R_B$
$= 102.5 \parallel 300 = \mathbf{76\ K}$

(iii) $Z_{out} \cong 1 / h_{oe} = 1 / 2 \times 10^{-5} = \mathbf{50\ K}$

(iv) $Z_{out}' = Z_{out} \parallel R_L = 50\text{ K} \parallel 2\text{ K} = \mathbf{1.92\ K}$

(v) A_i = transistor current gain = h_{fe} = **100**

Even when R_E is taken into account, gain of transistor *alone* would remain the same. The *circuit* current gain is

$$A_i' = \frac{h_{fe}\ R_B}{R_B + Z_{in\,(base)}} = \frac{100 \times 300}{300 + 102.5} = \mathbf{74.5}$$

Fig. 32.32

*Please refer to β-rule given in Art. 32.2.

(*vi*) Voltage gain of transistor *alone* is

$$A_v = -h_{fe}\, R_L / h_{ie} = -100 \times 2 / 1.5 = \mathbf{-133.3}$$

Voltage gain when R_E is taken into account is

$$A_v = \frac{-h_{fe} R_L}{h_{ie} + (1 + h_{fe})\, R_E}$$

$$\cong -\frac{R_L}{R_E} = \frac{-2}{1} = \mathbf{-2}$$

It is so because $h_{ie} \ll (1+h_{fe})\, R_E$ and $h_{fe} \gg 1$.

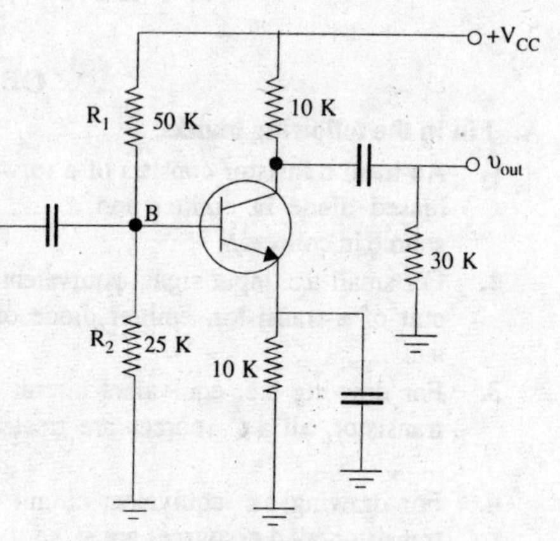

Fig. 32.33

Example 32.13. *The transistor of Fig. 32.33 has the following set of h-parameters:*

$h_{ie} = 2$ *K*, $h_{fe} = 100$,

$h_{re} = 5 \times 10^{-4}$, $h_{oe} = 2.5 \times 10^{-5}$ *seimens.*

Find the voltage gain and the a.c. input impedance of the stage.

Solution. Using exact formulas given in Art. 32.12, we have

$$Z_{in(base)} = Z_{in} = h_{ie} - \frac{h_{fe}\, h_{re}}{h_{oe} + 1/r_L}$$

Now, collector load r_L = 10 K || 30 K = 7.5 K

$$\therefore \quad Z_{in(base)} = 2000 - \frac{100 \times 5 \times 10^{-4}}{2.5 \times 10^{-5} + 1/7.5 \times 10^3} = 2000 - 316 = \mathbf{1684\ \Omega}$$

The a.c. input impedance of the stage *i.e.*, impedance when looking into point *B* is

$$Z_{in(stage)} = Z_{in(base)} \,||\, R_1 \,||\, R_2 = 1.684 \,||\, 50 \,||\, 25 = \mathbf{1.53\ K}$$

$$A_v = \frac{-h_{fe}}{Z_{in\,(base)}\,(h_{oe} + 1/r_L)}$$

Now, r_L = 10 K || 30 K = 7.5 K

$$\therefore \quad A_v = \frac{-100}{1684\,(2.5 \times 10^{-5} + 1/7500)} = \mathbf{375}$$

Obviously, R_E does not come into the picture because it has been a.c. grounded by the capacitor.

HIGHLIGHTS

1. The d.c. equivalent circuit of an ideal transistor consists of a forward-biased diode in emitter and a current source in collector.
2. The small-signal a.c. equivalent circuit of an ideal transistor consists of an a.c. resistance in emitter and a current source in collector.
3. Voltage gain of a small-signal transistor amplifier is $A_v = r_c / r_e'$ where r_c is the a.c. load as seen by the collector.
4. The T-model of a transistor utilizes *r*-parameters.
5. The four *h*-parameters of a transistor are: h_i, h_f, h_r and h_o. Generally, h_i is the largest and h_o is the smallest (almost negligible).
6. The approximate hybrid formulae are:

1. $$R_{in} = h_i - \frac{h_r\, h_f}{h_o + 1/r_L} \qquad \text{where } r_L \text{ is a.c. load resistance}$$

$$\cong h_i$$

2. $$R_{out} = \frac{h_i + R_S}{(h_i + R_S)\, h_o - h_f\, h_r}$$

3. $$A_i = \frac{h_f}{1 + h_o\, r_L} \cong h_f$$

4. $A_v = \dfrac{-h_f r_L}{h_i + \Delta h \,.\, r_L}$.

OBJECTIVE TESTS—32

A. Fill in the following blanks.

1. An ideal transistor consists of a forward-biased diode in emitter and a.............. source in collector.
2. For small a.c. input signal equivalent circuit of a transistor, emitter diode offers a.c. ....
3. For drawing d.c. equivalent circuit of a transistor, all a.c. sources are treated as ....
4. For drawing a.c. equivalent circuit of a transistor, all d.c. sources are ....
5. The *r*-parameters of a transistor are measured under .... circuit conditions.
6. The *h*-parameters of a transistor are .... signal parameters.
7. The letter '*h*' of *h*-parameters has come from the word ....
8. The parameter h_f of a transistor represents forward current gain with output ....

B. Answer True or False.

1. The ideal d.c. equivalent circuit of a transistor consists of a diode and a current source.
2. The a.c. equivalent circuit of a transistor consists of a resistance and current source.
3. For finding d.c. equivalent circuit of a transistor, all capacitors are treated as shorts.
4. The a.c. equivalent circuit of a transistor is found by treating all coupling capacitors as shorts.
5. The approximate voltage gain of a transistor is given by the ratio of output a.c. resistance to input a.c. resistance.
6. The emitter resistor of a transistor circuit can be a.c. grounded by connecting a suitable capacitor across it.
7. The T-model of a transistor is not much used these days.
8. The parameter h_r of a transistor gives the voltage ratio with input shorted.

C. Multiple Choice Questions.

1. The voltage gain of a well-designed single-stage CB amplifier is essentially determined by a.c. collector load and
   (*a*) emitter resistor R_E
   (*b*) a.c. alpha
   (*c*) input resistance of emitter diode
   (*d*) a.c. beta
2. In a single-stage CB amplifier, a smaller load resistance R_L will produce
   (*a*) high current gain
   (*b*) low voltage gain
   (*c*) better frequency response
   (*d*) higher power gain
3. The output signal from a single-stage CE amplifier is increased when
   (*a*) its a.c. load is decreased
   (*b*) input signal resistance is increased
   (*c*) base resistor is decreased
   (*d*) input signal resistance is small
4. The *h*-parameters of a transistor are called hybrid because they
   (*a*) are obtained from different characteristics
   (*b*) are mixed with other parameters
   (*c*) apply to circuits contained in a black box
   (*d*) are defined by using both open and short-circuit termination
5. The smallest of the four *h*-parameters of a transistor is
   (*a*) h_i (*b*) h_f
   (*c*) h_o (*d*) h_r
6. Typical value of h_{ie} for a transistor is
   (*a*) 1 K (*b*) 25 K
   (*c*) 50 Ω (*d*) 100 K

ANSWERS

A. 1. current 2. resistance 3. shorts 4. shorts 5. open 6. small 7. hybrid 8. shorted

B. 1. T 2. T 3. F 4. T 5. T 6. T 7. T 8. F

C. 1. *c* 2. *b* 3. *d* 4. *d* 5. *c* 6. *a*

33 TRANSISTOR AMPLIFIERS

33.1. Classification of Amplifiers

Linear amplifiers are classified according to their mode of operation *i.e.*, the way they operate according to a predetermined set of values. Various amplifier descriptions are based on the following factors :

1. **As based on its input**
   (*a*) small-signal amplifier (*b*) large-signal amplifier
2. **As based on its output**
   (*a*) voltage amplifier (*b*) power amplifier
3. **As based on its frequency response**
   (*a*) audio-frequency (AF) amplifier (*b*) intermediate-frequency (IF) amplifier
   (*c*) radio-frequency (RF) amplifier
4. **As based on its biasing conditions**
   (*a*) class-A (*b*) class-AB (*c*) class-B (*d*) class-C
5. **As based on transistor configuration**
   (*a*) common-base (CB) amplifier (*b*) common-emitter (CE) amplifier
   (*c*) common-collector (CC) amplifier

The description : small-signal, class-A, CE, voltage amplifier means that input signal is small, biasing condition is class-A, transistor configuration is common-emitter and its output concerns voltage amplification.

We will first take up the basic working of a single-stage amplifier *i.e.*, an amplifier having one amplifying element connected in CB, CE and CC configuration.

33.2. Common Base (CB) Amplifier

Both Figs. 33.1 and 33.2 show the circuit of a single-stage CB amplifier using NPN transistor. As seen, input a.c. signal is injected into the emitter-base circuit and output is taken from the collector-base circuit. The *E/B* junction is forward-biased by V_{EE} whereas *C/B* junction is reverse-biased by V_{CC}. The *Q*-point or d.c. working conditions are determined by the batteries along with resistors R_E and R_C.

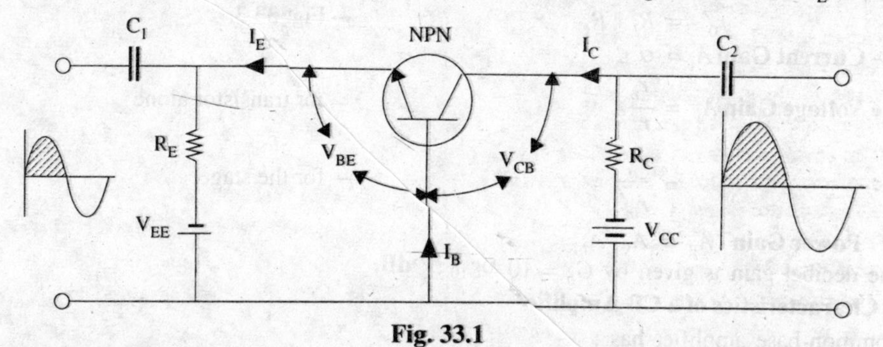

Fig. 33.1

When no signal is applied to the input circuit, the output just *sits at the Q-point* so that there is no output signal. Let us now see what happens when we apply an a.c. signal to the *E/B* junction *via* a coupling capacitor C_1 (which is assumed to offer no reactance to the signal).

Circuit Operation

When positive half-cycle of the signal is applied, then

1. forward bias is **decreased** because V_{BE} is already negative with respect to the ground as per biasing rule of Art. 30.2.
2. consequently, I_B is **decreased.**
3. I_E and hence I_C are **decreased** (because they are both nearly β times the base current).
4. the drop $I_C R_C$ is **decreased.**
5. hence, V_{CB} is **increased** because

$$V_{CC} = V_{CB} + I_C R_C \quad \text{or} \quad V_{CB} = V_{CC} - I_C R_C$$

It means that a positive output half-cycle is produced.

Since a *positive-going* input signal produces a *positive-going* output signal, there is no phase reversal between the two.

Voltage amplification in this circuit is possible by reason of relative input and output circuitry rather than current gain (α) which is always less than unity. The input circuit has low resistance whereas output circuit has very large resistance. Although changes in the input and output currents are the same, the a.c. drop across R_L is very large. Hence, changes in V_{CB} (which is the output voltage) are much larger than changes in input a.c. signal. Hence, the voltage amplification.

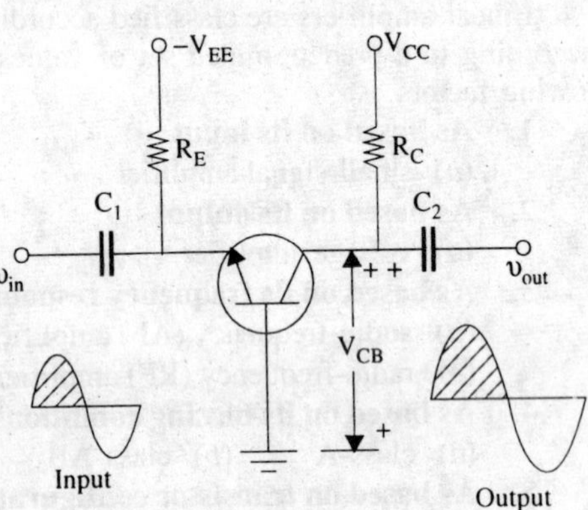

Fig. 33.2

33.3. Various Gains of a CB Amplifier

1. Input Resistance

The a.c. input resistance of the *transistor alone* is given by the emitter junction resistance

$$r_e' = \frac{25 \text{ mV}}{I_E} \quad \text{or} \quad \frac{50 \text{ mV}}{I_E}$$

As seen from the a.c. equivalent circuit (Fig. 33.3), the input resistance of the *stage* is

$$r_{in}' = r_e' \parallel R_E$$

2. Output Resistance

$$r_0 = R_C \qquad \text{— Fig. 33.2}$$

If a load resistance R_L is connected across output terminals, then output resistance of the stage is

$$r_0' = R_L \parallel R_C \qquad \text{— Fig. 33.3}$$

3. **Current Gain** $A_i = \alpha$

4. **Voltage Gain** $A_v = \dfrac{r_o}{r_{in}}$ — for transistor alone

$$= \frac{r_o'}{r_{in}'} \qquad \text{— for the stage}$$

5. **Power Gain** $A_p = A_v \cdot A_i$

The decibel gain is given by $G_p = 10 \log_{10} A_p$ **dB.**

33.4. Characteristics of a CB Amplifier

Common-base amplifier has :

1. very low input resistance (30–150 Ω);
2. very high output resistance (upto 500 K);
3. a current gain α < 1;
4. large voltage gain of about 1500;
5. power gain of upto 30 dB;

6. no phase reversal between input and output voltages.

Uses

One of the important uses of a CB amplifier is in matching a low-impedance circuit to a high-impedance circuit.

It also has high stability of collector current with temperature changes.

Example 33.1. *For the single-stage CB amplifier shown in Fig. 33.3 (a), find (a) stage input resistance (b) stage output resistance (c) current gain (d) voltage gain of the stage (e) stage power gain in dB.*

Assume $\alpha = 1$. *Neglect* V_{BE} *and use* $r_e' = 25\ mV / I_E$.

(Basic Electronics, Mumbai Univ. 1987)

Solution. The a.c. equivalent circuit is shown in Fig. 33.3 (*b*).

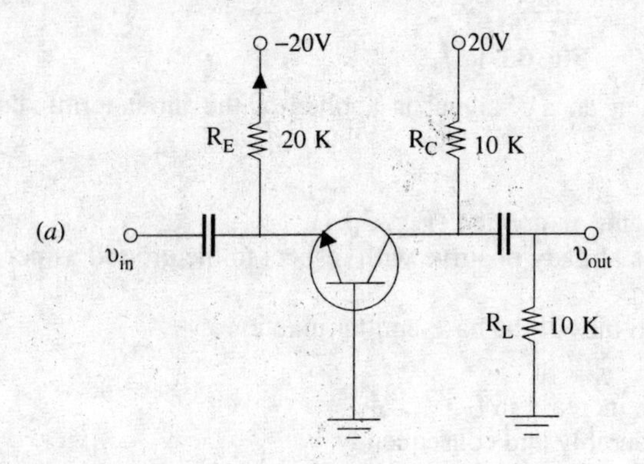

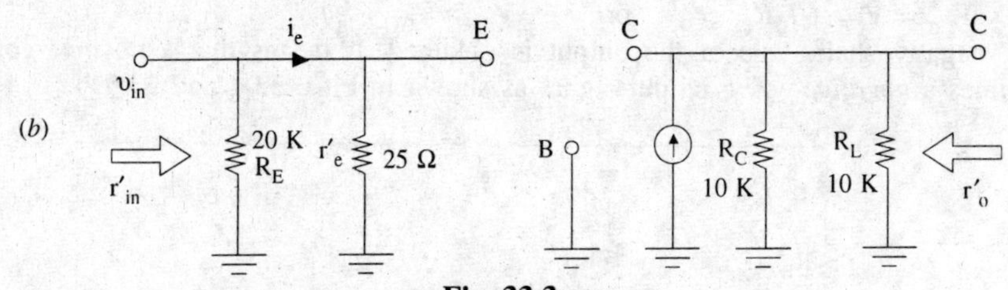

Fig. 33.3

Here, $I_E = 20\ V / 20\ K = 1\ mA$, $r_e' = 25 / 1 = 25\ \Omega$

(*a*) $r_{in}' = r_e' \parallel R_E = 25\ \Omega \parallel 20\ K \cong \mathbf{25\ \Omega}$

(*b*) $r_o' = R_C \parallel R_L = 10\ K \parallel 10\ K = \mathbf{5\ K}$

(*c*) $A_i = \alpha = \mathbf{1}$

(*d*) $$A_v = \frac{r_o'}{r_{in}'} = \frac{5\ K}{25\ \Omega} = \mathbf{200}$$

(*e*) $A_p = A_v \cdot A_i = 200 \times 1 = \mathbf{200}$

$G_p = 10 \log^{A_p}\ dB = 10 \log_{10}^{200} = \mathbf{23\ dB}$

33.5. Common Emitter (CE) Amplifier

Figs. 33.4 and 33.5 show the circuit of a single-stage CE amplifier using an NPN transistor. Here, base is the driven element. The input signal is injected into the base-emitter circuit whereas output signal is taken out from the collector-emitter circuit. The *E/B* junction is forward-biased by V_{BB} and *C/B* junction is reverse-biased by V_{CC} (in fact, same battery V_{CC} can provide d.c. power for both base and collector as in Fig. 33.5). The *Q*-point or working condition is determined by V_{CC} together with R_B and R_C.

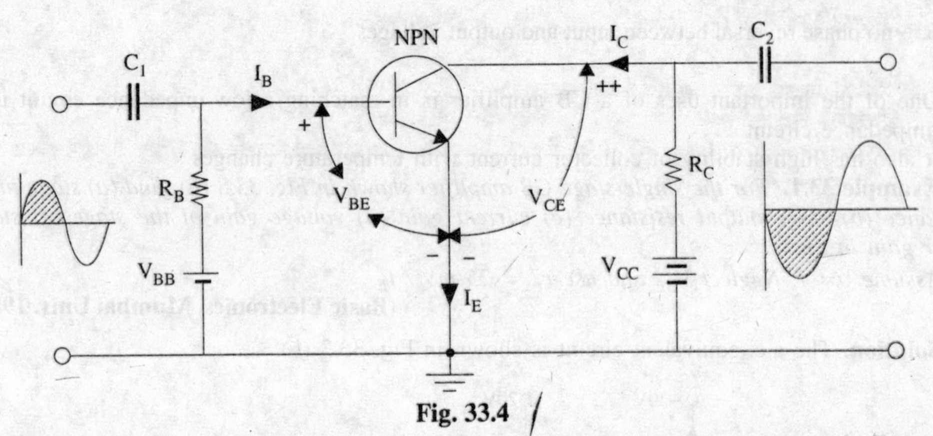

Fig. 33.4

Now, let us see what happens when an a.c. signal is applied at the input terminals of the circuit.

Circuit Operation

When positive half-cycle of the signal is applied (Fig. 33.4).

1. V_{BE} is **increased** because it is already positive with respect to the ground as per biasing rule of Art. 30.2.
2. It leads to **increase** in forward bias of the base-emitter junction.
3. I_B is **increased** somewhat.
4. I_C is **increased** by β times the increase in I_B.
5. drop $I_C R_C$ is **increased** considerably and consequently.
6. V_{CE} which represents the output voltage is **decreased.**

Now $\quad V_{CC} = V_{CE} + I_C R_C \qquad$ or $\qquad V_{CE} = V_{CC} - I_C R_C$

Hence, negative half-cycle of the output is obtained. It means that a *positive-going* input signal becomes a *negative-going* output signal as shown in Figs. 33.4 and 33.5.

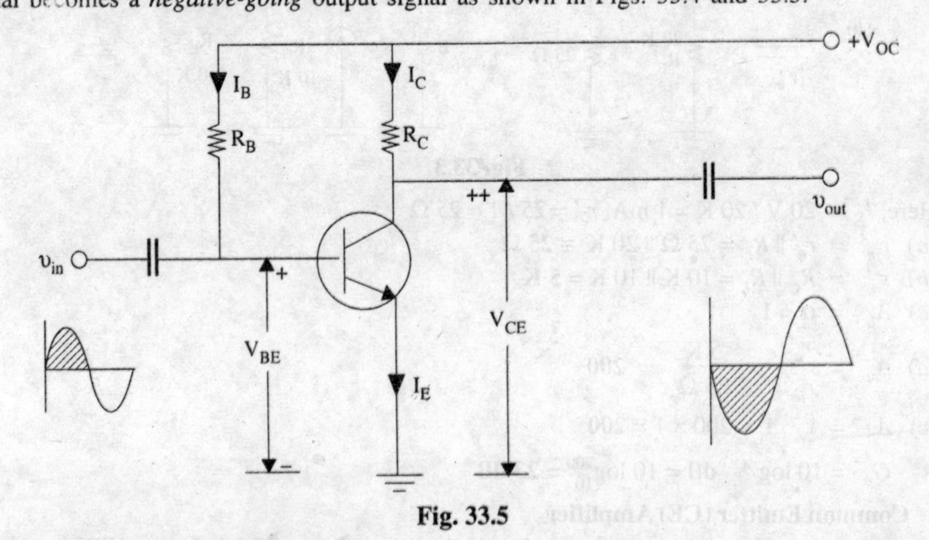

Fig. 33.5

33.6. Various Gains of a CE Amplifier

The a.c. equivalent of the given circuit (Fig. 33.5) is similar to the one shown in Fig. 33.6 (*b*).

1. **Input Resistance**

When viewed from base, a.c. resistance of the emitter junction is $\beta r_e'$. As seen from Fig. 33.6 (*b*), *circuit* or *stage* input resistance is

$$r_{in}' = R_B \parallel \beta r_e' \qquad \text{— remember } \beta\text{-rule (Art. 32.2)}$$
$$\cong \beta r_e' \qquad \text{— when } R_B \gg \beta r_e'$$

It is called input resistance of the stage *i.e.* $r_{in(stage)}$.

2. **Output Resistance** $r_o = R_C$ — Fig. 33.5

However, if a load resistor R_L is connected across the output terminals (Fig. 33.6), then

$$r_o' = R_C \parallel R_L = r_L$$

3. **Current Gain** $A_i = \beta$
4. **Voltage Gain :**

$$A_v = \beta \frac{r_o'}{r_{in}'} \cong \beta \cdot \frac{r_o'}{\beta r_e'} = \frac{r_o'}{r_e'} \qquad \text{— if } R_B \gg \beta r_e'$$

It is the *stage* voltage gain.

5. **Power Gain**

$$A_p = A_v \cdot A_i = \beta \frac{r_o'}{r_e'}; \qquad G_p = 10 \log_{10}{}^{Ap} \text{ dB}$$

33.7. Characteristics of a CE Amplifier

A CE transistor amplifier has the following characteristics :

1. it has moderately low input resistance (1 K to 2 K);
2. its output resistance is moderately large (50 K or so);
3. its current gain (β) is high (50–300)
4. it has very high voltage gain of the order of 1500 or so;
5. it produces very high power gain of the order of 10,000 or 40 dB;
6. it produces *phase reversal* of input signal *i.e.,* input and output signal are 180° out of phase with each other.

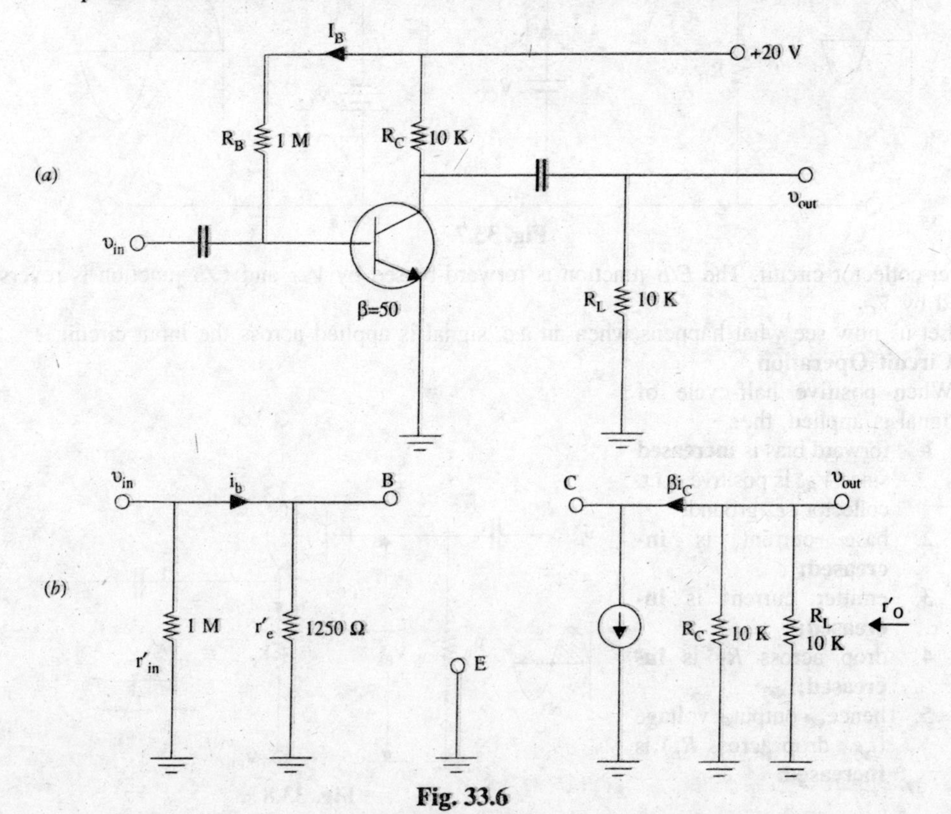

Fig. 33.6

Uses

Most of the transistor amplifiers are of CE type because of large gains in voltage, current and power. Moreover, its input and output impedance characteristics are suitable for many applications.

Example 33.2. *For the single-stage CE amplifier circuit shown in Fig. 33.6 (a), calculate (a) r_{in} (b) r_0' (c) A_i (d) A_v and (e) G_p .*

Take transistor $\beta = 50$. Neglect V_{BE} and take $r_e' = 25\ mV / I_E$.

Solution. $I_B = 20\ V / 1\ M, 20\ \mu A; I_C = \beta I_B = 50 \times 20 = 1\ mA$

$r_e' = 25 / 1 = 25\ \Omega\ ; \beta r_e' = 50 \times 25 = 1250\ \Omega$

The a.c. equivalent circuit is shown in Fig. 33.6 (*b*)

(*a*) $r_{in}' = 1\ M \parallel 1250\ \Omega \cong$ **$1250\ \Omega$**

Obviously, it is the input resistance of the *stage* and not that of the *transistor alone.*

(*b*) $r_o' = R_C \parallel R_L \doteq 10\ K \parallel 10\ K =$ **$5\ K$**

(*c*) $A_i = \beta =$ **50**

(*d*) $A_v = r_o' / r_e' = 5\ K / 25\ \Omega =$ **200**

(*e*) $G_p = 10 \log_{10}^{200} =$ **23 dB**

33.8. Common Collector (CC) Amplifier

Figs. 33.7 and 33.8 show the circuit of a single-stage CC amplifier using an NPN transistor. The input signal is injected into the base-collector circuit and output signal is taken out from the

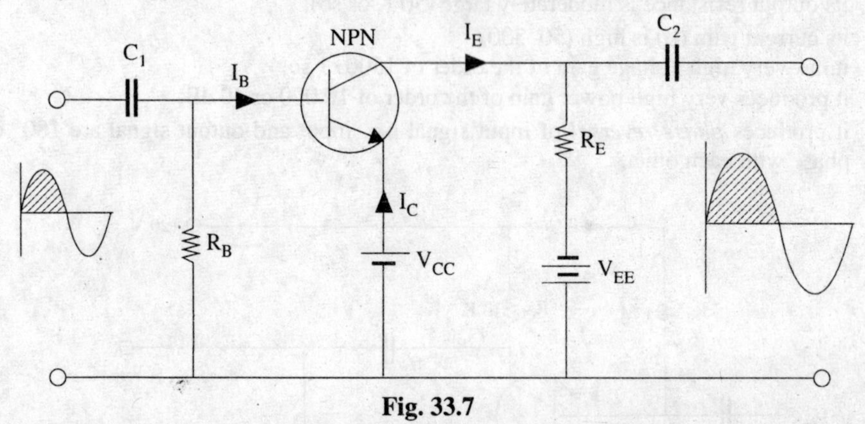

Fig. 33.7

emitter-collector circuit. The *E/B* junction is forward-biased by V_{EE} and *C/B* junction is reverse-biased by V_{CC}.

Let us now see what happens when an a.c. signal is applied across the input circuit.

Circuit Operation

When positive half-cycle of the signal is applied, then

1. forward bias is **increased** since V_{BE} is positive w.r.t. collector *i.e.,* ground;
2. base current is **increased;**
3. emitter current is **increased;**
4. drop across R_E is **increased;**
5. hence, output voltage (*i.e.,* drop across R_E) is **increased.**

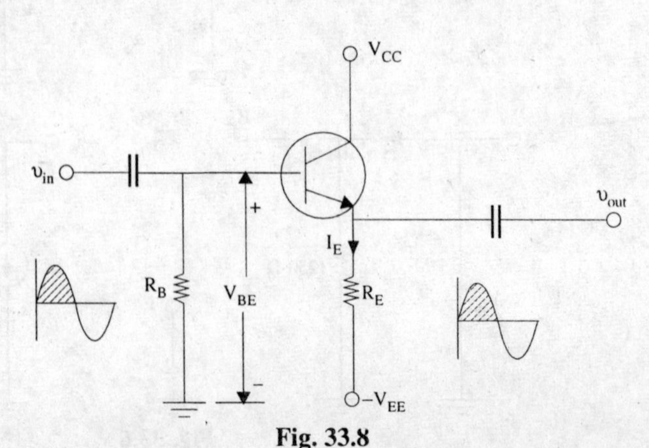

Fig. 33.8

Consequently, we get positive half-cycle of the output.

It means that a *positive-going* input signal results in a *positive-going* output signal and, consequently, the input and output signals are in phase with each other as shown in Fig. 33.8.

33.9. Various Gains of a CC Amplifier

The a.c. equivalent circuit of the CC amplifier (Fig. 33.8) is given in Fig. 33.9.

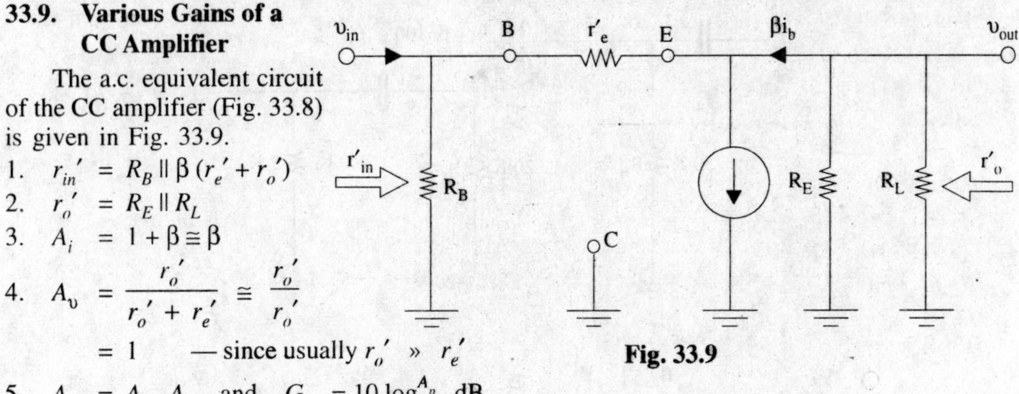

Fig. 33.9

1. $r_{in}' = R_B \parallel \beta\,(r_e' + r_o')$
2. $r_o' = R_E \parallel R_L$
3. $A_i = 1 + \beta \cong \beta$
4. $A_\upsilon = \dfrac{r_o'}{r_o' + r_e'} \cong \dfrac{r_o'}{r_o'}$

   $= 1$ — since usually $r_o' \gg r_e'$
5. $A_p = A_v \cdot A_i$ and $G_p = 10 \log_{10}^{A_p}$ dB

33.10. Characteristics of a CC Amplifier

A CC amplifier has the following characteristics:

1. high input impedance (20 – 500 K);
2. low output impedance (50 Ω – 1 kΩ);
3. high current gain of (1 + β *i.e.*, (50 – 500);
4. voltage gain of less than 1;
5. power gain of 10 to 20 dB;
6. no phase reversal of the input signal.

33.11. Uses

The CC amplifiers are used for the following purposes:

1. for impedance matching *i.e.*, for connecting a circuit having high output impedance to one having low input impedance;
2. for circuit isolation;
3. as a two-way amplifier since it can pass a signal in either direction;
4. for switching circuits.

33.12. Phase Reversal in Amplifiers

From the fore-going discussion of different amplifiers, it is seen that

1. CB amplifier does not change the phase of the input a.c. signal. As seen from Fig. 33.10, the input and output signals are in phase.
2. The CE amplifier inverts signal *i.e.*, it causes a phase reversal of 180° in the signal.
3. The CC amplifier does not change the phase of the input signal *i.e.*, the input and output signals are in phase.

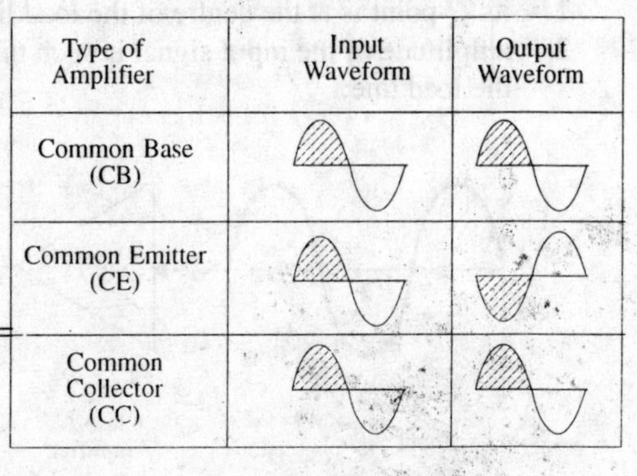

Type of Amplifier	Input Waveform	Output Waveform
Common Base (CB)		
Common Emitter (CE)		
Common Collector (CC)		

Fig. 33.10

Example 33.3. *For the CC amplifier circuit of Fig. 33.11, compute (i) r_{in}' (ii) r_o' (iii) A_v and (iv) A_p. Take transistor β = 100. Neglect V_{BE} and use $r_e' = 25\ mV/I_E$.*

(Electronics-1, MS Univ. Vadodara, 1991)

Solution. I_E 20 V / 20 K = 1 mA ; $r_e' = 25/1 =$ 25 Ω.

The a.c. equivalent circuit is shown in Fig. 33.11 (*b*)

(*i*) $r_{in}' = R_B \parallel \beta\,(r_e' + r_o') \cong R_B \parallel \beta r_o'$

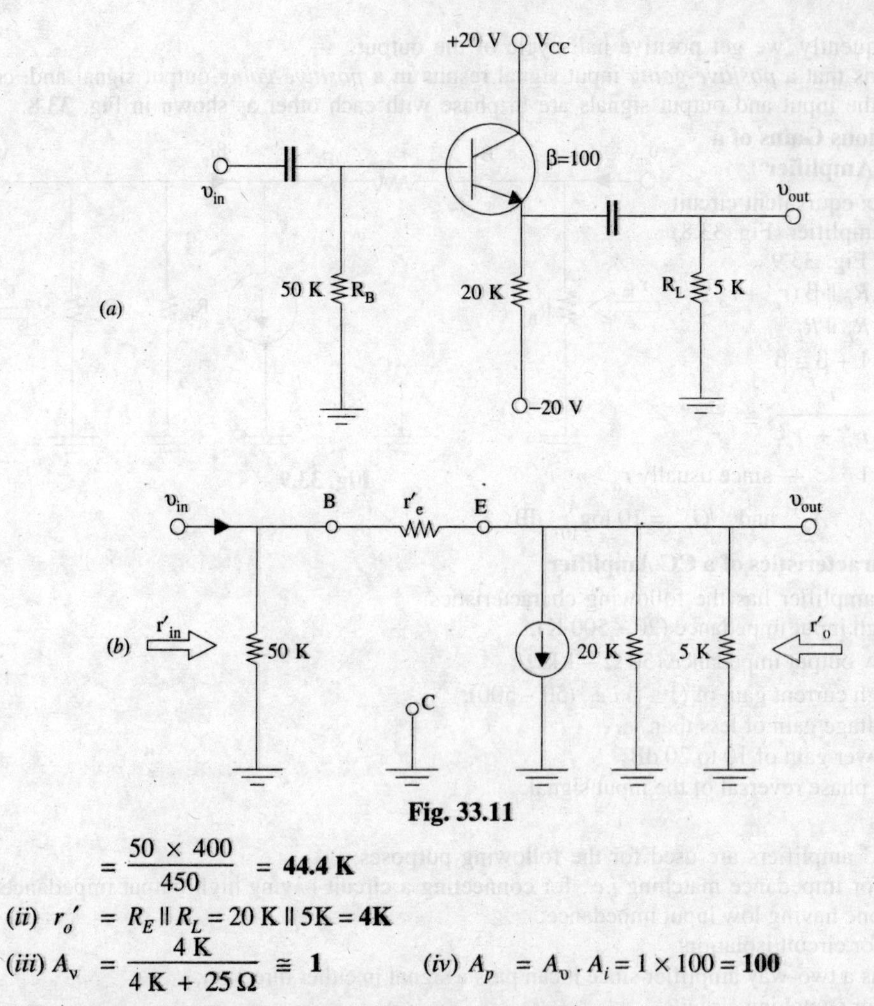

Fig. 33.11

$$= \frac{50 \times 400}{450} = \mathbf{44.4\ K}$$

(ii) $r_o' = R_E \parallel R_L = 20\text{ K} \parallel 5\text{K} = \mathbf{4K}$

(iii) $A_v = \dfrac{4\text{ K}}{4\text{ K} + 25\,\Omega} \cong \mathbf{1}$ (iv) $A_p = A_\upsilon \,.\, A_i = 1 \times 100 = \mathbf{100}$

33.13. Amplifier Classification Based on Biasing Conditions

This classification is based on the amount of transistor bias and the amplitude of the input signal. It takes into account the portion of the cycle for which the transistor conducts. The three main classifications are : (i) Class A (ii) Class B and (iii) Class C.

33.14. Class-A Amplifier

In this amplifier, the transistor is so biased that

1. its Q-point is at the centre of the load line;
2. amplitude of the input signal is such that the transistor operates over the linear portion of the load line;

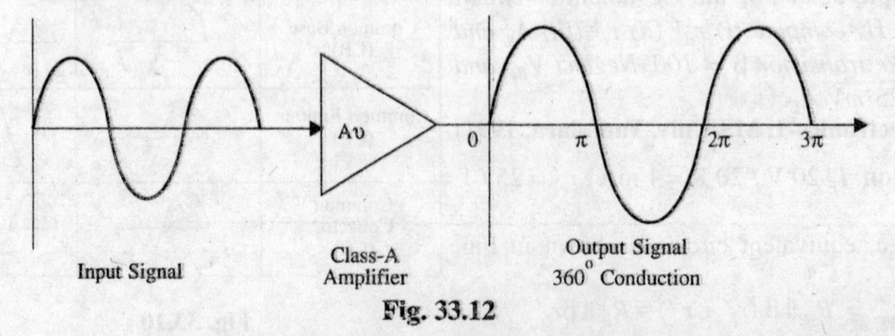

Fig. 33.12

3. output current flows during the entire cycle of the input signal;
4. its conduction angle is 360° as shown in Fig. 33.12.

The above points have been shown graphically in Fig. 33.13 where CE output characteristics have been taken. It is so because CE configuration is widely used in Electronics.

Here, i_b represents the input signal whereas i_c is the output signal. Since the transistor remains FR-biased throughout the input cycle, its output current flows for the entire cycle.

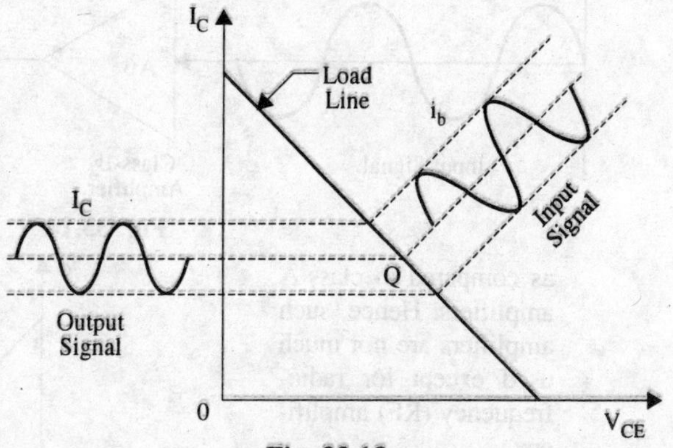

Fig. 33.13

Characteristics

1. Since the transistor operates over the *linear portion* of the load line, the output waveform is exactly similar to the input waveform. Hence, class-A amplifiers are used where high-fidelity and distortion-free output is required as in radio receivers and television sets.
2. Since its operation is restricted only over a small central region of the load line, this amplifier is meant only for amplifying input signals of *small amplitude*. Large signals will shift the *Q*-point into non-linear regions near saturation or cut-off and thus produce distortion.
3. Due to the limitation of the input signal amplitude, a.c. power output per active device (*i.e.*, transistor) is small.
4. The overall efficiency of the amplifier circuit is

$$= \frac{\text{a.c. power delivered to the load}}{\text{total power delivered by d.c. supply}}$$

$$= \frac{\text{average a.c. power output}}{\text{average d.c. power input}}$$

The maximum possible overall efficiency of a class-A amplifier with resistive load is 25%.

5. The collector efficiency of a transistor is defined as

$$= \frac{\text{average a.c. power output}}{\text{average d.c. power input to transistor}}$$

The maximum possible collector efficiency of a class-A amplifier with resistive load is 50%.

6. In case an output transformer is used, the maximum possible overall efficiency and maximum possible collector efficiency for a class-A amplifier are both 50%.
7. Since under zero-signal condition, there is no a.c. output power, it means that all the power given to the transistor is wasted as heat. Hence, the transistor dissipates maximum power under *zero-signal condition.*

33.15. Class-B Amplifier

In this amplifier, the transistor is so biased that

1. the *Q*-point lies at the cut-off point (Fig. 33.15);
2. there is no output current when input current is zero;
3. it conducts only for half-cycle of the input *i.e.*, its conduction angle is only 180° (Fig. 33.14).

The above facts have been shown graphically on the CE characteristics in Fig. 33.15. It is seen that output consists of only positive half-cycles, the negative half-cycles having been suppressed when transistor becomes biased below cut-off.

Characteristics

1. since negative half-cycles are totally absent from the output, the signal distortion is high

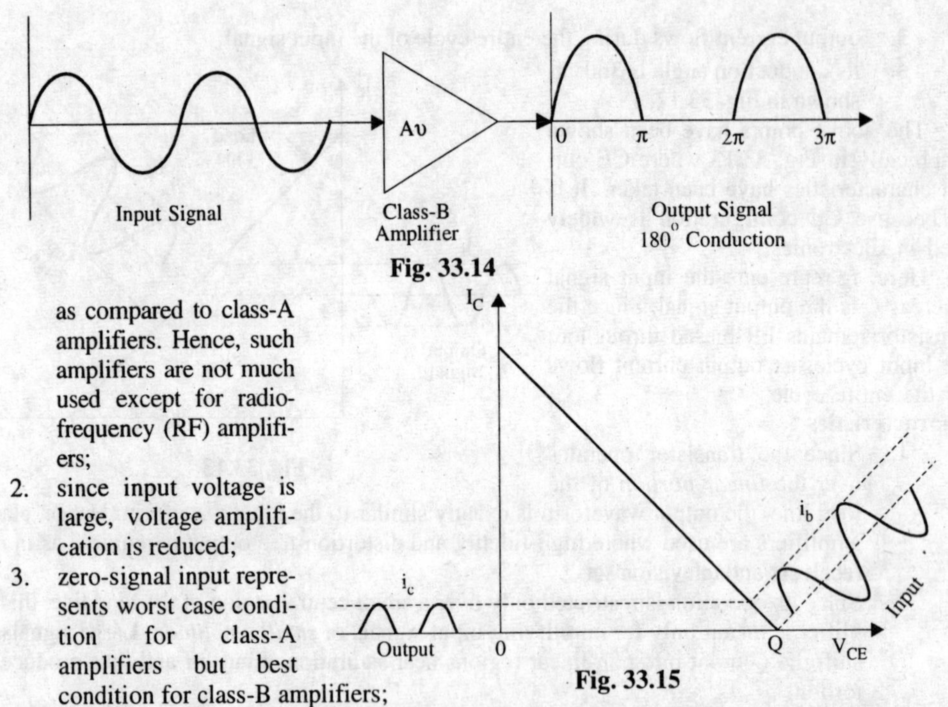

Fig. 33.14

Fig. 33.15

as compared to class-A amplifiers. Hence, such amplifiers are not much used except for radio-frequency (RF) amplifiers;

2. since input voltage is large, voltage amplification is reduced;
3. zero-signal input represents worst case condition for class-A amplifiers but best condition for class-B amplifiers;
4. in class-B amplifiers, transistor dissipates more power as signal increases but opposite is the case in class-A amplifiers;
5. average current in class-B operation is less than in class-A, hence power dissipated is less. Consequently, maximum circuit (or overall) efficiency of a class-B amplifier is 78.5%.

33.16. Class-C Amplifier

In this amplifier, the transistor is biased much beyond cut-off (Fig. 33.17). Hence,

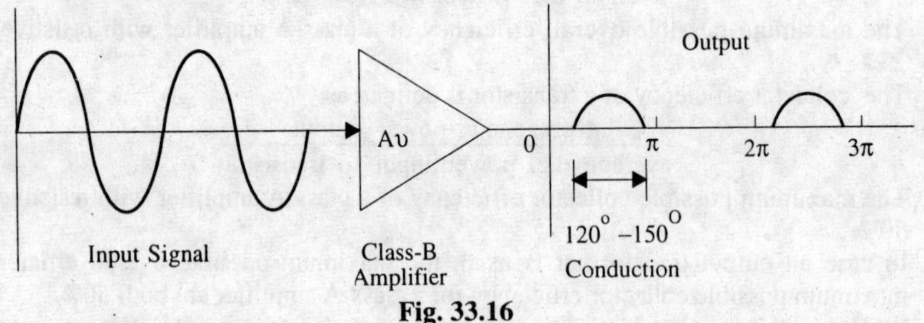

Fig. 33.16

1. output current flows only during a *part of the positive half-cycle* of the input signal. The conduction angle varies from 120° – 150° (Fig. 33.16);
2. there is no output current flow during *any part of the negative half-cycle* of the input signal;
3. output signal has hardly any resemblance with the input signal. It consists of *short pulses only.*

The above facts have been shown on the V_{CE}/I_C characteristic of Fig. 33.17.

Characteristics

1. Since output current flows for almost two-thirds of each half-cycle, power loss in such amplifiers is the least. Hence, they have very high efficiency of about 85 – 90%.
2. Since output signal is much different from the input signal, class-C amplifiers suffer from very high distortion. Hence, such amplifiers are mainly used in oscillators for radio-frequency amplification where high efficiency is very essential regardless of distortion.

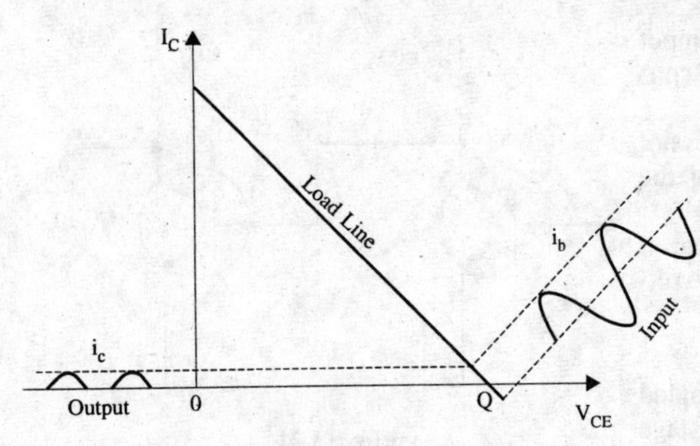

Fig. 33.17

33.17. Amplifier Coupling

All amplifiers need some coupling network. Even a single stage amplifier has to be coupled to the input and output devices. In the case of multi-stage systems, there is *interstage* coupling. The type of coupling used determines the characteristics of the cascaded amplifier. In fact, amplifiers are classified according to the coupling network used.

The four basic methods of coupling are:

1. **Resistance-capacitance (RC) Coupling**

It is also known as capacitive coupling and is shown in Fig. 33.18. Amplifiers using this coupling are known as RC-coupled amplifiers. Here, RC coupling network consists of two resistors R_{C_1} and R_B and one capacitor *C*. The connecting link between the two stages is *C*. The function of RC coupling network is two fold:

(*a*) to pass a.c. signal from one stage to the next;

(*b*) to block the passage of d.c. voltages from one stage to the next.

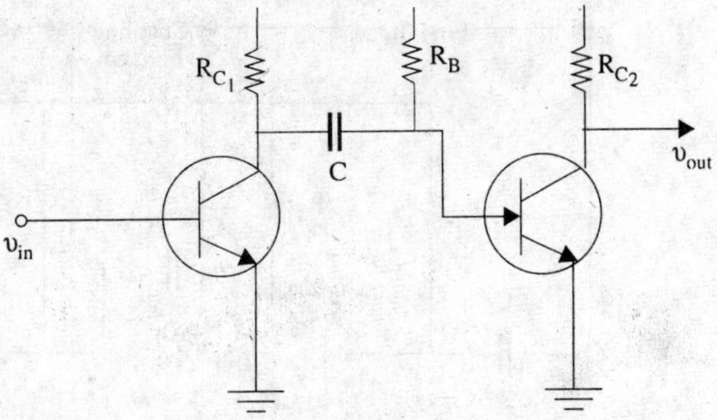

Fig. 33.18

2. **Impedance Coupling or Inductive Coupling**

It is also known as choke-capacitance coupling and is shown in Fig. 33.19. Amplifiers using this coupling are known as impedance-coupled amplifiers. Here, the coupling network consists of L_1, *C* and R_B. The impedance of the coupling coil depends on (*i*) its inductance and (*ii*) signal frequency.

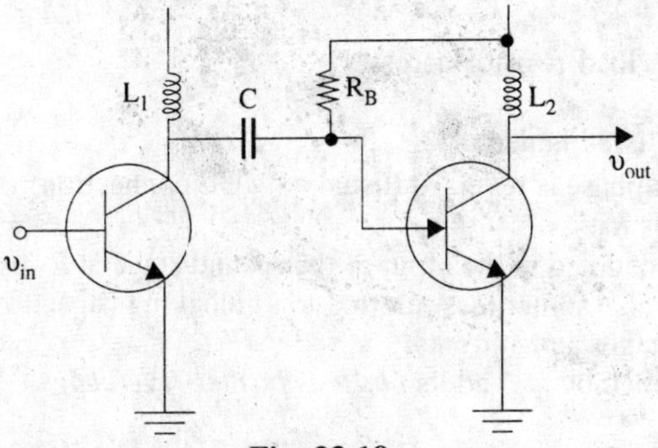

Fig. 33.19

3. **Trasformer Coupling**

It is shown in Fig. 33.20. Since secondary of the coupling transformer conveys the a.c. component of the signal directly to the base of the second stage, there is no need for a *coupling capacitor.* Moreover, the secondary winding also provides a base return path hence, there is no *need for a base resistance.* Amplifiers using this coupling are called transformer-coupled amplifiers.

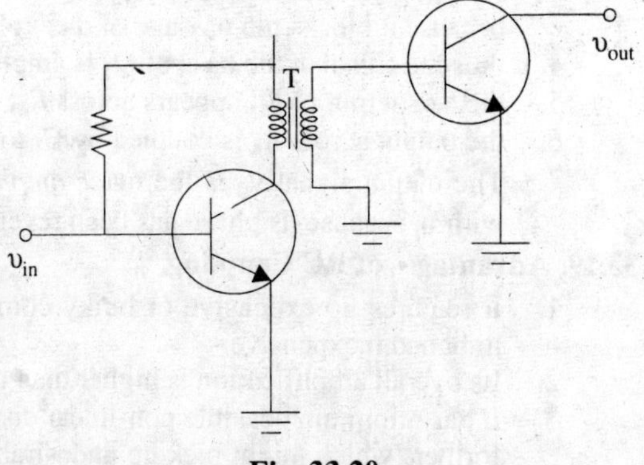

Fig. 33.20

4. **Direct Coupling**

It is shown in Fig. 33.21. This coupling is used where it is desirable to connect the load directly in series with the output terminal of the active circuit element. The examples of such load devices are (*i*) headphones (*ii*) loud-speakers (*iii*) d.c. meters (*iv*) *d.c.* relays and (*v*) input circuit of a transistor etc. Of course, direct coupling is permissible only when

(*i*) d.c. component of the output does not disturb the normal operation of the load device;

(*iii*) device resistance is so low that it does not appreciably reduce the voltage at the electrodes.

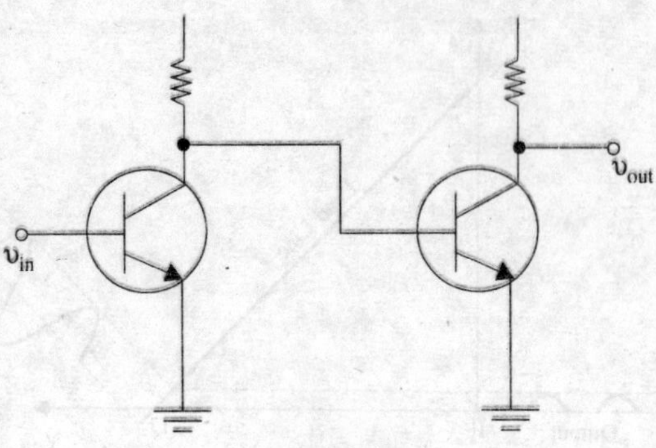

Fig. 33.21

33.18. RC-coupled Two-stage Amplifier

Fig. 33.22 shows a two-stage RC-coupled amplifier which consists of two single-stage transistor amplifiers using the CE configuration. The two resistors R_2 and R_3 and capacitor C_2 form the coupling network. R_2 is collector load of Q_1 and R_4 is that of Q_2. Capacitor C_1 couples the input signal whereas C_3 couples out the output

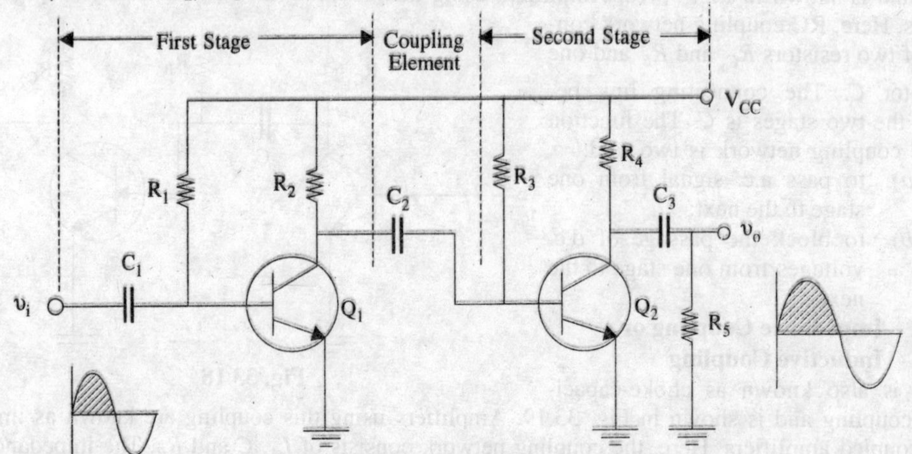

Fig. 33.22

signal. R_1 and R_2 provide d.c. base bias. R_5 is the load resistor across Q_2 .

Circuit Operation

The brief description of the circuit operation is as under:

1. the input signal v_i is amplified by Q_1. Its phase is reversed (usual with CE connection);
2. the amplified output of Q_1 appears across R_2;
3. the output of the first stage across R_2 is coupled to the input of the second stage at R_3 by coupling capacitor C_2. This capacitor is also sometimes referred to as blocking capacitor because it blocks the passage of d.c. voltages and currents;
4. thus the signal at the base of Q_1 is amplified by Q_2 and its *phase is further reversed;*
5. the a.c. output of Q_2 appears across R_4;
6. the output across R_4 is coupled by C_3 to load resistor R_5;
7. The output signal v_0 is the *twice-amplified replica of the input signal* υ_i. It is in phase with υ_i because its phase has been reversed twice.

33.19. Advantages of RC Coupling

1. It requires no expensive or bulky components and no adjustments. Hence, it is small, light and inexpensive.
2. Its overall amplification is higher than that of the other couplings.
3. It has minimum possible non-linear distortion because it does not use any coils or transformers which might pick up undesirable signals. Hence, there are no magnetic fields to interfere with the signal.

4. As shown in Fig. 33.23, it has a very flat frequency versus gain curve *i.e.*, it gives uniform voltage amplification over a wide range from a few hertz to a few megahertz because resistor values are independent of frequency changes.

As seen from Fig. 33.23, amplifier gain falls off at very low as well as at very high frequencies, the fall in gain (called *roll-off*) is due to capacitive reactance of the coupling capacitor between the two stages. The high-frequency roll-off is due to output capacitance of the first stage, input capacitance of the second stage and the stray capacitance.

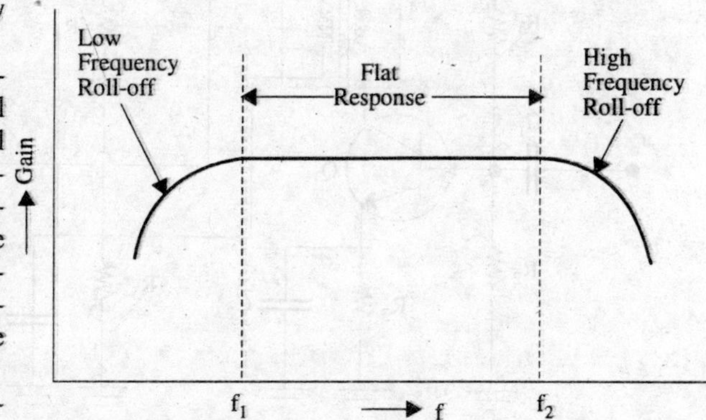

Fig. 33.23

The only drawback of this coupling is that due to large drop across collector load resistors, the collectors work at relatively small voltages unless higher supply voltage is used to overcome this large drop.

33.20. Impedance-coupled Two-stage Amplifier

The circuit is shown in Fig. 33.24. The coupling network consists of L, C_2 and R_3. The only basic difference between this circuit and the one shown in Fig. 33.22 is that inductor *L has replaced the resistor* R_2.

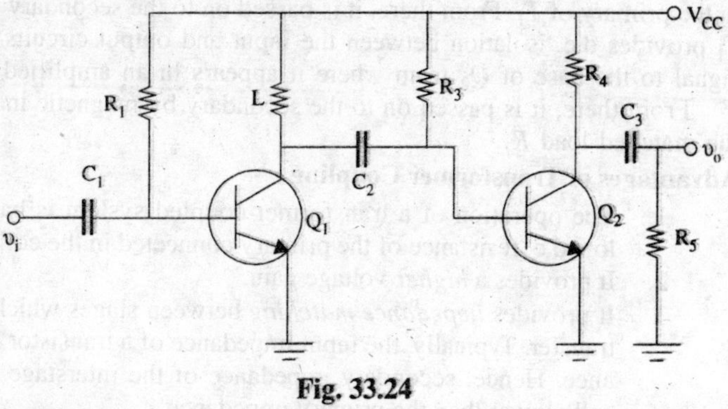

Fig. 33.24

Advantages and Disadvantage

The biggest advantage of this coupling is that there is hardly any d.c. drop across L so that low collector supply voltages can be used.

However, it has many disadvantages:

1. It is larger, heavier and costlier than RC coupling.
2. In order to prevent the magnetic field of the coupling inductor from affecting the signal, the inductor turns are wound on a closed core and are also shielded.
3. Since inductor impedance depends on frequency, the frequency characteristics of this coupling are not as good as those of RC coupling. The flat part of the frequency versus gain curve is small (Fig. 33.25).

At low frequencies, the gain is low due to large capacitance offered by the coupling capacitor just as in RC-coupled amplifiers. The gain *increases with frequency* till it levels off at the middle frequencies of the audio range.

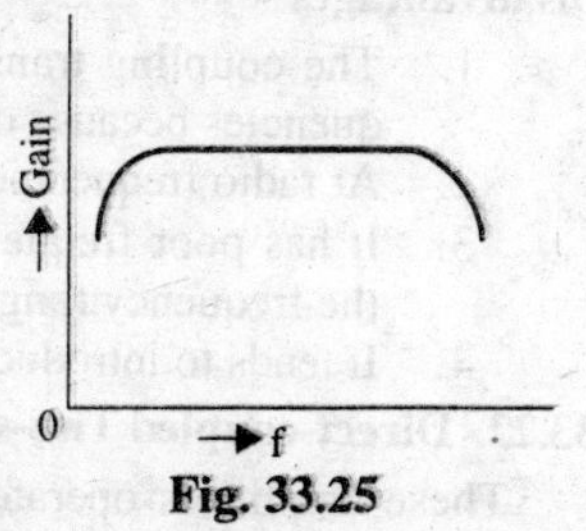

At relatively high frequencies, gain drops off again because of the increased reactance.

Fig. 33.25

Hence, impedance coupling is rarely used beyond audio range.

33.21. Transformer-coupled Two-stage Amplifier

The circuit for such a cascaded amplifier is shown in Fig. 33.26. T_1 is the coupling transformer whereas T_2 is the output transformer. C_1 is the input coupling capacitor whereas C_2, C_3 and C_4

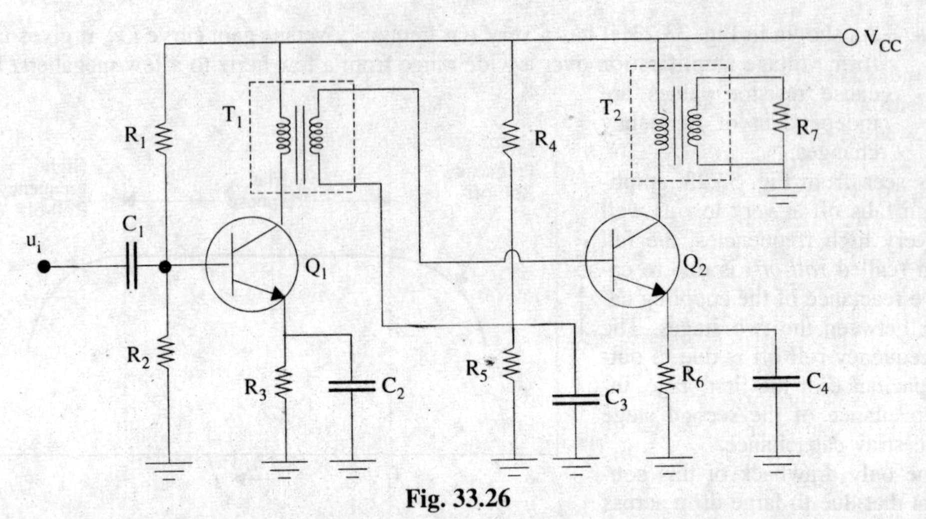

Fig. 33.26

are the bypass capacitors. Resistors R_1 and R_2 as well as R_4 and R_5 form voltage-divider circuit whereas R_3 and R_6 are the emitter-stabilizing resistors.

Circuit Operation

When input signal is coupled through C_1 to the base of Q_1, it appears in an amplified form in the primary of T_1. From there, it is passed on to the secondary by magnetic induction. Moreover, T_1 provides d.c. isolation between the input and output circuits. The secondary of T_1 applies the signal to the base of Q_2 from where it appears in an amplified form in the primary of T_2.

From there, it is passed on to the secondary by magnetic induction and finally appears across the matched load R_7.

Advantages of Transformer Coupling

1. The operation of a transformer-coupled system is basically *more efficient* because of low d.c. resistance of the primary connected in the collector circuit.
2. It provides a *higher* voltage gain.
3. It provides *impedance matching* between stages which is desirable for maximum power transfer. Typically, the input impedance of a transistor stage is less than its output impedance. Hence, secondary impedance of the interstage (or coupling) transformer is typically lower than the primary impedance.

This coupling is effective when the final amplifier output is fed to a low-impedance load. For example, the impedance of a typical loud-speaker varies from 4 Ω to 16 Ω whereas output impedance of a transistor stage is several hundred ohms. Use of an output audio transformer can avoid the bad effects of such a mismatch.

Disadvantages

1. The coupling transformer is costly and bulky particularly when operated at audio frequencies because of its heavy iron core.
2. At radio frequencies, the inductance and winding capacitance present lot of problems.
3. It has poor frequency response because the transformer is frequency sensitive. Hence, the frequency range of the tranformer-coupled amplifiers is limited.
4. It tends to introduce 'hum' in the output.

33.22. Direct-coupled Two-stage Amplifier

These amplifiers operate without the use of frequency-sensitive components like capacitors, inductors and transformers etc. They are especially suited for amplifying:

1. a.c. signals with frequencies as low as a fraction of a hertz;
2. changes in d.c. voltages.

Fig. 33.27 shows the circuit of such an amplifier which uses similar transistors each connected in the CE mode. Both stages employ direct coupling (*i*) collector of Q_1 is connected directly to the base of Q_2 and (*ii*) load resistor R_2 is connected directly to the collector of Q_2. The resistor R_1 establishes the forward bias of Q_1 and also indirectly that of Q_2.

Any signal current at the base of Q_1 is amplified β_1 times and appears at the collector of Q_1 and becomes base signal for Q_2. Hence, it is further amplified β_2 times by Q_2. Obviously, signal current gain of the cascaded amplifier is

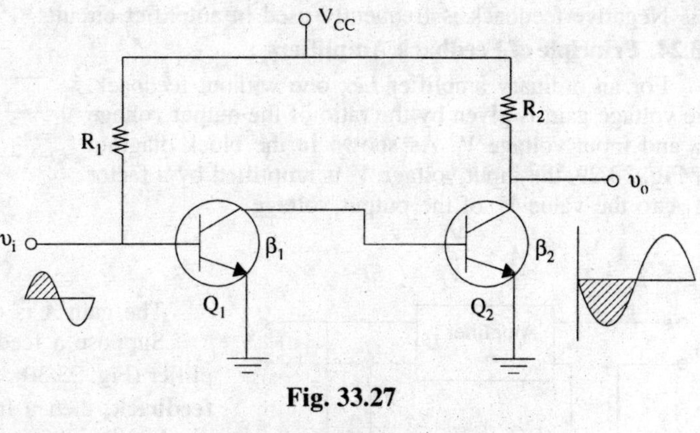

Fig. 33.27

$$A_i = \beta_1 \times \beta_2 = \beta^2 \quad \text{— if transistors are identical}$$

Advantages

1. The circuit arrangement is very simple since it uses minimum number of components.
2. It is quite inexpensive.
3. It has the outstanding ability to amplify direct current (*i.e.*, as d.c. amplifier) and low-frequency signals.
4. It has no coupling or by-pass capacitors to cause a drop in gain at low frequencies. As seen in Fig. 33.28, the frequency-response curve is flat upto upper cut-off frequency determined by stray wiring capacitance and internal transistor capacitance.

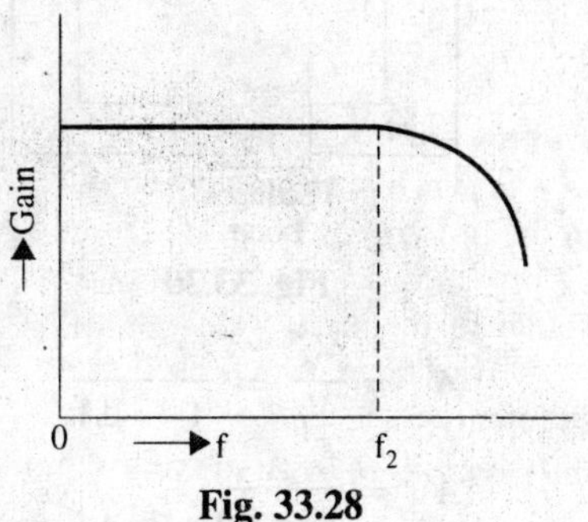

Fig. 33.28

Disadvantages

1. It cannot amplify high-frequency signals.
2. It has poor temperature stability.

It is due to the fact that any variation in base current (due to temperature changes) in one stage is amplified in the following stage (or stages) thereby shifting the *Q*-point. However, stability can be improved by using emitter-stabilizing resistors.

Applications

Some of the applications of direct-coupled amplifiers are in:

1. regulator circuits of electronic power supplies,
2. pulse amplifiers,
3. differential amplifiers,
4. computer circuitry,
5. electronic instruments.

33.23. Feedback Amplifiers

A feedback amplifier is that in which *a fraction* of the amplified output is feedback to the input circuit. This partial dependence of amplifier input on its output helps to control the output. A feedback amplifier consists of two part : an amplifier and a feedback circuit.

(*i*) Positive feedback

If the feedback voltage (or current) is so applied as to increase the input voltage (*i.e.*, it is in phase with it), then it is called *positive feedback.* Other names for it are : *regenerative* or *direct* feedback.

Since positive feedback produces excessive distortion, it is seldom used in amplifiers. However, because it increases the power of the original signal, it is used in *oscillator circuits.*

(*ii*) Negative feedback

If the feedback voltage (or current) is so applied as to reduce the amplifier input (*i.e.*, it is 180° out of phase with it), then it is called *negative feedback.* Other names for it are : *degenerative* or *inverse* feedback.

Negative feedback is frequently used in amplifier circuits.

33.24. Principle of Feedback Amplifiers

For an ordinary amplifier *i.e.*, one without feedback, the voltage gain is given by the ratio of the output voltage V_0 and input voltage V_i. As shown in the block diagram of Fig. 33.29, the input voltage V_i is amplified by a factor of A to the value V_o of the output voltage

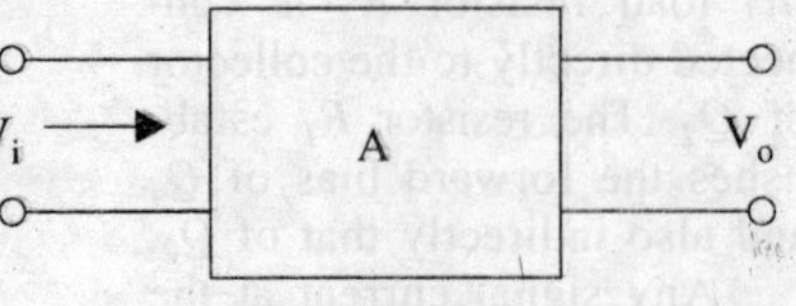

Fig. 33.29

$$\therefore \quad A = \frac{V_o}{V_i}$$

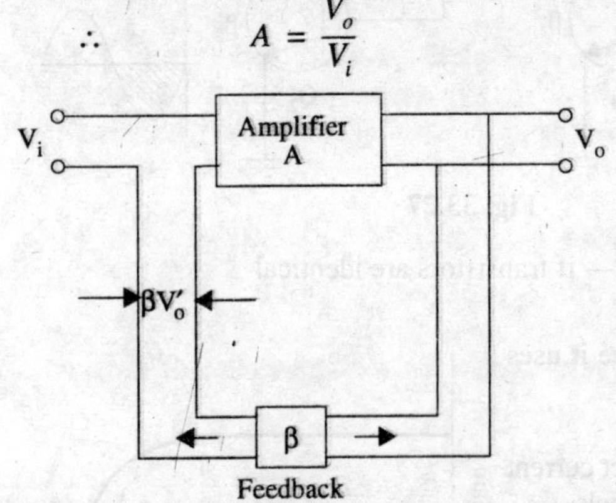

Fig. 33.30

The gain A is often called **open-loop** gain.

Suppose a feedback loop is added to the amplifier (Fig. 33.30). If V_o' is the output voltage **with feedback**, then a fraction β of this voltage is applied to the input voltage which, therefore, becomes $(V_i \pm \beta V_o')$ depending on whether the feedback voltage is in phase or antiphase with it. Assuming positive feedback, the input voltage will become $(V_i + \beta V_o')$. When amplified A times, it becomes $A(V_i + \beta V_o')$.

$$\therefore \quad A(V_i \pm \beta V_o') = V_o' \quad \text{or} \quad V_o'(1 - \beta A) = AV_i$$

The amplifier gain A' **with feedback** is given by

$$A' = \frac{V_o'}{V_i} = \frac{A}{1 - \beta A} \qquad \text{— from above}$$

$$\therefore \quad A' = \frac{A}{1 - \beta A} \qquad \text{— positive feedback}$$

$$\text{and} \quad A' = \frac{A}{1 - (-\beta A)} = \frac{A}{1 + \beta A} \qquad \text{— negative feedback}$$

The term 'βA' is called **feedback factor** whereas β is known as **feedback ratio**. The expression $(1 \pm \beta A)$ *is called* **loop gain.** The amplifier gain A' with feedback is also referred to as **closed loop gain** because it is the gain obtained after the feedback loop is closed.

(a) Negative Feedback

The amplifier gain with negative feedback is given by

$$A' = \frac{A}{1 + \beta A}$$

Obviously, $A' < A$ because $|1 + \beta A| > 1$. Suppose, $A = 90$ and $\beta = 1/10 = 0.1$. Then, gain *without* feedback is 90 and with feedback is

$$A' = \frac{A}{1 + \beta A} = \frac{90}{1 + 0.1 \times 90} = 9$$

As seen, negative feedback *reduces the amplifier gain.* That is why it is called *degenerative* feedback. A lot of voltage gain is sacrificed due to negative feedback. when $|\beta A| \gg 1$,

$$A' \cong \frac{A}{\beta A} = \frac{1}{\beta}$$

It means that A' depends only on β. But it is very stable because it is not affected by changes in temperature, device parameters, supply voltage and from the ageing of circuit components etc. Since resistors can be selected very precisely with almost zero temperature-coefficient of resistance, it is possible to achieve a highly precise and stable gain with negative feedback.

(b) Positive Feedback

The amplifier gain with positive feedback is given by

$$A' = \frac{A}{1 - \beta A}$$

Since $| 1 - \beta A | < 1$, $A' > A$

Suppose gain without feedback is 90 and $\beta = 1/100 = 0.01$, then gain with positive feedback is

$$A' = \frac{90}{1 - (0.01 \times 90)} = 900$$

Since positive feedback *increases* the amplifier gain, it is called *regenerative* feedback. If $\beta A = 1$, then mathematically, the gain becomes infinite which simply means that there is an output without any input. However, electrically speaking, this cannot happen. What actually happens is that the amplifier becomes an oscillator which supplies its own input. In fact, the two important and necessary conditions for circuit oscillation are:

1. the feedback must be positive,
2. feedback factor must be unity *i.e.*, $| \beta A | = 1$.

33.25. Advantages of Negative Feedback

The numerous advantages of negative feedback outweigh its only disadvantage of reduced gain.

Among the advantages are :

1. higher fidelity *i.e.,* more linear operation,
2. highly stabilized gain,
3. increased bandwidth *i.e.,* improved frequency response,
4. less distortion,
5. reduced noise,
6. input and output impedances can be modified as desired.

Example 33.4. *In the series-parallel (SP) feedback amplifier of Fig. 33.31, calculate*

(a) open-loop gain of the amplifier *(b) gain of the feedback network*

(c) closed-loop gain of the amplifier *(d) sacrifice factor, S*

(Electronic Engg. - I, Osmania Univ. 1990)

Solution. (*a*) Since 1 mV goes into the amplifier and 10 V come out

$$\therefore \quad A = \frac{10\text{ V}}{1\text{ mV}} = \mathbf{10{,}000}$$

(*b*) The feedback network is being driven by the output voltage of 10 V.

$\therefore$ gain of feedback network

$$= \frac{\text{output}}{\text{input}} = \frac{250\text{ mV}}{10\text{ V}} = \mathbf{0.025}$$

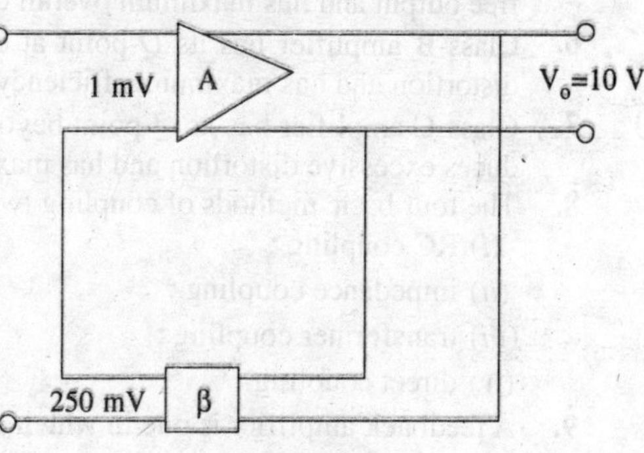

Fig. 33.31

(*c*) So far as the feedback amplifier is concerned, input is (250 + 1) = 251 mV and final output is 10 V. Hence, gain *with feedback* is

$$A' = \frac{10\text{ V}}{251\text{ mV}} = \mathbf{40}$$

(*d*) The sacrifice factor is given by

$$S = \frac{A}{A'} = \frac{10{,}000}{40} = \mathbf{250}$$

By sacrificing so much voltage gain, we have improved many other amplifier quantities (Art. 33.25).

Example 33.5. *Calculate the gain of a negative feedback amplifier whose gain without feedback is 1000 and $\beta = 1/10$.*

Solution. Since $| \beta A | > 1$, the closed-loop gain is

$$A' = \frac{1}{\beta} = \frac{1}{1/10} = \mathbf{10.}$$

Example. 33.6. *An amplifier has a gain of 360° ∠ 0°. When negative feedback is applied, the gain reduces to 240° ∠ 0°. Determine the feedback ratio.* **(Electronics, A.M.Ae.S.I. Dec. 1994)**

Solution. The amplifier gain with negative feedback is given by the expression

$$A' = \frac{A}{1 + \beta A}$$

$$\therefore \quad 240 \angle 0 = \frac{360 \angle 0}{1 + \beta \times 360 \angle 0}, \quad \beta = \mathbf{\frac{1}{720}}$$

Example 33.7. *In a negative feedback amplifier, A = 100, β = 0.04 and V_i = 50 mV. Find*
(a) gain with feedback *(b) output voltage*
(c) feedback factor *(d) feedback voltage*

(Basic Electronics, Pune Univ. 1993)

Solution. (*a*) $A' = \frac{A}{1 + \beta A} = \frac{100}{1 + 0.04 \times 100} = \mathbf{20}$

(*b*) $V_o' = A'V_i = 20 \times 50\ \text{mV} = \mathbf{1\ V}$

(*c*) feedback factor $= \beta A = 0.04 \times 100 = \mathbf{4}$

(*d*) feedback voltage $= \beta V_o' = 0.04 \times 1 = \mathbf{0.04\ V}$

HIGHTLIGHTS

1. CB and CC amplifiers do not invert the input signal whereas CE amplifier does.
2. CB circuit has highest input impedance but lowest output impedance.
3. CE circuit has the highest power gain.
4. CC circuit has highest current gain and input impedance but lowest power gain and output impedance.
5. Class-A amplifier has a centred *Q*-point, a conduction angle of 360°, gives distortion-free output and has maximum overall efficiency of 25%.
6. Class-B amplifier has its *Q*-point at cut-off, conduction angle of 180°, produces large distortion and has maximum efficiency of about 78.5%.
7. Class-C amplifier has its *Q*-point beyond cut-off, conduction angle of 120° – 150°, produces excessive distortion and has maximum efficiency of about 85 – 90%.
8. The four basic methods of coupling two amplifiers are :
   (*i*) RC coupling ;
   (*ii*) impedance coupling ;
   (*iii*) transformer coupling ;
   (*iv*) direct coupling.
9. A feedback amplifier is one in which a fraction of the amplifier output is feedback to the input circuit.
10. Positive feedback is used in oscillator circuits whereas negative feedback is employed in amplifier circuits.
11. Negative feedback amplifiers have highly stabilized gain, less distortion, high fidelity and increased bandwidth.

OBJECTIVE TESTS—33

A. Fill in the following blanks.

1. A CB amplifier produces no phase ...... in the input signal.
2. A CE amplifier produces a phase reversal of ...... degrees in the input signal.
3. In a CC amplifier, the input and output signals are in ...... with each other.
4. As compared to other amplifiers, a CE amplifier has the ...... power gain.
5. The maximum possible overall efficiency

of a class-A amplifier is ...... per cent.

6. The class-C amplifier produces high ...... in the output signal.
7. Negative feedback is commonly used in ...... circuits.
8. The ratio of the amplifier gain without feedback and with feedback is called ...... factor.

B. Answer True or False.

1. A CC amplifier has highest current gain but lowest power gain.
2. A class-B amplifier conducts only when there is input signal.
3. In amplifiers, high efficiency and high distortion go together.
4. Class-A amplifiers operate with centred *Q*-point and large input signals.
5. Class-C operation is most efficient because it hardly operates during the input cycle.
6. An RC-coupled amplifier has a very flat frequency response curve in the audic range.
7. Negative feedback always reduces the amplifier gain.
8. Negative feedback in amplifiers provides gain stability and higher fidelity.

C. Multiple Choice Questions.

1. A CE amplifier is characterised by
   (*a*) low voltage gain
   (*b*) moderate power gain
   (*c*) signal phase reversal
   (*d*) very high output impedance
2. In a CC amplifier, voltage gain
   (*a*) cannot exceed unity
   (*b*) depends on output impedance
   (*c*) is dependent on input signal
   (*d*) is always constant
3. The main use of a class-C amplifier is
   (*a*) as an RF amplifier.
   (*b*) as stereo amplifier
   (*c*) in communication sound equipment
   (*d*) as distortion generator
4. Direct coupling in amplifiers is especially suited for amplifying.
   (*a*) high frequency a.c. signals
   (*b*) changes in d.c. voltage
   (*c*) high-level voltage
   (*d*) sinusoidal signals
5. Feedback in an amplifier always helps to
   (*a*) control its output
   (*b*) increase its gain
   (*c*) decrease its input impedance
   (*d*) stabilize its gain
6. Closed-loop gain of a feedback amplifier is the gain obtained when
   (*a*) its output terminals are closed
   (*b*) negative feedback is applied
   (*c*) feedback loop is closed
   (*d*) feedback factor exceeds unity

ANSWERS

A. 1. reversal 2. 180 3. phase 4. highest 5. 25 6. distortion 7. amplifier 8. sacrifice

B. 1. T 2. T 3. T 4. F 5. T 6. T 7. T 8. T

C. 1. *c* 2. *a* 3. *a* 4. *b* 5. *a* 6. *c*

34 FIELD EFFECT TRANSISTORS

34.1. What is a FET?

The acronym 'FET' stands for **field effect transistor**. It is a three-terminal, unipolar, solid-state device in which current is controlled *by an electric field* as is done in vacuum tube. Broadly speaking, there are two types of FETs:

(*a*) junction field effect transistor (JFET)

(*b*) metal-oxide semi-conductor FET (MOSFET)

It is also called insulated-gate FET (IGFET). It may be further sub-divided into:

(*i*) depletion-enhancement MOSFET *i.e.*, DE MOSFET

(*ii*) enhancement-only MOSFET *i.e.*, E-only MOSFET

Both of these can be either P-channel or N-channel devices.

The FET family tree is shown in Fig. 34.1.

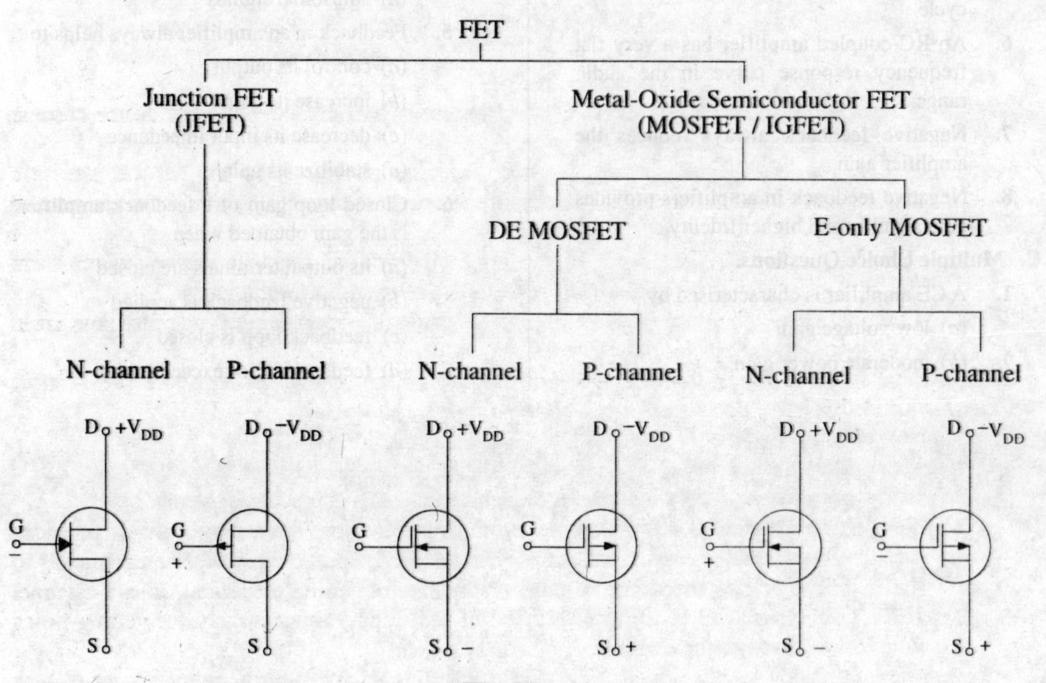

Fig. 34.1

34.2. Junction FET (JFET)

(*a*) Basic Construction

As shown in Fig. 34.2, it can be fabricated with either an N-channel or P-channel though N-channel is generally preferred. For fabricating N-channel JFET, first a narrow bar of N-type semi-conductor material is taken and then two P-type junctions are diffused on opposite sides of its middle part [Fig. 34.2 (*a*)]. These junctions form two P-N diodes or **gates** and the area between these gates is called **channel**. The two P-regions are internally connected and a single lead is brought out which is called **gate terminal**. Ohmic contacts (direct electrical connections) are made at the two ends of the bar, one lead is called **source terminal** *S* and the other **drain terminal** *D*. When potential difference is established between drain and source, current flows along the length of the 'bar' through the channel located between the two P-regions. The current consists of **only majority carriers** which, in the present case, are electrons. P-channel JFET is similar in construction except that it uses P-type bar and two N-type junctions. The majority carriers are holes which flow through the channel located between the two N-regions or gates.

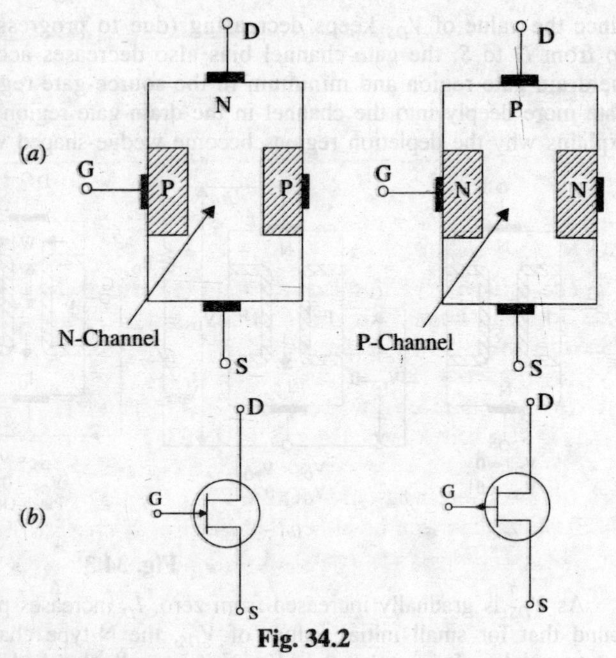

Fig. 34.2

Following FET notation is worth remembering:

1. **Source**. It is the terminal through which majority carriers *enter* the bar. Since carriers originate from it, it is called the source.
2. **Drain.** It is the terminal through which majority carriers *leave* the bar *i.e.*, they are drained out from this terminal. The drain-to-source voltage V_{DS} drives the drain current I_D.
3. **Gate.** These are two internally-connected heavily-doped impurity regions which form two P-N junctions. The gate-source voltage V_{GS} reverse-biases the gates.
4. **Channel.** It is the space between two gates through which majority carriers pass from source-to-drain when V_{DS} is applied.

Schematic symbols for N-channel and P-channel JFET, are shown in Fig. 34.2 (*b*). It must be kept in mind that **gate arrow always points to N-type material.**

(*b*) Theory of Operation

While discussing the theory of operation of a JFET, it should be kept in mind that

1. Gates are always reverse-biased. Hence, gate current I_G is practically zero.
2. The source terminal is always connected to that end of the drain supply which provides the necessary charge carriers. In an N-channel JFET, source terminal *S* is connected to the negative end of the drain voltage supply (for obtaining electrons). In a P-channel JFET, *S* is connected to the positive end of the drain voltage supply for getting holes which flow through the channel.

Let us now consider an *N-channel* JFET and discuss its working when either V_{GS} or V_{DS} or both are changed.

(*i*) When $V_{GS} = 0$ and $V_{DS} = 0$

In this case, drain current $I_D = 0$ because $V_{DS} = 0$. The depletion regions around the P-junctions are of equal thickness and symmetrical as shown in Fig. 34.3 (*a*).

(*ii*) When $V_{GS} = 0$ and V_{DS} is increased from zero

For this purpose, the JFET is connected to the V_{DD} supply as shown in Fig. 34.3 (*b*). The electrons (which are the majority carriers) flow from *S* to *D* whereas conventional drain current I_D flows through the channel from *D* to *S*. Now, the gate-to channel bias at any point along the channel is = $|V_{DS}| + |V_{GS}|$ *i.e.*, the numerical sum of the two voltages. In the present case, external bias $V_{GS} = 0$. Hence, gate-channel reverse bias is *provided by V_{DS} alone.*

Since the value of V_{DS} keeps decreasing (due to progressive drop along the channel) as we go from D to S, the gate-channel bias also decreases accordingly. It has maximum value in the drain-gate region and minimum in the source-gate region. Hence, depletion regions penetrate more deeply into the channel in the drain-gate region than in the source-gate region. This explains why the depletion regions become wedge-shaped when V_{DS} is applied [Fig. 34.3 (*b*)].

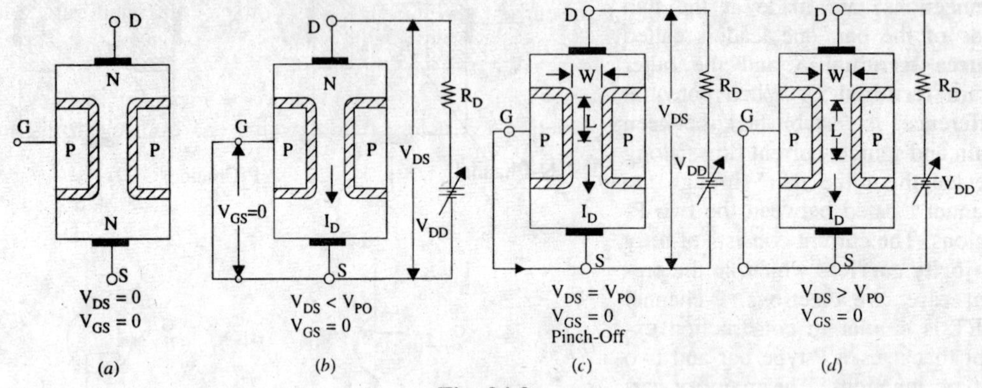

Fig. 34.3

As V_{DS} is gradually increased from zero, I_D increases proportionally as per Ohm's law. It is found that for small initial values of V_{DS}, the N-type channel material acts like a resistor of constant value. It is so because V_{DS} being small, the depletion regions are not large enough to have any significant effect on channel cross-section and, hence, its resistance. Consequently, I_D increases linearly as V_{DS} is increased from zero onwards (Fig. 34.6).

This ohmic relationship between V_{DS} and I_D continues till V_{DS} reaches a certain critical value called *pinch-off* voltage V_{PO} when drain current becomes constant at its maximum value called I_{DSS}. The *SS* in I_{DSS} indicates that the gate is shorted to source to make sure that $V_{GS} = 0$. This current is also known as *zero-gate-voltage drain current*. It is seen from Fig. 34.3 (*c*) that under pinch-off conditions, separation between the depletion regions near the drain end reaches a minimum value *W*. It should, however, be carefully noted that *pinch-off does not mean 'current-off'*. In fact, I_D is maximum at pinch-off.

When V_{DS} is increased beyond V_{PO}, I_D remains constant at its maximum value I_{DSS} upto a certain point. It is due to the fact that further increase in V_{DS} (beyond V_{PO}) causes more of the channel on the source end to reach the minimum width as shown in Fig. 34.3 (*d*). It means that channel width does not increase, instead its length *L* increases. As more of the channel reaches the minimum width, the resistance of the channel *increases at the same rate at which* V_{DS} *increases*. In other words, increase in V_{DS} is neutralized by increase in R_{DS}. Consequently, I_D (= V_{DS} / R_{DS}) remains unchanged even though V_{DS} is increased. Ultimately, a certain value of V_{DS} (called V_{DSO}) is reached when JFET breaks down and I_D increases to an excessive value as seen from drain characteristic of Fig. 34.6.

(*iii*) When V_{DS} = 0 and V_{GS} is decreased from zero

In this case, as V_{GS} is made more and more negative, the gate reverse bias increases which increases the thickness of the depletion regions. As negative value of V_{GS} is increased, a stage comes when the two depletion regions touch each other as shown in Fig. 34.4. In this condition, the channel is said to be *cut-off*. This value of V_{GS} which outs off the channel and hence the drain current is called $V_{GS(off)}$.*

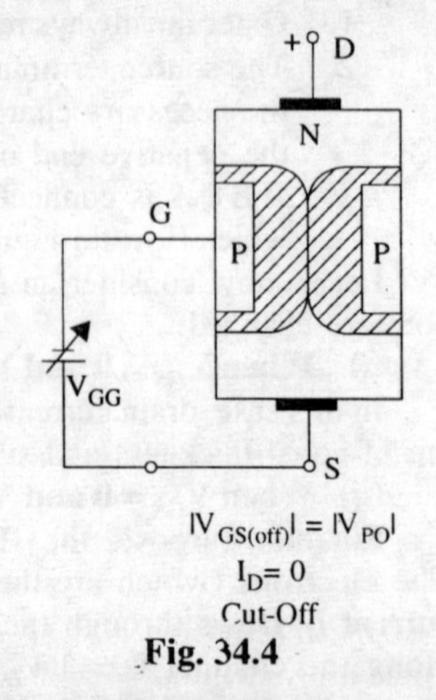

Fig. 34.4

It may be noted that $V_{GS(off)} = -V_{PO}$ or $|V_{PO}| = |V_{GS(off)}|$.

*It has negative value for an N-channel JFET but a positive value for a P-channel JFET.

As seen from Fig. 34.7 because V_{PO} = 4 V, $V_{GS(off)}$ = – 4 *V*. Obviously, their absolute values are equal.

***(iv)* When V_{GS} is negative and V_{DS} is increased**

As seen from Fig. 34.7, as V_{GS} is made more and more negative, values of V_P as well as breakdown voltage V_B are decreased.

Since gate voltage controls the drain current, JFET is called a voltage-controlled device. A P-channel JFET operates exactly in the same manner as an N-channel JFET except that current carriers are holes and polarities of both V_{DD} and V_{GS} are reversed.

Since only one type of majority carrier (either electron or hole) is used in JFETs, they are called *unipolar devices* unlike bipolar junction transistors (BJTs) which use both electrons and holes as carriers (and, hence, are called *bipolar devices*).

34.3. Static Characteristics of a JFET

We will consider the following two characteristics:

***(i)* drain characteristic**

It gives relation between I_D and V_{DS} for different values of V_{GS} (which is called *running variable*).

***(ii)* transfer characteristic**

It gives relation between I_D and V_{GS} for different values of V_{DS}.

We will analyse these characteristics for an N-channel JFET connected in the common-source mode as shown in Fig. 34.5. We will first consider the drain characteristic when V_{GS} = 0 and then when V_{GS} has any negative value upto $V_{GS(off)}$.

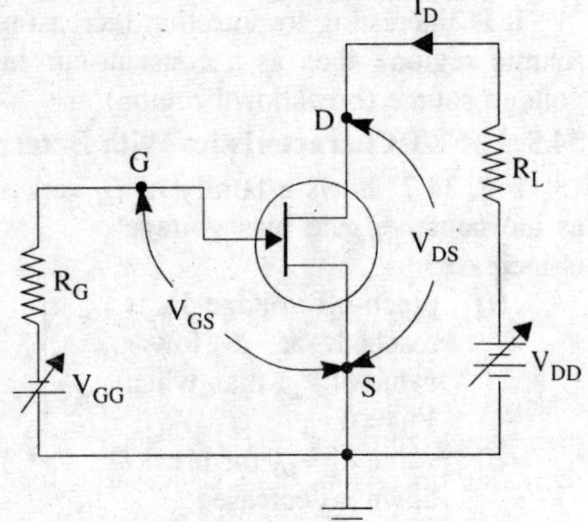

Fig. 34.5

34.4. JFET Drain Characteristic with V_{GS} = 0

Such a characteristic is shown in Fig. 34.6 and has been already discussed briefly in Art. 34.2. It can be sub-divided into following four regions:

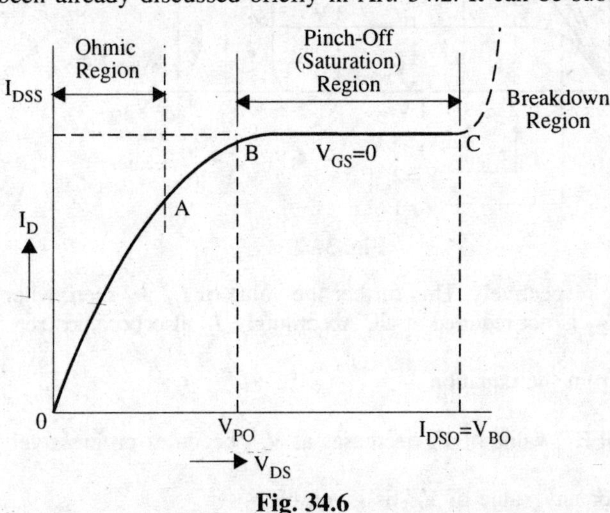

Fig. 34.6

1. Ohmic-Region OA

This part of the characteristic is linear indicating that for low values of V_{DS}, current varies directly with voltage following Ohm's Law. It means that JFET behaves like an ordinary resistor till point *A* (called *knee*) is reached.

2. Curve AB

In this region, I_D increases at inverse square-law rate upto point *B* which is called *pinch-off point*. This progressive fall in the rate of increase of I_D is caused by the square law increase in the depletion region at each gate upto point *B* where the two regions are closest without touching each other. The drain-to-source voltage V_{DS} corresponding to point *B* is called *pinch-off voltage* V_{PO} (the subscript *O* indicates that V_{GS} = 0). But it is essential to remember that '*pinch-off' does not mean 'current-off*'.

3. Pinch-off Region BC

It is also known as **saturation region** or **'amplifier' region.** Here, JFET operates as a **constant-current** device because I_D is relatively independent of V_{DS}. It is due to the fact that as V_{DS} increases, channel resistance also increases proportionally thereby keeping I_D practically constant at I_{DSS}. It should also be noted that the reverse bias required by the gate-channel junction is supplied entirely by V_{DS} (Art. 34.2) and none by the external bias because $V_{GS} = 0$.

Drain current in this region is given by Shockley's equation

$$I_D = I_{DSS}\left(1 - \frac{V_{GS}}{V_{PO}}\right)^2 = I_{DSS}\left(1 - \frac{V_{GS}}{V_{GS(off)}}\right)^2$$

It is the normal operating region of the JFET when used as an amplifier.

4. Breakdown Region

If V_{DS} is increased beyond its value corresponding to point *C* (called *avalanche breakdown voltage*), JFET enters the breakdown region where I_D increases to an excessive value. This happens because the reverse-biased gate-channel P-N junction undergoes avalanche breakdown when even small changes in V_{DS} produce very large changes in I_D.

It is interesting to note that increasing values of V_{DS} make a JFET behave first as a resistor (ohmic region), then as a constant-current source (pinch-off region) and finally, as a constant-voltage source (breakdown region).

34.5. JFET Characteristics With External Bias

Fig. 34.7 shows a family of I_D versus V_{DS} curves for different values of V_{GS}. It is seen that as the negative gate bias voltage is increased:

(*i*) pinch-off voltage V_P is reached at a lower value of V_{DS} than when $V_{GS} = 0$,

(*ii*) value of V_{DS} for breakdown is decreased.

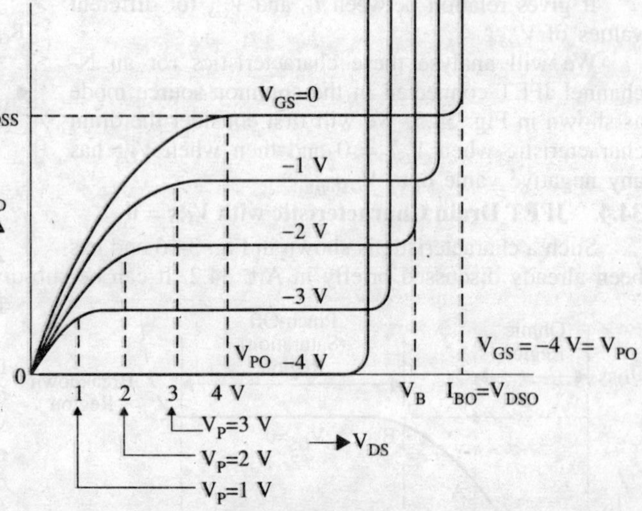

Fig. 34.7

It is seen that with $V_{GS} = 0$, I_D saturates at I_{DSS} and the characteristic shows $V_{PO} = 4$ V. When an external bias of –1 V is applied, gate-channel junction still requires – 4V to achieve pinch-off (remember $|V_{GS}| = |V_{PO}|$). It means that a value of $V_{DS} = 3$V is now required instead of the previous 4V. Consequently, the value of I_D is reduced. Similarly, when V_{GS} is –2 V and –3V, pinch-off is achieved with V_{DS} equal to 2V and 1 V respectively. This further the value of I_D. As seen, when $V_{GS} = -4$ V (*i.e.*, numerically = V_{PO}), V_{DS} is not required at all. Accordingly, I_D also becomes zero.

The above fact could be verified from the equation

$$V_{DS(P)} = V_P = V_{PO} + V_{GS}$$

Since V_{PO} is constant for a given JFET, value of V_P decreases as V_{GS} becomes progressively more negative.

The value of breakdown voltage for any value of V_{GS} is given by

$$V_B = V_{DSO} + V_{GS}$$

As seen, as V_{GS} becomes more and more negative, V_B keeps decreasing. Obviously, when $V_{GS} = 0$, $V_B = V_{BC} = V_{DSO}$.

34.6. Transfer Characteristic

It is a plot of I_D versus V_{GS} for a constant value of V_{DS} and is shown in Fig. 34.8. It is similar to the transconductance characteristic of a vacuum tube or a transistor. It is seen that when $V_{GS} = 0$, $I_D = I_{DSS}$ and when $I_D = 0$, $V_{GS} = V_{PO}$. The transfer characteristic approximately follows the equation

$$I_D = I_{DSS}\left(1-\frac{V_{GS}}{V_{PO}}\right)^2 = I_{DSS}\left(1-\frac{V_{GS}}{V_{GS(off)}}\right)^2$$

The above equation can be written as

$$V_{GS} = V_{GS(off)}\left(1-\sqrt{\frac{I_D}{I_{DSS}}}\right)$$

This characteristic can be obtained from the drain characteristics by reading off V_{GS} and I_{DSS} values for different values of V_{DS}.

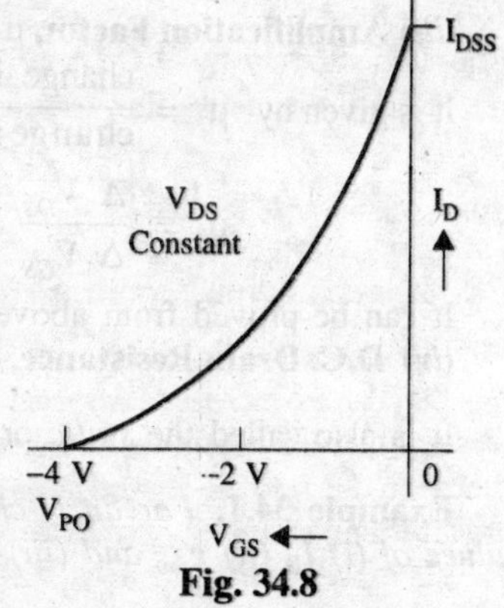

Fig. 34.8

34.7. Small Signal JFET Parameters

The various parameters of a JFET can be obtained from its two characteristics. The main parameters of a JFET when connected in common-source mode are as under:

(*i*) AC Drain Resistance r_d

It is the a.c. resistance between drain and source terminals when JFET is operating *in the pinch-off region*. It is given by

$$r_d = \frac{\text{change in } V_{DS}}{\text{change in } I_D} \quad \ldots V_{GS} \text{ constant} \quad \text{or} \quad r_d = \left.\frac{\Delta V_{DS}}{\Delta I_D}\right| V_{GS}$$

An alternative name is *dynamic drain resistance*. It is given by the slope of the drain characteristic in the pinch-off region. It is sometimes written as r_{ds} emphasizing the fact that it is the resistance from drain to source. Since r_d is usually the output resistance of a JFET, it may also be expressed as an output admittance y_{os}. Obviously, $y_{os} = 1 / r_d$. It has a very high value.

(*ii*) Transconductance, g_m

It is simply the slope of transfer characteristic.

$$g_m = \frac{\text{change in } I_D}{\text{change in } V_{GS}} \quad \ldots V_{DS} \text{ constant or} \quad g_m = \left.\frac{\Delta I_D}{\Delta V_{GS}}\right| V_{DS}$$

Its unit is siemens (*S*) earlier called *mho*. It is also called *forward transconductance* (g_{fs}) or *forward transadmittance* y_{fs}.

The transconductance measured at I_{DSS} is written as g_{mo}.

Mathematical Expression for g_m

The Shockley equation* is $I_D = I_{DSS}\left(1-\dfrac{V_{GS}}{V_P}\right)^2$

Differentiating both sides, we have

$$\frac{dI_D}{dI_{DSS}} = 2\,I_{DSS}\left(1-\frac{V_{GS}}{V_P}\right)\left(-\frac{1}{V_P}\right)$$

or $$g_m = -\frac{2\,I_{DSS}}{V_P}\left(1-\frac{V_{GS}}{V_P}\right)$$

When $V_{GS}=0$, $g_m = g_{mo}$ $\quad\therefore\quad g_{mo} = \dfrac{2\,I_{DSS}}{V_{PO}}$

From the above two equations, we have

$$g_m = g_{mo}\left(1-\frac{V_{GS}}{V_P}\right) = g_{mo}\sqrt{\frac{I_D}{I_{DSS}}}$$

---

*Because of the squared term in the equation, JFET and MOSFET are referred to as square law devices.

(*iii*) Amplification Factor, μ

It is given by $\mu = \dfrac{\text{change in } V_{DS}}{\text{change in } V_{GS}}$ — I_D constant

or $\mu = \dfrac{\Delta V_{DS}}{\Delta V_{GS}}\bigg|\, I_D$

It can be proved from above that $\mu = g_m \times r_d = g_{fs} \times r_d$

(*iv*) D.C. Drain Resistance, R_{DS}

It is also called the *static* or *ohmic* resistance of the channel. It is given by $R_{DS} = \dfrac{V_{DS}}{I_D}$.

Example 34.1. *For an N-channel JFET, I_{DSS} =8.7 mA, V_P = –3 V, V_{GS} = – IV. Find the values of (i) I_D (ii) g_{mo} and (iii) g_m.* **(Basic Electronics, Bombay Univ. 1989)**

Solution. (*i*) $I_D = I_{DSS}\left(1 - \dfrac{V_{GS}}{V_P}\right)^2 = 8.7\left(1 - \dfrac{-1\text{ V}}{-3\text{ V}}\right) =$ **3.87 mA**

(*ii*) $g_{mo} = \dfrac{-2I_{DSS}}{V_P} = \dfrac{-2 \times 8.7}{-3} =$ **5.8 mS**

(*iii*) $g_m = g_{mo}\left(1 - \dfrac{V_{GS}}{V_P}\right) = 5.8\left(1 - \dfrac{-1}{-3}\right) =$ **3.87 mS**

34.8. D.C. Biasing of a JFET

A JFET may be biased by using either

1. a separate power source V_{GG} as shown in Fig. 34.9 (*a*) or
2. some form of self-bias as shown in Fig. 34.9 (*b*) or
3. source bias as in Fig. 34.9 (*c*) or
4. voltage divider bias as in Fig. 34.9 (*d*).

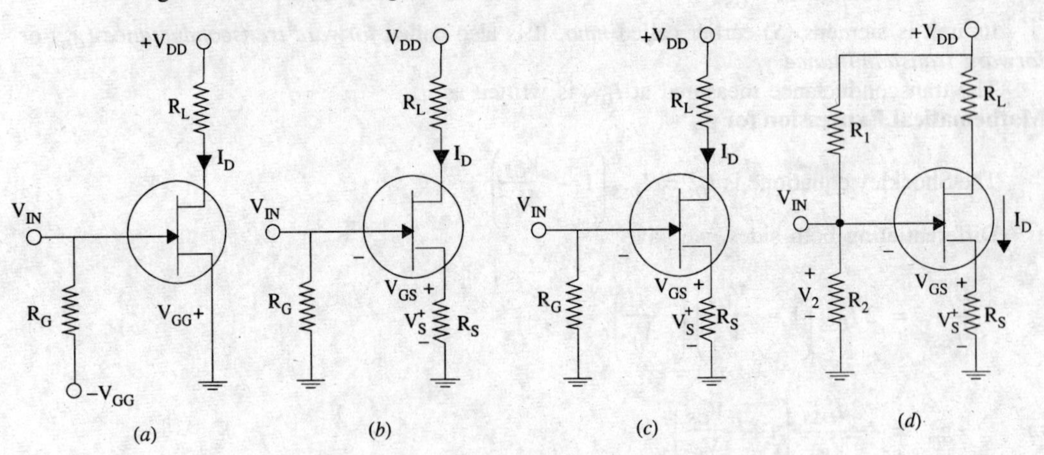

Fig. 34.9

The circuit of Fig. 34.9 (*b*) is called *self-bias circuit* because the V_{GS} bias is obtained from the flow of JFET's own drain current I_D through R_S.

∴ $V_S = I_D R_S$ and $V_{GS} = -I_D R_S$

The gate is kept at this much negative potential with respect to the ground.

The addition of R_G in Fig. 34.9 (*b*) does not upset this *d.c.* bias for the simple reason that no gate current flows through it (the gate leakage current is almost zero). Hence, gate is essentially at *d.c. ground*. Without R_G, gate would be kept 'floating' which could collect charge and ultimately cut-off the JFET.

The resistance R_G additionally serves the purpose of avoiding short-circuiting of the a.c. input voltage, v_{in}. Moreover, in case leakage current is not totally negligible, R_G would provide it an escape route. Otherwise, the leakage current would build up static charge (voltage) at the gate which could change the bias or even destroy the JFET.

Fig. 34.9 (*c*) shows the *source* bias circuit which employs a self-bias resistor R_S to obtain V_{GS}. Here, $V_{SS} = I_D R_S + V_{GS}$ or $V_{GS} = V_{SS} - I_D R_S$.

Fig. 34.9 (*d*) shows the familiar voltage divider bias. In this case, $V_2 = V_{GS} + I_D R_S$ or $V_{GS} = V_2 - I_D R_S$.

Since, $V_2 = V_{DD}\dfrac{R_2}{R_1 + R_2}$ $\quad\therefore\quad V_{GS} = V_{DD}\dfrac{R_2}{R_1 + R_2} - I_D R_S$

Example 34.2 *Find the values of V_{DS} and V_{GS} in Fig. 34.10 for $I_D = 4$ mA.*

(Applied Electronics-I, Punjab Univ. 1991)

Solution. $V_S = I_D R_S = 4 \times 10^{-3} \times 500 = 2.0$ V

$V_D = V_{DD} - I_D R_L = 12 - 4 \times 1.5 = 6$ V

$V_{DS} = V_D - V_S = 6 - 2 = 4$ V

Since $V_G = 0$, $V_{GS} = V_G - V_S$

$= 0 - 2.0 = \mathbf{-2.0\ V}$

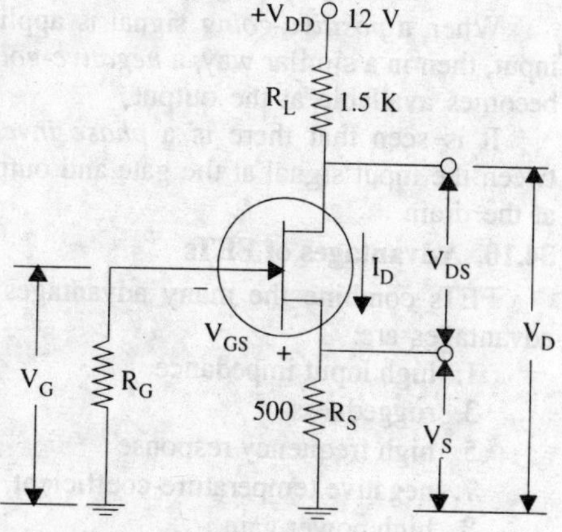

Fig. 34.10

Example 34.3. *Determine the quiescent value of V_{GS}, I_D and V_{DS} for the JFET circuit of Fig. 34.11 given that $I_{DSS} = 10$ mA, $R_S = 5$ K and $V_P = -5$ V.*

(Electronic Devices & Circuits, Pune Univ. 1988)

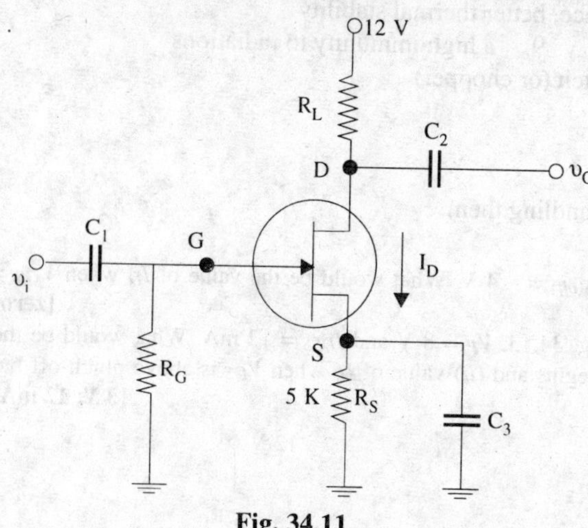

Fig. 34.11

Solution. Since $I_S \equiv I_D$,

$V_{GS} = -I_D R_S = -5000\ I_D$

Now, $I_D = I_{DSS}\left(1 - \dfrac{V_{GS}}{V_P}\right)^2$

$= 10 \times 10^{-3}\left(1 - \dfrac{V_{GS}}{-5}\right)^2$

$= 10 \times 10^{-3}\ (1 + 0.2\ V_{GS})^2$

Substituting this value in the above equation, we get

$V_{GS} = -5000\ (10 \times 10^{-3})\ (1 + 0.2\ V_{GS})^2$

Expanding and rearranging the above, we have

$2\ V_{GS}^2 + 21\ V_{GS} + 50 = 0$

$\therefore\quad V_{GS} = -3.65$ V or -6.85 V

Rejecting the higher value because it is more than V_P, we have $V_{GS} = \mathbf{-3.65\ V}$

$\therefore\quad -3.65 = -5000\ I_D \quad\therefore\quad I_D = \mathbf{0.73\ mA}$

$V_D = V_{DD} - I_D R_L = 12 - 0.73 \times 2 = 10.54$ V

$V_S = I_D R_S = 0.73 \times 5 = 3.65$ V

$\therefore\quad V_{DS} = V_D - V_S = 10.54 - 3.65 = \mathbf{6.89\ V}$

34.9. Common Source JFET Amplifier

A simple circuit for such an amplifier is shown in Fig. 34.12. Here, R_G serves the purpose of providing leakage path to the gate current, R_S develops gate bias, C_3 provides a.c. ground to the input signal and R_L acts as drain load.

Working

When *negative-going* signal is applied to the input

1. gate bias is *increased*
2. depletion regions are *widened*
3. channel resistance is *increased*
4. I_D is *decreased*
5. drop across R_L is *decreased*
6. consequently, a *positive-going* signal becomes available at the output through C_2 as shown in Fig. 34.12.

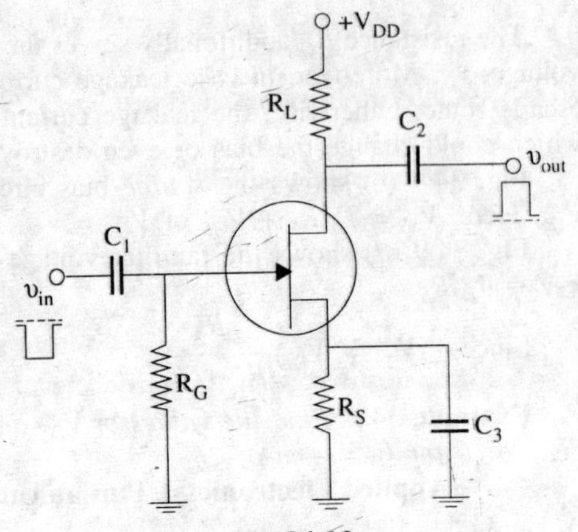

Fig. 34.12

When a *positive-going* signal is applied at the input, then in a similar way, a *negative-going* signal becomes available at the output.

It is seen that there is a *phase inversion* between the input signal at the gate and output signal at the drain.

34.10. Advantages of FETs

FETs combine the many advantages of both BJTs and vacuum tubes. Some of their main advantages are:

1. high input impedance
2. small size
3. ruggedness
4. long life
5. high frequency response
6. low noise
7. negative temperature-coefficient, hence, better thermal stability
8. high power gain
9. a high immunity to radiations
10. no offset voltage when used as a switch (or chopper)
11. square law characteristics.

The only disadvantages are :

1. small gain-bandwidth product
2. greater susceptibility to damage in handling them.

Tutorial Problems No. 34.1

1. For a particular *N*-channel JFET, $V_{GS(off)} = -4$ V. What would be the value of I_D when $V_{GS} = -6$ V? **[zero]**
2. For the *N*-channel JFET shown in Fig. 34.13, $V_P = 8$ V and $I_{DSS} = 12$ mA. What would be the value of (*i*) V_{DS} at which pinch-off begins and (*ii*) value of I_D when V_{DS} is above pinch-off but below the breakdown voltage? **[3 V; 12 mA]**

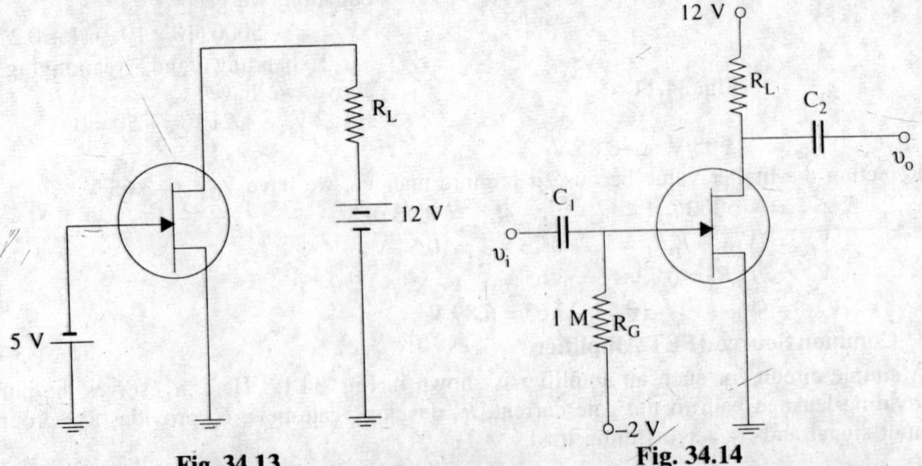

Fig. 34.13 Fig. 34.14

3. The data sheet of a JFET indicates that it has

$I_{DSS} = 15$ mA and $V_{GS(off)} = -5$ V.

Calculate the value of I_D when V_{GS} is (*i*) 0 (*ii*) –1 V and (*iii*) – 4 V.

[(*i*) 15 mA (*ii*) 9.6 mA (*iii*) 0.6 mA]

4. The data sheet of a JFET gives the following information:

$I_{DSS} = 20$ mA, $V_{GS}(off) = -8$V and $g_{mo} = 4000$ μS

Calculate the value of I_D and g_m for $V_{GS} = -4$V. **[5mA; 2000 μS]**

5. For a JFET, $I_{DSS} = 5$ mA and $g_{mo} = 4000$ μS.

Calculate (*i*) $V_{GS(off)}$ and (*ii*) g_m at mid-point bias. **[(*i*) –5 V (*ii*) 3000 μS]**

6. At a certain point on the transfer characteristics of an *N*-channel JFET, following values are read : $I_{DSS} = 8.4$ mA, $V_{GS} = -0.5$ V and $V_P = -3.0$ V.

Calculate (*i*) g_{mo} and (*ii*) g_m at that point. **[(*i*) 5600 μS (*ii*) 4670 μS]**

7. For the JFET circuit of Fig. 34.14, $I_{DSS} = 9$ mA and $V_P = -3$ V.

Find the value of R_L for setting the value of V_{DS} at 7 V. **[5 K]**

(**Hint** : $V_{GS} = -2$ V as $I_G = 0$)

34.11. MOSFET or IGFET

It could be further sub-divided as follows:

(*i*) Depletion-enchancement MOSFET or DE MOSFET

This MOSFET is so-called because it can be operated in both *depletion* mode and *enhancement* mode by changing the polarity of V_{GS}. When negative gate-to-source voltage is applied, the *N*-channel DE MOSFET operates in the depletion mode [Fig. 34.16 (*a*)]. However, with positive gate voltage,

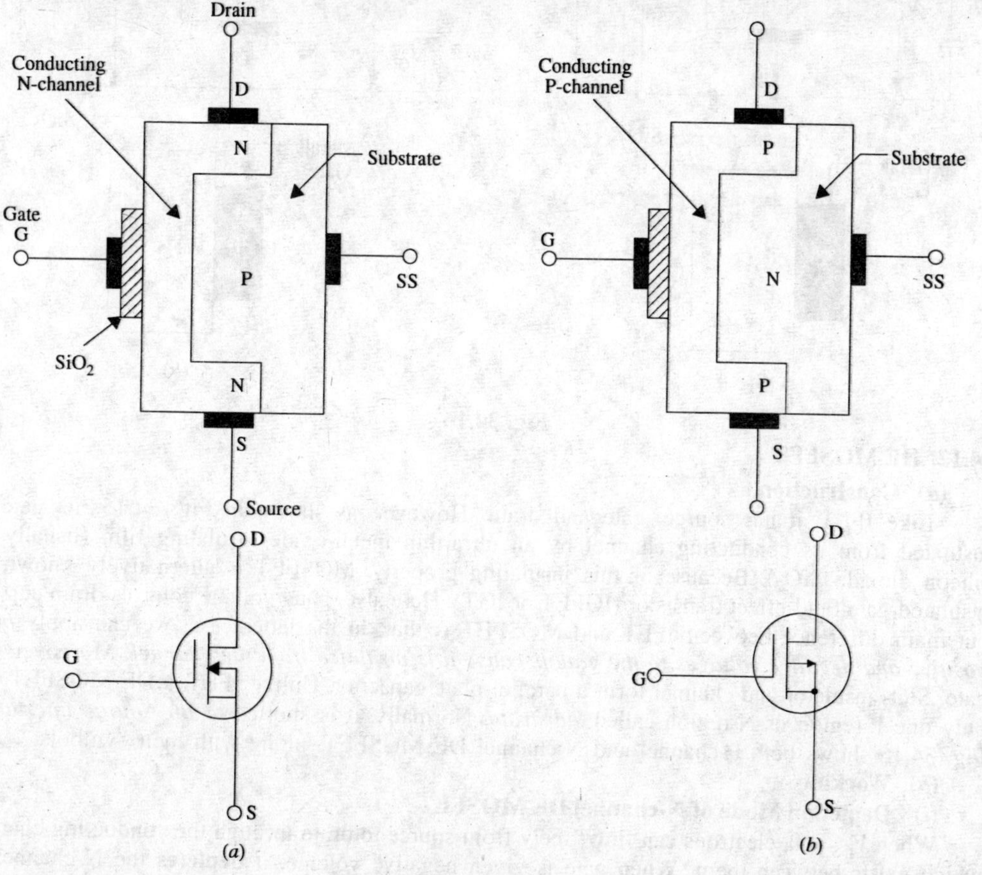

Fig. 34.15

it operates in the enhancement mode [Fig. 34.16 (*b*)]. Since a channel exists between drain and source, I_D flows even when $V_{GS} = 0$. That is why DE MOSFET is known as **normally-ON** MOSFET.

(*ii*) Enhancement-only MOSFET

As its name indicates, this MOSFET operates *only* in the enhancement mode and has no depletion mode. It works with *large positive* gate voltages only. It differs in construction from the DE MOSFET in that structurally *there exists no channel between the drain and source* (Fig. 34.20). Hence, it does not conduct when $V_{GS} = 0$. That is why it is called **normally-OFF** MOSFET.

In a DE MOSFET, I_D flows even when $V_{GS} = 0$. It operates in depletion mode with negative values of V_{GS}. As V_{GS} is made more negative, I_D decreases till it ceases when $V_{GS} = V_{GS(off)}$. It works in enhancement mode when V_{GS} is positive as shown in Fig. 34.16 (*b*).

In the case of E-only MOSFET, I_D flows only when V_{GS} exceeds $V_{GS(th)}$ as shown in Fig. 34.22 (*c*).

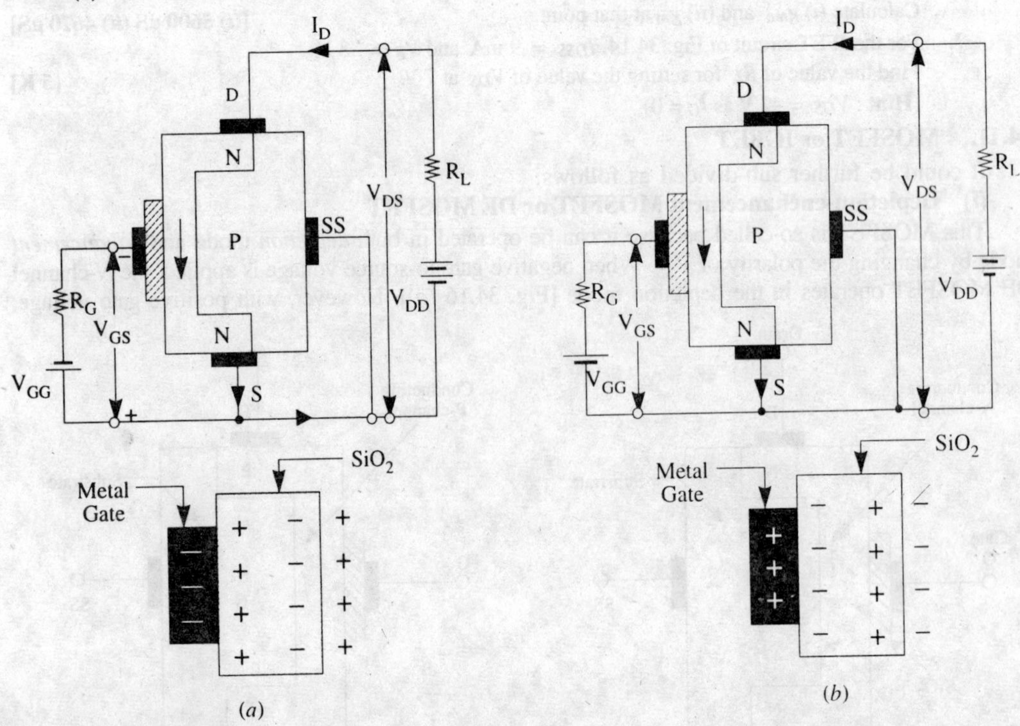

Fig. 34.16

34.12. DE MOSFET

(*a*) Construction

Like JFET, it has source, gate and drain. However, as shown in Fig. 34.15, its gate is insulated from its conducting channel by an ultra-thin metal-oxide insulating film (usually of silicon dioxide SiO_2). Because of this insulating property, MOSFET is alternatively known as insulated-gate field-effect transistor (IGFET or IGT). Here also, gate, voltage controls drain current but main difference between JFET and MOSFET is that, in the latter case, we can apply *both positive and negative voltages to the gate because it is insulated from the channel*. Moreover, the gate, SiO_2 insulator and channel form a parallel-plate capacitor. Unlike JFET, a DE MOSFET has only one P-region or N-region called *substrate*. Normally, it is shorted *to the source internally*. Fig. 34.15 shows both P-channel and N-channel DE MOSFETs along with their symbols.

(*b*) Working

(*i*) Depletion Mode of N-channel DE MOSFET

When $V_{GS} = 0$, electrons can flow freely from source to drain through the conducting channel which exists between them. When gate is given negative voltage, it **depletes** the N-channel of its electrons by inducing positive charge in it as shown in Fig. 34.16 (*a*). Greater the negative voltage on the gate, greater is the reduction in the number of electrons in the channel and, consequently, lesser its conductivity. In fact, too much negative gate voltage called $V_{GS(off)}$ can cut-off the channel. Hence, with negative gate voltage, a DE MOSFET behaves like a JFET.

For obvious reasons, negative-gate operation of a DE MOSFET is called *its depletion mode operation.*

(*ii*) Enhancement Mode of N-channel DE MOSFET

The circuit connections are shown in Fig. 34.16 (*b*). Again, drain current flows from source to drain even with zero gate bias. When positive voltage is applied to the gate, the input gate capacitor is able to create free electrons in the channel which increases I_D. As seen from the enlarged view of the gate capacitor in Fig. 34.16 (*b*), free electrons are induced in the channel *by capacitor action.* These electrons are added to those already existing there. This increased number of electrons *increases* or **enhances** the conductivity of the channel. As positive gate voltage is increased, the number of induced electrons is increased, so, conductivity of the source-to-drain channel is increased and, consequently, increasing amount of current flows between the terminals. That is why, positive gate operation of a DE MOSFET is known as its **enhancement mode** operration.

Since gate current in both modes is negligibly small, input resistance of a MOSFET is incredibly high varying from 10^{10} Ω to 10^{14} Ω. In fact, MOSFET input current is the leakage current of the capacitor unlike the input current for JFET which is the leakage current of a reverse-biased P-N junction.

34.13. Schematic Symbols for a DE MOSFET

Schematic symbol of an N-channel normally-ON or DE MOSFET is shown in Fig. 34.17. The gate looks like a metal plate. The arrow is on the substrate and **towards the N-channel**. When *SS* is connected to an external load, we have a 4-terminal device as shown in Fig. 34.17 (*a*) but when it is internally shorted to *S*, we get a 3-terminal device as shown in Fig. 34.17 (*b*). Fig. 34.17 (*c*) shows the symbol for a P-channel DE MOSFET.

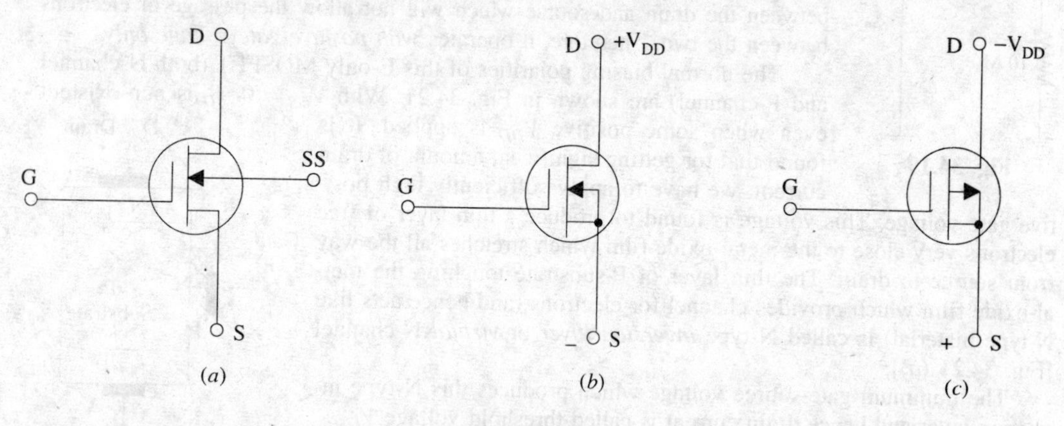

Fig. 34.17

34.14. Static Characteristics of a DE MOSFET

In Fig. 34.18 are shown the drain current and transfer characteristics of a common-source N-channel DE MOSFET for V_{GS} varying from + 2 V to $V_{GS(off)}$.

It acts in the enhancement mode when gate is *positive* with respect to source and in the depletion mode when gate is *negative*. As usual, $V_{GS(off)}$ represents the gate-source voltage which cuts off the source-to-drain current. The transfer characteristic is shown in Fig. 34.18 (*b*). For a given V_{DS}, I_D flows even when $V_{GS} = 0$. However, keeping V_{DS} constant, as V_{GS} is made more negative, I_D keeps decreasing till it becomes zero at $V_{GS} = V_{GS(off)}$. When used in the enhancement mode, I_D increases as V_{DS} is increased positively.

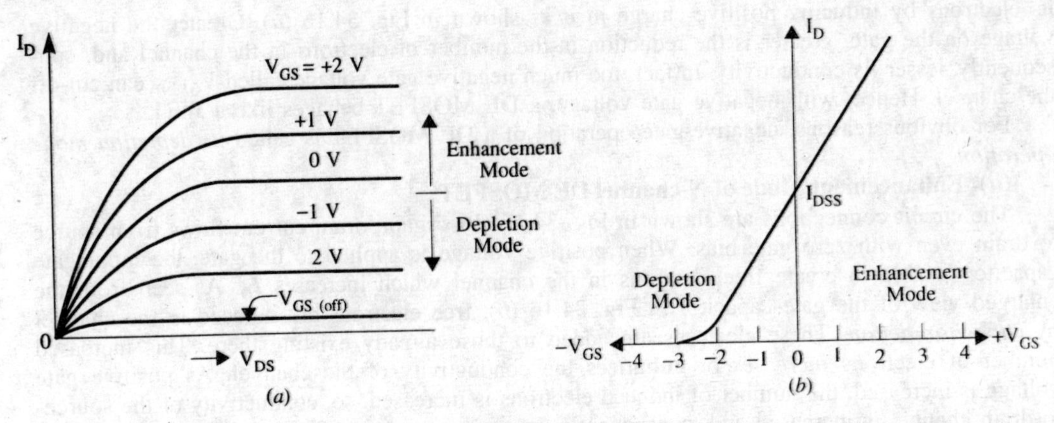

Fig. 34.18

Example 34.4. *For the N-channel zero-biased DE MOSFET circuit of Fig. 33.19, calculate V_{DS} if I_{DSS} = 10 mA and $V_{GS(off)}$ = − 6 V.* **(Electronics-I, Bangalore Univ. 1988)**

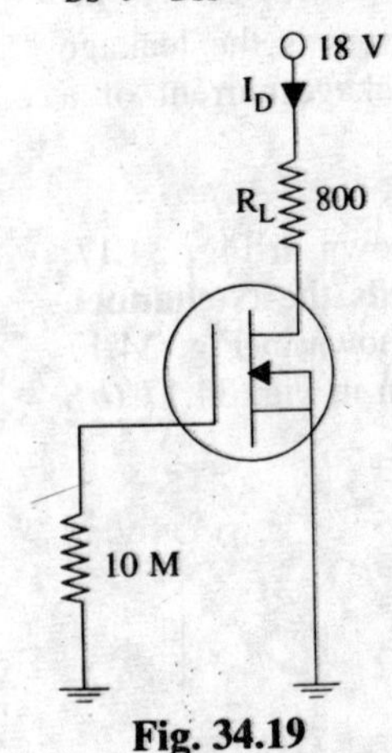

Fig. 34.19

Solution. Since $I_D = I_{DSS} = 10$ mA

$$\therefore \quad V_{DS} = V_{DD} - I_{DSS} R_L$$

$$= 18 - 10 \times 10^{-3} \times 800 = \mathbf{10\ V}$$

34.15. Enhancement-only N-channel MOSFET

This N-channel MOSFET (also called *NMOS*) finds wide application in digital circuitry. As shown in Fig. 34.20 in this NMOS, the P-type substrate extends all the way to the metal-oxide layer. Structurally, *there exists no channel between the source and the drain.* Hence, an NMOS can never operate with a negative gate voltage because it will induce *positive* charge in the space between the drain and source which will not allow the passage of electrons between the two. Therefore, it operates *with positive gate voltage only.*

The normal biasing polarities of this E-only MOSFET (both N-channel and P-channel) are shown in Fig. 34.21. With $V_{GS} = 0$, I_D is non-existent even when some positive V_{DD} is applied. It is found that for getting significant amount of drain current, we have to apply sufficiently high positive gate voltage. This voltage is found to produce a thin layer of free electrons very close to the metal-oxide film which stretches all the way from source to drain. The thin layer of P-substrate touching the metal-oxide film which provides channel for electrons (and hence acts like N-type material) is called N-type *inversion layer* or *virtual* N-channel [Fig. 34.21 (*a*)].

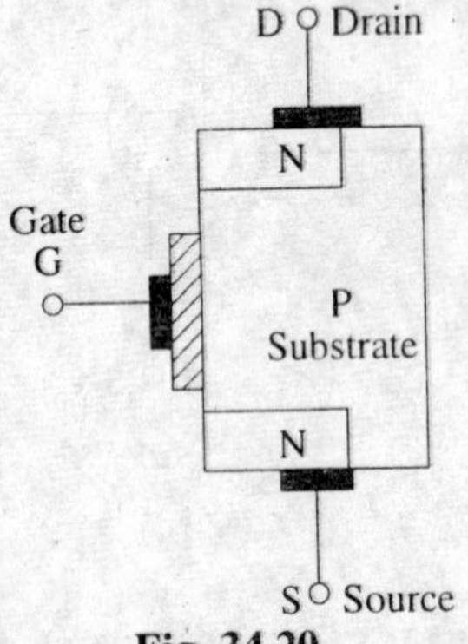

Fig. 34.20

The minimum gate-source voltage which produces this N-type inversion layer and hence drain current is called threshold voltage $V_{GS(th)}$ as shown in Fig. 34.22 (*c*). When $V_{GS} < V_{GS(th)}$, $I_D = 0$. Drain current starts only when $V_{GS} > V_{GS(th)}$. For a given V_{DS}, as V_{GS} is increased, virtual channel deepens and I_D increases. The value of I_D is given by $I_D = K(V_{GS} - V_{GS(th)})^2$ where K is a constant which depends on the particular MOSFET. Its value can be determined from the data sheet by taking the specified value of I_D called $I_{D(ON)}$ at the given value of V_{GS} and then substituting the values in the above equation. Fig. 34.22 (*a*) shows the schematic symbol for an E-only N-channel MOSFET whereas Fig. 34.22 (*b*) shows its typical drain characteristics. As usual, arrow on the substrate points to the N-type material and the vertical line (representing channel) is broken as a reminder of the normally-OFF condition.

A P-channel E-only MOSFET (PMOS) is constructed like NOMS except that all the P- and N-regions are interchanged. It operates with *negative* gate voltage only.

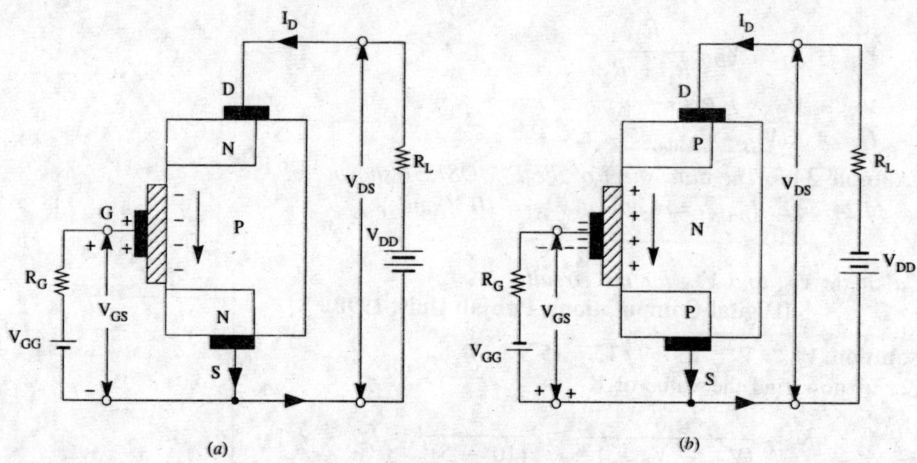

Fig. 34.21

Differentiating the above-given drain current equation respect to V_{GS}, we get

$$\frac{dI_D}{dV_{GS}} = g_m = 2\,K\,(V_{GS} - V_{GS(th)})$$

Transfer Characteristic

It is shown in Fig. 34.22 (*c*). I_D flows only when V_{GS} exceeds threshold voltage $V_{GS(th)}$. This MOSFET does not have an I_{DSS} parameter as do the JFET and DE MOSFET.

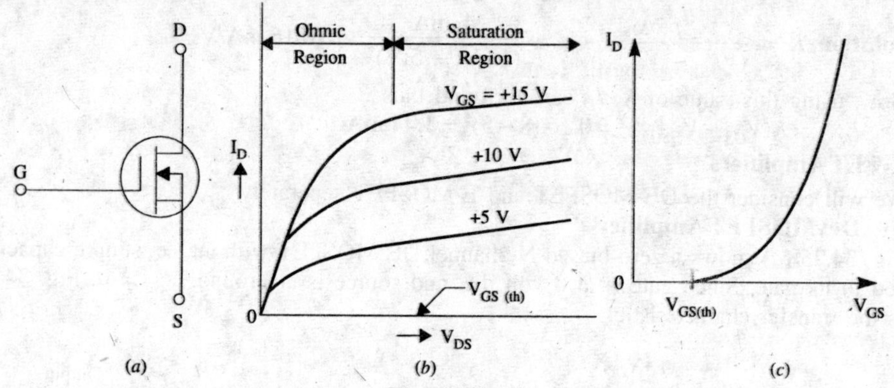

Fig. 34.22

34.16. Biasing E-only MOSFET

As stated earlier, enhancement-only MOSFET must have a V_{GS} greater than $V_{GS(th)}$. Fig. 34.23 shows two methods of biasing an N-channel E-MOSFET. In either case, the purpose is to make the gate voltage more positive than the source by an amount exceeding $V_{GS(th)}$. Fig. 34.23 (*a*) shows drain-feedback bias whereas Fig. 34.23 (*b*) shows voltage-divider bias.

Considering the circuit of Fig. 34.23 (*b*), we have

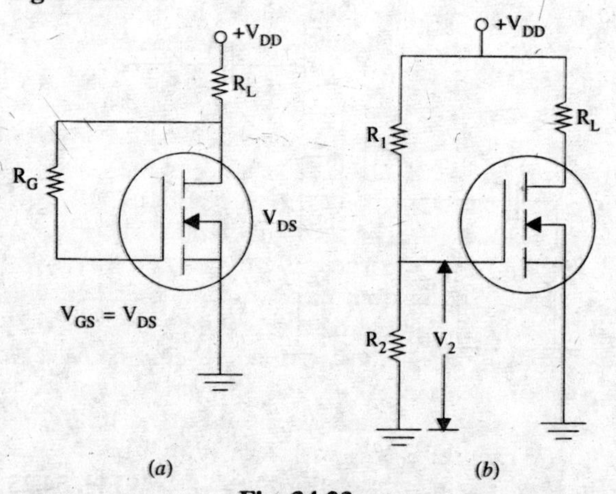

Fig. 34.23

$$V_{GS} = V_2 = V_{DD}\frac{R_2}{R_1 + R_2}$$

and $\quad V_{DS} = V_{DD} - I_D R_L$

where $\quad I_D = K(V_{GS} - V_{GS(th)})^2.$

Example 34.5. *The data sheet of the E-MOSFET shown in Fig. 34.24 gives $I_{D(ON)} = 4$ mA at $V_{GS} = 10$ V and $V_{GS(th)} = 5$ V.*

Calculate V_{GS} and V_{DS} for the circuit.

(Digital Computations, Punjab Univ. 1990)

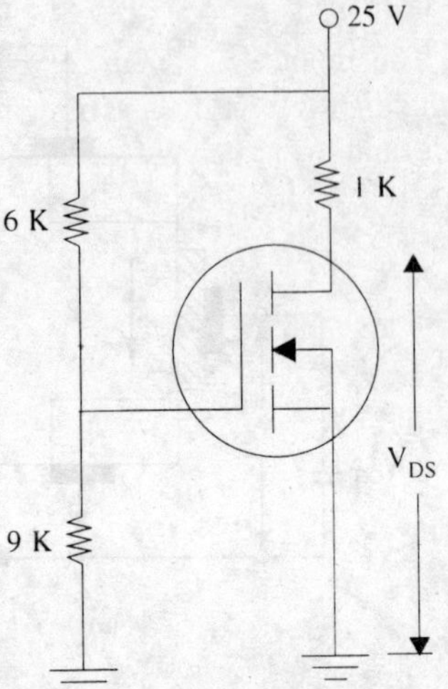

Solution. $V_{GS} = V_2 = 25 \times 9 / 15 = 15$ V

Let us now find the value of K.

$$K = \frac{I_{D(ON)}}{(V_{GS} - V_{GS(th)})^2} = \frac{4}{(10 - 5)^2}$$

$$= 0.16 \text{ mA/V}^2$$

Value of I_D for $V_{GS} = 15$ V is given by

$$I_D = K(V_{GS} - V_{GS(th)})^2 = 0.16\,(15 - 5)^2 = 16 \text{ mA}$$

Now, $\quad V_{DS} = V_{DD} - I_D R_L = 25 - 16 \times 1 = \mathbf{9\ V}$

Fig. 34.24

Example 34.6. *An N-channel E-MOSFET has the following parameters:*

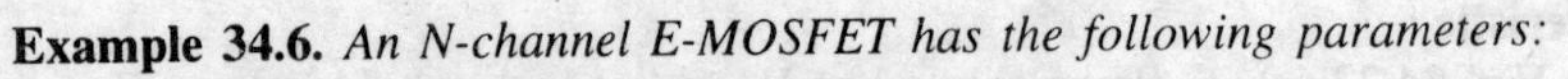

$$I_{D(ON)} = 4 \text{ mA at } V_{GS} = 10 \text{ V and } V_{GS(th)} = 5 \text{ V}$$

Calculate its drain current for $V_{GS} = 8$ V.

Solution. $K = \dfrac{I_{D(ON)}}{(V_{GS} - V_{GS(th)})^2} = \dfrac{4 \text{ mA}}{(10 \text{ V} - 5 \text{ V})^2} = 0.16 \text{ mA/V}^2$

Now, using this value of K, I_D can be found thus:

$$I_D = K(V_{GS} - V_{GS(th)})^2 = 0.16\,(8 - 5)^2 = \mathbf{1.44\ mA}$$

34.17. FET Amplifiers

We will consider the DE MOSFET and E-MOSFET separately.

(i) DE-MOSFET Amplifier

Fig. 34.25 (*a*) shows a zero-biased N-channel DE MOSFET with an a.c. source capacitively-coupled to its gate. Since gate is at 0 volt d.c. and source is at ground, $V_{GS} = 0$. Fig. 34.25 (*b*) shows the transfer characteristic.

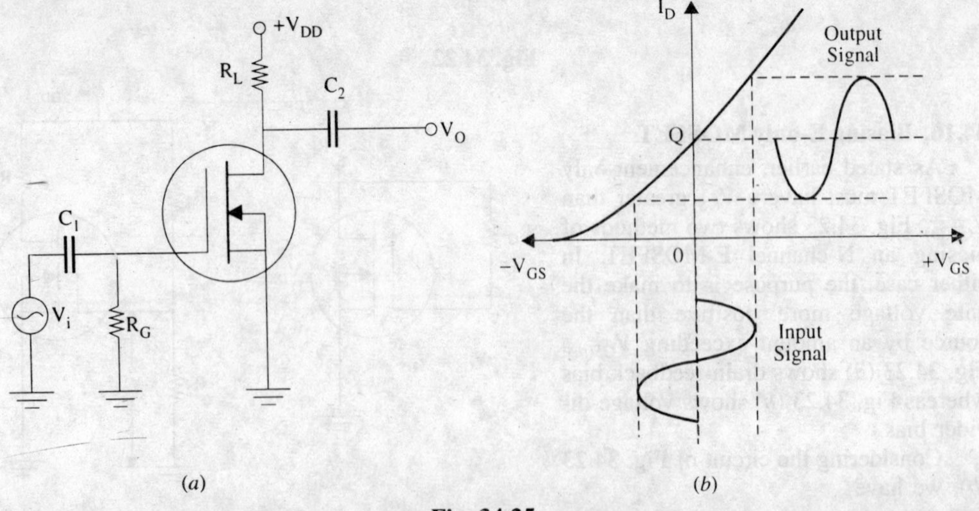

Fig. 34.25

The input a.c. signal voltage υ_i causes υ_{GS} to swing above and below its zero value thus producing a swing in I_D as shown in Fig. 34.25 (*b*). The negative swing in υ_{GS} produces delpetion mode and I_D is decreased. A positive swing in υ_{GS} produces enhancement mode so that I_D is increased. This leads to a large swing in drop across R_L and can be taken out via C_2 as υ_0.

(*ii*) E-MOSFET Amplifier

In Fig. 34.26 (*a*) is shown a voltage-divider biased N-channel E-MOSFET having an a.c. signal source coupled to its gate. The gate is biased with a positive voltage such that V_{GS} exceeds $V_{GS(th)}$.

As shown in Fig. 34.26 (*b*), the signal voltage produces a swing in V_{GS} below and above its Q-point value. This, in turn, causes a swing in I_D and hence in I_DR_L which gives rise to υ_o.

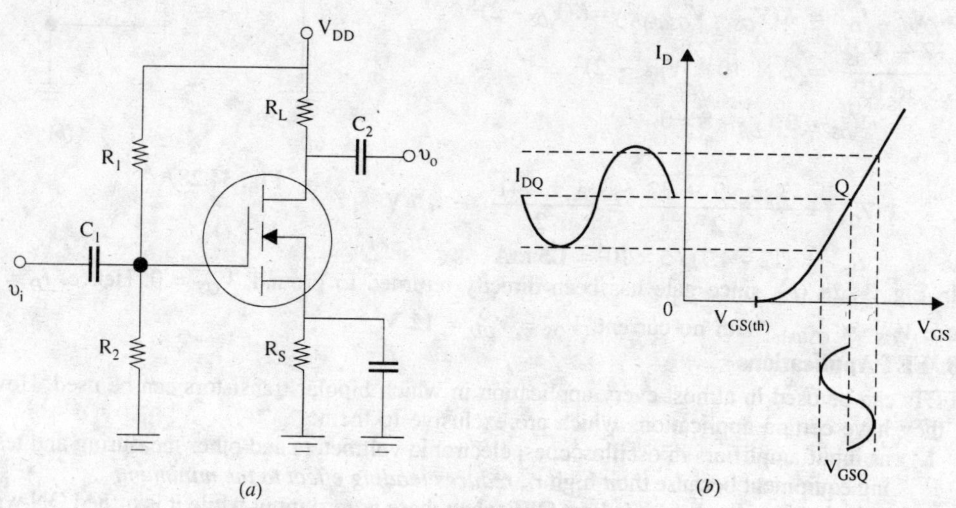

Fig. 34.26

Example 34.7. *The N-channel E-MOSFET used in the common-source amplifier of Fig. 33-27 has the following parameters:*

$I_{D(ON)}$ = 4 mA at V_{DS} = 10 V,

$V_{GS(th)}$ = 4 V and g_m = 5000 μS.

Calculate V_{GS}, I_D, V_{DS} *and* v_o.

(Elect. and Electronic Engg., Annamalai Univ. 1992)

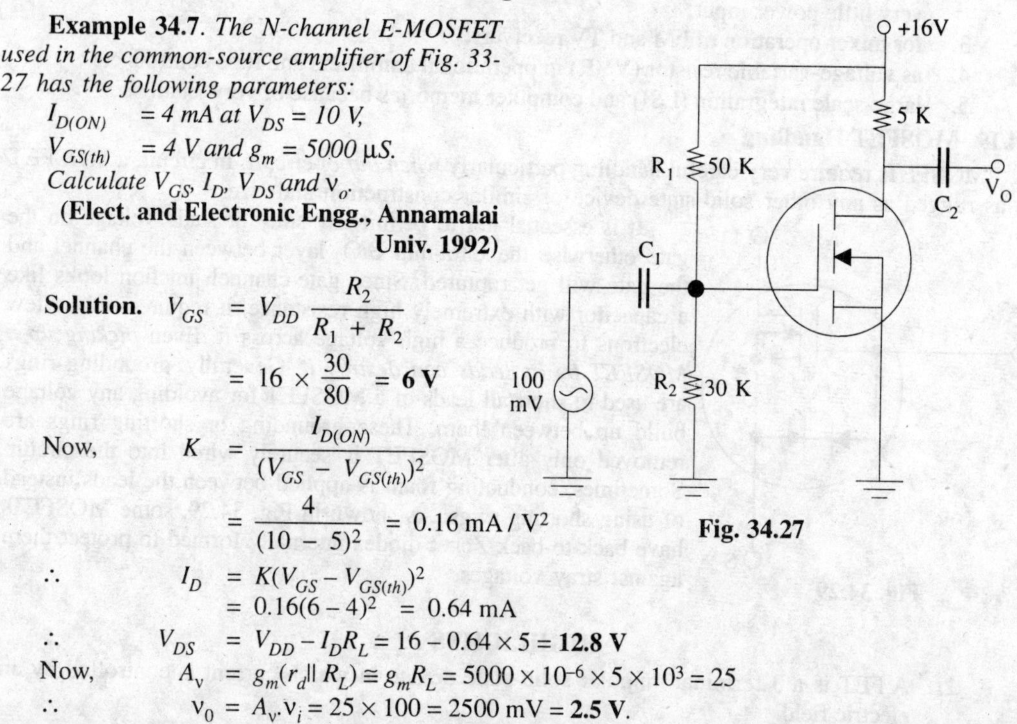

Fig. 34.27

Solution. $V_{GS} = V_{DD}\dfrac{R_2}{R_1 + R_2}$

$= 16 \times \dfrac{30}{80} = \mathbf{6\ V}$

Now, $K = \dfrac{I_{D(ON)}}{(V_{GS} - V_{GS(th)})^2}$

$= \dfrac{4}{(10-5)^2} = 0.16\ \text{mA/V}^2$

$\therefore \quad I_D = K(V_{GS} - V_{GS(th)})^2$

$= 0.16(6-4)^2 = 0.64$ mA

$\therefore \quad V_{DS} = V_{DD} - I_DR_L = 16 - 0.64 \times 5 = \mathbf{12.8\ V}$

Now, $A_v = g_m(r_d \| R_L) \cong g_mR_L = 5000 \times 10^{-6} \times 5 \times 10^3 = 25$

$\therefore \quad \upsilon_0 = A_v.v_i = 25 \times 100 = 2500\ \text{mV} = \mathbf{2.5\ V}.$

Example 34.8. *The parameters of the enhancement-only NMOS shown in Fig. 34.28 are $V_{GS(th)} = 2$ V and $K = 2 \times 10^{-4}$ A / V^2. Calculate the values of I_D and V_{DS}.*

(Electronic Circuits, Mysore Univ. 1992)

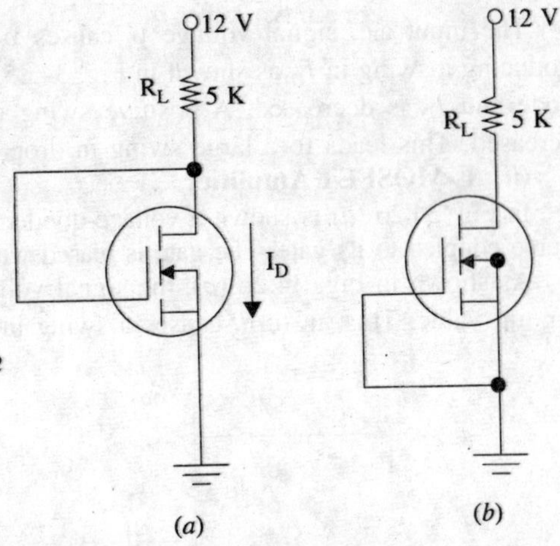

Fig. 34.28

Solution. Since drain is directly returned to gate, $V_{GS} = V_{DS}$. As seen from the figure

$$V_{DS} = V_{DD} - I_D R_L = 12 - 5 \times 10^3 I_D$$

$$\therefore \quad I_D = (12 - V_{DS}) / 5 \times 10^3 \text{ A}$$

$$\text{Now,} \quad I_D = K(V_{GS} - V_{GS(th)})^2 = K(V_{DS} - 2)^2$$

$$\text{or} \quad \frac{12 - V_{DS}}{5 \times 10^3} = 2 \times 10^{-4} (V_{DS} - 2)^2$$

$$\text{or} \quad V_{DS}^2 - 3 V_{DS} - 8 = 0$$

$$\therefore \quad V_{DS} = \frac{3 \pm \sqrt{9 + 32}}{2} = \frac{3 \pm \sqrt{41}}{2} = \mathbf{4.7\ V}$$

$$\therefore \quad I_D = (12 - 4.7) / 5 \times 10^3 = \mathbf{1.5\ mA}$$

In Fig. 34.28 (*b*), since gate has been directly returned to ground, $V_{GS} = 0$. Hence, $I_D = 0$ because $V_{GS} \ll _{GS(th)}$. With no current, $V_{DS} = V_{DD} =$ **12 V.**

33.18. FET Applications

FETs can be used in almost every application in which bipolar transistors can be used. However, they have certain applications which are exclusive to them:

1. as input amplifiers in oscilloscopes, electronic voltmeters and other measuring and testing equipment because their high r_{in} *reduces loading effect to the minimum;*
2. in logic circuits where it is kept OFF when there is zero input while it is turned ON with very little power input;
3. for mixer operation of FM and TV receivers;
4. as voltage-variable resistor (VVR) in operational amplifiers and tone controls etc ;
5. large-scale integration (LSI) and computer memories because of very small size.

34.19. MOSFET Handling

MOSFETs require very careful handling particularly *when out of circuit*. In circuit, a MOSFET is as rugged as any other solid-state device of similar construction and size.

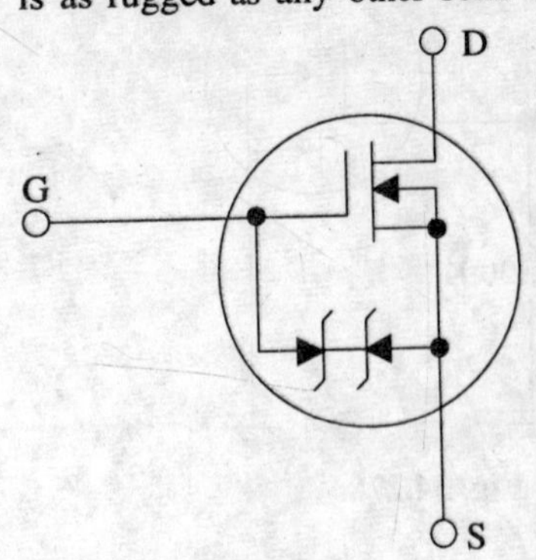

Fig. 34.29

It is essential not to permit any stray or static voltage on the gate otherwise the ultra-thin SiO_2 layer between the channel and the gate will get ruptured. Since gate-channel junction looks like a capacitor with extremely high resistance, it requires only a few electrons to produce a high voltage across it. Even *picking up a MOSFET by its leads can destroy it.* Generally, grounding rings are used to short all leads of a MOSFET for avoiding any voltage build up between them. These grounding or shorting rings are removed only after MOSFET is securely wired into the circuit. Sometimes, conducting foam is applied between the leads instead of using shorting rings. As shown in Fig. 34.29, some MOSFETs have back-to-back Zener diodes internally formed to protect them against stray voltages.

HIGHLIGHTS

1. A FET is a 3-terminal unipolar solid-state device in which current is controlled by an electric field.

2. There are two types of FETs (*i*) JFET and (*ii*) MOSFET.
3. MOSFET can be sub-divided into (*i*) DE MOSFET and (*ii*) E-MOSFET.
4. A JFET consists of a narrow bar of either an N-type or P-type semi-conductor material with two internally-connected gates on its either side. Its drain current flows through the channel situated between the two gates.
5. A JFET can behave like a resistor or as a constant-current source as well as a constant-voltage source.
6. DE MOSFET can work both in the depletion mode and enhancement mode. It can be operated with positive and negative gate voltages. There is a drain current even when $V_{GS} = 0$.
7. In an E-MOSFET, there exists no channel between the source and the drain. Hence, there is no drain current when $V_{GS} = 0$. Drain current cannot start unless V_{GS} has a minimum value called $V_{GS(th)}$.
8. Special applications of FETs are as input amplifiers in a oscilloscopes, electronic voltmeters, in logic circuits for mixer operation of FM and TV receivers, as voltage-variable resistors and in computer memories.

OBJECTIVE TESTS—34

A. Fill in the following blanks.

1. In a JFET, the gate-channel junctions are .... biased.
2. A JFET is a .... controlled device.
3. Unlike a BJT, JFET is a .... device.
4. The .... current of a JFET is practically zero.
5. The input resistance of an ideal JFET is ....
6. In a JFET, drain current is .... when $V_{GS} = 0$.
7. In a MOSFET, gate and channel are .... from each other.
8. An E-only MOSFET works only with .... gate voltages.

B. Answer True or False.

1. A FET is a unipolar device.
2. High input impedance of a JFET is due to its reverse-biased gate.
3. A JFET does not conduct with zero gate bias voltage.
4. A JFET can operate with zero gate bias.
5. In a JFET, pinch-off means current-off.
6. Negative-gate operation of an N-channel MOSFET is called its depletion mode operation.
7. With negative gate voltage, an N-channel MOSFET behaves like an N-channel JFET.
8. Even picking up a MOSFET by hand can destory it.

C. Multiple Choice Items.

1. A FET consists of a
   (*a*) source (*b*) drain
   (*c*) gate (*d*) all of the above
2. FETs have similar properties to
   (*a*) PNP transistors
   (*b*) NPN transistors
   (*c*) thermionic valves
   (*d*) uni-junction transistors
3. After V_{DS} reaches pinch-off value V_p in a JFET, drain current I_D becomes
   (*a*) zero (*b*) low
   (*c*) saturated (*d*) reversed
4. In a JFET, drain current is maximum when V_{GS} is
   (*a*) zero (*b*) negative
   (*c*) positive (*d*) equal to V_p
5. For the operation of enhancement-only N-channel MOSFET, value of gate voltage has to be
   (*a*) high positive (*b*) high negative
   (*c*) low positive (*d*) zero
6. The main factor which makes a MOSFET likely to break-down during normal handling is its
   (*a*) very low gate capacitance
   (*b*) high leakage current
   (*c*) high input resistance
   (*d*) both (*a*) and (*c*)

ANSWERS

A. 1. reverse 2. voltage 3. unipolar 4. gate 5. infinity 6. maximum 7. insulated 8. positive

B. 1. T 2. T 3. F 4. T 5. F 6. T 7. T 8. T

C. 1. *d* 2. *b* 3. *c* 4. *a* 5. *a* 6. *d*

35 THYRISTORS

35.1. What is a Thyristor?

The term *thyristor* usually refers to a family of four-layer solid-state devices having turn-ON characteristics that can be externally controlled by either current or voltage. They are also referred to as breakdown devices because their working depends on avalanche breakdown.

Thyristors have only two states: OFF and ON. We will now briefly discuss the following most popular thyristors:

1. SCR 2. GCS (GTO) 3. SCS 4. Triac 5. Diac and 6. UJT*

35.2. Silicon Controlled Rectifier (SCR)

It is one of the prominent members of the thyristor family. It is a four-layer or PNPN device. Basically, *it is a rectifier with a control element.* In fact, it consists of *three diodes* connected back-to-back with a gate connection. It is widely used as a switching device in power control applications. It can control loads by switching current OFF and ON up to many thousand times a second. It can switch ON for variable lengths of time, thereby delivering selected amount of power to the load. Hence, it possesses the advantages of a rheostat and a switch with none of their disadvantages.

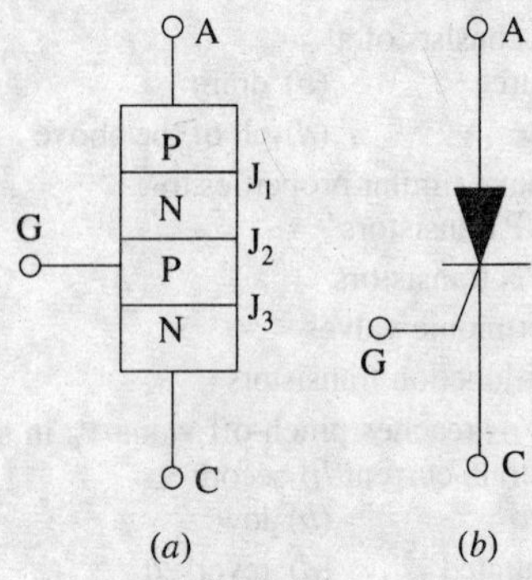

Fig. 35.1

(a) Construction

As shown in Fig. 35.1 (*a*), it is a three-junction, three-terminal, four-layer transistor, the layers being alternately of P-type and N-type silicon. The three junctions are marked J_1, J_2 and J_3 whereas the three terminals are : anode (*A*), cathode (*C*) and gate (*G*) which is connected to the *inner* P-type layer. The function of the gate is to control the firing of SCR. The schematic symbol is shown in Fig. 35.1 (*b*).

SCRs come in all shapes and sizes (Fig. 35.2). Smaller ones look like transistors and can handle currents upto nearly 10 A. Larger ones come in less familiar packages and can handle anode currents upto 2000 A.

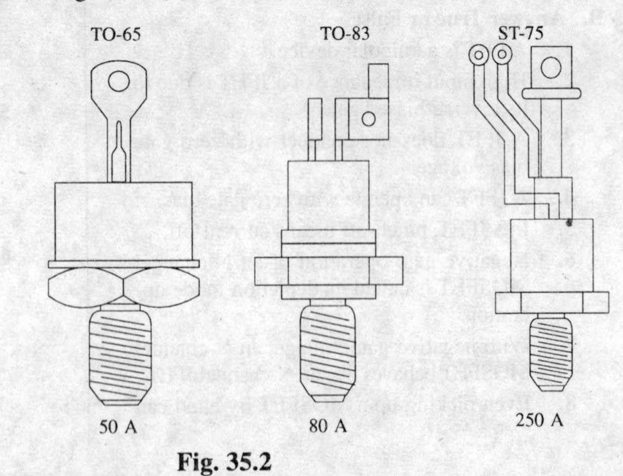

Fig. 35.2

(b) Biasing

With the polarity of V as shown in Fig. 35.3 (*a*), the junctions J_1 and J_3 become forward-biased whereas J_2 is re-

*Strictly speaking, UJT is not a thyristor because it is not a 4-layer device although it operates in two states (ON and OFF).

verse-biased. Hence, no current (except leakage current) can flow through the SCR.

In Fig. 35.3 (*b*), polarity of V has been reversed. It is seen that, now, junctions J_1 and J_3 become reverse-biased and only J_2 is forward-biased. Again, there is no flow of current through the SCR.

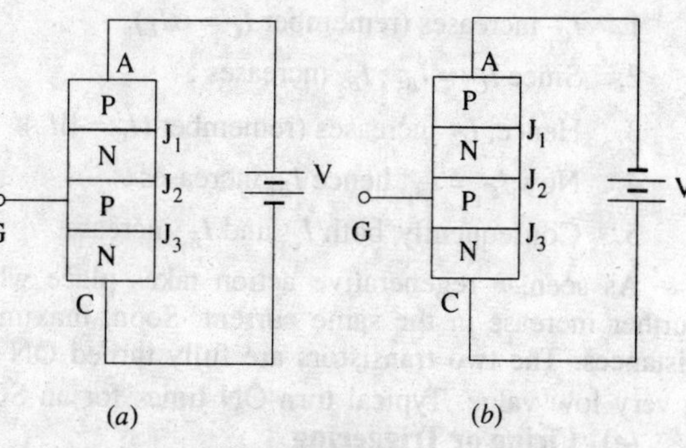

Fig. 35.3

(*c*) Operation

In Fig. 35.3 (*a*) above, current flow is blocked due to reverse-biased junction J_2. However, when anode voltage is increased, a certain critical value called forward break-over voltage V_{BO} is reached when J_2 breaks down and SCR switches suddenly to a highly conducting state. Under this condition, SCR offers very little forward resistance (0.01 Ω – 1.0 Ω) so that voltage across it drops to a low value (about 1 V) as shown in Fig. 35.4 and current is limited only by the power supply and the load resistance. Current keeps flowing indefinitely until the circuit is opened briefly or current falls below the holding current, I_H (also called latching current). Value of V_{BO} is determined by the magnitude of the gate current. Higher the gate current, lower the breakover voltage.

With supply connection as in Fig. 35.3 (*b*), the current through the SCR is blocked by the two reverse-biased junctions J_1 and J_3. When V is increased, a stage comes when Zener breakdown occurs which may destroy the SCR (Fig. 35.4). Hence, it is seen that SCR is a unidirectional device unlike triac *which is bi-directional.*

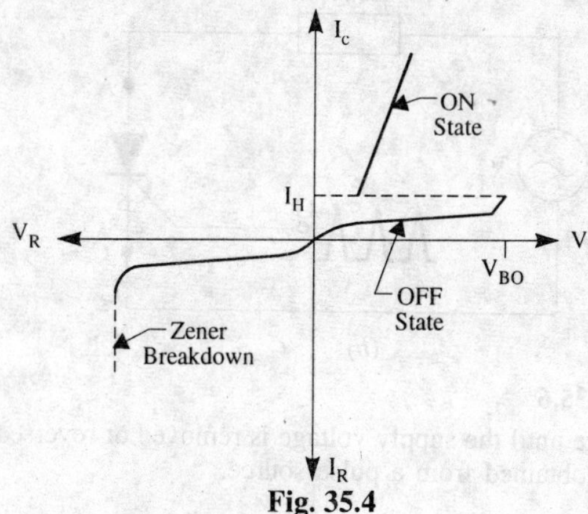

Fig. 35.4

(*d*) Two Transistor Analogy

The basic operation of an SCR can be described by using two-transistor analogy, For this purpose, SCR is split into two 3-layer transistor structures as shown in Fig. 35.5 (*a*). As seen, transistor Q_1 is a PNP transistor whereas Q_2 is an NPN device interconnected together. It will also be noted from Fig. 35.5 (*b*) that

(*i*) collector current of Q_1 is also the base current of Q_2; and

(*ii*) base current of Q_1 is also the collector current of Q_2.

Suppose that the supply voltage across terminals *A* and *C* is such that the reverse-biased junction J_2 starts breaking down. Then, current through the device begins to rise. It means that I_{E_1} begins to increase. Then

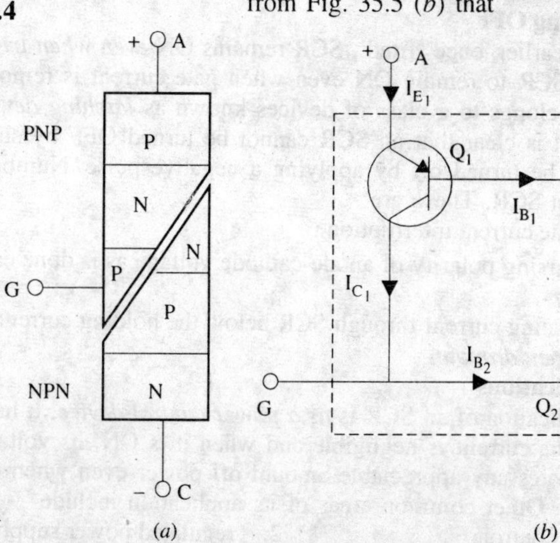

Fig. 35.5

1. I_{C_1} increases (remember $I_C = \alpha I_E$).
2. Since $I_{C_1} = I_{B_2}$, I_{B_2} increases .
3. Hence, I_{C_2} increases (remember ($I_C = \beta I_B$).
4. Now, $I_{C_2} = I_{B_1}$, hence I_{B_1} increases .
5. Consequently, both I_{C_1} and I_{E_1} increase.

As seen, a regenerative action takes place whereby an initial increase in current produces further increase in the same current. Soon, maximum current is reached limited by external resistances. The two transistors are fully turned ON and voltage across the two transistors falls to a very low value. Typical turn-ON times for an SCR are 0.1 to 1.0 μS.

(e) Firing or Triggering

Usually, SCR is operated with an anode voltage *slightly less* than the forward breakover voltage V_{BO} and is triggered into conduction by a low-power gate pulse. Once switched ON, gate has no further control on the device current. Gate signals can be

(a) d.c. firing signals — Fig. 35.6 (a)
(b) pulse signals — Fig. 35.6 (b)

In Fig. 35.6 (a) with S open, SCR does not conduct and the lamp is out. When S is closed momentarily, a positive voltage is applied to the gate which forward-biases the centre PN junction.

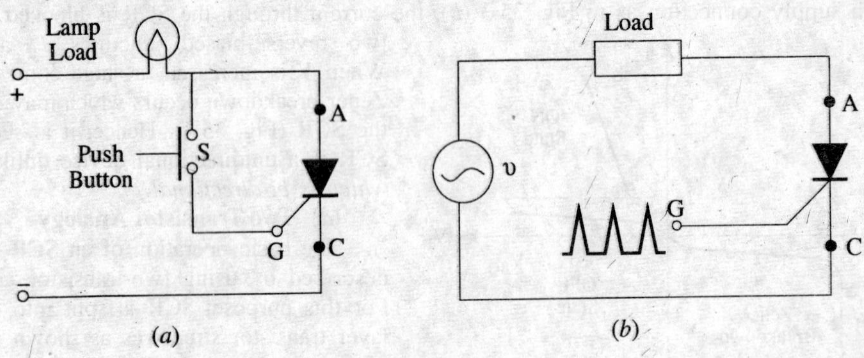

Fig. 35.6

As a result, SCR is pulsed in the conducting state until the supply voltage is removed or reversed. Fig. 35.6 (b) shows triggering by timed pulses obtained from a pulse source.

(f) Turning OFF

As stated earlier, once 'fired', SCR remains ON *even when triggering pulse is removed*. This ability of the SCR to remain ON even when gate current is removed is referred to as **latching.** In fact, SCR belongs to a class of devices known as *latching devices.*

By now, it is clear that an SCR cannot be turned OFF by simply removing the gate pulse. Neither can it be turned off by applying a negative pulse. Number of techniques are employed to turn OFF an SCR. These are

1. anode current interruption ;
2. reversing polarity of anode-cathode voltage as is done each half cycle by υ in Fig. 35.6 (b) ;
3. reducing current through SCR below the holding current I_H (Fig. 35.4). It is called *low-current dropout.*

(g) Applications

Main application of an SCR is *as a power control device*. It has been shown above that when SCR is OFF, its current is negligible and when it is ON, its voltage is negligible. Consequently, it never dissipates any appreciable amount off power even when controlling substantial amounts of load power. Other common areas of its application include

1. relay controls
2. regulated power supplies

3. static switches
4. motor controls
5. inverters
6. battery chargers
7. heater controls
8. phase control etc.

SCRs have been designed to control powers upto 10 MW with individual ratings as high as 2000 A at 1.8 kV. Its frequency range application has been extended to about 50 kHz.

35.3. Half-wave Power Control

SCRs are commonly used for the control of a.c. power for lamp dimmers, electric heaters and motors etc. A half-wave variable-resistance phase-control circuit is shown in Fig. 35.7. For this circuit, gate triggering current is derived from the supply itself. Here, R_L is the resistance of the load (say, a heating element), R_1 is current-limiting resistor because it limits the gate current during the positive half-cycles of the supply. Potentiometer R_2 sets the trigger level for the SCR. By adjusting R_2, the SCR can be made to trigger at any point on the positive half-cycle of the a.c. waveform from 0° to 90° as shown in Fig. 35.7. If R_2 is adjusted to a very low value (point *A*), the SCR will trigger almost immediately at the commencement of the positive half-cycle of the input thus giving a conduction angle of 180°. It means that the current will flow for the entire positive half-cycle thereby delivering maximum power to the load. If value of R_2 is further increased by moving the contact to point *B*, the SCR triggering is delayed by 45° so that conduction angle is reduced to 135°. Hence, relatively less power is delivered to R_L. If R_2 is further increased by moving the contact to point *C*, SCR triggering is further delayed by 90° due to decrease in gate current I_G. In this case, still less power is delivered to the load. From the above description, it is clear that by adjusting R_2, firing angle of the SCR can be changed and so can be the power delivered to the load Fig. 34.8 shows a full-wave controlled rectifier arrangement which supplies higher average current. The two SCRs are fed by a centre-tapped transformer and supply the same load R_L. The SCRs are triggered one on each alternate half-cycle by the same trigger circuit. Each of the two SCRs remains conducting unless its current falls below the holding current I_H. The average d.c. voltage supplied by the above rectifier is

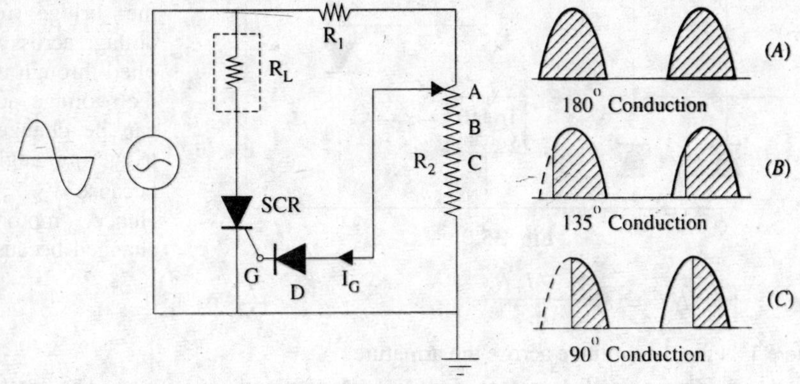

Fig. 35.7

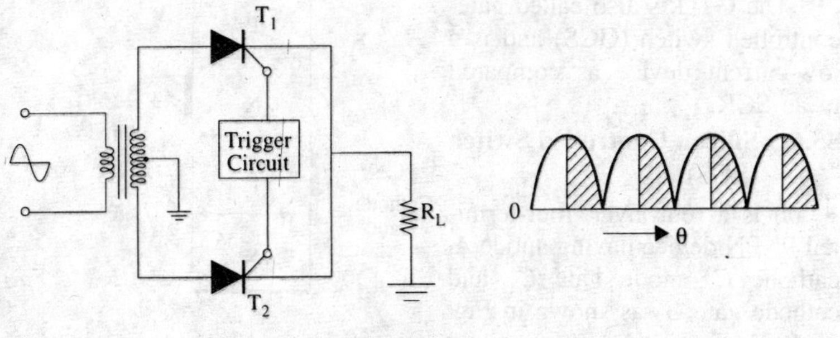

Fig. 35.8

$$V_{dc} = \frac{V_m}{\pi}(1 + \cos\theta)$$

where V_m = maximum voltage across each thyristor

θ = delay angle

35.4. D.C. Motor Speed Control

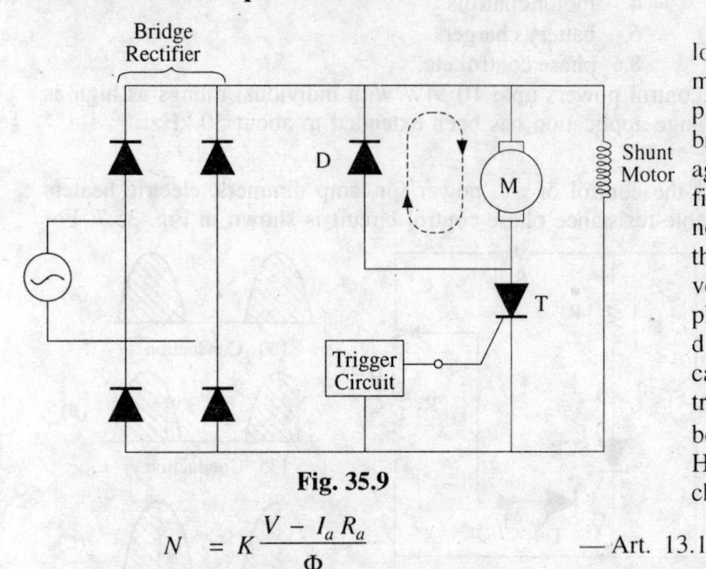

Fig. 35.9

Fig. 35.9 illustrates the open-loop* armature speed control method of a d.c. shunt motor supplied from an a.c. source. The bridge rectifier converts a.c. voltage into d.c. voltage. The shunt field winding is directly connected across the d.c. output of the bridge rectifier. However, voltage across the armature is applied through the thyristor *T*. The d.c. voltage across the armature can be changed by varying the triggering angle of the thyristor because $V_{dc} = V_m (1 + \cos\theta)$. Hence, motor speed can be changed because it is given by

$$N = K\frac{V - I_a R_a}{\Phi} \qquad \text{—Art. 13.1}$$

where *V* is the d.c. voltage across the armature.

As θ increases, *V* decreases and *N* decreases and *vice versa.* The free-wheeling diode *D* connected across the motor provides a circulating current path (shown dotted) for the energy stored in the inductance of the armature winding at the time thyristor *T* turns OFF. Without *D*, current will flow through *T* and bridge rectifier thus prohibiting *T* from turning OFF.

35.5. Gate Turn-off Switch

The gate-turn-off switch (GTO) is nothing else but a gate-turn-off SCR. It is exactly similar to an SCR in construction but the difference is that, unlike an SCR, it can be turned OFF by applying a negative gate signal.

As with an SCR, GTO will not conduct when forward-biased until a positive pulse is applied to its gate. But unlike an SCR, a GTO can be turned OFF by a negative gate pulse. The ratio of the anode current in the ON state to the negative gate current required to turn-off the GTO is called the turn-off current gain.

The GTO is also called gate-controlled switch (GCS) and is a low-current device as compared to an SCR.

35.6. Silicon Controlled Switch (SCS)

It is a four-layer, four-terminal PNPN device having anode *A*, cathode *C*, anode gate G_1 and cathode gate G_2 as shown in Fig. 35.10. In fact, *it is a low-current SCR with two gate terminals***. The two transistor equivalent cir-

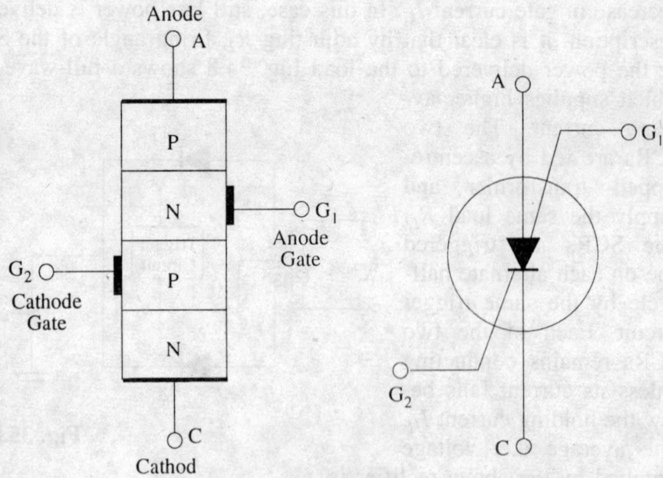

Fig. 35.10

*In the closed-loop method, a tachogenerator is used which provides voltage to the comparator depending on the shaft speed of the motor.

**It is sometimes called tetrode thyristor because it has four terminals, one for each layer.

cuit is shown Fig. 35.11.

Switching ON and OFF

The device may be switched ON or OFF by suitable pulse applied at either gate. As seen from Fig. 35.11, a negative pulse is required at anode gate G_1 to turn the device ON whereas positive pulse is needed to turn it OFF as explained below.

Similarly, at cathode gate G_2, a negative pulse is required to switch the device OFF and a positive pulse to turn it ON.

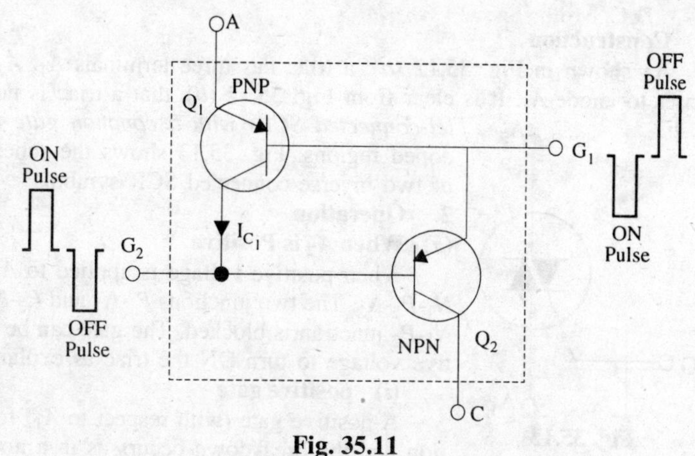

Fig. 35.11

As seen from Fig. 35.11 when a negative pulse is applied to G_1, if forward-biases Q_1 (being PNP) which is turned ON. The resulting heavy collector current I_C being the base current of Q_2 turns it ON. Hence, SCS is switched ON. A positive pulse at G_1 will reverse-bias *E/B* junction of Q_1 thereby switching the SCS OFF.

The SCS is a versatile device because it can be operated in many different ways. For example, it can be used as an SCR when its anode gate is not connected.

V/I Characteristics

The *V/I* characteristics of an SCS are essentially the same as **those of the SCR** (Fig. 35.4).

As compared to an SCR, an SCS has *much reduced turn-OFF time*. Moreover, it has higher control and triggering sensitivity and a more predictable firing situation.

Applications

The more common areas of SCS application are as under:

1. in counters, registers and timing circuits of computers;
2. pulse generators;
3. voltage sensors;
4. oscillators etc.

35.7. Triac

It is a *3-terminal*, *5-layer*, *bi-directional device* which can be triggered into conduction by both *positive and negative* voltages at its anodes and with both *positive and negative* triggering pulses at its gate. It behaves like two SCRs *connected in parallel, upside down with respect to each other*. That is, the anode of one is tied to the cathode of the other and their gates are directly tied together. Hence, anode and gate voltages applied in either direction will fire a triac because they would fire at least one of the two SCRs which are in opposite directions.

Since a triac responds to both positive and negative voltages at the anode, the concept of cathode used for SCR is dropped. Instead, the two electrodes are called anodes A_1 and A_2.

Triac is called bi-directional device because, unlike an SCR, it can allow current to flow in both directions.

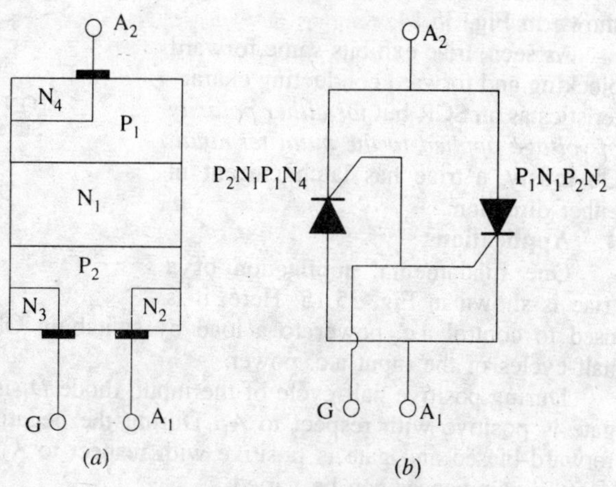

Fig. 35.12

1. Construction

As shown in Fig. 35.12 (*a*), a triac has three terminals A_1, A_2, and G. As seen, gate G is closer to anode A_1. It is clear from Fig. 35.12 (*b*), that a triac is nothing but *two inverse parallel-connected SCRs with a common gate terminal.* As seen, it has **six** doped regions. Fig. 35.13 shows the schematic symbol which consists of two inverse-connected SCR symbols.

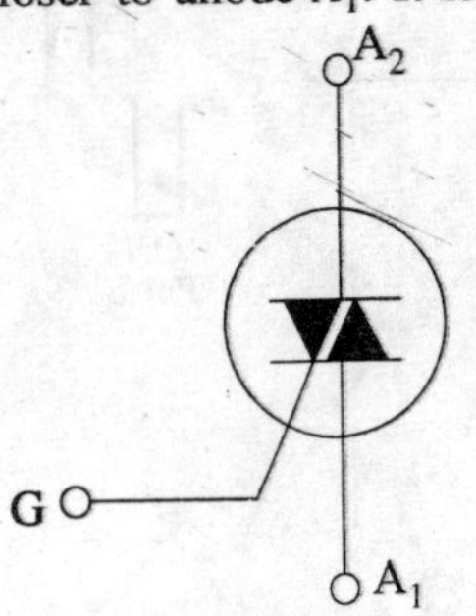

Fig. 35.13

2. Operation

(*a*) When A_2 is Positive

When positive voltage is applied to A_2, path of current flow is P_1-N_1-P_2-N_2. The two junctions P_1-N_1 and P_2-N_2 are forward-biased whereas N_1-P_2 junction is blocked. The gate can be given either positive or negative voltage to turn ON the triac as explained below.

(*i*) positive gate

A positive gate (with respect to A_1) forward-biased the P_2-N_2 junction and the breakdown occurs as in a normal SCR.

(*ii*) negative gate

A negative gate forward-biases the P_2-N_3 junction and current carriers injected into P_2 turn on the triac.

(*b*) When A_1 is Positive

When positive voltage is applied to anode A_1, path of current flow is P_2-N_1-P_1-N_4. The two junctions P_2-N_1 and P_1-N_4 are forward-biased whereas junction N_1-P_1 is blocked. Conduction can be achieved by giving either positive or negative voltage to G as explained below.

(*i*) positive gate

A positive gate (with respect to A_1) injects current carriers by forward-biasing P_2-N_2 junction and thus initiates conduction.

(*ii*) negative gate

A negative gate injects current carriers by forward-biasing P_2-N_3 junction thereby triggering conduction.

It is seen that there are four triac-triggering modes, two each for the two anodes.

Low-current dropout is the only way to open a triac.

3. V/I Characteristics

Typical characteristics of a triac are shown in Fig. 35.14.

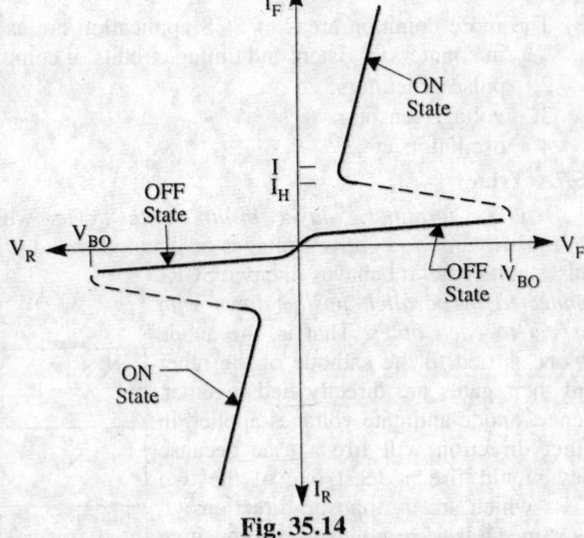

Fig. 35.14

As seen, triac exhibits same forward blocking and forward conducting characteristics as an SCR but *for either polarity of voltage applied to the main terminal.* Obviously, a triac has latch current in either direction.

4. Applications

One fundamental application of a triac is shown in Fig. 35.15. Here, it is used to control a.c. power to a load by switching ON and OFF during positive and negative half-cycles of the input a.c. power.

During positive half-cycle of the input, diode D_1 is forward-biased, D_2 is reverse-biased and gate is positive with respect to A_1. During the negative half-cycle, D_1 is reverse-biased, D_2 is forward-biased and gate is positive with respect to A_2. By adjusting R, the point at which conduction commences can be varied.

Diac-triac combination for a.c. load power control is shown in Fig. 35.16. Firing control of diac is achieved by adjusting *R*.

Other applications of a triac include

1. as static switch to turn a.c. power OFF and ON ;
2. for minimizing radio interference;
3. for light control;
4. for motor speed control etc.

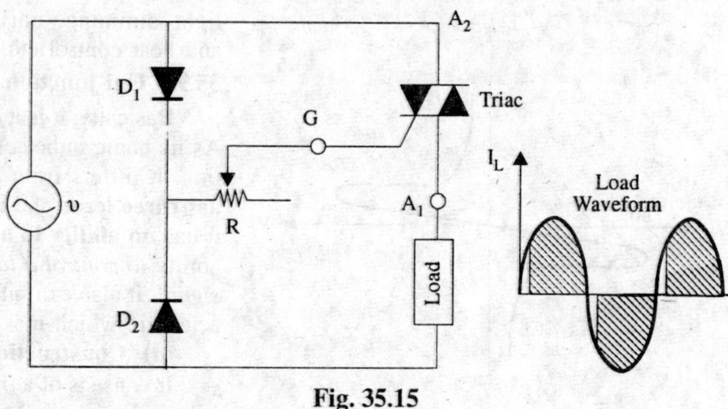

Fig. 35.15

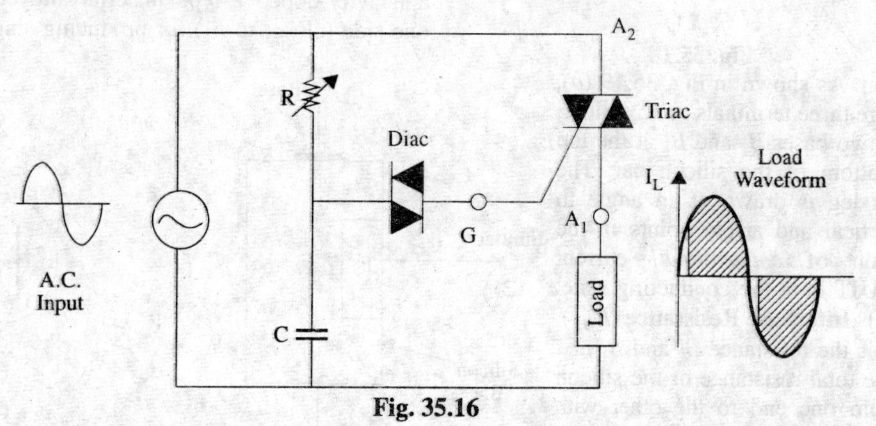

Fig. 35.16

The only disadvantage of a triac is that it takes comparatively longer time to recover to OFF state. Hence, its use is limited to a.c. supply frequencies of upto 400 Hz.

35.8. Diac

To put it simply, a diac is nothing else but a triac without its gate terminal as shown in Fig. 35.17 (*a*). Its equivalent circuit is a pair of inverted four-layer diodes. Its schematic symbol is shown in Fig. 35.17 (*b*). As seen, *it can breakdown in either direction.*

(*a*) Working

When anode A_1 is positive, the current path is P_2-N_2-P_1-N_1. Similarly, when A_2 is positive, the current flow path is P_1-N_2-P_2-N_3. Diac is designed to trigger triacs or provide protection against over-voltages.

The operation of a diac can best be explained by imagining it as **two diodes connected in series**. Voltage applied across it in either direction turns ON one diode, reverse-biasing the other. Hence, it can be switched from OFF to ON state for either polarity of the applied voltage.

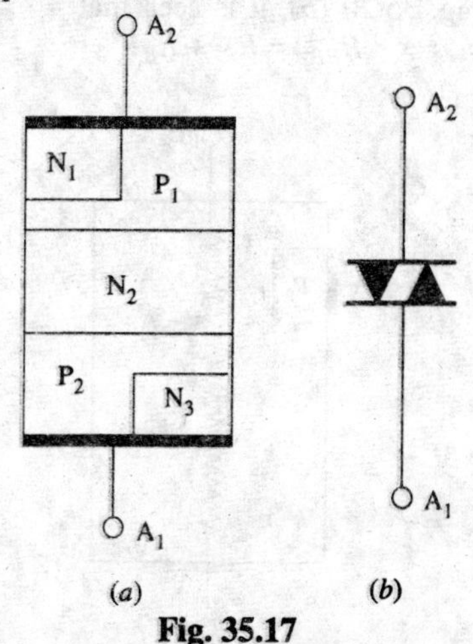

Fig. 35.17

(*b*) V/I Characteristics

The characteristic curve of a typical diac is shown in Fig. 35.18. It resembles the letter *Z* since diac breaks down in either direction.

(*c*) Applications

As stated above, diac has symmetrical bi-directional switching characteristics. Because of this feature, diacs are frequently used as triggering devices in triac phase control circuits used

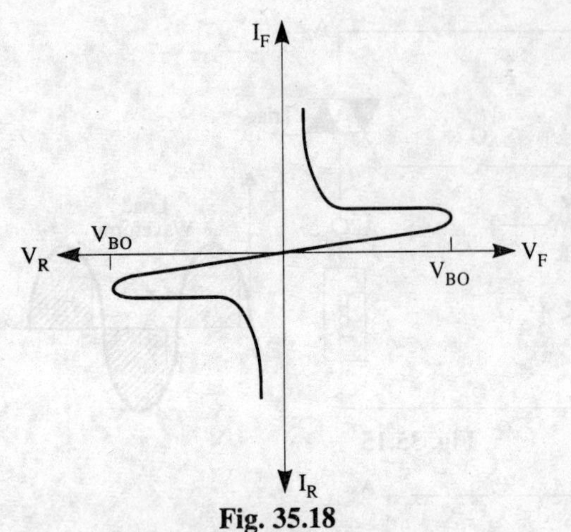

Fig. 35.18

light dimming, universal motor speed control and heat control etc.

35.9. Uni-junction Transistor

Basically, it is a three-terminal silicon diode. As its name indicates, it has only one P-N junction. It differs from an ordinary diode in that it has **three** leads and it differs from a FET in that it has **no ability to amplify**. However, it has the ability *to control a large a.c. power with a small signal.* It also exhibits a negative resistance characteristic which makes it useful as an oscillator.

(*a*) Construction

It consists of a *lightly-doped* silicon bar with a heavily-doped P-type material alloyed to its one side (closer to B_2) for producing single P-N junction. As shown in Fig. 35.19 (*a*), there are three terminals : one emitter, E and two bases B_2 and B_1 at the top and bottom of the silicon bar. The emitter leg is drawn at an angle to the vertical and arrow points in the direction of *conventional* current when UJT is in the conducting state.

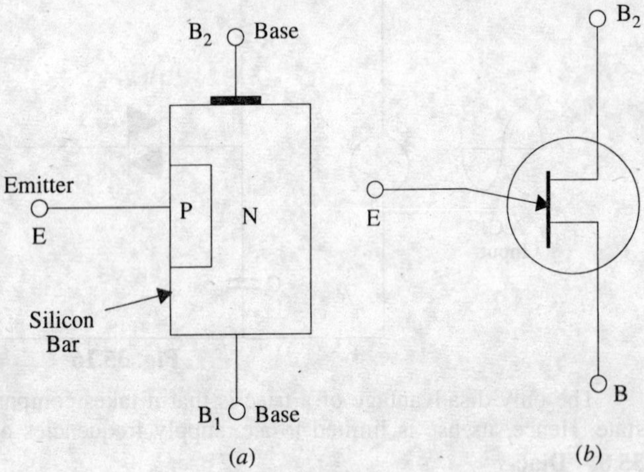

Fig. 35.19

(*b*) Interbase Resistance (R_{BB})

It is the resistance B_2 and B_1 *i.e.*, it is the total resistance of the silicon bar from one end to the other with emitter terminal open [Fig. 35.20 (*a*)].

From the equivalent circuit of Fig. 35.20 (*b*), it is seen that

$$R_{BB} = R_{B_2} + R_{B_1}$$

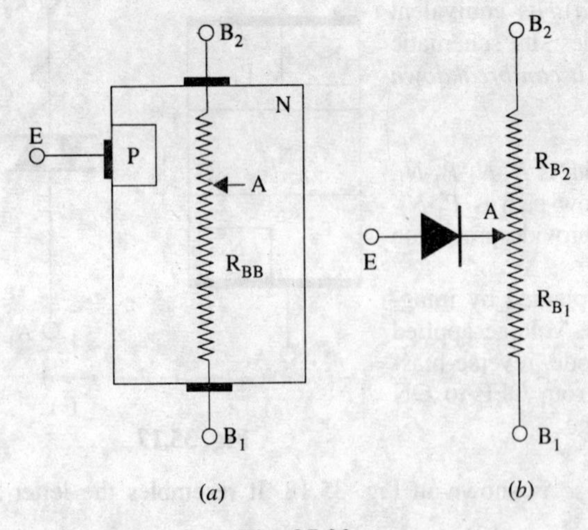

Fig. 35.20

It should also be noted that point A is such that $R_{B_1} > R_{B_2}$. Usually, $R_{B_1} = 60\%$ of R_{BB}.

(*c*) Intrinsic Stand-off Ratio

As seen from Fig. 35.21 (*a*), when a battery of 30 V is applied across B_2B_1, there is a progressive fall of voltage over R_{BB} provided E is open. It is obvious from Fig. 35.21 (*b*) that emitter acts as a voltage-divider tap on fixed resistance R_{BB}.

With emitter open, $I_1 = I_2$. The interbase current is given by Ohm's Law

$$I_1 = I_2 \frac{V_{BB}}{R_{BB}}$$

For example, if V_{BB} = 30 V and R_{BB} = 20 K, $I_1 = I_2$ = 2 mA.

It may be noted that part of V_{BB} is dropped over R_{B_2} and part on R_{B_1}. Let us call the voltage drop across R_{B_1} as V_A. Using simple voltage divider relationship

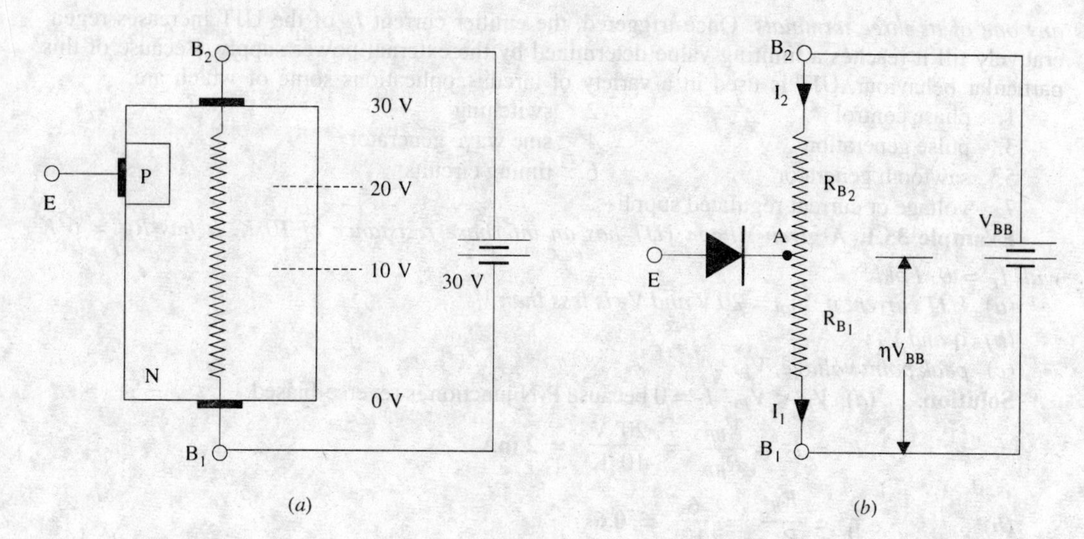

Fig. 35.21

$$V_A = V_{BB} \frac{R_{B_1}}{R_{B_1} + R_{B_2}}$$

The voltage division factor is given a special symbol (η) and the name of **'intrinsic stand-off ratio'.**

$$\eta = \frac{R_{B_1}}{R_{B_1} + R_{B_2}}$$

$$\therefore \qquad V_A = \eta\, V_{BB}$$

The intrinsic stand-off ratio is the property of the UJT and is always less than unity (0.5 to 0.85). If V_{BB} = 30 V and η = 0.6, then potential of point *A* with respect to point B_1 = 0.6 × 30 = 18 V. The remaining 12 V drop across R_{B_2}.

(*d*) Operation

When V_{BB} is switched on, V_A is developed and *reverse-biases the junction.* If V_B is the barrier voltage of the P-N junction, then total reverse bias voltage is

$$= V_A + V_B = \eta V_{BB} + V_B$$

Value of V_B for Si is 0.7 V.

It is obvious that emitter junction will not become forward-biased unless its applied voltage V_E exceeds (η V_{BB} + V_B). This value of V_E is called *peak-point voltage* V_P. When $V_E = V_P$, emitter (peak current) I_P starts to flow through R_{B_1} to ground (*i.e.*, B_1). The UJT is then said to have been *fired* or turned ON. Due to the flow of I_E (=I_P) through R_{B_1}, number of charge carriers in R_{B_1} is increased which *reduces its resistance*. As η depends on R_{B_1}, *its value is also decreased.*

Hence, we find that as V_E and hence I_E increases (beyond I_P), R_{B_1} decreases, η decreases and V_A decreases. This decrease in V_A causes more emitter current to flow which causes a further reduction in R_{B_1}, η and V_A. Obviously, the process is regenerative. V_A as well as V_E quickly drop as I_E increases. Since V_E decreases when I_E increases, the UJT *possess negative resistance.*

It is seen that only terminals *E* and B_1 are the active terminals whereas B_2 is the bias terminal *i.e.*, it is meant only for applying external voltage across the UJT.

Generally, UJT is triggered into conduction by applying a suitable positive pulse at its emitter. It can be brought back to OFF state by applying a negative trigger pulse.

(*e*) Applications

One unique property of UJT is that it can be triggered by (or an output can be taken from) *any one of its three terminals.* Once triggered, the emitter current I_E of the UJT increases regeneratively till it reaches a limiting value determined by the external power supply. Because of this particular behaviour, UJT is used in a variety of circuit applications some of which are:

1. phase control
2. switching
3. pulse generation
4. sine wave generator
5. sawtooth generator
6. timing circuits
7. voltage or current regulated supplies.

Example 35.1. *A given silicon UJT has an interbase resistance of 10 K. It has $R_{B_1} = 6$ K with $I_E = 0$. Find*

(a) UJT current if $V_{BB} = 20$ V and V_E is less than V_P;

(b) η and V_A;

(c) peak point voltage, V_P.

Solution. (*a*) $V_E < V_P$, $I_E = 0$ because P-N junction is reverse-biased.

$$\therefore \quad I_1 = I_2 = \frac{V_{BB}}{R_{BB}} = \frac{20 \text{ V}}{10 \text{ K}} = \mathbf{2\ mA}$$

(*b*)
$$\eta = \frac{R_{B_1}}{R_{BB}} = \frac{6}{10} = \mathbf{0.6}$$

HIGHLIGHTS

1. SCR is a rectifier with a control element. It is widely used as a switching device in power control applications. It can switch ON for variable lengths of time, thereby delivering selected amount of power to the load. It, in fact, possesses the advantages of a rheostat and a switch with none of their disadvantages.
2. Gate-turn-off switch (GTO) or gate-controlled switch (GCS) is identical to an SCR in construction but the difference is that it can be turned OFF by a negative gate pulse whereas an SCR cannot be.
3. Silicon-controlled switch (SCS) is a low-current SCR with two gate terminals. It can be switched ON or OFF by a suitable pulse applied at either gate.
4. Triac is a 3-terminal bi-directional device having two anodes and one gate. It can be triggered into conduction by both positive and negative voltages at its anodes and with both positive and negative triggering pulses at its gate. Unlike SCR, it allows current flow in both directions.
5. Diac is just a triac without its gate terminal. It can breakdown in either direction.
6. Unijunction transistor is a three-terminal silicon diode having one P-N junction. It possesses negative resistance characteristic which makes it useful as an oscillator.

OBJECTIVE TESTS—35

A. Fill in the following blanks.

1. Basically, an SCR is a rectifier with a ...... element.
2. An SCR has the advantages of a ...... and a switch with none of their disadvantages.
3. Once 'fired', an SCR remains ON even when triggering pulse is ....
4. A GTO can be turned OFF by applying a ...... gate signal.
5. A silicon-controlled switch is a low-current SCR with...... gate terminals.
6. A triac is called a ...... device because it allows current flow in both directions.
7. A diac is nothing else but a triac without its ...... terminal.
8. A UJT possesses ...... resistance.

B. Answer True or False.

1. An SCR is a unidirectional device unlike triac which is bi-directional
2. A GTO can, like an SCR, be turned OFF by a negative pulse.
3. An SCS can be operated as an SCR.
4. A triac has latch current in either direction.
5. A triac has six doped regions.
6. A diac can be imagined to consist of two diodes connected in parallel.

7. A unijunction transistor is just like an ordinary transistor but with one P-N junction.
8. A UJT can be triggered by any one of its three terminals.

C. Multiple Choice Questions.

1. In an SCR, the function of the gate is to
   (*a*) switch it OFF
   (*b*) control its firing
   (*c*) make it unidirectional
   (*d*) reduce forward breakdown voltage
2. The silicon-controlled switch (SCS)
   (*a*) cannot be used as an SCR
   (*b*) is a high-current SCR
   (*c*) cannot be switched ON by a negative pulse
   (*d*) is a four-layer four-terminal device
3. A triac can be triggered into conduction by
   (*a*) only positive voltage at either anode
   (*b*) positive or negative voltage at either anode
   (*c*) positive or negative voltage at gate
   (*d*) both (*b*) and (*c*)
4. A diac is nothing else but a / an
   (*a*) triac without its gate terminal
   (*b*) GTO with two gates
   (*c*) SCS with one gate
   (*d*) transistor with one junction.
5. A unijunction transistor has
   (*a*) anode, cathode and a gate
   (*b*) two bases and one emitter
   (*c*) two anodes and one gate
   (*d*) anode, cathode and two gates
6. A UJT has R_{BB} = 10 K and R_{B_2} = 4 K. Its intrinsic stand-off ratio is
   (*a*) 0.6 (*b*) 0.4
   (*c*) 2.5 (*d*) 5 / 3

ANSWERS

A. 1. control 2. rheostat 3. removed 4. negative 5. two 6. bi-directional 7. gate 8. negative

B. 1. T 2. F 3. T 4. T 5. T 6. F 7. F 8. T

C. 1. *a* 2. *d* 3. *d* 4. *a* 5. *b* 6. *a*

36 DIGITAL ELECTRONICS

36.1. Introduction

Digital electronics is concerned with two-state switching type circuits in which signals are in the form of electrical pulses as shown in Fig. 36.1. The outputs and inputs of digital devices involve only two levels of voltage referred to as 'high' and 'low'. In digital systems, the information being operated on is usually present in binary form. In fact, digital electronics is the world of logical 0 s and 1 s.

Digital circuits are used in hand-held calculators, electronic watches, computers, in control systems (for domestic appliances such as washing machines, for robots and for production processes in industry), TV and other games and, increasingly, in electronic telecommunication.

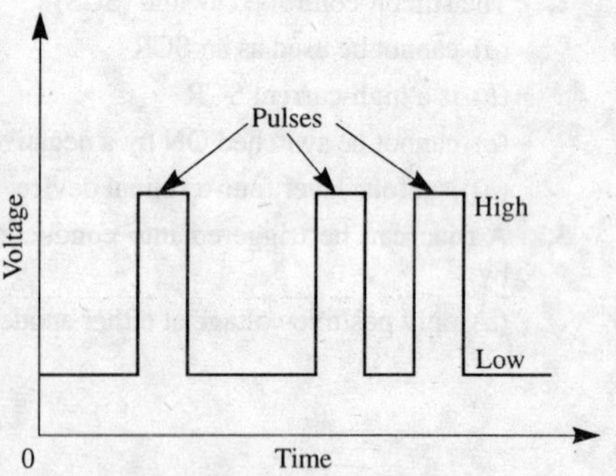

Fig. 36.1

Digital circuits are also used in digital multi-meters (DMMs) and digital capacitance meters and frequency meters etc.

Digital electronic circuits are being increasingly used in modern automobiles. There is a microprocessor in the ignition circuit to precisely control many ignition and fuel system variables. There is also digital tachometer and a digital speedometer.

Domestic appliances such as microwave ovens, washers and dryers have sophisticated microprocessor-controlled digital circuitry.

36.2. Why Use Digital Circuits?

Digital electronics and circuits have become very popular because of some distinct advantages they have over analog systems.

1. Digital systems can be fabricated into integrated circuit (IC) chips. These ICs can be used to form digital circuits with few external components.
2. Information can be stored for short periods or indefinitely.
3. Data can be used for precise calculations.
4. Digital systems can be designed more easily using suitable logical families.
5. Digital systems can be programmed and they show a certain manner of 'intelligence'.

However, digital circuitry has following disadvantages as compared to analog systems.

1. Most real-world events are analog in nature.
2. Analog processing is usually simpler and faster.

However, digital circuits are appearing in more and more products primarily because of the availability of low-cost reliable digital ICs. Other reasons for the growing popularity are accuracy, added stability, computer compatibility, ease of use and simplicity of design.

36.3. Numbers used in Digital Electronics

Digital electronic devices do not use the familiar decimal system of numbers but instead use binary number system which consists of 0 s and 1 s. Other number systems in use are octal and hexadecimal. We will discuss these number systems one-by-one.

36.4. Decimal Number System

This system has a base of 10 and is a *position value system* (meaning that value of a digit depends on its *position*). It has the following characteristics:

(*i*) Base or Radix

It is defined as the number of different digits which can occur in each position in the number system.

This system has ten unique digits *i.e.,* 0, 1, 2, 3, 4, 5, 6, 7, 8, 9. Anyone of these may be used in each position of the number. Hence, base of this system is 10.

(*ii*) Position Value

The absolute value of each digit is fixed but its *position value* (or place-value or weight) is determined by its position in the overall number. For example, position value of 3 in 3000 is not the same as in 300. Also, position value of each 5 in the number 5555 is different as shown in Fig. 36.2.

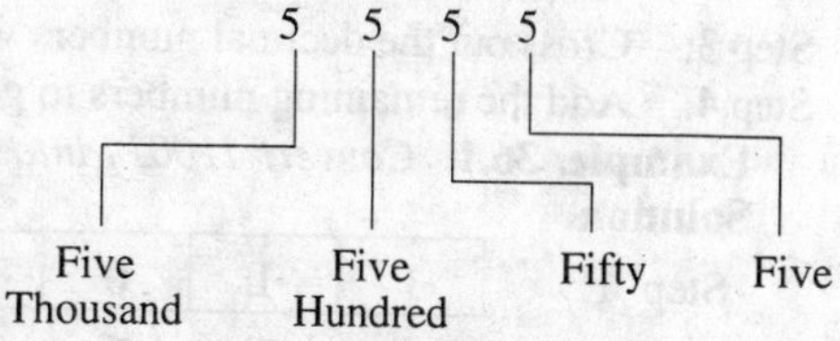

5 5 5 5

Five Thousand | Five Hundred | Fifty | Five

Fig. 36.2

Similarly, the number 2573 can be broken down as follows:

$$2573 = 2 \times 10^3 + 5 \times 10^2 + 7 \times 10^1 + 3 \times 10^0$$

It will be noted that in this number, 3 is the least signficant digit (LSD) whereas 2 is the most significant digit (MSD). Again, the number 2573.469 can be written as

$$2573.469 = 2 \times 10^3 + 5 \times 10^2 + 7 \times 10^1 + 3 \times 10^0 + 4 \times 10^{-1} + 6 \times 10^{-2} + 9 \times 10^{-3}$$

It is seen that position values are found by raising the base of the number system (10 in this case) to the power of the position. Also, powers are numbered to the left of the decimal point starting with zero and to the right of the decimal point starting with −1.

36.5. Binary Number System

Like decimal system, it has a radix and it also uses the same type of position value system.

(*i*) Radix

Its radix is 2 because it uses two digits 0 and 1 (the two binary digits are contracted to *bits*). All binary numbers consist of a string of 0 s and 1 s. Examples are : 10, 101 and 1011 which are read as one-zero, one-zero-one and one-zero-one-one to avoid confusion with decimal numbers. Another way to avoid confusion is to add a subscript of 10 for decimal numbers and of 2 for binary numbers as illustrated below:

10_{10}, 101_{10}, 5632_{10} .... decimal numbers

10_2, 101_2, 11001_2 .... binary numbers

(*ii*) Position Value

In binary system, position value of each bit corresponds to some power of 2. In each binary number, the value increases by powers of 2 starting with 0 to the left of the binary point and decreases to the right of the binary point starting with the power of −1. The position value (or weight) of each bit in the binary number 1101. 011 is shown in Fig. 36.3.

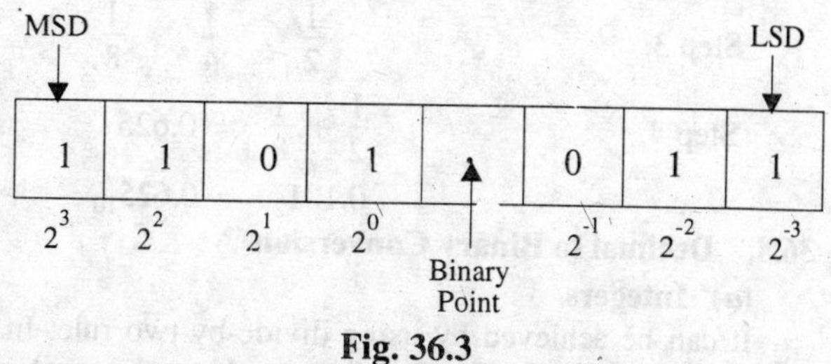

Fig. 36.3

As seen, the fourth bit to the left of the binary point is MSD and the third bit to the right of the binary point is LSD.

The decimal equivalent of the above binary number may be found as under:

$$1101.011_2 = (1 \times 2^3) + (1 \times 2^2) + (0 \times 2^1) + (1 \times 2^0) + (0 \times 2^{-1}) + (1 \times 2^{-2}) + (1 \times 2^{-3})$$

$$= 8 + 4 + 0 + 1 + 0 + 1/4 + 1/8 = 13.375_{10}$$

Binary numbers are used extensively by all digital systems primarily due to the nature of electronics itself. The bit one may be represented by a saturated (fully-conducting) transistor, a light turned ON, a relay energised or a magnet magnetized in a particular direction. The bit 0, on the other hand, can be represented by a cut-off transistor, a light turned OFF, a relay de-energised or a magnet magnetized in the opposite direction. In all such cases, there are only two values which a device can assume.

36.6. Binary to Decimal Conversion

Following procedure should be adopted for converting a given binary integer (whole number) into its equivalent decimal number.

Step 1. Write the binary number *i.e.*, all its bits in a row.

Step 2. Directly under the bits and starting from right to left, write 2^0, 2^1, 2^2, 2^3, 2^4 etc., or 1, 2, 4, 8, 16 etc.

Step 3. Cross out the decimal numbers which lie under 0 bits.

Step 4. Add the remaining numbers to get the decimal equivalent.

Example. 36.1. *Convert 11001_2 into its equivalent decimal number.*

Solution.

Step 1.	1	1	0	0	1
Step 2.	2^4	2^3	2^2	2^1	2^0
Step 3.	16	8	~~4~~	~~2~~	1

Step 4. $16 + 8 + 1 = 25$

$\therefore \quad 11001_2 = 25_{10}$

36.7. Binary Fractions

Here, procedure is the same as for binary integers except that the following weights are used for different bit positions:

•	2^{-1}	2^{-2}	2^{-3}	2^{-4}	....
	$\frac{1}{2}$	$\frac{1}{4}$	$\frac{1}{8}$	$\frac{1}{16}$	....

Binary point

Example 36.2. *Convert the binary fraction 0.101 into its decimal equivalent.*

Solution. The following four steps would be used for this purpose.

Step 1.	0	•	1	0	1
Step 2.			$\frac{1}{2}$	$\frac{1}{4}$	$\frac{1}{8}$
Step 3.			$\frac{1}{2}$	~~$\frac{1}{4}$~~	$\frac{1}{8}$

Step 4. $\frac{1}{2} + \frac{1}{8} = 0.625$

$\therefore \quad 0.101_2 = 0.625_{10}$

36.8. Decimal to Binary Conversion

(*a*) Integers

It can be achieved by using divide-by-two rule. In this method, we progressively divide the given decimal number by 2 and write down the remainder after each division. These remainders taken in the *reverse* order (*i.e.*, from bottom to top) form the required binary number. As an example, let us convert the decimal number 37 into its binary equivalent.

$37 \div 2 = 18$ with a remainder of 1 ↑
$18 \div 2 = 9$ with a remainder of 0
$9 \div 2 = 4$ with a remainder of 1
$4 \div 2 = 2$ with a remainder of 0
$2 \div 2 = 1$ with a remainder of 0
$1 \div 2 = 0$ with a remainder of 1

$\therefore \quad 37_{10} = 100101_2$

It may also be put in the following form:

$37 \div 2 = 18 + 1$
$18 \div 2 = 9 + 0$
$9 \div 2 = 4 + 1$
$4 \div 2 = 2 + 0$
$2 \div 2 = 1 + 0$
$1 \div 2 = 0 + 1$

$\therefore$ decimal 37 = 1 0 0 1 0 1

(*b*) Fractions

In this case, multiply-by-2 rule is used *i.e.*, we multiply each bit by 2 and record the carry in the integer position. These carries taken in the forward (top to bottom) direction give the required binary fraction. Let us convert the decimal fraction 0.8125_{10} into its binary equivalent.

$0.8125 \times 2 = 1.625 = 0.625$ with a carry of 1
$0.625 \times 2 = 1.25 = 0.25$ with a carry of 1
$0.25 \times 2 = 0.5 = 0.5$ with a carry of 0
$0.5 \times 2 = 1.0 = 0$ with a carry of 1

$\therefore \quad 0.8125_{10} = 0.1101_2$

It may be noted that we have to add the binary points from our side.

36.9. Binary Operations

Now, we will consider the following four binary operations:

1. addition 2. substraction 3. multiplication 4. division.

Addition is the most important of these four operations. In fact, by using 'complements', subtraction can be reduced to addition. Similarly, multiplication is nothing but repeated addition and, finally, division is but repeated substraction.

36.10. Binary Addition

The rules for binary addition are as under:

(*i*) $0 + 0 = 0$ (*ii*) $0 + 1 = 1$
(*iii*) $1 + 0 = 1$ (*iv*) $1 + 1 = 0$ with a carry of 1.
or $= 10_2$

36.11. Binary Substraction

The four rules for binary substraction are as under:

(*i*) $0 - 0 = 0$ (*ii*) $1 - 0 = 1$
(*iii*) $1 - 1 = 0$ (*iv*) $0 - 1 = 1$ with a borrow of 1 from the next column of the minuend.
or $10 - 1 = 1$

36.12. Binary Multiplication

The four rules for binary multiplication are as under:

(*i*) $0 \times 0 = 0$ (*ii*) $0 \times 1 = 0$
(*iii*) $1 \times 0 = 0$ (*iv*) $1 \times 1 = 1$

36.13. Binary Division

It is similar to the division in the decimal system. The two rules are

(*i*) $0 \div 1 = 0$ (*ii*) $1 \div 1 = 1$

36.14. Octal Number System

(*i*) Radix or base

It has a base of 8 which means that it has 8 distinct counting digits 0, 1, 2, 3, 4, 5, 6 and 7.

These digits 0 through 7 have exactly the same physical meaning as in decimal system.

For counting beyond 7, 2-digit combinations are formed taking the second digit followed by the first, then the second followed by the second and so on. Hence, after 7 the next octal number is 10 (second digit followed by first), then 11 (second digit followed by second) and so on. Hence, different octal numbers are:

0, 1, 2, 3, 4, 5, 6, 7,
10, 11, 12, 13, 14, 15, 16, 17,
20, 21, 22, 23, 24, 25, 26, 27,
30, 31, 32, 33, ...

(*ii*) Position Value

The position value for each digit is given by different powers of 8 as shown below

$\leftarrow 8^3 \quad 8^2 \quad 8^1 \quad 8^0 \quad . \quad 8^{-1} \quad 8^{-2} \quad 8^{-3} \rightarrow$

↑
octal point

36.15. Octal to Decimal Conversion

Procedure is exactly the same as given in Art. 36.6 except that we will use digit of 8 rather than 2. Procedure for converting octal number 206.104_8 is under :

2	0	6	•	1	0	4
8^2	8^1	8^0		8^{-1}	8^{-2}	8^{-3}

$$\therefore \quad 206.104_8 = 2 \times 8^2 + 0 \times 8^1 + 6 \times 8^0 + 1 \times 8^{-1} + 0 \times 8^{-2} + 4 \times 8^{-3}$$

$$= 128 + 6 + \frac{1}{8} + \frac{1}{28} = \left(134 \frac{17}{128}\right)_{10}$$

36.16. Decimal to Octal Conversion

The same method is used as in Art. 36.8 except that 8 is the multiplying factor for integers and is dividing factor for fractions.

The octal equivalent of the decimal number 175_{10} can be found as under:

$$175 \div 8 = 21 + 7$$
$$21 \div 8 = 2 + 5$$
$$2 \div 8 = 0 + 2$$
$$\therefore \quad 175_{10} = 2\ 5\ 7_8$$

Now, take decimal fraction 0.15. Its octal equivalent can be found as under :

$0.15 \times 8 = 1.20 = 0.20$ with a carry of 1
$0.20 \times 8 = 1.60 = 0.60$ with a carry of 1
$0.60 \times 8 = 4.80 = 0.80$ with a carry of 4

$\therefore \quad 0.15_{10} \equiv 114_8$

As seen, here carries have been taken in the forward direction *i.e.*, from top to bottom.

36.17. Hexadecimal Number System

The characteristics of this system are as under:

(*i*) It has a base of 16. Hence, it uses sixteen distinct counting digits 0 through 9 and A through F as detailed below:

0, 1, 2, 3, 4, 5, 6, 7, 8, 9, A, B, C, D, E, F

(*ii*) Place value for each digit is in ascending power of 16 for integers and descending power of 16 for fractions.

The main use of this system is in connection with byte-organised machines. It is used for specifying addresses of different binary numbers stored in the computer memory.

The advantage of hexadecimal system lies in its ability to convert directly from a 4-bit binary number. For example, hexadecimal F stands for the 4-bit binary number 1111. Similarly, hexadecimal number A6 represents the 8-bit binary number 10100110. Hexadecimal notation is widely used in microprocessor-based systems to represent 8-, 16-, or 32- bit binary numbers. The following table gives the equivalent numbers in the decimal, binary and hexadecimal systems

Table 36.1

Decimal	Binary	Hexadecimal
0	0000	0
1	0001	1
2	0010	2
3	0011	3

36.18. Binary Logic Gates

The basic building block of any digital circuit is a logic gate. Since logic gates operate with binary numbers, they are called *binary logic gates*. They can be constructed by using simple switches, relays, transistors and diodes or ICs. Because of their easy availability and low cost, ICs are widely used to construct digital circuits.

A logic gate has one output but one or more inputs. The output signal appears only for certain combinations of input signals. They implement the hardware logic function based on the Boolean algebra.

A unique characteristic of the Boolean algebra is that variables used in it can assume only one of the two values *i.e.*, either 0 or 1.

36.19. Positive and Negative Logic

In computing systems, the number symbols 0 and 1 represent the two possible states of a circuit or device. The two states may be referred to as ON and OFF, CLOSED and OPEN, HIGH and LOW, TRUE and FALSE depending on the circumstances. The main point is that they must be symbolized by two opposite conditions.

Hence, in positive logic, a 1 represents

(*i*) an ON circuit (*ii*) a CLOSED switch

(*iii*) a HIGH voltage (*iv*) a TRUE statement

Consequently, a 0 represents

(*i*) an OFF circuit (*ii*) an OPEN switch

(*iii*) a LOW voltage (*iv*) a FALSE statement

In negative logic, just opposite conditions prevail.

Unless stated otherwise, we will be using only positive logic in this chapter.

36.20. The OR Gate

The electronic symbol for a two-input OR gate is shown in Fig. 36.4 (*a*) and its equivalent switching circuit in Fig. 36.4 (*b*). The two inputs have been marked as *A* and *B* and the output is *C*. As per Boolean algebra, the three variables *A*, *B* and *C* can have only one of the two values *i.e.*, either 0 or 1.

Logic Operation

The OR gate has an output of 1 when either *A* or *B* or both are 1.

In other words, it is any-or-all gate because an output occurs when any or all the inputs are present.

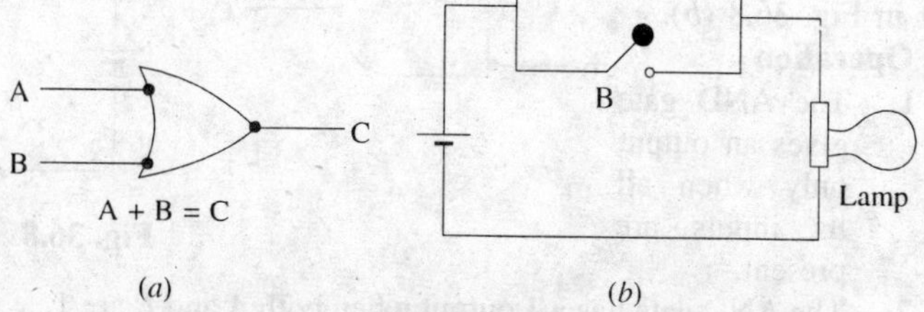

A + B = C

(*a*) (*b*)

Fig. 36.4

As seen from Fig. 36.4 (*b*), the lamp will light up (logic 1) when either switch *A* or *B* or both are closed.

Obviously, the output will be 0 if and only if both its inputs are 0. In terms of the switching conditions, it means that lamp would be OFF (logic 0) only when both switches A and B are OFF.

It will be seen that an OR gate is equivalent to a parallel circuit in its logic function.

The OR function is represented by the Boolean equation.

$$A + B = C$$

It may be noted that the plus (+) sign is the Boolean symbol for OR.

The logic operation of OR gate can be summarised with the help of the truth table given in Fig. 36.5.

A truth table is defined as a table *which gives the output state* ***for all possible input combinations.*** The table shown in Fig. 36.5 for an OR gate gives all outputs for all possible AB inputs of 00, 01, 10 and 11.

Table 36.2

A	B	C
0	0	0
0	1	1
1	0	1
1	1	1

Fig. 36.5

Another point worth remembering is that the above OR gate is called inclusive OR gate because it includes the case when both inputs are true. It may also be noted that OR gate performs logical addition according to Boolean algebra.

36.21. Exclusive OR Gate

Its electronic symbol is shown in Fig. 36.6 (*a*) and equivalent switching circuit in Fig. 36.6 (*b*). In this gate, output is 1 if any of its input but not both, is 1. In other words, it has an output 1 when its input are *different*. The output is 0 only when inputs *are the same.*

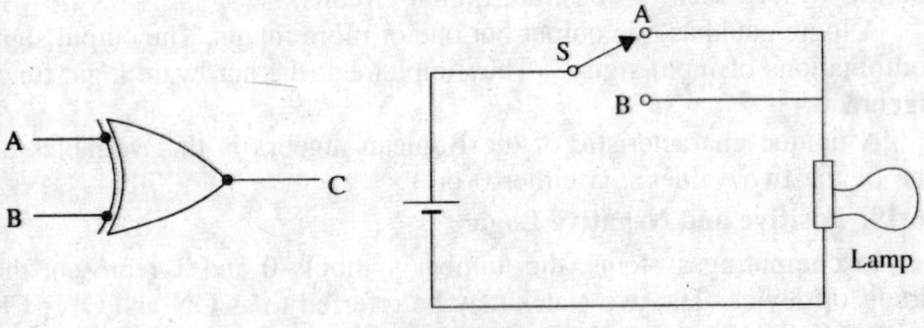

Fig. 36.6

To put it a bit differently, this logic gate has output when inputs are either all 0 or all 1.

This gate represents the Boolean equation

$$A \oplus B = C$$

The circle around plus (+) sign is worth noting.

The truth table for a 2-input exclusive OR (XOR) gate is given in table.

Table 36.3

A	B	C
0	0	0
0	1	1
1	0	1
1	1	0

Fig. 36.7

36.22. The AND Gate

The logic symbol for a 2-input AND gate is shown in Fig. 36.8 (*a*) and its equivalent switching circuit in Fig. 36.8 (*b*).

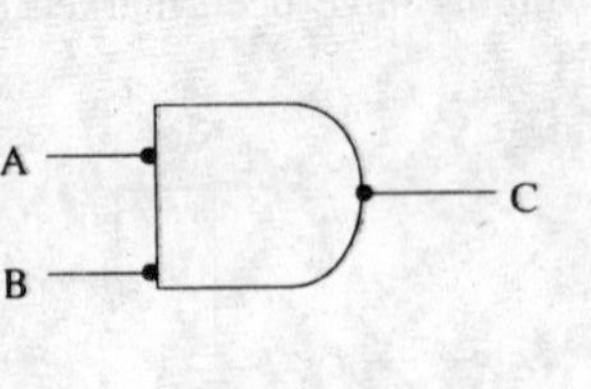

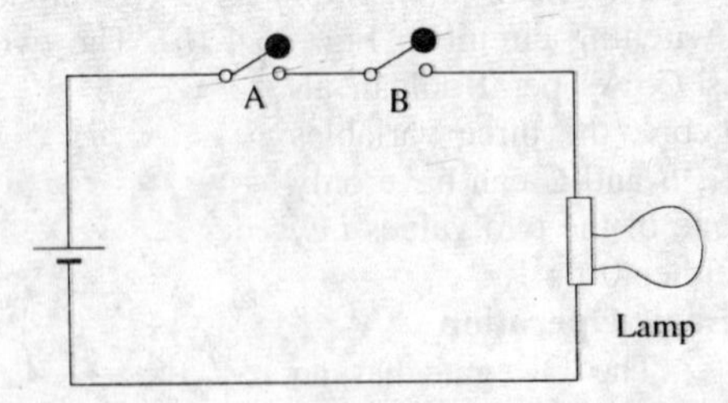

Fig. 36.8

Logic Operation

1. The AND gate gives an output only when all its inputs are present.
2. The AND gate has a 1 output when both A and B are 1.

Hence, this gate is an all-or-nothing gate whose output occurs only when all its inputs are present.

3. In True / False terminology, the output of an AND gate is true only if *all its inputs are true*. Its output would be false if *any* of its inputs is false.

The AND gate represents the Boolean equation.

$$A \times B = C$$

or $$A \,.\, B = C$$

or $$AB = C$$

The AND gate performs logical multiplication.

As seen from Fig. 36.8 (*b*), the lamp would be ON only when both switches *A* and *B* are closed. Even when one switch is open, the lamp would be OFF. Obviously, the performance of an AND gate is similar to a series switching circuit. The truth table of this gate is shown in Fig. 36.9.

Table 36.4

A	B	C
0	0	0
0	1	0
1	0	0
1	1	1

Fig. 36.9

36.23. The NOT Gate

It is so-called because its output is NOT the same as its input. It is also called an inverter because it inverts the input signal. It has one input and one output as shown in Fig. 36.10 (*a*). All it does is to invert (or complement) the input as seen from its truth table of Fig. 36.10 (*b*).

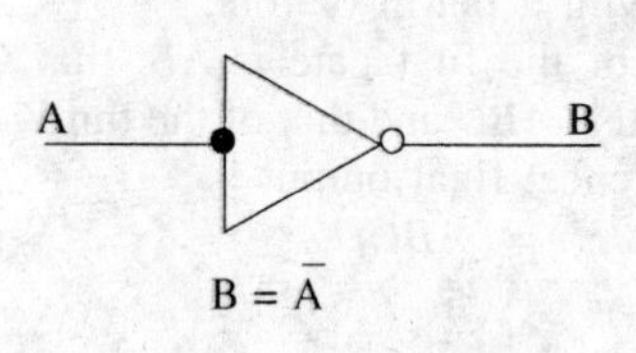

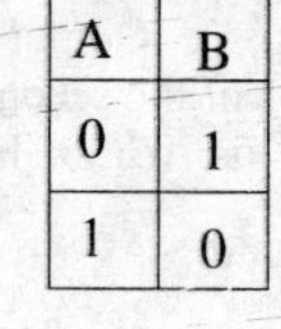

A	B
0	1
1	0

Fig. 36.10

The schematic symbol for inversion is a small circle as shown in Fig. 36.10 (*a*). The logical symbol for inversion or negation is a bar over the function to indicate the opposite state. For example, $\overline{A}$ means not–*A*. Similarly, $\overline{(A + B)}$ means the complement of $(A + B)$.

The operation of a NOT gate can also be defined as under:

$$\overline{0} = 1 \quad \text{or} \quad \overline{1} = 0$$

Double complementation gives the original value as shown in Fig. 36.11.

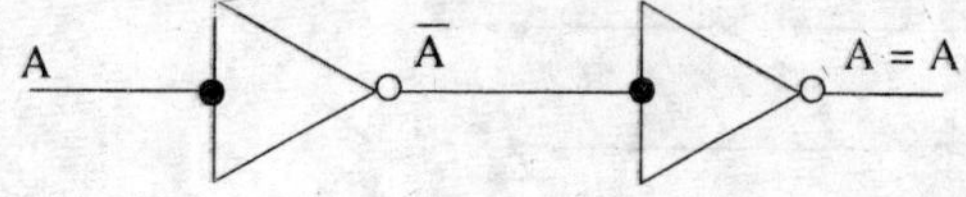

Fig. 36.11

For example

$$\overline{\overline{1}} = \overline{0} = 1 \quad \text{or} \quad \overline{\overline{0}} = \overline{1} = 0$$

Example 36.3. *Find the Boolean equation for the output D of Fig. 36.12 (a). Evaluate D when*

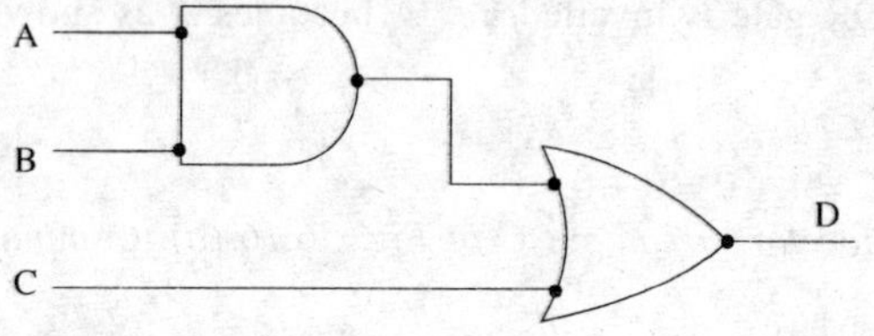

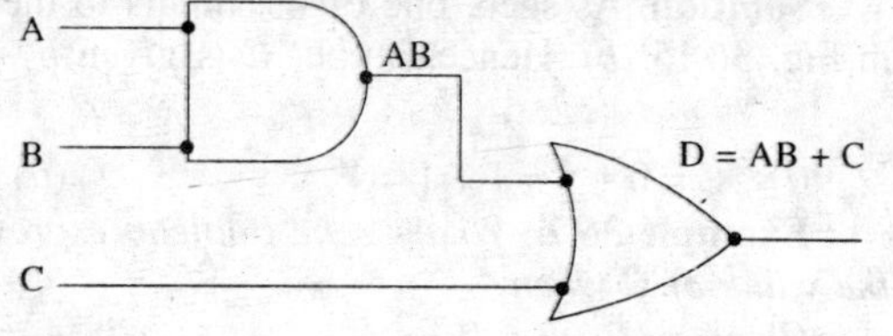

Fig. 36.12

(i) $A = 0$ $\quad B = 1$ $\quad C = 1$

(ii) $A = 1$ $\quad B = 1$ $\quad C = 1$

Solution. The output of the AND gate is *AB*. It then becomes one of the inputs for the 2-input OR gate. When *AB* is ORed with *C*, we get $(AB + C)$.

$\therefore \qquad D = AB + C$ ....Fig. 36.12 (*b*)

(*i*) $D = 0.1 + 1 = 0 + 1 = 1$ ...Art. 36.10

(*ii*) $D = 1.1 + 1 = 1 + 1 = 1$ ...Art. 36.10

Example 36.4. *Find the Boolean expression for the output of Fig. 35.13 (a) and evaluate it when*

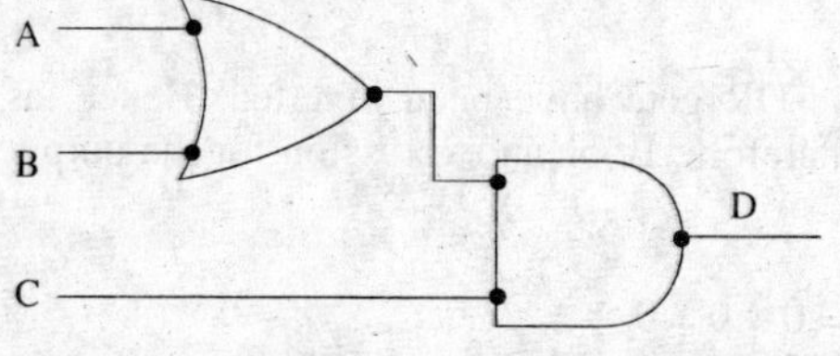

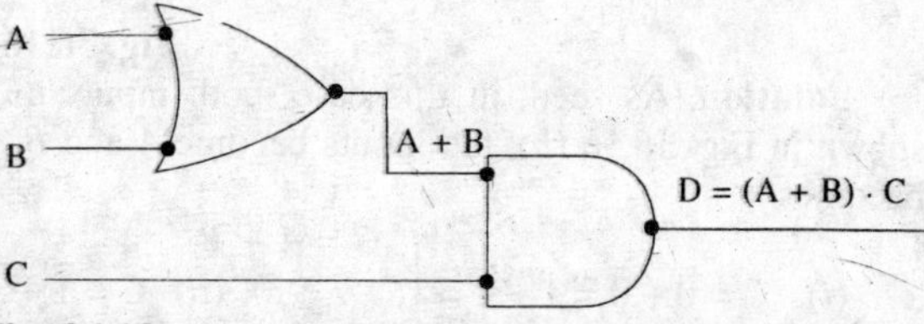

Fig. 36.13

(*i*) $A = 0, \quad B = 1, \quad C = 1$

(*ii*) $A = 1, \quad B = 1, \quad C = 0$

Solution. The output of the OR gate is $(A + B)$. Afterwards, it becomes the input of the AND gate. When ANDed with C, it becomes $(A + B)\ C$.

$$D = (A + B)C \qquad \text{...Fig. 36.13 } (b)$$

(*i*) $D = (0 + 1).1 = 1.1 = 1$ (*ii*) $D = (1 + 1).0 = 1.0 = 0$

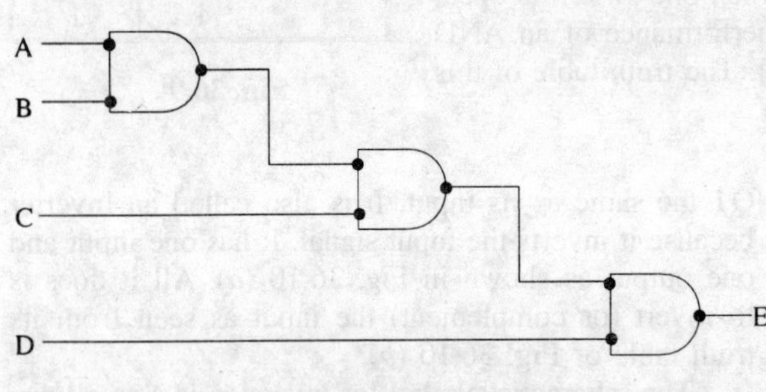

Fig. 36.14

Example 36.5 *Find the Boolean expression for the output of the Fig. 36.14 and compute its value when A = B = C = 1 and D = 0.*

Solution. The circuit is made up of three AND gates. Obviously, it is equivalent to a single 4-input AND gate *i.e.*, an AND gate with a fan-in of four.

Output of the first gate is AB, that of the second is ABC and that of the third is $ABCD$. Hence, final output is

$$E = ABCD$$

Substituting the given values, we get

$$E = 1.1.1.0 = 1.1.0 = 1.0 = 0$$

Example 36.6. *Find the Boolean expression for the output C of Fig. 36.15 (a) and compute its value when*

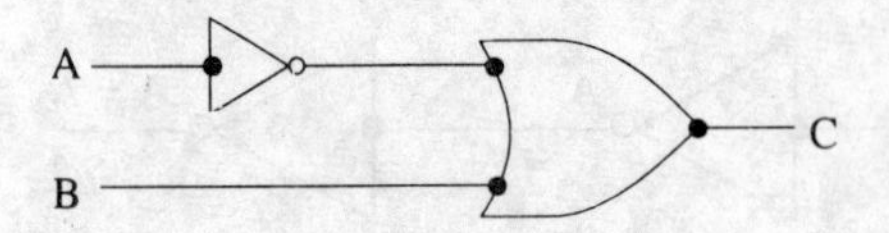

Fig. 36.15

(*i*) $A = 0, B = 1$ (*ii*) $A = 1, B = 0$

Solution. As seen, one of the inputs to the OR gate is inverted *i.e.*, A becomes $\bar{A}$ as shown in Fig. 36.15 (*b*). Hence, output C is given by

$$C = \bar{A} + B$$

(*i*) $C = \bar{0} + 1 = 1 + 1 = 1$ (*ii*) $C = \bar{1} + 0 = 0 + 0 = 0$

Example 36.7. *What is the Boolean expression for the output C of Fig. 36.16 (a). Compute the value of C when*

(*i*) $A = 0, \quad B = 0$ (*ii*) $A = 1, \quad B = 1$

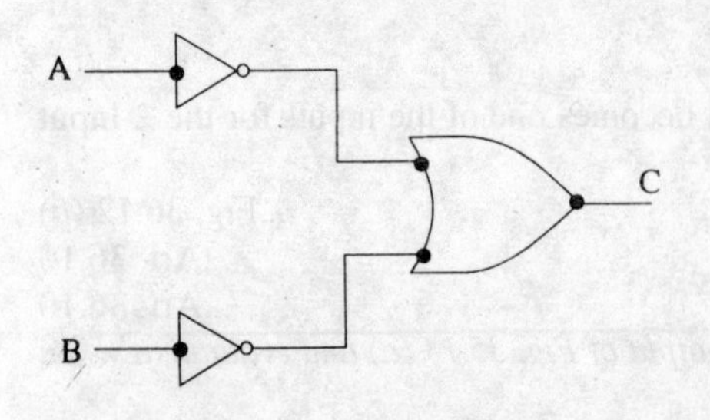

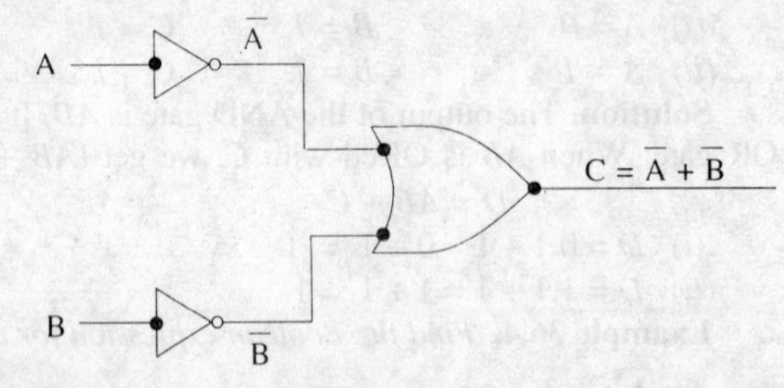

Fig. 36.16

Solution. As seen, in this case, both inputs to the OR gate have been inverted. Hence, as shown in Fig. 36.16 (*b*), the inputs become $\bar{A}$ and $\bar{B}$. Therefore, Boolean expression for the output becomes

$$C = \bar{A} + \bar{B}$$

(*i*) $C = \bar{0} + \bar{0} = 1 + 1 = 1,$ (*ii*) $C = \bar{1} + \bar{1} = 0 + 0 = 0$

36.24. Non-inverting Buffer / Driver

It has no logical purpose but is used to supply greater drive current at its output than is normal for a regular gate. Since, normal digital ICs have limited drive current capabilities, the non-inverting buffer/driver assumes importance when interfacing ICs with other devices such as LEDs and lamps etc. Logical symbol of a non-inverting buffer is shown in Fig. 36.17.

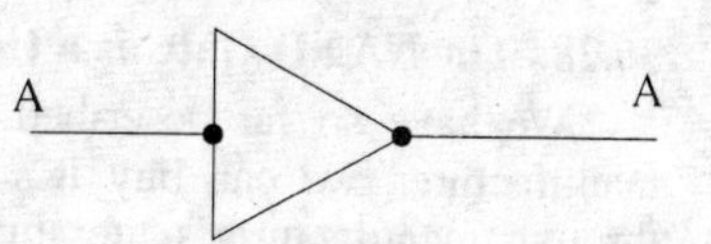

Fig. 36.17

36.25. The NOR Gate

It is, in fact, a NOT–OR gate. It can be made out of an OR gate by connecting an inverter in its output as shown in Fig. 36.18.

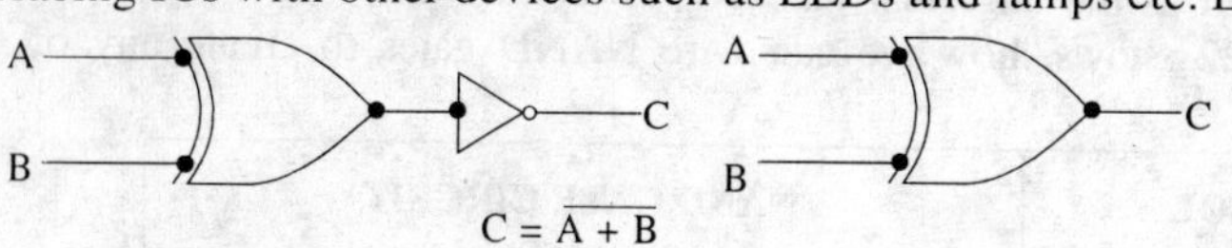

Fig. 36.18

The NOR gate represents the following Boolean equation.

$$C = \overline{A + B}$$

A NOR function is just the reverse of the OR function

Logic Operation

A NOR gate will have an output of 1 only when *all its inputs are 0*. Obviously, if any input is 1, the output will be 0. Alternatively, in an NOR gate, the output is true only when all inputs are false. The truth table for a 2-input NOR gate is shown in Fig. 36.19.

Table 36.5

A	B	C
0	0	1
0	1	0
1	0	0
1	1	0

Fig. 36.19

36.26. The Exclusive NOR Gate

The term exclusive NOR gate is often shortened to XNOR gate. In fact, it can be made out of a NOR gate by connecting an inverter in its output as shown in Fig. 36.20 (*a*). As seen, the logical symbol for this gate is the same as for the XOR gate but with the added invert bubble on the output side.

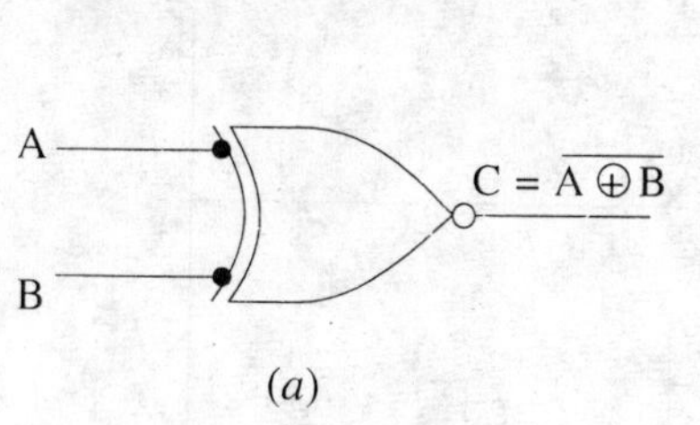

(*a*)

Table 36.6

A	B	C
0	0	1
0	1	0
1	0	0
1	1	1

(*b*)

Fig. 36.20

The truth table for the XNOR gate is given in Fig. 36.20 (*b*). It will be seen that this gate has an output 1 when its both inputs are either 0 or 1. It means that both of its inputs should be at the *same* logic level of either 0 or 1. Obviously, it produces no output if its two inputs are at the *opposite* logic level.

36.27. The NAND Gate

It is, in fact, a NOT-AND gate. It can be obtained by connecting a NOT gate in the output of an AND gate as shown in Fig. 36.21. Its output is represented by the Boolean equation.

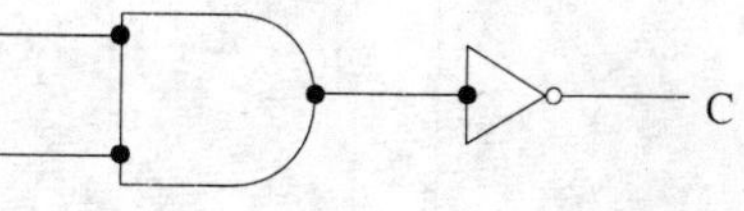

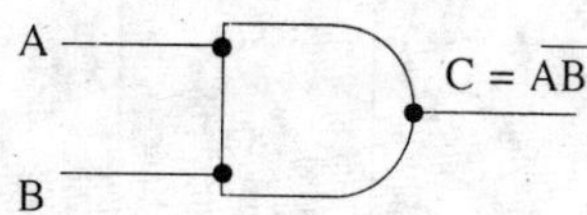

Fig. 36.21

$$C = \overline{AB}$$

Logic Operation

This gate gives an output of 1 if its both inputs are not 1. In other words, it gives an output of 1 if either *A* or *B* or both are 0. The truth table for a 2-input NAND gate is given in Fig. 36.22. It is just the opposite of the truth table for an AND gate.

Table 36.7

A	B	C
0	0	1
0	1	1
1	0	1
1	1	0

Fig. 36.22

36.28. The NAND Gate as a Universal Gate

We have so far described seven types of gating circuits and their characteristics. From a manufacturer, we can buy ICs that perform any of these seven basic functions. When looking through manufacturer's literature, it is found that NAND gates are more widely available than many other type of gates. The reason for this is that the NAND gate can be used to make other types of gates. Hence, it is known as the universal gate.

The table given in Fig. 36.22 shows, how we can wire NAND gates to create any of the other basic logic functions.

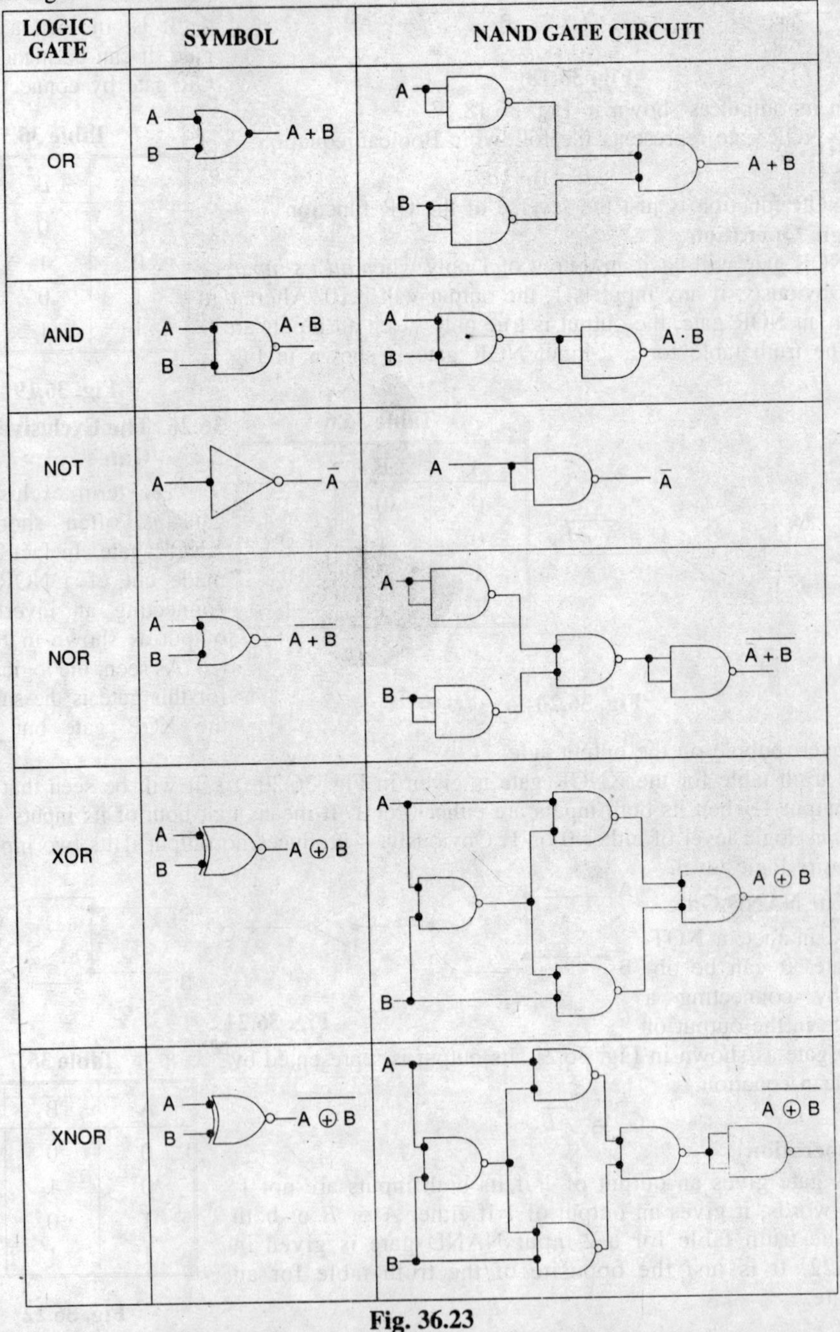

Fig. 36.23

36.29. Summary of the Gates

OR : output 1, unless all inputs 0.
XOR : output 1, when inputs different.
AND : output 1, only when all inputs 1.
NOT : output always opposite of input.
NOR : output 1, only when all inputs 0.
XNOR : output 1, when inputs at same logic level.
NAND : output 1, unless all inputs 1.

Example 36.8. *Convert the Boolean expression (AB + C) into a logic circuit using different logic gates.*

Solution. In such cases, it is best to start with the output and work towards the input. As seen, C has been ORed with AB. Hence, the output gate must be a 2-input OR gate as shown in Fig. 36.24 (*a*).

Now, term AB is an AND function. Hence, we need an AND gate with inputs A and B. The complete logic circuit is shown in Fig. 36.24 (*b*).

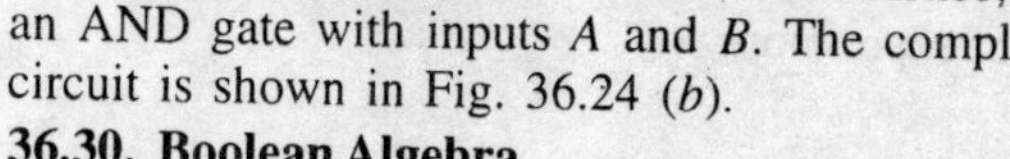

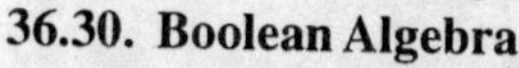

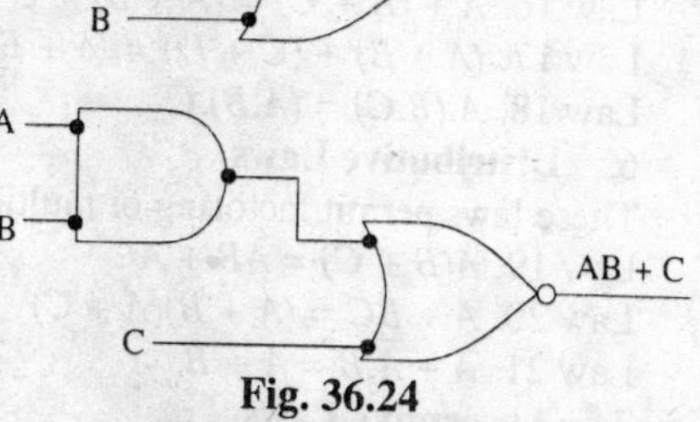

Fig. 36.24

36.30. Boolean Algebra

Boolean algebra is the algebra of logic presently applied to the operation of computer devices. It is a mathematical system of logic in which the truth functions are expressed as symbols which are then manipulated to arrive at a conclusion. This Boolean algebra is not the ordinary numerical algebra we know from our high school days but a totally new system called *logic algebra.* For example, in Boolean algebra, $A + A = A$ and not $2A$ as is the case in ordinary algebra.

Because of its logical nature, Boolean algebra is ideal for the design and analysis of logic circuits used in computers.

36.31. Laws of Boolean Algebra

This algebra has its own set of fundamental laws which are necessary for manipulating different Boolean expressions.

1. **OR Laws**

These are as under:

Law 1. $A + 0 \quad = A$

Law 2. $A + 1 \quad = 1$

Law 3. $A + A \quad = A$

Law 4. $A + \overline{A} \quad = 1$

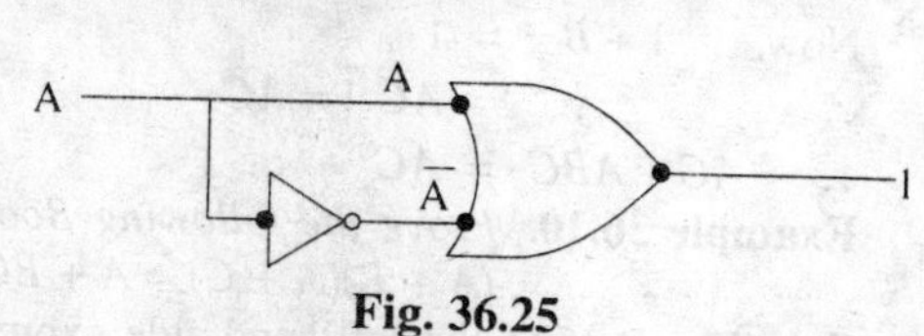

Fig. 36.25

The expression given in Law. 4 can be understood with the help of Fig. 36.25. Consider the following two possibilities.

(*i*) when $A = 0, \overline{A} = 1$

$\therefore \quad A + \overline{A} = 0 + 1 = 1$

(*ii*) when $A = 1, \quad \overline{A} = 0$

$\therefore \quad A + \overline{A} = 1 + 0 \quad = 1$

2. **AND Laws**

Law 5. $A.0 = 0;$ Law 6. $A.1 = A$

Law 7. $A.A = A;$ Law 8. $A.\overline{A} = 0$

The expression for Law 8 can be easily understood with the help of the logic circuit of Fig. 36.26. Consider the following two possibilities.

(*i*) when $A = 0, \overline{A} = 1$

$\therefore \quad A.\overline{A} = 0.1 = 0$

(*ii*) when $A = 1, \overline{A} = 0$

$\therefore \quad A.\overline{A} = 1.0 = 0$

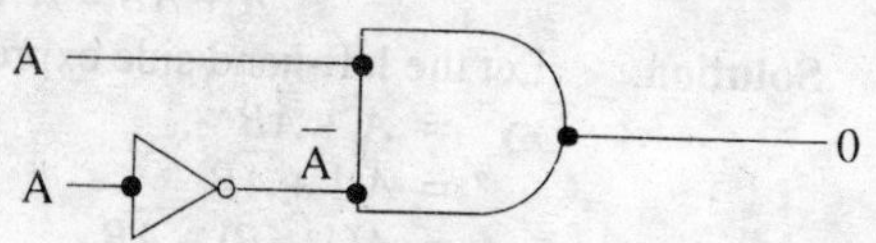

Fig. 36.26

3. **Laws of Complementation**

Law 9. $\overline{0} = 1$; Law 10. $\overline{1} = 0$

Law 11. if $A = 0$, then $\overline{A} = 1$;

Law 12. if $A = 1$, then $\overline{A} = 0$; Law 13. $\overline{\overline{A}} = A$

4. **Commutative Laws**

These laws allow change in the position of variables in OR and AND expressions.

Law 14. $A + B = B + A$; Law 15. $A.B = B.A$

These two laws express the fact that the order in which a combination of terms is performed does not affect the final result of the combination.

5. **Associated Laws**

These laws allow removal of brackets from logical expressions and regrouping of variables.

Law 16. $A + (B + C) = (A + B) + C$

Law 17. $(A + B) + (C + D) = A + B + C + D$

Law 18. $A.(B.C) = (A.B).C$

6. **Distributive Laws**

These laws permit factoring or multiplying out of expressions.

Law 19. $A(B + C) = AB + AC$

Law 20. $A + BC = (A + B)(A + C)$

Law 21. $A + \overline{A}.B = A + B$

7. **Absorptive Laws**

Law 22. $A + AB = A$

Law 23. $A.(A + B) = A$

Law 24. $A.(\overline{A} + B) = B$

The above laws can be used to prove any given Boolean identity and to simplify complicated expressions.

Example 36.9. *Prove the following Boolean identity:*

$$AC + ABC = AC$$

Solution. Taking the left-hand side expression as y, we get

$$y = AC + ABC = AC(1 + B) \quad \text{... Law 2}$$

Now, $1 + B = 1$ ... Law 6

$\therefore \quad y = AC.1 = AC$

$\therefore \quad AC + ABC = AC$

Example 36.10. *Prove the following Boolean identity*

$$(A + B)(A + C) = A + BC$$

Solution. Putting the left-hand side expression equal to y, we get

$$\begin{aligned}
y &= (A + B)(A + C) \\
&= AA + AC + AB + BC && \text{... Law 19} \\
&= A + AC + AB + BC && \text{... Law 7} \\
&= A + AB + AC + BC \\
&= A(1 + B) + AC + BC && \text{... Law 19} \\
&= A + AC + BC && \text{... Law 2} \\
&= A(1 + C) + BC && \text{... Law 19} \\
&= A + BC && \text{... Law 2}
\end{aligned}$$

$\therefore \quad (A + B)(A + C) = A + BC$

Example 36.11. *Prove the following Boolean identity*

$$A + \overline{A}B = A + B$$

Solution. Let the left-hand side expression be put equal to y.

$$\begin{aligned}
y &= A + \overline{A}B \\
&= A.1 + \overline{A}B && \text{... Law 6} \\
&= A(1 + B) + \overline{A}B && \text{... Law 2 and 14} \\
&= A.1 + AB + \overline{A}B && \text{... Law 19}
\end{aligned}$$

$$= A + BA + B\overline{A} \quad \text{... Law 6 and 15}$$
$$= A + B(A + \overline{A}) \quad \text{... Law 19}$$
$$= A + B.1 \quad \text{... Law 4}$$
$$= A + B \quad \text{... Law 6}$$
$$\therefore \quad A + \overline{A}B = A + B$$

36.32. DE MORGAN's Theorems

These two theorems or rules are a great help in simplifying complicated logical expressions. The theorems can be stated as under:

Law 25. $\overline{A + B} = \overline{A}.\overline{B}$

Law 26. $\overline{A.B} = \overline{A} + \overline{B}$

The first statement says that *the complement of a sum equals the product of complements.* The second statement says that *the complement of a product equals the sum of the complements.* In fact, it allows transformation from a sum-of-products form to a product-of-sum form.

As seen from the above two laws, the procedure required for taking out an expression from under a NOT sign is as follows:

1. complement the given expression *i.e.*, remove the overall NOT sign,
2. change all ANDs to ORs and all ORs to ANDs,
3. complement or negate all individual variables.

As an illustration, take the following example

$$\overline{A + BC} = A + BC \quad \text{... step 1}$$
$$= A(B + C) \quad \text{... step 2}$$
$$= \overline{A}(\overline{B} + \overline{C}) \quad \text{... step 3}$$

Next, consider this example.

$$\overline{(\overline{A} + B + \overline{C})\,(\overline{A} + B + C)} = (\overline{A} + B + \overline{C})(\overline{A} + B + C) \quad \text{.... step 1}$$
$$= \overline{A}\,B\,\overline{C} + \overline{A}\,B\,C \quad \text{... step 2}$$
$$= \overline{\overline{A}}\;\overline{B}\,\overline{\overline{C}} + \overline{\overline{A}}\;\overline{B}\,\overline{C} \quad \text{... step 3}$$
$$= A\,\overline{B}\,C + A\,\overline{B}\,\overline{C}$$

This process is called *demorganisation*. It should, however, be noted that opposite procedure would be followed in bringing an expression under the NOT sign.

Let us bring the expression $A + B + C$ under the NOT sign.

$$\overline{A} + \overline{B} + \overline{C} = \overline{\overline{A}} + \overline{\overline{B}} + \overline{\overline{C}} \quad \text{... step 3}$$
$$= A + B + C$$
$$= ABC \quad \text{... step 2}$$
$$= \overline{ABC} \quad \text{... step 1}$$

OBJECTIVE TESTS—36

A. Fill in the blanks with the most appropriate word(s) or numerical value(s).

1. Digital electronics is concerned with circuits in which signals are in the form of electrical ....
2. In digital systems, the information being operated on is usually present in .... form.
3. The radix of binary number system is ....
4. The divide-by-two-rule is used for decimal to .... conversion.
5. The decimal fraction 0.75 corresponds to the binary fraction of ....
6. The hexadecimal number D3 represents ....
7. The basic building block of any digital circuit is a .... gate.
8. A logic gate has .... output but one or more inputs.
9. In positive logic, a 1 represents a .... voltage.

B. Answer True or False.

1. Digital circuits are used in computers.

2. In digital circuits, information can be stored indefinitely.
3. Most real-world events are digital in nature.
4. In binary number system, the place value of a bit increases or decreases by powers of 2.
5. In 1110, the 0 carries the heighest weight.
6. All binary numbers consist of a string of 1 s and 0 s.
7. The divide-by-two method is used for decimal to binary convertion only.
8. According to binary addition, 1 + 1 = 0.
9. In octal number system, after 7 the next number is 10.
10. The main use of hexadecimal system is for byte-organised machines.
11. Binary logic gates implement the hardware logic function based on the Boolean algebra.
12. In XOR gate, output is 1 when inputs are similar.
13. In NOR gate, output is 1 when all inputs are 0.
14. The NAND gate is called the universal gate.
15. In Boolean algebra, $A + A = 2A$.
16. According to absorptive laws of Boolean algebra,

$$A + AB = A$$

C. Multiple Choice Items.

1. The digital systems usually operate on .... system.
   (*a*) binary (*b*) decimal
   (*c*) octal (*d*) hexadecimal
2. After counting 0, 1, 10, 11, the next binary number is
   (*a*) 12 (*b*) 100
   (*c*) 101 (*d*) 110
3. The binary addition of 1 + 1 + 1 gives
   (*a*) 111 (*b*) 10
   (*c*) 110 (*d*) 11
4. In octal number system, the number after 7 is
   (*a*) 8 (*b*) 10
   (*c*) 15 (*d*) 20
5. A logic gate is an electronic circuit which
   (*a*) makes logic decisions
   (*b*) allows electron flow only in one direction
   (*c*) works on binary algebra
   (*d*) alternates between 0 and 1 values
6. In positive logic, logic state 1 corresponds to
   (*a*) positive voltage
   (*b*) higher voltage level
   (*c*) zero voltage
   (*d*) lower voltage level
7. In negative logic, the logic state 1 corresponds to
   (*a*) negative voltage
   (*b*) zero voltage
   (*c*) more negative voltage
   (*d*) lower voltage level
8. The output of a 2-input OR gate is 0 only when its
   (*a*) both inputs are 0
   (*b*) either input is 1
   (*c*) both inputs are 1
   (*d*) either input is 0
9. When an input electrical signal $A = 10100$ is applied to a NOT gate, its output signal is
   (*a*) 01011 (*b*) 10101
   (*c*) 10100 (*d*) 00101
10. The only function of a NOT gate is to
    (*a*) stop a signal
    (*b*) recomplement a signal
    (*c*) invert an input signal
    (*d*) act as universal gate
11. A NOR gate is ON only when all its inputs are
    (*a*) ON (*b*) positive
    (*c*) high (*d*) OFF
12. Boolean algebra is essentially based on
    (*a*) symbols (*b*) logic
    (*c*) high (*d*) OFF
13. According to Absorptive Laws of Boolean algebra, $A.(A + B)$ equals
    (*a*) B (*b*) AB
    (*c*) $A + B$ (*d*) A

ANSWERS

A. Fill in the blanks

1. pulses 2. binary 3. 2 4. binary, octal 5. 0.11 6. 11010011
7. logic 8. one 9. higher

B. True or False

1. T 2. T 3. F 4. T 5. F 6. T 7. F 8. F 9. T 10. T
11. T 12. F 13. T 14. T 15. F 16. T

C. Multiple Choice

1. *a* 2. *b* 3. *d* 4. *b* 5. *a* 6. *b* 7. *d* 8. *a* 9. *a* 10. *c*
11. *d* 12. *b* 13. *d*

37 SINE WAVE OSCILLATORS

37.1. Function of an Oscillator

An oscillator is a circuit which converts electric energy at d.c. (zero frequency) to electric energy at frequency varying from a few Hz to gigahertz (GHz). Stated simply, it is a source of alternating voltage or current. An electronic oscillator is, essentially, a feedback amplifier that supplies its own input *i.e.*, it requires not external signal to initiate or maintain the energy conversion process.

The basic difference between an amplifier and an oscillator is shown in Fig. 37.1.

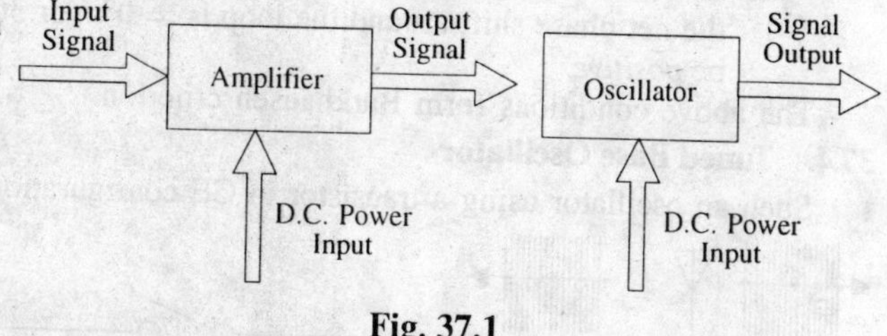

Fig. 37.1

37.2. Classification of Oscillator Circuits

Oscillator circuits may be broadly divided into two groups :

(*a*) sinusoidal (harmonic) oscillators–such circuits are used as electronic timing and control circuits in TV, radar, oscilloscope and industrial control equipment.

(*b*) non-sinusoidal (relaxation) oscillators–they produce output which has square or rectangular or sawtooth waveform.

Sine wave oscillators may be further subdivided into:

(*i*) LC feedback oscillators such as Hartley, Colpitts and Clapp etc. These are used at higher frequencies and their output frequency is adjustable.

(*ii*) RC phase-shift oscillators such as Wienbridge oscillator. They work at relatively low frequencies.

(*iii*) Negative-resistance oscillators such as tunnel diode oscillator.

(*iv*) Crystal controlled oscillators–they are used where stable and accurate output is required. The output frequency is determined by the crystal and is not adjustable.

However, only LC feedback and RC phase-shift oscillators would be considered in this book.

37.3 Essential of a Feedback LC Oscillator

The essential components of a feedback oscillator shown in Fig. 37.2 are:

1. a resonator which consists of an LC circuit. It is also known as frequency determining network (FDN) or tank circuit;
2. an amplifier whose function is to am-

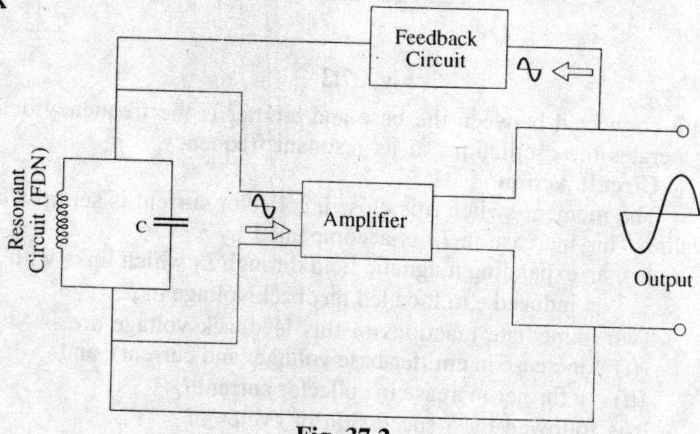

Fig. 37.2

plify the oscillations produced by the resonator;

3. a positive feedback network (PFN) whose function is to transfer part of the output energy to the resonant LC circuit in proper phase. The amount of energy fed back is sufficient to meet I^2R losses in the LC circuit.

The gain of an amplifier with positive feedback is

$$A' = \frac{A}{1 - \beta A}$$

If the circuit is to oscillate, the gain must be infinite.

$$\therefore \quad |1 - \beta A| = 0 \quad \text{or} \quad |\beta A| = 1;$$

In general, the essential condition for maintaining oscillations and for finding the vlaue of frequency is

$$\beta A = 1 + j0 \quad \text{or} \quad \beta A \angle \phi = 1 \angle 0°$$

It means that

(*i*) the open loop gain $|\beta A| = 1$;

(*ii*) the net phase shift around the loop is = 0° (or 360°). In other words, feedback should be *positive*.

The above conditions form Barkhausen criterion.

37.4. Tuned Base Oscillator

Such an oscillator using a transistor in CE configuration is shown in Fig. 37.3. Resistors R_1, R_2 and R_3 determine the d.c. bias of the circuit. The mutually-coupled coils L_1 and L forming primary and secondary of an RF transformer provide the required feedback between the collector and base circuits. The amount of feedback depends on the coefficient of coupling between the two coils. The CE-connected transistor itself provides a phase shift of 180° between its input and output circuits. The transformer provides another 180° phase shift thus producing a total phase shift of 360° (or 0°) which is an essential condition for producing oscillations.

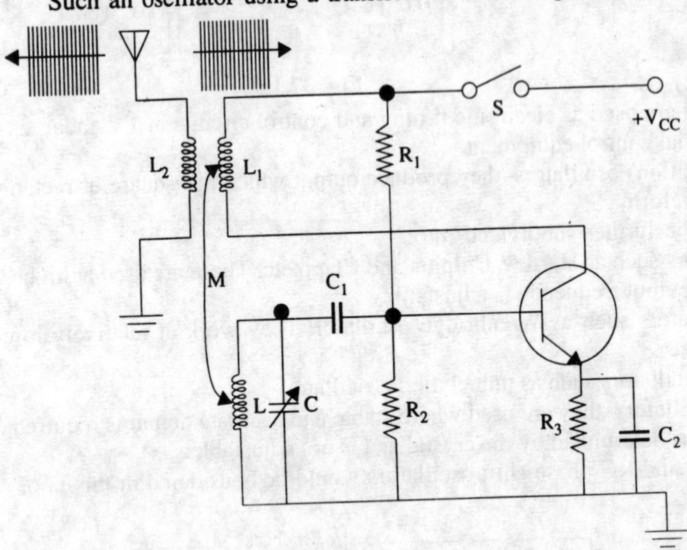

Fig. 37.3

The parallel-tuned LC circuit connected between the base and emitter is the frequency determining network (FDN) *i.e.*, it generates the oscillations at its resonant frequency.

Circuit Action

The moment switch *S* is closed, collector current is set up which tends to rise to its quiescent value. This increase in I_C is accompanied by :

1. an expanding magnetic field through L_1 which links with *L*; and
2. an induced e.m.f. called feedback voltage in *L*.

Two immediate reactions of this feedback voltage are :

(*i*) increase in emitter-base voltage (and current); and

(*ii*) a further increase in collector current I_C.

It is followed by a succession of cycles of

1. an increase in feedback voltage;
2. an increase in emitter-base voltage; and

3. an increase in I_C until saturation is reached.

Meanwhile, C gets charged. As soon as I_C ceases to increase, magnetic field of L_1 ceases to expand and thus no longer induces feedback voltage in L. Having been charged to a maximum value, C starts to discharge through L. However, decrease in voltage across C causes the following sequence of reactions:

1. a decrease in emitter-base bias,
2. a decrease in I_C,
3. a collapsing magnetic field in L_1,
4. an induced feedback voltage in L though in the *opposite* direction,
5. further decrease in emitter-base bias ,

and so on till I_C reaches its cut-off value.

During this time, the capacitor having lost its original charge, again becomes fully charged though with *opposite* polarity. Transistor being in cut-off, the capacitor will again begin to discharge through L. Since polarity of capacitor charge is opposite to that when transistor was in saturation, the sequence of reactions now will be

1. an increase in emitter-base bias,
2. an increase in I_C,
3. an expanding magnetic field in L_1,
4. an induced feedback voltage in L,
5. a further increase in emitter-base bias,

and so on till I_C increases to its saturation value.

This cycle of operation keeps repeating so long as enough energy is supplied to meet losses in the LC circuit.

The output can be taken out by means of a third winding L_2 magnetically coupled to L_1. It has approximately the same waveform as collector current.

The frequency of oscillation is given by

$$f_o = 1 / 2\pi \sqrt{LC}$$

37.5. Tuned Collector Oscillator

Such an oscillator using a transistor in CE configuration is shown in Fig. 37.4.

(*i*) Frequency Determining Network (FDN)

It is made up of a variable capacitor C_1 and a coil L_1 which forms primary winding of a step-down transformer. The combination of L_1 and C_1 forms an oscillatory tank circuit.

(*ii*) Positive Feedback

Feedback between the collector-emitter circuit and base-emitter circuit is provided by the transformer secondary winding L_2 which is mutually-coupled to L_1. As far as a.c. signals are concerned, L_2 is connected to emitter via low-reactance capacitors C_2 and C_3.

Since transistor is connected in CE configuration, it provides a phase shift of 180° between its input and output circuits. Another phase shift of 180° is provided by the transformer thus producing a total phase shift of 360° between the output and input voltages resulting in positive feedback between the two.

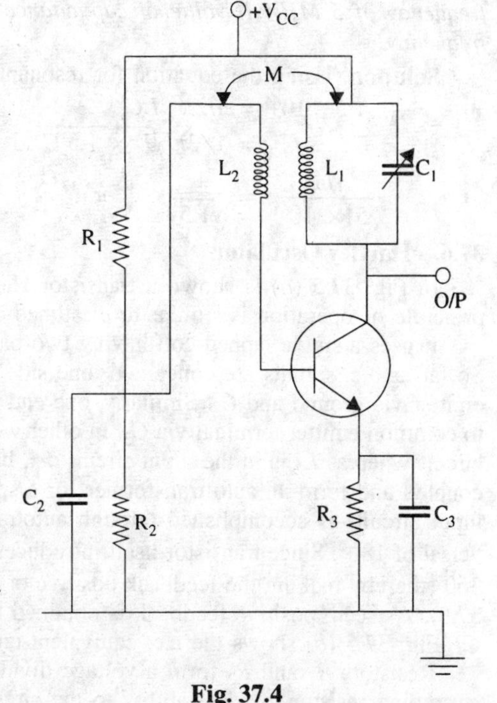

Fig. 37.4

(*iii*) Amplifying Action

The transistor amplifier provides sufficient gain for oscillator action to take place.

The resistors R_1, R_2 and R_3 are used to d.c. bias the transistor.

(*iv*) Working

When the supply is first switched on, a transient current is developed in the tuned L_1-C_1 circuit as the collector current rises to its quiescent value. This transient current initiates natural oscillations in the tank circuit. These natural oscillations induce a small e.m.f. into L_2 by mutual induction which causes corresponding variations in base current. These variations in I_B are amplified β times and appear in the collector circuit. Part of this amplified energy is used to meet losses taking place in the oscillatory circuit and the balance is radiated out in the form of electromagnetic waves.

The frequency of oscillatory currents is almost equal to the resonant frequency of the tuned circuit.

$$\therefore \quad f_0 = \frac{1}{2\pi\sqrt{L_1 C_1}}$$

Example 37.1 *A tuned-collector oscillator has a fixed inductance of 100 μH and has to be tunable over the frequency band of 500 kHz to 1500 kHz. Find the range of variable capacitor to be used.*

Solution. Resonant frequency is given by

$$f_0 = 1/2\pi\sqrt{LC} \quad \text{or} \quad C = 1/4\pi^2 f_0^2 L$$

where L and C refer to the tank circuit.

When f_0 = 500 kHz

$$C = 1/4\pi^2 \times (500 \times 10^3)^2 \times 100 \times 10^{-6} = \mathbf{1015\ pF}$$

When f_0 = 1500 kHz

$$C = 1015 / (1500/500)^2 = \mathbf{133\ pF.}$$

Hence, capacitor range required is **113 – 1015 pF.**

Example 37.2 *The resonant circuit of a tuned-collector transistor oscillator has a resonant frequency of 5 MHz. If value of capacitance is increased by 50%, calculate the new resonant frequency.* **(Electronics-I, Bangalore Univ. 1994)**

Solution. Using the equation for resonant frequency, we have

$$5 \times 10^6 = 1/2\pi\sqrt{LC}$$

$$f_0 = 1/2\pi\sqrt{L \times 1.5\ C}$$

$$\therefore \quad \frac{f_0}{5 \times 10^6} = \frac{1}{\sqrt{1.5}} \quad \text{or} \quad f_o = \mathbf{4.08\ MHz.}$$

37.6. Hartley Oscillator

In Fig. 37.5 (*a*) is shown a transistor Hartley oscillator using CE configuration. Its general principle of operation is similar to the tuned-collector oscillator discussed in Art. 37.5.

It uses a single tapped-coil having two parts marked L_1 and L_2 instead of two separate coils. So far as a.c. signals are concerned, one side of L_2 is connected to base via C_1 and the other to emitter via ground and C_3. Similarly, one end of L_1 is connected to collector via C_2 and the other to common emitter terminal via C_3. In other words, L_1 is in the output circuit *i.e.*, collector-emitter circuit whereas L_2 is in the input circuit *i.e.*, base-emitter circuit. These two parts are inductively-coupled and form an auto-transformer or a split-tank inductor. Feedback between the output and input circuits is accomplished through autotransformer action which also introduces a phase reversal of 180°. Since transistor itself introduces a phase shift of 180°, the total phase shift becomes 360° thereby making the feedback positive or regenerative which is essential for oscillations (Art. 37.3). As seen, positive feedback is obtained from the tank circuit and is coupled to the base via C_1. Fig. 37.5 (*b*) shows the a.c. equivalent circuit of Hartley oscillator.

Resistors R_1 and R_2 form a voltage divider for providing the base bias and R_3 is an emitter swamping resistor to add stability to the circuit. Capacitor C_3 provides a.c. ground thereby preventing any degeneration while still providing temperature stabilisation. The radio-frequency choke (RFC) provides collector load resistor which is necessary for amplifier action and inductance which is necessary to keep a.c. currents out of d.c. supply V_{CC}.

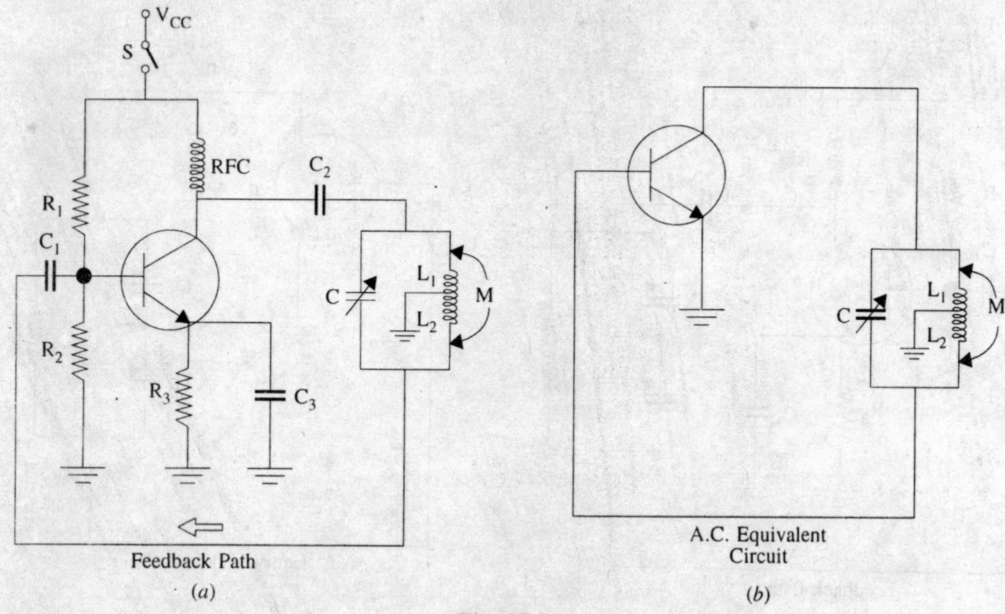

Fig. 37.5

When V_{CC} is first switched on through S, an initial bias is established by R_1-R_2 and oscillations are produced because of positive feedback from the LC tank circuit (L_1 and L_2 constitute L). The frequency of oscillation is given by

$$f_o = \frac{1}{2\pi\sqrt{LC}} \qquad \text{where} \quad L = L_1 + L_2 + 2M$$

The output from the tank may be taken out by means of another coil coupled either to L_1 or L_2.

37.7. Colpitts Oscillator

This oscillator is essentially the same as Hartley oscillator except for one difference. Colpitts oscillator uses tapped capacitance whereas Hartley oscillator uses tapped inductance.* Fig. 37.6 (*a*) shows the complete circuit with its power source and d.c. biasing circuit whereas Fig. 37.6 (*b*) shows its equivalent a.c. circuit. The two series capacitors C_1 and C_2 form the voltage divider used for providing the feedback voltage (the voltage drop across C_2 constitutes the feedback voltage).

The tank circuit consists of two ganged capacitors C_1 and C_2 and a single fixed coil. The frequency of oscillation (which does not depend on mutual inductance) is given by

$$f_o = \frac{1}{2\pi\sqrt{LC}} \quad \text{where} \quad C = \frac{C_1 C_2}{C_1 + C_2}$$

Transistor itself produces a phase shift of 180°. Another phase shift of 180° is provided by the capacitive feedback thus giving a total phase shift of 360° between the emitter-base and collector-base circuits.

Resistors R_1 and R_2 form a voltage divider across V_{CC} for providing base bias and R_3 is for emitter stabilisation. The RFC provides collector load necessary for amplifier action and its inductance keeps a.c. currents out of the d.c. supply. Capacitor C_3 is a by-pass capacitor whereas C_4 conveys feedback from the collector-to-base circuit.

When S is closed, a sudden surge of I_C shock-excites the tank circuit into oscillations which are sustained by feedback and the amplifying action of the transistor.

---

*As an aid to memory, remember that Hartley begins with *H* for Henry *i.e.*, coil and Colpitts begins with *C* for Capacitor.

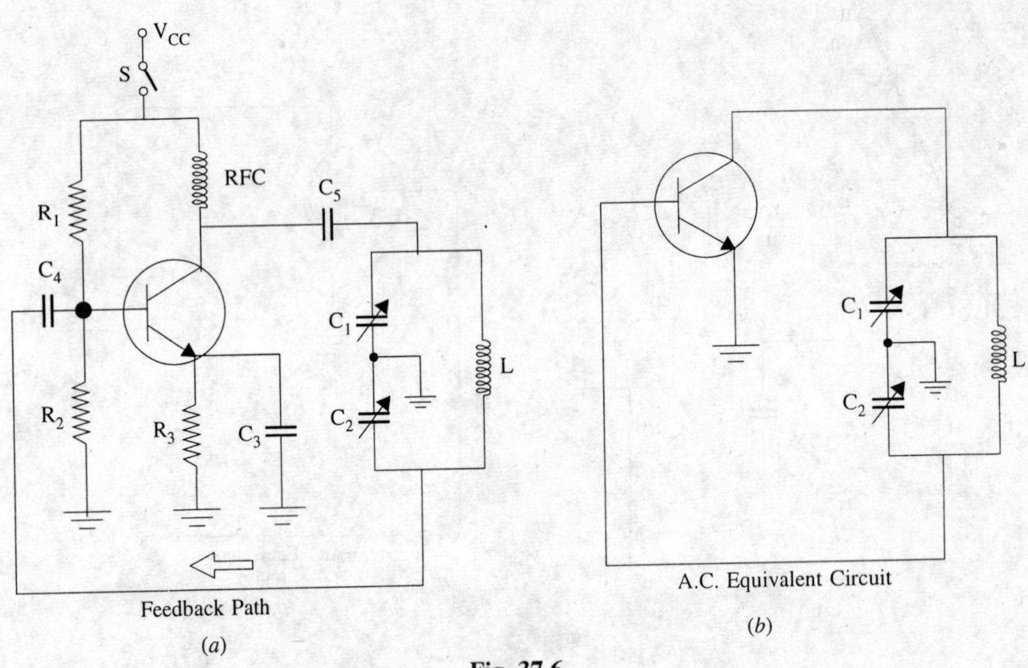

Fig. 37.6

Colpitts oscillator is widely used in commercial signal generators upto 1 MHz. Frequency of oscillation is varied by gang-tuning the two capacitors C_1 and C_2.

37.8. Clapp Oscillator

It is a variation of Colpitts oscillator and is shown in Fig. 37.7 (*a*). It differs from Colpitts oscillator in respect of capacitor C_3 only which is joined in series with the tank inductor.

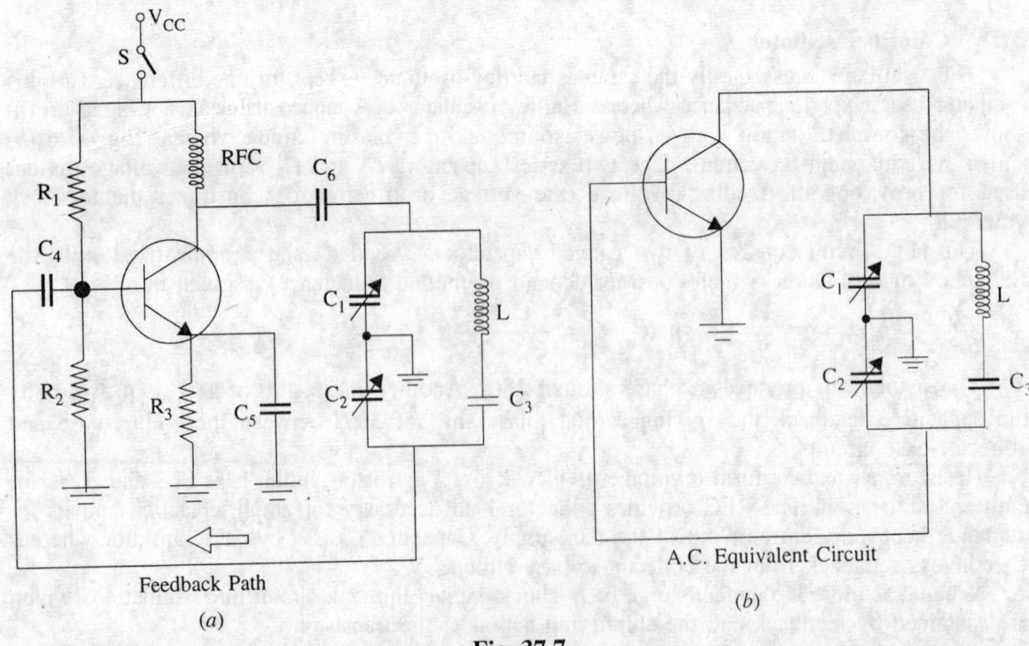

Fig. 37.7

Addition of C_3 (*i*) improves frequency stability and (*ii*) eliminates the effect of transistor's parameters on the operation of the circuit. The working of this circuit is the same as that of the

Colpitts oscillator.

The frequency of oscillation is given by

$$f_o = \frac{1}{2\pi\sqrt{LC}}$$

where $1/C = 1/C_1 + 1/C_2 + 1/C_3$

37.9. Crystal Controlled Oscillator

Fig. 37.8 shows the use of a crystal to stabilise the frequency of a tuned-collector oscillator which has a crystal (usually quartz) in the feed-back circuit.

The LC tank circuit has a frequency of oscillation $f_o = 1/2\pi\sqrt{LC}$. The circuit is adjusted to have a frequency near about the desired operating frequency but the exact frequency is set by the crystal and stabilized by the crystal. For example, if natural frequency of vibration of the crystal is 27 MHz, the LC circuit is made to resonate at this frequency.

As usual, resistors R_1, R_2 and R_3 provide a voltage-divider stabilised d.c. bias circuit. Capacitor C_1 by-passes R_3 in order to maintain large voltage gain. RFC coil L_1 prevents a.c. signals from entering d.c. line whereas R_c is the required d.c. load of the collector. The coupling capacitor C_2 has negligible impedance at the operating frequency but prevents any d.c. link between collector and base. Due to extreme stability of crystal oscillations, such oscillators are widely used in communication transmitters and receivers where frequency stability is of prime importance.

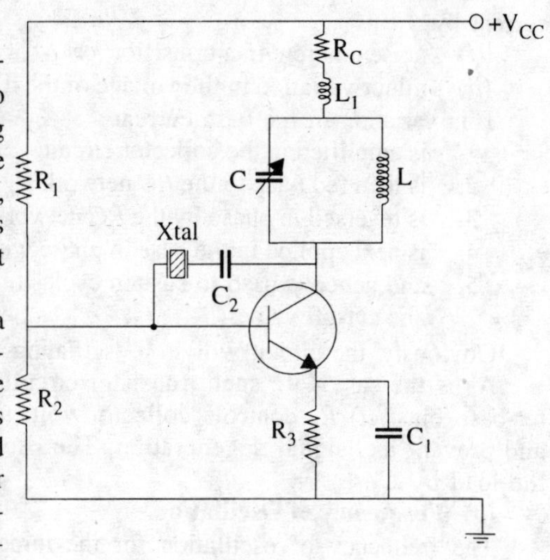

Fig. 37.8

37.10. Phase Shift Principle

Tuned circuits are *not an essential requirement for oscillation*. What is essential is that there should be a 180° phase shift around the feedback network* and loop gain should be equal to unity. The 180° phase shift in the feedback signal can be achieved by using a suitable *RC* network consisting of three or four *R-C* sections. The sine wave oscillators which use *R-C* feedback network are called **phase-shift oscillators.**

37.11. Phase-shift Oscillator

Fig. 37.9 shows a transistor phase-shift oscillator which uses a three-section R-C feedback network for producing a total phase shift of 180° (*i.e.*, 60° per section) in the sig-

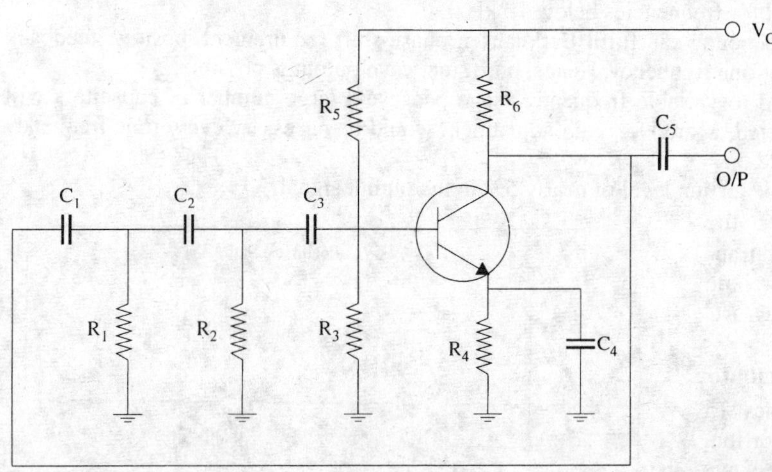

Fig. 37.9

*Total phase shift required is 360°. However, the balance of 180° is provided by the active device of the amplifier itself.

nal fed back to the base. Since CE amplifier produces a phase reversal of the input signal, total phase shift becomes 360° or 0° which is essential for **regeneration and hence for sustained oscillations.**

Values of *R* and *C* are so selected that each *RC* section produces a phase advance of 60°. Addition of a fourth section improves oscillator stability. It is found that phase shift of 180° occurs **only at one frequency** which becomes the oscillator frequency.

(*a*) Circuit Action

The circuit is set into oscillations by any random or chance variation caused in the base current by

(*i*) noise inherent in a transistor; or

(*ii*) minor variation in the voltage of the d.c. source.

This variation in the base current

1. is amplified in the collector circuit;
2. is then fed back to the *RC* network $R_1 C_1$, $R_2 C_2$ and R_3C_3;
3. is reversed in phase by the *RC* network;
4. is next applied to the base in phase with initial change in base current;
5. and hence is used to sustain cycles of variations in collector current between saturation and cut-off values.

Obviously, the circuit will stop oscillating the moment phase shift differs from 180°.

As is the case with such transistor circuits (*i*) voltage divider $R_5 - R_3$ provides d.c. emitter-base bias (*ii*) R_6 controls collector voltage and (*iii*) R_4, C_4 provide temperature stability and prevent a.c. signal degeneration. The oscillator output voltage is capacitively coupled to the load by C_5.

(*b*) Frequency of Oscillation

The frequency of oscillation for the three-section *RC* oscillator when the three *R* and *C* components are equal is roughly given by

$$f_o = \frac{1}{2\pi\sqrt{6}\,.\,RC}\ \text{Hz} = \frac{0.065}{RC}\ \text{Hz}$$

Moreover, it is found that the value of $\beta = 1/29$. It means that amplifier gain must be more than 29 for oscillator operation.

(*c*) Advantages and Disadvantages

1. Since they do not require any bylky and expensive high-value inductors, such oscillators are well-suited for frequencies below 10 kHz.
2. Since only frequency can fulfil Barkhausen phase-shift requirement, positive feedback occurs only for one frequency. Hence, pure sine wave output is possible.
3. It is not suited to variable frequency usage because a large number of capacitors will have to be varied. Moreover, gain adjustment would be necessary every time frequency change is made.
4. It produces a distortion level of nearly 5% in the output signal.
5. It necessitates the use of a high β transistor to overcome losses in the *RC* network.

37.12. Wien Bridge Oscillator

It is a low-frequency (5 Hz – 500 kHz) low-distortion, tunable, high-purity sine wave generator often used in laboratory work. As shown in the block diagram of Fig. 37.10

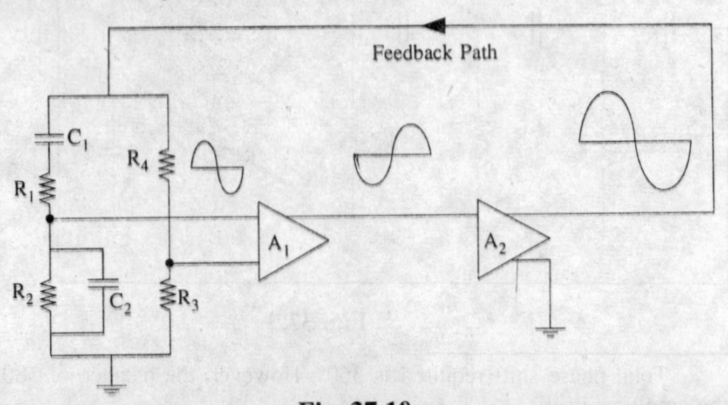

Fig. 37.10

and Fig. 37.11, this oscillator uses two CE-connected *RC*-coupled transistor amplifiers and one *RC*-bridge (called *Wien bridge*) network to provide feedback. Here, Q_1 serves as amplifier-oscillator and Q_2 provides phase reversal and additional amplification. The bridge circuit is used to control the phase of the feedback signal at Q_1.

(*a*) Phase Shift Principle

Any input signal at the base of Q_1 appears in the amplified but phase-reversed form across collector resistor R_6 (Fig. 37.11). It is further inverted by Q_2 in order to provide a total phase

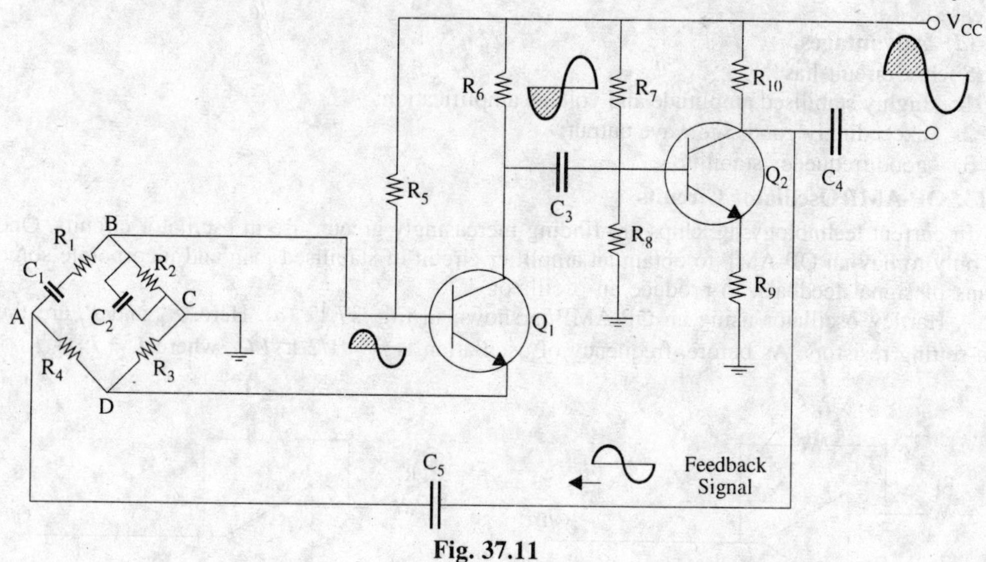

Fig. 37.11

reversal of 360° for positive feedback. Obviously, the signal at R_{10} is an amplified replica of the input signal at Q_1 and is of the same phase since it has been inverted twice. We could feed this signal back to the base of Q_1 directly to provide regeneration needed for oscillator operation. But because Q_1 will amplify signals over a wide range of frequencies, direct coupling would result in poor frequency stability. By adding the Wien bridge, oscillator becomes sensitive to a signal of only *one particular frequency*. Hence, we get an oscillator of good frequency stability.

(*b*) Bridge Circuit Principle

It is found that the Wien bridge would become balanced at that signal frequency for which phase shift is exactly 0° (or 360°).

The balance conditions are

$$\frac{R_4}{R_3} = \frac{R_1}{R_2} + \frac{C_2}{C_1}$$

and $$\omega_o = \frac{1}{\sqrt{R_1 C_1 R_2 C_2}} \quad \text{or} \quad f_0 = \frac{1}{2\pi\sqrt{R_1 C_1 R_2 C_2}}$$

If $R_1 = R_2 = R$ and $C_1 = C_2 = C$, then

$$f_o = \frac{1}{2\pi RC} \quad \text{and} \quad \frac{R_4}{R_3} = 2$$

(*c*) Circuit Action

Any random change in base current of Q_1 can start oscillations. Suppose, the base current of Q_1 is increased due to some reason. It is equivalent to applying a positive-going signal to Q_1. Following sequence of events will take place:

1. an amplified but phase-reversed signal will appear at the collector of Q_1;
2. a still further amplified and twice phase-reversed signal will appear at the collector of Q_2. Having been inverted twice, this output signal will be in phase with the input signal at Q_1;

3. a part of the output signal at Q_2 is fed back to the input points of the bridge circuit (points *A* and *C*). A part of this feedback signal is applied to emitter resistor R_3 where it produces degenerative effect. Similarly, a part of the feedback signal is applied across base-bias resistor R_2 where it produces regenerative effect.

At the rated frequency f_o, effect of *regeneration is made slightly more than that of degeneration* in order to maintain continuous oscillations.

By replacing R_3 with a thermistor, amplitude stability of the oscillator output voltage can be increased.

(*d*) Advantages

Such a circuit has

1. highly stabilised amplitude and voltage amplification;
2. exceedingly good sine wave output;
3. good frequency stability.

37.13. OP-AMP Oscillator Circuits

In current technology, IC chips are finding increasingly greater use in oscillator circuits. One has only to buy an OP-AMP to obtain an amplifier circuit of stabilised gain and incorporate some means of signal feedback to produce an oscillator.

A Hartley oscillator using an OP-AMP is shown in Fig. 37.12 (*a*). Here, R_1 and R_f are the gain-setting resistors. As before, frequency of oscillation is $f_0 = 1/2\,\pi\sqrt{LC}$ where $L = L_1 + L_2 +$

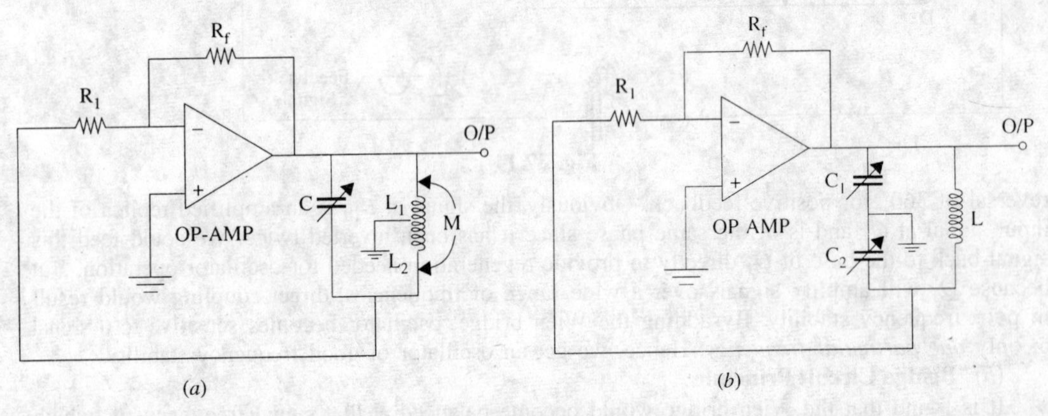

Fig. 37.12

2*M*. Fig. 37.12 (*b*) shows Colpitts oscillator using an OP-AMP. As before, its frequency of oscillation is given by $f_0 = 1/2\pi\sqrt{LC}$ where $C = C_1C_2/(C_1 + C_2)$.

A phase-shift oscillator using OP-AMP inverting amplifier and an *RC* phase-shift network which provides 180° phase shift is shown in Fig. 37.13.

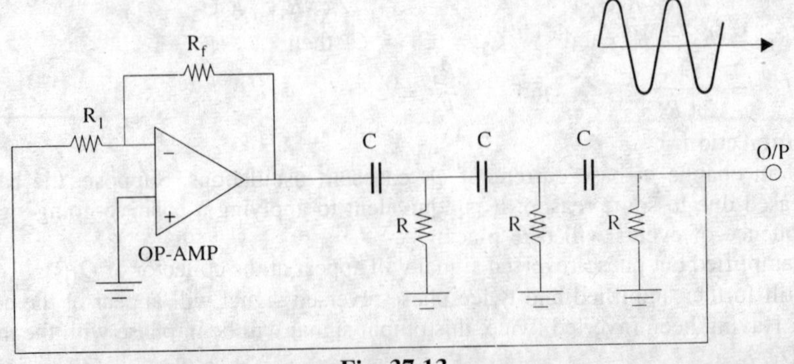

Fig. 37.13

The frequency of oscillation is given by $f_o = 1/2\pi\sqrt{6}\,RC$. For successful operation, the amplifier gain must be more than 29.

Fig. 37.14 shows a modified Colpitts oscillator using a crystal in resonant parallel mode. The crystal replaces the *LC* tuned circuit. The two capacitors C_1 and C_2 form a voltage divider which shunt the crystal but have no effect on frequency of operation. The voltage across C_2 is fed back to the input.

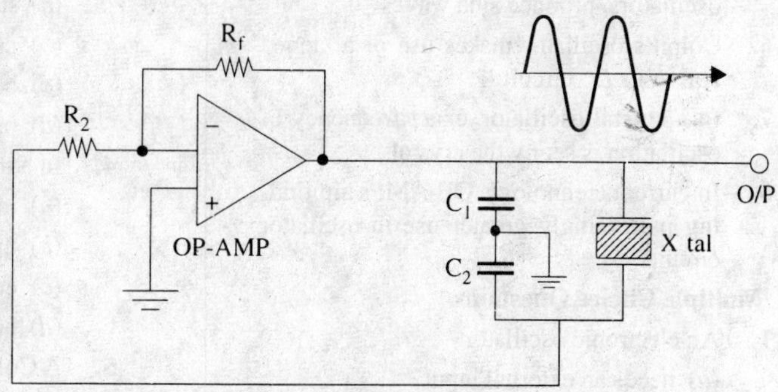

Fig. 37.14

HIGHLIGHTS

1. The essential condition for maintaining oscillations in an oscillator is

$$\beta A = 1 + j0 = 1 \angle 0°$$

The above conditions constitute Barkhausen criterion.

It means that

(*i*) open loop gain $|\beta A| = 1$

(*ii*) net phase shift around the loop is 0° or 360°.

2. In a tuned-base oscillator circuit, the oscillatory tank circuit is connected across the base and emitter terminals.
3. In a tuned-collector oscillator, the oscillator tank circuit is connected across the collector and emitter terminals.
4. Hartley oscillator is similar to tuned-collector oscillator except that it uses a tapped coil oscillator tank circuit.
5. Colpitts oscillator is essentially the same as Hartley oscillator except that it uses tapped capacitance instead of tapped inductance.
6. Clapp oscillator is variation of Colpitts oscillator.
7. Crystal-controlled oscillator is a tuned-collector oscillator which uses a crystal not only for frequency stabilization but for feedback purposes as well.
8. In an *RC* phase-shift oscillator, three RC sections are used to produce a phase difference of 180° in the feedback voltage.
9. Wien bridge oscillator is a low-frequency (5 Hz – 500 kHz) phase-shift oscillator.

OBJECTIVE TESTS—37

A. Fill in the following blanks.

1. Unlike an amplifier, an oscillator requires no .... signal.
2. Positive feedback is also known as .... feedback.
3. For a circuit to oscillate, its gain must be ....
4. In a tuned-collector oscillator, *LC* circuit is connected across collector and .... terminals.
5. Hartley oscillator uses a single .... coil.
6. Colpitts oscillator uses tapped ....
7. Colpitts oscillator is a variation of .... oscillator.
8. Wien-bridge oscillator is a high-purity ....-wave generator.

B. Answer True or False.

1. All sine wave oscillators must satisfy Barkhausen criterion.
2. Negative feedback is essential for a circuit to oscillate.
3. An oscillator circuit essentially consists of an amplifer, feedback network and a resonant circuit.
4. An oscillator produces an a.c. output with no externally applied input.
5. Relaxation oscillators, like sinusoidal

oscillators, produce sine waves.

6. Colpitts oscillator makes use of a tapped coil in its *LC* circuit.
7. In a crystal oscillator, exact frequency of oscillation is set by the crystal.
8. In current technology, OP-AMPs are finding increasingly greater use in oscillator circuits.

C. Multiple Choice Questions.

1. An electronic oscillator
   (*a*) needs an external input
   (*b*) provides its own input
   (*c*) is nothing but an amplifier
   (*d*) is just a d.c. / a.c. converter
2. For sustaining oscillations in an oscillator circuit
   (*a*) feedback factor should be unity
   (*b*) phase shift should be 0°
   (*c*) feedback should be negative
   (*d*) both (*a*) and (*b*)
3. If Barkhausen criterion is not fulfilled by an oscillator circuit, it will
   (*a*) stop oscillating
   (*b*) produce damped waves continuously
   (*c*) become an amplifier
   (*d*) produce high-frequency whistles
4. In a transistor Hartley oscillator
   (*a*) inductive feedback is used
   (*b*) untapped coil is used
   (*c*) entire coil is in the output circuit
   (*d*) no capacitor is used
5. A Colpitts oscillator uses
   (*a*) tapped coil
   (*b*) inductive feedback
   (*c*) tapped capacitance
   (*d*) no tuned *LC* circuit
6. In *RC* phase-shift oscillator circuits
   (*a*) there is no need for feedback
   (*b*) feedback factor is less than unity
   (*c*) pure sine wave output is possible
   (*d*) transistor parameters determine oscillation frequency

ANSWERS

A. 1. input 2. regenerative 3. infinite 4. emitter 5. tapped 6. capacitors 7. Colpitts 8. sine

B. 1. T 2. F 3. T 4. T 5. T 6. F 7. T 8. T

C. 1. *b* 2. *d* 3. *a* 4. *a* 5. *c* 6. *c*

38 ANALOG AND DIGITAL COMMUNICATION

38.1. Communication Systems

By communication is generally meant "electronic communication" which began with the opening of a commercial telegraph in London in 1839. More's telegraph was started over a distance of 65 km in 1844 in America. Telegraph grew very rapidly thereafter leading to the laying of submarine cables. Telegraphs sent coded signals whereas telephone (invented by Bell in 1874) could transmit speech itself. From then on, telephonic communication grew very rapidly which led to the development of better and better switching systems.

Radio telegraphy became possible in 1894 with Marconi's spark transmitter which made use of a new medium for electromagnetic communication. However, radio and television had to wait for the invention of diode by Fleming in 1901 and the triode by De Forest in 1907. In 1934, microwaves of frequency 1.7 gigahertz (GHz) were used for communication in a link across the English channel. By about 1960, microwave became standard carrier for long distance telephone communications throughout the world. But, now-a-days, optical fibre is being increasingly used for the purpose.

(*a*) Analog Communication System

In this system, the entire analog signal or message is processed to convey information to the user or receiver. An analog signal (Fig. 38.1) has one of a continuum of possible amplitudes at any given time. When the signal or message to be sent over a communication system is analog, we refer to the system as analog.

(*b*) Digital Communication System

It is a system which processes only digital signals *i.e.*, electrical pulses to convey information (Fig. 38.2).

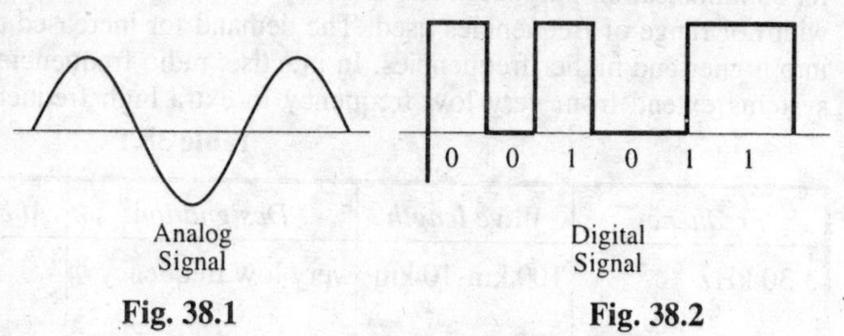

Fig. 38.1 **Fig. 38.2**

Digital systems can also convey analog messages provided they are first converted to digital form with the help of analog-to-digital (A/D) converter. Most of the modern communication systems are digital instead of analog.

38.2. Advantages of Digital System

Digital communication systems are preferred over analog systems for the following reasons:

1. Digital systems interface with computers easily.
2. Discrete data can be efficiently processed.
3. They are very reliable and give high performance with low cost.
4. They are very flexible and can handle both digital and analog messages with the help of A/D converters.
5. They provide privacy of message to users with the help of readily available a security techniques.

However, there are some disadvantages of such a system as given under:

1. Digital system needs larger bandwidth than an equivalent analog system.
2. Digital system is comparatively more complex.

However, the recent trend in designing communication systems is to make as much of the system digital as possible. In fact, digital systems exist that can simultaneously transmit audio, television and digital data over the same channel.

38.3. Elements of a Communication Systems

All electronic communication systems are concerned with the flow of information. Each system essentially consists of three elements as shown in Fig. 38.3.

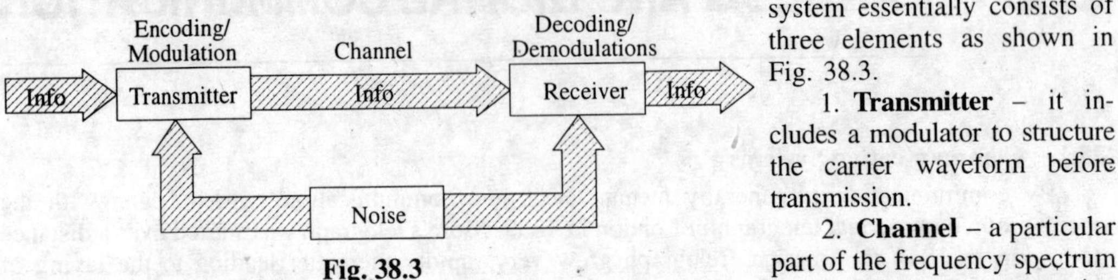

Fig. 38.3

1. **Transmitter** – it includes a modulator to structure the carrier waveform before transmission.

2. **Channel** – a particular part of the frequency spectrum being used in the propogating medium through which communication occurs. The medium could be a vacuum, air, water, earth, wire, waveguide, optical fibre *i.e.*, anything which can support wave motion.

3. **Receiver** – it includes a demodulator which recovers the message from the received signals.

In each of the above three stages, there is a possibility of the information being lost or being corrupted by signal distortion, noise attenuation and equipment malfunction. The receiver is designed to minimize the effects of the noise on the recovered information at the destination.

38.4. Electromagnetic Spectrum

Radiowaves, infrared waves, visible light, ultra-violet light, X-rays etc., all are electromagnetic waves which differ in frequency or wavelength since $f = c/\lambda$ where c is the velocity of light which equals 3×10^8 m/s. The region of the electromagnetic spectrum below the ultra-violet region is used for communication purposes. The capacity of a communication system is proportional to the bandwidth of range of frequencies used. The demand for increased capacity has driven communication into higher and higher frequencies. In practise, radio frequencies used by different communication systems extend from very low frequency to extra high frequencies as tabulated below:

Table 38.1

Frequency	*Wave length*	*Designation*	*Abbrviations*	*Uses*
3-30 kHz	100 km-10 km	very low frequency	VLF	long-range communication, maritime mobile radio, navigational aids.
30 - 300 kHz	10 km - 1 km	low frequency	LF	long-wave radio and communications, aeronautical radio navigation.
300 kHz - 3 MHz	1 km - 100 m	medium frequency	MF	medium-wave, local and distant radio broadcast.
3 - 30 MHz	100 m - 10 m	high frequency	HF	short-wave broadcast, amateur and citizens band.
30 - 300 MHz	10 m - 1 m	very high frequency	VHF	television, FM radio, navigational aids, meteorology devices, data links.

(Contd.)

300 MHz - 3 GHz	1 m - 10 cm	ultra high frequency	UHF	television, communication data links, radar, satellite communication.
3 - 30 GHz	10 cm - 1 cm	super high frequency	SHF	radar, communication satellites, microwave telephone and television links.
30 - 300 GHz	10 mm - 1 mm	extra high frequency	EHF	radar communications, specialised equipment.

38.5. Radio Wave Propogation

Generally speaking, radio wave propogation is divided into three broad ranges.

(*i*) Ground Wave or Surface Wave or Direct

Here, the signal propogates along the earth's surface or through it.

(*ii*) Sky Wave

Here, the signal propogates *via* reflections from the ionosphere which is a layer of charged particles in the upper atmosphere.

(*iii*) Space Wave or Line-of-sight

Here, the wave propagates like optical or light signals in straight lines and, hence requires that the transmitting and the receiving stations should be approximately within sight of each other.

These three modes of propagation are illustrated in Fig. 38.4.

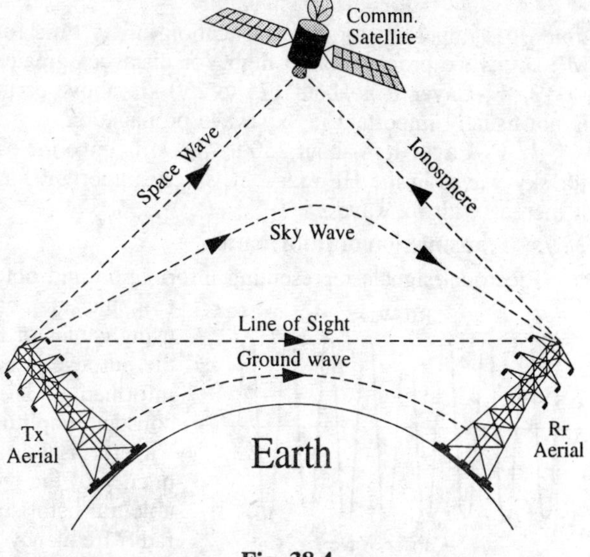

Fig. 38.4

From 10 to 500 kHz (VLF and LF bands), ground wave is normally used for medium and long-range communications. From 500 kHz to 3 MHz (MF band), the range of ground wave varies from 25 km to 350 km. Sky wave transmission is excellent at night for ranges up to several thousand kilometres. However, in the day time, the sky wave transmission becomes erratic. From 3 to 30 MHz (HF band), the range of ground waves decreases rapidly and the sky wave transmission becomes somewhat unpredictable. From 30 to 300 MHz (VHF band), neither ground waves nor sky waves are much usable. Hence, space wave transmission (line-of-sight) is used. From 30 MHz to 300 GHz (UHF, SHF, EHF bands), line-of-sight or space wave transmission is used almost exclusively.

38.6. The Ionosphere

The ionosphere consists of many layers of ionized air which surround the earth and extend from 50 km to 600 km above its surface. The important properties of ionosphere affecting wave propagation are absorption and reflection. Different layers of regions of the ionosphere affect wave propagation differently. Following are the principal layers of the ionosphere. (Fig. 38.5)

1. D-Layer It is 50 to 90 km above the earth's surface usually in the day time. It reflects VLF and LF, absorbs MF and weakens HF.

2. E-Layer It is about 110 km above the earth's surface and is usually most intense in day

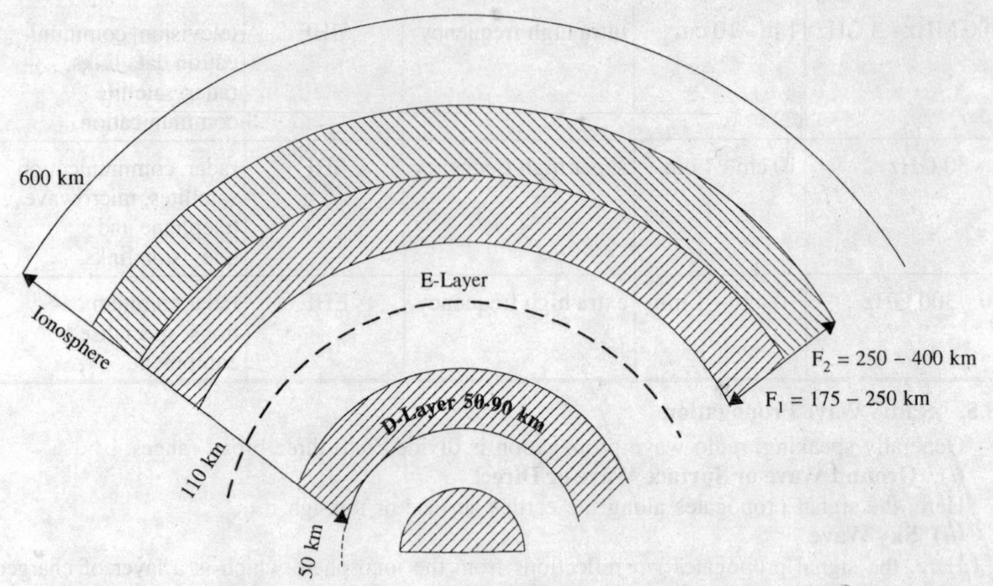

Fig. 38.5

time. It is important for HF propagation in day time for distances greater than 1600 km. Allows MF sky wave propagation at night for distances greater than 160 km.

3. F_1-Layer It is about 175 to 250 km above earth's surface and exists only in day time. It is not usually important in sky wave propagation.

4. F_2-Layer It is about 250 to 400 km above the earth's surface and is the principal reflector for sky waves in the HF range. It is very important for long distance communications. At night, it merges with F_1 waves.

38.7. Transmission of Information

Electrical signals representing information and obtained from microphones, a TV camera, a computer etc., can be sent from one place to a another by using cables or radio waves. The radio wave required for the purpose is called "carrier wave" because it carries the information. It is a very high-frequency wave having a constant amplitude as shown in Fig. 38.6. It has sinusoidal waveform and its frequency is much higher than the frequency of the information signal. It is produced by a suitable transistorised oscillator (Chapter 37). It is often called radio frequency (RF) wave.

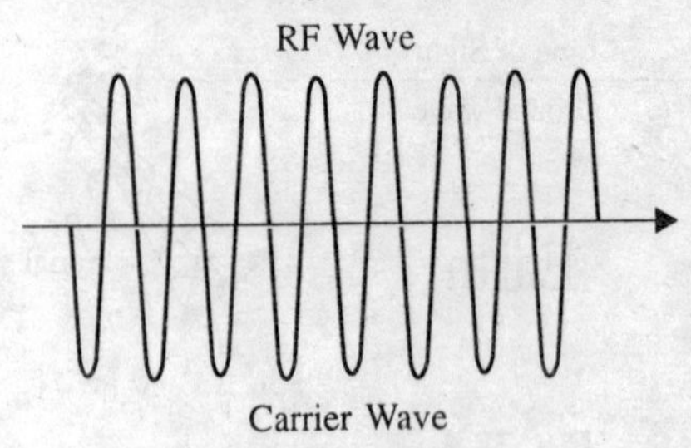

Fig. 38.6

38.8. Bandwidth

This term is used in the following two ways:

(*i*) **Bandwidth of a Signal.** It gives the range of frequencies, a signal occupies. For human speech, it is about 3 kHz (*i.e.*, 300 to 3400 Hz), for high quality music, it is about 20 kHz and for TV signals, it is about 8 MHz.

(*ii*) **Bandwidth of a Channel.** It is the range of frequencies a communication channel can accommodate. Greater the bandwidth, higher the information carrying capacity of the channel. High-frequency transmission channels have greater bandwidth than lower frequency channels. For example, the MF waveband (300 kHz–3 MHz) can carry 270 signals of 10 KHz bandwidth. However, VHF band (30 MHz–300 MHz) can accommodate 2700 such signals.

38.9. Radio Broadcasting

Let us see how radio broadcasting stations (whether AM or FM) broadcast speech or music etc., from their broadcasting studios.

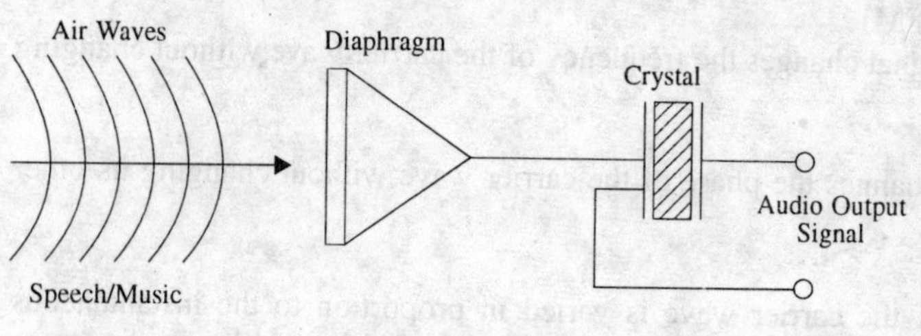

Fig. 38.7

First, the speech or music which consists of a series of compressions and rarefactions is converted into a varying electric current with the help of a suitable microphone (Fig. 38.7). Since the frequency of variations of this current lies in the audio range (20 Hz–20 kHz), hence it is known as audio frequency (AF) signal. This AF signal which is to be broadcast is then superimposed on the RF carrier waves which carry it through space to distant places. The radio frequency carrier wave is produced by a suitable oscillator as shown in Fig. 38.8.

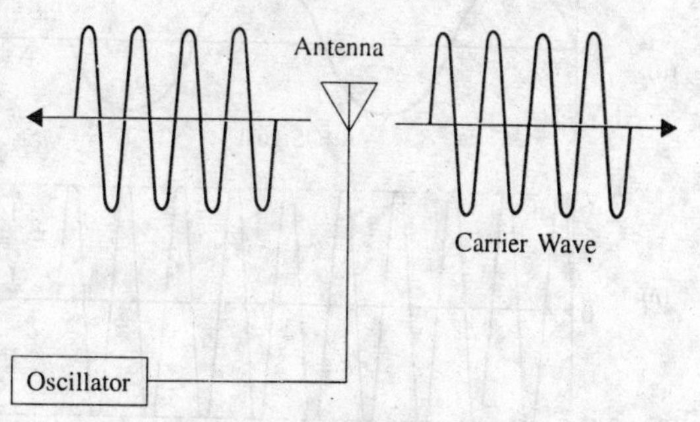

Fig. 38.8

The process by which the information or intelligence (in the form of AF signal) is impressed on the carrier wave is known as *modulation* (or encoding). This modulated carrier wave is radiated out from the antenna of the transmitter as shown in Fig. 38.8. After travelling long distance, this modulated carrier wave is picked up by the aerial of the receiver. The receiver has the arrangement to separate the AF signal from carrier wave. This process by which the RF wave and AF wave are separated is known as *demodulation* or *decoding*. It is also referred to as detection. When the AF signal is passed through a suitable speaker, the original sound is produced.

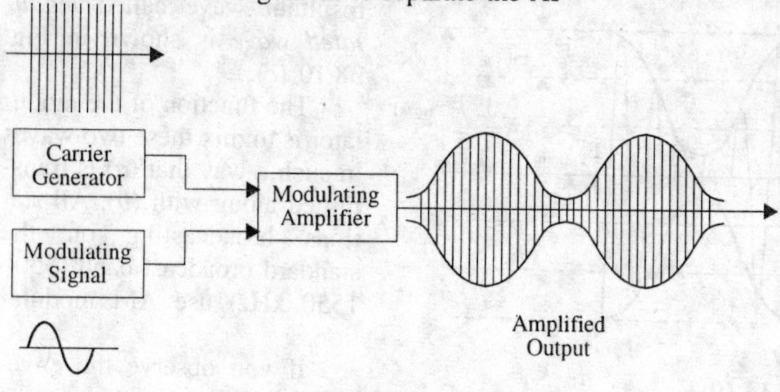

Fig. 38.9

38.10. Modulation

It is the process of combining a low-frequency signal with a radio-frequency carrier wave (Fig. 38.9). The low frequency signal is also called the *modulating signal* and the resultant wave produced is called *modulated wave.*

During modulation, some characteristic of the carrier wave is varied in time with the modulating signal and is accomplished by combining the two. Fig. 38.9 shows a carrier wave whose amplitude is made to vary at the audio signal rate. It is called *amplitude-modulated (AM) wave.*

38.11. Types of Modulation

The mathematical expression for a sinusoidal carrier wave is

$$e = E_c \sin (\omega_c t + \phi) = E_c \sin (2\pi f_c + \phi)$$

Obviously, the waveform can be varied by any of its following three factors or parameters:

1. E_c — the amplitude 2. f_c — the frequency 3. ϕ — the phase

Accordingly, there are three types of sine wave modulations known as:

1. Amplitude Modulation (AM)

Here, the information or AF signal changes the amplitude of the carrier wave without changing its frequency or phase.

2. Frequency Modulation (FM)

In this case, the information signal changes the frequency of the carrier wave without changing its amplitude or phase.

3. Phase Modulation (PM)

Here, the information signal changes the phase of the carrier wave without changing its other two parameters.

38.12. Amplitude Modulation

In this case, the amplitude of the carrier wave is varied in proportion to the instantaneous amplitude of the information signal or AF signal. Obviously, the amplitude (and hence the intensity) of the carrier wave is changed but not its frequency. Greater the amplitude of the AF signal, greater the fluctuations in the amplitude of the carrier wave.

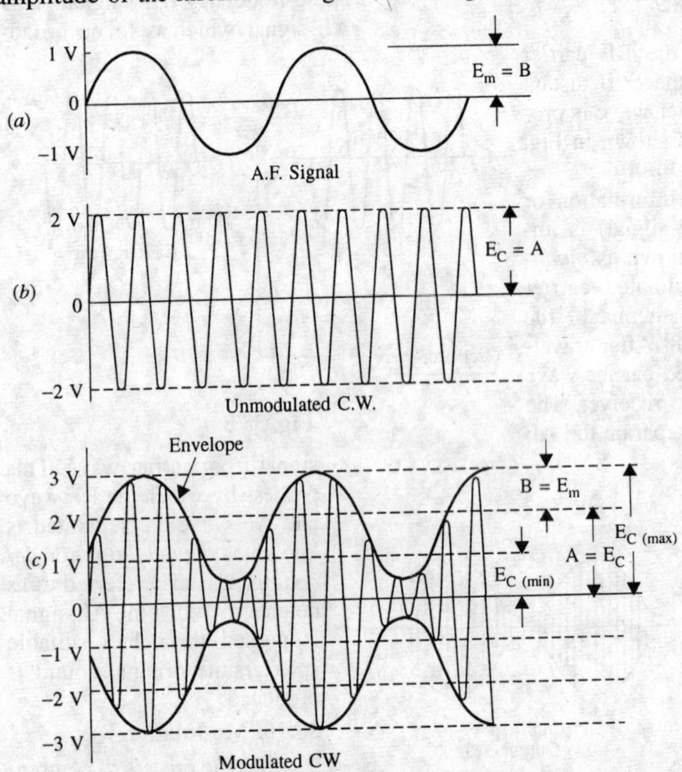

Fig. 38.10

The process of amplitude modulation is shown graphically in Fig. 38.10. For the sake of simplicity, the AF signal has been assumed sinusoidal [Fig. 38.10 (*a*)]. The carrier wave by which it is desired to transmit the AF signal is shown in Fig. 38.10 (*b*). The resultant wave called *modulated wave* is shown in Fig. 38.10 (*c*).

The function of the modulator is to mix these two waves in such a way that (*a*) is transmitted along with (*b*). All stations broadcasting on the standard broadcast band (550–1550 kHz) use AM modulation.

If you observe the envelope of the modulated carrier wave, you will realize that it is an exact replica of the AF signal wave.

In summary

(*i*) fluctuations in the amplitude of the carrier wave depend on the signal amplitude;

(*ii*) rate at which these fluctuations take place depends on the frequency of the audio signal.

38.13. Percent Modulation

When a carrier wave is not modulated by an audio signal, the percent of modulation is zero. In general, percent modulates indicates the degree to which the AF signal modulates the carrier wave. Percent modulation will be defined in the following ways:

(*i*)
$$m = \frac{\text{peak value of the signal wave}}{\text{peak value of the carrier wave}} \times 100$$
$$= \frac{\text{signal amplitude}}{\text{carrier amplitude}} \times 100 = \frac{E_m}{E_c} = \frac{B}{A} \times 100$$
$$= \frac{1}{2} \times 100 = 50\%$$

The ratio $E_m/E_c = B/A$ expressed as a fraction is called *modulation index.*

(*ii*) Modulation may also be defined in terms of the values referred to the modulated carrier wave only. With reference to Fig. 38.10.

$$m = \frac{E_{c(max)} - E_{c(min)}}{E_{c(max)} + E_{c(min)}} \times 100 = \frac{A - B}{A + B} \times 100$$

where $E_{c(max)}$ and $E_{c(min)}$ are the maximum and minimum values of the amplitude of the modulated carrier wave.

As seen from Fig. 38.10,

$$m = \frac{3 - 1}{3 + 1} \times 100 = 50\%$$

In general, it may be kept in mind that if the positive peak of the audio signal causes the amplitude of the carrier wave to increase by 50% of its unmodulated value, the carrier is 50% modulated Fig. 38.11 (*a*). Similarly, the negative peak of the audio signal will cause the amplitude of the carrier wave to decrease by 50% of its unmodulated wave.

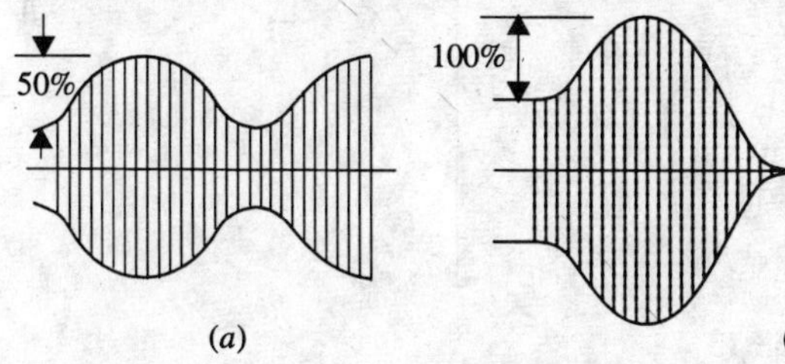

(*a*) (*b*)

Fig. 38.11

If, however, positive peak of the audio signal causes the amplitude of the carrier wave to increase by 100% of its unmodulated value. The carrier wave is 100% modulated Fig. 38.11 (*b*).

Fig. 38.12 shows a modulated wave with different degrees of modulation. As before, both the signal and carrier waves are assumed to be sine waves. Smallest value of m = 0 *i.e.*, when amplitude of the modulating signal is zero. It means that $m = 0$ for an unmodulated carrier wave. Maximum value of $m = 1$ when $B = A$. Value of m can vary anywhere from 0 to 100% without introducing distortion. Maximum undistorted power of a radio transmitter is obtained when m = 100%. If m is less than 100%, power output is reduced though the power content of the carrier is not. Modulation in excess of 100% produces severe distortion and interference (called *splatter*) in the transmitter output.

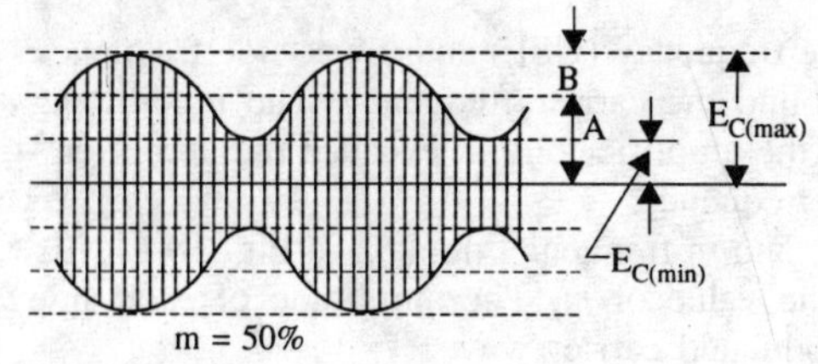

(*a*)

B
A
m = 100%

(*b*)

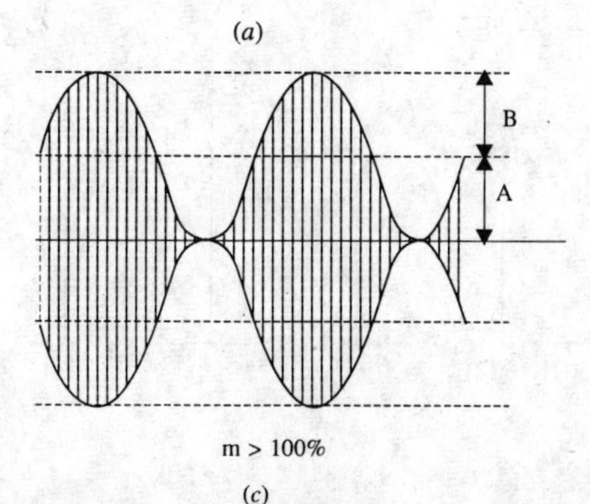

(*c*)

Fig. 38.12

38.14. Upper and Lower Side Frequencies

An unmodulated carrier wave consists of only one single-frequency component having a frequency of f_c. When it is combined with a modulating signal of frequency f_m, heterodyning action takes place. As a result, to additional frequencies called *side frequencies* are produced. The AM wave is found to consist of three-frequency components (Fig. 38.13).

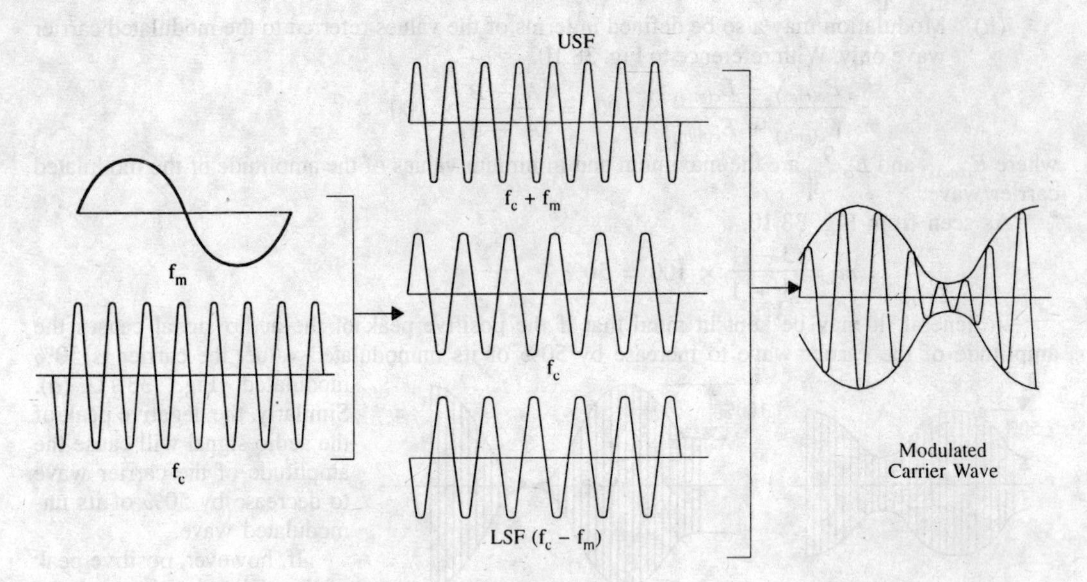

Fig. 38.13

1. the original carrier frequency components, f_c.
2. a higher frequency component ($f_c + f_m$).
   It is called the *sum component*.
3. a lower frequency component ($f_c - f_m$).
   It is called the *difference component*.

The two new frequencies are called the upper-side frequency (USF) and lower-side frequency (LSF) respectively and are symmetrically located around the carrier frequency. The modulating frequency remains unchanged but does not appear in the amplifier output because the amplifier's load presents practically zero impedance to this low frequency.

These are shown in time domain in Fig. 38.14 (*a*) and in frequency domain in Fig. 38.14 (*b*). The amplitude of the side frequencies depends on the value of *m*. The amplitude of each side frequency = $mA/2$ where *A* is the amplitude of unmodulated carrier wave.

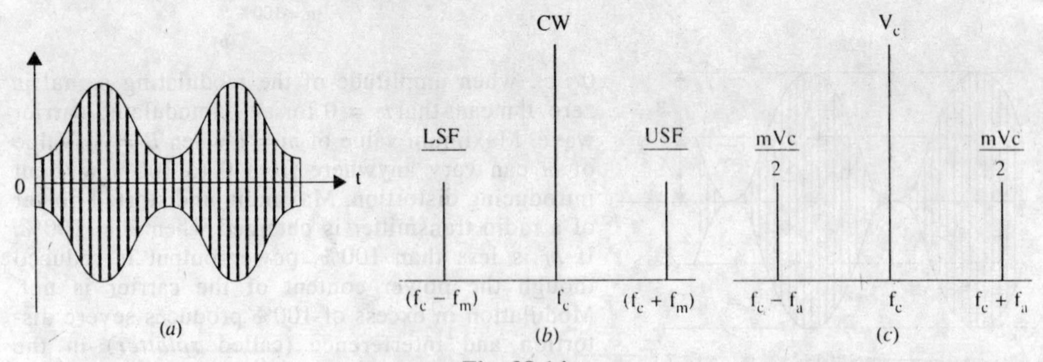

Fig. 38.14

The frequency-versus-voltage spectrum of an AM wave is shown in Fig. 38.14 (*c*)

Example. 38.1. *A 10 MHz sinusoidal carrier wave of amplitude 10 mV is modulated by a 5 kHz sinusoidal audio signal wave of amplitude 6 mV. Find the frequency components of the resultant modulated wave and their amplitudes.* **(Electronics & Commun. Pune Univ. 1992)**

Solution. Here, f_c = 10 MHz and f_m = 5 kHz = 0.005 MHz. The modulated carrier contains the following frequencies:

1. original carrier wave of frequency
$f_c = 10$ MHz
2. USF of frequency = 10 + 0.005
= 10.005 MHz
3. LSF of frequency = 10 − 0.005
= 9.995 MHz

The frequency spectrum is shown in Fig. 38.15.

Here, $m = \frac{B}{A} = \frac{6}{10} = 0.6$

Amplitude of LSF = USF = mA/2 = 0.6 × 10/2 = 3 mV as shown.

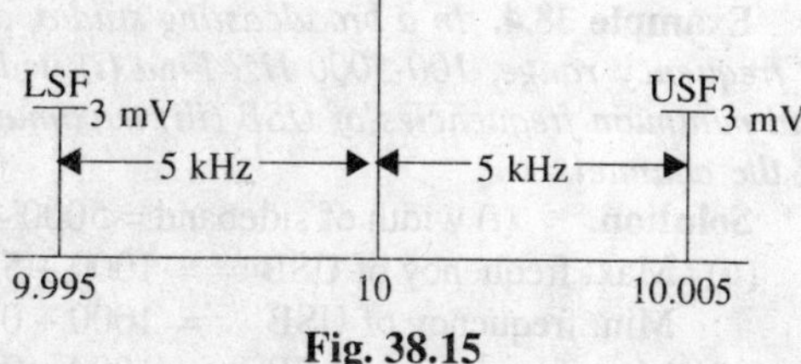

Fig. 38.15

38.15. Upper and Lower Sidebands

In Art. 38.14, it was assumed that the modulating signal was composed of one frequency component only. However, in a broadcasting station, the modulating signal is the human voice (or music) which contains waves with a frequency range of 20 – 4000 Hz. Each of these waves has its own LSF and USF. When combined together, they give rise to an upper-side band (USB) and a lower-side band (LSB) as shown in Fig. 38.16. The USB, in fact, contains all sum components of the signal and carrier frequency whereas LSB contains their difference components.

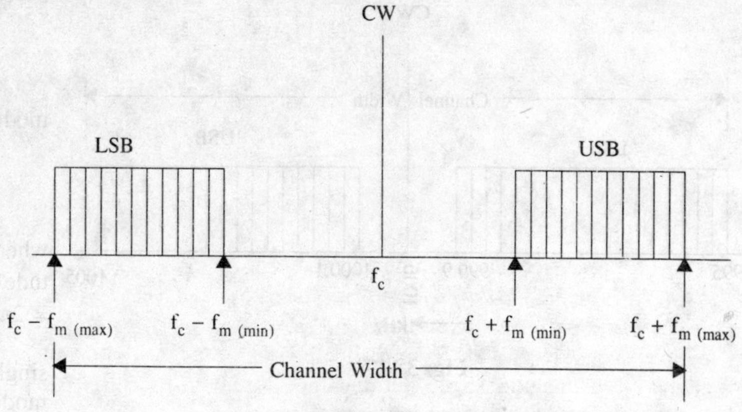

Fig. 38.16

The channel width (or bandwidth) is given by the difference between extreme frequencies *i.e.,* between maximum frequency of USB and minimum frequency of LSB. As seen

Channel width = 2 × maximum frequency of modulating signal
$= 2 \times f_{m(max)}$

Example. 38.2. *An audio signal given by 15 sin 2π (2000 t) amplitude-modulates a sinusoidal carrier wave 60 sin 2π (100,000) t.*

Determine: (a) modulation index (b) percent modulation (c) frequencies of signal and carrier (d) frequency spectrum of the modulated wave.

Solution. Here, $B = 15$ and $A = 60$

(*a*) $\text{M.I.} = \frac{B}{A} = \frac{15}{60} = 0.25$

(*b*) m = M.I. × 100 = 0.25 × 100 = 25%

(*c*) f_m = 2000 Hz — by inspection of the given equation
f_c = 100,000 Hz — by inspection of the given equation

(*d*) the three frequencies present in the modulated CW are

(*i*) 100,000 Hz = 100 kHz (*ii*) 100,000 + 2000 = 102,000 Hz = 102 kHz (*iii*) 100,000 – 2000 = 98,000 Hz = 98 kHz.

Example 38.3. *A bandwidth of 15 MHz is available for AM transmission. If the maximum audio signal frequency used for modulating the carrier is not to exceed 15 kHz, how many stations can broadcast within this band simultaneously without interfering with each other?*

Solution. BW required by each station

$$= 2 \times f_{m(max)} = 2 \times 15 = 30 \text{ kHz}$$

Hence, the number of stations which can broadcast within this frequency band without interfering with one another is

$$15 \text{ MHz} / 30 \text{ kHz} = 500$$

Example 38.4. *In a broadcasting studio, a 1000 kHz carrier is modulated by an audio signal of frequency range, 100-5000 Hz. Find (i) width or frequency range of side-bands (ii) maximum and minimum frequencies of USB (iii) maximum and minimum frequencies of LSB and (iv) width of the channel.*

Solution. (*i*) width of sideband = 5000 – 100 = 4900 Hz

(*ii*) Max. frequency of USB = 1000 + 5 = 1005 kHz
Min. frequency of USB = 1000 + 0.1 = 1000.1 kHz

(*iii*) Max. frequency of LSB = 1000 – 0.1 = 999.9 kHz
Min. frequency of LSB = 1000 – 5 = 995 kHz

(*iv*) Width of the channel = 1005 – 995 = 10 kHz

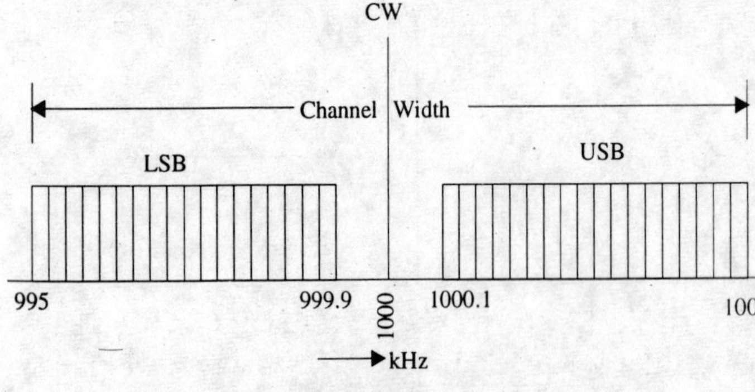

Fig. 38.17

38.16. Mathematical Analysis of a Modulated Carrier Wave

The equation of an unmodulated carrier wave is

$$e_c = E_c \sin 2\pi f_c t$$
$$= A \sin 2\pi f_c t$$
$$= A \sin \omega t$$

where A is the constant amplitude of the carrier wave and $\omega = 2\pi f_c$.

Let the equation of the single-frequency sinusoidal modulating signal be

$$e_m = E_m \sin 2\pi f_m t = B \sin 2\pi f_m t = B \sin pt \text{ – where } p = 2\pi f_m$$

As seen from Fig. 38.18, the amplitude of the modulated carrier wave at any instant is

$$= A + e_m \qquad (\because A \text{ is constant})$$
$$= (A + B \sin pt)$$

Hence, its instantaneous value is given by

$$e = (A + B \sin pt) \sin \omega t$$
$$= A \sin \omega t + B \sin \omega t \times \sin pt$$
$$= A \sin \omega t + \frac{B}{2} \times 2 \sin \omega t \times \sin pt$$
$$= A \sin \omega t + \frac{B}{2} [\cos (\omega - p)t - \cos (\omega + p)t]^*$$
$$= A \sin \omega t + \frac{B}{2} \cos (\omega - p)t - \frac{B}{2} \cos (\omega + p)t$$
$$= A \sin 2\pi f_c t + \frac{B}{2} \cos 2\pi (f_c - f_m)t - \frac{B}{2} \cos 2\pi (f_c + f_m)t$$

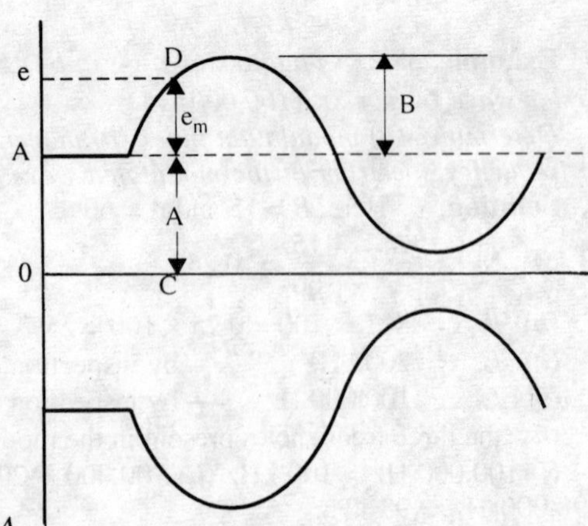

Fig. 38.18

As seen from Art. 38.13, $m = B/A$ or $B = mA$

*2 sin A, sin B = cos (A – B) – cos (A + B).

As seen Art. 38.13, m = B/A or B = mA

$$\therefore \quad e \quad A \sin 2\pi f_c t + \left(\frac{mA}{2}\right) \cos 2\pi (f_c - f_m)t - \left(\frac{mA}{2}\right) \cos 2\pi (f_c + f_m)\, t$$

It is seen that the modulated wave contains three components:

(*i*) $A \sin 2\pi f_c t$ — the original carrier wave

(*ii*) $\frac{mA}{2} \cos 2\pi (f_c + f_m)t$ — upper side frequency

(*iii*) $\frac{mA}{2} \cos 2\pi (f_c - f_m)t$ — lower side frequency

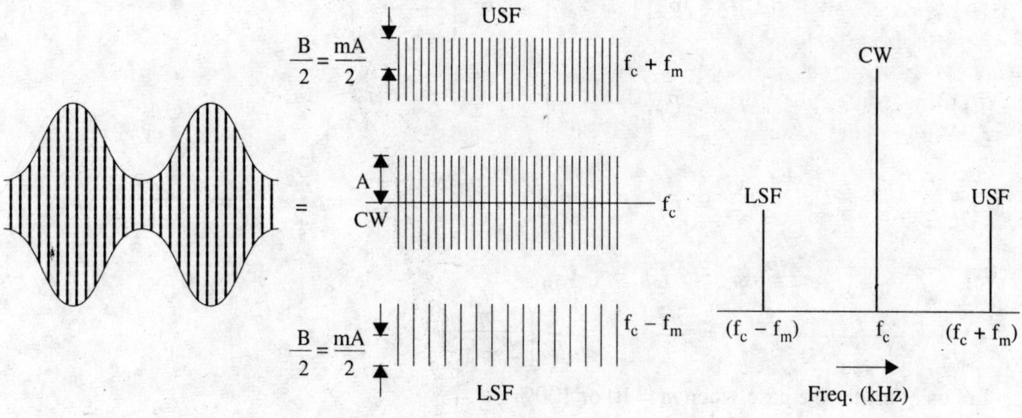

Fig. 38.19

These three frequencies are not a mathematical fiction but they actually exists. In fact, with the help of a narrow-band filter, we can separate side frequencies from the carrier wave.

Example 38.5. *The tuned circuit of the oscillator in an AM transmitter uses a 40 μH coil and a 1 nF capacitor. If the carrier wave produced by the oscillator is modulated by audio frequencies upto 10 kHz, calculate the frequency band occupied by the side bands and channel width.*

Solution. The resonant frequency is given by

$$f_c = \frac{1}{2\pi \sqrt{LC}} = \frac{1}{2\pi \sqrt{40 \times 10^{-6} \times 1 \times 10^{-9}}} = 796 \text{ kHz}$$

The frequency range occupied by the side-bands is from 786 kHz to 806 kHz. The channel width

$$= 2 \times 10 = 20 \text{ kHz}$$

38.17. Power Relations in an AM Wave

As discussed in Art. 38.16, a modulated carrier wave consists of the following three components:

1. original carrier wave of amplitude A,
2. USF wave of amplitude (mA/2),
3. LSF wave of amplitude (mA/2).

Now, power radiated out by a wave through an antenna is proportional to (amplitude)^2.

Carrier power, $P_c \propto A^2 = KA^2$

$$\text{USB power, } P_{USB} \propto \left(\frac{B}{2}\right)^2 = \frac{KB^2}{4}$$

$$\text{LSB power, } P_{LSB} \propto \left(\frac{B}{2}\right)^2 = \frac{KB^2}{4}$$

Total sideband power $= P_{SB} = 2 \times \frac{KB^2}{4} = \frac{KB^2}{2}$

Total power radiated out from the antenna is

$$P_T = P_C + P_{SB} = KA^2 + \frac{KB^2}{2}$$

Substituting $B = mA$, we get

$$P_T = KA^2 + \frac{K}{2}(mA)^2 = KA^2\left(1 + \frac{m^2}{2}\right)$$

Now, $P_C = KA^2$

(i) $\therefore$ $$P_T = P_C\left(1 + \frac{m^2}{2}\right) = P_C\left(\frac{2 + m^2}{2}\right)$$

(ii) $$P_C = P_T\left(\frac{2}{2 + m^2}\right)$$

(iii) $$P_{SB} = P_T - P_C = P_C\left(1 + \frac{m^2}{2}\right) - P_C = \frac{m^2}{2}P_C = \left(\frac{m^2}{2 + m^2}\right)P_T$$

(iv) $$P_{USB} = P_{LSB} = \frac{1}{2}P_{SB}$$

$$= \frac{m^2}{4}P_C = \frac{1}{2}\left(\frac{m^2}{2 + m^2}\right)P_T$$

Let us consider the case when $m = 10$ or 100%

(i) $P_T = 1.5\,P_C = 1.5 \times$ carrier power

(ii) $P_C = \frac{2}{3}P_T = \frac{2}{3} \times$ total power radiated

(iii) $P_{USB} = P_{LSB} = \frac{1}{4}P_C = 25\%$ of the carrier power

(iv) $P_{SB} = \frac{1}{2}P_C$

$= 50\%$ of carrier power

(v) $$\frac{P_{USB}}{P_T} = \frac{0.25\,P_C}{1.5\,P_C} = \frac{1}{6}$$

It means that single side-band contains 1/6th of the total power radiated out by the transmitter. That is why single-side-band (SSB) transmission is more power efficient.

38.18. Modulation Efficiency

It should be noted that it is only the power caused by the information signal or message contributes towards message quality in the receiver. Hence, it is called *useful power*. The carrier's power although important to the receiver's ability to recover the message, is not useful power in the sense that it carries no useful information. Modulation efficiency of the transmitted signal is defined as the ratio of useful power to the total power.

$$\eta_{AM} = \frac{\text{signal power}}{\text{carrier power + signal power}} = \frac{P_m}{P_C + P_m}$$

If A is the amplitude of the carrier wave, then $P_C \propto A^2$. If $f(t)$ represent the message signal, then

$$\eta_{AM} = \frac{f^2(t)}{A^2 + f^2(t)}$$

If $f(t)$ is a square wave of peak value A, then maximum value of modulation efficiency is

$$\eta_{AM} = \frac{A^2}{A^2 + A^2} = 0.5 \text{ or } 50\%$$

However, if $f(t)$ is a sinusoid, then $\eta_{AM} \leq 33.3\%$ although for practical messages such as voice or music, efficiency is less than 33.3%.

Example 38.6. *In AM modulation, the equation of the modulating signal is given by $f(t) = A_m \cos \omega_m t$. If the amplitude of the carrier wave is A, find the modulation efficiency if there is no over-modulation.*

Solution. We will first find the average squared value of the signal.

$$\overline{f^2(t)} = \frac{1}{T_m} \int_{-Tm/2}^{Tm/2} A_m^2 \cos^2(\omega_m t)\, dt$$

$$= \frac{A_m^2}{8\pi} \int_{-2\pi}^{2\pi} (1 + \cos x)\, dx = \frac{A_m^2}{2}$$

Hence, modulation efficiency is given by

$$\eta_{AM} = \frac{A_m^2 / 2}{A^2 + A_m^2 / 2} = \frac{(A_m / A)^2}{2 + (A_m / A)^2}$$

For no over-modulation, $A_m \leq A$, hence $\eta_{AM} \leq 1/3$ or 33.3%.

Example 38.7. *A radio telephone transmitter using AM has an unmodulated carrier output power of 20 kW and can be modulated to a maximum depth of 80% by sinusoidal modulation. Find the value to which unmodulated carrier power may be increased without resulting in over-loading if maximum permitted modulation index is restricted to 60%.*

(Electronics, A.M.Ae.S.I. Dec., 1994)

Solution. $P_T = P_C\left(\frac{2 + m^2}{2}\right) = 20\left(\frac{2 + 0.8^2}{2}\right) = 26.4 \text{ kW}$

Now then $m = 0.6$, P_T is still 26.4 kW. Hence, the new value of carrier power is given by the relation

$$26.4 = P_C\left(\frac{2 + 0.6}{2}\right)^2 \qquad \therefore\ P_C = 22.4 \text{ kW}$$

Example 38.8. *The total power content of an AM wave is 1500 W. For a 100 per cent modulation, determine*

(i) power transmitted by carrier;

(ii) power transmitted by each side-band.

Solution. $P_T = 1500$ W, $P_C = ?$ $P_{USB} = ?$ $P_{LSB} = ?$

(i) $$P_C = P_T\left(\frac{2}{2 + m^2}\right) = P_T\left(\frac{2}{2 + 1^2}\right)$$

$$= \frac{2}{3} P_T = \frac{2}{3} \times 1500 = 1000 \text{ W}$$

(ii) $$P_{USB} = P_{LSB} = \frac{m^2}{4} P_C$$

$$= \frac{1^2}{4} \times 1000 = 250 \text{ W}$$

As seen, $P_{USB} = \frac{1}{6} P_T$.

Example 38.9. *The total power content of an AM wave is 2.64 kW at a modulation factor of 80%. Determine the power content of*

(i) carrier *(ii) each side-band.*

Solution. (*i*) $P_C = P_T\left(\frac{2}{2 + m^2}\right)$

$$= 2.64 \times 10^3\left(\frac{2}{2 + 0.8^2}\right) = 2000 \text{ W}$$

(*ii*) $P_{USB} = P_{LSB} = \frac{m^2}{4}P_C$

$$= \frac{0.8^2}{4} \times 2000 = 320 \text{ W}$$

Example 38.10. *A certain transmitter radiates 9 kW with carrier unmodulated and 10.125 kW when carrier is sinusoidally modulated. Calculate modulation index. If another sine wave corresponding to 40% modulation is transmitted simultaneously, determine total radiated power.*

(Electronics, A.M.Ae.S.I. June, 1994)

Solution. The relation between the total power radiated and the carrier power is given by the equation.

$$P_T = P_C\left(\frac{2 + m^2}{m}\right) = 9\left(\frac{2 + m^2}{2}\right)$$

$\therefore$ $m = 0.5$ and $M = 50\%$

When another sine wave is used, the total radiated power is given by the relation

$$P_T = 9\left(1 + \frac{0.5^2}{2} + \frac{0.4^2}{2}\right)$$

$$= 10.845 \text{ kW}$$

Example 38.11. *A spectrum analysis of an AM signal shows :*

(i) 50 volts at 1000 kHz; (ii) 20 volts at 1010 kHz; (iii) 20 volts at 990 kHz.

Calculate the carrier frequency, USB, LSB, modulating frequency, modulation index, percentage of modulation, radiated power and power content of side-bands. Assume carrier power to be 5 kW.

(Electronics, A.M.Ae.S.I. Dec., 1993)

Solution. As seen from the given data, 50 V corresponds to carrier wave and 20 V represents upper and lower side frequency voltage.

Obviously, carrier frequency = 1000 kHz, USF = 1010 kHz. LSF = 990 kHz.

If V is the carrier voltage, then voltage of each side frequencies = $mV_c / 2$

$\therefore$ $20 = m \times 50/2$, $m = 0.8$ or $M = 80\%$

Total radiated power is given by

$$P_T = P_C\left(\frac{2 + m^2}{2}\right) = 5\left(\frac{2 + 0.8^2}{2}\right) = 6.6 \text{ kW}$$

Combined power content of the two side-bands

$$= P_T - P_C$$

$$= 6.6 - 5 = 1.6 \text{ kW}$$

Alternatively, $P = P\left(\frac{m^2}{2 + m^2}\right) = 6.6 \times \frac{0.64^2}{2.64} = 1.6 \text{ kW}$

Again power content of each side-band is

$$= \frac{m^2 P_C}{4} = \frac{0.8^2 \times 5}{4} = 0.8 \text{ kW}$$

Hence power content of both side-bands = $2 \times 0.8 = 1.6$ kW.

38.19. Forms of Amplitude Modulation

As shown in Art. 38.14, one carrier and two sidebands are produced in AM generation. It is found that it is not necessary to transmit all these signals to enable the receiver to reconstruct the original signal. Accordingly, we may attenuate or altogether remove the carrier or any one of the side-bands without affecting the communication process. The advantages would be

1. less transmitted power and 2. less bandwidth required.

The different suppressed component systems are:

***(a)* DSB-SC**

It stands for double-side-band suppressed carrier system [Fig. 38.20 (*a*)]. Here, carrier component is suppressed thereby saving enormous amount of power. As seen from Art. 38.18, carrier signal contains 66.7 per cent of the total transmitted power for $m = 1$. Hence, power saving amounts to 66.7% at 100% modulation.

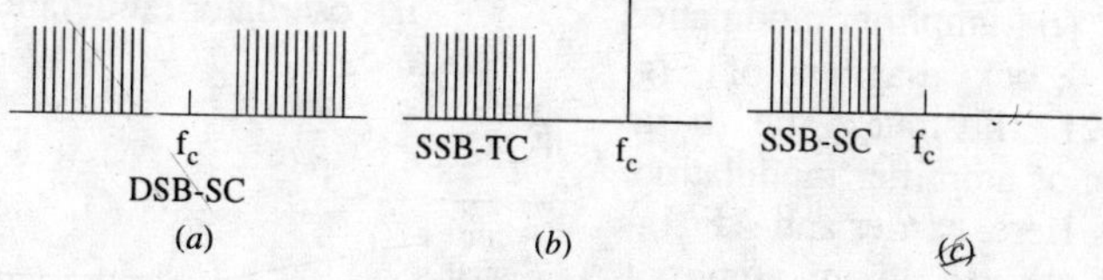

Fig. 38.20

***(b)* SSB-TC**

As shown in Fig. 38.20 (*b*), in this case, one side-band is suppresed but the other sideband and carrier are transmitted. It is called *single-side-band* transmitted carrier system. For $m = 1$, power saved is 1/6 of the total transmitted power.

***(c)* SSB-SC**

This is the most dramatic suppression of all because it suppresses one side-band and the carrier and transmits only the remaining side-band as shown in Fig. 38.20 (*c*). In the standard or double-side-band full-carrier (DSB-FC) AM, carrier conveys no information but contains maximum power. Since the two side-bands are exact images of each other, they carry the same audio information. Hence, all information is available in one side-band only. Obviously, carrier is superfluous and one sideband is redundant. Hence, one sideband and the carrier can be discarded with no loss of information. The result is SSB signal. The advantages of SSB-SC system are as follows:

1. total saving of 83.3% in transmitted power (66.7% due to suppression of carrier wave and 16.6% due to suppression of one side-band). Hence, power is conserved in an SSB transmitter;
2. bandwidth required is reduced by half *i.e.,* 50%. Hence, twice as many channels can be multiplexed in a given frequency range;
3. the size of power supply required is very small. This fact assumes vital importance particularly in a spacecraft;
4. since the SSB signal has narrower bandwidth, a narrower passband is permissible within the receiver, thereby limiting the noise pick up.

However, the main reason for wide-spread use of DSB-FC (rather than SSB-SC) transmission in broadcasting is the relative simplicity of its modulating equipment.

Example 38.12. *In an AM wave, calculate the power saving when the carrier and one side-band are suppressed corresponding to*

(i) $m = 1$ *(ii)* $m = 0.5$ **(Electronics, A.M.Ae.S.I. Dec., 1994)**

Solution. (*i*) When , $m = 1$

$$P_T = P_C\left(1 + \frac{m^2}{2}\right) = 1.5\,P_C$$

$$P_{LSB} = P_{USB} = \frac{m^2}{4}P_C = 0.25\,P_C$$

$$\therefore \quad \text{saving} = P_T - P_{SUB} = 1.5\,P_C - 0.25\,P_C = 1.25\,P_C$$

$$\therefore \quad \%\ \text{saving} = \frac{1.25\,P_C}{1.5\,P_C} \times 100 = 83.3\%$$

(*ii*) When, $m = 0.5$

$$P_T = P_C\left(1 + \frac{m^2}{2}\right) = P_C\left(1 + \frac{0.5^2}{2}\right) = 1.125\,P_C$$

$$P_{LSB} = P_{USB} = \frac{m^2}{4}P_C = \frac{0.5^2}{4}P_C = 0.0625\ P_C$$

$$\therefore \quad \%\ \text{saving} = \frac{1.125\ P_C - 0.0625\ P_C}{1.125\ P_C} \times 100 = 94.4\%$$

38.20. Methods of Amplitude Modulation

There are two methods of achieving amplitude modulation:

(i) amplifier modulation *(ii)* oscillator modulation

Block diagram of Fig. 38.21 illustrates the basic idea of amplifier modulation.

Here, carrier and AF signal are fed to an amplifier and the result is an AM output. The modulation process takes place in the active device used in the amplifier.

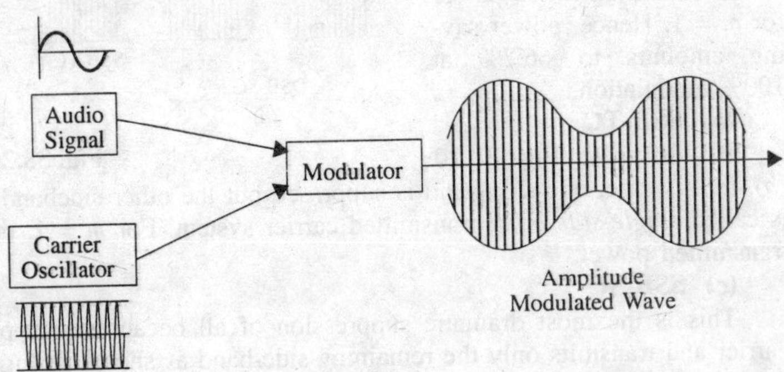

Fig. 38.21

38.21. Block Diagram of an AM Transmitter

Fig. 38.22 shows the block diagram of a typical transmitter. The carrier wave is supplied by a crystal-controlled oscillator at the carrier frequency. It is followed by a tuned buffer amplifier and an RF and output amplifier.

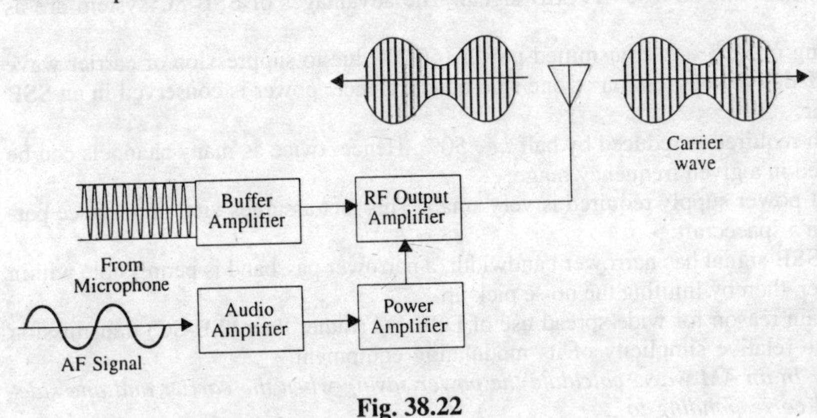

Fig. 38.22

The source of AF signal is a microphone. The audio signal is amplified by a low-level audio amplifier and, finally, by a power amplifier. It is then combined with the carrier wave to produce a modulated carrier wave which is ultimately radiated out in the free space by the transmitter antenna as shown.

38.22. Frequency Modulation

FM radio has become a very popular method of radio communication. It is capable of transmitting relatively high audio sound within the allotted spectrum band. Also, the general acceptance of stereophonic sound has been encouraged by FM transmission of dual channels of sound by multiplex systems.

As the name shows, in this modulation, it is only the frequency of the carrier which is changed and not its amplitude.

The amount of change in frequency is determined by the amplitude of the modulating signal whereas rate of change is determined by the frequency of the modulating signal. As shown in Fig. 38.23, in an FM carrier, information (or intelligence) is carried as variations in its frequency. As seen, frequency of the modulated carrier increases as the signal amplitude increases but decreases as the signal amplitude decreases. It is at its highest frequency (point H) when the signal

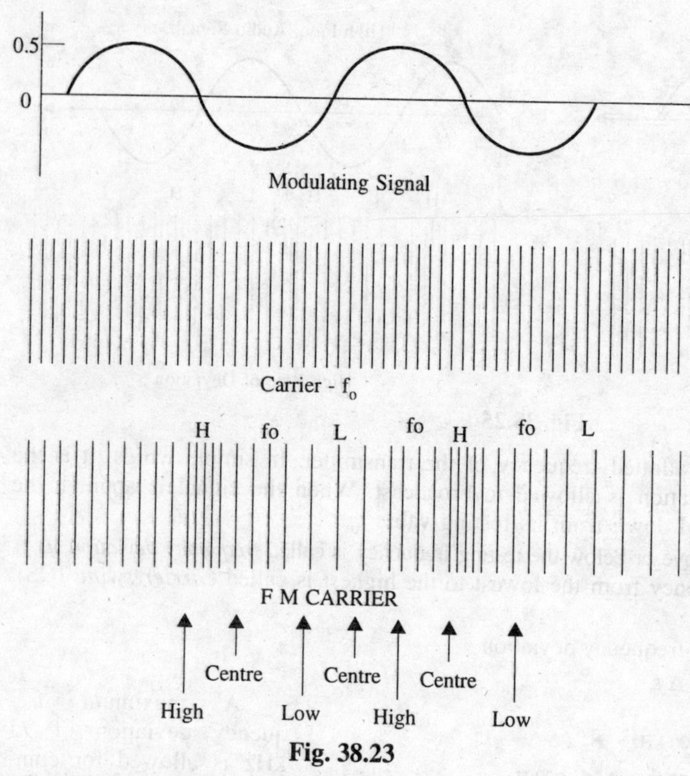

Fig. 38.23

amplitude is at its maximum positive value and is at its lowest frequency (point L) when signal amplitude is zero, the carrier frequency is at its normal frequency f_0 (also called resting or centre frequency).

The fact that amount of change in frequency depends on signal amplitude is illustrated in Fig. 38.24 where R stands for resting frequency. Here, signal amplitude is almost double of that in Fig. 38.23 though its frequency is the same. This louder signal causes greater frequency change in modulated carrier as indicated by increased bunching and spreading of the waves as compared with relatively weaker signal of Fig. 38.23.

The rate at which frequency shift takes place depends on the signal frequency as shown in Fig. 38.25. For example, if the modulating signal is 1 kHz, then the modulated carrier will swing between its maximum frequency and lowest frequency 1000 times per second. If $f_m = 2$ kHz, the rate of frequency swing would be twice as fast.

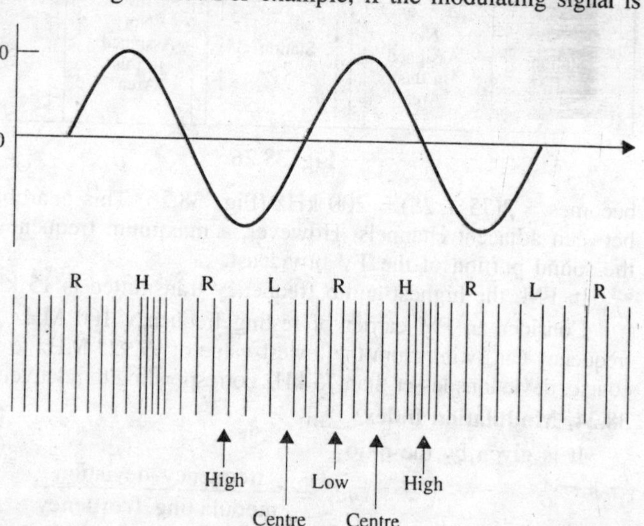

Fig. 38.24

In short, we have established two important points about the nature of frequency modulation:

(*i*) the amount of frequency deviation (or shift or variation) depends on the amplitude (loudness) of the audio signal. Louder the sound, greater the frequency deviation and *vice versa*. However, for the purposes of FM broadcasts, it has been internationally agreed to restrict maximum deviation to 75 kHz on each side of the centre frequency for sounds of maximum loudness. Sounds of lesser loudness are permitted proportionately less frequency deviation.

(*ii*) the rate of frequency deviation depends on the signal frequency.

38.23. Frequency Deviation and Carrier Swing

The frequency of an FM transmitter without signal input is called the *resting frequency* or

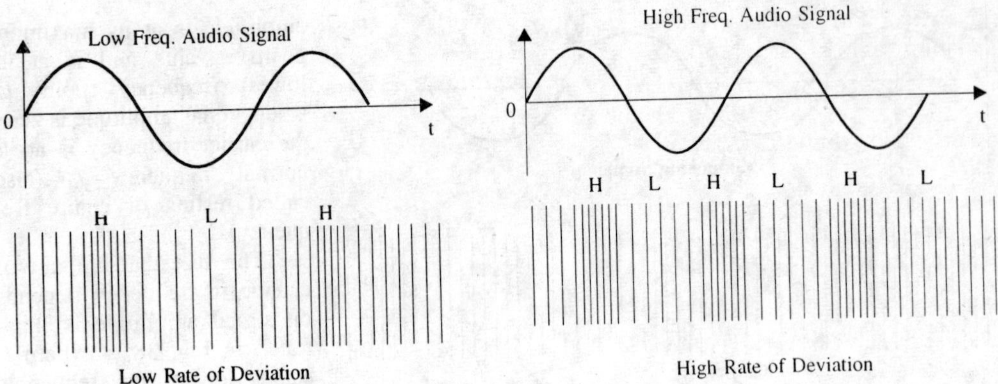

Fig. 38.25

centre frequency (f_0) and is the allotted frequency of the transmitter. In simple words, it is the carrier frequency on which a station is allowed to broadcast. When the signal is applied, the carrier frequency deviates up and down from its resting value f_0.

This change or shift either above or below the resting frequency is called *frequency deviation* (Δf).

The total variation in frequency from the lowest to the highest is called *carrier swing* (CS). Obviously,

$$\text{carrier swing} = 2 \times \text{frequency deviation}$$

$$CS = 2 \times \Delta f$$

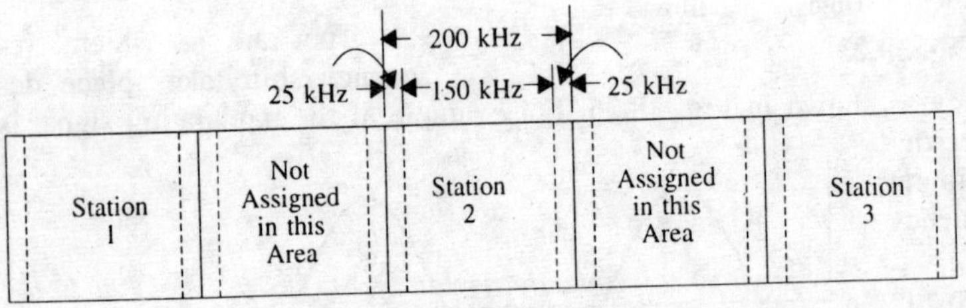

Fig. 38.26

A maximum frequency deviation of 75 kHz is allowed for commercial FM broadcast stations in the 88 to 168 MHz VHF band. Hence, FM channel width is 2 × 75 = 150 kHz. Allowing a 25 kHz guard band on either side. The channel width becomes = 2(75 + 25) = 200 kHz (Fig. 38.26) This guardband is meant to prevent interference between adjacent channels. However, a maximum frequency deviation of 25 kHz is allowed in the sound portion of the TV broadcast.

In FM, the highest audio frequency transmitted is 15 kHz.

Consider an FM carrier of resting frequency 100 MHz. Since $(\Delta f)_{max}$ = 75 kHz, the carrier frequency can swing from the lowest value of 99.925 MHz to the highest value of 100.075 MHz. Of course, deviations lesser than 75 kHz corresponding to relatively softer sounds are always permissible.

38.24. Modulation Index

It is given by the ratio

$$m_f = \frac{\text{frequency deviation}}{\text{modula ting frequency}}$$

Unlike amplitude modulation, this modulation index can be greater than unity. By knowing the value of m_f, we can calculate the number of significant sidebands and the bandwidth of the FM signal.

38.25. Deviation Ratio

It is the worst-case modulation index in which maximum permitted frequency deviation and maximum permitted audio frequency are used.

$$\therefore \quad \text{deviation ratio} = \frac{(\Delta f)_{max}}{f_{m\,(max)}}$$

Now, for FM broadcast stations, $(\Delta f)_{max}$ = 75 kHz and maximum permitted frequency of modulating audio signal is 15 kHz.

$$\therefore \quad \text{deviation ratio} = \frac{75 \text{ kHz}}{15 \text{ kHz}} = 5$$

For sound portion of commercial TV

$$\text{deviation ratio} = \frac{25 \text{ kHz}}{15 \text{ kHz}} = 1.67$$

38.26. Percent Modulation

When applied to FM, this term has slightly different meaning than when applied to AM. In FM, it is given by the ratio of actual frequency deviation to the maximum allowed frequency deviation.

$$m = \frac{(\Delta f)_{actual}}{(\Delta f)_{max}}$$

Obviously, 100% modulation corresponds to the case when actual deviation equals the maximum allowable frequency deviation. If, in some case, actual deviation is 50 kHz, then

$$m = \frac{50}{75} = \frac{2}{3} = 0.667$$
$$= 66.7\%$$

Value of $m = 0$ corresponds to zero deviation *i.e.*, unmodulated carrier wave. It is seen from the above equation $m \propto (\Delta f)_{actual}$. It means that when frequency deviation (*i.e.*, signal loudness) is doubled, modulation is doubled.

Example 38.13. *What is the modulation index of an FM carrier having a carrier swing of 100 kHz and a modulating signal of 5 kHz ?*

Solution $\quad CS = 2 \times \Delta f$

$$\Delta f = \frac{CS}{2} = \frac{100}{2} = 50 \text{ kHz}$$

$$m_f = \frac{\Delta f}{f_m} = \frac{50}{5} = 10$$

Example 38.14. *An FM transmission has a frequency deviation of 18.75 kHz. Calculate percent modulation if it is broadcast*

(i) in the 88-108 MHz band *(ii) as a portion of a TV broadcast.*

Solution (*i*) For this transmission band,

$$(\Delta f)_{max} = 75 \text{ kHz}$$

$$m = \frac{18.75}{75} \times 100 = 25\%$$

(*ii*) In this case, $(\Delta f)_{max}$ = 25 kHz

$$m = \frac{18.75}{25} \times 100 = 75\%$$

Example 38.15. *An FM signal has a resting frequency of 105 MHz and highest frequency of 105.03 MHz when modulated by a signal of frequency 5 kHz. Determine*

(i) frequency deviation (ii) carrier swing (iii) modulation index

(iv) percent modulation (v) lowest frequency by the FM wave.

Solution (*i*) Δf = 105.03 – 105 = 0.03 MHz = 30 kHz

(*ii*) $CS = 2 \times \Delta f = 2 \times 30 = 60$ kHz

(*iii*) $m_f = \frac{30}{5} = 6$

(*iv*) $m = \frac{30}{75} \times 100 = 40\%$

(*v*) lowest frequency = 105 – 0.03 = 104.87 MHz.

38.27. FM Side-bands

In FM, when a carrier is modulated, a number of side-bands are formed*. Though theoretically their number is infinite, their strength becomes negligible after a few side-bands. They lie on both sides of the centre frequency spaced f_m apart as shown in Fig. 38.27. Side-bands at equal distances from f_0 have equal amplitudes. If f_0 is the centre frequency and f_m the frequency of the modulating signal, then FM carrier contains the following frequencies:

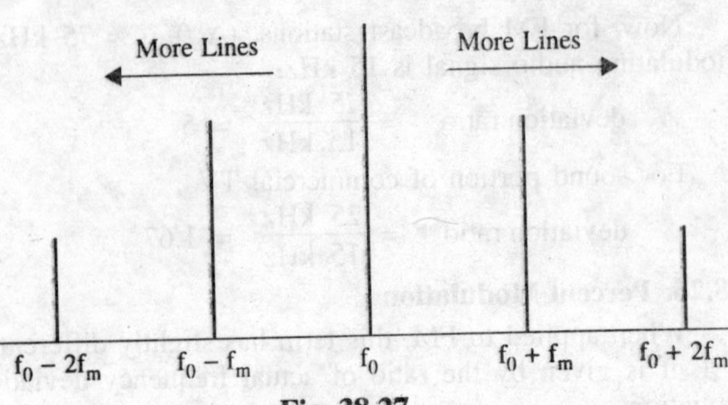

Fig. 38.27

(*i*) f_0 (*ii*) $f_0 \pm f_m$ (*iii*) $f_0 \pm 2f_m$ (*iv*) $f_0 \pm 3f_m$ and so on

The bandwidth occupied by the spectrum is

$$BW = 2\,n\,f_m$$

where n is the highest order of the significant sideband.

Another approximate expression for spectrum bandwidth is

$$BW = 2\,(1 + m_f)\,f_m$$

Now, $m_f = \dfrac{\Delta f}{f_m}$, hence

$$BW = 2\,(\Delta f + f_m)$$

This expression is based on the assumption that sidebands having amplitudes less than 5% of the unmodulated carrier wave are negligible or when m_f is at least 6.

Example 38.16. *A 5 kHz audio signal is used to frequency-modulate a 100 MHz carrier causing a frequency deviation of 20 kHz. Determine*

(i) modulation index, *(ii) bandwidth of the FM signal.*

Solution (*i*) $m_f = \dfrac{\Delta f}{f_m} = \dfrac{20}{5} = 4$

$$BW = 14\,f_m = 14 \times 5 = 70 \text{ kHz}.$$

Note. We cannot use the alternative expression for BW given in Art. 38.17 above because $m_f < 6$.

Example 38.17. *In an FM circuit, the modulation index is 10 and the highest modulation frequency is 20 kHz. What is the approximate bandwidth of the resultant FM signal?*

Solution. Since the value of m_f is more than 6, we will use the expression

$$BW = 2\,(\Delta f + f_m)$$

Now, $m_f = \dfrac{\Delta f}{f_m}$

or $10 = \dfrac{\Delta f}{20}$ $\therefore$ $\Delta f = 200$ kHz

$\therefore$ $BW = 2\,(200 + 20) = 440$ kHz.

38.28. Mathematical Expression for FM Wave

The unmodulated carrier is given by

$$e_c = A \sin 2\pi f_0 t$$

The modulating signal frequency is given by

$$e_m = B \sin 2\pi f_m t$$

The modulated carrier frequency f swings around the resting frequency f_0, thus

$$f = f_0 + \Delta f \times \sin 2\pi f_m t$$

Hence, equation for the frequency-modulated wave becomes

*An AM signal has only two side frequencies for each modulating frequency (Art. 38.14).

$$e = A \sin 2\pi ft$$
$$= A \sin [2\pi (f_0 + \Delta f \times \sin 2\pi f_m t)t]$$
$$= A \sin (2\pi f_0 t + \frac{\Delta f}{f_m} \cos 2\pi f_m t)$$
$$= A \sin (2\pi f_0 t + m_f \cos 2\pi f_m t)$$

38.29. Multiplexing

Multiplexing involves sending several different information signals along the same communication channel so that they do not interfere with each other. Two methods are briefly given here.

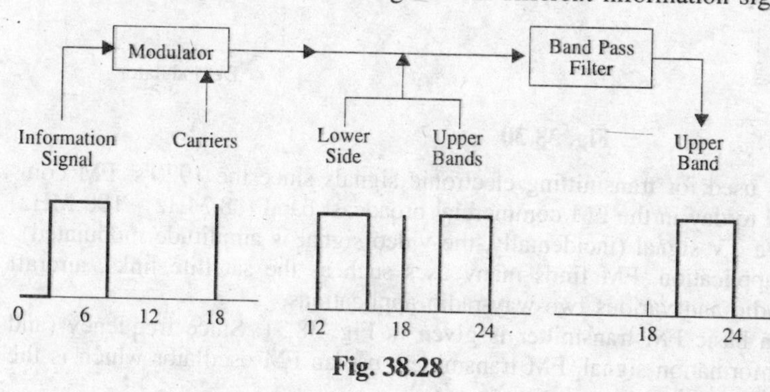

Fig. 38.28

(*a*) Frequency Division

In this method the analog audio signals modulates carriers of different frequencies which are then transmitted. Of course, it requires a greater bandwidth for accommodating all the signals.

The principal of frequency division is shown in Fig. 38.29 for cable transmission. The audio signal contains frequencies up to 6 kHz and the carrier frequency 18 kHz. Each sideband contains all the information and hence only one is really necessary. Here, a band pass filter is used to let only one sideband *i.e.*, upperside band to pass through.

The multiplexing of three 6 kHz wide information signals *a*, *b* and *c* using carriers of 18, 24 and 30 kHz is shown in Fig. 38.29.

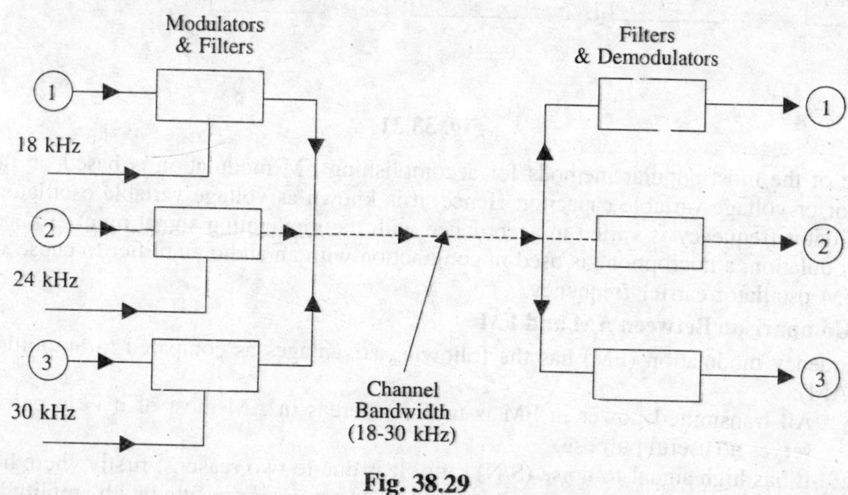

Fig. 38.29

(*b*) Time Division

Time division is shown in Fig. 38.30 for three signals. A multiplexer (*i.e.*, an electronic switch) samples each signal in turn (8000 time per second for speech).

The encoder converts the samples into a stream of pulses representing the level of each signal. The decoder and a demultiplexer which are synchronised with the multiplexer and encoder reverse the operation at the receiving end.

38.30. FM Transmission

Frequency modulation was invented by Major E.H. Armstrong, the inventor of superhetero-

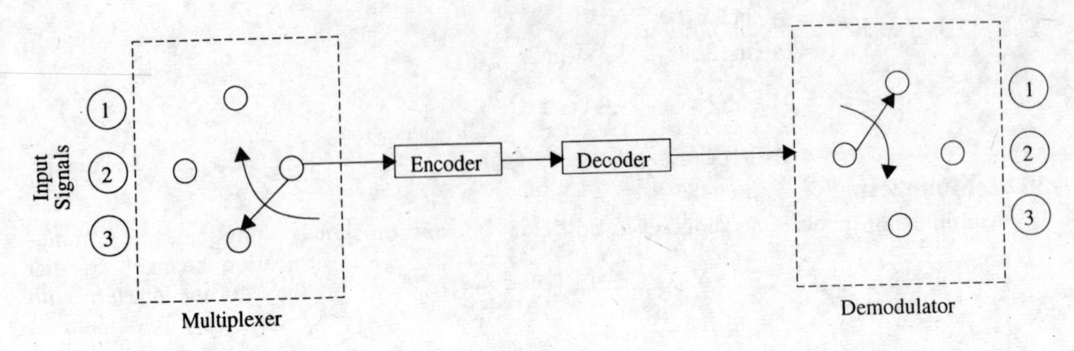

Fig. 38.30

dyne receiver. FM has been used for transmitting electronic signals since the 1920's. FM communication systems are used to-day in the FM commercial broadcast band (88 MHz – 108 MHz) and in the audio portion of a TV signal (incidentally, the video signal is amplitude modulated).

Apart from broadcast application, FM finds many uses such as the satellite links, aircraft altimetry, radars, amateur radio and various two-way-radio applications.

The block diagram of a basic FM transmitter is given in Fig. 38.31. Since frequency (and not amplitude) carries the information signal, FM transmitters use an FM oscillator which is the heart of an FM transmitter.

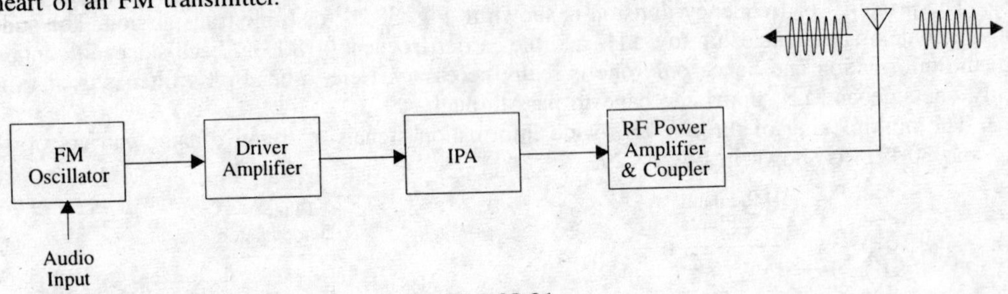

Fig. 38.31

One of the most popular methods for accomplishing FM modulation is based on the use of a varactor or voltage variable capacitor. Hence, it is known as voltage variable oscillator (VVO). The oscillator frequency is varied in accordance with the modulating signal input. To accomplish voice modulation, a microphone is used in conjunction with an audio amplifier to cause variations in the FM oscillator carrier frequency.

38.31. Comparison Between AM and FM

Frequency modulation (FM) has the following advantages as compared to amplitude modulation (AM):

1. All transmitted power in FM is useful whereas in AM most of it is in carrier which serves no useful purpose.
2. It has high signal-to-noise (S/N) ratio. It is due to two reasons: firstly, there happens to be less noise at VHF band and secondly, FM receivers are fitted with amplitude limiters which remove amplitude variations caused by noise.
3. Due to 'guard-band' there is hardly any adjacent-channel interference.
4. Since only transmitter frequency is modulated in FM, only fraction of a watt of audio power is required to produce 100% modulation as compared to high power required in AM.

However, FM has the following disadvantages:

1. It requires much wider channel — almost upto 20 times as large as needed by AM.
2. It requires complex and expensive transmitting and receiving equipment.

3. Since FM reception is limited to only line-of-sight, area of reception for FM is much smaller than for AM.

38.32. The Four Fields of FM

There are four major areas of application for FM transmission.

1. First use is in FM broadcast band 88-108 MHz with 200 kHz channels in which commercial FM stations broadcast programmes to their listeners.
2. Second use is in TV. Though video signal is amplitude-modulated, sound is transmitted by a separate transmitter which is frequency-modulated.
3. Third use is in the mobile or emergency services which transmit voice frequencies (20–4000 Hz) only.
4. Fourth use is in the amateur bands where again only voice frequencies are transmitted.

OBJECTIVE TESTS—38

A. Fill in the blanks with appropriate word(s) or numerical value(s).

1. Microwaves of 1.7 GHz frequency were used for communication in the year ....

2. Digital systems can also convey .... messages with the help of A/D converters.

3. One of the disadvantages of digital communication system is that it needs a larger ....

4. The region of the electromagnetic spectrum .... the ultra-violet region is used for communication purposes.

5. The range of frequencies occupied by a signal is called its ....

6. The range of frequencies a communication channel can accommodate is called its .... bandwidth.

7. Transmitted electromagnetic waves may be classified as ground waves and .... waves

8. The ionosphere extends from 50 km to .... km above Earth's surface.

9. In AM transmission, the .... of the carrier wave is changed.

10. When information is added to a carrier wave to vary its frequency, the process is called frequency ....

11. An AM wave has .... component frequencies.

12. In AM transmission, the frequency of the modulating signal is .... than the carrier wave frequency.

13. The smallest value of percent modulation in an AM wave is ....

14. In an AM wave, the amplitude of the side-frequencies depends on the value of .... modulation.

15. Single-side band (SSB) transmission is more .... efficient.

16. In double-side band full-carrier AM, carrier conveys no information but contains .... power.

17. In a single-side band suppressed carrier (SSB-SC) transmission system, there is a total saving of .... percent in transmitted power.

18. FM transmission is limited to almost line-of- ....

19. The process of sending two or more signals on one carrier wave is called ....

20. All the transmitted power in an FM signal is ....

B. Answer True or False.

1. Electromagnetic communication systems use a high frequency carrier wave which is modulated before transmission.

2. The capacity of a communication system is proportional to its bandwidth.

3. AM radio is a line-of-sight system.

4. An AM wave has two component frequencies.

5. AM broadcast band has higher frequencies than FM broadcast band.

6. Multiplexing is often used to send a stereo signal over FM waves.

7. Signals over 1500 kHz tend to be absorbed by the Earth.

8. Low-frequency transmission is much better at night than during day.

9. Video portion of a television signal is frequency-modulated.

10. All spark-created noise is amplitude-modulated.

11. FM was invented by Major E.H. Armstrong.

12. Noise places a limit on the performance of any communication system.

13. FM has noise-immunity advantage.

14. The number of sidebands in FM trnsmission is finite.

15. In an AM signal, two-thirds of the power is in the carrier and one-thirds in the two sidebands for 100 per cent modulation.

16. Multiplexing can be achieved either by frequency division or time division.

C. Multiple Choice Items.

1. The three essential elements of a communication system are............

(*a*) a transmitter (*b*) a channel

(*c*) a receiver (*d*) all of the above

2. Which region of the electromagnetic spectrum is used for communication purposes?

(*a*) infrared (*b*) ultra-violet

(*c*) visible (*d*) below ultra-violet

3. AM radio operates on wavelengths roughly between 10 m and ....

(*a*) 100 m (*b*) 1 km

(*c*) 2 km (*d*) 5 km

4. If a 1 MHz carrier wave is frequency-modulated by 1 kHz audio signal, the third upper sideband will be produced at .... kHz.

(*a*) 1.001 (*b*) 0.999

(*c*) 1.003 (*c*) 0.997

5. The maximum amplitude of the 100% amplitude-modulated carrier wave is .... that of the unmodulated wave.

(*a*) one-half (*b*) one-third

(*c*) twice (*d*) thrice

6. If a 500 kHz carrier wave is amplitude-modulated with a 2 kHz audio signal, the bandwidth of the transmitted signal will be .... kHz.

(*a*) 2 (*b*) 4

(*c*) 502 (*d*) 498

7. In AM transmission, the upper and lower sidebands are essentially .... images of each other.

(*a*) inverted (*b*) erect

(*c*) similar (*d*) mirror

8. If an AM receiver were to pick up only the carrier and exclude the sidebands, the only information available would be whether ....

(*a*) the carrier were on or off

(*b*) the lower or upper sidebands were missing

(*c*) there were two or more sidebands

(*d*) the bandwidth were wide or narrow

9. The layer of the ionosphere which is closest to the Earth is called .... layer

(*a*) A (*b*) D

(*c*) E (*d*) F

10. The maximum permissible value of the modulation index for an AM carrier wave is ....

(*a*) 0.5 (*b*) 1

(*c*) 1.5 (*d*) 2

11. The largest possible value of modulation efficiency of an AM system is .... per cent.

(*a*) 25 (*b*) 50

(*c*) 75 (*d*) 100

12. Frequency modulation had become very popular for ....

(*a*) hi-fi broadcasting

(*b*) TV-audio

(*c*) VHF-UHF communication

(*d*) all of the above

13. As compared to AM transmission, FM transmission has the advantages of ....

(*a*) higher fidelity

(*b*) noise immunity

(*c*) easy transmission of multiple signals

(*d*) all of the above

14. The frequency modulation technique used in many modern receivers was invented in 1933 by ....

(*a*) Armstrong (*b*) Marconi

(*c*) De Forest (*d*) Fleming

15. In order to avoid distortion of the signal in an amplitude-modulated carrier wave, its modulation index should lie between ....

(*a*) 0 and 0.5 (*b*) 0 and 1

(*c*) 0 and 1.5 (*d*) 1 and 2

16. FM requires upto .... times more bandwidth than AM.

(*a*) 5 (*b*) 10

(*c*) 20 (*d*) 50

17. The frequency modulation is used in
(*a*) satellite links
(*b*) aircraft altimetry
(*c*) radars
(*d*) all of the above

18. The video portion of a telephone signal is modulated
(*a*) pulse (*b*) amplitude
(*c*) frequency (*d*) phase

ANSWERS

A. Fill in the blanks

1. 1934 2. analog 3. bandwidth 4. below 5. bandwidth 6. channel
7. sky 8. 600 9. amplitude 10. modulation 11. three 12. lower
13. zero 14. percent 15. power 16. maximum 17. 83.3 18. sight
19. multiplexing 20. useful

B. True or False

1. T 2. T 3. F 4. F 5. F 6. T 7. T 8. T 9. F 10. T
11. T 12. T 13. T 14. F 15. T 16. T

C. Multiple Choice

1. *d* 2. *d* 3. *c* 4. *c* 5. *c* 6. *b* 7. *d* 8. *a* 9. *c* 10. *b*
11. *b* 12. *d* 13. *d* 14. *a* 15. *b* 16. *c* 17. *d* 18. *b*

39 VACUUM TUBES AND GAS VALVES

39.1 Electrons

Some elementary ideas about the atomic structure of matter have already been given in Chapter 1. As explained there, electricity is composed of extremely small negative charges called *electrons*. An electrically neutral atom of matter consists of a positively charged nucleus surrounded by orbital electrons whose number is such as to give an equal negative charge, so that the *resultant charge on the entire atom is zero*. In bad conductors of electricity, these electrons are comparatively more rigidly attached to their nuclei, hence it is very difficult to detach them away from the nuclei of atoms of such materials. But in the so-called good conductors of electricity like metals, a small portion of electrons is found free in the sense that electrons can easily pass from one atom to another. In other words, these electrons are not firmly attached to their parent nuclei. These free electrons form a kind of electron *gas* within a conductor and, in fact, it is their circulation around a closed circuit that constitutes the ordinary electric current.

The electrons constituting electron gas within metallic conductors are in a state of constant agitation and keep on colliding with each other and with other atoms, thereby executing very erratic and chaotic motion.

When such a free electron is darting around well within the interior of the metal, the resultant attractive force acting on it due to various positive ions lying all round it, is zero because it is pulled equally from all sides. Now, let us suppose that the electron is moved towards the metal surface. As it approaches the surface, the number of positive ions lying behind it increases and of those lying in front of it decreases till when it finally reaches the boundary, all ions lie behind it and none in front of it. Hence, during this journey of the electron towards the boundary, the backward pull on it goes on increasing till it is maximum at the surface beyond which it tapers off.

It is clear that as the electron is moved towards the surface of the metal, work has to be done because the electorn is being moved against the united backward pulling force of all ions lying behind the electron in motion.

As soon as electron tries to pass out of the interior of the metal through the boundary surface, it is immediately pulled back by the united force of all the ions which it is leaving behind, whereas before this, it was immersed in their force fields. Therefore, the outgoing electron experiences a potential energy barrier to its escape, this barrier being known as work function (ϕ) of the material.

39.2. Methods of Producing Electronic Emission

In the light of the above discussion, it may be guessed that if an electron in the electron gas of a metal be given sufficient energy and if this speeded-up electron approaches metal surface perpendicularly, then it will escape from the metal surface.

Ordinarily, none of the electrons in the electronic gas has kinetic energy equal to or greater than the amount necessary to overcome the "potential energy barrier". Hence, they cannot leave the surface and are pulled back every time they try to approach the surface.

There are many ways in which electronic emission can be brought about. In all these methods, energy, in one form or the other, is imparted to the free electrons, in a quantity which gives them kinetic energy sufficient enough to overcome potential energy barrier represented by the work function of the metal. Some of these ways are discussed below:

1. By heating the metals sufficiently high, thermal energy may be imparted to electrons to

enable them to escape from the metal surface. It is known as *thermionic emission* and is employed in vacuum tubes.

2. Electrons can also be pulled out by the application of a strong electrostatic field on the surface of the metal. Such an emission is known as *field emission.*
3. A beam of high-velocity electrons may strike a metal surface and so eject electrons out of it. This process is known as *secondary emission.*
4. When the surfaces of certain metals are illuminated by a beam of light, electrons are ejected out of these metals by the photons incident on their surface. Such an electronic emission is known as *photo-electric emission* and is used in photo-electric cells.

39.3. Thermionic Emission

Many formulae for the thermionic emission of electrons have been developed. But the formula employed today is the Richardson-Dushman equation

$$J_S = AT^2 e^{-b/T} \text{ A/m}^2$$

where J_S = emission current / m^2 of the emitter
T = temperature in °K
A = a constant
e = natural logarithmic base
b = a constant for the given metal.

As explained earlier, normally the free electrons in a metal do not have enough energy to escape from the confines of their metal boundaries or to surmount the potential energy barrier. For doing so, they must be given additional energy. In a thermionic cathode, the additional boost of energy is given to the electrons through the medium of heat. As the metal is heated more and more, the average kinetic energy of the electrons increases. If the temperature becomes high enough, then some of the electrons may get sufficient energy so as to overcome the potential-energy barrier and may escape. At high temperatures, the electrons are literally being 'boiled' off from the metal surface. This is analogous to the boiling of water.

From practical point of view, the work function of the metal used in a thermionic emitter, should be as low as possible to permit operation at the lowest possible temperature. This is very much desirable for increasing the efficiency of emission *i.e.*, to increase the milliamperes of emitted current per watt of heating power supplied to the emitter.

39.4. Cathodes

A cathode is an essential part of an electron tube because it emits electrons which are essential for tube operation. When heat energy is given to the meterial of the cathode, electrons are released. Different forms of cathodes can be distinguished from each other by the manner in which they are heated. In general, cathodes may be classified into two categories:

1. *directly-heated cathodes or filamentary cathodes;*
2. *indirectly-heated cathodes or heater cathodes.*

The construction of an indirectly-heated cathode is shown in Fig. 39.1. It consists of a thin metal sleeve (of nickel or konel) which is coated on the outside with electron-emitting oxides. Within the sleeve is a heater wire of tungsten coated with a layer, about 0.5 mm thick of the oxides of aluminium and beryllium and electrically insulated from the sleeve. The purpose of the heater wire is merely to heat the sleeve and the sleeve coating by radiation and conduction to an electron-emitting temperature. It is to be noted that useful electronic emission does not take place from the heater wire. The heater wire usually operates at a temperature of about 1000°K and maintains the sleeve at about 800°K which is sufficient for copious emission from the oxide coating on the metal sleeve.

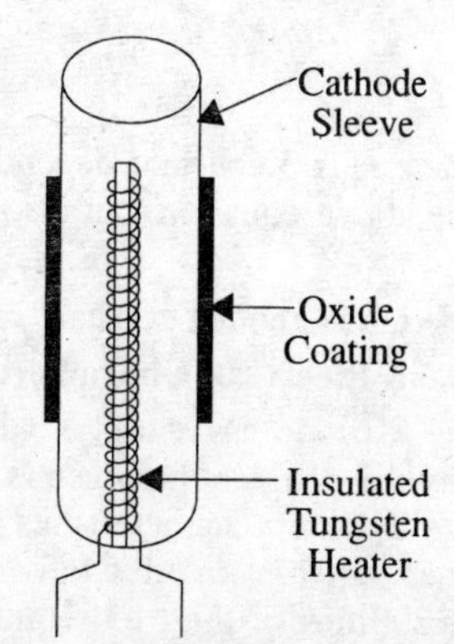

Fig. 39.1

39.5. Triode–Physical Characteristics

A typical cut-away sketch of a high-vaccum triode has been shown in Fig. 39.2. As its name implies, a triode contains three active electrodes. The cathode is located at the centre of the tube and is surrounded by a grid which is, in turn, surrounded by an anode. Its symbol is shown in Fig. 39.3 and it is the same as for a diode except for the addition of a dotted line between the anode and the cathode representing the grid.

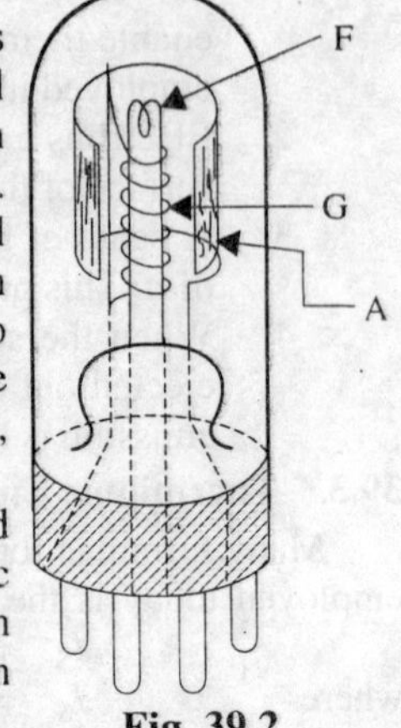

Fig. 39.2

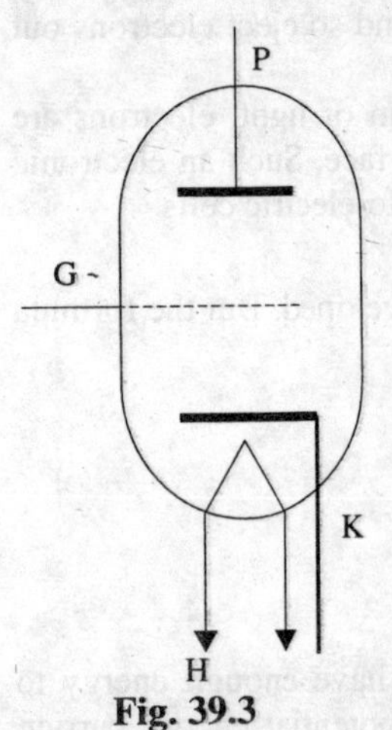

Fig. 39.3

The grid may be in the form of a helix, a squirrel cage or it has apertures or perforations through which electrons can freely pass from cathode to anode. The shape of the anode is influenced by the shape of cathode and grid and is usually round, elliptical or even rectangular.

The primary function of the additional third electrode *i.e.*, grid is to serve as an electrostatic screen and thus partially shield the cathode from the electrostatic field of the anode. The operation of the grid can be explained thus:

First, let is be pointed out that grid is directly situated in the path of the electronic beam and secondly, it is nearer to cathode which is the source of electrons. Because of this advantageous position, it exerts a very sensitive and powerful control on the flow of electrons which constitutes plate current. When the grid is given some negative potential with respect to the cathode, it repels back or retards the flow of electrons towards the plate, thereby reducing the plate current. Because of its advantageous position as pointed out, even with a small negative potential, it is capable of exerting greater controlling effect on plate current than anode itself even with a much larger positive potential. Because of this current controlling property of the grid, it is often referred to as *control* grid. As the negative potential of the grid (called *grid bias*) is increased, I_p decreases continuously till ultimately it is reduced to zero. Grid-bias corresponding to zero plate current is known as *cut-off* potential of the grid. Its value can be found by putting $I_p = 0$ in the triode equation given below. If on the other hand, it is made positive relative to cathode, it helps the anode in pulling the electrons up by partially neutralizing the negative space charge. Hence, I_p is increased. It is found that small changes in grid voltage produce such changes in plate current which can be produced by comparatively large changes in plate voltage. It is because of this that a triode possesses that valuable property of amplification. If a change of 2 volts on the grid produces the same change in plate current as produced by a change of 50 volts in plate voltage, then the amplification factor (μ) is 50/2 = 25. It means that the grid (because of its advantageous positioning) is 25 times more effective than anode in controlling the plate current.

From the above discussion it is obvious that the plate current depends not only on cathode temperature and plate voltage V_p but on grid voltage V_g also. In the region where plate current is principally limited by space-charge, the expression for plate current is given by

$$I_p = K\left(V_g + \frac{V_p}{\mu}\right)^{3/2}$$

where K is a constant depending upon the geometry of the tube and μ is the amplification factor. The above equation can also be put in an alternate form

$$I_p = A(V_p + \mu V_g)^{3/2}$$

where A is another constant equal to $K / \mu^{3/2}$.

39.6. Electrical Characteristics of a Triode

A triode has static as well as dynamic characteristics. The static characteristics give relation between different parameters of a triode when different d.c. potentials are applied to its electrodes. The dynamic characteristics are the values obtained with a.c. voltage applied to the control grid when other electrodes have d.c. potentials. Hence this characteristic indicates the performance capabilities of the triode under actual working conditions.

The three static characteristics of a triode are :

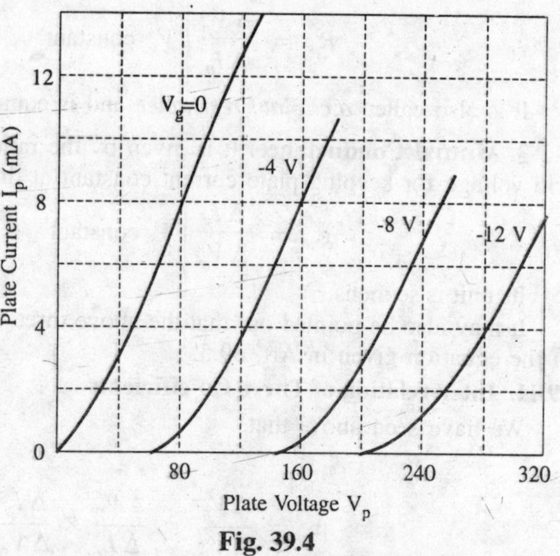

Fig. 39.4

1. **V_p/I_p characteristic.** It is also called plate or anode characteristic.
2. **V_g/I_p characteristic.** It is also called transfer or mutual characteristic.
3. **V_g/V_p characteristic.** It is also called constant-current characteristic.

39.7. Plate Characteristic of a Triode

For obtaining this characteristic, d.c. grid voltage V_g is kept constant at a given d.c. value whereas plate voltage V_p is changed and corresponding values of I_p are noted. The results so obtained are plotted in Fig. 39.4 which are typical for a triode.

39.8. Transfer Characteristic

This characteristic shows variation of I_p with V_g with fixed values of V_p. Different characteristics corresponding to different plate potentials are shown in Fig. 39.5.

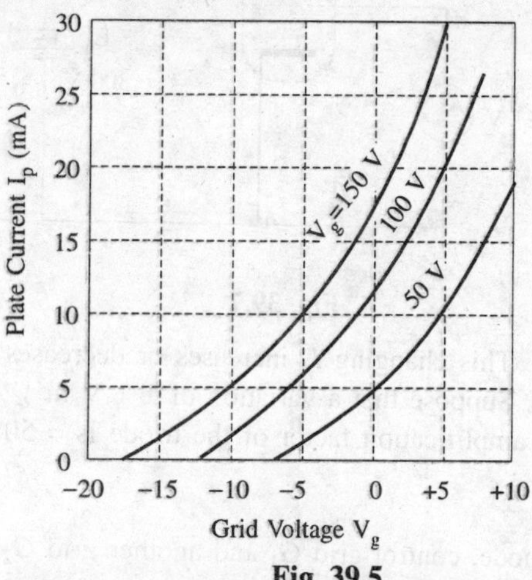

Fig. 39.5

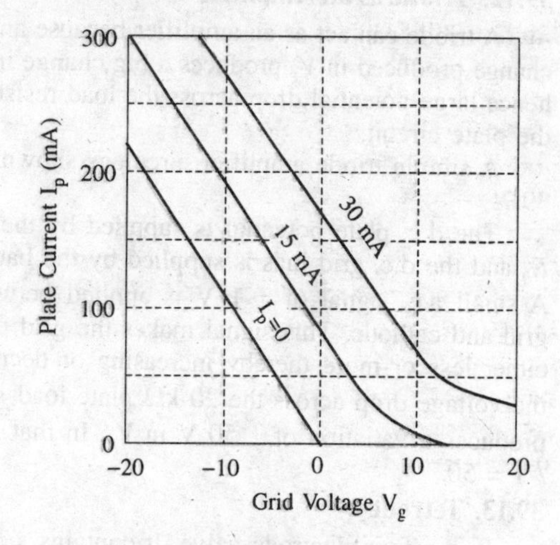

Fig. 39.6

39.9. Constant-current Characteristic

It shows the relation between plate voltage and grid voltage for keeping plate current constant at a given value (Fig. 39.6).

39.10. Triode Co-efficient

Three co-efficients or parameters of a triode are as under:

1. **Amplification Factor** (μ). It is given by the ratio of the change in plate voltage to the change in grid voltage for keeping the plate current constant.

$$\therefore \qquad \mu = \left.\frac{\Delta V_p}{\Delta V_g}\right| I_p \text{ constant.}$$

It is a dimensionless number and is represented by the slope of the constant-current characteristic.

2. **Plate Resistance.** It is given by the ratio of the change in plate voltage to the change in plate current for a constant grid voltage.

$$\therefore \qquad R_p = \frac{\Delta V_p}{\Delta I_p}\Bigg| V_g \text{ constant.}$$

It is also called *a.c. plate resistance* and is equal to the reciprocal of the plate characteristic.

3. **Mutual Conductance.** It is given by the ratio of change in plate current to the change in grid voltage for keeping plate current constant at the given value.

$$\therefore \qquad g_m = \frac{\Delta I_p}{\Delta V_g}\Bigg| V_p \text{ constant}$$

Its unit is seimens.

It may also be pointed out that the above three co-efficients are, in fact, the partial derivative of the equation given in Art. 39.5.

39.11. Inter-relation of Three Co-efficients

We have seen above that,

$$\mu = \frac{\Delta V_p}{\Delta V_g} = \frac{\Delta V_p}{\Delta I_p} \times \frac{\Delta I_p}{\Delta V_g} = R_p \times g_m$$

or $$\mu = R_p \times g_m$$

39.12. Triode as an Amplifier

A triode can act as an amplifier because any small change produced in V_g produces a big change in I_p and hence large potential drop across the load resistance in the plate circuit.

A simple triode amplifier circuit is shown in Fig. 39.7.

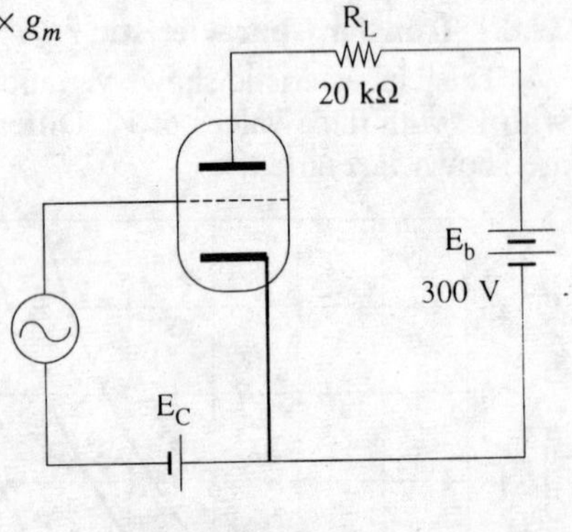

The d.c. plate potential is supplied by the battery E_b and the d.c. grid bias is supplied by the battery E_c. A small a.c. signal of ± 1 V is applied between the grid and cathode. This signal makes the grid potential either less or more thereby increasing or decreasing I_p. This changing I_p increases or decreases the voltage drop across the 20 kΩ plate load resistance. Suppose that a variation of ± 1 V in V_g produces a variation of ± 50 V in V_p. In that case, the amplification factor of the triode is = 50 / 1 = 50.

R

Fig. 39.7

39.13. Tetrode

It is a four-electrode valve. It contains a plate, cathode, control grid G_1 and another grid G_2

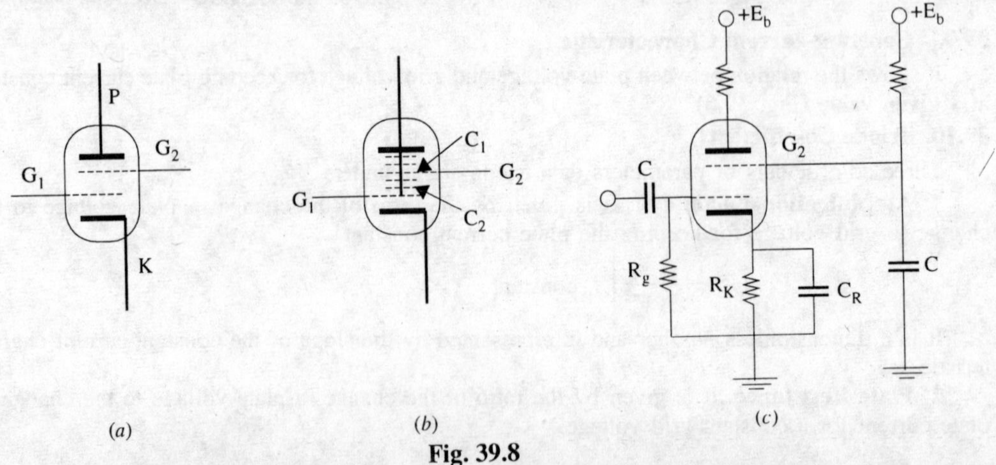

Fig. 39.8

known as screen grid as shown in Fig. 39.8 (*a*). It is called screen grid because its main purpose is to screen or shield the plate from the control grid in order to reduce grid-plate inter-electrode capacitance C_{gp}. As shown in Fig. 39.8 (*c*) screen grid is grounded so far as a.c. potentials are concerned with the help of the by-bass capacitor *C* which by offering very low reactance to high frequency currents acts as a short-circuit for them. G_2 has to be an a.c. ground if it is to act as an effective shield. The by-pass capacitor keeps G_2 and cathode at the same a.c. potential. However, G_2 is given large d.c. potential which is only slightly less than V_p.

As seen from Fig. 39.8 (*b*), screen grid breaks up the control grid-plate capacitance into two capacitances marked C_1 and C_2 which are connected in series. Hence, their combined capacitance is reduced. In fact, its value is reduced to about 0.01 pF. Another advantage of adding the screen grid is that it makes I_p almost independent of V_p thereby increasing the amplification factor to about 800 or so.

39.14. Tetrode Characteristics

The V_p/I_p characteristic of a triode has been shown in Fig. 39.9. For plotting this curve, screen grid G_2 is given a large but fixed d.c. potential from *B*-battery [Fig. 39.8 (*c*)] and control grid G_1 has a fixed bias. Then, V_p is gradually increased in steps and corresponding values of I_p noted.

(*i*) It is seen from Fig. 39.9 that starting from zero, as V_p is increased, I_p keeps on increasing up to point *A*. It is so because screen grid voltage helps this small V_p to attract more electrons most of which pass straight through the comparatively large spaces between the screen grid wires. Very few of them are retained by G_2 itself.

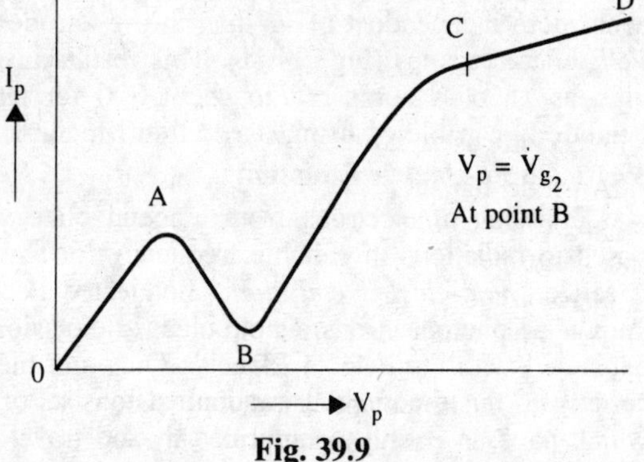

Fig. 39.9

(*ii*) As V_p is further increased between points *A* and *B*, I_p starts *decreasing.* Over this region, tetrode is said to possess *negative* resistance because, contrary to Ohm's law, current decreases with increase in applied voltage. This has been explained in terms of *secondary emission* from the plate. In the region *AB*, electrons starting from cathode, gather up large speed and hence energy as they approach the plate. As they strike the plate, they dislodge electrons from it. These 'splashed' secondary electrons are attracted towards the screen grid whose fixed d.c. potential V_{g_2} is *as yet* higher than the plate potential. Consequently, I_p is decreased upto point *B* where V_p becomes equal to V_{g_2}.

(*iii*) As V_p is further increased (portion *BC*) and begins to exceed the value of screen-grid voltage, the force of attraction exerted by the plate on the secondary electrons becomes

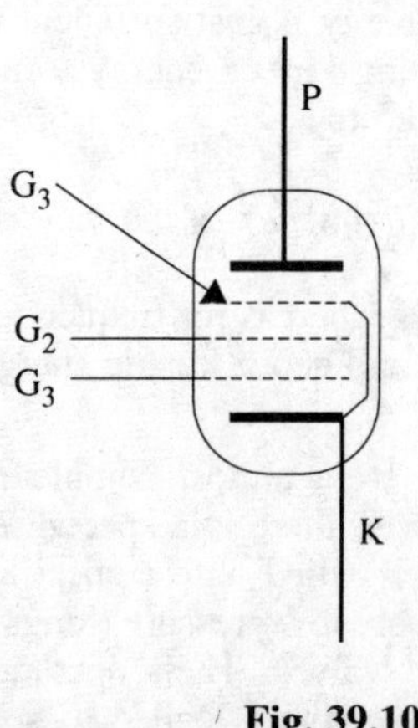

Fig. 39.10

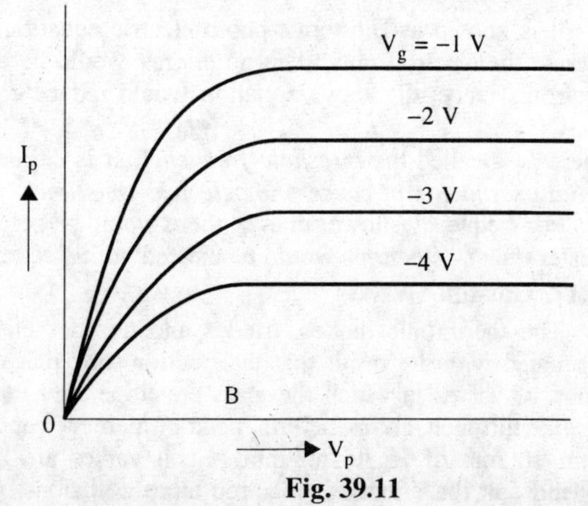

Fig. 39.11

greater than that exerted by screen-grid with the result that the secondary electrons return to it. Consequently, I_P rise sharply. Since now V_p exceeds the value of screen-grid voltage, plate collects all the electrons coming from the cathode.

(*iv*) From point *C* onwards, I_p remains practically constant. Over the region *CD*, V_p hardly affects I_p due to the screening effect of the screen grid G_2.

39.15. Pentode

When one more grid is added to a tetrode, it becomes pentode. This additional grid is known as *suppressor grid* . Its main job is to suppress the bad effects of secondary emission which takes place in a tetrode. For this purpose, suppressor grid G_3 is joined (internally) to the cathode so that it is at negative potential with respect to the plate. Hence, it repels the secondary electrons emitted by the plate and drives them back to the plate itself. Thus, while not eliminating secondary emission from the plate, the suppressor grid does eliminate the bad effects of secondary emission. Consequently, I_p increases smoothly with increasing V_p from zero upto maximum value for each control grid voltage V_{g_1} (Fig. 39.11).

Another point worth noting is that suppressor grid provides further shielding between the plate and control grid and thus further reduces the grid-to-plate capacitance C_{gp}. This makes I_p even more independent of V_p than in the tetrode as demonstrated by the further flattening of the V_p/I_p characteristics (Fig. 39.11). This results in making the amplification factor of a pentode as high as 1500 as compared to about 800 for tetrodes and about 50 for triodes. Consequently, pentode is capable of being used at still higher frequencies.

39.16. Photo-electric Emission

Emission of electrons from a metal plate when exposed to radiations of suitable frequency (or wavelength) is called *photo-electric emission.* The ejected electrons are known as photoelectrons. Photo-electric emission can be explained with the help of Planck's Quantum theory. According to this theory, every radiation consists of *photons* which have an energy content of '*hf*' and travel with the velocity of light. Einstein proposed in 1905 that when a photon falls on a metal surface (Fig. 39.12). It is completely absorbed and in the process, gives its energy to a single electron of the metal. This energy is utilized for two purposes:

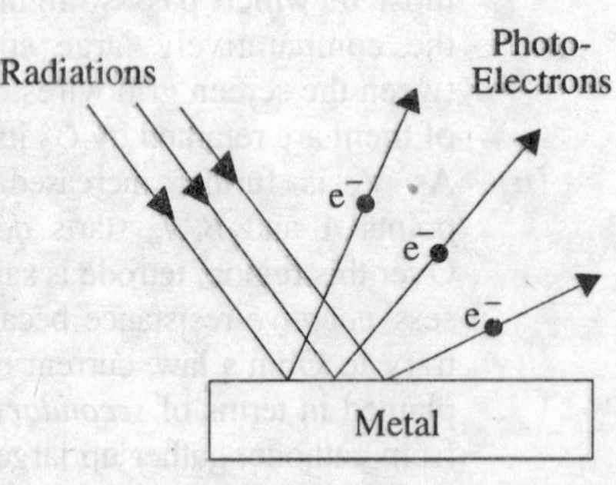

Fig. 39.12

(*i*) partly for dislodging the electron from its orbit in the atom. This energy is known as photo-electric *work function* and is represented by W_0; and

(*ii*) partly for giving kinetic energy of $1/2\ mv^2$ to the electron

$$\therefore \quad hf = W_0 + \frac{1}{2}mv^2$$

It is known as Einstein's photo-electric equation. In case, photon energy is just sufficient to liberate the electron only, then no energy would be available for imparting kinetic energy to the electron. Hence, the above equation would reduce equation would reduce to

$$hf_0 = W_0$$

where f_0 is called the *threshold frequency.* It is defined as the *minimum frequency of the incident radiation which can cause photoelectric emission.*

For frequencies lower than f_0, there would be no emission of electrons whereas for frequencies greater than f_0, electrons would be ejected with a certain definite velocity and hence kinetic energy.

39.17. Gas-filled Valves

The thermionic diodes, triodes and tetrodes etc., discussed so far were almost completely evacuated with the result that their action took place in vacuum. However, there is a specialised group of valves in which the glass envelope after evacuation is purposely filled with a specified gas like nitrogen, argon, helium, neon or mercury vapours at a comparatively low pressure (ranging from 10 mm of Hg to 50 mm). Such valves are known as gas-filled valves. Their operation depends on the effects of electron-atom collisions. Usually, a gas-filled valve is called a 'soft' valve whereas an evacuated one is known as 'hard' valve.

The construction of gas-filled valves is almost similar to that of high-vacuum valves except that their cathodes, grids and anodes are usually larger in order to carry heavier currents.

Gas-filled valves are classified into two groups depending on the type of electron emission employed:

(*a*) *Cold cathode type* – in which electorn emission is obtained from a cold-cathode.

(*i*) a two-element cold-cathode vlave is called a *glow tube* because its action depends on the glow discharge through the gas.

(*ii*) a three-element cold-cathode valve is known as *grid-glow tube*.

(*b*) *Hot cathode (or thermionic) type*–in which electron emission is obtained from oxide-coated heated cathode just as in the conventional thermionic vacuum valves.

(*i*) a two-element hot-cathode valve is known as *phanotron* and its action depends on the arc discharge through the gas.

(*ii*) a three-element hot-cathode valve is known *as thyratron* and its operation also depends on the arc discharge.

The schematic symbols for the different types of gas-filled valves are shown in Fig. 39.13. A small dot put anywhere in the envelope indicates that the valve is gas-filled.

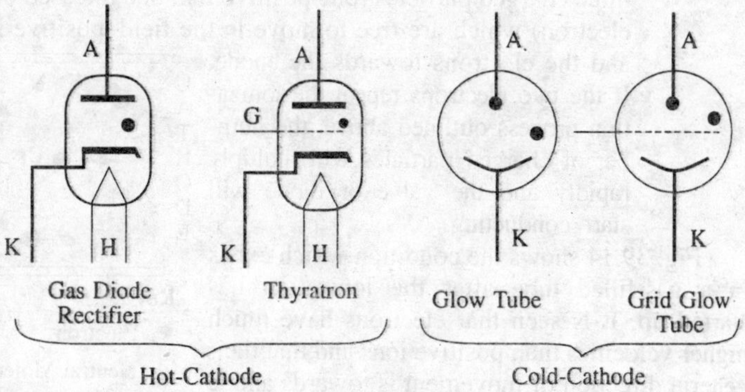

Fig. 39.13

Cold-cathode diodes (also called glow tubes) are used as

(*i*) indicating and testing lamps; and

(*ii*) as voltage regulators (for which role they are called VR tubes).

These tubes are available in various sizes ranging from 1/25 to 3 W capacity.

Low-pressure hot-cathode diodes are usually designed for use as rectifiers in preference to high vacuum diodes because they have low tube drop and tube loss.

High-pressure hot-cathode diodes (with gas pressure of 50 mm of Hg) are known as Tunger or rectigon tubes and are extensively used in chargers for storage batteries.

Electrically speaking, a hot cathode triode or thyratron acts as a switch because of its trigger characteristics. Hence, it is put to diverse uses such as:

(*i*) for arcless switching and relay applications as in the control of d.c. motors and d.c. power regulating systems.

(*ii*) in relaxation type sweep-circuit generators or saw-tooth oscillators for TV and radar applications.

(*iii*) in controlled rectifiers used for electric welding and lighting control installations in theatres etc.

39.18. Mechanism of Gaseous Conduction

Consider the case of a gas-filled diode either of the cold-cathode or the hot-cathode type. In the former type of diode, there are no emitted electrons present in the gas although some electrons are usually produced by cosmic rays or by radiations from a radioactive material lying in the neighbourhood. In the hot-cathode type of diode, plenty of electrons are emitted by the cathode.

When a potential difference is applied between the anode and cathode of such a diode, an electric field is set up between the two. Under the influence of this field, the free electrons start moving towards the anode with increasing speed and hence increasing kinetic energy. During their flight, the electrons collide with the gas atoms (or molecules). Now, when an electron bumps into a gas atom, three things can happen depending on the energy of the electron at the time of its impact:

(*i*) if the electron is moving *slowly,* it will produce an elastic collision with conservation of total momentum. Since the electron is much lighter than the gas atom, it simply bounces off without producing any change in the gas atom except a slight change in its motion.

(*ii*) if the electron is moving *rapidly*, it may use some or all of its kinetic energy in raising the outermost orbital electron of the gas atom from its normal orbit to a higher-energy orbit. This action is called *excitation* and the atom is said to be in an excited state. The atom remains in excited state for about 10^{-8} second after which it returns to its normal or unexcited state. In so doing, it gives out the energy previously taken from the colliding electron in the form of radiations (which may be in the visible or invisible region of the spectrum).

(*iii*) if the electron is moving *very rapidly* (and hence more energetically), it may knock off the orbital electron completely out of the atom instead of merely raising it to a higher energy orbit. In that case, the gas atom becomes a positive ion (because of having lost one electron) and the process is known as *ionisation* by *collision*. As seen, ionisation of a gas atom produces one positive ion and one free electron. Altogether, now there are three charged particles (one positive ion, one released electron and the original colliding electron) which are free to move in the field–positive ion moving towards the cathode and the electrons towards the anode. If the two electrons repeat the ionization process outlined above, the number of charged particles will multiply rapidly and the valve or diode will start conducting.

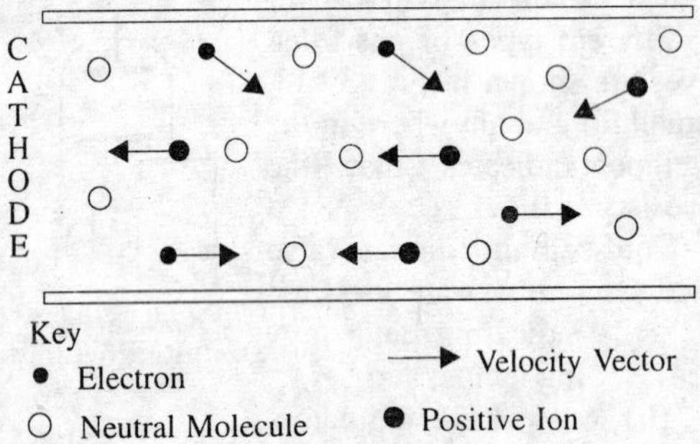

Fig. 39.14 shows the condition which exists in a gas-filled tube after the ionization has started up. It is seen that electrons have much higher velocities than positive ions and that their general direction of movement is towards anode whereas the positive ions drift towards the cathode. The neutral gas molecules have random motion.

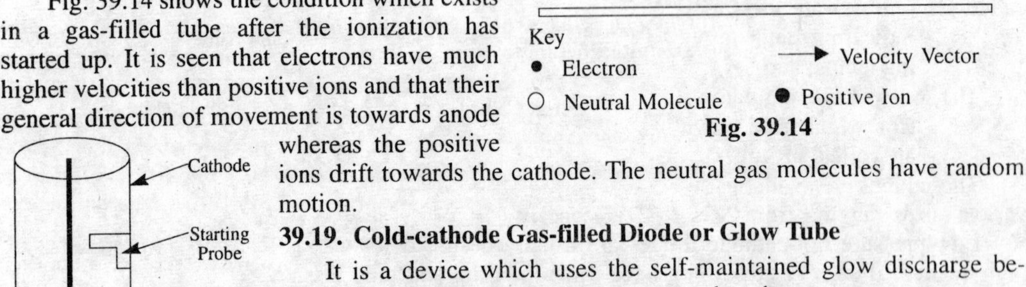

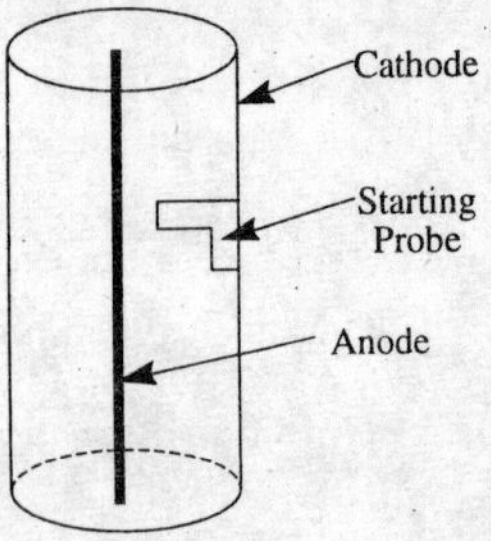

Fig. 39.14

Cathode
Starting Probe
Anode

Fig. 39.15

39.19. Cold-cathode Gas-filled Diode or Glow Tube

It is a device which uses the self-maintained glow discharge between two cold electrodes in a gas at reduced pressure.

(*a*) Construction. Fig. 39.15 shows the construction of one such simple tube used commercially. It consists of a central anode wire coaxial with a surrounding cathode. Both the anode wire and cathode are made of nickel, with the inner surface of the cathode being oxide-coated. The gases commonly used in such tubes are neon, argon and helium.

There is a 'starting probe' which consists of a metal piece attached to the cathode and directed towards and near to the anode. Its purpose is to reduce the breaking or firing potential of the tube.

(*b*) Characteristic. The volt-ampere (V_A/I_A) characteristic of such a tube may be found with the help of the circuit diagram shown in Fig. 39.16 (*a*).

As seen, the glow tube is connected to a high-voltage battery B through a current limiting resistance R and an indicating ammeter A. The voltmeter V_1 reads anode voltage V_A whereas V_2 measures the supply voltage applied across both R and the tube.

When anode voltage V_A is gradually increased from zero with the help of potentiometer slider S, no measurable current is found to flow through the tube at first because A reads zero. However, soon, a certain value of V_A is reached when there is a sudden flow of current. This value of the anode voltage is known as *breakdown* or *sparking voltage* V_S* (and is also known as striking or

*Its value depends on many factors such as gas pressure and geometrical arrangement of electrodes. The function of the starting probe is to reduce the breakdown voltage. Sparking takes place first to the starting probe and then spreads to the rest of the cathode.

ionizing potential). After the breakdown occurs, the voltage across the tube (*i.e.*, reading of V_1) falls to a lower value V_M and a glow appears in the tube. This voltage V_M is known as *maintenance* voltage or sometimes the *operating* or *burning* voltage. Like V_S, its value also depends on the type of gas used, its pressure and the nature of the electrodes. Its value varies from 60 V to several hundred volts for different diodes. The colour of the luminous glow depends on the nature of the gas, being pink for neon and blue for argon.

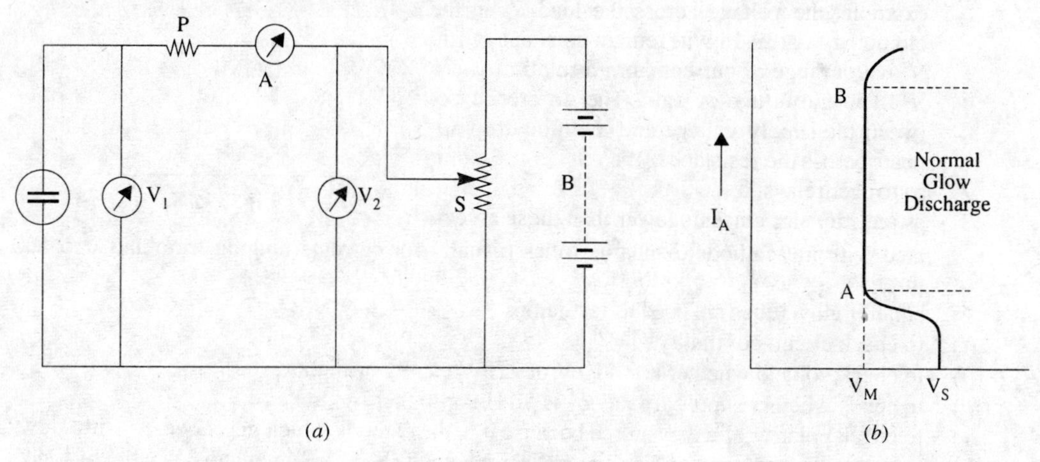

Fig. 39.16

It is seen from Fig. 39.16 (*b*) that while working over the normal glow region *i.e.*, over the portion *AB* of the V_A/I_A characteristic, the voltage drop across the tube remains practically constant and is independent of the current flowing through the tube.

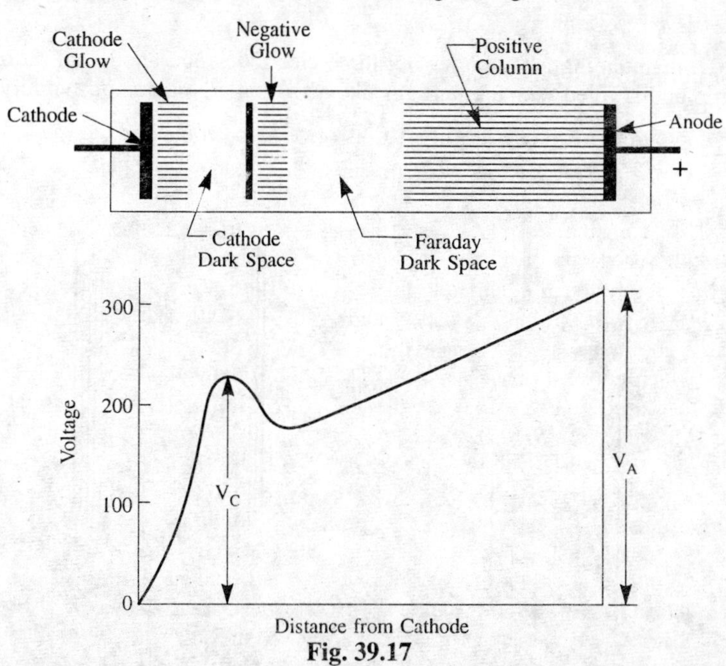

Fig. 39.17

Mechanism of operation of such a tube is exactly like that explained in Art. 39.18. The initial start of the operation depends on the presence of a free electron within the tube, such an electron may be released by ionization due to collision between a gas molecule and cosmic rays or by radiations from a neighbouring radioactive material or photo-electric emission. With the application of a potential difference between the anode and cathode of the tube, the electron drifts towards the anode and eventually ionizes the gas molecules by collision. As a consequence, the number of charged particles increases to three. If the field is strong enough, the resulting cumulative ionization may continue till breakdown occurs. Once breakdown occurs, the potential distribution within the tube is completely modified and most of the region of the discharge (extending from negative glow to positive column) becomes virtually equipotential or force-free because it contains as much positive charge as the negative one. As seen from Fig. 39.20, most of the potential drop occurs in the very narrow region near the cathode. Normal values of cathode-fall potential V_C range from 60 V (a potassium surface and helium gas) to about 350 V.

(c) Uses

1. The principal use of glow tubes is as voltage regulators (or stabilisers *i.e.*, as VR tubes. For example, the voltage across the load R_L in the circuit of Fig. 39.18 will remain constant at 150 V over a range of current from 5 to 50 mA, if a VR-150 glow tube is used. The difference between the supply voltage and the tube drop appears across the resistance R*;

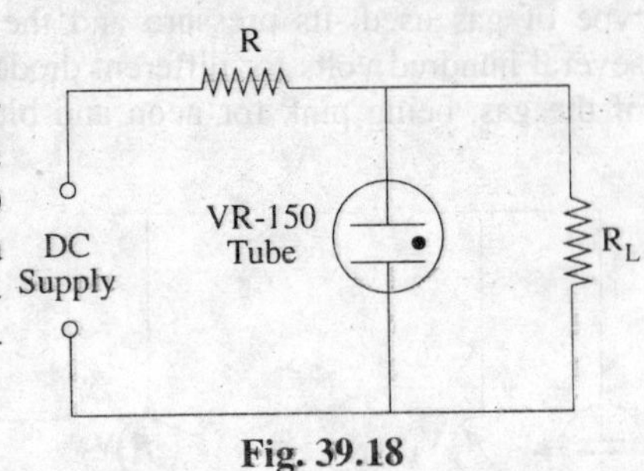

Fig. 39.18

2. as trigger relays;
3. as rectifiers for currents lower than those associated with hot-cathode discharge tubes (usually for currents ranging from 1 μA to 100 mA);
4. smaller glow tubes are used as test lamps
   (*i*) to check circuit continuity,
   (*ii*) to check voltage whether it is 110 V or 220 V etc. by brilliancy,
   (*iii*) to check whether supply frequency is 50 Hz or higher by flicker,
   (*iv*) to check polarity of a d.c. source because it is the cathode which gets covered with glow,
   (*v*) to detect the presence of a strong radio-frequency (R.F.) field which is capable of ionizing the gas without direct connection to the tube,
   (*vi*) as pilot lamps to indicate that a circuit is energised,
   (*vii*) as dim-light sources for exit markers, for fire-station markers and for location markers is general.

39.20. Thyratron

It is a low-voltage non-self-maintaining arc tube having three electrodes *i.e.*, anode, an indirectly-heated cathode and a grid and is filled with an inert gas like neon and argon etc., or mercury vapours at low pressure.

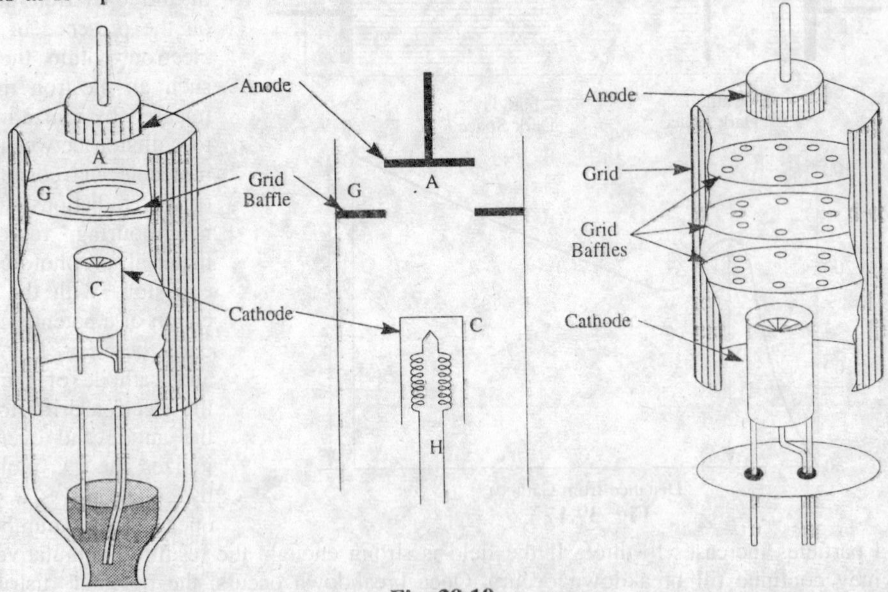

Fig. 39.19

*Of late, glow tubes have been replaced by Zener diodes.

(*a*) Construction. The electrode structure of one type of a gas tube or thyratron* is as shown in Fig. 39.19. It consists of an anode and cathode which are approximately planar. The grid is a cylinder which *surrounds* both the anode and the cathode, a baffle or a series of baffles having small holes being inserted between the anode and the cathode. Obviously, the grid acts as a complete electrostatic shield between anode and cathode except for the holes in the grid baffles. Hence, even a small negative grid potential is sufficient to neutralize the effect of a large anode potential and thereby prevent the arc from being initiated. However, once the arc has been struck between the anode and the cathode, the grid has no control over the anode current. In other words, grid is used only to *start* conduction of anode current by ionization; it cannot be employed to control anode current or to stop it.

(*b*) Working. When grid voltage is sufficiently negative, the electrons emitted from the cathode are prohibited by the grid to go to the anode even when it has large positive potential. As grid potential is made less negative, an occasional electron from the cathode passes through the hole in the grid baffle but still does not have enough energy to produce ionization. As the negative grid voltage is gradually reduced further, the electrons acquire more speed and hence energy and a potential called *critical grid voltage* is reached, when ionization occurs and the anode current changes rapidly from zero to a large value. At the same time, an arc discharge appears inside the valve and the anode-to-cathode voltage falls to about 15 to 20 V.

Once 'firing' has taken place, grid has no control over anode current (because during conduction, its negative charge is neutralized by the positive ions that collect on it). Obviously, once fired, the thyratron behaves like a closed switch. To stop anode current and to re-establish grid control, it is necessary to reduce the anode voltage below the maintenance voltage or to open the anode circuit. When discharge ceases, de-ionization is generally complete in about 0.1 to 1 millisecond.

(*c*) Grid-control Characteristic. It is the curve which shows relation between the anode voltage and the critical grid voltage at which the valve strikes. It is also known as *starting characteristic* or *breakdown characteristic* and may be found with the help of the circuit shown in Fig. 39.20.

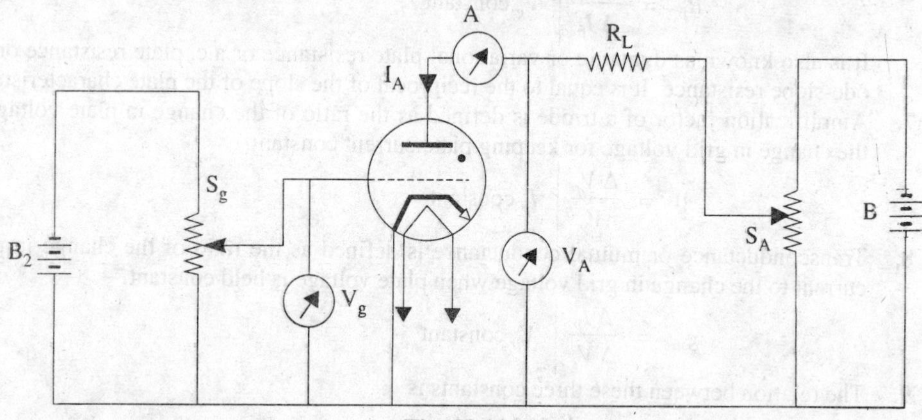

Fig. 39.20

The grid can be given any negative voltage with the help of the slider S_g and its value read on the voltmeter V_g. Similarly, anode voltage can be varied with the help of S_A and its value read on V_A. Ammeter *A* reads anode current I_A.

Suppose we adjust V_A=100 V when grid is at –10 V. The anode current is found to be zero because this grid bias does not permit the emitted electrons to go to the anode. When grid voltage is gradually reduced to about –2 V, it is found that all of a sudden I_A achieves large value due to ionization by the high-velocity electrons. The value of this critical voltage is noted.

If the test is repeated for higher anode voltages, somewhat larger critical grid voltages would

*Which in Greek means a 'door'. Originally, it was the trade name of G.E.C. (U.S.A.).

be obtained. Fig. 36.21 shows a typical grid-control characteristic of a thyratron. The slope of this characteristic gives (what is known as) control ratio which is defined as = $\Delta V_A / \Delta V_g$.

(d) Uses

(i) Thyratrons are used for arcless switching and relay applications as in the control of d.c. motors etc.

(ii) The small argon-filled thyratron is generally used in saw-tooth sweep generators for TV and radar equipment.

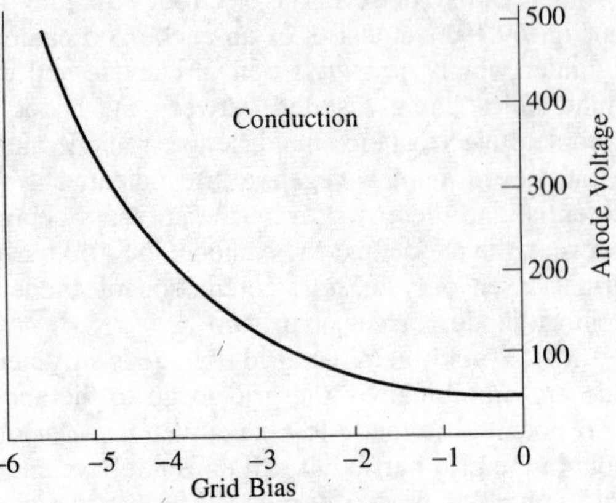

Fig. 39.21

HIGHLIGHTS

1. Thermionic emission, as given by Richardson-Dushman equation is given by

$$J_S = AT^2 e^{-b/T} \text{ A/m}^2$$

2. A diode valve consists of two electrodes–one positive called anode or plate and the other negative called cathode.
3. A diode is mostly used as a rectifier *i.e.,* for converting alternating currents into unidirectional currents.
4. The plate current for a diode is given by Child's Three-halves Law

$$I_p = KV_p^{3/2}$$

5. A triode consists of a plate (or anode), cathode and a grid. Its plate current is given by

$$I_p = A(V_p + \mu V_g)^{3/2}$$

6. The plate resistance of a triode is defined as the ratio of change in plate voltage to change in plate current when grid potential is held constant

$$\therefore \quad R_p = \left.\frac{\Delta V_p}{\Delta I_p}\right| V_g \text{ constant}$$

It is also known as dynamic or variational plate resistance or a.c. plate resistance or anode-slope resistance. It is equal to the reciprocal of the slope of the plate characteristic.

7. Amplification factor of a triode is defined as the ratio of the change in plate voltage to the change in grid voltage for keeping plate current constant.

$$\therefore \quad \mu = \left.\frac{\Delta V_p}{V_g}\right| I_p \text{ constant}$$

8. Transconductance or mutual conductance is defined as the ratio of the change in plate current to the change in grid voltage when plate voltage is held constant.

$$\therefore \quad g_m = \left.\frac{\Delta I_p}{\Delta V_g}\right| V_p \text{ constant}$$

9. The relation between these three constants is

$$\mu = R_p \times g_m$$

10. The voltage amplification of a triode is given by

$$A = \frac{-\mu}{1 + (R_p / R_L)}$$

11. The power output to the load resistance is

$$P_0 = \frac{\mu^2 E_g^2 R_L}{(R_L + R_p)^2}$$

where R_L = load resistance

R_p = plate resistance

E_g = grid input voltage

12. Einstein's photo-electric equation is

$$hf = W_0 + \frac{1}{2} mv_2$$

OBJECTIVE TESTS—39

A. Fill in the following blanks.

1. Working of vaccum tubes depends on .... emission.

2. The plate current of a vacuum diode is given by .... three-halve's power law.

3. A vacuum triode consists of an anode, cathode and a ....

4. The amplification factor of a triode is given by the product of plate resistance and ....

5. The slope of constant-current characteristic gives the .... factor of a triode.

6. Tetrode is a .... electrode valve.

7. Pentode has one anode, one cathode and .... grids.

8. Once 'fired', a thyratron behaves like a .... switch.

B. Answer True or False.

1. Vacuum tubes utilize field emission.

2. Tungsten filamentary cathodes are particularly suited for high power transmitting tubes.

3. The grid in a triode controls its plate current.

4. In a triode valve, cut-off grid potential varies with plate voltage.

5. Amplification factor of a triode is given by the slope of its transfer characteristic.

6. The a.c. equivalent circuit of a triode consists of a voltage generator and a series resistance.

7. Tetrode characteristic has a region of negative resistance due to secondary emission.

8. Working of gas valves depends on ionization by collision.

C. Multiple Choice Questions.

1. In vacuum tubes, indirectly-heated cathodes are preferred to directly-heated cathodes because they

(*a*) provide higher electron emission

(*b*) minimize the introduction of hum from a.c. heater supply

(*c*) work at higher temperatures

(*d*) are more rugged

2. A triode has a plate resistance of 6 K and a transconductance of 1 mS. Its amplification factor is

(*a*) 6000 (*b*) 60,000

(*c*) 6 (*d*) 1/6

3. The amplification factor of a pentode is much higher than that of a triode due to the presence of .... in it.

(*a*) control grid (*b*) space charge

(*c*) screen grid (*d*) suppressor grid

4. Working of a glow tube depends on

(*a*) thermionic emission

(*b*) arc discharge

(*c*) ionization by collision

(*d*) cathode glow

5. A thyratron can be changed from ON to OFF state by

(*a*) making grid more negative

(*b*) opening the anode

(*c*) reducing anode voltage below maintenance voltage

(*d*) either (*b*) or (*c*)

ANSWERS

A. 1. thermionic 2. Child's 3. grid 4. transconductance 5. amplification 6. four 7. three 8. closed

B. 1. F 2. T 3. T 4. T 5. F 6. T 7. T 8. T

C. 1. *b* 2. *c* 3. *d* 4. *c* 5. *d*

40 ELECTRON BALLISTICS

40.1. Introduction

We will discuss the motion of a charged particle (usually, an electron or an electron beam) in an electric field as well as in a magnetic field.

40.2. Uniform Electric Field: Zero Initial Velocity

Let there be two large plane parallel plates A and B situated in vaccum at a distance of d metre from each other and having a potential difference of V volt between them (Fig. 40.1). Obviously, there will be a uniform electric field of strength $E = V/d$ volt/metre between the two plates. An electron placed at plate A will be attracted towards the positively-charged plate B. If free to move, the electron will be accelerated towards plate B along X-axis as shown.

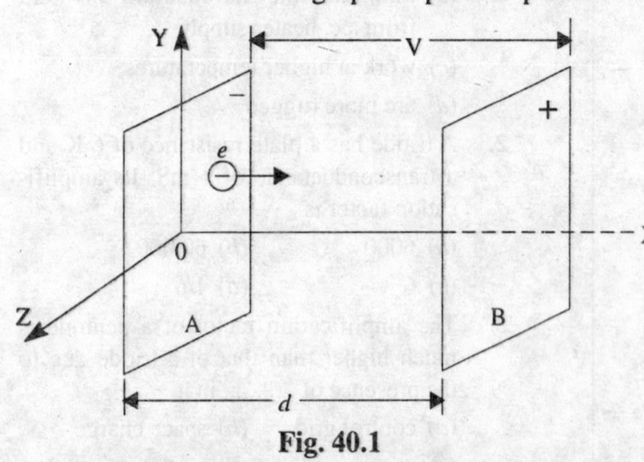

Fig. 40.1

Since the negatively-charged electron is situated in an electric field, it is acted upon by a force given by

$$F = eE$$

Also, $$F = ma$$

where a is the acceleration along X-axis, m the mass of the electron in kg and e its charge in coulomb.

$$\therefore \quad ma = eE$$

$$\therefore \quad a = \frac{eE}{m} = \frac{e}{m} \times \frac{V}{d} \text{ m/sec}^2$$

(*i*) Velocity of the electron at any time t is given by

$$v = at = \frac{e}{m}\frac{V}{d}\, t \text{ metre/second (use, } v = u + at) \qquad ...(i)$$

(*ii*) Distance travelled by the electron during that time is

$$x = \frac{1}{2}at^2 = \frac{1}{2}\frac{e}{m}\frac{V}{d}t^2 \text{ metre} \left(S = ut + \frac{1}{2}at \right) \qquad ...(ii)$$

(*iii*) Eliminating t from the above two equations, we get

$$v = \sqrt{2\frac{e}{m}\frac{V}{d} \times} \text{ metre/second} \qquad ...(iii)$$

Now, $e/m = 1.6 \times 10^{-19} / 9.1 \times 10^{-31} = 1.76 \times 10^{11}$ C/kg

$$\therefore \quad v = \sqrt{2 \times 1.76 \times 10^{11} \frac{V}{d} x}$$

$$= 5.93 \times 10^5 \sqrt{\frac{V}{d} x} \text{ m/s}$$

$$= 5.93 \times 10^5 \sqrt{E\, x} \text{ m/s}$$

(*iv*) The impact velocity v_i of the electron as it strikes plate *B* can be found by putting $x = d$ in Eq. (*iii*) above

$$v_L = \sqrt{2\frac{e}{m}\frac{V}{d}d} = \sqrt{\frac{2\,eV}{m}} \text{ metre/second}$$

$$= 5.93 \times 10^5\sqrt{V} \text{ metre/second*}$$

(*v*) The time of transit from plate *A* to plate *B* may be found by putting $x = d$ in Eq. (*ii*) above

$$t^2 = \frac{2\,mxd}{eV} = \frac{2\,md^2}{eV}$$

$$\therefore \quad t = \sqrt{\frac{2md^2}{eV}} = \sqrt{\frac{4d^2}{2\,eV/m}} \text{ second}$$

or

$$t = \frac{2d}{v_i} \text{ second}$$

Note: In the case of such uniformly-accelerated motion, transit time may also be found by dividing the distance with the average speed.

$$\text{Average speed} = \frac{0 + v_i}{2} = \frac{v_i}{2}$$

$$\therefore \quad t = \frac{d}{v_i/2} = \frac{2d}{v_i} \text{ second.}$$

Example 40.1. *Calculate the time taken by an electron which has been accelerated through a potential difference of 1000 V to traverse a distance of 2 cm.*

Given : $e = 1.6 \times 10^{-19}$ C and $m = 9.1 \times 10^{-31}$ kg.

Solution. Impact velocity acquired by the electron

$$v_i = \sqrt{\frac{2\,eV}{m}} = 5.93 \times 10^5\sqrt{V}$$

$$= 5.93 \times 10^5\sqrt{10^3} = 1.875 \times 10^7 \text{ m/s}$$

$$\text{Time,} \quad t = \frac{2d}{v_i} = \frac{2 \times 0.02}{1.875 \times 10^7} = 2.13 \times 10^{-9} \text{ s}$$

Example 40.2. *A potential gradient of 3×10^6 V/m is maintained between two horizontal parallel plates 1 cm apart. An electron starts from rest at the negative plate and travels under the influence of the potential gradient to the positive plate. Given that mass of electrons is 9.1×10^{-31} kg and the charge 1.6×10^{-19} coulomb, calculate (i) the force acting on the electron (ii) the ratio of electric force to gravitational force (iii) the acceleration (iv) the time taken to reach the positive plate.* **(Electrical Science, AMIE Winter, 1993)**

Solution. We are given that $E = 3 \times 10^6$ V/m, $d = 1$ cm

(*i*) $F = eE = 1.6 \times 10^{-19} \times 3 \times 10^6 = 4.8 \times 10^{-13}$ N

(*ii*) Gravitational force acting on the electron

$= mg = 9.1 \times 10^{-31} \times 9.81 = 8.93 \times 10^{-30}$ N

Ratio of the electric force to the gravitational force

$= 4.8 \times 10^{-13} / 8.93 \times 10^{-30} = 5.37 \times 10^{16}$

(*iii*) Acceleration $= \dfrac{eE}{m} = \dfrac{1.6 \times 10^{-19} \times 3 \times 10^6}{9.1 \times 10^{-31}} = 5.27 \times 10^{17}$ m/s^2

(*iv*) As given in Art. 40.2, the impact velocity of the electron is given by $v_i = 5.93 \times 10^5\sqrt{V}$.

Now, $E = V/d$ or $V = Ed = 3 \times 10^6 \times 1 \times 10^{-2} = 3 \times 10^4$ V/m

$\therefore \quad v_i = 5.93 \times 10^5 \times \sqrt{3 \times 10^4} = 10.27 \times 10^7$ m/s

*It is seen that if an electron is accelerated through 1 V, its impact velocity is 5.93×10^5 m/second or 593 km/s! Despite this tremendous velocity, it possesses very little kinetic energy because if its extremely small mass.

Hence, time taken by the electron to reach the positive plate is

$$t = \frac{2d}{v_i} = \frac{2 \times 10^{-2}}{10.27 \times 10^7} = 0.973 \times 10^{-10} \text{ second}$$

40.3. Uniform Electric Field: Initial Velocity in the Direction of Field

In this case, the acceleration of the electron is the same as in Art. 40.2 and is, as before, given by

$$a = \frac{e}{m}\frac{V}{d} = 1.76 \times 10^{11} \frac{V}{d} \text{ m/s}^2$$

The other quantities are as follows:

(*i*) the velocity at any time is given by

$$v = u + at = u + \frac{e}{m}\frac{V}{d}t \quad \text{...(i)}$$

(*ii*) the distance travelled by the electron in time t is

$$x = ut + \frac{1}{2}at^2$$

$$= ut + \frac{1}{2}\frac{e}{m}\frac{V}{d}t^2 \quad \text{...(ii)}$$

(*iii*) the electron velocity at any distance x from the negative plate can be found by eliminating t from Eq. (*i*) and (*ii*). The result is

$$v = \sqrt{u^2 + 2\frac{e}{m}\frac{V}{d}x} \quad \text{...(iii)}$$

(*iv*) the impact velocity or velocity of arrival at the positive plate B (Fig. 40.2) can be found by putting $x = d$ in Eq. (*iii*) above. The result so obtained is

$$v_i = \sqrt{u^2 + 2\frac{e}{m} \cdot V} = \sqrt{u^2 + v_1'^2}$$

where $v_i'^2$ is the impact velocity when the electron has no initial velocity (Art. 40.2).

(*v*) the transit time can be found by dividing the distance d by the average velocity.

$$\therefore \quad t = \frac{d}{(v_i + v_i')/2} = \frac{2d}{(v_i + v_i')}$$

Example 40.3. *In a parallel-plate electronic diode, the anode is at 250 V with respect to the cathode which is 4 mm away from it. An electron is emitted from cathode with an initial velocity towards cathode of 2×10^6 m/s. Calculate:*

(i) the arrival velocity of the electron at anode

(ii) time of transit

(iii) the velocity and the travel time when the electron is half way through

(iv) the velocity and distance travelled by the electron after 0.5×10^{-9} s.

Solution. (*i*) $v_i = \sqrt{u^2 + 2\frac{e}{m}\text{V}} = \sqrt{u^2 + 2 \times 1.76 \times 10^{11}\ \text{V}}$

$= \sqrt{(2 \times 10^6)^2 + 2 \times 1.76 \times 10^{11} \times 250}$

$= \mathbf{9.59 \times 10^6}$ **m/s**

(*ii*) Average velocity $= \dfrac{2 \times 10^6 + 9.59 \times 10^6}{2} = \dfrac{11.59 \times 10^6}{2}$ m/s

$\therefore$ Transit time $= \dfrac{2 \times 4 \times 10^{-3}}{11.59 \times 10^6} = \mathbf{6.91 \times 10^{-10}}$ **s**

(*iii*) $v = \sqrt{u^2 + 2\frac{e}{m}\frac{V}{d}x}$

At mid-point, voltage $V = 250/2 = 125$ V and $x = 0.02$ m

$\therefore \quad v = \sqrt{(2 \times 10^6)^2 + 2 \times 1.76 \times 10^{11} \times 125/2} = 4.9 \times 10^6$ m/s

Average velocity of electron upto the mid-point is

$$= (2 \times 10^6 + 4.9 \times 10^6)/ 2 = 6.9 \times 10^6 / 2 = 3.45 \times 10^6 \text{ m/s}$$

$\therefore$ time of travel upto mid-point is

$$t = 0.02 / 3.45 \times 10^6 = \mathbf{5.9 \times 10^{-9}\ s}$$

(*iv*) $\quad v = u + \dfrac{e}{m}\dfrac{V}{d}t$

Here, $\quad t = 0.5 \times 10^{-9}$ s. Hence,

$$v = 2 \times 10^6 + 1.76 \times 10^{11} \times 250 \times 0.5 \times 10^{-9} / 4 \times 10^{-3} = \mathbf{7.5 \times 10^6\ m/s}$$

$$x = ut + \frac{1}{2}\frac{e}{m}\frac{V}{d}t^2$$

$$= 2 \times 10^6 \times 0.5 \times 10^{-9} + \frac{1}{2} \times 1.76 \times 10^{11} \times 250 \times (0.5 \times 10^{-9})^2 / 4 \times 10^{-3}$$

$$= 2.375 \times 10^{-3} \text{ m} = \mathbf{2.375\ mm}$$

40.4. Uniform Electric Field: Initial Velocity Perpendicular to the Field

Let an electron having an initial velocity of u along X-axis enter at point 0 the space between two plane parallel plates A and B where an electric field E exists along the Y-axis as shown in Fig. 40.2. While moving between the two plates, the electron experiences a vertical acceleration along Y-axis but none along X-axis. It is worth emphasizing that since there is no force along X-axis, the electron velocity remains constant along this direction. The axial distance travelled by the electron during time t is

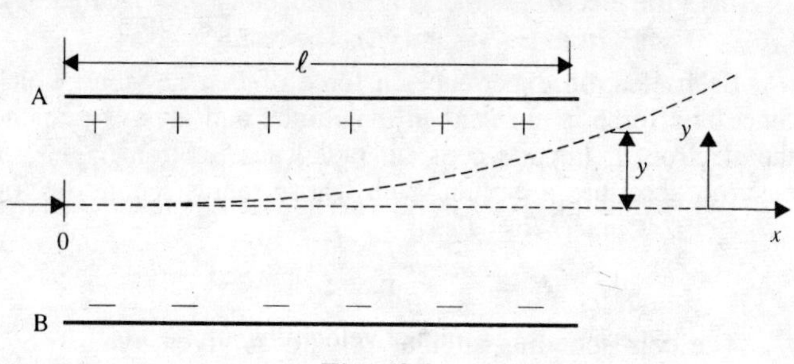

Fig. 40.2

$$x = ut \qquad \text{....(}i\text{)}$$

There is no initial electron velocity along Y-axis but as the electron moves between the plates, its velocity along Y-axis keeps on increasing.

$$a_y = \frac{F}{m} = \frac{eE}{m} = \frac{e}{m}\frac{V}{d}$$

The velocity and displacement along Y-axis after time t are given by

$$v_y = 0 + a_y t = a_y t \qquad (v = u + at)$$

$$y = \tfrac{1}{2} a_y t^2 \qquad \text{....(}ii\text{)} \qquad (s = ut + \tfrac{1}{2} at^2)$$

Substituting value of t from Eq. (*i*) in Eq. (*ii*), we get

$$y = \frac{1}{2} a_y \left(\frac{x}{u}\right)^2 = \left(\frac{1}{2}\frac{a_y}{u^2}\right) x^2$$

It shows that the electron moves along a parabolic path in the region between the two plates.

40.5. Force on an Electron Moving in a Magnetic Field

Suppose an electron moves a distance of l metres in a transverse magnetic field of flux density B with a velocity of v. Motion of an electron constitutes a negative current flow. The direction of the conventional current is opposite to the motion of the electron. Now, an electron charge of e coulomb travels a distance of l metres in l/v second. Hence, current which is the rate of flow of charge, is given by

$$I = \frac{e}{l/v} = \frac{ev}{l} \text{ ampere}$$

Force experienced by a current-carrying conductor when placed in a magnetic field is

$$F = BIl$$

$$F = B \cdot \frac{ev}{l} \cdot l = Bev \text{ newton}$$

Direction of this force may be found by applying Fleming's left-hand rule.*

40.6. Deflection of a Moving Electron in a Transverse Magnetic Field

In Fig. 40.3 (*a*) is shown an electron gun which injects a beam of electrons with an initial velocity of *u* at right angles to a magnetic field which is very extensive or very strong. The field is perpendicular to the plane of the paper and is directed downwards (as shown by crosses).

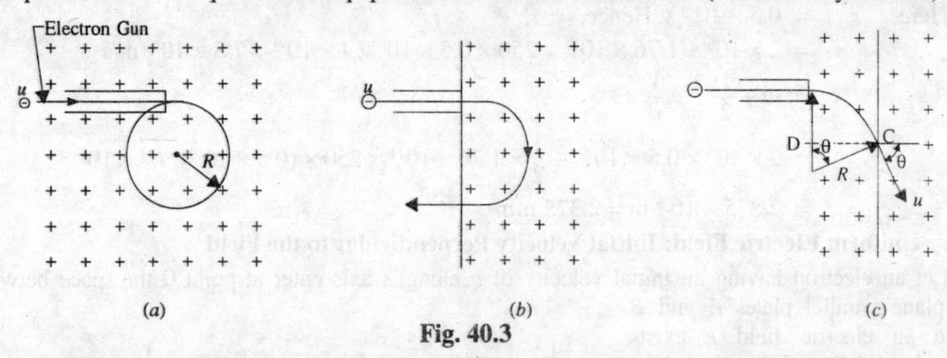

Fig. 40.3

Each electron experiences a force of *Beu* newtons which is at right angles both to *u* and *B*. Since this force is constant in magnitude and always perpendicular to the direction of motion of the electron, it imparts constant radial acceleration ($= u^2 / R$) to the electron. Consequently, the electron describes a circular path whose radius *R* may be found from the equation

$$m.u^2 / R = Beu$$

$$\therefore \quad R = \frac{m}{e} \frac{u}{B} \text{ metre} \quad \text{...}(i)$$

The corresponding angular velocity is given by

$$\omega = \frac{u}{R} = \frac{eB}{R} \text{ rad/s}$$

The time for one complete revolution *i.e.*, the time-period is given by

$$T = \frac{2\pi}{\omega} = \frac{2\pi}{B} \cdot \frac{m}{e}$$

Now, for an electron, $e/m = 1.76 \times 10^{11}$ C / kg, hence

$$T = \frac{2\pi}{1.76 \times 10^{11} \times B} = \frac{3.57 \times 10^{-11}}{B} \text{ s}$$

If the initial electron velocity *u* is obtained by an accelerating anode potential *V*, then

$$u = \sqrt{2\ eV/m}$$

Substituting this value in Eq. (*i*), we get

$$R = \frac{1}{B} \cdot \sqrt{\frac{2m}{e} \cdot V} = \frac{3.37 \times 10^{-6} \sqrt{V}}{B} \text{ m}$$

Note: (*i*) There is no change in the kinetic energy of the electron when it executes a circular path because its velocity remains constant *i.e.*, equal to the initial velocity.

(*ii*) It is seen from Eq. (*i*) above that *R* is directly proportional to the speed of the electron but the time-period *T* and angular velocity are independent of speed. It simply means that a faster-moving electron executes larger circles in the same time in which a slower-moving electron executes its smaller circles. This important fact forms the basis of operation of many devices such as a cyclotron and magnetic focussing apparatus.

* When applying this rule, the middle finger must be pointed in a direction opposite to the direction of motion of the electron *i.e.*, in the direction of the conventional current (and not electronic current).

(*iii*) If there is a charged particle (other than the electron) of charge Q, then

$$R = \frac{m}{Q} \cdot \frac{u}{B} \text{ metre}$$

(*iv*) If the magnetic field is not sufficiently extensive or strong, the electron executes a curved path as shown in Fig. 40.3 (*b*). Fig. 40.3 (*c*) shows the case when an electron passes through a magnetic field which is confined over a small region. The electron emerges out of the field at an angle θ to its original direction given by

$$\sin\theta = \frac{l}{R} = \frac{Bl}{u}\frac{e}{m} \qquad \text{(from right-angled } ACD\text{)}$$

Obviously, the exit velocity of the electron is the same as its initial velocity u.

Example 40.4. *An electron is accelerated by a p.d. of 12 kV. It then enters a uniform magnetic field of 10^{-3} T applied perpendicular to its path. Find the radius of the path of the electron after entering the magnetic field. (Assume: mass of the electron, $m = 9 \times 10^{-31}$ kg and charge of the electron, $e = 1.6 \times 10^{-19}$ C).* **(Electrical Science, AMIE Summer, 1993)**

Solution. $R = \frac{1}{B}\sqrt{\frac{2\,mV}{e}}$

Substituting the given values, we get

$$R = \frac{1}{10.3}\sqrt{\frac{2 \times 9 \times 10^{-31} \times 12 \times 10^{3}}{1.6 \times 10^{-19}}} = 0.369 \text{ m} = 36.9 \text{ cm.}$$

OBJECTIVE TESTS—40

A. Fill in the blanks with appropriate word(s) or numerical value(s).

1. The acceleration of an electron moving in a uniform electric field is directly proportional to its accelerating voltage and ....
2. The impact velocity of an electron accelerated by a uniform electric field is proportional to............... root of the accelerating voltage.
3. The impact velocity of an electron accelerated through a potential difference of 1 volt is .... km / s.
4. When an electron travels through a transverse uniform electric field it moves along a .... path.
5. When an electron is injected in a strong magnetic field with an initial velocity perpendicular to the field, it describes a .... path.
6. The radius of the circular path describe by an electron in a transverse magnetic field is inversely proportional to the............ density of the the magnetic field.
7. There is .... change in the kinetic energy of an electron describing a circular path in a magnetic field.
8. The time-period of a circularly revolving electron in a transverse magnetic field is independent of its initial ....

B. Answer True or False.

1. The acceleration of an electron in a uniform electric field depends directly on its mass.
2. The charge of an electron determines its acceleration in a uniform electric field.
3. The value of the *e/m* ratio of an electron is 1.76×10^{10} C / kg.
4. The impact velocity of an electron accelerated through a potential difference of 100 volt is nearly 6000 km/s.
5. When an electron moving with a certain velocity enters a transverse uniform electric field it describes a circular path.
6. The direction of a force acting on an electron moving in a uniform magnetic field can be found with the help of Fleming's rule.
7. An electron entering a transverse uniform magnetic field describes a circular path.
8. The time-period of an electron revolving circularly in a uniform transverse magnetic field is solely determined by its initial speed.

C. Multiple Choice Items.

1. The acceleration of an electron in a uniform electric field depends
   (*a*) directly on its charge
   (*b*) indirectly on its mass
   (*c*) directly on field strength

(*d*) all of the above

2. An electron is accelerated between two conducting plates *A* and *B* having a potential difference of 100 V. If initial velocity of the electron is zero, it will strike plate *B* with a velocity of nearly .... km / s.

(*a*) 100 (*b*) 1000
(*c*) 3000 (*d*) 6000

3. The time taken by an electron which has been accelerated through a potential difference of 10,000 V to travel a distance of 296.5 m is .... second.

(*a*) 10^{-4} (*b*) 10^{-5}
(*c*) 10^{-6} (*d*) 10^{-8}

4. An electron with a certain initial velocity enters a transverse uniform field. The path followed by the electron would be

(*a*) parabolic (*b*) straight line
(*c*) circular (*d*) elliptical

5. The force experienced by an electron in a transverse magnetic field depends on

(*a*) flux density
(*b*) velocity
(*c*) electronic charge
(*d*) all of the above

6. An electron with a certain initial velocity moves in a transverse magnetic field. The radius of the circular path described by it

(*a*) depends directly on flux density
(*b*) depends inversely on its initial velocity
(*c*) is independent of its charge
(*d*) depends on electron accelerating voltage

7. The time-period of an electron rotating circularly in a transverse magnetic field is independent of

(*a*) its speed
(*b*) flux density
(*c*) electronic charge
(*d*) electronic mass

8. The velocity *v* of a free electron, whose mass is *m* and possesses charge *e* in an electric field produced due to potential difference V volts between two metal plates is given by m / s

(*a*) $v = \sqrt{(2\ Ve/m)}$ m/s
(*b*) $v = 2\ Ve / m$ m/s
(*c*) $v = Ve / m$ m/s
(*d*) $v = (Ve / m)^2$ m/s

(Elect. Engg. A.M.Ae.S.I. Dec., 1994)

ANSWERS

A. Fill in the blanks
1. charge 2. square 3. 593 4. parabolic 5. circular 6. flux 7. no 8. speed

B. True or False
1. F 2. T 3. F 4. T 5. F 6. T 7. T 8. F

C. Multiple Choice
1. *d* 2. *d* 3. *b* 4. *a* 5. *d* 6. *d* 7. *a* 8. *a*

41 ILLUMINATION

41.1. Production of Light

The different methods of producing light by electricity may, in a broad sense, be divided into three groups:

1. By temperature incandescence. In this method, an electric current is passed through a filament of thin wire placed in vacuum or an inert gas. The current generates enough heat to raise the temperature of the filament to luminosity.
   Incandescent tungsten filament lamps are examples of this type and since their output depends on the temperature of their filaments, they are known as temperature radiators.
2. By establishing an arc between two carbon electrodes. The source of light, in their case, is the incandescent electrode.
3. Discharge Lamps—in these lamps, gas or vapour is made luminous by an electric discharge through them. The colour and intensity of ligh *i.e.*, candle-power emitted depends on the nature of the gas or vapour only. It should be particularly noted that these discharge lamps are luminiscent-light lamps and do not depend on temperature for higher efficiencies. In this respect, they differ radically from incandescent lamps whose efficiency is dependent on temperature. Mercury-vapour lamp, sodium-vapour lamp, neon gas lamp and fluorescent lamps are examples of light sources based on discharge through gases and vapours.

Discharge lamps can be further subdivided into two types:

(*a*) those in which the colour of the light is the same as produced by the discharge through the gas or vapour.
Examples are : neon gas lamps, mercury vapour lamps and sodium vapour lamps.

(*b*) those which use the phenomenon of flourescence. In their case, the discharge through the vapour produces invisible ultra-violet rays (especially of λ = 2537 Å) which cause flourescence in certain internally-coated materials known as **phosphors**. These phosphors absorb these incident ultra-violet rays and then re-radiate them at longer wavelengths of the visible spectrum. The colour of the flourescent light depends on the type of phosphor used.

The wavelength of red light is $\lambda = 0.78\ \mu m$ (or 780 nm or 7800×10^{-10} m) and that of violet light is $0.39\ \mu m$ (or 390 nm or 3900×10^{-10} m).

41.2. Definitions

1. **Steradian (sr.)** It is the unit of solid angle which is defined as the angle subtended by a certain surface at a given point. As shown in Fig. 41.1, if A is the area of the spherical surface, then solid angle

$$\omega = \frac{A}{r^2} \text{ steradian}$$

If $A = 1\ m^2$, $r = 1$ m, then $\omega = 1$ steradian (sr).

Hence, steradian may be defined as the angle subtended at the centre of a sphere of radius 1 m by a part of its surface having an area of 1 m^2. Obviously, solid angle subtended at the centre by whole of the spherical surface of a radius r is $= 4\pi r^2 / r^2 = 4\pi r$ sr.

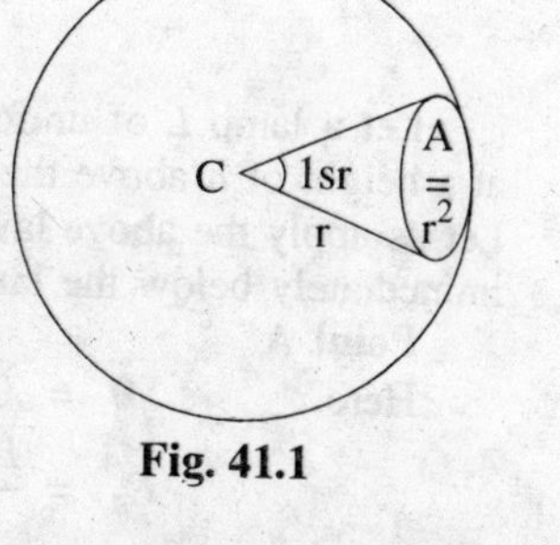

Fig. 41.1

2. **Candela (cd)**. It is the unit of luminous intensity of a *source*.

A light source has a luminous intensity of one candela if it emits 1 lumen (lm) per steradian. As shown in Fig. 41.2, the source has a luminous intensity of one candela. Since it emits a flux of one lumen in a solid angle of one steradian, total luminous flux emitted by it alround is

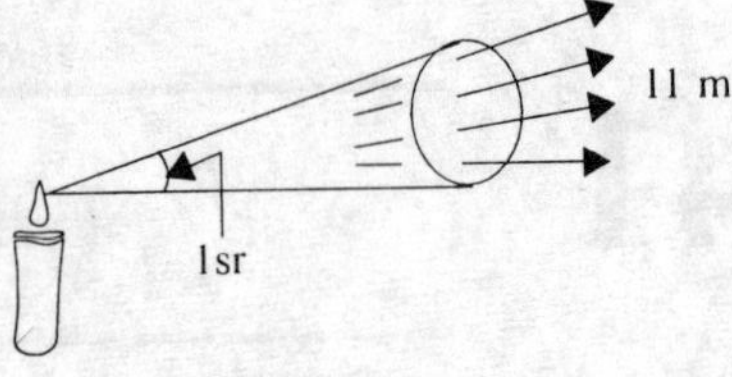

Fig. 41.2

$$= 4\pi \times 1 = 4\pi \text{ lumen}$$

It is so because solid angle subtended by the surrounding space at a point is 4π steradian.

3. **Luminous Flux (F or Φ)**. It is the total light energy radiated out per second from a luminous source.

Its unit is *lumen* (lm). Since, it is a rate of flow of light energy, it is a sort of *power* unit.

One lumen is equal to the flux given out in a solid angle of one steradian by a source of one candela.

4. **Luminous Intensity (I)**. It is defined as the luminous flux radiated out per unit solid angle by a light source.

If $d\Phi$ is the luminous flux radiated out by a source within a solid angle of $d\omega$ in any particular direction, then

$$I = \frac{d\Phi}{d\omega} \text{ lm/sr or cd}$$

It may be noted that if a source has a luminous intensity of I candela, then total flux radiated out by it in the space around it is

$$= 4\pi \times I \text{ lumen}$$

5. **Illuminance or Illumination (*E*)**. When luminous flux falls on a surface, it is said to be illuminated.

The illumination of a surface is measured by the normal flux per unit area received by it. If $d\Phi$ is the luminous flux incident normally on an area dA, then

$$E = \frac{d\Phi}{dA} \text{ lm/m}^2 \text{ or lux}$$

In general, $$E = \frac{\Phi}{A} \text{ lux}$$

41.3. Laws of Illuminance for Point Sources

The illuminance (E) of a *surface* depends on the following factors. The light source is assumed to be a point source or is otherwise sufficiently away from the surface to be regarded as such.

(*i*) E is directly proportional to the luminous intensity (I) of the *source i.e.*, $E \propto I$.

(*ii*) E is *inversely* proportional to the square of the distance between the source and the surface, *i.e.*, $E \propto 1/r^2$.

(*iii*) E is *directly* proportional to the cosine of the angle made by the normal to the illuminated surface with the direction of the incident flux.

Combining the above three factors, we have

$$E = \frac{I \cos\theta}{r^2}$$

Let a lamp L of uniform luminous intensity I be suspended at a height of h above the working plane as shown in Fig. 41.3. Let us apply the above law to calculate illuminance E at point A immediately below the lamp and at point B.

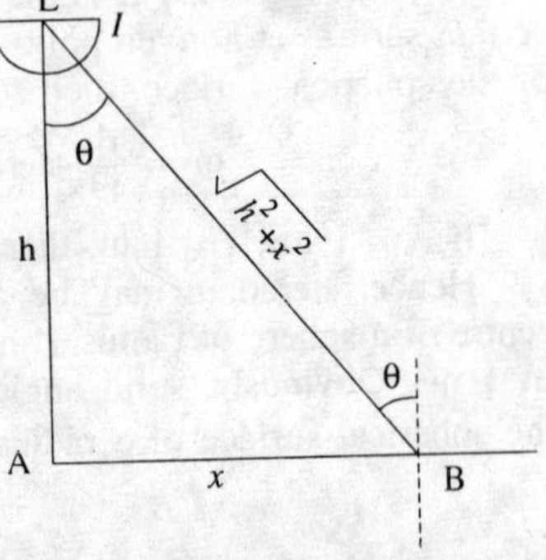

Fig. 41.3

Point A

Here, $\theta = 0, \cos\theta = 1$

$$\therefore \quad E_A = \frac{I\cos\theta}{r^2} = \frac{1}{h^2} \quad (\because r = h)$$

Point B

Here, $\cos\theta = \dfrac{h}{LB} = \dfrac{h}{\sqrt{h^2 + x^2}}$

$\therefore \quad E_B = \dfrac{I}{LB^2} \times \dfrac{h}{LB} = I \cdot \dfrac{h}{LB^3}$

We may also express E_B in terms of E_A as under:

$$E_B = I\frac{h}{LB^3} = \frac{I}{h^2} \cdot \frac{h^3}{LB^3} = E_A \cos^3\theta$$

Example 41.1. *An incandescent lamp having a luminous intensity of 60 cd in all directions gives an illumination of 15 lux at the surface of a table vertically below it.*

(a) What distance is the lamp from the table?

(b) What illumination would be given if 60 cd lamp is replaced by a similar 100 cd lamp?

Solution. (*a*) $E = I / h^2$ or $15 = 60 / h^2$; h = **2 m**

(*b*) Since $E_1 \propto I_1$ and $E_2 \propto I_2$

$\therefore \quad \dfrac{E_2}{E_1} = \dfrac{I_2}{I_1}$ or $\dfrac{E_2}{15} = \dfrac{100}{60}$ or E_2 = **25 lux**

Example 41.2. *A lamp giving out 1260 lumen in all directions is suspended 8 m above the working plane. Calculate the illumination at a point on the working plan 6 m away from the foot of the lamp.*

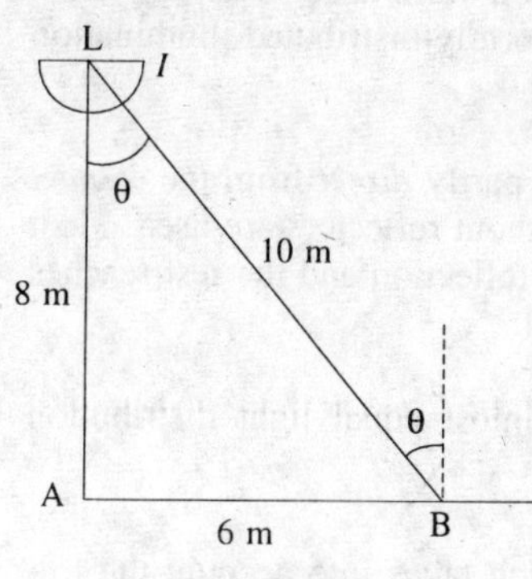

Fig. 41.4

Solution. $\Phi = 4\pi I$ lumen

$1260 = 4\pi I$ or $I = 100$ cd

As seen from Fig. 41.4,

$LB = \sqrt{8^2 + 6^2} = 10$m

$\cos\theta = LA / LB = 8 / 10 = 0.8$

Now, $E = I \cos\theta / r^2$

$= 100 \times 0.8 / 10^2 =$ **0.8 lux**

Example 41.3. *A certain lamp, giving equal illuminance below the horizontal, hangs from the ceiling of a room. The illuminance received on a small horizontal screen lying on a bench vertically below the lamp is 60 lux. When the screen is moved a distance of 1.2 m along the bench, the illumination is 30.72 lux.*

Calculate the luminous intensity of the lamp and its vertical distance from the bench.

(Utilis. of Elect. Energy Haryana, 1992)

Solution. As seen from Fig. 41.5,

$E_1 = I / h^2$ or $I = E_1 h^2$ ...(*i*)

$E_2 = I\cos\theta / d^2 = Ih / d^3$

or $I = E_2 d^3 / h$ ... (*ii*)

From (*i*) and (*ii*), we get

$E_1 h^2 = E_2 d^3 / h$ or $E_2 / E_1 = h^3 / d^3$

Now, $h / d = \cos\theta$

$\therefore \quad \cos^3\theta = E_2 / E_1$

or $\cos\theta = (30.72 / 60)^{1/3} = 0.8$

$\therefore \quad \theta = 36.8°$

Now, $\tan\theta = 1.2 / h$

$\therefore \quad h = 1.2 / \tan 36.8° =$ **1.6 m**

Substituting this value in Eq. (*i*), we have

$I = 60 \times 1.6^2 =$ **154 candela**

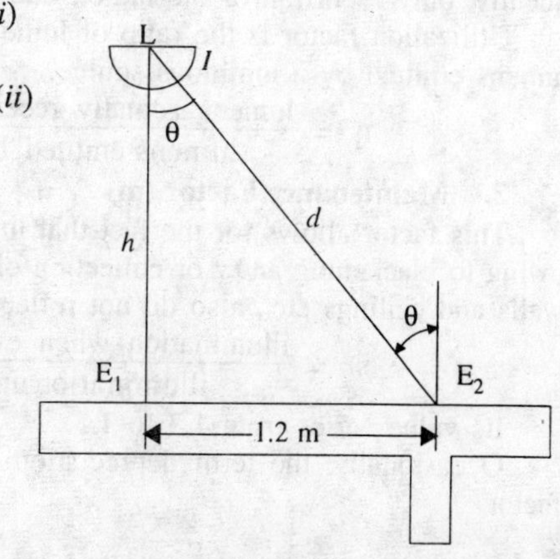

Fig. 41.5

41.4. Practical Lighting Schemes

The simple inverse-square-law calculations are applicable only to point sources of light or where there are no reflecting or absorbing surfaces.

For indoor illumination, we have to consider not only the light received directly from the source itself but also that reflected from walls and ceilings etc.

Different lighting schemes are classified as under:

1. Direct Lighting

In this case, light from the source falls directly on the surface or object to be illuminated. Most of the light is directed in the lower hemisphere with the help of shades, globes and reflectors of various types. It is essential to keep the lamps and fittings clean otherwise decrease in effective illumination due to dirty bulbs or reflectors may amount to about 20 to 25% in offices and homes. This type of lighting is liable to cause glare and hard shadows.

2. Indirect Lighting

In this case, light does not reach the working plane *directly* from the source but indirectly by diffuse reflection. Lamps and tubes are placed either behind the cornice or in suspended opaque bowls. In this way, maximum light is thrown upwards on the ceiling from where it is distributed all over the room by diffuse reflection. Indirect lighting provides shadowless illumination which is very suitable for drawing offices, composing rooms and in workshops.

3. Semi-direct Lighting

This system utilizes luminaries which send most of the light downwards directly on the working plane though a considerable amount reaches the ceiling and walls also. Such a system is best suited to rooms with high ceilings where a high level of uniformly-distributed illumination is desirable.

4. Semi-indirect Lighting

In this system, light is partly received by *diffuse* reflection and partly *direct* from the source. Instead of using opaque bowls with reflectors, *translucent* bowls without reflectors are used. Most of the light is, as before, directed upwards to the ceiling for diffuse reflection and the rest reaches the working plane directly except for some absorption by the bowl.

5. General Diffusing System

In this system, such luminaries are employed which have almost equal light distribution downwards and upwards.

41.5. Design of Lighting Schemes

Practical lighting schemes are based on the *lumen method* which takes into account the following factors:

1. Utilization Factor (η)

It is well-known that part of the lumen output of a source is lost in the fittings. Some output is directed towards the walls and ceiling where it is partly absorbed and partly reflected. Consequently, only a portion of the lumen output reaches the working plane.

Utilization factor is the ratio of lumens actually received by a particular surface to the total lumens emitted by a luminous source.

$$\therefore \quad \eta = \frac{\text{lumens actually received on working plane}}{\text{lumens emitted by the light source}}$$

2. Maintenance Factor (m)

This factor allows for the fact that luminous intensity of all lamps or luminaries deteriorates owing to blackening and / or collection of dust or dirt on the globes and reflectors etc. Similarly, walls and ceilings etc., also do not reflect as much light as when they are clean.

$$m = \frac{\text{illumination when everything is perfectly clean}}{\text{illumination under actual conditions}}$$

Its value varies from 1.3 to 1.5.

Occasionally, the term depreciation factor (p) is used which is reciprocal of maintenance factor.

$$\text{depreciation factor} = \frac{1}{\text{maintenance factor}} \quad \text{or} \quad p = 1/m$$

3. Space-height Ratio

It is given by

$$\frac{\text{horizontal distance between two lamps}}{\text{mounting height of lamps}}$$

This ratio depends on the nature of the polar curve when *used along with its reflector.* For reflectors normally used in indoor lighting, the value of this ratio lies between 1 and 2.

41.6. Calculations Based on Lumen Method

Taking into account the above given factors, total number of lumens required for a certain illuminated surface are

$$\text{total flux required } \Phi = \frac{E \times A}{\eta \times p} = \frac{E \times A}{\eta \times 1/m} = \frac{mEA}{\eta}$$

If a group of light units is given, then illuminance produced on a given surface is

$$E = \frac{\Phi \times \eta \times p}{A} = \frac{\Phi \times \eta}{m \times A}$$

Example 41.4. *A workshop measures 10 m × 12 m and is lighted by 20 lamps of 100 W each. Taking depreciation factor of 0.75, coefficient of utilisation 0.6 and efficiency of lamps as 15 lm/W, find the illumination on the working plane.*

Solution. $A = 10 \times 12 = 120 \text{ m}^2, \; p = 0.75$

$\eta = 0.6, \Phi = (20 \times 100) \times 15 = 30{,}000 \text{ lm}$

$E = ?$

Now, $\Phi = \dfrac{EA}{\eta p}$ or $30{,}000 = \dfrac{E \times 120}{0.75 \times 0.6}$

$\therefore$ $E = \mathbf{112.5 \text{ lux}}$

Example 41.5. *A light assembly shop, 15 m long, 9 m wide and 3 m upto trusses is to be illuminated to a level of 200 lux. The utilisation and depreciation factors are respectively 0.9 and 0.8. Using tungsten lamps and dispersive metallic reflectors, draw their spacing lay-out. Assume a lamp efficiency of 13 lm/W and space/height ratio of unity.*

(Utilis. of Energy., Gujarat Univ., 1993)

Solution. $E = 200 \text{ lux}, A = 15 \times 9 = 135 \text{ m}^2$

$\eta = 0.9, p = 0.8, \Phi = ?$

$$\Phi = \frac{200 \times 135}{0.9 \times 0.8} = 37{,}500 \text{ lm}$$

$S/H = 1$ or $S/3 = 1$ or $S = 3 \text{ m}$

Obviously, 5 lamps can be mounted along the length and 3 along the width of the assembly shop as shown in the layout diagram of Fig. 41.6. Total number of lamps required is 5 × 3 = 15.

If W is the wattage of each lamp, then total wattage is $15 \times W$. Taking lamp efficiency into account, total lumens output

$= 15 \times W \times 13$ lumens

$\therefore \; 15 \times 13 \times W = 37{,}500$

or $W = 192.3 \text{ watt} \cong \mathbf{200 \text{ watt.}}$

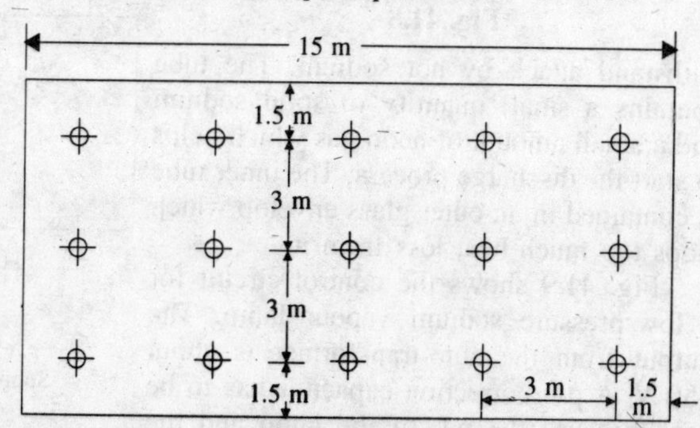

Fig. 41.6

41.7. Tungsten Filament Lamp

The basic components of a tungsten-filament lamp are shown in the Fig. 41.7.

It consists of a hollow glass bulb containing a glass stem which carries a tungsten filament. The filament is either single-or double-coiled. The glass bulb itself is cemented into a brass cap having two pins and two contact points as shown.

The filament lamps are of two main types (*i*) vacuum type (*ii*) gas-filled type.

In the vacuum type lamp, the filament operates in a vacuum in the glass bulb. The filament temperature can achieve maximum temperature of 2000°C because of which lamp efficiency is poor.

In the gas-filled type lamps, the bulb is filled with an inert gas such as nitrogen or argon. This enables the operating temperature to reach upto 2500°C. It increases the luminous efficiency of the bulb tremendously. In fact, the bulb is usually so bright that it is given an opaque coating internally.

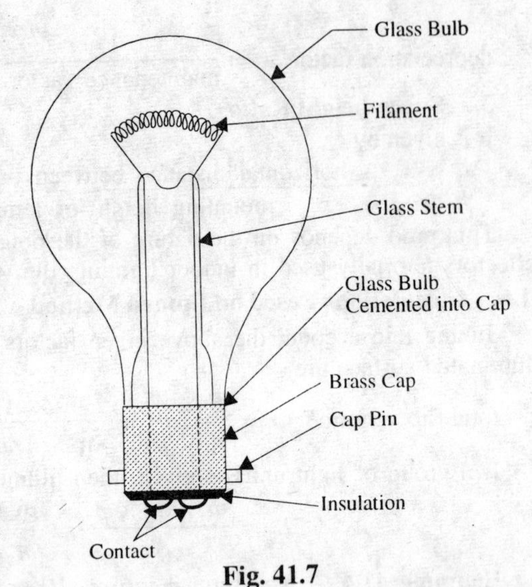

Fig. 41.7

The efficiency of a tungsten lamp depends on many factors including its size and age but tends to be around 12 lm / W for a 100 W lamp. The colour of its light is mostly red and yellow. In its basic form, this type of lamp is used where high level of illumination is not required.

Other lamps of the filament type include (*i*) tubular strip lights (*ii*) oven lamps (*iii*) infrared heating lamps (*iv*) spot and floodlights and (*v*) tungsten-halogen lamps.

41.8. Discharge Lamps

In such lamps, ionization of a gas is used to produce light. Since high voltage is used for this purpose, special precaution has to be taken with regards to their high-voltage circutry. Typical discharge lamps include (*i*) decorative neon signs (*ii*) fluorescent lighting (*iii*) mercury-vapour lamps and (*iv*) sodium-vapour lamps.

41.9. Sodium Vapour Lamp

Two types of sodium vapour lamps are available which work at high vapour pressure and low vapour pressure respectively.

(*a*) Low-pressure Sodium-vapour Lamp

As shown in Fig. 41.8 the lamp consists of a U-shaped double-thickness glass tube the inner wall of which is made of low-silica glass which can withstand attack by hot sodium. The tube contains a small quantity of solid sodium and a small amount of neon gas which helps to start the discharge process. The inner tube is contained in an outer glass envelop which stops too much heat loss from it.

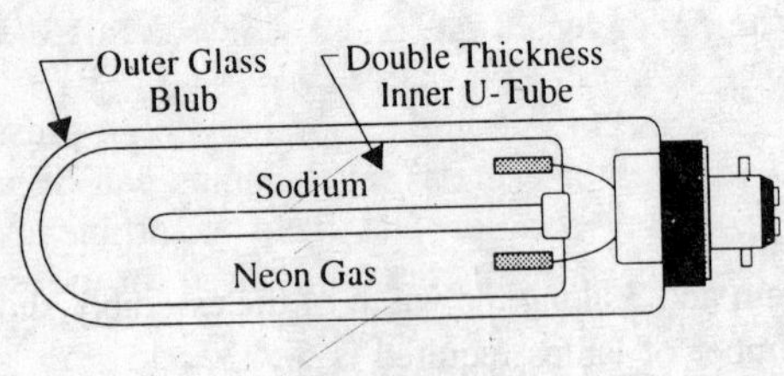

Fig. 41.8

Fig. 41.9 shows the control circuit for a low-pressure sodium vapour lamp. The output from the auto-transformer is about 450 V. A p.f. correction capacitor has to be used because the p.f. of the lamp and the

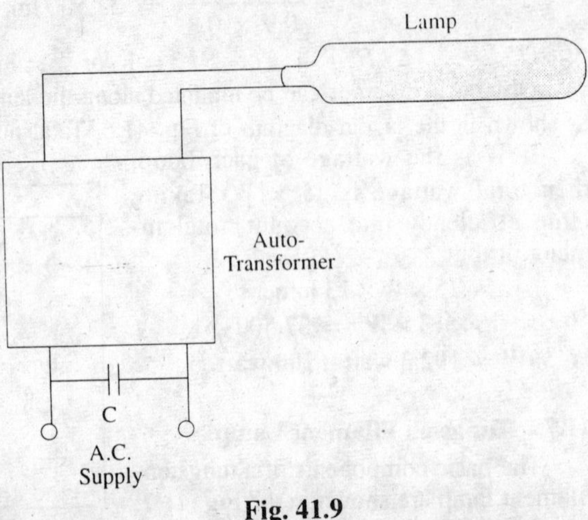

Fig. 41.9

transformer is as low as 0.3 lagging. The recommended burning position of the lamp is horizontal ± 20°. This factor ensures that hot sodium does not collect at one end of the tube in sufficient quantities to attack and damage it. Several minutes are necessary after switching on before the discharge comes to the full but the lamp may be restarted immediately after stoppage. The light output from the lamp is almost pure yellow which distorts the surrounding colours. Hence, such lamps are useful for street lighting only. Their efficiency is as high as 140 lm/W.

(*b*) High-pressure Sodium-vapour Lamp

This lamp differs from other discharge lamps in the fact that its discharge tube is made of compressed aluminium oxide which can withstand the intense chemical activity of the sodium vapours at high temperature and pressure. The lamp may be mounted in any position. Its luminous efficiency is about 100 lm / W and the colour of its light is golden white which produces little surrounding colour distortion. Hence, such a lamp is suitable for many applications including shopping centres, car parks, sports ground and dockyards etc.

(*c*) High-pressure Mercury-vapour Lamp

The basic components of this lamp are shown in Fig. 41.10. It consists of a quartz tube containing mercury vapour at high pressure and a small amount of argon gas which helps in starting the discharge. There are three electrodes two main and one auxiliary. The auxiliary electrode is used for starting the discharge. The initial discharge takes place in the argon gas between the auxiliary electrode and the main electrodes close to it. This causes the main electrode to heat up and then the main discharge between the main two electrodes takes place. The lamp requires 4 to 5 minutes to attain its full brilliancy.

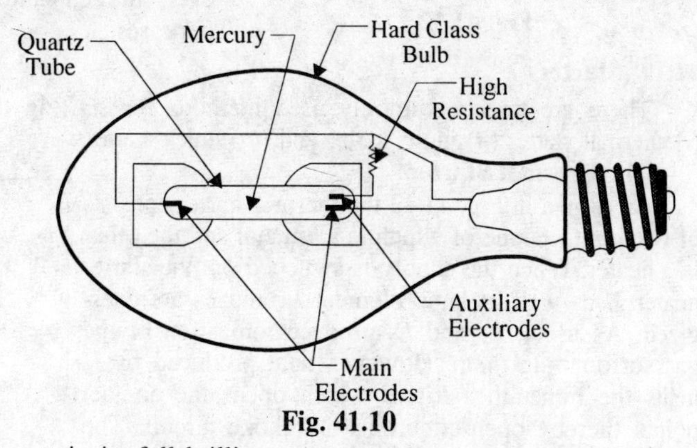

Fig. 41.10

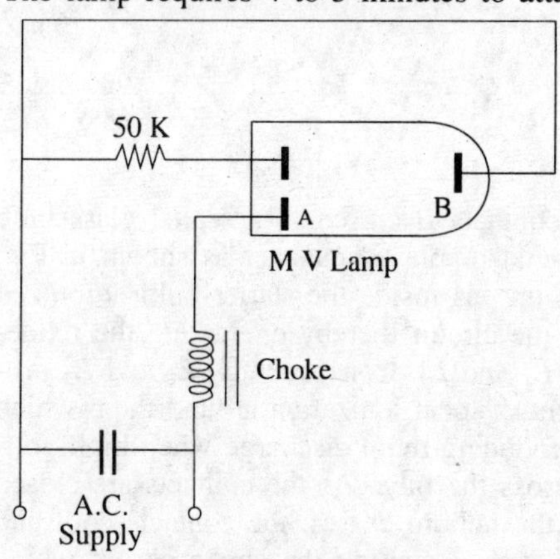

Fig. 41.11

Fig. 41.11 shows the control circuit for a high-pressure mercury-vapour lamp which are available in many types. As seen, a power correction capacitor has been used for improving the p.f. of the choke and the lamp. The unconnected p.f. is 0.6 which is raised with the help of a suitable capacitor.

The colour of the light given by high-pressure mercury lamp is blue-green. Such lamps can be used in any mounting position and have an efficiency of about 50 lm/W. They are commonly used for industrial and street lighting, commercial and display lighting.

(*d*) Low-pressure Mercury-vapour Lamp

Such a lamp is popularly known as fluorescent lamp or fluorescent tube because it consists of a glass tube the interior of which is coated with a fluorescent phosphor. The tube is filled with mercury vapour at low-pressure and a small quantity or argon gas which helps in starting. There is an oxide-coated filament at each end of the tube. Discharge takes place through the mercury vapours when a high voltage is applied across the two ends of the tube. The circuit diagram for a single flourescent tube is shown in Fig. 41.12

Practical Operation

When supply is switched on, the circuit is completed via the choke lamp element L_1, the bimetallic starter switch, lamp element L_2 and the neutral. The oxide-coated lamp elements L_1

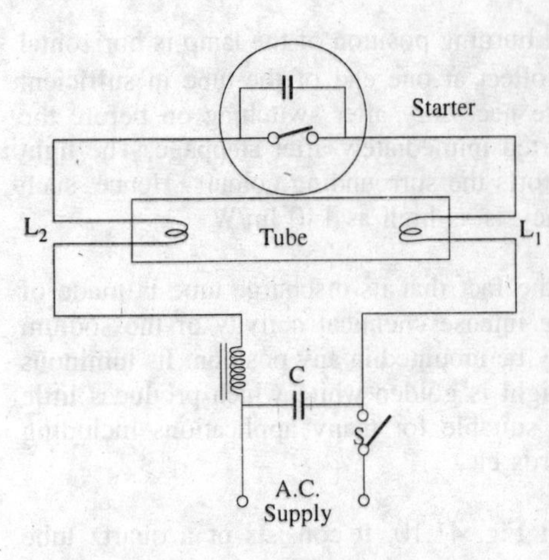

Fig. 41.12

and L_2 are heated up and their oxide coatings emit some electrons which ionize the gas in small space around them, (this helps the main ionization process). Soon after the bimetallic starter contacts open-up on being heated thereby opening circuit the highly inducting choke. This sudden breaking of the choke circuit induces high voltage across the breaking contacts and hence across L_1 and L_2. Consequently, the gas inside the tube gets ionized and emits *ultra-violet* light. This ultra-violet light excites the fluorescent powder coating and makes it to emit visible light. The choke limits the current to a pre-determined value. As seen from the circuit diagram, there are two capacitors used in the circuit. The capacitor connected across the starter contacts is meant for suppressing radio interference. However, the capacitor C is meant for p.f. correction purposes.

41.10. Starters

There are three commonly used methods for starting the discharge in a fluorescent tube. (*i*) thermal starter (*ii*) glow starter and (*iii*) quick starter.

(*i*) Thermal Starter

As shown in Fig. 41.13 the thermal starter consists of two contacts one of which is a bimetal and the other is a heater. When the supply is switched on, the starter heater H as well as lamp element L_1 and L_2 are energized. As usual, L_1 and L_2 produce ionization of the gas surrounding them. However, heat produced by H heats the bimetallic strip B which opens the contact points thereby open-circuiting the choke circuit. The high voltage so produced starts the main discharge and the tube glows up. The capacitor inside the starter is for the suppression of radio interference whereas the other is for p.f. correction purposes.

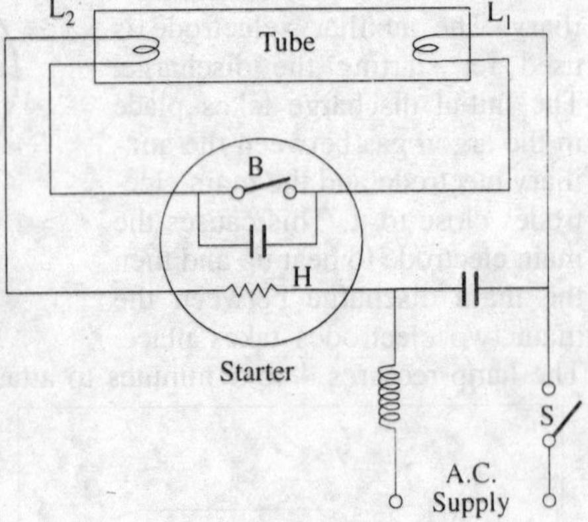

Fig. 41.13

(*ii*) Glow Starter

It is the most popular of all the methods of starting the discharge. It consists of a pair of open bimetallic contacts B closed in a sealed glass bulb filled with helium gas. This assembly is housed in a metal or plastic canister as shown in Fig. 41.14. Initially, when the supply is switched on the helium gas inside the starter bulb is ionized and hence heats up the bimetallic contacts which close the circuit thereby energizing the oxide-coated filaments L_1 and L_2. Consequently, L_1 and L_2 produce small amount of local ionization around them which is essential for producing main discharge when high voltage is applied across the tube. As the contacts are closed the discharge in the helium ceases, the contacts cool and open up thereby open–circuiting the choke circuit which produces high voltage required for the main discharge.

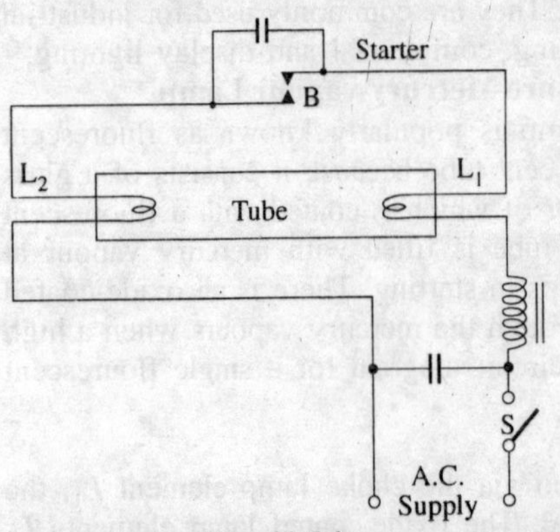

Fig. 41.14

(*iii*) Quick Starter / Instant Starter

In this case, the starting is achieved by the use of an auto transformer and an earthed metal strip located very close to the tube as shown in Fig. 41.15. When the supply is switched on, the mains voltage appears across the ends of the tube and the small part of the winding at each end of the transformer energizes the filaments L_1 and L_2 which heat up and produce local ionization. The difference in potential between the electrodes and the earthed strip causes ionization which spreads along the fluorescent tube.

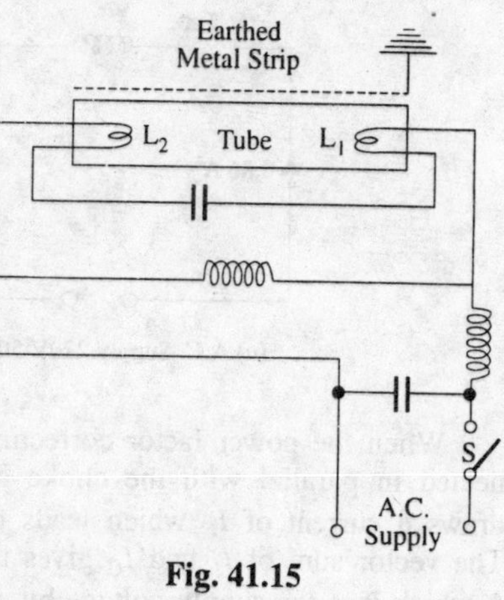

Fig. 41.15

41.9. Stroboscopic Effect

While a fluorescent lamp is in operation, the light may flicker. Under some circumstances, a rotating machine wheel may appear to slow down or even reverse its direction of rotation. This is called *stroboscopic* effect. With 50 Hz supply frequency, the lamp will be extinguished 100 times per second. Hence, any machinery part revolving at 100 r.p.s. will appear to be stationary. Since it produces undesirable results, it has to be removed. Following two methods can be used to remove this stroboscopic effect.

(*i*) Balancing the Light Loads

If a large lighting load is installed in a three-phase installation where there is some rotating machinery, the stroboscopic effect can be overcome by connecting alternate group of lights to a different phase as shown in Fig. 41.16.

This also has the advantage of balancing the light.

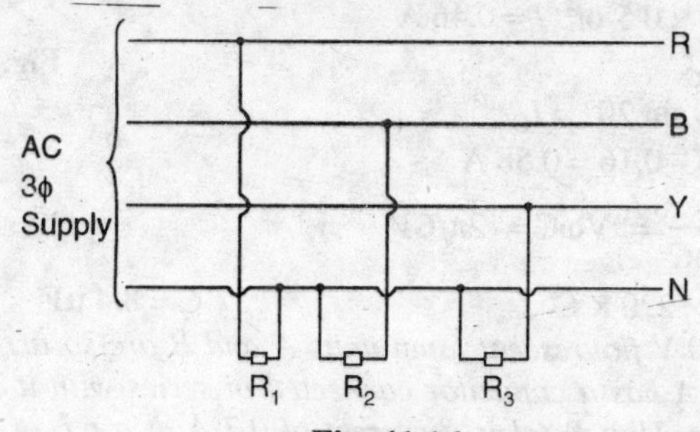

Fig. 41.16

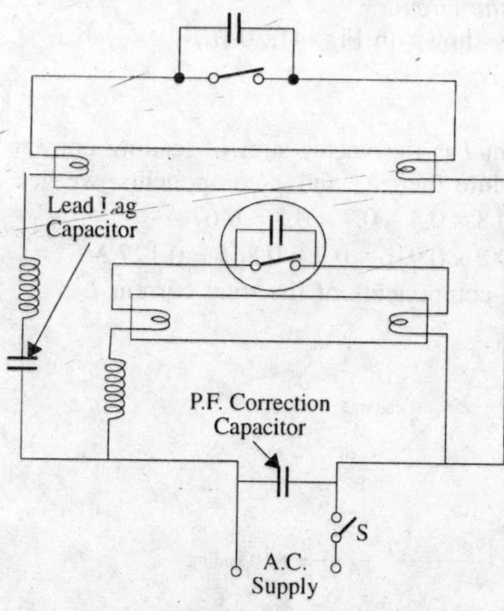

Fig. 41.17

(*ii*) The Lead-Lag Circuit

In this method, a capacitor is connected in series with every alternate lamp in a row (Fig. 41.17).

The value of the capacitor is such that it gives an overall leading p.f. to the lamp unit concerned. This means that any pair of lamps will have a lagging and leading power factor. This results in the cancellation of the resultant flicker.

Example 41.6. *The power factor correction capacitor in a 220 V, 50 Hz fluorescent light unit has broken down and needs replacing. It is found that without the capacitor, the unit takes a supply current of 0.86 A at a lagging power factor of 0.5. The values quoted on the original capacitor are unreadable except for the information that the working power factor of the unit should be 0.94. Compute the value of the capacitor required.*

Solution. The diagrams for the fluorescent light unit with and without the power-correction capacitor are shown in Fig. 41.18.

Without the capacitor, p.f. = cos θ = 0.5, so that θ = 60°. Hence, the vector for the current $I_L = 0.86\ A$ is drawn at a lagging angle of 60° with the voltage vector as shown in Fig. 41.19.

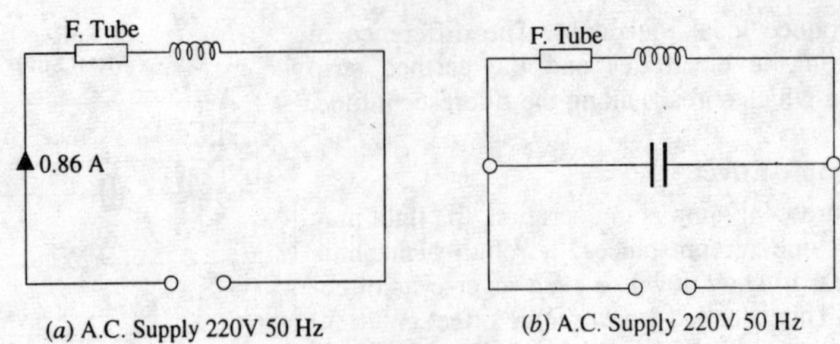

(a) A.C. Supply 220V 50 Hz (b) A.C. Supply 220V 50 Hz

Fig. 41.18

When the power factor correcting capacitor is connected in parallel with the choke and lamp circuit, it draws a current of I_C which leads the voltage by 90°. The vector sum of I_L and I_C gives the resultant current I. which lags the supply voltage by an angle ϕ such that $\cos \phi = 0.94$ or $\phi = 20°$ as shown in Fig. 41.19.

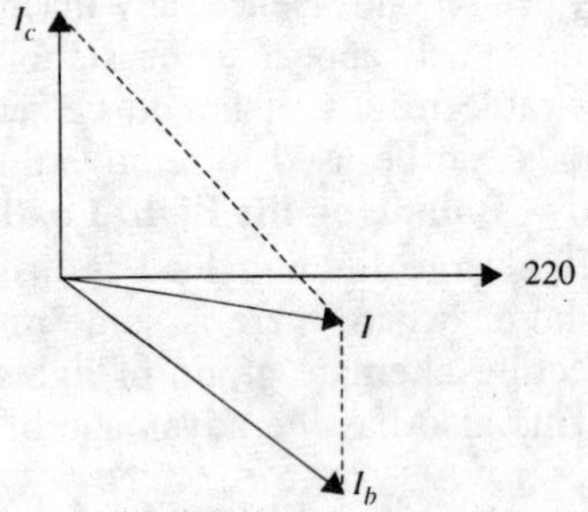

It is obvious from the vector diagram that

$$I \cos \phi = I_L \sin \theta \qquad \text{(each} - AB)$$

$$\therefore \quad I \times 0.94 = 0.86 \times 0.5 \text{ or } I = 0.46 \text{ A}$$

Aslo, $I_L \sin \theta - I \sin \phi = I_C$

$$\therefore \quad 0.86 \times \sin 60° - 0.46 \sin 20° = I_C$$

$$\therefore \quad I_C = 0.74 - 0.16 = 0.58 \text{ A}$$

Fig. 41.19

$$\text{Now,} \quad I_C = \frac{V}{1/\omega C} = \text{V}\omega\text{C} = 2\pi fCV$$

$$\text{or,} \quad 0.58 = 2\pi \times 50 \times 220 \times C \qquad \therefore \quad C = \mathbf{8.4\ \mu F}$$

Example. 41.7. *Two 220-V flourescent lamp units A and B are so arranged as to overcome the stroboscopic effect. Unit A has a capacitor connected in series with it and takes a current of 0.8 A at a p.f. of 0.4 leading. Unit B takes a current of 0.7 A at a p.f. of 0.5 lagging Calculate the total current taken and the overall power factor of the circuit.*

Solution. The two units are connected in parallel as shown in Fig. 41.20 (*a*)

Unit *A*. $\cos \phi_A = 0.4, \phi_A = 66.4°$ (lead)

Unit *B*. $\cos \phi_B = 0.5, \phi_B = 60°$ (lag)

As seen from Fig. 41.20 (*b*), the total circuit current I is the vector sum of leading current I_A and lagging current I_B. Resolving these two currents into their *X*- and *Y*-components, we get

Total *X*-component $= I_A \cos \phi_A + I_B \cos \phi_B = 0.8 \times 0.4 + 0.7 \times 0.5 = 0.67$ A

Total *Y*-component $= I_A \text{ sing } \phi_A - I_B \sin \phi_B = 0.8 \times 0.916 - 0.7 \times 0.866 = 0.127$ A

Obviously, the above values represent the *X*- and *Y*-components of the total current I.

$$I = \sqrt{0.67^2 + 0.127^2} = \mathbf{0.68\ A}$$

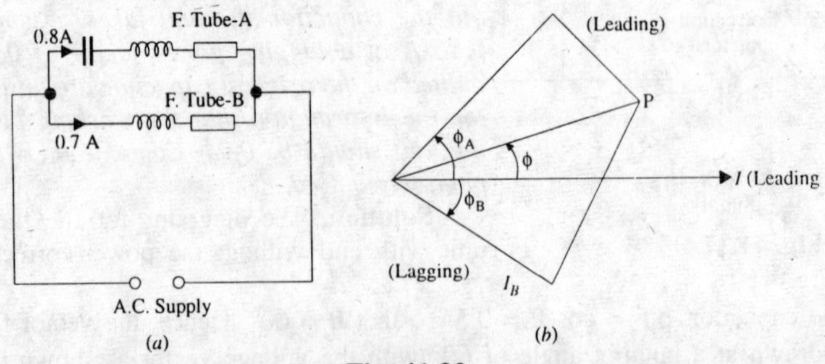

(a) (b)

Fig. 41.20

Combined p.f. $\cos \phi$ = total X-component / total current
= 0.67 / 0.68 = **0.985 leading.**

HIGHLIGHTS

1. Candela is the unit of luminous intensity of a source.
2. Luminous intensity of a light source is given by luminous flux radiated out by it per unit solid angle.
3. Illuminance of a surface is measured by the normal flux per unit area received by it.
4. Laws of illuminance are given by the expression

$$E = \frac{I \cos \theta}{r^2}$$

5. For indoor lighting, illuminance formula is

$$E = \frac{\Phi \times \eta}{m \times A} \quad \text{or} \quad \frac{\Phi \times \eta \times p}{A}$$

where η = utilization factor
m = maintenance factor
p = depreciation factor

OBJECTIVE TESTS—41

A. Fill in the following blanks.

1. Sodium vapour lamp is an example of .... lamp.
2. Candela is the unit of .... intensity of a source.
3. The unit of illumination is ....
4. In indirect lighting scheme, light reaches the working plane by .... reflection.
5. Depreciation factor of a luminary is the reciprocal of .... factor.
6. Lumen is the unit of ....

B. Answer True or False.

1. Tungsten filament lamps work on the principle of temperature incandescence.
2. Steradian is the unit of luminous flux.
3. A light source of one candela emits a total luminous flux of 4π lumens.
4. Flourescent lamps are an example of discharge lamps.
5. Candela is the same as lumen / steradian.
6. Direct lighting scheme is ideal for workshops.

C. Multiple Choice Questions

1. The unit of luminous intensity of a light source is
(*a*) lux (*b*) candela
(*c*) steradian (*d*) lumen
2. A 100 candela light source radiates out a total flux of .... lumen.
(*a*) 0.1257 (*b*) 1.257
(*c*) 12.57 (*d*) 1257
3. A flux of 50 lumen falls perpendicularly on a surface of area 100 m^2. The surface illumination is .... lux.
(*a*) 0.5 (*b*) 5000
(*c*) 2 (*d*) 150
4. Direct lighting scheme for indoor illumination
(*a*) is ideal for drawing offices
(*b*) is best suited to rooms with high ceilings.
(*c*) produces hard shadows
(*d*) is often used for composing rooms
5. Practical lighting schemes based on lumen method take into account
(*a*) utilization factor
(*b*) maintenance factor
(*c*) space / height ratio
(*d*) all the above
6. A point light source giving out 1257 lumen in all directions is hung 8 m above a flat surface. The illumination at a point 6 m away from the point vertically below the source would be .... lux.
(*a*) 126 (*b*) 12.6
(*c*) 0.8 (*d*) 1.

ANSWERS

A. 1. discharge 2. luminous 3. lux. 4. diffuse 5. maintenance 6. flux
B. 1. T 2. F 3. T 4. F 5. T 6. F
C. 1. *b* 2. *d* 3. *a* 4. *c* 5. *d* 6. *c*

INDEX

A

B

C

D

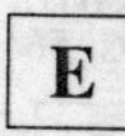

E

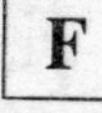

F

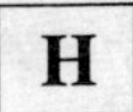

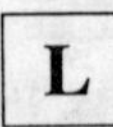

T

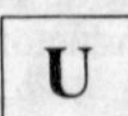

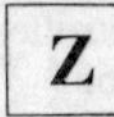